DYNAMICS OF PHYSICAL SYSTEMS

ROBERT H. CANNON, JR.
Charles Lee Powell Professor Emeritus
Stanford University

DOVER PUBLICATIONS, INC.
Mineola, New York

To *Philip*
Douglas
Beverly
Frederick
David
Joseph
James
and *Dorothea*

Copyright

Copyright © 1967 by Robert H. Cannon, Jr.
All rights reserved.

Bibliographical Note

This Dover edition, first published in 2003, is an unabridged republication of the work originally published by McGraw-Hill, Inc., New York, in 1967.

Library of Congress Cataloging-in-Publication Data

Cannon, Robert H.
    Dynamics of physical systems / Robert H. Cannon, Jr.
       p. cm.
    Originally published: New York : McGraw-Hill, 1967.
    Includes bibliographical references and index.
    ISBN 0-486-42865-6 (pbk.)
    1. Systems engineering. I. Title.

TA168.C33 2003
620'.001'171—dc21

                                                              2003043996

Manufactured in the United States of America
Dover Publications, Inc., 31 East 2nd Street, Mineola, N.Y. 11501

FOREWORD

The engineering subject of dynamic systems has evolved as the keystone element of modern system engineering and, as a result, has gradually moved toward a central role in curricula in all engineering disciplines. This evolution has been accompanied by a spate of textbooks which have attempted to use slight modifications of conventional subjects (circuit theory, vibrations, and control) in order to familiarize the student with the potency of the fundamental concepts and the breadth of applicability of the viewpoints included within dynamic systems.

Professor Cannon's excellent new book is a major departure from these earlier attempts in several ways of particular significance to the engineering educator:

1. Emphasis is placed on the engineering concepts (modeling, dynamics, feedback, and stability, as examples), rather than on the mechanistic analysis procedures designed to yield routine answers to programmable problems.

2. The emphasis on modeling permits development of appreciation for the breadth of dynamic systems without resorting to analog solutions or depending on dexterity in electric-circuit formulation and analysis.

3. The unique breadth is realized successfully also through a truly remarkable series of examples and problems—examples which develop physical insight in the best traditions of modern engineering, provide excitement and motivation at a level ranking this as a classic book in its field, and lead the student and reader into richer and richer technical ground (up to Cannon's intriguing work on the balancing of inverted sticks and the control of unstable systems).

The significance of these items is particularly important today, as engineering education attempts to prepare students both for a rapidly changing technology and for our national dedication to the application of technology to the solution of social problems in urban, educational, and biomedical engineering areas. The fundamental concepts which Dr. Cannon develops constitute the common language of engineering, regardless of the area of application, and indeed represent a large share of the contribution of engineering to the intellectual world.

For example, the concept of dynamics as presented in this book is the basis for understanding not only of physical devices, but also of systems in such fields as management and transportation. Indeed, much of the recent progress in these areas has been based upon first principles of dynamics. Dr. Cannon lays the foundations for such extensions as he moves naturally from simple mechanical and electrical configurations into a novel discussion of distributed-parameter and compound systems.

Finally, the book is unusual in its applicability to a wide variety of particular courses and groups of students, ranging from sophomores to graduate students. The flexibility is achieved by focussing on fundamental concepts and physical insight, with the mathematical techniques used to illustrate applications of the ideas to problems of varying depth.

Thus, Professor Cannon's work appropriately appears at a time when the concepts and techniques of dynamic-systems engineering have just evolved from a period of major development and when these concepts are providing the basis for the extension of engineering toward new demands and challenges. At this critical epoch, the publication of a masterful development of a fundamental subject is indeed a most welcome enrichment of the technological literature. I believe it will have a major impact.

John G. Truxal
Provost, Polytechnic Institute of Brooklyn

Brooklyn, N.Y.
May 8, 1967

PREFACE

The subject of dynamic behavior is central to much of technology. In recent years the ascendance of system engineering to a pivotal position has engendered a unifying trend among the traditional dynamics disciplines from which it has grown, particularly classical dynamics, electrical circuit theory, mechanical vibrations, and automatic control.

At the universities the trend is being fostered by the development, integral to the core curriculum, of a course to introduce the system-dynamics viewpoint and build upon it, beginning at a reasonably early point. At the professional level, men whose careers have lead into system engineering are pursuing additional study to extend their earlier dynamics training to the expanded domain of systems.

This book has been prepared as a comprehensive text and reference for a first study of system dynamics, with emphasis on the development of physical insight. It treats carefully the modeling of physical systems in several media, and derivation of their differential equations of motion (Part A). Then it determines the physical behavior those equations connote: the free and forced motions of, first, elementary systems (Part B), and then compound "systems of systems" (Parts C-E). Dynamic stability and natural behavior are given comprehensive linear treatment in Part C, response to continuing and abrupt forcing inputs in Part D and, at a fundamental level, analysis and synthesis of feedback control systems in Part E.

The book is designed to provide foundation training and perspective to students planning to pursue any of the various dynamic subjects: mechanical vibrations, electrical circuits, electromechanics, systems theory and automatic control, classical dynamics (including gyro behavior, satellite motions, etc.), vehicle guidance and control, thermal and fluid systems and, less directly, plasma, chemical, biological, and organizational system dynamics.

There are two pervading objectives: (1) to develop understanding of the basic *concepts* of dynamic behavior, (2) to develop facility with the *analytical techniques* for predicting and assessing dynamic behavior.

The book is carefully packaged to serve equally well at each of several levels. The typical user will employ perhaps half the book for his own course of study (as described later in this preface), entering the book where previous training suggests, proceeding as far as requirements prescribe, and omitting optional chapters of less immediate interest. Thus although the book is quite long, it is in effect several shorter, integrated books.

There is, moreover, from beginning to end, a closely knit core that carries basic concepts forward, on a building-block basis, to systems of increasing complexity, and to analytical techniques of increasing sophistication. The pace is deliberately slower in the first half of the book, and the steps between concepts smaller than in the second half. Hence, while the nominal prerequisites are a year of college physics and calculus, readers who have also studied mechanics and circuits may take full advantage thereof by using Chapter 2 for review and proceeding with dispatch through Part B.

There is also a selected deployment of "bridges" to adjacent disciplines such as nonlinear mechanics, state-model analysis, and vibrations of distributed-parameter systems.

A major goal has been to choose and present examples and problems so that physical insight is developed at every turn. There are over 1,000 problems. Some of them form a chapter-to-chapter sequence in which a given physical system is considered from an increasingly sophisticated viewpoint. Others provide experience with a rather wide variety of physical devices (as may be gauged by perusing the index). At the same time, in the core development the examples, while having physical variety, are physically simple, extensions of the concepts to more advanced systems being demonstrated in optional sections. Answers are given to odd-numbered problems.

Learning by images has been stressed, with special attention to the figures and their placement. For reference utility, sections on such techniques as Routh's method, root-locus plotting, partial fraction expansion, and Lagrange's method, have been written to be self-contained, and are given a special index inside the front cover. (They are also identified by informal "Procedure" numbers.)

This book has been developed in a continuing, intensive effort over a period of some fourteen years, while teaching system dynamics at UCLA, MIT, and Stanford University, and consulting in inertial guidance and automatic flight control at several aerospace firms. Classes of ME, EE, Aeronautics and Astronautics, and IE majors have participated, class levels ranging from sophomores to graduate students.

The book's building-block development of the subject and its use in specific courses of study are described below.

SUBJECT DEVELOPMENT

Part A presents a systematic procedure for obtaining the equations of motion of physical systems, first describing a physical model, then apply-

PREFACE

ing the laws of physics and of interelement relations in systematic sequence. Simple mechanical and electrical systems are treated in a core chapter. Electromechanical, heat-conduction, fluid, and three-dimensional mechanical systems are also treated with care and rigor, for optional use.

Part B introduces the elementary motions—exponentials and sine waves—which comprise the total response of any linear system, however complicated. The mathematical simplicity attending first-order systems is exploited to introduce the concepts of natural motion, forced response, the characteristic equation, the pervasive role of function e^{st}, impulse response, and convolution. Then natural undamped vibrations and finally the behavior of damped, second-order systems are studied in detail. s-plane representation is featured. "Frequency-response" methods of analysis (including Bode's plotting methods) and nonlinear topics are also presented (optionally).

Part C is concerned with the natural characteristics of *compound* systems—their possible natural motions, dynamic stability, and modes of coupled oscillation. Since these characteristics are unrelated to external forcing or disturbing influences, they are studied first by themselves. The behavior of roots in the s plane is introduced naturally in Part B, and the root-locus method of Evans is developed readily in Part C and employed naturally thereafter. (It is treated more elaborately in Part E.)

Against this background, Part D considers the *forced* response of compound systems, with e^{st} the basic function. The transfer-function viewpoint is used first for continuing inputs, particularly sine waves, and then (with the Laplace transform) to obtain total response to abruptly applied inputs, via partial fraction expansion.

Part E provides the basis for a first short course in automatic control. It presents rigorously the concepts of control system dynamics—the philosophy and benefits of feedback, the precise description and measurement of system performance, and the algebra of feedback coupling—and develops capability for analysis and synthesis using root-locus methods.

A feature of the book is the point at which the Laplace transform is introduced. The Laplace method is without peer for handling the response of compound linear systems to abruptly applied inputs and to initial conditions, but it is unnecessarily cumbersome for studying natural-motion characteristics and stability. A common practice is to present the Laplace transform procedure at the beginning of the course, and then to treat a variety of problems by adapting them to the Laplace mold and "turning the crank." This author believes that such a procedure all too often stifles seriously the development of physical judgment. Accordingly, in this book the method is not introduced until the middle of Part D, where it is really useful in finding total system response, and at which point the student should appreciate its implications more fully. A comprehensive 25-page table of Laplace transform pairs is included, as Appendix J. (Readers already familiar with the Laplace transform method can use it readily throughout the book.)

The spectrum of
COURSE SYLLABI
*for which this book is designed**

	1	2	3	4	5
Typical course title	System dynamics	System dynamics	Mechanical vibrations	Dynamics and control	Reference for Flight Control and Guidance
Level	Soph.	Jr.-Sr.	Jr.-Sr.	Sr.-Grad.	
Prerequisites	Physics, Calculus	Introductory EE, Mechanics (dynamics)	Mechanics (dynamics)	Networks or vibrations	
Course length	1 sem. / 2 sem.	1 sem. / 2 sem.	1 sem.	1 sem.	
Part A	1, 2 / 1, 2, 3, 4	1, (2)† / 1, (2), 3, 4, 5	1, (2)	1, (2)	(5)
Part B	6, 7, 8 / 6, 7, 8, 9	6, (7), (8) / 6, (7), (8), 9, 10	(6), 7, 8, 9, 10		(11)
Part C	11 / 11, 12	11 / 11, 12, 13	11, 12, 13	11	
Part D	14 / 14, 15, 16, 17, 18, 19	14 / 14, 15, 16, 17, 18, 19	14, 15, 16	14, 16, 17, 18, 19	(17), (18), (19)
Part E		16, 17, 18, 19		20, 21, 22	(20), (21), (22)

Optional chapters‡
3, 4, 5
9, 10
12, 13
15

* Refer to chapter titles in Table of Contents.
† Parentheses imply use for rapid review.
‡ One may bypass these chapters without loss in continuity.

PREFACE   xi

The book culminates in a series of case studies based on interesting engineering experiences with dynamic systems in general (Chapter 19) and with automatic control systems in particular (Chapter 22). Included are the electromechanical drive (e.g., for an audio speaker or vibration tester), the hydraulic servo, the classical two-axis gyroscope, the remote-indicator servo, the aircraft autopilot, and a system for balancing sticks upside down.

For the instructor planning a specific university course, some suggested syllabi are offered in the following section of this preface.

USE IN SPECIFIC ENGINEERING COURSES

Among the universities there is considerable variation in the point in the curriculum at which system dynamics is introduced, early fostering of the system viewpoint being desirable, but solid mastry of physical principles being an essential prerequisite. Similarly, professional engineers take up the subject from a variety of backgrounds. Hence, for instructors laying out a course syllabus, and for career engineers mapping a course for self-study, the scope and level of subject matter required may vary considerably with the previous training and experience of the student. This book is designed to accommodate, in a fundamental and comprehensive way, the variation in level at which it may be used, and thus to facilitate the continuing evolution of curricula.

Providing such versatility has dictated a rather long book, but it is packaged with care, and the core of it is contained in eleven of the first nineteen chapters. (Please refer to Contents, pages xv–xx.) Eight other chapters, marked ⊙, are included to provide additional perspective or depth in special areas, as desired, and any of them may be omitted (or interchanged) without loss of continuity.

As noted earlier, the text pace is deliberately slower in Part B than in Parts C-E.

Some typical course syllabi are tabulated on page x. For example, at Stanford University this book is used in two courses. The first course (column 2) at the junior-senior level follows core courses in mechanics (dynamics) and electrical circuits, and provides the foundation for subsequent courses in automatic control, network theory, advanced systems theory, and advanced vibrations. (In the ME curriculum, this course replaces an earlier introductory vibrations course.) Chapters 1 and 2 on equations of motion (mechanical and electrical) are used for review, Chapters 6–8 on elementary motions are covered briskly, and Chapters 11 on systems stability and 14 through 19 on forced response of compound systems are studied deliberately, dwelling finally on the case studies of Chapter 19, wherein all of the techniques of modeling and system analysis are brought to bear, in concert, to predict the total behavior of interesting real-world systems.

For a course at the sophomore level (column 1), the instructor would proceed more slowly through Chapters 1 and 2, 6 through 8, 11 and 14, covering the later parts of the book (as above) in a second semester.

The book is arranged so that it may be used, as it has been, for the traditional introductory vibrations course (column 3). The first half of Chapter 2 and mechanical sections of Chapters 6 through 16 are used with special emphasis on natural modes of motion and eigenvectors, which are given (augmented by the s-plane viewpoint) in Chapters 12 for two-degree-of-freedom systems and 13 for many-degree-of-freedom and distributed-parameter systems. (The sections on music in Chapter 13 are included to give further insight, motivation and enjoyment, rather than technological import.) The traditional introductory vibration subject matter is thus covered thoroughly, and with the broader perspective and terminology of dynamic systems in general.

For a course at the senior-graduate level (column 4), following mechanical vibrations and/or network dynamics courses, Parts A and B are used for reference only, and the course begins with Chapter 11, proceeding in depth through general system response (Parts C and D), and ending with automatic control (Part E) as a special, very important application of system dynamics.

The second course at Stanford is a senior-graduate course in automatic control (column 4), wherein Chapters 11 and 17 through 19 are used for review and reference, and the main part of the course is based on Part E.

At the advanced graduate level, the author has found the material in this book a most useful supplement for courses in flight control and inertial guidance (column 5), where the class members come with widely differing undergraduate backgrounds. Chapter 5 provides a concise reference for deriving equations of motion of three-dimensional systems, Chapters 11 and 17–19 review the basis for determining and efficiently displaying system behavior, and Chapters 20–22 present primary tools and concepts for control-system synthesis. The advanced courses may thereby proceed more rapidly, in an orderly way, from a common base.

The optional chapters (last column) may be introduced as desired, or used by the student later. For example, where frequency response will be of interest, Chapter 9 for simple systems and 15 for compound systems would be included. Similarly, readers with a special interest in thermal or fluid systems may find Chapter 3 a suitable summary of the fundamental physical considerations in deriving approximate models for them, and will find examples and problems throughout the book featuring such systems.

For the instructor or reader who finds the network approach to mechanical problems a helpful bridge to his previous experience, Sec. 4.5 (with Table V) presents the method as a formal procedure. However, as pointed out there, formal analysis by analog is *not* recommended as a way of life. Rather, the student is urged always to contemplate systems in their own physical terms.

ACKNOWLEDGMENTS

First thanks are to four good friends who first taught me the fundamentals, the intrigue, and the sheer fun of dynamics: Professor J. P. Den Hartog, J. G. Baker, W. D. Mullins, Jr., and W. R. Evans.

Professors J. G. Truxal and A. E. Bryson reviewed the entire manuscript in detail and made comprehensive, penetrating, and encouraging comments. Several other colleagues generously reviewed all or part and made many valuable suggestions: Professors G. F. Franklin, W. W. Harman, R. E. Scott, M. J. Rabins, T. Kailath, B. Roth, D. D. Boyden, R. N. Clark, Dr. D. B. DeBra, W. R. Evans, and J. M. Slater.

The complete manuscript was typed by Cheryl Sampson, without whose great skill, care, amazing speed, and unfailing cheerfulness, the book would never have been finished. Preliminary sections were typed by a series of fine young ladies: Donna Villarinho, Pauline Harris, Pauline Eckman, and Katherine Mendez.

PREFACE

The talent and craftsmanship of artists Felix Cooper and David Strassman are abundantly evident. Technical proofreading by Barry Likeness and Brent Silver was important. Mr. Likeness also prepared the Laplace transform table and answers to the odd-numbered problems, a prodigious task.

No one who has not written a technical book can possibly understand the depth of gratitude with which I acknowledge the support of my wife and children. The years of encouragement and sacrifice have been unstinting and inspiring. This is their book. It is dedicated to them.

Robert H. Cannon, Jr.

Washington, D.C.
April 1, 1967

METHODS AND RELATIONSHIPS

Physical modeling, classes of approximations (Table I) 18
Dimensions and units (Table II) 22

A: EQUATIONS OF MOTION FOR PHYSICAL SYSTEMS

"Checklists," Procedures A
- A-e   Electrical 61, 75
- A-em   Electromechanical 82
- A-f   Fluid 104
- A-h   Heat conduction. 97
- A-m   Mechanical 35, 74, 151
- A-mL   Lagrange 168

Dynamic equilibrium 40, 152
    Vector differentiation, rotating frame 149, 714
Kirchhoff's laws. 64

B: DYNAMIC RESPONSE OF ELEMENTARY SYSTEMS

Basic form of response (Procedure B-1). 182
Convolution integral 221
Euler's equation (complex numbers) 237
First-order system 183
Second-order system
    Characteristics (s-plane) 248, 252, 387
    Natural-response coefficients 255, 259, 264
    Sketching natural response (Procedure B-3) 260
    Step response 286, 288
    Unit-impulse response 293
    Frequency response (Bode plot). 336, 353
State-space analysis. 360

C: NATURAL BEHAVIOR OF COMPOUND SYSTEMS

Characteristics and stability
    Root-locus method (Procedure C-2) 401, 646
    Root-locus sketching rules (Procedures C-3, E-2). . . . 405, 652
    Routh's method (Procedure C-4) 410
    Rayleigh's method (Procedure C-5) 483
Modes of natural motion (eigenvectors) 428, 460
    Distributed-parameter systems 469, 471

D: TOTAL RESPONSE OF COMPOUND SYSTEMS

Bode plotting. 513
Fourier series (complex). 533
Laplace transform (one-sided) 552
Laplace transform method (Procedure D-1) 537
    Short table of $\mathcal{L}$-transform pairs (Table X) 540
    Long table of $\mathcal{L}$-transform pairs (Appendix J). 731
Partial fraction expansion (Procedure D-2) 595
Total investigation, Procedure D-3 (contains Procedure D-2) . . . 599

E: FUNDAMENTALS OF CONTROL-SYSTEM ANALYSIS

Basic algebraic relations 639
Root-locus method (Procedure C-2) (basic construction) . . . 401, 646
Root-locus sketching (Procedures C-3, E-2) 405, 652

CONTENTS

1 DYNAMIC INVESTIGATION — 1

    1.1   The scope of dynamic investigation — 3
    1.2   The stages of a dynamic investigation — 4
    1.3   The block diagram: a conceptual tool — 8
    1.4   Stage I. Physical modeling: from actual system to physical model — 10
    1.5   Dimensions and units — 19

PART A: EQUATIONS OF MOTION FOR PHYSICAL SYSTEMS

2 EQUATIONS OF MOTION FOR SIMPLE PHYSICAL SYSTEMS: MECHANICAL, ELECTRICAL, AND ELECTROMECHANICAL — 31

    2.1   Stage II. Equations of motion: from physical model to mathematical model — 32
    2.2   One-dimensional mechanical systems — 34
    2.3   Mechanical energy and power — 53
    2.4   Gear trains and levers — 54
    2.5   Motion in two and three dimensions — 57
    2.6   Simple electrical systems — 57
    2.7   A recapitulation of Procedure A — 74
    2.8   Amplifiers and transformers — 75
    2.9   Simple electromechanical systems — 79
   ⊙2.10 Electromechanical elements: an empirical sampling — 85

⊙3 EQUATIONS OF MOTION FOR SIMPLE HEAT-CONDUCTION AND FLUID SYSTEMS — 94

    3.1   Simple heat conduction — 94
    3.2   Simple fluid systems — 102

⊙4 ANALOGIES — 121

    4.1   Analogies between physical media — 121
    4.2   The electrical analog of mechanical systems — 125

⊙ Optional chapter or section, may be omitted without loss in continuity.

4.3	Classification of dynamic system elements	132
4.4	The benefits and limitations of analysis by analog	133
4.5	The network approach to analysis	137

⊙5 EQUATIONS OF MOTION FOR MECHANICAL SYSTEMS IN TWO AND THREE DIMENSIONS 143

5.1	Geometry of motion in two and three dimensions	143
5.2	Rotating reference frames	149
5.3	Dynamic equilibrium for rigid body in general motion	152
5.4	Equations of motion for systems of rigid bodies: examples	156
5.5	Advantages of the D'Alembert method. The gyro	159
5.6	Energy methods	163
5.7	Lagrange's method	167
5.8	Lagrange's method for conservative systems	169
5.9	Lagrange's method for nonconservative systems	173
5.10	The relative advantages of Lagrange's method	178

PART B: DYNAMIC RESPONSE OF ELEMENTARY SYSTEMS

Introduction 180

6 FIRST-ORDER SYSTEMS 182

6.1	First-order systems	183
6.2	Natural (unforced) motion	186
6.3	Forced motion	193
6.4	Linearity and superposition	197
6.5	Initial conditions	200
⊙6.6	Special case: the pure integrator	204
⊙6.7	Special case: resonance	207
6.8	Response to a very short impulse	210
⊙6.9	Initial conditions involving sudden change	218
⊙6.10	Generalization to an arbitrary input: convolution	220

7 UNDAMPED SECOND-ORDER SYSTEMS: FREE VIBRATIONS 225

7.1	Physical vibrations	226
7.2	Complex numbers	236
7.3	Mathematical operations with complex numbers	238
7.4	Complex-vector (phasor) representation of a sine wave	242

8 DAMPED SECOND-ORDER SYSTEMS 244

8.1	Second-order systems	245
8.2	Natural motion	246
8.3	Dynamic characteristics and the s plane	248
8.4	Initial conditions: $1 < \zeta$	255
8.5	Initial conditions: $\zeta < 1$	257
⊙8.6	Sketching time response when $\zeta < 1$	259
8.7	Initial conditions: $\zeta = 1$	263

8.8	Forced motion alone	266
8.9	Transfer functions and pole-zero diagrams	273
⊙8.10	Impedance and admittance	280
8.11	Total response to abrupt disturbances	282
⊙8.12	Transient and steady state	290
⊙8.13	Impulse response of certain second-order systems	291
⊙8.14	Response of a second-order system by convolution	298
⊙8.15	Simulation: the analog computer	299

⊙9 FORCED OSCILLATIONS OF ELEMENTARY SYSTEMS 309

9.1	The nature of sinusoidal response	311
⊙9.2	Operation at a single frequency: the impedance viewpoint	313
9.3	Frequency response	314
9.4	Computing frequency response	321
9.5	Logarithmic scales for plotting frequency response	322
9.6	Forced oscillation of first-order systems: the plotting techniques of Bode	324
9.7	Forced oscillation of undamped second-order systems: resonance	330
9.8	Forced oscillation of damped second-order systems	335
⊙9.9	Techniques for plotting second-order frequency response	337
9.10	Obtaining frequency response from the s plane	342
9.11	Seismic instruments	345
9.12	Frequency response of RLC circuits	352

⊙10 NATURAL MOTIONS OF NONLINEAR SYSTEMS AND TIME-VARYING SYSTEMS 355

10.1	Methods of linear approximation	356
10.2	State-space analysis	360
10.3	Large motions of pendulum with damping	367
10.4	Piecewise-linear elements	369
10.5	Linear equations with time-varying coefficients	370

PART C: NATURAL BEHAVIOR OF COMPOUND SYSTEMS

Introduction 374

11 DYNAMIC STABILITY 376

11.1	The concept of stability	377
11.2	The elementary second-order system	379
11.3	Locus of roots by graphical construction	384
11.4	Damping as a variable	386
11.5	Coupled pairs of first-order systems	392
11.6	Feedback systems	393
11.7	Third-order systems	396
11.8	The root-locus method of Evans	401
⊙11.9	An introduction to root-locus sketching	404
11.10	The method of Routh: third-order example	406
11.11	Routh's method: general case	409

	11.12	Special case of a zero term in the first column	415
	11.13	Special case of a zero row	416

⊙12 COUPLED MODES OF NATURAL MOTION: TWO DEGREES OF FREEDOM — 419

	12.1	Forms of physical coupling	420
	12.2	Coupled equations of motion	423
	12.3	A simple vibrating system of coupled members	426
	12.4	The effect of coupling strength	431
	⊙12.5	Beat generation	437
	12.6	Inertial coupling	442
	12.7	Normal coordinates	445
	12.8	General case, with damping: eigenvalues and eigenvectors	447

⊙13 COUPLED MODES OF NATURAL MOTION: MANY DEGREES OF FREEDOM — 457

	13.1	Many degrees of freedom	458
	⊙13.2	Distributed-parameter systems: equations of motion	462
	⊙13.3	Natural motions of a class of one-dimensional distributed-parameter systems	467
	⊙13.4	Brief description of general distributed-parameter systems	472
	⊙13.5	Certain musical instruments	475
	⊙13.6	A note on the musical scale	480
	13.7	Rayleigh's method	482

PART D: TOTAL RESPONSE OF COMPOUND SYSTEMS

Introduction — 490

14 e^{st} AND TRANSFER FUNCTIONS — 492

	14.1	Review of the transfer-function concept	493
	14.2	Transferred response of subsystems in cascade	495
	14.3	Graphical evaluation of transfer functions from the system pole-zero diagram	497
	14.4	Transferred response of systems with general coupling	500
	14.5	Matrix representation and standard form	502
	14.6	Matrix description of eigenvector calculation	509

⊙15 FORCED OSCILLATIONS OF COMPOUND SYSTEMS — 510

	15.1	Frequency response of subsystems in cascade	511
	15.2	Frequency response of systems with two-way coupling	514
	15.3	Resonance in coupled systems	518
	15.4	Design for a single frequency: the vibration absorber	525

16 RESPONSE TO PERIODIC FUNCTIONS: FOURIER ANALYSIS — 529

	16.1	Real Fourier series	529
	16.2	Complex Fourier series	533
	16.3	Spectral representation	534

CONTENTS xix

17 THE LAPLACE TRANSFORM METHOD 535

17.1 Demonstration of the Laplace transform method 537
⊙17.2 Evolution of the Laplace transform 544
17.3 Application of the Laplace transform: the one-sided Laplace transform 549
17.4 Summary of basic Laplace transform relations 554
17.5 Derivation of common Laplace transform pairs 555
17.6 Transfer functions from Laplace transformation 559
17.7 Total response by the Laplace transform method 562
17.8 The final-value theorem and the initial-value theorem 567
17.9 A system's response to an impulse and its $\mathcal{L}$ transform 572
17.10 Initial conditions and impulse response: a physical interpretation 572
17.11 Equations of motion in standard form: state variables 576
17.12 Convolution and the Laplace transform 580

18 FROM LAPLACE TRANSFORM TO TIME RESPONSE BY PARTIAL FRACTION EXPANSION 584

18.1 Formulation of the task 584
18.2 Partial fraction expansion: case one 585
18.3 Special handling of complex conjugate poles 588
18.4 Use of the s-plane pole-zero array to compute response coefficients 589
18.5 The case of repeated poles: case two 592
18.6 Summary of Procedure D-2: partial fraction expansion 595

19 COMPLETE SYSTEM ANALYSIS: SOME CASE STUDIES 597

19.1 A fluid clutch 600
19.2 An electromechanical shaker 605
19.3 A thermal quenching operation 610
19.4 An aircraft hydraulic servo 614
19.5 The two-axis gyroscope 617

PART E: FUNDAMENTALS OF CONTROL-SYSTEM ANALYSIS

Introduction 628

20 FEEDBACK CONTROL 630

20.1 The philosophy of feedback control 631
20.2 Performance objectives 635
20.3 The sequence of control-system analysis 637
20.4 Review of dynamic coupling 638
20.5 The algebra of loop closing 639

21 EVANS' ROOT-LOCUS METHOD 644

21.1 The basic principle 645
21.2 Root-locus sketching procedure 651
21.3 Sketching rules for 180° loci 653

21.4	Rule 1: Real-axis segments	655
21.5	Rule 2: Asymptotes	657
21.6	Rule 3: Directions of departure and arrival	660
21.7	Rule 4: Breakaway from the real axis	664
⊙21.8	Rule 4 (continued): The saddle-point concept	669
21.9	Rule 5: Routh and Evans	673
21.10	Rule 6: Fixed centroid	677
21.11	A typical construction of root loci	678
21.12	Summary	682

22 SOME CASE STUDIES IN AUTOMATIC CONTROL — 683

22.1	Analysis of an electromechanical remote-indicator servo	683
22.2	Synthesis of indicator servo using network compensation to improve performance	691
22.3	Roll-control autopilot: a multiloop system	696
22.4	Control of an unstable mechanical system: the stick balancer	703

APPENDICES

A	Physical conversion factors to eight significant figures	711
B	Vector dot product and cross product	712
C	Vector differentiation in a rotating reference frame	714
D	Newton's laws of motion	716
E	Angular momentum and its rate of change for a rigid body; moments of inertia	718
F	Fluid friction for flow through long tubes and pipes	721
G	Duals of electrical networks	723
H	Determinants and Cramer's rule	725
I	Computation with a Spirule	727
J	Table of Laplace transform pairs	731

Problems	757
Answers to odd-numbered problems	865
Selected references	881
Index	885
About the Author	905

DYNAMICS OF
PHYSICAL SYSTEMS

Chapter 1

DYNAMIC INVESTIGATION

It is, at last, the day and the hour for launch. The forty-story gantry yawns slowly open and rolls back, leaving its sometime tenant to stand alone, tall, white, cloaked now only in a veil of liquid-oxygen vapor.

Minutes hence the huge vehicle will thunder skyward on a white-hot cone of fire, and in an hour its forebody will be pursuing an unerring path to a distant planet. But at this moment it stands silent in the spotlight of a hundred arc lamps, mixed with the early rays of dawn, while a thousand test signals are checked and rechecked by the men and machines who attend it.

The loudspeaker intones: "All systems are go." It is a commonplace phrase by now, but its connotation is almost unbelievable! Twenty months ago this vehicle—this collection of hundreds of intricate systems and subsystems—was no more than a concept, a bundle of specifications. Today every system will work, and in perfectly matched, intimate harmony with all the others.

When the start button is pushed, high-capacity pumps will pour fuel into the ravenous rocket chambers. The enormous fuel tanks will be emptied in one minute flat; but flow will be precisely uniform, and the last drop will be delivered exactly on schedule.

Quick, sure hydraulic pistons will swivel and aim the huge rocket engines, as the automatic pilot solves the problem of balancing a long, limber reed on one end.

Precise inertial instruments—each an intricate dynamic system in its own right—will sense the vehicle's path; and a miniature, high-capacity computer will solve the trajectory equations and generate continuous path-correction signals. Two dozen radio and television channels will handle instantly the voluminous communication between the vehicle and its ground base, and a world-wide radar tracking network will monitor its path through the sky.

In the vehicle's nose a myriad of delicate instrument systems are ready to measure, to observe, to photograph, to record and report their findings. Additional systems provide the essential environment for these instruments. A sophisticated air-conditioning system controls temperature and humidity. Carefully designed shock-mounting systems will isolate the delicate instruments from the tremendous vibration of the rocket engines. If there are human passengers, an elaborate complex of life-support systems will operate as well.

Few of the hundreds of systems in this vehicle existed until this decade. Some are based on technology that did not exist four years ago. Many will be making their first flight today. All have been developed specifically for this vehicle, and matched meticulously to one another.

How could so many interrelated systems be developed so rapidly at the same time? How can we be sure each will respond and perform as it must through the abruptly changing sequence of thunderous launch, searing air, and cold, empty space?

There are several underlying answers to these questions. The first is that we have come to understand some of nature's laws, and to know that they are perfectly dependable. This we call *science*.

Another answer is that we have learned how to use nature's laws to build systems of our own to perform tasks we wish done. This we call *engineering*.

A third answer is part of the second: We have learned to predict the dynamic behavior of systems not yet built. This is called *dynamic analysis*. When used in careful support of design and testing programs, dynamic analysis is the key to telescoping the development time of new systems, the key to confidence that they will work properly together in a strange new environment.

A fourth answer is in turn part of the third: Nature is orderly and systematic, and the dynamic behavior of large, intricate compound systems is found to be made up of elementary behavior patterns which can be discerned and studied one by one. The process of discernment is sometimes an involved one; but it is straightforward and it can be accomplished by repeated (and astute) application of relatively elementary analytical techniques; and thereon hangs our ability to contemplate very involved systems of systems, and to predict their behavior with confidence.

Thus, in the development of a space-vehicle system—or of a television network, a power complex, or any other compound dynamic system—a program of *dynamic investigation* is the central cord that threads together the myriad of physical systems at their inception. It writes the script from which specifications are established, preliminary designs are evolved, "breadboard" models are constructed, early tests are conceived, performed, and analyzed, and final design decisions are made. This is the script against which the final performance of each system will be measured, first alone and then in concert with its teammates. This is the script by which several hundred complex systems have been developed in months, have been tested and integrated together and tested again and again. And now this system of systems is ready to be launched.

This process of dynamic investigation—its fundamental concepts, its basic building blocks, and its applications—is the subject of this book.

1.1 THE SCOPE OF DYNAMIC INVESTIGATION

Dynamics is the study of how things change with time, and of the forces that cause them to do so. It is an intriguing discipline. Moreover, in this transilient era of space travel, instant communication, and pervasive automation, it is a pivotal discipline: Analysis of the dynamic behavior of physical systems has become a keystone to modern technology.

The motion of a space vehicle, for example, must be thoroughly understood, and its precision control correctly provided for early in the design of the vehicle, many months before it is actually launched. The design of an atomic power plant is predicated in part on the *predicted dynamic response* of the plant to sudden changes in load. The circuit design and component selection for a high-fidelity radio receiver system are based on calculations of how the electronic section will combine with the speakers, in their enclosures, to produce *dynamic response* that will match to the desired degree the original sound at the broadcasting station.

In these engineering problems and countless others the first requirement is to predict, before construction, the dynamic behavior a physical system will have—its natural motions when disturbed, and its response to commands and stimuli. More, perhaps, than any other field, the study of dynamic behavior links the engineering disciplines.

In a larger sense, the field of dynamics extends well beyond the realm of physical phenomena. In the field of biology, the response of the eye pupil to a sudden change in light intensity, or of the hand to motor commands from the brain, and the transient adjustment of the body's energy balance to the trauma of major surgery are exciting subjects for dynamic analysis. In economics, the response of a banking system to fluctuations in market activity, of an industry to variations in consumer demand, and, more broadly, the dynamic behavior of the entire economy, have become the subject of dynamic analysis of increasing penetration and importance. Even the phenomena of "group dynamics"—the collective dynamic behavior of teams of individuals having specific tasks to perform—are being studied quantitatively with useful results.

The twofold objective of this book is to develop familiarity with the elementary *concepts* of dynamic behavior, and to develop proficiency with the *techniques* of linear dynamic analysis.

Consideration is confined to physical systems in the interest of efficient exposition, because physical phenomena are more familiar and because their behavior is more easily analyzed. Once a degree of proficiency and insight has been attained for one medium, the extension to others will be found to follow by analogy (and to be most intriguing).

Moreover, the similarity in dynamic behavior of different physical systems is accompanied by a striking consistency in the *pattern of analytical investigation* by which that behavior can most effectively be studied. There are certain broad stages through which an investigation nearly always proceeds; at each stage a small number of basic concepts and analytical methods will be found to be the keys to efficient, successful prediction of dynamic behavior, whatever the physical medium or the particular arrangement.

This "universality" of the concepts and methods of dynamic analysis is a remarkable thing; and it is, of course, a very fortunate thing for people who enjoy moving to new fields of study, and for people responsible for the development of large systems of systems in which numerous physical media interact dynamically. In particular, it will add substantial breadth to our study, in this book, of the process of dynamic investigation.

We begin our study with an overview of the process.

1.2 THE STAGES OF A DYNAMIC INVESTIGATION

The objective of a dynamic investigation is to understand and predict the dynamic behavior of a given system and sometimes to improve upon it.

The detailed tasks in a program of dynamic investigation depend, of course, on the physical media involved, the size and complexity of the system, the stringency with which it must perform, and so on. But whatever the particular physical system under study, the procedure for analytical investigation usually incorporates each of the following four stages:

> I. Specify the system to be studied and imagine a simple PHYSICAL MODEL whose behavior will match sufficiently closely the behavior of the actual system.
> II. Derive a mathematical model to represent the physical model; that is, write the differential EQUATIONS OF MOTION† of the physical model.
> III. Study the DYNAMIC BEHAVIOR of the mathematical model, by solving the differential equations of motion.
> IV. Make DESIGN DECISIONS; i.e., choose the physical parameters of the system, and/or augment the system, so that it will behave as desired.

The stages of a dynamic investigation

---

† Throughout this book the word "motion" will be used in a general context to connote the change of any physical variable—mechanical displacement and velocity, electrical voltage and current, pressure, temperature, etc.

In Stages I, II, and III the emphasis is on analysis, while in Stage IV it is on synthesis.

To illustrate the stages let us single out one of the hundreds of dynamic systems that make up a rocket vehicle, and focus attention on the dynamic investigation that has accompanied its development during the months preceding launch day. Consider, for example, the rather straightforward but vital problem of designing a suitable mounting for a "payload" of delicate scientific instruments. Figure 1.1 shows a panel containing such a payload.

The vibration of the vehicle is so severe during rocket firing that if the panel were mounted rigidly to the vehicle structure the instruments would fail to function properly. A simple cushioning arrangement to protect the instruments from this vibration must be designed. Rubber cups, such as those depicted in Fig. 1.1a, are considered by the dynamic analyst as candidates to furnish the necessary cushioning effect. His next task is to specify the size and stiffness of the cups to achieve the required degree of isolation in the face of expected vehicle vibration.

In *Stage I* the analyst contrives an imaginary model of the system he wishes to study—a model essentially like the real system, but much easier to analyze mathematically. For the real system of Fig. 1.1a he imagines (and sketches) a simplified model such as that shown in Fig. 1.1b, where the vehicle structure has been replaced by a completely rigid member, the instrument panel by another, and the rubber cups by linear springs plus viscous dashpots. In this physical model, distortions of the vehicle frame and motions of the instrument parts have been ignored, and the rather nonlinear rubber cups have been replaced by the simpler linear spring-and-dashpot model. In his studies the analyst decides to consider only motions in one direction at a time (e.g., vertical). The symmetry of the mount indicates that this is reasonable, and one-dimensional motion is much easier to study.

To complete the model, the analyst inspects the boundaries of his chosen system, Fig. 1.1a, and judges that the motions of the vehicle structure segments will be affected negligibly by motions of the instrument panel. Accordingly in the model, Fig. 1.1b, he considers x_1 a *prescribed input* to the system.

Next, in *Stage II*, the analyst applies the appropriate physical laws—in this case Newton's laws of motion—to obtain a set of "dynamic equations of motion" which, for this model, will be a set of ordinary differential equations in the variables x_1 and x_2.

In *Stage III*, the equations of motion are solved to obtain explicit expressions for the time variation of x_2 as a function of input motion x_1. Further, expressions are obtained for the actual excursions and accelerations which would be produced by the "worst inputs," $x_1(t)$, that the rocket vehicle is expected to impose on the panel. A major task is to estimate accurately the maximum magnitudes of inputs of various types—sudden shock, steady vibration, and so on.

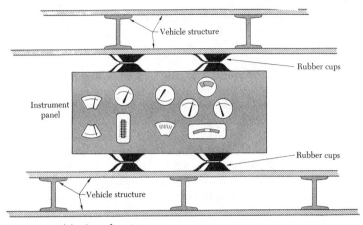

(a) *Actual system*

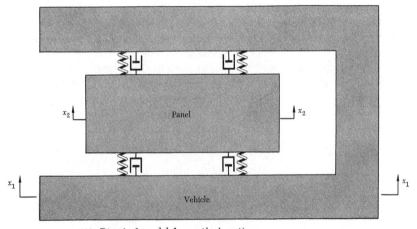

(b) *Physical model for vertical motion*

Fig. 1.1 *A vibration isolation system and its model*

Finally, in *Stage IV*, the calculated behavior from Stage III is studied to determine what combinations of spring and dashpot characteristics will do the job—will insure that the resulting panel motions are small enough not to impair instrument performance.

When the final stage has been completed, and a preliminary design established, the entire analysis will be repeated with some of the simplifications omitted, to get a more precise estimate of the performance to be expected.

The analyst will also arrange to have a hardware model of the panel constructed, using actual rubber shock mounts, to give him added experience

[Sec. 1.2] THE STAGES OF A DYNAMIC INVESTIGATION 7

by allowing him to check his imagined model against physical reality. Eventually an actual instrument panel will be vibration-tested to make sure the design is acceptable.

Clearly, without an astute initial analysis, costly experimental time could be wasted in a trial-and-error approach to solving even this rather elementary problem in dynamic design.

The four stages of dynamic investigation are depicted in a more general way in Fig. 1.2. Note particularly the "feedback" feature in the procedure: that errors in the analysis may be detected by comparison with actual test behavior, and corrections then made ("fed back") to the original physical model to make it more realistic. The application of design specifications to the *actual* system is also in the nature of a "feedback" process.

In the five parts of this book we shall describe in greater detail, and with numerous examples, the analytical procedures involved in each of the above stages. In Part A we shall study the process of deriving the equations of motion of physical systems, first imagining a physical model, and then applying physical laws to the model to obtain the governing equations (Stages I and II). We shall study simple mechanical, electrical, electromechanical, fluid, and thermal systems, and will note the detailed analogy between their mathematical descriptions.

In Parts B, C, and D of this book we shall be concerned with determin-

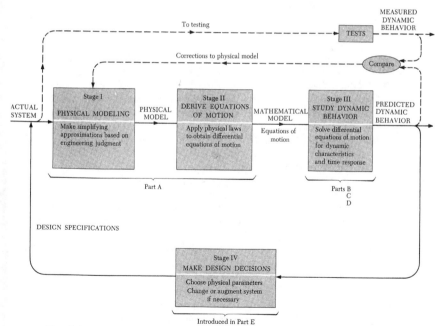

Fig. 1.2 *The stages of a dynamic investigation*

ing the dynamic behavior of systems (Stage III), once their equations of motion have been derived.

Finally, in Part E we shall introduce the synthesis of systems (Stage IV) and, in particular, of automatic-control systems.

1.3 THE BLOCK DIAGRAM: A CONCEPTUAL TOOL

Block diagrams, for portraying a sequence of events or interrelationships, have important technical, as well as organizational, applications. In working with the physical model (Stage I) or the mathematical model (Stage II) of a complicated system, it is sometimes convenient to indicate the interrelation between subsystems in a block diagram. This is especially easy when the interaction is in one direction only.

For example, the shock-mount in Fig. 1.1 was considered to have no effect on the vehicle. The whole shock-mount "subsystem" might be represented by a single block in a block diagram, Fig. 1.3, that indicates the propagation

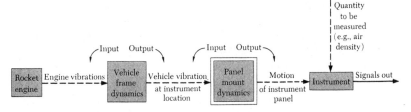

Fig. 1.3 *A block diagram to represent the transmission of engine vibration*

of vibration from the rocket engine through the vehicle frame to the instrument mount, and thence to the separate instruments where it contaminates their output signals. Such a diagram dramatizes the sequence of cause and effect.

Other systems involve more elaborate interactions, like the "automatic pilot" system in Fig. 1.4, which steers a rocket vehicle in response to radioed commands from the ground. The actual system is indicated in Fig. 1.4a. The direction of vehicle flight is to be controlled by rotating the rocket engine so that a lateral component of its thrust will then rotate the vehicle to point in the desired direction. (The system shown controls vehicle direction about only one axis of the vehicle; similar systems would control the other axes.)

The interconnection of the various parts of the autopilot is indicated in the block diagram of Fig. 1.4b. In the following description of the system's operation, we shall see that the sequence of cause and effect is easier to follow in Fig. 1.4b than in Fig. 1.4a.

The autopilot computer receives a heading-command signal from the radio receiver. The computer is kept informed about the present state of motion of the vehicle, and about the position of the engine, by means of signals from gyros, accelerometers, and other sensors of vehicle motion, plus a simple angle "pickoff" on the engine gimbal. The computer uses all of this information in deciding what action to take next to carry out the radio command.

If the computer decides to move the engine it sends a signal to the hydraulic valve opener, an electromagnetic device that opens the valve and ports oil into one end or the other of the big hydraulic cylinder. The flow of oil forces the piston to move, and this moves the engine. (The oil supply is not shown.) Using the gimbal pickoff, the computer makes sure the engine has moved the amount it was told to, and then sends a signal to close the valve.

With the engine in its new position the engine-thrust line of action misses the center of the vehicle (by an amount proportional to angle ϵ) and thus produces a change in moment which rotates the vehicle. The computer keeps track of vehicle rotation via gyroscopic instruments, and continually adjusts the engine (via the hydraulic system) until the vehicle flight path matches that commanded.

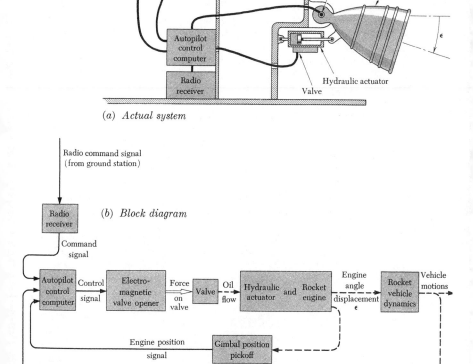

Fig. 1.4 *Block diagram representation of the interrelation between the subsystems of an automatic pilot for a rocket vehicle*

The entire system shown in Fig. 1.4 is, of course, only part of a larger and more complicated system. The path of the vehicle through the sky is monitored by ground radar installations and is studied by a large complex of computers and people who make changes in the radio command signals. The rocket engine is fed by an intricate fuel system, and its thrust is controlled by another part of the autopilot. Each of the components in Fig. 1.4a is furnished electric power from an elaborate power system, and so on.

Moreover, when we contemplate a dynamic analysis of just the system shown in Fig. 1.4 we realize that we have many physical models to conceive and many equations of motion to derive and solve—Newtonian equations for motion of the vehicle itself, network equations for the radio and the computer, fluid-dynamic equations for the hydraulic and fuel systems, an electromagnetic equation for the valve opener, mechanical equations for each gyro (a very sophisticated dynamic system in itself), and so on.

The block diagram of Fig. 1.4b helps us to *partition* the analysis task into subsystems which can be readily studied one at a time. We carefully choose these partitions, or blocks, so that the *signals crossing their boundaries* are easy to identify and to measure.† Then each subsystem—the radio and computer networks, the hydraulic system, the vehicle dynamics, etc.—can be studied by itself until it is understood and is well described by a mathematical model.

Eventually, of course, the system is to be studied *in toto*. Again, we shall find that the block diagram is a most helpful tool. In each block we can write the *mathematical model* we have derived for the *dynamic behavior* of the physical subsystem represented; and then the block diagram will help indicate how these mathematical models are to be combined to predict behavior of the whole system.

In this book we shall approach the study of complex dynamic systems as we would the study of the blocks in a diagram—first one block at a time, and then joined into systems. In Part B we shall study the dynamic behavior of individual blocks ("building blocks") from typical systems—a single rotor with damping, individual "RC" and "RLC" networks, a single lumped-mass thermal capacitance, etc. Then in Parts C, D, and E we shall study systems in which a number of such "building blocks" are interconnected; and we shall learn how to predict the behavior of the entire system. The behavior of systems using feedback, such as the autopilot of Fig. 1.4, will be the special concern of Part E.

Problems 1.1 through 1.4

1.4 STAGE I. PHYSICAL MODELING: FROM ACTUAL SYSTEM TO PHYSICAL MODEL

By a "physical model" we mean an imaginary physical system which resembles an actual system *in its salient features* but which is simpler

---

† The same philosophy has become the natural basis for partitioning engineering organizations. (Some interesting working language has developed. For example, the word *interface* emerged first as a noun and more recently as a verb, as, "Have you interfaced with the telemetry group?")

(more "ideal"), and is thereby more amenable to analytical studies. Invariably the top-notch engineer will be distinguished by his ability to imagine simplified physical models of actual systems which prove to be sound, and which permit him to predict behavior accurately and swiftly and thence to make valid decisions.

Making approximations

As we mentally construct a physical model for a system we presume numerous approximations; and to be unequivocal we make a sketch of our imagined system, making sure that all approximations are precisely noted, and that conditions at boundaries of our model properly represent the interaction between the real system and its environment.

In preparing to study the vibration mount in Fig. 1.1a, for example, the analyst defines as his "system" the instrument panel plus short segments of the vehicle structure to which it is mounted. Then he begins to sketch a physical model, in Fig. 1.1b. He assumes the aircraft structure to be rigid in the vicinity of the mounting. He knows that in fact the structure is not quite rigid, but he perceives intuitively that slight nonrigidity in the structure will not appreciably affect the motion of the instrument panel, and since the problem is so much easier with a rigid structure he makes that approximation. He makes a number of other approximations before beginning analysis of the system, and indicates them in the sketch of Fig. 1.1b.

This is not to say that a system will ever be designed and built on the basis of the first set of simplifying assumptions, which will be (and should be) quite gross. On the contrary, a number of refined analyses will follow in which key assumptions will be checked for their validity. But initial comparisons can be made and gross conclusions can be reached much more rapidly, and at a great saving in expense, if the engineer knows intuitively which assumptions will lead to major mathematical simplification and at the same time will appropriately represent the true system. Indeed, it may be just as poor engineering practice to adopt a physical model which is more complicated than necessary as it is to adopt one which is an oversimplification, because of the waste of time and energy that will attend elaborate analysis of unnecessarily complicated models.

The astuteness with which approximations are made at the outset of an investigation is, in fact, the very crux of engineering analysis. The ability to make shrewd and viable approximations which greatly simplify the system and still lead to a rapid, reasonably accurate prediction of its behavior is the hallmark of every successful engineer. This ability involves a special form of carefully developed intuition known as "engineering judgment."

Engineering judgment cannot really be learned with pencil and paper alone, although thorough analytical experience is an essential requisite. The engineer is well advised to develop at every opportunity his ability to make astute engineering approximations by working with actual physical systems in a test environment as much as possible, and as early as possible. Meanwhile, we can develop here an appreciation for which kinds of assumptions produce important analytical simplification.

The following kinds of engineering approximations are useful and appropriate in varying degrees in most problems:

(a) Neglecting small effects
(b) Assuming that the environment surrounding the system is unaltered by the system
(c) Replacing "distributed" characteristics by similar "lumped" characteristics
(d) Assuming simple linear cause-and-effect relationships between physical variables
(e) Assuming that physical parameters do not change with time
(f) Neglecting uncertainty and noise

Let us see how each approximation simplifies and speeds up the subsequent mathematical analysis.

(a) Neglecting small effects

Neglecting small effects in the physical model simplifies the subsequent mathematics principally by reducing the number of variables, and hence the number and complexity of the equations of motion. Normally the complexity of an analysis balloons in geometric proportion to the number of different "moving parts" it has. For this reason, astute judgment regarding which effects to neglect in a given analysis contributes greatly to the usefulness of the analysis.

For example, the network of Fig. 1.5 is a model of an actual circuit built by connecting two inductors, a capacitor, and two resistors to a source of current as shown. In the model these elements are assumed to be *purely* inductive, capacitive, and resistive, respectively. That is, the small inductance possessed by a real resistor has been neglected in the model, etc. Obviously the analysis of this network of a few elements will be much simpler than for a more exact model having more elements.

In any problem a myriad of "small effects" will be neglected practically

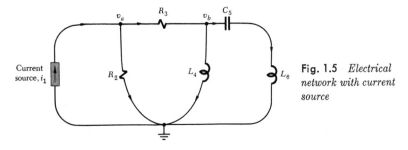

Fig. 1.5 *Electrical network with current source*

[Sec. 1.4] STAGE I: FROM ACTUAL SYSTEM TO PHYSICAL MODEL

without thought. For example, in analyzing the motion of an airplane we are unlikely to consider the fact that sunlight exerts a force on its wings, or that the earth's magnetic field can exert a torque on it if current is running through its electrical system, or that there will be a gravitational torque when it is banked because one wing is a few feet closer to the earth than the other. All of these forces are entirely minute, compared to the enormous aerodynamic forces acting all over the airplane and the major gravitational force acting through its center of gravity.

It is to be emphasized, however, that "small effects" are neglected on a *relative* basis. Thus the forces mentioned above—solar pressure, earth's magnetic field, and gravity gradient—which were so tiny as not even to come to mind in the airplane example, are the dominating influences on rotation of a space vehicle! Gravity gradient may be used to cause the vehicle always to point toward the earth like a pendulum, and either solar pressure or the earth's magnetic field may overturn the vehicle if care is not exercised. To ignore these effects in the space-vehicle problem would lead to grossly incorrect results; but to include them in the airplane problem would lead to needless complexity which could make the analysis hopelessly involved and impair its usefulness.

Engineering judgment is of the essence.

(b) Independent environment

Another common assumption is that the environment surrounding a physical system is unaffected by the behavior of the system. Referring again to Fig. 1.1, we assumed that the given motion x_1 of the vehicle frame was independent of the response motion x_2 of the instrument panel. This seems to be a valid assumption if the vehicle is very rigid, and if its mass is large compared to the panel's. However, it is an assumption one would want to check at the end of the first analysis to be sure that it is in fact valid.

In electrical problems it is common to assume that current or voltage is supplied to a section of a system from a "source" whose output will be unaffected by behavior "downstream" in the network, as in Fig. 1.5 where the "current source" is actually another network whose elements are such that i_1 is unaffected by the behavior of voltages v_a and v_b "downstream" from i_1.

When the input to a system *is* affected by the response of the system, the situation becomes much more complex. Thus, the assumption of an independent environment simplifies the analysis in the same way as the neglecting of small physical effects: by making the resulting mathematical expressions less complicated.

(c) Lumped characteristics

The approximation of distributed physical characteristics by a lumped-element model affords analytical simplification in a more fundamental way. It leads not just to less complexity of the differential equations of motion but, indeed, to equations that are amenable to a much simpler *method* of solution. Specifically, distributed physical elements must be represented by *partial* differential equations, which are in general very difficult to solve, while lumped elements can be represented by *ordinary* differential equations, which are relatively simple to solve.

A single-lump model may be adequate. Consider, for example, the cantilever spring with a large mass on the end shown in Fig. 1.6a. If the end mass is much greater than that of the spring itself, it is common to consider the spring to be massless (Fig. 1.6b), which leads to a very simple analytical solution for the motion of the beam. On the other hand, if the distributed mass of the spring must be considered, then the equations of motion are much more difficult to solve because they are partial differential equations, involving variation of y not only with time but also with location: $y = y(x,t)$. The resulting motions are much more elaborate, involving many bending mode configurations of the cantilever. Thus, if the distributed mass can be represented by a lumped one the problem is much simpler. (The numerical value assigned to the lump must be chosen astutely.)

A several-lump model may be advisable. When the dimensions of a distributed-mass system are such that the assumption of a single mass is, in

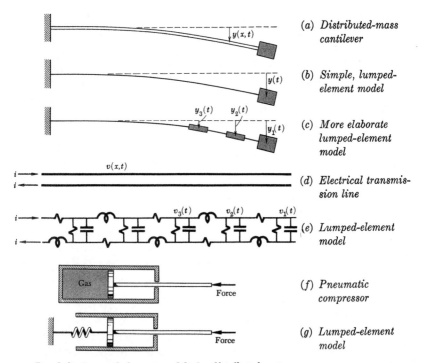

(a) *Distributed-mass cantilever*

(b) *Simple, lumped-element model*

(c) *More elaborate lumped-element model*

(d) *Electrical transmission line*

(e) *Lumped-element model*

(f) *Pneumatic compressor*

(g) *Lumped-element model*

Fig. 1.6 *Lumped-element models for distributed systems*

fact, a rather poor approximation, a closer model is often used in which the distributed mass is represented by a number of lumped masses, separated by massless springs (Fig. 1.6c). Such a model can still be represented by *ordinary* differential equations [in the several variables $y_1(t)$, $y_2(t)$, etc.] rather than by partial differential equations. Moreover, by increasing the number of masses used in the model, the approximation may be made as close as required. As the number of masses is increased, of course, the subsequent analysis rapidly becomes more laborious. But, since the equations are still *ordinary* differential equations, the solution remains straightforward.†

Figure 1.6d represents a long electrical transmission line having resistance, inductance, and capacitance distributed continuously all along its length. Figure 1.6e shows a lumped model for analysis in which the distributed properties are approximated by lumped elements at discrete points along the line. (Again, numerical values for the model elements must be properly chosen.)

Another example of parameter lumping is shown in Fig. 1.6f and g, where the gas within a cylinder is represented simply by an extensible (massless) spring. That is, the distributed mass of the gas has been neglected compared to the mass of the piston. This assumption will be valid only when low-speed operation is expected, since wave motions within the gas will involve both its mass and its "springiness." Again, intuition and experience will be involved in making the initial approximation, which will be checked numerically at a later time.

Partial differential equations will be described and studied briefly in Chapter 13. Elsewhere in this book, physical elements will be assumed to be lumped, and hence only ordinary differential equations will be encountered.

(d) Linearity

An ordinary linear differential equation is one having the form

$$A_n \frac{d^n x}{dt^n} + A_{n-1} \frac{d^{n-1} x}{dt^{n-1}} + \cdots + A_2 \frac{d^2 x}{dt^2} + A_1 \frac{dx}{dt} + A_0 x$$
$$+ B_m \frac{d^m y}{dt^m} + B_{m-1} \frac{d^{m-1} y}{dt^{m-1}} + \cdots + B_2 \frac{d^2 y}{dt^2} + B_1 \frac{dy}{dt} + B_0 y$$
$$+ \cdots = f(t) \quad (1.1)$$

Variables x, y, etc., are functions of only the independent variable [t in (1.1)], and the coefficients (the A's, B's, etc.) may vary with t but not with x, y, etc. Term $f(t)$ may vary with t in any manner, but may not involve x, y, etc. Note that no products of variables or their derivatives may be present, such as xy, x^2, $x\dot{x}$, $x\dot{y}$. An example of an ordinary linear differential equation is

$$3 \frac{d^2 x}{dt^2} + 14 \frac{dx}{dt} + (\sin 4t) x = 5 t^3 \quad (1.2)$$

† In special cases the partial differential equation may be easy to solve. Then, of course, the lumped-parameter approximation would not be needed.

16   DYNAMIC INVESTIGATION   [Chap. 1]

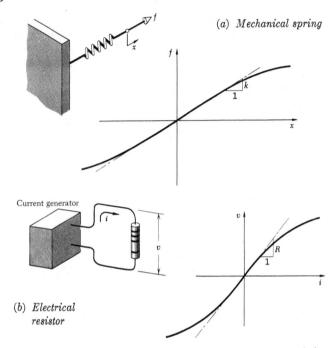

Fig. 1.7 *Linear approximation of physical-element characteristics*

If the coefficients (the A's, B's, etc.) are constant, the equation is said to be time invariant, or to have constant coefficients. Most of this book will involve linear differential equations with constant coefficients, which we shall call *l.c.c. equations*.

Often a nonlinear system can be described approximately by linear differential equations in which terms like xy and x^2 are dropped, any A which varies with x or y is replaced by some average A which varies only with t or is constant, etc.

Linear approximations are tremendously helpful in simplifying the mathematical analysis of a system's dynamic behavior, as we shall begin to see very soon.

In the first place, nonlinear differential equations are seldom amenable to analysis at all, usually requiring a computer for their solution (or a laborious graphical procedure).

But there is a more fundamental difference: When a linear equation with either constant or time-varying coefficients has been solved once, the solution is general, holding for all magnitudes of motion. More precisely stated: solutions to linear equations may be superposed on one another. By contrast, one solution to a nonlinear equation holds only for the particular magnitude of motion for which it was obtained and for no others, so that a whole family of solutions—very numerous in some situations—must be computed before the system's behavior can be said to be understood. Hence, the behavior of a linear model can be understood much more comprehensively and with a small fraction of the effort required to study a nonlinear model.

Thus, even with the amazing computational facilities now available for solving large sets of nonlinear equations, the role of the linear model in furnishing early, comprehensive insight into the dynamic behavior of a new system is a vital one.

Few physical elements display truly linear characteristics. For example, the relation between force on a spring and displacement of the spring is always nonlinear to some degree, as illustrated in Fig. 1.7a. The relation between current through a resistor and voltage drop across it also deviates from a straight-line relation, as in Fig. 1.7b. However, if in each case the relation is "reasonably" linear, then it will be found that the system behavior will be very close to that obtained by assuming an ideal, linear physical element, and the analytical simplification is so enormous that we make linear assumptions wherever we can possibly do so in good conscience.

(e) Constant parameters

If the physical parameters of a system vary with time, the equations of motion will also be more difficult to solve (although not nearly so difficult, in general, as when nonlinear relationships prevail). Suppose, for example, that the length of the cantilever spring in Fig. 1.6a is being varied with time, as indicated in Fig. 1.8. (This occurs in certain instrument applications, for example.) Then the resulting differential equation of motion will contain a time-varying coefficient (the spring constant of the cantilever), and will be considerably more difficult to solve. However, the solutions, once obtained, will be much more broadly applicable than for a nonlinear system, because they will hold for both small and large vibrations.

If the length ℓ of the cantilever is varying at a sufficiently slow rate we can obtain a good approximation to its motion by adopting a *physical model* in which ℓ is constant. We can then solve for the motion much more easily, doing so for each of several values of ℓ. From these several solutions we will obtain a good understanding of the motion of the actual system.

Physical problems are often simplified by the adoption of a model in which all the physical parameters are constant. Electrical apparatus in which values of resistance vary slowly with temperature is studied by means of a model in which resistance is constant. The motions of an airplane

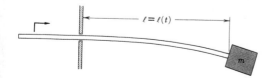

Fig. 1.8 *System with time-variable spring*

diving from a high altitude, where the air is rare and the controls are relatively ineffective, to a low altitude, where the elevator, rudder, and aileron are extremely effective, is studied first in terms of a model in which the airplane flies at various *constant* altitudes, so that control-surface effectiveness will not appear as a time variable.

(f) Neglecting uncertainty and noise

In real systems we are uncertain, in varying degrees, about the values of parameters, about measurements, and about expected inputs and disturbances. Further, the latter contain random components, called noise, which can influence system behavior. Sometimes disturbances are wholly random.

In elementary dynamic studies it is common to neglect such uncertainties and noise, and to proceed as if all quantities have definite values that are known precisely. This is called a *deterministic* approach; it simplifies analysis by avoiding the need for statistical treatment.

The assumption that numerical values are known can be examined, at the end of a first analysis, by checking the effect on system behavior of several values for the key parameters.

The study of stochastic processes—dynamic effects of uncertainty and response to random disturbances—is an important part of advanced study.

Summary

The exact equations of motion of a physical system will, in general, be a set of partial, nonlinear differential equations. We have described six common kinds of approximation that beget physical models more amenable to mathematical analysis. These are listed in Table I, together with

TABLE I

Approximations used in physical modeling

Approximation	*Mathematical simplification*
(a) Neglect small effects	Reduces number and complexity of differential equations
(b) Assume environment independent of system motions	Same as (a)
(c) Replace distributed characteristics with appropriate lumped elements	Leads to *ordinary* (rather than partial) differential equations
(d) Assume linear relationships	Makes equations linear, allows superposition of solutions
(e) Assume constant parameters	Leads to constant coefficients in differential equations
(f) Neglect uncertainty and noise	Avoids statistical treatment

the nature of the simplification they afford. If approximations (c), (d), and (e) can all be made, the resulting equations of motion will be ordinary, linear differential equations with constant coefficients, and the subsequent analysis will be much simpler—by orders of magnitude in some cases.

> Problems 1.5 through 1.10

1.5 DIMENSIONS AND UNITS

The study of dynamics is an altogether quantitative one. Measurement is implicit, and the magnitude and timing of motions are the heart of the problem. It is therefore mandatory that the dimensions of quantities and their units of measure be understood precisely.

Dimensions

The physical world with which we shall be concerned in this book—the world of mechanics, electricity, fluids, and heat—can be described quantitatively in terms of just five dimensions: length ℓ, force f,† charge q, temperature T, and, of course, time τ.

Some of the variables with which we work are described by only one of these dimensions—as displacement of a point (ℓ), charge on a body (q), period of a motion (τ). Others require combinations. For example, velocity has the dimensions of length per unit time $(\ell \tau^{-1})$; current has the dimensions of quantity of charge per unit time $(q\tau^{-1})$; pressure has the dimensions of force per unit area $(f\ell^{-2})$. Work has the fundamental dimensions of force times distance $(f\ell)$; and so, therefore, do *all* forms of energy.

Units of measure

In order to describe the *amount* of anything we must have standard units into which it can be subdivided, units that can be counted. Thus a distance is measured by counting the number of standard meters or feet into which it can be divided; and time is measured by counting seconds (or hours or centuries).

Sometimes the units we choose, like the variables they measure, involve only one of the primary dimensions. Meters have the dimension of length ℓ; coulombs have the dimension of charge q. Other units may involve several dimensions. For example, the units of work (e.g., foot-pounds) involve

---

† The choice of force or of mass as a primary dimension is arbitrary. Physicists commonly prefer to choose mass, with force a derived quantity. In our work here, however, we are dealing so much with the relations between force and geometric quantities—displacement and velocity, as well as acceleration—and force equilibrium plays such a dominant role, that for convenience and to be systematic we elect to consider force as the primary dimension. Then mass is a derived quantity, the mass of a body being defined as the ratio between a force applied to the body and the accompanying acceleration of the body. The units of mass are thus force/acceleration, or $f\ell^{-1}\tau^2$.

the dimensions $f\ell$, while voltage is a unit with dimensions $f\ell q^{-1}$, work per unit charge.

Often it is convenient to define new units of measure as combinations of others. Thus electrical current is almost always measured in amperes, which are equivalent to coulombs per second and have the fundamental dimensions $q\tau^{-1}$. Electrical potential difference is measured in volts (with the units $f\ell q^{-1}$, as noted above).

Table II lists many of the variables used in studying the dynamic behavior of physical systems. The fundamental dimensions, and also typical units, are given for each variable. The British system and both metric systems—the mks and cgs systems—are represented. Some convenient conversion factors are also given. Some simple problems are offered to familiarize the reader with Table II.†

Dimensional checking

Whenever an important equation or expression has been reached in the course of an analytical investigation—a geometric relation, an equilibrium equation, a final time response, or the like—it is always a good idea to pause briefly to check the equation dimensionally. If the separate terms to be added do not all have the same dimensions, then a mistake has been made (probably in the algebraic manipulation leading up to this point) and can be found and corrected before more time is spent carrying forth an incorrect study.

Problems 1.11 through 1.13

† The standard scientific notation $N \times 10^m$, e.g., 3.1×10^4, is used in Table II. Elsewhere in this book we shall employ the more concise "slant" notation, letting $N\backslash^m$ mean $N \times 10^m$. For example, $3.1\backslash^4$ (read "3.1 slant 4") means 3.1×10^4.

TABLE II

Dimensions and units

TABLE II

Dimensions and units

The primary dimensions are boxed. Three common, self-consistent sets of units are tabulated. Vector quantities are shown boldface. Commonly used conversion formulas are given in brackets. The factors are repeated, to eight significant figures, in Appendix A.

PHYSICAL MEDIUM	QUANTITY	DIMENSIONS	UNITS		
			mks	*cgs*	*British*[a] *engineering units*
All	Time t	$\boxed{\tau}$	second (sec)	second (sec)	second (sec)
Mechanical	Length ℓ	$\boxed{\ell}$	meter (m)	centimeter (cm)	foot (ft) $\left[\begin{smallmatrix}(ft)\\ \ell\end{smallmatrix}/ 3.281 \atop (ft/m) = \begin{smallmatrix}(m)\\ \ell\end{smallmatrix}\right]$
	Force[b] $\mathbf{f}$	$\boxed{f}$	newton (n)	dyne (dn) or kilodyne (kd) $\left[\begin{smallmatrix}(n)\\ \mathbf{f}\end{smallmatrix} = \begin{smallmatrix}(dn)\\ \mathbf{f}\end{smallmatrix} / {10^5 \atop (dn/n)}\right]$	pound (lb)[c] $\left[\begin{smallmatrix}(lb)\\ \mathbf{f}\end{smallmatrix} \times {(n/lb) \atop 4.448} = \begin{smallmatrix}(n)\\ \mathbf{f}\end{smallmatrix}\right]$
	Moment $\mathbf{M}$	$f\ell$	n m	dn cm $\left[\begin{smallmatrix}(n\ m)\\ \mathbf{M}\end{smallmatrix} = \begin{smallmatrix}(dn\ cm)\\ \mathbf{M}\end{smallmatrix} / {10^7 \atop (dn\ cm/n\ m)}\right]$	ft lb $\left[\begin{smallmatrix}(ft\ lb)\\ \mathbf{M}\end{smallmatrix} \times {(n\ m/ft\ lb) \atop 1.356} = \begin{smallmatrix}(n\ m)\\ \mathbf{M}\end{smallmatrix}\right]$
	Displacement $\mathbf{x}$ or $\mathbf{r}$	ℓ	m	cm	ft $\left[\begin{smallmatrix}(ft)\\ \mathbf{x}\end{smallmatrix} / 3.281 \atop (ft/m) = \begin{smallmatrix}(m)\\ \mathbf{x}\end{smallmatrix}\right]$

		m/sec	cm/sec	ft/sec
Velocity **v** or **u**	$\ell\tau^{-1}$			
Acceleration **a**	$\ell\tau^{-2}$	m/sec²	cm/sec²	ft/sec²
Angular displacement θ	(dimensionless)^d	radian (rad) $\left[\dfrac{(\text{deg})}{\theta} / \dfrac{57.30}{(\text{deg/rad})}\right]$	radian (rad) $\left[\dfrac{(\text{rad})}{\theta}\right]$	radian (rad) $\left[\dfrac{(\text{rev})}{\theta} \times \dfrac{(\text{rad/rev})}{2\pi} = \dfrac{(\text{rad})}{\theta}\right]$
Angular velocity Ω	τ^{-1}	rad/sec	rad/sec	rad/sec
Angular acceleration $\dot{\Omega}$	τ^{-2}	rad/sec²	rad/sec²	rad/sec²
Translational spring constant k	$\dfrac{f}{\ell} = f\ell^{-1}$	n/m	dn/cm	lb/ft
Translational damper b	$\dfrac{f}{\ell/\tau} = f\ell^{-1}\tau$	n/(m/sec)	dn/(cm/sec)	lb/(ft/sec)
Mass m	$\dfrac{f}{\ell/\tau^2} = f\ell^{-1}\tau^2$	n/(m/sec²) or kilograms (kg) $\left[\dfrac{[\text{n/(m/sec}^2)]}{m} = \dfrac{(\text{kg})}{m}\right]$	dn/(cm/sec²) or grams (gm) $\left[\dfrac{[\text{dn/(cm/sec}^2)]}{m} = \dfrac{(\text{gm})}{m}\right]$	lb/(ft/sec²) or slugs $\left[\dfrac{[\text{lb/(ft/sec}^2)]}{m} = \dfrac{(\text{slugs})}{m}\right]$ $\left[\dfrac{[\text{lb/(ft/sec}^2)](\text{kg/slug})}{m \times 14.59} = \dfrac{(\text{kg})}{m}\right]$
Torsional spring k	$\dfrac{f\ell}{1} = f\ell$	n m/rad	dn cm/rad	ft lb/rad
Torsional damper b	$\dfrac{f\ell}{1/\tau} = f\ell\tau$	n m/(rad/sec)	dn cm/(rad/sec)	ft lb/(rad/sec)
Moment of inertia J	$\dfrac{f\ell}{1/\tau^2} = f\ell\tau^2$	n m/(rad/sec²)	dn cm/(rad/sec²)	ft lb/(rad/sec²)

TABLE II (Continued)

PHYSICAL MEDIUM	QUANTITY	DIMENSIONS	UNITS mks	cgs	British[a] engineering units
Mechanical (*continued*)	Momentum (translational or "linear") $\mathbf{p} = m\mathbf{v}$	$\dfrac{f}{\ell/\tau^2} \cdot \dfrac{\ell}{\tau} = f\tau$	n sec	dn sec	lb sec
	Angular momentum $\mathbf{H}$ (moment of momentum)	$f\tau\ell$	n m sec	dn cm sec	lb ft sec
	Acceleration of gravity g Standard[f] value g_0	$\ell\tau^{-2}$	m/sec^2 $9.81\ m/sec^2$	cm/sec^2 $981\ cm/sec^2$	ft/sec^2 $32.2\ ft/sec^2$
	Weight $W = mg$ Weight at "standard" location $W = mg_0$	f	n $\left[\dfrac{(n)}{W} = \dfrac{(kg)}{m} \times \dfrac{(m/sec^2)}{9.81}\right]$	dn $\left[\dfrac{(dn)}{W} = \dfrac{(gm)}{m} \times \dfrac{(cm/sec^2)}{981}\right]$	lb $\left[\dfrac{(lb)}{W} = \dfrac{(slugs)}{m} \times \dfrac{(ft/sec^2)}{32.2}\right]$
Fluid	Pressure p	$\dfrac{f}{\ell^2} = f\ell^{-2}$	n/m^2	dn/cm^2	lb/ft^2
	Mass flow rate w	$\dfrac{f/\ell\tau^{-2}}{\tau} = f\tau\ell^{-1}$	$\dfrac{n/(m/sec^2)}{sec} = n\ sec/m$ or kg/sec	$\dfrac{dn/(cm/sec^2)}{sec} = dn\ sec/cm$ or gm/sec	$\dfrac{lb/(ft/sec^2)}{sec} = lb\ sec/ft$ or slugs/sec
	Mass density ρ	$\dfrac{f/\ell\tau^{-2}}{\ell^3} = f\tau^2\ell^{-4}$	$n\ sec^2/m^4$ or kg/m^3	$dn\ sec^2/cm^4$ or gm/cm^3	$lb\ sec^2/ft^4$ or $slugs/ft^3$
	Viscosity μ $\left(\dfrac{stress}{velocity\ gradient}\right)$	$\dfrac{f/\ell^2}{(\ell/\tau)/\ell} = f\ell^{-2}\tau$	$\dfrac{(n/m^2)}{(m/sec)/m} = n\ sec/m^2$	Poise $\dfrac{(dn/cm^2)}{(cm/sec)/cm} = dn\ sec/cm^2$	$\dfrac{lb/ft^2}{(ft/sec)/ft} = lb\ sec/ft^2$

TABLE II (Continued)

PHYSICAL MEDIUM	QUANTITY	DIMENSIONS	Common combination unit	UNITS mks	cgs	British engineering units
Electrical and electromechanical	Charge q	q	coulomb (cl)	coulomb (cl)	coulomb (cl)	coulomb (cl)
	Electric potential emf ε; voltage v	$\dfrac{f\ell}{q} = f\ell q^{-1}$	volt (v)	n m/cl $\left[v = \dfrac{\text{n m/cl}}{v}\right]$	dn cm/cl	ft lb/cl
	Current i	$\dfrac{q}{t} = q\tau^{-1}$	ampere (amp) $\left[i = \dfrac{\text{amps}}{i}\right]$	cl/sec flowing through designated cross-sectional area $\left[i = \dfrac{\text{cl/sec}}{i}\right]$		
	Resistance R	$\dfrac{f\ell q^{-1}}{q\tau^{-1}} = f\ell\tau q^{-2}$	ohm v/amp $\left[R = \dfrac{\text{v/amp}}{R} = \dfrac{\text{(ohms)}}{R}\right]$	n m sec/cl² $\left[R = \dfrac{\text{n m sec/cl}^2}{R}\right]$	dn cm sec/cl²	ft lb sec/cl²
	Capacitance C	$\dfrac{q}{f\ell/q} = q^2 f^{-1}\ell^{-1}$	farad cl/v $\left[C = \dfrac{\text{cl/v}}{C} = \dfrac{\text{(farads)}}{C}\right]$	cl²/n m $\left[C = \dfrac{\text{cl}^2/\text{n m}}{C}\right]$	cl²/dn cm	cl²/ft lb
	Inductance L	$\dfrac{(f\ell/q)}{q\tau^{-1}/\tau} = f\ell\tau^2 q^{-2}$	henry (h) v/(amp/sec) $\left[L = \dfrac{\text{v/(amp/sec)}}{L} = \dfrac{\text{(henrys)}}{L}\right]$	n m sec²/cl² $\left[L = \dfrac{\text{n m sec}^2/\text{cl}^2}{L}\right]$	dn cm sec²/cl²	ft lb sec²/cl²

TABLE II (Continued)

PHYSICAL MEDIUM	QUANTITY	DIMENSIONS	Common combination unit	UNITS mks	cgs	British engineering units
Electrical and electromechanical (*continued*)	Electric field **E**	$\dfrac{f}{q} = fq^{-1}$		n/cl (v/m)	dn/cl (v/cm)	lb/cl (v/ft)
	Magnetic induction **B**	$\dfrac{f}{q\ell/\tau} = fr q^{-1}\ell^{-1}$	gauss weber/m² v sec/m² $\left[\dfrac{\text{(gauss)}}{\mathbf{B}} \Big/ 10^4 \dfrac{\text{gauss}}{[\text{web}/\text{m}^2]}\right] = \dfrac{(\text{web}/\text{m}^2)}{\mathbf{B}}$	n sec/cl m $\left[\dfrac{(\text{n sec/cl m})}{\mathbf{B}}\right]$	dn sec/cl cm	lb sec/cl ft

TABLE II (Continued)

PHYSICAL MEDIUM	QUANTITY	DIMENSIONS	UNITS		
			mks	cgs	British engineering units
Thermal	Temperature	T	degrees Kelvin (°K) or degrees centigrade (°C)		degrees Fahrenheit absolute or degrees Fahrenheit (°F) $\left[T - 32 \right)^{(°F)} \times \frac{5}{9} = T^{(°C)}$
	Amount of heat Q	$f\ell$	n m or joule (jl) or kilogram-calorie (kcal) $\left[\overset{(n\,m)}{Q} = \overset{(jl)}{Q} = \overset{(kcal)}{Q}/2.389 \times 10^{-4} \right]_{(kcal/jl)}$	dn cm or erg or calorie (cal) $\left[\overset{(dn\,cm)}{Q} = \overset{(ergs)}{Q} = \overset{(cal)}{Q}/2.389 \times 10^{-8} \right]_{(cal/erg)}$	ft lb or British Thermal Unit (Btu) $\left[\overset{(Btu)}{Q} \times 777.9 \overset{(ft\,lb/Btu)}{=} \overset{(ft\,lb)}{Q} \right]$
	Heat flow rate q	$\dfrac{f\ell}{\tau} = f\ell\tau^{-1}$	n m/sec or jl/sec or kcal/sec	dn cm/sec or cal/sec	ft lb/sec or Btu/sec
	Heat capacity[a] C	$\dfrac{f\ell}{T} = f\ell T^{-1}$	n m/°K or jl/°K or kcal/°K	dn cm/°K or cal/°K	ft lb/°F or Btu/°F
	Specific heat[a] c	$\dfrac{f\ell/T}{f/(t\tau^{-2})} = \ell^2 T^{-1} \tau^{-2}$	m²/sec²°K or (jl/°K)/kg or (kcal/°K)/kg	cm²/sec²°K or (cal/°K)/gm	ft²/sec²°F or (Btu/°F)/lb
	Thermal resistance R	$\dfrac{T}{f\ell/\tau} = T f^{-1} \ell^{-1} \tau$	°K/(n m/sec) or °K/(jl/sec) or °K/(kcal/sec)	°K/(dn cm/sec) or °K/(cal/sec)	°F/(ft lb/sec) or °F/(Btu/sec)

TABLE II (Continued)

PHYSICAL MEDIUM	QUANTITY	DIMENSIONS	UNITS		
			mks	cgs	British engineering units
All	Work W and energy U, T, (also, heat Q and all other forms of energy)	fl	n m or jl $\left[\frac{(\text{n m})}{W} = \frac{(\text{jl})}{W}\right]$	dn cm or erg $\left[\frac{(\text{dn cm})}{W} = \frac{(\text{erg})}{W}\right]$	ft lb
	Power P (rate of doing work)	$fl\tau^{-1}$	n m/sec or watt (w) $\left[\frac{(\text{n m/sec})}{P} = \frac{(\text{w})}{P}\right]$	dn cm/sec or watt (w) $\left[\frac{(\text{dn cm/sec})}{P} \Big/ \frac{10^7}{(\text{dn cm/n m})} = \frac{(\text{w})}{P}\right]$ $\left[\frac{(\text{w})}{P} = \frac{(\text{hp})}{P} \times \frac{(\text{w/hp})}{745.7}\right]$	ft lb/sec or horsepower (hp) $\left[\frac{(\text{ft lb/sec})}{P} \Big/ \left(\frac{550}{\frac{\text{ft lb/sec}}{\text{hp}}}\right) = \frac{(\text{hp})}{P}\right]$

[a] It is noted with appreciation that these will soon be known as "Old British" units, as Great Britain converts to the metric system.
[b] The selection of force, rather than mass, as a primary dimension is discussed in the footnote on p. 19.
[c] In certain cases the word "pound" is used to denote a unit of mass. In this book, however, it will denote only force.
[d] Dimensionless ratio of arc length to radius.
[e] In certain cases the word "kilogram" is used to denote force. In this book, however, it will denote only mass.
[f] The value of g varies with location, of course. Its value comes from the law of gravitation: $f = Gm_1 \int_{\text{earth}} dm_2/R^2$, where f is the mass attraction on any test mass m_1, dm_2 is a segment of the earth, and $G = 6.670 \times 10^{-11}$ n m²/kg²; and dm_2/R^2 must be integrated over the earth's volume. Then g is defined by $g = f/m_1 = G\int_{\text{earth}} dm_2/R^2$. At altitude h above the earth's surface, g is given closely by: $g = g_0(R_0/R)^2$, with $R_0 = 6.378 \times 10^6$ m, $R = R_0 + h$.
[g] Conditions of measurement must be specified; e.g., at constant pressure or at constant volume.

PART A

EQUATIONS OF MOTION FOR PHYSICAL SYSTEMS

Chapter 2. *Equations of Motion for Simple Physical Systems: Mechanical, Electrical, and Electromechanical*

Chapter 3. *Equations of Motion for Simple Heat-Conduction and Fluid Systems*

Chapter 4. *Analogies*

Chapter 5. *Equations of Motion for Mechanical Systems in Two and Three Dimensions*

Chapter 2

EQUATIONS OF MOTION
FOR SIMPLE PHYSICAL SYSTEMS:
MECHANICAL, ELECTRICAL,
AND ELECTROMECHANICAL

In Chapters 2, 3, and 4 we want to develop the ability to write the equations of motion for models of simple physical systems. We wish also to indicate the similarities among the various physical media. Five representative physical media will be discussed: mechanical, electrical, and electromechanical (Chapter 2), and fluid and heat conduction (Chapter 3—an optional chapter). In each case we shall first discuss physical models (Stage I), and then derive the equations of motion (Stage II).

The principal feature that distinguishes the study of one physical medium from that of another is, of course, the specific cause-and-effect phenomena exhibited by its elements. In addition, the kinds of approximations made in physical modeling will be different; the details of selecting variables and writing equilibrium relations will also depend on the medium. There is much in common, however, in the process of deriving equations of motion, as we shall note at the outset, and repeatedly in our studies.

In Chapter 4 (optional) we discuss the mathematical analogies between the equations of motion from the different physical media, and the corresponding physical analogies implied. The network approach is also presented.

In Chapter 5 (optional) the derivation of equations of motion for mechanical systems is treated in greater depth, with the consideration of two- and three-dimensional systems of rigid bodies.

Chapters 3, 4, and 5 are included in Part A to supply important perspective and added capability. However, study of the simple systems treated in Chapter 2 will be adequate preparation for the study of *dynamic response*

(Stage III) pursued in most of this book, beginning in Part B. Thus, in a first use of the book the reader may wish to omit Chapters 3 through 5, and proceed at once from Chapter 2 to the study of dynamic behavior which begins in Part B, Chapter 6 (returning again to Part A at a later time).

2.1 STAGE II. EQUATIONS OF MOTION: FROM PHYSICAL MODEL TO MATHEMATICAL MODEL

Before considering physical systems of a specific kind (mechanical, electrical, etc.), it is worth-while to discuss in general terms the process of obtaining equations of motion.†

The focal point in deriving the equations of motion, for a given physical system model, is the writing of *equilibrium* relations to describe the *balance*— of forces, of flow rates, of energy—which must exist for the system and for its subsystems; or the writing of system *compatibility* relations to describe how motions of the system elements are interrelated because of the way they are *interconnected*. These are the inter-element or *system* relations.

Two further considerations are involved in deriving equations of motion. One is the selection of precise *physical variables* (velocity, voltage, pressure, flow rate, etc.) with which to describe the instantaneous *state* of a system, and in terms of which to study its behavior. The final basic consideration is for the natural *physical laws* which the individual elements of the system obey—mechanical relations between force and motion, electrical relations between current and voltage, electromechanical relations between force and magnetic field, thermodynamic relations between temperature, pressure, internal energy, etc. These are properly called *constitutive* physical relations: They concern only the individual elements, or constituents, of the system. We shall usually call them physical relations, for short.

Finally, when (1) variable selection, (2) equilibrium or compatibility, and (3) physical laws have been considered individually, the resulting relations are *combined* algebraically into a compact set of equations of motion.

Physical variables may be classified conveniently as either "through" or "across" variables. Through variables measure the transmission of something *through* an element, as electric current through a resistor, fluid through a duct, Fig. 2.1c, or force through a spring. Across variables measure a difference in state between the ends of an element, as the voltage drop *across* a resistor, pressure drop between the ends of a duct, or the difference in velocity between the ends of a dashpot, Fig. 1.1.‡

---

† As noted in the footnote on p. 4, the word "motion" will be used throughout this book in a general context to connote the change of any physical variable—mechanical displacement and velocity, electrical voltage and current, pressure, temperature, etc.
‡ Across variables are called *two-point variables* because they are defined by the difference between quantities at two points, the terminals of an element. Through variables are one-point variables.

Equilibrium relations are always relations among *through* variables, and are sometimes called also *vertex* or *node* or *continuity* or *flow* relations, as Kirchhoff's Current Law (at an electrical node), continuity of fluid flow, or equilibrium of forces meeting at a point.

Compatibility relations are always relations among *across* variables, as Kirchhoff's Voltage Law around a circuit, pressure drop across all the connected stages of a fluid system, or geometric compatibility in a mechanical system. They are sometimes called also *path* or *loop* or *connectedness* relations.

Physical relations are relations *between* the through and the across variables of each individual physical element, as $f = kx$ for a spring, or $i = (1/R)v$ for a resistor, Fig. 1.7.

To summarize, in deriving equations of motion we must consider individually:

•*Definition of (through and across) variables*
•*System relations of equilibrium or compatibility*
•*Physical relations for each element*

Considerations in deriving equations of motion

and then we must *combine relations*.

For each particular physical medium we shall find it convenient to divide these considerations into a checklist of steps—a pattern of procedure.

Two further general comments, about equilibrium relations and about physical laws, are important at this point:

A system must be defined before equilibrium relations can be written. Unless the physical *boundaries* of a system are *clearly specified*, any "equilibrium relations" we may write are meaningless. We shall confront this principle over and over in every physical medium.

In mechanical systems we draw an isolated "free-body diagram," as in Fig. 2.1a; *then* we can state the equilibrium of forces acting on the boundaries of the isolated free-body subsystem. In electrical networks we must identify isolated junctions (as node b in Fig. 2.1b) before we can write Kirchhoff's Current Law. In fluid dynamics and in thermodynamics a precise "control surface" is imagined, like the one shown (———-———-———) in Fig. 2.1c, and at its boundaries are carefully measured the inward or outward flow of energy, momentum, and matter.

Always we define a subsystem first; then we write equilibrium relations for that subsystem.

Physical phenomonological relations are purely empirical. It is emphasized that the physical relations between force and displacement in a spring, between voltage and current in a resistor, and so forth, are not deduced from any basic principles. They are, rather, empirical relations observed by experiment. Thus, the fact that the extension of a spring is nearly linearly

34 EQUATIONS OF MOTION: SIMPLE PHYSICAL SYSTEMS [Chap. 2]

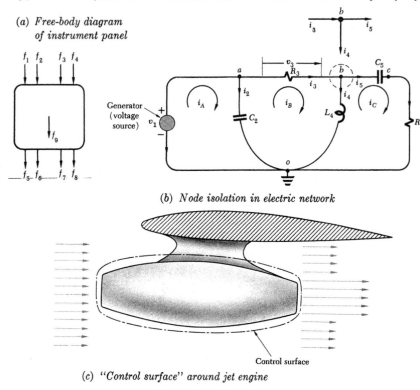

(a) *Free-body diagram of instrument panel*

(b) *Node isolation in electric network*

(c) *"Control surface" around jet engine*

Fig. 2.1 To write equilibrium relations, subsystems must be defined

proportional to the force applied to it over a fairly large range is a purely experimental result. (It is a happy one for the solving of differential equations involving spring characteristics.) Fortunately many of the cause-and-effect relations in nature turn out to be nearly linear, within useful ranges of operation.

In obtaining the differential equations for a given physical system, we make use of a storehouse of known physical cause-and-effect relations for individual elements—a storehouse built up through innumerable physical experiments, performed (sometimes at great expense) by many research teams, because the results are so important.

2.2 ONE-DIMENSIONAL MECHANICAL SYSTEMS

STAGE I. PHYSICAL MODELS

In this section we want to study in some detail how to derive equations of motion for mechanical systems. To concentrate attention on the method, we begin with simple systems that move in only one dimension. Useful models of many important mechanical systems fall in this category—for

[Sec. 2.2] ONE-DIMENSIONAL MECHANICAL SYSTEMS 35

example, the vertical-motion-only model of the instrument mount in Fig. 1.1b (p. 6). As another example, in Fig. 2.7 (p. 48), a one-dimensional model is used to good advantage in a first analysis of motions of an aircraft landing on the deck of a carrier.

Two- and three-dimensional models are considered in Chapter 5.

STAGE II. EQUATIONS OF MOTION

Analytical procedure

When the physical model of a system consists of mechanical elements only, the accompanying procedure† is an efficient way of making sure that all three considerations of Sec. 2.1 have been taken into account.

> (1) **Geometry.** Picture system in an *arbitrary configuration* (with respect to a reference configuration), then *define coordinates* and their positive directions. Note geometric *identities;* note relations implied by geometric *constraints*. Write system geometric *compatibility* relation, if advantageous.
> (2) **Force equilibrium.**
>   (a) Write *force* balance relations.
>   (i) Draw free-body diagram.
>   (ii) Write equations of equilibrium of all the forces acting on the free body.
> or (b) Write *energy* balance relations.
>   (i) Define *system envelope*.
>   (ii) Invoke conservation of energy for system.
> (3) **Physical force-geometry relations.** Write these for the individual elements.

Procedure A-m
Mechanical equations of motion

Then **combine relations.** Be sure there are as many independent equations as there are unknowns. Then insert relations from (3) and (1) into (2), for example, to reduce the number of unknowns.

In more concise form, Procedure A-m can be stated as follows‡ for the case when Step (2a) is selected rather than (2b):

> (1) **Geometry**
> (2) **Forces**
> (3) **Force-geometry relations**

Procedure A-m
Mechanical

Then **combine,** inserting (1) and (3) into (2).

† Informal "procedures" are offered at various points throughout the book as convenient guides or checklists. They are indexed inside the front cover.
‡ This form for mechanical systems is introduced in S. H. Crandall and N. C. Dahl, "The Mechanics of Solids," McGraw-Hill Book Company, New York, 1957.

Mechanical forces are through variables, and geometric motions are across variables. Thus, comparing Procedure A-m with the considerations listed on p. 33, we see that item (1) of Procedure A-m is concerned with state-of-geometric-motion (across) variables and with compatibility relations among them as well; whereas item (2) concerns force (through) variables and equilibrium relations among them.

We shall first discuss the three kinds of relations; then we shall demonstrate Procedure A-m with examples and suggested problems.

(1) Geometry

The selection of dynamic variables is governed principally by two questions: "Which variables interest us in this problem?" and "Which variables will be easiest to use in the analysis?"

Configuration. Usually the first variables to define for a mechanical system are position coordinates to describe its configuration. As the system moves, the coordinates vary; and the time variation of coordinates constitutes the description of the system's dynamic behavior.

In defining position coordinates, two configurations are involved: a *reference* configuration, and an *arbitrary* configuration. These should be sketched, perhaps superposed as in Fig. 2.2a. Sometimes a reference mark

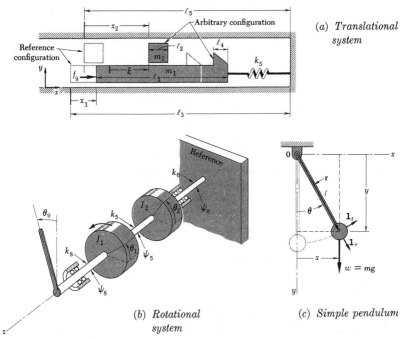

Fig. 2.2 *To define coordinates, the system is sketched in an arbitrary configuration*

[Sec. 2.2] ONE-DIMENSIONAL MECHANICAL SYSTEMS 37

"painted" on a reference frame, as in Fig. 2.2a, is adequate to identify the reference configuration. From these sketches we can define precisely a set of coordinates—displacements of the system—which describe completely the system's configuration (the orientation of all of its parts). In Fig. 2.2a any two of coordinates x_1, x_2, and ξ will serve. In Fig. 2.2b three must be known; for example θ_9, θ_1, θ_2. (Name some other sets using the twist angles ψ in the shafts between the disks.) In Fig. 2.2c a single coordinate, e.g. angle θ, specifies the configuration.

We emphasize that by *arbitrary configuration* we mean that *every* part of the system is displaced arbitrarily from its reference location. Without such an arbitrary configuration we cannot define coordinates with full generality.

Positive directions. Let us agree that whenever a coordinate has the direction shown in our arbitrary-configuration sketch we shall say it has a *positive* value. Thus, when blocks m_1 and m_2 in Fig. 2.2a are to the right of their painted reference positions, then x_1 and x_2 are positive; if m_1 is to the left of its reference, then x_1 is negative; and if m_2 is to the left of its reference, then x_2 is negative. Similarly, in Fig. 2.2b, when a disk is rotated counterclockwise about the z axis we call the angular displacement θ positive; otherwise it is negative.

Note that we are at liberty to draw our arbitrary configuration any way we like (e.g., we could have shown m_1 displaced to the left and m_2 to the right) and thus to define either direction as the positive one. Once we make our choice, however, we must follow it *consistently* thereafter in all subsequent analysis of the system.

Degrees of freedom. The number of degrees of freedom a mechanical system has is defined traditionally as the minimum number of geometrically independent coordinates required to describe its configuration completely. The system in Fig. 2.2a has two degrees of freedom, the system in Fig. 2.2b has three, and the system in Fig. 2.2c has only one.

Geometric identities. Alternative sets of coordinates are always related by identities. For example, if in Fig. 2.2a we select x_1 and x_2 as independent variables, then ξ is given by the identity

$$\xi \equiv x_2 - x_1$$

Often we don't write these explicitly; but we recognize them and use them implicitly. (Having chosen coordinates x_1 and x_2 in Fig. 2.2a, we might never use the symbol ξ, for example.)

Geometric constraint relations. Usually the parts of a physical system have some path constraint upon where they may move (without something breaking). For example, in Fig. 2.2a the blocks cannot move vertically, so that $y_1 = 0$ and $y_2 = 0$ for all time. Other constraints in Fig. 2.2a are that $x_1 \leq \ell_3 - \ell_1$ (due to the wall), and $x_2 \leq \ell_5 - (\ell_2 + \ell_4)$ (due to the lug).

Alternatively, a geometric variable may be constrained to move in a definite way versus time, as was x_1 in Fig. 1.1b (p. 6). (This often constitutes the "input" to the system.) In Fig. 2.2b, θ_9 vs. t might be prescribed.

Sometimes, because of the physical constraints, the most convenient coordinates may not be simple rectangular quantities. In Fig. 2.2c, for example, the pendulum is constrained to move in a circular path, and the most convenient coordinate is the angle θ through which the pendulum has swung. The variables θ, x, and y are related by

$$x = \ell \sin \theta \qquad y = \ell \cos \theta$$

Velocity and acceleration. For motion in one dimension, velocity is merely the time derivative of position†

$$v = \dot{x}$$

and acceleration is the time derivative of velocity

$$a = \dot{v} = \ddot{x}$$

Similarly, angular velocity and acceleration are the time derivatives of the angular position coordinate

$$\Omega = \dot{\theta}$$
$$\dot{\Omega} = \ddot{\theta}$$

It is *essential* to note that when the unknown velocity and acceleration of a point are expressed in terms of a position coordinate, *the sign convention for positive velocity and acceleration must be in the same direction as it is for positive displacement.* For example, in Fig. 2.2a, x_1 is taken positive to the right, hence $v_1 = \dot{x}_1$ and $a_1 = \ddot{x}_1$ must also be taken positive to the right.

> This is sometimes a point of confusion when the point is displaced to the right, but is known to be traveling to the left. But this merely means that, for the instant, x happens to have a positive *numerical* value while $\dot{x}$ happens to have a negative *numerical* value. The fundamental sign convention always holds: If x is considered positive when it is to the right, then $\dot{x}$ is also considered positive when it is to the right. All the differential calculus is based on this premise.

If velocity v is measured with respect to a fixed reference, as $\dot{x}_1$ is in Fig. 2.2a, then v is absolute. Otherwise it is relative (as $\dot{\xi}$ in Fig. 2.2a). Similarly, if a is measured with respect to a nonaccelerating reference, then a is absolute; otherwise it is relative.

The state of a system. What we mean by the state of a mechanical system may be introduced by analogy with configuration: The state of a system is often defined by an independent set of *position coordinates, plus their derivatives.* Thus, state implies *configuration plus velocity.* Configuration tells only where the system *is*, but state tells both where it is and how fast (and in what direction) it is going; and as expressed by Professor L. Zadeh,

† For general two- and three-dimensional motion the situation may be much more complicated, as discussed in Chapter 5.
  We shall be using the shorthand dot notation, invented by Newton, throughout this book: $\dot{x}$ means dx/dt.

[Sec. 2.2] ONE-DIMENSIONAL MECHANICAL SYSTEMS

"the state of a system at a given time, plus its differential equations of motion and inputs, will determine its configuration for all future time." The state of the system in Fig. 2.2a is given (at any instant) by the values of x_1, $\dot{x}_1$, x_2, and $\dot{x}_2$. The state of the system in Fig. 2.2c is given, for example, by θ and $\dot{\theta}$. The number of independent coordinates of a system is its number of degrees of freedom, while the number of independent state variables is its *order*, as we shall show precisely later.

> Although *state* comes from the Latin *status*, a standing or position, a more helpful image for our purposes is that of a stopping or freezing of action, as when a moving-picture film is stopped on one frame, but the instantaneous speed and direction of motion are revealed by blurring of the picture.

The state of a system at a given instant is related to the energy stored in the elements of the system at that instant. Thus, one straightforward set of mechanical state variables consists of the velocity of each independent mass plus the distortion of each spring (or set of springs, if grouped) plus the gravity-field displacement of each independent weight.

Precise mathematical definitions of a system's state are suggested by its equations of motion, as we shall see later (Sec. 17.11).

System geometric compatibility. In a system of connected mechanical elements we may inspect, along an imaginary closed line through all the connections and back to the starting point, the state of motion of each element, and note that, *because of the connectedness*, the individual motions must add up to zero: The motions of the parts must be *compatible* with the motion of the whole.

In Fig. 2.2b, for example, we may write, obviously,

$$\theta_9 - \psi_8 - \psi_5 - \psi_6 = 0$$

This is the equation of geometric compatibility for the system. We have traversed an imaginary path from the reference out directly to the end of the shaft and then back through the shaft, noting all twist angles along the way.

We may obtain a similar relation for the rates of rotation, either by traversing our path again or by differentiating the above expression:

$$\dot{\theta}_9 - \dot{\psi}_8 - \dot{\psi}_5 - \dot{\psi}_6 = 0$$

Often in static mechanical systems, and sometimes in dynamic systems, the system geometric compatibility relation is the focus for deriving system equations; but usually in dynamics, force equilibrium is a more useful focus.

In nearly all real mechanical problems, however, considerable attention must be devoted to geometric relations, whether the system compatibility relation per se is used or not. (This is particularly true in three-dimensional problems.) Thus, (1) geometry and (2) forces are given "equal billing" in Procedure A-m because typically they involve equal levels of attention.

Problems 2.1 through 2.17

(2) Force equilibrium

(i) *Defining a system: The free-body diagram.* We speak of a *system* as being in *equilibrium*. Without a system there can be no "equilibrium"—the word is meaningless. So the first task is to define precisely what system— what collection of matter—we shall discuss.

In mechanics our systems consist of "free bodies" which we draw as if severed from their environment; but we show carefully and completely in place of the environment all the forces it exerts on our system—our free body.

Figure 2.3a, for example, shows free-body diagrams of the blocks in Fig. 2.2a. On each block we have shown the forces that hold it in equilibrium: normal and friction forces from the adjacent block and from the floor, the force exerted by the compressed spring, the force of gravity, external force f_9, and an "inertial" force (shown dashed), which we shall discuss presently.

We now have a *second set of arbitrary choices* to make: We can choose either direction as the positive direction for each force. (As with the choice of positive directions for motion, once we make the choice for a force we must follow it consistently thereafter.) As it turns out, it is convenient to choose to make the positive direction for each force opposite to that for the corresponding motion, and we shall normally do so in this book, purely for convenience.† The other choice is perfectly valid in any case (and many people prefer it).

By Newton's Third Law,‡ the forces exerted by block m_2 on block m_1 are equal and opposite to those exerted by block m_1 on block m_2, and this is noted in labeling forces f_4 and f_8.

(ii) *Writing equations of force equilibrium: The focal relations.* We adopt D'Alembert's view of dynamic equilibrium, whereby, for a *system* of rigid bodies, Newton's Second Law is written in the following form:

$$\text{Equations of dynamic equilibrium} \quad \boxed{\begin{array}{ll} \Sigma f^* = 0 & \text{in any direction} \\ \Sigma M^* = 0 & \text{about any axis} \end{array}} \quad \begin{array}{l}(2.1)\\(2.2)\end{array}$$

in which the asterisk indicates that inertial forces and moments are included.§ That is, the equilibrium relations are identical to those for a statics problem. This will be explained further on p. 45 [Eqs. (2.9)–(2.12)]. *Throughout this book "equilibrium" will always mean "dynamic equilibrium."*

For example, (2.1) applied in turn to each of the free bodies in Fig. 2.3a gives, for the $-x$ direction,

$$\begin{array}{ll} \text{Block } m_1: & \Sigma f^* = 0 = f_1 + f_3 - f_4 + f_5 - f_9 \\ \text{Block } m_2: & \Sigma f^* = 0 = f_2 + f_4 \end{array} \quad (2.3)$$

---

† This choice is convenient because reaction forces oppose motions. This choice thus makes the signs come out positive in the physical relations, Step (3), which makes it a little easier to keep track of the signs.
‡ Newton's Laws of Motion may be found in Appendix D.
§ Equation (2.1) is read "sigma f star equals zero," etc.

[Sec. 2.2] ONE-DIMENSIONAL MECHANICAL SYSTEMS 41

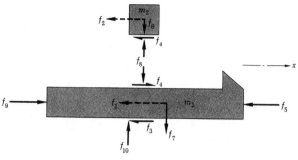

(a) *Translational system*

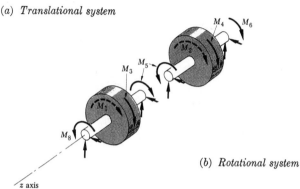

(b) *Rotational system*

Fig. 2.3 *Free-body diagrams for the systems of Fig. 2.2a and b*

These are the dynamic force-equilibrium equations for the system of Fig. 2.2a.

For the two free bodies of Fig. 2.3b, application of (2.2) about the $-z$ axis gives†

$$\text{Disk } J_1: \quad \Sigma M^* = 0 = M_1 + M_3 + M_5 - M_8$$
$$\text{Disk } J_2: \quad \Sigma M^* = 0 = M_2 + M_4 - M_5 + M_6 \quad (2.4)$$

These are the dynamic force-equilibrium equations for the system of Fig. 2.2b.

Problems 2.18 through 2.28

(3) Physical force-geometry relations

In mechanical systems three important classes of physical relations are force-displacement, force-velocity, and force-acceleration relations.

† We shall sometimes use the notation $\overset{\frown}{\Sigma M^*}$ or $\overset{\frown}{\Sigma M^*}$ to indicate the direction we are taking as positive when summing moments in a plane-motion problem. Thus in (2.4) we would write: $\overset{\frown}{\Sigma M^*} = \ldots$.

42   EQUATIONS OF MOTION: SIMPLE PHYSICAL SYSTEMS   [Chap. 2]

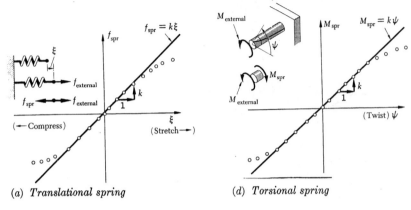

(a) *Translational spring* (d) *Torsional spring*

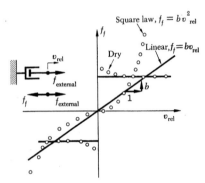
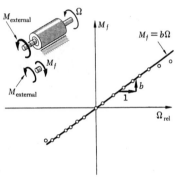

(b) *Translational friction (damper)* (e) *Torsional friction (damper)*

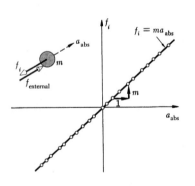
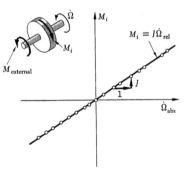

(c) *Mass under translational acceleration* (f) *Inertia under angular acceleration*

Fig. 2.4 *Measured physical relations for lumped mechanical elements, with linear approximations*

[Sec. 2.2]   ONE-DIMENSIONAL MECHANICAL SYSTEMS   43

Force-displacement relations. A displacement-opposing force may be produced by distortion, or by displacement in a force field (e.g., the gravity field). For example, an external force is required to compress a spring. Therefore, when a spring is connected to another element (as k_5 is connected to m_1 in Fig. 2.2a) and the element is then displaced, the force exerted *by* the spring *on* the other element is directed opposite to the displacement. (The spring tries to return to its neutral state.)

The force in a stretched or compressed spring (as in Fig. 2.2a) or the torque in a twisted shaft (Fig. 2.2b) is often found, experimentally, to be approximately proportional to displacement, as shown in Fig. 2.4a and d. The small circles represent typical plotted data points, and the straight line represents a linear approximation. Constant k has the dimensions of force/displacement: f/ℓ; or of moment/angular displacement: $f\ell$. (A radian is a dimensionless ratio.)

The restoring torque on a displaced pendulum (Fig. 2.2c) is due to its reorientation in the gravity field. For small angles, it is nearly linearly proportional to displacement: $M \cong (mg\ell)\theta = k\theta$. The exact expression, of course, is $M = (mg\ell) \sin \theta = k \sin \theta$, again opposing displacement. (What are the dimensions of $mg\ell$?)

Force-velocity relations. It is usually some sort of friction that produces a force opposing relative velocity (which makes the phenomenon dissipative†). In Fig. 2.2a rubbing of block 2 on block 1 and of block 1 on the ground will produce, respectively, the forces f_4 and f_3 in Fig. 2.3a, which are related to the relative velocities involved. (That is, f_4 is related to the velocity of block 2 relative to block 1, and f_3 is related to the velocity of block 1 relative to the ground.) On the rotors of Fig. 2.2b, bearing friction and other effects may produce moments M_3 and M_4 shown in Fig. 2.3b opposing the angular velocity of the rotor relative to its case.

These friction relations are not likely to be very linear. Typical measured data will resemble one of the three curves in Fig. 2.4b. The linear one is found only when there is laminar fluid flow (flow in which the fluid shears smoothly in layers, without turbulent eddies), which is generally attributable to high viscosity. Hence the name "viscous damping." For dry friction the simpler horizontal line may be a good approximation. The approximately parabolic relation in Fig. 2.4b (sometimes called "square-law damping") accompanies turbulent flow. When this exists we shall often use a linear approximation (solid line). This approximation turns out to be surprisingly useful. Constant b has the dimensions of force/velocity, $f/\ell\tau^{-1}$ (Fig. 2.4b), or of moment/angular velocity, $f\ell/\tau^{-1}$ (Fig. 2.4e).

Note that the friction force due to a relative velocity is always in the direction to oppose that velocity (e.g., f_4 in Fig. 2.3a and f_f in Fig. 2.4b).

Force-acceleration relations. The most precise measurements made (Fig. 2.4c) indicate that the relation between force on a mass and the resulting

† This is discussed in Sec. 2.3.

absolute acceleration† of the mass is perfectly linear (so long as relativistic velocities are not involved). This is a remarkable and extremely important fact. If the relation were nonlinear, dynamics and technology generally would doubtless be far behind their present state. The proportionality constant m has, of course, the dimensions of force/acceleration: $f/\ell\tau^{-2}$. (Refer to Sec. 1.5.)

Like the force-displacement and the force-velocity relations described above, the force-acceleration relation (Newton's law) is a purely empirical physical relation, based exclusively on experimental measurements.

Newton's law and D'Alembert's principle. Newton's Second Law is given as follows for a mass particle (refer to Appendix D):

$$\Sigma f = m a_{\text{abs}} \quad \text{in any direction} \tag{2.5}$$

in which Σf is the sum of all "real" forces acting in a given direction, and a_{abs} is the resulting absolute acceleration in that direction. Equation (2.5) is readily extended to apply verbatim to a rigid body in translation, with a_{abs} the acceleration of the body's mass center.

Newton's law can be further extended, for a *rigid body in pure rotation about a fixed axis*, Fig. 2.4f:

$$\Sigma M = J\dot{\Omega} \quad \text{fixed axis} \tag{2.6}$$

in which ΣM and angular velocity Ω are taken about the fixed axis,‡ and J is a mass property of the body, called its *moment of inertia*. J is defined generally by§

$$\text{Moment of inertia} \quad \boxed{J \triangleq \int_m r^2\, dm} \tag{2.7}$$

in which dm is an element of mass, r is distance from the given axis to dm, and integration is performed over the body. For a uniform disk symmetrical about its mass center, for example, the integral gives

$$J = \tfrac{1}{2} m r_o^2 \tag{2.8}$$

in which m is the disk's mass and r_o its outer radius. The dimensions of J are moment/angular acceleration: $f\ell/\tau^{-2}$.

---

† Formally, absolute acceleration is acceleration with respect to a reference frame that is not itself accelerating. For many problems a fixed point on the earth will do, the small centripetal acceleration due to the earth's rotation being relatively negligible. (This effect is evaluated in Example 5.3.)

‡ In this book, for clarity, we shall use the symbol Ω for angular velocity, reserving the symbol ω always for frequency, as in $\sin \omega t$.

§ The symbol $\triangleq$ means "is defined to be equal to."

[Sec. 2.2] ONE-DIMENSIONAL MECHANICAL SYSTEMS

The further extension of Newton's Second Law to a rigid body in arbitrary general motion is described in Sec. 5.3.

D'Alembert pointed out, in a rule known as *D'Alembert's principle*, that (2.5) can be written in the alternative form

$$\Sigma f^* \triangleq \Sigma f + f_i = 0 \qquad \text{in any direction} \tag{2.9}$$

[same as (2.1)] with f_i defined, for a rigid body, by

$$f_i \triangleq m a_{c,\text{abs}} \qquad \text{directed opposite to } a_{c,\text{abs}} \tag{2.10}$$

in which a_c is acceleration of the body's mass center.

This suggestion of D'Alembert's has far-reaching implications in dynamic analysis. At this point, however, we note only that D'Alembert's principle allows us to cast the physical relation for "acceleration force" in the same role as those for force due to displacement and velocity. It is this feature that makes the D'Alembert approach so convenient in systems analysis.

For a *rigid body in pure rotation about a fixed axis through its mass center*, D'Alembert's principle can be applied to (2.6) to write the equilibrium relation

$$\Sigma M^* \triangleq \Sigma M + M_i = 0 \tag{2.11}$$

with M_i defined by

$$M_i \triangleq J \dot{\Omega}_{\text{abs}} \qquad \text{directed opposite to } \dot{\Omega}_{\text{abs}} \tag{2.12}$$

These relations are useful in studying rotors, turbines, gears, pulleys, and so forth. Equations (2.11) and (2.12) actually represent a special case of (2.2), which has much greater generality, as we shall see in Secs. 5.3 and 5.4.

We shall refer to Eqs. (2.1) and (2.2) as *equations of dynamic equilibrium*. We shall refer to the quantities f_i and M_i as *inertial force* and *inertial moment*, respectively.

It was with this form of the equation of force equilibrium in mind that we showed the inertial forces f_1 and f_2 in Fig. 2.3a and the inertial moments M_1 and M_2 in Fig. 2.3b. These are body forces, much like gravitational forces. It should perhaps be noted explicitly that D'Alembert's principle is entirely rigorous, and that the concept of inertial forces is as legitimate as Newton's Second Law.[†]

We elect to use the D'Alembert method in this book because it is convenient and systematic (as noted above), because (as we shall see in Secs. 5.3, 5.4, and 5.5) it provides important analytical simplification in more advanced problems, and because physical intuition is strengthened by thinking of inertial forces opposing acceleration whenever it tries to occur. The

---

† Some people like to draw each inertial force in red, as a reminder that they are dealing with a force of a different color.

reader will certainly have little difficulty substituting the Newtonian method if he prefers.

Summary. The principal physical relations for mechanical elements are as shown in the typical experimental plots of Fig. 2.4. These can be represented by the following linear approximations:

Translational elastic spring	$f = k\xi$ opposing ξ	(2.13)
Translational viscous damping	$f = bv_{rel}$ opposing v_{rel}	(2.14)
Translational inertial mass	$f = ma_{abs}$ opposing a_{abs}	(2.15)
Rotational elastic spring	$M = k\psi$ opposing ψ	(2.16)
Rotational viscous damping	$M = b\Omega_{rel}$ opposing Ω_{rel}	(2.17)
Rotational inertial mass	$M = J\dot{\Omega}_{abs}$ opposing $\dot{\Omega}_{abs}$	(2.18)

Relations (2.15) and (2.18) are exact.† The others are approximate, their accuracy depending on the device, as indicated in Fig. 2.4.

We shall now demonstrate the concepts discussed above with three very simple examples and with a number of suggested problems (Problems 2.29 through 2.41). Other, more advanced examples will be found in Chapter 5 and in other sections throughout this book.

Example 2.1. A "torque motor." Figure 2.5 shows a physical model of a simple system that consists of a rotor having moment of inertia J, mounted in bearings and restrained by a flexible shaft. The rotor is driven by a torque applied electromechanically via windings not shown. (More detail is given in Fig. 6.1, p. 184.) Such a device is used, for example, to control the opening of a hydraulic valve. Suppose that a moment is applied which is some specified function of time, $M_1(t)$.‡ Derive the equations of motion for the system.

† Excluding relativistic effects (which can be neglected for velocities that are small compared with the velocity of light—186,000 miles per second).
‡ The symbol $M_1(t)$, read "M_1 of t," may be thought of conveniently as the plot of M_1 versus t.

Fig. 2.5 *Physical model of a torque motor*

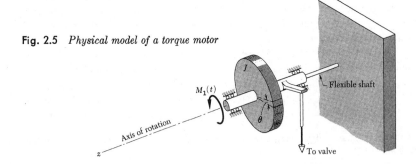

[Sec. 2.2] ONE-DIMENSIONAL MECHANICAL SYSTEMS 47

STAGE I. PHYSICAL MODEL. We assume, as implied in the problem statement, that $M_1(t)$ does not depend on the motion of the rotor. We assume that the bearings are very good, so that there is only rotary motion about the bearing axis. For small motions the restoring torque exerted by the twisted shaft on the rotor will be very nearly proportional to the angle of twist; in our model we assume that this proportionality is exact. Finally, we assume that all of the resisting force from the valve plus the bearing friction torque simply depends linearly on (and, of course, opposes) the rotor angular velocity. (Tests show that this is approximately true and, as we have discussed, linear approximation greatly facilitates the analysis in Stage III.)

STAGE II. EQUATIONS OF MOTION. We follow Procedure A-m.

(1) *Geometry.* In Fig. 2.5 we "picture the system in an arbitrary configuration" and define coordinate θ, positive when it is counterclockwise (as viewed from the z axis). The derivatives of θ ($\dot\theta$, angular velocity, and $\ddot\theta$, angular acceleration) will then be positive when they are counterclockwise.

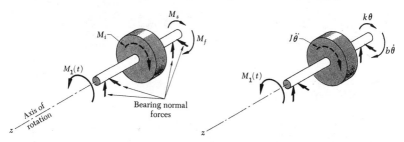

(a) *With unknown moments*   (b) *With moments represented by their physical-relation equivalents*

Fig. 2.6 *Free-body diagram of torque-motor rotor*

(2) *Forces (equilibrium).* (i) Before we can write force-equilibrium relations we must select a subsystem and *draw a free-body diagram.* We select the rigid rotor alone, replacing the bearings and the limber shaft by the forces and moments they exert *on* the rotor, Fig. 2.6a.

We elect (arbitrarily) to call the moments positive in the directions indicated (i.e., opposing θ). (ii) Now we can write an equation of dynamic equilibrium (2.2), summing moments about the $-z$ axis:

$$\Sigma M^* = 0 = M_i + M_f + M_s - \dot M_1(t) \tag{2.19}$$

(3) *Physical force-geometry relations.* Our physical model assumes linearity throughout, so that M_s, M_f, and M_i are given by (2.16), (2.17), and (2.18), respectively:

$$M_s = k\theta \qquad M_f = b\dot\theta \qquad M_i = J\ddot\theta \tag{2.20}$$

[Is it clear to you that the signs in (2.20) are all positive because of the choices of (1) the direction in which motion is taken positive in Fig. 2.5, *and* (2) the directions in which M_s, M_f, and M_i are taken positive in Fig. 2.6a? For example,

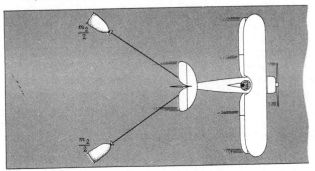

(a) *Actual arrangement*

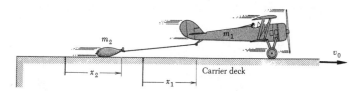

(b) *Side view*

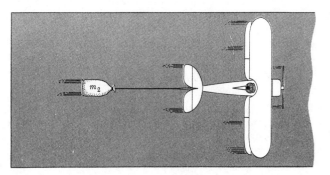

(c) *Simpler arrangement for first analysis*

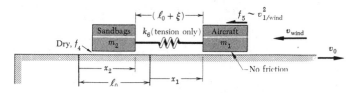

(d) *Physical model (in arbitrary configuration)*

Fig. 2.7 *Early carrier-landing aircraft-arresting scheme*

[Sec. 2.2] ONE-DIMENSIONAL MECHANICAL SYSTEMS 49

if θ is positive, then will M_s be positive or negative—which way will the spring try to turn the rotor?]

Combining relations [entering (2.20) into (2.19)] we obtain the equation of motion

$$J\ddot{\theta} + b\dot{\theta} + k\theta - M_1(t) = 0 \qquad (2.21)$$

Steps (2) and (3) may be further condensed by indicating the physical relations directly on the free-body diagram, as in Fig. 2.6b, and then simply writing (2.21) directly by inspection of the diagram. This makes it even easier to check the signs against intuition.

Show that each term in (2.21) has the dimensions of moment.

In Part B we shall be solving (2.21) to determine the time behavior (Stage III), $\theta(t)$, of the rotor in Fig. 2.5 under many circumstances. (In fact, this particular system is a much-used example throughout Part B.)

Example 2.2. An arresting system for airplanes. On early aircraft carriers, landing aircraft were stopped in the manner illustrated in Fig. 2.7. As the aircraft tail touched down, it hooked an arresting wire which was connected at each end to a large sandbag. The aircraft's kinetic energy was then dissipated as the sandbags were dragged across the deck. Derive a set of equations of motion suitable for a first preliminary analysis of the aircraft's forward motion after it has hooked the arresting wire.

STAGE I. PHYSICAL MODEL. Although there are two sandbags, and the angle between the wire and the fore-and-aft deck line varies during the landing, we make a first study of the dynamics based on a slightly simpler model in which there is a single sandbag and a straight wire, as shown in Fig. 2.7c. We do not, however, omit the stretchability of the wire, which appears intuitively to play an important role in the dynamic behavior.† We assume that the friction between sandbag and deck is probably of the "dry" variety, Fig. 2.4b. We neglect friction between the aircraft and the deck (the pilot does not use his brakes in a carrier landing). Wind drag on the aircraft will vary as the square of the velocity of the aircraft relative to the air.

In our first analysis we neglect rotational motion of the aircraft, and we also assume that the carrier is steaming in a straight line with constant velocity. We indicate our various assumptions in the simplified diagram of Fig. 2.7d. [Will this model be valid if the cable is not always in tension? See (2.24) below.] This is a *one-dimensional* model because it concerns motion only in one direction, the x direction.

STAGE II. EQUATIONS OF MOTION. We now derive the equations of motion for our model, following Procedure A-m.

(1) *Geometry.* Let the constant velocity of the ship be v_0. Let the location of the aircraft and sandbag be specified with respect to the deck by coordinates x_1 and x_2, respectively, as they are carefully defined in Fig. 2.7b or d. The reference configuration is that at touchdown, when the tail has just hooked the wire, but has not stretched it yet. The velocities of the two masses will be given (for

---

† In the model, should the spring constant k_6 of the single wire be the same as that of the two wires it has replaced? (Explain on geometrical grounds.)

this simple one-dimensional situation) by the geometric identities

$$\text{Absolute velocities} \begin{cases} v_{1,\text{abs}} \equiv \dot{x}_1 + v_0 \\ v_{2,\text{abs}} \equiv \dot{x}_2 + v_0 \end{cases}$$

$$\text{Relative velocities} \begin{cases} v_{1/\text{wind}} \equiv \dot{x}_1 + v_0 + v_{\text{wind}} \\ v_{2/\text{deck}} \equiv \dot{x}_2 \end{cases} \quad (2.22)$$

and their absolute† accelerations will be given by the geometric identities

$$\text{Absolute accelerations} \begin{cases} a_{1,\text{abs}} \equiv \dfrac{d}{dt}(\dot{x}_1 + v_0) \equiv \ddot{x}_1 \\ a_{2,\text{abs}} \equiv \dfrac{d}{dt}(\dot{x}_2 + v_0) \equiv \ddot{x}_2 \end{cases} \quad (2.23)$$

because v_0 is constant. Because of our choice of reference configuration, the amount of stretch in the cable is given by the identity

$$\xi \equiv x_1 - x_2 \qquad x_2 < x_1 \quad (2.24)$$

Identity (2.24) holds only for $x_2 < x_1$. Otherwise, the cable is not stretched, but collapses into some indeterminate shape, and exerts no force on either the aircraft or the sandbag.

(2) *Force equilibrium.* (i) We must first select and "isolate" free bodies. It is convenient to isolate the two masses (airplane and sandbag), replacing their connection with the rest of the world by the forces the world exerts on them, as shown in Fig. 2.8a.

(ii) We elect to use D'Alembert's form (2.1) for force equilibrium, and thus we show the inertial reaction forces f_1 and f_2 on the free-body diagram. We write the equations of equilibrium in the form (2.1), each in the $-x$ direction

$$\begin{array}{ll} \text{For the aircraft:} & \Sigma f^* = f_1 + f_5 + f_6 = 0 \\ \text{For the sandbag:} & \Sigma f^* = f_2 + f_4 - f_6 = 0 \end{array} \quad (2.25)$$

(Since we are considering only motion in the horizontal direction, we do not need the equilibrium relations for the vertical direction.)

(3) *Force-geometry physical relations.* From (2.13), Fig. 2.4b, and (2.15) [note the signs as defined (arbitrarily) in the free-body diagram]

$$\begin{array}{ll} f_2 = m_2 a_{2,\text{abs}} & f_1 = m_1 a_{1,\text{abs}} \\ f_4 = f_{40}\, \text{sgn}\,(v_{2/\text{deck}}) & \\ f_6 = \begin{cases} k_6 \xi & \text{if} \quad x_2 < x_1 \\ 0 & \text{if} \quad x_2 > x_1 \end{cases} & f_5 = \beta_5 (v_{1/\text{wind}})^2 \end{array} \quad (2.26)$$

in which sgn( ) means $+1$ if the sign of ( ) is positive and -1 if the sign of ( ) is negative.‡ [What definition for sgn(0) is appropriate for this problem?] ξ is, of course, the stretch in the cable (not its total length). Do you agree with the signs in (2.26)? (Refer to the free-body diagrams.)

Combine relations. Counting, we find that we have 14 equations in 14 unknowns (v_0 is known). To eliminate some variables we first enter (2.22), (2.23),

---

† We adopt a point on the earth's surface as a nonaccelerating reference for this problem. (See footnote, p. 44.) Wind velocity, v_{wind}, is taken positive in the direction drawn because a carrier heads into the wind to land aircraft.

‡ The abbreviation sgn is read "signum."

[Sec. 2.2] ONE-DIMENSIONAL MECHANICAL SYSTEMS 51

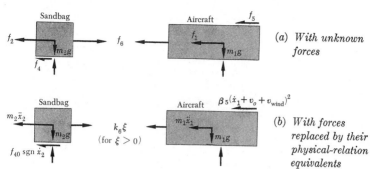

Fig. 2.8 *Free-body diagrams of aircraft and sandbag*

and (2.24) into (2.26), and then enter (2.26) into the equilibrium equations (2.25). The result is

$$m_1\ddot{x}_1 + \beta_5(\dot{x}_1 + v_0 + v_{wind})^2 + k_6(x_1 - x_2) = 0 \qquad x_2 < x_1$$
$$m_2\ddot{x}_2 + f_{40} \operatorname{sgn} \dot{x}_2 - k_6(x_1 - x_2) = 0 \qquad x_2 < x_1 \quad (2.27)$$

These are the equations of motion of the system. (Are the dimensions consistent? Consider some special cases—e.g., no spring, $k = 0$; or very stiff spring, $k \to \infty$. Do the equations make sense for the special cases? What are the equations for $x_2 > x_1$?)

Notice that the equilibrium equations are the focal point for combining the results of the individual steps.

A useful alternative to the above procedure is to display relations (2.26) on the free-body diagrams, as in Fig. 2.8b. The equations of motion (2.27) can then be written directly from the free-body diagram.

Careful contemplation of a sketch such as Fig. 2.8b yields additional physical insight. For example, give careful thought to each of the signs in Fig. 2.8b. Does every one make sense physically?

If the carrier were not moving at constant velocity, but had also a substantial acceleration, a_0, what changes would be made in relations (2.22), (2.23), and (2.24)? In Fig. 2.8b?

Contrive and sketch a more elaborate physical model of the above situation, in which flexibility of the aircraft landing gear and a two-sandbag arrangement are included. Discuss the geometric relations required to analyze this model.

In Stage III we shall be solving Eqs. (2.27) simultaneously to obtain x_1 versus time, the motion of the aircraft as it moves down the flight deck. We shall be interested to learn, for example, whether $\dot{x}_1$ reaches zero before x_1 reaches L, the total distance to the end of the deck.

Problems 2.29 through 2.41

As the final mechanical example in this section we consider a system in which geometric compatibility is a more convenient focal point than force equilibrium for deriving the equations of motion. (This situation is less common in dynamics.)

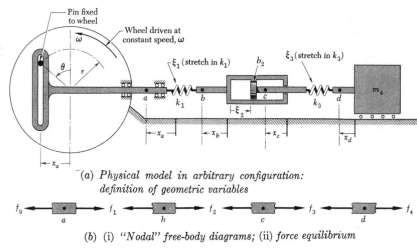

(a) *Physical model in arbitrary configuration: definition of geometric variables*

(b) (i) *"Nodal" free-body diagrams;* (ii) *force equilibrium*

Fig. 2.9 *Scotch yoke mechanism*

Example 2.3. A force transmission system. Figure 2.9 shows a Scotch yoke mechanism for driving machine elements. The yoke has the special virtue of delivering a motion that is exactly sinusoidal. It happens that in this test arrangement the input motion is transmitted through a viscous dashpot. Moreover, the connecting links are not completely rigid, but stretch and compress slightly in proportion to the force in them. Guides permit horizontal motion only.

For design purposes it is desired to find, versus time, the force f_3 in connecting rod k_3. Write the differential equation of motion from which f_3 can be computed.

STAGE I. PHYSICAL MODEL. The lumping of elements in our physical model into massless extensible rods, a rigid mass, and a massless dashpot is indicated in Fig. 2.9a. We assume that the connecting rods behave as stiff linear springs, and that the dashpot has a linear force-velocity characteristic. We assume further that friction forces other than the dashpot force can be neglected.

STAGE II. EQUATIONS OF MOTION. In this problem it will be convenient to use geometric compatibility as the focal relation.

(1) *Geometry.* The definitions of several coordinates, θ, x_a, ..., x_d, and ξ_2 are given by the arbitrary-configuration sketch of Fig. 2.9a. Let ξ_1 and ξ_3 be the *stretch* in springs k_1 and k_3, and let all ξ's (as well as x's and θ) be zero in the reference configuration. (How many degrees of freedom has the system? Write identities relating the ξ's and x's. Let the springs have unstretched lengths ℓ_1 and ℓ_2.)

Geometric compatibility for the entire system requires that the sum of the stretches in the individual elements must equal the total stretch in the system:

$$\xi_1 + \xi_2 + \xi_3 = x_a - x_d \tag{2.28}$$

We note also that x_a is constrained by the yoke:

$$x_a = r \sin \theta = r \sin \omega t$$

(2) *Force equilibrium.* (i) *Free-body diagrams:* In this problem it is propitious to consider free bodies that are not masses (since there is only one mass in the system). Instead we select small "massless" points at the connection between each pair of elements. These are labeled a, b, c, d in Fig. 2.9a. Free-body diagrams for these† are shown in Fig. 2.9b, from which (ii) *force equilibrium* gives

$$f_1 = f_2 = f_3 = f_4 \tag{2.29}$$

in which each of f_1, f_2, f_3 is taken positive when its member is in tension.

(3) *Force-geometry physical relations.* Using inverted versions of (2.13), (2.14), (2.15) (and appropriate identities):

$$\xi_1 = \frac{f_1}{k_1} \qquad \xi_2 = \frac{\int f_2 \, dt}{b_2} \qquad \xi_3 = \frac{f_3}{k_3} \qquad x_d = x_4 = \frac{\iint f_4 \, dt}{m_4} \tag{2.30}$$

Combine relations. Entering (2.30) into (2.28), and using (2.29) to express all quantities in terms of f_3, in which we are interested, we obtain

$$\frac{f_3}{k_1} + \frac{\int f_3 \, dt}{b_2} + \frac{f_3}{k_3} + \frac{\iint f_3 \, dt}{m_4} = r \sin \omega t \tag{2.31}$$

which is the equation of motion for the system. In Stage III we are to solve it for the value of f_3 versus time.

It should be noted that the time response may also be calculated (with considerably more labor) from the equations of motion obtained by substituting the physical relations directly into the equations of force equilibrium, (2.29).

Problems 2.42 through 2.46

2.3 MECHANICAL ENERGY AND POWER

For future reference we note that a mechanical system can store energy as potential energy in a spring or a displaced pendulum, or as kinetic energy in a moving mass; and it can "dissipate" energy (convert it into heat) through damping devices.

Potential-energy storage depends on a change in system configuration; and the amount of energy stored ΔU equals the work ΔW required to change

† For simple one-dimensional systems, it may be found convenient to write force equilibrium for free-body diagrams of massless points (or nodes) as a general routine, and some people prefer this because of the similarity to writing Kirchhoff's Current Law at a network node. However, for two- and three-dimensional systems like Fig. 2.11 (p. 57), in which forces and moments are distributed over rigid bodies, the node method cannot be used. (Occasionally, artificial "equivalent" one-dimensional systems can be contrived whose equations of motion have the same form as a given three-dimensional system, and for which node equations can be written; but the process of contriving the equivalent system is seldom worth-while, and is a diversion from the actual physical system.)

the configuration. The calculation has the form

$$\Delta U = \Delta W = \int \mathbf{f} \cdot d\mathbf{x} \tag{2.32}$$

($\mathbf{f}$ is force and $d\mathbf{x}$ is displacement), in which the dot product indicates that only the component of force along the direction of displacement does work (see Appennix B). For a linear spring, $f = k\xi$, and the total energy stored is

$$U = \tfrac{1}{2}k\xi^2 \tag{2.33}$$

where ξ is the *total* amount the spring is stretched or compressed.

For a mass in a gravity field, the change in potential energy is simply the weight times the change in altitude:

$$\Delta U = mg\,\Delta h \tag{2.34}$$

Kinetic-energy storage depends on velocity. A mass m in *pure translation* at speed v has kinetic energy T, given by

$$T = \tfrac{1}{2}mv^2 \tag{2.35}$$

A mass in *pure rotation* at angular velocity Ω has

$$T = \tfrac{1}{2}J\Omega^2 \tag{2.36}$$

in which J is the moment of inertia about the axis of rotation, (2.7). (The general, three-dimensional case is discussed in Sec. 5.6.)

Power is the time rate of doing work. Thus, the power dissipated in a frictional device is given by

$$P = \mathbf{f} \cdot \mathbf{v} \tag{2.37}$$

The power used in compressing a spring is given by $P = f\,d\xi/dt = k\xi\dot{\xi} = \dot{U}$, the rate of change of potential energy [from (2.33)]. Show that the power used to accelerate a mass in a straight line is the rate of change of kinetic energy.

For a linear damper $f = bv$ (along v), and

$$P = bv^2 \tag{2.38}$$

which is, of course, the rate at which energy is being dissipated in the device. What are the dimensions of power? (Refer to Table II, p. 28.)

2.4 GEAR TRAINS AND LEVERS

An ideal gear train or lever is a device that transmits energy from one part of a mechanical system to another in such a way that both the force level and the motion "level" are altered, but the power level is not. Real gears and levers, of course, have inertia and friction (and often appreciable compliance as well), but these are neglected in the idealized devices. (Some more realistic models of gears and levers are suggested for study in Problems 2.47 through 2.54.)

[Sec. 2.4] GEAR TRAINS AND LEVERS

Consider the ideal (massless, frictionless, rigid) gear train in Fig. 2.10a. It is transmitting both torque and velocity. The ratio of its radii is called N:

$$N \triangleq \frac{r_2}{r_1} \qquad (2.39)$$

(N is also the ratio of the number of teeth on the two gears, of course.)

From *geometric compatibility* the surface speeds must be the same (to avoid slipping). Therefore

$$\Omega_2 r_2 = \Omega_1 r_1$$

or

$$\boxed{\frac{\Omega_2}{\Omega_1} = \frac{r_1}{r_2} = \frac{1}{N}} \qquad (2.40a)$$

From *force equilibrium* on the free bodies in Fig. 2.10b (where f_3 is the force exerted on one another by the gear teeth), we have

GEAR TRAIN:

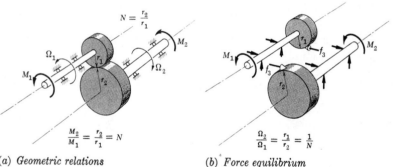

(a) Geometric relations   (b) Force equilibrium

LEVER:

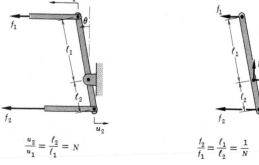

(c) Geometric relations   (d) Force equilibrium

Fig. 2.10 *Ideal gear train and lever*

$$M_1 = f_3 r_1$$
$$M_2 = f_3 r_2$$
$$\boxed{\frac{M_2}{M_1} = \frac{r_2}{r_1} = N} \qquad (2.40b)$$

[With the definitions of positive directions indicated in Fig. 2.10a, all signs are plus in (2.40a) and (2.40b).] Equations (2.40) are the governing relations for the gear train. From (2.40a) and (2.40b), an ideal gear train transmits power undiminished:

$$P = M_1 \Omega_1 = M_2 \Omega_2 \qquad \text{for the ideal gear train} \qquad (2.41)$$

Levers transmit translational motion and force in the same way that gears transmit rotational motion and torque. Consider the ideal, massless lever in Fig. 2.10c. The reader may readily show that its "input–output" relations are

From Fig. 2.10c: $\qquad \theta = \dfrac{u_1}{\ell_1} = \dfrac{u_2}{\ell_2} \qquad$ for small θ

From Fig. 2.10d: $\qquad f_1 \ell_1 = f_2 \ell_2$

or, with

$$N \triangleq \frac{\ell_2}{\ell_1} \qquad (2.42)$$

$$\boxed{\frac{u_2}{u_1} = \frac{\ell_2}{\ell_1} = N} \qquad (2.43a)$$

$$\boxed{\frac{f_2}{f_1} = \frac{\ell_1}{\ell_2} = \frac{1}{N}} \qquad (2.43b)$$

Equations (2.43) are the governing relations for a lever. Note that in the drawings $N < 1$ for the lever, but $N > 1$ for the gear train, so that $u_2 < u_1$ and $f_2 > f_1$ in both cases.

Like the ideal gear train, the ideal lever converts both motion and force to new levels, while transmitting power unchanged. That is, from (2.43),

$$P = f_2 u_2 = f_1 u_1 \qquad \text{for the ideal lever} \qquad (2.44)$$

Because they convert mechanical energy from one force and motion level to another, gear trains and levers are sometimes called *mechanical-mechanical energy converters*.

Problems 2.47 through 2.54

2.5 MOTION IN TWO AND THREE DIMENSIONS

We note again that, for simplicity, we have considered here a very restricted class of mechanical problems; namely, one-dimensional ones.

More generally, as in Fig. 2.11 for example, motion may be simultaneously translational and rotational in three-dimensional space; and individual forces will not converge at a single point (vertex), but may be distributed

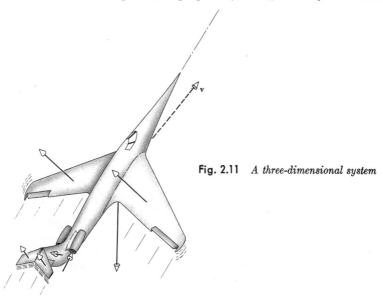

Fig. 2.11   *A three-dimensional system*

about the system, intersecting at many points, or at none, and producing moments about any axis.

The geometric description of such motions may become very involved, but the fundamental force-equilibrium relations (2.1) and (2.2) still apply, when used properly.

Three-dimensional motion is the subject of Chapter 5.

2.6 SIMPLE ELECTRICAL SYSTEMS

Electrical devices are familiar to us in our everyday activities. A light bulb, for example, is essentially a pure resistor, as are electric heaters, toasters, electric irons, and cigarette lighters. When one of these devices is connected to a voltage source—for example, a flashlight bulb to a battery—current flows through the resistance, and energy is converted to light and heat.

From studying the inside of a radio, or perhaps building one, we are familiar also with other electrical elements used in circuits, such as capacitors (sometimes called "condensers"), inductors, vacuum tubes, and transistors.

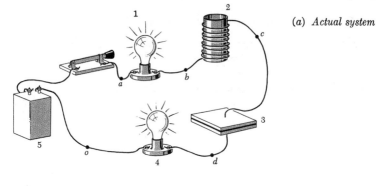

(a) *Actual system*

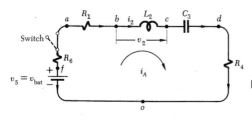

(b) *Physical model (network)*

Fig. 2.12 *A simple electrical circuit and its model*

When electrical elements are connected together in certain ways, the resulting "circuits" may be used to process and amplify electrical signals in communication, power, and control equipment. Very simple circuits are shown in Figs. 2.12 and 2.13. Electrical circuits are a vital part of nearly every modern technological system, from tiny hearing aids to vast electrical power complexes; from simple thermostats to intricate space-vehicle guidance systems.

In all electrical systems the signals which a circuit receives and processes and transmits are the electrical currents that run through its branches and the (time-varying) voltages that exist at the terminals of its elements. These currents and voltages are the quantities in terms of which we study the *dynamic behavior* of an electrical circuit, just as we study the behavior of mechanical systems in terms of the geometric displacements of elements and the forces exerted between them.

Currents and voltages are manifestations of the behavior of charged particles, which repel or attract one another statically according to Coulomb's law:†

$$\mathbf{f} \cong \frac{q_1 q_2}{r_{12}^2} \mathbf{1}_r \tag{2.45}$$

and which exert additional forces on one another when they are in motion, as Ampére, Faraday, Henry, and others discovered. The fundamental laws of

† Vector $\mathbf{f}$ is the force of repulsion on particle 1, q_1 and q_2 are the charges on the two particles, r_{12} is the distance between them, and $\mathbf{1}_r$ is a unit vector pointing at particle 1 from particle 2.

[Sec. 2.6] SIMPLE ELECTRICAL SYSTEMS

electrical behavior are summarized in Maxwell's equations (just as mechanical behavior is described by Newton's laws).

Quantity of *charge* is measured in coulombs. (The charge of one electron is -1.6×10^{-19} coulomb.) We have taken charge as a primary dimension, q, in Table II, p. 25.

Physically, an electrical *current* consists of a stream of charged particles (usually electrons) flowing through a conductor. The current is precisely defined as the total net quantity of charge passing any cross section of the conductor *per unit of time*. Current is a through variable.

Current is measured in *amperes*. A flow of 1 coulomb per second is defined as 1 ampere. (The fundamental dimensions are $q\tau^{-1}$, Table II, p. 25.)

Physically an electrical *voltage* difference connotes an electric potential difference between two points: It implies that a certain amount of work must be done to move a charged particle from one of the points to the other. Voltage difference is an across variable.

Voltage is measured in *volts*. The electric potential difference $v_a - v_b$ between two points a and b is $+1$ volt if 1 newton-meter (10^7 dyne-centimeters) of work is required to move a particle having 1 coulomb of charge from point b to point a. (That is, the voltage difference is the difference in potential energy *per unit charge*. The fundamental dimensions are $f\ell q^{-1}$.)

Electric potential difference (voltage) is perfectly analogous to gravitational potential difference: The gravitational potential difference $U_a - U_b$ between points a and b is unity if 1 newton-meter of work is required to move a mass of 1 kilogram from point b to point a.

Consider, for example, the circuit in Fig. 2.12a. After the switch has been closed, a current of electrons flows through the elements of the system producing different voltage levels at the various junctions (nodes) between elements. Unfortunately, we cannot "see" an electrical current or voltage in the same sense that we can see the displacement of a mechanical device. We can, however, readily measure electrical current and voltage, and we can develop a physical image of their behavior, using, for example, the analogy to the flow of water in a pipe suggested by Benjamin Franklin, on the basis of which we can study and predict the behavior of electrical circuits quite satisfactorily. In the hydraulic analogy, the flow of electrons is likened to the flow of water, and hydraulic pressure is likened to voltage. Then in a hydraulic circuit of elements (connected by pipe of negligible resistance) the pressure (voltage) at various points along the hydraulic circuit will depend upon the pressure drop across each of the elements, which in turn depends upon the flow rate (current) through the elements.

Because the elements of an electrical circuit remain physically fixed (rather than changing their form and shape as the elements of a mechanical system do) the visualization and mathematical handling of circuit dynamics is, in fact, generally much simpler than the corresponding study for mechanical systems.

STAGE I. PHYSICAL MODEL

It is a great simplification, in circuit analysis, to consider as physical models only idealized circuits made up of lumped ideal electrical elements, each of which is characterized as either a pure resistance, pure capacitance, or pure inductance. We can on the one hand state categorically that real physical elements are *never* pure. All resistors display a certain amount of capacitive and inductive behavior; all capacitors have a certain amount of resistance, and so forth.

On the other hand, we can state just as strongly that most of the physical elements used to build circuits have *predominantly* the characteristic for which they were designed (either resistance, capacitance, or inductance) *almost* to the exclusion of the other characteristics. Indeed, in most cases a degree of purity can be achieved far beyond the requirements of a given system. For example, if a resistor is to be used in a circuit in which all of the signals of interest will be sinusoidal in the 0 to 200 cycles per second range, then the fact that it has capacitive properties that become manifest at frequencies above 10,000 cycles per second will not impair its usefulness as a pure resistor; and in contriving a model of the circuit for analysis the error we make by assuming the element to be a pure resistor will be entirely negligible.

At the same time, hi-fi builders will testify that when dealing with very-high-frequency equipment, great care must be taken (e.g., isolation of certain components) to ensure that the capacitive properties of "resistors" do not distort the signals passing through the system. In cases where this effect is in fact not negligible we may still develop a satisfactory model of "pure" elements by representing each real resistor as a combination of an ideal resistor plus an ideal capacitor. For purposes of analysis the problem is simply to identify lumps, in the real circuit, which can be represented by ideal elements in a model.

At this point let us introduce the term "electrical *network*" to connote the *physical model* of a real electrical circuit. In the network or model we permit only ideal lumped elements—principally resistors, inductors, and capacitors—together with ideal source elements. For example, the actual circuit of physical components in Fig. 2.12a is represented by the network of ideal lumped elements shown in Fig. 2.12b. A more complicated network is shown in Fig. 2.13. (The term "network" derives from the fishnet appearance of drawings like Fig. 2.13.)

For network models that represent only part of a larger system we want to be sure we have properly represented the situation at all *boundaries* of the system envelope—the "interfaces" between our system and the outside world—where voltages or currents may be specified, and through which electrical power may be flowing. Commonly, inputs from the outside world are represented in a network as idealized "sources." (We shall discuss sources presently.)

[Sec. 2.6] SIMPLE ELECTRICAL SYSTEMS 61

STAGE II. EQUATIONS OF MOTION

Analytical procedure

When the physical model of a system consists of electrical elements only, the considerations listed in Sec. 2.1 can be written in the following more specific terms:

> (1) **Network variables.**† *Define variables*—voltages and currents. Note *identities*. Note *constraints* imposed by sources.
> (2) **Equilibrium.** *Apply Kirchhoff's Current Law* (KCL) (node analysis); or, **Compatibility.** *Apply Kirchhoff's Voltage Law* (KVL) (loop analysis).‡
> (3) **Physical voltage-current relations.** Write these for the individual elements.

Procedure A-e
Electrical equations of motion

Then *combine relations* [inserting (1) and (3) into (2), for example].

As with Procedure A-m, the equilibrium or compatibility relations are the focus in deriving equations of motion.

Again, as with Procedure A-m,§ the steps can be summarized more tersely:

> (1) **Network variables**
> (2) **Kirchhoff's laws**
> (3) **Voltage-current relations**

Procedure A-e
Electrical

Then *combine relations.*
We now discuss each item in turn.

(1) Network variables

The topology of electrical networks is usually simple and routine, so that often it is not even considered explicitly, and Step (1) is not listed.

† This step is commonly not used explicitly, but is merely implied.
‡ Some people prefer to use Kirchhoff's Current Law exclusively (which is always sufficient). Others elect to write either KCL or KVL, depending on which is more convenient. It is never necessary, and seldom convenient, to use both laws in the same problem. (The useful abbreviations KCL and KVL are introduced in B. Friedland, O. Wing, and R. Ash, "Principles of Linear Networks," McGraw-Hill Book Company, New York, 1961.)
§ Procedure A-e is seen to follow exactly the form of the list of considerations in Sec. 2.1. Procedure A-e is thus not completely analogous to Procedure A-m, which takes across variables and compatibility in Step (1) and through variables and equilibrium in Step (2). The reason why the convenient subdivision for electrical systems differs slightly from that for mechanical systems is that *either* KCL *or* KVL is used in an electrical problem, but seldom both; while geometry and forces usually require equal levels of attention in a mechanical problem.

But whether variables are considered explicitly or only implicitly, precision in their definition is essential.

The *variables* in electrical networks are voltages and currents; and the *state* of an electrical system is specified by an appropriate set of voltages and/or currents and their derivatives. Both node voltages (v_b, v_c, etc., in Fig. 2.12b) and voltage drops across elements ($v_2 \equiv v_b - v_c$) are of interest. Similarly, both branch currents (i_2) and loop currents (i_A) are of interest.

To ensure clarity in discussing network geometry, the parts of a network, such as those in Fig. 2.12b and 2.13, are given names. Each physical

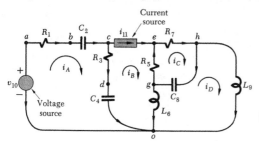

Fig. 2.13 *Network of a more general nature*

element in the network is called a *branch*; each junction where elements are fastened together is called a *node*; and any closed path (beginning at a node and tracing around through any set of elements back to the same node again) is called a *loop*.

We can contribute further to the simplicity of handling network variables by adopting a consistent nomenclature to use in all of our studies of electrical networks. The following set of conventions is suggested. (The conventions are quite arbitrary, but are very useful.)

The network elements—resistors, inductors, capacitors, sources, etc.—have two terminals (one at each end). Label each element with a number, as in Fig. 2.12b and Fig. 2.13, and label with a small letter the junctions (nodes) where the terminals of elements are connected together (a, b, etc.). On a wire of each element draw an arrow; and agree (arbitrarily) that current will always be considered positive in the direction of the arrow,† and (also arbitrarily) that voltage *drop* will also be considered positive in the direction of the arrow. The current in an element and the voltage drop across it are designated with the element's own number, as i_6 and v_6 in Fig. 2.13. (Node voltages have letter subscripts, as v_d.) With this nomenclature the physical relation for each individual element will involve an i, a v, and a parameter, all with the same number subscript, as $v_6 = L_6(di_6/dt)$ in Fig. 2.13.

Voltage *identities* are always quite obvious, as $v_2 \equiv v_b - v_c$ in Fig. 2.13.

Finally, label the area inside each loop of a network with a capital letter, and use the letter as a subscript for the loop current of that loop, as i_B in Fig.

---

† As in the geometry of mechanical systems, the arrow does not imply that current always flows in the direction of the arrow, but only that *when* current flows in the direction of the arrow we call it positive and when it flows in the opposite direction we call it negative.

[Sec. 2.6] SIMPLE ELECTRICAL SYSTEMS 63

2.13. Current identities are also obvious. In Fig. 2.13, for example, we have $i_3 \equiv i_A - i_B$.

The utility of the above set of conventions will become clear as we proceed.†

A voltage or current in a network can be *constrained* to have a prescribed value by, respectively, a voltage or a current *source* (such as the battery in Fig. 2.12, whereby $v_5 = v_{\text{bat}}$).

Number of independent variables. Let b be the total number of branches and n_{total} the total number of nodes in a network. Then the possible network variables include: n_{total} node voltages, plus b branch currents, plus b voltage drops through elements, plus a number of loop currents equal to the number of closed paths in the network. These are, of course, not all independent variables. The following rules are helpful in establishing sets of independent variables:

(a) The number of independent voltages is equal to the number of independent nodes n, which is given by

$$n = n_{\text{total}} - 1 \tag{2.46}$$

(b) The number of independent currents is equal to the number of independent loops ℓ, which is given by

$$\ell = b - n \tag{2.47}$$

The two common sets of independent variables are:

> (a) A set of n *node voltages* (one node is taken as a reference).
> (b) The set of ℓ *mesh currents*, for a planar network.‡

The above relations can be proved using a network concept known as a "tree." (This method of proof is given by Guillemin.§) In Fig. 2.13 there are $b = 11$ branches, $n = 7$ independent nodes, and $\ell = 4$ independent loops.

It is important to know how many independent variables there are in a given network, because this tells us how many independent equations of motion (Kirchhoff's-law equations) we must obtain. Thus, if the node method is used, there will be n independent equations of motion; if the mesh method is used, there will be ℓ. (When an independent variable is given, versus time, e.g., is constrained by a source, then the number of unknowns and the number of independent equations may be reduced accordingly. This is demonstrated in Example 2.5.)

† As an added feature, readers familiar with network duals will find this nomenclature helpful in their construction: Capital letters transform to small ones, and vice versa.
‡ A mesh is a loop which has no smaller loops within it. Loops A, B, C, D in Fig. 2.13 are all meshes; but loop c, e, h, g, o, c, for example, is not. A planar network is one that can be drawn on a plane without branch crossings.
§ E. A. Guillemin, "Introductory Circuit Theory," John Wiley & Sons, New York, 1953.

State variables. The number of variables required to prescribe the state of an electrical system—both its "configuration" *and* its "velocity" (refer to p. 38)—is in general larger than the number of independent variables. For example, it can be shown that the state of a circuit is specified at a given instant when the energy stored in every independent capacitor and inductor is known. Thus, a straightforward set of state variables consists of the (instantaneous) voltage drop across every capacitor (or set, for grouped capacitors), plus the current in every inductor (or set). Alternatively, an appropriate set of voltages plus their derivatives, or currents plus their derivatives, may be chosen.

For example, the circuit of Fig. 2.12 has only one independent current, e.g., loop current i_A; but two state variables are required, e.g., i_2 plus v_3, because both L_2 and C_3 store energy. Alternative state variables might be i_A plus its first derivative.

We shall further pursue formally the concept of state variables in Sec. 17.11.

Problems 2.55 through 2.60

(2) Kirchhoff's laws: The electrical equilibrium and compatibility relations

The equilibrium and compatibility relations in electric networks are quite straightforward. A network is a simple, one-dimensional thing in which currents always meet at a point (unlike forces, which may be distributed over a free body) and loops are usually trivial to trace. The relations are, simply:

Equilibrium: Kirchhoff's Current Law (KCL)

$$\boxed{\sum_{\text{node}} i = 0} \qquad (2.48)$$

The sum of all currents entering a node must be zero at every instant.

Compatibility: Kirchhoff's Voltage Law (KVL)

$$\boxed{\sum_{\substack{\text{closed} \\ \text{path}}} v = 0} \qquad (2.49)$$

The sum of all voltage drops around any closed path must be zero at every instant.

KCL is an equilibrium (or "vertex") relation among through variables.

KVL is a compatibility (or "path") relation among across variables.

[Sec. 2.6] SIMPLE ELECTRICAL SYSTEMS

Kirchhoff's Current Law is simply a statement of the conservation of *charge:* Since charge cannot be stored at an electrical junction the amount coming in must equal the amount going out at every instant. This is of course analogous to the continuity relation in fluid flow, which is based on the conservation of *mass*. This very helpful image is illustrated in Fig. 2.14.

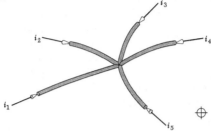

Fig. 2.14 *Current entering a junction (node). Kirchhoff's Current Law: At every instant the amount leaving must equal the amount entering. (Obviously, at any given instant, some i's must have negative values.)*

If the picture is compared with segments of pipe in which water is flowing, it is clear that some of the flow rates $i_1, \ldots, i_5$ will have to be negative, such that the same amount of water is leaving as entering, simply because there is no provision at the junction (node) for storing any water. Kirchhoff's Current Law simply states that electrical current flowing into a node behaves exactly in the same way.

As an example, consider node h in Fig. 2.13, for which KCL gives

$$\sum_h i = 0 = i_7 - i_8 - i_9$$

Kirchhoff's Voltage Law is axiomatic: If at any instant we traverse a closed path in a network we must arrive at the starting voltage when we

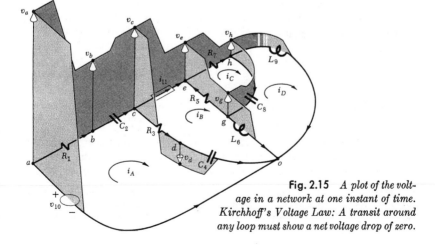

Fig. 2.15 *A plot of the voltage in a network at one instant of time. Kirchhoff's Voltage Law: A transit around any loop must show a net voltage drop of zero.*

return to the starting point. An easy way to visualize this law is to draw a three-dimensional picture of a circuit in which the network is drawn on a plane and the voltage at each point in the network is plotted vertically above the plane, as shown in Fig. 2.15 for the network of Fig. 2.13. In this picture we imagine a set of vertical partitions above the network whose altitude at every point represents the voltage of the network below. The picture is changing continuously with time, of course, so that Fig. 2.15 should be thought of as a "snapshot" taken at a particular instant.

Imagine now that we make an instantaneous excursion around a closed loop in this network. Suppose for example we start at point e and proceed around the loop e, h, g, e. As we go from one node to another we experience a change in altitude (voltage) either up or down; but so long as we return to our starting point e *we must return also to the starting altitude*. That is

$$v_{eh} + v_{hg} + v_{ge} = 0$$
or
$$v_7 + v_8 + v_5 = 0 \qquad (2.50)$$

and this is the essence of Kirchhoff's Voltage Law. (Guillemin likens this to performing a surveying transit in mountainous terrain: Large variations in altitude may be recorded from point to point, but the net change must be zero when the transit returns to its starting point.)

For a given network all of the KCL and KVL relations are not independent. The most straightforward way to obtain an independent set of relations is either to:

> (a) Write Kirchhoff's Current Law for each of the *nodes* in the network but one, or
> (b) Write Kirchhoff's Voltage Law around each of the independent *loops* (e.g., the meshes, for a planar network) corresponding to the independent currents selected in Step (1).

[Refer to relations (2.46) and (2.47), p. 63.] Method (a) is known as analysis on a *node basis* (or simply *nodal* analysis), while method (b) is known as analysis on a *loop basis*.

Problems 2.61 through 2.66

(3) Physical voltage-current relations for electrical elements

Ideal passive elements: Capacitor, resistor, inductor. The physical relations for these elements in a network are taken to be linear; a network is by definition a physical model in which all elements are ideal, lumped elements, and at this point we take them to be linear as well, Fig. 2.16:[†]

[†] Relations (2.51), (2.52), and (2.53) are based on the assumption that the voltage-current relations are linear, but notice that the development of the equations prior to this point has *not* presumed linearity; and indeed Procedure A-e does not depend upon linearity, but is general.

Sec. 2.6] SIMPLE ELECTRICAL SYSTEMS

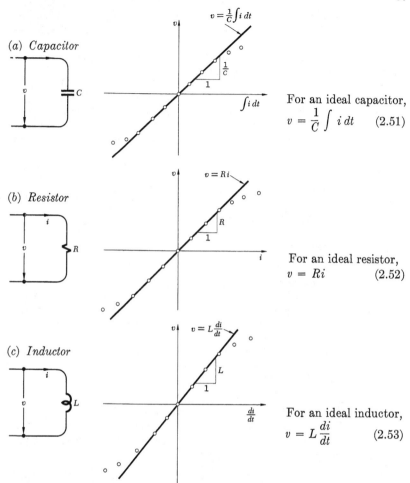

Fig. 2.16 *Physical relations for passive electrical elements*

For an ideal capacitor,
$$v = \frac{1}{C} \int i\, dt \qquad (2.51)$$

For an ideal resistor,
$$v = Ri \qquad (2.52)$$

For an ideal inductor,
$$v = L\frac{di}{dt} \qquad (2.53)$$

In Fig. 2.16, and in (2.51), (2.52), (2.53), v is the voltage drop across any branch, and i is the current in that branch; and the voltage drop and current are taken to be positive in the same direction. Relation (2.52) is, of course, the well-known *Ohm's Law*.

The physical phenomena underlying the relations of Fig. 2.16 are described in elementary physics texts.† A brief review is given below as a reminder of the physical nature of electrical behavior.

† See, for example, D. Halliday and R. Resnick, "Physics," John Wiley & Sons, New York, 1960. Capacitors are described in Secs. 30-1 to 30-3, resistors in 31-1 to 31-4, and inductors in 35-1 to 35-5 and 36-1 and 36-2.

Capacitors. Physically a capacitor consists of two thin metallic plates separated by an insulator (dielectric material). Sometimes the plates are large and rigid with air or vacuum between them, as in the tuning condenser of a radio. More commonly in communications equipment capacitors are made from thin metal foil sandwiched between a sort of waxed paper.

The operation of a capacitor can be visualized from Fig. 2.17, where one is connected across the terminals of a current source. As the (specified) current, or stream of charged particles (electrons) flows into one plate of the capacitor an excess of charge builds up. This charge induces an equal and opposite charge on the other plate, simply by repelling free electrons from across the gap.

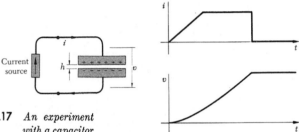

Fig. 2.17 *An experiment with a capacitor*

(Electrons do not flow through the insulator; the action is at a distance across the gap.) The repelled electrons in the second plate flow out through the other terminal of the capacitor and back to the current source.

Thus, at any time after the terminals have been connected to the current source, total charge on the capacitor will be equal to the integral of the flow of current up to that time:

$$q = \int i \, dt \tag{2.54}$$

The voltage across the capacitor will be proportional to the charge, the constant of proportionality depending upon the dimensions of the plates, the spacing between them, and the dielectric property. For a given element the proportionality constant is known as the capacitance C of the element:

$$v = \frac{q}{C} \tag{2.55}$$

Equation (2.51) is obtained by combining (2.54) and (2.55).

The dimensions of C are given in Table II, p. 25. The value of C can be calculated, for a given physical configuration (e.g., two plates), from Coulomb's law (2.45) for two particles, by suitable integration over the areas involved. For a pair of closely spaced parallel plates C is found to be

$$C = \epsilon \frac{A}{h} \tag{2.56}$$

where A is the area of the plates, h is the gap between them, and ϵ is a property (called the "dielectric constant") of the medium separating them.

By differentiating (2.51) we can of course obtain the alternative relation

$$i = C \frac{dv}{dt} \tag{2.57}$$

[Sec. 2.6] SIMPLE ELECTRICAL SYSTEMS

which states that the instantaneous current flowing through an ideal capacitor is proportional to the rate of change of voltage across its terminals.

It is important to notice that an ideal capacitor, such as those just described, contains no mechanism for dissipating energy, i.e., for converting electrical energy into heat. (In any real capacitor, to be sure, some minute current will flow through the insulating material between the plates; but it is possible to manufacture these elements so that such "leakage" is practically negligible. There will also be a tiny amount of resistance in the lead-in wires to the capacitor, but this too can be minimized.) Thus, the ideal capacitor is a *conservative* physical element, conserving electrical energy in the same way that a spring or moving mass conserves mechanical energy [in the potential (2.33) and kinetic (2.35) forms respectively]. Electrical energy is stored in the capacitor in the form of charge, and this stored energy can be recovered completely. For example, when a capacitor is used in a circuit, electrical charge that flows into it during part of the time usually flows out again (into the rest of the circuit) during another part of the time.

In setting up a particular network problem for dynamic study it is obviously essential to know the value of the charge or the voltage on every capacitor in the system at some time, since each capacitor is an integrator, (2.51), and integration requires knowledge of the "initial condition" from which the integration is carried forth.

Resistors. A resistor consists of a long, very thin metallic conductor or wire, packaged compactly with its terminals protruding for connection to other parts of the circuit. An approximate physical description of what happens when a voltage (electrical potential difference) is applied between the two ends of a resistor is that swarms of electrons are driven through the resistor by the resulting electrical field, and as they travel along they "rub" on the conductor with a sort of viscous friction which produces heat in the conductor and restricts the electrons to a particular velocity. Using this physical image we can deduce that the magnitude of the current flowing will be proportional to the strength of the applied field and the cross-sectional area of the conducting metal, and will be inversely proportional to the length of the path through which the electrons must travel:

$$i \sim v \frac{A}{\ell} \tag{2.58}$$

which is the basis for (2.52), which defines resistance R. Further, the rate of heat generation (by the "rubbing") will be equal to the power consumed:

$$P = vi\dagger \tag{2.59}$$

A more correct physical explanation (which predicts the same gross behavior) is that the electrons travel largely in free space within the conductor, during which time they do not travel at constant velocity, but are continuously accelerated by the electric field. However, the electrons collide frequently with the fixed particles of the material, and the collisions stop the electrons (briefly) and increase the magnitude of vibration of the fixed particles, thus

---

† This is perfectly analogous to mechanical power, the time rate of doing work, which is given by force times velocity: $P = fu$. The fundamental dimensions must be the same also—$fl\tau^{-1}$ (see Table II).

effecting an increase in the heat energy in the material. After each collision the electrons accelerate again, and thus they have a certain average velocity on which the current is based.

Inductors. An inductor is constructed in the form of a coil, and has the behavior property that there always exists a voltage between its terminals proportional (ideally) to the rate of change of current through it: Fig. 2.16c and Eq. (2.53). Physically, the voltage drop exists because the magnetic field produced by the current in each turn of the coil links all the other turns; and when this magnetic field has a rate of change (due to changing current) an electromotive force is produced in all the other turns, and thereby a potential difference is established between the ends of the coil. Expression (2.53) can of course be integrated to give the alternative form (L is called the *inductance*):

$$i = \frac{1}{L} \int v \, dt \tag{2.60}$$

It is important to notice that an ideal inductor, like an ideal capacitor, contains no mechanism for dissipating energy. In any real inductor there will of course be some resistance, and hence some energy dissipation. However, such losses can be made extremely small. The ideal inductor is thus a *conservative* physical element. Electrical energy is stored in the inductor in the form of magnetic flux, and this stored energy can be recovered completely. Thus, when an inductor is used in a circuit, an electrical current that flows into it during part of a time interval usually flows out again (into the rest of the circuit) during another part of the interval.

In setting up a particular network problem for dynamic study, it is essential to know the value of the current in every inductor in the system at some time (for example, at an initial time when the dynamic response is to begin), because each inductor is an integrator, (2.60), and integration requires knowledge of the "initial condition" from which the integration is carried forth.

Mutual inductance. When two coils are in close proximity, changing current in one can of course produce a voltage drop in the other. In Fig. 2.13, for example, if inductor L_6 were physically near to L_9 then the total voltage drop v_9 in inductor L_9 would be given by

$$v_9 = L_9 \frac{di_9}{dt} \pm M_{96} \frac{di_6}{dt} \tag{2.61a}$$

in which either the $+$ or $-$ sign would apply, depending on the relative directions of the windings. By the same token, the voltage drop v_6 would be given by

$$v_6 = L_6 \frac{di_6}{dt} \pm M_{69} \frac{di_9}{dt} \tag{2.61b}$$

in which M_{69} will always equal $+M_{96}$.

Ideal active elements: Sources and isolators. In addition to the three passive elements, several other elements appear in ideal networks, including "sources" and "isolators." Ideal sources serve to describe, in the physical network model, interactions with the outside world.

A *voltage source* is assumed (in the ideal physical model) to be capable of imposing a prescribed voltage difference (vs. time) between two nodes,

[Sec. 2.6] SIMPLE ELECTRICAL SYSTEMS 71

regardless of the current flow. The battery in Fig. 2.12 is a special form of voltage source. Obviously, the validity of this assumption is much better in some situations than in others. Consider the extreme case of a short circuit across a voltage source, for example. (If necessary, the accuracy of representing a battery as a voltage source may be enhanced by assuming an additional small resistance in series with the battery to represent the fact that when current is being drawn the voltage at the battery terminals is reduced proportionately. Resistance R_6 was added in Fig. 2.12b for this purpose.)

Similarly, a *current source* is assumed (in the ideal physical model) to be capable of imposing a prescribed current in its branch of a network regardless of the voltage difference between the ends of the branch. Again, the validity of the assumption is sometimes very good—e.g., when a powerful current amplifier drives a network having elements of reasonable physical characteristics.

Isolators are discussed in Sec. 2.8.

> **Problems 2.67 through 2.73**

Example 2.4. A simple circuit. To demonstrate the foregoing ideas, consider now the simple electrical system shown in Fig. 2.12a. (The purpose for this particular assemblage of elements is not given.) We are asked to find, versus time, the voltage drop in R_4 following closing of the switch.

Figure 2.12b shows a *physical model* (Stage I) of the system, a *network* in which the light bulbs are represented as ideal resistors (having no capacitive or inductive characteristics), and the capacitor and inductor are also represented as ideal elements. The battery is represented as an ideal voltage source plus a resistor R_6. The network of Fig. 2.12b is repeated in Fig. 2.18.

To write the equations of motion (Stage II) we follow Procedure A-e.

(1) *Network variables.* We select the voltage at node o as the reference (arbitrarily), and define the voltage at each of the other nodes with respect to the reference: v_a, v_b, v_c, v_d, v_f. The voltage drops in the branches then become additional or alternative variables: v_1, v_2, v_3, v_4, v_6. (Recall the convention that the direction of the arrows in the branches indicates the direction in which voltage drop is considered positive. The direction for each arrow is chosen quite arbitrarily.)

The current in each of the elements is also a variable: i_1, i_2, i_3, i_4, i_5, i_6, positive in the direction of the arrows. In addition, we define a loop current i_A as the current that flows around the closed loop in Fig. 2.18. We choose (arbitrarily) a direction to call positive, and indicate this with a loop arrow.

The variable we are interested in is v_4.

The voltage drops through the elements are obviously related to the voltages at the junctions by *identity relations*, such as $v_2 \equiv v_b - v_c$. In this example each of the element currents is *identical* to the loop current: $i_1 \equiv i_A$, $i_2 \equiv i_A$, etc. Because of the voltage source, node voltage v_f is *constrained* to have a specified value: $v_f = v_{bat}$.

All of the above definitions and relations are more or less self-evident, and are seldom written down.

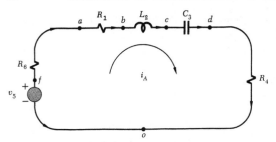

Fig. 2.18 *The network of Fig. 2.12*

(2) *Kirchhoff's laws.* Because there is only one loop, we choose the loop basis for analysis and write Kirchhoff's Voltage Law (clockwise around the loop):

$$\text{KVL:} \sum_{\text{loop}} v = 0 = v_1 + v_2 + v_3 + v_4 - v_5 + v_6 \quad (2.62)$$

in which the sign convention of p. 62 is followed (v_5 is written with a negative sign because its arrow is counterclockwise).

(3) *Physical voltage-current relations.* In our idealized model we presumed that the elements are all "ideal" and linear, so that physical relations (2.51), (2.52), and (2.53) apply. For the elements in Fig. 2.18, they are

$$\begin{aligned}
v_1 &= R_1 i_1 = R_1 i_A \\
v_2 &= L_2 \frac{di_2}{dt} = L_2 \frac{di_A}{dt} \\
v_3 &= \frac{1}{C_3} \int i_3 \, dt = \frac{1}{C_3} \int i_A \, dt \\
v_4 &= R_4 i_4 = R_4 i_A \\
v_6 &= R_6 i_6 = R_6 i_A
\end{aligned} \quad (2.63)$$

The second equality in each case comes from the current identities.

Finally, we *combine* the results by substituting relations (2.63) into the KVL equation (2.62)†:

$$0 = + R_1 i_A + L_2 \frac{di_A}{dt} + \frac{1}{C_3} \int i_A \, dt + R_4 i_A - v_{\text{bat}} + R_6 i_A \quad (2.64)$$

which is the equation of motion of the system in terms of the single unknown i_A. In Stage III it is to be solved for i_A vs. time. Then, from (2.63), the desired solution $v_4 = R_4 i_A$ versus time may be obtained. (See Problem 8.52.)

The use of the KCL equations is slightly less convenient in this example, since it leads to a set of three equations of motion instead of just one. (See Problem 2.74.)

Streamlined procedure

The demonstration given in Example 2.4 (above) occupied considerable space because we explained and wrote out each step and thought process in detail as we proceeded. In practice it is unnecessary to go through all the

† Notice that the Kirchhoff's-law equation is the natural focal point for combining the steps.

[Sec. 2.6] SIMPLE ELECTRICAL SYSTEMS

detail explicitly in writing the equations of motion for a simple network. A streamlined procedure is used, wherein Steps (2) and (3) are combined, and Step (1) may not even be mentioned.

Thus, with a glance at Fig. 2.18 the experienced analyst will simply write KVL, Step (2), including physical relations (3) as he goes, and invoking identities and constraints as needed:

$$0 = + R_1 i_A + L_2 \frac{di_A}{dt} + \frac{1}{C_3} \int i_A \, dt + R_4 i_A - v_{\text{bat}} + R_6 i_A$$

To demonstrate that writing equations of motion for electrical systems is normally quick and routine, consider another, more complicated example in which we shall proceed about as we would in practice, combining some steps and doing others mentally.

Example 2.5. A multinode, multiloop network. Write the equations of motion (Stage II) for the network (physical model) of Fig. 2.19 preparatory to finding the time history of voltage v_7. The time history of current i_1 is prescribed.

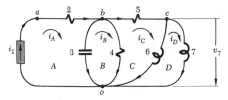

Fig. 2.19 *A more involved network*

(1) *Network variables.* We first label nodes and select positive directions (arbitrarily) as indicated in the drawing (following the conventions of p. 62). The available variables—voltages and currents—are clear from the drawing. We take point o as the arbitrary voltage datum: $v_o \triangleq 0$. The *constraint* is that i_1 is specified. The various *identities* are obvious. For example

$$v_2 \equiv v_a - v_b \quad \text{and} \quad i_2 \equiv i_A \equiv i_1$$

The quantity in which we are interested is $v_7 \equiv v_c$.

(2) *Kirchhoff's laws.* We select the node method (because there are only three independent nodes, and because the desired quantity is a node voltage, v_c).

KCL at node b: $\quad i_2 - i_3 - i_4 - i_5 = 0$
KCL at node c: $\quad i_5 - i_6 - i_7 = 0$
KCL at node a: $\quad i_1 - i_2 = 0$

(3) *Physical relations.* We insert these in KCL (using identities as needed), and combine the third KCL into the first, because i_1 is given:

$$i_1(t) - C_3 \dot{v}_b - \frac{1}{R_4} v_b - \frac{1}{R_5}(v_b - v_c) = 0$$
$$\frac{1}{R_5}(v_b - v_c) - \frac{1}{L_6} \int v_c \, dt - \frac{1}{L_7} \int v_c \, dt = 0$$

(2.65)

These are the equations of motion. There are two equations in the two unknowns v_b and v_c [$i_1(t)$ being prescribed]. In Stage III we would solve them simultaneously to obtain v_c versus time.

Note that every term in (2.65) has the dimensions of current.

In Problem 2.82 the effect of mutual coupling between L_6 and L_7 is to be studied.

Problems 2.74 through 2.83

2.7 A RECAPITULATION OF PROCEDURE A

It may be helpful at this point to review and compare Procedure A-m and Procedure A-e, for deriving equations of motion, with the set of considerations on which they are based (p. 33).

We study the dynamic behavior of systems in terms of carefully defined through and across *variables*. For a collective system of elements, there are *system relations*—equilibrium relations among through variables and compatibility relations among across variables. Finally, for each element in the system, there is a *"constitutive" physical relation* between its through and across variable. These are the considerations in deriving equations of motion.

The steps of Procedure A-m account for these considerations as indicated in the following diagram:

Variables	(2) Through variables **forces**	(1) Across variables **geometry**	
System relations	Equilibrium among through variables $\Sigma f^* = 0$ $\Sigma M^* = 0$	Compatibility among across variables $\Sigma x = 0$ $\Sigma \theta = 0$, etc.	**Procedure A-m** *Mechanical equations of motion*
Constitutive relations	(3) Physical force-geometry relations between through and across variables $f = kx$ $f = b\dot{x}$ $f = m\ddot{x}$		

[Sec. 2.8] AMPLIFIERS AND TRANSFORMERS

Similarly, the steps of Procedure A-e account for the considerations as indicated in the following diagram:

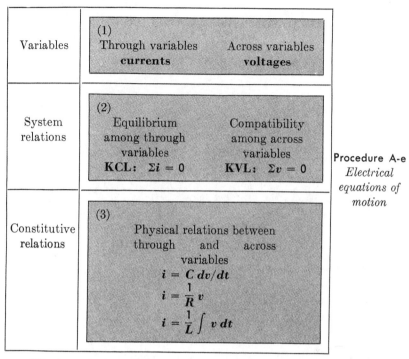

We shall find a similar procedure helpful in writing the equations of motion for electromechanical systems in Sec. 2.9, another for heat-conduction systems in Sec. 3.1, and another for fluid systems in Sec. 3.2. In each case, the steps furnish a systematic way of accounting for all the basic considerations of p. 33.

2.8 AMPLIFIERS AND TRANSFORMERS

Amplifiers: Electrical isolators

Amplifiers are electronic subsystems that produce output signals having some predetermined relation to the input signals they receive. An amplifier is like a *valve*, whereby a low-power signal can be used to control a much higher level of power.

A typical application is the *voltage-isolation amplifier*, which is designed to accept a voltage input and to produce a voltage output that is faithfully *proportional* to the input. This end is accomplished by making electrical power available to the amplifier as required, and by using feedback techniques to ensure that the output follows the input precisely and is not

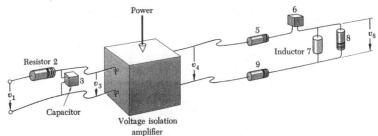

(a) *Voltage-isolation amplifier in a circuit*

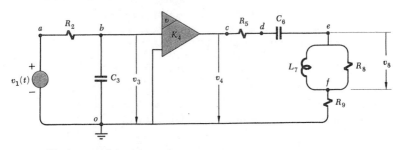

(b) *A network model*

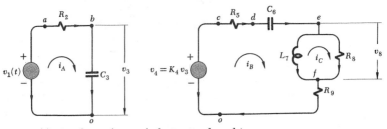

(c) *An alternative, equivalent network model*

Fig. 2.20 *Network models for a voltage-isolation amplifier*

affected by elements connected to it, as the original signal would have been. The amplifier thus serves to *isolate* a low-power signal from the network it is to drive. The voltage-isolation amplifier serves as a *voltage source* which has the time variation of the input voltage. This type of amplifier is a key element in communication systems and in computers.

A voltage-isolation amplifier is depicted in operation in a circuit in Fig. 2.20a. The input to the system is a low-power voltage, v_1 (modified slightly by the "filter" network preceding the amplifier). The output of the amplifier, voltage v_4, has available a much higher power level, and is guaranteed to be directly proportional to input voltage v_3, regardless of what circuitry may be connected to the output terminals of the amplifier. That is,

[Sec. 2.8] AMPLIFIERS AND TRANSFORMERS 77

the presence of resistors R_5, R_8, R_9, capacitor C_6, and inductor L_7, in Fig. 2.20a, has no effect on the value of voltage v_4, which will at every instant be equal to some constant times voltage v_3.

Figure 2.20b shows a network model of the circuit of Fig. 2.20a, based on the above discussion of an ideal voltage-isolation amplifier. Because the amplifier output acts as a voltage source, and because the elements "downstream" from the amplifier have no effect on voltage v_3, the amplifier effectively divides the network into two separate parts in cascade, as shown in Fig. 2.20c. This equivalent, two-part network is convenient to use in analyzing the behavior of circuits involving voltage-isolation amplifiers, because the first part can be studied independently of the second part.

Another kind of amplifier, which we shall call (somewhat arbitrarily) a *current amplifier*, is so designed that its output is a *current* proportional to the input signal. (The input signal is usually a voltage.) Again, the amplifier has sufficient power and control that its output current is unaffected by whatever circuitry is connected to its output. Such an amplifier acts as a *current source* which has the time variation of its input signal. This kind of amplifier is useful, for example, for driving electric motors (see Fig. 2.25, p. 85), where it is desired to apply to the motor a driving torque which is accurately proportional to a given signal. (The signal might come, typically, from a computer, as in an automatic-machine-tool application or in the rudder-positioning servo of an automatic pilot.)

If the voltage-isolation amplifier in Fig. 2.20a is replaced by a *current amplifier*, then the corresponding ideal network models will be as shown in Fig. 2.21a and b. The amplifier still separates the network into two parts in cascade; now, however, the input to the second part is a *current source*.

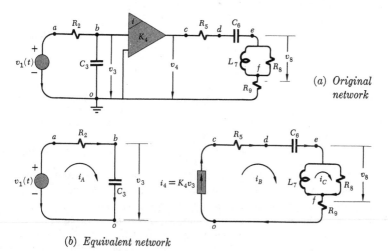

Fig. 2.21 *Network models for a current amplifier*

There are many other kinds of amplifiers, designed for various specific purposes. In this book, however, we shall concern ourselves only with ideal versions of the two types mentioned, the voltage-isolation amplifier and the current amplifier.

Transformers

The ideal transformer is a device that transmits a-c power from one part of an electrical system to another in such a way that both the voltage level and the current level are altered, but the power is not.

Transformers consist of two coils wound on the same core, one of which (the "primary") is connected to the input circuit, and the other (the "secondary") to the output circuit.

Transformers are shown symbolically as in Fig. 2.22, where the dots indicate whether the output current and voltage drop have the same sense

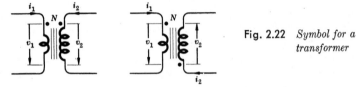

Fig. 2.22 *Symbol for a transformer*

(dots together) or opposite sense (dots not together) as the input current and voltage.

If the output coil has N times as many windings as the input coil

$$N = \frac{n_2}{n_1} \tag{2.66a}$$

then the currents and voltage drops are related by

$$\boxed{\frac{i_2}{i_1} = \frac{1}{N}} \tag{2.66b}$$

$$\boxed{\frac{v_2}{v_1} = N} \tag{2.66c}$$

The power going out of an ideal transformer is identical to the power going in:

$$P = i_2 v_2 = i_1 v_1 \tag{2.67}$$

Because they convert electrical energy from one current and voltage level to another, transformers are sometimes called *electrical-electrical energy converters*. Obviously, there is an analogy between a transformer and a gear train or lever (mechanical-mechanical energy converter); see pages 55 and 56. Transformers are usable only with a-c (alternating sinusoidal) signals. In typical control and communication circuitry, however, the signals of interest are superimposed on a high-frequency sinusoidal "carrier" signal.

Then (2.66) can be applied to the signals, and the carrier ignored.

Problems 2.84 through 2.90

2.9 SIMPLE ELECTROMECHANICAL SYSTEMS

There have been developed a large number and wide variety of physical devices to exploit various electromechanical phenomena for the general purpose of converting electrical signals to mechanical motions, and vice versa. The devices that come most readily to mind are those involving electromagnetic-mechanical phenomena. Devices such as the speaker and microphone, the electric motor, the electric generator, the galvanometer, and the tachometer are in this category.

In the speaker shown in Fig. 2.23, for example, the current in the coil is varied (via an amplifier), and this produces a varying magnetic force on

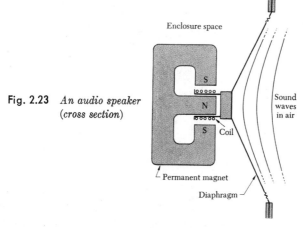

Fig. 2.23 *An audio speaker (cross section)*

the coil because of its location in a magnetic field. The coil is embedded in the diaphragm so that this varying force drives the diaphragm mechanically back and forth, generating pressure waves in the air, and thus sound in the room. The current in the coil is varied according to some input signal (from a radio receiver, for example) and thereby the speaker is caused to generate sound—music, speech, etc. (The exact mechanical response of the speaker to a particular time variation in the current will depend on a number of things, of course, such as the mass and elastic properties of the diaphragm, the manner in which air around the speaker is enclosed, etc.)

For an inductive microphone the pattern of cause and effect is very similar to that of the speaker, but in reverse. Sound waves (e.g. from a human speaker's voice) impinge upon the diaphragm of the microphone and cause it to vibrate in a particular way, depending upon the character of the sound waves. These vibrations of the diaphragm move the coil back and forth in a

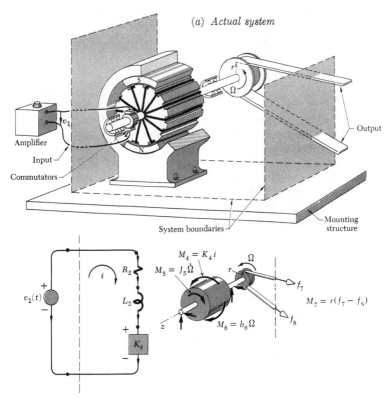

Fig. 2.24 *An electric motor*

magnetic field, and this motion induces a current in the coil. The current is measured and amplified electrically for transmission (through a radio broadcast system or into a tape recorder, for example). Sometimes (when high fidelity is not required) the same device is used both as microphone and speaker, as in office "intercom" systems.

Another widely used electromechanical device is the electric motor, a form of which is shown in Fig. 2.24. Electric motors come in all sizes, from tiny D'Arsonval galvanometers used to measure milliamps of current to huge electric locomotives. As with the speaker, the basic action involves the force produced because current-carrying wire segments are in the magnetic field of permanent magnets. In the motor it is necessary to reverse the current in each wire segment as it passes from the vicinity of the south magnetic pole to the vicinity of the north magnetic pole, in order that the forces produced always torque the rotor in the same direction. This reversal is effected by means of the commutators indicated. The torque thus produced can be controlled by varying the current input to the wires, as indi-

[Sec. 2.9] SIMPLE ELECTROMECHANICAL SYSTEMS 81

cated in the figure. In Fig. 2.24 the motor is delivering power by means of a belt drive.

As in the case of the speaker, the inverse of the motor is also very useful. In this case the inverse device is called an *electric generator*. In Fig. 2.24 mechanical power could be delivered *to* the rotor by means of the belt (which now might be connected to a steam turbine, for example) and electric power would then be generated in the wires due to their velocity through the magnetic field. This power would be delivered through the commutators to an electrical power-distribution system.

Interesting electromechanical phenomena are utilized in many other devices. Among these phenomena are the production of mechanical force by an electric field, as in the electrostatically supported gyroscope; the variation in resistance with position, as in the potentiometer and in the strain gauge; the variation in capacitance with mechanical configuration, as in the tuning "condenser" and in the capacitance microphone; the variation in magnetic coefficient with position, as in the variable-reluctance phonograph pickup; and the piezoelectrical effect, wherein a voltage is produced in proportion to the distortion of certain materials or, conversely, distortion is produced by application of a voltage. A number of these physical phenomena will be described in Sec. 2.10.

Analytical procedure

We wish to consider an orderly procedure for deriving the equations of motion for electromechanical systems, just as we did for mechanical systems in Sec. 2.2 and for electrical systems in Sec. 2.6.

STAGE I. PHYSICAL MODELS

In Stage I of the analysis we shall, as always, isolate the system to be studied and contrive a physical model for it. We first define carefully the boundaries of the system, specifying what is included and what is not. We then inspect the boundaries and note precisely the interactions that take place between the system and its environment. Then we sketch a simple physical model whose behavior we judge will be sufficiently close (for our purpose) to that of the actual system.

Suppose, for example, we want to study the dynamic behavior of the motor in Fig. 2.24a when a specified, time-varying voltage[†] is applied to its rotor windings, and when some specified time-varying load is applied to the output drive.

We might specify, as the system to be studied, everything contained within the dotted envelope indicated in Fig. 2.24a. At one boundary of this

---

[†] As we shall find later, better control of the motor will result if a controlled current input to the rotor winding (a "current source") can be used. However, sometimes what we have to work with is a battery and rheostat, for example, and then we are interested in the situation postulated here: a controlled voltage input.

system we find the input, an electrical voltage, and at another boundary we find the output, mechanical drive forces on the belt. At the third boundary, the point of attachment of the motor housing to the fixed mounting structure, there will be a set of static forces (which do no work on the system). For the mechanical part of the physical model we might assume that bearing friction and windage combine to produce a torque on the rotor which is simply proportional (and opposite) to rotor angular velocity as indicated in the physical model of Fig. 2.24b.

A reasonable physical model for the electrical part of this system consists of representing the armature coil, through which current $i(t)$ flows, as a resistor and an inductance plus an additional voltage drop proportional to angular velocity of the coil through the stationary magnetic field [Eq. (2.71) below]. These circuit elements are shown in Fig. 2.24b as R_2, L_3, K_4, respectively.

The torque on the rotor will be simply proportional to the current through the armature coil [Eq. (2.70) below], to a very good approximation.

STAGE II. EQUATIONS OF MOTION

In electromechanical systems we will have, in general, both electrical circuits and mechanical elements, coupled through some physical relationship. We therefore find that the appropriate procedure involves a *combination* of *both* the mechanical and electrical checklists, Procedure A-m and Procedure A-e. This combined procedure may be conveniently remembered in the following concise form:†

> (1) **Variables and geometry**
> (2) **Kirchhoff's laws and force equilibrium**
> (3) **Physical relations:**
>   *Voltage-current*
>   *Force-geometry*
>   *Electromechanical*

Procedure A-em
Electromechanical

Then **combine relations** (to reduce the number of unknowns).

(1) *Variables and geometry.* Both electrical and mechanical variables are to be defined (and positive directions specified); constraints and identity relations are to be noted. A few new variables may be employed in studying an electromechanical system—notably magnetic field strength and electric field strength—but most of the variables are simply the electrical and mechanical variables of the separate components. Recall that current and force are "through" variables, and voltage and velocity are "across" variables.

† A more systematic listing would be (1) KVL and geometry, (2) KCL and force equilibrium. (Refer to Sec. 2.1.) The author finds the above more convenient in practice.

[Sec. 2.9] SIMPLE ELECTROMECHANICAL SYSTEMS

(2) *Kirchhoff's laws and force equilibrium.* These may consist simply of KCL or KVL for the electrical subsystems plus force equilibrium equations for mechanical subsystems.†

(3) *Electromechanical physical relations.* These are, as we have said, the thing that is completely new. An introductory survey of them is given in Sec. 2.10.

For illustration we discuss here a typical and important set: those physical relations pertaining to a current-carrying wire in a magnetic field, as in the speaker or microphone of Fig. 2.23 and the motor or generator of Fig. 2.24. For a given arrangement (coil size, magnetic-core shape and material, etc.), the governing relations are that there is a force f or moment M proportional to current i; and there is a voltage drop v—or more properly, an electromotive force, called *back emf*, that produces a voltage drop—proportional to velocity u‡ or angular velocity Ω:

$$f = Ki \tag{2.68}$$
$$v = Ku \tag{2.69}$$

$$M = Ki \tag{2.70}$$
$$v = K\Omega \tag{2.71}$$

If appropriate units are used (e.g., newtons, meters, and seconds), the K's for a given device are numerically equal, as indicated. [What are the primary dimensions of K in (2.68), (2.69)? In (2.70), (2.71)? Refer to Table II, p. 25.] Notice that these devices are electrical-mechanical energy converters, with no power loss ($fu = iv$ and $M\Omega = iv$). In each case, constant K gives the conversion from through variable to through variable, and from across variable to across variable.

Example 2.6. An electric motor. Derive the equations of motion for the physical model, Fig. 2.24b, of the electric motor in Fig. 2.24a, from which the response $\Omega(t)$ of the rotor to changes in voltage $v_1(t)$ can be calculated.

(1) *Variables and geometry.* For the network part of the system we choose current i as the independent variable and plan to use the loop basis for analyzing the network. For the mechanical part, the quantity of interest is rotor angular velocity Ω, which we therefore select as the mechanical independent variable.

The electrical *constraint* is that v_1 is specified as a function of time.

---

† Sometimes it is desirable to use energy relations, rather than a force balance, as the electromechanical equilibrium equations. The philosophy and techniques are described in texts on electromechanical energy conversion, e.g., H. H. Skilling, "Electromechanics," John Wiley & Sons, New York, 1962; D. C. White and H. H. Woodson, "Electromechanical Energy Conversion," John Wiley & Sons, New York, 1959. Special application to systems studies is presented in W. W. Harman and D. W. Lytle, "Electrical and Mechanical Networks," McGraw-Hill Book Company, New York, 1962.

‡ When both voltage and velocity are involved in the same problem we shall use v for voltage and u for velocity.

(2a) *Kirchhoff's Voltage Law.*

$$\Sigma v = 0 = v_2 + v_3 + v_4 - v_1 \tag{2.72}$$

(2b) *Force equilibrium*

(i) A free-body diagram of the rotor is shown in Fig. 2.24b. The moments acting about the z axis are magnetic torque M_4, a friction torque M_6, inertial torque M_5, and the load torque $M_7 \triangleq r(f_7 - f_8)$, applied through the belt.

(ii) Summing moments about the $+\cdot z$ axis [from (2.2)]:

$$\Sigma M^* = 0 = +M_4 - M_5 - M_6 - r(f_7 - f_8) \tag{2.73}$$

(3) *Physical relations.* The voltage-current relations for the electrical elements are [(2.52) and (2.53)]:

$$v_2 = R_2 i \qquad v_3 = L_3 \frac{di}{dt} \tag{2.74}$$

The force-geometry relations are [(2.17) and (2.18)]:

$$M_5 = J_5 \dot{\Omega} \qquad M_6 = b_6 \Omega \tag{2.75}$$

$M_7 = r(f_7 - f_8)$ is specified as a function of time.

And finally, the *electromechanical physical relations* are [(2.71) and (2.70)]:†

$$v_4 = K_4 \Omega \qquad M_4 = K_4 i \tag{2.76}$$

Combining the above physical relations into the equilibrium expressions, (2.72) and (2.73), we obtain the following equations of motion for the system:

$$\left. \begin{array}{l} R_2 i + L_3 \dfrac{di}{dt} + K_4 \Omega = v_1 \\ K_4 i - J_5 \dot{\Omega} - b_6 \Omega = M_7 \end{array} \right\} \tag{2.77}$$

In Stage III these will be solved simultaneously to obtain Ω versus time for any given input v_1 versus time.

Electric motors are often driven by current amplifiers; then a very good physical model for the system may be as shown in Fig. 2.25. In such a case the analysis is much easier, simply because variable i is now constrained in terms of applied voltage v_1 by the direct relation

$$i = K_1 v_1$$

(That is, although there will still be voltage drops in the electrical loop, these are no longer of concern because they do not affect the value of i.) Then the

---

† Constant K_4 is the same in both expressions only if consistent units are used (e.g., newtons, meters, seconds).

[Sec. 2.10] ELECTROMECHANICAL ELEMENTS: A SAMPLING

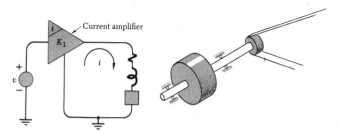

Fig. 2.25 *Electric motor driven by a current source*

total equations of motion required to relate Ω to v_1 are simply:

$$\left.\begin{aligned} i &= K_1 v_1 \\ K_4 i - J_5 \dot{\Omega} - b_6 \Omega &= M_7 \end{aligned}\right\} \qquad (2.78)$$

in which the second equation is the same as in (2.77), but the first equation is much simpler. As one would assume, the determination of time behavior of the rotor in the second case is correspondingly much simpler than in the first case.†

⊙2.10 ELECTROMECHANICAL ELEMENTS: AN EMPIRICAL SAMPLING

There is a wide variety of electrical-mechanical devices, utilizing a broad spectrum of operating principles and applications. Two examples are shown in Figs. 2.23 and 2.24; several more are shown in Figs. 2.26, 2.27, and 2.28.

For the purpose of analyzing their use in a system, the various electromechanical devices are merely elements in the system, just as resistors or capacitors or masses or springs are. We can regard them as elements on which we have made many input-output experimental measurements to obtain and plot an empirical curve, like those in Fig. 2.4 for mechanical systems, and Fig. 2.16 for electrical systems. When we are satisfied that the curve represents correctly the physical input-output behavior of an element (for the circumstances it will experience in operation in our system), the curve is all we need to complete Step (3) in writing the equations of motion. (If the curve can be approximated by a straight line, we are pleased at our good luck.)

For example, Figs. 2.27a and b show the basic pair of experimental curves for the dynamic microphone and the speaker. One plot is of emf (electromotive

---

† As noted on p. 32, the reader may wish to skip at once from here to Part B, where the dynamic response of elementary systems is discussed. Chapter 2 (the present chapter) is adequate preparation for Part B. Later (e.g., while studying Part C or D on the dynamic behavior of compound systems), he may wish to return to Chapters 3 through 5 to study the derivation of equations of motion for such systems (particularly to Chapter 5 on mechanical systems in two and three dimensions).

⊙ Optional: This section may be omitted without loss of continuity. Other optional sections and chapters will be so designated throughout the book.

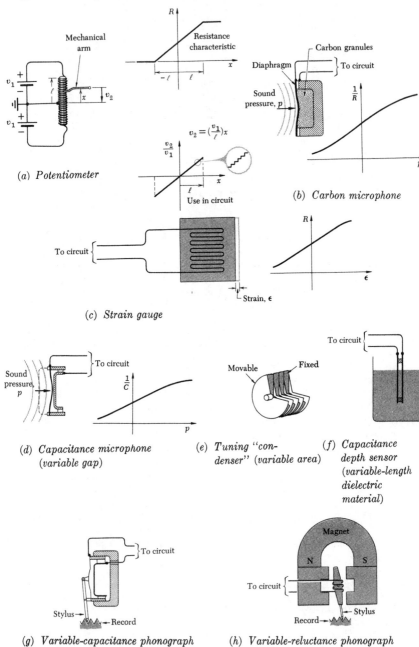

Fig. 2.26 Electrochemical modulation devices

force) versus velocity and the other is of force versus current. The slopes of these curves are the values of K in (2.68) and (2.69). (As we noted earlier, it can be shown from more fundamental empirical laws that the two slopes are the same in a consistent set of units.)

The problem of cataloging electromechanical physical relations

In the study of purely mechanical systems the physical relations (between force and motion) are made to fall neatly into three categories: force-displacement, force-velocity, and force-acceleration. Similarly, in electrical systems the physical relations between current and voltage also fall into three categories: voltage-charge (integral of current), voltage-current, and voltage-rate-of-change-of-current.

When we seek a similarly simple classification for electromechanical relations we find our efforts less rewarding, because there are so many more possibilities. For one thing, we are now dealing, in effect, with four mechanical quantities (force, displacement, velocity, and acceleration) plus four electrical quantities (e.g., current, voltage plus its derivative and its integral). Thus there may be, in principle, as many as 16 *new* relations. Therefore classification of electromechanical devices on the basis of their input-output relations is not very helpful, as a practical matter.

Another consideration in cataloging input-output relations is that many of the mechanical-to-electrical devices operate by changing or *modulating* a parameter—resistance, capacitance, or inductance—rather than by producing a voltage or a current directly. To see the difference between these two kinds of devices, compare the dynamic microphone of Fig. 2.27a with the carbon microphone of Fig. 2.26b. In both cases the input is sound pressure, which deflects a diaphragm, and the output is an electrical signal, which is related to the input pressure. But in the dynamic microphone the diaphragm motion *produces electrical emf directly* (in the coil); whereas in the carbon microphone, motion of the diaphragm merely *changes the resistance* of the device. This change in resistance is made available as an electrical signal by connecting the resistor in a circuit having a voltage source. Thus in one case the mechanical input produces an electrical signal by *direct energy conversion*, while in the other case an existing electrical signal is *modulated* by the mechanical input.

It is therefore sometimes convenient to classify electromechanical devices first as either modulation devices or energy-conversion devices, as we have done in Figs. 2.26 and 2.27.†

Then (and more fundamentally) both modulation and energy-conversion devices are most conveniently classified and studied in terms of the basic physical phenomena employed, and the electromagnetic physical laws that underlie them.

Modulation devices

There are a host of modulation devices. Whenever we find (or can arrange for) mechanical motion or force to alter the value of R in a resistor, C in a

---

† It should be noted that capacitive and inductive modulation devices do, in fact, accomplish also some energy conversion. But in devices designed for the purpose of modulating a signal, the energy conversion can generally be safely neglected compared to the magnitude of the large external signal being modulated.

capacitor, or L in an inductor, we can exploit the situation by putting that element in a circuit with a constant voltage (or current) source. Then the output from the circuit will vary with the mechanical input, and we have a motion or force sensor.

One of the simplest and most straightforward of such devices is the potentiometer, which may be connected as indicated in Fig. 2.26a. In this device the mechanical displacement to be measured is simply made to vary the effective length of the resistor by moving the wiper arm along the tightly wound coil of resisting wire. The resulting input-output relation for the subsystem shown is plotted in Fig. 2.26a, the governing relation being

$$v = \left(\frac{v_1}{\ell}\right) x \tag{2.79}$$

If angular, rather than linear, motion is to be measured, the resisting coil may be wound around a circular ring. (Note that there is, in a potentiometer signal, an inherent granularity equal to the distance between wires, as indicated by the magnified inset in Fig. 2.26a.)

More generally, the relations for modulation devices are of the form

$$v = cx \quad \text{or} \quad v = c\theta \quad \text{or} \quad v = cp \text{ (pressure sensor)} \ldots \tag{2.80}$$

in which c is the input-output "constant" of the device.

Other variable-resistance devices include the carbon microphone of Fig. 2.26b and the strain gauge, Fig. 2.26c. In the carbon microphone the mechanical quantity is sound pressure, which compresses the carbon granules in the microphone, changing their resistance. A strain gage (Fig. 2.26c) is a long strand of wire laid tightly back and forth in a postage-stamp-sized patch. The patch is cemented firmly to the structural member of interest—e.g., the skin of an aircraft wing. When the member is strained the wire is stretched or compressed, which changes its resistance. Precision measurements can be made with the strain gauge as a member of a Wheatstone bridge (refer to any introductory physics text). Strain gauges have played a vital role in experimental research in structural dynamics.

Several typical *capacitance-modulating devices* are shown in Fig. 2.26d through g. Each of these uses a parallel-plate capacitor, and force or motion varies the geometry and thence the capacitance according to Eq. (2.56), p. 68. In the capacitance microphone, Fig. 2.26d, the plates are clamped and the input is pressure which forces the plates together, changing gap h, and thus C. In the capacitance phonograph pickup, Fig. 2.26g, one plate is only slightly restrained so that displacement of the phonograph needle changes gap h. In the radio tuning condenser, Fig. 2.26e, the gap is constant, but mechanical displacement changes the area A of the plates that are in proximity. The liquid depth sensor, Fig. 2.26f, exploits the difference in the dielectric constants ϵ of air and the liquid: as depth increases, average ϵ, and hence C, changes.

A common inductive modulation device is the variable-reluctance phonograph pickup, Fig. 2.26h, in which (in principle) mechanical displacement changes the configuration of the iron core in an induction coil, and hence the L of the coil.

[Sec. 2.10]   ELECTROMECHANICAL ELEMENTS: A SAMPLING   89

Most of the above modulation devices do not have nice linear input-output plots like Fig. 2.26a; and therefore they are generally used only over a small portion of their total range of mechanical freedom, the portion wherein their characteristic may be approximated by a linear relation.

Energy-conversion devices

In electrical-mechanical energy-conversion devices, mechanical motion is used to *generate* electrical signals, as in the dynamic microphone and the generator, Fig. 2.27a or c; or electrical power is used to produce mechanical motion, as in the loudspeaker and the motor, Fig. 2.27b or d. These *electromagnetic* devices are the most common electrical-mechanical energy converters. They were discussed in Sec. 2.9, and their governing physical relations were given as (2.68) to (2.71).

Notice that for this class of energy converter the signal transformations are simultaneously

$$\left. \begin{array}{rcl} \text{current} & \rightarrow & \text{force} \\ \text{emf(voltage)} & \leftarrow & \text{velocity} \end{array} \right\}$$

That is, energy converters are two-way devices, while modulators are essentially one-way.

The basic physical phenomenon in the devices of Fig. 2.27a through e, and in the cathode ray tube of Fig. 2.28b as well, is that a particle having charge q and velocity $\mathbf{u}$ through a magnetic field of strength $\mathbf{B}$ experiences a force $\mathbf{f}$ given by

$$\mathbf{f} = q\mathbf{u} \times \mathbf{B} \tag{2.81}$$

In Fig. 2.28b, force $\mathbf{f}$ acting on individual electrons (charge $q = -1.6 \times 10^{-19}$ coulomb) diverts their path as shown.

In the devices of Fig. 2.27a through e, phenomenon (2.81) is manifested in two relations: (1) a current-carrying element $d\boldsymbol{\ell}$ of wire† in a magnetic field experiences a lateral force given by

$$d\mathbf{f} = i\,d\boldsymbol{\ell} \times \mathbf{B} \tag{2.82}$$

which results in a corresponding net force $\mathbf{f} = \int_l d\mathbf{f}$ on the whole winding;‡ and (2) when an element $d\boldsymbol{\ell}$ of conducting wire has lateral velocity $\mathbf{u}$ through a magnetic field $\mathbf{B}$, the resulting forces $\mathbf{f}$ on free charge along the wire produce a cumulative longitudinal *electromotive force* ϵ given by

$$d\epsilon = (\mathbf{u} \times \mathbf{B}) \cdot d\boldsymbol{\ell} \tag{2.83}$$

which results in a corresponding voltage $v = \int_l d\epsilon$ between the terminals of the coil. Relations (2.82) and (2.83) are the genesis for Eqs. (2.68) through (2.71).

---

† Recall that electric current i is the rate of flow of charge.
‡ The force is enhanced many times by the magnetic properties of the soft-iron core of the windings.

As a point of interest, in the devices of Fig. 2.27a through d a through variable is transformed to a through variable, and an across to an across variable. Compare this electrical-to-mechanical form of energy converter, Eqs. (2.68) to (2.71), with the mechanical-to-mechanical converters, Eqs. (2.40), (2.43), and the electrical-to-electrical converter, Eqs. (2.66). Note that K in (2.68) to (2.71) plays a role analogous to that of N in (2.40), (2.43), and (2.66).

Another energy-conversion device—one designed for one-way utilization —is the D'Arsonval meter, shown in Fig. 2.27e, in which current *in* produces, finally, displacement *out* (current produces a force which moves the meter against a spring). The operating principle is the same as for a motor.

Electrostatic energy-conversion devices are also used. Most have approximately a parallel-plate-capacitor configuration with variable geometry, as in Fig. 2.27f. The governing equations are†

$$f = \frac{q^2}{2\epsilon A} = \frac{1}{2}\frac{C}{h}v^2 = \frac{1}{2}\frac{\epsilon A}{h^2}v^2 \qquad (2.84)$$

where f is the attractive force between the plates, A is their area, ϵ is the dielectric constant of the separating medium, q is the charge (taken equal and opposite on the two plates), v is the voltage difference, C is capacitance, and h is the distance between the plates. In a typical device, both displacement and force vary as v is varied, and the analysis is somewhat complicated. However, the energy exchange is again two-way, electrical-to-mechanical and mechanical-to-electrical, the relations having the forms

force ↔ voltage
displacement ↔ voltage

Note that here a mechanical through variable is transformed to an electrical across variable, and vice versa. (The relations are nonlinear, but may be approximated as linear changes from a nominal operating point.) Figure 2.27f shows an electric suspension system for a spherical gyro rotor. The plates and rotor are in a high vacuum. The nominal plate voltage is 10,000 volts, and this is varied automatically (by three control loops, one of which is shown) to maintain the rotor centered between the three pairs of plates (two pairs are shown).

An interesting limiting case of the energy converter is the *cathode ray tube* (crt), Fig. 2.28, in which the mechanical-to-electrical effect is quite negligible. The electrostatic version of a crt is shown in Fig. 2.28a. A stream of electrons is pointed at the screen by an electron gun. The stream passes between a pair of electrostatic plates. A voltage difference v across the plates produces an electrostatic field between them, which causes the electrons to accelerate (in a short segment of a circular path) during the short time they travel between the plates. This effectively deflects the electron beam by an amount proportional to the voltage v. The electrons have a negligible effect on v, so we have essentially a one-way relation

voltage → displacement

† These equations can all be derived directly from Coulomb's law for two charged particles, (2.45), by integrating over the plate areas and using the definition of capacitance C from (2.55). For parallel plates, C is given by (2.56).

[Sec. 2.10] ELECTROMECHANICAL ELEMENTS: A SAMPLING   91

(The energy conversion is there, nevertheless, with the imperceptible work required to displace the electrons coming from the electric circuit.)

In the electromagnetic version of the crt, Fig. 2.28b, the electron stream travels through a short electromagnetic field created by input current, so that the overall relation is

$$\text{current} \rightarrow \text{displacement}$$

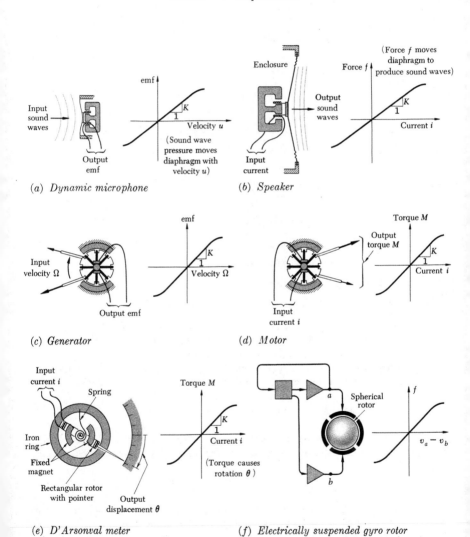

Fig. 2.27 Electromechanical energy-conversion devices

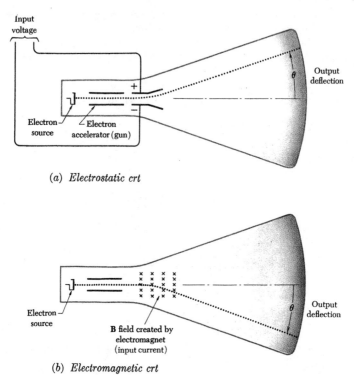

(a) *Electrostatic crt*

(b) *Electromagnetic crt*

Fig. 2.28 *The cathode ray tube (crt)*

Another physical phenomenon that is exploited for electrical-mechanical energy conversion is the piezoelectric effect. It is found that crystals of certain materials generate a voltage between their faces when they are distorted and, conversely, that application of a voltage across the crystal faces produces a distortion of the crystal. Thus there is the *two-way* conversion

$$\text{displacement} \leftrightarrow \text{voltage}$$

Since the crystals are also elastic, distortion of a crystal can be produced by (and can produce) a pressure on the crystal, so that the two-way conversion

$$\text{force} \leftrightarrow \text{voltage}$$

also is available. The first application gives conversion between across variables; the second, between a through variable in one medium and an across variable in the other. The crystal phonograph pickup and the crystal microphone are common applications.

Fundamental laws

In general, the gross empirical relations described above can, of course, be explained and understood in terms of basic empirical relations at a much more fundamental level, namely, Maxwell's equations. These fundamental laws of the behavior of charged particles are presented in elementary physics courses, and are pursued in depth in subsequent physics courses in electromagnetic theory and in engineering courses in electrical energy conversion.[†]

Problems 2.91 through 2.98

---

† See, for example, D. Halliday and R. Resnick, "Physics," John Wiley and Sons, New York, 1960, Chaps. 26-39.

⊙Chapter 3

EQUATIONS OF MOTION FOR SIMPLE HEAT-CONDUCTION AND FLUID SYSTEMS

In this chapter we continue our study of deriving the equations of motion for physical systems with a discussion of some very simple heat-conduction systems and fluid systems.

Fluid systems and heat-conduction systems are inherently distributed in space. However, in many important problems it is possible to identify discrete lumps, regions throughout which the properties—such as pressure, density, temperature, velocity—may be considered approximately uniform. The resulting lumped models greatly facilitate analysis.

Fluid systems are, in addition, usually inherently nonlinear. Many important problems, however, involve small changes—*perturbations*—from a steady-flow condition; and these can often be studied usefully on a linearized basis.

3.1 SIMPLE HEAT CONDUCTION

The nature of heat conduction

A number of heat-conduction systems are indicated in Fig. 3.1: the quenching of a steel ingot, the transfer of heat from a rocket engine, the cooling of an automobile engine, and the removal of heat from an instrument. The reader will doubtless think of many others.

⊙ Optional chapter, may be omitted without loss in continuity.

[Sec. 3.1] SIMPLE HEAT CONDUCTION

Heat flows from body a to body b at a rate, q, proportional to the temperature difference between body a and body b

$$q = \frac{1}{R}(T_a - T_b) \qquad (3.1)$$

where R is the *resistance* to heat flow (which depends on the thermal conductivity and geometry of the material separating body b from body a). Thus, heat-flow rate q is the through variable and temperature T is the across variable.

Heat flow is a form of energy transfer. Energy that flows into a body as heat may be stored in the body as internal energy (molecular kinetic

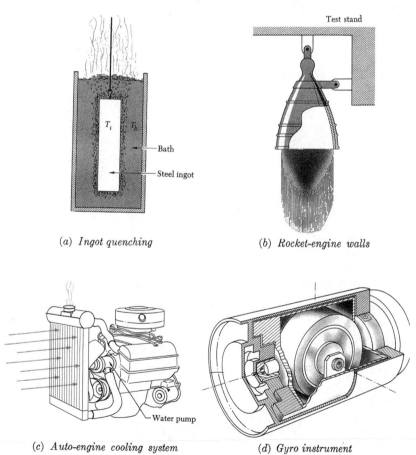

(a) *Ingot quenching*   (b) *Rocket-engine walls*

(c) *Auto-engine cooling system*   (d) *Gyro instrument*

Fig. 3.1 *Some heat-conduction systems*

energy), typically with an accompanying increase in the temperature of the body. When no work is involved, and for the simple case of a lumped model, with no phase change (e.g., from liquid to gas), the rate of change of a body's temperature indicates directly the rate at which heat energy is flowing into it and being stored:

$$\boxed{\frac{dT}{dt} = \frac{1}{C} q_{\text{net}}} \qquad \text{No work}\dagger \qquad (3.2)$$

C is the thermal storage *capacity* of the body (which depends on its material and mass). The dimensions of heat Q are those of energy or work: $f\ell$. Thus the dimensions of heat-flow rate q are $f\ell\tau^{-1}$.

Equations (3.1) and (3.2) are the principal relations used in simple heat-conduction problems. They are discussed in more detail below: (3.1) is an empirical physical law (constitutive relation), while (3.2) is a special case of the First Law of Thermodynamics, the dynamic equilibrium relation.

Analytical procedure

STAGE I. PHYSICAL MODEL. An exact physical model for any heat-conduction system will lead inevitably to partial differential equations of motion; because, in general, when heat is flowing, as for example in the ingot shown in Fig. 3.1a and again in Fig. 3.2, temperature is distributed in a

† Work done on a system will also raise its temperature (as Count Rumford discovered in about 1795, while boring cannon).

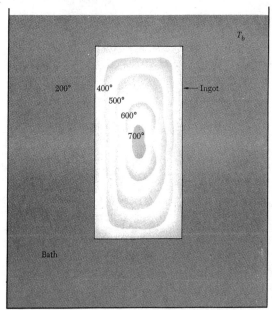

Fig. 3.2 *Temperature distribution in the ingot of Fig. 3.1a*

[Sec. 3.1] SIMPLE HEAT CONDUCTION

continuous manner throughout the body. This occurs because both thermal resistance and thermal storage capacity are distributed throughout the body. For analytical purposes it is very helpful to represent the actual system approximately by a lumped-element model, wherein temperature is considered uniform in lumps having thermal capacity (e.g., the inside of the ingot); and resistance to flow is considered concentrated in other lumps (the outside coating on the ingot and the steam layer surrounding it) which have negligible thermal capacity. The accuracy of the model may be made as high as desired by dividing the conducting medium into sufficiently many lumps.

STAGE II. EQUATIONS OF MOTION. For a heat-conduction system it is convenient to list the considerations of Sec. 2.1 in the following more specific terms:

---

(1) *Variables*
(2) *Equilibrium:*
   (i) Define system
   (ii) Write heat-flow balance
   or, *compatibility* of temperature distribution
(3) *Physical heat-temperature relations*

Procedure A-h
Heat conduction

Then **combine** relations. (This grouping is exactly analogous to Procedure A-e.)

(1) Variables

In heat transmission the principal variables selected will be heat-flow rate q as the through or flow variable, and temperature T as the across variable. In a distributed-parameter model, q and T will be continuous functions of location; in a lumped-parameter model, q will be uniform through the boundaries chosen as resistances, and T will be uniform everywhere in the heat-capacitance lumps. Besides T, other thermodynamic properties may be involved as variables, including density ρ and internal energy per unit mass u. *Constraints* may be that temperature is prescribed at certain places (usually boundaries) in the system by external reservoirs (as T_b in Fig. 3.1a); or that heat flow is prescribed at certain places in the system (as from the motor inside the case in Fig. 3.1d).

(2) Equilibrium†

(i) We first define subsystems—prescribed collections of matter—and (ii) for each we write the First Law of Thermodynamics (conservation of energy).

---

† It must be noted that, traditionally, equilibrium in a thermodynamic system connotes *static* equilibrium. Our use of the word here to mean dynamic equilibrium is thus unconventional, thermodynamically, but consistent with our procedure in other physical media.

The general statement of the First Law of Thermodynamics is commonly given (for a system) in the form

$$dU = dQ - d\mathcal{W}$$

which may be written (for a lumped model)

$$\rho \mathcal{V}\, du = q_{net}\, dt - d\mathcal{W}$$

in which $dQ = q_{net}\, dt$ is the amount of heat transferred into the system, q_{net} being the net *rate* of heat flow into the system and dt an increment of time; $d\mathcal{W}$ is the work done *by* the system, and $dU = \rho \mathcal{V}\, du$ is the increase in the system's internal energy. U is total internal energy, ρ is mass density, $\mathcal{V}$ is total volume of the system, and u is internal energy per unit mass. (In the lumped model, ρ is considered uniform throughout $\mathcal{V}$.)

In pure heat-transmission problems there is no work done by the system, and the First Law can be written as follows:

$$(\rho \mathcal{V})\, du = q_{net}\, dt \qquad \text{No work}$$

or

$$(\rho \mathcal{V}) \frac{du}{dt} = q_{net} \qquad \text{No work} \qquad (3.3)$$

Relation (3.2) is obtained from (3.3) by using physical relation (3.4), given subsequently:

$$C \frac{dT}{dt} = q_{net} \qquad \text{No work} \qquad \begin{matrix}(3.2)\\ \text{[repeated]}\end{matrix}$$

in which thermal capacitance $C \triangleq c\rho\mathcal{V}$ (c being defined below).

Temperature *compatibility* is used less frequently in heat-conduction studies.

(3) Physical relations

In heat conduction two principal constitutive physical relations are important. The first is the empirical observation (3.1), that the rate of heat flow through a given material is proportional to the temperature drop across it:

$$q = \frac{1}{R}(T_a - T_b) \qquad \begin{matrix}(3.1)\\ \text{[repeated]}\end{matrix}$$

in which heat flow is from body a to body b, and resistance R is given,[†] for the case of conduction through a solid interface (as plane 4 in Fig. 3.3), by

$$\frac{1}{R} = \frac{kA}{\ell}$$

[†] The symbol R, for thermal resistance in (3.1), is used to emphasize the perfect and useful analogy of a one-dimensional lumped heat-conduction system to an electric network. Temperature T is analogous to voltage and heat-flow rate q is analogous to electric current (not to charge). The analogy will be developed completely in Chapter 4.

[Sec. 3.1] SIMPLE HEAT CONDUCTION

where k is the thermal conductivity of the material separating a and b, and A and ℓ are the cross-sectional area and length of the heat-transmission path. (kA/ℓ is called the thermal conductance.) For the interface between a fluid and a solid, a heat-transfer coefficient h is defined, and R is given by

$$\frac{1}{R} = hA$$

The second constitutive physical relation is the state relation that, under given conditions, change in temperature is proportional to change in internal energy per unit mass:

$$dT = \frac{1}{c} du \tag{3.4}$$

c is called the *specific heat* of the material.

The value of c depends on how other state variables are constrained during the change. For example, if specific volume ($v \triangleq 1/\rho$)† is held constant (with no phase change), c is the "specific heat at constant volume," c_v. Formally, c is defined by partial derivatives, e.g.

$$c_v \triangleq \left(\frac{\partial u}{\partial T}\right)_v$$

Combine relations

When the equilibrium and physical relations have been written for the separate subsystems, they are then combined, the latter commonly being entered into the former. That is, the equilibrium relations are the convenient focus, as they are in other media. As we have noted, in heat-conduction problems relations (3.3) and (3.4) are usually combined into (3.2) at the outset.

Determine, from physical reasoning, what the dimensions of C, c, u, and R, must be. Then refer to Table II, pp. 27 and 28.

Example 3.1. Heat conduction from a rocket engine. To demonstrate the above ideas, consider the problem of determining, versus time, the temperature in the outer insulating layer of a rocket engine during a one-minute firing. The whole engine is represented in Fig. 3.1b and a small segment of its case is represented in Fig. 3.3a. It is known that the temperature inside the chamber will rise almost instantly to the (approximately) steady value T_i. The outside temperature is T_o. The conductivities of the inner lining (k_2) and outer coating (k_6) are known, as are the conductance and thermal capacity of the two intermediate layers, an inner layer 3, having specific heat c_3 and high conductivity k_3, and the outer insulating layer 5, having specific heat c_5 and low conductivity k_5. The conductivity k_4 of the interface between layers 3 and 5 is also known (approximately).

† This is the only place in this book where v is used to mean specific volume. (Everywhere else it means either voltage or velocity, as we have noted.)

PHYSICAL MODEL (STAGE I). We wish to approximate the actual system of Fig. 3.3a by a lumped physical model. That is, even though the temperature will actually vary in a continuous way through the several layers, as indicated (solid plot) in Fig. 3.3a—because all the layers have *both distributed heat capacity and distributed resistance to flow*—we wish, in our model, to lump all the capacity in the two thick layers 3 and 5 (where most of it is), and to lump the resistance in the other three thin layers—the lining 2, the outer coating 6, and the interface 4, between 3 and 5—pretending that among them they contain all the thermal resistance. (This means that part of the resistance of 3 will be lumped with the lining and part with the interface, and so forth.) Then the temperature is considered uniform in 3 and in 5, with all the temperature drop occurring in the other, resistive, layers. This approximation to the temperature distribution is shown by the heavy dashed line in Fig. 3.3a, at an intermediate time. We assume also, for this study, that there is no heat flow along the walls; specifically, in our model segment no heat flows out the top or out the bottom—only horizontally.

EQUATIONS OF MOTION (STAGE II). We follow Procedure A-h.

(1) *Variables.* Define T_a, T_b, T_c, T_d, and heat-flow rates q_2, q_4, q_6, as indicated in Fig. 3.3a. (The temperature-drop relation $\Delta T_4 \equiv T_b - T_c$ is one of three obvious identities.) Temperatures T_a and T_d are *constrained* to have the given constant values T_i and T_o, respectively.

(2) *Equilibrium.* (i) Define as isolated subsystems, Fig. 3.3b, segments of b and c having area A. The other boundaries of the subsystems will be four arbitrary planes perpendicular to the engine walls, as shown in Fig. 3.3b. By symmetry it was assumed in the physical model that heat flows only in the direction perpendicular to the walls (i.e., out through the walls, but not along them). Then, using the First Law of Thermodynamics, with work equal to zero, in the form (3.2)

$$\text{For } b: \quad q_2 - q_4 = C_3 \dot{T}_b$$
$$\text{For } c: \quad q_4 - q_6 = C_5 \dot{T}_c \quad (3.5)$$

in which $C_3 = (\rho_3 A \ell_3) c_3$ and $C_5 = (\rho_5 A \ell_5) c_5$.

(3) *Physical relations.* By (3.1), and using temperature identities as needed:

$$q_2 = \frac{1}{R_2} (T_i - T_b)$$
$$q_4 = \frac{1}{R_4} (T_b - T_c) \quad (3.6)$$
$$q_6 = \frac{1}{R_6} (T_c - T_o)$$

in which $1/R_2 = k_2 A/\ell_2$, etc. (see p. 98).

Combine relations. We have now five independent equations and five unknowns (two temperatures and three heat-flow rates). As usual, it is convenient to use the equilibrium relations (3.5) as the focal point for combining:

[Sec. 3.1] SIMPLE HEAT CONDUCTION

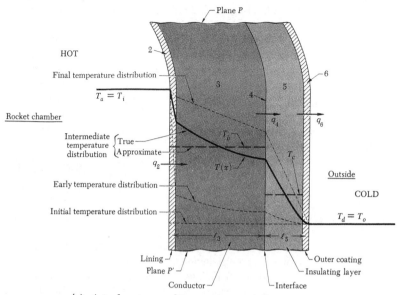

(a) *Actual system, and temperature variation*

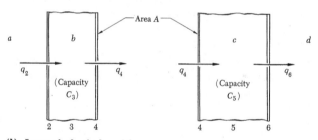

(b) *Lumped physical model*

Fig. 3.3 *Heat conduction from a rocket engine*

$$\frac{1}{R_2}(T_i - T_b) - \frac{1}{R_4}(T_b - T_c) = C_3 \dot{T}_b$$
$$\frac{1}{R_4}(T_b - T_c) - \frac{1}{R_6}(T_c - T_o) = C_5 \dot{T}_c \tag{3.7}$$

which are the equations of motion of the system in the unknowns T_b and T_c. (Do they check dimensionally?) From them we can determine the time history of T_b and T_c, as we shall in Chapter 19 (Stage III).

T_b and T_c are of course the temperatures in our model. They represent an approximation to the average temperatures in layers 3 and 5 of the actual system. For each time t, we can use these values of T_b and T_c to help us estimate (e.g., draw a plot of) the actual temperature distribution, as is indicated by the sets of lightly dashed curves in Fig. 3.3a.

Some additional familiarity with heat conduction can be gained from the suggested problems.

It is emphasized that this section treats heat transfer only in an introductory and approximate way. The important mechanisms of convection, radiation, fluid transport, and phase change are omitted (or grossly approximated, Problems 3.9–3.10), as are distributed systems and two- and three-dimensional systems. To pursue the subject further, see books on heat transmission, such as *Introduction to Heat and Mass Transfer*, E. R. G. Eckert and J. F. Gross, McGraw-Hill, 1963. (Others are listed in the bibliography.)

Problems 3.1 through 3.13

3.2 SIMPLE FLUID SYSTEMS

Some examples of interesting fluid systems are shown in Fig. 3.4. Figure 3.4a is a pneumatic bellows device used in the computer portion of an all-fluid control system. Figure 3.4b is part of the water-supply system for a town. Figure 3.4c is a hydraulic servo, used, for example, to position a rocket engine as in Fig. 1.4. Figure 3.4d is an aircraft in flight. Figure 3.4e shows a rocket vehicle during the boost phase of its journey.

Analytical procedure

STAGE I. PHYSICAL MODEL. The dynamics of fluids—liquids and gases—is fundamentally an extremely complicated thing to study because the elements of fluid are generally free to roam anywhere within regions of three-dimensional space. Progress is made in understanding fluid dynamics because some people with marvelous physical insight have been able to contrive *physical models* which were simple enough to analyze, and yet which have represented many of the important problems in fluid flow (e.g., the lift on aircraft, the force on turbine blades, the sound paths in a concert hall, and so forth) well enough to predict behavior with high accuracy. In fluid systems it is in Stage I (physical modeling) that, to a large degree the possibility and usefulness of analysis are determined.

It may be reasonable to make one or several of the following common approximations in establishing a physical model for a given fluid system: (a) that large elements of the fluid may be "lumped" (assumed to have uniform properties throughout), (b) that the flow is nearly steady (non-time-varying), the dynamic behavior consisting of small perturbations from steady flow, (c) that the fluid is incompressible, (d) that inertial forces may be neglected, (e) that there is no friction, and (f) that there is no exchange of heat.

In the present section we consider fluid dynamics on general grounds, and then note specific approximations which may make possible a linearized,

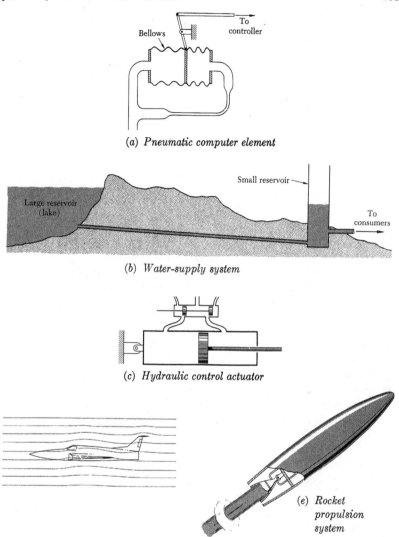

(a) *Pneumatic computer element*

(b) *Water-supply system*

(c) *Hydraulic control actuator*

(e) *Rocket propulsion system*

Fig. 3.4 *Some fluid systems*

lumped analysis. Our treatment here is necessarily cursory. Our purpose is to indicate the *approach* to studying non-steady-flow problems; the examples chosen are comparatively simple. The bibliography lists some comprehensive texts on the subject.

STAGE II. EQUATIONS OF MOTION. In fluid systems a slight modification of Procedure A-m provides a convenient outline for deriving the equations of motion:

> (1) **Variables and geometry**
> (2) **Equilibrium**
>   (i) Define closed or open system
>   (ii) Continuity
>   Force and momentum
>   Energy
> (3) **Physical relations**

Procedure A-f
Fluid systems

Then *combine relations.*

We consider each subdivision in turn.

(1) Variables and geometry

As in other media, the first step in studying the behavior of a fluid system is to select variables with which to define the system's behavior (and to note identities and constraints among them).

In fluid problems the variables include geometric quantities—coordinates, velocity v, mass flow rate w, and sometimes elevation h—and the properties pressure p, density ρ, and (sometimes) temperature T. (Additional properties are employed in thermodynamics.)

In the *selection of variables* we might, as did Lagrange, identify each particle of fluid and follow its behavior as time runs on. This is how we deal with rigid mechanical members, and fluid problems are sometimes studied this way also. A more widely used approach, due to Euler, is to consider the situation at each *point in space*, and to describe the situation in terms of the properties of whatever particle happens to occupy that point at a given time. This greatly simplifies the definition of mechanical coordinates.

In fluid mechanics we are sometimes interested in small perturbations from a steady-flow condition. In those cases we may define variables as *changes* from a large constant value (e.g., total pressure = $p_0 + p$, etc.). Then we shall study the dynamic behavior in terms of the small changes, which will lead to important simplification, particularly linearization.

Through and across variables. Sometimes we can assume a physical model for a fluid system which is not only lumped, but in which the only variables considered are (change in) pressure drop *across* the lumped elements and mass flow rate *through* them. Then these two variables are naturally classified as the across and through (or "flow") variables, respectively, which is convenient. In studying hydraulic systems using storage in gravity reservoirs (as in Fig. 3.4b), liquid height h may be the most useful choice for the across variable.

The *geometric relations between across variables* include the *constraints* of rigid walls and the obvious *identity* relations between liquid depth and volume in a vessel, or between the density of a (compressible) fluid and the volume that the fluid is constrained to occupy. Similarly, *flow* may be constrained to a prescribed value, as by a constant-displacement liquid pump, and so forth.

(2) Equilibrium

(i) *Systems.* Equilibrium relations can be written only for precisely defined systems. Two types are commonly used in fluid mechanics: the "closed system" and the "open system."

A *closed system* is any specified collection of matter, such as we show in a free-body diagram.

An *open system* is defined using an imaginary envelope in space, called a *control surface*. (The concept is illustrated in Fig. 2.1c, p. 34, for example.) Matter may enter and leave the envelope. At each instant the system is all the matter within the envelope at that instant.

(ii) *Equilibrium relations.* There are three important equilibrium relations in fluid mechanics.

Continuity. The first equilibrium relation in a fluid-flow problem is the "equation of continuity," which is merely a statement of the principle of conservation of matter: For a given open system (envelope in space), such as a reservoir, the difference between the rate of inflow and the rate of outflow of matter must be equal to the rate of increase of matter within the envelope

$$\boxed{\frac{dm}{dt} = w_{\text{in}} - w_{\text{out}}} \qquad (3.8)$$

in which m is the total mass of fluid inside the envelope, and w_{in} and w_{out} are the mass-flow rates. For the special case when the open system has no storage capacity (e.g., a junction of pipes), (3.8) resembles Kirchhoff's Current Law (refer to Fig. 2.14).

Force equilibrium. For a closed system (specified collection of fluid), just as for any mechanical free-body diagram:

$$\Sigma f^*_{\text{any direction}} = 0 \qquad (3.9)$$

$$\Sigma M^*_{\text{any axis}} = 0 \qquad (3.10)$$

That is, D'Alembert Eqs. (2.1) and (2.2), p. 40, hold for a free body of fluid just as for a system of rigid bodies. The relations are useful, of course, only when the motion is such that inertial force $\mathbf{f}_i$ and inertial moment $\mathbf{M}_i$ are readily identified, as when all of a lump of fluid translates as a rigid body.†

---

† More generally, (3.9) (or Newton's Second Law) must be written for each mass particle. The resulting fundamental set of nonlinear, partial differential equations for the general motion of a fluid are known as the Navier-Stokes equations. Certain integrations of them for special circumstances are widely used. For example, Euler's equation presumes that friction effects are negligible. Bernoulli's formula for hydraulics presumes, in addition, incompressibility and steady state. Unfortunately, space prohibits inclusion of the fundamental-particle approach in this book.

Conservation of momentum and energy. These powerful tools are derived, for an open system, from Newton's Second Law and from the First Law of Thermodynamics, respectively. We shall not require them for the simple problems we study here, and their derivation is outside the scope of this book.

(3) Physical relations

The constitutive physical relations of fluid mechanics include (among other things) the relations between the properties pressure and density, and the relations between flow rate and pressure drop, and between flow rate and friction forces. For large motions, these are likely to be very nonlinear relations. Sometimes we can consider *small changes* from a steady-flow condition; and then a linear relation between the changes may be a reasonable approximation.

Pressure-density relation. The simplest pressure-density relation is that of incompressibility, wherein density is independent of pressure. This approximation normally leads to accurate results for liquids (as in the hydraulic systems of Fig. 3.4b and c). It may also be good for the flow of gases in certain cases. (It predicts accurately the lift on an airplane wing, Fig. 3.1d, in low-speed, steady flow, for example.)

More generally, the best model for studying gas behavior may be that for behavior of an ideal gas in the absence of heat transfer—the "ideal gas law"

$$p = k\rho^\gamma \tag{3.11}$$

in which γ is a constant. For air $\gamma = 1.4$. (For a given system, k is obtained from one known condition.) Relation (3.11) is illustrated in Fig. 3.5.

The pressure-density relation in perturbation studies. In those cases where we are interested in the dynamics of small changes from steady flow, we can derive approximate linear relations by replacing p by $p_0 + p$ and ρ by $\rho_0 + \rho$, in (3.11):

$$\begin{aligned}(p_0 + p) &= k(\rho_0 + \rho)^\gamma \\ &= k\rho_0^\gamma \left(1 + \frac{\rho}{\rho_0}\right)^\gamma\end{aligned} \tag{3.12}$$

in which p_0 and ρ_0 are steady-state values, and p and ρ are now *small increments* (*perturbations*) from the steady state. Equation (3.12) may be expanded in a binomial series:

$$p_0 + p = k\rho_0^\gamma \left[1 + \gamma \frac{\rho}{\rho_0} + \frac{\gamma(\gamma-1)}{2!}\left(\frac{\rho}{\rho_0}\right)^2 + \cdots\right] \tag{3.13}$$

and then product terms dropped (as being very small), giving (with $p_0 = k\rho_0^\gamma$):

$$p \cong (k\gamma\rho_0^{\gamma-1})\rho \tag{3.14}$$

as the desired linear (approximate) relation. What we have done, in effect, is to use the *local slope* from the pressure-density plot, Fig. 3.5. Show this by differentiating (3.11) to obtain $\left(\dfrac{dp}{d\rho}\right)_{\rho=\rho_0}$ and substituting into $p = \left(\dfrac{dp}{d\rho}\right)\rho$.

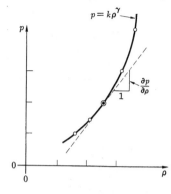

Fig. 3.5 *The pressure-density physical relation for an ideal gas in the absence of heat transfer*

Fluid capacitance. In fluid systems there are often elements—pressure vessels, water-storage tanks, etc.—that store fluid part of the time for delivery at another time. Several are shown in Fig. 3.6. These elements play the same dynamic role in fluid systems as across-variable energy-storage elements do in mechanical and electrical systems (i.e., mechanical mass and electrical capacitance). For each element, the appropriate mathematical description (including an expression for capacitance) will be obtained when we write continuity relation (3.8) for the element as a subsystem.

Fluid resistance and fluid friction. It is useful to consider separately fluid flow through short constrictions, such as those in Fig. 3.7a through d, and flow in long tubes or pipes, Fig. 3.7e, or through long channels between plates, and so forth. In the former case there is negligible mass of fluid within the element at a given instant, and body forces—gravity and inertia —can always be ignored. Then a direct relation between flow and pressure drop can be given in the form

$$w = \frac{1}{R}(p_1 - p_2)^{1/\alpha} \qquad (3.15)$$

in which R is simply a "resistance." In long tubes or channels, it may not be possible to ignore body forces on the fluid within the element. Then it is necessary to consider friction forces per se, along with the other forces, and to write an equilibrium relation, (3.9) or (3.10). Friction force is a function of fluid velocity in the form

$$f_f = bv^\alpha \qquad (3.16)$$

and the ensuing analysis is a little more elaborate. The value of α in (3.15) and (3.16) is generally either 1 or (approximately) 2, depending on whether

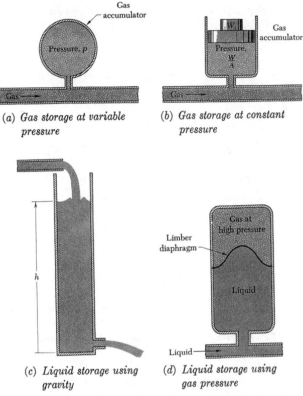

Fig. 3.6 *Dynamic storage of fluid: elements having capacitance*

viscous friction is dominant or negligible, respectively. There are fundamental reasons for this which are most clearly seen physically for flow in long tubes, which we shall discuss presently. First we consider short constrictions.

Fluid resistance in short constrictions. When there is a pressure difference between the ends of a short constriction such as those in Fig. 3.7a through d, there results a flow which is generally related to the pressure drop (approximately) in one of two ways [compare with (3.15)]:

| Either | $w = c_\alpha \sqrt{p_1 - p_2}$ | Most short constrictions | (3.17) |
| or | $w = c(p_1 - p_2)$ | Porous plug | (3.18) |

If we can neglect viscous friction and compressibility, which are almost always excellent approximations for typical flow speeds in short constrictions, then (3.17) can be derived rigorously from energy and continuity considerations.

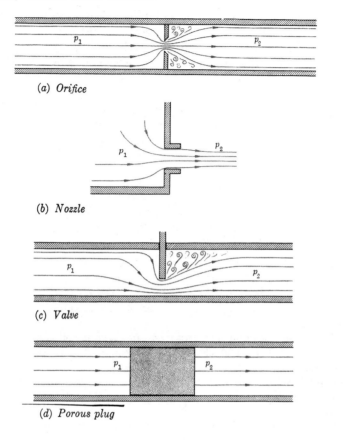

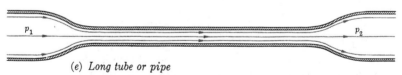

Fig. 3.7 *Resistance to fluid flow*

The porous plug is an exception: Viscosity dominates the flow through the pores, and relation (3.18) governs.

Fluid friction in long tubes or pipes. Either of two quite different flow patterns can exist in a tube, depending on the parameters of flow. Specifically, the kind of flow depends on the value of a dimensionless number called Reynolds number, defined by:

$$\mathcal{R} = \frac{vD\rho}{\mu} \tag{3.19}$$

in which v is average velocity ($v = w/\rho A$), D is pipe diameter (or channel width, etc.), ρ is fluid mass per unit volume [dimensions $(f/\ell\tau^{-2})/\ell^3 = f\tau^2\ell^{-4}$], and μ is fluid viscosity—shear stress/velocity gradient (dimensions $(f/\ell^2)/[(\ell/\tau)/\ell] = f\tau\ell^{-2}$). Some values of μ and ρ are given in Table III.

TABLE III
Typical values of fluid density and viscosity
(approximate values, accurate to 5%)

Fluid	Density ρ ($f\tau^2\ell^{-4}$)	Viscosity μ ($f\tau\ell^{-2}$)	$\dfrac{\mu}{\rho}$ ($\ell^2\tau^{-1}$)
Water at 73°F	1 dn sec²/cm⁴ 1×10^3 n sec²/m⁴ 2 lb sec²/ft⁴	1×10^{-2} dn sec/cm² 1×10^{-3} n sec/m² 2×10^{-5} lb sec/ft²	1×10^{-2} cm²/sec 1×10^{-6} m²/sec 1×10^{-5} ft²/sec
Air at 73°F and atmospheric pressure	1.2×10^{-3} dn sec²/cm⁴ 1.2 n sec²/m⁴ 2.3×10^{-3} lb sec²/ft⁴	1.8×10^{-4} dn sec/cm² 1.8×10^{-5} n sec/m² 3.8×10^{-7} lb sec/ft²	0.15 cm²/sec 0.15×10^{-4} m²/sec 1.7×10^{-4} ft²/sec

As an example, water flowing at 10 ft/sec in a conduit of one-foot diameter has a value of $\mathcal{R} \cong 1{,}000{,}000$. Air at atmospheric pressure in the same circumstances has $\mathcal{R} \cong 50{,}000$.

The following rules of thumb will provide a reasonable model for friction force versus pressure drop in many fluid problems:

If $\mathcal{R} < 1100$, flow is *laminar*, friction force is linearly proportional to velocity:

$$\boxed{f_f = bv} \quad \mathcal{R} < 1100 \tag{3.20}$$

with $b = 8\pi\ell\mu$ (if v is average velocity and ℓ is pipe length). Note that f_f is independent of density in the laminar regime. If, in addition, fluid *body forces can be ignored*, then this leads to a flow relation like (3.18):

$$\boxed{w = \frac{1}{R}(p_1 - p_2)} \quad \mathcal{R} < 1100 \tag{3.21}$$

If $\mathcal{R} > 3500$, flow is *turbulent*, friction force varies as a power of velocity:

$$\boxed{f_f = b_\alpha v^\alpha} \quad \mathcal{R} > 3500 \tag{3.22}$$

[Sec. 3.2] SIMPLE FLUID SYSTEMS   111

in which $\alpha = 1.75$ for smooth tubes, and $\alpha = 2$ for high-$\Re$ flow in rough pipes (e.g., for $\Re > 10^5$). (See Appendix F.) If, in addition, *body forces can be ignored*, then (3.22) leads to an approximate flow relation like (3.17):

$$\boxed{w = c_\alpha(p_1 - p_2)^{1/\alpha}} \qquad \Re > 3500 \qquad (3.23)$$

The constants b, R, b_α, c_α, and α depend, of course, on geometry, fluid medium, etc. Data for their calculation is given in Appendix F. (What are the dimensions of each?) Turbulent flow is more common in engineering. The physical behavior leading to these results is discussed briefly below.

Reynolds number is a measure of the relative importance of inertial forces and viscous forces in the microscopic flow patterns. When $\Re$ is low—small inertial forces (e.g., small ρv), large viscous forces (large μ)—the flow pattern is laminar:

In *laminar flow* the fluid moves in concentric cylindrical layers or lamina (of infinitesimal thickness) as indicated in Fig. 3.8a. Each layer slides over the

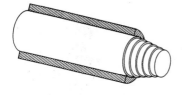

Fig. 3.8 *Regimes of fluid flow*

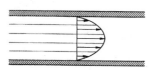

(a) *Laminar flow in a pipe*
(*low Reynolds number*)

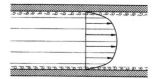

(b) *Turbulent flow in a pipe*
(*high Reynolds number*)

next (like layers of paper in a roll when you push your finger in the center), producing a shear stress linearly proportional to relative velocity; and the stress between the outermost layer and the wall, the resistance to flow, is simply linearly proportional to velocity. Relation (3.20) can be calculated directly and rigorously from this concept, which is due first to Newton (including its mathematical description). The velocity distribution is found to be exactly parabolic, as indicated in Fig. 3.8a.

For large values of $\Re$ the inertial forces (within the tiny elements of fluid) dominate the flow pattern, and the flow is turbulent:

In *turbulent flow* through a pipe, Fig. 3.8b, a reasonably good image is that most of the fluid moves approximately as a rigid cylinder, rolling through the tube on a turbulent outer boundary layer (which is very thin). Forces are transmitted from the wall through the boundary layer in a complicated way, and the force varies approximately as $v^{1.75}$ or v^2. The velocity distribution is nearly even across the pipe, except at the boundaries.

As a fundamental matter, in laminar flow, resistance to flow depends only on viscosity, while in turbulent flow, resistance is due predominantly to local accelerations in the boundary layer, to the extent that resistance depends only on fluid mass and is virtually independent of viscosity.

Fluid resistance in perturbation studies. When we are interested in the dynamics of small changes from steady flow, we can obtain approximate linear relations between flow and pressure drop or friction force, even when a nonlinear relation, (3.17), (3.22), or (3.23), is in effect. The perturbation technique is used. We demonstrate for (3.17), which we rewrite as follows:

$$w_0 + w = c_\alpha \sqrt{(p_{01} + p_1) - (p_{02} + p_2)}$$

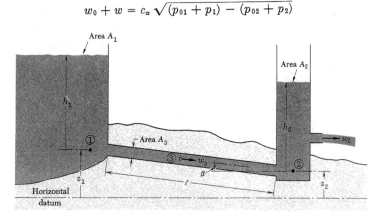

(a) *System in arbitrary configuration*

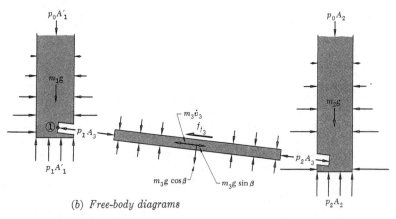

(b) *Free-body diagrams*

Fig. 3.9 *The dynamics of a water-supply system*

[Sec. 3.2] SIMPLE FLUID SYSTEMS

in which w is now the change in mass rate of flow from a steady-flow value w_0, and p_1 and p_2 are changes from the steady values p_{10} and p_{20} of pressure at the upstream and downstream ends of the restrictive element. Again, we express $w_0 + w$ in a series:

$$w_0 + w = c_\alpha \sqrt{p_{01} - p_{02}} \left[1 + \frac{1}{2} \frac{p_1 - p_2}{p_{01} - p_{02}} - \frac{1}{8} \left(\frac{p_1 - p_2}{p_{01} - p_{02}} \right)^2 + \cdots \right] \quad (3.24)$$

and approximate flow rate by dropping products of variables as being, relatively, very small. After subtracting the steady-state component, $w_0 = c_\alpha \sqrt{p_{01} - p_{02}}$, we have, for the *change* w in flow rate,

$$w \cong \frac{1}{2} c_\alpha \sqrt{p_{01} - p_{02}} \frac{p_1 - p_2}{p_{01} - p_{02}}$$

or

$$\boxed{w \cong \frac{1}{R} (p_1 - p_2)} \quad (3.25)$$

in which $R = 2(p_{01} - p_{02})/w_0 = 2 \sqrt{p_{01} - p_{02}}/c_\alpha$.

Linear approximations like (3.14) and (3.25) for small perturbations from steady flow are often the key to linear modeling of fluid systems.

Relation between h and p in a reservoir. In hydraulics problems where storage of fluid is in gravity reservoirs, Fig. 3.6c, pressure in a reservoir (at any point where velocity is negligible) is simply proportional to depth, by hydrostatics:

$$p = \rho g h \quad (3.26)$$

Then it is often convenient to write the equations of motion in terms of the heights h in the reservoirs in the system [by substituting for p from (3.26)], the heights being more obvious quantities to measure. For example, (3.23) may be written:

$$\boxed{w = c_\alpha'(h_1 - h_2)^{1/\alpha}} \quad (3.27)$$

with $c_\alpha' = (\rho g)^{1/\alpha} c_\alpha$; and (3.21) and (3.25) may be written

$$\boxed{w = \frac{1}{R'} (h_1 - h_2)} \quad (3.28)$$

with $R' = R/\rho g$.

Example 3.2. A hydraulic reservoir system. The system shown in Fig. 3.9 represents two large hydraulic reservoirs (part of a residential water system for a town) connected by a long uniform line.

We are interested in the equations of motion of the system, from which to determine what the time history of the heights of the two reservoirs will be for any given time history of the usage rate, $w_2(t)$. (Recall that we are using w to mean mass-flow rate.)

STAGE I. PHYSICAL MODEL. We assume a lumped model having three lumps: the two reservoirs and the connecting pipe. The reservoirs are so large compared to the pipe diameter that fluid velocity in a reservoir is neglected everywhere. Compressibility is neglected.

STAGE II. EQUATIONS OF MOTION.

(1) *Variables and geometry.* We choose as variables h_1, h_2 (defined in Fig. 3.9a), pressures p_1 and p_2 at points 1 and 2 respectively, flow rates w_2 out of reservoir 2 and w_3 in the connecting line (and therefore out of reservoir 1). Average velocity in the pipe is given by $v_3 \equiv w_3/\rho A_3$. The total mass of fluid in the line is $m_3 = \rho A_3 \ell$.

(2) *Equilibrium—continuity.* (i) Consider each reservoir as an open system. (Sketch appropriate control surfaces in Fig. 3.9a.) Then (ii)

$$\text{For reservoir 1:} \quad \rho A_1 \dot{h}_1 = -w_3$$
$$\text{For reservoir 2:} \quad \rho A_2 \dot{h}_2 = w_3 - w_2$$

Force equilibrium. (i) Figure 3.9b shows a free-body diagram of the fluid in the pipe. (ii) Applying (3.9) along the pipe:

$$\Sigma f^* = 0 = p_1 A_3 - p_2 A_3 - f_{f3} - m_3 \dot{v}_3 + m_3 g \frac{z_1 - z_2}{\ell} \quad (3.29)$$

(Note that z_1 and z_2 are the elevations of points 1 and 2 above an arbitrary horizontal datum.)

(3) *Physical relations.* We estimate that flow in the pipe will be turbulent, and we therefore use expression (3.22) (friction coefficient b_α to be measured or calculated using data such as that in Appendix F):

$$f_{f3} = b_\alpha v_3^\alpha$$

in which we take α to be 2 (turbulent flow in a rough pipe). (When we have finished Stage III of the whole problem we must then calculate w_3 and v_3, and thence $\mathfrak{R}$, to see if our guess of the $\mathfrak{R}$ regime was correct.) Finally, for the two reservoirs, (3.26) gives:

$$p_1 = \rho g h_1 \qquad p_2 = \rho g h_2 \quad (3.30)$$

Combine relations. We have seven equations in the seven unknowns v_3, w_3, h_1, h_2, p_1, p_2, and f_{f3} (plus w_2 which is prescribed). Combining to eliminate four of these, we obtain

$$\rho A_1 \dot{h}_1 \qquad\qquad\qquad + w_3 = 0 \quad (3.31)$$
$$\rho A_2 \dot{h}_2 \qquad\qquad\qquad - w_3 = -w_2 \quad (3.32)$$
$$-(h_1 - h_2) + \frac{\ell}{g\rho A_3} \dot{w}_3 + \frac{b_\alpha}{g(\rho A_3)^{\alpha+1}} w_3^\alpha = (z_1 - z_2) \quad (3.33)$$

These are the equations of motion which are to be solved in Stage III for the system's time behavior—i.e., for h_1 and h_2 versus time. Is each equation dimen-

[Sec. 3.2] SIMPLE FLUID SYSTEMS 115

sionally consistent? Note that, as usual, the equilibrium relations have been the convenient focus for combining relations.

If, as is probably the case, inertial force $\ell\dot{w}_3$ is not important, then (3.33) can be solved for w_3 and the result entered into (3.31) and (3.32). Then we have the following equations of motion for the system of Fig. 3.9a:

$$\left.\begin{array}{l}\rho A_1 \dot{h}_1 + C[(h_1 + z_1) - (h_2 + z_2)]^{1/\alpha} = 0 \\ \rho A_2 \dot{h}_2 - C[(h_1 + z_1) - (h_2 + z_2)]^{1/\alpha} = -w_2\end{array}\right\} \quad (3.34)$$

What is C? [The same result could have been obtained by using (3.23), with (3.26), directly in (3.31) and (3.32).] Do all terms check dimensionally? Consider some special cases (e.g., $A_2 = 0$).

With $\alpha \neq 1$ the above sets of equations are nonlinear, so that in Stage III an exact time solution would have to be obtained by use of machine computation. Alternatively, an approximate solution could be obtained using small perturbation from a steady-flow condition, as suggested in Problem 3.14. That is, as

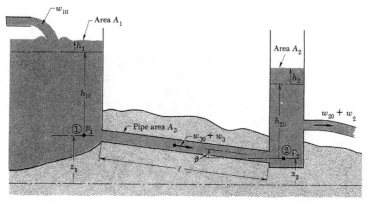

Fig. 3.10 *The dynamics of a water-supply system studied in terms of small changes from steady flow*

shown in Fig. 3.10, assume that there is a steady flow w_{10} into reservoir 1, and that there results a steady-flow condition indicated by $w_{30} = w_{20} = w_{10}$ and accompanied by nominal reservoir heights h_{10} and h_{20}. We are then interested in small changes from the steady condition, indicated by w_2, w_3, h_1, and h_2 in Fig. 3.10. Before carrying out this problem it may be helpful to study Example 3.3, below, in which the perturbation technique is demonstrated.

Example 3.3. A bellows element for a pneumatic computer. Various pneumatic devices are commonly used to perform dynamic operations on measured signals in fluid systems. That is, they are used as computing elements.

An example is shown in Fig. 3.11. In this case it is desired to have a mechanical displacement θ which will indicate promptly any rates of change in gas pressure at point 1, but which will indicate null when p_1 is constant at any value (regardless of that value). The arrangement shown performs that function, as may be determined qualitatively by inspection. We wish now to derive the equations of motion of the device, from which its dynamic behavior can be predicted.

STAGE I. PHYSICAL MODEL. We assume that there is no resistance to flow except in R. Further, we postulate that flow in R is laminar, so that resistance to flow is proportional to flow rate. (If D is 0.03 in., what maximum velocity, v_2, does this imply?) At the end of Stage III we must calculate $\mathfrak{R}_{max}$ from $v_2(t)$

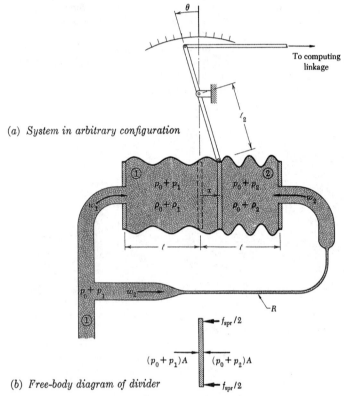

(a) *System in arbitrary configuration*

(b) *Free-body diagram of divider*

Fig. 3.11 A *pneumatic computer element*

to be sure that our laminar-flow assumption is viable. The mass of the bellows, including the center divider, is to be neglected. Intuition indicates that fluid compressibility may play a strong role and must not be neglected. The bellows is assumed to have the characteristic of a *linear spring*. We decide to neglect the flow of heat. Fluid properties are considered homogeneous throughout each chamber; further, they are taken homogeneous in the lines, and compressibility and inertial forces are neglected in the lines. Thus, we identify two fluid lumps in the system: chamber 1 with its adjacent line, and chamber 2, with its adjacent line.

[Sec. 3.2] SIMPLE FLUID SYSTEMS 117

STAGE II. EQUATIONS OF MOTION. We follow Procedure A-f.

(1) *Variables and geometry.* The system is shown in an arbitrary configuration in Fig. 3.11a. We define each variable in terms of a steady-state value (e.g., p_0) plus a small change (e.g., p) from that steady-state value. Thus the variables are as shown in Fig. 3.11a, in which p_1 is a given input to the system—it is *constrained* to a prescribed value. (Note that pressure $p_0 + p_1$ is taken to be uniform throughout chamber 1 and the line approaching it, in accordance with the physical model.)

The output θ to the computing linkage is given by the geometric identity:

$$\theta \equiv \frac{x}{\ell_2} \tag{3.35}$$

(2) *Equilibrium relations. Continuity.* (i) Consider each chamber as an open system. (Again, each system may be thought of as defined by an imaginary control surface.) Then (ii) the rate of increase in the amount of fluid in each chamber is equal to the flow rate into it.

For chamber 1: $\quad A\ell\dot{\rho}_1 + A\rho_0\dot{x} = w_1 \tag{3.36}$
For chamber 2: $\quad A\ell\dot{\rho}_2 - A\rho_0\dot{x} = w_2 \tag{3.37}$

[More precisely, note that the amount m_1 of fluid in chamber 1, for example, is $m_1 = A(\rho_0 + \rho)(\ell + x)$. Differentiate to obtain dm_1/dt, drop products of variables—e.g., $\dot{\rho}_1 x$ is very small compared to $\dot{\rho}_1 \ell$—and set $dm_1/dt = w_1$.]

Force equilibrium. (i) Taking the divider as a free body, Fig. 3.11b, we find that (ii) Eq. (2.1) gives

$$(p_1 - p_2)A = f_{\mathrm{spr}} \tag{3.38}$$

(3) *Physical relations.*

Mechanical spring (bellows): $\quad f_{\mathrm{spr}} = k_3 x \tag{3.39}$

Flow through tube R using (3.21):† $\quad w_2 = \frac{1}{R}(p_1 - p_2) \tag{3.40}$

No heat flows, so we assume the ideal gas law, (3.14):

$$p_1 = (k\gamma\rho_0^{\gamma-1})\rho_1 = K_1\rho_1 \tag{3.41}$$
$$p_2 = (k\gamma\rho_0^{\gamma-1})\rho_2 = K_2\rho_2 \tag{3.42}$$

Combine relations. We see that we now have seven equations in seven unknowns (p_1 is prescribed). w_1 appears only in (3.36). Since w_1 and ρ_1 are not of prime interest, we drop (3.36) and (3.41) from the set. The remaining equations can be conveniently combined by first differentiating (3.42) with respect to time:

$$\dot{p}_2 = K_2\dot{\rho}_2$$

and then eliminating $\dot{\rho}_2$ and w_2 from (3.37):

$$\left(\frac{A\ell}{K_2}\right)\dot{p}_2 - A\rho_0\dot{x} = \frac{1}{R}(p_1 - p_2) \tag{3.43a}$$

† As noted above, the propriety of this linear assumption must be checked at the end of the problem by computing $\mathfrak{R}_{\max}$.

Entering (3.39) into (3.38) gives:

$$(p_1 - p_2)A = k_3 x \qquad (3.43b)$$

Again, the equilibrium relations—continuity and force equilibrium—are the focal relations. These are two equations in the two unknowns p_2 and x and the known (given) input p_1. From them (in Stage III) we can learn the response x to changes in p_1. (See Problems 6.62 and 9.18.)

Do the dimensions in Eqs. (3.43) check? If R becomes very small, is the resulting effect on the equations as expected? What other special-case checks can you think of?

Example 3.4. A hydraulic controller. One method of steering a rocket vehicle is to swivel the whole rocket engine to control the direction of its thrust, as in Fig. 1.4. Rocket engines are massive, and they must be moved very rapidly to stabilize a vehicle on its course. To control quickly the large forces involved, the engines are moved by a hydraulic system, as shown in Fig. 3.12a. A small, easily moved valve admits oil under high pressure into a cylinder containing a large piston, so that a large hydraulic force (pressure times piston area) is established to move the engine. (Consider the analogy to a high-gain electrical isolation amplifier, Fig. 2.20.) The parts in Fig. 3.12 are not to scale, of course.

We want to write the equations of motion governing the response θ of the engine due to a motion x of the valve. Later (pp. 614, 615, 616) we shall study more of the system including a signal feedback from the engine position to close the valve when the engine reaches its desired new position.

Measured data on the valve-flow characteristics are given in the plot of Fig. 3.12b.

STAGE I. PHYSICAL MODEL. For this first analysis we assume that the solid parts of the engine are rigid, that the liquid hydraulic fluid is *incompressible*, and that it flows through the valve into one end of the piston and out the other at a rate given by the linear approximation indicated in Fig. 3.12b.

A separate and equally important part of the physical model is that the pressure is assumed always *uniform* throughout each of the chambers s, e, 1, and 2. That is, all the fluid in a given chamber is considered as "one lump." The resulting lumped-system model will lead to ordinary differential equations of motion.

STAGE II. EQUATIONS OF MOTION. We follow Procedure A-f.

(1) *Variables and Geometry.* The variables are defined in Fig. 3.12a, where the system is shown in an arbitrary configuration. The supply and exhaust pressures p_s and p_e are constrained to be constant, and valve displacement x is prescribed. For small θ, the geometric identity

$$y \equiv \ell_1 \theta \qquad (3.44)$$

is valid. (For larger θ a more accurate, nonlinear relation would be used.)

(2) *Equilibrium—continuity.* From (3.8), with chambers 1 and 2 as the open systems:

$$A \rho_0 \dot{y} = w_1 = w_2 \qquad (3.45)$$

[Sec. 3.2] SIMPLE FLUID SYSTEMS 119

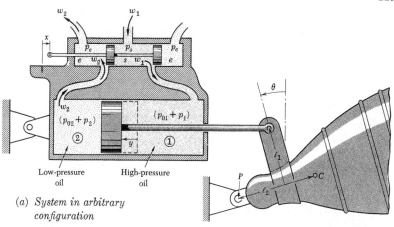

(a) System in arbitrary configuration

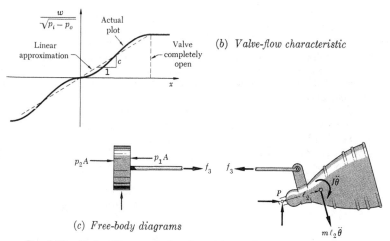

(b) Valve-flow characteristic

(c) Free-body diagrams

Fig. 3.12 *Hydraulic control of rocket-engine angle*

(That is, the piston must move with the incompressible fluid. ρ_0 is the constant density).

Force equilibrium. (i) From the free-body diagrams of Fig. 3.12c, (ii)

For the piston† [Eq. (2.1)]: $\quad \Sigma f^* = 0 = (p_1 - p_2)A - f_3$
For the engine‡ [Eq. (2.2)]: $\quad \Sigma M^* = 0 = (J + m\ell_2^2)\ddot{\theta} - f_3 \ell_1$ $\quad$ (3.46)

in which the mass of the piston is neglected, J is the moment of inertia of the

† The slight decrease in piston area due to the piston rod has been neglected here. It is easy to include.
‡ Application of (2.2) to rotation about a point other than the mass center is discussed fully in Sec. 5.3.

engine about its mass center C, and θ is considered small. (How must the second equation be altered if point P is accelerating?) Also, it has been tacitly assumed that nominally there is no load, so that $p_{01} = p_{02} = (p_s - p_e)/2$.†

(3) *Physical relations.* From the valve-flow data:

$$w_1 = c \sqrt{p_s - p_{01} - p_1}\, x \qquad (3.47)$$

which is nonlinear. In keeping with our perturbation approach, we expand in a binomial series:

$$w_1 = c \sqrt{p_s - p_{01}} \left[1 - \frac{1}{2} \frac{p_1}{p_s - p_{01}} - \frac{1}{8} \left(\frac{-p_1}{p_s - p_{01}} \right)^2 - \cdots \right] x$$

and then drop the products of variables, such as $p_1 x$, $p_1^2 x$, etc.:

$$w_1 = c \sqrt{p_s - p_{01}}\, x \qquad (3.48)$$

Combine relations. We now notice that the three equations (3.44), (3.45), and (3.48) contain only the three unknowns y, θ, w_1, plus the prescribed x. Combining:

$$A \rho_0 \ell_1 \dot\theta = c \sqrt{p_s - p_{01}}\, x \qquad (3.49)$$

or *Approximate behavior of hydraulic controller* $\boxed{\ell_1 \dot\theta = Cx}$ $\qquad (3.50)$

That is, to first order, the motion of the piston and engine is governed entirely by simple continuity considerations, without regard for changes in hydraulic pressure. This relation is used in Stage III to calculate system time behavior. (It is employed on p. 615, for example.)

The chamber pressures certainly have to change to push the inertia load. The value of pressure difference $p_1 - p_2$ is given by (3.46). But, to first order, the pressures do not, in turn, mitigate the direct relation (3.50) between valve opening and piston velocity.

The more interesting case when compressibility cannot be neglected is the subject of Problem 3.23.

Problems 3.14 through 3.25

---

† The last equality is deduced from symmetry. Physically, when the valve is in neutral there is normally leakage around the valve, from supply chamber s directly into exhaust chambers e. Half the accompanying pressure drop occurs between s and 1 or 2, and the other half between 1 or 2 and e.

ⓒChapter 4
ANALOGIES

The equations of motion and the behavior of systems involving various physical media are found to be analogous, and these analogies can be exploited to carry a firm analytical foundation in one medium over into improved understanding of behavior in another.

The classification of variables as through and across variables is accompanied by a useful classification of physical elements, Table IV.

The electrical analogy for thermal, fluid, and mechanical systems is developed formally, and demonstrated with a number of examples.†

The approach is then presented (for those interested) of representing very simple physical systems in these media by "network" diagrams, which can then be analyzed conveniently, using the techniques already available from electrical-network theory. The symbolism used in this procedure is displayed in Table V.

Some discussion is given of the disadvantages and pitfalls of exclusive reliance on analog and network techniques in analyzing physical systems.

We begin the discussion of analogies by considering, in Sec. 4.1, those between heat flow and electricity and between fluid flow and electricity, where the obvious similarity between the flow of heat, of a fluid, and of electric charge has strong intuitive appeal. (Section 4.1 may be omitted without loss of continuity, however, by the reader interested only in the mechanical and electrical media.)

4.1 ANALOGIES BETWEEN PHYSICAL MEDIA

We may well suspect from the discussions of Chapters 2 and 3 that there are strong analogies between the equations of motion for lumped physical models of systems in the different physical media. The set of considerations (Sec. 2.1) in deriving the equations of motion is the same,

† Electrical duals—pairs of networks which are analogous to one another in a precise way—are treated in Appendix G.

and the equations of motion have the same form, each physical parameter in one medium having a counterpart that plays the analogous role in another.

The existence, and indeed the complete correspondence, of such analogies in some cases, is most perfectly illustrated in the case of conduction of electricity and conduction of heat in a lumped-system model.

The electrical analog of heat conduction

Dynamic studies of these two media are exactly analogous, as follows.

In the *physical model (Stage I)*, thermal resistance and storage capacity are considered lumped in discrete elements. Then:

Thermal capacitance C is analogous to electrical capacitance C.

Thermal resistance R is analogous to electrical resistance R.

In writing the *equations of motion (Stage II)*, Procedures A-h (p. 97) and A-e (p. 61) are followed.

(1) *Variables.*

Through variables: Heat flow q is analogous to flow of electrical current i (not to charge†).

Across variables: Temperature T is analogous to voltage v.

The constraints are:

Prescribed temperature T at a given boundary is analogous to prescribed voltage v at a given node. (The environment is a temperature "source.")

Prescribed heat flow q through a given layer is analogous to prescribed current i through a given branch. (The environment is a heat-flow source.)

The identities are also analogous.

(2) *Equilibrium.* The First Law of Thermodynamics, with no work, in the form (3.2) applied at a junction where no heat is stored, is analogous to Kirchhoff's Current Law.

Compatibility. Temperature compatibility around a path is analogous to Kirchhoff's Voltage Law.

(3) *Physical relations.* The empirical relations for heat flow are analogous to those for flow of electrical current as follows:

$$\text{Resistance:} \qquad q = \frac{1}{R}(T_a - T_b) \qquad i = \frac{1}{R}(v_a - v_b) \qquad (4.1)$$

$$\text{Capacitance:} \qquad q = C\dot{T} \qquad i = C\dot{v} \qquad (4.2)$$

(There is no thermal analog for inductance.)

Finally, the heat-flow electrical analogy is enhanced by the well-known fact that materials which are good conductors of electricity are usually also good conductors of heat.

---

† This accidental nomenclature mismatch—q being used for heat-flow rate and for electric charge—is a little annoying from the analogy viewpoint, but the nomenclature is thoroughly established in both fields.

[Sec. 4.1] ANALOGIES BETWEEN PHYSICAL MEDIA 123

As an example, consider again the heat-conducting system of Fig. 3.3 (the rocket engine wall), the lumped model for which is repeated in Fig. 4.1a. It is a simple matter, using the above analogous quantities, to construct an electrical network, Fig. 4.1b, which is perfectly analogous in every way to the thermal system model of Fig. 4.1a. (Thermal, rather than electrical labels are used in Fig. 4.1b, to emphasize the point.) Note that all the T's (like all the v's in a network) are measured with respect to the same reference T_r. Equations of motion (3.7) of Example 3.1 can be derived by applying "KCL" to Fig. 4.1b.

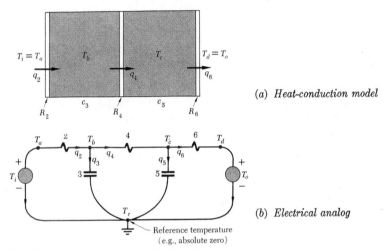

(a) *Heat-conduction model*

(b) *Electrical analog*

Fig. 4.1 *Electrical analog of a heat-conduction system*

We can see at once two potentially useful applications of the heat-flow electrical analogy: (1) the expert in one of the fields (heat transfer or electricity) can learn the other more easily by referring often to the field with which he is familiar; (2) it is possible to construct a real electric circuit whose dynamic behavior will be the same as that of a given heat-flow model. Electric circuits are much easier to build than thermal systems; and for very complicated models it may be easier and cheaper (and more instructive!) to study behavior by experimenting with a small electric network than by compiling and sifting numerous digital-computer solutions to the equations of motion.

Problems 4.1 through 4.12

The electrical analogs of fluid systems

STAGE I. PHYSICAL MODEL. As with thermal systems, only fluid-system models which are lumped are amenable to electrical-analog representation.

Further, electrical analogs are commonly made only for fluid systems in which body inertial forces can be ignored—i.e., systems in which resistance to fluid flow can be represented in the form (3.17), (3.18), (3.21), (3.23), (3.25), (3.27), or (3.28). [When the resistance relation is linear, (3.18), (3.21), (3.25), or (3.28), the analog is of course simpler.]

For such physical models the analogy has been used a great deal in the form of analog-computer studies of fluid systems. Moreover, in the inverse sense, it has played an important historical role, having been first noted by Benjamin Franklin, who used it to help him develop his understanding of the nature of electricity.

For the special physical model described above there are only two kinds of fluid elements, resistive elements and storage elements. The resistive elements may be lines or orifices. The storage elements are either gravity tanks, as in Fig. 3.6c, or pressure tanks, as in Fig. 3.6a, b, d.

STAGE II. EQUATIONS OF MOTION. For the above special cases, the fluid-flow electrical analogy is rather similar to the heat-flow electrical analogy. Compare Procedures A-f (p. 104) and A-e (p. 61).

(1) *Variables and geometry.* The analogous variables are as follows:

Through variables: Fluid-flow rate w is analogous to electrical current i.

Across variables: Pressure p is analogous to voltage v; or, for liquid storage in a gravity reservoir, reservoir height h is analogous to voltage v.

(2) *Equilibrium.* Continuity of flow is analogous to Kirchhoff's Current Law.

(3) *Physical relations.*

Resistance. Relations (3.17), (3.18), (3.21), (3.23), (3.25), (3.27), and (3.28) for flow through fluid resistance, e.g.,

$$w = c_\alpha (p_1 - p_2)^{1/\alpha} \qquad (3.23)$$
[repeated]

or†
$$w = c_\alpha' (h_1 - h_2)^{1/\alpha} \qquad (3.27)$$
[repeated]

are analogous to Ohm's law:

$$i = \frac{1}{R} v$$

when $\alpha = 1$. In cases when $\alpha \neq 1$, an electrical element must be employed which exhibits the desired behavior:

$$i = cv^{1/\alpha}$$

Such elements can, in fact, be made without too much difficulty.

† Recall that form (3.27) is derived by using the hydrostatic expression (3.26) for liquid systems employing gravity storage.

[Sec. 4.2] THE ELECTRICAL ANALOG OF MECHANICAL SYSTEMS 125

Capacitance. When relations for storage of fluid under pressure or under gravity (refer to Fig. 3.6) can be written in the form

$$\dot{p} = \frac{1}{C} w_{\text{net}} \qquad (4.3)$$

or

$$\dot{h} = \frac{1}{C} w_{\text{net}} \qquad (4.4)$$

then they are analogous to the relation for storage of electric charge in a capacitor:

$$\dot{v} = \frac{1}{C} i \qquad (4.5)$$

To illustrate, note that the continuity equilibrium relations in Example 3.2 (p. 114) have precisely the form (4.4).

Demonstration of the electrical analogy for hydraulic systems is offered in Problems 4.13 through 4.20.

Problems 4.13 through 4.20

4.2 THE ELECTRICAL ANALOG OF MECHANICAL SYSTEMS

The most important analogy in the development of system dynamics has been the mechanical-electrical analogy. The reason is that many of the important contributions to the field of dynamics of mechanical systems, particularly automatic control systems, were made by electrical engineers who were able to carry over their experience with electrical networks and feedback amplifiers to the mechanical medium. (Early development of the servos for aiming radar antennas is a case in point.)

The mechanical-electrical analogy is not so self-evident as the heat-electrical or hydraulic-electrical ones, where there is an obvious correspondence between the flow of heat, of fluid, and of electric current. There is not, in rigid mechanical systems, a "flow" of anything, in the same sense. The potential to produce "flow"—like voltage, pressure, or temperature—likewise is not obvious in an analogous way in rigid mechanical systems. The mechanical quantity which is usually taken to be analogous to current is force. It takes some imagination to think about force flowing. (The phrase "through variable" was coined to help the cause.)

There is, however, a firm and rigorous mathematical basis for the analogy between the behavior of lumped, linear mechanical systems and of electrical networks. The fundamental basis for analogy is that the equations of motion for such mechanical systems and for electrical systems are identical in form. Thus a network can always be drawn whose equations of motion are identical in form to those of the original mechanical system. Then, by studying the node voltages in the network, the motions of the original mechanical system can be predicted.

Working from the equations of motion, two analogies are actually possible: one in which velocity is analogous to voltage, and force to current, and another in which force is analogous to voltage, and velocity to current.

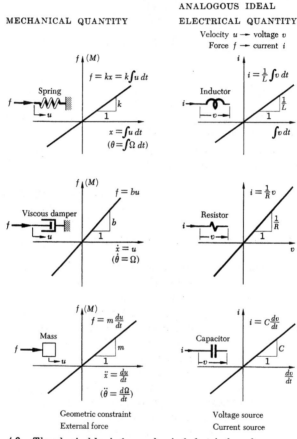

Fig. 4.2 *The physical basis for mechanical-electrical analogy*

The former is preferred pedagogically, and is presented here. It is called the "mobility analog," and it is preferred because it has across variables analogous to one another, and through variables analogous to one another. (The other, the force-voltage analogy, is the subject of Problem 4.36.)

(1) Variables

Through variables: Force f is analogous to current i
Across variables: Velocity u is analogous to voltage v.

(2) Equilibrium

Force equilibrium: For a free body $\begin{cases} \Sigma f^* = 0 & (2.1) \text{ [repeated]} \\ \Sigma M^* = 0 & (2.2) \text{ [repeated]} \end{cases}$

is analogous to

KCL: For a node $\quad \Sigma i = 0 \quad\quad (2.48)$ [repeated]

Geometric compatibility around a loop is analogous to KVL.

(3) Physical relations

The underlying physical basis for successful analogy between one-dimensional mechanical systems and electrical networks is the perfectly parallel structure of the sets of physical relations of Figs. 2.4 (p. 42) and 2.16 (p. 67). The relations are repeated, side-by-side, in Fig. 4.2.

The first column shows the force-geometry relations for an ideal mechanical spring, viscous damper, and mass. Beside it in column 2 are shown the analogous relations between current and voltage for an ideal electrical inductor, resistor, and capacitor. The one-to-one correspondence is evident.

Demonstrations of the network analog for mechanical systems

Analogous networks for models of two mechanical systems are shown in Figs. 4.3 and 4.4. We shall first indicate that the networks are indeed perfectly analogous to the mechanical models in every respect by deriving the equations of motion involved. Then we shall discuss briefly a procedure for constructing the electrical analogs of simple, one-dimensional mechanical systems by inspection.

Example 4.1. A simple system. We begin with the very simple mechanical system model shown in the upper left-hand corner of Fig. 4.3, which involves each of the three kinds of mechanical elements. To construct an electrical analog for this system we will be representing each of the mechanical elements by its electrical counterpart as indicated in the second column of Fig. 4.2. That is, the mass is to be replaced by a capacitor, the spring by an inductor, and each of the viscous dampers by an electrical resistance. Finally, the externally applied force on the mechanical mass is to be represented by a current source in the analogous network.

The key to *interconnection* of the electrical elements, so that they will properly represent the mechanical system, is the fact that node voltage v must be analogous to velocity u. Since all four mechanical elements see velocity u, all four of the corresponding electrical elements must be connected to the node, as shown in the right-hand drawing of Fig. 4.3. Moreover, since applied force

f_5 acts on mass m_1, the current source i_5 in the analog must act on the (ungrounded) terminal of capacitor C_1, as shown.

To prove that the network is analogous to the mechanical system, we simply derive the equations of motion for each, following a combination of Procedures A-m and A-e.

(1) *(Across) variables and geometry.* For the mechanical system choose velocity u as the independent variable; and for the network, node voltage v, as indicated. Identities and constraints are indicated in Fig. 4.3, Step (1).

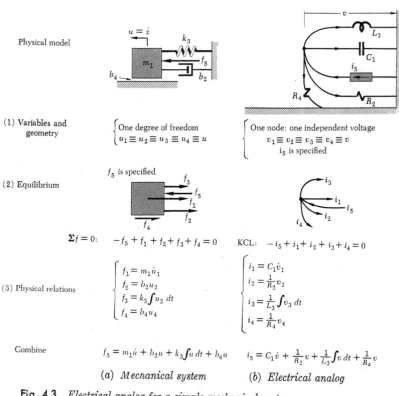

(a) *Mechanical system*   (b) *Electrical analog*

Fig. 4.3 *Electrical analog for a simple mechanical system*

Consider carefully the meaning of the various arrows in Fig. 4.3. (As usual, we draw the voltage arrow in the direction voltage *drop* is taken positive—see p. 62.)

(2) *Equilibrium relations.* Force balance Eq. (2.2) is written for the mechanical free-body diagram shown; Kirchhoff's Current Law is written for the isolated node a. The results are stated in Fig. 4.3, Step (2). Again, note carefully the directions of the arrows.

(3) *Physical relations.* The force-motion relations are given in column 1, and the voltage-current relations in column 2.

[Sec. 4.2] THE ELECTRICAL ANALOG OF MECHANICAL SYSTEMS 129

Finally, *combining* the results of Steps (2) and (3) yields the two equations of motion (using the equilibrium relations as the focus):

$$f_5 = m_1 \dot{u} + b_2 u + k_3 \int u \, dt + b_4 u \tag{4.6}$$

$$i_5 = C_1 \dot{v} + \frac{1}{R_2} v + \frac{1}{L_3} \int v \, dt + \frac{1}{R_4} v \tag{4.7}$$

It is seen that these equations are identical in form, the correspondence between terms being as given in Fig. 4.2.

Example 4.2. A more complex system. The analog for the physical model of a more complicated mechanical system—and one which is of a much more general nature—is shown in Fig. 4.4. Let us begin by postulating that the analog

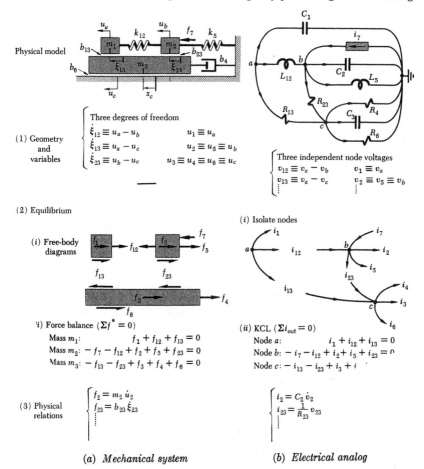

(a) *Mechanical system*   (b) *Electrical analog*

Fig. 4.4 *Electrical analog for a three-degree-of-freedom (one-dimensional) mechanical system*

is correct, and prove this by writing the equations of motion. Later we shall indicate the synthesis philosophy by which the analog was contrived.

(1) *(Across) variables and geometry.* The geometry in this system is more interesting than that in Fig. 4.3. There are three degrees of freedom, and moreover, in addition to the absolute velocities of the masses, some relative velocities will be required. Geometric identities are indicated in Fig. 4.4, Step (1).

The equations of motion for the network are to be written in terms of the three independent node voltages, and the corresponding identity relations for voltage drops in the elements are as indicated in Fig. 4.4, Step (1).

(2) *Equilibrium relations.* Newton's law (2.1) is written for each of the three free-body diagrams, Fig. 4.4, Step (2). Kirchhoff's Current Law is written for the three corresponding isolated nodes. (Explain why the signs all match, but f_{12} is to the left on m_2 while i_{12} is to the right at node b.)

(3) *Physical relations.* These are indicated in Fig. 4.4, Step (3).

Finally, the relations of Steps (1), (2), and (3) are *combined* to obtain the equations of motion for the system. For example, the second of the three equations of motion is as follows:

For the mechanical system:
$$f_7 = -k_{12} \int (u_a - u_b) \, dt + m_2 \dot{u}_b + b_{23}(u_b - u_c) + k_5 \int u_b \, dt \qquad (4.8)$$

For the network:
$$i_7 = -\frac{1}{L_{12}} \int (v_a - v_b) \, dt + C_2 \dot{v}_b + \frac{1}{R_{23}}(v_b - v_c) + \frac{1}{L_5} \int v_b \, dt \qquad (4.9)$$

Construction procedure. A few comments about the procedure for building up the electrical analog for a mechanical system may be helpful. The steps by which the analog in column 2 of Fig. 4.4 was constructed are shown in column 2 of Fig. 4.5. An easy place to begin is with the masses m_1, m_2, m_3, which are to be represented by capacitors C_1, C_2, and C_3. Because the physical relation for a mass (Newton's law) involves motion with respect to inertial space, all the capacitors which represent masses *must* be tied to ground, as shown in Fig. 4.5b.

Next, we note that the node voltage with respect to ground on each capacitor will represent a velocity with respect to inertial space of the corresponding mass. Therefore these are the nodes a, b, c, the voltages of which will correspond to the velocities u_a, v_b, and u_c. (Recall that the arrows in the network represent the direction of positive voltage *drop*, corresponding to velocity drop.)

Finally, in Fig. 4.5c the other components are connected between the nodes in an arrangement which parallels the connections between the masses in the mechanical system. For example, resistor R_4 connects the node of capacitor C_3 (representing mass m_3) directly to ground, while inductor L_{12} (representing spring k_{12}) connects together the nodes of capacitors C_1 and C_2 (masses m_1 and m_2). In this way all of the elements are added to the network as shown in Fig. 4.5c.

As an added convenience, the signs in the equations of motion of the network can all be made to come out the same as those of the mechanical

[Sec. 4.2] THE ELECTRICAL ANALOG OF MECHANICAL SYSTEMS 131

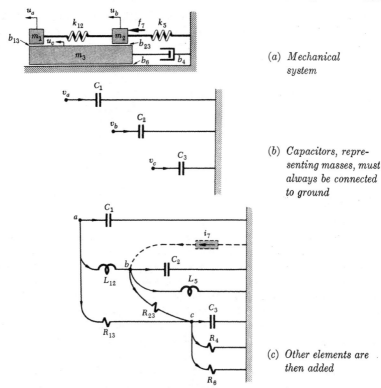

(a) *Mechanical system*

(b) *Capacitors, representing masses, must always be connected to ground*

(c) *Other elements are then added*

Fig. 4.5 *Procedure for constructing electrical analogs*

system by choosing the (arbitrary) direction of the branch arrows in the network to correspond to the (also arbitrary) direction chosen for the force arrows in the mechanical free-body diagrams [Fig. 4.4, Step (2)].

When the elements have been properly arranged and connected together, the last step is to insert the current source representing applied force f_7. Because the applied force acts on mass m_2, the current source is applied to the terminal of capacitor C_2 that is not connected to ground.

Electrical analogy for rotary mechanical systems. The preceding discussion of electrical analogies applies verbatim whether the mechanical system involved is translational in one dimension, as in Fig. 4.3a, or is rotational in one dimension, as in Fig. 4.6.

The system in Fig. 4.6 has the equation of motion

$$M_5 = J_1\dot{\Omega} + b_2\Omega + k_3\int\Omega\,dt \tag{4.10}$$

Comparing (4.10) with (4.6), it is clear that the appropriate correspondence in Fig. 4.2 is for rotary moments of inertia J to occupy the same position in the table as linear inertias m; for rotary spring constants and damping constants to occupy the same position as their linear counterparts; and for torque M and angular displacement θ to occupy, respectively, the same positions as force f and linear displacement x. These are all indicated in parentheses in Fig. 4.2.

Fig. 4.6 *A rotary mechanical system*

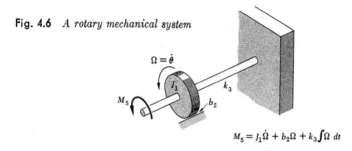

$$M_5 = J_1 \dot{\Omega} + b_2 \Omega + k_3 \int \Omega \, dt$$

For mechanical motions in two or three dimensions, the development of electrical analogs may be considerably more sophisticated, and will not be treated here. Usually it is necessary (and highly advisable!) first to derive the mechanical equations of motion completely in mechanical terms (e.g., using Procedure A-m). Then an electrical network having the same equations of motion can often be contrived. As a practical matter the electrical analog approach is nearly always too cumbersome to be useful in two- and three-dimensional problems. The analog computer, however, is tremendously useful, as noted on p. 133. Mechanization of the analog computer is performed directly from the mechanical equations of motion, without any regard for element-to-element electrical-to-mechanical correspondence. The process will be studied in Sec. 8.15.

Problems 4.21 through 4.38

4.3 CLASSIFICATION OF DYNAMIC SYSTEM ELEMENTS

In each of the foregoing discussions of the analogy between one physical medium and another we have begun by classifying the analogous quantities—variables, physical elements and parameters, and governing relations. By way of summary it is useful to tabulate the classification for the several simple physical systems considered here. Such a classification is presented in Table IV.[†]

It is to be emphasized that the physical models considered are all simple ones: All elements are represented by lumped models, only one-dimensional systems are considered, and nearly all relations are taken to be linear. In fluid systems, body forces (gravity and inertia) are also neglected.

[†] The classification of Table IV was suggested, in part, by class notes of Professor R. E. Kronauer of Harvard University.

It is also emphasized that Table IV is by no means comprehensive: Many interesting dynamic media are not represented.

The first two "rows" in Table IV—items (1) Variables, and (2) Equilibrium and Compatibility—merely tabulate concepts already developed, beginning in Sec. 2.1. These are the *system* relations.

(3) Physical relations have been subdivided into energy-storage and energy-dissipation elements, sources, and energy converters. Energy-storage elements† are classified as "T-type" or "A-type" according to whether storage is achieved by virtue of a through or an across variable. Similarly, sources are either "T-type" or "A-type" according to whether they constrain a through or an across variable, respectively.

4.4 THE BENEFITS AND LIMITATIONS OF ANALYSIS BY ANALOG

Many benefits derive, of course, from the existence of physical analogies. For one thing, additional insight into the behavior of a system often accrues from consideration of its analog. In fact, the analytical techniques used in studying the automatic control of mechanical systems (Part E), which have been extremely important in the design of such systems, were carried over in large part from their development in the context of electrical feedback-amplifier design; and the transfer function and impedance methods (Part D) developed originally for electrical-circuit analysis are an important tool in the analysis of mechanical vibrations, as a result of the analogy between the two systems.

Moreover, the development of analog computers, based on the analogous behavior of electrical systems to other physical systems, has widespread importance in the synthesis of dynamic systems of all kinds. The importance of simulation on analog computers in the rapid development of highly sophisticated automatic control and guidance systems for air and space vehicles, for example, can hardly be overestimated. (Analog-computer mechanization will be introduced in Sec. 8.15.)

Analogies are detrimental, however, when they entice us into stopping our thinking about new physical phenomena, or when they cause us to force a system arbitrarily into a mold of analysis which it does not fit. Unfortunately, the discovery that mechanical systems can be represented mathematically by electrical networks has been regarded as a panacea by many inexperienced engineers who find electrical circuits much more pleasant to think about than mechanical systems. With as little thought as possible, they draw a network to represent every mechanical system with which they are confronted in the course of an engineering project, and then proceed to analyze the network exhaustively with no thought for the actual system with which they should be concerned, obtaining detailed answers without understanding the physical significance of those answers. The use of the

† The adjectives "T-type" and "A-type," applied to elements, were introduced in class notes of Professors J. L. Shearer, H. H. Richardson, and A. T. Murphy at MIT.

TABLE IV
A classification of relations for simple physical system models

CLASSIFICATION		PHYSICAL MEDIUM			
		Electrical	Mechanical	Heat conduction	Fluid
(1) **Variables**	Through variable	Current i	Force f (Moment M)	Heat-flow rate q	Flow rate w
	Across variable	Voltage drop v (Note: arrow indicates direction of voltage drop.)	Velocity u (Angular velocity Ω)	Temperature T	Pressure p or Liquid height h
(2) **Equilibrium** relations (among through variables)		KCL: $\Sigma i_{net} = 0$	Force equilibrium (Newton's law, à la D'Alembert), $\Sigma f^* = 0$ $\Sigma M^* = 0$	First Law of Thermodynamics $C\dfrac{dT}{dt} = q_{net\,in}$, no work, no phase change	Continuity $w_{net\,in} = \dfrac{dm}{dt}$ Force equilibrium $\Sigma f^* = 0$
Compatibility relations (among across variables)		KVL: $\Sigma v_{loop} = 0$	$\Sigma u_{loop} = 0$ $\Sigma \Omega_{loop} = 0$	$\Sigma T_{loop} = 0$	$\Sigma p_{loop} = 0$
(3) **Constitutive physical relations** Passive energy-storage elements	"T-type" (storage via a through variable)	Inductor $i = \dfrac{1}{L}\int v\,dt$	Spring $f_{spr} = k\xi$ $f_{spr} = k\int u\,dt$ $M_{spr} = k\psi$ $M_{spr} = k\int \Omega\,dt$		
	"A-type" (storage via an across variable)	Capacitor $i = C\dot v$	Mass $f_i = m\dot u$ $M_i = J\dot\Omega$	Heat capacity $q_{net} = C\dot T$ (derived from the First Law)	Gas storage $w = C\dot p$ Liquid storage $w = C\dot h$

TABLE IV (Continued)

CLASSIFICATION	PHYSICAL MEDIUM				
	Electrical	Mechanical	Heat conduction	Fluid	
(3) Constitutive physical relations (continued) Passive energy-dissipation elements	Resistor $i = \frac{1}{R} v$	Damper $f_f = bu$ $M_f = b\Omega$	Heat resistance $q = \frac{1}{R}(T_1 - T_2)$	Fluid resistance $w = \frac{1}{R}(p_1 - p_2)^{1/\alpha}$	
Sources "T-type" (Constraining a through variable)	$i = i(t)$ prescribed Current source	$f = f(t)$ prescribed Force source	$q = q(t)$ prescribed Heat-flow source	$w = w(t)$ prescribed Mass-flow source	
"A-type" (Constraining an across variable)	$v = v(t)$ prescribed Voltage source	$u = u(t)$ prescribed Velocity source	$T = T(t)$ prescribed Temperature source	$p = p(t)$ prescribed Pressure source	
Isolators	$v_2 = Kv_1$ (independent of load) Amplifier			$y = cx$ (independent of load) Hydraulic integrating amplifier	

TABLE IV (Continued)

CLASSIFICATION	PHYSICAL MEDIUM		
	Electrical-electrical	Electrical-mechanical	Mechanical-mechanical
Energy-conversion elements	Transformer $v_2 = Nv_1$ $i_2 = \dfrac{1}{N} i_1$ (N is the turns ratio)	Motor or generator $v = K\Omega$ $M = Ki$	Gear train $\Omega_2 = \dfrac{1}{N} \Omega_1$ $M_2 = NM_1$ (N is the ratio of radii: $N \triangleq \dfrac{r_2}{r_1}$)

analog technique to avoid thinking about the actual physical situation is *not* to be regarded as a benefit of the analog: Its widespread use has seriously impeded the development of engineering judgment in too many young men.

Philosophically, analogies are especially useful when they help give us insight into the behavior of a system which is new to us by showing us its similarity to systems that we have previously come to understand rather well. The analyst is well advised, however, to study each kind of physical system in its own terms, and to teach himself to be equally at home with each of the physical media with which he is working. Only in this way will strong physical intuition and good judgment develop at each step of the analysis, so that sound approximations can be made, and so that the physical consequences of the analytical results can be recognized at once, and in depth.

4.5 THE NETWORK APPROACH TO ANALYSIS

Pursuit of the technique of analysis by analog leads to one further step which is popular with some analysts: the representation of physical systems by a network.

This is what we have always done for electrical systems, using the symbols we have listed again in the first column of Table IV. To extend the method to other systems, additional symbols have been developed, as shown in Table V. (Table V is merely a copy of part of Table IV, with the new symbols added.) Note that in column 2, because Newton's Second Law holds only for motion with respect to inertial space, one terminal of each mass symbol must always be connected to "ground."

The procedure for network analysis is, then, the following:

(1) **Construct network.**
(2) **Write node or loop equations** (i.e., KCL or KVL).
(3) **Include physical relations.**

Procedure A-n
Network approach

(1) *Constructing a network.* The entire burden of the method is, obviously, borne in this first step. If the physical system is sufficiently simple that a network can be constructed without actually writing any equations of motion, then this step is relatively straightforward. For example, using the symbols of Table V, Fig. 4.7b may be constructed to represent the mechanical system of Fig. 4.4, which is shown again in Fig. 4.7a. The nodes a, b, and c may be thought of as representing the points of connection between elements (as they did in Example 2.3, p. 52). Node a, for example, is the point where m_1 connects to k_{12}; it has velocity u_a. (Notice that the arrows in Fig. 4.7b indicate the direction of force, and also of velocity *drop*, in accordance with our standard network convention.)

(2) *Writing node and loop equations.* This consists of writing "KCL" or "KVL" routinely for the network contrived in (1). For example, KCL for nodes a, b, and c in Fig. 4.7 may be quickly written. This is left as an exercise. (Compare the results with the equilibrium relation in column 1 of Fig. 4.4.)

Alternatively, electrical-network specialists prefer to manipulate their networks a bit, replacing them with equivalent networks which are simpler, before applying KCL or KVL. It is this opportunity to take advantage of their manipulative skill that makes the mechanical-network approach attractive to electrical specialists.

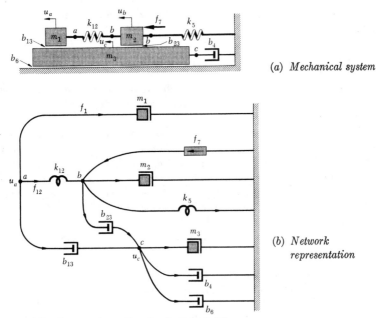

Fig. 4.7 Constructing a "mechanical network"

(3) *Including physical relations.* Table V gives the relation implied by the network symbol for each element in the network. The relations are inserted routinely into the equilibrium equations from Step (2). As with electrical networks, items (2) and (3) are commonly combined and done in one step.

As another demonstration of the network approach, Fig. 4.8 shows a "network" to represent the carrier landing of an aircraft described in Example 2.2, p. 49. Some nonlinear electrical elements have had to be added to the repertoire of Table V. From Fig. 4.8 the equations of motion for the physical model can be written by straightforward application of KCL, as the reader may verify.

[Sec. 4.5] THE NETWORK APPROACH TO ANALYSIS 139

It is certainly enjoyable and academically challenging to synthesize diagrams such as Fig. 4.8, and additional insight is often to be gained in the process. Moreover, as noted, there is the chance to apply network manipulation tricks to obtain the equations of motion more quickly.

However, it must be admitted that, compared with Fig. 2.7 (p. 48), Fig. 4.8 does very little by way of bringing to mind the image of an aircraft skidding across the deck of a ship. Moreover, it was necessary to sketch the physical model and to derive parts of the equations of motion before the network could even be drawn. Thus, the network approach really involved

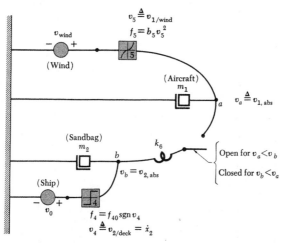

Fig. 4.8 *Mechanical network to represent an aircraft landing on a carrier*

extra steps—steps in the direction of detachment from the actual physical system. Therefore, while we note, as we have, the aesthetic appeal of such mechanical networks, we shall nevertheless not employ them much in this book.

Rather, to keep as close as possible to physical reality *throughout the analysis*, we shall always draw physical models to look as much like the real system as possible—not like a model in some other medium.

For the same reason, we shall not try to force the process of deriving equations of motion in one medium into the topological mold of another. (Summing forces at a "mechanical node," for example, is rather awkward and unhelpful in many of the more interesting mechanical systems—e.g., Fig. 2.11.)

Problems 4.39 through 4.49

TABLE V
Elements for "Network" Representation of Simple Physical Systems

CLASSIFICATION		PHYSICAL MEDIUM			
		Electrical	Mechanical (one-dimensional)	Thermal	Fluid
Passive energy-storage elements	"T-type"	Inductor $\quad i = \frac{1}{L}\int v\,dt$	Spring $\quad f = k\int u\,dt$ Note: arrow indicates direction in which velocity *drop* is positive.		
	"A-type"	Capacitor $\quad i = C\dot{v}$	Mass $\quad f = m\dot{u}$	Heat-storage capacity $\quad q = C\Delta\dot{T}$	$\begin{cases} w = C\dot{p} \\ \text{Gas storage} \end{cases}$ $\begin{cases} w = C\dot{h} \\ \text{Liquid storage} \end{cases}$
Passive energy-dissipation elements		Resistor $\quad i = \frac{1}{R}v$	Damper $\quad f = bu$ Note: arrow indicates direction in which velocity *drop* is positive.	Resistance $\quad q = \frac{1}{R}\Delta T$ Note: arrow indicates direction in which temperature *drop* is positive.	Resistance $\quad w = \frac{1}{R}\Delta p$ or $w = \frac{1}{R'}\Delta h$ Note: arrow indicates direction in which pressure or elevation *drop* is positive.

TABLE V (Continued)

CLASSIFICATION	PHYSICAL MEDIUM			
	Electrical	Mechanical (one-dimensional)	Thermal	Fluid
Sources (Constraints) "T-type"	$i = i(t)$ prescribed Current source	$f = f(t)$ prescribed Force source	$q = q(t)$ prescribed Flow-rate source	$w = w(t)$ prescribed Flow-rate source
"A-type"	$v = v(t)$ prescribed Voltage source	$u = u(t)$ prescribed Velocity source	$T = T(t)$ prescribed "Infinite" fixed-temperature reservoir	$p = p(t)$ prescribed "Infinitely" large constant-pressure reservoir

TABLE V (Continued)

CLASSIFICATION	PHYSICAL MEDIUM		
	Electrical-electrical	*Electrical-mechanical*	*Mechanical-mechanical*
Energy-conversion elements	$v_2 = Nv_1$ $\quad i_2 = \frac{1}{N}i_1$ (N is the turns ratio) Network symbol[a] Physical arrangement — Transformer	$v = K\Omega$ $\quad M = Ki$ Network symbol Physical arrangement — Motor or generator	$\Omega_2 = \frac{1}{N}\Omega_1$ $M_2 = NM_1$ (N is the ratio of radii: $N \triangleq \frac{r_2}{r_1}$) Network symbol Physical arrangement — Gear train

[a] Suggested by Kronauer (see footnote, p. 132).

Chapter 5

EQUATIONS OF MOTION FOR MECHANICAL SYSTEMS IN TWO AND THREE DIMENSIONS

The purpose of this chapter is to extend the study of mechanical systems that was begun in Chapter 2, considering now motion in two and three dimensions.

A principal topic will be the geometry of three-dimensional motion. Another will be force-acceleration physical relations in two and three dimensions, for systems of several rigid bodies.

Finally, in Secs. 5.6 through 5.10 we present the energy approach to deriving equations of motion, and the method of Lagrange. These sections may readily be studied without first reading Secs. 5.1 through 5.5, if desired.

The present chapter is intended to serve primarily as a summary of some important concepts, rather than as an exposition of them. More detailed presentations are, of course, given in texts on dynamics.

5.1 GEOMETRY OF MOTION IN TWO AND THREE DIMENSIONS

Mechanical systems may have motion in one, two, or three dimensions of space, which makes them much more interesting to study.

In two-dimensional motion, called *plane motion*, the translational part of the motion is confined to a plane, and the rotational part takes place only about axes perpendicular to that plane, as with a rolling cylinder or the pitching motions of an airplane.

In three-dimensional motion the mass center of a rigid body may translate in all three dimensions of space, or the body may rotate about an axis in any direction, or both (as a gyroscope, or an airplane in a general maneuver).

Configuration

To describe the configuration of a system we must define coordinates which are easily visualized (for example, quantities which might be measured with a ruler and protractor). An arbitrary-configuration sketch is essential. Some exercises in defining configuration coordinates are suggested in Problems 5.1 through 5.15.

Degrees of freedom. As we stated in Sec. 2.2, the number of degrees of freedom a system has is the number of geometrically independent coordinates required to specify its configuration.

That is, a degree of freedom is a position coordinate the system can change—a way it can displace—without changing any other coordinate in the set of independent position coordinates.† For example, an unconstrained

† We shall use the words "translation" and "location" to imply linear displacement, "orientation" to imply angular displacement, and "position" or "configuration" to imply either or both.

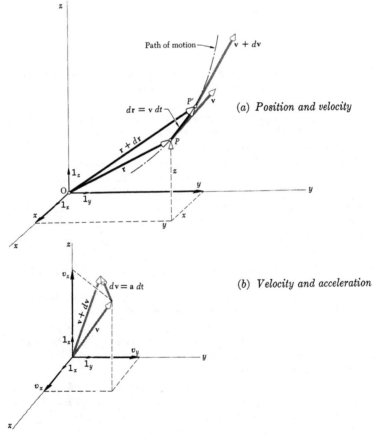

Fig. 5.1 *Defining position, velocity, and acceleration*

[Sec. 5.1] GEOMETRY IN TWO AND THREE DIMENSIONS 145

rigid body has six degrees of freedom: one point in it (e.g., its mass center) can be located with three linear coordinates, and its orientation about that point can be specified with three angular coordinates; and these coordinates may be so chosen that they are all *geometrically independent*—any one can change without a change in the others.

Usually when geometric constraints exist they reduce both the number of coordinates c required to specify configuration and (therefore) the number of degrees of freedom d. Examples are a point mass ($d = c = 3$) being constrained to the end of a simple pendulum like Fig. 2.2c ($d = c = 1$), and a rigid sphere ($d = c = 6$) being confined to motion on a plane, but free to slip ($d = c = 5$). Such constraints are called *holonomic constraints*.

Sometimes, however, constraints introduce geometric relations (typically velocity relations) which make it impossible to change a coordinate without changing some other coordinate, but which do *not* reduce the number of coordinates required to prescribe system configuration. Such constraints are called *nonholonomic constraints*. The classic example of nonholonomic constraint is a sphere or disk rolling, without slipping, on a plane. The number of position coordinates required to describe completely its configuration is $c = 5$. However, because of the rolling constraint, there is one geometric relation ($n = 1$) *between the rates of change* of the position coordinates (see Problem 5.16). Therefore, the system has $c - n = 5 - 1 = 4$ degrees of freedom.

Thus, in general, if c position coordinates are required (necessary and sufficient) to define the configuration of a system, but there are n geometric relations between them (n ways in which they are not independent, because of nonholonomic constraints), then *the number of degrees of freedom d—the number of position coordinates which can be altered independently*—is

$$\text{Degrees of freedom} \quad \boxed{d = c - n}$$

Problems 5.1 through 5.20.

Velocity and acceleration

The translational velocity of a point—such as the mass center of a rigid body—is the rate of change (derivative) of its location. Location of a point can be designated by a vector **r** (drawn from some reference) as in Fig. 5.1a. (Point P might be the mass center of the airplane in Fig. 2.11, for example.) Then velocity is defined as the rate of change of **r** (note the vector construction of $d\mathbf{r}$ in Fig. 5.1a). That is, the velocity vector is the ratio of change in **r** to time elapsed, in the limit as time elapsed becomes infinitesimally short:

$$\mathbf{v} \triangleq \frac{d\mathbf{r}}{dt} \tag{5.1}$$

(Conversely, during infinitesimal time dt, point P moves along **v** through a distance $d\mathbf{r}$ given by $d\mathbf{r} = \mathbf{v}\, dt$.) Similarly, acceleration is defined as the rate of change of velocity (the vector construction of $d\mathbf{v}$ is given in Fig. 5.1b, where

v and **v** + d**v** are compared):

$$\mathbf{a} \triangleq \frac{d\mathbf{v}}{dt} \tag{5.2}$$

The *"frame of reference,"* with respect to which these quantities are measured, must be considered precisely. A reference frame is, in general, a rigid set of three noncoplanar lines with respect to which the locations of points are to be measured. (Commonly, the three lines are mutually orthogonal, like the lines at the corner of a room.)

The simplest reference frame to consider is one which is neither translating nor rotating. In some situations, however, it is more convenient to use a reference frame which is translating, or rotating, or both (a reference frame painted on a moving aircraft, for example).

Stationary, nonrotating reference frame. When the reference frame is not rotating, then (and only then) the derivative of a position vector **r**,

$$\mathbf{r} = \mathbf{1}_x x + \mathbf{1}_y y + \mathbf{1}_z z$$

is given by:

$$\boxed{\mathbf{v} = \frac{d\mathbf{r}}{dt} = \mathbf{1}_x \dot{x} + \mathbf{1}_y \dot{y} + \mathbf{1}_z \dot{z}} \quad \text{Nonrotating reference frame} \tag{5.3}$$

where $\mathbf{1}_x$, $\mathbf{1}_y$, $\mathbf{1}_z$ are unit vectors in the x, y, and z directions, as in Fig. 5.1a.† (When the reference frame *is* rotating, the unit vectors also have rates of change, and $d\mathbf{r}/dt$ is much more involved. This is discussed in Sec. 5.2.)

The velocity vector **v** can also be resolved into components in the $\mathbf{1}_x$, $\mathbf{1}_y$, and $\mathbf{1}_z$ directions, as shown in Fig. 5.1b:

$$\mathbf{v} = \mathbf{1}_x v_x + \mathbf{1}_y v_y + \mathbf{1}_z v_z \tag{5.4}$$

By comparing (5.3) and (5.4) we see that, for a nonrotating reference frame,

$$v_x = \dot{x} \qquad v_y = \dot{y} \qquad v_z = \dot{z}$$

If the reference frame has zero velocity, then **v**, the rate of change of **r**, is the *absolute* velocity.‡

Again, if the reference frame is not rotating, the differentiation of **v** is simple because the unit vectors do not change. The derivative is, for a *nonrotating reference frame*,

$$\mathbf{a} = \frac{d\mathbf{v}}{dt} = \mathbf{1}_x \dot{v}_x + \mathbf{1}_y \dot{v}_y + \mathbf{1}_z \dot{v}_z \tag{5.5}$$

---

† We shall always use orthogonal right-handed coordinate frames wherein the cross-product relation $\mathbf{1}_x \times \mathbf{1}_y = \mathbf{1}_z$ holds. (Cross product is defined in Appendix B.)

‡ Precisely, absolute motion means motion with respect to "inertial space." Inertial space, in turn, is defined as a reference frame with respect to which Newton's laws of motion hold. As we have noted, a reference frame fixed to the earth is often a reasonable approximation. At other times the earth's motion cannot be neglected; then we may use instead a reference frame which we imagine fixed in the sun or in the "fixed" stars.

[Sec. 5.1]   GEOMETRY IN TWO AND THREE DIMENSIONS

Or, using (5.3), we have

$$\mathbf{a} = \boxed{\frac{d\mathbf{v}}{dt} = \mathbf{1}_x\ddot{x} + \mathbf{1}_y\ddot{y} + \mathbf{1}_z\ddot{z}} \qquad \text{Nonrotating reference frame} \quad (5.6)$$

If the reference frame is not accelerating, then the rate of change of the vector **v** is the *absolute* acceleration.

Relative motion: translating nonrotating reference frame. If $\mathbf{v}^{P/O}$ is the velocity of point P relative to a reference point O, which, in turn, has absolute velocity $\mathbf{v}^O$, then the absolute velocity of P is given by

$$\boxed{\mathbf{v}^P = \mathbf{v}^{P/O} + \mathbf{v}^O} \qquad (5.7)$$

If O is a point in a nonrotating reference frame, then all points in the frame have velocity $\mathbf{v}^O$. Thus, if the components of the velocity of P relative to the O frame are

$$\mathbf{v}^{P/O} = \mathbf{1}_x\dot{x} + \mathbf{1}_y\dot{y} + \mathbf{1}_z\dot{z}$$

then the components of absolute velocity $\mathbf{v}^P$ are given by

$$\mathbf{v}^P = \mathbf{1}_x(\dot{x} + v_x^O) + \mathbf{1}_y(\dot{y} + v_y^O) + \mathbf{1}_z(\dot{z} + v_z^O) \qquad \text{Nonrotating reference frame} \quad (5.8)$$

in which v_x^O, v_y^O, v_z^O are the components of $\mathbf{v}^O$.

Similarly, if $\mathbf{a}^O$ is the absolute acceleration of point O, then the absolute acceleration of P is given by:

$$\boxed{\mathbf{a}^P = \mathbf{a}^{P/O} + \mathbf{a}^O} \qquad (5.9)$$

and, again, in terms of the components of acceleration of the nonrotating reference frame,

$$\mathbf{a}^P = \mathbf{1}_x(\ddot{x} + a_x^O) + \mathbf{1}_y(\ddot{y} + a_y^O) + \mathbf{1}_z(\ddot{z} + a_z^O) \qquad \text{Nonrotating reference frame} \quad (5.10)$$

If the reference frame is rotating, the expressions for $\mathbf{v}^P$ and $\mathbf{a}^P$ are much more involved. They are given in Sec. 5.2.

Angular velocity and acceleration

The angular velocity and angular acceleration of a rigid body are also vector quantities. They obey the laws of vector addition. Thus, for example, if $\mathbf{\Omega}^{c/b}$ is the angular velocity of body c relative to body b, and $\mathbf{\Omega}^b$ is the absolute angular velocity of body b, then the absolute angular velocity of body c is

$$\mathbf{\Omega}^c = \mathbf{\Omega}^{c/b} + \mathbf{\Omega}^b \qquad (5.11)$$

For the special case of *plane motion* (which includes one-dimensional motion, of course), these vectors are all normal to the fixed plane of motion at all times: That is, they *do not change their direction in space*. Then, and only then in general, the angular velocity can be obtained by simple scalar differentiation of an angle, and angular acceleration, in turn, can be obtained by scalar differentiation of angular velocity. That is, the relations for Ω and $\dot{\Omega}$ on p. 38 hold also for plane motion.

More generally, the rule for differentiation of a vector given in Sec. 5.2 must be employed.

Example 5.1. Cylinder on wedge. Express the absolute velocity and acceleration of the mass center C of the cylinder in Fig. 5.2 if the cylinder rolls on the wedge without slipping, and the wedge slides on the floor.

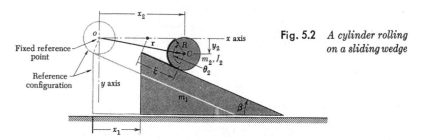

Fig. 5.2 *A cylinder rolling on a sliding wedge*

We first picture the system in an arbitrary configuration (shaded), displaced with respect to a reference configuration (gray outline).† Then various coordinates can be drawn and labeled. We perceive, however, that the system has only two degrees of freedom—that only *two* coordinates are required to specify completely its configuration (and the two are independent). For example, we can imagine getting the system to the arbitrary configuration by first sliding the block laterally by x_1 and then rolling the cylinder down the block a distance $\xi = R\theta_2$.

Selecting, as independent variables, the coordinates x_1 and θ_2, for example, we can describe the location of point C in terms of vector $\mathbf{r}$, from fixed reference point O to C. We write the components of $\mathbf{r}$ in the fixed, nonrotating reference frame xy:

$$\mathbf{r} = \mathbf{1}_x(x_2) + \mathbf{1}_y(y_2)$$
$$= \mathbf{1}_x(x_1 + R\theta_2 \cos \beta) + \mathbf{1}_y(R\theta_2 \sin \beta) \tag{5.12}$$

in which the appropriate identity and constraint relations were used—e.g., $x_2 \equiv x_1 + \xi \cos \beta$, where $\xi = R\theta_2$ (note "tread mark" on cylinder and wedge), and so forth.

Then, using (5.4), the velocity of C is given by:

$$\mathbf{v} = \dot{\mathbf{r}} = \mathbf{1}_x(\dot{x}_1 + R\dot{\theta}_2 \cos \beta) + \mathbf{1}_y(R\dot{\theta}_2 \sin \beta) \tag{5.13}$$

and the acceleration of C is given [from (5.6)] by:

$$\mathbf{a} = \dot{\mathbf{v}} = \ddot{\mathbf{r}} = \mathbf{1}_x(\ddot{x}_1 + R\ddot{\theta}_2 \cos \beta) + \mathbf{1}_y(R\ddot{\theta}_2 \sin \beta) \tag{5.14}$$

† Our choice of sign—of x_1 positive to the right, for example—is entirely arbitrary. Actually, intuition tells us the wedge will move to the left; so when we finally solve the equations of motion for $x_1(t)$ we shall expect it to have negative numerical values.

[Sec. 5.2] ROTATING REFERENCE FRAMES

Because **v** and **a** are measured with respect to a fixed reference point (point O), they are the *absolute* velocity and acceleration of point C.

Problems 5.21 through 5.28

5.2 ROTATING REFERENCE FRAMES

The geometry of studying motion in rotating reference frames is greatly facilitated by the following theorem.

Let 1_x, 1_y, 1_z be unit vectors along the axes of a rotating reference frame α (e.g., one painted on an aircraft). Let the reference frame have angular velocity $\mathbf{\Omega}^\alpha$. Let **A** be any vector whose xyz components (α-frame components) are given:

$$\mathbf{A} = 1_x A_x + 1_y A_y + 1_z A_z \tag{5.15}$$

A might be the velocity of the aircraft, for example, or its angular momentum.

Then the rate of change of **A** can be written:

$$\boxed{\dot{\mathbf{A}} = \overset{\alpha}{\dot{\mathbf{A}}} + \mathbf{\Omega}^\alpha \times \mathbf{A}} \tag{5.16}$$

in which $\dot{\mathbf{A}}$ is the rate of change of **A** as seen from a nonrotating reference frame, and $\overset{\alpha}{\dot{\mathbf{A}}}$ is the rate of change as seen from the α frame (xyz frame). That is, $\overset{\alpha}{\dot{\mathbf{A}}}$ is given by

$$\overset{\alpha}{\dot{\mathbf{A}}} \triangleq 1_x \dot{A}_x + 1_y \dot{A}_y + 1_z \dot{A}_z$$

(the unit vectors being viewed as constants). A proof of (5.16) is given in Appendix C. (The cross product is defined in Appendix B.)

Expression (5.16) can be used to find **v** from **r**, and **a** from **v**, when **r** and **v** are given by their components in a rotating reference frame, as illustrated in Example 5.2, and in Problems 5.30 through 5.37.

Expression (5.16) is also used to study angular motion in rotating reference frames, Example 5.7 and Problems 5.38, 5.39.

Example 5.2. Simple pendulum. Find the velocity and acceleration of the bob of a simple pendulum. For convenience, consider a reference frame β imbedded in the pendulum arm as shown in Fig. 5.3, with axes RTz. (Axis z is outward from the page.) The angular velocity of this frame is $\mathbf{\Omega}^\beta = 1_z \dot{\theta}$. The position of the bob is given by $\mathbf{r} = 1_R \ell$. Its velocity is therefore given by [from (5.16)]:

$$\begin{aligned}\mathbf{v} = \dot{\mathbf{r}} &= \overset{\beta}{\dot{\mathbf{r}}} + \mathbf{\Omega}^\beta \times \mathbf{r} \\ &= 0 + 1_z \dot{\theta} \times 1_R \ell \\ &= 1_T \ell \dot{\theta}\end{aligned} \tag{5.17}$$

150   EQUATIONS OF MOTION: TWO AND THREE DIMENSIONS   [Chap. 5]

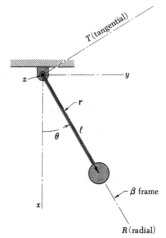

Fig. 5.3 *Simple pendulum*

[Note that $\overset{\beta}{\dot{\mathbf{r}}} = \mathbf{1}_x(d\ell/dt)$ is zero because ℓ is constant.] Similarly, the bob's acceleration is given by

$$\begin{aligned}\mathbf{a} = \dot{\mathbf{v}} &= \overset{\beta}{\dot{\mathbf{v}}} + \boldsymbol{\Omega}^\beta \times \mathbf{v} \\ &= \mathbf{1}_T(\ell\ddot{\theta}) + \mathbf{1}_z\dot{\theta} \times \mathbf{1}_T\ell\dot{\theta} \\ &= \mathbf{1}_T\ell\ddot{\theta} + \mathbf{1}_R(-\ell\dot{\theta}^2)\end{aligned} \qquad (5.18)$$

The tangential and radial components of $\mathbf{v}$ and $\mathbf{a}$ are familiar.

Problem 5.29 suggests that the reader perform the same derivation, using a nonrotating reference frame.

Example 5.3. Motion of a vehicle on the earth. Find the absolute velocity and acceleration of an auto traveling north at latitude 30° if its velocity and acceleration with respect to the highway are $u = 50$ mph and $a = 0.01$ g, respectively. Consider the effect of earth rotation, but neglect motion of the earth's center around the sun. Note the identities

$$\dot{\lambda} \equiv \frac{u}{R} \qquad \ddot{\lambda} \equiv \frac{a}{R} \qquad (5.19)$$

which hold for travel north. (λ is latitude.)

Consider a reference frame α fixed in the auto. (This is often done for vehicles.) The axes xyz of this frame are taken to be local north, west, and vertical, respectively, as shown in Fig. 5.4. The angular velocity of the α frame is [from (5.11)]

$$\begin{aligned}\boldsymbol{\Omega}^\alpha &= \boldsymbol{\Omega}^{\alpha/e} + \boldsymbol{\Omega}^e \\ &= \mathbf{1}_y\dot{\lambda} + \mathbf{1}_N\Omega_e\end{aligned} \qquad (5.20)$$

or $\qquad = \mathbf{1}_x(\Omega_e \cos \lambda) + \mathbf{1}_y\dot{\lambda} + \mathbf{1}_z(\Omega_e \sin \lambda) \qquad (5.21)$

The position of the auto is conveniently defined by a vector from the earth's center:

$$\mathbf{r} = \mathbf{1}_z R \qquad (5.22)$$

[Sec. 5.3] DYNAMIC EQUILIBRIUM: RIGID BODY, GENERAL MOTION 151

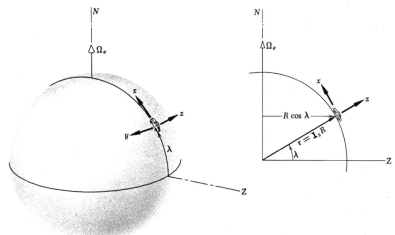

Fig. 5.4 *Auto traveling north*

Then, applying (5.16), we obtain

$$\mathbf{v} = \dot{\mathbf{r}} = \overset{\alpha}{\mathbf{r}} + \Omega^\alpha \times \mathbf{r}$$
$$= 0 + [\mathbf{1}_x(R\dot\lambda) + \mathbf{1}_y(-R\Omega_e \cos \lambda)] \quad (5.23)$$

and $\mathbf{a} = \dot{\mathbf{v}} = \overset{\alpha}{\mathbf{v}} + \Omega^\alpha \times \mathbf{v}$

$$= \mathbf{1}_x R\ddot\lambda + \mathbf{1}_y(R\Omega_e\dot\lambda \sin \lambda)$$
$$+ [\mathbf{1}_x(R\Omega_e^2 \sin \lambda \cos \lambda) + \mathbf{1}_y(R\Omega_e\dot\lambda \sin \lambda)$$
$$+ \mathbf{1}_z(-R\dot\lambda^2 - R\Omega_e^2 \cos^2 \lambda)] \quad (5.24)$$

Arranging in more familiar form, and employing (5.19), we obtain

$$\mathbf{v} = \underset{\substack{\text{relative}\\\text{to earth}}}{\mathbf{1}_x(u)} + \underset{\substack{\text{local velocity of}\\\text{earth (tangential)}}}{\mathbf{1}_y[-(R\cos\lambda)\Omega_e]} \quad (5.25)$$

$$\mathbf{a} = \underset{\substack{\text{relative}\\\text{to earth}}}{\mathbf{1}_x(a)} + \underset{\text{Coriolis}}{\mathbf{1}_y(2\Omega_e u \sin \lambda)} + \underset{\substack{\text{centripetal}\\\text{due to}\\\text{traveling}}}{\mathbf{1}_z\left(-\frac{u^2}{R}\right)} + \underset{\substack{\text{local acceleration of}\\\text{earth (centripetal)}}}{\mathbf{1}_Z[-(R\cos\lambda)\Omega_e^2]} \quad (5.26)$$

where $\mathbf{1}_Z$ is normal to the earth's axis, and through the auto, while $\mathbf{1}_z$ is local vertical. The individual terms are identified. (Acceleration terms of the form $2\Omega v$ were first discovered by G. Coriolis and bear his name.)

For the numerical values given (take $R = 4000$ miles),

$$\mathbf{v} = [\mathbf{1}_x(73.3) - \mathbf{1}_y(1331)] \text{ ft/sec} \quad (5.27)$$
$$\mathbf{a} = [\mathbf{1}_x(0.01) + \mathbf{1}_y(0.000166) - \mathbf{1}_z(0.00000798) - \mathbf{1}_Z(0.00300)]\,g \quad (5.28)$$

Problems 5.29 through 5.39

5.3 DYNAMIC EQUILIBRIUM FOR RIGID BODY IN GENERAL MOTION

When Newton's Second Law is extended to give the equations of motion for a rigid body translating and rotating in the most general way in three dimensions (an aircraft or space vehicle, for example), the general result is the independent pair of vector equations

$$\mathbf{f} = m\dot{\mathbf{v}}_c \qquad (5.29)$$
$$\mathbf{M}_c = \dot{\mathbf{H}}_c \qquad \text{About mass center} \qquad (5.30)$$

in which $\mathbf{f}$ is the vector resultant of all forces acting on the body, $\dot{\mathbf{v}}_c$ is the acceleration of the mass center of the body, $\mathbf{M}_c$ is the resultant of all moments acting on the body, and $\mathbf{H}_c$ is the total angular momentum of the body; and both $\mathbf{M}_c$ and $\mathbf{H}_c$ are measured with respect to the mass center of the body.†

Equations (5.29) and (5.30) are really remarkable in that they say, in particular, that the motion of a force free rigid body can be separated dynamically into two independent parts: the translational motion of the mass center of the body, and its angular motion about the mass center. The independence of these two motions leads to great simplification of the general equations of motion of a rigid body.

In the most general case the scalar components of the quantity $\mathbf{H}_c$ are 36 in number (Appendix E).

Equations (5.29) and (5.30) can be written in D'Alembert form:

$$\mathbf{f}^* \triangleq \mathbf{f} + \mathbf{f}_i = 0 \qquad (5.31)$$
$$\mathbf{M}^* \triangleq \mathbf{M} + \mathbf{M}_i = 0 \qquad (5.32)$$

in which‡

$$\mathbf{f}_i \triangleq -m\dot{\mathbf{v}}_c \qquad (5.33)$$
$$\mathbf{M}_i \triangleq -\dot{\mathbf{H}}_c \qquad \text{About mass center} \qquad (5.34)$$

It is convenient to write (5.31) and (5.32) in terms of scalar components

$$\Sigma f^*_{\text{any direction}} = 0 \qquad (5.35)$$
$$\Sigma M^*_{\text{any axis}} = 0 \qquad (5.36)$$

These equations say the following:

> Once the inertial force $\mathbf{f}_i$ and the inertial moment $\mathbf{M}_i$ have been included on the free-body diagram of a rigid body, equilibrium relations can be written as for a statics problem.

Relations (2.1) and (2.2) for one-dimensional motion (p. 40) are, of course, special cases of (5.35) and (5.36), respectively.

---
† Equation (5.30) holds also with respect to a fixed point and to certain other kinematically singular points.
‡ The minus signs indicate vectorially that the inertial forces oppose the acceleration—are directed opposite from it.

[Sec. 5.3] DYNAMIC EQUILIBRIUM: RIGID BODY, GENERAL MOTION 153

We shall be demonstrating the use of the above relations first in an example involving a single rigid body (Example 5.4).

Plane motion. For plane (two-dimensional) motion a *major* simplification occurs in the equations of motion because we are concerned only with axes of rotation and angular-momentum vectors in a single direction, namely, perpendicular to the plane of motion. In this case, of the 36 possible terms in the quantity $\dot{\mathbf{H}}_c$, only *one* remains! To wit:

$$-\mathbf{M}_i = \dot{\mathbf{H}}_c = \mathbf{1}_z(J_z\dot{\Omega}_z) \tag{5.37}$$

in which $\dot{\mathbf{H}}_c$ is the rate of change of angular momentum, which is now perpendicular to the plane of motion, J_z is the moment of inertia about an axis normal to the plane of motion and passing through the mass center of the body [refer to (2.7), p. 44], and Ω_z is the angular velocity of the body, again about an axis perpendicular to the plane of motion. (For a rigid body the magnitude of the angular velocity does not depend upon the location of the axis of rotation.)

Equation (5.37) is printed with some misgivings for the reason that it is too inviting to nonexperts in dynamics to apply this equation universally to all rigid-body motion. Indeed, the approach to dynamics is all too often something like the following: "Motion is always of one of two types, either translational or rotational. For translational motion the equation is $f = ma$, and for rotational motion the equation is $M = J\alpha$." Stated without qualification these descriptions of dynamic equilibrium are false. First, motion is *not* "always either translational or rotational" but is generally a combination of both kinds of motion. Second, the equation $\mathbf{f} = m\mathbf{a}$ is a vector equation, and it applies to a rigid body only when $\mathbf{a}$ is the acceleration of its mass center. Finally, the equation $M = J\alpha$ is true only in the case of plane motion, and applies only when the quantities involved are measured with respect to the mass center† or, in the very special case of rotation about a fixed axis, to quantities measured about the fixed axis.

Three-dimensional motion. The value of $\mathbf{H}_c$ for a rigid body having general motion is most simply expressed in terms of the body's moments of inertia about its three principal axes‡ through its mass center (say, x_p, y_p, z_p). Then

$$\mathbf{H}_c = \mathbf{1}_{x_p}(J_{x_p}\Omega_{x_p}) + \mathbf{1}_{y_p}(J_{y_p}\Omega_{y_p}) + \mathbf{1}_{z_p}(J_{z_p}\Omega_{z_p}) \tag{5.38}$$

For each axis, moment of inertia J is defined by (2.7) on p. 44. The value of $\dot{\mathbf{H}}_c$ must then be found by differentiating $\mathbf{H}_c$, using (5.16), for example.

---

† Or certain other kinematically singular points.

‡ The principal axes of a body are its axes of inertial symmetry. Every body has, through its mass center, one set of three, and they are always orthogonal. (In special cases there may be more. To wit: for a spherically symmetrical body—a sphere, a cube, or the like—every axis is a principal axis; for an axially symmetrical body—a cylinder, a square disk, or the like—the axis of symmetry and all axes perpendicular to it are principal axes.) Moments of inertia are defined in Appendix E. The subject is treated, for example, in G. W. Housner and D. E. Hudson, "Applied Mechanics-Dynamics," 2nd ed., chap. 7, D. Van Nostrand Company, Inc., Princeton, N.J., 1959.

This is demonstrated in Example 5.7. When **H** must be expressed in non-principal axes, each of its components contains three terms, which leads to additional complication. (See Appendix E.)

Example 5.4. Air-cushion vehicle. Air-cushion vehicles are used to simulate (in two dimensions) the frictionless environment of space. Astronauts riding on them can practice working with tools, for example (a difficult task when there is nothing to push against). Jets can be used to maneuver the vehicles.

Figure 5.5 is a top view of an air-cushion vehicle which is free to roam about over the test floor. Write its equations of motion if only the one jet is firing, applying a prescribed (time-varying) force $f_1(t)$.

As a *physical model* (Stage I), assume that the only horizontal force, besides f_1 and inertial forces, is a friction force f_4 due to motion over the floor, concentrated at an off-center point P (due to nonsymmetry in the air cushion).

To write the equations of motion (Stage II) we use Procedure A-m as an outline:

(1) *Geometry.* Let the reference configuration be as shown in Fig. 5.5a by the gray outline (painted on the floor) at the origin of the fixed xy reference frame. Then, with the system in the arbitrary configuration shown (solid), define the coordinates x, y, and θ. Are these all independent? (How many degrees of freedom has the system?) The position, velocity, and acceleration of the mass center are given by:

$$\begin{aligned} \mathbf{r}^C &= \mathbf{1}_x x + \mathbf{1}_y y \\ \mathbf{v}^C &= \mathbf{1}_x \dot{x} + \mathbf{1}_y \dot{y} \\ \mathbf{a}^C &= \mathbf{1}_x \ddot{x} + \mathbf{1}_y \ddot{y} \end{aligned} \qquad (5.39)$$

because the reference frame is not rotating. (Refer to pp. 146 and 147.)

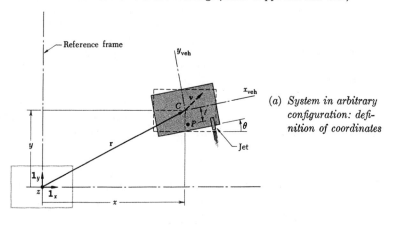

(a) *System in arbitrary configuration: definition of coordinates*

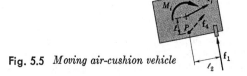

(b) *Free-body diagram*

Fig. 5.5 *Moving air-cushion vehicle*

[Sec. 5.3] DYNAMIC EQUILIBRIUM: RIGID BODY, GENERAL MOTION 155

The velocity of point P is given, from (5.7), by:

$$\mathbf{v}^P = \mathbf{v}^{P/C} + \mathbf{v}^C$$

in which $\mathbf{v}^{P/C}$, the velocity of P relative to C, is

$$\mathbf{v}^{P/C} = \mathbf{1}_{x_{\text{veh}}}(\ell\dot\theta)$$
$$\mathbf{v}^{P/C} = \mathbf{1}_x(\ell\dot\theta \cos\theta) + \mathbf{1}_y(\ell\dot\theta \sin\theta)$$

so that the velocity of P with respect to the floor is, using (5.7),

$$\mathbf{v}^P = \mathbf{1}_x(\dot x + \ell\dot\theta \cos\theta) + \mathbf{1}_y(\dot y + \ell\dot\theta \sin\theta) \tag{5.40}$$

(2) *Force equilibrium.* (i) A free-body diagram is shown in Fig. 5.5b. The forces are $\mathbf{f}_1$ (from the jet), $\mathbf{f}_4$ (from friction), and $\mathbf{f}_i$ and M_i (the inertial force and moment). $\mathbf{f}_4$ and $\mathbf{f}_i$ are shown in arbitrary directions because we do not know their directions yet.

(ii) In this problem we shall need all three of the plane-motion equations of equilibrium. Writing (5.35) in the x and y directions,† and (5.36) about $-z_{\text{veh}}$, we have

$$\Sigma f_x^* = 0 = f_{ix} + f_{4x} - f_1 \sin\theta$$
$$\Sigma f_y^* = 0 = f_{iy} + f_{4y} + f_1 \cos\theta \tag{5.41}$$
$$\Sigma (M_C^*)_{-z} = 0 = M_i - f_1\ell_2 - \ell f_{4x}\cos\theta - \ell f_{4y}\sin\theta$$

(3) *Physical force-geometry relations.* These are [from Stage I, above, and (5.33) and (5.34)]

$$\mathbf{f}_4 = \mathbf{1}_x(-bv_x^P) + \mathbf{1}_y(-bv_y^P)$$
$$\mathbf{f}_i = -m\mathbf{a}_C = \mathbf{1}_x(-ma_x^C) + \mathbf{1}_y(-ma_y^C) \tag{5.42}$$
$$M_i = J\ddot\theta \quad \text{(directed opposite to } \ddot\theta, \text{ as shown in Fig. 5.5}b\text{)}$$

Combine relations. In counting equations we note that each vector equation could have been written as two scalar equations, e.g.,

$$v_x^P = \dot x + \ell\dot\theta \cos\theta$$
$$v_y^P = \dot y + \ell\dot\theta \sin\theta$$

so, counting each vector equation as two equations, we have 12 equations and 12 unknowns.‡ We combine these, entering (5.39) and (5.40) into (5.42), and the result into the equations of equilibrium (5.41), thus eliminating the a's, v's, and f's (except f_1, which is given), to obtain:

$$\left.\begin{aligned}0 &= -m\ddot x - b(\dot x + \ell\dot\theta \cos\theta) - f_1 \sin\theta\\0 &= -m\ddot y - b(\dot y + \ell\dot\theta \sin\theta) + f_1 \cos\theta\\0 &= J\ddot\theta - f_1\ell_2 + \ell b(\dot x + \ell\dot\theta \cos\theta)\cos\theta + \ell b(\dot y + \ell\dot\theta \sin\theta)\sin\theta\end{aligned}\right\} \tag{5.43}$$

The last equation can be simplified as follows:

$$0 = J\ddot\theta - f_1\ell_2 + b\ell(\dot x \cos\theta + \dot y \sin\theta) + b\ell^2\dot\theta$$

(Identify the last term physically.)

---

† We could alternatively have written (5.35) along x_{veh} and y_{veh}. In fact, doing so is more convenient for studying vehicles moving in three dimensions.

‡ We did not include $r_x^C, r_y^C, v_x^C, v_y^C, v_x^{C/P}, v_y^{C/P}$, which are already replaced in subsequent equations.

156 EQUATIONS OF MOTION: TWO AND THREE DIMENSIONS   [Chap. 5]

These are the differential equations of motion which the physical model of the free air-cushion vehicle must obey. When we solve them simultaneously (Stage III) we shall learn what motions—translation $x(t)$, $y(t)$, and rotation $\theta(t)$—the vehicle will have in response to a given jet-firing pattern, $f_1(t)$. Do equations (5.43) check dimensionally? (See Prob. 19.27.)

Consider the special case when $\ell = 0$. Does each equation take a reasonable form in this case? What happens to the "coupling" between equations?

Problems 5.40 through 5.56

5.4 EQUATIONS OF MOTION FOR SYSTEMS OF RIGID BODIES: EXAMPLES

The equations of dynamic equilibrium, e.g., equations (5.35) and (5.36), apply not only to single rigid bodies (as they were used in Sec. 5.3) but to any *system* of rigid bodies as well. All we need do is add an inertial force $\mathbf{f}_i$ and an inertial moment $\mathbf{M}_i$ to *each* rigid body. Then by D'Alembert's principle, the system of rigid bodies is in dynamic equilibrium, and (5.35) and (5.36) apply to the entire system! That is, again, we can write equilibrium relations as for a problem in statics.

It is in dealing with *systems* of rigid bodies that the power of D'Alembert's method is most valuable.

Example 5.5. Bucket and pulley. Write the equation of motion for the pulley and bucket system of Fig. 5.6, if the ball bearings are very good, and the pulley is accurately centered. Let the bucket have vertical motion only.

STAGE I. PHYSICAL MODEL. The problem statement almost defines an ideal model. Assume that the pulley is perfectly centered, that there is *no* friction in the system, and that the rope does not stretch.

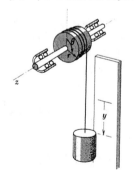
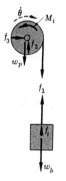

(a) *Physical model*

(b) *Free-body diagram and force equilibrium: pulley alone and bucket alone*

(c) *Alternative free-body diagram: pulley and bucket taken together*

Fig. 5.6 *A system of rigid bodies in dynamic equilibrium*

[Sec. 5.4] SYSTEMS OF RIGID BODIES: EXAMPLES 157

STAGE II. EQUATIONS OF MOTION. We follow Procedure A-m.

(1) *Geometry.* In Fig. 5.6a the system is shown in an arbitrary configuration which either coordinate, y or θ, describes completely. (The system has one degree of freedom.) They are related by the identity $y \equiv r\theta$. We have plane motion, and y is measured in a stationary, nonrotating reference frame, so the velocity and acceleration of the bucket are obtained simply by differentiating y: $\mathbf{v} = \mathbf{1}_y \dot{y}$, $\mathbf{a} = \mathbf{1}_y \ddot{y}$, and the angular velocity and angular acceleration of the pulley are likewise obtained by differentiating θ: $\mathbf{\Omega} = \mathbf{1}_z \dot{\theta}$, $\dot{\mathbf{\Omega}} = \mathbf{1}_z \ddot{\theta} = \mathbf{1}_z (\ddot{y}/r)$.

(2) *Force equilibrium.* Two free-body diagrams are shown in Fig. 5.6b, one for the pulley and one for the bucket. When the free-body diagrams are drawn, each point of isolation must have the appropriate forces applied, as shown.

Using the D'Alembert point of view, we include an inertial force f_i on the bucket and an inertial moment M_i on the pulley about its mass center (which happens to be a fixed axis). The positive directions for f_i and M_i have been chosen (arbitrarily) in the directions opposite to y and θ, respectively.

Equations of equilibrium can now be written, as in statics, for *any* free-body diagram (any *system* of rigid bodies): (5.35) can be written in any desired direction, and (5.36) about any desired axis.

At least two force-equilibrium relations must be written for the free-body diagrams of Fig. 5.6b, and they will contain unknown force f_1, in which we are not interested.

An alternative, and more efficient, free-body diagram is shown in Fig. 5.6c. Here, by a shrewd application of (5.35), (5.36), we can write a single equilibrium relation, namely,

$$\Sigma M^*_{\text{pulley center}} = 0 = M_i + r(f_i - w_B) \tag{5.44}$$

which will be adequate. Unknown forces f_1, f_2, f_3 have all been avoided.

(3) *Force-geometry physical relations.* By our assumptions, we have only inertial forces to consider:

$$\begin{aligned} f_i &= m\ddot{y} \\ M_i &= J\ddot{\theta} \end{aligned} \tag{5.45}$$

Finally, *combining relations*, we obtain the following equation of motion for the system:

$$0 = J\frac{\ddot{y}}{r} + r(m\ddot{y} - w_B) \tag{5.46}$$

Note that the simplification of a single equilibrium relation is possible in this problem *only* with the application of D'Alembert's principle.

This equation is readily checked for dimensional consistency by noting that each term has the dimensions of a moment.

As a special limiting case, note that if the inertia of the wheel is zero, we have simply the equation of motion for a freely falling mass:

$$m\ddot{y} = w_B$$

while if $m = 0$ (and hence $w_B = 0$) we have just a pure rotary inertia:

$$J\ddot{\theta} = 0$$

Example 5.6. Rolling cylinder on sliding wedge. Write the equations of motion for the cylinder and wedge of Fig. 5.2 if the cylinder rolls without slipping, and the wedge slides on the floor without friction, and is driven by a force $f_5(t)$.

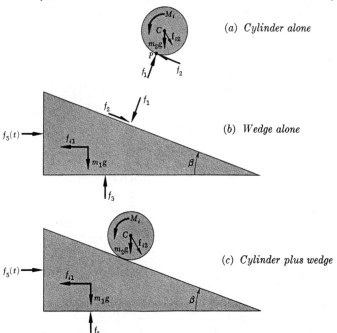

Fig. 5.7 *Free-body diagrams for the system of Fig. 5.2*

(1) *Geometry.* From Example 5.1,

$$\begin{aligned} \mathbf{r} &= 1_x(x_1 + R\theta_2 \cos\beta) + 1_y(R\theta_2 \sin\beta) \\ \mathbf{v}^C &= \dot{\mathbf{r}} = 1_x(\dot{x}_1 + R\dot{\theta}_2 \cos\beta) + 1_y(R\dot{\theta}_2 \sin\beta) \\ \dot{\mathbf{v}}^C &= 1_x(\ddot{x}_1 + R\ddot{\theta}_2 \cos\beta) + 1_y(R\ddot{\theta}_2 \sin\beta) \end{aligned} \qquad (5.47)$$

(The xy reference frame is stationary and nonrotating.)

(2) *Equilibrium.* (i) Three possible *free-body diagrams* are shown in Fig. 5.7a, b, c. (ii) *Equilibrium equations.* The inclusion of f_i and M_i on both rigid bodies makes (5.35) and (5.36) valid for *any* of the systems—a, b, or c in Fig. 5.7. There are only two independent variables, e.g., x_1 and θ_2; so in principle we need only two equilibrium equations. Moreover, we are again not interested in forces f_1, f_2, or f_3. With a little thought we see that by exploiting the versatility of (5.35) and (5.36) we can, in fact, write two independent equilibrium equations which do not contain f_1, f_2, or f_3. One is $\Sigma M_P{}^* = 0$ for free body a;† the other is $\Sigma f_x{}^*$ for free body c.‡

---

† Note that it would have been incorrect to apply Newton's Second Law in the form (5.30) to take moments about point P, for point P is not the mass center, and it is accelerating. In the form (5.35), derived from D'Alembert's principle, however, the equation is correct.
‡ Note that Newton's Second Law in the form (5.29) and (5.30) applies only to single rigid bodies. Using that form, a minimum of four equations must be written, and forces f_1 and f_2 eliminated from among them; while with the D'Alembert form (5.35), (5.36), the final equations can be written directly, as we have shown.

$$\Sigma \overset{\frown}{M}_P{}^* = 0 = M_i - f_{i2x}R\cos\beta - (f_{i2y} + m_2 g)R\sin\beta$$
$$\Sigma f_x{}^* = 0 = f_5(t) - f_{i1} + f_{i2x} \tag{5.48}$$

(3) *Physical relations.* From (5.33) and (5.34), using (5.47),

$$f_{i1} = m_1 \ddot{x}_1 \text{ directed as shown}$$
$$\mathbf{f}_{i2} = -m_2 \dot{\mathbf{v}}^c = 1_x[-m_2(\ddot{x}_1 + R\ddot{\theta}_2 \cos\beta)] + 1_y(-m_2 R\ddot{\theta}_2 \sin\beta)$$
$$= 1_x[f_{i2x}] + 1_y(f_{i2y})$$
$$M_i = J\ddot{\theta}_2$$

Combining relations, we have

$$0 = J\ddot{\theta}_2 + m_2(\ddot{x}_1 + R\ddot{\theta}_2 \cos\beta)R\cos\beta + m_2(R\ddot{\theta}_2 \sin\beta - g)R\sin\beta$$
$$0 = f_5(t) - m_1 \ddot{x}_1 - m_2(\ddot{x}_1 + R\ddot{\theta}_2 \cos\beta) \tag{5.49}$$

which are the equations of motion. Some additional combining gives the first of Eqs. (5.49) a simpler form:

$$0 = (J + m_2 R^2)\ddot{\theta}_2 + (m_2 R \cos\beta)\ddot{x}_1 - m_2 gR \sin\beta$$

Problems 5.57 through 5.68

5.5 ADVANTAGES OF THE D'ALEMBERT METHOD. THE GYRO

As the preceding examples demonstrate, there are sometimes two advantages to the D'Alembert approach when systems of rigid bodies are involved:

(1) The procedure is *more systematic* because inertial forces are cast in the same form as spring forces and friction forces, and need no special treatment.
(2) The procedure is *more versatile,* and therefore *more efficient,* because (a) free-body diagrams can be drawn of *systems* of rigid bodies taken together, so that internal forces need not be considered, and (b) moments can be taken about any axis.

In addition, physical intuition is reinforced by the concept of inertial reaction forces opposing acceleration whenever it tries to occur.

For these reasons the D'Alembert approach (5.35), (5.36) will be used for most of the examples in this book. But the reader will certainly find little difficulty in substituting a Newtonian derivation whenever he chooses.

We submit one further, rather advanced example, in which rotating reference frames play a helpful role. (Refer again to Sec. 5.2.)

Example 5.7. The two-axis gyro. Gyros for sensing angular motion are the heart of airplane automatic pilots, rocket-vehicle launch-guidance systems, space-vehicle attitude-control systems, ship's gyrocompasses, and submarine inertial autonavigators. A physical model for a typical gyro instrument is shown in Fig. 5.8. The rotor is driven at constant speed n with respect to inner gimbal g.

160 EQUATIONS OF MOTION: TWO AND THREE DIMENSIONS [Chap. 5]

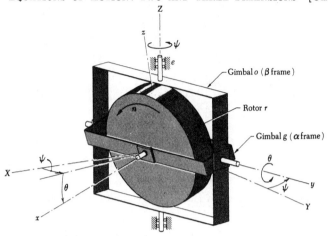

(a) *System in an arbitrary configuration*

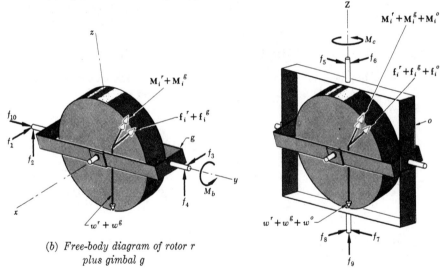

(b) *Free-body diagram of rotor r plus gimbal g*

Fig. 5.8 *Two-axis gyro*

(c) *Free-body diagram of rotor r plus gimbal g plus gimbal o*

Angle θ defines the angular position of the inner gimbal g with respect to the outer gimbal o; and angle ψ defines the angular position of the outer gimbal with respect to the earth e (or with respect to the vehicle to which the gyro is mounted). (Gimbal angles θ and ψ are measured with electrical pickoffs in many applications.)

Derive the equations of motion for study of rapid transient motions of the system.

To further define the *physical model* (Stage I) we assume that the rotor is perfectly balanced and symmetrical, and that the bearings are rigid and perfectly centered. We assume that a moment M_b is exerted about the y axis, between gimbals g and o, and that a moment M_c is exerted about the Z axis on the outer gimbal by the fixed base. In this analysis of high-speed transient motions of the rotor we take the earth (or vehicle) to be an inertially fixed reference.†

To derive the *equations of motion* (Stage II) we follow Procedure A-m, p. 35.

(1) *Geometry.* Figure 5.8a shows the system in an arbitrary configuration (with respect to fixed reference XYZ), and coordinates θ and ψ are defined. The system has two degrees of freedom (the speed n of the rotor with respect to gimbal g being prescribed as constant).

It will be convenient to employ both the α reference frame (fixed in the inner gimbal g) and the β frame (fixed in the outer gimbal o). The angular velocities of these frames are [using (5.11)]

$$\mathbf{\Omega}^\beta = \mathbf{1}_Z \dot\psi \tag{5.50}$$
$$\mathbf{\Omega}^\alpha = \mathbf{\Omega}^{\alpha/\beta} + \mathbf{\Omega}^\beta$$
$$= \mathbf{1}_y(\dot\theta) + \mathbf{1}_Z(\dot\psi)$$
or
$$\mathbf{\Omega}^\alpha = \mathbf{1}_x(-\dot\psi s\theta) + \mathbf{1}_y(\dot\theta) + \mathbf{1}_z(\dot\psi c\theta) \tag{5.51}$$

in which the last expression is obtained by decomposing $\mathbf{1}_Z\dot\psi$ into its components along the x and z axes, the shorthand $s\theta \triangleq \sin\theta$ and $c\theta \triangleq \cos\theta$ being employed. Finally, the angular velocity of the rotor is

$$\mathbf{\Omega}^r = \mathbf{\Omega}^{r/\alpha} + \mathbf{\Omega}^\alpha$$
$$= \mathbf{1}_x n + \mathbf{1}_y \dot\theta + \mathbf{1}_Z \dot\psi$$
or
$$\mathbf{\Omega}^r = \mathbf{1}_x(n - \dot\psi s\theta) + \mathbf{1}_y(\dot\theta) + \mathbf{1}_z(\dot\psi c\theta) \tag{5.52}$$

(2) *Force equilibrium.* (i) Select subsystems. Since there are only two independent variables, θ and ψ, we expect that two equations of motion will be sufficient to describe the motion. The key to writing them directly is the shrewd choice of free bodies. The pair of free-body diagrams which do the job are shown in Fig. 5.8b, the rotor plus inner gimbal, and 5.8c, the entire system.

(ii) Write force-balance equations. For the free-body diagram of Fig. 5.8b there is just one equilibrium equation that can be written which does not involve the bearing forces $f_1, \ldots, f_4$, and that is Eq. (5.36) written about the y axis:

$$\text{System } g + r \qquad (\Sigma \mathbf{M}^*)_y = 0 = (\mathbf{M}_i{}^r + \mathbf{M}_i{}^g)_y + M_b \tag{5.53}$$

(Note that the assumption of perfect balance in Stage I implies that each of the forces $\mathbf{f}_i{}^r$, $\mathbf{f}_i{}^g$, w^r, w^g, has zero moment about the y axis.)

Similarly, for the free-body diagram of Fig. 5.8c there is just one equilibrium equation that can be written which does not involve the bearing forces $f_5, \ldots, f_8$, and that is Eq. (5.36) written about the Z axis:

$$\text{System } g + r + o \qquad (\Sigma \mathbf{M}^*)_Z = 0 = (\mathbf{M}_i{}^r + \mathbf{M}_i{}^g + \mathbf{M}_i{}^o)_Z + M_c \tag{5.54}$$

† By contrast, if we were interested in analyzing the gyro as a gyrocompass, the motion of the earth would, of course, be of primary importance.

which is more convenient to write in the form

$$0 = (\mathbf{M}_i{}^r + \mathbf{M}_i{}^g)_z \cos\theta - (\mathbf{M}_i{}^r + \mathbf{M}_i{}^g)_x \sin\theta + (\mathbf{M}_i{}^o)_Z + M_c \quad (5.55)$$

as we shall see.

(3) *Physical relations.* Using (5.37) and (5.16), we can write expressions for the inertial moments $\mathbf{M}_i$, shown in the free-body diagrams of Fig. 5.8b and c:

$$\begin{aligned}
-\mathbf{M}_i{}^r &= \dot{\mathbf{H}}^r = \overset{\alpha}{\dot{\mathbf{H}}}{}^r + \mathbf{\Omega}^\alpha \times \mathbf{H}^r \\
-\mathbf{M}_i{}^g &= \dot{\mathbf{H}}^g = \overset{\alpha}{\dot{\mathbf{H}}}{}^g + \mathbf{\Omega}^\alpha \times \mathbf{H}^g \\
-\mathbf{M}_i{}^o &= \dot{\mathbf{H}}^o = \overset{\beta}{\dot{\mathbf{H}}}{}^o + \mathbf{\Omega}^\beta \times \mathbf{H}^o
\end{aligned} \quad (5.56)$$

in which differentiation in the α and β frames, respectively, has been selected because $\mathbf{H}^r$ and $\mathbf{H}^g$ are most simply expressed in the α frame (in which their moments of inertia are constant), while $\mathbf{H}^o$ is most simply expressed in the β frame (in which its moments of inertia are constant). Axes x, y, z, are (by symmetry) principal axes of both r and g, and thus, using (5.38), (5.52), and (5.51), we obtain

$$\left.\begin{aligned}
\mathbf{H}^r &= \mathbf{1}_x[J_x{}^r(n - \dot\psi s\theta)] + \mathbf{1}_y[J_y{}^r\dot\theta] + \mathbf{1}_z[J_z{}^r\dot\psi c\theta] \\
\mathbf{H}^g &= \mathbf{1}_x[J_x{}^g(-\dot\psi s\theta)] \phantom{{}+n} + \mathbf{1}_y[J_y{}^g\dot\theta] + \mathbf{1}_z[J_z{}^g\dot\psi c\theta]
\end{aligned}\right\} \quad (5.57)$$

Similarly, Z is a principal axis for o:

$$\mathbf{H}^o = \phantom{\mathbf{1}_x[J_x{}^g(-\dot\psi s\theta)] + \mathbf{1}_y[J_y{}^g\dot\theta] + {}} \mathbf{1}_Z[J_Z{}^o\dot\psi]$$

We define the following, for convenience in handling the nomenclature:

$h \triangleq J_x{}^r n$ (this is the rotor's constant "spin momentum" with respect to the gimbal g)

$J_x \triangleq J_x{}^r + J_x{}^g \qquad J_y \triangleq J_y{}^r + J_y{}^g \qquad J_z \triangleq J_z{}^r + J_z{}^g$

Then, substituting (5.57) into (5.56), and computing only the components required by (5.53) and (5.55), gives

$$-(\mathbf{M}_i{}^r + \mathbf{M}_i{}^g)_y = J_y\ddot\theta + h\dot\psi\cos\theta + (J_z - J_x)\dot\psi^2\sin\theta\cos\theta \quad (5.58)$$

$$\begin{aligned}-(\mathbf{M}_i{}^r + \mathbf{M}_i{}^g)_z\cos\theta &= J_z(\ddot\psi\cos^2\theta - \dot\psi\dot\theta\sin\theta\cos\theta) - h\dot\theta\cos\theta \\ &\quad + (J_x - J_y)\dot\psi\dot\theta\sin\theta\cos\theta \end{aligned} \quad (5.59)$$

$$\begin{aligned}-(\mathbf{M}_i{}^r + \mathbf{M}_i{}^g)_x\sin\theta &= -J_x(\ddot\psi\sin^2\theta + \dot\psi\dot\theta\cos\theta\sin\theta) \\ &\quad + (J_z - J_y)\dot\psi\dot\theta\cos\theta\sin\theta \end{aligned} \quad (5.60)$$

Combine terms. Substituting (5.58), (5.59), and (5.60) into (5.53) and (5.55), respectively, as required, gives:

$$J_y\ddot\theta + h\dot\psi\cos\theta + (J_z - J_x)\dot\psi^2\sin\theta\cos\theta = M_b \quad (5.61)$$

$$(J_Z{}^o + J_z\cos^2\theta + J_x\sin^2\theta)\ddot\psi - h\dot\theta\cos\theta$$
$$\qquad + 2(J_x - J_z)\dot\psi\dot\theta\sin\theta\cos\theta = M_c \quad (5.62)$$

which are the exact equations of motion for the ideal two-axis gyro.

Linear approximation for small motions. In many applications of interest it may be assumed that the motions ψ and θ will be very small (the order of milliradians or arc seconds). Further, the spin speed n will be large compared to $\dot\psi$. This permits dropping the $J\dot\psi^2\sin\theta$ and $J\dot\psi\dot\theta\sin\theta$ terms in (5.61) and (5.62), compared with the terms in $h\dot{\psi} = Jn\dot{\psi}$ and $h\dot{\theta} = Jn\dot{\theta}$. Then, using the usual small-angle approximations, $\sin\theta = \theta$ and $\cos\theta = 1$, we obtain the following linear equations:

Two-axis gyro equations
for small motions

$$J_y\ddot{\theta} + h\dot{\psi} = M_b \quad (5.63)$$
$$J_Z\ddot{\psi} - h\dot{\theta} = M_c \quad (5.64)$$

in which $J_Z \triangleq J_{Z}{}^{o} + J_z$. These linear equations for a two-axis gyro are very useful in making linear calculations of gyro behavior, as we shall demonstrate in Sec. 19.5. Usually M_b and M_c are made as small as possible (very good bearings are used), and consist primarily of light viscous damping, $M_b = -b\dot{\theta}$, and $M_c = -c\dot{\psi}$.

The single-axis gyro. In many gyro applications the outer gimbal is fastened rigidly to a vehicle. Then only the inner gimbal is free to move, and the instrument is called a single-axis gyro. For this instrument, only Eq. (5.63) is required, the motion of ψ now being a prescribed input motion:

$$J_y\ddot{\theta} = M_b - h\dot{\psi} \quad (5.65)$$

Several versions of the single-axis gyro can be made, depending on the arrangements made for M_b. For example, if a spring and viscous damper are provided, so that $M_b = -k\theta - b\dot{\theta} + M_u$ (refer again to Fig. 5.8a), then the resulting instrument is called a *rate gyro*, and has the equation of motion:

Single-axis gyro

$$J\ddot{\theta} + b\dot{\theta} + k\theta = -h\dot{\psi} + M_u \quad (5.66)$$

in which M_u is uncertainty torque due to unbalance, stray magnetic field, and other causes. An actual single-axis gyro is shown in Fig. 19.13a.

Versions of the single-axis gyro will be studied in Problems 19.18 and 19.19.

Problems 5.69 through 5.70

5.6 ENERGY METHODS†

As is indicated in Procedure A-m, the *equilibrium relations* required [Step (2)] in deriving the dynamic equations of motion for a system can be obtained in two alternative ways: either by writing *force-balance* relations (as we have been doing up to now) or by writing *energy-balance* relations. Indeed, in the absence of heat exchange, one method is derivable from the other.

More fundamentally, energy methods are based on the Law of Conservation of Energy—the First Law of Thermodynamics—an empirical law

† Sections 5.6 through 5.10 may be studied independently of Secs. 5.1 through 5.5.

for which a violation has never been observed. It states that the increase in the total energy within a system is equal to the net amount of energy which has entered the system:

$$\Delta E_{\text{syst}} = \Delta E_{\text{net in}} = \Delta E_{\text{in}} - \Delta E_{\text{out}} \\ = Q_{\text{in}} + W_{\text{in}} \qquad (5.67)$$

in which Q_{in} is the net heat transferred into the system and W_{in} is the net work done on the system. Before applying this law it is, of course, essential first to *define the system* precisely.

For a purely mechanical system the energy within the system is in the form of either kinetic or potential energy; and energy enters and leaves the system only in the form of mechanical work,† and thus (5.67) becomes:

$$\Delta(T + U) = W_{\text{in}} \qquad (5.68)$$

in which T is the kinetic energy of the system,‡ U is the potential energy of the system, and W is the net work done *on* the system by all the external forces acting. For each force the work is given by

$$W \triangleq \int \mathbf{f} \cdot d\mathbf{x} \qquad (5.69)$$

in which $d\mathbf{x}$ is the displacement of the point of action of external force $\mathbf{f}$. (The dot product indicates, of course, that only that component of $\mathbf{f}$ which is along the direction of $d\mathbf{x}$ does work on the system. See Appendix B.)

For reference, we note the general expression for the kinetic energy of a rigid body:

$$\boxed{T = \tfrac{1}{2}mv_c^2 + \tfrac{1}{2}\boldsymbol{\Omega} \cdot \mathbf{H}_c} \qquad (5.70)$$

where $\boldsymbol{\Omega}$ is the body's absolute angular velocity and $\mathbf{H}_c$ is defined by (5.38) (subscript c refers to the body's center of mass); T is a scalar quantity—it does not have any direction. For the special case of *plane motion*, $\boldsymbol{\Omega} = \mathbf{1}_n \Omega$, $\mathbf{H}_c = \mathbf{1}_n J_c \Omega$ (where $\mathbf{1}_n$ is normal to the plane of motion), and

$$T = \tfrac{1}{2}mv_c^2 + \tfrac{1}{2}J_c\Omega^2 \qquad (5.71)$$

[Moment of inertia J_c is defined by (2.7).]

Energy dissipation through friction. In all real systems—and therefore in many of the models we study—there is friction which "dissipates" energy. It is convenient (and correct, if boundaries are properly chosen) to account for this by computing the *work* done by friction, and assuming that the conversion of this work to heat takes place outside the system. (We shall do just this in Sec. 5.9—see Fig. 5.15.) The salient point is that friction work

---

† Precisely, then, we mean by a "purely mechanical system" that there is no flow of heat through the boundaries of the system and no energy stored by the system in the form of heat during the process we are considering.

‡ The use of T for both kinetic energy and temperature is unfortunate, but is well established in both cases.

[Sec. 5.6] ENERGY METHODS 165

may not be reconverted to mechanical energy, and must therefore be counted as energy lost from the system.

As a simple example of the application of (5.68) to deriving equations of motion, consider the mass and spring shown in Fig. 5.9a. This is a model of a real physical system. In the model it is assumed that there is no friction and that the spring is linear. We now apply the energy version of Procedure A-m.

(1) *Geometry.* This involves merely choosing as the coordinate the displacement x of the mass from its rest position (spring unstretched). The absolute velocity of the mass is $v_{abs} = \dot{x}$.

(2) *Equilibrium.* (i) We *define*, as our *system*, the mass plus the spring, i.e., all matter included in the bounding envelope shown in Fig. 5.9a. (ii) Then we note that the only *outside* forces acting on the system (Fig. 5.9b) are f_1 from the wall, f_2 from the floor, and the force of gravity w. But none of these forces does any work, because none of them moves along its line of action. Therefore

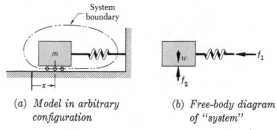

(a) *Model in arbitrary configuration*

(b) *Free-body diagram of "system"*

Fig. 5.9 *A simple mass-spring system*

the work done on the given system is zero, and (5.68) becomes

$$\Delta(T + U) = 0$$

from which
$$T + U = \text{const.} \tag{5.72}$$

that is, the total energy *stored* in the system is *conserved*. The system of Fig. 5.9a is therefore known as a *conservative system*.

(3) *Physical (geometry-energy) relations.* Using (5.71) and the well-known expression for potential energy stored in a linear spring [see (2.33)], we obtain

$$T = \tfrac{1}{2}m\dot{x}^2 \qquad U = \tfrac{1}{2}kx^2 \tag{5.73}$$

and inserting (5.73) into (5.68), we obtain

$$\tfrac{1}{2}m\dot{x}^2 + \tfrac{1}{2}kx^2 = 0 \tag{5.74}$$

This is a form of the equation of motion of the system in Fig. 5.9a. We can convert it to the more familiar form by differentiating the entire equation with respect to time:

$$\frac{d}{dt}[T + U] = 0 \tag{5.75}$$

$$m\ddot{x}\dot{x} + kx\dot{x} = 0 \tag{5.76}$$

then, dividing through by $\dot{x}$, we obtain

$$m\ddot{x} + kx = 0 \qquad (5.77)$$

For conservative, one-degree-of-freedom systems the above procedure is rather convenient.

> Problems 5.71 through 5.75

Use of the energy principle to derive the equations of motion in the above example (or more generally, in any conservative, one-degree-of-freedom system) is unusually simple. In less special cases, however, the procedure just used does not produce the desired results: If the system has more than one degree of freedom, the steps from (5.76) and (5.77) cannot be performed; and if the system is not conservative, the form (5.72) cannot be used at all.

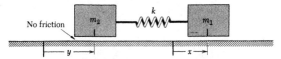

Fig. 5.10 *A conservative two-degree-of-freedom system*

To illustrate the former difficulty, consider the system whose physical model is shown in Fig. 5.10.

Since this is a conservative system, we can use relation (5.72):

$$\tfrac{1}{2}m_1\dot{x}^2 + \tfrac{1}{2}m_2\dot{y}^2 + \tfrac{1}{2}k(x - y)^2 = \text{const.}$$

Next we try differentiating this equation [as we did in getting from (5.74) to (5.77)]:

$$m_1\ddot{x}\dot{x} + m_2\ddot{y}\dot{y} + k(x - y)(\dot{x} - \dot{y}) = 0 \qquad (5.78)$$

We now find, however, that we have a nonlinear equation and that we cannot divide out velocity, as we could in going from (5.76) to (5.77), because there are now two different velocity terms involved, $\dot{x}$ and $\dot{y}$. Some more subtle approach will have to be used.

About 1780 J. L. Lagrange developed a procedure for writing equations of motion which is based on energy methods, but which can be used in the general, many-degree-of-freedom case. Lagrange's method accounts for the work done by all of the forces acting on a system, whether they are conservative or not; and it avoids the difficulty of separation of variables by employing a procedure of partial differentiation.

A derivation of Lagrange's method, which will not be given here, is based on Newton's Second Law and on the dynamic equilibrium concept

[Sec. 5.7] LAGRANGE'S METHOD

of D'Alembert, and employs a differential version of (5.68) in which incremental "virtual displacements" are imagined.†

Lagrange's method will be presented and demonstrated, beginning in Sec. 5.7.

There are some situations in which Lagrange's method is more convenient to use than the Newton-D'Alembert approach, as we discuss in Sec. 5.10. Some people (often those who deal mostly with conservative systems) prefer to use Lagrange's method exclusively.

In addition to its other possible advantages, Lagrange's method has the substantial virtue that it provides for an almost completely independent check on the derivation of the equations of motion, which is a most useful thing to have when dealing with complicated systems.

5.7 LAGRANGE'S METHOD

Lagrange's method is based on a form of the energy-balance relation known as Lagrange's equations, which (for a precisely defined system) are written in terms of the total kinetic energy T of the system, the total potential energy U of the system, and a set of *independent coordinates* chosen to describe the configuration of the system. The coordinates are denoted (purely for convenience) by the symbols $q_1, q_2, q_3, \ldots$.

For example, in Fig. 5.10 we would probably choose as a set of independent coordinates‡

$$q_1 \triangleq x \qquad q_2 \triangleq y$$

In Fig. 5.2 we might select

$$q_1 \triangleq x_1 \qquad q_2 \triangleq \theta_2$$

(Name some alternative sets.) For Fig. 5.5 we might choose

$$q_1 \triangleq x \qquad q_2 \triangleq y \qquad q_3 \triangleq \theta$$

and so on.

For a system having n independent coordinates $q_1, \ldots, q_n$, Lagrange's equations are

$$\frac{d}{dt}\left(\frac{\partial T}{\partial \dot{q}_1}\right) - \frac{\partial T}{\partial q_1} + \frac{\partial U}{\partial q_1} = Q_1$$

$$\frac{d}{dt}\left(\frac{\partial T}{\partial \dot{q}_2}\right) - \frac{\partial T}{\partial q_2} + \frac{\partial U}{\partial q_2} = Q_2$$

$$\cdot \qquad \cdot \qquad \cdot$$
$$\cdot \qquad \cdot \qquad \cdot$$

$$\frac{d}{dt}\left(\frac{\partial T}{\partial \dot{q}_n}\right) - \frac{\partial T}{\partial q_n} + \frac{\partial U}{\partial q_n} = Q_n$$

† An elementary derivation is given in, for example, G. W. Housner and D. E. Hudson, *op. cit.*, Sec. 9.2.
‡ Recall that $\triangleq$ means "is defined to equal."

or, more compactly,

Lagrange's equations

$$\frac{d}{dt}\left(\frac{\partial T}{\partial \dot{q}_i}\right) - \frac{\partial T}{\partial q_i} + \frac{\partial U}{\partial q_i} = Q_i \qquad i = 1 \text{ through } n \qquad (5.79)$$

in which the Q_i's are special combinations of the external forces which will be discussed in Sec. 5.9. (They are called "generalized forces." They exist only for nonconservative systems.)

As indicated, any set of geometrically independent coordinates† can be used in Lagrange's equations, but one cannot use a set of coordinates which are not geometrically independent, such as x_2, ξ, θ_2 in Fig. 5.2, or x and y in Fig. 5.3, for example.

Sometimes, for convenience, a special function, called "the Lagrangian" L of a system is defined as

$$L \triangleq T - U$$

Then Lagrange's equations can be written

$$\frac{d}{dt}\left(\frac{\partial L}{\partial \dot{q}_i}\right) - \frac{\partial L}{\partial q_i} = Q_i \qquad i = 1 \text{ through } n \qquad (5.80)$$

(Notice that, since U can never depend on $\dot{q}$, we have $\partial U/\partial \dot{q} \equiv 0$.)

Lagrange's equations have the very important property that they are not vector equations: only magnitudes are involved. This property is part of the basis for the preference for Lagrange's equations in some circumstances.

An orderly procedure for deriving equations of motion by Lagrange's method is the following concise form of Procedure A-m (p. 35):

(1) **Geometry**
(2) **Equilibrium: Lagrange's equations**
(3) **Energy-geometry relations**

Procedure A-mL
Lagrange

Then **combine relations,** inserting (1) and (3) into (2). We shall demonstrate this procedure with examples and problems.

To simplify the presentation we consider first, in Sec. 5.8, the special (but important) case of "conservative systems." For such systems all the Q_i's are zero. In Sec. 5.9 the general case will be considered.

† Sometimes sets of independent coordinates are called *generalized coordinates*. Any independent set of coordinates—i.e., any set which is sufficient to describe completely a system's configuration—may be termed a set of generalized coordinates. The term is used to imply that they need not be rectangular ("Cartesian") coordinates like x and y in Fig. 2.2c or like x_1 and x_2 in Fig. 2.2a, but may also be angles like θ in Fig. 2.2c or θ_1, θ_2, θ_3 in Fig. 2.2b. The generalized coordinate of a runner in the indoor mile might simply be the number of yards he has run. The requirement is that the generalized coordinates constitute a geometrically independent set, as x_1, x_2 or x_1, ξ or x_2, ξ in Fig. 2.2a; θ_1, θ_2, θ_3 or ψ_a, ψ_b, ψ_c in Fig. 2.2b; θ or x or y in Fig. 2.2c.

5.8 LAGRANGE'S METHOD FOR CONSERVATIVE SYSTEMS

A conservative mechanical system is one in which the total mechanical energy in the system (potential plus kinetic energy) remains constant; that is, no work is done on the system by external forces, and no mechanical energy is "dissipated" (e.g., converted into heat). Figure 5.11 shows the idealized physical models for four mechanical systems: Each physical model is a conservative system. By comparison, the system of Fig. 5.15 is not conservative, both because of the dissipation of energy and because an external force acts on the system as it moves, thereby doing work on the system and changing its total energy. Many additional examples of conservative mechanical systems are shown, for example, in Fig. 7.2, p. 230. (Figure 7.2g, Fig. 7.2h, and Fig. 7.2i do not represent conservative systems.)

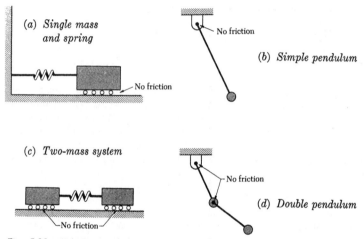

Fig. 5.11 *Typical conservative systems*

For the special case of a conservative system, all the Q's in (5.79) are zero, and thus *Lagrange's equations for a conservative system* are as follows:

$$\frac{d}{dt}\left(\frac{\partial T}{\partial \dot{q}_i}\right) - \frac{\partial T}{\partial q_i} + \frac{\partial U}{\partial q_i} = 0 \qquad i = 1 \text{ through } n \qquad (5.81)$$

We now demonstrate the use of (5.81) with some examples.

Example 5.8. Mass-spring system. Derive the equations of motion for the simple mass-spring system shown in Fig. 5.12a. Suppose that in our physical model (Stage I) we assume that there is no friction acting on the mass and that the spring is linear. To derive the equation of motion (Stage II) for this model we follow Procedure A-mL, above.

(1) *Geometry.* We describe the configuration of the system with coordinate x, and we let $x = 0$ when the spring is unstretched.

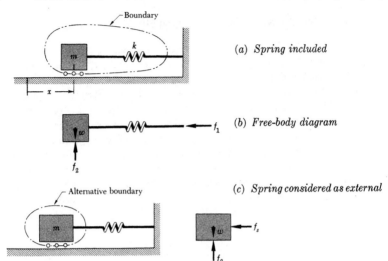

Fig. 5.12   *Alternative choices for a system boundary*

(2) *Lagrange's equation.* We define our system by the boundary shown, which includes both the mass and the spring. In a free-body diagram of this system, Fig. 5.12b, we notice (as we did on p. 165) that none of the forces acting on the system—f_1 from the wall, f_2 from the floor, and gravity w—does any work on the system. Since there is also no energy dissipation, the situation is conservative, and thus we can write Lagrange's equation in the form (5.81):

$$\frac{d}{dt}\left(\frac{\partial T}{\partial \dot{x}}\right) - \frac{\partial T}{\partial x} + \frac{\partial U}{\partial x} = 0 \tag{5.82}$$

(3) *Energy-geometry relations.* We consider the system in an arbitrary state in which it has both an arbitrary displacement x and an arbitrary velocity $\dot{x}$. Then, from relations (5.71) and (2.33), we have the following expressions for the kinetic and potential energy of the system, respectively,

$$\left. \begin{array}{l} T = \tfrac{1}{2}m|\mathbf{v}|^2 = \tfrac{1}{2}m\dot{x}^2 \\ U = \tfrac{1}{2}kx^2 \end{array} \right\} \tag{5.83}$$

Combining results of Steps (1), (2), and (3) [i.e., performing the various differentiations indicated by (5.82)],

$$\frac{\partial T}{\partial \dot{x}} = m\dot{x} \qquad -\frac{\partial T}{\partial x} = 0$$

$$\frac{d}{dt}\left(\frac{\partial T}{\partial \dot{x}}\right) = m\ddot{x} \qquad \frac{\partial U}{\partial x} = kx$$

and inserting in (5.82), we obtain

$$m\ddot{x} + kx = 0 \tag{5.84}$$

which is the well-known equation of motion for this system.

[Sec. 5.8] LAGRANGE'S METHOD: CONSERVATIVE SYSTEMS

Lagrange's method seems a bit cumbersome for such a simple system. Its real advantage appears for systems that involve several degrees of freedom, or complicated geometry, or both.

We note in passing that we may alternatively define the boundary of our system [in (2) above] so that it does not include the spring, as shown in Fig. 5.12c. In this case there is an external force f_s, which does work on the system as it moves, and therefore (5.81) does not apply. We can still obtain the equation of motion by using (5.79), as we shall demonstrate in Sec. 5.9.

Example 5.9. Simple pendulum. Write the equation of motion for a simple pendulum, Fig. 5.13.

For a physical model we assume that there is no friction in the hinge. We also assume simple plane motion of the pendulum. Finally, we assume, for simplicity, that the arm is weightless and the entire mass of the pendulum therefore concentrated in the bob. For this physical model we now write the equation of motion, following the Lagrange Procedure A-mL.

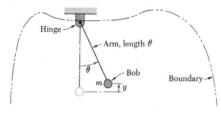

Fig. 5.13 *Simple pendulum, with system defined to include gravity field*

(1) *Geometry.* We depict the system in an arbitrary configuration, Fig. 5.13, and select angle θ as the independent coordinate ($q = \theta$). Then the altitude y of the bob above its lowest position is given by

$$y = \ell(1 - \cos \theta) \tag{5.85}$$

and the magnitude of the bob's velocity is

$$|\mathbf{v}| = \ell \dot{\theta} \tag{5.86}$$

($|\mathbf{v}|$ means magnitude of $\mathbf{v}$.) An important advantage of Lagrange's method is that it requires only the *speed*—the magnitude of $\mathbf{v}$, but not its direction nor its derivative.

(2) *Lagrange's equation.* We choose to include within our system the entire pendulum and also the entire gravity field of the earth, as indicated by the boundary shown in Fig. 5.13. The only external force acting on this system is at the wall, and since this force does not move, it does no work on the system. Since we have assumed no friction we have a conservative system, and we can write Lagrange's equation in the form (5.81):

$$\frac{d}{dt}\left(\frac{\partial T}{\partial \dot{\theta}}\right) - \frac{\partial T}{\partial \theta} + \frac{\partial U}{\partial \theta} = 0 \tag{5.87}$$

(3) Energy-geometry relations. With the system in the arbitrary state indicated, the kinetic and potential energies are, respectively, [Eqs. (5.70), (2.34)],

$$T = \tfrac{1}{2}m|\mathbf{v}|^2 = \tfrac{1}{2}m(\ell\dot\theta)^2 \tag{5.88}$$

and
$$U = mgy = mg\ell(1 - \cos\theta) \tag{5.89}$$

Combining (1), (2), and (3) [entering relations into (5.87)], we obtain

$$\frac{\partial T}{\partial \dot\theta} = m\ell^2\dot\theta \qquad -\frac{\partial T}{\partial \theta} = 0$$

$$\frac{d}{dt}\left(\frac{\partial T}{\partial \dot\theta}\right) = m\ell^2\ddot\theta \qquad \frac{\partial U}{\partial \theta} = mg\ell\sin\theta$$

and thus
$$m\ell^2\ddot\theta + mg\ell\sin\theta = 0 \tag{5.90}$$

which is well known to be the correct result. Compare the derivation here with that in Problem 5.48, where the Newton-D'Alembert method was used.

Notice that in the preceding example we did not have to calculate the velocity as a vector, and that we did not have to find the acceleration at all. This advantage is one of the important attractions of the Lagrange method.

Example 5.10. A two-degree-of-freedom system. Write the equations of motion for the simple two-mass system shown in Fig. 5.14.

For a physical model we assume that no friction acts on either mass and that the connecting spring is linear, within the motions that we shall consider.

(1) *Geometry.* This is a two-degree-of-freedom system (although it is only one-dimensional). Therefore any two independent coordinates can be used, e.g., coordinates x and y or x and ξ, where ξ is the stretch of the spring. We select x and y:

$$q_1 = x \qquad q_2 = y$$

such that when the spring is unstretched, $x = y$.

The absolute velocities of masses m_1 and m_2 have the magnitudes

$$|\mathbf{v}_1| = \dot x \qquad |\mathbf{v}_2| = \dot y \tag{5.91}$$

(2) *Lagrange's equations.* We define the system we shall consider by the boundary shown in Fig. 5.14. The only forces acting on this system are normal forces which do no work, and since there is no energy dissipation, we can use form (5.81) for the Lagrange equations. In terms of the independent coordinates x and y these are

$$\frac{d}{dt}\left(\frac{\partial T}{\partial \dot x}\right) - \frac{\partial T}{\partial x} + \frac{\partial U}{\partial x} = 0 \qquad \frac{d}{dt}\left(\frac{\partial T}{\partial \dot y}\right) - \frac{\partial T}{\partial y} + \frac{\partial U}{\partial y} = 0 \tag{5.92}$$

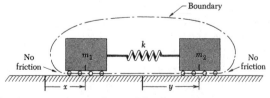

Fig. 5.14 *A conservative two-degree-of-freedom system*

[Sec. 5.9] LAGRANGE'S METHOD: NONCONSERVATIVE SYSTEMS 173

(3) *Energy-geometry relations.* For the system defined above, the total kinetic and potential energies are, respectively,

and
$$T = \tfrac{1}{2}m_1\dot{x}^2 + \tfrac{1}{2}m_2\dot{y}^2 \tag{5.93}$$
$$U = \tfrac{1}{2}k\xi^2 = \tfrac{1}{2}k(y-x)^2 \tag{5.94}$$

Combining (1) through (3) [entering relations into (5.92)], we obtain

$$\frac{\partial T}{\partial \dot{x}} = m_1\dot{x} \qquad\qquad \frac{\partial T}{\partial \dot{y}} = m_2\dot{y}$$

$$\frac{d}{dt}\left(\frac{\partial T}{\partial \dot{x}}\right) = m_1\ddot{x} \qquad\qquad \frac{d}{dt}\left(\frac{\partial T}{\partial \dot{y}}\right) = m_2\ddot{y}$$
$$-\frac{\partial T}{\partial x} = 0 \qquad\qquad -\frac{\partial T}{\partial y} = 0$$
$$\frac{\partial U}{\partial x} = k(y-x)(-1) \qquad\qquad \frac{\partial U}{\partial y} = k(y-x)(+1)$$

and thus

$$m_1\ddot{x} - k(y-x) = 0 \qquad\qquad m_2\ddot{y} + k(y-x) = 0 \tag{5.95}$$

Problems 5.76 through 5.89

5.9 LAGRANGE'S METHOD FOR NONCONSERVATIVE SYSTEMS

External forces. When external forces act on a system they can do work, and this must be accounted for in any energy balance in general, and in Lagrange's equations in particular. Such work is represented by the Q_i's on the right-hand side of (5.79):

$$\frac{d}{dt}\left(\frac{\partial T}{\partial \dot{q}_i}\right) - \frac{\partial T}{\partial q_i} + \frac{\partial U}{\partial q_i} = Q_i \qquad i = 1 \text{ through } n \qquad \begin{array}{c}(5.79)\\ \text{[repeated]}\end{array}$$

The Q_i's are defined as follows. Consider the system in an arbitrary configuration at an arbitrary time in its motion. Imagine that the system is suddenly frozen there, and that then a very small displacement, δq_i, called a "virtual displacement," of just *one* of the independent coordinates is permitted.† Then, from a free-body diagram of the total system (as defined by a boundary envelope):

> Calculate the work, δW_i, done on the system by all external forces during the displacement δq_i. Then Q_i, the *"generalized force"* (corresponding to this coordinate), is defined by
> $$Q_i \triangleq \frac{\delta W_i}{\delta q_i} \tag{5.96}$$

† The process of considering small, imaginary deviations from the actual situation is an application of variational principles. It is clearly essential that the coordinates chosen be geometrically independent.

174 EQUATIONS OF MOTION: TWO AND THREE DIMENSIONS [Chap. 5]

Quantity δW_i is known as the "virtual work" done during displacement δq_i. In Example 5.11 this concept will be demonstrated for a system with an external driving force.

Friction. One straightforward way (which works very well) of extending Lagrange's method to systems containing dissipative elements is to define the system boundaries so that all such elements are excluded from "the system" for which Lagrange's equations are written. Then such elements exert "external forces" on the system and thereby they make their appearance in the equations of motion via Q_i's. (Note that this is in strict accord with our discussion of the First Law of Thermodynamics on p. 164.) This method for analyzing systems with damping will also be demonstrated in Example 5.11, which follows.

Example 5.11. A two-mass system with damping and external force. In this physical model we include damping, but for convenience of demonstration, we concentrate it in two dashpots, Fig. 5.15a, labeled b_2 and b_4. We assume that there is otherwise no friction in the system. There is an external force acting on mass m_1, as shown in Fig. 5.15.

To obtain the equations of motion (Stage II) by Lagrange's method we follow Procedure A-mL.

(1) *Geometry.* Like Fig. 5.14, this is a two-degree-of-freedom (one-dimensional) system. The configuration of the system can be determined by two independent coordinates, and for these we choose $q_1 = x$ and $q_2 = y$.

Define: Compression of k_1, $\xi_1 \equiv x - y$
           Compression of k_3, $\xi_3 \equiv y$ (5.97)

(2) *Lagrange's equations.* (i) Define the system. We define the boundary of the system as shown in Fig. 5.15a. Notice that this boundary carefully excludes the dissipative dashpot elements, whose influence on the system we shall consider as that of external forces. The boundary includes in the system the springs, which are conservative elements. (ii) Write Lagrange's equations.

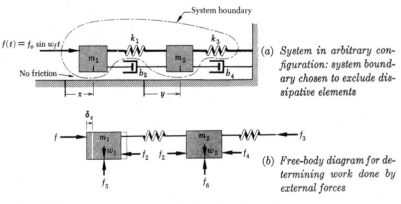

(a) *System in arbitrary configuration: system boundary chosen to exclude dissipative elements*

(b) *Free-body diagram for determining work done by external forces*

Fig. 5.15 *A nonconservative two-degree-of-freedom system*

[Sec. 5.9] LAGRANGE'S METHOD: NONCONSERVATIVE SYSTEMS 175

Since external forces act on our system, we must use Lagrange's equations in the form (5.79):

$$\frac{d}{dt}\left(\frac{\partial T}{\partial \dot{x}}\right) - \frac{\partial T}{\partial x} + \frac{\partial U}{\partial x} = Q_x \qquad \frac{d}{dt}\left(\frac{\partial T}{\partial \dot{y}}\right) - \frac{\partial T}{\partial y} + \frac{\partial U}{\partial y} = Q_y \quad (5.98)$$

To obtain Q_x and Q_y we use a free-body diagram of the system in an arbitrary configuration, Fig. 5.15b. Consider now a small "virtual" displacement δx of mass m_1 (with no displacement of mass m_2). The work done by all external forces during displacement δx is

$$\delta W_x = (f - f_2)\, \delta x$$

(No other force does any work because no other force has a component of displacement along its line of action during δx.) Thus, by (5.96), the value of Q_x is

$$Q_x \triangleq \frac{\delta W_x}{\delta x} = f - f_2 \quad (5.99)$$

Next we hold mass m_1 fixed and make an arbitrary displacement δy of mass m_2. The work done on the system by external forces during this displacement is

$$\delta W_y = (f_2 - f_4)\, \delta y$$

(Note again that force f_3 does no work because it does not move.) Applying (5.96) again, we obtain

$$Q_y \triangleq \frac{\delta W_y}{\delta y} = f_2 - f_4 \quad (5.100)$$

(3) *Energy-geometry relations.* The kinetic energy and potential energy of the system defined by the boundary are, respectively,

$$T = \tfrac{1}{2}m_1|\mathbf{v}_1|^2 + \tfrac{1}{2}m_2|\mathbf{v}_2|^2 \quad (5.101)$$
$$= \tfrac{1}{2}m_1\dot{x}^2 + \tfrac{1}{2}m_2\dot{y}^2$$

and
$$U = \tfrac{1}{2}k_1\xi_1{}^2 + \tfrac{1}{2}k_3\xi_3{}^2 \quad (5.102)$$

We have also the force-geometry relations,

$$f_2 = b_2\dot{\xi}_1 \qquad f_4 = b_4\dot{\xi}_3$$

Combining the results of the above steps [entering relations into (5.98)]:

$\dfrac{\partial T}{\partial \dot{x}} = m_1\dot{x}$	$\dfrac{\partial T}{\partial \dot{y}} = m_2\dot{y}$
$\dfrac{d}{dt}\left(\dfrac{\partial T}{\partial \dot{x}}\right) = m_1\ddot{x}$	$\dfrac{d}{dt}\left(\dfrac{\partial T}{\partial \dot{y}}\right) = m_2\ddot{y}$
$-\dfrac{\partial T}{\partial x} = 0$	$-\dfrac{\partial T}{\partial y} = 0$
$\dfrac{\partial U}{\partial x} = k_1(x - y)$	$\dfrac{\partial U}{\partial y} = k_1(x - y)(-1) + k_3 y$
$m_1\ddot{x} + k_1(x - y) = f - b_2(\dot{x} - \dot{y})$	$m_2\ddot{y} - k_1(x - y) + k_3 y = b_2(\dot{x} - \dot{y}) - b_4\dot{y}$

$$(5.103)$$

176   EQUATIONS OF MOTION: TWO AND THREE DIMENSIONS   [Chap. 5]

These two equations of motion are to be solved simultaneously, in Stage III, for x and y versus time.

For the particular system in Example 5.11 the equations of motion can be more easily derived by Procedure A-m, as the reader may verify.

It may be noted also that the system in Example 5.11 could have been defined with one or both springs excluded, and the correct equations of motion obtained by considering the spring forces as also external to the system. In this case the spring forces would do work on the system because their point of attachment moves during the virtual displacements δx and δy, and thus these forces enter into the equations of motion, in this case, via the Q's rather than via the potential-energy expression. (The reader may wish to repeat the above example with the springs excluded from the defined system to show that the same answer is obtained.) Since we can represent spring forces and gravity forces as potential-energy terms, it is usually more convenient to do so, particularly if no generalized forces are otherwise required (i.e., if the system is conservative).

Another technique which is sometimes used to accommodate friction forces within the framework of Lagrange's equations is to define an artificial function, called a "dissipation function" F, in such a way that the appropriate Q will be obtained automatically by taking $\partial F/\partial \dot{q}$, in the same way that $\partial U/\partial q$ produces the proper spring term. A description of this method can be found in texts on advanced dynamics.

Example 5.12. Rolling cylinder on sliding wedge. Use Lagrange's method to obtain the equations of motion of the system of Figs. 5.2 and 5.7. Compare both the results and the analysis with those of Example 5.6.

(1) *Geometry.* We take as independent (generalized) coordinates:

$$q_1 = x_1 \\ q_2 = \theta_2 \tag{5.104}$$

From Example 5.1,

$$v_1 = \dot{x}_1 \\ \mathbf{r} = \mathbf{1}_x(x_1 + R\theta_2 \cos \beta) + \mathbf{1}_y(R\theta_2 \sin \beta) \\ \mathbf{v}^C = \dot{\mathbf{r}} = \mathbf{1}_x(\dot{x}_1 + R\dot{\theta}_2 \cos \beta) + \mathbf{1}_y(R\dot{\theta}_2 \sin \beta) \\ y_2 = R\theta_2 \sin \beta \tag{5.105}$$

(The xy reference frame is fixed and nonrotating.)

(2) *Lagrange's equations.* (i) Define system. We include the block plus the cylinder plus the gravity field in our system. From Fig. 5.7c, the only forces acting on this system are f_3, which does no work, and $f_5(t)$, which does some. (The friction force f_2 does none; there is no energy dissipation in ideal rolling friction.)

(ii) Write Lagrange's equations. We must use form (5.79):

$$\frac{d}{dt}\left(\frac{\partial T}{\partial \dot{x}_1}\right) - \frac{\partial T}{\partial x_1} + \frac{\partial U}{\partial x_1} = Q_1 \qquad \frac{d}{dt}\left(\frac{\partial T}{\partial \dot{\theta}_2}\right) - \frac{\partial T}{\partial \theta_2} + \frac{\partial U}{\partial \theta_2} = Q_2 \tag{5.106}$$

[Sec. 5.9] LAGRANGE'S METHOD: NONCONSERVATIVE SYSTEMS

To obtain Q_1 and Q_2 let the system in Fig. 5.7c have first a small (virtual) displacement δx_1 and note that the virtual work done on the system by all external forces is simply

$$\delta W_1 = f_5 \, \delta x_1$$

and thus
$$Q_1 \triangleq \frac{\delta W_1}{\delta x_1} = f_5 \tag{5.107}$$

Next, let the system have virtual displacement $\delta \theta_2$, and note that no work is done by forces external to the system. (We have included gravity within our system.) Therefore

$$Q_2 = 0 \tag{5.108}$$

(3) *Energy-geometry relations.* From (5.71) and (2.34):

$$\begin{aligned} T &= \tfrac{1}{2} m_1 v_1{}^2 + \tfrac{1}{2} m_2 v^{C2} + \tfrac{1}{2} J \Omega^2 \\ U &= -m_2 g y_2 \end{aligned} \tag{5.109}$$

Combine relations. Equations (5.105) are first entered into (5.109):

$$T = \tfrac{1}{2} m_1 (\dot{x}_1{}^2) + \tfrac{1}{2} m_2 [(\dot{x}_1 + R \dot{\theta}_2 \cos \beta)^2 + (R \dot{\theta}_2 \sin \beta)^2] + \tfrac{1}{2} J \dot{\theta}_2{}^2 \tag{5.110}$$
$$U = -m_2 g (R \theta_2 \sin \beta) \tag{5.111}$$

These are then entered into (5.106), along with (5.107) and (5.108):

$\dfrac{\partial T}{\partial \dot{x}_1} = m_1 \dot{x}_1$
$\qquad + m_2 (\dot{x}_1 + R \dot{\theta}_2 \cos \beta)$

$\dfrac{d}{dt}\left(\dfrac{\partial T}{\partial \dot{x}_1}\right) = m_1 \ddot{x}_1$
$\qquad + m_2 (\ddot{x}_1 + R \ddot{\theta}_2 \sin \beta)$

$-\dfrac{\partial T}{\partial x_1} = 0$

$\dfrac{\partial U}{\partial x_1} = 0$

$m_1 \ddot{x}_1 + m_2 (\ddot{x}_1 + R \ddot{\theta}_2 \cos \beta) = f_5$

which are identical to (5.49).

$\dfrac{\partial T}{\partial \dot{\theta}_2} = m_2 [(\dot{x}_1 + R \dot{\theta}_2 \cos \beta) R \cos \beta$
$\qquad + (R \dot{\theta}_2 \sin \beta) R \sin \beta] + J \dot{\theta}_2$

$\dfrac{d}{dt}\left(\dfrac{\partial T}{\partial \dot{\theta}_2}\right) = m_2 [R \cos \beta \ddot{x}_1 + R^2 \ddot{\theta}_2] + J \ddot{\theta}_2$

$-\dfrac{\partial T}{\partial \theta_2} = 0$

$\dfrac{\partial U}{\partial \theta_2} = -m_2 g R \sin \beta$

$(J + m_2 R^2) \ddot{\theta}_2 + (m_2 R \cos \beta) \ddot{x}_1$
$\qquad - m_2 g R \sin \beta = 0 \tag{5.112}$

Problems 5.90 through 5.99

In comparing the analysis of Examples 5.6 and 5.12, we note that the Lagrange procedure (used in the latter) requires less geometry and more algebra. Specifically, only velocities are required in (5.105) [Step (1) of Example 5.12], while acceleration is required in (5.47) of Example 5.6. Also in Example 5.6, moment-taking in Fig. 5.7a required care and geometric visualization. The trade-off in Example 5.12 was for the extra algebra between Eqs. (5.111) and (5.112).

5.10 THE RELATIVE ADVANTAGES OF LAGRANGE'S METHOD

We can now list some of the advantages of Lagrange's method:

(1) The amount of geometric reasoning required may be substantially less because only velocity, and not acceleration, is required. (Further, the sign problem is easier because the square of velocity is used.)
(2) The method deals essentially with scalar, rather than vector relations.
(3) For conservative systems, forces need not be considered, per se.
(4) The method avoids consideration of internal forces within the system.
(5) The required number of equations are delivered automatically (once independent coordinates are correctly chosen).

It should be remarked that item (4) can also be achieved in a force-equilibrium procedure, Procedure A-m, by skillful use of D'Alembert's principle, as we demonstrated in Examples 5.5, 5.6, and 5.7.

If both methods are used, an almost independent check is thereby available which may be of considerable value in complicated problems. (It should be mentioned that the equations of motion obtained by Lagrange's method and by the Newton-D'Alembert method are not always identical, but simple combinations of the equations obtained by one method will always yield those produced by the other, if both sets are correct.)

The principal disadvantages of Lagrange's method are the following:

(1) It proceeds in a routine way without indicating physical cause and effect until the final step, and thus the development of and benefit from physical intuition are severely inhibited.
(2) The amount of algebra involved may be substantially greater than in a force-equilibrium procedure, particularly if the system is not conservative.

The choice of whether to use, in Procedure A-m, a force-balance procedure or the method of Lagrange is often a matter of individual taste. Those who prefer routine algebraic manipulation to geometric and mechanical reasoning are naturally attracted to Lagrange's method. The situation was well put by Lagrange himself in his *Mécanique Analytique* (1788), as follows:

> The methods which I present here require neither constructions nor reasoning of geometrical or mechanical nature, but only algebraic operations proceeding after a regular and uniform plan. Those who love analysis will view with pleasure mechanics being made a branch of it and will be grateful to me for having thus extended its domain.

PART B

DYNAMIC RESPONSE
OF ELEMENTARY SYSTEMS

Chapter 6. *First-Order Systems*

Chapter 7. *Undamped Second-Order Systems: Free Vibrations*

Chapter 8. *Damped Second-Order Systems*

Chapter 9. *Forced Oscillations of Elementary Systems*

Chapter 10. *Natural Motions of Nonlinear Systems and Time-Varying Systems*

In Part A we discussed the first two of the four stages of a dynamic investigation (p. 33): Stage I, imagining an astute physical model for a real system, and Stage II, deriving a mathematical model—a set of equations of motion—for the physical model.

We now come to the heart of the matter (and the principal concern of this book): Stage III, determining how the physical model will behave—what motions† it will have. In general we do this by solving completely the differential equations of motion; although we can often learn all we need to from partial solutions.

As we have indicated in Sec. 1.3, when the equations of motion are linear with constant coefficients a great simplification occurs. Then the solutions to compound, complicated sets of differential equations are made up of the sum of only a few simple kinds of response. In Part B these simple response patterns will be studied, per se, for elementary systems. Then in Parts C and D we shall consider more elaborate, compound systems and look for the same response patterns.

Specifically, we shall find that when a linear, constant-coefficient dynamic system is disturbed by some forcing function the resulting motion is the sum of two distinct components:

(1) A *forced response* which resembles in character the forcing function, and
(2) A *natural motion* whose character depends only on the physical characteristics of the system itself, and not upon the forcing function.

In formal mathematical parlance these are known as (1) the particular solution, and (2) the homogeneous or complementary solution.

Further, the *natural motion* of a linear, constant-coefficient system will always be made up of some combination of two elementary motion patterns, the exponential decay and the sinusoidal motion. In Part B each type of natural motion is first investigated thoroughly in its pure form for very simple dynamic systems—the exponential decay for first-order systems (Chapter 6) and the pure sine wave for undamped second-order systems (Chapter 7). Finally, the damped second-order system is investigated (Chapter 8), and it is found that its natural motion may be an exponentially damped sine wave; that is, the *product* of the two elementary types of natural motion.

Investigation of these basic elements forms the core for all of our future studies of linear systems because, as we shall see in subsequent chapters, all of the possible natural motions of any linear system, however complicated,

---

† As we stated in Part A, the term "motion" is used throughout this book in the broad sense of "change in a physical variable," so that "equations of motion" may be written in terms not only of geometrical displacement, but also of voltage, temperature, pressure, etc., as the case may be.

can always be computed by superposing the responses of simple first-order and damped second-order systems.

In Chapter 7 the complex-number concept is developed, whereby a function of the form e^{st} may be used to describe mathematically either of the basic types of motion (the exponential or the sine wave) by letting s be a complex number.

Although the complex-number technique is introduced in the study of natural motions, it proves also to be the key to the study of forced response, because most common disturbing functions—including steps, exponentials, and damped sinusoids—can be represented in the form e^{st} when s is allowed to be complex.

> In fact, even completely arbitrary disturbing functions can be represented mathematically in the form e^{st} by the methods of convolution and the Laplace transformation.
>
> The Laplace technique is, of course, also a convenient method for solving differential equations with simple forcing functions. It constitutes a powerful alternative to the procedure of assuming a solution of the form e^{st}. In this book, however, formal introduction of the Laplace transform is deferred because it is not needed for the simple business at hand, and we wish to get on with the physical study at once. Later, in Part D, when it will be helpful—and its physical significance more fully appreciated—the Laplace method will be introduced as a generalization of the method of assuming e^{st} solutions.
>
> Meanwhile, readers already familiar with the Laplace method will have no difficulty in applying it to obtain the results in the text.

Forced response to general disturbances of the form e^{st} is discussed for first- and second-order systems in Chapters 6 and 8, respectively. Finally, the particular, but very important, case of sinusoidal disturbances is discussed in greater detail in optional Chapter 9, and a number of applications in mechanical vibration and electric circuitry are analyzed.

To help put the study of linear systems in perspective, a brief introduction to methods of analyzing nonlinear systems is given in optional Chapter 10. Techniques for linear approximation are included.

Throughout the studies of Stage III it is important to consider continuously the physical connotations of each step. In particular, the final calculated motion should always be scrutinized on the basis of physical understanding of the actual system, to be certain that it is physically reasonable, both qualitatively and quantitatively.

Chapter 6

FIRST-ORDER SYSTEMS

In this chapter we introduce the following concepts† concerning the response of physical systems that are represented by linear, ordinary differential equations with constant coefficients—l.c.c. equations:‡

> (1) **Superposition** of time responses is valid.
> (2) The **total response** will consist of (the linear sum of) *two distinct parts*, a "forced motion" and a "natural motion"
>
> $$\theta = \theta_f + \theta_n \qquad (6.1)$$
>
> (Or, in mathematical parlance, the solution to the differential equation will consist of the particular solution plus the homogeneous or complementary solution.)
> (3) The **forced motion** will have the same character as the "forcing function" (system input, or disturbance). Its magnitude will be proportional to the magnitude of the forcing function.

Procedure B-1
Dynamic response of linear systems

† In each chapter certain basic concepts will be introduced for the first time. At the beginning of each chapter we shall state very tersely the new basic concepts. The reader may then watch for them as they emerge in the development of the chapter.

‡ We shall use the abbreviation "l.c.c. equation" to denote a linear, ordinary differential equation with constant coefficients, i.e., one having form (1.1) with constant A's and B's; e.g.,

$$A_n \frac{d^n x}{dt^n} + A_{n-1} \frac{d^{n-1} x}{dt^{n-1}} + \cdots + A_1 \frac{dx}{dt} + A_0 x = f(t)$$

in which the A's are constant. By an l.c.c. *system* we mean a physical model whose equations of motion are l.c.c. equations. (Such systems are often called also "l.t.i."—linear time-invariant—systems.)

[Sec. 6.1] FIRST-ORDER SYSTEMS

> (4) The **natural motion** will always be of the form
> $$\theta_n = \Theta e^{st} \qquad (6.2)$$
> in which Θ and s are constants, and
> (a) s will depend only on the physical system, and not on how it is forced or disturbed (s tells the *character* of the natural motion, e.g., how rapidly it is damped);
> (b) the value of Θ in (6.2) will depend on the initial state of the system and on the forcing function.
> (5) The **characteristic equation** is an algebraic expression in s, obtained by substituting (6.2) into the unforced ("homogeneous") differential equation of motion. The *roots* of the characteristic equation are the values of s which make (6.2) a correct solution to that differential equation.

(6) *"Resonance"* occurs when the input to a system is like one of its natural motions.

(7) An *impulse* is an input of very short duration (zero in the limit) but having a finite area (amplitude times duration). A *unit impulse* has unit area.

(8) *Unit-impulse response.* The dynamic character of a system can be represented by its *response to a unit impulse*, usually denoted $h(t)$, which is like its response to an initial state.

(9) *Convolution.* The response of a system to any arbitrary input $x(t)$ can be obtained by the *convolution* of $x(t)$ with $h(t)$.

Each of the above basic concepts applies, in fact, to an l.c.c. system of any order. In the present chapter, to make development of the concepts just as simple and unencumbered as possible, only first-order systems will be discussed.

6.1 FIRST-ORDER SYSTEMS

A system that can store energy in only one form and location is called a "first-order" dynamic system because the mathematical equation describing its motion can be written in terms of a single variable and its first derivative only. Several examples are shown in Fig. 6.1—a single mass moving against friction (Fig. 6.1a and g), a single electrical capacitance with resistors (b), a single inductance with resistors (c), a single mechanical spring with friction (d), a single thermal capacitance with thermal resistance (e). In each case the resistor or friction dissipates energy, and in each case, as we shall see, the system will return naturally to a position of static equilibrium. The hydraulic servo in Fig. 6.1f is also mathematically a first-order system

184 FIRST-ORDER SYSTEMS [Chap. 6]

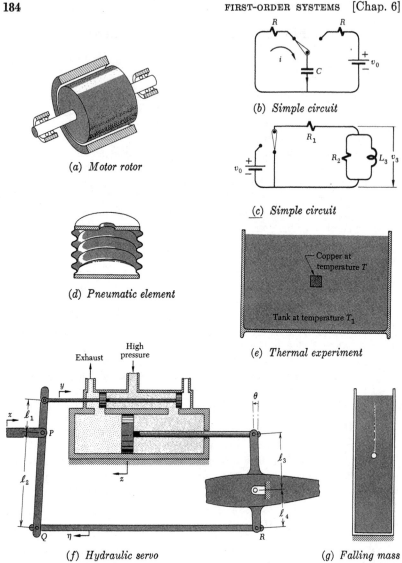

Fig. 6.1 *Many physical systems are essentially first order*

having the same behavior as the others, although its physical arrangement is different.

Let us study the typical physical system in Fig. 6.2a, which consists of a spring-restrained rotor, subject to heavy viscous damping and driven by an applied torque M (M is produced by current in the armature windings on the rotor). Figure 6.2 is a more detailed picture of the system modeled in

[Sec. 6.1] FIRST-ORDER SYSTEMS 185

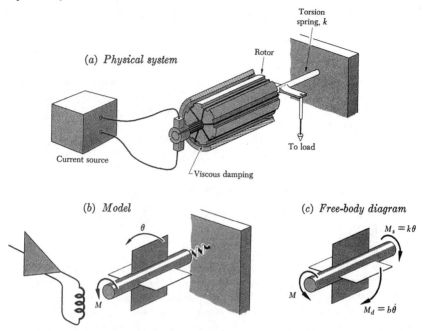

Fig. 6.2 *First-order model of a "torque motor" (when inertia may be neglected)*

Fig. 2.5. Such an element is used, for example, to control precisely the opening of small, high-performance hydraulic valves. It is sometimes called a "torque motor." Torque M is controlled electrically, and is presumed to be independent of motion of the rotor. (This is often a very good approximation.)

Let us suppose further that, for this example, the relative magnitudes of the mass, viscosity, and spring stiffness involved are such that the inertial (acceleration) moment may clearly be neglected, compared to the damping and spring moments. (This is also a good approximation for many real physical systems.) For this particular situation we may then assume the physical system to be represented by the *first-order physical model* of Fig. 6.2b—a spring with damping. This is one of the simplest possible dynamic systems, and is chosen so that the steps in solution can be illustrated with a minimum of mathematical complexity.

In Fig. 6.2b the only energy-storage element is the spring (the damper dissipates energy). We therefore expect that the equation of motion will be first order.

To be systematic in our analysis, we state precisely (Stage I) our physical model (Fig. 6.2b): a massless element with linear damping and spring, driven by a time-varying moment M which is unaffected by motion of the element. Then we derive the equation of motion (Stage II), using

Procedure A-m. The result, from Example 2.1, with $J = 0$ (or from Fig. 6.2c) is

Equation of motion for a first-order system
$$b\dot{\theta} + k\theta = M \quad (6.3)$$

Next we turn to the question (Stage III) of what system behavior is implied by (6.3). We want to know θ as a function of time. We would like a plot of θ versus t. It would also be convenient to have an analytical expression for $\theta(t)$.†

We consider first, in Sec. 6.2, the special case of unforced motion, $M = 0$. Forced motion is considered in Sec. 6.3.

6.2 NATURAL (UNFORCED) MOTION

Consider the special case in which the disturbing torque M is absent in (6.3):
$$b\dot{\theta} + k\theta = 0 \quad (6.4)$$

[Equation (6.4) is the *homogeneous* equation that corresponds to (6.3).]

Mathematical solutions to (6.4) are well known, but it is instructive first to deduce, by reasoning physically, the motions the element must have.

At any instant of time there are only two forces acting on the element: spring torque $k\theta$ (opposing displacement θ) and damping torque $b\dot{\theta}$ (opposing velocity $\dot{\theta}$). These torques must therefore be equal and opposite at each instant:
$$b\dot{\theta} = -k\theta$$

The spring will always attempt to return the element to the position where the spring force is zero. The velocity with which it does so will be limited by the damping torque, so that at any instant the angular velocity of the element will be proportional to the amount by which the spring is twisted at that instant. When the element finally reaches the neutral position it will have zero velocity, and hence will come to rest without overshooting the neutral position.

> The motion of the system can be plotted stepwise, using only the knowledge that the rate of return toward neutral is always proportional to displacement from neutral. Thus, if the system is released from a displacement θ_0, Fig. 6.3a, it will have instantly a relative velocity proportional to θ_0:
> $$\dot{\theta}(0) = -C\theta(0) = -C\theta_0$$
> The proportionality constant C is k/b, because in general [from (6.4)], $\dot{\theta} = -(k/b)\theta$, and thus when $t = 0$, we have $\dot{\theta}(0) = -(k/b)\theta_0$. If the element continued at this velocity, it would reach neutral in b/k seconds, as shown. As

---

† The symbol $\theta(t)$, pronounced "theta of t," is shorthand for "θ at time t." Thus $\theta(4)$ means "θ at $t = 4$," $\theta(0)$ means "θ at $t = 0$," and $\theta(t)$ means "θ at $t = t$." A useful image for $\theta(t)$ is the time plot of θ versus t.

[Sec. 6.2] NATURAL (UNFORCED) MOTION

the displacement decreases, however, the slope also decreases: at point (1), for example, we must make the slope much less, $\dot{\theta} = -(k/b)\theta$. As before, however, if the element continued at constant velocity from point (1), it would again reach neutral in b/k seconds. A fairly accurate stepwise plot can actually be carried out on this basis, using small increments of t (much smaller than b/k) and changing slope each increment. We shall call b/k the "time constant" of the system and use τ to denote it: $\tau = b/k$.

To obtain a mathematical solution to Eq. (6.4), we note that θ must be a quantity whose derivative is always proportional to the quantity itself. The exponential function e^{st} has this property. Therefore, following the classical mathematical routine, we assume that the solution is

$$\theta = \Theta e^{st} \qquad (6.2)$$
[repeated]

and this solution is introduced into the differential equation of motion (6.4) to see whether values of constants Θ and s exist for which the equation is satisfied:

$$bs\Theta e^{st} + k\Theta e^{st} = 0$$

(Note that θ is time variable, but Θ is a constant.)

The characteristic equation

This equation can be satisfied in either of two ways. One way is for Θ to be zero. This is a correct, but pretty uninteresting solution (it implies no motion at all). If Θ does not equal zero, then the common term Θe^{st} is seen to cancel out of the equation, leaving a simple algebraic expression,

$$bs + k = 0 \qquad (6.5)$$

which is known as the *characteristic equation* of the system. From simple algebra this equation is satisfied when $s = -k/b$. The unforced solution to differential equation (6.4), therefore, has the form $\theta = \Theta e^{(-k/b)t}$. Equation (6.5) is known as the characteristic equation because *its root*, $s = -k/b$, *determines completely the dynamic character of natural motions* of the system. (This is clear from Fig. 6.3.)†

The value that Θ must have can be established from the position of the system at time $t = 0$ (the so-called "initial condition" of the system). In the present case the equation is satisfied if $\Theta = \theta_0$. That is, the unforced motion is

Natural response of a first-order system $\qquad \boxed{\theta = \theta_0 e^{(-k/b)t}} \qquad (6.6)$

Note that the amplitude and rate given by this mathematical solution

† The roots of a system's characteristic equation are called, formally, the system's natural dynamic characteristics, or its *eigenvalues*. Eigen is the German word meaning "one's own characteristic . . . ."

agree with the values deduced from physical reasoning; e.g., at $t = 0$, we have $\theta = \theta_0$ and $\dot\theta = -(k/b)\theta_0$.

The solution is plotted in Fig. 6.3a. It substantiates further our physical reasoning. Note that if the spring is extremely stiff (large k), Fig. 6.3a shows that the system returns quickly toward $\theta = 0$, while if it is soft, the system returns slowly. Conversely, if the damping is heavy (large b), the return is slow, while if damping is light, the return is rapid.

Plotting time response

It is convenient to note the amount by which θ has "decayed" at the end of one time constant, $\tau = b/k$. This is calculated by evaluating the solution at time $t = \tau = b/k$:

$$\theta(\tau) = \theta_0 e^{-1} = \frac{1}{e}\theta_0 = 0.368\theta_0$$

This result can be used to make an *accurate sketch* of an exponential function quite rapidly, as demonstrated in Fig. 6.3b. Beginning at the point $t = 0$, $\theta = \theta_0$, an asymptote is drawn which intersects the horizontal axis τ seconds later. A point is plotted above the τ intersection a distance 0.37θ. Beginning at this point, another line is drawn to intersect the axis another τ seconds later, and a point plotted at $0.37 \times (0.37\theta_0) = 0.14\theta_0$.

The process is repeated as many times as desired. The exponential curve is then drawn passing through, and tangent to, the upper end of each "asymptote." The magnitudes at the end of successive time constants can be read on the LL00 or LL/3 scale of a slide rule. The first five values of θ/θ_0 are (approximately) 0.37, 0.14, 0.05, 0.02, and 0.007, for $t/\tau = 1, 2, 3, 4$, and 5, respectively. An alternative procedure is to draw the asymptotes from points marked on the θ axis ($t = 0$ axis). These are shown dotted in Fig. 6.3b. The first four points in this case are (approximately) at $\theta/\theta_0 = 1, 0.73, 0.4, 0.2$, and 0.1. (It is easy and convenient to memorize these two sets of values.)

It is often convenient to plot the motion to a semi-log scale, as shown in Fig. 6.3c. (Amplitude is plotted to a log scale, but time to a linear scale.) The exponential function then becomes a straight line, and the motion can be followed accurately for as many time constants as desired. [On a linear scale the trace is rather indistinguishable from zero after four or five or six time constants (Fig. 6.3b), and in some practical cases this is not sufficient.] This technique is especially helpful in the inverse problem of determining time constants of simple systems from experimental data. The data are plotted to a semi-log scale, and the system's time constants may be estimated with the help of straight-line asymptotes sketched through the data. (See Problem 6.4.)

We emphasize that τ in Fig. 6.3 depends only on the physical parameters of the system. It is entirely independent of initial position and of any disturbing forces which may be applied to the system. Hereafter, we shall refer to such free motion as the "natural" motion of a system.

Some other common first-order physical systems are shown in Fig. 6.1. Two of these are discussed in examples, and the others are topics for problems.

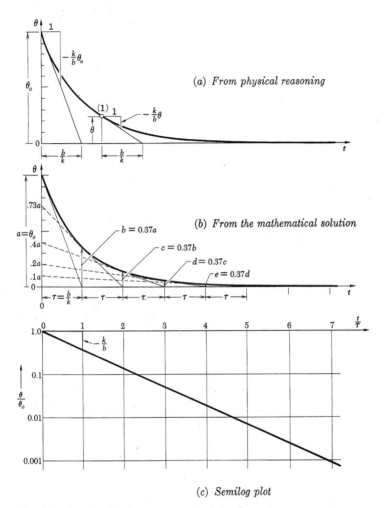

(a) *From physical reasoning*

(b) *From the mathematical solution*

(c) *Semilog plot*

Fig. 6.3 *Constructing the response of a first-order system*

Example 6.1. A free rotor. The rotor in Fig. 6.1a is free to turn on its bearings, but is subject to a retarding torque proportional to its speed Ω, owing to friction and to magnetic forces produced by back emf in its windings. If the total of such torques may be represented by $M_d = b\Omega$ opposing Ω, plot the coast-down transient from an initial speed Ω_0. Given: moment of inertia $J = 3$ (in.-lb)/(rad/sec^2), $b = 0.15$ (in.-lb)/(rad/sec), $\Omega_0 = 1200$ rpm. [This value of J implies, for example, a rotor weighing 23 lb and having a radius of gyration of 7 in.; refer to (2.7):

$$J = m\rho^2 = \frac{w}{g}\rho^2 = \frac{23(7)^2 \;\text{lb in.}^2}{32 \times 12 \;\text{(in./sec}^2)} = 3.0 \text{ in.-lb/(rad/sec}^2)]$$

STAGE I. PHYSICAL MODEL. A simple linear model is described in the problem statement.

STAGE II. EQUATION OF MOTION.

(1) *Geometry.* We are interested in the rotor's speed, Ω, but not in the angle through which it has turned. We thus choose Ω as the independent variable,† positive counterclockwise.

(2) *Force equilibrium*, and (3) *physical (force-geometry) relations*. These two steps are readily combined by indicating the force-geometry relations on

Fig. 6.4 *Free-body diagram of rotor in Fig. 6.1a*

a free-body diagram of the rotor, $M_d = b\Omega$ opposing Ω, and $M_i = J\dot{\Omega}$ opposing $\dot{\Omega}$, Fig. 6.4. The result is

$$\Sigma M^* = 0 = J\dot{\Omega} + b\Omega - M \qquad (6.7)$$

in which, in this problem, there is no external driving torque, $M = 0$:

$$J\dot{\Omega} + b\Omega = 0 \qquad (a)$$

STAGE III. DYNAMIC BEHAVIOR. Following (6.2) we assume that the solution to (a) is

$$\Omega = \Omega e^{st} \qquad (b)$$

Substitution into (a) yields the characteristic equation

$$Js + b = 0 \qquad (c)$$

(provided that Ω is not zero). From (c), solution (b) is correct, provided that $s = -b/J$:

$$\Omega = \Omega e^{-(b/J)t} \qquad (d)$$

Finally, Ω is evaluated by noting that at $t = 0$, $\Omega = \Omega_0$:

$$\Omega(0) = \Omega e^0$$

so that Ω must equal Ω_0, and the complete solution is

$$\Omega = \Omega_0 e^{-t/(J/b)} \qquad (e)$$

or, with the numerical values given,

$$\Omega = 1200 e^{-t/20} \text{ rpm} \qquad (f)$$

which we have plotted in Fig. 6.5, using the method of Fig. 6.3b. The rotor slows down to $1200/e$ rpm in 20 seconds.

The solution shows that if J is increased, the time to slow down will be increased, while if b is increased, the time will be shortened, which is to be expected on physical grounds.

† See footnote, p. 44.

[Sec. 6.2] NATURAL (UNFORCED) MOTION 191

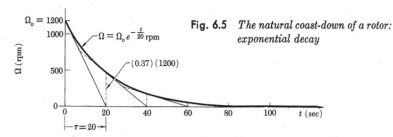

Fig. 6.5 *The natural coast-down of a rotor: exponential decay*

As a limiting-case check, note that if $b = 0$ (no damping), then

$$\Omega(t) = \Omega_0 e^{0t} = \Omega_0 \qquad (g)$$

that is, the rotor continues at its initial speed.

Suggested experiment. Find several rotors having bearings of varying quality, and record and plot their coast-down transients (speed versus time). Rotor speed may be measured conveniently and accurately with a stroboscope. (A semi-log plot, like Fig. 6.3c, will make it especially easy to see how nearly exponential the coast-downs are—i.e., how good the viscous-damping model is.)

Example 6.2. The natural motion of a first-order network. The capacitor in Fig. 6.1b is initially connected to the battery for a long time. Then the switch is suddenly moved to the position shown. Find the resulting time history of current flow.

For convenience, let $t = 0$ be the instant of switching.

STAGE I. PHYSICAL MODEL. This is given by Fig. 6.1b. (A network *is* a physical model, in which the capacitance of wires, the internal resistance of a capacitor, and so on, are neglected or lumped. Refer to Sec. 2.6.)

STAGE II. EQUATION OF MOTION. (Procedure A-e, p. 61, is used as a checklist.)

(1) *Variables.* Choose i as the independent variable: $i_R \equiv i_C \equiv i$.
(2) *Equilibrium.* KVL after switching:†

$$v_R + v_C = 0 \qquad 0 < t$$

(3) *Physical (voltage-current) relations.* We combine these into KVL. After switching:

$$Ri + \frac{1}{C} \int i\, dt = 0 \qquad 0 < t \qquad (a)$$

STAGE III. DYNAMIC BEHAVIOR. Following (6.2), we assume that the solution to this equation of motion is

$$i = Ie^{st} \qquad 0 < t \qquad (b)$$

† The condition $0 < t$ can of course also be written $t > 0$. We elect, arbitrarily, to make a custom of placing the larger quantity always to the right in inequalities, thereby achieving a spatial, as well as algebraic, image of relative magnitude.

Substituting in (*a*), and canceling the common term Ie^{st}, we obtain the characteristic equation

$$R + \frac{1}{Cs} = 0 \qquad (c)$$

from which, for (*b*) to be a correct solution to (*a*), we must have

$$s = -\frac{1}{RC} \quad \text{and thus} \quad i = Ie^{-t/RC} \quad 0 < t \qquad (d)$$

To establish the value of I we need to know i at $t = 0$, just *after* the switch was thrown. This is not a trivial matter, and we must be careful.

The key to determining the state of the system at "$t = 0$, just after the switch is thrown" (more formally referred to as "$t = 0^+$") is the behavior of the energy-storage element, the capacitor. The voltage across a capacitor will not change instantaneously, and therefore $v_C(0^+)$ just after the switch is thrown will be the same as $v_C(0^-)$ just before, and the current $i(0^+)$ will have to accommodate this. Before switching, the capacitor had been connected to the battery for a long time, and thus $v_C(0^-) = v_0$.

In the given circuit, after switching, v_C is also the voltage across the resistor, and thus, in general, $i = -v_C/R$, and in particular just after switching $i(0^+) = -v_0/R$. Since (*d*) will be true at $t = 0^+$ only if $I = i(0^+)$ [because $i(0^+) = Ie^0 = I$], we have

$$i = -\frac{v_0}{R} e^{-t/RC} \qquad (e)$$

The capacitor discharges exponentially through the resistor.

Suggestion: Let $v_0 = 50$ volts, $R = 2000$ ohms, $C = 10^{-5}$ farad, and plot (*e*).

Note that increasing either the resistance or the capacitance will increase the time required for the capacitor to discharge. In the limit if $R \to 0$, discharge is instantaneous, while if $R \to \infty$, the capacitor holds its charge indefinitely, both of which results agree with physical reasoning.

The problem of establishing "initial conditions," which arose in Example 6.2, will be treated more generally in Sec. 6.5. (The problem was also present in Example 6.1, but was simple because the chosen variable Ω itself was known at $t = 0^+$.)

Stability

There is the possibility that the solution to a characteristic equation such as (6.5) will turn out to be a positive real number—that s will be positive. What will that mean?

The characteristic equation came from assuming that the time response was [from (6.2)]

$$\theta = \Theta e^{st}$$

Therefore a positive real value for s ($s = +0.2$, say) would connote a *growing* exponential motion, such as Fig. 6.6, where $\theta = \Theta_0 e^{+0.2t} = \Theta_0 e^{+t/5}$. When the *natural motion* of a system grows without bound, we say that the system is *unstable*. Thus we see that (at least for a first-order system)

[Sec. 6.3] FORCED MOTION 193

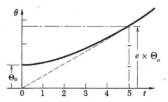

Fig. 6.6 *Natural motion of an unstable, first-order system: exponential growth*

if a root s of a system's characteristic equation is a positive real number, the system will be unstable.

As a singular case, when $s = 0$ the natural motion neither grows nor decays, but remains at its initial amplitude: $\theta = \Theta e^0 = \Theta$. We shall call such a system "neutrally stable."

The subject of stability will be discussed in depth in Chapter 11, where we shall learn that the statement in italics above is true for an l.c.c. system of any order.

Problems 6.1 through 6.15

6.3 FORCED MOTION

Response to a sudden step input

Let us now consider what happens when an external torque M is applied to the element of Fig. 6.2. As a first example, suppose that M is simply a constant torque, suddenly applied when $t = 0$, as shown in Fig. 6.7.

From simple physical reasoning it is clear that a long time after the step change in moment, the element will be stationary at a new position where the spring torque is just great enough to oppose M (the damping torque being zero when the element is stationary). If, after equilibrium is established at this new neutral position, the rotor were then displaced back to its original (pre-M) position and released, the response (shown in Fig. 6.7) would be like that found in Fig. 6.3: the rotor would proceed to neutral exponentially, with time constant $\tau = b/k$.

The formal *mathematical development* by which this same conclusion is reached illustrates a *pattern of solution* that is quite general. As before, a solution to the equation of motion is guessed which has the necessary properties indicated by physical reasoning. The assumed solution is then introduced into the differential equation of motion to determine for what values of its parameters it is a correct solution.

The equation of motion is, from (6.3) and Fig. 6.7,

$$b\dot\theta + k\theta = M = M_0 u(t) \tag{6.8}$$

The factor $u(t)$ is used to represent precisely the fact that M_0 is applied suddenly at $t = 0$. That is, $u(t)$ is the symbol for the "unit step function," which has, by definition, the values

$$\text{Unit step} \qquad u(t) = \begin{cases} 0 & \text{for } t < 0 \\ 1 & \text{for } 0 < t \end{cases} \qquad (6.9)$$

From physical reasoning we have deduced that the solution to this equation must contain both a constant (to represent the constant, final displacement) and an exponential term (to represent the transient). Such a solution is assumed:

$$\theta = \theta_f + \theta_n$$
$$\theta = \Theta_0 + \Theta e^{st} \qquad \text{for } 0 < t$$

(That is, the forced motion $\theta_f = \Theta_0$ plus the natural motion $\theta_n = \Theta e^{st}$ are assumed.) This is substituted into Eq. (6.8):

$$k\Theta_0 + (bs + k)\Theta e^{st} = M_0 \qquad \text{for } 0 < t \qquad (6.10)$$

Since Eq. (6.10) must be correct at any time $0 < t$, it must, in particular, be correct when t is very large. When t is very large, however, e^{st} will be zero (assuming that the motion doesn't diverge, i.e., that s is a negative number). Therefore the rest of the equation must be zero, independent of the portion involving e^{st}:

$$k\Theta_0 = M_0 \qquad (6.11)$$

Furthermore, if this is true, then the portion of the equation that *is* multiplied by e^{st} must also be zero by itself:

$$(bs + k)\Theta e^{st} = 0 \qquad (6.12)$$

Thus it is legitimate to divide Eq. (6.10) into two separate equations, (6.11) and (6.12), and each equation may be solved separately for the parameters involved in it.

Equation (6.12) is recognized immediately as the equation for *unforced motion*, as in Fig. 6.3. It contains the characteristic equation,

$$bs + k = 0$$

and results in the usual transient solution for the *natural* motion of the system:

$$\theta_n = \Theta e^{-(k/b)t} \qquad 0 < t$$

(where Θ is yet to be determined).

The *forced portion* of the solution is obtained from (6.11):

$$\theta_f = \Theta_0 = \frac{M_0}{k} \qquad 0 < t$$

[In mathematical parlance the solution to (6.12)—θ_n—is known as the "complementary" or "homogeneous" solution, and the solution to (6.12) —θ_f—is known as the "particular" solution.]

The *constant* Θ in the natural part of the solution must be just large enough so that the total solution satisfies the initial displacement of the

[Sec. 6.3] FORCED MOTION

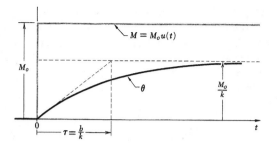

Fig. 6.7 *Response of a first-order system to a step disturbance*

element, which in this case is zero. That is, at $t = 0^+$ ("just after" $t = 0$) the rotor has not yet been displaced,† and we can write

$$\theta(0^+) = 0 = \Theta_0 + \Theta e^0 = \Theta_0 + \Theta$$

from which

$$\Theta = -\Theta_0 = -\frac{M_0}{k}$$

The total solution, then, is

$$\theta = \frac{M_0}{k}\left(1 - e^{-(k/b)t}\right) \qquad 0 < t \qquad (6.13)$$

of which the first term describes the forced motion of the element and the second term describes the transient natural motion that is superposed upon it.

The response given by (6.13) is plotted in Fig. 6.7.

If the initial position of the element is not zero, $\theta(0) \neq 0$, an additional transient motion, $\theta = \theta(0)e^{-(k/b)t}$, will be produced (as in Fig. 6.3) *independent of*, and superposed upon, the other motions, so that the total motion is then

Total response of a first-order system after a step input

$$\theta = \left\{\frac{M_0}{k} + \left[\theta(0) - \frac{M_0}{k}\right]e^{-(k/b)t}\right\} \qquad 0 < t \qquad (6.14)$$

That is, we utilize *the validity of superposition for linear systems*. We find separately (a) the total forced motion (6.13) for zero initial conditions and (b) the total unforced motion for a given initial condition. Then we simply add (a) and (b) to obtain the total forced motion with the given initial condition. Linearity and superposition are discussed more formally in Sec. 6.4.

Response to a suddenly applied exponential input

Consider next that an external disturbing torque described by the relation

$$M = M_0 e^{-\sigma_f t} u(t) \qquad (6.15)$$

† Because that would imply an infinite velocity, which cannot be imposed upon the damper by any finite disturbance. This is discussed carefully in Sec. 6.5.

is applied to the system in Fig. 6.2. This forcing function, shown in Fig. 6.8, is of a more general nature than the "step" of Fig. 6.7, and will reveal some additional facts about the process of solution. At the same time, however, this function is sufficiently simple that no special mathematics will be required to handle it. (A sine wave can be represented by a similar function, but with complex parameters, as we shall see in Sec. 7.2.)

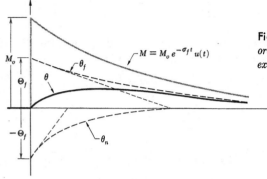

Fig. 6.8 *Response of a first-order system to an abrupt exponential disturbance*

As with the step, we assume that the total motion consists of a forced plus a natural component:

$$\theta = \theta_f + \theta_n \qquad (6.1)$$
[repeated]

We concentrate attention first on the *forced* component and assume that it has the same mathematical form as the forcing function. That is, we try a solution of the form

$$\theta_f = \Theta_f e^{-\sigma_f t} \qquad 0 < t$$

in the differential equation (6.3):

$$-b\sigma_f \Theta_f e^{-\sigma_f t} + k\Theta_f e^{-\sigma_f t} = M = M_0 e^{-\sigma_f t} \qquad 0 < t$$
or
$$(-b\sigma_f + k)\Theta_f e^{-\sigma_f t} = M_0 e^{-\sigma_f t} \qquad 0 < t$$

In this case, we see that our assumed solution is correct, provided that Θ_f has the value

$$\Theta_f = \frac{M_0}{-b\sigma_f + k} \qquad (6.16)$$

This response is plotted dashed in Fig. 6.8.

The *natural motion* in Fig. 6.8 will exhibit the characteristic time constant of the element [just as it did in Eq. (6.14)] and will have a magnitude just great enough to satisfy the initial condition. Thus the total motion is

[Sec. 6.4] LINEARITY AND SUPERPOSITION

given by

$$\theta = \underbrace{\left(\frac{M_0}{-b\sigma_f + k}\right)e^{-\sigma_f t}}_{\text{Forced motion, }\theta_f} + \underbrace{\left[\theta(0) - \frac{M_0}{-b\sigma_f + k}\right]e^{-(k/b)t}}_{\text{Natural motion, }\theta_n} \qquad 0 < t \quad (6.17)$$

The above results are plotted in Fig. 6.8 for the case $\theta(0) = 0$. [Check Eq. (6.17) to be sure that at $t = 0$, $\theta = \theta(0)$.]

The pattern of forced response

From these two simple examples of forced motion† we can deduce the following general *pattern of solution*.

> When a forcing function is applied to an l.c.c. dynamic system, the response of the system will consist of (1) a **forced response** θ_f that is a modification of the *input* signal and (2) a **natural motion** θ_n having dynamic characteristics determined by the system's own *characteristic equation,* and having an amplitude just great enough so that when it is combined with the forced response, the *initial conditions* are satisfied:
> $$\theta = \theta_f + \theta_n$$

From Procedure B-1

A formal mathematical statement of Procedure B-1 (which can be proved formally) is the following: *The complete solution* to a linear ordinary differential equation of order r will consist of *any* particular solution that can be found $[\theta_f(t)]$ plus all of the r homogeneous solutions $[\theta_n(t)]$, provided that the coefficients of the differential equation are continuous. (Obviously, an l.c.c. equation‡ is a satisfying special case: The coefficients are constant and therefore continuous.) If the *state*—the zeroth through $(r - 1)$st derivative—is known at some instant, the proportions of the homogeneous solutions are to be so chosen that they match the state at that instant.

6.4 LINEARITY AND SUPERPOSITION

The above pattern of solution involves computing the motion of the system for each of several special conditions, and then assuming that the total motion will be the *linear sum* of the separate motions [as we have done in (6.14) and (6.17), for example]. License to carry out such *superposition* comes from the *linear property* assumed for each element of the

---
† These two examples do not, of course, constitute complete coverage of forced motion, but other forms of forcing function will be more easily handled later, after some additional techniques have been developed. In particular, the important response to a sinusoidal disturbance will be discussed thoroughly in Chap. 9.
‡ See footnote, p. 182.

system—that if, for example, the spring twist θ is doubled, the spring torque will also be doubled. Because this is true, equation of motion (6.8) has the form (1.1). This assumption of linear elements, and the attendant license to superpose solutions, is essential to the entire line of analytical development we are pursuing.

More formally, we say that if a system is linear, then (1) its elements obey the law of proportionality, illustrated in Fig. 6.9a, whereby the output

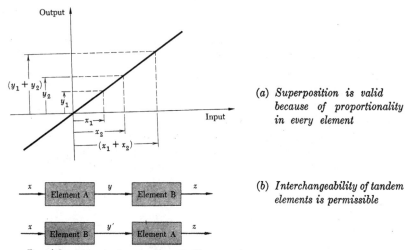

(a) *Superposition is valid because of proportionality in every element*

(b) *Interchangeability of tandem elements is permissible*

Fig. 6.9 *Important properties of a linear system*

of each element is *linearly proportional* to the input (the input-output relation is a straight line), and (2) (therefore) the entire system obeys the rule, illustrated in Fig. 6.10, that if input $x_1(t)$ produces output $y_1(t)$ and input $x_2(t)$ produces output $y_2(t)$, then the input $C_1 x_1(t) + C_2 x_2(t)$ will produce the output† $C_1 y_1(t) + C_2 y_2(t)$, where C_1 and C_2 are arbitrary constants. In the illustration of Fig. 6.10 the system is a very simple one, and C_1 and C_2 each are unity; but the principle holds for assemblages of linear elements, however complicated, with C_1 and C_2 arbitrary. For a system of lumped elements, conditions (1) and (2) imply a differential equation of the form (1.1).

Conversely, if any element in a system is nonlinear, having a characteristic like Fig. 6.11, for example, then the output produced by two inputs together will *not* be the sum of the outputs produced by the separate inputs, and may not even resemble them.

Another important property of a linear system is that any two of its

---

† Assuming that the initial conditions are zero. [If they are not, the system's natural response to initial conditions is also to be simply superposed, as in (6.17).]

[Sec. 6.4] LINEARITY AND SUPERPOSITION 199

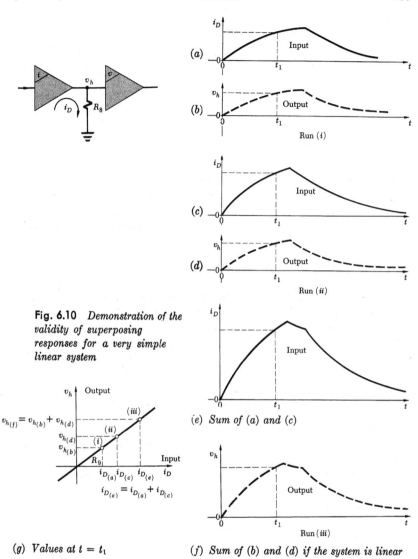

Fig. 6.10 *Demonstration of the validity of superposing responses for a very simple linear system*

(g) *Values at $t = t_1$*

elements that are in tandem, as in Fig. 6.9b, can be interchanged without affecting the overall output. That is, for a given x in Fig. 6.9b the same z results for either arrangement. This property is demonstrated in Problem 6.27.

For linear systems, then, *we can use the power of superposition in two ways:* (i) *to decompose complicated inputs into their simple elements, and* (ii)

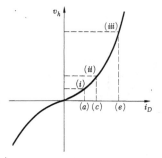

Fig. 6.11 *A nonlinear resistor, for which superposition is not valid*

to decompose complicated systems into their simple elements. In each case the simple elements can be studied separately and the results then superposed.

Problems 6.16 through 6.27

6.5 INITIAL CONDITIONS

Each of the examples in Sec. 6.3 concerns the dynamic response of a physical system following an abrupt disturbance. In each case we have had to consider carefully the "initial" state of the physical system before its complete time response could be determined. Specifically, we chose the *magnitude* of the *natural* motion so that the *total* motion would be correct at "$t = 0$." (That is, for convenience, we took our time origin at the instant the disturbance began.)

Actually, this situation needs to be examined in general with a little more precision, with particular attention to the definition of "$t = 0$." There are really three such times in a typical problem, which we shall designate (rather arbitrarily) as $t = 0^-$, $t = 0$, and $t = 0^+$. The definitions of these times will be clear from the following discussion.

Definition of $t = 0$. Let us choose to call $t = 0$ the exact time when a disturbance (e.g., a step) is applied, as shown in Fig. 6.12.

Definition of $t = 0^-$. The "initial" state of the system is given as

$$\theta = 0 \qquad \dot{\theta} = 0$$

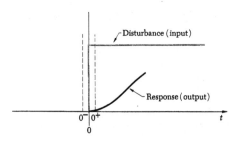

Fig. 6.12 *The meaning of $t = 0$, $t = 0^+$, and $t = 0^-$*

[Sec. 6.5] INITIAL CONDITIONS

but this means the state of the system *before* the step disturbance was applied. It is conventional to refer to this time more precisely as "$t = 0^-$," and to define this as a time before, but infinitesimally close to $t = 0$. This is indicated in Fig. 6.12.

Definition of $t = 0^+$. Finally, the time response we calculate, e.g., Eq. (6.14), holds only for $0 < t$. It does not hold for $t < 0$; and specifically, it also does not hold for $t = 0$. We therefore refer to the earliest time for which our solution holds as "$t = 0^+$," which we define as a time after, but infinitesimally close to $t = 0$. (Thus the specification on p. 195, for example, that the various expressions hold for $0 < t$, may also be written as $0^+ < t$.)

Now the crux of the initial-condition problem is this: "Given the state of the system at $t = 0^-$, what components of the system's state can be specified at $t = 0^+$?" In terms of Fig. 6.7, for example, given $\theta(0^-) = 0$, what is $\theta(0^+)$? We need to answer this question in order to determine the magnitude of the natural part of the time response.†

Variables that do not change instantaneously

The answer to the above question often lies in a reluctance to move on the part of certain of the physical elements involved, and the answer is fortunately almost always clear, once the problem is considered carefully. We have in fact solved the problem of initial conditions for particular cases in the discussion of Sec. 6.3. Let us look again at what we did.

In Fig. 6.7 we noted that "the displacement of a rotor restrained by viscous damping cannot be changed instantaneously." We therefore concluded that, despite the sudden application of a large (but finite) moment at $t = 0$, the value of θ at $t = 0^+$ would be the same as its value at $t = 0^-$:

$$\theta(0^+) = \theta(0^-)$$

and this established the initial condition for us.

Similarly, in electrical systems (e.g., Example 6.2) we may make use of the fact that, despite the possible sudden change in current flowing through a capacitor, the voltage across the capacitor cannot change instantaneously, so that we may write $v(0^+) = v(0^-)$; or the current in an inductor cannot change instantaneously so that $i(0^+) = i(0^-)$. Such relations will normally establish the initial conditions in electrical problems.

Thus, in general, we establish the initial conditions on the basis of various energy-storage (or dissipation) elements whose associated physical variable cannot be changed abruptly by finite disturbances (because the amount of stored energy cannot be changed abruptly). A list of some common elements of this kind, with their physical variables, is given in Table VI. The last two rows represent a useful generalization.

---

† By way of a preview, we shall find in Chapter 17 that when we employ the one-sided Laplace transform this problem does not arise. Initial conditions at $t = 0^-$ may be inserted into the routine solution without further thought.

In general, also, to find the motion of a *first-order system*, the initial value of the independent variable (associated with the energy storage) must be known, but the initial value of its derivative need *not*. (Note the abrupt change in $\dot{\theta}$ between $t = 0^-$ and $t = 0^+$ in Fig. 6.7, for example.)

TABLE VI

Some physical variables that cannot be changed instantaneously by any finite disturbance†

Variable	Element	Cannot be changed instantaneously by any finite
Displacement $\begin{Bmatrix} x \\ \theta \end{Bmatrix}$ of	Mechanical element having viscous restraint or inertia, or both	Force or torque input $\begin{Bmatrix} f \\ M \end{Bmatrix}$
Velocity $\begin{Bmatrix} \dot{x} \\ \dot{\theta} \end{Bmatrix}$ of	Mass or rotary inertia	Force or torque input $\begin{Bmatrix} f \\ M \end{Bmatrix}$
Voltage v across	Capacitor	Current input i
Current i through	Inductor	Voltage input v
Through variable in	T-type storage element‡	Across-variable input
Across variable of	A-type storage element‡	Through-variable input

† Finite as seen by the element in column 2.
‡ Refer to Table IV, p. 134.

Once we have determined which variables will not change between $t = 0^-$ and $t = 0^+$, we have only to establish the values of those variables at $t = 0^-$. Sometimes they are given to us explicitly, as in Example 6.1. In other cases (e.g., Example 6.2 and Example 6.4 below) they are not, and we must analyze the situation prior to $t = 0$ to determine the needed values at $t = 0^-$. Usually this pre-$t = 0$ analysis is a static (rather than dynamic) one and is therefore relatively simple, as it was in Example 6.2, and as it will be found in Example 6.4.

The case of instantaneously changing variables

There are important situations in which a variable *does* change between $t = 0^-$ and $t = 0^+$. The following are two cases frequently encountered: (*i*) the input is a very short impulse (considered to occur within the interval $0 < t < 0^+$); (*ii*) the input commences abruptly (as a step), and the system differentiates the input. When the solution requires knowledge of the value, at $t = 0^+$, of a variable which changes between $t = 0^-$ and $t = 0^+$, then, of course, the magnitude of the change must be calculated. Discussion of how to make such calculations is deferred until Sec. 6.9, following the study of response to an impulse in Sec. 6.8.

[Sec. 6.5] INITIAL CONDITIONS   203

Summary of initial-conditions procedure

We may summarize Sec. 6.5 in the following procedure for establishing initial conditions at $t = 0^+$:

> (1) (a) Determine which variables will not change between $t = 0^-$ and $t = 0^+$.
> (b) Find values for these variables at $t = 0^-$ by inspection or by static analysis.
> (2) If necessary, compute variable changes between $t = 0^-$ and $t = 0^+$, e.g., using knowledge of the system's impulse response.

Procedure B-2
Initial conditions

We close this section with two examples which demonstrate item 1. (Item 2 is demonstrated in Example 6.8.)

Later, in Part D, we shall find that proper employment of the Laplace transform obviates the whole problem of initial conditions. Values of variables at $t = 0^-$ are inserted routinely into the Laplace expressions, and the situation at $t = 0^+$ never comes up. It is valuable, however, first to consider the problem stepwise as we are doing here; we shall then have much more insight into the Laplace results.

Example 6.3. Stopping a sliding box. A box, of mass m, is riding on a railroad flatcar, Fig. 6.13. The flatcar moves at a (practically) constant velocity of

Fig. 6.13 *A box on a flatcar*

3 feet per second for a long time until suddenly it strikes an earth embankment and stops abruptly. What is the velocity, $u(0^+)$, of the box just after the car hits the embankment? Assume any sort of friction between box and flatcar.

Since the flatcar has been moving at constant velocity for a long time, transient motions of the box relative to the flatcar will have vanished prior to impact (given any sort of friction), and the velocity of the box will be the same as that of the car:

$$u(0^-) = 3 \text{ fps}$$

Then, since no "infinite" force is exerted *on the box* during impact, its velocity will not change instantaneously (Table VI), and

$$u(0^+) = u(0^-) = 3 \text{ fps}$$

Example 6.4. Opening a switch. The switch in the network of Fig. 6.14 has been closed for a long time. It is opened suddenly at $t = 0$. Find the voltage drop, $v_3(0^+)$, across the capacitor, and the current, $i_6(0^+)$, in the inductor just after the switch is opened. (Refer to Procedure B-2, item 1.)

(a) From Table VI, $v_3(0^+) = v_3(0^-)$, and $i_6(0^+) = i_6(0^-)$.

(b) To find $v_3(0^-)$ and $i_6(0^-)$ we write the "steady-state" equilibrium equations for $t = 0^-$. By "steady state" we mean that currents and voltages have no time rates of change—no transient motions remain in the network. At $t = 0^-$ the switch is still closed.

(1) *Variables.* Identities: $\quad i_1 \equiv i_A; \quad i_4 \equiv i_6 \equiv i_B; \quad i_2 \equiv i_3 \equiv i_A - i_B$.

(2) *Equilibrium.*

$$\text{KVL} \begin{cases} \text{Loop A:} & v_1 + v_2 + v_3 = v_0 \\ \text{Loop B:} & -v_3 - v_2 + v_4 + v_6 = 0 \end{cases}$$

(3) *Physical relations.*

$$i_3 = C_3 \frac{dv_3}{dt} = 0; \quad v_6 = L_6 \frac{di_6}{dt} = 0 \text{ at } t = 0^-, \text{ by definition of steady state.}$$

(Other physical relations are standard.)

Fig. 6.14 *Network with switch*

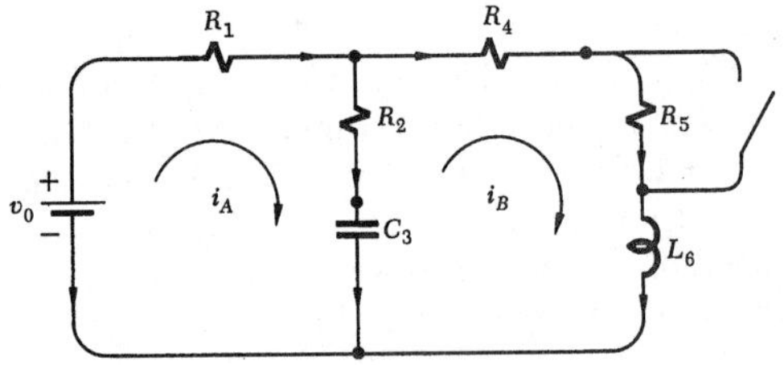

Combining results, we obtain

$$\left. \begin{array}{ll} & i_A = i_B = i_6 \\ \text{Loop A:} & i_6 R_1 + 0 + v_3 = v_0 \\ \text{Loop B:} & -v_3 - 0 + i_6 R_4 + 0 = 0 \end{array} \right\} \quad t = 0^-$$

Solving simultaneously for v_3 and i_6 yields

$$v_3(0^-) = \frac{R_4}{R_1 + R_4} v_0 = v_3(0^+)$$

$$i_6(0^-) = \frac{v_0}{R_1 + R_4} = i_6(0^+)$$

The reader may have been able to arrive at this result more quickly by intuitive reasoning (e.g., by noting that at $t = 0^-$, C_3 is "open" and R_1 and R_4 form a voltage divider for v_0). But should intuition not suffice, the above procedure is straightforward and seldom laborious.

Problems 6.28 through 6.33

°6.6 SPECIAL CASE: THE PURE INTEGRATOR

Many physical devices have the property that they perform almost a pure integration of their input signal. Each such device is a special case of a first-order system. Suppose, for example, that friction on the rotor in

[Sec. 6.6] SPECIAL CASE: THE PURE INTEGRATOR

Fig. 6.1a could be made truly zero.† Then Eq. (6.7) of Example 6.1 would be

$$J\dot{\Omega} = M \tag{6.18}$$

which can be integrated in general to produce

$$\Omega = \frac{1}{J} \int M \, dt \tag{6.19}$$

That is, the speed of the rotor now gives a direct measure of the *integral* of the applied torque.

Another physical device that integrates is the circuit shown in Fig. 6.15. (Recall from p. 77 that an ideal current amplifier converts an input volt-

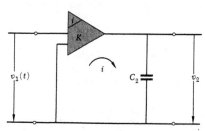

Fig. 6.15 *An electrical circuit that integrates. (Similar circuits are used in analog computers.)*

age to a proportional output current, independent of the output load.) The equation of motion for the circuit in Fig. 6.15 is

$$C_2 \frac{dv_2}{dt} = i = K v_1 \tag{6.20}$$

which, again, can be integrated exactly to give

$$v_2 = \frac{K}{C_2} \int v_1 \, dt \tag{6.21}$$

Circuits based on the principle of Fig. 6.15 are used as integrating elements in analog computers.

A third example of a physical integrating device is the hydraulic system of Fig. 6.1f with the "feedback" link QR removed and with point Q fixed. This system is the subject of Problem 6.38.

An excellent spot check on the response calculated for any first-order system is to consider the special case of setting the appropriate physical constant equal to zero to make the system a pure integrator. For example expression (6.17) gives the response of the system of Fig. 6.2, whose equation of motion is

$$b\dot{\theta} + k\theta = M \tag{6.3}$$
[repeated]

to the input $M = M_0 e^{-\sigma_f t} u(t)$.

---

† Several advanced techniques have, in fact, been developed for making bearings having extremely low friction.

Now if $k = 0$, then the "forced" part of (6.17) becomes

$$\theta = \frac{M_0}{-b\sigma_f} e^{-\sigma_f t} \qquad 0 < t \qquad (6.22)$$

which is indeed the integral of the input torque. (The reader may wish to ponder the significance of the system in Fig. 6.2 when both k and J are zero. It is easiest to consider this as a limiting case, in which b is large so that k and J *approach* zero in their relative importance.)

Natural motion of a pure integrator

If the system of Fig. 6.2 is a pure integrator, that is, if $k = 0$, then its natural response following an initial displacement is given by (6.6) with $k = 0$:

$$\theta = \theta_0 e^{0t} = \theta_0 \qquad 0 < t \qquad (6.23)$$

That is, the system does not move; whatever initial displacement it has simply persists. This is pretty obvious from a physical consideration of Fig. 6.2: If there is no spring, then (in the absence of disturbance) there is no reason for the rotor to return or to move at all.

Similarly, if the rotor in Fig. 6.1a has no viscous friction, and is given an initial velocity, it will simply continue to rotate at that velocity indefinitely:

$$\Omega = \Omega_0$$

as can be seen from Eq. (e) of Example 6.1 with $b = 0$. We can, in fact, make the following general statement: *The natural (unforced) behavior of a pure integrator is simply to retain its initial condition.*

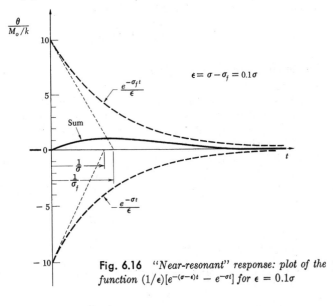

Fig. 6.16 *"Near-resonant"* response: plot of the function $(1/\epsilon)[e^{-(\sigma-\epsilon)t} - e^{-\sigma t}]$ for $\epsilon = 0.1\sigma$

⊙6.7 SPECIAL CASE: RESONANCE

Let us return again to the discussion on p. 195, of the response of a first-order system to an input of the form

$$M = M_0 e^{-\sigma_f t} u(t)$$

that is, an input of the same form as the natural motion of a first-order system, Eq. (6.6). The response in such a case was found to be given by Eq. (6.17), which we repeat here (for the case of zero initial θ). For brevity, let $\sigma = k/b$. Then

$$\theta = \frac{M_0}{b} \frac{e^{-\sigma_f t} - e^{-\sigma t}}{-\sigma_f + \sigma} \qquad 0 < t \qquad (6.24)$$

This is the total response to the abrupt input when $\theta(0) = 0$.

We now raise the question "what happens if $\sigma_f = \sigma$; that is, what happens *when the system is forced with a time function which looks just like its own natural motion.*" This special situation we shall call *resonance*.†

What we find is that, as a matter of fact, nothing unusual happens to the response of a first-order system when σ_f is made equal to σ. The response is well-behaved‡ and not much different from the response when $\sigma_f = 0.9\sigma$, or $\sigma_f = 1.1\sigma$, as we now show.

At first glance we might conclude, because the denominator goes to 0 in (6.24) when $\sigma_f \to \sigma$, that θ will therefore go to ∞. However, when we look more carefully we see that the numerator also goes to 0; that is, (6.24) reduces to an indeterminate form. There are several methods for resolving such an indeterminate form. Perhaps the most meaningful, physically, is to approach the problem from the viewpoint that σ_f is different by a small amount from σ:

$$\sigma_f = \sigma - \epsilon$$

and then investigate the limiting case $\epsilon \to 0$.

For a small but finite ϵ, Eq. (6.24) becomes

$$\theta = \frac{M_0}{b\epsilon} \left(e^{-\sigma_f t} - e^{-\sigma t} \right)$$

which is plotted in Fig. 6.16 for $\epsilon = 0.1\sigma$. That is, to obtain a numerical or graphical solution for this case we must find the time-varying difference between two very large quantities. In the limit the two quantities each become infinitely large, although their difference remains finite.

As an alternative to subtracting large numbers from each other, it is much more accurate to rewrite (6.24) in a different form, using a series

---

† The term "resonance" is used much more commonly in connection with the response of a second-order system to sinusoidal inputs at its natural frequency. See p. 272.
‡ This is not the case for resonance in an undamped, second-order system, which produces quite singular behavior.

expansion for the exponential function at the appropriate point. The algebra involved follows.

To obtain our result in a general form, let us begin with the equation of motion in the following form:

$$\dot{y} + \sigma y = x(t) \qquad (6.25)$$

Then if $x(t) = Xe^{-\sigma_f t}u(t)$, the time response will be

$$y = X\frac{e^{-\sigma_f t} - e^{-\sigma t}}{-\sigma_f + \sigma} \qquad 0 < t$$

Let $\sigma_f = \sigma - \epsilon$. Then

$$\frac{e^{-\sigma_f t} - e^{-\sigma t}}{-\sigma_f + \sigma} = \frac{e^{-\sigma_f t} - e^{-(\sigma_f + \epsilon)t}}{\epsilon}$$

$$= \frac{e^{-\sigma_f t}(1 - e^{-\epsilon t})}{\epsilon}$$

$$= \frac{e^{-\sigma_f t}\left[1 - \left(1 - \epsilon t + \frac{\epsilon^2 t^2}{2!} - \frac{\epsilon^3 t^3}{3!} + \cdots\right)\right]}{\epsilon}$$

$$= te^{-\sigma_f t}\left(1 - \frac{\epsilon t}{2!} + \frac{\epsilon^2 t^2}{3!} - \cdots\right) \qquad (6.26)$$

Then, for $\epsilon = 0$ the correct response is readily obtained, and is certainly well behaved. [It can be shown that as $t \to \infty$, the expression in (   ) approaches unity.]

Exact resonance. To summarize, if a first-order system having the equation of motion

$$\dot{y} + \sigma y = x(t) \qquad (6.25)$$
[repeated]

has an abrupt input with exactly the natural time constant of the system,

$$x(t) = Xe^{-\sigma t}u(t)$$

then the response will be

$$y(t) = Xte^{-\sigma t} + y(0^+)e^{-\sigma t} \qquad 0 < t \qquad (6.27)$$

This response is plotted in Fig. 6.17, for $y(0^+) = 0$.

For the specific system of Fig. 6.2 the response to a resonant input is [by comparison of (6.25) with (6.3)]

$$\theta = \frac{M_0}{b}te^{-\sigma t}u(t)$$

[Sec. 6.7] SPECIAL CASE: RESONANCE

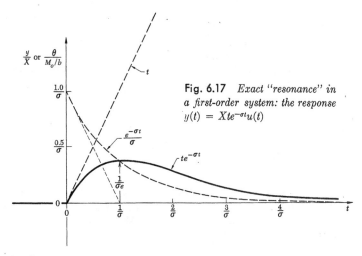

Fig. 6.17 Exact "resonance" in a first-order system: the response $y(t) = Xte^{-\sigma t}u(t)$

This is also indicated in Fig. 6.17. (The solid curve is obtained by multiplying σ times the product of the dashed curves.)

Near resonance

As a practical matter we should not expect to see any abrupt change in the *character* of the response of a physical system when the input to the system is changed only slightly in character, as from $\sigma_f = \sigma - \epsilon$ to $\sigma_f = \sigma$.

As a practical matter also, even when σ_f is slightly different from σ, there is considerable difficulty in making a time plot from an expression like (6.24), as we found in Fig. 6.16, because of the necessity for subtracting very large numbers from each other. It is therefore useful to develop an alternative basis for plotting the response of a first-order system which is disturbed *near* resonance.

For this purpose we return to Eq. (6.26) and then, instead of allowing ϵ to vanish, we merely allow it to become quite small, and drop all but first-order terms in ϵ. The result is

$$y(t) \cong Xte^{-\sigma_f t}\left(1 - \frac{\epsilon t}{2!}\right) \qquad 0 < t \qquad (6.28)$$

From this expression we notice two things: (1) if ϵ is quite small, then the plot for $\epsilon = 0$ is a good approximation for the response of the system, and (2) a simple way to get a more accurate plot is simply to correct the resonant response by multiplying the plot by $(1 - \epsilon t/2!)$. If ϵ is small, say $\epsilon = 0.1\sigma$, this is an easier and much more accurate method of plotting than that used in Fig. 6.16. Notice that at the maximum-amplitude point ($t = 1/\sigma$ in Fig. 6.17) the correction is only

$$\frac{1}{2}\frac{\epsilon}{\sigma}$$

For example, if $\epsilon = 0.1\sigma$, the correction is only 5 percent. [The percent correction increases with time, of course, but the response itself becomes small anyway after a few time constants ($1/\sigma$'s) and is usually of less interest.]

The reader is invited to study this correction further in Problem 6.35.

Resonance in a pure integrator

As we learned in Sec. 6.6, a pure integrator $\dot{y} = x(t)$ has the characteristic equation $s = 0$ and therefore the natural motion $y = Ye^{0t}$. Resonance for such a system occurs when the input is of the form $x(t) = Xe^{0t}u(t)$. That is, the "resonant input" to an integrator is a step.

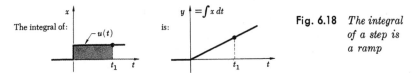

Fig. 6.18 *The integral of a step is a ramp*

The response of an integrator to a step is intuitively obvious, Fig. 6.18. That is, the value of $\int x\, dt$ at time t_1 is the area under the x plot up to that time.

It is instructive to obtain this same answer as a special case of (6.27), with $\sigma = 0$:

$$y = Xtu(t)$$

Problems 6.34 through 6.45

6.8 RESPONSE TO A VERY SHORT IMPULSE

We have considered the natural motion of first-order systems and their response to several typical disturbances or inputs. We now want to consider response to a special kind of input known as an impulse—an input which has substantial amplitude but which is of short duration, as shown in Fig. 6.19. The "magnitude" of an impulse is defined as its area in a time plot: $A = x_0 \Delta t$ in Fig. 6.19.

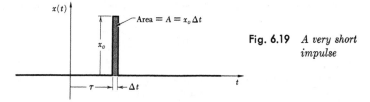

Fig. 6.19 *A very short impulse*

The response to an impulse has a very special place in the study of physical system dynamics, not only because impulsive inputs occur commonly in physical problems, but even more because they provide (1) an elegant link between the study of natural and of forced motion, and (2) a means for finding response to completely arbitrary inputs.

We shall consider the impulse response of a system, first from the point of view of natural response, and then (in Sec. 6.10) from the point of view of forced response. We shall find on the one hand that the response of a first-order system to an impulse at $t = 0$ is the same as its *natural*

[Sec. 6.8] RESPONSE TO A VERY SHORT IMPULSE

motion following an initial condition. On the other hand, we shall find that the *forced* response to an arbitrary input can be computed as the response to a series of impulses.

The unit impulse

A unit impulse—denoted by $\delta(t - \tau)$ [and in particular, if it occurs at $t = 0$, by $\delta(t)$]—is basically a mathematical device obtained by reducing the width of a real impulse until it occupies a time which is very short compared to characteristic time constants of the system being studied, but at the same time maintaining the magnitude (area) of the impulse unity. Formally, the unit impulse is defined by

$$\delta(t - \tau) = \lim_{\Delta t \to 0} a[u(t - \tau) - u(t - \tau + \Delta t)] \quad \text{with } a\Delta t = 1 \quad (6.29)$$

The concept is illustrated in Fig. 6.20, which shows the response of the system of Fig. 6.2 to an applied unit moment impulse, $M_1 \Delta t = 1$.

When the impulse is of long duration, Fig. 6.20a, the element has time to reach its steady-state position, M_1/k, before the moment M_1 is removed. As the duration of the impulse is shortened, Fig. 6.20b, M_1 is increased so that the magnitude of the impulse (area under the M plot) continues to equal unity. Now the impulse is terminated while the element still has quite high velocity.

Finally, in Fig. 6.20c, the duration of the impulse is reduced until it is very short compared to the time constant, b/k, of the element, so that during the impulse the element moves essentially at the constant (initial) velocity M_1/b. The position of the element when the moment is removed is then velocity $\times \Delta t = (M_1/b) \Delta t$. But $M_1 \Delta t = 1$, by definition, so that the position at the end of the impulse is just $1/b$, *independent of the exact duration of the impulse* (provided that it is very short).

After the impulse is over, the system is then unforced, and it exhibits its usual natural motion from the initial condition $\theta(0^+) = 1/b$:

$$\theta(t) = \frac{1}{b} e^{-(k/b)t} \qquad 0 < t \quad (6.30)$$

That is, from Fig. 6.20 we can make the following statement:

> *The response of a first-order system to a unit impulse is identical with its natural motion from an initial position* [of magnitude $\theta(0^+) = 1/b$ in this case]. *The impulse may be considered to generate the initial condition* in such a short time that it has no other effect on the system. Physically, the element is "jarred" to an initial position by the impulse.

This is a result of major importance, because it permits us to regard natural motions as forced motions if we wish. This result also means that *a linear system can be characterized by its unit-impulse response, just as it can be characterized by its equation of motion.*

It is important to recognize a singular property of the impulse as a forcing function: *the response to an impulse does not have separate "forced" and "natural" components; there is only the total response.*

> Mathematically speaking, what we are saying is that if a system has a first-order equation of motion which can be written in the form
> $$\dot{y} + \sigma y = x(t) \tag{6.31}$$
> then the response of the system to an impulse $x(t)$ of magnitude A,
> $$x(t) = A\,\delta(t) \tag{6.32}$$
> will always be the same as its response to the initial state $y(0^+) = A$:
> $$y = Ae^{-\sigma t}u(t) \tag{6.33}$$

The *unit-impulse* response of a system is sometimes given the special symbol $h(t)$. This symbol brings forth an image of the whole time response of the system, vs. t, following a unit impulse. Thus, Fig. 6.21a shows the image represented by symbol "$h_2(t)$" for the system of Fig. 6.2.

Special case: the value of $y(0^+)$ is given by

$$\boxed{y(0^+) = A} \tag{6.34}$$

This result is sometimes quite useful in establishing initial conditions when there is an abrupt input and the system differentiates it.

More generally, for a first-order system represented by (6.31), the unit-impulse response is

$$h(t) = e^{-\sigma t}u(t) \tag{6.35}$$

[That is, (6.35) is the response for $A = 1$ in (6.33).]

It is also sometimes convenient to represent elementary systems, like the one in Fig. 6.2, as building blocks in larger compound systems, as in Fig. 6.21b. In such "block diagrams"† the dynamic properties of the individual subsystems can be represented in a number of ways. The differential equation of motion of each could be written in its block. Alternatively, the impulse response $h(t)$ of each subsystem can be used, as shown in Fig. 6.21b. (There are some other useful ways, which we shall learn later.)

So far we have considered only the response to impulses occurring at $t = 0$. If an impulse occurs at some other time, $t = \tau$, say, this involves nothing new, of course; the response of the system is the same, but is delayed by τ seconds, as indicated in Fig. 6.22.‡

The analytical expressions, (6.30), (6.32), and (6.33), can therefore be made to represent response to an impulse at $t = \tau$ merely by replacing

---

† We introduced the block-diagram idea in Sec. 1.3 (see Figs. 1.3 and 1.4).
‡ Here, τ is not the time constant of a system.

[Sec. 6.8] RESPONSE TO A VERY SHORT IMPULSE 213

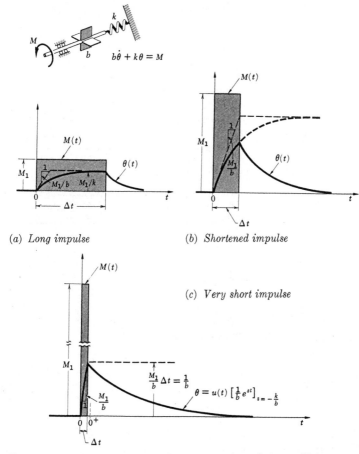

(a) Long impulse

(b) Shortened impulse

(c) Very short impulse

Fig. 6.20 *Evolution of unit-impulse response: in each figure, $M_1 \Delta t = 1$*

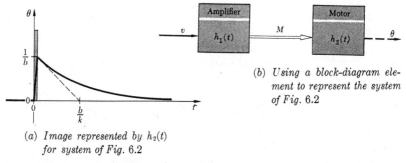

(a) Image represented by $h_2(t)$ for system of Fig. 6.2

(b) Using a block-diagram element to represent the system of Fig. 6.2

Fig. 6.21 *Use of the symbol $h(t)$, representing system response to a unit impulse*

t by $(t - \tau)$ everywhere in the response:

$$\theta(t) = \frac{1}{b} e^{-\frac{k}{b}(t-\tau)} \qquad \tau < t$$

or

$$\theta(t) = \frac{1}{b} e^{-\frac{k}{b}(t-\tau)} u(t - \tau) \tag{6.36}$$

Impulse response of any linear system
Input: $\quad v(t) = A\delta(t - \tau)$ (6.37)
Response: $\quad y = Ah(t - \tau)$ (6.38)

in which, for a first-order system, from (6.35),

Unit-impulse response of a first-order system
$$h(t - \tau) = e^{-\sigma(t-\tau)} u(t - \tau) \tag{6.39}$$

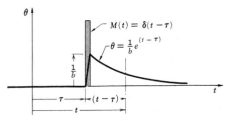

Fig. 6.22 *Response to a unit impulse at $t = \tau$*

Fig. 6.23 *Impulses are often represented by vertical arrows*

There is one more thing we need to say to complete our formal description of the unit impulse, and this is that we so define $t = 0^+$ (somewhat arbitrarily) that a unit impulse $\delta(t)$ which occurs at time $t = 0$ shall be completed before time $t = 0^+$. (Refer to Sec. 6.5.) This is indicated in Fig. 6.20c, for example. The utility of this custom is that then *motion after $t = 0^+$ is exactly the same for response to an impulse as it is for response following an initial condition.*

Impulses are often represented by vertical arrows, as in Fig. 6.23, to emphasize that, in the mathematical limit, they have "zero" thickness and "infinite" height [per Eq. (6.29)].

Example 6.5. Impulse response of a damped rotor. Find the response of the rotor of Fig. 6.1a to a torque impulse of magnitude 10 in.-lb-sec, if the rotor inertia is $J = 3$ (in. lb)/(rad/sec²) and the viscous damping restraint is $b = 0.15$ (in. lb)/(rad/sec), as they were in Example 6.1. What magnitude of impulse would be required to produce the time response plotted in Fig. 6.5 [i.e., to "generate" the initial condition $\theta(0^+) = 1200$ rpm]?

[Sec. 6.8] RESPONSE TO A VERY SHORT IMPULSE 215

STAGES I AND II. The equation of motion, (6.7), for a physical model of this system was derived in Example 6.1:

$$J\dot{\Omega} + b\Omega = M \qquad (6.7)$$
[repeated]

In the present example we are given

$$M(t) = (M_{max} \Delta t) \delta(t) = 10 \delta(t)$$

STAGE III. To find the time response, it is convenient to put (6.7) in the form (6.31) and (6.32):

$$\dot{\Omega} + \frac{b}{J} \Omega = \frac{M_{max} \Delta t}{J} \delta(t)$$

and then to apply (6.33):

$$\Omega = \frac{M_{max} \Delta t}{J} e^{-(b/J)t} u(t) \qquad (6.40)$$

or, with the numerical values given,

$$\Omega = \frac{10}{3} e^{-t/20} u(t) \text{ rad/sec}$$

(Do the units check?)

To answer the second question above, we note that the initial condition in Example 6.1 was

$$\Omega(0^+) = 1200 \text{ rpm} = 126 \text{ rad/sec}$$

which, from (6.34), can be established with an impulse of magnitude

$$\frac{M_{max} \Delta t}{J} = 126$$

$$(M_{max} \Delta t) = 126J = 378 \text{ in.-lb-sec}$$

This is a pretty substantial "jar" to give the rotor, even though the impulse can be applied over a period of 2 or 3 seconds without violating the spirit of an impulse, the system time constant being 20 seconds. That is, the "impulse" could consist of $M_{max} = 189$ in.-lb, $\Delta t = 2$ sec, for example.

Example 6.6. A telegraph circuit. A simplified physical model of a telegrapher's key is shown in Fig. 6.24. (Coil L_2 represents the solenoid which actuates the mechanical "clicker"; a more realistic model would include the electromechanical portion of the system.) For the network shown, if the telegrapher sends a series of "dots" by depressing the key for 0.05 sec, find the current in L_2 that results. Let $R_1 = 10$ ohms, $L_2 = 1$ henry, $R_3 = 10,000$ ohms.

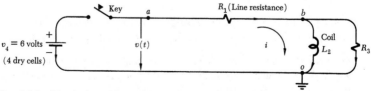

Fig. 6.24 *Physical model of a telegraph circuit*

Starting with the given physical model (network) and neglecting R_3† we derive the equations of motion (Procedure A-e):

(1) *Variables:* $i_1 \equiv i_2 \equiv i$
(2) *KVL:* $v_1 + v_2 = v(t)$
(3) *Physical relations* (insert in KVL):

$$R_1 i + L_2 \frac{di}{dt} = v(t) \tag{a}$$

For one impulse, taken at $t = 0$,

$$v(t) = (v_4 \, \Delta t) \, \delta(t) \tag{b}$$

Rewriting (a) in the form of (6.31), we have

$$\frac{di}{dt} + \frac{R_1}{L_2} i = \frac{v_4 \, \Delta t}{L_2} \delta(t) \tag{c}$$

Then the response, by comparison with (6.33), is

$$i = \frac{v_4 \, \Delta t}{L_2} e^{-(R_1/L_2)t} u(t) \tag{d}$$

where the residual current just before a "dot" is neglected; or, with numerical values,

$$i = \frac{6 \times 0.05}{1} e^{-t/0.1} u(t) = 0.3 e^{-t/0.1} u(t) \text{ amp} \tag{e}$$

after a "dot" sent at $t = 0$. This is the same as the response to an initial current $i(0^+) = 0.3$ amp in the coil. For well-spaced dots, the same response will follow each one. For closely spaced dots, responses must be superposed (see Problem 6.55).

Impulse response of a pure integrator

If the input to an integrator is $x(t) = A \, \delta(t) = (x_0 \, \Delta t) \, \delta(t)$, as in Fig. 6.25, then we expect the output of the integrator to be simply the area

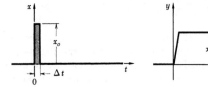

Fig. 6.25 *The response of an integrator to an impulse is a step output*

under this curve:

$$y = \int_0^t A \, \delta(t) \, dt = A u(t) = (x_0 \, \Delta t) u(t) \tag{6.41}$$

That is, *an impulse input to an integrator produces a step output.*

This result is also seen as a special case of (6.33) with $\sigma = 0$.

Example 6.7. Impulse response of a frictionless, free rotor. Repeat Example 6.5 assuming the bearings of the rotor in Fig. 6.1a to be frictionless (i.e., assuming that the rotor is a pure integrator).

† The effect of doing so is to be studied in Problem 6.47.

[Sec. 6.8] RESPONSE TO A VERY SHORT IMPULSE 217

The answer is available at once from (6.40) with $b = 0$:

$$\Omega = \frac{M_{\max} \Delta t}{J} u(t) = \frac{10}{3} u(t) \text{ rad/sec}$$

The derivative of a step or other abrupt function

It is useful to consider the inverse of the preceding discussion, and to note that, since the integral of an impulse is a step, the derivative of a step must be an impulse. From (6.41) we can write

$$\frac{d}{dt}[Au(t)] = A\,\delta(t) \qquad (6.42)$$

As a matter of fact, the derivative of *any* abruptly commencing function must involve an impulse. For example, the derivative of the function $Ae^{-\sigma t}u(t)$ is found to be

$$\frac{d}{dt}[Ae^{-\sigma t}u(t)] = A[\delta(t) - \sigma e^{-\sigma t}u(t)] \qquad (6.43)$$

as can be proved by integrating, for example:

$$\int_{-\infty}^{t} A\,\delta(t)\,dt - \int_{-\infty}^{t} A\sigma e^{-\sigma t}u(t)\,dt = Au(t) - A(1 - e^{-\sigma t})u(t)$$

The most convincing way to see (6.43) is to sketch the time function $e^{-\sigma t}u(t)$ and beneath it sketch its derivative as given by (6.43). Then it is obvious that the $\delta(t)$ term in (6.43) corresponds to the infinite slope of $e^{-\sigma t}u(t)$ at $t = 0$.

Is an impulse a finite disturbance?

It is important to point out that an impulse, unlike the "finite" disturbances of Sec. 6.5, will indeed change "instantaneously" the velocity of a mass, the current in an inductor, or the like, in the spirit of our mathematical definition of an impulse. In Example 6.5 the angular velocity of the rotor [of moment-of-inertia 3 in.-lb/(rad/sec^2)] was changed "instantaneously" from $\Omega(0^-) = 0$ to $\Omega(0^+) = \frac{10}{3}$ rad/sec. In Example 6.6 the current in coil L_2 was changed "instantaneously" from $i(0^-) = 0$ to $i(0^+) = 0.3$ amp.

Mathematically, the reason this can happen is that an ideal impulse has finite area but "zero" duration (Fig. 6.20c) (it occurs entirely within the interval $0 < t < 0^+$). It must therefore have "infinite" height.

In practice impulses have only finite height, but they also have finite—albeit short—duration, so that they do have some time in which to impart velocity to masses, charge to capacitors, and so forth. We simply consider that they do so between $t = 0$ and $t = 0^+$.

The point is that, on one ground or another, impulses are specifically exempted from the category of "finite disturbances" in the spirit of Table VI, p. 202. So, for exactly the same reasons, are the derivatives of steps.

Problems 6.46 through 6.58

⊙6.9 INITIAL CONDITIONS INVOLVING SUDDEN CHANGE

We can now return to item 2 of Procedure B-2 (p. 203): the question of how to calculate the value of a variable at $t = 0^+$ when the variable changes "instantaneously" between $t = 0^-$ and $t = 0^+$. The usual reason for wishing to do so is that the value of the variable at $t = 0^+$ is needed as an initial condition in its time solution.

Our treatment here will not be comprehensive. We shall, however, consider the two most common cases of sudden change: (a) the input is an impulse, and (b) the input is abrupt and the system differentiates it. These two situations are readily handled as follows:

(a) The amount by which an impulse causes a first-order system to change between $t = 0^-$ and $t = 0^+$ is obtained from (6.34).

(b) The amount by which the system changes between $t = 0^-$ and $t = 0^+$ because it has differentiated an abrupt input is obtained from an expression like (6.43).

These are demonstrated in Examples 6.5 through 6.8.

More generally, of course, we may be interested also in the effect on initial conditions at $t = 0^+$ of higher derivatives of abrupt functions, or in other singular functions—the derivative of an impulse, for example. We elect to defer consideration of those situations to Part D, where we shall find that they are handled neatly and routinely when the Laplace transform method is employed. Here in Part B we shall be content with demonstrating, in Example 6.8, the solution for the first derivative of an abrupt function, to indicate how derivatives may be handled.

We now consider a system which differentiates its input.

Example 6.8. Step response of a differentiating system. A physical model is shown in Fig. 6.26 of a system in which mass is neglected and small motions assumed. (The units of b_1 and b_2 are different, b_1 being rotary damping and b_2 translational.)

The input to the system, motion $x(t)$, is an abrupt function:

$$x = x_0 u(t) e^{-\sigma_f t} \qquad (6.44)$$

Find the response if the initial position of the arm is $\theta(0^-) = \theta_0$.

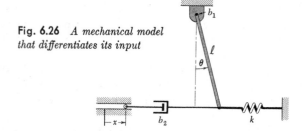

Fig. 6.26 *A mechanical model that differentiates its input*

[Sec. 6.9] INITIAL CONDITIONS INVOLVING SUDDEN CHANGE

STAGE II. EQUATION OF MOTION. The reader may show that (for small values of θ) the equation of motion for the above physical model is

$$(b_1 + b_2 \ell^2)\dot\theta + k\ell^2 \theta = \ell b_2 \dot x(t) \qquad (6.45)$$

STAGE III. RESPONSE. After $t = 0^+$ this can be written as [letting $b = (b_1/\ell) + b_2 \ell$]

$$\dot\theta + \frac{k\ell}{b}\theta = \frac{b_2}{b}(-\sigma_f x_0 e^{-\sigma_f t}) \qquad 0^+ < t \qquad (6.46)$$

where discontinuity at $t = 0$ has been avoided by considering only $0^+ < t$.

The solution to (6.46) is (Procedure B-2)

$$\theta = \theta_f + \theta_n$$

in which θ_f has the form of the forcing function:

$$\theta_f = \Theta_f e^{-\sigma_f t} \qquad 0^+ < t$$

and Θ_f is readily shown (by substitution) to be

$$\Theta_f = -\frac{b_2}{b}\frac{x_0 \sigma_f}{-\sigma_f + (k\ell/b)} \qquad (6.47)$$

The form of the natural solution is readily found, and thence the total motion:

$$\theta = -\frac{b_2}{b}\frac{x_0 \sigma_f}{-\sigma_f + (k\ell/b)} e^{-\sigma_f t} + \Theta_n e^{-(k\ell/b)t} \qquad 0^+ < t \qquad (6.48)$$

To find Θ_n we must first find $\theta(0^+)$. We know $\theta(0^-)$, but there is an abrupt change in θ between $t = 0^-$ and $t = 0^+$ because the system differentiates an abrupt input. From (6.43), the impulsive part of the input is $b_2 x_0 \delta(t)$. Then, using (6.34), we see that this will give a change in θ, between $t = 0^-$ and $t = 0^+$, of

$$\theta(0^+) - \theta(0^-) = \frac{b_2 x_0}{b} \qquad (6.49)$$

and thus the total value of $\theta(0^+)$ is

$$\theta(0^+) = \theta_0 + \frac{b_2}{b}x_0$$

This can now be used in (6.48) at $t = 0^+$ to calculate Θ_n:

$$\theta(0^+) = -\frac{b_2}{b}\frac{x_0 \sigma_f}{-\sigma_f + (k\ell/b)} + \Theta_n = \theta_0 + \frac{b_2}{b}x_0$$

Then the total motion is

$$\theta = -\frac{b_2}{b}\frac{\sigma_f x_0}{-\sigma_f + (k\ell/b)} e^{-\sigma_f t} + \left(+\frac{b_2}{b}\frac{x_0 \sigma_f}{-\sigma_f + (k\ell/b)} + \theta_0 + \frac{b_2}{b}x_0\right) e^{-(k\ell/b)t}$$
$$0^+ < t \qquad (6.50)$$

It is evident that there are two possible approaches to finding the response when a system which differentiates its input is disturbed by an abrupt function:

(i) obtain the solution for $0^+ < t$ to avoid discontinuities at $t = 0$, and calculate initial conditions for $t = 0^+$;

(ii) obtain the solution for $0^- < t$ including discontinuities, and use the given initial conditions at $t = 0^-$.

In Example 6.8 we used approach (i). The solution can be checked by using method (ii), which is suggested in Problem 6.59.

> Problems 6.59 through 6.62

⊙6.10 GENERALIZATION TO AN ARBITRARY INPUT: CONVOLUTION

We have, in the present chapter, attempted to develop a fairly complete understanding of the response of first-order systems to initial conditions and to certain kinds of inputs or disturbances, namely, those of the form

$$Xe^{st}$$

in which s may equal 0 (constant input), or s may equal $-\sigma_f$ (exponential decay). Actually, response to these and other inputs of the form e^{st} (see Chapter 8) constitute remarkably good coverage of situations of physical interest, and lead to quite thorough understanding of the dynamic behavior of simple systems.

However, it is of philosophical interest to know that the response of a linear system to *any input whatever* can be obtained by extension of the methods of the present chapter, using a process called convolution. We present the process here in connection with first-order systems.

Convolution is a somewhat sophisticated procedure based on a simple concept. The concept is that an arbitrary input, like Fig. 6.27a, can be thought of as the sum of a sequence of impulses (two of which are shaded in Fig. 6.27a), and the response of a *linear* system to the whole input can be found as the superposition of its response to the individual impulses. [The response to one impulse is shown in Fig. 6.27b. It was calculated in (6.38).]

Analytical definition

The remarkable thing is that this concept is also simple to state analytically. The magnitude (area) of a shaded impulse of width $d\tau$ occurring at time $t = \tau$ is

$$A = x(\tau)\, d\tau \tag{6.51}$$

and the input function $x(t)$, for this impulse alone, is

$$x(t) = A\, \delta(t - \tau) = [x(\tau)\, d\tau]\, \delta(t - \tau) \tag{6.52}$$

The corresponding response of a first-order system represented by the equation

$$\dot{y} + \sigma y = x(t) \tag{6.31}$$
[repeated]

to this particular impulse is, by (6.38) and (6.39), as follows:

[Sec. 6.10] GENERALIZATION TO ARBITRARY INPUT: CONVOLUTION 221

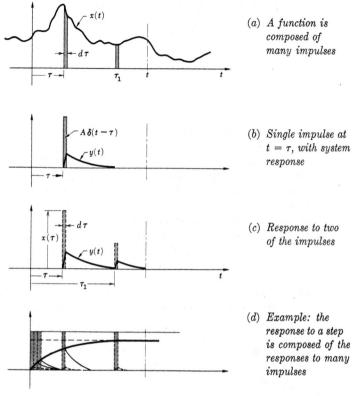

Fig. 6.27 *Why convolution works*

For a single impulse at $t = \tau$:

$$y(t) = Ah(t - \tau) \\ = [x(\tau)\, d\tau][e^{-\sigma(t-\tau)}u(t - \tau)] \quad (6.53)$$

This response is shown in Fig. 6.27b.

If we consider now another of the impulses which make up the total input—say, the one at $t = \tau_1$, $A = x(\tau_1)\, d\tau$—then the response to this one can simply be added to the other as in Fig. 6.27c.

To find, at a given time t, the total response due to the entire input, it is merely necessary to add up the individual responses to *all* of the individual impulses up to time t; that is, to integrate [refer to (6.53)]:

For all impulses together:
$$y = \int_{\tau=-\infty}^{\tau=t} [x(\tau)\, d\tau]h(t - \tau) \\ \text{or } y = x(t) * h(t)$$
(6.54)

Expression (6.54) is known as the "convolution integral," and has many uses in mathematics. The * notation is convenient. It should be clear from the symmetry of the integral that convolution is commutative. That is, $h(t) * x(t) = x(t) * h(t)$.

For the first-order system we have found that [from (6.39)]:

$$h(t - \tau) = e^{-\sigma(t-\tau)}u(t - \tau) \qquad (6.39)$$
[repeated]

and this is the expression we would insert in (6.54) for a first-order system.

Moreover, Equation (6.54) is true in fact, for *any* linear system whose impulse response $h(t - \tau)$ is known, and represents not only a key philosophical concept but a most important tool for finding the time response of linear systems. When teamed with the computational capacity of a modern digital computer, it is in fact a tool of tremendous power.

To demonstrate the application of (6.54) we use it to calculate—in Examples 6.9 and 6.10, and Problems 6.63 through 6.72—the response of a first-order system to a number of inputs. In some cases we can readily check the results against calculations we have made previously.

Example 6.9. Response of restrained rotor by convolution. Use convolution to find the response of the system model in Fig. 6.2b to the moment input:

$$M = M_0 e^{-\sigma_f t} u(t)$$

The equation of motion of the system is [from (6.3)]

$$b\dot{\theta} + k\theta = M$$

or, in the form of (6.31),

$$\dot{\theta} + \frac{k}{b}\theta = \frac{M}{b}$$

We know that, for this first-order system, the unit impulse response is, by (6.39):

$$h(t - \tau) = e^{-(k/b)(t-\tau)}u(t - \tau)$$

The response to the given input, M/b, is therefore, by (6.54),

$$\theta = \int_{\tau=-\infty}^{\tau=t} \left[\frac{M_0}{b} e^{-\sigma_f \tau} u(\tau) \, d\tau \right] e^{-(k/b)(t-\tau)} u(t - \tau)$$

$$= \frac{M_0}{b} e^{-(k/b)t} \int_{\tau=0}^{\tau=t} e^{[-\sigma_f + (k/b)]\tau} u(t - \tau) \, d\tau$$

$$= \frac{M_0}{b} e^{-(k/b)t} \left[\frac{u(t-\tau) e^{[-\sigma_f + (k/b)]\tau}}{-\sigma_f + (k/b)} \right]_{\tau=0}^{\tau=t}$$

$$= \frac{M_0}{b} e^{-(k/b)t} \left[\frac{e^{[-\sigma_f + (k/b)]t} - 1}{-\sigma_f + (k/b)} \right] \qquad 0 < t$$

$$= \frac{M_0}{b} u(t) \left(\frac{e^{-\sigma_f t} - e^{-(k/b)t}}{-\sigma_f + (k/b)} \right)$$

which checks with (6.17) for $\theta(0) = 0$.

[Sec. 6.10] GENERALIZATION TO ARBITRARY INPUT: CONVOLUTION 223

Example 6.10. Response to a triangular input. If a moment having the time history shown in Fig. 6.28 is applied to the system of Fig. 6.2b, compute the system response during the time interval $T < t < 2T$. (System response for other time intervals is the subject of Problem 6.71.)

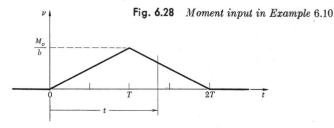

Fig. 6.28 *Moment input in Example* 6.10

The given input is simple to describe graphically, but is awkward to handle by the analysis methods of Secs. 6.3 through 6.6. Convolution will be found to be simpler.

The function shown in Fig. 6.28 can be written [for use with (6.31) and (6.39)]:

$$x(t) = \frac{M(t)}{b} = \begin{cases} \frac{M_0}{b}\left(\frac{t}{T}\right) & 0 < t < T \\ \frac{M_0}{b}\left(2 - \frac{t}{T}\right) & T < t < 2T \\ 0 & 2T < t \end{cases} \quad (a)$$

or, using the unit-step notation,

$$x(t) = \frac{M(t)}{b} = \frac{M_0}{b}\left\{\frac{t}{T}[u(t) - u(t-T)] + \left(2 - \frac{t}{T}\right)[u(t-T) - u(t-2T)]\right\} \quad (b)$$

To apply convolution to the problem, we are to substitute (a) or (b) into the following expression, from (6.54)

$$\theta = \int_0^t \frac{M(\tau)}{b} h(t-\tau)\, d\tau \quad (c)$$

For the particular time interval requested, $T < t < 2T$, this can be written simply

$$\theta = \int_0^T \frac{M_0}{b}\left(\frac{\tau}{T}\right) h(t-\tau)\, d\tau + \int_T^t \frac{M_0}{b}\left(2 - \frac{\tau}{T}\right) h(t-\tau)\, d\tau \quad (d)$$

Recalling from Example 6.9 that, for this system, $h(t-\tau) = e^{-(k/b)(t-\tau)} u(t-\tau)$, and performing the indicated integrations, we obtain finally

$$\theta = \frac{M_0}{k}\left(\frac{b/k}{T}\right)\left\{\left[1 + \left(\frac{k}{b}T - 1\right)e^{(k/b)T}\right]e^{-(k/b)t} \right. \\ \left. + \left[\frac{2T}{b/k}(1 - e^{-(k/b)(t-T)}) - \left(\frac{k}{b}t - 1\right) + \left(\frac{k}{b}T - 1\right)e^{-(k/b)(t-T)}\right]\right\} \quad (e)$$

in which the first [ ] term is the response to the total moment impulse delivered between times $t = 0$ and $t = T$, and the second [ ] term is the response to the total moment impulse delivered between time $t = T$ and $t = t$. The terms of (e) may be collected conveniently as follows:

$$\frac{M_0}{k} \left\{ \frac{b/k}{T} (1 - 2e^{-(k/b)(t-T)} + e^{-(k/b)t}) + \left(2 - \frac{t}{T}\right) \right\} \qquad T < t < 2T \qquad (f)$$

Graphical interpretation

Further insight into why convolution works can be gained by a little further graphical development. As we have seen, what convolution does is to consider that at a given time t the total response due to input $x(t)$ is made up

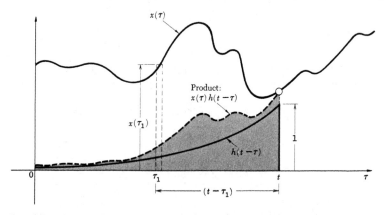

Fig. 6.29 *Graphical meaning of convolution. The convolution integral determines the area under the product of the two plots*

of the response to many individual impulses (at various times τ) having width $d\tau$ and height $x(\tau)$. The response to one such impulse depends upon two things (Fig. 6.29): (1) *the height* $x(\tau)$ *of the impulse*, and (2) *how long ago* $(t - \tau)$ it occurred. The convolution integral adds up all these individual responses, and it takes into account both factors in the manner shown in Fig. 6.29, which is plotted versus τ instead of versus t.

Figure 6.29 shows plots of $x(\tau)$ and of $h(t - \tau)$ for a given time t. [The impulse response, $h(t - \tau)$ runs "backward" on the τ plot.] What the convolution integral does is *first* to multiply these two plots together to obtain their product (dotted plot). Thus the contribution of the "impulse" shown at a particular time τ_1 is taken care of on both counts: (1) its height, $x(\tau_1)$, and (2) its age, $(t - t_1)$. (As Evans puts it, convolution asks each impulse two questions: its height and its age.) *Second*, after the two curves have been multiplied together, the convolution integral simply finds the area under their product (shaded area), which is the response of the system to all of the impulses up to time t.

Problems 6.63 through 6.72

Chapter 7

UNDAMPED SECOND-ORDER SYSTEMS: FREE VIBRATIONS

In Chapter 6 we dealt with motions which decay exponentially with time, slowly approaching zero without overshoot, as in Fig. 6.3. In the present chapter we shall introduce a quite different-appearing motion, the sinusoidal oscillation shown in Fig. 7.1. Such vibratory motion is exhibited by a pendulum swinging back and forth, by a violin string, by a mass bouncing on the end of a spring, by the current in an oscillator circuit, and by a host of other systems (see Fig. 7.2).

Vibratory motion is the second basic building block we shall use in studying the dynamic behavior of linear systems. In the present chapter we shall consider the special case of free vibration without damping, which is a purely sinusoidal motion like that in Fig. 7.1. In Chapter 8 we shall consider damped vibrations (Fig. 8.1d) and both free and forced motions.† In Chapter 9 we shall study the response when sinusoidal inputs are applied to simple systems.

New concepts

The equation of free motion for an undamped, second-order system has the form

$$\boxed{\ddot{\theta} + \omega^2 \theta = 0} \qquad (7.1)$$

---

† Readers who prefer to start with the general case may read Chapter 8 before Chapter 7 without loss in continuity.

226 UNDAMPED SECOND-ORDER SYSTEMS: FREE VIBRATIONS [Chap. 7]

In this chapter the following new concepts will appear:

(1) The natural motion of a system having two different forms of energy storage and no damping is an undamped, sinusoidal oscillation:

$$\theta = \theta_m \cos(\omega t - \psi) \tag{7.2}$$

(2) Sinusoidal motion can be represented in the form

$$\theta = Ce^{st}$$

by considering s as a complex number.

(3) The basis for the manipulation of complex numbers is Euler's equation:

$$\textit{Euler's equation} \quad \boxed{e^{j\phi} = \cos\phi + j\sin\phi} \tag{7.3}$$

$$j \triangleq \sqrt{-1}$$

7.1 PHYSICAL VIBRATIONS

A system having two separate energy-storage elements is called a second-order system. A common system of this class is shown in Fig. 7.1a, in which energy is stored both in the spring (potential energy) and in the rotating mass (kinetic energy).

The motion of a system containing two energy-storage elements can be described by a second-order differential equation—i.e., an equation in one variable and its first two derivatives. We consider here the case of natural (unforced) motion. Forced motion of undamped, second-order systems will be discussed as a special case in Chapters 8 and 9.

Natural (unforced) motion

The system modeled (Stage I) in Fig. 7.1a is a special case of that in Fig. 2.5: damping is negligible and there is no disturbing moment.

We can obtain its equation of motion (Stage II) quickly from Procedure A-m:

(1) *Geometry.* θ is arbitrarily taken positive in the direction shown. $\dot{\theta}$ and $\ddot{\theta}$ are *necessarily* taken positive in the same direction.

(2) *Force equilibrium,* and (3) *physical force-geometry relations.* (i) A free-body diagram of the rotor is shown in Fig. 7.1b. Only two moments act on this free body, a spring torque, $M_s = k\theta$, which opposes θ (as shown), and a D'Alembert inertial torque $M_i = J\ddot{\theta}$, which opposes $\ddot{\theta}$ (as shown). (ii) By equilibrium relation (2.2) these sum to zero. Taking moments to be positive in the clockwise direction, we have

$$\begin{aligned}\Sigma M^* = M_i + M_s &= 0 \\ J\ddot{\theta} + k\theta &= 0\end{aligned} \tag{7.4}$$

[Sec. 7.1] PHYSICAL VIBRATIONS 227

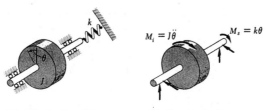

(a) Physical model   (b) Free-body diagram

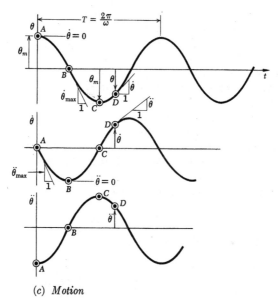

(c) Motion

Fig. 7.1 *The natural motion of an undamped second-order system is oscillatory*

Confusion sometimes arises about the signs in (7.4) because one feels that "the inertial torque must oppose the spring torque." Indeed, if we were to replace $k\theta$ and $J\ddot{\theta}$ by their *numerical values* at any *particular instant*, they would be found (for this system) always to have opposite signs. But $k\theta$ and $J\ddot{\theta}$ are general; what they say (for Fig. 7.1) is

If θ is positive, then the spring torque is clockwise; if $\ddot{\theta}$ is positive, then the inertial torque is clockwise.

They *also* say (with equal precision)

When θ is negative, the spring torque is counterclockwise; when $\ddot{\theta}$ is negative, the inertial torque is counterclockwise.

When we *solve* the equation of motion (below), *then* we shall find out when (timewise) the numerical value of θ is positive and when it is negative, and when the numerical value of $\ddot\theta$ is positive and when it is negative.

We now turn to the question (Stage III) of what motions the system may have.

The pattern of motion is deduced physically in Fig. 7.1c. To start the motion, let the element be rotated an amount θ_m (point A, Fig. 7.1c) and released with no initial velocity. The spring immediately attempts to rotate the element toward the position of zero spring force. Since the element has inertia, however, it will not assume a velocity instantly, but will accelerate toward the neutral position. By the time the rotor has reached neutral (point B), it has built up a high velocity, and thus it has momentum which carries it through neutral. The spring force is zero at B, but as soon as the element passes neutral the spring force begins to build up in the opposite direction, decelerating the mass until at point C its velocity has been reduced to zero again. Thus, the system oscillates back and forth between the extreme positions $\theta = +\theta_m$ and $\theta = -\theta_m$.

To obtain a mathematical solution to Eq. (7.4), we note from the physically deduced plots of displacement, velocity, and acceleration (Fig. 7.1c) that θ must be a quantity which is oscillatory and nondiminishing, and whose second derivative is always proportional to itself. Sine and cosine waves are well known to satisfy these conditions; and so the trial relation

$$\theta = \theta_m \cos \omega t \tag{7.5}$$

is substituted into (7.4):

$$J(-\theta_m \omega^2 \cos \omega t) + k(\theta_m \cos \omega t) = 0$$
or
$$-J\omega^2 \theta_m + k\theta_m = 0$$

where the first term is the maximum acceleration torque, and the second term is the maximum spring torque. We conclude that $\theta_m \cos \omega t$ is a solution to (7.4) (for any θ_m), provided that

$$\omega = \sqrt{\frac{k}{J}} \tag{7.6}$$

ω is known as the *natural frequency* of the element, and θ_m is, of course, the maximum excursion in Fig. 7.1. The *period* of the oscillation (time between peaks in Fig. 7.1c) is given by $T = 2\pi/\omega$.

Pure sinusoidal motion, as given by (7.5) and plotted in Fig. 7.1c, is called *simple harmonic motion*. Undamped systems having such natural motion, like that in Fig. 7.1a, are called *harmonic oscillators*.

Before studying some further examples, we note that the solution (7.5) does not quite represent the general case of unforced motion of the system in Fig. 7.1a, because it corresponds to the special case in which the

[Sec. 7.1] PHYSICAL VIBRATIONS

system was released with zero velocity, as in Fig. 7.1c. More generally, the unforced motion may have both initial velocity and initial displacement. The form of the solution required in the general case is

$$\theta = \theta_m \cos(\omega t - \psi) \qquad (7.2)$$
[repeated]

where the additional constant ψ, together with θ_m, can accommodate any initial state. For example, if $\dot\theta(0)$ has a value but $\theta(0)$ does not, then the solution is

$$\theta = \theta_m \cos\left(\omega t - \frac{\pi}{2}\right) = \theta_m \sin \omega t$$

in which θ_m will be $\dot\theta(0)/\omega$.

Initial conditions will be studied further in Chapter 8.

Energy analysis

Undamped second-order systems are elegant to analyze on an energy basis because they are so-called "conservative" systems—the energy of such systems is constant because there are no dissipative forces acting (refer to Sec. 5.8). Consider Fig. 7.1. At point A the system is not moving, and so has no kinetic energy. However, it has potential energy stored in the wound-up spring. At point B the spring is unwound, hence the system has no potential energy; the kinetic energy at B must therefore be equal to the potential energy the system had at A. Finally, at point C the velocity is again zero, which means that the spring must have been wound up by the same amount it was at A because the potential energy of the system at C is the same as at A. Thus it is seen that energy is converted from potential to kinetic and back again, and the maximum deflection of the spring is the same, cycle after cycle.

We can use the energy analysis to check result (7.6) by writing the expression for the total energy of the system:

$$\begin{aligned}\text{Total energy} &= (\text{K.E.})_{\max} = (\text{P.E.})_{\max} \\ &= \tfrac{1}{2} J \dot\theta_{\max}{}^2 = \tfrac{1}{2} k \theta_{\max}{}^2 \\ &= \tfrac{1}{2} J (-\omega \theta_m)^2 = \tfrac{1}{2} k (\theta_m)^2\end{aligned}$$

The last expression requires that

$$J(-\omega)^2 = k \qquad \text{or} \quad \omega^2 = \frac{k}{J}$$

Lord Rayleigh used this energy principle to develop an important method for finding simply the approximate natural frequencies of complicated dynamic systems. Rayleigh's method is presented in Sec. 13.7. Further, the use of energy relations to derive complete equations of motion is the foundation of the powerful method of Lagrange, as we discussed in Secs. 5.6 through 5.10.

Some additional examples of harmonic oscillators—undamped second-order systems—are shown in Fig. 7.2.

230 UNDAMPED SECOND-ORDER SYSTEMS: FREE VIBRATIONS [Chap. 7]

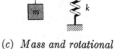

(a) *Mass and spring*

(b) *Mass and set of springs*

(c) *Mass and rotational inertia with spring*

(d) *Free piston on air column*

(e) *Floating buoy*

(f) *Mercury in a U tube*

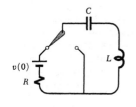
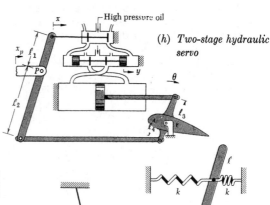

(g) *LC circuit*

(h) *Two-stage hydraulic servo*

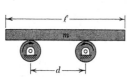

(i) *Plank on rollers (dry friction)*

(j) *Simple pendulum*

(k) *Inverted pendulum with restoring springs*

[Sec. 7.1] PHYSICAL VIBRATIONS

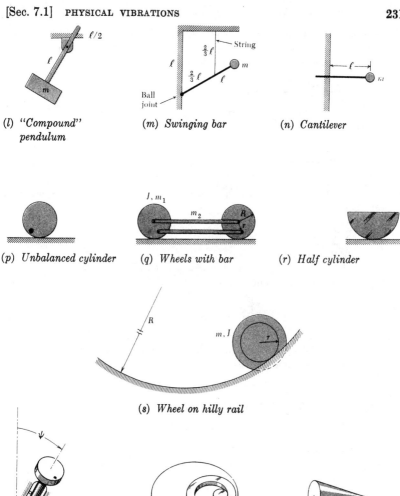

(l) "Compound" pendulum

(m) Swinging bar

(n) Cantilever

(p) Unbalanced cylinder

(q) Wheels with bar

(r) Half cylinder

(s) Wheel on hilly rail

(t) Unbalanced, canted wheel and axle

(u) Constrained disk on rotating table

(v) Unbalanced cone

Fig. 7.2 Some undamped second-order systems

232 UNDAMPED SECOND-ORDER SYSTEMS: FREE VIBRATIONS [Chap. 7]

Example 7.1. An oscillator circuit. In the network shown in Fig. 7.3, the switch has been closed for a long time. Then, at $t = 0$, it is opened. Describe the resulting variation in the current i_1 in the inductor.

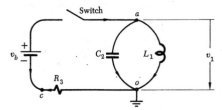

Fig. 7.3 *An oscillator circuit*

After $t = 0$ the switch is open, and we can write the equations of motion as follows:

(1) *Variables.* $v_1 \equiv v_2 \equiv v_a$ (identity)
(2) *Equilibrium.* KCL at node a: $i_1 + i_2 = 0$ $0 < t$ \hfill (7.7)
(3) *Physical relations.* $v_1 = L_1\, di_1/dt$ \hfill (7.8)
$i_2 = C_2\, dv_2/dt = C_2\, d/dt[L_1(di_1/dt)]$

Combining, we obtain

$$\boxed{i_1 + C_2 L_1 \frac{d^2 i_1}{dt^2} = 0} \tag{7.9}$$

This differential equation of motion has the same form as (7.1), and so we try a solution of the same form as (7.2):

$$i_1 = i_{1m} \cos(\omega t - \psi) \tag{7.10}$$

Substituting (7.10) into (7.9), we obtain

$$i_{1m}[1 - L_1 C_2 \omega^2] \cos(\omega t - \psi) = 0$$

from which the solution can be correct, provided that

$$\omega = \sqrt{\frac{1}{L_1 C_2}} \tag{7.11}$$

Next we find i_{1m} and ψ from consideration of the initial conditions. Using Table VI, p. 202, we note that neither i_1 nor v_2 will change suddenly:

$$i_1(0^+) = i_1(0^-) \quad \text{and} \quad v_2(0^+) = v_1(0^+) = v_2(0^-) \tag{7.12}$$

We need next to find $i_1(0^-)$ and $v_2(0^-)$.

At $t = 0^-$ the switch is still closed, all transient currents have died out (because of the resistor), and the steady current $i_1(0^-) = v_3(0^-)/R_3$ will be flowing. There will be no voltage drop across the inductor $[(di_1/dt) = 0]$ and hence $v_1(0^-) = v_2(0^-) = 0$. Then KVL before switching gives

$$0 + v_3(0^-) = v_b \quad t < 0 \tag{7.13}$$

Combining these relations, we obtain

$$i_1(0^+) = \frac{v_b}{R_3} \quad v_1(0^+) = 0 \tag{7.14}$$

[Sec. 7.1] PHYSICAL VIBRATIONS

These may be substituted into (7.10) and (7.8), both taken at $t = 0^+$:

$$i_1(0^+) = i_{1m} \cos(-\psi) = \frac{v_b}{R_3}$$

$$v_1(0^+) = L_1 \frac{di_1}{dt}(0^+) = -L_1 i_{1m} \omega \sin(-\psi) = 0$$

[The last term is obtained by differentiating (7.10).] From this the constants are found: $\psi = 0$ and $i_{1m} = \dfrac{v_b}{R_3}$, and thus we see that the current has the variation

$$i_1 = \frac{v_b}{R_3} \cos \omega t \qquad \omega = \sqrt{\frac{1}{L_1 C_2}}$$

which is the same pattern θ has in Fig. 7.1c.

From (7.7), $i_2 = -i_1$ at all times; thus the current flows around the isolated LC network first clockwise and then counterclockwise, charging the capacitor first one way and then the other. Energy is stored alternately in the electric field of the capacitor and in the magnetic field of the inductor.

Example 7.2. Simple pendulum. Find the approximate motion of a simple pendulum that has been released with zero velocity from a small angle. As a physical model (Stage I), assume that there is no friction and the string is massless. (See Fig. 7.4.)

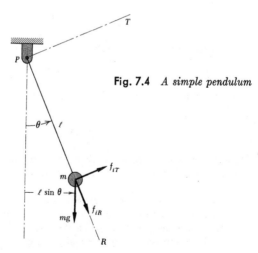

Fig. 7.4 *A simple pendulum*

The equation of motion (Stage II) for this system model is derived as follows:

(1) *Geometry.* From (5.18) (Example 5.2, p. 150), the acceleration of mass m at an arbitrary time is

$$\mathbf{a} = \mathbf{1}_T(\ell \ddot{\theta}) + \mathbf{1}_R(-\ell \dot{\theta}^2) \tag{5.18}$$
[repeated]

234 UNDAMPED SECOND-ORDER SYSTEMS: FREE VIBRATIONS [Chap. 7]

(2) *Force equilibrium.* We write (2.2) about point P:†

$$\Sigma \widehat{M}_P = 0 = -\ell f_{i_T} + (\ell \sin \theta) mg = 0$$

(3) *Physical relations.* We need only $f_{i_T} = -ma_T = -m\ell\ddot{\theta}$.

Combining, we obtain

$$m\ell^2\ddot{\theta} + (\ell \sin \theta) mg = 0 \qquad (7.15)$$

where the first term is recognized as the inertial moment $J\ddot{\theta}$, and the second is the gravity-restoring moment. A more convenient form is

Simple pendulum
$$\boxed{\ddot{\theta} + \frac{g}{\ell} \sin \theta = 0} \qquad (7.16)$$

This is a nonlinear differential equation for which an exact solution is quite complicated. However, for small values of θ—say less than 11.5° or 0.2 rad—we can approximate $\sin \theta$ by the first term in its series expansion:

$$\sin \theta = \theta - \frac{\theta^3}{3!} + \frac{\theta^5}{5!} - \frac{\theta^7}{7!} + \cdots$$

with an error of less than 0.2 percent! (The reader may verify this.) With this approximation, (7.16) becomes a linear differential equation

$$\ddot{\theta} + \frac{g}{\ell} \theta = 0 \qquad (7.17)$$

which looks just like (7.1). Its solution is

$$\theta = \theta_0 \cos \omega t$$

where the natural frequency ω is, by analogy with (7.6),

$$\omega = \sqrt{\frac{g}{\ell}} \qquad (7.18)$$

and θ_0 is the initial displacement. [Note that the pendulum inertia will not abide a sudden change in θ or $\dot{\theta}$, and therefore $\theta(0^+) = \theta(0^-) = \theta_0$; and $\dot{\theta}(0^+) = \dot{\theta}(0^-) = 0$.]

Result (7.18) is, of course, very easy (and instructive) to check experimentally. The fact that the frequency of small motions of a simple pendulum depends only on its length (and not on the mass of its bob, for example) has been the basis for precision clocks for many centuries.

Problems 7.1 through 7.19

Suggested experiments. Construct apparatus like several of the examples in Fig. 7.2, such as (a), (f), (j), (m), (r), or, if the necessary

† The convention $\Sigma \widehat{M}_P$ is defined in the footnote on p. 41.

elements are available, (c), (d), (i), (k), or (s). In each case measure the mass involved and the static force-displacement relation [e.g., spring constant in (a), (c), and (k), friction coefficient in (i), and so on]. Then measure the natural frequency and compare with the calculated value. Try to use several values of a parameter in each case. That is, use several pendulum lengths in (j), different pivot locations in (k), different amounts of liquid in (f), and so on.

Representing sine waves by rotating vectors

It is often convenient in analyzing oscillatory systems to make use of the fact that a sine wave is generated by the projection of a rotating arrow—a vector—as shown in Fig. 7.5a where an angle, ωt, on the circular plot corresponds to an amount of time on the time plot. In this way the amplitude and phase relations between several sinusoidal motions (of the same frequency) can be represented by a simple "vector" picture instead of by plotting each sine wave in detail. This is demonstrated in Fig. 7.5b where vector picture 1 gives as much information about the relation between

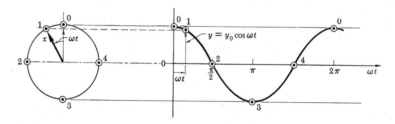

(a) A cosine wave is generated by a rotating vector

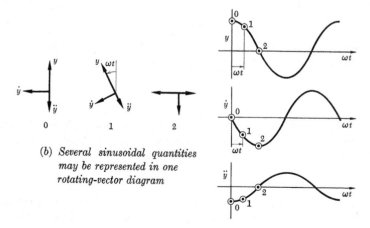

(b) Several sinusoidal quantities may be represented in one rotating-vector diagram

Fig. 7.5 *Representing sine waves with rotating "vectors"*

angular position, velocity, and acceleration as the combination of three sine waves plotted point by point.

When combined with complex-number theory (to be presented in Sec. 7.2), the vector sine-wave representation becomes a rigorous mathematical tool of considerable utility, as we shall see.

R 7.2 COMPLEX NUMBERS

In Chapter 6 and in Sec. 7.1, the two fundamental types of linear response—the exponential decay and the sinusoidal oscillation—have been encountered in their pure form in the natural motions of first-order systems (Fig. 6.3) and undamped second-order systems (Fig. 7.1), respectively.

The two types of motion appear, superficially, to be quite different. However, they are encountered in combination in most dynamic systems, so that further analysis would be enormously facilitated if the two types of motion could be represented by the same form of mathematical function.

This can, in fact, be accomplished, thanks to complex-number theory! Specifically, the function e^{st} can be extended to describe not only exponential motion but sinusoidal motion as well.

We take time at this point to develop carefully the complex-number relationships, so that we can benefit in our subsequent studies from the much greater efficiency they afford. We begin by returning to Eq. (7.4),

$$J\ddot{\theta} + k\theta = 0 \qquad (7.4)$$
[repeated]

and assuming that its solution can be written in the form $\theta = Ce^{st}$. We then set about to determine what values of C and s, if any, will make it a correct solution. When $\theta = Ce^{st}$ is substituted into (7.4), a characteristic equation is obtained, similar to Eq. (6.5):

$$Js^2 + k = 0 \qquad (7.19)$$

Next, solution of this characteristic equation gives the values that s must have to satisfy the differential equation of motion:

$$s = \pm \sqrt{-\frac{k}{J}} \quad \text{or} \quad s = \pm \sqrt{-1}\sqrt{\frac{k}{J}} \qquad (7.20)$$

Now, s is found to contain $\sqrt{-1}$. We are thus faced with the problem of reconciling the function e^{st} with the known sinusoidal motion of the system—i.e., *the problem is to make e^{st} look like a sine wave.*

Euler's equation

This can be accomplished by performing a series expansion on each of the functions of interest, namely, e^{ϕ}, cos ϕ, and sin ϕ (where $\phi = st$ for

---

R It is expected that this section will be review material for many readers and can be read quickly.

[Sec. 7.2] COMPLEX NUMBERS

the case of interest):

$$e^{\phi} = 1 + \phi + \frac{\phi^2}{2!} + \frac{\phi^3}{3!} + \frac{\phi^4}{4!} + \frac{\phi^5}{5!} + \cdots$$

$$\cos \phi = 1 \qquad - \frac{\phi^2}{2!} \qquad + \frac{\phi^4}{4!} - \cdots$$

$$\sin \phi = \qquad \phi \qquad - \frac{\phi^3}{3!} \qquad + \frac{\phi^5}{5!} - \cdots$$

Then by multiplying the exponent of e by $\sqrt{-1}$ and also by multiplying each term in the sine expansion by $\sqrt{-1}$, the three functions are simply related by the following important equation, known as Euler's equation:

Euler's equation
$$\boxed{e^{\sqrt{-1}\,\phi} = \cos \phi + \sqrt{-1} \sin \phi} \qquad [(7.3) \text{ repeated}] \quad (7.21)$$

With this interpretation of $\sqrt{-1}$, the assumed solution to (7.4) can now be converted to the expected sinusoidal motion. We find that two constants, C_1 and C_2, are required: Both the positive and negative values of (7.20) are included to account for the two constants of integration necessary in a second-order equation.† That is, there are *two initial conditions* which must be accommodated, initial position and initial velocity:

$$\theta = Ce^{st} = C_1 e^{\sqrt{-1}\sqrt{\frac{k}{J}}\,t} + C_2 e^{-\sqrt{-1}\sqrt{\frac{k}{J}}\,t} \qquad (7.22a)$$

$$= C_1 \left(\cos \sqrt{\frac{k}{J}}\,t + \sqrt{-1} \sin \sqrt{\frac{k}{J}}\,t\right)$$

$$+ C_2 \left(\cos \sqrt{\frac{k}{J}}\,t - \sqrt{-1} \sin \sqrt{\frac{k}{J}}\,t\right) \qquad (7.22b)$$

$$= (C_1 + C_2) \cos \sqrt{\frac{k}{J}}\,t + \sqrt{-1}\,(C_1 - C_2) \sin \sqrt{\frac{k}{J}}\,t \qquad (7.22c)$$

or $\quad \theta = C_3 \cos \left(\sqrt{\frac{k}{J}}\,t - \psi\right) \qquad (7.23a)$

We note that C_3 and ψ in (7.23a) are combinations of C_1 and C_2. The reader may verify that

$$C_3 = \sqrt{(C_1 + C_2)^2 + (C_1 - C_2)^2} \qquad \text{and} \qquad \psi = \tan^{-1} \frac{C_1 - C_2}{C_1 + C_2} \qquad (7.23b)$$

by expanding $C_3 \cos \left(\sqrt{\frac{k}{J}}\,t - \psi\right)$ and using other appropriate identities.

Two other extremely useful and widely used relations, based on Eq. (7.21), are

$$\cos \varphi = \frac{1}{2}\left(e^{j\varphi} + e^{-j\varphi}\right) \qquad (7.24a)$$

$$\sin \varphi = \frac{1}{2j}\left(e^{j\varphi} - e^{-j\varphi}\right) \qquad (7.24b)$$

[These are readily verified by substituting from (7.21).]

† In general, as noted on p. 197, for an rth order system r constants will be required.

The complex plane

A simple method of visualizing Euler's equation (7.21) is to plot ϕ as an angle and label its sine and cosine as in Fig. 7.6. This way of looking at the relation is so useful that a coordinate system has been universally adopted to accommodate it. The coordinate system is known as the "complex plane," and the components of functions plotted in it are known as "complex numbers." Also, $\sqrt{-1}$ is given the symbol j for convenience.

Since it takes two numbers to describe a quantity in the complex plane, the quantities can be thought of as mathematical vectors. They are

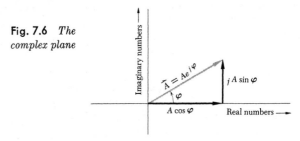

Fig. 7.6 *The complex plane*

called "complex vectors" or "phasors" to distinguish them from physical vectors (representing forces, velocities, and so on).† It will be seen that the addition of complex vectors is similar to the addition of physical vectors, but the multiplication process is different. Manipulations of complex numbers and complex vectors (phasors) are all based on Euler's fundamental equation, (7.21).

R 7.3 MATHEMATICAL OPERATIONS WITH COMPLEX NUMBERS

The various mathematical operations involving complex numbers can be most easily understood in terms of geometric manipulation of the complex vector (phasor) of Fig. 7.6.

The various operations are depicted in Fig. 7.7. Complex-vector *addition* is demonstrated in Fig. 7.7a which shows the addition of two complex vectors $\vec{A}$ and $\vec{B}$. For addition, the vectors are most conveniently represented in the form

$$\vec{A} = A \cos \phi_A + jA \sin \phi_A$$
$$\vec{B} = B \cos \phi_B + jB \sin \phi_B$$

in which we have let A mean $|\vec{A}|$, and so on. The sum of the two vectors is simply obtained by adding separately their real and imaginary parts:

$$\vec{A} + \vec{B} = (A \cos \phi_A + B \cos \phi_B) + j(A \sin \phi_A + B \sin \phi_B) \quad (7.25)$$

† As we shall use it, the word *phasor* is synonymous with *mathematical complex vector*, and we shall use the two terms interchangeably. Some people use only phasor for this, reserving the word vector exclusively for physical quantities.

R See footnote, p. 236.

[Sec. 7.3] MATHEMATICAL OPERATIONS WITH COMPLEX NUMBERS 239

This is easily seen in Fig. 7.7a.

Multiplication is most easily accomplished with the vector in the form†
$\vec{A} = Ae^{j\phi_A}$:

$$(\vec{A})(\vec{B}) = (Ae^{j\phi_A})(Be^{j\phi_B}) = ABe^{j(\phi_A+\phi_B)} \tag{7.26a}$$

That is, *the amplitude of the product is equal to the product of the amplitudes of the individual vectors, and the phase angle of the product is equal to the sum of the individual phase angles.* Multiplication is demonstrated in Fig. 7.7b.

Division. By application of (7.26a), division is performed by dividing magnitudes and *subtracting* phase angles:

$$\frac{\vec{A}}{\vec{B}} = \frac{Ae^{j\phi_A}}{Be^{j\phi_B}} = \frac{A}{B}e^{j(\phi_A-\phi_B)} \tag{7.26b}$$

When many complex numbers are to be multiplied and divided, it is common to write the vectors as $\vec{A} = Ae^{j\phi_A} = A\underline{/\phi_A}$. Then:

$$\boxed{\frac{(\vec{A})(\vec{B})(\vec{C})}{(\vec{D})(\vec{E})} = \frac{ABC}{DE}\underline{/\phi_A + \phi_B + \phi_C - \phi_D - \phi_E}} \tag{7.26c}$$

In studying dynamic response, and particularly stability, this is the most frequently used operation.

Differentiation. In performing differentiation of a complex vector it must be remembered that, in general, its amplitude as well as its phase angle may be a function of time, and thus the expression $Ae^{j\phi}$ is the product of two variables, and the proper derivative is

$$\frac{d\vec{A}}{dt} = \frac{dA}{dt}e^{j\phi} + A\frac{d}{dt}e^{j\phi} = (\dot{A} + Aj\dot{\phi})e^{j\phi} \tag{7.27a}$$

Differentiation in this general case is illustrated in Fig. 7.7c.

With $\vec{A}$ in the alternative form $\vec{A} = A_x + jA_y$, we have

$$\frac{d\vec{A}}{dt} = \frac{dA_x}{dt} + j\frac{dA_y}{dt} \tag{7.27b}$$

The reader may show that (7.27a) and (7.27b) are identical (Problem 7.32).

A special situation which often arises when we deal with time vectors is that the magnitude of the complex number is constant and only its phase angle is variable: $\vec{A} = A_0 e^{j\phi}$, and

$$\frac{d\vec{A}}{dt} = A_0 j\dot{\phi}e^{j\phi} = j\dot{\phi}\vec{A} \tag{7.28}$$

† Alternatively, the form $\vec{A} = A\cos\phi_A + jA\sin\phi_A$ could again be used, but the algebraic manipulation is much more tedious.

In such cases each successive derivative of the number involves merely a multiplication by $j\dot\phi$ and therefore is represented by a complex vector which is rotated 90° in the positive direction from the original vector, as illustrated in Fig. 7.7d. It is left as an important exercise (Problem 7.29) to show that

$$\frac{d}{dt}(\text{Re }[\quad]) = \text{Re}\frac{d}{dt}[\quad]$$

Examples of complex-number manipulation

Example 7.3. Find the sum $\vec{A} + \vec{B}$, when

(a) $\quad\quad\quad \vec{A} = 4 + j1 \quad\quad \vec{B} = 2 + j3$

(b) $\quad\quad\quad \vec{A} = 7e^{j\frac{\pi}{6}} \quad\quad \vec{B} = 5e^{j\frac{\pi}{4}}$

Solution for (a) is immediate: Adding real and imaginary parts gives

(a) $\quad\quad\quad\quad\quad \vec{A} + \vec{B} = 6 + j4$

To do part (b) algebraically, the vectors are first converted to Cartesian form:

(b) $\quad\quad \vec{A} = 7\cos\frac{\pi}{6} + j7\sin\frac{\pi}{6} = 6.06 + j3.50$

$$\vec{B} = 5\cos\frac{\pi}{4} + j5\sin\frac{\pi}{4} = 3.54 + j3.54$$

$$\vec{A} + \vec{B} = 9.60 + j7.04$$

Addition can also be done graphically—using the vectors in *either* form—and this is strongly recommended when graph paper or a scale and protractor are at hand.

Example 7.4. Find the product $(\vec{A})(\vec{B})$ for each set in Example 7.3. In this case the solution to (b) is immediate:

$$(\vec{A})(\vec{B}) = (7e^{j\frac{\pi}{6}})(5e^{j\frac{\pi}{4}}) = 35e^{j\left(\frac{\pi}{6}+\frac{\pi}{4}\right)} = 35e^{j\frac{5}{12}\pi}$$

To find an algebraic solution to (a), the vectors are put in exponential form:

$$\vec{A} = \sqrt{4^2 + 1^2}\, e^{j\tan^{-1}(1/4)}$$
$$= 4.12 e^{j14.0°}$$
$$\vec{B} = 3.60 e^{j56.3°}$$
$$(\vec{A})(\vec{B}) = 14.8 e^{j70.3°}$$

Again, a graphical method could have been used. A plastic device known as a "Spirule"† greatly facilitates multiplying vectors graphically. The device incorporates a protractor for adding phase angles with a logarithmic spiral for multiplying magnitudes mechanically. (See Appendix I.)

† Available commercially from bookstores or the Spirule Company, 9728 El Venado, Whittier, California.

[Sec. 7.3] MATHEMATICAL OPERATIONS WITH COMPLEX NUMBERS 241

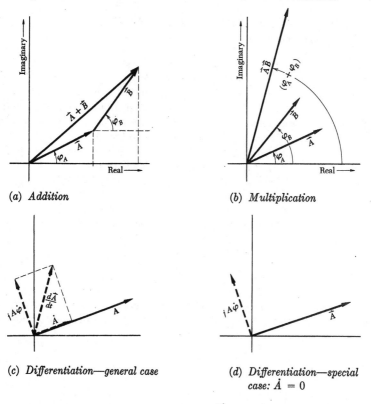

(a) *Addition*

(b) *Multiplication*

(c) *Differentiation—general case*

(d) *Differentiation—special case:* $\dot{A} = 0$

Fig. 7.7 *Operations with complex numbers:* $\vec{A} = Ae^{j\phi} = A\cos\phi + jA\sin\phi$

Example 7.5. Find the time derivative of a complex vector, $\vec{A}$, which has constant length and rotates at constant speed ω. If it is given that $\vec{A} = A_0 e^{j\omega t}$ then, by (7.28), we have

$$\dot{\vec{A}} = j\omega A_0 e^{j\omega t} = j\omega \vec{A} \qquad (7.29)$$

That is, the vector is rotated ahead 90° (and multiplied by ω). In Sec. 7.4 this vector is used to form a rigorous representation of a sine wave, as suggested by Fig. 7.5.

Example 7.6. Find the time derivative of the vector $\vec{A} = A_0 e^{-\sigma t} e^{j\omega t}$. This may be written $\vec{A} = A_0 e^{(-\sigma + j\omega)t}$ from which we obtain at once

$$\dot{\vec{A}} = (-\sigma + j\omega)\vec{A} \qquad (7.30)$$

Problems 7.20 through 7.32

7.4 COMPLEX–VECTOR (PHASOR) REPRESENTATION OF A SINE WAVE

It was demonstrated in Fig. 7.5 that a considerable convenience and saving of labor can be realized by using a rotating arrow, instead of a time plot, to represent a sinusoidally varying quantity. This concept can be put on a rigorous mathematical basis by applying Euler's equation (7.21).

First, the complex vector $e^{+j(\omega t - \psi)}$ is drawn in the complex plane as shown in Fig. 7.8a. This vector rotates counterclockwise with angular velocity ω. Next, the conjugate of this vector is also drawn: $e^{-j(\omega t - \psi)}$. This vector rotates clockwise; that is, it rotates with velocity $-\omega$.

When these two vectors are put in the form of Eq. (7.21),

$$e^{j(\omega t - \psi)} = \cos(\omega t - \psi) + j\sin(\omega t - \psi)$$
$$e^{-j(\omega t - \psi)} = \cos(\omega t - \psi) - j\sin(\omega t - \psi)$$

and added, the result is seen to be a real quantity:

$$e^{j(\omega t - \psi)} + e^{-j(\omega t - \psi)} = 2\cos(\omega t - \psi) \qquad (7.31)$$

This can be seen in Fig. 7.8a where the imaginary components of the two vectors cancel, leaving only the real components, whose sum grows and shrinks along the real axis sinusoidally with time.

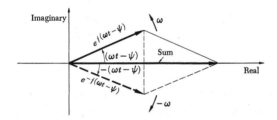

(a) *The sum of a conjugate pair of rotating complex vectors is real*

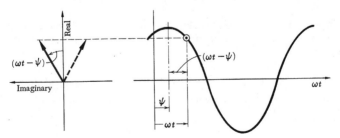

(b) *A sine wave is generated by projection from a rotating complex vector*

Fig. 7.8 *Rotating complex vectors (phasors)*

From this we conclude that a cosine wave can be represented in the exponential form

$$A \cos(\omega t - \psi) = \tfrac{1}{2} A [e^{j(\omega t - \psi)} + e^{-j(\omega t - \psi)}] \qquad (7.32)$$

Just as correctly, only the real part of the right-hand side need be considered,

$$A \cos(\omega t - \psi) = \tfrac{1}{2} \operatorname{Re} A [e^{j(\omega t - \psi)} + e^{-j(\omega t - \psi)}]$$

since the left-hand side has no imaginary part. Or, finally, from the geometry of Fig. 7.8a, we can say†

$$\boxed{A \cos(\omega t - \psi) = \operatorname{Re} [A e^{j(\omega t - \psi)}]} \qquad (7.33)$$

That is, the real-axis projection of a rotating complex vector (phasor) is a sine wave in the manner of Eq. (7.33).

The use of this relation is illustrated in Fig. 7.8b, which is constructed by projecting points from the complex plane on the left to the time plot on the right. Note that it is necessary to tilt the complex plane 90°—with its real axis vertical and its imaginary axis to the left—in order to produce a plot with the time axis horizontal, as it conventionally is. Figure 7.8b illustrates the meaning of the quantity ψ: It is a constant which allows expression $\cos(\omega t - \psi)$ to represent a cosine wave having its maximum at any chosen time. Specifically, the maximum will occur at $t = \psi/\omega$. In the rotating-vector (phasor) picture, $-\psi$ is the angle of the vector at $t = 0$; ψ is called a "phase angle."

Equation (7.33) is very important in the analysis of complicated dynamic systems. For example, if a system is being forced by a sinusoidal disturbance of the form $M_0 \cos \omega_f t$, the system analysis can be carried out using the disturbing function $M_0 e^{j\omega_f t}$—which is very much easier to manipulate—and then simply taking the real part of the answer, which it is allowable to do by Eq. (7.33). Equation (7.33) will be used many times in the remainder of this book as a basis for streamlining the algebra associated with oscillatory motions—motions which will henceforth play a central role in all that we study of the dynamic behavior of systems.

† We sometimes denote = Re [ ] by $\stackrel{r}{=}$.

Chapter 8

DAMPED SECOND-ORDER SYSTEMS

In this chapter we shall study linear, damped second-order systems (systems having two separate energy-storage elements plus a mechanism for energy dissipation). The governing differential equation has the form

Equation of motion $\quad \boxed{\ddot{\theta} + 2\sigma\dot{\theta} + \omega_0^2 \theta = \alpha(t)} \quad$ (8.1)

We shall find that

(1) The *natural motion* of such systems consists either of
 (a) the sum of two first-order responses

$$\theta = C_1 e^{-\sigma_1 t} + C_2 e^{-\sigma_2 t} \qquad 0 < t \qquad (8.2a)$$

with σ_1 and σ_2 given by (8.6a), or
 (b) the *product* of a first-order response and a sinusoid, that is, a damped sinusoid,

$$\theta = \theta_m e^{-\sigma t} \cos(\omega t - \psi) \qquad 0 < t \qquad (8.2b)$$

with ω given by (8.5). In (8.2), constants C_1, C_2, θ_m, and ψ depend upon the initial state, and are given by (8.10) or (8.11b).

(2) *The s plane* furnishes a most convenient way to display the natural characteristics (eigenvalues) of a dynamic system (the roots of its characteristic equation).

(3) Several of the most important *forcing functions* can be represented in the form $x(t) = Xe^{st}$, with X a (complex) constant.

(4) The forced response of an l.c.c. system† to the above input will be of the same form, for example, $y = Ye^{st}$. The ratio Y/X—the proportion

† See footnote, p. 182.

[Sec. 8.1] SECOND-ORDER SYSTEMS 245

of such an input that a given system *transfers* to its output—is called the system's *transfer function*. (The factors of the denominator and numerator of a transfer function are known, respectively, as its *poles* and *zeros*.)

In addition, all of the basic concepts introduced in Chapters 6 and 7 are shown in the present chapter to apply also to damped second-order systems. In particular, Procedure B-1 applies verbatim.

Simulation using an analog computer, an important analytical tool, is introduced at the end of the chapter.

8.1 SECOND-ORDER SYSTEMS

In Chapter 6 we found that first-order systems (like those in Fig. 6.1 and Fig. 6.2b) have a natural motion whose time history is a simple exponential (Fig. 6.3). In Chapter 7 we found that the natural motion of an undamped second-order system (like those in Fig. 7.1a and Fig. 7.2) is a simple sinusoid (Fig. 7.1c). We also learned how either type of motion—the exponential or the sine wave—can be represented in the form Ce^{st} by allowing s to be a complex number.

In Chapter 6 we also found that if a system is disturbed by a forcing function of the form e^{st}, then its response will also be of the form e^{st} — although with different magnitude.

In the present chapter we shall find that the natural motion of a lightly damped second-order system, like that in Fig. 8.1a, is a combination of *both* exponential and sinusoidal motion, which can be conveniently represented using complex numbers. We shall also see how forcing functions of the form e^{st} produce, as before, response of the same form.

As a model for study we refer again to the system of Fig. 2.5, and we now consider the case in which all three components—inertia, spring, and damper—produce torques of important magnitude. Complex-number techniques will be employed in the general pattern [Eq. (6.1)] to study first natural motions and then forced response.

We can quickly obtain the equation of motion (Stage II)† for the physical model (Stage I) of Fig. 8.1a:

(1) *Geometry:* Take θ positive counterclockwise. (Then $\dot\theta$ and $\ddot\theta$ are also positive counterclockwise.)
(2) *Equilibrium*
(3) *Physical relations* } From the free-body diagram of Fig. 8.1b,

$$J\ddot\theta + b\dot\theta + k\theta = M \tag{8.1'}$$

(A detailed discussion of signs was given on p. 227.)

Note that (8.1') contains the terms of both (6.3) and (7.4).

† A more detailed derivation was given in Example 2.1, p. 46.

We now proceed to a study of the possible motions implied by (8.1'), Stage III.

8.2 NATURAL MOTION

Consider first the case of unforced motion: $M = 0$ in (8.1'). The response of the system to initial displacement can be deduced qualitatively by considering special cases. If the inertia of the system is quite small, the spring energy will be dissipated in the damper and the response will be a simple exponential decay like that of Fig. 8.1c (which is repeated from Fig. 6.3). If, on the other hand, the damping is quite small, the system will respond with a simple sinusoidal oscillation like that of Fig. 8.1e (repeated from Fig. 7.1). When the inertia, damping, and spring forces are about equal, the response will be something between the responses of Fig. 8.1c and Fig. 8.1e, and (if the damping does not dissipate all of the energy in the spring before the mass reaches neutral) a damped oscillation will result, as shown in Fig. 8.1d.

Because of the inertia, neither θ nor $\dot\theta$ can change instantly. Therefore both the initial position and the initial velocity of the system must be specified before the response can be calculated, and the mathematical expression for θ must satisfy both initial conditions.

From the above discussion we expect that the formal mathematical solution for θ must be capable of representing either a damped oscillation (for small damping) of the form $Ce^{-\sigma t} \cos \omega t$, or an exponential decay (for large damping) of the form $Ce^{-\sigma t}$. To obtain a universal solution, we take advantage of the complex-number method of representing a cosine wave, and assume an expression of the form

$$\theta = \Theta e^{st} \tag{8.2'}$$

which can represent either kind of motion, depending on the relative values of J, b, and k (s being, in general, a complex number: $s = -\sigma \pm j\omega$).

The values that Θ and s must have, for a given system with given starting conditions, are obtained, as usual, by substituting the assumed solution (8.2') into the homogeneous part of the differential equation (8.1'):

$$J(\Theta s^2 e^{st}) + b(\Theta s e^{st}) + k(\Theta e^{st}) = 0$$

As usual also, each term in the equation contains the common factor Θe^{st}, which can be canceled out,† leaving the *characteristic equation* of the system:

$$Js^2 + bs + k = 0 \tag{8.3}$$

As before, the characteristic equation contains s but does not contain

† It is assumed that $\Theta \neq 0$. (The solution $\Theta = 0$ is, of course, also a possible—but uninteresting—one.)

[Sec. 8.2] NATURAL MOTION 247

Θ. It therefore represents completely the dynamic characteristics of the system but does not contain any information about starting conditions.

The roots s of the characteristic equation are

$$s = -\frac{b}{2J} \pm \sqrt{\left(\frac{b}{2J}\right)^2 - \frac{k}{J}} \quad \text{for } \frac{k}{J} \leq \left(\frac{b}{2J}\right)^2 \tag{8.4a}$$

or

$$s = -\frac{b}{2J} \pm j\sqrt{\frac{k}{J} - \left(\frac{b}{2J}\right)^2} \quad \text{for } \left(\frac{b}{2J}\right)^2 \leq \frac{k}{J} \tag{8.4b}$$

The roots are given in two forms, depending on the relative magnitudes of b, J, and k, because it is advantageous to maintain the square-root expression, $\sqrt{(b/2J)^2 - (k/J)}$, in proper form. This is done by factoring out $\sqrt{-1}$ when it occurs and giving it the usual symbol, j.

The roots of a system's characteristic equation are called the *eigenvalues* or *characteristics* of the system (see footnote, p. 188).†

† They are sometimes also called *complex natural frequencies*.

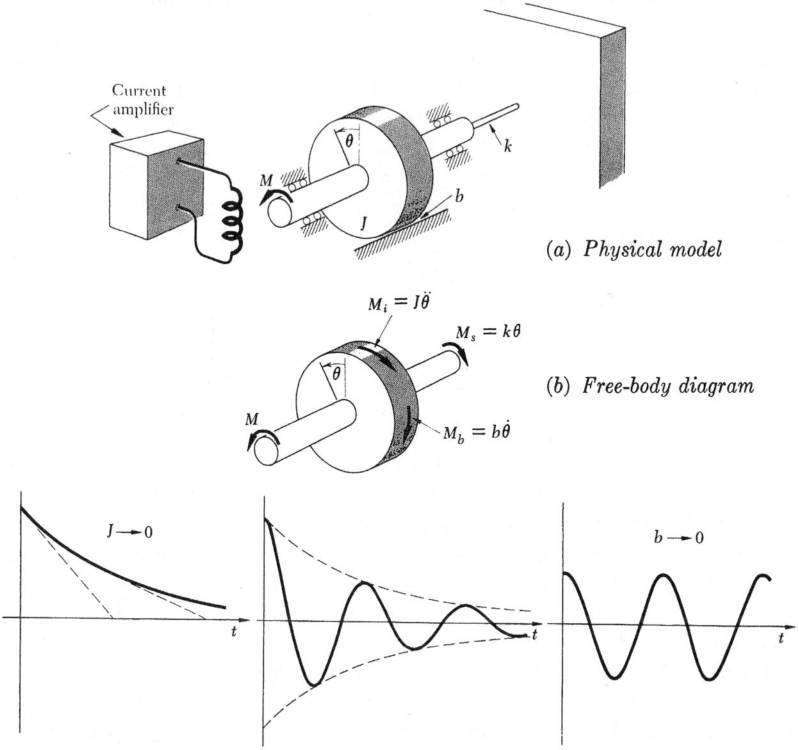

(a) *Physical model*

(b) *Free-body diagram*

(c) *Pure exponential* (from Fig. 6.3)   (d) *Damped sine*   (e) *Pure sine* (from Fig. 7.1)

Fig. 8.1 *The natural motion of a damped, second-order system*

8.3 DYNAMIC CHARACTERISTICS AND THE s PLANE

Expressions (8.4), for the roots (eigenvalues) of a second-order system, occur so often, and must be manipulated so much, that a universal set of shorthand symbols has been adopted for them:

Shorthand definitions
$$\sigma \triangleq \frac{b}{2J} \qquad \omega_0 \triangleq \sqrt{\frac{k}{J}}$$
$$\omega \triangleq \sqrt{\omega_0^2 - \sigma^2}$$
$$\zeta \triangleq \frac{\sigma}{\omega_0}$$
(8.5)

Each of these symbols will be found in the discussion which follows to have a clear physical meaning. [For a given physical system, the combinations of physical constants denoted by σ and ω_0 will always be clear from the equation of motion in the form (8.1). See, for example, the constants following Eq. (8.12), p. 262.]

Definitions (8.5) make it convenient to express the equation of motion (8.1) (or any other l.c.c. second-order differential equation) in the form

$$\ddot{\theta} + 2\sigma\dot{\theta} + \omega_0^2\theta = \alpha(t)$$
(8.1) [repeated]

The roots (8.4) of the corresponding characteristic equation (8.3) can be written compactly:

Characteristic roots
$$s = -\sigma_1, -\sigma_2$$
$$\sigma_1 = \sigma\left(1 - \sqrt{1 - \frac{1}{\zeta^2}}\right)$$
$$\sigma_2 = \sigma\left(1 + \sqrt{1 - \frac{1}{\zeta^2}}\right)$$
$\quad 1 < \zeta \quad$ (8.6a)

$s = -\sigma, -\sigma \qquad \zeta = 1$ (8.6b)
$s = -\sigma \pm j\omega \qquad \zeta < 1$ (8.6c)

in which the necessary definitions are given by (8.5).

Substitution of these eigenvalues into the assumed solution $\theta = \Theta e^{st}$, yields the following expressions for motion θ (Fig. 8.1):†

Natural response
$\theta = C_1 e^{-\sigma_1 t} + C_2 e^{-\sigma_2 t} \qquad 1 < \zeta$ (8.7a)
$\theta = e^{-\sigma t}(C_1 e^{j\omega t} + C_2 e^{-j\omega t}) = C_3 e^{-\sigma t}\cos(\omega t - \psi) \qquad \zeta < 1$ (8.7c)

depending on the amount of damping. The C's, like Θ, are constants which may be complex. In each case the two values of s are both used, with separate coefficients C, so that the two separate initial conditions (displacement and velocity) can be taken care of mathematically. (Formally: the homogeneous solution to a linear differential equation of order r must

† The case $\zeta = 1$ [result (8.7b)] requires special development, Sec. 8.7.

[Sec. 8.3] DYNAMIC CHARACTERISTICS AND THE s PLANE 249

contain r constants, C.) Before evaluating these constants, however, the dynamic characteristics implied by the values of s will be investigated more thoroughly. The condition $\zeta = 1$ will be discussed as a special case in Sec. 8.7.

The s plane

The values of s given in (8.6) are plotted in Fig. 8.2 in a special complex plane known as the s *plane*. When s-plane pictures are interpreted in terms of the transient motions to which they correspond, they are found to indicate, in a simple way, all of the dynamic characteristics of the system. For this reason, simple s-plane plots like those of Fig. 8.2 will be used as the

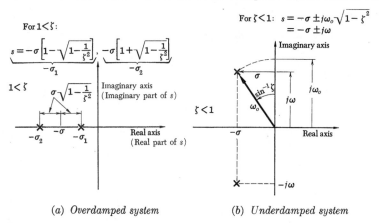

(a) *Overdamped system* (b) *Underdamped system*

Fig. 8.2 *Plotting system characteristics in the s plane*

starting point for nearly all of the dynamic analysis which follows in this book.

For the case of an "underdamped" but stable system, $0 < \zeta < 1$, the system's characteristics are located in the s plane at $s = -\sigma \pm j\omega$, Fig. 8.2b.† Then, from (8.5), the length of the complex s vector (phasor) is ω_0, and the angle between that vector and the j axis is the angle whose sine is ζ.

The correspondence between the s plane and the time response is illustrated in Fig. 8.3 for a given set of numerical values.

The quantity ω is, of course, the *frequency of natural oscillation* in rad/sec [from (8.7c)], and thus the period of natural oscillation T, the time between zero crossings in the same direction, (and the time between maxima of the same sign) is given by

$$T = \frac{2\pi}{\omega} \qquad (8.8)$$

We call ω the *damped natural frequency*, to distinguish it from ω_0, the

† The axes of the s plane are often labeled σ and $j\omega$. However, this is not consistent, signwise, with our definition of σ as a positive coefficient in the general second-order equation of motion. In this book, σ is simply a number related to s by: Re$[s] = -\sigma$. (Similarly, ω is related to s by: Im$[s] = \pm\omega$.)

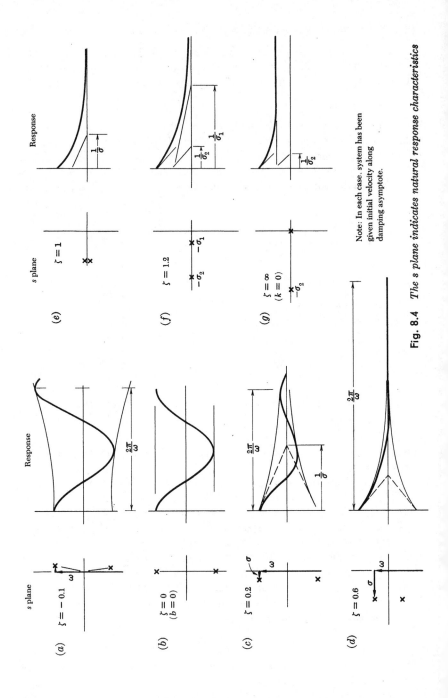

Fig. 8.4 *The s plane indicates natural response characteristics*

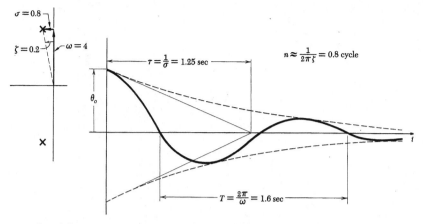

Fig. 8.3 *Natural motion from s-plane characteristics*

undamped natural frequency. (The damped natural frequency in cycles per second is given by $f = 1/T = \omega/2\pi$.)

The quantity ω_0 is the *frequency* at which the same system would oscillate *if damping were absent* (i.e., if the viscous force, Fig. 8.1, were removed).

The quantity $1/\sigma$ indicates the *time* required for the motion *to damp* to $(1/e)$th its original value. This time

$$\tau = \frac{1}{\sigma} \tag{8.9}$$

is known as the *damping time constant* of the system.

The quantity ζ implies the relative damping of the system, i.e., the rate of damping with respect to rate of oscillation; ζ is commonly called the *damping ratio*. In fact, for small values of ζ, ζ indicates explicitly the *number of cycles n to damp* to $(1/e)$th:

$$n \approx \frac{1}{2\pi\zeta}$$

(That is, n = number of cycles to damp to $\frac{1}{e} = \frac{\text{time to damp to } 1/e}{\text{time for one cycle}} = \frac{\tau}{T} = \frac{1/\sigma}{2\pi/\omega} \approx \frac{1}{2\pi\zeta}$, for $\zeta \ll 1$.)

Further comparison between s-plane plots and the corresponding transient responses is given in Fig. 8.4 for various values of ζ, from which it is seen that ζ is an excellent index of system stability.

In particular, if ζ is negative (Fig. 8.4a), the system is unstable—the motion grows without bound (refer to p. 192). When ζ is zero, the system is just neutrally stable (Fig. 8.4b) (the motion neither growing nor decaying); and as ζ is increased toward 1, the relative damping of the system

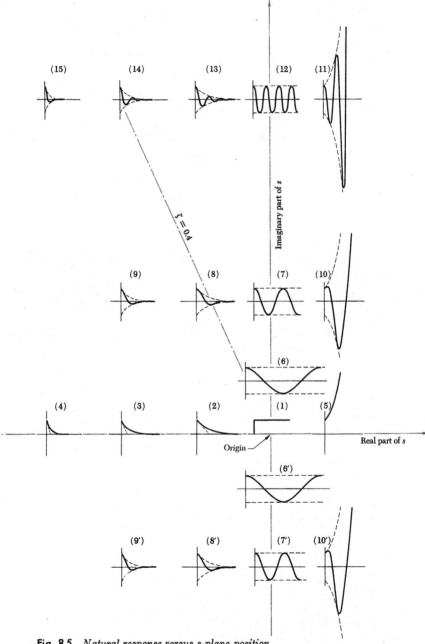

Fig. 8.5 *Natural response versus s-plane position*

[Sec. 8.3] DYNAMIC CHARACTERISTICS AND THE s PLANE 253

increases (Fig. 8.4c, d, and e). For values of ζ greater than 1 the system will not overshoot neutral, without larger initial velocity.

When ζ is greater than 1, the two roots of (8.3) are, from (8.6a):

$$s_1 = -\sigma_1 = -\sigma\left(1 - \sqrt{1 - \frac{1}{\zeta^2}}\right)$$

and

$$s_2 = -\sigma_2 = -\sigma\left(1 + \sqrt{1 - \frac{1}{\zeta^2}}\right)$$

Both of these roots (eigenvalues) lie on the negative real axis in the s plane (Fig. 8.2a) and equidistant from point $-\sigma$ [as defined in (8.5)]. The corresponding response, shown in Fig. 8.4f, is the sum of two simple exponential decays—one fast and one slow.

For the special case that ζ goes to infinity, one of the roots lies at the origin, corresponding to an infinite time constant. This means that if the system is displaced to some position and released with zero velocity, it will simply stay put, a fact which is physically clear from Fig. 8.1, where $\zeta = \infty$ corresponds to removing the spring. The response of the same system when released with a (negative) initial velocity is illustrated in Fig. 8.4g. Physically, the initial kinetic energy of the inertia is dissipated by the damper before the system reaches neutral (for the initial conditions shown).

In Fig. 8.4a through d, each curve has been drawn with an initial velocity such that the motion starts out along the damping asymptote. This particular set of initial conditions results in a cosine motion involving no initial phase. The effects of initial conditions on the response are investigated in Secs. 8.4 through 8.7.

The development of a strong mental correlation between the s-plane picture of dynamic characteristics (eigenvalues) and the corresponding character of the motion is of paramount importance in the study of dynamic behavior. All of Part C, for example, will deal with the dynamic character of compound systems as revealed in the s plane.

An aid to developing this correlation is the plot of Fig. 8.5, in which miniature pictures of time response are spotted on the s plane, each at the coordinates of its characteristics.[†] On the real axis the motion is always a pure exponential: the farther from the origin, the faster the response [(1), (2), (3), (4)]; in the right half-plane the motion grows unstably [(5)].

On the imaginary axis the motion is always an undamped oscillation: the farther from the origin, the higher the frequency [(1), (6), (7), (12)].

A constant distance from the real axis [e.g., (12), (13), (14), (15)] means a constant frequency ω, but variable decay time [or, in the right half-plane, (11), growth time].

A constant distance from the imaginary axis [e.g., (3), (9), (14)] means a constant decay time $1/\sigma$, but variable frequency.

† This display was introduced by W. W. Harman in 1954. See W. W. Harman and D. W. Lytle, "Electrical and Mechanical Networks," McGraw-Hill Book Company, New York, 1962.

Along a constant ζ line [(8), (14)] the number of *cycles* to damp is constant, but, as in general, the farther from the origin, the faster the whole response.

The picture is, of course, symmetrical with respect to the real axis [(9)(9'), (8)(8'), and so on] because complex roots come in conjugate pairs.

In Secs. 8.4, 8.5, and 8.7 we shall be discussing the natural time response of systems having, respectively, roots on the real axis [Points (1), (2), (3), (4)], roots off the real axis, but not in the right half-plane [(6), (7), (8), (9), (12), (13), (14), (15)], and double roots on the real axis. In Sec. 8.6 the sketching of oscillatory time response is discussed. Some examples of unstable behavior [Points (5), (10), (11)] are described in Secs. 11.2 and 11.4.

Example 8.1. The locus of roots. Plot in the s plane the roots of (8.3) if the system in Fig. 8.1 has the following sets of physical constants.

Case	J [(in lb)/(rad/sec²)]	b [(in lb)/(rad/sec)]	k (in lb/rad)
(a)	3	0	192
(b)	3	4.8	192
(c)	3	24	192
(d)	3	36	192
(e)	3	48	192
(f)	3	72	192
(g)	3	−4.8	192

In each case state whether the natural motion will be oscillatory. If it will, give its period and damping time constant.

Using (8.5), (8.6), (8.8) and (8.9), we can construct Table VII.

TABLE VII

Calculation of the roots of Eq. (8.3) for given numerical values

Case	σ (sec⁻¹)	ω_0 (rad/sec)	ζ	ω (rad/sec)	$T = \dfrac{2\pi}{\omega}$ (sec)	$\tau = \dfrac{1}{\sigma}$ (sec)
(a)	0	8	0	8	0.785	0
(b)	0.8	8	0.1	7.96	0.79	1.25
(c)	4	8	0.5	6.92	0.907	0.25
(d)	6	8	0.75	5.29	1.19	0.17
(e)	8	8	1	...†		0.125, 0.125
(f)	12	8	1.5	...†		0.327, 0.048
(g)	−0.8	8	−0.1	7.96	0.79	−1.25‡

† Not oscillatory. Values in the τ column are $\tau = 1/\sigma_1, 1/\sigma_2$, from (8.6a).
‡ Time constant for *growth* (unstable).

[Sec. 8.4] INITIAL CONDITIONS: $1 < \zeta$

Since ω_0 is found to be constant (the undamped natural frequency is constant because k and J are constant), the roots lie on a circle in the s plane for $\zeta < 1$. They are readily plotted from Table VII, as are the roots for $1 \leqslant \zeta$. Points (a), (b), (c), (d) will be seen to form a circle about the origin.

Problems 8.1 through 8.14

8.4 INITIAL CONDITIONS: $1 < \zeta$

When ζ for a second-order system is greater than unity, the system is said to be "supercritically damped," or "overdamped." Then the natural motion, from (8.6), consists of two exponential decays, each like those studied in Chapter 6:

$$\theta = C_1 e^{-\sigma_1 t} + C_2 e^{-\sigma_2 t} \qquad 0 < t \qquad (8.7a)$$
[repeated]

with the values of σ_1 and σ_2 given in terms of the physical parameters of the system by (8.6a):

$$\sigma_1 = \sigma \left[1 - \sqrt{1 - \frac{1}{\zeta^2}} \right]$$
$$\sigma_2 = \sigma \left[1 + \sqrt{1 - \frac{1}{\zeta^2}} \right] \qquad (8.6a)$$
[repeated]

Following the classical routine, constants C_1 and C_2 are evaluated by setting the general solution (8.7a), and its derivative—the initial velocity $\dot{\theta}$—equal to their known values at some instant. Usually the "instant" chosen is $t = 0$ and the "known values" are the "initial conditions," i.e., the state of the system at $t = 0$. This gives the following two simultaneous expressions for C_1 and C_2:

$$\theta = \theta(0) = C_1 + C_2$$
$$\dot{\theta} = \dot{\theta}(0) = -\sigma_1 C_1 - \sigma_2 C_2$$

with the result that

Natural-response coefficients, $1 < \zeta$

$$\boxed{C_1 = \frac{\sigma_2 \theta(0) + \dot{\theta}(0)}{\sigma_2 - \sigma_1} \qquad C_2 = \frac{-\sigma_1 \theta(0) - \dot{\theta}(0)}{\sigma_2 - \sigma_1}} \qquad (8.10)$$

The unforced motion, for $1 < \zeta$, is given by substituting (8.10) and (8.6a) into (8.7a).

The response θ in this case is merely the sum of two exponential decays, one fast (corresponding to σ_2 in the s-plane plot of Fig. 8.6a) and the other slow (corresponding to σ_1). The time plot, Fig. 8.6b, is constructed by the technique of Fig. 6.3. It is seen that the relative amplitudes of the two

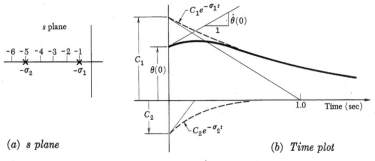

Fig. 8.6 *Natural (unforced) response from general initial conditions of an overdamped $(1 < \zeta)$ second-order system*

terms (8.10) are just such that the initial position and velocity are both satisfied. As a special case, for example, when the initial *position* is zero, the two exponentials will begin at equal and opposite amplitudes. When initial *velocity* is zero, they will begin with equal and opposite slopes. Figure 8.6b illustrates the general case when both $\theta(0)$ and $\dot{\theta}(0)$ have values.

Example 8.2. Overdamped response. The physical constants of the system shown in Fig. 8.7a have the relationships $b/m = 5$, $k/m = 4$. The spring may be taken as linear and the friction as proportional to velocity; x is measured from the zero-spring-strain position. Plot the time response from the following initial conditions:

(i) $x(0) = 2$ in. $\quad \dot{x}(0) = 0$
(ii) $x(0) = 0$ $\quad \dot{x}(0) = 2$ in./sec
(iii) $x(0) = 2$ in. $\quad \dot{x}(0) = 2$ in./sec

The physical model (Stage I) is given in the problem statement. The equation of motion (Stage II) for this physical model is (Problem 2.19):

$$m\ddot{x} + b\dot{x} + kx = 0$$

or, in the form of (8.1)

$$\ddot{x} + \frac{b}{m}\dot{x} + \frac{k}{m}x = 0$$
$$(\ddot{\theta} + 2\sigma\dot{\theta} + \omega_0^2\theta = 0)$$

From the given physical constants [using (8.5)], we have

$$\sigma = 2.5 \quad \omega_0 = 2 \quad \zeta = 1.25$$

From (8.6a), we have

$$\sigma_{1,2} = 2.5\left(1 \mp \sqrt{1 - \frac{1}{1.56}}\right)$$
$$\sigma_1 = 1.00 \text{ sec}^{-1}$$
$$\sigma_2 = 4.00 \text{ sec}^{-1}$$

[Sec. 8.5] INITIAL CONDITIONS: $\zeta < 1$

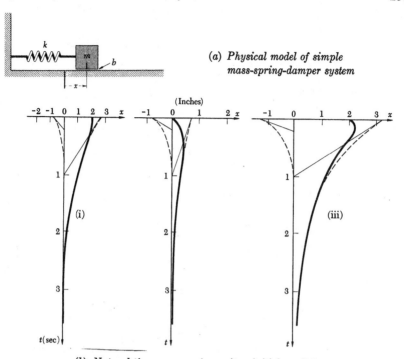

(a) *Physical model of simple mass-spring-damper system*

(b) *Natural time response for various initial conditions*

Fig. 8.7 *Natural motion of an overdamped, second-order system*

The given initial conditions lead to the following constants [from (8.10)]:

	C_1	C_2
(i)	2.67	−0.67
(ii)	0.67	−0.67
(iii)	3.33	−1.33

the response being given by (8.7a):

$$x = C_1 e^{-t} + C_2 e^{-t/0.25}$$

The responses are plotted in Fig. 8.7b, with time running downward (as it does on many recorders) so that the horizontal ordinates correspond to the horizontal displacement x. As a check, note that (iii) may be obtained by superposing (i) and (ii).

Problems 8.15 through 8.17

8.5 INITIAL CONDITIONS: $\zeta < 1$

We now turn to the problem of finding the magnitude and phase of the natural motion of a "subcritically damped" or "underdamped" system,

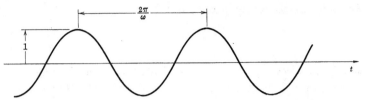

Step (1) (*a*) *and* (*b*): *Plot a cosine of unit height*

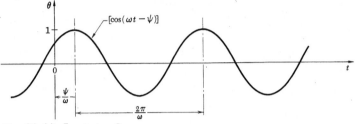

Step (1) (*c*): *Locate* $t = 0$

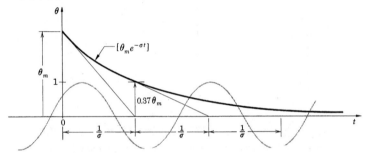

Step (2): *Plot* $[\theta_m e^{-\sigma t}]$

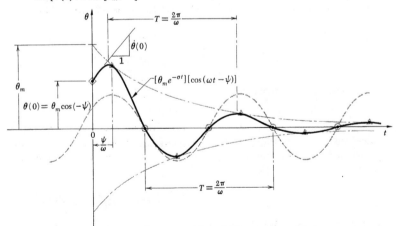

Step (3): *Multiply plots of Steps* (1) *and* (2)

Fig. 8.8 *Using Procedure B-3 to plot a damped cosine* ($\zeta < 1$): $\theta = [\theta_m e^{-\sigma t}][\cos(\omega t - \psi)]$

[Sec. 8.6] SKETCHING TIME RESPONSE WHEN $\zeta < 1$

$\zeta < 1$. We want to find C_1 and C_2 or C_3 and ψ in (8.7c) in terms of the initial conditions $\theta(0)$ and $\dot\theta(0)$. We do so by considering $\theta(t)$ [given by (8.7c)] and $\dot\theta(t)$ at $t = 0$:

$$\theta(t) = C_1 e^{(-\sigma+j\omega)t} + C_2 e^{(-\sigma-j\omega)t}$$
$$\dot\theta(t) = C_1(-\sigma + j\omega)e^{(-\sigma+j\omega)t} + C_2(-\sigma - j\omega)e^{(-\sigma-j\omega)t}$$
$$\theta(0) = C_1 + C_2$$
$$\dot\theta(0) = C_1(-\sigma + j\omega) + C_2(-\sigma - j\omega)$$

Simultaneous solution of the above $t = 0$ equations gives expressions for C_1 and C_2 which are simple combinations of dynamic parameters σ and ω, and the initial conditions:

$$C_1 = \theta(0)\left(\frac{\sigma + j\omega}{2j\omega}\right) + \dot\theta(0)\left(\frac{1}{2j\omega}\right)$$
$$C_2 = \theta(0)\left(\frac{-\sigma + j\omega}{2j\omega}\right) + \dot\theta(0)\left(\frac{-1}{2j\omega}\right)$$

When these constants are substituted back into (8.7c), the result includes a complex combination of terms in $e^{j\omega t}$ and $e^{-j\omega t}$. It is known from physical reasoning (Fig. 8.1) that the resultant must contain only sinusoidal terms with real coefficients. The complex terms are unscrambled by the use of Euler's equation (7.21), with the following result (see Problem 8.18):

$$\theta = e^{-\sigma t}\left\{\theta(0)\cos\omega t + \left[\frac{\sigma}{\omega}\theta(0) + \frac{\dot\theta(0)}{\omega}\right]\sin\omega t\right\} \qquad (8.11a)$$

or, alternatively,

Natural response, $\zeta < 1$

with
$$\theta = \theta_m e^{-\sigma t}\cos(\omega t - \psi)$$
$$\theta_m = \sqrt{\left[\frac{\sigma}{\omega}\theta(0) + \frac{\dot\theta(0)}{\omega}\right]^2 + [\theta(0)]^2} \qquad (8.11b)$$
$$\psi = \tan^{-1}\left[\frac{\sigma}{\omega} + \frac{\dot\theta(0)/\omega}{\theta(0)}\right]$$

Result (8.11b) can also be obtained without the use of complex numbers, simply by considering $\theta = \theta_m e^{-\sigma t}\cos(\omega t - \psi)$ and its derivative at $t = 0$. (See Problem 8.19.) Note that the use of a phase angle, ψ, has allowed us to write the sum of a cosine and a sine wave, (8.11a), as a single cosine wave, (8.11b). The significance of ψ is discussed on p. 243.

°8.6 SKETCHING TIME RESPONSE WHEN $\zeta < 1$

It is worthwhile to develop techniques for sketching rapidly, but accurately, the time response of a second-order system whose initial conditions are known, not only because one is occasionally called upon to do so, but also because one develops thereby a strong mental correlation between the system parameters and their effect upon system behavior.

The most convenient basis for constructing the time plot is Eq. (8.11b). Procedure B-3, which follows, is a technique for rapid construction as demonstrated in Fig. 8.8. This technique is based on plotting separately the two components of (8.11b) indicated by the square brackets in (8.11c):

$$\theta = [\theta_m e^{-\sigma t}][\cos(\omega t - \psi)] \qquad (8.11c)$$

Procedure B-3
Sketching a damped sinusoid
(Refer to Fig. 8.8)

(1) **Sketch** $[\cos(\omega t - \psi)]$. The easy thing is simply to:
  (a) Plot a cosine wave to convenient dimensions [Fig. 8.8, Step (1) (a) and (b)].
  (b) Decree that its amplitude is 1 and that the time between peaks is $2\pi/\omega$ [not $2\pi/\omega_0$; refer to (8.11c)].
  (c) Then locate the $t = 0$ axis at the proper distance, ψ/ω, to the left of a peak.
(2) **Plot** $[\theta_m e^{-\sigma t}]$. For this we use verbatim the technique of Fig. 6.3b, p. 187.
(3) **Multiply** *the plots together, using easy points.*
  (a) Plot the mirror image of the $\theta_m e^{-\sigma t}$ curve.
  (b) Mark zero crossings as points ○ and peaks of the cosine curve as points △. (The product is *on* the $\theta_m e^{-\sigma t}$ plot because $\cos(\omega t - \psi) = 1$ there.)
  (c) Find accurately the point at $t = 0$, point ◇:
      $\theta(0) = \theta_m \cos(-\psi)$.
  (d) Sketch in the rest of the plot.

Notice that the maxima of the function θ do *not* occur just at the maxima of $\cos(\omega t - \psi)$, i.e., at points △. They occur slightly earlier than points △. [It is still accurate, however, to judge the period of a damped sine wave by measuring the time between peaks (see Problem 8.20). The time between zero crossings is also an accurate measure, as shown.]

As a final check, the initial values $\theta(0)$ and $\dot{\theta}(0)$ should always be measured in the finished plot, to make sure they are the given values.

Example 8.3. Underdamped response. The system of Fig. 8.1 has the following physical constants: $b/J = 5 \text{ sec}^{-1}$, $k/J = 81 \text{ sec}^{-2}$. The system is given an initial velocity of 1.7 rad/sec but has no initial displacement. Plot the resulting time response.

From the given data, and using (8.5), we have

$$\sigma = 2.5 \qquad \omega_0 = 9 \qquad \omega = 8.6 \qquad \tau = \frac{1}{\sigma} = 0.4 \qquad T = \frac{2\pi}{\omega} = 0.73$$

From (8.9b), we have

$$\theta_m = \sqrt{\left[\frac{\dot{\theta}(0)}{\omega}\right]^2} = \frac{1.7}{8.6} = 0.2$$

$$\psi = \tan^{-1}(\infty) = \frac{\pi}{2}$$

$$\theta = [0.2 e^{-t/0.4}][\sin 8.6 t]$$

[Sec. 8.6] SKETCHING TIME RESPONSE WHEN $\zeta < 1$ 261

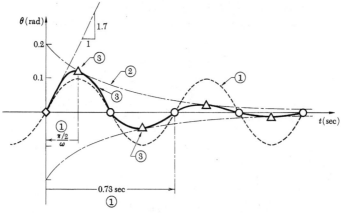

Fig. 8.9 *Natural response of the system in Fig. 8.1, with $\theta(0) = 0$ and $\zeta < 1$*

This is plotted in Fig. 8.9. (The steps of Procedure B-3 are labeled ①, ②, ③.) Note in the plot that $\theta(0) = 0$, $\dot{\theta}(0) = 1.7$ rad/sec.

Example 8.4. RLC network. Consider the network (ideal physical model) shown in Fig. 8.10. The switch is connected to the battery for a long time. Then, at $t = 0$, it is switched to ground. Plot the subsequent time history of voltage v_3 across the capacitor if $R_2/L_1 = 100$ sec^{-1} and $1/L_1C_3 = 40{,}000$ sec^{-2}.

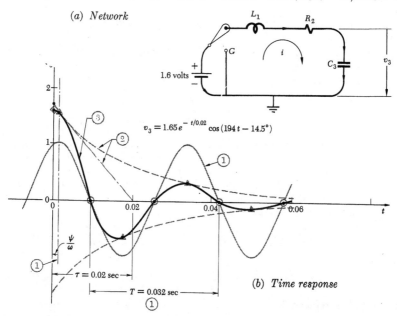

Fig. 8.10 *Natural response of an underdamped RLC network after grounding its input*

First (Stage II) we write the equation of motion.
(1) *Variables.* Single loop; choose i as independent variable. (Identity: $i_1 \equiv i_2 \equiv i_3 \equiv i$)
(2) *Equilibrium.* KVL after switching:
$$v_1 + v_2 + v_3 = 0$$
(3) *Physical relations.* $\quad v_1 = L_1 \dfrac{di}{dt} \quad\quad v_2 = R_2 i$
$$v_3 = \frac{1}{C_3}\int i\, dt \quad \text{or} \quad i = C_3 \dot{v}_3$$

Combining, we obtain
$$L_1 \frac{di}{dt} + R_2 i + \frac{1}{C_3}\int i\, dt = 0 \quad\quad 0 < $$
or, since we are interested in v_3,
$$L_1 C_3 \ddot{v}_3 + R_2 C_3 \dot{v}_3 + v_3 = 0$$

For convenience we write this in the form of (8.1):
$$\ddot{v}_3 + \frac{R_2}{L_1}\dot{v}_3 + \frac{1}{L_1 C_3} v_3 = 0$$
$$(\ddot{\theta} + 2\sigma\dot{\theta} + \omega_0^2 \theta = 0) \tag{8.12}$$

Then, from the given numbers, using (8.5), (8.8), and (8.9), we obtain

$$\sigma = \frac{R_2}{2L_1} = 50 \quad\quad \omega_0 = \frac{1}{\sqrt{L_1 C_3}} = 200 \quad\quad \omega = 194 \quad\quad \tau = 0.02 \quad\quad T = 0.032$$

We consider next the initial conditions. Before switching, there is no current (transients have died out), and KVL gives
$$0 + 0 + v_3(0^-) = 1.6 \text{ volts}$$
This capacitor voltage will not change instantly (unless subjected to an impulse, which it is not here); i.e., $v_3(0^+) = v_3(0^-)$. Since the system is second-order, we require $\dot{v}_3(0^+)$ as well as $v_3(0^+)$. We obtain $\dot{v}_3(0^+)$ from the physical relation $\dot{v}_3 = (1/C_3)i$, and the facts that (1) $i_1(0^-) \equiv i(0^-) = 0$, and (2) the current in an inductor will not change instantaneously: $i_1(0^+) = i_1(0^-)$. Thus $\dot{v}_3(0^+) = 0$. Then from (8.11b) we obtain

$$v_m = v_3(0^-)\sqrt{\left(\frac{\sigma}{\omega}\right)^2 + 1} = 1.6\sqrt{\left(\frac{200}{194}\right)^2} = 1.65$$

$$\psi = \tan^{-1}\left(\frac{\sigma}{\omega}\right) = \tan^{-1}(0.26) = 14.5° \quad\text{or}\quad 0.25 \text{ rad}$$

The response is
$$v_3 = 1.65 e^{-t/0.02} \cos(194t - 14.5°) \tag{8.13}$$

[Sec. 8.7] INITIAL CONDITIONS: $\zeta = 1$

We have plotted this response in Fig. 8.10b, using Procedure B-3. (The plotting steps are numbered.) Note that in the plot $v_3(0) = 1.6$, $\dot{v}_3(0) = 0$.

Problems 8.18 through 8.24

8.7 INITIAL CONDITIONS: $\zeta = 1$

When damping is exactly "critical," $\zeta = 1$, we have a special case which requires some special handling mathematically. We hasten to state, however, that the physical response of the system is not noticeably different when $\zeta = 1$ from what it is when $\zeta = 0.95$ or $\zeta = 1.05$. Nothing singular happens *physically* just because ζ is exactly equal to 1—only mathematically.

Mathematically, from either (8.10) or (8.11) we can see that the solution is in indeterminate form when $\zeta = 1$. [That is, in (8.11) $\zeta = 1$ means $\omega = 0$ and therefore $\theta_m = \infty$, and the indeterminate form is $0 \times \infty$; while in (8.10) $\sigma_1 = \sigma_2$, and the indeterminate form is $0/0$.] Mathematically, we could consider approaching $\zeta = 1$ as a limiting case either of (8.10) or of (8.11). It is a little simpler to work from (8.10), and besides, we have already worked out the limiting process involved there in connection with "resonance" in first-order systems (Sec. 6.7).

We begin with the differential equation of motion in the form

$$\ddot{\theta} + 2\sigma\dot{\theta} + \omega_0^2 \theta = 0$$

and we assume that ζ is slightly greater than 1, so that the solution is given by (8.10). Now in (8.10) let

$$\left. \begin{array}{l} \sigma_1 = \sigma(1 - \epsilon) \\ \sigma_2 = \sigma(1 + \epsilon) \end{array} \right\} \quad (8.14)$$

which implies [using a little algebra on (8.6)] that

$$\zeta = \frac{1}{\sqrt{1-\epsilon^2}} \approx \left(1 + \frac{1}{2!}\epsilon^2 + \cdots \right) \quad (8.15)$$

Substituting (8.14) into (8.10), and rearranging, we obtain

$$\theta = \theta(0) \left[\frac{\sigma(1+\epsilon)e^{-\sigma(1-\epsilon)t} - \sigma(1-\epsilon)e^{-\sigma(1+\epsilon)t}}{2\epsilon\sigma} \right] + \dot{\theta}(0) \left[\frac{e^{-\sigma(1-\epsilon)t} - e^{-\sigma(1+\epsilon)t}}{2\epsilon\sigma} \right]$$

or, letting

$$e^{\sigma\epsilon t} = 1 + \sigma\epsilon t + \frac{\sigma^2\epsilon^2 t^2}{2!} + \cdots$$

and

$$e^{-\sigma\epsilon t} = 1 - \sigma\epsilon t + \frac{\sigma^2\epsilon^2 t^2}{2!} - \cdots$$

we obtain

$$\theta = \theta(0)\frac{e^{-\sigma t}}{2\epsilon}\left[(1+\epsilon)\left(1+\sigma\epsilon t+\frac{\sigma^2\epsilon^2 t^2}{2!}+\frac{\sigma^3\epsilon^3 t^3}{3!}+\cdots\right)\right.$$
$$\left.-(1-\epsilon)\left(1-\sigma\epsilon t+\frac{\sigma^2\epsilon^2 t^2}{2!}-\frac{\sigma^3\epsilon^3 t^3}{3!}+\cdots\right)\right]$$
$$+\dot{\theta}(0)\frac{e^{-\sigma t}}{2\sigma\epsilon}\left[\left(1+\sigma\epsilon t+\frac{\sigma^2\epsilon^2 t^2}{2!}+\frac{\sigma^3\epsilon^3 t^3}{3!}+\cdots\right)-\left(1-\sigma\epsilon t+\frac{\sigma^2\epsilon^2 t^2}{2!}\right.\right.$$
$$\left.\left.-\frac{\sigma^3\epsilon^3 t^3}{3!}+\cdots\right)\right]$$

$$\theta \approx \theta(0)e^{-\sigma t}\left[1+\sigma t+\epsilon\frac{\sigma^2 t^2}{2!}+\epsilon^2\frac{\sigma^2 t^2}{2!}\left(1+\frac{\sigma t}{3}\right)+\cdots\right]$$
$$+\dot{\theta}(0)te^{-\sigma t}\left[1+\epsilon^2\frac{\sigma^2 t^2}{3!}\right] \quad (8.16)$$

Then, taking the limit as $\epsilon \to 0$, we have

Natural response for $\zeta = 1$
$$\boxed{\theta = \theta(0)(1+\sigma t)e^{-\sigma t}+\dot{\theta}(0)te^{-\sigma t}} \quad (8.17)$$

[As one check, notice that when $t = 0$ in (8.17), we have $\theta = \theta(0)$. Show also, by differentiating (8.17), that when $t = 0$, $\dot{\theta} = \dot{\theta}(0)$.]

A plot of (8.17) for arbitrary values of $\theta(0)$ and $\dot{\theta}(0)$ is given in Fig. 8.11.

As an example of critically damped response, let us find for what value of b/J the system in Example 8.3 will be exactly critically damped ($\zeta = 1$), and repeat Example 8.3 for that case.

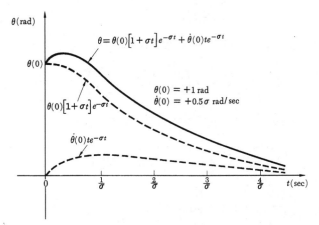

Fig. 8.11 *Natural response from general initial conditions of a critically damped ($\zeta = 1$) second-order system*

[Sec. 8.7] INITIAL CONDITIONS: $\zeta = 1$

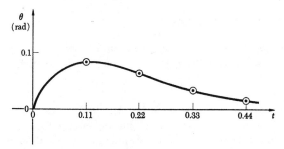

Fig. 8.12 *The natural motion of a critically damped system ($\zeta = 1$) having initial velocity only*

For $\zeta = 1$, we require $\sigma = \omega_0$, or $b/J = 2\sigma = 2\omega_0 = 18$. Then the other constants are

$$\sigma = 9 \qquad \omega_0 = 9 \qquad \omega = 0 \qquad \tau = 0.11$$

We were given the initial conditions $\theta(0) = 0$, $\dot{\theta}(0) = 2$ rad/sec; hence the response is (8.17)

$$\theta = 2te^{-t/0.11}$$

which is plotted in Fig. 8.12.

We note in passing, for future reference, that it is possible to generate a function of the form

$$Cte^{-\sigma t} \tag{8.18}$$

(as shown in Fig. 8.12) by releasing, with initial velocity, a second-order system having equal roots.

Computing response for $\zeta \approx 1$

When ζ is not exactly equal to 1, but is nearly so, plotting the response (8.6) will consist of plotting the small difference between two large-valued curves, as it did in Sec. 6.7. As in that section, it is easier, and in fact more accurate, to obtain the plot by first making a plot for equal roots (8.17) and then correcting this by the small error due to the fact that the roots are not quite equal. The correction is indicated from (8.16):

$$\theta = \theta_{\zeta=1} + \dot{\theta}(0)\epsilon \frac{\sigma^2 t^2}{2!} e^{-\sigma t} + \dot{\theta}(0)\epsilon^2 \frac{\sigma^2 t^2}{3!} te^{-\sigma t} \tag{8.19}$$

It is often quite acceptable simply to ignore the correction. [The error involved in doing so is readily evaluated from (8.19).]

Problems 8.25 through 8.28

8.8 FORCED MOTION ALONE

To find the response of the second-order system of Fig. 8.1 to an applied moment $M(t)$, we begin with the complete Eq. (8.1'):

$$J\ddot{\theta} + b\dot{\theta} + k\theta = M \qquad (8.20)$$

In accordance with Procedure B-1 (p. 182) we can, for convenience, investigate first the forced portion of the response, without regard for any accompanying natural motions. (Sometimes this is really all we are interested in.)† Later (in Sec. 8.11) we shall look at the combination of forced plus natural motions accompanying, for example, sudden application of a forcing function.

Actually, we can at this point handle readily only a few of the many possible forcing functions to which a system may be subjected; but they are the important few! Specifically, we shall find that we can calculate promptly the response to the first four functions in Fig. 8.13, because they can all be represented in the form $M = \mathsf{M}e^{st}$ in which M is a constant. Handling, for example, the remaining functions in Fig. 8.13 requires using convolution or techniques we have not yet discussed. We defer their study until later (Part D), and concentrate attention on the first four functions.

As it turns out, these particular functions—especially the undamped sinusoid—are the key to a broad segment of the problems in elementary dynamic response. In each case the value of $M(t)$ in (8.20) can be written in the form $M = \mathsf{M}e^{st}$ as follows:

$$\left. \begin{array}{lll} \text{General} & M = \mathsf{M}e^{st} & \\ \text{Constant} & M = \mathsf{M} & (s = 0) \\ \text{Exponential} & M = \mathsf{M}e^{-\sigma_f t} & (s = -\sigma_f) \\ \text{Sinusoid}\ddagger & M \stackrel{r}{=} \mathsf{M}e^{j\omega_f t} & (s = j\omega_f) \\ \text{Damped sinusoid}\ddagger & M \stackrel{r}{=} \mathsf{M}e^{(-\sigma_f + j\omega_f)t} & (s = -\sigma_f + j\omega_f) \end{array} \right\} \quad (8.21)$$

The solution is therefore most efficiently obtained in the general form.

Some additional flexibility is available by allowing the constant M to be complex. Then if $M(t)$ has a phase angle, it can be incorporated in the constant M, which is then a phasor (footnote, p. 238). This is demonstrated in Example 8.5.

We have already seen, for first-order systems (p. 197), that if the input has the form e^{st}, then the output will have the same form (provided that the system is l.c.c.—linear, with constant coefficients). We therefore assume:

$$\theta_f = \Theta e^{st}$$

† As when the forcing function has been present a long time before the start of the problem—i.e., before $t = 0$—so that, as Morse puts it, the steady-state motion has forgotten how it started. P. M. Morse, "Vibration and Sound," McGraw-Hill Book Company, New York, 1948.

‡ The notation $\stackrel{r}{=}$ means "equals the real part of."

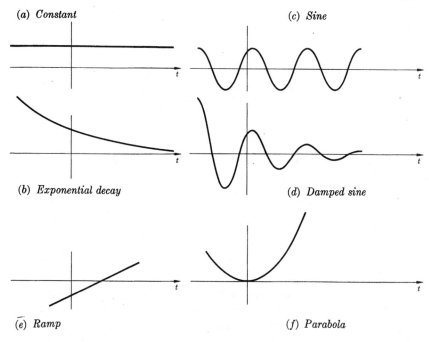

(a) *Constant*

(b) *Exponential decay*

(c) *Sine*

(d) *Damped sine*

(e) *Ramp*

(f) *Parabola*

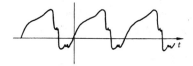

(g) *Periodic*

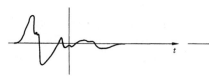

(h) *Random, nonperiodic*

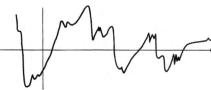

(i) *Random, nonperiodic*

Fig. 8.13 *Types of continuing time functions*

in which Θ is again a (complex) constant, and substitute this into (8.20):

$$Js^2\Theta e^{st} + bs\Theta e^{st} + k\Theta e^{st} = Me^{st}$$

or

$$(Js^2 + bs + k)\Theta = M$$

from which

$$\Theta = \frac{M}{Js^2 + bs + k} \tag{8.22a}$$

That is, if $M = \vec{M}e^{\vec{\alpha}t}$, then

$$\theta_f = \left[\frac{Me^{st}}{Js^2 + bs + k}\right]_{s=\vec{\alpha}} \quad (8.22b)$$

For any one of the functions (8.21), the value of Θ can be obtained by letting s have the appropriate value. For example, for the exponential input $M = M_0 e^{-\sigma_f t}$, $s = -\sigma_f$ (and $\mathsf{M} = M_0$), and

$$\Theta = \frac{M_0}{J\sigma_f{}^2 - b\sigma_f + k}$$

Notice that when s is complex, the constant Θ is also complex. Thus, when the element is driven by an undamped sinusoid, for example, $M = M_0 \cos \omega_f t \stackrel{r}{=} M_0 e^{j\omega_f t}$, then $s = j\omega_f$, $\mathsf{M} = M_0$ and

$$\Theta = \frac{M_0}{-J\omega_f{}^2 + jb\omega_f + k} \quad (8.23a)$$

and
$$\theta_f \stackrel{r}{=} \left[\frac{M_0}{-J\omega_f{}^2 + jb\omega_f + k}\right]e^{j\omega_f t} \quad (8.23b)$$

[Using (7.24a), we have let $M_0 \cos \omega_f t = M_0(e^{j\omega_f t} + e^{-j\omega_f t})/2$, then solved half the problem. This general result is to be proved in Problem 8.28.]

The forced motion for each case in (8.21) except $s = 0$ is plotted in Fig. 8.14. Note that in each case the magnitude of the response is proportional to the forcing function, and that the decay time constant and the frequency of the response are the same as those of the forcing function.

The case $s = j\omega_f$—the simple undamped sine-wave input—is so important, in nature and in engineering, that we shall devote a whole chapter to it, Chapter 9. It is also the subject of Examples 8.6 and 8.7 in the present section.

The last case—the most general of the four—is calculated in detail in Example 8.5.

Phase angle

Recall that for natural vibrations, Fig. 7.8b, we defined a phase angle ψ to indicate the time $t = \psi/\omega$ at which the cosine-wave peak occurs. Thus we write $\theta_n = \theta_{n\,max} \cos(\omega t - \psi)$ (so that $\theta_n = \theta_{n\,max}$ when $\omega t - \psi = 0$, and the sign is convenient).

For forced motion, it is common to define phase angle ψ_f to indicate the time by which a cosine peak *precedes* (or "leads") time 0. Thus in Fig. 8.14b we write $\theta_f = \theta_{f\,max} \cos(\omega_f t + \psi_f)$ and also $M = M_0 \cos(\omega_f t + \psi_o)$. The angle $(\psi_f - \psi_o)$ by which the output *leads* the input is then given directly and systematically from the transfer function, (8.23b):

$$\psi_f - \psi_0 = \underline{/M_o/(-J\omega_f{}^2 + jb\omega_f + k)}$$

This is why ψ_f is chosen to represent lead, rather than lag. (Commonly in physical systems, $\psi_f - \psi_o$ comes out negative when numerical values are substi-

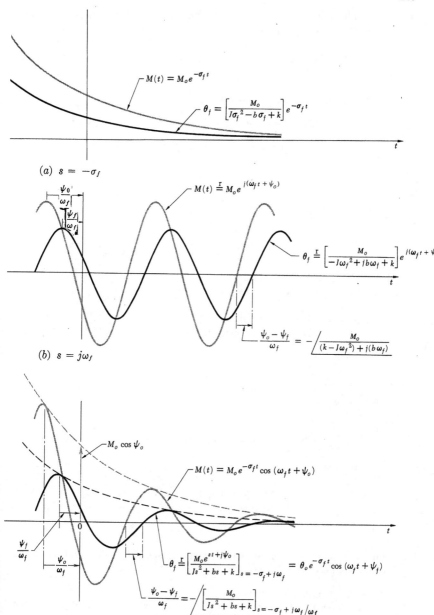

Fig. 8.14 Forced response alone of a second-order system, when the input has the general form $M(t) = \mathsf{M}e^{st}$

tuted, as indeed it does in Fig. 8.14c: the output actually lags the input. However, the choice of sign we have made is still the more systematic.)

Example 8.5. Second-order system response to a damped sine input. The system of Fig. 8.1 is driven by a forcing function like Fig. 8.13d. Find the forced response.

The forcing function can be written

$$M(t) = M_0 e^{-\sigma_f t} \cos(\omega_f t + \psi_0) \stackrel{r}{=} M_0 e^{(-\sigma_f + j\omega_f)t + j\psi_0} = \mathsf{M} e^{(-\sigma_f + j\omega_f)t}$$

with M_0 and ψ_0 identified as in Fig. 8.14c and $\mathsf{M} = M_0 e^{j\psi_0}$. Then, from (8.23b),

$$\theta_f \stackrel{r}{=} \left[\frac{M_0}{J(-\sigma_f + j\omega_f)^2 + b(-\sigma_f + j\omega_f) + k} \right] e^{(-\sigma_f + j\omega_f)t + j\psi_0}$$
$$= \theta_0 e^{-\sigma_f t} \cos(\omega_f t + \psi_f)$$

with
$$\theta_0 = \left| \frac{M_0}{J(-\sigma_f + j\omega_f)^2 + b(-\sigma_f + j\omega_f) + k} \right|$$
$$\psi_f = \psi_0 + \bigg/ \frac{M_0}{J(-\sigma_f + j\omega_f)^2 + b(-\sigma_f + j\omega_f) + k}$$

This result is plotted in Fig. 8.14c for specific numerical values of the constants. Details are included in Problem 8.29.

Example 8.6. Vibration isolation. The simple model (Stage I) shown in Fig. 8.15 represents an instrument package that is isolated from the main

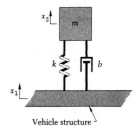

Fig. 8.15 *Simple model for vibration isolation*

structure of a vehicle. (This is a simplified model for the system in Fig. 1.1.) Motions of the vehicle structure can be represented by a continuous function,

$$x_1(t) = X_1 e^{st}$$

which is sometimes exponential and sometimes sinusoidal. Obtain an expression for the response x_2 of the instrument package.

STAGE II. EQUATIONS OF MOTION

(1) *Geometry.* Choose x_2 so that $x_2 = 0$ when $x_1 = 0$ and m is in static equilibrium under gravity. Let δ equal the static deflection under gravity. Then:

Spring compression $= x_1 - x_2 + \delta$
Rate of dashpot compression $= \dot{x}_1 - \dot{x}_2$
Absolute acceleration of $m = \ddot{x}_2$

[Sec. 8.8] FORCED MOTION ALONE 271

(2) *Force equilibrium.*

Static: $mg - f_{k\text{ static}} = 0$

Dynamic: $f_i + mg - f_b - f_k = 0$

(3) *Physical relations.* These are simply inserted into the equation of motion—note that $f_{k\text{ static}} = k\delta = mg$

$$m\ddot{x}_2 - b(\dot{x}_1 - \dot{x}_2) - k(x_1 - x_2) = 0 \tag{8.24}$$

The weight and the static spring compression have canceled in (8.24), and thus gravity will not affect the dynamic behavior.

STAGE III. DYNAMIC BEHAVIOR

We assume that motions x_1 and x_2 are both of the form Xe^{st}. Substituting this assumption in the equation of motion we obtain

$$(ms^2 + bs + k)X_2 = (bs + k)X_1$$

which leads to the relation

$$X_2 = \frac{bs + k}{ms^2 + bs + k} X_1 \tag{8.25}$$

Consider first that the input is exponential, with a time constant $1/\sigma_f$:

$$x_1 = X_1 e^{-\sigma_f t}$$

Then the forced part of response x_2 motion is given by

$$x_2 = X_2 e^{st} = \left[\frac{bs + k}{ms^2 + bs + k} X_1 e^{st}\right]_{s=-\sigma_f}$$

$$= \frac{(-b\sigma_f + k)X_1}{m\sigma_f^2 - b\sigma_f + k} e^{-\sigma_f t}$$

Consider next that the input is a continuing sine wave,

$$x_1 = x_{1m} \cos \omega_f t \stackrel{r}{=} X_1 e^{j\omega_f t} \qquad X_1 = x_{1m}$$

Then the forced response is given by

$$x_2 = X_2 e^{st} \stackrel{r}{=} \left[\frac{bs + k}{ms^2 + bs + k} X_1 e^{st}\right]_{s=j\omega_f}$$

$$\stackrel{r}{=} \frac{bj\omega_f + k}{-m\omega_f^2 + bj\omega_f + k} x_{1m} e^{j\omega_f t}$$

$$x_2 = \left|\frac{bj\omega_f + k}{-m\omega_f^2 + bj\omega_f + k}\right| x_{1m} \cos(\omega_f t + \psi) \tag{8.26}$$

$$\psi = \bigg/\frac{bj\omega_f + k}{-m\omega_f^2 + bj\omega_f + k}$$

Resonance

The results (8.26) of Example 8.6 indicate precisely how the response of the given system to a sinusoidal disturbance depends on the frequency of the disturbance. Typically, if damping b is relatively small, there will be a range of values of ω_f over which the output motion x_2 will actually be larger than the input motion x_1.

In particular, if the disturbing frequency happens to be the same as the undamped natural frequency of the system, $\omega_f = \sqrt{k/m}$, then the first and third terms in the denominator of (8.26) cancel, and only damping b prevents the output from being "infinitely" large. This value of ω_f is called the *resonant frequency*, and the phenomenon of a large output for a small input near the resonant frequency is called *resonance*.

One easy way to experience the phenomenon is with a weight hung on the end of a chain of rubber bands, as in Fig. 8.16 (which duplicates closely the system of Fig. 8.15, with small damping being furnished by the rubber). First let the weight oscillate naturally to find its natural frequency. Then oscillate the upper end of the rubber band with your fingers at the same frequency and *small* amplitude. With a little trial, the frequency will be found for which the output oscillation builds up to a large amplitude, even though the input amplitude is negligible. [It will also be noticed that the *velocity* of the weight is approximately in phase with the *displacement* of your fingers. Does this correspond to result (8.26) for b very small?]

The physical explanation is that at the resonant frequency the motion adjusts so that the small energy furnished in each cycle by the input is mostly stored in the spring and mass (while a little is dissipated in the damping and none is put back into your fingers).

Fig. 8.16
A resonance experiment

Resonance is discussed in much more detail in Sec. 9.7. The above experiment is treated in Example 9.3.

Example 8.7. RLC Network. The ideal RLC network of Example 8.4 is driven by a sinusoidal voltage source having frequency ω_f, Fig. 8.17. Find the forced part of the response of current i (the continuing response after the transient natural motions have died out). Find also the response of voltage v_5 across R and C.

We augment the equation of motion obtained in Example 8.4 by including the driving voltage v_1:

$$L\frac{di}{dt} + Ri + \frac{1}{C}\int i\, dt = v_1 \qquad (8.27)$$

(The subscripts shown on L, R, and C in Example 8.4 are dropped here.)

Next, let us consider the general form of input $v_1 = V_1 e^{st}$ and assume the corresponding response $i = I e^{st}$. Substituting these in (8.27) gives the relation

$$I = \frac{1}{Ls + R + (1/Cs)} V_1 \tag{8.28}$$

Then for the special case at hand, namely,

$$v_1 = v_{1m} \cos(\omega_f t + \psi_1) \stackrel{r}{=} v_{1m} e^{j\psi_1} e^{j\omega_f t} \stackrel{r}{=} V_1 e^{st}$$

with $V_1 = v_{1m} e^{j\psi_1}$ and $s = j\omega_f$, the solution is $i \stackrel{r}{=} I e^{j\omega_f t}$
in which phasor I is given by $I = \left[\dfrac{1}{Ls + R + (1/Cs)} \right]_{s = j\omega_f} V_1$

Thus $\quad i \stackrel{r}{=} I e^{j\omega_f t} \stackrel{r}{=} \dfrac{1}{Lj\omega_f + R + (1/Cj\omega_f)} v_{1m} e^{j\psi_1} e^{j\omega_f t} \quad (8.29)$

That is, I has a phase angle as well as a magnitude which depend on frequency (and which are different from V_1's phase and magnitude).

To find $v_5(t)$ we first write the identity $v_5 \equiv v_2 + v_3$ and the physical relations

$$v_2 = Ri \qquad v_3 = \frac{1}{C} \int i \, dt$$

and combine:

$$v_5 = Ri + \frac{1}{C} \int i \, dt$$

Then assuming $v_5 = V_5 e^{st}$, we obtain

$$\begin{aligned} V_5 &= \left(R + \frac{1}{Cs} \right) I \\ &= \frac{R + (1/Cs)}{Ls + R + (1/Cs)} V_1 \end{aligned} \tag{8.30}$$

Finally, as in (8.29), we let $s = j\omega_f$:

$$V_5 = \left[\frac{R + (1/Cs)}{Ls + R + (1/Cs)} \right]_{s = j\omega_f} V_1 \tag{8.31}$$

and thus the time response is given by

$$v_5 \stackrel{r}{=} V_5 e^{j\omega_f t} \stackrel{r}{=} \left[\frac{R + (1/Cj\omega_f)}{Lj\omega_f + R + (1/Cj\omega_f)} \right] v_{1m} e^{j\psi_1} e^{j\omega_f t} \tag{8.32}$$

For what value of ω_f will resonance be experienced by the system of Fig. 8.17? What will be the ratio of $v_{5\max}/v_{1\max}$ at that frequency?

Problems 8.29 through 8.34

8.9 TRANSFER FUNCTIONS AND POLE-ZERO DIAGRAMS

The concept

Whenever we assume the variables in a system to have behavior of the form $x = X e^{st}$, we find that we are concerned with the ratios of constants

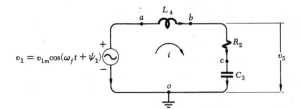

Fig. 8.17 *Network driven by sinusoidal voltage*

X. The ratio between any two such constants is known as a transfer function.

In the demonstration of Sec. 8.8, for example, the input has the form $M = \mathsf{M}e^{st}$, and the output is assumed to have the form $\theta = \Theta e^{st}$. The transfer function from M to θ is then defined as the ratio of the constants Θ and M, which is, from (8.22),

$$\frac{\Theta}{\mathsf{M}} = \frac{1}{Js^2 + bs + k} \tag{8.33}$$

The transfer function indicates how much of the e^{st} signal is *transferred* from the input to the output.

Similarly, in Example 8.6 the transfer function from input motion x_1 to output motion x_2 is, from (8.25),

$$\frac{X_2}{X_1} = \frac{bs + k}{ms^2 + bs + k} \tag{8.34}$$

Again, in Example 8.7 we considered an input voltage of the form $v_1 = V_1 e^{st}$ which produced a current of the form $i = I e^{st}$. The transfer function from v_1 to i is then, from (8.28),

$$\frac{I}{V_1} = \frac{(1/L)s}{s^2 + (R/L)s + (1/LC)} \tag{8.35}$$

The constants, like X, Θ, M, V, I, may in general be complex numbers (phasors) (as M is in Sec. 8.8).

It is useful (and proper) to think of transfer functions as *cause-and-effect* relations. In Fig. 8.1, M is the physical "cause" or "input" which produces motion, and θ is the "effect" or "output" in which we are interested. The transfer function (8.33) gives the ratio of effect to cause, Θ/M.

In Example 8.6, input motion x_1 is the cause, output motion x_2 the effect, and transfer function (8.34) is the ratio X_2/X_1.

In Example 8.7, input voltage v_1 is the cause, and two effects were considered, current i and output voltage v_5. Correspondingly, there are two transfer functions (ratio of effect to cause). One is (8.35), and the other is obtainable from (8.30). In another study of the same circuit we might encounter a transfer function which is the reciprocal of one of these. For example, suppose we ask what voltage v_1 results when we drive the network in Example 8.7 with a current source i_1, as in Fig. 8.18. Now the roles have changed, and i_1 is the "cause" ("input") and v_1 the "effect" ("output").

[Sec. 8.9] TRANSFER FUNCTIONS AND POLE-ZERO DIAGRAMS 275

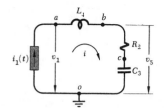

Fig. 8.18 *Network of Fig. 8.17 driven by a current source*

The equation of motion is still (8.27), but the transfer function of interest is†

$$\frac{V_1}{I_1} = Ls + R + \frac{1}{Cs} \quad (8.36)$$

In mechanical systems this role inversion is less common, but may occasionally be of interest. For example, in Fig. 8.1 we might (using a powerful driving amplifier and tight feedback system) impose upon the rotor a given displacement pattern $\theta(t)$ and then ask what torque M must exist at the input shaft as a result. The transfer function would then be the reciprocal of (8.33):

$$\frac{M}{\Theta} = Js^2 + bs + k \quad (8.37)$$

Other examples are noted in Problems 8.36 and 8.37.

Block representation

A block representation is often used to emphasize the cause-effect relations in physical systems (as in Fig. 1.4). The transfer function lends itself most elegantly to this kind of representation. Thus in Sec. 8.8 we can show the relation between moment M (the cause) and response θ (the effect), as in Fig. 8.19a.

In Example 8.6 the relation between input x_1 and output x_2 is shown by the block in Fig. 8.19b. In Example 8.7, with a voltage source v_1, the relation between "cause" v_1 and "effect" i is shown by the block in Fig. 8.19c; while with the current source (Fig. 8.18) we take i_1 to be the "cause" and v_1 to be the "effect," and show the relation in the block form of Fig. 8.19d.

In the discussion of simple systems like those described above, transfer functions serve principally to lend convenience and organization. But further, in the analysis of compound systems they constitute a really important tool, because they help us partition the compound system into elementary subsystems whose response is easily found. The use of transfer functions to study compound systems is the subject of Chapter 14.

It is of central importance to note that the denominator (or, in certain

---

† Note that here for element 1 (the current source) we have (for the first time in this book) not followed our convention of defining voltage *drop* (v_1) to be positive in the same direction that current (i_1) is positive. We have done so out of deference to well-established custom regarding "impedance" (Sec. 8.10). There should be no confusion, the directions of v_1 and i_1 being clearly labeled on the diagrams.

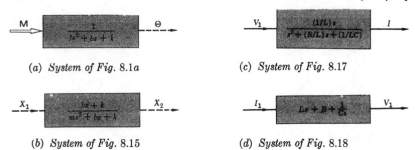

(a) System of Fig. 8.1a

(b) System of Fig. 8.15

(c) System of Fig. 8.17

(d) System of Fig. 8.18

Fig. 8.19 Block-diagram representation of system dynamics using transfer functions

cases, the numerator) of a transfer function contains the terms of the characteristic equation of the system. That is, the characteristic equation has the form

$$D(s) = 0$$

and the denominator† of a transfer function contains $D(s)$. [Let us call $D(s)$ the system's *characteristic function*.] This is evident, for example, in expressions (8.33), (8.34), (8.35), (8.36), and (8.37). It is easy to see why this would be true in general. If, in each case, we return to the derivation of the transfer function, we find that its denominator (or else its numerator) invariably began at the homogeneous side of the equation of motion (the "left-hand side") and as we know, the characteristic equation for an l.c.c. system is always obtained simply by substituting Ce^{st} into the homogeneous part of the equation of motion. Later, in Part D, we shall find that *for any linear system with constant coefficients, the denominator† of any transfer function will contain the characteristic function of the system.*

This statement is awfully important, of course, because the characteristic equation—and therefore the denominator† of any transfer function—tells us all the possible kinds of natural motion the system can have. That is, we may *factor* the denominator,† and its factors are the roots of the characteristic equation, and therefore the system's natural characteristics (its eigenvalues).

Poles and zeros

For any algebraic function of the form

$$G(s) = \frac{N(s)}{D(s)} \stackrel{eg}{=} \frac{s^2 + 9s + 20}{s^3 + 6s^2 + 11s + 6} \tag{8.38}$$

there may be certain values of s which make the function zero:

$$G(s) = 0$$

† Or, in certain cases, the numerator. Transfer function (8.36) illustrates this.

[Sec. 8.9] TRANSFER FUNCTIONS AND POLE-ZERO DIAGRAMS

These are called the *zeros* of the function. For the example given, the zeros are $s = -4$, and $s = -5$, as the reader may readily verify.

Similarly, there may be certain values of s which make the function infinite. These are called *poles* of the function. For the example given, the poles are $s = -1$, $s = -2$, $s = -3$, as the reader may again verify. That is, the sample function (8.38) can be written in the following factored form:

$$G(s) = \frac{(s+4)(s+5)}{(s+1)(s+2)(s+3)} \qquad (8.39)$$

[which, again, is readily verified by multiplying out the factors to obtain (8.38)].

If the function in question is a transfer function, then the denominator† contains the characteristic function, and therefore the poles‡ of the transfer function are the roots of the system characteristic equation. Thus the poles‡ of a transfer function tell what natural motions the system can exhibit.

Thus, the transfer function between two variables of a system tells us two things:

> (1) The transfer function tells how much of the *input* variable will be *transferred* to the output variable.
> (2) The poles‡ of the transfer function tell what kind of *natural motions* the system can have (regardless of how the natural motion is stimulated).

Obviously, a transfer function which has been factored, like (8.39), is ready-made for plotting in the s plane (refer to p. 249), and this is a most useful thing to do. For transfer function (8.39) all the poles and zeros are on the real axis. Plotting zeros as $\bigcirc$ and poles as $\times$, the s-plane picture is Fig. 8.20a.

When a transfer function contains complex poles or zeros (or both), the s-plane picture is more interesting. For example, suppose the values of physical constants in Example 8.6 are such that the transfer function (8.34) for the system has these numerical values:

$$\frac{X_2}{X_1} = \frac{b}{m} \frac{(s+8.5)}{(s+1-j4)(s+1+j4)}$$

Then the poles are at $s = -1 + j4$ and $s = -1 - j4$, and there is a zero at $s = -8.5$. The s-plane plot looks like Fig. 8.20b. More generally, from Eq. (8.1) and Fig. 8.2b, we can plot the pole-zero diagram for this particular transfer function with literal labels as in Fig. 8.20c, if the system is underdamped.

† Or, in certain cases, the numerator.
‡ Or, in the special cases, the zeros.

If the system in Example 8.7 happens to be underdamped, the pole-zero plot is as shown in Fig. 8.20d.

The overdamped case is left as an exercise, Problem 8.35.

In the same circuit the transfer function from v_1 to v_5 is, from (8.30),

$$\frac{V_5}{V_1} = \frac{R}{L} \frac{s + (1/RC)}{s^2 + (R/L)s + (1/LC)} \tag{8.40}$$

For this function the pole-zero diagram, Fig. 8.20e, looks qualitatively just like the one for the mechanical system of Example 8.6, Fig. 8.20c.

(a) $G(s) = \dfrac{(s+4)(s+5)}{(s+1)(s+2)(s+3)}$ [Eq. (8.39)]

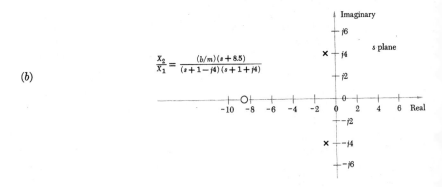

(b) $\dfrac{X_2}{X_1} = \dfrac{(b/m)(s+8.5)}{(s+1-j4)(s+1+j4)}$

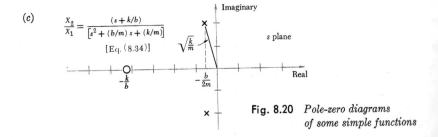

(c) $\dfrac{X_2}{X_1} = \dfrac{(s+k/b)}{[s^2 + (b/m)s + (k/m)]}$ [Eq. (8.34)]

Fig. 8.20 *Pole-zero diagrams of some simple functions*

[Sec. 8.9] TRANSFER FUNCTIONS AND POLE-ZERO DIAGRAMS 279

[Refer to Eq. (8.34).] Note that the poles are the same as in Fig. 8.20d because the characteristic equation and the system natural motions do not depend upon which outputs we happen to be observing.

For the mechanical system discussed in Sec. 8.8, the transfer function is given by (8.33), and the pole-zero diagram, Fig. 8.20f, happens to have only poles.

The location of poles tells us at once the kind of natural behavior to expect (refer to Fig. 8.5, p. 252). The complete pole-zero array indicates the system response to inputs, as we shall come to learn later.

Problems 8.35 through 8.45

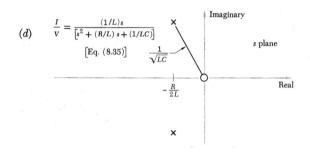

(d) $\dfrac{I}{V} = \dfrac{(1/L)s}{[s^2 + (R/L)s + (1/LC)]}$ [Eq. (8.35)]

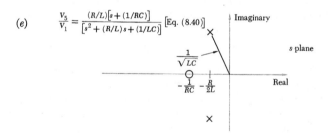

(e) $\dfrac{V_5}{V_1} = \dfrac{(R/L)[s + (1/RC)]}{[s^2 + (R/L)s + (1/LC)]}$ [Eq. (8.40)]

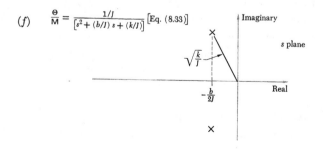

(f) $\dfrac{\Theta}{M} = \dfrac{1/J}{[s^2 + (b/J)s + (k/J)]}$ [Eq. (8.33)]

○8.10 IMPEDANCE AND ADMITTANCE

The name *impedance* is often given to certain transfer functions of certain systems (usually with the expectation that the input is to be an undamped sinusoid, $s = j\omega_f$).

Most commonly, the term refers to a transfer function *from current to voltage* in an electrical system when the signals are undamped sine waves. That is, *electrical impedance* is the ratio of a voltage phasor (complex vector) to a current phasor. More generally, any transfer function between a through variable† as input and an across variable as output may be called an impedance—for example, X/F.

The impedances of a network are defined in terms of an implied current source. Precisely, an impedance Z_{pq} of a network is the transfer function from a current i_q inserted by a source in one of its branches to the resulting voltage v_p across one pair of its nodes, with no other sources present. (The node may or may not be across the current source.)

The impedance most often used is that for a single pair of terminals in a network (i.e., for v and i measured at the same terminals). In Fig. 8.18, for example, the voltage v_1 produced by a current source i_1 in branch 1 leads to the transfer function (8.36), and this particular impedance is known as the *driving-point impedance* of the circuit,

$$Z_{11} = \frac{V_1}{I_1} = Ls + R + \frac{1}{Cs} \tag{8.41}$$

(because the voltage is measured at the terminals to which the driving current is applied). The same applied current source also produces a voltage v_5 across node pair bo for which, from (8.30), the transfer function is

$$Z_{51} = \frac{V_5}{I_1} = R + \frac{1}{Cs} \tag{8.42}$$

and this ratio of voltage to current is known as a *transfer impedance*, because it is the voltage at one place due to current at another.

The concept of electrical impedance was first developed to study operation at a single frequency, and is thought of as a sort of generalized resistance. That is, just as a resistance R *resists* the flow of steady current, so the impedance Z *impedes* the flow of *alternating* current. For the network in Fig. 8.18, the total impedance to sinusoidal current flow is the sum of the three terms in (8.36). The out-of-phase components are sometimes lumped in a single term called the *reactance* X:

$$Z = R + jX$$

In (8.36) the reactance is

$$X = \left(L\omega_f - \frac{1}{C\omega_f}\right)$$

Reactance, like resistance, is measured in ohms.

† Refer to Table IV, p. 134, for definitions of through and across variables.

[Sec. 8.10] IMPEDANCE AND ADMITTANCE 281

Electrical *admittance* is a converse of impedance. It is a measure of the degree to which a circuit *admits* the flow of (alternating) current, and it is always the ratio of a current phasor to a voltage phasor. More generally, any transfer function between an across variable as input and a through variable as output may be called an admittance.

Electrical admittances are defined in terms of an applied voltage source. Precisely, an admittance Y_{pq} of a network is the transfer function from a voltage v_q impressed by a source across one pair of its terminals to the resulting current i_p in one of its branches, with no other sources present.

The admittance most often used is that for a single pair of terminals in a network (i.e., for v and i measured at the same terminals). In Example 8.7 the current i_1 produced by a voltage source in branch 1 (i.e., across node pair oa) leads to the transfer function (8.35), and this particular admittance is known as the *driving-point admittance* of the network (because the current is measured in the same branch that contains the driving-voltage source):

$$Y_{11} = \frac{I_1}{V_1} = \frac{1}{Ls + R + (1/Cs)}$$

or, for the common case $s = j\omega_f$,

$$Y_{11} = \frac{1}{Lj\omega_f + R + (1/Cj\omega_f)}$$

(8.43)

Any other admittance—i.e., the current in any other branch due to a voltage source in branch 1—will be the same as (8.43) for a simple one-loop network like Example 8.7, because $i_1 \equiv i_2 \equiv i_3 \equiv i$. In simple systems an admittance is often simply the reciprocal of an impedance (e.g., $Y_{11} = 1/Z_{11}$ in Fig. 8.18); but this is not true in general (e.g., $Y_{51} \neq 1/Z_{51}$ in Fig. 8.18).

The special transfer functions impedance and admittance are often handy in analyzing the response of electrical networks to sinusoidal inputs, as we shall see in Chap. 9.

Notice that the transfer function (8.40) implied by (8.30) is a ratio of two voltage phasors, and is therefore neither an impedance (ratio of voltage phasor to current phasor) nor an admittance (ratio of current phasor to voltage phasor).

Poles and zeros. We pointed out on p. 277 that either the poles or (in certain cases) the zeros of any transfer function written for a physical system will always be the roots of the system's characteristic equation.

Impedances and admittances are transfer functions, and therefore either the poles or the zeros of every impedance and admittance will always be the roots of the characteristic equation of the system to which they pertain.

In the present chapter we have studied only two simple kinds of networks, one-loop networks and one-node-pair networks, such as those shown in Fig. 8.21. Others are considered in the problems.

Broader usage. The term impedance is also applied to mechanical systems by workers in the fields of vibrations, acoustics, and sound repro-

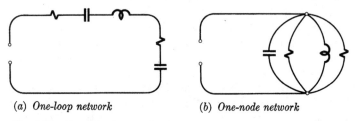

(a) One-loop network   (b) One-node network

Fig. 8.21 *Two simple second-order networks*

duction, where it refers to transfer functions F/U or M/Ω between applied force f and linear velocity u, or between applied torque M and rotational velocity Ω.

The evolution and use of the impedance idea will be discussed in more detail for simple systems in Sec. 9.2.

Problems 8.46 through 8.50

8.11 TOTAL RESPONSE TO ABRUPT DISTURBANCES

For some time after the sudden application of a disturbance to a dynamic system, the response of the system contains not only the so-called forced response, but also natural motions of the system, as we demonstrated for the first-order system in Sec. 6.2. If the system is well-damped, then the natural motions die out after a short time (e.g., several times the natural time constant of the system for most practical purposes), leaving the "forced" part of the response, as in Fig. 8.23. In other cases the system may be undamped, or lightly damped, and then the natural motions engendered by the disturbance persist for a long time.

When the natural motion is well damped, it is often acceptable to approximate the behavior as the forced response alone. Sometimes, however, the first few moments are all-important for practical reasons, and the total motion must be calculated.

An interesting case in point involved an early long-range missile which was carried to altitude and to supersonic speed mounted "piggy-back" on a booster rocket, as shown in Fig. 8.22. When the rocket fuel had been exhausted, the empty rocket case was separated from the missile. During separation the rocket's shockwave slapped the missile suddenly, causing it to vibrate in its various bending modes. Inside the missile, the guidance package was mounted on a spring suspension (as in Fig. 1.1) to protect it from the vibration climate of the missile. For reasons noted in Problem 8.64, it was necessary to make the natural frequency of the shock mount quite high, with the result that it was close to that of the missile. The shock mount then received an input consisting of a suddenly commenced large sinusoidal motion of its base. This transient input turned out to be the most crucial in the problem of designing the shock mount.

[Sec. 8.11] TOTAL RESPONSE TO ABRUPT DISTURBANCES

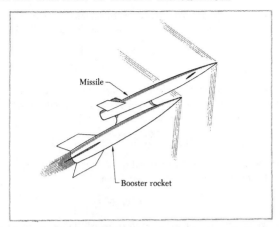

Fig. 8.22 *Launch configuration for an early long-range missile*

In the present section through Sec. 8.13 we shall carry out six detailed calculations—Examples 8.8 through 8.13—of the total response of various second-order systems to different abruptly commencing inputs. We may be somewhat dismayed at the degree of complication we find in these calculations, but total response *is* rather complicated—albeit straightforward.

For the practicing engineer much of the annoyance may be removed by the availability of computers for performing the calculations, once the problem is properly set up. For example, use of the analog computer is described at some length in Sec. 8.15.

Even purely analytical solution for this class of problem is made more routine and systematic—and sometimes less complicated as well—by the Laplace transform method, which we shall study in Chapter 17, *before* we approach the problem of finding the total response of *compound* systems. (Chapters 11 through 13 and 14 through 16 are concerned, respectively, with the natural motion alone and the forced motion alone of compound systems.)

Meanwhile, it is important to go through the following examples in order to develop insight into, and familiarity with, the mechanism of total response, so that future studies using a computer or the Laplace transform method can be undertaken with understanding and authority.

In the following examples the reader is urged to watch for the overall pattern of solution, as outlined in Procedure B-1 (p. 182), of finding separately the forced motion and the *form* of the natural motions, and then of combining these into the total motion by adjusting the *amount* of the natural motions to match the initial conditions.

Example 8.8. Second-order-system response to a step. Find the response of the system in Fig. 8.1 to the sudden application of a step moment at $t = 0$:

$$M(t) = \mathsf{M} u(t)$$

Assume (to make the problem more definite) that the natural damping of the system is less than critical ($\zeta < 1$). The equation of motion is (8.1')

$$J\ddot{\theta} + b\dot{\theta} + k\theta = M \qquad (8.1')$$
[repeated]

In order to demonstrate the general approach, we shall follow a procedure which is more elaborate than necessary for this special input. First we write the input in the equivalent form

$$M(t) = \mathsf{M}e^{0t}u(t) = [\mathsf{M}e^{st}]_{s=0}u(t)$$

Next, following the method in Procedure B-1, we assume that the total motion consists of a forced and a natural component:

$$\theta = \theta_f + \theta_n \qquad (8.44)$$

in which we assume the forced motion to be of the form

$$\theta_f = [\Theta_f e^{st}]_{s=0} \qquad 0^+ < t \qquad (8.45)$$

and, following Sec. 8.1, we assume the natural motion to be of the form

$$\theta_n = \Theta_n e^{st} \qquad 0^+ < t \qquad (8.46)$$

As usual, Θ_f and Θ_n are both constants (which may be complex numbers).

Substituting these assumed solutions into the equation of motion yields—as in Chapter 6—two separate equations, one from θ_f:

$$[Js^2 + bs + k]_{s=0}\Theta_f = \mathsf{M} \qquad 0^+ < t \qquad (8.47)$$

and the other (the characteristic equation) from θ_n:

$$Js^2 + bs + k = 0 \qquad (8.48)$$

The first of these, (8.47), leads to the transfer function:

$$\frac{\Theta_f}{\mathsf{M}} = \frac{1}{Js^2 + bs + k} \qquad (8.49)$$

and to the value Θ_f must have:

$$\Theta_f = \left[\frac{1}{Js^2 + bs + k}\right]_{s=0}\mathsf{M} = \frac{\mathsf{M}}{k} \qquad (8.50)$$

[which, for this problem, could have been obtained directly from (8.22) by substituting $s = 0$].

The second equation, (8.48), is the characteristic equation, the roots of which are the values of s corresponding to natural motion. For this particular case these roots are available from earlier calculation, (8.4a), and the natural motion is available from (8.11) (with $C_3 = \Theta_n$):

$$\theta_n = [C_3 e^{-\sigma t} \cos(\omega t - \psi)]u(t) \qquad (8.11b)$$
[repeated]

[Sec. 8.11] TOTAL RESPONSE TO ABRUPT DISTURBANCES 285

The total motion is given, then [from (8.45) and (8.46)], by

$$\theta = \left[\frac{M}{k} + C_3 e^{-\sigma t} \cos(\omega t - \psi)\right] \qquad 0^+ < t \qquad (8.51)$$

in which C_3 and ψ are to be obtained from the initial conditions, $\theta(0^+)$ and $\dot\theta(0^+)$. It is given that $\theta(0^-) = 0$ and $\dot\theta(0^-) = 0$. The input is finite, and therefore inertia J cannot suddenly change its velocity or displacement (Table VI). Thus $\theta(0^+) = \theta(0^-) = 0$, $\dot\theta(0^+) = \dot\theta(0^-) = 0$, and

$$\theta(0^+) = 0 = \frac{M}{k} + C_3 \cos(-\psi)$$

$$\dot\theta(0^+) = 0 = 0 + C_3[-\sigma \cos(-\psi) - \omega \sin(-\psi)]$$

from which
$$C_3 = -\frac{M_0}{k}\frac{1}{\cos(-\psi)}$$

and
$$\psi = \tan^{-1}\frac{\sigma}{\omega} = \sin^{-1}\zeta$$

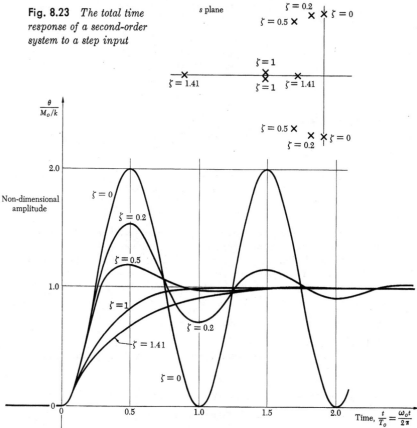

Fig. 8.23 *The total time response of a second-order system to a step input*

The total motion is therefore given by

Step response of second-order system

$$\theta = \frac{M_0}{k}\left[1 - \frac{1}{\cos\psi}e^{-\sigma t}\cos(\omega t - \psi)\right] \qquad 0^+ < t \qquad (8.52)$$

since $\cos\psi = \cos(-\psi)$ $\psi = \sin^{-1}\zeta$

A plot of the above response is given in Fig. 8.23 for a range of values of ζ. [For $1 < \zeta$, the calculation of initial conditions is slightly different, as we showed in Sec. 8.4, Eqs. (8.10).] Detailed calculation of several of the responses is left as an important problem for the reader (Problems 8.51 and 8.54).

Example 8.9. Mass-spring system with damping. A mass hangs from a movable support by a linear spring, Fig. 8.24. The system is initially in static equilibrium. Suddenly the support is given the motion

$$x(t) = x_0 u(t) \cos\omega_f t \qquad (8.53)$$

Find the resulting vertical motion y of the mass. The system is known to be underdamped.

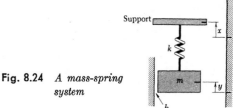

Fig. 8.24 *A mass-spring system*

It is left as a problem for the reader (Problem 8.53) to show that, with suitable modeling, the system equation of motion is

$$m\ddot{y} + b\dot{y} + ky = kx \qquad (8.54)$$

in which y is taken to be 0 when x is 0 and the mass hangs statically on the spring. That is, the presence of gravity does not influence the dynamic behavior of the system, but only determines the static equilibrium position from which motion starts (as in Example 8.6).

For convenience, cast (8.53) in the equivalent form†

$$x(t) \stackrel{r}{=} x_0 u(t) e^{j\omega_f t} \qquad (a)$$

or
$$x(t) = Xe^{st} \qquad (b)$$

with $X = x_0$, $s = j\omega_f$, $0^+ < t$.

By (6.1) we expect the total response to be made up of two terms,

$$y = y_f + y_n \qquad (c)$$

Forced component. We expect y_f to have the form

$$y_f \stackrel{r}{=} Y_f[e^{st}]_{s=j\omega_f} \qquad 0^+ < t \qquad (d)$$

† Recall that $\stackrel{r}{=}$ means "equals the real part of."

[Sec. 8.11] TOTAL RESPONSE TO ABRUPT DISTURBANCES 287

To find Y_f we substitute (d) and (b) into (8.54):

$$(ms^2 + bs + k)Y_f e^{st} = kXe^{st} \qquad s = j\omega_f, \qquad 0^+ < t \qquad (e)$$

This leads to the transfer function

$$\frac{Y_f}{X} = \frac{k}{ms^2 + bs + k} \tag{8.55}$$

and to the following value for Y_f

$$Y_f = \left[\frac{k}{ms^2 + bs + k}\right]_{s=j\omega_f} x_0 \tag{8.56}$$

We see that Y_f is a complex constant, which can be written

$$\vec{Y}_f = |\vec{Y}_f| e^{j/\vec{Y}_f} \tag{f}$$

Natural component. We expect y_n to have the form

$$y_n = Y_n e^{st} \tag{g}$$

which, substituted into (8.54), yields the characteristic equation:

$$ms^2 + bs + k = 0 \tag{8.57}$$

whose roots are the eigenvalues

$$s = -\frac{b}{2m} \pm j\sqrt{\frac{k}{m} - \left(\frac{b}{2m}\right)^2} \tag{8.58}$$

$$= -\sigma \pm j\omega$$

where the complex form is used because we were told that the system is underdamped. This makes (g), if we use (8.7c),

$$\begin{aligned} y_n &= C_3 e^{-\sigma t} \cos(\omega t - \psi) \\ \text{or} \quad y_n &\stackrel{r}{=} C_3 e^{-\sigma t} e^{j(\omega t - \psi)} \end{aligned} \tag{h}$$

in which C_3 and ψ are to be such as to satisfy the initial conditions for the *total motion*.

Summarizing the results to this point, we have

$$y \stackrel{r}{=} |\vec{Y}_f| e^{j/\vec{Y}_f} e^{j\omega_f t} + C_3 e^{-\sigma t} e^{j(\omega t - \psi)} \tag{i}$$

or†

$$y = |\vec{Y}_f| \cos(\omega_f t + \underline{/\vec{Y}_f}) + C_3 e^{-\sigma t} \cos(\omega t - \psi) \tag{j}$$

with $\vec{Y}_f$ given by (8.56) and with C_3 and ψ to be chosen so that

$$\begin{aligned} y(0^+) &= y(0^-) = 0 \\ \dot{y}(0^+) &= \dot{y}(0^-) = 0 \end{aligned}$$

[We have a finite disturbance to a mass and thus, by Table VI, p. 202, $y(0^+) = y(0^-)$ and $\dot{y}(0^+) = \dot{y}(0^-)$.]

† The form (j) is more convenient than form (i) for accounting for initial conditions.

The last step is to determine C_3 and ψ from (j) and its derivative, with $t = 0$. (We must differentiate first, then set $t = 0$.) Thus

$$y(0^+) = 0 = |\vec{Y}_f| \cos \underline{/\vec{Y}_f} + C_3 \cos(-\psi)$$
$$\dot{y}(0^+) = 0 = -\omega_f |\vec{Y}_f| \sin \underline{/\vec{Y}_f} + C_3[-\sigma \cos(-\psi) - \omega \sin(-\psi)]$$
(k)

After performing a little simple algebra with relations (k), we find the values of C_3 and ψ to be

$$C_3 = -|\vec{Y}_f| \frac{\cos \underline{/\vec{Y}_f}}{\cos \psi}$$

$$\tan \psi = -\frac{\omega_f}{\omega} \tan \underline{/\vec{Y}_f} + \frac{\sigma}{\omega}$$
(l)

so that (j) becomes

Total response
$$y = |\vec{Y}_f| \left[\cos(\omega_f t + \underline{/\vec{Y}_f}) - \frac{\cos \underline{/\vec{Y}_f}}{\cos \psi} e^{-\sigma t} \cos(\omega t - \psi) \right]$$
(8.59)

with $\vec{Y}_f$ given by (8.56), and ψ by (l).

This motion is plotted in Fig. 8.25 for the following set of numerical values:

$$\omega_f = 3.93 \qquad \sqrt{\frac{k}{m}} = 31.5 \qquad \frac{b}{2m} = 1.67$$

for which $\left|\frac{\vec{Y}_f}{x_o}\right| = 1.012 \qquad \underline{/\vec{Y}_f} = -0.76° \qquad \psi = 3.5°$

Example 8.10. Undamped mass-spring system. If the system in Example 8.9 has no damping, find the response $y(t)$ of the mass when the movable support is given the motion

$$x(t) = x_0 e^{-\sigma_f t} u(t)$$
(8.60)

(which can be written $x = X e^{st}$ with $X = x_0$, $s = -\sigma_f$, $0 < t$).

Without damping, the equation of motion is

$$\ddot{y} + \frac{k}{m} y = \frac{k}{m} x$$
(8.61)

Again, we expect a motion with two terms

$$y = y_f + y_n$$
(a)

and we expect the *forced component* y_f to look like (8.60):

$$y_f = Y_f [e^{st}]_{s=-\sigma_f} \qquad 0 < t$$
(b)

To find Y_f, substitute (b) in (8.61):

$$\left(s^2 + \frac{k}{m}\right) Y_f e^{st} = \frac{k}{m} X e^{st} \qquad \text{with } s = -\sigma_f, X = x_0, 0 < t$$
(c)

which leads to the transfer function

$$\frac{Y_f}{X} = \frac{k/m}{s^2 + (k/m)}$$
(8.62)

[Sec. 8.11] TOTAL RESPONSE TO ABRUPT DISTURBANCES 289

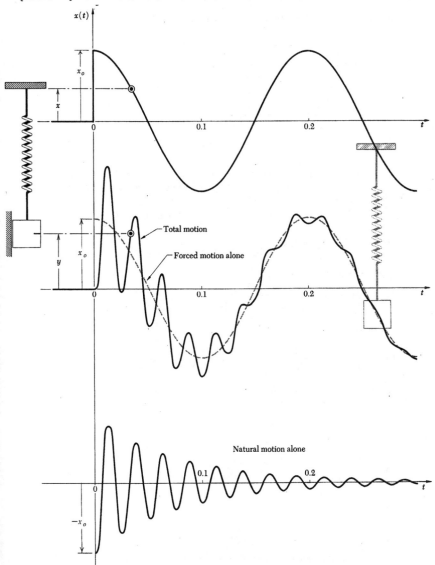

Fig. 8.25 *The total response of a second-order system to an abrupt sinusoidal input*

and to the following value for Y_f:

$$Y_f = \left[\frac{k/m}{s^2 + (k/m)}\right]_{s=-\sigma_f} x_0 \qquad (d)$$

For the *natural component* we assume

$$y_n = Y_n e^{st} \qquad 0 < t \qquad (e)$$

and substitute in (8.61) to obtain the characteristic equation:

$$s^2 + \frac{k}{m} = 0 \tag{8.63}$$

whose roots are the eigenvalues

$$s = \pm j \sqrt{\frac{k}{m}} \tag{8.64}$$

The total solution to this point is thus

$$y = \frac{x_o}{[\sigma_f^2/(k/m)] + 1} e^{-\sigma_f t} + C_3 \cos\left(\sqrt{\frac{k}{m}}\, t - \psi\right) \qquad 0 < t \tag{i}$$

To evaluate C_3 and ψ we consider initial conditions: Both y and $\dot{y}$ are to be zero at $t = 0^-$, and therefore also at $t = 0^+$ (because of the inertia of m). Thus

$$\begin{aligned} y(0^+) = 0 &= \frac{x_o}{[\sigma_f^2/(k/m)] + 1} + C_3 \cos(-\psi) \\ \dot{y}(0^+) = 0 &= \frac{-\sigma_f x_o}{[\sigma_f^2/(k/m)] + 1} - C_3 \sqrt{\frac{k}{m}} \sin(-\psi) \end{aligned} \tag{j}$$

After some algebra, these yield

$$C_3 = -\frac{x_0/\cos\psi}{[\sigma_f^2/(k/m)] + 1} \qquad \psi = -\tan^{-1}\frac{\sigma_f}{\sqrt{k/m}} \tag{k}$$

and thus the total response is

$$y = \frac{x_o}{1 + [\sigma_f^2/(k/m)]} \left[e^{-\sigma_f t} - \frac{1}{\cos\psi} \cos\left(\sqrt{\frac{k}{m}}\, t - \psi\right) \right] \tag{8.65}$$

$$\psi = -\tan^{-1}\frac{\sigma_f}{\sqrt{k/m}}$$

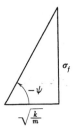

The reader may wish to check (8.65) at $t = 0$.
Response (8.65) is plotted in Fig. 8.26.

⚬8.12 TRANSIENT AND STEADY STATE

The terms "transient" and "steady state" are often used to distinguish the parts of a dynamic response. The obvious meanings of the terms are as follows:

Transient: Motion which dies out.
Steady state: Motion which persists indefinitely.

Thus in Fig. 8.25 the natural motion of the element has died out after a cycle or so of the forced motion, the latter persisting indefinitely. This common situation has led to the custom of referring to the natural motion as *"the* transient," and to the forced motion as *"the* steady state."

This custom is not accurate, in general, and sometimes leads to inaccurate thinking. It is true that natural motion is *usually* transient, and forced motion is *commonly* steady, as they were in Fig. 8.25; but it is by no means always so.

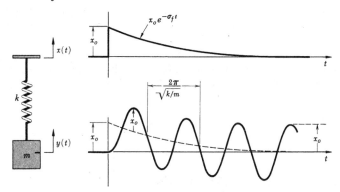

Fig. 8.26 *A case of forced motion that is transient, accompanied by natural motion that is "steady state"*

Consider, for example, the situation in Example 8.10, an undamped mass-spring system, Fig. 8.26, the upper end of which is suddenly given the disturbance motion $x(t) = u(t)x_0 e^{-\sigma_f t}$. The response $y(t)$ contains a forced-motion component $y_f = u(t)y_0 e^{-\sigma_f t}$. But long after both the disturbance and the forced response have died out, the mass continues to display a *natural* oscillation (at its own natural frequency). Clearly, this natural motion is not transient. Moreover, the forced motion is not steady.

The student is encouraged to use the four words—forced motion, natural motion, steady state, transient—with accuracy.

Problems 8.51 through 8.64

⊙8.13 IMPULSE RESPONSE OF CERTAIN SECOND-ORDER SYSTEMS

As we noted in Secs. 6.8 through 6.10, the impulse response of a linear system is of great interest because (1) it allows us to view natural motion as a special case of forced motion, and (2) it makes it possible, through convolution, to calculate the response of the system to any arbitrary input.

To obtain the response of a second-order system to a unit impulse, it is instructive to follow the same limiting procedure we used in Sec. 6.8 for a first-order system. The steps in the limiting process are illustrated in Fig. 8.27. The physical model for this illustration is the spring-restrained rotor with damping of Fig. 8.1a, p. 246. The equation of motion is (8.1′), and the transfer function is (8.33), p. 274.

In Fig. 8.27a a step input has been applied to the system; so the system begins to respond as it did in Fig. 8.23. However, the step is terminated after a time Δt, after which the system exhibits a damped oscillation about neutral. The magnitude of the impulse applied is $M_0 \, \Delta t$.

In Fig. 8.27b a larger input is applied for a shorter time, the total area of the impulse, $M_0 \, \Delta t$, being kept the same, however. In this case the impulse is terminated before the system has time to be displaced very far, but it has reached substantial velocity, and has therefore a substantial response, as shown.

In the limiting case of Fig. 8.27c the duration of the impulse Δt, has been reduced to "zero," but the area $M_0 \, \Delta t$ has been kept the same by making M_0 correspondingly very large. Now the system has no time to

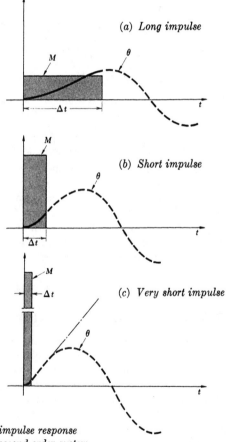

Fig. 8.27 *The impulse response of a second-order system*

[Sec. 8.13] IMPULSE RESPONSE: CERTAIN SECOND-ORDER SYSTEMS

displace, and its displacement at $t = 0^+$, just after the impulse, is zero. However, a finite impulse has been applied to the mass, and it has achieved a finite velocity at $t = 0^+$.

To find the velocity of the system at $t = 0^+$ we can notice that, since $\theta = 0$ throughout the time Δt, the term $k\theta$ in (8.1′) is zero, so that *during this interval* we can consider the equivalent equation of motion

$$J\dot{\Omega} + b\Omega = M$$

or
$$\dot{\Omega} + \frac{b}{J}\Omega = \frac{M}{J} = \frac{M_0 \Delta t}{J}\delta(t) \qquad 0 < t < 0^+ \qquad (8.66)$$

We have already found the response for this exact situation on p. 215. The time response is given by (6.40):

$$\Omega = \frac{M_0 \Delta t}{J} e^{-(b/J)t} \qquad 0 < t$$

We are interested in this response only at $t = 0^+$, because Eq. (8.66) does not hold after that. At $t = 0^+$ we have, then,

$$\dot{\theta}(0^+) = \Omega(0^+) = \frac{M_0 \Delta t}{J} \qquad (8.67)$$
$$\theta(0^+) = 0$$

After $t = 0^+$ we have simply the free (unforced) motion of a second-order system with initial conditions (8.67).

To summarize: for a second-order system which has an equation of motion of the form

$$\ddot{y} + 2\sigma\dot{y} + \omega_0^2 y = \propto(t)$$

the response to a *unit impulse* $\propto(t) = \delta(t)$, is [from (8.10), (8.17) and (8.11a)]

$$y(t) = h(t) = \frac{e^{-\sigma_1 t} - e^{-\sigma_2 t}}{\sigma_2 - \sigma_1} \qquad 1 < \zeta \qquad (a)$$
$$y(t) = h(t) = te^{-\sigma t} \qquad \zeta = 1 \qquad (b) \qquad (8.68)$$
$$y(t) = h(t) = \frac{1}{\omega}e^{-\sigma t}\sin \omega t \qquad \zeta < 1 \qquad (c)$$

---

in which
$$\zeta = \frac{\sigma}{\omega_0}$$
$$\sigma_1 = \sigma\left(1 - \sqrt{1 - \frac{1}{\zeta^2}}\right)$$
$$\sigma_2 = \sigma\left(1 + \sqrt{1 - \frac{1}{\zeta^2}}\right)$$
$$\omega = \sqrt{\omega_0^2 - \sigma^2}$$

and $h(t)$ is a standard symbol for a system's unit impulse response.

As an interesting special case, the values of $h(0^+)$ and $\dot{h}(0^+)$ are given by

$$h(0^+) = 0 \qquad \dot{h}(0^+) = 1 \tag{8.68'}$$

regardless of ζ; i.e., the result is the same for (8.68a), (8.68b), or (8.68c).

Example 8.11. Isolated test pad. An isolated test pad (on which to test delicate instruments) may be represented by the model shown in Fig. 8.28. This is a steel pad mounted on relatively soft springs, and designed to have an undamped natural frequency of only 0.5 cps. (What will be the static deflection of the springs under the mass?) The light damping of the springs is represented in the model by the dashpot.

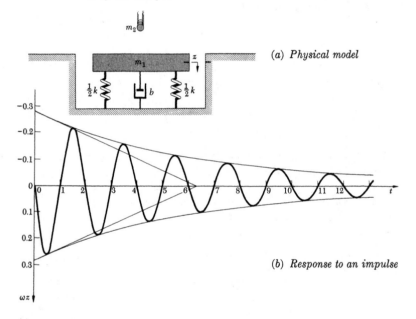

Fig. 8.28 *The impulse response of an isolated test pad*

A steel ball of mass m_2 is dropped onto the center of the pad from a height d, and then caught (by a bystander) on the first bounce. Find the resulting motion of the pad. Plot the response if numerical values are as follows:

$$\zeta = 0.05 \qquad \frac{m_2}{m_1} = 0.01 \qquad d = 3 \text{ ft}$$

STAGE I. PHYSICAL MODEL. The model for the test pad is as shown. For a first analysis we assume, in addition, that the impact of the ball is perfectly elastic, and occurs during a negligibly short time. We consider only vertical motion (which by symmetry will be uncoupled from rotary and lateral motions).

[Sec. 8.13] IMPULSE RESPONSE: CERTAIN SECOND-ORDER SYSTEMS 295

STAGE II. EQUATIONS OF MOTION. We choose as our system the pad m_1 only. From numerous previous analyses, the equation of motion of mass m_1 will be

$$m_1 \ddot{z} + b\dot{z} + kz = f(t) \tag{8.69}$$

where z is displacement from equilibrium† (we take z positive downward) and $f(t)$ in this case is the force applied to the pad by the ball.

With an elastic impact, the momentum of the ball is changed from $m_2 v$ downward (at $t = 0^-$) to $m_2 v$ upward (at $t = 0^+$), or a total change of $2m_2 v$, where v is the velocity of the ball just before impact. This change must be produced by an impulse $\mathcal{I}$ exerted on the ball by the pad, having magnitude

$$\mathcal{I} = \int_{t=0^-}^{t=0^+} f(t) \, dt = 2m_2 v \tag{8.70}$$

where $f(t)$ acting upward on the ball is the same as the $f(t)$ in Eq. (8.69) acting downward on the table. That is, $f(t)$ is an impulse of area $A = 2m_2 v$:

$$f(t) = A \, \delta(t) = 2m_2 v \, \delta(t) \tag{8.71}$$

[and thus $\int f(t) \, dt = 2m_2 v$].

The velocity of m_2 after falling distance d is, of course,

$$v = \sqrt{2gd}$$

STAGE III. RESPONSE. The response of the system (8.69) to impulse (8.71) is, from (8.68),

$$z = \frac{2m_2 \sqrt{2gd}}{m_1 \omega} e^{-\sigma t} \sin \omega t \tag{8.72}$$

where $\qquad \sigma = \dfrac{b}{2m_1} \qquad \omega = \sqrt{\omega_0^2 - \sigma^2} \qquad \omega_0 = \sqrt{\dfrac{k}{m_1}}$

[Form (8.68c) was used because ζ is given less than 1. Check the units of (8.72).]
For the numerical values given,

$$\sigma = \zeta \omega_0 = 0.157 \quad \left(\frac{1}{\sigma} = 6.4 \text{ sec}\right)$$

$$\omega = 3.14 \qquad T = \frac{2\pi}{\omega} = 2.0 \text{ sec}$$

$$v = \sqrt{2gd} = 14 \text{ ft/sec}$$

and thus $\qquad z = 0.089 e^{-t/6.4} \sin 3.14 t \text{ ft} \tag{8.73}$

This response is plotted in Fig. 8.28b.

Example 8.12. System with zero. To see the influence of zeros in a transfer function, repeat Example 8.9 if there is an ideal dashpot b_2, as well as a spring, connecting the mass to the support above it, Fig. 8.29. Let the input be a suddenly applied exponential,

$$x = x_0 u(t) e^{-\sigma_f t} \tag{8.74}$$

Let the initial value of x and $\dot{x}$ be zero, and let $\zeta < 0$.

† The force of gravity is opposed statically by the equilibrium spring deflection, and these two terms are dropped from the equation of motion.

296 DAMPED SECOND-ORDER SYSTEMS [Chap. 8]

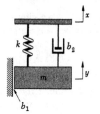

Fig. 8.29 *A system having a zero*

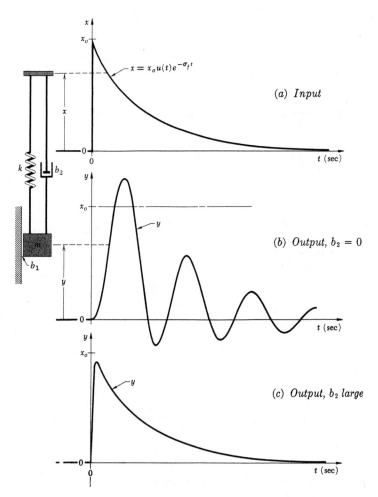

(a) *Input* $x = x_o u(t) e^{-\sigma_f t}$

(b) *Output, $b_2 = 0$*

(c) *Output, b_2 large*

Fig. 8.30 *Response, to an abrupt input, of a system having a zero*

[Sec. 8.13] IMPULSE RESPONSE: CERTAIN SECOND-ORDER SYSTEMS   297

STAGE II. It is left as an exercise to show that the equation of motion is

$$m\ddot{y} + (b_1 + b_2)\dot{y} + ky = b_2\dot{x} + kx \tag{8.75}$$

STAGE III. The total motion is given as the sum of

$$y = y_f + y_n \tag{8.76}$$

We can find y_f conveniently from the transfer function from X to Y, which is obtained by assuming $x = Xe^{st}$ and $y = Ye^{st}$ and substituting in (8.75), per Secs. 8.8 and 8.9. The transfer function can be written thus:

$$\frac{Y}{X} = \frac{b_2/m[s + (k/b_2)]}{s^2 + [(b_1 + b_2)/m]s + (k/m)} \tag{8.77}$$

where it is understood that we are interested in the case $X = x_0$, $s = -\sigma_f$, $0^+ < t$, from (8.74). The forced part of the motion is then

$$y_f = Y_f e^{-\sigma_f t} \qquad 0^+ < t \tag{8.78}$$

in which, letting $b = b_1 + b_2$,

$$Y_f = \frac{b_2/m[-\sigma_f + (k/b_2)]x_0}{\sigma_f^2 - (b/m)\sigma_f + (k/m)} \tag{8.79}$$

The advantages of confining our attention to $0^+ < t$ were discussed on p. 219. The natural motion has, of course, the same form as in Example 8.9:

$$y_n = Ce^{-\sigma t} \cos(\omega t - \psi) \qquad 0^+ < t \tag{8.80}$$

with
$$\sigma = -\frac{b}{2m} \qquad \omega = \sqrt{\frac{k}{m} - \left(\frac{b}{2m}\right)^2}$$

and C and ψ so chosen that the total motion has the proper initial conditions at $t = 0^+$, as discussed in Sec. 6.5.

Combining (8.80) with (8.79), we obtain the total motion:

$$y = Y_f e^{-\sigma_f t} + Ce^{-\sigma t} \cos(\omega t - \psi) \qquad 0^+ < t \tag{8.81}$$

Its derivative is

$$\dot{y} = -\sigma_f Y_f e^{-\sigma_f t} + Ce^{-\sigma t}[-\sigma \cos(\omega t - \psi) - \omega \sin(\omega t - \psi)] \qquad 0^+ < t \tag{8.82}$$

Initial conditions at $t = 0^+$. To determine the values of y and $\dot{y}$ at $t = 0^+$, we note first that, because the system differentiates an abrupt input at $t = 0$, y and $\dot{y}$ may change between $t = 0^-$ and $t = 0^+$ (Case 2 of Procedure B-2 page 203). (Physically, the abrupt displacement of x at $t = 0$ implies infinite velocity across the dashpot, and therefore "infinite" force on m, during the time interval $0^- < t < 0^+$. This force impulse imparts a finite velocity to m.) To find out how much they change, we (a) note that the derivative of the input is given by (6.43):

$$\frac{d}{dt}[x_0 u(t)e^{-\sigma_f t}] = x_0[\delta(t) - \sigma_f e^{-\sigma_f t} u(t)] \qquad \begin{matrix}(6.43)\\ \text{[repeated]}\end{matrix}$$

and (b) note that the response of the system at $t = 0^+$ will be, for the first term in (6.43), from (8.68'),

$$y(0^+) = 0 \qquad \dot{y}(0^+) = \frac{b_2}{m}\dot{x}_0$$

[where coefficient b_2/m occurs in front of $\dot{x}$, after (8.75) is divided by m], and for the second term, from Table VI, p. 202,

$$y(0^+) = 0 \qquad \dot{y}(0^+) = 0$$

These are then the values we shall use as initial conditions in determining the constants for the natural part of the solution.

Evaluation of constants: magnitude of natural motion. At $t = 0^+$ we have, therefore,

$$\left.\begin{aligned} y(0^+) &= 0 = Y_f + C\cos(-\psi) \\ \dot{y}(0^+) &= \frac{b_2}{m}x_0 = -\sigma_f Y_f + C[-\sigma\cos(-\psi) - \omega\sin(-\psi)] \end{aligned}\right\} \quad (8.83)$$

from which

$$\left.\begin{aligned} C &= \frac{-Y_f}{\cos(-\psi)} \\ \psi &= -\tan^{-1}\left[\frac{\sigma_f - \sigma}{\omega} + \frac{b_2}{m\omega}\frac{x_0}{Y_f}\right] \end{aligned}\right\} \quad (8.84)$$

The total motion is given, then, by

$$y = Y_f\left[e^{-\sigma_f t} - \frac{1}{\cos(-\psi)}e^{-\sigma t}\cos(\omega t - \psi)\right] \quad (8.85)$$

with Y_f given by (8.79) and ψ by (8.84).

Consider now the effect on the solution (8.85) of the zero in the transfer function (8.77). If, first, b_2 is zero, the response is as shown in Fig. 8.30b. If b_2 is quite large, the response will be as in Fig. 8.30c: while x makes its initial jump the large dashpot force imparts a high upward velocity to the mass, but the mass does not undergo an instantaneous displacement.

The important effect of system zeros on forced response is clear from Example 8.12 and Fig. 8.30. So, also, is the possibility that they may cause considerable complication. We shall not consider more-complicated systems until we have the Laplace transform method at our disposal in Chapter 17.

Problems 8.65 through 8.71

○8.14 RESPONSE OF A SECOND-ORDER SYSTEM BY CONVOLUTION

The procedure for using convolution to find the response of a system to any arbitrary input is outlined in Sec. 6.10. The fundamental expression is

$$y = \int_{\tau=-\infty}^{\tau=t} x(\tau)h(t-\tau)\,d\tau \qquad (6.37)$$
[repeated]

The procedure described in Sec. 6.10 is perfectly general and can be used for any linear system whose impulse response $h(t)$ is known. We demonstrate the procedure further in a second-order example below.

Example 8.13. Convolution calculation. Use convolution to find the response requested in Example 8.10.

The equation of motion is (8.61)

$$\ddot{y} + \frac{k}{m} y = \frac{k}{m} x(t)$$

in which the input $x(t)$ is given by (8.60)

$$\frac{k}{m} x(t) = \frac{k}{m} x_0 e^{-\sigma_f t} u(t)$$

From (8.68) the impulse response of the second-order system (with $\sigma = 0$) is

$$h(t) = \frac{1}{\omega} \sin \omega t \, u(t)$$

(in which $\omega = \sqrt{k/m}$).

Inserting the above quantities into the general convolution expression, (6.54), yields the following result:

$$y(t) = \int_{\tau = -\infty}^{t} [\omega^2 x_0 e^{-\sigma_f \tau} u(\tau)] \left[\frac{1}{\omega} \sin \omega(t - \tau) u(t - \tau) \right] d\tau$$

$$= \omega^2 x_0 \int_{\tau = 0}^{t} e^{-\sigma_f \tau} \left[\frac{e^{j\omega(t-\tau)} - e^{-j\omega(t-\tau)}}{2j\omega} \right] d\tau$$

which is integrated and regrouped to use Euler's equation, (7.3), yielding

$$y(t) = \omega^2 x_0 \left[\frac{e^{-\sigma_f t}}{(\sigma_f^2 + \omega^2)} + \frac{\sigma_f}{\omega(\sigma_f^2 + \omega^2)} \sin \omega t - \frac{1}{(\sigma_f^2 + \omega^2)} \cos \omega t \right]$$

$$y(t) = \frac{x_0}{(\sigma_f^2/\omega^2) + 1} \left[e^{-\sigma_f t} - \frac{1}{\cos \psi} \cos(\omega t - \psi) \right] \quad (8.86)$$

$$\psi = -\tan^{-1} \frac{\sigma_f}{\omega} \qquad \omega = \sqrt{\frac{k}{m}}$$

which is the same as the solution (8.65) obtained by assuming motions of the form e^{st} and solving for the various coefficients.

Problems 8.72 through 8.78

○8.15 SIMULATION: THE ANALOG COMPUTER

If an analog computer is available, it is worthwhile at this point to do some experiments with it, to become more familiar with the ideas introduced in Chapters 6, 7, and 8.

An analog computer is a collection of specially designed electronic and electromechanical components which can be connected together to form a dynamic system whose voltages obey the same equation of motion as the physical *model* which is to be simulated.

Analog computers are used extensively to study the behavior of physical models of mechanical, chemical, fluid, and heat-transfer systems, and

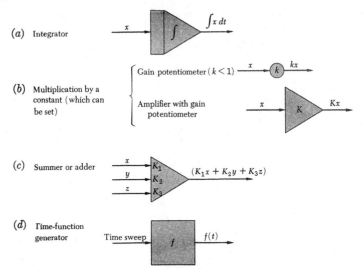

Fig. 8.31 *The elements of a simple, linear analog computer*

many others, as a step in predicting quickly and efficiently the behavior of the corresponding real systems.

Indeed, the analog computer itself constitutes another physical model for the same equations of motion, the dynamic behavior of its components and the time variation of its voltages being exactly analogous to the corresponding mechanical (or other) components and variables of the system being simulated. (Early analog computers were constructed by mechanizing directly the electrical analog for the mechanical system of interest—one capacitor representing each mass, one inductor each spring, etc. Modern general-purpose analog equipment is mechanized instead from the differential equations of motion, which provides much greater flexibility. The principle is the same, however, and one-to-one correspondence is readily established between the simulated and actual physical variables.)

Analog simulation has played a particularly significant role in the development of aeronautical and space-vehicle control systems, first in predicting the behavior of new vehicles before they were built, then in providing a means for testing control components in the dynamic context of actual flight (the simulator "standing in" for the dynamics of the still-unfinished vehicle), and finally in studying actual flight test results.

An outstanding example of the latter role was the detective work performed by simulation following a disastrous crash of the then-quite-new F100 during a flight test in 1951. The aircraft was thrown out of control by a subtle but strong cross-coupling effect, and a spectacular violent maneuver followed, which tore the wings from the airplane and killed the test pilot. It was necessary to ground all F100 aircraft pending determination of measures which would

[Sec. 8.15] SIMULATION: THE ANALOG COMPUTER

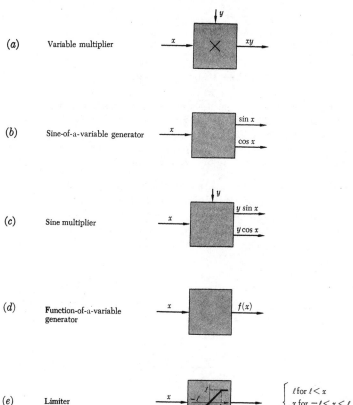

(a) Variable multiplier

(b) Sine-of-a-variable generator

(c) Sine multiplier

(d) Function-of-a-variable generator

(e) Limiter
$$\begin{cases} \ell \text{ for } \ell < x \\ x \text{ for } -\ell < x < \ell \\ -\ell \text{ for } x < -\ell \end{cases}$$

Fig. 8.32 *Nonlinear components used in more elaborate computers*

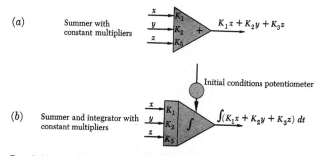

(a) Summer with constant multipliers

(b) Summer and integrator with constant multipliers

Fig. 8.33 *Common combination components used in analog computers*

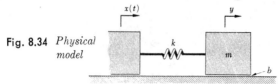

Fig. 8.34 *Physical model*

positively prevent a recurrence. Astute nonlinear, six-degree-of-freedom analog simulation (directed by Frederick Frankel of Autonetics) duplicated exactly the test records of the fatal maneuver and played a key role in establishing preventive measures which could be positively relied upon, so that the squadrons of F100s could be returned to operational status.

In our discussion here we are interested in simulating the elementary linear systems we have been studying in Chapters 6, 7, and 8. We consider, therefore, only quite simple analog computer arrangements.

It takes relatively little space to explain the procedure for setting up and operating an analog computer. We shall do that first and then proceed to some experiments.

Analog computer setup

While the design and development of an analog computer is a fairly sophisticated business, it is relatively easy to operate one. We shall there-

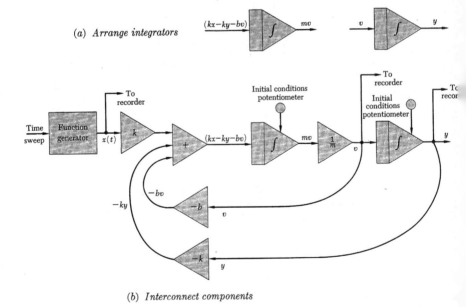

Fig. 8.35 *Steps in arranging an analog computer to simulate the system model of Fig. 8.34, from Eqs. (8.89)*

[Sec. 8.15] SIMULATION: THE ANALOG COMPUTER

fore assume here that we already have one, and that it consists of a set of "black boxes," each of whose characteristics are simple and well known.

The basic elements of a simple, linear analog computer are shown in Fig. 8.31, with the symbols commonly used to represent them. There are many types of function generator, from slow mechanical curve-followers to electronic circuits driven by a sawtooth voltage.

In addition, more elaborate computers have a number of nonlinear elements such as those in Fig. 8.32. Most of the devices which perform the operations shown are all electric, but some (e.g., the variable multiplier, the sine generator, and the function-of-a-variable generator) are usually electromechanical. Often the same device performs two of the operations—for example, multiplying one variable by the sine of another.

Differentiation is not performed on analog computers (for the practical reason that any noise present in a signal is greatly amplified, relative to other signals present, by differentiation). Instead, the differential equations are cast in integral form [see Eqs. (8.89)], so that integrators will be the only dynamic components required.

In this brief discussion we confine our attention to the linear components shown in Fig. 8.31, with which we can mechanize linear differential equations. These are usually used in the combinations shown in Fig. 8.33. With each integrator there is also provision for setting the initial value of the output, as indicated.

The *first step* in analog simulation is to *write the equations of motion* of the physical model to be simulated in the form of a set of first-order differential equations. This is referred to formally as *"standard form."*

Suppose, for example, we wish to simulate the system of Example 8.9 (p. 286). The physical model is shown again in Fig. 8.34, for convenience. We begin with Eq. (8.54) from Example 8.9:

$$m\ddot{y} + b\dot{y} + ky = kx \tag{8.87}$$

Next we define a new variable, $v = \dot{y}$. Then we can write (8.87) in the following alternative form, which contains only first derivatives:

$$\begin{aligned} m\dot{v} &= kx - ky - bv \\ \dot{y} &= v \end{aligned} \tag{8.88}$$

Or, alternatively,
$$\begin{aligned} v &= \frac{1}{m} \int (kx - ky - bv)\, dt \\ y &= \int v\, dt \end{aligned} \tag{8.89}$$

Variables y and v are a set of *state variables* for the system.

The *second step* in simulation is to *arrange* a group of analog *computer components* in such a way that they will solve the desired equations—in this case, Eqs. (8.89). It is most systematic to begin with the integrators, of which we need two, as shown in Fig. 8.35a. We see that we need one summer and four amplifiers (with the required gains) to obtain the signals we are to enter into the two integrators. We also need a function generator to generate $x(t)$. These are added to the two integrators, as shown in Fig. 8.35b. The last move, of course, is to connect the outputs v and y of the integrators to the proper amplifier inputs. The gains k (two places), b, and $1/m$ can then be set to the desired values, and the system is ready to operate.†

A vital part of simulation is, of course, observing and recording the results. One much-used device is the pen recorder, Fig. 8.36, which plots the various variables versus time as paper is moved past the pens.‡ In the present problem we might typically elect to have input x and variables v and y plotted in this way.

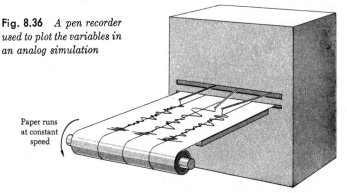

Fig. 8.36 *A pen recorder used to plot the variables in an analog simulation*

Paper runs at constant speed

When a pen recorder is used, the problem may be run at various time scales. For example, the problem may be run in "real time." This means that in Fig. 8.35, for example, the scaling is such that the voltages v and x in the simulator vary at the same speed as the actual velocity and displacement of the mass in Fig. 8.34. Real-time simulation is especially attractive (and necessary) when the simulator is to be operated in conjunction with actual hardware. But any other scale—say, 100 times as fast as real time, or one-tenth as fast as real time—may be chosen, within the dynamic limits of the computer elements and the recorder pens. In addition, various paper speeds may be selected, to adjust the distance that corresponds to one second of time on the paper, and thereby compress or spread out the response trace to a convenient shape. These ideas are most readily seen through a little experience with an analog computer and pen recorder.

An alternative, and also very useful display method is the crt (cathode-ray tube, see p. 92). This is most useful in high-speed analog computers which are scaled to solve a whole problem in 0.01 sec or so. (That is, the

---

† It should be remarked that the simulation described here could be set up with fewer components by utilizing the combination components of Fig. 8.33. Also, potentiometers alone would probably be used in place of the k, b, and $1/m$ amplifiers, needed amplification being provided by the adders and integrators. Finally, the amplifiers in most analog computers effect a sign reversal, which must be accommodated in the setup.

‡ A further discussion of the pen recorder is given in Secs. 9.1 and 9.3.

[Sec. 8.15] SIMULATION: THE ANALOG COMPUTER 305

computer time scale is, typically, thousands of times as fast as real time.) The computer resets automatically after each run and repeats the same problem over and over, producing perhaps 50 or 100 solutions per second. The variables of interest, e.g., voltage y in Fig. 8.35, are connected to the vertical plates of the crt; and, to plot t horizontally, the horizontal plates are given a sawtooth voltage which sweeps the beam horizontally once per solution (the sawtooth being synchronized with the problem reset switch). The crt screen is persistent, so that the whole time solution will appear on the scope simultaneously and continuously.

The crt can, of course, be used also with slower (e.g., real time) computation; but then the beginning of a response plot fades before the end of the problem is reached, and the display is less convenient.

Still another kind of recorder of great usefulness and versatility is the so-called "x-y plotter" or input-output table," which can plot (on nonmoving paper) one of the variables against another—for example, y versus x, or v versus y in Fig. 8.35. (The latter combination is called a phase-plane plot. See Sec. 10.2.) A crt can obviously also be used in this way simply by connecting the horizontal plates to one variable and the vertical plates to the other.

The *third step* in simulation is *operation* of the analog computer. The function generator is set up to produce the required function [e.g., $x(t) = u(t) \cos \omega_f t$ in Example 8.9], and the initial-condition potentiometers on each integrator are set to "clamp" voltages v and y at their specified initial values. Then the OPERATE switch is thrown. This starts the function generator, and also releases voltages v and y, whose subsequent motion is then determined by Eqs. (8.89). After the problem has run as long as desired, the recorder is stopped and the OPERATE switch returned to RESET. Then the values of v and y are again clamped at their specified initial values and the system is ready for the next run.

Some suggested experiments with analog simulation

The analog simulation of Fig. 8.35 may be used to make all of the following experiments.

Natural response, experiment 1. Duplicate (approximately) many of the traces in Figs. 8.4 and 8.5 by removing the input ($x = 0$) and setting the k, b, and $1/m$ pots so as to produce various values of $\sigma = b/2m$ and $\omega_0 = k/m$. Set initial condition $y(0)$ equal to some convenient positive value, and set $v(0) = 0$. Check the values of σ and ω by making measurements on the recorded time plot. (Selecting an appropriate recording paper speed or crt sweep speed can contribute to the ease of making measurements.)

The instructional value of this set of runs may be enhanced by following a systematic procedure. For example, runs (1), (6), (7), and (12) of Fig. 8.5 may be made by removing the term $-bv$ from the summer and then varying k. (On a crt high-repetition-rate display, the whole time plot can be seen to change simultaneously as the k gain pot is turned.) Note that the sign of the b amplifier must be changed to produce the plots in the right half-plane of Fig. 8.5.

Natural response, experiment 2. Reproduce the response traces shown in Fig. 8.7 for a system having $1 < \zeta$.

Natural response, experiment 3. Notice that ζ can be varied without changing ω_0 by operating the b pot. Use this fact to reproduce the upper dotted plot in Fig. 8.11 [$\theta(0) = 0$, $\zeta = 1$] and two other plots for $\zeta = 1.05$ and $\zeta = 0.95$. Then, with ζ set equal to 1, reproduce first the other dotted plot in Fig. 8.11, and then the solid curve, by altering the initial conditions—i.e., first $\theta(0) = 0$; and then $\theta(0) = 1$, $\dot{\theta}(0) = 0.5\sigma$.

Total response, experiment 1. Reproduce all the plots in Fig. 8.23 by arranging for the function generator in Fig. 8.35 to produce a step. Set the various values of ζ (one at a time) by adjusting the b pot and leaving the k pot fixed. Note that both $v(0)$ and $y(0)$—or $\dot{\theta}(0)$ and $\theta(0)$, as they happen to be called—are zero in Fig. 8.23.

Total response, experiment 2. Reproduce Fig. 8.26 by arranging for the function generator to generate an abruptly started exponential curve, and by disconnecting the output of the b amplifier in Fig. 8.35. Other interesting plots may be obtained by setting values of b other than zero.

Total response, experiment 3. Animate Example 8.9 by arranging for the function generator to produce an abruptly started sine wave, and reproducing Eq. (8.59). The effect of the initial condition can be readily added, to show that the result is simply a superposition of the natural motion due to initial conditions only (sans forcing functions) superposed on the total motion when initial conditions are zero.

Total response, experiment 4. Use the analog computer to produce the two plots of Fig. 8.30. The computer arrangement must be altered slightly by the addition of the term $b_2(\dot{x} - \dot{y})$. The function generator is to introduce the signal $x = x_0 u(t) e^{-\sigma_1 t}$. Initial conditions are zero. The two different plots are produced by using two different values of b_2.

Nonlinear system, experiment 5. If a component is available, like the second one indicated in Fig. 8.32, which will generate the sine of a variable, then the complete nonlinear equation of motion for a simple pendulum making large natural motions can be set up. The equation was derived in Example 7.2 [Eq. (7.15)]. Here we add damping:

$$m\ell^2 \ddot{\theta} + b\dot{\theta} + mg\ell \sin \theta = 0$$

Put the equation in standard form and work out the proper analog-computer arrangement. (You will find that the choice of scales for the analog components in a nonlinear problem requires much more careful planning than for a linear problem.) Then time traces for small and large motions can be obtained and compared with Fig. 10.5 for $b = 0$, which gives a graphical solution. If an "input-output" plotter (p. 305) is also available, the interesting patterns of Fig. 10.6 can also be generated.

[Sec. 8.15] SIMULATION: THE ANALOG COMPUTER

Some closing comments about analog simulation

It should be clear at this point that the analog computer can simulate any given *physical model* of an actual system, as distinguished from simulating the actual system itself. The computer merely solves the equations of motion as derived for the physical model. Therefore the simulation is only as true to life as the physical model on which it is based.

A physical model contrived for analog simulation can, however, reasonably be a much closer approximation to the real thing—including nonlinearities, higher-order terms, sinusoidal functions, multiplications, and so on—than a physical model contrived for hand calculations. This is so because the analog computer can solve quickly much more difficult, and much higher-order differential equations than we are willing to contemplate for hand calculation.

It is this power to determine readily the behavior of systems of higher order, and containing nonlinearities, together with the convenience to the user of watching voltages behave exactly as the corresponding physical variables do, and of seeing at once the result of changing any parameter, that makes the analog computer so useful as an exploratory and design tool in dynamics.

We mentioned earlier the technique of using an analog simulation to represent part of a system for the purpose of testing another part. An example is the testing of a jet and reaction-wheel attitude control system for a satellite vehicle. These systems must be tested dynamically on the ground because the cost of orbital tests is so prohibitively high (the cost of the rocket booster alone is several million dollars). The technique, called "fixed-base simulation," is to mount all the control components and sensors on special fixtures so that their response can be monitored. The vehicle itself is represented by the analog simulator. The satellite's sensors are stimulated artificially and feed signals to the satellite's own computer, which calls for appropriate control action. The actual control components are actuated (wheels torqued, jets fired), and the resulting reaction torques are measured and sent to the simulator. The simulator then computes the response the vehicle would have, and the sensor inputs are altered accordingly. Thus the complete system can be tested dynamically, as a system, on the ground.

Astronauts practice many hundreds of hours on simulators to become familiar with the dynamic behavior of their vehicle before they make a flight. (These simulators become highly elaborate, with extensive visual cues and even motion cues being furnished the astronaut by the simulator.)

Digital computers

We wish not to leave the subject of computer simulation without also mentioning the *digital computer*, the important counterpart and coworker of the analog computer.

Functionally, the digital computer solves a set of differential equations by first casting them in its own terms, and then integrating them step by step, using sufficiently small increments and sufficiently elaborate iteration techniques to produce any desired degree of accuracy.

By its nature, the digital computer does not provide the one-to-one association between its behavior and that of the physical system being studied that an analog computer does. Moreover, operation of a digital computer is necessarily more formal, and rapid intuitive exploration cannot be pursued with the same casualness as on an analog computer. But the tremendous capacity, versatility, and speed of large modern digital computers, as well as their inherent accuracy, make them a tool of profound importance in dynamics.

Often we do our initial exploratory studies of a new system on an analog-computer simulation in which we make rather gross approximations. Later, when we want to know the behavior of our final configuration with great accuracy, we turn to the digital computer.

We shall not discuss further the rather extensive subject of digital computers, but their availability and capability must be borne in mind at all times if we are to maintain perspective in our further study of dynamic analysis.

Finally, a technique known as "hybrid" simulation—employing a combination of analog and digital computing equipment—is often an optimum procedure. For example, in studying a rocket-vehicle control system design, the vehicle dynamics would be simulated on an analog computer and the digital controller on a digital computer. The processing of signals for transmission from one computer to the other is called "A to D conversion."

⊙Chapter 9

FORCED OSCILLATIONS
OF ELEMENTARY SYSTEMS

The steady-state response of a dynamic system to an undamped sinusoidal disturbance has already been discussed, for first- and second-order systems, as a special case of response to a disturbance of the form e^{st}; but the occurrence of this particular form of response is so widespread and so important that we want to consider it in much greater detail. Common important examples include electrical-power-transmission and home-utility equipment, mechanical vibrations in machinery, communication and broadcasting equipment, and both the electrical circuitry and the electromechanical portions (microphones and speakers) of sound-reproduction systems†. The response of mechanical systems to sinusoidal mechanical disturbances is commonly called *forced vibration*, while the response of a system to sinusoidal electrical inputs over a range of frequencies is often called the *frequency response* of the system.

Frequency response plays an important role in dynamics, not only because sinusoidal signals dominate many systems, as noted above, but also because the frequency response of any system is usually easy to measure accurately, and provides a valid and convenient measure of the dynamic range of the system and, further, because frequency response is amenable to versatile analytical methods.

Techniques for analyzing sinusoidal response have been highly developed, particularly in the communication and servomechanism field,

---

† Indeed, sinusoidal motion seems to be one of nature's favorites, occurring in many natural forms and at a very wide range of frequencies, from the pendulous swing of a vine (at perhaps one cycle per second) upward to the atomic vibration in a crystal (a few megacycles per second), and downward to the precession of the equinoxes (once every 26,000 years).

and will be useful in the studies of system dynamics ahead. In particular, they will furnish a helpful intuitive background for the study of Fourier series, transfer functions of systems, and the forced response of systems generally, which we shall undertake in Part D. These techniques also form the basis for an important segment of the methods of analysis of feedback control systems.

These frequency-response techniques are *not* related to the studies of the natural behavior of systems in Part C, so that the present chapter can as well be deferred until after Part C if desired.

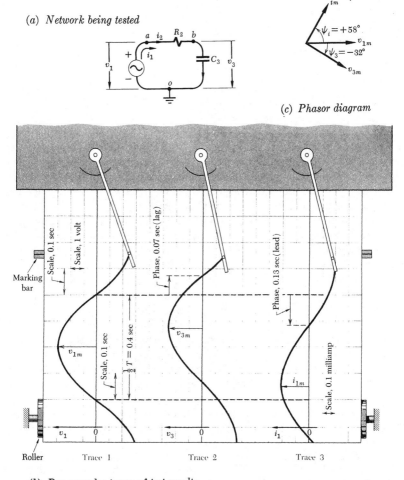

(a) *Network being tested*

(c) *Phasor diagram*

(b) *Pen-recorder traces of test results*

Fig. 9.1 *Experimental measurement of sinusoidal response*

[Sec. 9.1] THE NATURE OF SINUSOIDAL RESPONSE

Concepts and techniques

There are no new concepts introduced in the present chapter. Rather, some concepts introduced earlier are examined and demonstrated in more detail. In particular, the concepts of the transfer function (Sec. 8.9) and impedance (Sec. 8.10), and of phase angle and amplitude ratio (Fig. 8.14b) will figure prominently in this chapter, as will the phenomenon of resonance.

The new techniques to be introduced include the phasor diagram; log plotting of frequency response, using asymptote methods developed by Bode; and the use of the s plane to calculate frequency response.

Several important physical systems will be studied for demonstration purposes, including electrical filters, the mechanical torque motor, mechanical shock mounts, seismic (inertial) instrumentation, and the many-purpose RLC network. These are all simple first- and second-order systems. Study of the frequency response of higher-order systems (by the same techniques and other ones) is the subject of Chapter 15.

9.1 THE NATURE OF SINUSOIDAL RESPONSE

As we have seen on several previous occasions, when any l.c.c. system is stimulated by a sinusoidal input, the system response (i.e., the time behavior of each of its physical variables—displacement, voltage, temperature, and the like) will also be sinusoidal, at the same frequency as the stimulus.

To dramatize this fact, let us step into the laboratory and witness a dynamic test demonstration. The system being tested is a simple RC circuit represented by the network in Fig. 9.1a. The input to the circuit, from a voltage source, is a pure sine wave of frequency 1.25 cycles per second:

$$v_1 = v_{1m} \cos \omega_f t \quad \text{with } \omega_f = 2\pi(1.25) \text{ rad/sec}$$

An operator is "looking" at this voltage with the aid (as it happens) of a pen recorder like the one shown in Fig. 8.36, p. 304. This useful device consists of a set of pens each of which follows faithfully the voltage applied to its controlling amplifier (over a relatively wide dynamic range). The pens mark on a sheet of ruled paper moved beneath them at a precisely constant speed by a motor-driven roller.† The operator runs the paper underneath the oscillating pens for a short time at a precisely known paper speed. Then by making measurements of the trace on the paper he can determine such

---

† Even though the pens are pivoted with rather short arms, they make a true orthogonal plot on the paper as a result of an ingenious arrangement: The pens mark the paper by burning it, and the burning can only occur where the pen intersects a straight knife edge underneath the paper (Fig. 9.1b). The pen-motor circuitry is so designed that the resulting orthogonal plot is proportional to the input voltage. In another model, linearity is achieved through a pen-linkage arrangement. A rather striking dynamic image of the pen-recording process may be simulated by placing a card over Fig. 9.1 with its lower edge along the roller edge and then moving the card slowly upward.

quantities as the magnitude of the input voltage and the period T of the sine wave.

In this particular test the operator is interested in two "output" quantities: the voltage v_3 across the capacitor and the current i_1 in the circuit. He has arranged for these quantities to be measured and connected to two other pens, as shown in Fig. 9.1b, so that he can also measure the amplitude of the two output sine waves, and can make a direct measurement of the "phase angle" between the input and each of the two outputs. From the burst of data shown in Fig. 9.1b, the operator makes the measurements shown in Table VIII. In the last two columns he has calculated the ratios of magnitudes and the "phase angles" in degrees. (The phase of v_1 is taken as 0, for convenience.)

TABLE VIII
Data measured from pen recorder at 1.25 cps

Trace	Period T (sec)	Quantity plotted	Magnitude	Phase difference (sec)	Magnitude ratio (amp/volt)	Phase angle = $\dfrac{\text{phase diff}}{T} \times 360°$
1	0.8	v_1	$v_{1m} = 3.0$ volts	...	...	...
2	0.8	v_3	$v_{3m} = 2.5$ volts	-0.07	$\dfrac{v_{3m}}{v_{1m}} = 0.83$	$\psi_3 = \dfrac{-0.07 \text{ sec}}{0.8 \text{ sec}} \times 360° = -32°$
3	0.8	i_1	$i_{1m} = 0.2$ amp	$+0.13$	$\dfrac{i_m}{v_{1m}} = 0.067$	$\psi_i = \dfrac{+0.13 \text{ sec}}{0.8 \text{ sec}} \times 360° = +58°$

The term phase angle comes from the fact that any sine wave can, as we have noted (Fig. 7.8), be regarded as generated by the "shadow" of a rotating arrow—a "complex vector," or phasor. To demonstrate the idea, phasors for the voltage input and for the two outputs have been sketched in Fig. 9.1c, using magnitudes and phase angles measured from the recorder paper, Table VIII. (What the operator actually measures is a time interval on the recorder paper, which he then compares to the time interval recorded for *one period* of the sine wave. The ratio, multiplied by 360°, gives the phase angle in degrees.)

What we are watching in this test is a system the input to which is the sine wave

$$v_1 = v_{1m} \cos \omega_f t \stackrel{r}{=} v_{1m} e^{j\omega_f t} = V_1 e^{j\omega_f t} \tag{9.1}$$

and the outputs from which are also sine waves of the form

$$i_1 = i_{1m} \cos (\omega_f t + \psi_i) \stackrel{r}{=} (i_{1m} e^{j\psi_i}) e^{j\omega_f t} = I_1 e^{j\omega_f t} \tag{9.2}$$
$$v_3 = v_{3m} \cos (\omega_f t + \psi_3) \stackrel{r}{=} (v_{3m} e^{j\psi_3}) e^{j\omega_f t} = V_3 e^{j\omega_f t} \tag{9.3}$$

the capital letters representing phasors, $V_1 = v_{1m}$, $I_1 = i_{1m} e^{j\psi_i}$, $V_3 = v_{3m} e^{j\psi_3}$. The operator is measuring the quantities v_{1m}, i_{1m}, v_{3m}, and ψ_i and ψ_3—the components of each phasor.

[Sec. 9.2] OPERATION AT A SINGLE FREQUENCY 313

To understand the test results of Fig. 9.1b we have only to analyze the linear network of Fig. 9.1a. The equation of motion (using, for example, KCL at node b) is $(v_1 - v_3)/R_2 = C_3 \dot{v}_3$, or

$$R_2 C_3 \dot{v}_3 + v_3 = v_1 \tag{9.4}$$

In finding i_1 we shall have use also for the physical relation

$$i_1 \equiv i_3 = C_3 \dot{v}_3 \tag{9.5}$$

We are interested only in the forced motion, and the input is of the general form e^{st}, so it is convenient to use the analytical procedure outlined in Sec. 8.8. Accordingly, we assume that the input is of the general form $v_1 = V_1 e^{st}$, and that the outputs are of the form $v_3 = V_3 e^{st}$ and $i = I e^{st}$. Substituting these in the above differential equations (and dropping subscripts on R and C, there being no ambiguity) leads to

$$\frac{V_3}{V_1} = \frac{1}{RCs + 1} \tag{9.6}$$

$$\frac{I_1}{V_1} = \frac{Cs}{RCs + 1} = Y_{11} \tag{9.7}$$

in which we let $s = j\omega_f$ because we are concerned only with sinusoidal motion at a specified frequency ω_f:

$$\frac{V_3}{V_1} = \frac{1}{RCj\omega_f + 1} \tag{9.8}$$

$$\frac{I_1}{V_1} = \frac{Cj\omega_f}{RCj\omega_f + 1} = Y_{11} \tag{9.9}$$

[Expressions (9.6) through (9.9) are transfer functions, and in addition (9.7) and (9.9) have the special name *admittance*, Y_{11} (Sec. 8.10).]

Note that the maximum *amplitude* v_{3m} of the sine plot of Fig. 9.1 is given by the *magnitude* of an algebraic expression: $v_{3m} = |1/(RCj\omega_f + 1)|v_{1m}$. We shall use the words *amplitude* and *magnitude* interchangeably to denote the peak value of a sine wave.

We can now get one check on our physical model, as represented by Eq. (9.4), by calculating the values of $V_3 = v_{3m}/\underline{\psi_3}$ and $I_1 = i_{1m}/\underline{\psi_i}$ from (9.8) and (9.9) for $\omega_f = 1.25$ cps $= 2\pi(1.25)$ rad/sec, and comparing them with the values measured by the operator and recorded in Table VIII. This the reader is urged to do, given that $R_2 = 25$ ohms, $C_3 = 0.00318$ farad.

◯9.2 OPERATION AT A SINGLE FREQUENCY: THE IMPEDANCE VIEWPOINT

Many common situations in electrical engineering involve operation at a single driving frequency. The electrical power system in private dwellings is the most familiar: The supply voltage and the current through all equipment in the house operate at the constant frequency of 60 cycles per second. (The constancy of this frequency is in fact, the basis for accurate, reliable electric clocks.) For studying electrical systems operating at a constant frequency, the impedance point of view introduced in Sec. 8.10 is a convenient one for making calculations.

To demonstrate, consider the electrical network of Fig. 9.1a, which we have just analyzed, and for which test data are given in Fig. 9.1b. (In this case operation is at 1.25 cps rather than 60 cps, but, of course, the principles are the same.)

The *driving-point impedance* (Sec. 8.10) is defined in terms of a hypothetical current source i_1, introduced in place of the voltage source:

$$Z_{11} = \frac{V_1}{I_1} = R + \frac{1}{Cs} = R + \frac{1}{Cj\omega_f} \tag{9.10}$$

Driving-point admittance (9.9) is defined in terms of an assumed voltage source (Fig. 9.1a). For this simple network, impedance (9.10) is merely the reciprocal of admittance (9.9). (This is not always so.)

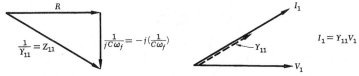

(a) *Constructing Z_{11} at a given frequency*   (b) *Obtaining I_1 from V_1 and Y_{11}*

Fig. 9.2 *Graphical construction of phasors*

For a given frequency these relationships can be shown completely in a phasor diagram (rotating "vector" diagram), such as Fig. 9.2, in which the V_1 phasor has been multiplied by the Y_{11} phasor to obtain the I_1 phasor, for example. (The rules for multiplying and dividing phasors are given in Sec. 7.3.)

In complicated circuits having numerous nodes and loops, the analysis of operation at a single frequency is often greatly facilitated by the use of the impedance viewpoint.

Problems 9.1 through 9.5

9.3 FREQUENCY RESPONSE

If we are interested in the response of the system to inputs of many frequencies, then we will want to repeat the test of Fig. 9.1 for a number of frequencies over the range of interest. The result of such testing of the circuit of Fig. 9.1 is shown in Fig. 9.3. Here the operator has repeated the test of Fig. 9.1 five times at four different frequencies. After each test he has changed the frequency of the input signal (but kept its amplitude the same). Then, after waiting for the transient motions to die out, he has

[Sec. 9.3] FREQUENCY RESPONSE 315

run the recorder for a few cycles and then stopped it. (After each run the operator has turned off the pens and moved the paper slightly to separate the runs.) The paper speed was the same for each run except the last (top of the page). The first run, at $\omega_f = 7.9$ (i.e., at 1.25 cps), is the same as the run in Fig. 9.1 (but at a slower paper speed).

Data measured from the time traces of Fig. 9.3, and from three others not shown, are tabulated in Table IX. (The data for 1.25 cps are the same as in Table VIII.) The maximum values v_{1m}, v_{3m}, i_{1m} are readily available from the traces shown. To obtain accurate measurements for the phase angles at the higher frequencies, it is necessary to rerun the tests at a much higher paper speed, as is demonstrated for one frequency in the run at the top of the page.

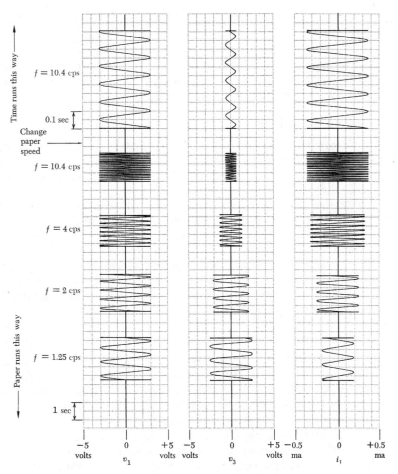

Fig. 9.3 *Recording sinusoidal response over a range of frequencies*

TABLE IX
Frequency response of circuit in Fig. 9.1a, as measured from data of Fig. 9.3

Period T	Frequency (cps)	(rad/sec)	v_{1m} (volts)	v_{3m} (volts)	$\dfrac{v_{3m}}{v_{1m}}$	$(\psi_3 - \psi_1)$ (deg) (Time lead/T)	i_{1m} (milli-amps)	$(\psi_i - \psi_1)$ (deg) (Time lead/T)
∞	0	0	3	3	1.0	0°	0	+90°
5	0.2	1.26	3	3	1.0	−6°	0.07	+79°
0.8	1.25	7.9	3	2.5	0.83	−32°	0.19	+58°
0.5	2	12.6	3	2.1	0.7	−45°	0.25	+45°
0.25	4	25	3	1.3	0.43	−63°	0.32	+27°
0.1	10	63	3	0.6	0.2	−79°	0.36	+11°
0.05	20	126	3	0.3	0.1	−84°	0.36	+6°

All the data in Table IX can be checked by a straightforward complex-number calculation from (9.8) and (9.9) (rules for manipulating complex numbers are given in Sec. 7.3), or by phasor calculation, Fig. 9.2.

Plotting measured data

It is common practice in plotting sinusoidal response data to plot separately amplitude ratio versus frequency and phase angle versus frequency. In Fig. 9.4 data from Table IX are plotted to linear amplitude and frequency scales.

It turns out to be much more convenient and more useful (as we shall see in Sec. 9.5) to plot frequency-response data to logarithmic scales, as demonstrated in Fig. 9.5. Forcing frequency is plotted to a log scale, as is amplitude (or amplitude ratio). The scale for phase angle is still linear.

A collection of numerical values such as Table IX, which indicates the response of a system over a range of forcing frequencies, is known as the "frequency response" of the system; and a plot of such data (e.g., Fig. 9.4 or Fig. 9.5) is called a frequency-response plot.

First-order examples

We shall now discuss the measured frequency response of two simple systems, and also some measurement techniques. Then in the following sections we shall discuss computation of predicted frequency response for first- and second-order systems, including the two discussed here.

The low-pass electrical filter. The simple circuit of Fig. 9.1 is a widely used one. Many times in physical systems an electrical signal is found to be polluted with an undesirably large portion of extraneous noise.

The problem of "cleaning up" the signal—that is, of reducing the ratio of noise amplitude to signal amplitude—is the center of attention of much

[Sec. 9.3] FREQUENCY RESPONSE   317

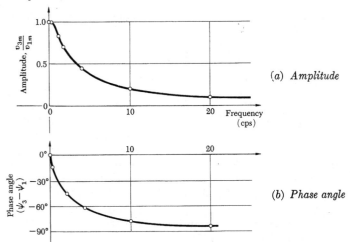

Fig. 9.4 *Plot of measured frequency response*

research in circuit analysis and information theory. (The problem is, of course, especially critical in the communication field.)

In most simple situations the problem is made relatively easy by the fact that the signal varies slowly, while most of the noise is varying rapidly. In such cases a "low-pass filter" is used—a network which attenuates all signals above a specific frequency; the higher the frequency, the greater the attentuation. The most common "low-pass filter" is simply the *RC* circuit

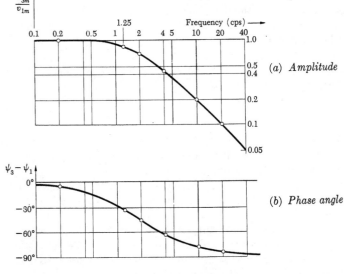

Fig. 9.5 *Logarithmic plot of measured frequency response*

of Fig. 9.1a, which is easy to assemble. The circuit is sometimes included between stages of amplification, as indicated in Fig. 9.6, in which case it is isolated from adjacent circuitry, so that its response will be exactly that of the model in Fig. 9.1a, which is plotted in Fig. 9.5.

It is evident from Fig. 9.5 that so long as the frequency of the input signal (v_1) is low (e.g., less than 0.1 cps), the output signal v_3 is a faithful transmission of it: That is, v_3 has the same amplitude and phase as v_1. But if the frequency of the input voltage is high (e.g., above 10 cps) the network greatly attenuates it, and the output voltage v_3 is much smaller. This behavior is understandable physically from Fig. 9.6: At low frequency there is ample time for charge on the capacitor to build up and down as required to follow v_1; at high frequency there is not. The values of R and C in a low-pass filter are so chosen that the frequency range of faithful reproduction includes all signals of interest but not higher frequencies. Then extraneous noise is greatly attenuated, and the output from the filter is a "cleaner" signal, as desired.

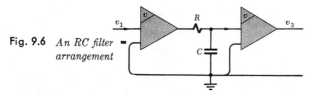

Fig. 9.6 *An RC filter arrangement*

A heavily damped torque motor. Consider again the idealized mechanical system model of Fig. 6.2 (p. 185), a heavily damped torque motor, whose dynamic behavior is entirely similar to that of the network of Fig. 9.1a. In this particular motor the damping and spring torque happen to be much larger than the inertia torque over the frequency range of interest, as the model indicates. Frequency-response tests are made on the motor, using a current source and measuring the output displacement. The results are very similar to Table VIII, as we expect by noting that the transfer function for the torque motor is, from (6.3),

$$\frac{\Theta}{M} = \frac{1}{[s/(k/b)] + 1} \tag{9.11}$$

which corresponds exactly to (9.6). Thus, as the operator increases the frequency of the input signal, the output magnitude begins to fall off with frequency after a certain point (perhaps at a different frequency than it did in Fig. 9.5), and the phase angle decreases with frequency in about the same manner as in Fig. 9.5 (again, perhaps with the whole plot shifted in frequency).

An added, interesting feature of the test of the torque motor is an acoustical hum which becomes audible at about $\omega_f = 100$ and which, of course, increases in pitch as the driving frequency is increased. Eventually it is noted that the loudness of the hum decreases with increasing frequency.

For electrical circuits it is so easy to measure the characteristics of

[Sec. 9.3] FREQUENCY RESPONSE 319

individual elements (e.g., resistors and capacitors) that the testing described in Sec. 9.1 is usually an unnecessary exercise. For mechanical systems, however, the separate elements (represented in the physical *model*) are not even well defined, so that frequency-response data often constitutes a vital contribution to the physical model, for subsequent analytical studies.

In Sec. 9.4 we shall calculate and plot the frequency response of the simple models of Fig. 9.1a and Fig. 6.2 to compare with and explain the tests we have just "witnessed." In the process we shall introduce some efficient techniques for making such calculations and for plotting the results.

Recording frequency-response measurements

A brief note about the available methods for recording signals over various ranges of frequencies may be of interest. The frequency range of sinusoidal signals to which a recorder responds accurately constitutes, of course, a valid and convenient measure of its dynamic range for signals of all types.

Pen recorder. The pen recorder, described above and on p. 304, is a high-quality tool of great utility for studies in the frequency range discussed. It was included in the demonstration of Sec. 9.1 because the significance of a pen record is so straightforward. (Circuits are usually tested using a cathode ray oscilloscope, as described below.)

The design of mechanical recorders is an interesting engineering problem which has been solved in an outstanding way by several firms. Pen recorders of the type described on p. 304 and having good response to 50 cycles per second are manufactured by the Sanborn Company of Waltham, Mass. A comparable recorder, using a special ink system and a rectilinear pen linkage, is made by the Brush Co. of Cleveland, Ohio. These models are used extensively in laboratory testing of dynamic behavior.

Mirror galvanometer. A different arrangement has been developed by the Consolidated Engineering Company of Pasadena, California, in which the pens are replaced by tiny mirrors which reflect beams of light onto photosensitive recorder paper. The mirrors respond faithfully out to 10,000 cps, and 24 of them in a single recorder produce time plots of 24 different quantities simultaneously on a roll of paper only 12 inches wide. These compact instruments are used widely in the dynamic flight testing of aircraft and rocket vehicles, for example.

The key element in either the pen or the mirror system is the D'Arsonval coil described on p. 90, which is driven by a current amplifier. These instruments are thus modern versions of the classical D'Arsonval galvanometer. The fundamental relations for their dynamic response are derived in Sec. 9.8.

As a pen or mirror is driven at higher and higher frequencies, a frequency is eventually reached, of course, above which the instrument will no longer respond. For a pen this frequency is typically about 50 cps, and for a mirror 10,000 cps. For operation at much higher frequencies the cathode-ray oscilloscope may be used for recording purposes.

The cathode-ray oscilloscope. Dynamic test voltages can also be projected on the screen of a cathode ray tube (crt), as shown in Fig. 9.7. (The physics of cathode ray tubes was described on p. 91 and in Fig. 2.28.) In Fig. 9.7a the electron beam is swept horizontally at constant velocity, the sweep being repeated over and over with any repetition rate the operator selects. The voltage to be measured (e.g., v_1 or v_3 in Fig. 9.1) is applied to the vertical plates

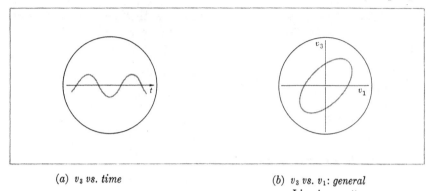

(a) v_3 vs. time (b) v_3 vs. v_1: general Lissajous pattern

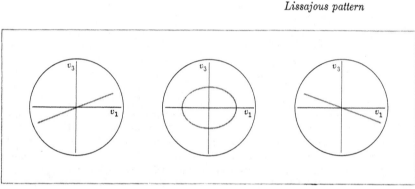

(c) Phase angle 0°   (d) Phase angle 90°   (e) Phase angle 180°

Fig. 9.7 *Frequency response patterns on a cathode-ray oscilloscope*

of the crt, so that if the operator adjusts the repetition rate to coincide with the frequency of the input signal (or a submultiple of it), a trace just like Fig. 9.1 is obtained. Some cathode ray tubes have two beams, so that two signals can be watched simultaneously. (A single beam may also be time-shared.) The operator establishes the time scale by knowing the repetition rate. If a permanent record is desired, the face of the crt may easily be photographed.

Since electrons have small inertia indeed, there is no "mechanical limit" to the response of a crt (although there are ultimately electrical circuit limits). Cathode ray tubes are available which will faithfully follow signals at frequencies of megacycles (millions of cycles per second).

An alternative common method of employing a crt, which is very useful for comparing the magnitude and phase of two signals, is to put one voltage (say, v_1) on the horizontal plates and the other (v_3) on the vertical plates. The electron beam then describes a closed figure, known as a Lissajous figure, the beam tracing around the figure once per cycle. From a Lissajous figure one can read the magnitudes of the two signals as indicated in Fig. 9.7b, and the phase angle between them can be read directly: If the signals are exactly in phase, the Lissajous pattern will collapse to a straight line in the first and third quadrants (Fig. 9.7c); if the signals are 90° out of phase, the Lissajous pattern will be a circle or horizontal ellipse (Fig. 9.7d); while if they are exactly 180° out of phase, the pattern will be a straight line in the second and fourth quadrants (Fig. 9.7e). For other phase angles, the general pattern is a tilted ellipse, as shown in Fig. 9.7b, and the phase angle can be read directly as the amount of opening of the ellipse (see Problem 9.11). This method of "plotting" data is especially useful when we are searching for the frequency at which the phase angle between two signals is 0°, 180°, and so on.

Problems 9.6 through 9.13

9.4 COMPUTING FREQUENCY RESPONSE

The predicted frequency response for a given linear system model can be computed in a number of ways, both algebraic and graphical. We shall describe several common methods. [For simple first-order systems like those of Fig. 9.1a and Fig. 6.2 the calculation is quite simple by any of the methods.]

In any case, one first obtains a complex expression like (9.8) or (9.11) relating the variables of interest. Then one evaluates this complex expression—for its magnitude and phase angle—for many values of $s = j\omega_f$, until the pattern is satisfactorily established. The various approaches are simply different ways of evaluating a complex algebraic expression.

For example, we can compute *algebraically* the values of magnitude ratio and phase angle for a given physical model, as we did on p. 313. We can repeat the calculation for a number of values of ω_f over the range of interest, entering our results in a table or plotting them. (A plot is helpful for interpolating between calculated values.)

Alternatively, we can avail ourselves of several useful *graphical* methods. One is the *phasor* construction of Fig. 9.2. A similar technique utilizes an *s-plane plot* of the poles and zeros. This method establishes an important link between the s plane and the frequency-response characterization of systems. It is described briefly below for first-order systems, and in Sec. 9.8 for second-order systems.

The most widely used method, however, is the logarithmic asymptote technique developed by *Bode*, which is presented in Sec. 9.6. The speed, accuracy, and simplicity of this construction have made it the "work-horse" method in the majority of problems.

Graphical calculation of frequency response from the s plane

This method is demonstrated in Fig. 9.8 for the ratio V_3/V_1 of expression (9.6). First the poles and zeros of the transfer function are plotted. (In this case there is only a pole, at $s = -1/RC$.) Next, with (9.6) in the form

$$\frac{V_3}{V_1} = \frac{1/RC}{(1/RC) + j\omega_f} \tag{9.12}$$

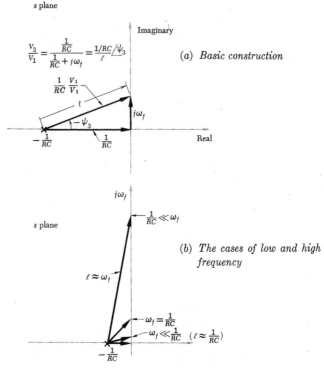

Fig. 9.8 *Obtaining magnitude and phase graphically from the s plane*

the real and imaginary parts of the denominator are readily plotted as complex vectors, and the amplitude and phase angle of the sum are measured geometrically. Note that since $1/RC$ is independent of frequency, the base leg (horizontal vector) in Fig. 9.8 is constant, while the vertical leg varies with frequency. The desired quantities are readily available from the figure in terms of ℓ and ψ_3 as follows:

$$\frac{v_{3m}}{v_{1m}} = \left|\frac{V_3}{V_1}\right| = \frac{1/RC}{\ell} \qquad \angle\frac{V_3}{V_1} = \psi_3$$

It is quite simple to mark off a number of values of ω_f on the vertical axis and then to measure the corresponding values of ℓ (with a ruler) and of ψ_3 (with a protractor). These can then be plotted, as in Figs. 9.4 and 9.5.

9.5 LOGARITHMIC SCALES FOR PLOTTING FREQUENCY RESPONSE

It was pointed out by H. W. Bode that if frequency response plots like Fig. 9.4 are constructed to logarithmic scales, then certain symmetries and asymptotic characteristics result which permit a great increase in

[Sec. 9.5] LOGARITHMIC SCALES FOR PLOTTING   323

the efficiency with which such plots can be constructed and used. In particular, the handling of systems of elements in series, and of feedback systems, is greatly facilitated. The principles involved and their application to simple systems will now be explained in Secs 9.5 and 9.6, and further in Chapter 15.

Log scales

It will be instructive to begin by constructing a logarithmic (or "log") scale. This can be done by using only the property that *equal distances along a log scale correspond to equal ratios of the quantity represented*. Thus we might assume some convenient length on the scale and assign to it the ratio of 2. We may then lay out this length any number of times, doubling the value after each interval, as shown in Fig. 9.9a. Each factor of 2 is known as an *octave*. It will be convenient to know also the location of factors of 10, which are known as *decades*. These are also easily found because it happens that the cube root of 2 is nearly five-fourths: $\sqrt[3]{2} = \tfrac{5}{4}$ (to closer than 1%). Therefore the distance between points 4 and 5 on the scale will be $\tfrac{1}{3}$ of an octave, as shown in Fig. 9.9b. (The distance corresponding to a ratio of $\tfrac{5}{4}$ is denoted u). In this way, points 5 and 10 are readily established, and one decade is found to consist of 10 units ($10u$), any three of which ($3u$) constitute an octave, as shown. Inexpensive one-quarter-inch quadrille paper is ideal for most Bode plotting, with one division = u.

Intermediate points can be easily estimated in Fig. 9.9. (For example, a point halfway between 1 and 2 will be $(2)^{\frac{1}{2}} = \sqrt{2}$ or 1.41.)

(*a*) *Octave plot: equal distances denote equal ratios*

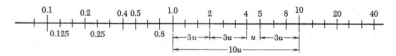

(*b*) *Decade plot*

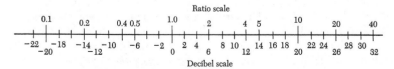

(*c*) *Comparison of ratio and decibel scales*

Fig. 9.9 *Scales for log plotting*

An alternative method of log plotting involves a scale known as the decibel scale (abbreviated db). One decibel has been arbitrarily defined as a distance of $\frac{1}{2}u$ in Fig. 9.9b, and the zero point has been taken at a ratio of unity. [The mathematical definition is: y(in db) $= 20 \log_{10} y$.] Thus the relations between a ratio scale and a db scale are as indicated in Fig. 9.9c.

Either scale can be used in logarithmic plots. The db scale is quite useful in communications work, where information is often obtained in tabular form; then it is convenient to multiply a number of quantities together by adding their corresponding db values (comparable to using a log table).

In dynamics, however, log scales are almost always used graphically (rather than algebraically), and there is no particular point in converting quantities into db for plotting, and then reconverting to ratios as part of the plot-reading process.† In dynamics it is the ratios themselves in which we are interested, and with which we want to work. Ratio scales will therefore be used throughout this book. (Additional advantages with regard to the slope of the asymptotes also attend the use of ratios instead of db, as will be seen.)

9.6 FORCED OSCILLATION OF FIRST-ORDER SYSTEMS: THE PLOTTING TECHNIQUES OF BODE

We can now replot Fig. 9.4, using logarithmic coordinates, as shown in Fig. 9.10c and d. We could do this point by point, as we did in Fig. 9.5. However, with the aid of Fig. 9.8 and Eq. (9.8) the plotting can be greatly speeded up by first constructing asymptotes, as we now demonstrate for the network of Fig. 9.1. (The network is shown again in Fig. 9.10a.)

Construction is based on the relation (9.8)

$$\frac{V_3}{V_1} = \frac{1}{j\omega_f/(1/RC) + 1} \qquad \begin{array}{c}(9.8)\\ \text{[repeated]}\end{array}$$

(We shall make a custom of including an s-plane picture like Fig. 9.10b with every plot of calculated frequency response, to display the poles and zeros of the corresponding transfer function.)

Plotting amplitude

Consider first the magnitude plot, Fig. 9.10c, and notice that at very low frequencies the length ℓ of the V_3/V_1 vector in Fig. 9.8b is not changing with frequency. In fact, from the geometry of Fig. 9.8, the length will change only slightly for all frequencies between zero and (about) $\omega_f/(1/RC) = 0.1$. The amplitude, therefore, will simply be a horizontal straight line up to about $\omega_f/(1/RC) = 0.1$. This result can, of course, be seen algebraically from (9.8), where the term $j\omega_f/(1/RC)$ in the denominator can be neglected for small values of $\omega_f/(1/RC)$, leaving $V_3/V_1 \approx \frac{1}{1}$. We therefore plot, as one asymptote in Fig. 9.10, a horizontal line at magnitude $V_3/V_1 = 1$.

---

† The whole process reminds one of the early custom wherein scientific papers were translated by their author into Latin for publication, and then retranslated into English by the reader. (See, for example, Appendix D.)

[Sec. 9.6] THE PLOTTING TECHNIQUES OF BODE

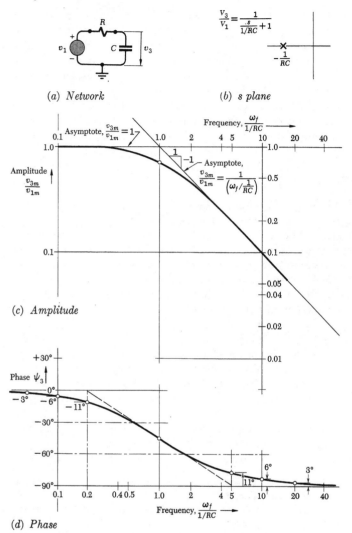

Fig. 9.10 *Calculated frequency response (Bode plot) for first-order system: asymptote construction*

At the other extreme, when $\omega_f/(1/RC)$ is very large, the term 1 in the denominator of (9.8) can be neglected, and thus the magnitude of V_3/V_1 becomes inversely proportional to frequency:

$$\frac{v_{3m}}{v_{1m}} \approx \left| \frac{1}{\omega_f/(1/RC)} \right| = \frac{1}{\omega_f/(1/RC)} \qquad (9.13)$$

A plot of (9.13) to a logarithmic scale, Fig. 9.10c, is simply a straight line which passes through the point (1,1) and has a slope of -1.† (The reader can readily verify this by plotting a few points or by indulging in a simple analytical proof.) From (9.18) or Fig. 9.8, we see that the magnitude of v_{3m}/v_{1m} coincides indistinguishably with this asymptote for frequencies greater than about $\omega_f/(1/RC) = 5$.

For values of $\omega_f/(1/RC)$ near 1 the magnitude can be most quickly obtained graphically from Fig. 9.8. As a practical matter, however, it is sufficient to note that when $\omega_f/(1/RC) = 1$, the magnitude is simply

$$\frac{v_{3m}}{v_{1m}} = \left| \frac{1}{j(1) + 1} \right| = \frac{1}{\sqrt{2}} \approx 0.7$$

which is plotted in Fig. 9.10c. The rest of the curve is faired in through the single plotted point and the two asymptotes.

The frequency $\omega_f = 1/RC$ is known as the "break frequency," or "corner frequency" for the first-order system, because it is the point at which the asymptotic plot breaks from a horizontal line to a line with a slope of -1. This will be a useful concept in plotting the response of numerous systems in series; a close approximation can be obtained by plotting the series of asymptotes. This will be demonstrated in Chapter 15.

Plotting phase angle

Points for the plot of phase angle versus frequency, Fig. 9.4 and Fig. 9.10d, can be most rapidly obtained graphically by actual measurement of the angle in Fig. 9.8. (The reader is invited to check a few points in Fig. 9.4 or Fig. 9.10d by actual graphical measurement, using a protractor or other device.)

Note, in particular, that for very low frequencies ψ_3 is near $0°$, while for very high frequencies it is near $-90°$. It is helpful, therefore, to construct asymptotes at $\psi_3 = 0°$ and $\psi_3 = -90°$, as shown in Fig. 9.10d.

Actual points on the ψ_3 plot can be obtained quickly and accurately for several special cases. For $\omega_f/(1/RC) = 1$ the phase angle is, of course, $-45°$ (see Fig. 9.8). For frequencies less than, say, $\omega_f/(1/RC) = 0.2$, the arc-tangent approximation is quite accurate. That is,

$$\psi_3 = -\tan^{-1}\frac{\omega_f}{1/RC} \approx -\frac{\omega_f}{1/RC}$$

and values of ψ_3 are readily calculated. For example,

for $\omega_f/(1/RC) =$	0.02	0.05	0.1	0.2
$\psi_3 =$	-0.02 rad	-0.05 rad	-0.1 rad	-0.2 rad
or $=$	$-1.1°$	$-3°$	$-6°$	$-11°$

† If a decibel scale had been used, the slope would be read as "minus six db per octave"— a rather cumbersome and unenlightening phrase compared to "minus one."

[Sec. 9.6] THE PLOTTING TECHNIQUES OF BODE 327

From the geometry of Fig. 9.8, a similar approximation can be used for frequencies *higher* than, say, $\omega_f/(1/RC) = 5$, as follows:

$$\psi_3 = -\tan^{-1}\frac{\omega_f}{1/RC} = -90° + \tan^{-1}\frac{1/RC}{\omega_f} = -90° + \frac{1}{\omega_f/(1/RC)}$$

so that if, for example,

$$\frac{\omega_f}{1/RC} = 5 \qquad 10 \qquad 20$$

then the *difference between* the $-90°$ asymptote and the actual angle ψ_3 is

	$\frac{1}{5}$ rad	$\frac{1}{10}$ rad	$\frac{1}{20}$ rad
or	$+11°$	$+6°$	$+3°$

It turns out that when phase angle is plotted versus the logarithm of frequency,† as it has been in Fig. 9.10d, then the phase angle plot exhibits symmetry about the frequency $\omega_f/(1/RC) = 1$. That is, at an equal distance (on the log chart) either above or below the frequency $\omega_f/(1/RC) = 1$, the phase angle differs from the asymptotes by the same amount [e.g., at $\omega_f/(1/RC) = \frac{1}{4}$, angle ψ_3 differs from 0° by the same amount that, at $\omega_f/(1/RC) = 4$, it differs from 90°]. The symmetry is clear in the figure.

As a result of this symmetry, it is possible—and sometimes quite convenient—to develop a slightly more elaborate asymptote method of constructing the entire phase-angle curve to accompany the asymptotes for the magnitude plot (Fig. 9.10c). One good asymptotic construction for phase angle consists of the 0° and 90° asymptotes, plus the dashed line shown in Fig. 9.10d. This line goes through the point $-45°$ at $\omega_f/(1/RC) = 1$, through 0° at $\omega_f/(1/RC) = 0.2$, and through 90° at $\omega_f/(1/RC) = 5$. It is found that the actual phase curve lies almost exactly on this asymptote between frequencies of $\omega_f/(1/RC) = \frac{1}{2}$ and $\omega_f/(1/RC) = 2$. Thus, with this asymptote, together with the arc-tangent points shown (3°, 6°, 11°, and so on), the entire phase curve can be faired in quite readily and accurately. The benefit of the asymptotic construction is fully realized when Bode plots are made for a series of elements, as we shall demonstrate in Chapter 15.

The reader should now check the data of Table IX (and Fig. 9.5) against Fig. 9.10, comparing the calculated frequency response with the "measured" values.

Example 9.1. A mechanical system. A physical system is represented by the model (Stage I) shown in Fig. 9.11: a mass m rides on a carriage. The motion of the carriage is known, and is sinusoidal: $x = x_m \cos \omega_f t$. Between the mass and the carriage is a viscous film, and thus the mass is coerced into following the carriage motion by a viscous force proportional to the *relative* velocity between mass and carriage, represented by damping coefficient b. Find the resulting motion $y(t)$ of the small mass with respect to a fixed reference.

STAGE II. EQUATIONS OF MOTION. The motions of carriage and mass are indicated by the geometry shown; the *given* displacement of the carriage relative

† On one-quarter-inch quadrille paper, one unit = 10° or = 30° are convenient scales.

to a fixed reference is denoted by x, while displacement of the mass with respect to a fixed reference is denoted by y. The reader may show that the equation of motion is then

$$m\ddot{y} - b(\dot{x} - \dot{y}) = 0 \tag{9.14}$$

from which the transfer function, from known motion x to unknown motion y, is (letting $x = Xe^{st}$, $y = Ye^{st}$)

$$\frac{Y}{X} = \frac{s}{s\{[s/(b/m)] + 1\}} \tag{9.15}$$

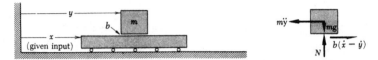

Fig. 9.11 *Forced vibration of a first-order mechanical system*

For the given input, $x = x_m \cos \omega_f t \stackrel{r}{=} x_m e^{j\omega_f t} = Xe^{j\omega_f t}$, and the assumed response $y \stackrel{r}{=} (y_m e^{j\psi_f})e^{j\omega t} = Ye^{j\omega_f t}$, we have

$$\frac{Y}{X} = \frac{1}{[j\omega_f/(b/m)] + 1} \tag{9.16}$$

which is identical in form to (9.8) [and to (9.11)], with the result that Fig. 9.10, with appropriately labeled coordinates, represents the system of Fig. 9.11, as well as those of Figs. 9.1a and 6.2. Relabel the coordinates of Fig. 9.10 appropriately.

Example 9.2. The high-pass electrical filter. A simple electrical network is isolated by amplifiers as shown in Fig. 9.12. The input to the system is a sinusoidal voltage: $v_1 = v_{1m} \cos \omega_f t$. Find the output voltage and plot the ratio of output to input voltage as a function of frequency ω_f.

Fig. 9.12 *A "high-pass" electrical filter*

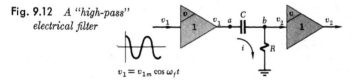

The dynamic equations of motion are obtained by applying Kirchhoff's Current Law at node b. (All of the current which flows through capacitor C flows also through the resistor R because none can flow into the isolating amplifier.) The result is

$$C(\dot{v}_1 - \dot{v}_2) = \frac{v_2}{R}$$

or
$$RC\dot{v}_2 + v_2 = RC\dot{v}_1$$

Thus, letting $v_1 = V_1 e^{st}$ and $v_2 = V_2 e^{st}$, we have the transfer function

$$\frac{V_2}{V_1} = \frac{s/(1/RC)}{[s/(1/RC)] + 1}$$

[Sec. 9.6] THE PLOTTING TECHNIQUES OF BODE

from which $v_2 = v_{2m} \cos(\omega_f t + \psi_f)$, with

$$\frac{v_{2m}}{v_{1m}} = \left| \frac{j[\omega_f/(1/RC)]}{j[\omega_f/(1/RC)] + 1} \right| \qquad \psi_f = \bigg/ \frac{j[\omega_f/(1/RC)]}{j[\omega_f/(1/RC)] + 1} \qquad (9.17)$$

From (9.17) we can obtain relations for the asymptotes on the frequency-response plot. For very low frequency the denominator of the magnitude expression is approximately 1, so that the magnitude asymptote is given by the expression:

$$\left(\frac{v_{2m}}{v_{1m}}\right)_{\text{LFasympt.}} = \left(\frac{\omega_f}{1/RC}\right)$$

and the phase angle asymptote is at $+90°$. At high frequency the denominator is approximately $j[\omega_f/(1/RC)]$, and thus the magnitude asymptote is

$$\left(\frac{v_{2m}}{v_{1m}}\right)_{\text{HFasympt.}} = 1$$

while the phase-angle asymptote is at $0°$. These asymptotes are plotted in

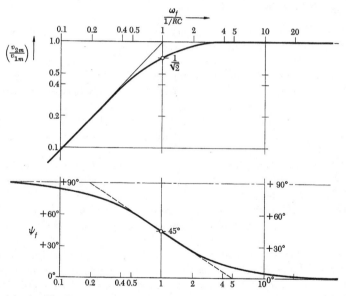

(a) *Amplitude and phase plot*

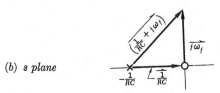

(b) *s plane*

Fig. 9.13 *Calculated frequency response (Bode plot) for high-pass filter of Fig. 9.12*

Fig. 9.13a. Before fairing in the final curves, it is helpful to make an s-plane picture, Fig. 9.13b, for the special case $\omega_f = 1/RC$. From this it is clear that when $\omega_f/(1/RC) = 1$, the magnitude is $v_{2m}/v_{1m} = 1/\sqrt{2}$, while the phase angle is $\psi_f = +90° - 45° = +45°$.

It is clear from Fig. 9.13a that this circuit earns its name by passing faithfully all signals of frequencies above $\omega_f \approx 10(1/RC)$, while attenuating low-frequency signals.

High-pass filters are used, typically, as a simple, cheap means of removing the d-c content from a signal. Notice also, from (9.17), that at low frequencies the output, v_2, is approximately proportional to the derivative, $\dot{v}_1$, of the input. This simple circuit is therefore much used as a computing element in spacecraft control systems, for example, where light weight and reliability are crucial.

Problems 9.14 through 9.21

9.7 FORCED OSCILLATION OF UNDAMPED SECOND-ORDER SYSTEMS: RESONANCE

A special case of widespread interest is the response of undamped second-order (mass-spring or LC-network) systems to sinusoidal excitation. The occurrences of such two-energy-storage systems, in which damping is negligible, are legion. A number are shown in Fig. 7.2.

Moreover, many important and often highly complex systems are made up of sets of many such subsystems. In particular, the majority of vibration problems—which dominated the field of dynamics in mechanical engineering for many years—involve systems of many subsystems like those in Fig. 7.1 or Fig. 7.2a or 7.2b.

To illustrate, Fig. 9.14a shows a model for a jet engine in which the sets of turbine blades and compressor blades are represented by disks connected by limber shafts. Torque is applied to the system by gas flowing through the turbine stages. (In a more detailed model, each of the disks would be replaced by a set of radial cantilever beams representing the blades.)

In Fig. 9.14b a gimbaled stable platform is isolated from its vibrating base by a shock mount. The gimbals are somewhat limber and, being metallic structures, have almost negligible inherent damping, and thus the system can be represented, for purposes of vibration analysis, by the mass-spring model shown.

Similarly, certain electrical networks, such as that shown in Fig. 9.14c, are built with negligible resistance. The electrical "high-Q" filter for selecting radio signals from a field of many signals differing only slightly in frequency consists of a number of such LC networks with nearly negligible damping, as do the circuits in the oscillators used by the radio stations to generate their signals.

Even distributed-mass systems like the rocket-vehicle frame shown in Fig. 9.14d, or analogous distributed-parameter systems such as electrical

[Sec. 9.7] UNDAMPED SECOND-ORDER SYSTEMS: RESONANCE 331

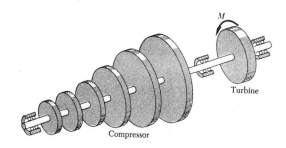

(a) *Vibration model of jet engine*

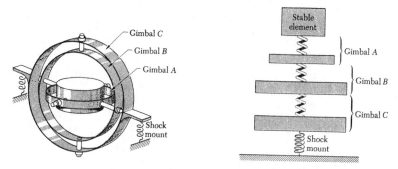

(b) *Stable platform and vibration model*

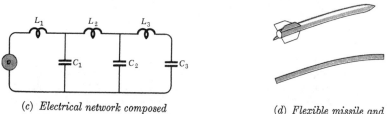

(c) *Electrical network composed of resonant elements*

(d) *Flexible missile and vibration model*

Fig. 9.14 *Systems made up of second-order subsystems*

transmission lines, can be represented by models consisting of discrete masses and springs, or capacitances and inductances, with very good analytical results. In Fig. 9.14d the rocket vehicle and its model are shown in the "first bending mode." (Modes of vibration are discussed in Chapters 12, 13.)

Second-order systems in which energy dissipation (damping or resistance) can be neglected are, of course, special cases of the more general second-order system *with* damping. Thus there is the choice of treating the general case first and then specializing the results obtained, or of treating the special case first and then the general. The latter course is prevalent in

Fig. 9.15 *Forced undamped second-order system*

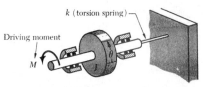

vibration texts. It will be followed here also because certain phenomena are more easily presented in the simpler context of the undamped model. However, the reader who is more inclined to the general approach is invited to reverse the order of reading the present section and the next, and to proceed immediately to the general case treated in Sec. 9.8 (without loss of continuity).

Consider the model shown in Fig. 9.15—a rotary mass J on a limber shaft of spring constant k, driven by a sinusoidal torque, $M = M_m \cos \omega_f t$. Its equation of motion, from (8.1') (with $b = 0$) is

$$J\ddot{\theta} + k\theta = M$$

and its steady-state forced response to the given input can be written, from (8.23') with $b = 0$,

$$\theta = \theta_m \cos \omega_f t = \frac{M_m}{k - J\omega_f^2} \cos \omega_f t \qquad (9.18)$$

or, using the common shorthand $\beta = \omega_f/\sqrt{k/J}$,

$$\boxed{\frac{\theta_m}{M_m/k} = \frac{1}{1 - \beta^2}} \qquad (9.19)$$

A simple linear plot of (9.19) is given in Fig. 9.16a. At low frequencies the motion simply follows along in phase with the disturbance and at an amplitude which is nearly independent of frequency: The amplitude depends merely on the spring stiffness, $\theta_m = M_m/k$. At very high frequencies the motion is 180° out of phase with the disturbance (regardless of frequency) and is much smaller: The inertia simply ignores high-frequency motions and refuses to follow them. (The spring force is of little consequence at high frequencies.)

But the most interesting point occurs where the disturbing frequency is just equal to the natural frequency of free motion of the system: $\omega_f = \sqrt{k/J}$. Then the motion of the system is infinitely large! (This result is, of course, evident from expression (9.18), whose denominator is zero when $J\omega_f^2 = k$, indicating infinite motion for any finite disturbance.) Physically we recall that, because there is no damping, if the system is set in motion it will vibrate indefinitely at its natural frequency with no loss of energy. Now if a disturbance is applied at exactly the natural frequency, then each cycle more energy is pumped into the system and its

[Sec. 9.7] UNDAMPED SECOND-ORDER SYSTEMS: RESONANCE

amplitude grows, finally, to an infinitely large value. In any real system, of course, the frictionless, linear-spring model no longer applies when the amplitude becomes very large.

This phenomenon—of large output in the presence of a small input—is known as *resonance*, as we discussed on p. 271, and the *undamped natural frequency* of a system (e.g., $\sqrt{k/J}$ for the system in Fig. 9.15) is often referred to as the *"resonant frequency"* of the system.

In effect, both magnitude and phase information are contained in Fig. 9.16a, which can be drawn only because the phase angle is always either 0° or 180° (except for the singular point at resonance). More generally, it is useful to plot the same result on separate magnitude and phase plots such as were used in Fig. 9.10. This is done (using logarithmic coordinates)

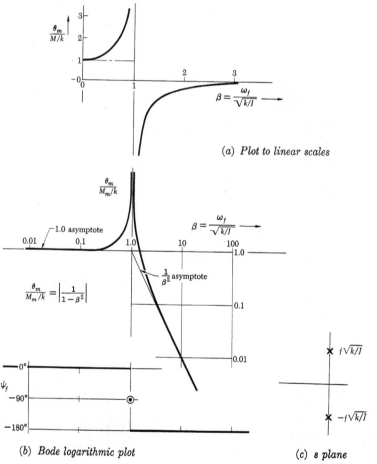

Fig. 9.16 *Sinusoidal response of undamped second-order system*

in Fig. 9.16b. The plot is most easily checked from expression (9.19). In particular, asymptotes can be drawn for very small and for very large frequencies. For small frequency—$\beta \ll 1$—the magnitude equals 1. For very large frequency—$1 \ll \beta$—the magnitude is given by $1/\beta^2$. Finally, when $\beta = 1$ (disturbance at the resonant frequency) the magnitude is infinite.

The phase curve is, of course, simply 0° for frequencies less than $\beta = 1$, and 180° for frequencies greater than $\beta = 1$. (It will be shown in the next section that when $\beta = 1$ the phase angle is $-90°$.)

The general transfer function for this system is (8.33) with $b = 0$:

$$\frac{\Theta}{M} = \frac{1}{J}\frac{1}{s^2 + (k/J)} = \frac{1}{J}\frac{1}{(s + j\sqrt{k/J})(s - j\sqrt{k/J})}$$

and this is displayed on the s plane, Fig. 9.16c, for reference.

Experimental example 9.3. Sinusoidal response of undamped system. Recall the experiment of Fig. 8.16. We wish to repeat it here more comprehensively.

Suspend a reasonably heavy mass from a good linear, fairly flexible coil spring as shown in Fig. 9.17. (Displacements x and y are zero at static equilibrium, and δ is the stretch in the spring at static equilibrium.) Perform the following experiments:

(a) Move the top end of the spring vertically up and down with a very slow sinusoidal motion. Notice that the mass follows the upper end of the spring at approximately the same amplitude ($y = x$ in the figure).

(b) Wiggle the top of the spring up and down very rapidly. Notice that the mass hardly moves ($y \approx 0$) and that its motion y is exactly (180°) out of phase with x.

(c) Find the frequency of motion of the upper end which causes the mass

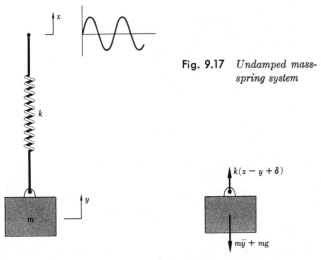

Fig. 9.17 Undamped mass-spring system

(a) Physical model   (b) Free-body diagram

[Sec. 9.8] DAMPED SECOND-ORDER SYSTEMS

to have a very large amplitude of motion—much larger than the motion x of the upper end of the spring. (This is best done by making motion x quite tiny and finding the frequency for which a tiny x produces a large y.) Notice that motion y of the mass is 90° out of phase with the input motion for this particular frequency.

(d) Hold the upper end of the spring fixed and let the mass vibrate up and down at its own natural frequency. Note that this is the same as the resonant frequency found in (c).

The reader may explain the experimental results in terms of Fig. 9.16 by first showing that the equation of motion (Stage II) for this model is [from Fig. 9.17b or (8.24)]

$$m\ddot{y} - k(x - y) = 0$$

and then showing that, since the input motion is $x = x_m \cos \omega_f t$, the output motion will be $y = y_m \cos \omega_f t$ with y_m/x_m given by

$$\frac{y_m}{x_m} = \frac{k}{-m\omega_f^2 + k} = \frac{1}{1 - \beta^2}$$

where $\beta = \omega_f/\sqrt{k/m}$, which is identical to (9.19), the relation plotted—magnitude and phase—in Fig. 9.16. (Now the coordinate of the magnitude curve is to be y_m/x_m.)

At very low frequencies [Experiment (a)] the ratio of $y_m/x_m = 1$, the stretch in the spring remains constant, and the mass follows the motion of the upper end of the spring exactly. At very high frequencies [Experiment (b)] the mass simply sits still while the upper end of the spring moves up and down. At $\beta = 1$ we have resonance [Experiments (c) and (d)].

Problems 9.22 through 9.29

9.8 FORCED OSCILLATION OF DAMPED SECOND-ORDER SYSTEMS

A typical second-order system is represented by the physical model in Fig. 9.18. It may be a torque motor (e.g., for positioning a hydraulic valve,

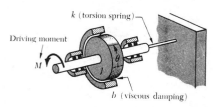

Fig. 9.18 *Forced damped second-order system*

p. 185), but not so heavily damped as the one in Fig. 6.2. Or it may be one of the recording pens or mirrors described on p. 319.† Its equation of motion is (8.1') (from p. 245)

$$J\ddot{\theta} + b\dot{\theta} + k\theta = M \qquad (8.1')$$
[repeated]

---

† For the recording pen, the spring may be partially supplied by electrical feedback proportional to θ. This does not change the equation of motion or the dynamic behavior.

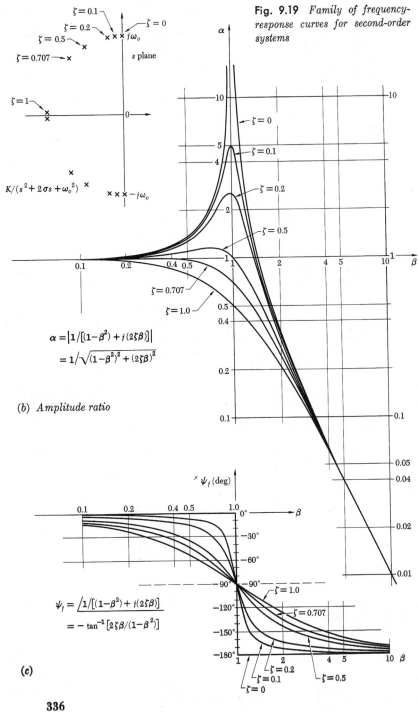

Fig. 9.19 Family of frequency-response curves for second-order systems

$$\alpha = |1/[(1-\beta^2) + j(2\zeta\beta)]|$$
$$= 1/\sqrt{(1-\beta^2)^2 + (2\zeta\beta)^2}$$

(b) Amplitude ratio

$$\psi_f = \underline{/1/[(1-\beta^2) + j(2\zeta\beta)]}$$
$$= -\tan^{-1}[2\zeta\beta/(1-\beta^2)]$$

(c)

336

[Sec. 9.9] PLOTTING SECOND-ORDER FREQUENCY RESPONSE

The corresponding transfer function, from M to θ, is (8.33)

$$\frac{\Theta}{\mathsf{M}} = \frac{1}{Js^2 + bs + k} \qquad (8.33)$$
[repeated]

As we showed in Sec. 8.8, if the system is disturbed by a torque $M = M_m \cos \omega_f t$, then its response will be given, from (8.23'), by

$$\theta = \theta_m \cos(\omega_f t + \psi_f)$$

with
$$\theta_m = \left| \frac{M_m}{-J\omega_f^2 + jb\omega_f + k} \right| = \frac{M_m}{\sqrt{(k - J\omega_f^2)^2 + (b\omega_f)^2}} \qquad (9.20)$$

$$\psi_f = -\tan^{-1} \frac{b\omega_f}{k - J\omega_f^2}$$

It will be convenient to convert (9.20) into nondimensional form by dividing through by k, and defining a dimensionless frequency ratio β and a dimensionless amplitude ratio α as follows (phase angle ψ_f is also given in terms of β):

$$\boxed{\begin{aligned} \alpha &= \frac{\theta_m}{M_m/k} = \left| \frac{1}{(1-\beta^2) + j2\zeta\beta} \right| = \frac{1}{\sqrt{(1-\beta^2)^2 + (2\zeta\beta)^2}} \\ \psi_f &= -\tan^{-1}\left(\frac{2\zeta\beta}{1-\beta^2}\right) \qquad \beta = \frac{\omega_f}{\sqrt{k/J}} \qquad \zeta = \frac{b/2J}{\sqrt{k/J}} \end{aligned}} \qquad (9.21)$$

Figure 9.19 gives the family of frequency-response curves for second-order systems, based on (9.21), for various values of ζ. Figure 9.19a gives the pole-zero diagram for the system as obtained from transfer function (8.33). Figures 9.19b and c plot α and ψ_f, respectively, versus β. Since the transfer function of interest for many second-order physical systems has precisely the form (8.33), the nondimensional display of Fig. 9.19 is much used in dynamic analysis and design.

The amplitude ratio at resonance ($\beta = 1$) is assigned the symbol Q (for "quality"; see p. 354). From (9.21) its value is

$$Resonance \boxed{\begin{aligned} \alpha_{\text{res.}} &= Q = \frac{1}{2\zeta} \\ \psi_{f\,\text{res.}} &= -90° \end{aligned}} \text{at } \beta = 1 \qquad (9.22)$$

⊙9.9 TECHNIQUES FOR PLOTTING SECOND-ORDER FREQUENCY RESPONSE

While Fig. 9.19 may supply ample information for dynamic analysis in many cases, it is often useful to be able to sketch rapidly the frequency response of a given second-order system from scratch (perhaps using only a quarter-inch quadrille pad), as when studying several systems connected

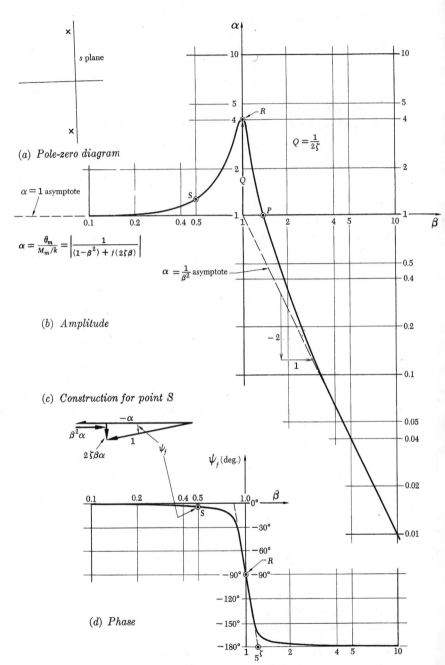

Fig. 9.20 *Sketching the frequency response (Bode plot) of a damped second-order system*

The use of asymptotes: amplitude plot

In plotting amplitude α versus frequency, Fig. 9.20b, two asymptotes can be used. From (9.21), the asymptote for very low frequency ($\beta \ll 1$) is a horizontal line at amplitude 1. The high-frequency asymptote ($1 \ll \beta$) is simply a line representing the function $1/\beta^2$, that is, a line having a slope of -2 and passing through unity at $\beta = 1$ (because at high frequency β^2 is much larger than the other terms in the denominator). These asymptotes are shown in Fig. 9.20b. (A third "asymptote" of interest is described in Problem 9.31.)

Next the resonant point R is plotted, from (9.22) (amplitude Q, phase $-90°$). (Point R is not quite the maximum of the response curve, as discussed below.)

Since we know that the amplitude curve will coincide with the low-frequency asymptote for $\beta \ll 1$, and with the high-frequency asymptote for $1 \ll \beta$, and since we know further that when $\beta = 1$ the magnitude is $Q = 1/2\zeta$, we can sketch in accurately much of the magnitude curve. In many practical problems design decisions can be made once the asymptotes plus point R are established.

Occasionally, however, as when several second-order systems are cascaded, it may be desirable to calculate intermediate points accurately. Some additional techniques for that purpose will now be described.

Unity amplitude

One additional point on the curve is quite easy to find, namely, the frequency β at which the amplitude curve crosses magnitude unity—point P in Fig. 9.20b. From (9.21), this value of β is obtained from $(1 - \beta^2)^2 + (2\zeta\beta)^2 = 1$, whose solutions are $\beta = 0$ (which we already knew), and $\beta = \sqrt{2 - 4\zeta^2}$ for which a few points are tabulated below:

ζ	β for $\alpha = 1$
0	$\sqrt{2} = 1.41$
0.1	$\sqrt{1.96} = 1.40$
0.125	$\sqrt{1.94} = 1.39$
0.5	$\sqrt{1} = 1.00$
Higher values	α always < 1 (for $0 < \beta$)

In particular, in Fig. 9.20b $\zeta = 0.125$, and point P occurs at $\beta = 1.39$. (For $0.5 < \zeta$, $\beta = \sqrt{2 - 4\zeta^2}$ is imaginary, which corresponds to the obvious fact, in Fig. 9.20b, that there is no unity crossing, except at $\beta = 0$.)

Resonance, Q and α_{max}

It is important to note that the *maximum* value of α (Fig. 9.20b) does not occur at the undamped natural frequency ("resonant frequency," $\beta = 1$) of the system but, in fact, near the *damped* natural frequency instead. The exact frequency of maximum α can be found by differentiating (α or $1/\alpha^2$) in (9.21) with respect to β and setting the result equal to zero. We find that β for maximum α is given by

$$\beta_{\alpha\,\max} = \sqrt{1 - 2\zeta^2}$$

(Recall that the damped natural frequency occurs at $\beta = \sqrt{1 - \zeta^2}$.) This result will be understood more clearly from the s-plane calculation of frequency response described in Sec. 9.10.

The actual magnitude at this frequency can be obtained directly from (9.21), and is found to be

$$\alpha_{\max} = \frac{1}{2\zeta\sqrt{1-\zeta^2}} = \frac{Q}{\sqrt{1-\zeta^2}}$$

which, *for small* ζ, is given approximately by

$$\alpha_{\max} \approx Q = \frac{1}{2\zeta}$$

(the value at $\beta = 1$).

To summarize: "resonance" or the "resonant frequency" is defined as $\beta = 1$, and the *resonant response* is given the special symbol "Q":

$$Q \triangleq (\alpha)_{\beta=1} = \frac{1}{2\zeta} \qquad (9.22)$$
[repeated]

This is not the largest value of α: $Q \neq \alpha_{\max}$. Instead, $\alpha_{\max}$ occurs at the lower frequency $\beta = \sqrt{1 - 2\zeta^2}$, and has the value

$$\alpha_{\max} = \frac{1}{2\zeta\sqrt{1-\zeta^2}} = \frac{Q}{\sqrt{1-\zeta^2}}$$

However, for small ζ's the two are nearly equal, while for large ζ's, Q and $\alpha_{\max}$ are usually not so important anyway.

[For the case at hand—$\zeta = 0.125$—the maximum value of α is 4.03, and it occurs at $\beta = \sqrt{1 - 2(0.125)^2} = 0.984$.]

Phase-angle plot

For the second-order function (9.21), with $\zeta < 1$, the phase-angle part of a Bode plot, Fig. 9.20d (i.e., phase angle to a linear scale versus frequency to a log scale), can be quite accurately sketched, using the following asymptotes:

For $\beta \ll 1$: slope = 0 $(\psi_t)_{\text{asympt.}} = 0°$
For $1 \ll \beta$: slope = 0 $(\psi_t)_{\text{asympt.}} = -180°$
For β near 1: asymptote passes through R, intersecting the other asymptotes at $\beta = e^{-\zeta\pi/2}$ and $e^{\zeta\pi/2}$

(Note that $e^{\pi/2} = 4.82 \approx 5$, so that the last two quantities are conveniently remembered as $1/5^\zeta$ and 5^ζ.)

The latter asymptote† can be shown to be tangent to the phase curve at R. Moreover, at its intersection with the two horizontal asymptotes the actual phase angle will differ from 0° or 180° by approximately the following value, depending on ζ:

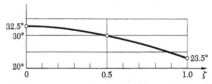

These points, together with the three asymptotes, normally give a sufficiently accurate phase plot as demonstrated in Fig. 9.20d.

When ζ is greater than 1, the system is the same as two first-order systems in tandem, and its frequency response can be plotted by extending the methods of Sec. 9.6, as we shall discuss in detail in Sec. 15.1.

Phasor construction

If additional intermediate points are required with precision, a phasor diagram may be used, as shown in Fig. 9.21. Figure 9.21a indicates the physical reasoning: The phasors (rotating arrows) represent the sinusoidal motions and torques involved in (8.1'), and the various torques must be in balance at any arbitrary time.

That is, if at some arbitrary time t the disturbing torque is represented by the phasor $\vec{M} = M_m e^{j\omega_f t}$, then the position $\vec{\theta} = \theta_m e^{j\omega_f t + \psi_f}$ of the rotor will be represented by another phasor as shown. Attached to the $\vec{\theta}$ phasor will be three torque phasors, representing the spring, damping, and inertial torques on the rotor, whose magnitude and orientation with respect to the $\vec{\theta}$ phasor are as shown. The magnitude of the $\vec{\theta}$ phasor, and its phase angle with respect to the applied torque $\vec{M}$, will therefore be such that the resultant of the three θ-dependent phasors exactly balances $\vec{M}$. These relations are clear from Fig. 9.21a, and in fact (9.20) can be obtained directly from Fig. 9.21a by simple geometry, as shown. The phasor diagram of Fig. 9.21a can also be converted to nondimensional form, as shown in Fig. 9.21b, by dividing each phasor by M_m and using α from (9.21). (The algebra is most easily seen by comparing the equations in Figs. 9.21a and b.)

Figure 9.21b, redrawn (for convenience) for $t = 0$ as in Fig. 9.21c, is helpful for calculating arbitrary points on the frequency-response plot. For example, the construction in Fig. 9.21c (shown also in Fig. 9.20c) happens to be for the particular frequency $\beta = 0.5$, and for $\zeta = 0.125$, point S in Fig. 9.20b. What is done, for convenience, is to choose an arbitrary length and call it α, then calculate the rest of the diagram, ending up, finally, with a length which represents 1 [steps ① through ④ in Fig. 9.21c]. The desired magnitude is the length ratio of phasor α to phasor 1, which is seen in this case to be about 1.25. This is plotted as point S in Fig. 9.20b. The construction also gives the phase angle,

† This construction is given in class notes by D. B. DeBra, Stanford University.

342 FORCED OSCILLATIONS OF ELEMENTARY SYSTEMS [Chap. 9]

as shown [step ⑤]. Other points along the curve can be found by the same construction.

The reader may wish to check some points in Fig. 9.20 (some Q's, for example, and some phase asymptotes) to be sure he understands the various techniques. Some check calculations are suggested in the problems.

Problems 9.30 through 9.36

9.10 OBTAINING FREQUENCY RESPONSE FROM THE s PLANE

Often the poles and zeros of a transfer function are available directly on the s plane as a result of preliminary graphical manipulation. In such cases it may be most convenient to compute frequency response directly from the s-plane poles and zeros. We demonstrate for (9.23) with σ, ω, and ω_0 defined as in (8.5):

$$\alpha = \frac{\theta_m}{M_m/k} = \left| \frac{k/J}{s^2 + (b/J)s + (k/J)} \right|_{s=j\omega_f}$$

$$= \left| \frac{\omega_0^2}{(s + \sigma - j\omega)(s + \sigma + j\omega)} \right|_{s=j\omega_f} \quad (9.23)$$

$$\psi_f = \bigg/\left[\frac{\omega_0^2}{(s + \sigma - j\omega)(s + \sigma + j\omega)} \right]_{s=j\omega_f}$$

Is it clear that expression (9.23) is the same as (9.20)? [Refer to (8.3)–(8.5).]

The poles for this function are plotted in Fig. 9.22 (for the particular case $\zeta = 0.35$). If we now select a frequency ω_f and plot the point $j\omega_f$ on the imaginary axis, as shown in Fig. 9.22, then the vector from the origin to point $j\omega_f$ is, of course, the vector $s = j\omega_f$, and the complex vectors labeled $\vec{A}$ and $\vec{B}$ in Fig. 9.22 are, respectively, the vectors† $(s + \sigma - j\omega)_{s=j\omega_f}$ and $(s + \sigma + j\omega)_{s=j\omega_f}$. But these are the vectors which appear in the denominator of (9.23). Thus we can write the nondimensional amplitude α and phase ψ_f as follows:

$$\alpha = \frac{k/J}{|\vec{A}| \cdot |\vec{B}|} = \frac{\omega_0^2}{|\vec{A}| \cdot |\vec{B}|}$$

and
$$\psi_f = -\underline{/\vec{A}} - \underline{/\vec{B}}$$

The most effective way to use the above relations is to make the calculations indicated—both magnitude and phase—with a Spirule.‡ If a Spirule is not available, then, of course, the magnitude can be calculated by measuring the lengths of the vectors $\vec{A}$ and $\vec{B}$ and performing a numerical multiplication, while phase angle can be calculated by measuring the appropriate angles with

---

† Again, these are mathematical vectors, not physical ones.
‡ A simple plastic device with a movable-arm memory. See Appendix I.

[Sec. 9.10] OBTAINING FREQUENCY RESPONSE FROM THE s PLANE 343

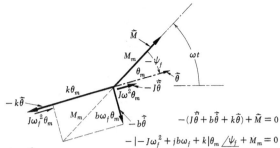

(a) Basic construction

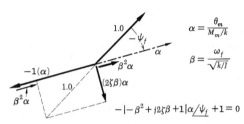

(b) Nondimensional form

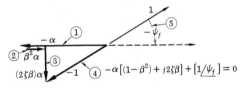

(c) Rapid construction for Bode plot

Fig. 9.21 Rotating phasor construction for forced oscillation of second-order system

a protractor. A close estimate of the ratios A/ω_0 and B/ω_0 can be made by eye from Fig. 9.22a.

Notice that the s plane shows again, Fig. 9.22b, that when ω_f is small compared with $\omega_0 = \sqrt{k/J}$, magnitude α is 1 and the net phase angle is near 0°, and when ω_f is large compared with ω_0 the magnitude falls off inversely with $\omega_f{}^2$ and the phase angle approaches 180°.

The important magnitude at resonance and the maximum magnitude can be estimated easily in Fig. 9.22c: Since vector $\vec{A}$ is varying much more with ω_f than $\vec{B}$ is, it is suggested from Fig. 9.22c that the minimum value of the denominator—and therefore the maximum value of α—will occur approximately when $\vec{A}$ is a minimum—that is, when the $j\omega_f$ vector falls nearly opposite the upper pole (i.e., near the *damped* natural frequency of the element, $j\omega$). The exact frequency at which $\alpha_{\max}$ occurs is given on p. 340.

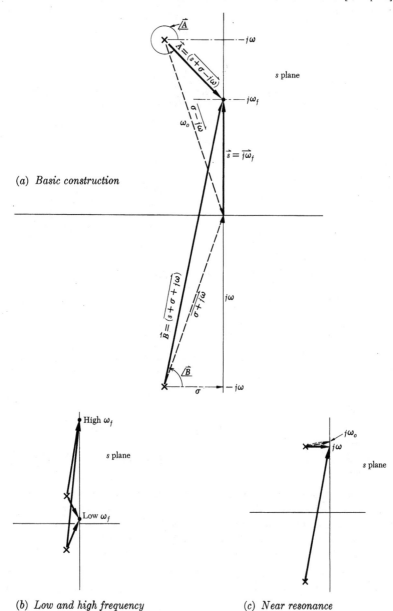

(a) Basic construction

(b) Low and high frequency (c) Near resonance

Fig. 9.22 *Frequency response from the s plane*

[Sec. 9.11] SEISMIC INSTRUMENTS 345

The reader is invited to show geometrically that when $\omega_f = \sqrt{k/J}$ (resonance), the value of ψ_f is always exactly $-90°$, regardless of ζ.

Problems 9.37 through 9.46

9.11 SEISMIC INSTRUMENTS†

A seismic instrument is one which senses and reports inertial motions of its case. Specifically, a *"seismograph"* indicates displacement of its case with respect to inertial space, while a *velocity meter* (or "velocimeter") indicates velocity, and an *accelerometer* indicates acceleration of its case, each with respect to inertial space (and with the effect of gravity superposed).

Accelerometer

An instrument for measuring the steady and vibratory acceleration of an aircraft or missile structure is depicted in Fig. 9.23.‡ In flutter tests, for example, many such instruments may be located along the wing of an aircraft to report simultaneously the acceleration at each station. (The signals from all the accelerometers are fed into a multichannel recording device, so that the complete mode of vibration can be seen at a glance from the aggregate of the traces.)

It will be worthwhile, for future reference, to derive the equation of motion of an accelerometer in detail. The result will be applicable also to all other seismic instruments.

As a physical model (Stage I) we assume that the spring and viscous damping are linear,§ and that the proof-mass guides are straight, so that cross-axis accelerations have no components along y.

To derive the equation of motion (Stage II) for the physical model we follow Procedure A-m (p. 35).

(1) *Geometry.* The system is shown in an arbitrary configuration in Fig. 9.23a. The case has (instantaneously) acceleration $a = \ddot{x}$, and is tilted upward by angle γ with respect to the local horizontal. Let $\ddot{y}$ be the absolute acceleration of the proof mass, and define motion of the mass relative to the case by

$$\xi \triangleq x - y$$

which is positive "backward," i.e., when $y < x$, as shown in Fig. 9.23a. The variable of interest is ξ, because that is what is reported as an electrical signal.

---

† The electrically inclined reader may prefer to study first Sec. 9.12 on RLC networks.
‡ An inductive pickoff is usually used instead of a potentiometer, to obtain better resolution and to avoid rubbing forces. (A potentiometer is shown here for graphic clarity.)
§ Many accelerometers are filled with the damping liquid. Then buoyant forces must be considered.

(2) *Force equilibrium* and (3) *Physical relations*. The forces acting on the proof mass are shown in a free-body diagram, Fig. 9.23b, from which, writing $\Sigma f^*_{\text{input axis}} = 0$, the equation of motion is

$$\boxed{-m\ddot{y} + b\dot{\xi} + k\xi - mg \sin \gamma = 0} \qquad (9.24)$$

where k is the combined spring constant.

We are now ready to study the response (Stage III) of the accelerometer. Since ξ is the variable of interest, we rewrite the equation of motion (9.24) in terms of $\xi \triangleq x - y$:

$$\underbrace{m(\ddot{\xi} - \ddot{x})}_{\text{inertial acceleration}} + \underbrace{b\dot{\xi}}_{\text{relative velocity}} + \underbrace{k\xi}_{\text{stretch in spring}} = \overbrace{mg \sin \gamma}^{\text{gravity}} \qquad (9.25)$$

If the acceleration of the case, $a = \ddot{x}$, is sinusoidal,

$$a = a_0 \cos \omega_f t \stackrel{r}{=} a_0[e^{st}]_{s=j\omega_f}$$

then we can solve as before for the electrical signal ξ, which will also be sinusoidal. Take $\gamma = 0$ initially. Then

$$\xi = \xi_m \cos (\omega_f t + \psi_f) \stackrel{r}{=} (\xi_m e^{j\psi_f})e^{j\omega_f t}$$

$$\frac{\xi_m}{a_0} = \frac{\xi_m}{(\ddot{x})_m} = \left| \left[\frac{m}{ms^2 + bs + k} \right]_{s=j\omega_f} \right| \qquad \psi_f = \angle \left[\frac{m}{ms^2 + bs + k} \right]_{s=j\omega_f}$$

or $\quad \dfrac{\xi_m}{a_0/\omega_0^2} = \left| \dfrac{1}{1 - \beta^2 + j2\zeta\beta} \right| \qquad \psi_f = \angle \dfrac{1}{1 - \beta^2 + j2\zeta\beta} \qquad (9.26)$

with $\qquad \omega_0 = \sqrt{k/m} \qquad \beta = \dfrac{\omega_f}{\omega_0} \qquad \zeta = \dfrac{b/2m}{\omega_0}$

which is (conveniently) exactly the quantity plotted in Fig. 9.19 and Fig. 9.20 if y is replaced by ξ. That is, Fig. 9.19 can be used not only for the forced-motion problem of Fig. 9.18 but, alternatively, to describe the dynamic performance of an accelerometer. Note, in particular, that for low-frequency motion the displacement of the seismic element with respect to its case represents faithfully the acceleration of the case for all motions of frequency lower than about $\omega_f = \frac{1}{4}\sqrt{k/m}$, ($\beta = \frac{1}{4}$). A properly designed accelerometer will therefore have a spring sufficiently stiff that $\sqrt{k/m}$ is at least four times as high as the frequency of the fastest signal to be measured.

Such design, of course, brings with it a noise problem, because all noise inputs to the instrument, up to the frequency $\sqrt{k/m}$, are reported along with the signal to be measured, which is often at a much lower frequency. (Moreover, *displacements* are reported by the accelerometer

[Sec. 9.11] SEISMIC INSTRUMENTS

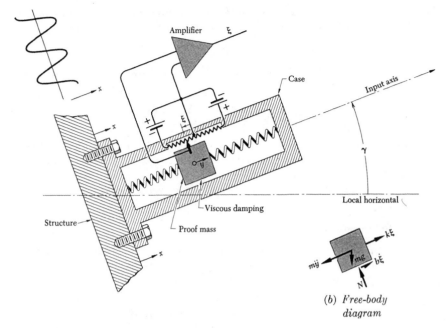

Fig. 9.23 *Accelerometer*

magnified by the square of their frequency. The apparently attractive technique of using an accelerometer to detect low-frequency displacements is therefore unsatisfactory in practice because small, extraneous, high-frequency displacements may completely swamp the signal of interest.) Accelerometers are usually built with fairly heavy damping—$\zeta = 0.5$ to 1—to avoid amplifying noise at the resonant frequency.

Accelerometers are used widely, not only to measure vibration, as in Fig. 9.23, but also, for example, as the primary precision sensor in inertial guidance systems. (The latter instruments are some orders of magnitude more precise than simple vibration-sensing instruments.)

The sensing of gravity

When an accelerometer is not oriented normal to the local gravity vector (i.e., when $\gamma \neq 0$), gravity† acts on the mass, of course, stretching the spring and producing a signal just as if there were an acceleration. Indeed, as a consequence of the equivalence principle, it is not possible to make an instrument which will sense only acceleration (but not gravity), or only gravity (but not acceleration).

From Fig. 9.23a, when the instrument is tilted upward $(0 < \gamma)$ the proof mass "hangs back" $(0 < \xi)$ and the electric output indicates a

† Or, less briefly, the gravitational mass attraction of the earth.

positive acceleration. Thus a constant tilt γ produces response identical to that of a constant acceleration $\ddot{x}$, as is clear from (9.25). That is,

$$\ddot{x}_{\text{indicated}} = g \sin \gamma \tag{9.27}$$

For example, a tilt of 1 arc second is interpreted by the instrument as an acceleration of

$$\frac{\ddot{x}_{\text{indicated}}}{g} = \sin\left(\frac{2\pi}{360° \times 60 \times 60}\right) = 0.485 \times 10^{-5} \text{ ``}g\text{'s''}$$

Accurate accelerometers are calibrated, in fact, with the use of a precision tilt table. Gravity, of course, produces a steady spring deflection, δ. Because accelerometers actually sense the *difference* between the acceleration force and the mass attraction force on their proof mass, they are sometimes called *specific force sensors*.

Seismograph

Consider next the seismograph for measuring displacements. A typical important application of the seismograph is to measure the ground displacement during earthquakes. An instrument for this purpose is depicted in

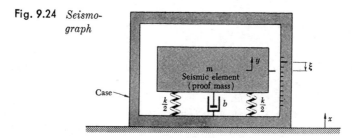

Fig. 9.24 *Seismograph*

Fig. 9.24.† Qualitatively, it is constructed exactly like an accelerometer; but its mass is much greater, and its springs are (relatively) much softer. For the seismograph, as for the accelerometer, what we can measure is relative displacement ξ (which we take downward from the position of static equilibrium). Then the total spring force is $k(\xi + \delta)$, and $k\delta$ cancels $mg \sin 90°$, and (9.24) becomes

$$m(\ddot{\xi} - \ddot{x}) + b\dot{\xi} + k\xi = 0$$
or
$$m\ddot{\xi} + b\dot{\xi} + k\xi = m\ddot{x} \tag{9.28}$$

Now, however, the input in which we are interested is x, rather than $\ddot{x}$. The corresponding transfer function is

$$\frac{\Xi}{X} = \frac{ms^2}{ms^2 + bs + k} \tag{9.29}$$

† The lateral suspension system is not shown. It is not involved in analysis of the vertical motion.

[Sec. 9.11] SEISMIC INSTRUMENTS 349

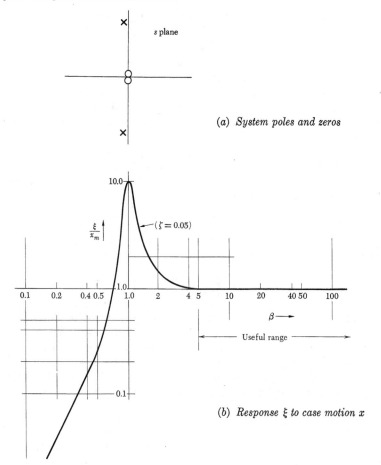

Fig. 9.25 *Frequency response of seismograph*

We therefore write the frequency response in the form

$$\frac{\xi_m}{x_m} = \left| \left[\frac{ms^2}{ms^2 + bs + k} \right]_{s=j\omega_f} \right|$$

$$\frac{\xi_m}{x_m} = \left| \frac{-\beta^2}{1 - \beta^2 + j2\zeta\beta} \right| \qquad \psi_f = \bigg/ \frac{-\beta^2}{1 - \beta^2 + j2\zeta\beta} \qquad (9.30)$$

The magnitude of this quantity is plotted in Fig. 9.25b, from which we see that the instrument reports displacements of its case faithfully only for motion of a frequency *higher* than the natural frequency ($\sqrt{k/m}$) of the instrument. Physically this is true because for very-low-frequency motions the seismic element follows the case up and down, and thus no stretch in the spring takes place and no relative motion is available to be measured. Above the natural frequency of the seismograph, the seismic element tends to remain fixed in space (as we saw also in Example 9.3), and thus all of the case motion is seen as a relative motion between case and element, and is readily measured. We want, therefore, to build our seismic instrument with a heavy mass and light spring—as light as static deflection limits and the elastic limit of the spring allow. Then we approach the ideal seismograph, in which the mass would be unattached to the base and would remain fixed in inertial space as the ground moved past it.

Notice that for the seismograph there is no problem from "extraneous" displacements, because the objective is to measure accurately all displacements of whatever frequency, and this the instrument does for all frequencies above its own low natural frequency.

Velocity meter

The velocity meter, by an analysis parallel to the preceding, is seen to have the frequency response

$$\frac{\xi_m}{v_m} = \frac{\xi_m}{(\dot{x})_m} = \left| \left[\frac{ms}{ms^2 + bs + k} \right]_{s=j\omega_f} \right| \tag{9.31}$$

If damping similar to that used in the accelerometer were used in the velocity meter, the system's pole-zero diagram would be as shown in Fig. 9.26a, the frequency response would look like Fig. 9.26b, and the instrument would have *no* frequency range in which it reported velocity accurately. The problem is solved by using very heavy damping, so that the poles of the denominator are real and widely separated, as in Fig. 9.26c. Then we have

$$\frac{\xi_m}{v_m} = \left| \left\{ \frac{ms}{m[s + (b/2m) + \sqrt{(b/2m)^2 - (k/m)}]} \right\} \right|$$
$$[s + (b/2m) - \sqrt{(b/2m)^2 - (k/m)}] \Big|_{s=j\omega_f}$$

or

$$\frac{\xi_m}{v_m/\sigma} = \left| \frac{j(\omega_f/\sigma)}{[j(\omega_f/\sigma) + (\sigma_2/\sigma)][j(\omega_f/\sigma) + (\sigma_1/\sigma)]} \right|$$

$$\psi_f = \angle \frac{j(\omega_f/\sigma)}{[j(\omega_f/\sigma) + (\sigma_2/\sigma)][j(\omega_f/\sigma) + (\sigma_1/\sigma)]}$$

where $\quad \sigma = \dfrac{b}{2m} \qquad \sigma_1 = \sigma\left(1 - \sqrt{1 - \dfrac{1}{\zeta^2}}\right)$

$\zeta = \dfrac{b/2m}{\sqrt{k/m}} \qquad \sigma_2 = \sigma\left(1 + \sqrt{1 - \dfrac{1}{\zeta^2}}\right)$

(9.32)

Then the frequency response of the instrument is as given in Fig. 9.26d for several ζ's.† It is seen that the instrument reports the magnitude of velocity accurately throughout the frequency range from about $2\sigma_1$ to

† This plot may be obtained graphically by multiplying two first-order plots together, as explained in detail in Sec. 15.1.

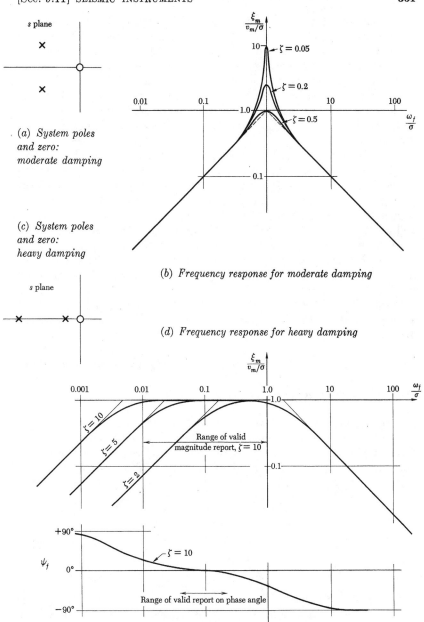

Fig. 9.26 Frequency response of velocity meter

$\frac{1}{2}\sigma_2$, but reports the proper phase angle only in a much narrower frequency band. For this reason velocity meters are not used when accurate phase data are required. A velocity meter (for magnitude only) should therefore be designed with the quantity $b/2m$ about twice the highest frequency of interest, and the quantity k/m low enough to make σ_1 about one-half the lowest frequency needed. (The purpose of spring k is to maintain the instrument somewhere near center, and avoid "bottoming." Thus in a practical design the spring is made just stiff enough to avoid bottoming in normal use, and no stiffer.)

In practical testing, if a large frequency range is to be investigated, it is found highly advantageous to use both an accelerometer and a velocity meter, in conjunction with either a seismograph or a direct displacement pickup (a wiper on a potentiometer or, for smaller motions, a variable-core inductance instrument). Then all three signals—displacement, velocity, and acceleration—will be available. The displacement signal will be found most usable at low frequencies; but at intermediate and high frequencies it tends to become so small that one is glad to switch over to the velocity meter and finally, at high frequencies, to the accelerometer. The reason for this is, of course, that for a given magnitude of motion, displacement is independent of frequency, velocity depends directly on frequency, and acceleration on frequency squared; thus for high-frequency motions the displacement will usually be very small.

Problems 9.47 through 9.53

9.12 FREQUENCY RESPONSE OF RLC CIRCUITS

Second-order networks are of major importance throughout nearly all branches of electrical engineering. The developments of Secs. 9.6 through 9.10 apply of course to electrical systems as well as to mechanical ones. (The derivations could as readily have been done using electrical models.)

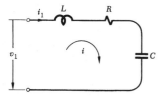

Fig. 9.27 *A common RLC network*

Just as second-order mechanical systems have two forms of energy storage, a spring and an inertia (plus damping which dissipates energy), so second-order electrical systems have two forms of energy storage, an inductor and a capacitor (plus resistors which dissipate energy): Hence, the common name "RLC circuit" for this type of electrical system.

As an example for discussion, the network in Fig. 9.27 represents a widely used RLC circuit. We have already analyzed this network in Exam-

[Sec. 9.12] FREQUENCY RESPONSE OF RLC CIRCUITS

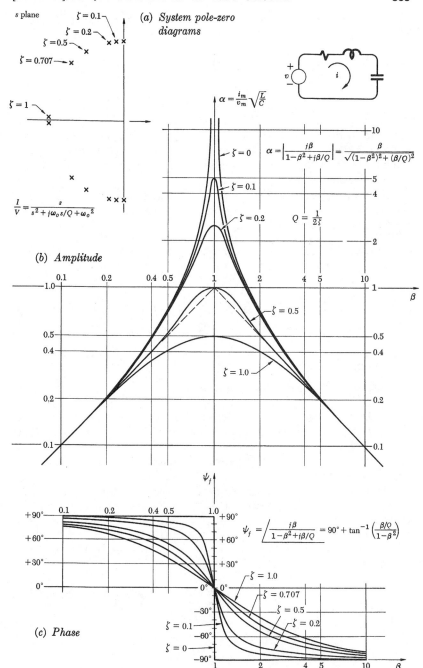

Fig. 9.28 *Frequency response of RLC network*

ple 8.7. Its transfer function I/V_1 is (8.35), and the parameters i_m and ψ_f of its frequency response, $i = i_m \cos(\omega_f t + \psi_f)$, may be written

$$i_m = \frac{v_m}{\sqrt{L/C}} \left| \frac{j\beta}{1 - \beta^2 + j\beta/Q} \right| \qquad \psi_f = \bigg/ \frac{j\beta}{1 - \beta^2 + j\beta/Q} \qquad (9.33)$$

where $\qquad \beta = \dfrac{\omega_f}{\sqrt{1/LC}} \qquad Q = \dfrac{\sqrt{L/C}}{R} = \dfrac{1}{2\zeta}$

The nondimensional quantities $i_m/(v_m/\sqrt{L/C})$ and ψ_f are plotted versus β in Fig. 9.28b and c, respectively, for various values of Q. As with Fig. 9.20, the plots of Fig. 9.28b are most readily sketched using asymptotes, as developed in the problem set.

In many applications (e.g., in the tuning circuit of a radio), it is desired that the amplitude at resonance—the "Q" of the network—be as large as possible; hence the symbol Q, for quality.

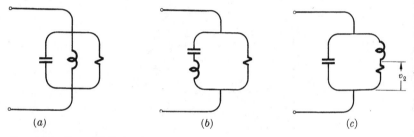

Fig. 9.29 *Some RLC circuit arrangements*

A useful property of log-magnitude plots, like Figs. 9.10a, 9.19a, 9.20a, and 9.28a, is that the reciprocal plot is simply the mirror image with respect to the unity-magnitude line. Thus if, in Fig. 9.28a, for example, we wish to have a plot of $v_m/i_m \sqrt{C/L}$ all we need do is relabel the vertical axis increasing downward from 1.0.

Similarly, since the phase of the reciprocal of a function like (9.35) or (9.36) is merely the negative of the phase of the original function, the phase plot of V/I is available at once if we merely relabel the vertical scale of Fig. 9.28b by changing all signs.

There are numerous other RLC circuits in use; some will be studied in the problem set.

Problems 9.54 through 9.63

Chapter 10

NATURAL MOTIONS OF NONLINEAR SYSTEMS AND TIME-VARYING SYSTEMS

This book is concerned primarily with the dynamic analysis of linear systems. While it is true that all real physical elements have characteristics that are nonlinear in some degree, it is also true that in a (fortunately) large majority of practical situations good predictions of physical behavior—good enough for making reliable design decisions—can be obtained by representing the actual physical elements by linear models with constant parameters.

It is fortunate that this is so, because linear analysis methods are much quicker and much more penetrating than those that must be used when nonlinear elements are present. What is more important, the results of analyzing a linear system are entirely comprehensive, applying for large or small motions and for all forms of forcing function and any initial conditions which do not violate the region of linearity, while the results of analyzing a nonlinear system under a particular set of circumstances may indicate almost nothing about the same system in other circumstances. Thus, many solutions may be required for a comprehensive understanding of the behavior of a nonlinear system.

It is possible to obtain solutions to nonlinear differential equations by tedious step-by-step numerical or graphical integration. However, this is no longer necessary. The advent of high-speed computing machines has made it possible to obtain large numbers of solutions in minutes instead of months. Even so, astute analytical study, and "hand" solution for a few cases, are important to basic understanding of the system's behavior. Thus analytical and graphical study now assume the complementary (and nontedious) role of "trail blazing" and providing philosophical organization.

The present chapter is included primarily to provide perspective: to show and evaluate methods of approximating nonlinear elements by linear models, and to give an introduction to the graphical state-space methods for analyzing nonlinear and time-varying systems. The treatment here is only introductory, and far from comprehensive. The methods of nonlinear analysis are given extensive treatment in the literature.†

Concepts and techniques

We summarize here the basic ideas and the analysis techniques that are introduced in this chapter.

(1) Each solution to a set of nonlinear differential equations is usually (a) difficult to obtain, and (b) usable only for one specific set of circumstances (forcing function and initial conditions).

(2) Even rather gross linear approximations are therefore very useful, at least in preliminary studies of a system. Two common techniques are described: (a) "equivalent linearization," and (b) perturbation.

(3) The concept of a "state space" is important in systems analysis generally, and for nonlinear systems especially. It involves (a) writing the equations of motion in "standard form," i.e., as a set of first-order differential equations. The set of variables used constitutes a set of *state variables:* they prescribe the state of the system at any instant. (b) A *state space* is conceived which has the state variables as its coordinates.

(4) Graphical techniques for tracing system motions in state space ("trajectories") often contribute greatly to solution of nonlinear problems. Some techniques are demonstrated, including "isoclines" and singular points.

(5) Piecewise-linear systems may often be studied best as linear for short intervals, with boundary conditions to be matched.

(6) The difficulty in obtaining one solution to linear differential equations with time-varying coefficients may be comparable with that for nonlinear systems, but one solution is far more useful because it can be applied for all amplitudes, superposition being valid.

10.1 METHODS OF LINEAR APPROXIMATION

There are a number of methods whereby nonlinear equations of motion can be approximated by linear equations, so that an estimate of the motions of the system represented can be obtained via the methods of linear analysis. Such methods are much quicker and more comprehensive than the methods which must be used to solve nonlinear differential equations, and thus there

† See Chihiro Hayashi, "Nonlinear Oscillations in Physical Systems," McGraw-Hill Book Company, New York, 1964, and Dunston Graham and Duane T. McRuer, "Analysis of Non-linear Control Systems," John Wiley & Sons, Inc., New York, 1961.

[Sec. 10.1] METHODS OF LINEAR APPROXIMATION

is great advantage in performing at least a preliminary linear analysis whenever a reasonably good linear approximation to a system can be made.

Two of the most important means for linearization are (a) the use of *"equivalent" linear relations*, and (b) *perturbation* techniques. Both methods can be demonstrated conveniently for the familiar problem of the simple pendulum, whose equation of motion was derived in Example 7.2:

$$m\ell^2 \ddot{\theta} + mg\ell \sin \theta = 0$$

or

$$\boxed{\ddot{\theta} + \frac{g}{\ell} \sin \theta = 0} \qquad (10.1)$$

Equivalent linearization

The relation that makes this equation nonlinear is the restoring torque, which depends in a nonlinear way upon displacement:

$$T = -mg\ell \sin \theta \qquad (10.2)$$

This relation is shown in Fig. 10.1

We have shown in Example 7.2 how this equation can be approximated for small motions about $\theta = 0$ by taking the first term in the series expansion for $\sin \theta$:

$$T = -mg\ell \left(\theta - \frac{\theta^3}{6} + \cdots \right) \approx -mg\ell\theta \qquad (10.3)$$

$$\ddot{\theta} + \frac{g}{\ell} \theta \approx 0 \qquad (10.4)$$

Fig. 10.1 shows that this approximation (line OL) is well justified for quite large values of θ (up to 25° or so).

If, however, we wish to consider larger motions about $\theta = 0$, say, $\theta = \pm 85°$, we shall come closer to the right answer with a linear equation in which $\sin \theta$ is replaced by 0.8θ:

$$\ddot{\theta} + 0.8 \frac{g}{\ell} \theta \approx 0 \qquad (10.5)$$

This linear relation is represented by Line OL' in Fig. 10.1, which is below the actual curve for small θ's and above it for large θ's. Line OL' is referred to as a *linear equivalent* to the actual curve. The technique of replacing nonlinear physical relations (such as the present one between torque and displacement) by approximate linear equivalents is a common method of linearization.

Perturbation

Consider, on the other hand, a case in which motion is about a point other than 0 in the figure. For example, suppose that we attach to the pendulum a torsional spring in such a way that the spring exerts no torque for the particular pendulum position θ_1, as depicted in Fig. 10.2a (θ_1 is the

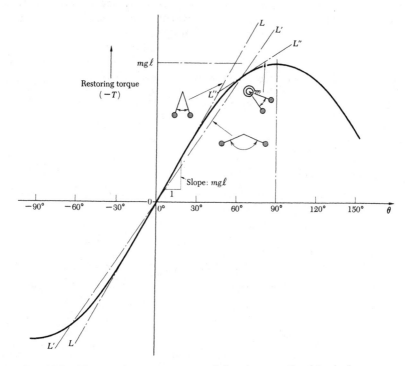

Fig. 10.1 *The restoring torque on a pendulum is proportional to sin θ*

"unwound" position of the spring). The equilibrium position of the pendulum will then be not $\theta = 0$, but some other position, $\theta = \theta_0$ (shown gray), such that the spring torque $k(\theta_1 - \theta_0)$ is just equal to the gravity torque, $mg\ell \sin \theta_0$. That is, the static equilibrium angle θ_0 is defined by:

$$mg\ell \sin \theta_0 - k(\theta_1 - \theta_0) = 0 \quad \text{(static equilibrium)} \quad (10.6)$$

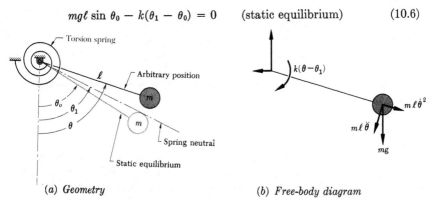

(a) *Geometry*   (b) *Free-body diagram*

Fig. 10.2 *Pendulum with torsional spring. (The neutral position is not vertical.)*

[Sec. 10.1] METHODS OF LINEAR APPROXIMATION

The equation of motion for this system is obtained from the free-body diagram in Fig. 10.2b, in which the pendulum is shown in an arbitrary position θ and moments are summed about the pivot:

$$m\ell^2\ddot{\theta} + mg\ell \sin \theta + k(\theta - \theta_1) = 0 \tag{10.7}$$

If we were to replace $\sin \theta$ in (10.7) by the "equivalent straight line" through angle θ_0, say, OL', in Fig. 10.1, we would obviously have a poor approximation to the actual relation for small motions about θ_0.

The perturbation technique is a better one in this case. It involves replacing the actual variable by a constant plus a small perturbation. In this case, we would replace θ by its static equilibrium value θ_0 plus a small displacement ϑ away from equilibrium,

$$\theta = \theta_0 + \vartheta \tag{10.8}$$

Then the sine of θ is obtained from the geometric identity for the sum of two angles:

$$\sin(\theta_0 + \vartheta) \equiv \sin\theta_0 \cos\vartheta + \cos\theta_0 \sin\vartheta \tag{10.9}$$

At this point we again employ the series expansions for $\sin \vartheta$ and $\cos \vartheta$:

$$\left.\begin{aligned}\sin \vartheta &= \vartheta - \frac{\vartheta^3}{6} + \cdots \\ \cos \vartheta &= 1 - \frac{\vartheta^2}{2} + \cdots\end{aligned}\right\} \tag{10.10}$$

and, dropping higher-order terms, we obtain, finally,

$$\sin(\theta_0 + \vartheta) \approx \sin\theta_0 + \vartheta \cos\theta_0 \tag{10.11}$$

This expression is now substituted into the equation of motion:

$$m\ell^2\ddot{\vartheta} + \underbrace{mg\ell(\sin\theta_0 + \vartheta \cos\theta_0)}\ + k(\vartheta + \underbrace{\theta_0 - \theta_1}) = 0 \tag{10.12}$$

The two terms marked $\smile$ sum to zero by the static equilibrium relation (10.6), leaving

$$m\ell^2\ddot{\vartheta} + (mg\ell \cos\theta_0 + k)\vartheta = 0$$

in which, from (10.6), $k = \dfrac{mg\ell \sin\theta_0}{(\theta_1 - \theta_0)}$ and thus the equation of motion may be written

$$\ddot{\vartheta} + \frac{g}{\ell}\left(\cos\theta_0 + \frac{\sin\theta_0}{\theta_1 - \theta_0}\right)\vartheta = 0 \tag{10.13}$$

a linear equation in ϑ which will be found to give a very accurate approximation to small motions about θ_0. The term $mg\ell \cos\theta_0$ in this expression is, of course, the slope $L''L''$ in Fig. 10.1.

More generally, the perturbation technique consists of performing, on the nonlinear physical relation involved (gravity-torque-versus-angle curve, for the pendulum problem), a series expansion about the mean value of interest (usually a static equilibrium point). This was the process we used for the original linearization for motions about $\theta = 0$, as well as for the motions about $\theta = \theta_0$. [(10.4) is the special case $\theta_0 = 0$ of (10.13).] It is also precisely the process we used to obtain linear equations of motion in fluid-flow problems, as on pp. 106 and 112. (A review of those pages in Sec. 3.2 is recommended.)

How closely an approximate linear solution approaches the motions of the actual system can only be found exactly by comparison with solutions to the exact equation. (For the case of large pendulum motions about $\theta = 0$, for example, this comparison can be made at the end of Sec. 10.2.) The more gross the approximation, of course, the more crude the expected linear solution. It is often striking, however, how closely such response features as period and rate of decay can be predicted by astute linear analysis of highly nonlinear systems.

Problems 10.1 through 10.5

10.2 STATE-SPACE ANALYSIS

A graphical method of dynamic analysis of considerable power has been developed from the simple idea of plotting velocity versus position in second-order problems. The method, for any order system, consists of first reducing the set of differential equations to an equivalent set which are all of first order, simply by defining appropriate additional variables. The result is a set of *state variables* (refer to p. 38). Then a state space, or "phase space," is conceived which has each of the state variables as one of its coordinates (with the independent variable—usually time—as a parameter). A considerable amount can then be learned about the motion by investigating slopes and singular points in such state space. This is exactly the procedure used to set up an analog computer (Sec. 8.15).

Moreover, techniques of graphical integration can then be applied to obtain the actual motions of the system, which are known as "trajectories" in the state space. The state-space (phase-space) technique is of great importance in analyzing the dynamic response of nonlinear systems because it facilitates graphical, stepwise integration and aids strikingly in visualization and classification of solutions. This is especially true when the system is only second order, in which case the "phase space" is only two-dimensional and is known as the *phase plane*.

To demonstrate the phase-space principle, the motion of the simple pendulum will be solved, first for small motions for which the equations are linear, and then for large motions for which they are not.

The state-space viewpoint has become central to advanced systems analysis, and the first-order form for the set of differential equations has

[Sec. 10.2] STATE-SPACE ANALYSIS

become known as "standard form." In dealing with nth-order dynamic systems, where n is large, the efficiency of the standard form, and the power of imagining behavior in an n-dimensional state space are especially valuable. (Standard-form analysis of higher-order systems is described in Sec. 14.5.)

Linear solution in the phase plane: simple pendulum

Consider again the simple pendulum without damping, Fig. 10.3. The equation of motion, for arbitrarily large motions θ, is (10.1)

$$\ddot{\theta} + \frac{g}{\ell}\sin\theta = 0 \qquad (10.1) \text{ [repeated]}$$

For small motions (e.g., $\theta < 10°$) the equation can be closely approximated by the linear equation:

$$\ddot{\theta} + \frac{g}{\ell}\theta = 0 \qquad (10.4) \text{ [repeated]}$$

as was discussed in Sec. 10.1. The solution to Eq. (10.4) is

$$\theta = \theta(0)\cos\sqrt{\frac{g}{\ell}}\,t + \frac{\dot{\theta}(0)}{\sqrt{g/\ell}}\sin\sqrt{\frac{g}{\ell}}\,t \qquad 0 < t$$

in which $\theta(0)$ is the initial angle and $\dot{\theta}(0)$ is the initial angular velocity of the pendulum. [This solution can be readily checked by substituting it into Eq. (10.4).]

It will be instructive to obtain the above solution by the alternative method of stepwise integration, using the phase-space method. The technique is to define new variables in such a way that expression (10.4) can be written as a set of equations of first order—i.e., in *standard form* (as we did on p. 303). As is often the case, one of the variables is conveniently taken as the original dependent variable:†

$$x \triangleq \theta \qquad (10.14)$$

The second variable is taken as a function of the derivative. In this case, it is convenient to form the nondimensional quantity:

$$y \triangleq \frac{\dot{\theta}}{\sqrt{g/\ell}} \qquad (10.15)$$

The quantities x and y are then the *state variables* for this problem. The original equation of motion (10.4) can now be written

$$\sqrt{\frac{g}{\ell}}\,\dot{y} + \frac{g}{\ell}x = 0$$

or
$$\dot{y} = -\sqrt{\frac{g}{\ell}}\,x \qquad (10.16a)$$

† Recall that the symbol $\triangleq$ means "is defined to equal."

Also, from the above definitions we can write

$$\dot{x} = \sqrt{\frac{g}{\ell}}\, y \qquad (10.16b)$$

Expressions (10.16a) and (10.16b) form a set (known as the "state equations") which can be solved graphically on a "phase plane" having coordinates x and y. (This is the phase space or state space for the special case of two variables.) Note that the state variables x and y completely prescribe the *state* of the system (configuration and velocity) at any instant.

The process of graphical integration consists of constructing in the phase plane all possible motions of the system, as dictated by its Eqs. (10.16). A useful tool in graphical integration is a construction known as an isocline, which is a line in the phase space at all points on which the trajectory has the same slope (isocline = "same inclination"). In the case of a linear system (such as the present one), the isoclines are straight lines whose equations are found by forming the derivative dy/dx thus:

$$\frac{dy}{dx} \equiv \frac{dy/dt}{dx/dt} = \frac{\dot{y}}{\dot{x}} \qquad (10.17)$$

In the present example, $\dot{y}/\dot{x}$ is available from the pair of equations (10.16):

$$\frac{dy}{dx} = \frac{\dot{y}}{\dot{x}} = -\frac{x}{y} \qquad (10.18)$$

Along an isocline, $dy/dx =$ constant, from which the isoclines have the equation $y = cx$, and the slope on any isocline is given, for this example, by

$$\frac{dy}{dx} = -\frac{1}{c} \qquad (10.19)$$

Some isoclines ($y = cx$) are drawn in Fig. 10.4b, and short lines having the proper slope ($1/c$) are distributed along each isocline. Note in particular that along the vertical axis ($y = \infty x$) the isoclines are horizontal ($dy/dx = 1/\infty$), while along the horizontal axis ($y = 0x$) they are vertical ($dy/dx = 1/0$).

We are now ready to begin constructing trajectories in the phase plane. We begin at any arbitrary point P representing initial conditions $x(0) = \theta(0)$ and $y(0) = \dot{\theta}(0)/\sqrt{g/\ell}$. We then simply begin to move parallel to the local slope line, continually correcting our heading in accordance with the local slope lines as we come to them. If the slope lines were infinitely close together, then we would trace an exact trajectory by this method. A quite close approximation can be made with a reasonable spacing of the isoclines.

We note that each trajectory we trace by this method is a circle about

[Sec. 10.2] STATE-SPACE ANALYSIS 363

Fig. 10.3 *A simple undamped pendulum*

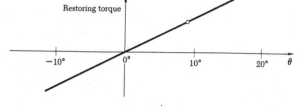

(a) *Restoring torque is taken linearly proportional to θ*

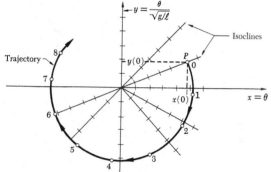

(b) *The phase-plane trajectories are circles*

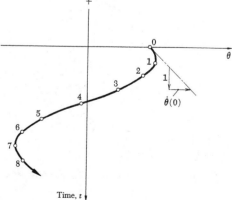

(c) *The plot of θ versus time is a sine wave*

Fig. 10.4 *Phase-plane solution for small motions of a simple undamped pendulum*

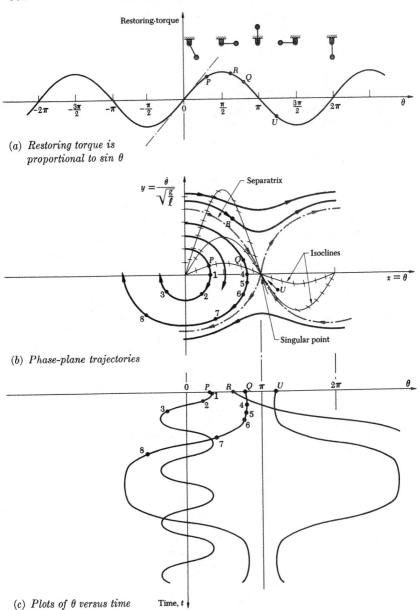

Fig. 10.5 *Phase-plane solutions for large motions of an undamped pendulum*

(a) *Restoring torque is proportional to sin θ*

(b) *Phase-plane trajectories*

(c) *Plots of θ versus time*

[Sec. 10.2] STATE-SPACE ANALYSIS

the origin. In the present—linear—problem, this could have been predicted alternatively from the exact solution:

$$\frac{dy}{dx} = -\frac{x}{y}$$

$$y\,dy = -x\,dx$$

$$\int_{y(0)}^{y} y\,dy = -\int_{x(0)}^{x} x\,dx$$

$$\frac{y^2}{2} - \frac{y(0)^2}{2} = -\frac{x^2}{2} + \frac{x(0)^2}{2}$$

$$y^2 + x^2 = y(0)^2 + x(0)^2 \tag{10.20}$$

which is the equation for a circle through point $x(0)$, $y(0)$.[†]

The time plot corresponding to this phase-plane plot is drawn at the bottom of Fig. 10.4. It is constructed by projecting the value of x. For this element the time plot is, of course, a simple sine wave as shown. Time pips $(1, 2, 3, 4, \ldots,)$ are distributed along both curves to indicate equal-time intervals.

The size of the circle in Fig. 10.4 depends only upon the initial conditions,

$$x(0) = \theta(0) \qquad y(0) = \dot{\theta}(0)/\sqrt{g/\ell}$$

Starting the swing of the pendulum with a larger angle or a larger initial velocity, or both, results in a circle of greater radius.

We must be careful, of course, not to employ the isoclines sketched in Fig. 10.4 to draw trajectories involving angles (values of x) larger than about 30°, because these isoclines hold only for the linear equation of motion on which they were based.

Nonlinear solution in the phase plane: simple pendulum

It is interesting to extend the above analysis to the nonlinear case: for large angles, the restoring force on the pendulum is related to the angle θ in accordance with the relationship drawn in Fig. 10.5a. Note that for small angles this relation is still a straight line, but for large angles it is sinusoidal in shape, and beyond $\theta = \pi$ the "restoring" force is negative: the pendulum tends to go on around to 360°. (When the angle is 360° the situation is, of course, identical with that at $\theta = 0°$.)

We choose again state variables $x \triangleq \theta$ and $y \triangleq \dot{\theta}/\sqrt{g/\ell}$. Then the state equations corresponding to nonlinear Eq. (10.1) are

$$\frac{\dot{y}}{\sqrt{g/\ell}} = -\sin x \qquad \text{[from equation of motion (10.1)]} \quad (10.21a)$$

$$\frac{\dot{x}}{\sqrt{g/\ell}} = y \qquad \text{[from definition of } y, \text{ (10.15)]} \quad (10.21b)$$

[†] Any undamped harmonic oscillator will, of course, trace this same circular pattern.

Obtaining the isoclines as before, we have

$$\frac{dy}{dx} = \frac{\dot{y}}{\dot{x}} = -\frac{\sin x}{y} \qquad (10.22)$$

from which, when $y = c \sin x$, we obtain

$$\frac{dy}{dx} = -\frac{1}{c} \qquad (10.23)$$

Thus the isocline is also a sine wave, as sketched in Fig. 10.5b.

The shape of the phase-plane trajectories now depends upon the magnitude of the motion (i.e., upon the starting conditions), which is the characteristic of nonlinear systems. For example, if the initial angle and velocity of the pendulum are small (point P), then the trajectory will be a circle, just as it was in Fig. 10.4. [In fact, near the origin ($x < \pi/6$) Fig. 10.5 is essentially identical to Fig. 10.4.] On the other hand, if the initial conditions are large (point Q), then the trajectories are found to be less and less circular. Moreover, the trajectory will not proceed at constant speed around 0 as it did from point P, but will move rapidly near the y axis, but slowly near the x axis, as the time pips indicate.

Ultimately, if the initial conditions lie beyond a certain boundary in the phase plane, for example, point R, then the trajectories are not closed curves at all, but continue on out the x axis. That is, θ increases indefinitely. Physically, of course, this is the case where the pendulum is given sufficient initial velocity that it continues to swing around and around.

The meaning of the three types of motion is more clear from the corresponding time plots constructed at the bottom of Fig. 10.5. For small initial conditions (point P) all the motions are pure sine waves of the same frequency, the amplitude depending upon initial conditions. (This is the standard linear solution.) For larger initial conditions—but still inside the boundary (point Q)—the motions are periodic about $\theta = 0$, but they are no longer sinusoidal: The pendulum moves rapidly through $\theta = 0$, but moves slowly through the top of its swing on either side.

For motions started *outside* the boundary, the pendulum moves always in the *same angular direction*, rapidly at the bottom of the swing, but more slowly at the top.

The boundary between motions of the second and third type is known as a *separatrix*. If motion of the pendulum is commenced exactly at some point on the separatrix, the pendulum will swing up to the upside-down position ($\theta = \pi$), where it will stop dead. It may then choose either to return whence it came, or to continue around. This particular point where the pendulum stops in an unstable position is known as a "singular point." Several other types of singular points are found in more general problems. Analysis in the vicinity of the singular points is often the key to understanding the pattern of motion.

From Fig. 10.5 we can evaluate precisely the accuracy of the equivalent

[Sec. 10.3] LARGE MOTIONS OF PENDULUM WITH DAMPING 367

linear approximation for a pendulum, for initial conditions $[\theta(0), \dot{\theta}(0)]$ of various magnitudes. For small initial conditions like point P the solutions are the same, the trajectories being circles and the time response a pure sine wave. For large initial conditions like Q or R, however, the exact solution exhibits the noncircular trajectories and nonsinusoidal motions shown, while the linear solution would continue to predict pure sinusoidal motion of one frequency.

The great advantage of the phase-plane method lies in the striking manner in which it provides at once visualization and clarification of all the possible motions of the nonlinear system.

10.3 LARGE MOTIONS OF PENDULUM WITH DAMPING

It is interesting to extend the above analysis to the pendulum with damping. Suppose, for example, that there is linear damping in the pivot of the pendulum analyzed above. Then, for small motions, the equation of motion is

$$\ddot{\theta} + \frac{b}{m\ell^2}\dot{\theta} + \frac{g}{\ell}\theta = 0 \tag{10.24}$$

The state variables are again chosen to be

$$x = \theta$$
$$y = \frac{\dot{\theta}}{\sqrt{g/\ell}} \tag{10.25}$$

and thus Eq. (10.24) can be written in terms of x and y as

$$\sqrt{\frac{g}{\ell}}\dot{y} + \frac{b\sqrt{g/\ell}}{m\ell^2}y + \frac{g}{\ell}x = 0$$

or
$$\frac{\dot{y}}{\sqrt{g/\ell}} = -(2\zeta y + x) \tag{10.26}$$

and
$$\dot{x} = \sqrt{g/\ell}\, y$$

where
$$\zeta \triangleq \frac{b/2m\ell^2}{\sqrt{g/\ell}}$$

From these the equation for an isocline is obtained:

$$\frac{dy}{dx} = \frac{\dot{y}}{\dot{x}} = -2\zeta - \frac{x}{y} = \text{const.} \tag{10.27}$$

or $\quad y = cx \quad$
$$\frac{dy}{dx} = -2\zeta - \frac{1}{c} \tag{10.28}$$

Some isoclines are plotted in Fig. 10.6. Note that the slope lines along the horizontal axis are still vertical, but the slope lines along the vertical axis are now tilted downward by the angle $\tan^{-1} 2\zeta$. That is, the greater the damping, the more the trajectories will be turned toward the origin. A few

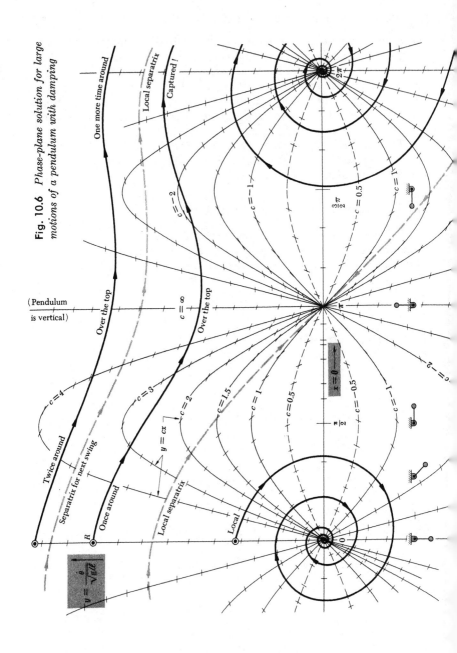

Fig. 10.6 Phase-plane solution for large motions of a pendulum with damping

small-θ trajectories are plotted in Fig. 10.6. They are logarithmic spirals into the origin.

For the nonlinear case of large motions, the damped equation of motion is

$$\frac{\ddot{\theta}}{g/\ell} + \frac{b}{mg\ell}\frac{\dot{\theta}}{\sqrt{g/\ell}} + \sin\theta = 0 \qquad (10.29)$$

With the state variables defined as before, the state equations are

$$\left. \begin{array}{l} \dfrac{\dot{y}}{\sqrt{g/\ell}} = -(2\zeta y + \sin x) \\[2mm] \dfrac{\dot{x}}{\sqrt{g/\ell}} = y \end{array} \right\} \qquad (10.30)$$

from the equation of motion, and from the definition of y.

Again, the equation for an isocline is obtained from the above pair:

$$\frac{dy}{dx} = \frac{\dot{y}}{\dot{x}} = -2\zeta - \frac{\sin x}{y} = \text{const.} \qquad (10.31)$$

or $y = c \sin x \qquad \dfrac{dy}{dx} = -2\zeta - \dfrac{1}{c} \qquad (10.32)$

Again the isoclines are sinewaves on the phase plane, but the slope lines corresponding to each are tilted clockwise by the damping. Some resulting trajectories are shown in Fig. 10.6.

It is fascinating to follow several of them. For example, consider motions started at point R in Fig. 10.6. The initial velocity in this case is great enough to carry the pendulum past the vertical on its first trip around. However, after another 360° its kinetic energy is further dissipated, and it cannot make it past the vertical on the second trip but falls back toward the neutral position, oscillates a few times, and then comes to rest. In phase-plane parlance, it might be said that the trajectory escaped the origin at $\theta = 0$, but was captured by another "origin" at $\theta = 2\pi$.

The separatrix, in this case, comes into the singular point at $\theta = \pi$ (pendulum upside down), whereupon the pendulum has the choice of proceeding either in the same direction to $\theta = 2\pi$, or back toward $\theta = 0$. Symmetry at the singular point is completed by the trajectory representing the pendulum approaching the vertical from the other side. It also, of course, has the option of going in either direction from its stop in the upside-down position.

10.4 PIECEWISE-LINEAR ELEMENTS

Another type of problem in which phase-plane analysis is particularly elegant is that of an element whose characteristics are piecewise linear, such as the spring shown in Fig. 10.7.

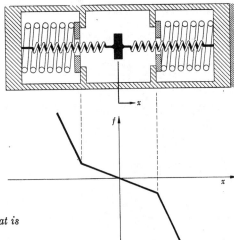

Fig. 10.7 *An element that is "piecewise linear"*

In this case one may actually proceed with the methods of linear analysis within the boundaries of each "piece," taking care that end conditions are all matched at each boundary (which usually involves trial and error). The phase plane is of great help in carrying out such analyses. Some examples are suggested in the problems.

Problems 10.6 through 10.15

10.5 LINEAR EQUATIONS WITH TIME-VARYING COEFFICIENTS

Before returning to the case of systems which are described by linear equations with constant coefficients, we call attention again, briefly, to systems that can be described by a set of differential equations which are linear but which have coefficients that vary with time.

Such equations are often difficult to solve but, as we have noted, their solutions are far more comprehensive than those of nonlinear equations for the important reason that *the solutions of linear equations can always be superposed.* That is, when we have obtained a solution to a linear equation for one set of amplitudes of the dependent variables, we have solved the equations for *all* amplitudes.

To illustrate the problem of time-varying coefficients, consider again the simple pendulum. Let us restrict the motion to quite small values of θ, so that the linear equation

$$\ddot{\theta} + \frac{g}{\ell}\theta = 0 \qquad (10.4)\ \text{[repeated]}$$

is correct. Suppose, however, that the pendulum is at the end of a string

[Sec. 10.5] TIME-VARYING COEFFICIENTS 371

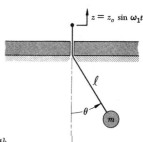

Fig. 10.8 *A pendulum with time-varying length*

and that we are varying the length of the string with time, as illustrated in Fig. 10.8. Then (10.4) is linear, but one of its coefficients is not constant:

$$\ell = \ell_0 + z(t) \tag{10.33}$$

The solution, even for this simple problem, is not easy. But we can guess some of its properties from experience. In particular, we can surmise that the equation is capable of exhibiting sinusoidal solutions which grow with time, as any child in a swing can demonstrate merely by "pumping." That is, the existence of time-varying coefficients leads to the possibility of forced motion without the usual "right-hand-side" forcing function. This phenomenon is known as *parametric excitation*, and has important physical applications. Unfortunately, space does not permit further discussion of it in this book.

The general case involves, of course, differential equations of motion which are nonlinear with time-varying coefficients. For example, if the pendulum of Fig. 10.8 has large-angle motions, nonlinear Eq. (10.1) applies, with coefficient ℓ a function of time, per (10.33). Computing machines—analog or digital—are used to study the behavior of such systems.

PART C

NATURAL BEHAVIOR OF COMPOUND SYSTEMS

Chapter 11. *Dynamic Stability*

Chapter 12. *Coupled Modes of Natural Motion: Two Degrees of Freedom*

Chapter 13. *Coupled Modes of Natural Motion: Many Degrees of Freedom*

In Part B we studied the dynamic behavior of elementary first- and second-order systems in considerable detail. In particular, we investigated the two elementary types of motion that are possible in a linear system—exponential motion and sinusoidal motion with exponentially varying magnitude—and we developed the complex-number and s-plane techniques for studying them efficiently. We found that when a simple system is isolated its natural motions can be represented by the roots of its characteristic equation, known as its *characteristics*, or *eigenvalues*.

We are now ready to begin investigating the behavior of *compound* dynamic systems, that is, groups of elementary subsystems which are so interconnected that they influence one another's behavior. Such coupling may result from any of numerous physical mechanisms.

In general, when simple systems are coupled together the motion of each is felt by the others. However, in many common systems only a few interactions are important, and sometimes these exist in one direction only. Special techniques (e.g., feedback analysis) for handling certain common cases have been highly developed.

Subsystems lose their identity

We shall find that a compound dynamic system has natural behavior characteristics which are, in general, distinct from those of any of its member subsystems. The members in a system interact with one another, and in so doing each member loses its own dynamic character. That is, the compound system is an entity unto itself; it has its own dynamic characteristics, which are related to, but may be quite different from those of any of its member subsystems alone. It is an important concept that when a small system becomes a part of a larger system it loses its own identity and takes on the dynamic characteristics of the larger system.

It is true, of course, that in a system in which the coupling between subsystems is light, each of the system characteristics may be easily identified with those of one of the members. For the system of Fig. 11.1a, for example, we might expect (and correctly so) that the two subsystems involved are so large, and the spring k_3 which couples them together so weak, that the natural frequencies of vibration of the complete system are almost exactly those of the member subsystems alone, as defined by parameters J_1, k_1, and J_2, k_2.

In Fig. 11.1b, however, the coupling spring is relatively strong, and now the frequencies of natural vibration are quite different from those of either subsystem alone.

An even stronger case in point is the simple feedback system of Fig. 11.14 (p. 398), made up of two dynamic subsystems, each of which exhibits stable, highly-damped, *non*oscillatory motions. But, as we shall see, when the subsystems are connected as shown, *the system* may exhibit *unstable* high-frequency natural oscillations which grow until the system fails structurally or until the motions encounter some sort of limit! That is, *the system* has natural motion which is *radically different* from any that its separate *members* are capable of by themselves: When they are connected in the system, all the members take part in the system's behavior, and lose their own individual characteristics.†

The preliminary study of natural behavior alone

The ultimate objective of our studies of system behavior is to predict the *response* of compound dynamic systems to any form of disturbance. Further, we would like to be able to see clearly how each physical parameter affects the behavior of the complete system, so that we can design the physical parameters at our disposal to have the most favorable values. In Part D we shall make considerable progress in this direction with the help of some powerful techniques of analysis.

But much can be learned about the *character of the motions* a system will exhibit without actually calculating its response to specific disturbances, and this will be our objective in Part C. For example, we can determine whether or not a system will be stable, whether or not its natural motions will be oscillatory and at what frequencies, and what are the rates of decay of all of its natural motions (oscillatory and exponential), without calculating the magnitudes of such motions or the response to any particular disturbance. We can do this because, as we know, these quantities—the *dynamic characteristics* of the system—are entirely independent of the initial conditions or the disturbances to which the system may be subjected (provided only that the system is linear). Moreover, it is usually much easier and quicker to determine the dynamic characteristics of a system than it is to determine the actual magnitudes of its motions; and much of the time the characteristics furnish ample information for the job at hand.

In Chapter 11, therefore, we shall deal only with the natural dynamic characteristics of compound systems. We shall begin by continuing the study—started in Fig. 8.5 (p. 252)—of the relation between the roots of the characteristic equation of a system and the natural motions it can have, with particular attention to the notion of dynamic stability.

In Chapter 12 we shall learn about another useful property of coupled, many-degree-of-freedom systems: that they have specific *modes of natural motion* associated with each natural characteristic; the system's members move in a definite relation to one another. Separate natural modes are easy to analyze; and complete natural motion may be studied simply by superposing natural modes, as we shall see for two-degree-of-freedom systems in Chapter 12, and for many-degree-of-freedom systems, including distributed systems, in Chapter 13. This will complete our study, in Part C, of natural motions alone.

Later, in Part D, we shall study the complete time response of compound systems to inputs and disturbances.

† The obvious sociological analogy is perhaps not altogether irrelevant.

Chapter 11

DYNAMIC STABILITY

This chapter introduces the following concepts and techniques pertaining to the natural behavior of compound dynamic systems which are linear with constant coefficients (l.c.c.):

(1) *Stability.* If the real part of any root s of the characteristic equation of a system is positive, then a natural motion of the system (given by Ce^{st}) grows with time and the system is *unstable*.
(2) *Compound systems.* When elementary systems are connected together to form compound systems, as in Fig. 11.1, they do not in general retain their own natural characteristics, but contribute to and take on the roots (eigenvalues) of the new system. The characteristic equation of a compound system may be found by Procedure C-1, p. 398.
(3) *s-plane plot.* An s-plane plot of a system's characteristic roots indicates succinctly the system's natural behavior, including both the existence (or absence) of stability and the degree of stability.
(4) *Locus of roots.* The locus of roots, plotted as a function of a physical parameter, gives an instantaneous view of that parameter's effect on natural behavior in general, and on stability in particular.
(5) *The root-locus method of Evans.* This important method, Procedure C-2, is a graphical procedure for plotting quickly the roots of characteristic equations for compound systems whose subsystems are already known. The procedure is stated formally in Sec. 11.8, and the most common sketching rules, Procedure C-3, are given in Sec. 11.9.
(6) *Routh's method.* Routh's method, Procedure C-4, tells, without solving for the roots, whether or not a characteristic equation represents a stable system. It does so by examining systematically the coefficients. If the characteristic equation is given in numerical form, Routh's method determines how many roots are unstable. More generally, for a characteristic equation in literal form the method gives constraints on

[Sec. 11.1] THE CONCEPT OF STABILITY 377

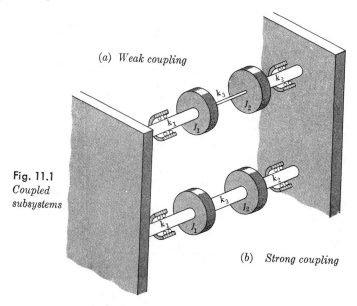

Fig. 11.1 Coupled subsystems

(a) Weak coupling

(b) Strong coupling

the coefficients necessary for stability. Routh's method is presented in its entirety in Secs. 11.11–11.13.

11.1 THE CONCEPT OF STABILITY

By the stability of a system we mean its tendency to seek a condition of static equilibrium after it has been disturbed. More precisely, if a system is in static equilibrium and is then given a small displacement from the equilibrium condition, it is considered stable if it returns to the equilibrium and unstable if it moves farther away. In addition, it is often convenient to establish arbitrary indexes of relative stability.

For a *nonlinear* system, stability must be defined with great care because, as we saw in Chapter 10 (sometimes dramatically), the character of the natural motions of a nonlinear system depends so much on the amplitude of the motions. Meticulous precision must be exercised, and such concepts as being "stable in the small," "stable in the large," and "asymptotically stable" are involved. Stated briefly, and without the necessary rigor, a real (generally nonlinear) system is considered to be "stable in the small" if for some small, carefully defined displacement from a static-equilibrium state its motions do not grow. It is "asymptotically stable in the small" if, further, it returns to that equilibrium state with time. A system is "stable in the large" if for "any displacement" its motions do not grow.

Because *the character of the natural motions of a linear system does not depend on their amplitude*, the problem of determining system stability is

(a) Second-order system

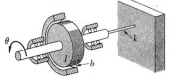

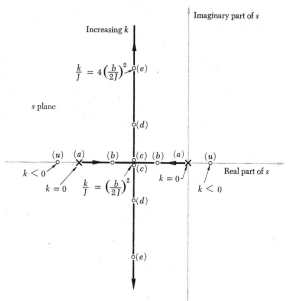

(b) Roots of the characteristic equation $s^2 + \dfrac{b}{J}s + \dfrac{k}{J} = 0$ plotted versus k

Fig. 11.2 s-plane plot of the characteristics of a second-order system

much simpler for linear systems. In Fig. 8.5 we saw that the dynamic stability of a linear system is intimately related to its characteristic equation. This comes about because the characteristic equation—an algebraic expression in s—is obtained by assuming motions to be of the form

$$\theta = \Theta e^{st} \tag{11.1}$$

and thus *the roots s of the characteristic equation determine the character of the motions.* Thus, *if the real part of any root of a characteristic equation is positive, the corresponding system is dynamically unstable.* From this we would expect to find—as we shall—that the most direct way to study the stability of a linear system is to observe the roots of its characteristic equation (its eigenvalues) in the s plane.

In our studies of system stability we not only want to see whether a given system is stable or not stable; we want also to have some measure of

[Sec. 11.2] THE ELEMENTARY SECOND-ORDER SYSTEM

the system's *degree of stability* and, further, to know exactly how stability depends upon any given parameter in which we may be interested.

The most straightforward indexes of system stability are (1) the *time* it takes for a given motion to decay and (2) the *number of oscillations* required for a motion to decay. In Chapters 6 and 8 we developed these two indexes for elementary first- and second-order systems: First, the rate of decay of motion is given precisely by the real parts of the roots of the characteristic equation, the real exponent $-\sigma$ in the assumed motion Ce^{st}; second, the number of oscillations which take place before a motion has decayed is indicated quite precisely by the index ζ (the "damping ratio"), as shown in Fig. 8.3. These quantities—σ and ζ—will be found still to be excellent indexes of system stability for complicated compound systems.

The easiest and most direct way to see how system stability depends on a given parameter is to plot the roots of its characteristic equation (its eigenvalues) in the s plane as a function of the parameter under study, as we did in Sec. 8.3. This technique will be demonstrated in Secs. 11.2 through 11.4 and 11.7 through 11.19.

11.2 THE ELEMENTARY SECOND-ORDER SYSTEM

Consider, for example, the second-order system shown in Fig. 11.2a, which was the major model used in Part B. Its equation of motion is (8.1′)

$$J\ddot{\theta} + b\dot{\theta} + k\theta = 0 \qquad (11.2)$$
[(8.1′) repeated]

Recall, from Chapter 8, that when the natural motions of this system were assumed to be of the form (11.1), $\theta = \Theta e^{st}$, then the characteristic equation of the system was found to be, from (8.3),

$$Js^2 + bs + k = 0 \qquad (11.3)$$

The roots of this characteristic equation are, from (8.4),

$$s = -\frac{b}{2J}\left[1 \pm \sqrt{1 - \frac{k/J}{[b/2J]^2}}\right] \qquad (11.4a)$$

or $$s = -\frac{b}{2J} \pm j\sqrt{\frac{k}{J} - \left[\frac{b}{2J}\right]^2} \qquad (11.4b)$$

We can see the effect on dynamic characteristics of any of the physical parameters in Fig. 11.2a by plotting roots (11.4) in the s plane as a function of the parameters of interest. This procedure was introduced in Sec. 8.3.

For example, the effect of varying spring constant k is shown in Fig. 11.2b. When k is 0, we see from Eq. (11.4a) that s has the values 0 and $-b/J$ [points (a) in Fig. 11.2b]. For small values of k we see that s has two real

380   DYNAMIC STABILITY [Chap. 11]

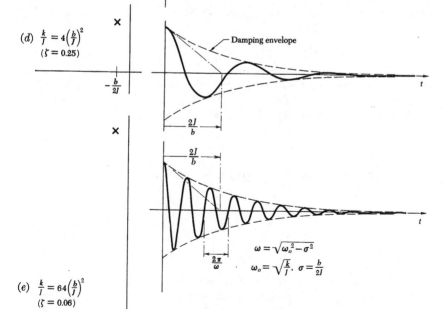

Fig. 11.3 *The s plane and natural response. Section I: varying k*

[Sec. 11.2] THE ELEMENTARY SECOND-ORDER SYSTEM 381

values [points (b), given by Eq. (11.4a)]. For the particular value of k given by

$$\frac{k}{J} = \left(\frac{b}{2J}\right)^2 \tag{11.5}$$

s has two equal roots: $s = -b/2J$ [points (c)]. For larger values of k, it is more convenient to use Eq. (11.4b). It is seen that for all larger values of k, the real part of s is the same: $-b/2J$. The imaginary part of s—corresponding to the frequency of oscillations of the element—increases with k as indicated by points (d) and (e) in Fig. 11.2b. (Figure 11.2b is to be plotted with numerical values in Problem 11.4.)

The physical meaning of variation in k can be seen precisely by correlating Fig. 11.2 with Fig. 11.3 (and more generally with Fig. 8.5). When $k = 0$, then motion is like Fig. 11.3a. For small values of k [points (b) in Fig. 11.2], it is like Fig. 11.3b. Points (c) correspond (approximately) to Fig. 11.3c, points (d) to Fig. 11.3d, and points (e) to Fig. 11.3e. For all values of k greater than that at points (c) in Fig. 11.2b [e.g., for points (d), (e), and higher], the *damping envelope* is the same; but as k increases the *frequency* of the oscillations increases, and thus for large values of k many closely spaced oscillations take place within the fixed damping envelope, as shown in Fig. 11.3e. (Note that the system has initially both displacement and velocity.)

From Fig. 11.2b it is clear that the system model of Fig. 11.2a will never become unstable, no matter how stiff the spring is made. In fact, its motions will always decay with the time constant $\tau = 2J/b$; but its *relative damping*—the amount the motion is *damped each cycle*—becomes smaller and smaller as k is increased, and thus in the limit, when k goes to ∞, ζ goes to 0. Refer to Fig. 8.5, p. 252.

Instability due to negative spring constant in physical systems

If, on the other hand, a "negative" spring were used in Fig. 11.2a, the system would be unstable. This is clear either from Eq. (11.4a) or from Fig. 11.2b, where the roots would be found in the positions marked

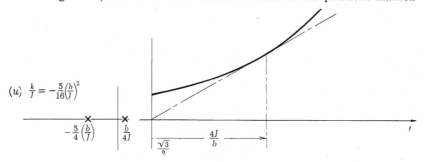

$(u) \quad \frac{k}{J} = -\frac{5}{16}\left(\frac{b}{J}\right)^2$

Fig. 11.3 *(continued)*

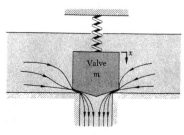

(a) A "negative spring"

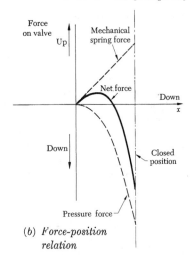

(b) Force-position relation

Fig. 11.4 *Model of a fuel injection valve: a second-order mechanical system that exhibits divergent instability*

(u). A number of mechanical spring configurations have actually been devised which have a negative constant, for use in special instrument applications. But more frequently a negative spring constant is encountered in nature when it is not wanted.

For example, fuel-injection valves, of which a model is shown in Fig. 11.4a, are sometimes known to behave unstably, popping closed when they are not supposed to. On close examination it is found that the valves display an exponential divergence corresponding to a real root in the right half-plane [point (u) in Fig. 11.2b].

The explanation for the phenomenon is, briefly, that the restoring force on the mass consists not only of the force in the physical spring, but also of a "Bernoulli" force due to pressure variation of the form $p \sim 1/v^2$ at the valve seat, caused by the rapid flow of fluid through the opening. This pressure force also varies with valve displacement, as plotted in Fig. 11.4b, and is seen to be such that when the valve is displaced slightly in the "closed" ($+x$) direction, the force in that direction is *increased*; that is, the pressure "spring constant" is negative (i.e., opposite that of Fig. 2.4a).

The *combined* spring constant of the valve—the sum of the physical and pressure characteristics—is shown as a solid curve in Fig. 11.4b. For small values of displacement x, the curve has a positive slope and the valve will oscillate stably; but for large displacements, the slope—and therefore the net spring constant k—is negative, and the valve rapidly *diverges* to the closed position, as predicted by point (u) in the right half of the s plane in Fig. 11.2b.

After the valve has closed, another mode of behavior ensues which is nonlinear and quite different: When the valve is completely closed the flow is stopped and the net pressure force returns to a steady lower value. Then

[Sec. 11.2] THE ELEMENTARY SECOND-ORDER SYSTEM 383

the spring (which is in tension) pulls the valve open again, flow starts, and the Bernoulli effect closes the valve. The cycle repeats as an unstable oscillation of limited amplitude.† Both phenomena—the divergence and the nonlinear oscillation—are readily demonstrated by holding the plug near the drain hole in a bathtub which is emptying. (The bather's fingers supply the physical spring in this experiment.)

A most important point to note is that unstable (growing) motions cannot occur in a system without an input of energy to the system. In Fig. 11.4 kinetic energy is being extracted from the flowing water.

Example 11.1. Inverted pendulum ("negative spring"). Plot the motion (up to 0.05 radian from the vertical) of the simple pendulum shown in Fig. 11.5a if its initial position is $\theta(0) = 0.01$ radian from the upside-down position, as shown, and its initial velocity is zero. Let $\ell = 10.7$ ft and $b/2m\ell^2 = 1.0$ (sec^{-1}).

STAGE I. As a physical model, assume the pendulum arm to be massless, and friction in the pivot to be viscous.

STAGE II. The equation of motion is written as follows.

(1) *Geometry.* The acceleration of m has been shown earlier [Eq. (5.18)] to have the tangential and radial components shown in Fig. 11.5a:

$$a_t = \ell\ddot{\theta} \qquad a_r = -\ell\dot{\theta}^2 \tag{11.6}$$

(2) *Equilibrium* and (3) *Physical relations.* The physical force-geometry relations are indicated by the forces and moments shown on the free-body diagram of Fig. 11.5b. Then, writing equilibrium relation (2.2), $\Sigma M^* = 0$, about the fixed point 0 yields the equation of motion:

$$\widehat{\Sigma M_0}^* = 0 = m\ell^2\ddot{\theta} + b\dot{\theta} - mg\ell \sin \theta \tag{11.7}$$

STAGE III. To study the behavior we first consider small values of θ, for which $\sin \theta \approx \theta$; this leads to the linear equation (dividing by $m\ell^2$)

$$\ddot{\theta} + \frac{b}{m\ell^2}\dot{\theta} - \frac{g}{\ell}\theta = 0 \tag{11.8}$$

which is identical in form to (11.2) but with a negative spring constant. This means that one root of its characteristic equation will be positive. The characteristic equation, obtained by assuming $\theta = \Theta e^{st}$, is

$$s^2 + \frac{b}{m\ell^2}s - \frac{g}{\ell} = 0 \tag{11.9}$$

and its roots are

$$s = -\frac{b}{2m\ell^2} + \sqrt{\left(\frac{b}{2m\ell^2}\right)^2 + \frac{g}{\ell}} \qquad s = -\frac{b}{2m\ell^2} - \sqrt{\left(\frac{b}{2m\ell^2}\right)^2 + \frac{g}{\ell}} \tag{11.10}$$

† As this example perhaps indicates, the study of stability of nonlinear systems is a difficult but fascinating one.

Since $\sqrt{(b/2m\ell^2)^2 + (g/\ell)}$ is greater than $b/2m\ell^2$, the first root will be positive. The time response will have the form

$$\theta = \Theta e^{st} = C_1 \exp\{[\sqrt{(b/2m\ell^2)^2 + (g/\ell)} - b/2m\ell^2]t\}$$
$$+ C_2 \exp\{-[\sqrt{(b/2m\ell^2)^2 + (g/\ell)} + b/2m\ell^2]t\}$$
$$= C_1 e^{+t/\tau_1} + C_2 e^{-t/\tau_2} \quad (11.11)$$

the first term of which diverges, the time constant of exponential divergence being

$$\tau_1 = \frac{1}{\sqrt{(b/2m\ell^2)^2 + (g/\ell)} - b/2m\ell^2} \quad (11.12)$$

or, from the numbers given,

$$\tau_1 \approx \frac{1}{\sqrt{1.0^2 + 3.0} - 1.0} = 1.0 \text{ sec}$$
$$\tau_2 \approx \frac{1}{\sqrt{1.0^2 + 3.0} + 1.0} = 0.33 \text{ sec} \quad (11.13)$$

and thus
$$\theta = \theta(0)(0.75 e^t + 0.25 e^{-t/0.33}) \quad (11.14)$$

The values of C_1 and C_2 were obtained from (8.10). This time response is plotted in Fig. 11.5c. It is seen to resemble, qualitatively, Fig. 11.3u (with $\sigma_1 = -1$).

Problems 11.1 through 11.9

11.3 LOCUS OF ROOTS BY GRAPHICAL CONSTRUCTION

The locus of roots in Fig. 11.2b can be plotted by a different process which will be found much more manageable when we deal with higher order systems. The process consists of rewriting Eq. (11.3) in the form

$$s\left(s + \frac{b}{J}\right) = -\frac{k}{J} \quad (11.15)$$

and making use of the fact that both $\vec{s}$ and $\overrightarrow{(s + b/J)}$ are vectors in the s plane† for any given value of s. When $k = 0$, these two vectors define points at $\vec{s} = 0$ and $\vec{s} = -b/J$. These are plotted as crosses ($\times$) in Fig. 11.6.

When k has the value corresponding to point (d) in Fig. 11.2b, the vector from the origin to point (d) is $\vec{s}$, as shown in Fig. 11.6. The vector $\overrightarrow{b/J}$ lies on the real axis, from point $-b/J$ to the origin. Thus the vector

† Again, these are mathematical, rather than physical, vectors. They obey the **rules** dictated by Euler's equation, as discussed in Sec. 7.2. Little arrows are added here over the quantities $\vec{s}$ and $\overrightarrow{(s + b/J)}$, simply as a reminder that the quantities are, indeed, vectors in the s plane.

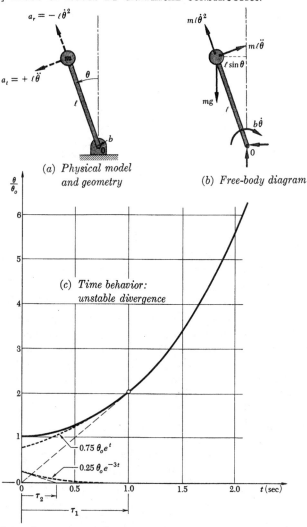

Fig. 11.5 *Behavior of an inverted pendulum*

from $-b/J$ to point (d) is the vector $\overrightarrow{(s + b/J)}$. By Eq. (11.15) the vector product $\overrightarrow{s(s + b/J)}$ must have a magnitude of $|k/J|$ and a phase angle of 180°. Therefore, multiplying the lengths of the two solid vectors in Fig. 11.6 will yield the value of $|k/J|$. The phase angle is checked by noting from geometry that $\phi_1 + \phi_2 = 180°$.

By using complex-vector geometry in this way, any points on the locus of roots in Fig. 11.2b can be found by trial and error. A point s is selected

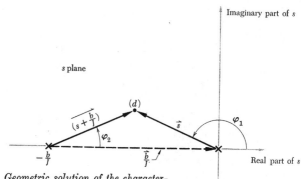

Fig. 11.6 *Geometric solution of the characteristic equation* $s\left(s + \dfrac{b}{J}\right) = -\dfrac{k}{J}$

for trial, and the vectors $\vec{s}$ and $\overrightarrow{(s + b/J)}$ are drawn to that point. Angles ϕ_1 and ϕ_2 are added. If their sum is not 180°, another point is selected. If their sum is 180°, then their lengths are measured and the product $|s||s + b/J|$ is computed. This is the value for k/J at this point, and the point is so labeled. Then another trial point is selected and the process repeated. This procedure is the basis of the root-locus method, Sec. 11.8.

For the second-order system this procedure is less direct than simply solving the characteristic equation explicitly in the form (11.4); but for higher-order systems the graphical trial-and-error procedure turns out to be much quicker than any analytical method.

11.4 DAMPING AS A VARIABLE

For a particular situation one might be more interested in the effect of damping constant b on system stability. A plot of roots s as a function of b is given in Fig. 11.7. Points (g) are for zero damping and indicate an

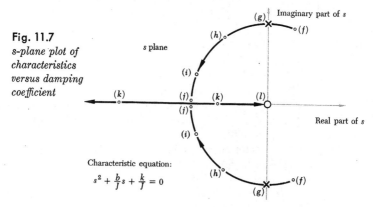

Fig. 11.7 *s-plane plot of characteristics versus damping coefficient*

Characteristic equation:
$$s^2 + \frac{b}{J}s + \frac{k}{J} = 0$$

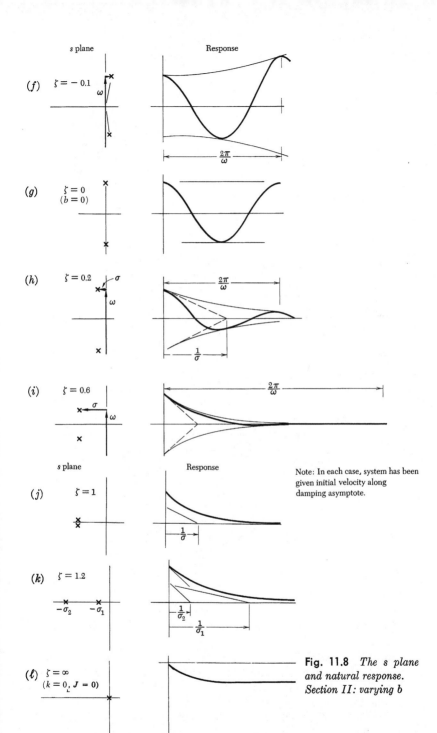

Fig. 11.8 *The s plane and natural response. Section II: varying b*

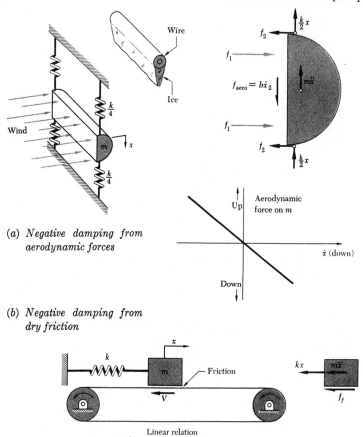

(a) *Negative damping from aerodynamic forces*

(b) *Negative damping from dry friction*

Fig. 11.9 Two second-order mechanical systems that exhibit oscillatory instability

undamped oscillation like that of Fig. 11.8g. (Fig. 11.8 is a continuation of Fig. 11.3.) For small damping, the roots are located between points (g) and (j)—e.g., points (h)—and the motion is like Fig. 11.8h. For heavy damping, the roots are on the real axis—e.g., points (k)—and the motion is like Fig. 11.8k. When the damping is infinitely large—point (ℓ) and Fig. 11.8ℓ—one root is at the origin, and the other at "$-\infty$." That is, the system

[Sec. 11.4] DAMPING AS A VARIABLE

is a simple integrator whose equation of natural motions is $b\dot{\theta} = 0$. (Physically, b is so large that k and J have negligible effect.) Figure 11.7 is to be plotted with numerical values in Problem 11.11.

Figure 11.7 can also be constructed by using the graphical technique of Fig. 11.6. This is accomplished by first writing the characteristic equation in the form

$$\frac{s^2 + (k/J)}{s} = -\frac{b}{J} \qquad (11.16)$$

so that the parameter b appears separate from the rest of the equation. When b is 0, Eq. (11.16) states that the roots are at $\pm j\sqrt{k/J}$—points (g) in Fig. 11.7. These are denoted by crosses (×). When b is infinite, Eq. (11.16) states that there is a root at $s = 0$. This root is denoted by a circle (○) at the origin in Fig. 11.7 and indicates, in another way, that as b is raised from its value at point (k) in Fig. 11.7 to a value of ∞ at point (ℓ), the extreme position of the root is at the origin (i.e., the locus cannot extend to the right of the origin).

Again, only when b is negative can the system be unstable, as indicated, for example, by points (f). Moreover, it is clear from Eq. (11.4) that *if b is negative, the system will always be unstable.* The motion corresponding to points (f) in Fig. 11.7 is plotted in Fig. 11.8f.

For the singular case $b = 0$ the roots are on the imaginary axis, points (g) in Fig. 11.7, and the motion, Fig. 11.8g, neither grows nor decays. *When a system has eigenvalues on the imaginary axis (and none in the right half-plane) we shall say it is "neutrally stable."*

Instability due to negative damping in physical systems

Negative damping is common in physical systems. Two examples are represented in Fig. 11.9a and Fig. 11.9b. Figure 11.9a is a (highly approximate) model for the "galloping transmission line" problem sometimes encountered in cold climates when ice-coated power lines are subjected to strong winds.[†] Figure 11.9b represents the dry-friction vibration which is so pleasant when produced by a skilled violinist, and so unpleasant when a machine tool begins to chatter.

In the model of Fig. 11.9a, air blowing across the flat plate of mass m is observed to produce a vertical oscillation of the plate on its springs. (In the transmission line the "flat plate" is a long wire on which ice has formed. The spring, of course, is furnished by tension in the wire.) Again, as in Fig. 11.4a, the explanation involves fluid flow. Here, however, there is no "negative spring constant"; the only force due to vertical *displacement* x is in the physical spring and is a stable restoring force. The aerodynamic flow is symmetrical and produces no force for any constant x. If the plate has a vertical *velocity*, however, the flow of air around it may be distorted in

---

[†] J. P. Den Hartog, "Mechanical Vibrations," Sec. 7.5, McGraw-Hill Book Company, New York, 1956. The demonstration model shown in Fig. 11.9a was developed by Professor Den Hartog.

(a) *An early stage in motion buildup (see Fig. 11.11)*

(b) *Catastrophic failure*

Fig. 11.10 *Dynamic instability on a large scale. The first bridge across the Tacoma Narrows at Puget Sound, Washington, was opened to traffic on July 1, 1940. The bridge became infamous because it oscillated whenever the wind blew. Finally after four months, on Nov. 7, 1940, a mild gale produced the results shown. (Reproduced from F. B. Farquharson, "Aerodynamic Stability of Suspension Bridges, with Special Reference to the Tacoma Narrows Bridge," Bull. '116, Part I, The Engineering Experiment Station, University of Washington, 1950.)*

[Sec. 11.4] DAMPING AS A VARIABLE 391

such a way as to produce an aerodynamic "lift" force *in the same direction as the velocity*. This constitutes a negative damping effect and leads to unstable oscillations, as depicted in Fig. 11.8(*f*). A similar, but more complicated, aerodynamic behavior is responsible for aerodynamic flutter of airplane wings, and was also the cause of the famous and spectacular failure of the Tacoma Narrows bridge,[†] Fig. 11.10.

In Fig. 11.9*b* the friction between the belt and the block is of the "dry" variety, for which the relation between friction force and relative velocity is as indicated by the solid curve. (Refer to Fig. 2.4*b*.) For comparison, the usual viscous friction relationship is indicated by the dashed line in Fig. 11.9*b*. Again, the *slope* of the force-velocity curve is effectively negative. That is, an increase in $\dot{x}$ leads to a decrease in the friction force opposing $\dot{x}$. (Recall, from Fig. 2.4, our convention that friction force is considered positive when it opposes velocity. The picture and the curve in Fig. 11.9*b* should be studied carefully to make sure that the directions of the forces and velocities involved are understood precisely.) Because of the dry friction, the effective damping constant *b*—the slope of the curve—is thus negative, the roots of the characteristic equation are at point (*f*) in Fig. 11.7, and the mass displays sinusoidal oscillations which grow with time, as in Fig. 11.8*f*.

Vibrations which are sustained, as in Figs. 11.9 and 11.10, by the energy from a steady (nonoscillating) force are known as "self-excited oscillations." As in Fig. 11.4, the unstable (growing) motions of each system in Fig. 11.9 and Fig. 11.10 can occur only because energy is being supplied to the system from an external source. In Fig. 11.9*a* and Fig. 11.10 the external source is the kinetic energy of the wind. In Fig. 11.9*b* it is the electrical power to the motor that drives the belt. The energy to sustain vibration of a violin string is, of course, supplied through the bow, by the musician's arm.

Example 11.2. Negative damping. The system of Fig. 11.9*a* has the following physical constants:

$m = 30{,}000$ dn/(cm/sec^2) $k = 30{,}000$ dn/cm

Vertical force due to wind, $f_{\text{aero}} = +b\dot{x}$ in the *same* direction as $\dot{x}$; with $b = 3000$ dn/(cm/sec)

Plot the motion of the system if it is released from the initial state: $x(0) = 2$ cm, $\dot{x}(0) = 0$.

We assume a linear physical model (Stage I) as implied by the physical constants given. To obtain the equations of motion (Stage II): (1) note that x and its derivatives are taken positive downward in Fig. 11.9*a*, (2) write vertical force equilibrium for the free-body diagram included in Fig. 11.9*a*, in which (3) the physical force-geometry relations (in the vertical direction) are included. The resulting equation of motion is

$$-\Sigma f_x{}^* = 0 = m\ddot{x} - b\dot{x} + kx = 0 \qquad (11.17)$$

[†] *Ibid.*, p. 307.

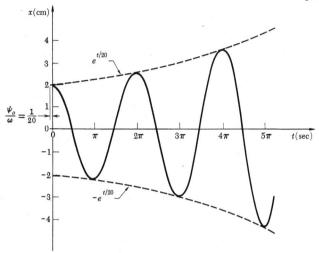

Fig. 11.11 *Natural time behavior of aerodynamic model in Fig. 11.9a: unstable oscillation*

Then, if we assume the solution $x = Xe^{st}$, the characteristic equation is

$$s^2 - \frac{b}{m} s + \frac{k}{m} = 0 \tag{11.18}$$

and the roots are

$$s = +\frac{b}{2m} \pm j \sqrt{\frac{k}{m} - \left(\frac{b}{2m}\right)^2} \tag{11.19}$$

and thus, from (8.11b), the time response is

$$x = \frac{x(0)}{\cos \psi} e^{+(b/2m)t} \cos (\omega t + \psi) u(t) \tag{11.20}$$

where $\omega = \sqrt{(k/m) - (b/2m)^2}$ and $\psi = \tan^{-1} (b/2m\omega)$

or, for the numbers given,

$$x = \frac{2}{\cos 3°} e^{+t/20} \cos (t + 3°) u(t) \text{ cm} \tag{11.21}$$

which is plotted in Fig. 11.11.

Problems 11.10 through 11.15

11.5 COUPLED PAIRS OF FIRST-ORDER SYSTEMS

Often two first-order systems are coupled together to produce a second-order system. Such systems (provided that they are l.c.c.) always have characteristic equations like (11.3) and therefore have characteristics which vary with system parameters in the manner of Fig. 11.2, or Fig. 11.7. Some systems of this sort are shown in Fig. 11.12.

In Fig. 11.12a two first-order mechanical systems like the model in Fig. 6.2b are connected via an electrical amplifier. Motion of the first subsystem is measured electrically, and the signal is amplified and used to drive the second subsystem. As in Part A, we use the triangular amplifier symbol to represent isolation and linear amplification (see p. 75). That is, the input signal enters the grid of an electronic tube (for example), after which it is amplified linearly and applies current to the windings of a torque motor; thus the torque applied to the second subsystem is linearly proportional to the displacement of the first subsystem, but the motion of the second subsystem in no way affects the first subsystem.

The system in Fig. 11.12b is similar to that of Fig. 11.12a except that the components are all electrical (and the amplifiers are "voltage amplifiers," furnishing isolation and giving a voltage out proportional to voltage in; see p. 76). Any input signal is modified in turn by each network element as it proceeds to the output point. The voltage at point a produces a voltage at point b, but the voltage at point b does not influence that at point a.

Figure 11.12c is also a combination of two first-order subsystems. (Although the mechanical subsystem has inertia, there is no spring, and it is the velocity that is measured by the tachometer and fed into the network.)

Figure 11.12d is similar to Fig. 11.12b, except that the middle isolation amplifier is missing. Now the voltage at point b *does* affect the voltage at point a, and the analysis is more involved. A similar degree of complexity exists in Fig. 11.12e in which two disks are coupled by the friction torque between them, because, again, the motion of each disk affects the other. Similarly, in Fig. 11.12f a change in either of the variables i or $\dot{x}$ affects the other.

Systems like Fig. 11.12d, e, and f, in which each subsystem may influence every other, involve dynamic coupling in a general way. Special cases like Fig. 11.12a, Fig. 11.12b and Fig. 11.12c, in which the motion of a subsystem influences *only* other subsystems that are *downstream* from it, are called *cascaded* or *tandem* systems. They are easier to analyze.

In the problems following Sec. 11.7, characteristic equations are to be obtained for the systems of Fig. 11.12 and system characteristics appraised.

11.6 FEEDBACK SYSTEMS

A class of dynamic systems known as "feedback" systems are of great interest; and many of them are only second (or even first) order. Several examples of such systems are given in Fig. 11.13. Commonly, feedback systems consist of a series of subsystems in cascade—perhaps only one or two (as in Fig. 11.13), or perhaps a great many—which, in addition, have *signal feedback* from the output of the last subsystem back into the input of the first. (More generally, feedback may also be used between any of the subsystems.)

Philosophically, when one subsystem is controlling another, feedback provides the controller with a measure of how well it is doing, and thus a basis for corrective action. Thus feedback systems are of central importance wherever the dynamic behavior of anything is to be controlled (electrons in a cathode ray tube, temperature in a building, the attitude of a satellite vehicle, and so on). The philosophy of feedback control, together with highly developed techniques for evaluating the complete performance of feedback systems, will be discussed in Part E. At this point we introduce the study of the *stability* and natural dynamic *characteristics* (eigenvalues) of feedback systems because it is natural to, and because they make ideal models from which to develop the basic principles for general higher-order dynamic systems, as we shall see.

The compound system in Fig. 11.13a is dynamically identical to the elementary system of Fig. 11.2a: In Fig. 11.13a, the spring has been replaced by an electrical feedback arrangement which (1) senses the rotation of the mass, and (2) causes a restoring torque proportional to displacement to be applied to the mass, in exactly the same way that the spring does in Fig. 11.2a. Thus if the feedback strength K in Fig. 11.13a is given the same magnitude as the spring constant k in Fig. 11.2a, then the dynamic behavior of the two systems will be exactly the same. In particular, *Fig. 11.2b represents the change in dynamic characteristics of the system in Fig. 11.13a as K is varied.*

The feedback system in Fig. 11.13a has been connected so that a positive displacement produces a negative (restoring) torque on the mass. This is called negative feedback, and is, of course, a necessary condition for stability. If positive feedback is used (i.e., if positive displacement calls for *positive* torque), then the roots of the characteristic equation will go to a position such as points (u) in Fig. 11.2b, and the system will display divergent instability.

The damping function in Fig. 11.2a can also be replaced by a feedback element, as shown in Fig. 11.13b, where a tachometer measures the angular velocity of the mass and applies a contrary torque. If the feedback "gain" in this system is given the same value b has in Fig. 11.2a, then again the systems will be dynamically identical. Thus the variation in dynamic characteristics as a function of b will be as shown in Fig. 11.7. Again, if the feedback is negative (as shown), the system is stable; but if the feedback is positive, the roots appear in the right half of the s plane [e.g., points (f) in Fig. 11.7], and the system is unstable.

The system in Fig. 11.13c is, dynamically, the simplest possible feedback system. It consists of a pure integrator† with position feedback, and the entire system equation of motion is thus only first order. (See Problem 11.21.) The system in Fig. 11.13c' is dynamically identical to that in Fig. 11.13c, the electrical feedback having been replaced by a mechanical one.

† The pure-integrator model of a hydraulic piston is derived on p. 120.

[Sec. 11.6] FEEDBACK SYSTEMS

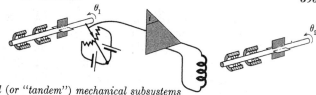

(a) *Cascaded (or "tandem") mechanical subsystems*

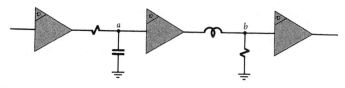

(b) *Cascaded electrical circuits*

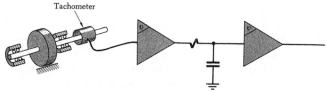

(c) *Mechanical and electrical subsystems in cascade*

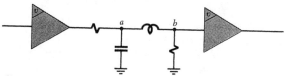

(d) *Two-way coupling of electrical subsystems*

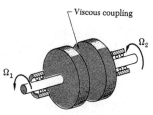

(e) *Two-way coupling of mechanical subsystems*

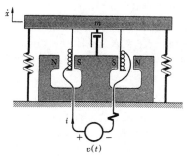

(f) *Two-way electrical-mechanical coupling*

Fig. 11.12 *Some coupled pairs of first-order systems*

In Fig. 11.13d the signal from the position pickoff is filtered through a first-order electrical network, and thus two first-order subsystems are involved. The characteristic equation for this system will be the same as for the element in Fig. 11.2a, even though physically the two systems look quite different. Further study is left as an exercise (Problem 11.23).

Another system of two first-order sybsystems is shown in Fig. 11.13e. In this case the system is also second order, but is somewhat different from the other examples. Specifically, the loci of roots are similar to those in Fig. 11.2b, except that they begin at two points on the real axis, neither of which is at the origin (see Problem 11.24).

The systems of Fig. 11.13 are to be studied further in the problem set that follows Sec. 11.7.

11.7 THIRD-ORDER SYSTEMS

Figure 11.14 is a modification of the system of Fig. 11.13a; the signal from the position pickoff is now filtered through a first-order electrical network. (Refer to the discussion on p. 316 on filtering noise in circuits.)

The first step in studying this system is, of course, to write its equations of motion. The equation for the mechanical subsystem is

$$J\ddot{\theta} + b\dot{\theta} = M \qquad (11.22a)$$

where M is the moment applied by the electromechanical torquer. This moment is proportional to voltage v_b: $M = K_2 v_b$, and v_b, in turn, is related to the voltage from the pickoff by the equation

$$RC\dot{v}_b + v_b = v_a \qquad (11.22b)$$

Finally, the pickoff voltage is simply proportional to position θ of the mechanical rotor:

$$v_a = -K_1 \theta \qquad (11.22c)$$

where $K_1 = K_3 K_4$ is the number of volts appearing at a per radian of rotation of θ (i. e., K_1 includes both the pickoff sensitivity K_3 and the amplifier gain K_4).

Obtaining the characteristic equation of a compound system

To obtain the characteristic equation for this entire system we extend the procedure we employed in Secs. 6.2 and 8.2: We now assume that *each of the variables*—θ, v_a, v_b, and M—varies according to the relation (11.1), Ce^{st}, *where s is the system characteristic, and is therefore the same for every element*. It is convenient to write these relations as $x = Xe^{st}$, using the capital letter to represent the constant value of the corresponding lower-case-letter variable (as has been our custom, beginning in Sec. 6.2). Thus, by substituting the assumed solutions

$$\theta = \Theta e^{st} \qquad v_a = V_a e^{st} \qquad v_b = V_b e^{st} \qquad M = M e^{st} \qquad (11.23)$$

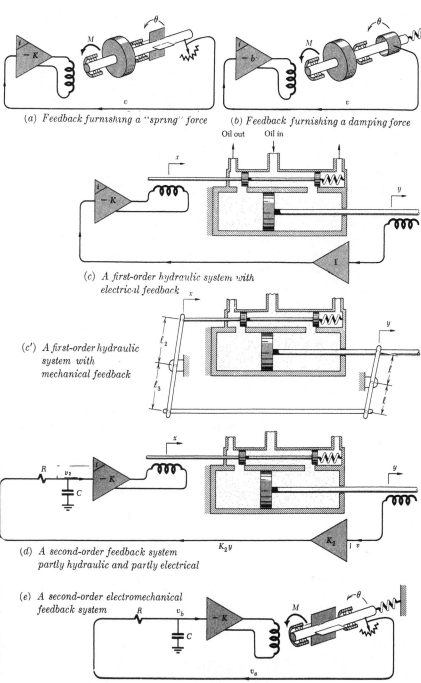

Fig. 11.13 Some simple systems having feedback coupling

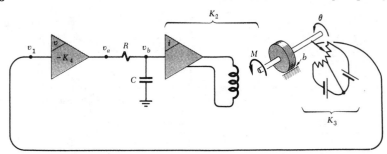

Fig. 11.14 *A third-order system*

into the equations of motion, (11.22), the following algebraic expressions are obtained:

$$(Js^2 + bs)\Theta = \mathsf{M}$$
$$\mathsf{M} = K_2 V_b$$
$$(RCs + 1)V_b = V_a$$
$$V_a = -K_1 \Theta$$
(11.24)

These can now be combined algebraically:

$$(Js^2 + bs)\Theta = K_2(RCs + 1)^{-1}(-K_1\Theta) \quad (11.25)$$

or, rearranging, and canceling Θ (i.e., ignoring the trivial solution $\Theta = 0$)

$$\boxed{(RCs + 1)(Js^2 + bs) + K_1 K_2 = 0} \quad (11.26)$$

which is the characteristic equation of the entire system. The roots of this characteristic equation are the exponents for the motion of *every* element in

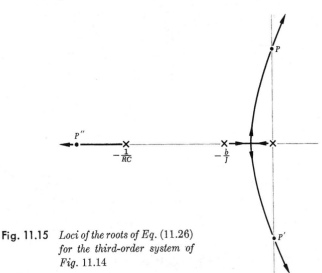

Fig. 11.15 *Loci of the roots of Eq. (11.26) for the third-order system of Fig. 11.14*

[Sec. 11.7] THIRD-ORDER SYSTEMS

the system, in accordance with the assumed motion (11.23). That is, *all the elements have natural motions with the same characteristics.*†

Before continuing, let us summarize and generalize the steps we have just taken as a formal procedure:

> To obtain the characteristic equation for a compound system represented by a *set* of l.c.c. differential equations of motion (with all external forcing functions zero):
> (1) Assume *every* variable to have motion of the form $x = Xe^{st}$ and substitute this *set* of solutions into the equations of motion.
> (2) Eliminate the constants X (and variable e^{st}) from the resulting set of algebraic equations to obtain a single equation in s, the characteristic equation.

Procedure C-1
Obtaining the characteristic equation for a compound system

Step (2) is most systematically carried out by setting equal to zero the *determinant* of the set of algebraic equations in s, as we shall see in more complicated systems.

A more penetrating understanding of the foregoing ideas, and of those that follow, can be gained by simulating the system of Fig. 11.14 on an analog computer and studying its natural motions as parameters are varied, following the procedures of Sec. 8.15.

Plotting system characteristics in the s plane

To see what the dynamic characteristics of the system in Fig. 11.14 are, we plot the roots of its characteristic equation (11.26)—its eigenvalues—in the s plane as a function of the physical parameter of interest, which is usually the overall feedback gain, $K = K_1 K_2$. This plot is given in Fig. 11.15. (The arrows indicate the direction of increasing K.) The method by which the plot is constructed will be described thoroughly in Secs. 11.8 and 11.9. Here we assume this has been done, and we study the completed plot to see what we can learn from it.

For small values of gain in Fig. 11.15 the locus of the roots emanating from points $s = 0$ and $s = -b/J$ looks very much like the locus in Fig. 11.2b. In addition, there is another root emanating from point $-1/RC$, which, for the relative magnitudes shown, connotes an unimportant, quickly decaying transient motion

For large values of K the locus in Fig. 11.15 is different in an important

† The *magnitudes* (and phase angles) of motions of the different members may, of course, have various values, depending on how the motions were started. (In particular, for certain starting conditions one or more of the magnitudes may be zero.)

way from the locus of Fig. 11.2b: In Fig. 11.15 the system *becomes unstable for all values of K larger than a certain value.* More precisely, for these large values of K the system will exhibit unstable oscillations like those of Fig. 11.8(f). The rate at which the oscillations diverge becomes greater as K is increased.†

Note that, taken separately, each of the subsystems in Fig. 11.14 is capable only of exponentially decaying natural motion; but the combined system is capable of natural oscillations which diverge!

A point of great interest is, of course, point P in Fig. 11.15, where the locus crosses the imaginary axis, i.e., where the system goes from a stable to an unstable condition. The value of K at this neutral stability point (as at any other point on the locus) can be calculated by multiplying together the magnitudes of complex vectors $\vec{s}$, $\overrightarrow{[s + (b/J)]}$, $\overrightarrow{[s + (1/RC)]}$.

Another powerful method for determining what values of system parameters will produce instability is the method of Routh. Unlike the root-locus method, however, Routh's method considers *only* the point of neutral stability (point P in Fig. 11.15). These two important methods, the root-locus and Routh methods, will be presented in the following sections.

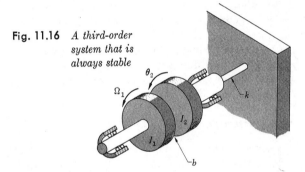

Fig. 11.16 *A third-order system that is always stable*

Third-order systems do not always have the capability of going unstable, of course. Consider, for example, the system shown in Fig. 11.16, with no external torque applied. The clutch damping b is taken to be viscous, and to be the only damping present. The reader may readily show that the characteristic equation for the system is

$$(J_1 s + b)(J_2 s^2 + bs + k) - b^2 s = 0$$

which can be rearranged to separate parameter b:

$$s\left(s^2 + \frac{k}{J_2}\right) + \left(\frac{J_2 + J_1}{J_2 J_1}\right) b \left(s^2 + \frac{k}{J_1 + J_2}\right) = 0$$

s and $[s^2 + (k/J_2)]$ representing the characteristics of disks 1 and 2, respec-

---

† The electrical power into the amplifier in Fig. 11.14 supplies the external energy required for these growing motions.

[Sec. 11.8] THE ROOT-LOCUS METHOD OF EVANS

tively, when $b = 0$ (no coupling). It can be shown that this system is stable for all positive values of b. The reason is, of course, that there is no external source of energy in Fig. 11.16 (like the wind in Fig. 11.9a or the amplifier power supply in Fig. 11.14). What can you deduce about this system's natural behavior when $b \to \infty$ (i.e., when the disks are cemented together)? See Problem 11.41.

Problems 11.16 through 11.26

11.8 THE ROOT-LOCUS METHOD OF EVANS†

This graphical method has affected control-system analysis rather profoundly since its introduction in 1948, because of the speed and directness with which it (1) solves the central problem of finding system roots and (2) displays them as a function of system parameters.

We have demonstrated the root-locus display in Figs. 11.2, 11.7, and 11.15, and we indicated the underlying construction in Fig. 11.6. Let us now state formally the procedure we followed, and the philosophy on which it is based. For clarity, we shall use the system of Fig. 11.14 as a model for discussion in this first study.

The root-locus method for solving for the roots of the characteristic equation of a system takes advantage of the fact that in most problems involving dynamic coupling the roots of the various subsystems are already known. For the system of Fig. 11.14, for example, the charactristic equation in the form (11.26) contains factors $[s + (1/RC)]$ and $s[s + (b/J)]$, which are the characteristics of the network and of the mechanical subsystem, respectively, both of which are already well known.

The root-locus method does not multiply these factors all together (as Routh would do), because then the identity of individual subsystems is lost. Instead, the characteristic equation (11.26) is put in the form

$$C(s) = s\left(s + \frac{1}{RC}\right)\left(s + \frac{b}{J}\right) = \frac{-K}{JRC} \quad (11.27)$$

in which the *left-hand side* of the equation is now totally factored.

Next, the roots of the left-hand side of (11.27) are plotted as X's (crosses) in the s plane, Fig. 11.17.‡ These are the roots of (11.27) for the special case $K = 0$. [More generally, $C(s)$ will contain a denominator, as (11.16) does. Then we plot the factors of the

Procedure C-2
Root-locus method: basic construction

† W. R. Evans, Graphical Analysis of Control Systems, *Trans. AIEE*, vol. 68, pp. 765–777, 1949. See also Evans' book "Control-System Dynamics," McGraw-Hill Book Company, New York, 1954.
‡ Figure 11.17 is the same locus as Fig. 11.15.

denominator as $\bigcirc$'s (circles†), as in Fig. 11.7. These correspond to $K = \infty$.] A root locus departs from each $\times$ as K increases from zero, Fig. 11.17. [A root locus arrives at each $\bigcirc$ as K approaches ∞, as in Fig. 11.7.]

The roots for other values of K are found in two parts:

① Find the loci (in the s plane) of all points s for which the total *phase angle* of the left-hand side of (11.27) is 180°. This satisfies the minus sign on the right-hand side of (11.27).‡

② Evaluate the *magnitude* of the left-hand side of (11.27) at each point of interest on the locus. This tells the magnitude of K/JRC, the right-hand side of (11.27).

Both Part ① and Part ② utilize the fact that each (complex) factor of the left-hand side of (11.27) is represented by a simple complex vector in Fig. 11.17. Consider, for example, arbitrary point Q. The vector from the origin to Q is simply $\vec{s}$. Similarly, the vector from $-b/J$ to Q is $\overrightarrow{[s + (b/J)]}$, because it is the sum of $\vec{s}$ and $\overrightarrow{b/J}$ (the vector from $-b/J$ to the origin). Finally, $\overrightarrow{[s + (1/RC)]}$ is drawn from $-1/RC$ to Q. Note that all vectors terminate at point Q.

Part ① is carried out by selecting likely points (e.g., point Q), and adding phase angles: $(\phi_1 + \phi_2 + \phi_3)$ to see whether they total 180°. If they do, Q is marked as a point on the "locus of roots." If they do not, another trial is made.

Procedure C-2 *(Cont.)*

Thus the method involves trial and error. It is, however, a rapid method, first because with some simple rules (given in Sec. 11.9, and in more detail in Chapter 21) and a little practice the loci can be estimated quite accurately, and second because each angle summation can be made very quickly. The second statement is true because, as shown, all the angles are available *at point* Q. Thus by centering a protractor with a movable arm at Q, the angles can be summed in a few seconds (without taking time to read any of them individually).

In practice, then, one would first construct the locus in Figure 11.17 approximately by freehand sketching and would then check perhaps two

† The pronunciation "crosses" and "circles" will circumvent a possible confusion, as discussed on p. 647.

‡ Obviously, if the sign on the right-hand side of (11.27) is plus instead of minus (as occasionally happens), we shall seek 0° loci instead of 180° loci.

[Sec. 11.8] THE ROOT-LOCUS METHOD OF EVANS

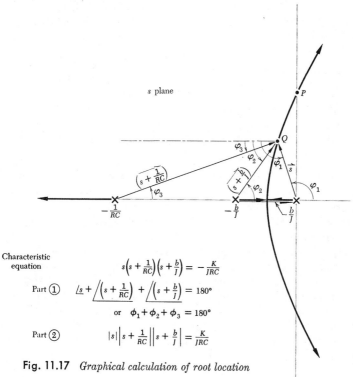

Fig. 11.17 *Graphical calculation of root location*

points—e.g., Q and P—with a protractor. Figure 11.17 can be constructed by an experienced person in less than 30 seconds.

> Part ② is carried out simply by multiplying together the *lengths* of the vectors in Fig. 11.17—either by measuring each length and multiplying with a slide rule, or by the faster method of using a logarithmic spiral on the protractor arm to accomplish measurement and multiplication simultaneously.†

As an example, let us compute the value of $|K/JRC|$ at point P in Fig. 11.17. (This point is, of course, especially interesting.) It happens that the case $1/RC = 3b/J$ was plotted in Fig. 11.17. Multiplication of the three vectors at point P gives:

$$|s|\left|s + \frac{1}{RC}\right|\left|s + \frac{b}{J}\right| \cong \left|1.7\frac{b}{J}\right|\left|2\frac{b}{J}\right|\left|3.5\frac{b}{J}\right| \cong 12\left(\frac{b}{J}\right)^3 = \frac{K}{JRC} \quad (11.28)$$

† Simple plastic devices for root-locus plotting, called "Spirules," are available commercially. Their use is explained in detail in Appendix I.

or $K/J = 4(b/J)^2$, which we shall find to agree with the value given by Routh in Eq. (11.45g) when $1/RC = 3b/J$. If K is made larger than this value, the system will oscillate unstably.

Root-locus method: general case

In general, the root-locus method is directly applicable whenever the characteristic equation can be written in the form

$$P(s) + KZ(s) = 0 \qquad (11.29)$$

where $P(s)$ and $Z(s)$ contain known factors (e.g., characteristics of individual subsystems) and K is the parameter whose effect on the system is under study. [Note that K need not be the "gain" of a control system. It might, for example, be the damper constant of a shock mount or the inductance in a network. In the latter cases some algebraic manipulation may be required to achieve the necessary form, (11.29).]

In nearly all control-system problems, and often in more general cases of dynamic coupling as well, the characteristic equation can be written in the form (11.29). In such problems the method often provides the fastest, most penetrating study available of system stability and behavior. For this reason, its effect on the control-system art in particular has been extremely important.

The simplest and most-used root locus techniques are presented in Sec. 11.9, which follows.

⊙11.9 AN INTRODUCTION TO ROOT-LOCUS SKETCHING

The root-locus method is the subject of Chapter 21, and the techniques for sketching loci are developed comprehensively and rather rigorously there.

The most-used techniques are quite readily learned, however, and the present point is a natural one for mastering them, if the reader so desires. We have therefore abstracted and abridged the simpler parts of Chapter 21 for presentation here, using equation numbers verbatim from Chapter 21.

Procedure

Consider a characteristic equation of the general form

$$(s + \sigma_1)(s + \sigma_2 - j\omega_2)(s + \sigma_2 + j\omega_2)(s + \sigma_4) \cdots (s + \sigma_n)$$
$$+ \ K(s + a_1)(s + a_2) \cdots (s + a_m) \ = \ 0 \qquad (21.1b)$$

which can be written more compactly in the form (11.29):

$$P(s) + KZ(s) = 0 \qquad (21.1a)$$

where the definitions of $P(s)$ and $Z(s)$ are clear. Rewrite the equation in

[Sec. 11.9] AN INTRODUCTION TO ROOT-LOCUS SKETCHING

the Evans form, which will be more convenient:†

$$C(s) = \frac{P(s)}{Z(s)} = \frac{(s+\sigma_1)(s+\sigma_2-j\omega_2)(s+\sigma_2+j\omega_2)(s+\sigma_4)\cdots(s+\sigma_n)}{(s+a_1)(s+a_2)\cdots(s+a_m)}$$
$$= -K = |K|\,\underline{/180°} \quad (21.2)$$

We call $C(s)$ the Evans function.

The loci of the roots of (21.1) may be sketched approximately in the s plane, using the rules given in Procedure C-3 (which are abstracted from Sec. 21.3).

Procedure C-3
Simple root-locus sketching rules for 180° loci

Special rule: **Symmetry.** The pattern of the set of loci is symmetrical with respect to the real axis.‡

Rule 1: **Real-axis segments.** There is a locus segment in each real-axis interval to the right of which the number of ×'s plus ○'s on the real axis is odd.

Rule 2: **Asymptotes.** As $K \to \infty$ in (21.2), $(n-m)$ loci will approach ∞ along asymptotes determined as follows:

(a) The asymptotes emanate from the centroid of the ×-○ array, ○'s being given negative weight. That is, the centroid is the location, $s = -c$, for which

$$c = \frac{(\sigma_1 + \sigma_2 + \cdots + \sigma_n) - (a_1 + a_2 + \cdots + a_m)}{(n-m)} \quad (21.6)$$

(b) The $(n-m)$ asymptotes have the directions

$$\phi = \pm(2r+1)\frac{180°}{(n-m)} \quad (21.7)$$

in which r is each integer, from 0 until ϕ repeats. The array of asymptotes is symmetrical with respect to the real axis.

Rule 3: **Directions of departure and arrival.** The direction in which a locus departs from a × at $s = -\vec{\beta}$ (or arrives at a ○ at $s = -\vec{\beta}$) is given by§

$$\left.\begin{array}{l}\phi_{\text{dep}} = 180° - \underline{/\left[\dfrac{1}{(s+\beta)}C(s)\right]_{s=-\vec{\beta}}} \\ \qquad\text{for departure from a ×} \\ (\text{or } \phi_{\text{arr}} = 180° + \underline{/[(s+\beta)C(s)]_{s=-\vec{\beta}}} \\ \qquad\text{for arrival at a ○})\end{array}\right\} \quad (21.8)$$

---

† As noted in Sec. 11.8, if the sign of K is negative, we shall write, instead, for the right-hand side of (21.2): "$\cdots = |K|\,\underline{/0°}$." For clarity, we present sketching rules for 180° loci only in Procedure C-3. The rules are readily rephrased for 0° loci (see Problem 11.49).
‡ Provided that the coefficients of the characteristic equation are real.
§ Provided there are not multiple ×'s (or ○'s). These are treated in Sec. 21.3.

> That is, if $[C'(s)]_{s=-\vec{\beta}}$ is the net phase angle, measured *at* $s = -\vec{\beta}$, of all terms in $C(s)$ *except* $s = -\vec{\beta}$, then ϕ is the *additional* angle required to make 180°.†

Examples

The sketching rules of Procedure C-3 are demonstrated, for a simple example, in Fig. 11.18. As the figure indicates, these simple rules are adequate for sketching the loci with surprising accuracy. The reader may develop considerable proficiency with root-locus sketching by working the problem set.

> Should additional examples seem desirable, a number are available in Chapter 21. These may be followed readily without reading the text in Chapter 21, which is concerned with deriving the given rules.
> Examples of Rule 1: Sec. 21.4 and Example 21.2.
> Examples of Rule 2: Examples 21.3 and 21.4.
> Examples of Rule 3: Examples 21.5 and 21.6.

When the rules are understood individually, then some rather interesting complete loci may be sketched, as suggested in the problems.

From working through the examples listed above and the problems given below, one can attain proficiency in *applying* the simple sketching rules. In Chapter 21 we shall learn *why* the rules work and how to augment them for more involved systems.

Problems 11.27 through 11.60

11.10 THE METHOD OF ROUTH: THIRD-ORDER EXAMPLE

Routh's method is a purely algebraic technique for investigating the stability of a linear system with constant coefficients. In general, it considers a characteristic equation of the form

$$s^n + A_{n-1}s^{n-1} + A_{n-2}s^{n-2} + \cdots + A_2s^2 + A_1s + A_0 = 0 \quad (11.30)$$

and states a set of inequalities between the A's necessary for the system represented to be stable.

To gain insight into the method, consider the special case of a third-order system whose characteristic equation has the form

$$s^3 + A_2s^2 + A_1s + A_0 = 0 \quad (11.31)$$

[This form would have resulted, for example, from multiplying out the

† A Spirule may be used to facilitate the angle summations in (21.8).

[Sec. 11.10] ROUTH'S METHOD: THIRD-ORDER EXAMPLE

Given the characteristic equation: $s(s + 4)(s + 3 - j2)(s + 3 + j2) + K(s + 2) = 0$
The Evans function is, from (21.2):

$$C(s) = \frac{s(s + 4)(s + 3 - j2)(s + 3 + j2)}{(s + 2)} = -K = K\underline{/180°}$$

Note that $n = 4$, $m = 1$.

Rule 1: Real-axis segments

Rule 2: Asymptotes

(a) Centroid:

$c = \dfrac{(0 + 4 + 3 + 3) - 2}{(4 - 1)}$

$= \dfrac{8}{3}$

(b) Directions [$(4 - 1) = 3$ of them]

$\phi = \pm(2r + 1)\dfrac{180°}{4 - 1}$

$= \pm 60°, \pm 180°$

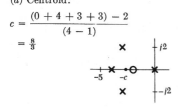

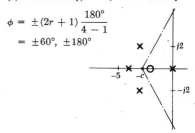

Rule 3: Direction of departure from × *at* $s = -3 + j2$

$\phi_{\text{dep}} = 180° - \underbrace{\left/\dfrac{s(s + 4)(s + 3 + j2)}{(s + 2)}\right.}_{\substack{\text{measured and added} \\ \text{by Spirule}}}$

$= 180° - [146.3° + 63.4° + 90° - 116.6°] = 3.3°$

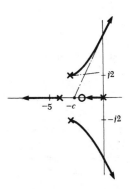

Complete root-locus plot. The upper locus was
obtained by fairing in between —·—
guides. The lower locus was drawn by
symmetry (Special rule).

Fig. 11.18 *Demonstration of root-locus sketching*

factors in Eq. (11.26).] We know that the roots of this equation must have either the form

$$s = -\sigma_1 \quad s = -\sigma_2 \quad s = -\sigma_3$$
or the form $\quad s = -\sigma_1 \quad s = -\sigma_2 \pm j\omega_2$ \hfill (11.32)

wherein any negative σ means instability.

At the point of *neutral* stability, (11.32) has the form

$$\begin{aligned} & s=0 & & s=-\sigma_2 & & s=-\sigma_3 \\ \text{or} \quad & s=0 & & s=-\sigma_2 \pm j\omega_2 & & \\ \text{or} \quad & s=-\sigma_1 & & s=\pm j\omega_2 & & \end{aligned}$$
(11.33)

The Routh method compares (11.33) and (11.31) by writing (11.31) in factored form:

$$s(s + \sigma_2)(s + \sigma_3) = 0$$
or $\quad s(s + \sigma_2 + j\omega_2)(s + \sigma_2 - j\omega_2) = 0$
or $\quad (s + \sigma_1)(s + j\omega_2)(s - j\omega_2) = 0$

and then multiplying the factors together:

$$s^3 + (\sigma_2 + \sigma_3)s^2 + (\sigma_2\sigma_3)s = 0$$
or $\quad s^3 + 2\sigma_2 s^2 + (\sigma_2{}^2 + \omega_2{}^2)s = 0$ \hfill (11.34)
or $\quad s^3 + (\sigma_1)s^2 + (\omega_2{}^2)s + \sigma_1\omega_2{}^2 = 0$

Then by comparing (11.34) with (11.31) it can be concluded that (11.31) represents a system which is neutrally stable either if

$$A_0 = 0$$
or if $\quad A_2 A_1 = A_0$ \hfill (11.35)

To determine the comparable relations for *non*-neutral stability, use the factors (11.32):

$$(s + \sigma_1)(s + \sigma_2)(s + \sigma_3) = 0$$
or $\quad (s + \sigma_1)(s + \sigma_2 + j\omega_2)(s + \sigma_2 - j\omega_2) = 0$

and again multiply the factors together:

$$s^3 + \underbrace{(\sigma_1 + \sigma_2 + \sigma_3)}_{A_2}s^2 + \underbrace{(\sigma_1\sigma_2 + \sigma_1\sigma_3 + \sigma_2\sigma_3)}_{A_1}s + \underbrace{\sigma_1\sigma_2\sigma_3}_{A_0} = 0$$
(11.36)
or $\quad s^3 + \overbrace{(\sigma_1 + 2\sigma_2)}s^2 + \overbrace{(2\sigma_1\sigma_2 + \sigma_2{}^2 + \omega_2{}^2)}s + \overbrace{\sigma_1(\sigma_2{}^2 + \omega_2{}^2)} = 0$

From these relations, if all σ's are positive (stable system), then A_0 is positive

[Sec. 11.11] ROUTH'S METHOD: GENERAL CASE

$(A_0 > 0)$, and (by comparison of $A_2 A_1$ to A_0) $A_2 A_1$ is *greater* than A_0. Thus

$$\boxed{\text{For stability:} \quad A_2 A_1 > A_0 > 0} \quad (11.37)$$

Moreover, from (11.36) we can make the additional statement that if *any* A is negative, then some σ must be negative, and the system is unstable:

$$\boxed{\text{For stability:} \quad \text{All } A\text{'s must be positive}} \quad (11.38)$$

Expressions (11.37) and (11.38) together constitute the *Routh Criteria for Stability*† for a third-order l.c.c. system.

For example, applying (11.37) and (11.38) to the system of Fig. 11.14, whose characteristic equation is (11.27),

$$s^3 + \left(\frac{1}{RC} + \frac{b}{J}\right) s^2 + \frac{1}{RC} \frac{b}{J} s + \frac{K}{JRC} = 0 \quad \begin{array}{c}(11.27)\\ \text{[repeated]}\end{array}$$

we can conclude that, for stability,

$$0 < \frac{K}{JRC} < \left(\frac{1}{RC} + \frac{b}{J}\right)\left(\frac{1}{RC} \frac{b}{J}\right) \quad (11.39)$$

The implications of this result are discussed in detail in Example 11.3.

Inequalities similar to (11.37) and (11.39) can be derived for higher-order systems. The reasoning is the same, but the mathematics becomes more involved. The results, for the general case, will now be given.

11.11 ROUTH'S METHOD: GENERAL CASE

Using reasoning analogous to the above, Routh deduced the general relations which must hold for the stability of a system represented by a linear characteristic equation of nth order. Routh read his results before the British Mathematical Society, beginning in 1874.‡

In most books the general rules deduced by Routh are stated without proof, as they will be here. Proofs are available in a few places,§ including the writings of Routh himself.¶

---

† In mathematical phraseology we have proved that (11.37) and (11.38) constitute *necessary* conditions for stability. [If all σ's are positive, then (11.37) and (11.38) will be true.] It can also be proved that, together, they constitute *sufficient* conditions.
‡ Adams Prize Essay, 1877.
§ E. A. Guillemin, "The Mathematics of Circuit Analysis," pp. 395–409, John Wiley & Sons, New York, 1956; see also W. R. Evans, "Control System Dynamics," Appendix D, McGraw-Hill Book Company, New York, 1954.
¶ R. J. Routh, "Dynamics of a System of Rigid Bodies," Part II, 6th rev. ed., arts. 290–307, Dover Publications, Inc., New York, 1955.

If a system can be represented by a characteristic equation of the form (11.30).†

$$s^n + A_{n-1}s^{n-1} + A_{n-2}s^{n-2} + \cdots \\ + A_2s^2 + A_1s + A_0 = 0 \quad (11.40)$$

the procedure for determining stability is as follows:

Step 1: **Negative coefficients.** Check for negative coefficients in the characteristic equation, (11.40). It can be shown that

$$\boxed{\text{For stability, all } A\text{'s must be positive.}} \quad (11.41)$$

[This result is derivable as a corollary of the general conditions, (11.44) of Step 5.]

Step 2: **Routhian array.** Form the following array:

$$\left.\begin{array}{c|ccccc}
s^n & 1 & A_{n-2} & A_{n-4} & A_{n-6} & A_{n-8} & \cdots \\
s^{n-1} & A_{n-1} & A_{n-3} & A_{n-5} & A_{n-7} & A_{n-9} & \cdots \\
s^{n-2} & b_1 & b_2 & b_3 & b_4 & \cdots \\
s^{n-3} & c_1 & c_2 & c_3 & \cdots \\
s^{n-4} & d_1 & d_2 & \cdots \\
\vdots \\
s^1 & j_1 \\
s^0 & k_1
\end{array}\right\} \quad (11.42)$$

Procedure C-4
Routh's method for determining stability

in which the first two lines are made up of *all* the coefficients in (11.40) taken alternately, and the remainder of the terms are calculated as follows:

Step 3: **Calculating terms.** The b's are calculated as:

$$\left.\begin{array}{l}
b_1 = -\dfrac{\begin{vmatrix} 1 & A_{n-2} \\ A_{n-1} & A_{n-3} \end{vmatrix}}{A_{n-1}} = \dfrac{A_{n-1}A_{n-2} - A_{n-3}}{A_{n-1}} \\[2ex]
b_2 = -\dfrac{\begin{vmatrix} 1 & A_{n-4} \\ A_{n-1} & A_{n-5} \end{vmatrix}}{A_{n-1}} \\[2ex]
b_3 = -\dfrac{\begin{vmatrix} 1 & A_{n-6} \\ A_{n-1} & A_{n-7} \end{vmatrix}}{A_{n-1}} \cdots
\end{array}\right\} \quad (11.43a)$$

The pattern is continued until the point is reached beyond which all the rest of the b's are zero.

[Sec. 11.11] ROUTH'S METHOD: GENERAL CASE

To calculate the c's, d's, and so on, the same pattern is followed, using, in each case, the two rows preceding. Thus

$$\left.\begin{array}{l} c_1 = -\dfrac{\begin{vmatrix} A_{n-1} & A_{n-3} \\ b_1 & b_2 \end{vmatrix}}{b_1} \\[2ex] c_2 = -\dfrac{\begin{vmatrix} A_{n-1} & A_{n-5} \\ b_1 & b_3 \end{vmatrix}}{b_1} \quad \cdots \\[2ex] d_1 = -\dfrac{\begin{vmatrix} b_1 & b_2 \\ c_1 & c_2 \end{vmatrix}}{c_1} \quad d_2 = -\dfrac{\begin{vmatrix} b_1 & b_3 \\ c_1 & c_3 \end{vmatrix}}{c_1} \quad \cdots \end{array}\right\} \quad (11.43b)$$

$\cdots \qquad \cdots$

Procedure C-4 (Cont.)

The procedure is followed down to the s^0 row. The array is found to be triangular, the last two rows having only one term each.

Step 4: **Simplification.** The terms in the array may be simplified numerically by multiplying or dividing any horizontal row by any *positive* number, at any stage of the calculation.

Step 5: **General conditions for stability.** Read down the first column in the array. The number of changes in sign encountered is equal to the number of roots of (11.40) with positive real parts. If there are any roots with positive real parts the system is, of course, unstable.‡ Thus

> For stability, all signs in the first column must be positive. $\qquad$ (11.44)

Some additional steps are necessary to handle the special cases of a *zero term* in the first column or a *zero row*. These will be discussed presently. First, however, the above general procedure will be illustrated with some examples.

† Assigning the value unity to coefficient A_n in (11.30) entails no loss of generality. A coefficient A_n may be added in (11.30) and in the upper left-hand corner of (11.42) and (11.43), if desired.

‡ If we are interested only in stability for a particular numerical example (and don't care how many unstable roots there are), we will stop calculating the array as soon as we encounter a change in sign in the first column.

Example 11.3. Stability of system in Fig. 11.14. Use Procedure C-4 to determine the conditions for stability for the system of Fig. 11.14. (This is a more formal presentation of the deductions of Sec. 11.10.)

The characteristic equation of the system has been derived, (11.26):

$$(RCs + 1)(Js^2 + bs) + K_1K_2 = 0 \qquad (11.26)$$
[repeated]

or, multiplying and combining terms in the form (11.40)

$$s^3 + \left(\frac{1}{RC} + \frac{b}{J}\right)s^2 + \frac{b}{JRC}s + \frac{K_1K_2}{JRC} = 0$$

Step 1. By (11.41) all coefficients must be positive, so that three conditions for stability are

$$0 < \frac{1}{RC} + \frac{b}{J} \qquad (11.45a)$$

$$0 < \frac{b}{JRC} \qquad (11.45b)$$

$$0 < \frac{K_1K_2}{JRC} \qquad (11.45c)$$

Steps 2 and 3. The Routhian array is formed, using (11.42) and (11.43):

s^3	1	$\dfrac{b}{JRC}$	0
s^2	$\left(\dfrac{1}{RC} + \dfrac{b}{J}\right)$	$\dfrac{K_1K_2}{JRC}$	0
s	$-\dfrac{\begin{vmatrix} 1 & \dfrac{b}{JRC} \\ \left(\dfrac{1}{RC} + \dfrac{b}{J}\right) & \dfrac{K_1K_2}{JRC} \end{vmatrix}}{\left(\dfrac{1}{RC} + \dfrac{b}{J}\right)}$	$-\dfrac{\begin{vmatrix} 1 & 0 \\ \left(\dfrac{1}{RC} + \dfrac{b}{J}\right) & 0 \end{vmatrix}}{\left(\dfrac{1}{RC} + \dfrac{b}{J}\right)} = 0$	
s^0	$\dfrac{K_1K_2}{JRC}$	0	

For the system to be stable, each of the terms in the first column must be positive [by (11.44)], and thus the general conditions for stability are

$$0 < \left(\frac{1}{RC} + \frac{b}{J}\right) \qquad (11.45d)$$

$$0 < \frac{b}{JRC}\left(\frac{1}{RC} + \frac{b}{J}\right) - \frac{K_1K_2}{JRC} \qquad (11.45e)$$

$$0 < \frac{K_1K_2}{JRC} \qquad (11.45f)$$

(11.45a) and (11.45c) are, of course, the same as (11.45d) and (11.45f). [Moreover, (11.45b) can be derived directly from (11.45d), (11.45e), and (11.45f).]

Results (11.45e) and (11.45f) may be written more compactly:

$$0 < \frac{K_1 K_2}{JRC} < \frac{b}{JRC}\left(\frac{1}{RC} + \frac{b}{J}\right) \quad (11.45g)$$

which is the same as (11.39). These results can be used to establish the value of $K = K_1 K_2$ at point P in Fig. 11.15 (see Problem 11.62).

The above example demonstrates an important feature of Routh's method: It can predict stability in terms either of general symbols (e.g., K, J, RC) or of numbers, while the root-locus method requires at least approximate numerical values. (But, of course, the root locus gives the dynamic characteristics completely, while Routh's method gives only the conditions for stability.)

It is very interesting and instructive to compare the Routh results with the root-locus analysis of Fig. 11.15, bearing in mind that Routh considers only points of neutral stability, i.e., points on the imaginary axis:

Inequalities (11.45b) and (11.45d) state that, for stability, both of the crosses—at $-b/J$ and at $-1/RC$—must be in the left half of the s plane in Fig. 11.17. (From a root-locus viewpoint it can be shown that, for this system, if either of these crosses is in the right half-plane, then at least one branch of the locus of roots will always remain in the right half-plane, and the system will always be unstable.)

Inequality (11.45f) states that if $K_1 K_2$ is negative, the system will be unstable. In Fig. 11.15 a negative $K_1 K_2$ would produce a locus proceeding to the *right* from the cross at the origin, confirming (11.45f).

Finally, inequality (11.45g) states that not only must $K_1 K_2 / JRC$ be positive, but it must not exceed the value $(b/JRC)[(1/RC) + (b/J)]$. This value of $K_1 K_2/JRC$ is shown in Eq. (11.28) to correspond precisely to point P in Fig. 11.17. That is, if $b[(1/RC) + (b/J)] < K_1 K_2$, the roots are beyond point P, and therefore in the right half of the s plane, confirming (11.45g). Routh's method can always be used to check root-locus results, but only where loci cross the imaginary axis.

Example 11.4. Negative coefficients. Determine whether or not a certain system is stable if its characteristic equation is

$$s^6 + 4s^5 + 7s^4 - 5s^3 + 2s^2 + 8s + 12 = 0$$

It is unnecessary in this case to form the Routhian array, because the characteristic equation contains a negative coefficient, and we know, by (11.41), that this indicates instability.

Example 11.5. Sixth-order system. Determine whether or not a certain system is unstable if its characteristic equation is

$$s^6 + 3s^5 + 10s^4 + 40s^3 + 84s^2 + 92s + 40 = 0$$

If the system is unstable, find the number of roots having positive real parts.

Step 1. All the coefficients are positive, hence Step 1 tells us nothing.

Step 2. The first two lines in the Routhian array are formed, following (11.42):

s^6	1	10	84	40	0
s^5	3	40	92	0	

Step 3. By (11.43) the coefficients of the third row are

$$b_1 = -\frac{\begin{vmatrix} 1 & 10 \\ 3 & 40 \end{vmatrix}}{3} \qquad b_2 = -\frac{\begin{vmatrix} 1 & 84 \\ 3 & 92 \end{vmatrix}}{3} \qquad b_3 = -\frac{\begin{vmatrix} 1 & 40 \\ 3 & 0 \end{vmatrix}}{3}$$

$$= -\tfrac{10}{3} \qquad\qquad = \tfrac{160}{3} \qquad\qquad = 40$$

Note that b_4 and subsequent terms are 0.

Step 4. This row can be simplified by multiplying by the positive number $\tfrac{3}{10}$:

$$\qquad -1 \qquad 16 \qquad 12 \qquad 0$$

The array is now

s^6	1	10	84	40	0
s^5	3	40	92	0	
s^4	−1	16	12	0	

Similarly, the coefficients of the fourth row are

$$c_1 = -\frac{\begin{vmatrix} 3 & 40 \\ -1 & 16 \end{vmatrix}}{-1} = +88 \qquad c_2 = -\frac{\begin{vmatrix} 3 & 92 \\ -1 & 12 \end{vmatrix}}{-1} = +128$$

while c_3 and subsequent terms are 0. Dividing by 8 (for convenience) and inserting in the array, we obtain

s^6	1	10	84	40	0
s^5	3	40	92	0	
s^4	−1	16	12	0	
s^3	+11	16	0		

The remaining terms are calculated by repeating (11.43), with the following final result:†

s^6	1	10	84	40	0
s^5	3	40	92	0	
s^4	−1	16	12	0	
s^3	+11	16	0		
s^2	$\tfrac{192}{11}$	12	0		
s^1	8.4	0			
s^0	12	0			

† Note the triangular form of the final array, and the pattern of the 0's which terminate the rows in pairs. This pattern always occurs (commencing with a pair of 0's when n—the order of the characteristic equation—is odd, and commencing with a single 0 when n is even, as it is in the present example).

Step 5. The general stability criterion, (11.44), can now be applied. Going down the first column we find two changes in sign, so we may conclude that two of the system's six roots have positive real parts.

The reader may prove that this is a correct conclusion by verifying that the roots of the original characteristic equation are

$$(s + 1)(s + 2)(s - 1 \pm j3)(s + 1 \pm j1) = 0$$

[where the convention $(s - 1 \pm j3) \triangleq (s - 1 + j3)(s - 1 - j3)$ is used].

Problems 11.61 through 11.63

11.12 SPECIAL CASE OF A ZERO TERM IN THE FIRST COLUMN

If the first term in a row is zero, but not all the other terms in the row are zero, then the array may still be completed by either one of the following steps:

> Step 6a: *Zero term in first column.* Replace the zero with a small positive quantity ϵ. Then complete the array as before, and apply stability criterion (11.44) by taking the limit as $\epsilon \to 0$.
> or
> Step 6b: *Zero term in first column.* Invert the characteristic equation by substituting $s = 1/x$. This will usually produce an alternative characteristic equation whose array has no zero term in the first column.

Procedure C-4 (*Cont.*)

Example 11.6. Zero term. How many unstable roots has the system whose characteristic equation is:

$$s^5 + 3s^4 + 2s^3 + 6s^2 + 6s + 9 = 0$$

Step 1. All the coefficients are positive, hence Step 1 yields no information.
Step 2. The first two lines of the Routhian array are

s^5	1	2	6
s^4	3	6	9

Dividing the second row by 3 and proceeding, we find that the third row is

s^3	0	3	0

Encountering zero in the first column, we proceed to Step 6.

Step 6a. Substitute ϵ—a small positive number—for 0 in the third row, and complete the array (Steps 3 and 4):

s^5	1	2	6
s^4	1	2	3
s^3	ϵ	3	0
s^2	$\dfrac{2\epsilon - 3}{\epsilon} \to -\dfrac{3}{\epsilon}$	3	0
s	$3 - 3\epsilon\left(\dfrac{\epsilon}{2\epsilon - 3}\right) \to +3$	0	
s^0	3	0	

Step 5. Now we apply (11.44) in the limit as $\epsilon \to 0$. The fourth term in column 1 goes to $-3/\epsilon$ in the limit, while the fifth term goes to $+3$. Then there are two sign changes in the first column, and two unstable roots. This completes the analysis by the method of Step 6a.

Step 6b. Alternative procedure. Replace s by $1/x$ in the characteristic equation:

$$\left(\frac{1}{x}\right)^5 + 3\left(\frac{1}{x}\right)^4 + 2\left(\frac{1}{x}\right)^3 + 6\left(\frac{1}{x}\right)^2 + 6\left(\frac{1}{x}\right) + 9 = 0$$

or, multiplying by x^5, we obtain

$$1 + 3x + 2x^2 + 6x^3 + 6x^4 + 9x^5 = 0$$

for which the Routhian array begins thus:†

x^5	9	6	3
x^4	6	2	1

Step 4. Divide first row by 3; complete the array:

x^5	3	2	1
x^4	6	2	1
x^3	1	$\frac{1}{2}$	0
x^2	-1	1	0
x	$+1\frac{1}{2}$	0	
x^0	1	0	

Step 5. Counting sign changes in the first column, we see at once that the system has two unstable roots.

11.13 SPECIAL CASE OF A ZERO ROW

The presence of a zero row indicates conjugate pairs of roots of the characteristic equation—i.e., an imaginary conjugate pair ($s = +j\omega, -j\omega$), or a real pair of equal magnitude, of which one is positive and the other

† See second footnote, p. 411.

[Sec. 11.13] SPECIAL CASE OF A ZERO ROW

negative ($s = +\sigma, -\sigma$), or, more generally, *a set of roots which are symmetrical with respect to the imaginary axis*, such as

$$s = +\sigma + j\omega \qquad s = +\sigma - j\omega \qquad s = -\sigma + j\omega \qquad s = -\sigma - j\omega$$

When this happens, the following procedure is followed:

Step 7: **Handling a zero row.** An auxiliary equation is formed from the preceding row. That is, if the row for s^m is a zero row, then the auxiliary equation is formed from the terms of the s^{m+1} row. If those terms are $k_1, k_2, \ldots$, then the auxiliary equation is

$$k_1 s^{m+1} + k_2 s^{m-1} + k_3 s^{m-3} + \cdots + \begin{Bmatrix} k_j s \\ \text{or} \\ k_j \end{Bmatrix} = 0 \quad (11.46)$$

(The last term depends on whether $m + 1$ is odd or even.)

The roots of the auxiliary equation are roots of the characteristic equation.

The auxiliary equation is then differentiated, and the coefficients of the differentiated equation are entered in the array in place of the zero row. The array is completed as before, and the stability criteria are the same.

Procedure C-4 (*Cont.*)

It is, of course, a highly useful thing that some roots of the characteristic equation are obtained free of charge for this special case.

Example 11.7. Zero row. Discuss the stability of the system whose characteristic equation is

$$s^5 + 5s^4 + 11s^3 + 23s^2 + 28s + 12 = 0$$

Step 1. There are no negative coefficients, hence no information.
Step 2. We begin with the Routhian array:

s^5	1	11	28
s^4	5	23	12
s^3	6.4	25.6	0
s^2	3	12	
s	0	0	

Encountering a zero row, we proceed to *Step 7*:
The *auxiliary equation* is, from (11.46),

$$3s^2 + 12 = 0$$

or
$$s^2 + 4 = 0$$

The roots of this equation—$s = \pm j2$—are roots of the characteristic equation. That is, the system will display, as one of its natural modes of motion, an undamped oscillation at $\omega = 2$.

Differentiating the auxiliary equation gives

$$2s + 0 = 0$$

The coefficients of this differentiated equation are entered in place of the zero row in the Routhian array, and the array completed:

s^5	1	11	28
s^4	5	23	12
s^3	6.4	25.6	0
s^2	3	12	0
s	2	0	
s^0	12	0	

Step 5. Since there are no sign changes in the first column, the characteristic equation has *no unstable roots*.

We conclude, therefore, that the system is characterized by three stable roots and a pair of *neutrally stable* oscillatory roots, namely, $s = \pm j2$.

An example of a characteristic equation which leads to a zero row, but which does not have roots on the imaginary axis, will be found in Problem 11.68.

Problems 11.64 through 11.81

Chapter 12

COUPLED MODES
OF NATURAL MOTION:
TWO DEGREES OF FREEDOM

In Chapter 11 we studied the *kinds* of natural behavior compound systems can have: their characteristics and, particularly, their stability. One notable finding was that all members of a coupled system have the *same* set of possible *kinds* of natural motion (characteristics), which are properties of the system. In the present chapter we shall study the *magnitudes* of such natural motions.

Basic concepts

(1) *Coupling: eigenvalues.* There are numerous physical mechanisms by which the subsystems of a compound system can be coupled together. As coupling becomes stronger, the *eigenvalues*, or *characteristics* of the system become less similar to those the member subsystems would have individually.
(2) *Natural modes.* Compound systems can move in natural modes wherein only one of the system's eigenvalues (one real root or one complex pair) is exhibited, the others being totally absent.
(3) *Eigenvectors.* In each natural mode, the motion of each subsystem has a specific magnitude and phase relation to that of the others. These relations, which describe the natural modes, are called *eigenvectors*. They involve total state, e.g., velocity as well as configuration, in general.
(4) *Superposition.* General natural motions of compound systems are most conveniently studied as the *superposition* of natural modes.
(5) *Special case: undamped vibration.* An undamped system having n degrees of freedom (and order $2n$) has n natural frequencies of vibration.

(These are the eigenvalues or roots of the system's characteristic equation.) To vibrate purely at one of its natural frequencies, the system must be released from a particular configuration, which it will then preserve cycle after cycle. This is known as vibrating in a *natural mode*, and the particular configuration corresponding to each natural frequency is known as a *natural mode shape*.

(6) *Beat phenomena*. These involve the very-low-frequency difference signal between two sinusoids of equal magnitude and nearly equal frequency. In undamped physical systems having two natural modes of nearly equal frequency, beats are generated when both modes are present in equal proportions.

12.1 FORMS OF PHYSICAL COUPLING

In Chapter 11 we studied some compound systems whose members were coupled together in a simple manner. In this section we shall look at dynamic coupling in general, to see in what forms it may arise and just how it may affect the behavior of the member subsystems in a system.

The system in Fig. 12.1a shows three dynamic subsystems connected in *cascade*.† In this arrangement no member is affected by the motion of members "downstream" from it (e.g., J_2 feels no effect from motions of J_3).‡ This is the simplest form of coupling.

Coupling is somewhat more complicated in feedback systems, where typically a signal feedback is used to make the motion of a particular downstream element felt at a particular place upstream. For example, in Fig. 12.1b members are still in cascade except that, in addition, a torque is applied to J_1 proportional to the displacement of J_3.

In general in a dynamic system, two-way interactions between subsystems may be more intimate and more numerous. For example, in the common system of Fig. 12.1c the subsystems are no longer in cascade: Not only does a displacement of mass J_1 produce a torque on J_2, but also a displacement of J_2 produces torque on J_1. The motions of J_2 and J_3 are similarly interrelated.

Coupling between the motions of members in a system can occur in various ways. In Fig. 12.1c, for example, coupling occurs because pairs of elements share a spring in common. That is, the motions θ_1 and θ_2 are coupled because spring k_{12} is twisted by both displacements θ_1 or θ_2. In this case the equation of motion for disk 1 will contain a term proportional to displacement θ_2, and vice versa.

Some other physical sources of coupling are indicated in Fig. 12.2.

---

† The genesis for the term "cascade" is the image of Fig. 14.1.

‡ As in Sec. 2.7, the symbol ─▷─ represents an idealized isolation amplifier producing a current output proportional to voltage in. Refer to p. 75.

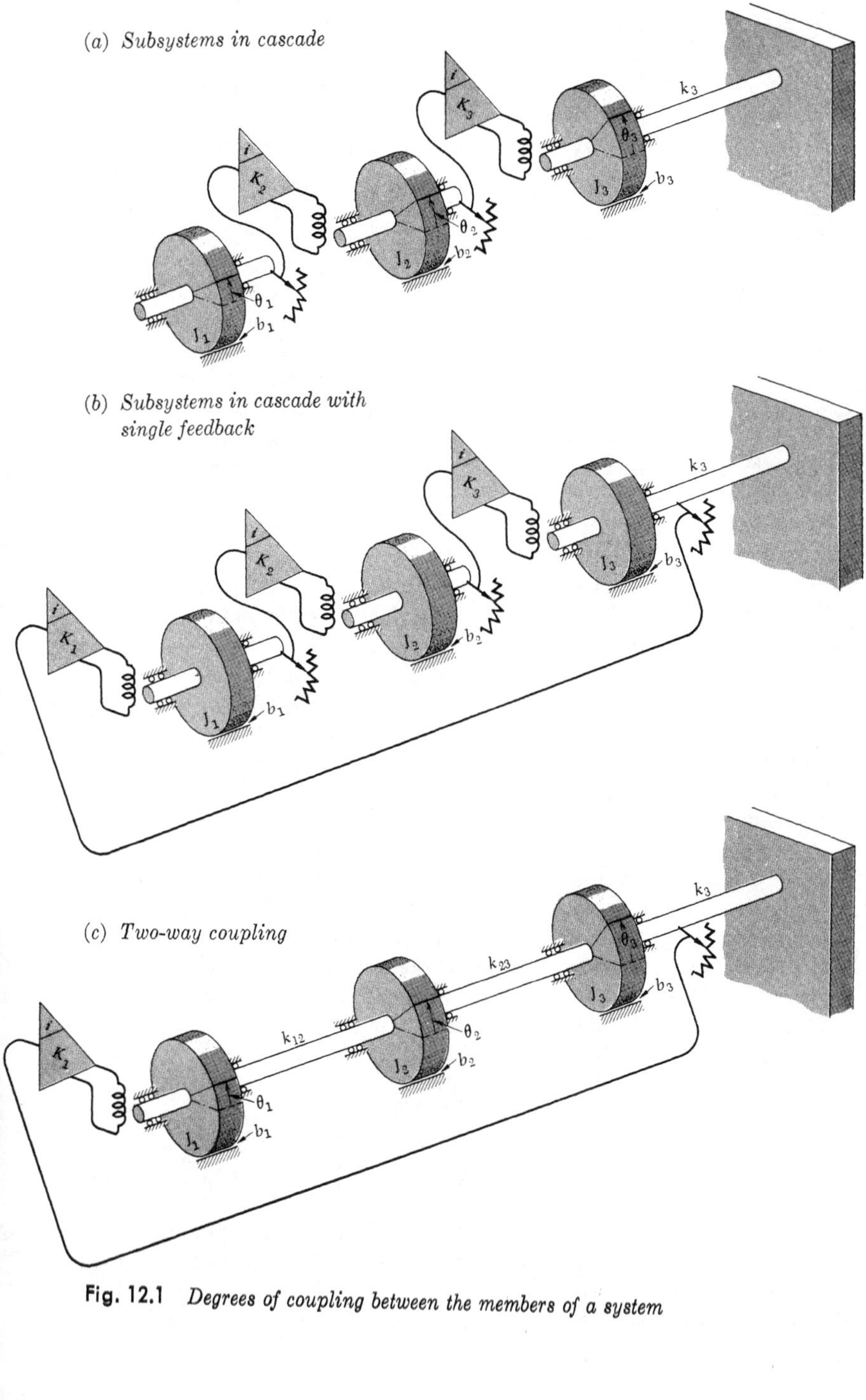

[Sec. 12.1] FORMS OF PHYSICAL COUPLING

(a) *Subsystems in cascade*

(b) *Subsystems in cascade with single feedback*

(c) *Two-way coupling*

Fig. 12.1 *Degrees of coupling between the members of a system*

Mechanical coupling is shown in Figs. 12.2a through 12.2d, while the other figures show electrical, electromechanical, and electrothermal coupling.

In Fig. 12.2a and Fig. 12.2b, the coupling is through common *damping*, so that a change in the *velocity* of one element produces a force on the other element. (This will result in coupling showing up in the equations of motion as *velocity* terms.)

In Fig. 12.2c and Fig. 12.2d the coupling is *inertial*. In Fig. 12.2c, if the coordinates of motion are chosen as x and θ, then it will be found that an *acceleration* in one of the coordinates produces a force or torque in the other. That is, if the pendulum has an acceleration $\ddot{\theta}$, then this will be accompanied by an inertial *force* on the pendulum which will be transmitted to the block, and so on. These relations will be studied carefully in subsequent sections.

Note that several kinds of mechanical coupling may exist simultaneously. In Fig. 12.2d, for example, we might take as coordinates the displacement x of the center of the bar and rotation θ about the center, and write the equations of motion as (1) a summation of vertical forces on the bar and (2) a summation of moments about the center of the bar. In this case we see that not only will a displacement x of the bar produce a moment (due to the springs) about the center, but also a velocity $\dot{x}$ will produce a moment (due to the dashpot), and an acceleration $\ddot{x}$ will produce a moment (because inertia m_1 does not equal m_2). Each of these moments represents coupling between x and θ.

As we might guess, the exact form of coupling will depend upon the coordinates we choose to use in writing the equations of motion. In particular, certain choices of coordinates will eliminate some (or even all) of the coupling terms. (Of course, the choice of coordinates can have no effect on the motions of the physical system; it can affect only the form of the mathematical relations we use to study those motions.)

Coupling occurs also in other physical media, of course. Coupling is most common, for example, in electrical networks where two loops (or two nodes) may share an element in common, as in Fig. 12.2e and Fig. 12.2f. Again, whether the coupling appears as a proportional or derivative quantity in the equations depends, for example, on whether the shared element is a resistor which affects the current itself, as in Fig. 12.2e, or an inductance which affects the derivative of current, as in Fig. 12.2f.

Figure 12.2g represents magnetic electromechanical coupling, as in an audio speaker. Here the coordinates chosen might be mechanical displacement and electrical current, then rate of change of mechanical displacement produces an electrical voltage drop, while electrical current, in turn, produces a mechanical force; thus, again, two-way coupling is found in the equations of motion.

Finally, in the thermocouple of Fig. 12.2h the coupling is essentially one-way, a change in temperature producing an electrical voltage, but the electrical current involved being so tiny as to produce no noticeable heat.

The mechanisms of coupling and their effects on system behavior will

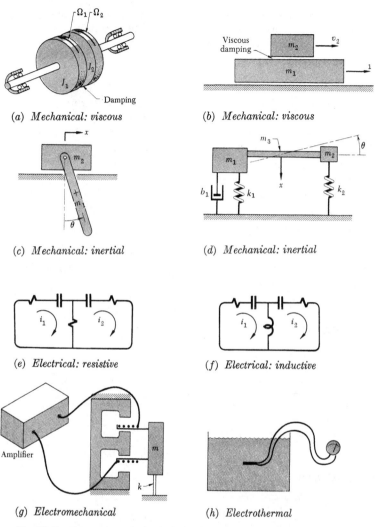

(a) Mechanical: viscous
(b) Mechanical: viscous
(c) Mechanical: inertial
(d) Mechanical: inertial
(e) Electrical: resistive
(f) Electrical: inductive
(g) Electromechanical
(h) Electrothermal

Fig. 12.2 *Some forms of physical coupling*

be better understood as we become more familiar with them, in the sections that follow, through writing equations of motion for various systems and studying the resulting behavior.

12.2 COUPLED EQUATIONS OF MOTION

It is especially easy to study the motions of cascaded subsystems like Fig. 12.1a, because (1) the motion of a given subsystem can be studied from its own differential equation wherein the motion of the preceding subsystem appears only as an input, and (2) the dynamic characteristics of the total system thus consist simply of the individual characteristics of the subsystems.

If, however, there is any coupling—even of simple feedback variety, as in Fig. 12.1b—then, as we have seen in Chapter 11, important changes take place in the dynamic behavior of the system: its natural motions no longer exhibit the dynamic characteristics of the individual elements, but instead take on new characteristics which can be quite different (see Figs. 11.14 and 11.15). The reasons for this change in characteristics can be seen directly by writing the equation of motion for each of the elements, and studying the interrelations between the equations.

The equations for the system in Fig. 12.1a are

$$\left.\begin{array}{l} J_1\ddot{\theta}_1 + b_1\dot{\theta}_1 = 0 \\ -K_2\theta_1 + J_2\ddot{\theta}_2 + b_2\dot{\theta}_2 = 0 \\ -K_3\theta_2 + J_3\ddot{\theta}_3 + b_3\dot{\theta}_3 + k_3\theta_3 = 0 \end{array}\right\} \quad (12.1)$$

These are more convenient to study if we first make the usual assumption that all motions will be of the form $\theta = \Theta e^{st}$:

$$\theta_1 = \Theta_1 e^{st} \qquad \theta_2 = \Theta_2 e^{st} \qquad \theta_3 = \Theta_3 e^{st} \quad (12.2)$$

and substitute in Eqs. (12.1):

$$\left.\begin{array}{l} (J_1 s^2 + b_1 s)\Theta_1 = 0 \\ -K_2\Theta_1 + (J_2 s^2 + b_2 s)\Theta_2 = 0 \\ -K_3\Theta_2 + (J_3 s^2 + b_3 s + k_3)\Theta_3 = 0 \end{array}\right\} \quad (12.3)$$

In like manner, the equations for the system in Fig. 12.1b are

$$\left.\begin{array}{l} (J_1 s^2 + b_1 s)\Theta_1 \hspace{3em} - K_1\Theta_3 = 0 \\ -K_2\Theta_1 + (J_2 s^2 + b_2 s)\Theta_2 = 0 \\ -K_3\Theta_2 + (J_3 s^2 + b_3 s + k_3)\Theta_3 = 0 \end{array}\right\} \quad (12.4)$$

and for the system in Fig. 12.1c,

$$\left.\begin{array}{l} (J_1 s^2 + b_1 s + k_{12})\Theta_1 \hspace{1em} - k_{12}\Theta_2 \hspace{1em} - K_1\Theta_3 = 0 \\ -k_{12}\Theta_1 + (J_2 s^2 + b_2 s + k_{12} + k_{23})\Theta_2 - k_{23}\Theta_3 = 0 \\ -k_{23}\Theta_2 + (J_3 s^2 + b_3 s + k_3 + k_{23})\Theta_3 = 0 \end{array}\right\} \quad (12.5)$$

From Eqs. (12.1) through (12.5) we see how additional coupling appears as an increase in the number of coupling terms in the array of equations. We expect this to make the equations increasingly difficult to solve. The solution to Eqs. (12.3) is quite simple: The first equation is solved by itself, then the second equation is solved from the first, and so on. But Eqs. (12.4) can no longer be solved so simply; in fact, as we showed in Sec. 11.7, sophisticated techniques may be required even for systems of a few elements. For feedback systems we shall show, in Part E, that the block-diagram technique makes an effective tool for manipulating the equations.

When the equations become more intimately coupled, as Eqs. (12.5)

[Sec. 12.2] COUPLED EQUATIONS OF MOTION

are, even the block-diagram approach becomes cumbersome, and usually the best procedure is to study the set of equations by determinant methods. For example, Eqs. (12.5) are a linear homogeneous set (i.e., they contain only terms in the variables Θ_1, Θ_2, Θ_3). All three equations can therefore be satisfied only if the determinant formed by the coefficients of the variables is zero:

$$\begin{vmatrix} (J_1 s^2 + b_1 s + k_{12}) & -k_{12} & -K_1 \\ -k_{12} & (J_2 s^2 + b_2 s + k_{12} + k_{23}) & -k_{23} \\ 0 & -k_{23} & (J_3 s^2 + b_3 s + k_3 + k_{23}) \end{vmatrix} = 0$$
(12.6)

When this determinant is multiplied out, we obtain the characteristic equation of the system, just as we always have:†

$$(J_1 s^2 + b_1 s + k_{12})[(J_2 s^2 + b_2 s + k_{12} + k_{23})(J_3 s^2 + b_3 s + k_3 + k_{23}) - k_{23}^2]$$
$$+ k_{12}[-k_{12}(J_3 s^2 + b_3 s + k_3 + k_{23}) - K_1 k_{23}] = 0 \quad (12.7)$$

The characteristics of natural motion of the system are then to be determined by finding the roots of (12.7)—which are, of course, the eigenvalues, the values of s which belong in the assumed solutions, (12.2), for natural motion.

But there is more to be learned from Eqs. (12.5). These appear to be three simultaneous equations in the four unknowns, s, Θ_1, Θ_2, and Θ_3. It is proper, however, to reduce these to three variables, by dividing all terms by Θ_1, for example. Then, once we have solved (12.7) for values of s, we can return to (12.5) and, for each value of s, solve for values Θ_2/Θ_1 and Θ_3/Θ_1. The physical meaning of this result is that *for each characteristic natural motion, the elements in the system move in a certain relationship with one another.* The *absolute* magnitude of this motion (e.g., the value of Θ_1) is arbitrary, because the system is linear; but the ratios between *relative* magnitudes—Θ_2/Θ_1 and Θ_3/Θ_1—are specifically determined by Eqs. (12.5).

This concept of relative magnitudes of motion is of special importance in systems with little or no damping, because then all of the motions will be undamped oscillations [the roots of the characteristic equation (12.7) are all imaginary] and the values of ratios like Θ_2/Θ_1 and Θ_3/Θ_1 represent the configurations or "mode shapes" in which these natural vibrations take place.

The ideas in the above general discussion will be developed in detail, and some numerical solutions for three and more degrees of freedom will be carried out in later sections. First, however, we shall treat a number of concepts that are more easily seen in the simpler context of two degrees of freedom.

Problems 12.1 through 12.8

---

† The reader unfamiliar with determinants can satisfy himself that (12.7) is correct merely by systematically eliminating two of the Θ's in (12.5). Determinants are discussed in Appendix H.

12.3 A SIMPLE VIBRATING SYSTEM OF COUPLED MEMBERS

Figure 12.3 shows the physical model of a simple system consisting of two rotary inertias connected to limber shafts. This system is a rather special case, not only because it involves only two degrees of freedom, but also because it involves no damping and is entirely symmetrical. We begin with this special case because from it three ideas of fundamental importance are especially easy to grasp: the ideas of *natural frequencies*, and of *natural modes of vibration* of a system of coupled subsystems, and the concept of *superposition* of natural modes of motion.

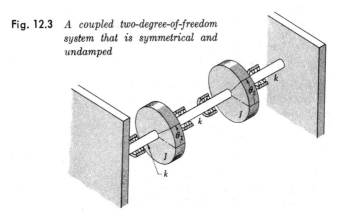

Fig. 12.3 *A coupled two-degree-of-freedom system that is symmetrical and undamped*

Later these same aspects of dynamic behavior will be investigated in the much more general context of unsymmetrical systems of many degrees of freedom with arbitrary amounts of damping. But the striking clarity with which the physical ideas can be seen from the system of Fig. 12.3 makes it well worth-while to dwell on this special case first.

We begin by writing the equations of motion for each of the disks in Fig. 12.3 (each is taken as a free body). A mark has been painted on the rim of each disk to indicate its displacement: θ_1 and θ_2. (The marks were painted when all shafts were untwisted, and were carefully aligned with each other.) The equations of motion, in terms of θ_1 and θ_2, are the following:

$$\begin{aligned} J\ddot{\theta}_1 + k\theta_1 + k(\theta_1 - \theta_2) = 0 &\quad \text{for the first disk} \\ J\ddot{\theta}_2 + k\theta_2 + k(\theta_2 - \theta_1) = 0 &\quad \text{for the second disk} \end{aligned} \quad (12.8)$$

To solve these equations we assume, as usual, that the solutions are

$$\theta_1 = \Theta_1 e^{st} \qquad \theta_2 = \Theta_2 e^{st} \quad (12.9)$$

Then Eqs. (12.8) become

$$\begin{aligned} (Js^2 + 2k)\Theta_1 - k\Theta_2 &= 0 \\ -k\Theta_1 + (Js^2 + 2k)\Theta_2 &= 0 \end{aligned} \quad (12.10)$$

[Sec. 12.3] SIMPLE VIBRATING SYSTEM OF COUPLED MEMBERS 427

Equations (12.10) are two simultaneous equations in three unknowns, s, Θ_1, and Θ_2. The first step in solving the equations is to reduce the number of unknowns by dividing all terms by Θ_1:

$$(Js^2 + 2k) - k\frac{\Theta_2}{\Theta_1} = 0$$
$$-k + (Js^2 + 2k)\frac{\Theta_2}{\Theta_1} = 0 \qquad (12.11)$$

We now have two equations in two unknowns: s and the ratio Θ_2/Θ_1. We shall solve first for s and then for Θ_2/Θ_1.

The direct way to solve for s is by cross-substitution between Eqs. (12.11):

$$\frac{\Theta_2}{\Theta_1} = \frac{Js^2 + 2k}{k} = \frac{k}{Js^2 + 2k} \qquad (12.12)$$

Then, by cross-multiplying, we obtain the *characteristic equation*:

$$(Js^2 + 2k)(Js^2 + 2k) - k^2 = 0 \qquad (12.13)$$

[A more systematic way to obtain the characteristic equation is to form the determinant of Eqs. (12.10), which procedure will be followed henceforth.] When Eq. (12.13) is expanded, and like terms combined, we find

$$s^4 + 4\frac{k}{J}s^2 + 3\left(\frac{k}{J}\right)^2 = 0 \qquad (12.14)$$

This equation is fourth-order in s; but it contains only even powers of s (because the system had no damping in it), hence it can be solved as a second-order equation in s^2 by the usual quadratic formula, with the result

$$s^2 = -2\frac{k}{J} \pm \sqrt{4\left(\frac{k}{J}\right)^2 - 3\left(\frac{k}{J}\right)^2} = (-2 \pm 1)\frac{k}{J}$$

$$\boxed{s = \pm j\sqrt{\frac{k}{J}} \qquad s = \pm j\sqrt{3}\sqrt{\frac{k}{J}}} \qquad (12.15)$$

These are, of course, the eigenvalues, the values of s which make Eqs. (12.9) correct solutions of (12.8). That is,

either $\theta_1 = \Theta_1 \exp\left(\pm j\sqrt{\frac{k}{J}}\,t\right)$

$\theta_2 = \Theta_2 \exp\left(\pm j\sqrt{\frac{k}{J}}\,t\right)$ First natural frequency

or $\theta_1 = \Theta_1 \exp\left(\pm j\sqrt{\frac{3k}{J}}\,t\right)$ (12.16)

$\theta_2 = \Theta_2 \exp\left(\pm j\sqrt{\frac{3k}{J}}\,t\right)$ Second natural frequency

The solutions represent natural motions which are undamped vibrations, either at frequency $\omega = \sqrt{k/J}$ or at frequency $\omega = \sqrt{3k/J}$.

To see what natural motions correspond to each of these natural frequencies, we substitute the above values of s into either of Eqs. (12.12); e.g.,

$$\frac{\Theta_2}{\Theta_1} = \frac{s^2}{k/J} + 2 \qquad (12.17)$$

with the result

$$\frac{\Theta_2}{\Theta_1} = -1 + 2 = +1 \quad \text{if } s = \pm j\sqrt{\frac{k}{J}} \quad \text{First mode}$$

or $\quad \dfrac{\Theta_2}{\Theta_1} = -3 + 2 = -1 \quad \text{if } s = \pm j\sqrt{\dfrac{3k}{J}} \quad \text{Second mode} \qquad (12.18)$

These results are summarized as follows:

$$\theta_1 = \Theta_1 \cos \omega t \quad \text{and} \quad \theta_2 = \Theta_2 \cos \omega t$$

with either $\quad \omega = \sqrt{\dfrac{k}{J}} \quad$ and $\quad \dfrac{\Theta_2}{\Theta_1} = +1 \quad$ First mode

or $\quad \omega = \sqrt{\dfrac{3k}{J}} \quad$ and $\quad \dfrac{\Theta_2}{\Theta_1} = -1 \quad$ Second mode $\qquad (12.19)$

The physical meaning of the above result is as follows (refer to Fig. 12.4):

(1) If the two disks are given identical initial displacements—e.g., 0.2 radian counterclockwise (Fig. 12.4a)—and released from that configuration, then they will vibrate back and forth together indefinitely at frequency $\sqrt{k/J}$, and thus there will never be any twist in the coupling shaft between them. Physically, this is an obvious result: since there is never any twist in the central shaft, it has no influence on the motion and each element does just what it

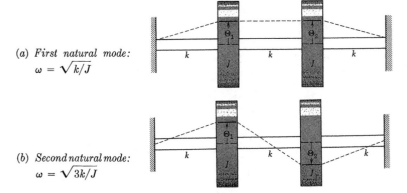

(a) *First natural mode:* $\omega = \sqrt{k/J}$

(b) *Second natural mode:* $\omega = \sqrt{3k/J}$

Fig. 12.4 *The two natural modes of vibration of the simple two-mass system of Fig. 12.3*

would have done if there were no coupling shaft (i.e., vibrates at frequency $\sqrt{k/J}$).

(2) If the two disks are rotated by equal but *opposite* initial angles (e.g., disk 1 by 0.2 radian counterclockwise, and disk 2 by 0.2 radian clockwise, as in Fig. 12.4b), and released, then the two disks will vibrate indefinitely, always at the same magnitude, but exactly 180° out of phase and at the higher frequency $\sqrt{3k/J}$. In this case the midpoint of the coupling shaft never rotates, and may be considered fixed. Then the frequency of vibration is quickly verified, because each disk is in effect vibrating against two springs, its own and half the coupling shaft. Its own shaft has stiffness k, and half the coupling shaft has stiffness $2k$, so the frequency is $\omega = \sqrt{3k/J}$.

The two modes of vibration of Fig. 12.4a and Fig. 12.4b are known as *natural modes* of motion of the system. Natural modes are of great importance in studying the vibration of systems because, as illustrated in the foregoing discussion, *vibration in a natural mode is like vibration of a single-degree-of-freedom system:* If an undamped system is once set in motion precisely in one of its natural modes, then it will oscillate indefinitely in that mode, at the corresponding natural frequency.

We might ask next what would happen if we should start a vibration of the system with initial displacements of the disks other than those specified in Fig. 12.4. What if, for example, we commence a vibration by displacing the first disk through an angle of 0.3 radian and the second disk through 0.1 radian in the same direction and then release the disks?

The answer is that if the system is released (as postulated) from a configuration other than a natural mode the resulting motion will consist of some combination of the two natural modes which will not be sinusoidal, and may not even be periodic (Fig. 12.5b). We can still study such motions in terms of natural modes, however, by making use of superposition (which is so vital in studying linear systems in general). That is, we look upon the specified initial configuration as consisting of the sum of a certain amount of each of the natural-mode configurations. This is shown graphically in Fig. 12.5a for the case postulated: $\theta_1(0) = 0.3$, $\theta_2(0) = 0.1$. By simple algebra, the amounts of each natural mode required to make up the prescribed total are 0.2 radian of mode 1 and 0.1 radian of mode 2. Thus the resulting motion will consist of a vibration at frequency $\sqrt{k/J}$ of magnitude 0.2 radian *superimposed* upon a vibration at frequency $\sqrt{3k/J}$ of magnitude 0.1 radian. The total motion is shown in Fig. 12.5b. Were we not privy to its simple natural-mode components, this motion would appear to be enigmatic indeed.

The above analysis of the simple system of Fig. 12.3 has demonstrated three fundamental concepts which play a key role in the analysis of more general, more complicated systems. They are as follows:

(1) A compound system will have natural motions whose characteristics or eigenvalues (*natural frequencies* in this case) are, in general, different from (but related to) the characteristics of the individual subsystems.

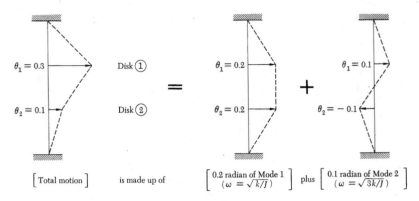

(a) *Superposition of natural modes*

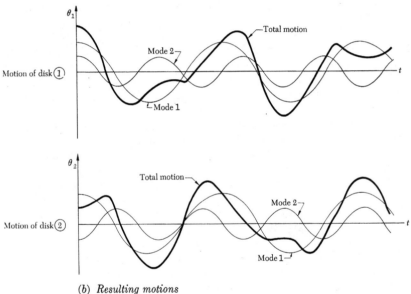

(b) *Resulting motions*

Fig. 12.5 *The natural-mode composition of arbitrary natural vibrations*

As in all linear systems, the characteristics of natural motion are obtained from the characteristic equation.

(2) Compound systems have *natural modes* of motion, in each of which each subsystem moves in definite relation to the others. In the absence of damping, as in the present example, the subsystems vibrate at a common frequency, in or out of phase with one another.

(3) Any given *arbitrary* natural motions of a system can be thought of as the *superposition* of specific amounts of its natural modes of motion.

12.4 THE EFFECT OF COUPLING STRENGTH

We now make the system of Fig. 12.3 somewhat more general, as shown in Fig. 12.6, by assuming inertias J_1 and J_2 and stiffnesses k_1 and k_2 to have different values, and by considering the stiffness k_3 of the coupling shaft as a parameter we wish to study. We are interested in the natural motions when coupling is extremely weak and when it is strong. (The results of our analysis in Sec. 12.3 will serve as a special case in the present study.)

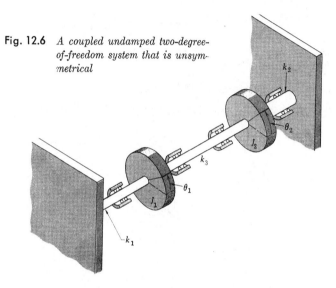

Fig. 12.6 *A coupled undamped two-degree-of-freedom system that is unsymmetrical*

By physical reasoning we can make some qualitative deductions about how stiffness k_3 will affect the natural motions of the system. For example, if the coupling shaft were absent ($k_3 = 0$) then each of the masses would vibrate, independently of the other, at its own natural frequency: J_1 at frequency $\omega_1 = \sqrt{k_1/J_1}$; J_2 at $\omega_2 = \sqrt{k_2/J_2}$. If, on the other hand, the coupling shaft were rigid ($k_3 \to \infty$), then the two masses would vibrate as one, at the frequency

$$\omega_c = \sqrt{\frac{k_1 + k_2}{J_1 + J_2}} \tag{12.20}$$

The problem at hand is to determine at what frequency the masses vibrate and with what relation to each other, when k_3 has some arbitrary intermediate value.

The equations of motion for the system in Fig. 12.6 are

$$\left. \begin{array}{ll} J_1 \ddot{\theta}_1 + k_1 \theta_1 + k_3(\theta_1 - \theta_2) = 0 & \text{for the first disk} \\ \text{and} \quad J_2 \ddot{\theta}_2 + k_2 \theta_2 + k_3(\theta_2 - \theta_1) = 0 & \text{for the second disk} \end{array} \right\} \tag{12.21}$$

or if we assume, as usual, that the motions are of the form $\theta_1 = \Theta_1 e^{st}$ and $\theta_2 = \Theta_2 e^{st}$, Eqs. (12.21) become

$$[J_1 s^2 + (k_1 + k_3)]\Theta_1 - k_3\Theta_2 = 0 \\ -k_3\Theta_1 + [J_2 s^2 + (k_2 + k_3)]\Theta_2 = 0 \quad (12.22)$$

Equations (12.22) are two simultaneous equations in three unknowns, s, Θ_1, and Θ_2. As in Sec. 12.3, the first step is to reduce the number of unknowns by dividing all terms by Θ_1:

$$[J_1 s^2 + (k_1 + k_3)] \quad\quad - k_3 \frac{\Theta_2}{\Theta_1} \quad\quad = 0 \\ -k_3 \quad\quad + [J_2 s^2 + (k_2 + k_3)]\frac{\Theta_2}{\Theta_1} = 0 \quad (12.23)$$

after which the pair of equations can be solved, first for *eigenvalues* s, and then for *mode shapes* Θ_2/Θ_1.

Natural frequencies from the characteristic equation

The systematic way to solve (12.22) or (12.23) for s is to set the determinant equal to 0:

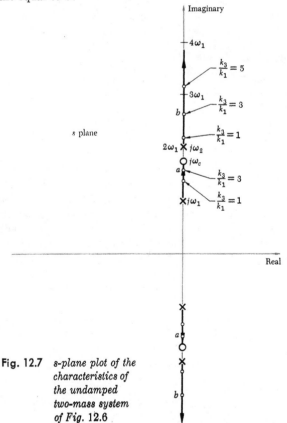

Fig. 12.7 *s-plane plot of the characteristics of the undamped two-mass system of Fig. 12.6*

[Sec. 12.4] THE EFFECT OF COUPLING STRENGTH

$$\begin{vmatrix} (J_1 s^2 + k_1 + k_3) & -k_3 \\ -k_3 & (J_2 s^2 + k_2 + k_3) \end{vmatrix} = 0 \quad (12.24)$$

The resulting *characteristic equation* is then (dividing by $J_1 J_2$)

$$\left[s^2 + \left(\frac{k_1 + k_3}{J_1}\right)\right]\left[s^2 + \left(\frac{k_2 + k_3}{J_2}\right)\right] - \frac{k_3^2}{J_1 J_2} = 0 \quad (12.25)$$

[The reader may verify that the same result is obtained by substitution from one of Eqs. (12.23) into the other.]

As a quick check, let k_3 be zero (no coupling) in (12.25). Then the roots are $s = \pm j\sqrt{k_1/J_1}$ and $s = \pm j\sqrt{k_2/J_2}$, indicating the expected undamped oscillations of each of the individual elements.

The reader is invited to show that if $k_3 \to \infty$ (rigid connection between J_1 and J_2), then (12.25) becomes $s^2 + [(k_1 + k_2)/(J_1 + J_2)] = 0$, indicating a single-degree-of-freedom vibration of the combined mass on the combined spring, at frequency (12.20).

For an arbitrary k_3, Eq. (12.25) is fourth-order in s; but it contains only even powers of s (because the system had no damping in it), and thus it can be solved as a second-order equation in s^2 by the usual quadratic formula, with the result

$$s^2 = -\frac{1}{2}\left[\frac{k_1 + k_3}{J_1} + \frac{k_2 + k_3}{J_2}\right]$$
$$\pm \sqrt{\frac{1}{4}\left[\frac{k_1 + k_3}{J_1} + \frac{k_2 + k_3}{J_2}\right]^2 - \frac{k_1 k_2 + k_1 k_3 + k_2 k_3}{J_1 J_2}} \quad (12.26)$$

[Verify that when $k_1 = k_2 = k_3$ and $J_1 = J_2$, Eq. (12.26) reduces to (12.15).]

Equation (12.26) is quite cumbersome, and it is hard to draw general conclusions from it algebraically. It is more convenient to study the implications of (12.26) as a locus of roots on the s plane, versus parameter k_3.†

We begin by putting (12.25) in the proper form, extracting coupling parameter k_3:

$$(s^2 + \omega_1^2)(s^2 + \omega_2^2) = -k_3 \left(\frac{J_1 + J_2}{J_1 J_2}\right)\left(s^2 + \frac{k_1 + k_2}{J_1 + J_2}\right) \quad (12.27)$$

$$\omega_1 = \sqrt{\frac{k_1}{J_1}} \qquad \omega_2 = \sqrt{\frac{k_2}{J_2}} \qquad \omega_c = \omega_{\text{combined}} = \sqrt{\frac{k_1 + k_2}{J_1 + J_2}}$$

When $k_3 = 0$, the roots of this characteristic equation are $s = \pm j\omega_1$ and $s = \pm j\omega_2$. These are plotted as X's in the s plane in Fig. 12.7. For very high values of k_3, only the right-hand side of (12.27) is significant, and the roots are simply $s = \pm j\sqrt{(k_1 + k_2)/(J_1 + J_2)}$, which are marked O in Fig. 12.7. As k_3 is varied, the locus of the roots of the characteristic equation is as shown: for small values of k_3, the roots are near the original

† The root-locus method—a convenience for the present no-damping investigation—will be found almost a necessity later when damping is present.

ones (×'s), but as k_3 is increased, they depart along the imaginary axis (because no damping is present), the higher-frequency roots becoming still higher, and the lower-frequency roots approaching the ○'s until finally, when k_3 is infinite (completely rigid), the lower root becomes exactly $s = \pm j\sqrt{(k_1 + k_2)/(J_1 + J_2)}$ (combined mass on combined spring), and the upper root becomes infinite: $s = \pm j\infty$.

Figure 12.7 is drawn for the particular case $J_2 = 2J_1$, $k_2 = 8k_1$, hence $\omega_2 = 2\omega_1$, and $\sqrt{(k_1 + k_2)/(J_1 + J_2)} = \sqrt{3}\omega_1$. [As a practical matter, the quick way to solve (12.27) is algebraically, taking advantage of the special fact that the roots will always be in the form $s^2 = -\omega^2$, because there is no damping. Then (12.27) becomes

$$-\frac{k_3}{k_1} = \frac{(\omega_1^2 - \omega^2)(\omega_2^2 - \omega^2)}{\omega_1^2(\omega_c^2 - \omega^2)} \left(\frac{J_2}{J_1 + J_2}\right)$$

which can be solved quickly for k_3/k_1 for each ω of interest. This algebraic simplification is not available when damping is present.]

Some values of parameter k_3/k_1 are marked in Fig. 12.7, corresponding to specific natural frequencies. These may be checked by the reader.

Natural modes of vibration

Having found the natural frequencies of the system from its characteristic equation (12.25), we now want to return to Eqs. (12.23) and find out about the relative-magnitude relation Θ_2/Θ_1. This can be done by direct substitution of results from Fig. 12.7 into (12.23). That is, with $s^2 = -\omega^2$, the first† of Eqs. (12.23) becomes

$$\frac{\Theta_2}{\Theta_1} = \frac{-\omega^2 + \omega_1^2}{k_3/J_1} + 1 \qquad (12.28)$$

where ω is to be obtained from Fig. 12.7 for any given value of k_3/K_1. For example, if—for the system of Fig. 12.7—$k_3/k_1 = 3$, then (from the figure) ω has the values 1.58 for the *first natural mode* of vibration, and 2.64 for the *second*. Moreover, k_3/J_1 has the value $k_3/J_1 = (k_3/k_1)(k_1/J_1) = 3\omega_1^2$. Then (12.28) gives the results

$$\frac{\Theta_2}{\Theta_1} = \frac{-1.58^2 + 1}{3} + 1 = +0.5 \qquad \text{for the first mode}$$

and $\qquad \dfrac{\Theta_2}{\Theta_1} = \dfrac{-2.64^2 + 1}{3} + 1 = -1.0 \qquad$ for the second mode

To summarize, if the system of Fig. 12.7 has the physical properties

$$J_2 = 2J_1 \qquad k_2 = 8k_1 \qquad k_3 = 3k_1$$

and if the disks are released in either of the configurations shown in Fig. 12.8, then the system will vibrate indefinitely in that configuration (mode), with the frequency indicated in Fig. 12.8.

† Either of Eqs. (12.23) may be used, as the reader may verify.

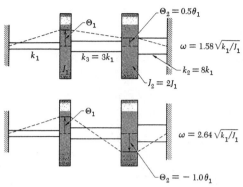

Fig. 12.8 *Natural modes of motion of the system of Fig.* 12.6

These results are shown also as one entry in the summary plot of Fig. 12.9.

Similarly, if $k_3/k_1 = 5$, the natural frequencies are (from Fig. 12.7)

$$\omega = 1.65\sqrt{\frac{k_1}{J_1}} \quad \text{for mode 1}$$

$$\omega = 3.15\sqrt{\frac{k_1}{J_1}} \quad \text{for mode 2}$$

and the corresponding mode shapes (entered also in Fig. 12.9) are

$$\frac{\Theta_2}{\Theta_1} = +0.66 \quad \text{for mode 1}$$

$$\frac{\Theta_2}{\Theta_1} = -0.77 \quad \text{for mode 2}$$

When the coupling shaft is extremely stiff, $k_3/k_1 \to \infty$, the first natural mode involves no twist of the coupling shaft, and thus

$$\Theta_2 = \Theta_1 \qquad \omega = \sqrt{\frac{k_1 + k_2}{J_1 + J_2}} = \omega_c$$

The second mode *does* involve twist of the coupling shaft, which renders stiffness k_1 and k_2 inconsequential, and involves J_1 and J_2 oscillating on shaft k_3 at extremely high (infinite, in the limit) frequency.† Since J_2 is twice as large as J_1, it moves only half as far: $\Theta_2/\Theta_1 = -0.5$. Again, the result is entered in Fig. 12.9.

From the above calculations, the variation in natural frequencies and mode shapes with coupling strength may be summarized as shown in Fig. 12.9.

† The frequency is $\omega = \sqrt{k_3\left(\dfrac{1}{J_1} + \dfrac{1}{J_2}\right)}$ which is infinite if k_3 is infinite. Refer to Problem 12.12.

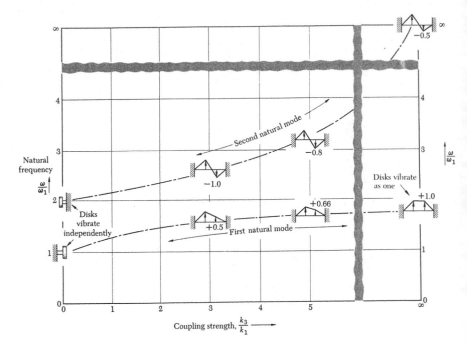

Fig. 12.9 *Natural modes of motion, versus coupling stiffness, of the undamped two-mass system of Fig.* 12.6 ($J_2 = 2J_1$, $k_2 = 8k_1$).

Special case of equal masses and equal springs

The system studied in this section (that shown in Fig. 12.6) has been assumed for generality to have unequal springs and unequal masses. If, as a special case, we let $k_2 = k_1$ and $J_2 = J_1$, then the variation, with k_3, of the natural frequencies of the system will be as shown in Fig. 12.10, which is a special case of Fig. 12.7.

When $k_3 = 0$ the two disks vibrate independently but at the same natural frequency, $\omega_1 = \omega_2 = \sqrt{k_1/J_1}$. This is indicated by the *pair* of ×'s at $s = j\omega_1$ (and another pair, of course, at $s = -j\omega_1$).

When $k_3 \to \infty$, one natural frequency is still at $s = j\omega_1$ (because the combined natural frequency is

$$\omega_c = \sqrt{(k_1 + k_2)/(J_1 + J_2)} = \sqrt{2k_1/2J_1} = \omega_1)$$

This is indicated by the ○ at $s = j\omega_1$. The other natural frequency, when $k_3 \to \infty$, is infinite. The two disks are vibrating out of phase, on their common, infinitely stiff shaft k_3.

For other values of k_3/k_1 the natural frequencies are as indicated: one is always at $s = \pm j\omega_1$ and the other is at a higher frequency which increases with k_3. The corresponding *natural modes* will, by symmetry, *always*

[Sec. 12.5] BEAT GENERATION

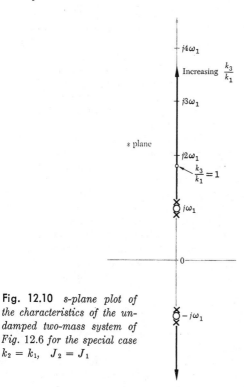

Fig. 12.10 *s-plane plot of the characteristics of the undamped two-mass system of Fig. 12.6 for the special case* $k_2 = k_1$, $J_2 = J_1$

involve *equal magnitudes* of displacement of the two disks: at the lower natural frequency the displacement will be in phase, while at the higher natural frequency they will be 180° out of phase, as depicted in Fig. 12.4.

In particular, when $k_3 = k_1$ (i.e., $k_3/J_1 = \omega_1^2$) we have the system of Fig. 12.3, whose higher natural frequency is $\sqrt{3k_1/J_1} = \sqrt{3}\,\omega_1$, as we found in Sec. 12.3, so that the natural frequencies, as well as the modes of motion, are as depicted in Fig. 12.4.

Before we proceed to include the generality of damping in our simple system, we shall discuss three other topics relating to undamped vibration. They are the phenomenon of *beat vibrations*, the occurrence of *inertial coupling*, and the concept of *normal coordinates*. These we discuss in Secs. 12.5, 12.6, and 12.7.

Problems 12.9 through 12.18

○12.5 BEAT GENERATION

An interesting special case of vibration of a two-mass system occurs when two subsystems having separately the *same natural frequency* are then *coupled very weakly* together. In this case the two natural modes of motion have almost exactly the same frequency, as indicated in Fig. 12.10 when $k_3/k_1 \ll 1$.

In this case the system can still be made to vibrate in either of its natural modes by releasing the masses with *exactly* the right mode shape. If the system is started in other than a natural mode shape, however, then the ensuing motions take on the interesting character shown in Fig. 12.11. If, for example, the system is released with one mass displaced and the other at zero (as shown), then the appropriate summation of the two modes (shown in detail in Fig. 12.11) results in a total motion of mass 1 which is nearly sinusoidal, at frequency $\omega \approx \omega_1 \approx \omega_2$, but which diminishes in amplitude as time goes on until, at one particular time, the motion of mass 1 is, instantaneously, zero.

Mass 2, meanwhile, has started with zero displacement, but its motion builds up steadily until it takes over the motion originally possessed by mass 1. Still later the motion will be seen to pass back to mass 1 again, and so on.

The name *beats* is given to the purely geometrical phenomenon of adding two sine waves of nearly equal frequency and obtaining a sum which is an amplitude-modulated sine wave (of their average frequency), being first strong and then weak as the sine waves first enforce and then cancel each other. The frequency of modulation is known as the "beat frequency," ω_B. The mathematical expression for the total motion is thus a product of sinusoids, e.g.,

$$\theta = \theta_0 \cos \frac{\omega_B}{2} t \cos \frac{\omega_1 + \omega_2}{2} t$$

That is, if $2\pi/\omega_B$ is the time between maxima in Fig. 12.11, then $4\pi/\omega_B$ is the period of the amplitude-modulation envelope.

In general, beats are usually *not* associated with coupling. They occur commonly, for example, in similar machines that are vibrating independently at nearly the same frequency. Passengers in twin-engine airplanes often hear the propeller noise grow loud, then soft, then loud again at a definite frequency (perhaps 2 or 3 cps), as the (high-frequency) propeller-generated sound waves first reinforce and then cancel one another. The same phenomenon occurs often in communications and is the basis for certain circuit designs.

The motion of Fig. 12.11 is a special occurrence of beats produced by coupling: The two natural frequencies of the coupled system are so close together that they add to produce beats.

A phasor (rotating vector) diagram accompanying the motion of mass 2 in Fig. 12.11 shows, again, the geometry of the motion. The mode-1 phasor rotates at frequency $\omega^{(1)}$, and the mode-2 phasor rotates at slightly higher frequency $\omega^{(2)}$; and thus it catches up with the mode-1 phasor periodically. The beat frequency (frequency of nulls or of maxima) is the difference between the two natural frequencies:

$$\omega_B = \omega^{(2)} - \omega^{(1)} \tag{12.29}$$

as can be seen from the phasor sequence or analytically from suitable trigonometric identities (Problem 12.19).

The transfer of motion from one mass to the other, depicted in Fig. 12.11, presents a rather dramatic demonstration of energy conservation. The kinetic energy initially possessed by mass 1 is transferred, a little each cycle, through the weak coupling spring to mass 2, and then back to mass 1 again.

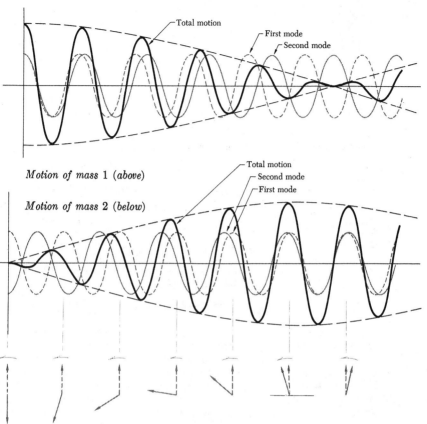

Fig. 12.11 *Beat vibrations in a coupled system*

Several other striking experiments are based on this phenomenon. Consider, for example, the two pendulums in Fig. 12.12a, which are coupled together by a very weak spring. The two natural modes of motion of this system—by symmetry and direct analogy with the results of Fig. 12.4—are (1) a mode in which the two pendulums swing back and forth parallel to one another, so that the spring is never stretched and the natural frequency is the same as for either pendulum alone; and (2) a mode in which the displacements of the pendulums are always exactly equal and opposite, so that the center of the coupling spring remains fixed, and the frequency is slightly higher than for mode (1) because of the stiffness of the spring.

To make the experiment, we keep one of the pendulums hanging straight down but displace the other through a small angle, and then release the system. Initially, the first pendulum swings back and forth in a motion which is almost sinusoidal, and the second hardly moves at all. Eventually, however, the swing of the second pendulum begins to build up, while that of the first dies down until, after quite a long time, $T = \pi/\omega_B$, the first mass is seen to be almost

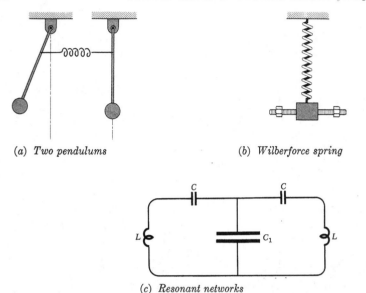

(a) *Two pendulums* (b) *Wilberforce spring*

(c) *Resonant networks*

Fig. 12.12 *Experiments with beat vibrations in weakly coupled systems*

completely stopped, while the second is now swinging with the amplitude originally held by the first. After another period π/ω_B, the motion has shifted back again to the first mass, and so on, as the kinetic energy of the system's vibration continues to shift back and forth from one pendulum to the other, a little each cycle, through the weak spring. The motion has the same character as that plotted in Fig. 12.11.

The device in Fig. 12.12b is known as a "Wilberforce spring." In this case the two natural modes of motion consist (almost) of a purely rotational motion of the mass about a vertical axis, and a purely translational vibration up and down. When the periods of these two motions are almost equal, the tiny amount of coupling, due to torsional stress in a linearly extended spring (and, conversely, to linear stress in a twisted spring) causes the kinetic energy in the system to be passed slowly back and forth between the purely rotational and purely vertical vibrations, in an amusing demonstration. (The adjustable nuts on the mass are for the purpose of tuning the system.)

Fig. 12.12c is an electrical version of Fig. 12.12a. Two resonant LC circuits are connected by the weak coupling of a large common capacitor, so that a current initiated in one loop will, after a time, be transferred entirely to the other loop, and then back again.

The discussion of beats to this point has been entirely descriptive. We now give it analytical substance with an example.

Example 12.1. Torsional beats. Find the time response of the system of Fig. 12.6 if $k_1 = k_2 = k$, $J_1 = J_2 = J$, and $k_3 = \epsilon^2 k$. Let $\omega_0 \triangleq \sqrt{k/J}$ and $\omega_3 \triangleq \sqrt{k_3/J}$, so that $\omega_3{}^2 = \epsilon^2 \omega_0{}^2$. Let $\theta_1(0) = 0.1$ rad., $\dot{\theta}_1(0) = 0$, $\theta_2(0) = 0$, $\dot{\theta}_2(0) = 0$.

[Sec. 12.5] BEAT GENERATION

The natural frequencies are given by (12.26). For the above special case they are

$$s^2 = -\tfrac{1}{2}(2\omega_0^2 + 2\omega_3^2) \pm \sqrt{\tfrac{1}{4}(2\omega_0^2 + 2\omega_3^2)^2 - (\omega_0^4 + 2\omega_0^2\omega_3^2)}$$
$$= -\omega_0^2[(1 + \epsilon^2) \pm \sqrt{\epsilon^4}]$$
$$s = \pm j\omega_0 \quad \text{and} \quad s = \pm j\omega_0 \sqrt{1 + 2\epsilon^2}$$

or $\quad \omega^{(1)} = \omega_0 \qquad \omega^{(2)} = \omega_0 \sqrt{1 + 2\epsilon^2}$

Suppose, for example, that $\epsilon^2 = 0.02$ (the coupling spring is 2% as strong as the others). Then the natural frequencies will be ω_0 and $1.02\omega_0$.

The natural modes of motion are, from (12.28),

$$\left(\frac{\Theta_2}{\Theta_1}\right)^{(1)} = +1 \qquad \left(\frac{\Theta_2}{\Theta_1}\right)^{(2)} = -1$$

which is also clear from symmetry (and independent of ϵ).

To summarize, for $\epsilon^2 = 0.02$, the natural motions of the disks in the first mode are

$$\theta_1 = \theta_{1\,\text{max}} \cos \omega_0 t \qquad \theta_2 = \theta_{1\,\text{max}} \cos \omega_0 t$$

and in the second natural mode they are

$$\theta_1 = \theta_{1\,\text{max}} \cos 1.02\omega_0 t \qquad \theta_2 = -\theta_{1\,\text{max}} \cos 1.02\omega_0 t$$

Now if we release the system in a configuration in which mass 1 is rotated by 0.1 radian, say, and mass 2 is not rotated, then the motion will consist of the *superposition* of the two natural modes in the following proportion:

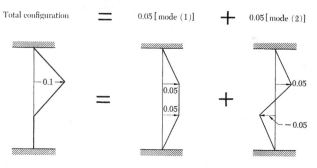

or $\quad \begin{cases} \theta_1 = 0.05 \cos \omega_0 t + 0.05 \cos 1.02\omega_0 t \\ \theta_2 = 0.05 \cos \omega_0 t - 0.05 \cos 1.02\omega_0 t \end{cases}$

The pattern of motion will be just as shown in Fig. 12.11, except that the energy will be transferred from mass 1 to mass 2 in

$$\tfrac{1}{2} T_B \times f_0 = \tfrac{1}{2} \frac{\omega_0}{\omega_B} = \tfrac{1}{2} \frac{\omega_0}{1.02\omega_0 - \omega_0} = \frac{1}{2(0.02)} = 25 \text{ cycles}$$

instead of in the five cycles shown in Fig. 12.11.

Problems 12.19 through 12.24

12.6 INERTIAL COUPLING

Some of the most interesting problems in dynamic coupling come about because a mass can move in more than one dimension. That is, a rigid body has six degrees of freedom—three in translation and three in rotation—and motion in one dimension can become coupled, through the inertias of the body, with motion in another dimension.

Inertial-coupling problems are especially intriguing when they involve motions in three dimensions, as with maneuvering aircraft, spinning tops and gyroscopes, or satellites (including a natural one, the moon) rocking about their mass centers because of the gradient in the earth's gravity field. However, the principle of inertial coupling can be well demonstrated in two dimensions (plane motion), as in the following example.

Consider the system in Fig. 12.13, which is a highly idealized model of an automobile on its suspension system and tires—the springs in Fig. 12.13 representing both. (A more realistic model would consider the mass of the tires and, of course, the damping of the shock absorbers.) Assume that there is only vertical motion, so that only two degrees of freedom need be considered: e.g., the extensions x_1 and x_2 of the two springs. (Other pairs of independent coordinates can, of course, be used to express the motions of

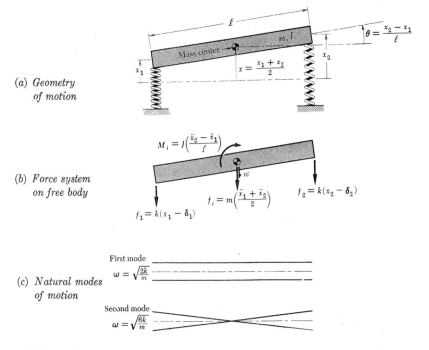

Fig. 12.13 *Analysis of inertial coupling (simplified model of an automobile)*

[Sec. 12.6] INERTIAL COUPLING

the system. In fact a judicious choice of coordinates can greatly simplify the analysis, as we shall see.)

For practice, let us derive the equations of motion of the physical model shown. We consider only motion in the plane of the paper, and we assume small θ, so that $\sin \theta \approx \theta$. We follow Procedure A-m (p. 35).

(1) *Geometry.* The system is shown in an arbitrary configuration in Fig. 12.13a. Choose x_1 and x_2 as the independent coordinates. For convenience, measure these from static equilibrium (where the springs support the weight of the car). Then in the arbitrary configuration the springs are stretched by $\xi_1 = x_1 - \delta_1$ and $\xi_2 = x_2 - \delta_2$, where the δ's are the compression in the springs under the car weight. The mass center has acceleration $a = (\ddot{x}_1 + \ddot{x}_2)/2$ and the angular acceleration is $\ddot{\theta} = (\ddot{x}_2 - \ddot{x}_1)/\ell$.

(2) *Force equilibrium.* A free-body diagram of the body is sketched in Fig. 12.13b. By D'Alembert's principle we can write equilibrium relations by summing forces in any direction or by taking moments about any point (Sec. 2.2). For example, summing moments about the right end gives

$$\overset{\curvearrowleft}{\Sigma M_2}{}^* = 0 = f_1 \ell + (f_i + w)\frac{\ell}{2} - M_i$$

Summing moments about the left end gives

$$\overset{\curvearrowright}{\Sigma M_1}{}^* = 0 = (f_i + w)\frac{\ell}{2} + f_2 \ell + M_i$$

(3) *Physical relations.* All of these—the force-displacement and force-acceleration relations—are noted in Fig. 12.13b. That is, $f_1 = k\xi_1$, $f_2 = k\xi_2$, $f_i = ma$, $M_i = J\ddot{\theta}$, the directions all being positive as drawn.

Combining steps by substituting (1) and (3) into (2) and rearranging, we obtain the following two equations in the unknowns x_1 and x_2:

$$\begin{aligned}\left(J + \frac{m\ell^2}{4}\right)\ddot{x}_1 + k\ell^2 x_1 + \left(-J + \frac{m\ell^2}{4}\right)\ddot{x}_2 &= 0 \\ \left(-J + \frac{m\ell^2}{4}\right)\ddot{x}_1 + \left(J + \frac{m\ell^2}{4}\right)\ddot{x}_2 + k\ell^2 x_2 &= 0\end{aligned} \quad (12.30)$$

(We have canceled the terms $k\delta_1 = k\delta_2 = w/2$.) Making the substitutions $x_1 = X_1 e^{st}$ and $x_2 = X_2 e^{st}$ results in the pair of algebraic equations

$$\begin{aligned}\left[\left(J + \frac{m\ell^2}{4}\right)s^2 + k\ell^2\right]X_1 + \left(-J + \frac{m\ell^2}{4}\right)s^2 X_2 &= 0 \\ \left(-J + \frac{m\ell^2}{4}\right)s^2 X_1 + \left[\left(J + \frac{m\ell^2}{4}\right)s^2 + k\ell^2\right]X_2 &= 0\end{aligned} \quad (12.31)$$

in which the coupling coefficients $[-J + (m\ell^2/4)]$, which couple the "x_1 equation" and the "x_2 equation," involve only inertia terms.†

† It should be noted, however, that other choices of coordinates could cause the coupling coefficients to involve spring constants, or even to vanish!

As usual, these equations are to be solved first for the natural frequencies of motion, and then for mode shapes X_2/X_1.

To permit specific calculation, suppose the value of J is $J = m\ell^2/12$. Then Eqs. (12.31) may be written

$$\left(s^2 + \frac{3k}{m}\right) X_1 + \left(\frac{s^2}{2}\right) X_2 = 0$$
$$\left(\frac{s^2}{2}\right) X_1 + \left(s^2 + \frac{3k}{m}\right) X_2 = 0 \qquad (12.32)$$

from which the characteristic equation is

$$\left(s^2 + \frac{3k}{m}\right)^2 - \left(\frac{s^2}{2}\right)^2 = 0$$

or
$$\frac{3}{4} s^4 + \frac{6k}{m} s^2 + \left(\frac{3k}{m}\right)^2 = 0 \qquad (12.33)$$

the roots of which are the eigenvalues

$$s = \pm j\sqrt{\frac{2k}{m}} \qquad s = \pm j\sqrt{\frac{6k}{m}} \qquad (12.34)$$

indicating undamped vibrations at those frequencies.

The eigenvalues are returned to either of Eqs. (12.32) to calculate the two mode shapes. From the first of Eqs. (12.32), the values of X_2/X_1 are

$$\frac{X_2}{X_1} = -\frac{\left(s^2 + \frac{3k}{m}\right)}{\frac{s^2}{2}} = \begin{cases} +1 & \text{for } \omega = \sqrt{\frac{2k}{m}} \\ -1 & \text{for } \omega = \sqrt{\frac{6k}{m}} \end{cases} \qquad (12.35)$$

These are plotted in Fig. 12.13c. Note that for the higher natural frequency, one point on the body, point P, remains fixed. This point is known as a node. Nodes are especially important for systems with many degrees of freedom, and particularly for distributed-mass systems such as a violin string. For the lower natural frequency the "node" is located at infinity: The mass moves parallel to itself. If the system were not symmetrical, both nodes would be located at finite points. (See Problem 12.28.)

Physically, the first mode of vibration in Fig. 12.13c consists simply, from (12.35), of the whole body moving up and down with no rotation. From symmetry this is what we should expect if we give the bar a purely translational vertical initial displacement: Since the equal springs have been stretched equal amounts, there will be no tendency for them to rotate the body. The natural frequency for this motion is simply that of mass m on two springs k: $\omega = \sqrt{2k/m}$, which checks the first of Eqs. (12.34).

The second mode of vibration, (12.35), is a purely rotational motion about the mass center, which we may instigate by giving the bar an initial angular displacement but no translation. Then, again by symmetry, the springs will have no tendency to induce translation motion. The frequency

[Sec. 12.7] NORMAL COORDINATES

of the resulting angular vibration is that of inertia $J = m\ell^2/12$ on two springs $k(\ell/2)^2$: $\omega = \sqrt{6k/m}$, which checks the second of Eqs. (12.34).

12.7 NORMAL COORDINATES

From the above results we may deduce an important general principle: If we had been foresighted enough to write the initial equations of motion in certain coordinates—x and θ in Fig. 12.13a— then we would have obtained an *uncoupled set of equations of motion*, one for the first natural mode, and another for the second natural mode. (That is, since x is the displacement of the node of the second mode, we deduce that second-mode motion will not involve coordinate x.) Let us prove this for the example at hand.

Using coordinates x and θ in Fig. 12.13a, and making the substitutions $x_1 = x - (\ell/2)\theta$ and $x_2 = x + (\ell/2)\theta$, we obtain from the free-body diagram of Fig. 12.13b the equations

$$\sum f^*_{\text{down}} = m\ddot{x} + k\left(x - \frac{\ell}{2}\theta\right) + k\left(x + \frac{\ell}{2}\theta\right) = 0$$

$$\sum{}^* \overset{\frown}{M}_{\text{c.m.}} = J\ddot{\theta} + \frac{\ell}{2}k\left(x + \frac{\ell}{2}\theta\right) - \frac{\ell}{2}k\left(x - \frac{\ell}{2}\theta\right) = 0$$

which, when simplified, yield

$$m\ddot{x} + 2kx = 0$$
$$J\ddot{\theta} + \frac{k\ell^2}{2}\theta = 0 \qquad (12.36)$$

The uncoupled Eqs. (12.36) are dynamically equivalent to the coupled set (12.32); but Eqs. (12.36) are, of course, much easier to work with. They are now just like two single-degree-of-freedom equations, yielding two separate characteristic equations:

$$s^2 + \frac{2k}{m} = 0$$
$$s^2 + \frac{k\ell^2}{2J} = 0 \qquad (12.37)$$

whose product is identical with (12.33). That is,

$$\left(s^2 + \frac{2k}{m}\right)\left(s^2 + \frac{k\ell^2}{2J}\right) = 0 \qquad (12.38)$$

is a factored version of (12.33) which leads instantly, of course, to roots (eigenvalues) (12.34) when $J = m\ell^2/12$.

The set of coordinates which makes possible the uncoupled form for the equation of motion, such as (12.36), is known as the set of *normal coordinates* for the system.

The concept of dealing with a higher-order system as a set of single-degree-of-freedom systems, by studying, one at a time, the natural modes of motion in normal coordinates, is of paramount importance in the study of complicated vibratory systems. It is particularly helpful in studying distributed-parameter systems, which have an infinite number of degrees of freedom, but which can usually be understood adequately by studying the first few natural modes of motion.

Further insight into the concept of finding normal coordinates in which natural modes of motion can be studied independently will be gained by working out Problem 12.28, which consists of an unsymmetrical version of the system in Fig. 12.13a. In that case two nodes are found, one for each natural mode of motion, and the proper normal coordinates are found to be the displacements of each of the nodes.

The normal coordinates for a system are not always single geometric quantities, as x and θ are in Fig. 12.13. More generally, *a normal coordinate is a measure of the amount of a natural mode that is present*. Thus, in Fig. 12.5 the normal coordinates would be defined as follows:

q_1 = the amount of mode 1 present
q_2 = the amount of mode 2 present

Then, from Fig. 12.5, we can write the relations

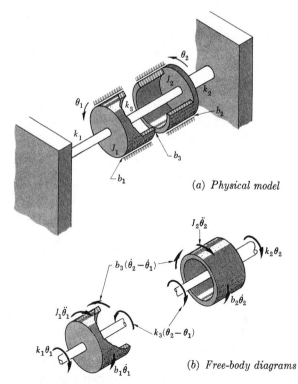

(a) *Physical model*

(b) *Free-body diagrams*

Fig. 12.14 *General two-degree-of-freedom system with coupling through both damping and spring*

[Sec. 12.8] GENERAL CASE: EIGENVALUES AND EIGENVECTORS

$$\begin{aligned}\theta_1 &= (1)q_1 + (1)q_2 \\ \theta_2 &= (1)q_1 + (-1)q_2\end{aligned} \quad (12.39)$$

These, in turn, can be solved simultaneously (e.g., using Cramer's rule†) for the inverse relations

$$\begin{aligned}q_1 &= (\tfrac{1}{2})\theta_1 + (\tfrac{1}{2})\theta_2 \\ q_2 &= (\tfrac{1}{2})\theta_1 + (-\tfrac{1}{2})\theta_2\end{aligned}$$

The above two sets of relations relate the normal coordinates q_1 and q_2 to the geometric coordinates θ_1 and θ_2. For example, in Fig. 12.5 we considered the case $q_1 = 0.2$, $q_2 = 0.1$, for which

$$\begin{aligned}\theta_1 &= (1)(0.2) + (1)(0.1) = 0.3 \\ \theta_2 &= (1)(0.2) + (-1)(0.1) = 0.1\end{aligned}$$

As we have noted, this concept is vital in the study of distributed systems, where complicated arbitrary motion can be represented adequately by superposing just a few of the "single-degree-of-freedom" motions, $q_1, q_2, \ldots$.

Problems 12.25 through 12.32

12.8 GENERAL CASE, WITH DAMPING: EIGENVALUES AND EIGENVECTORS

If damping is present in the general two-mass system of Fig. 12.6, then the analysis becomes considerably more difficult because the characteristic equation is found to contain all powers of s up to s^4, and the roots of a quartic are much more difficult to extract than the roots of a quadratic.

In general, damping may act on either of the masses individually, and in addition there may be damping between the masses in the system. The general situation is shown in Fig. 12.14a.

Denoting θ_1 and θ_2 as independent coordinates, we write the equations of motion from the free-body diagrams of the individual disks, Fig. 12.14b, on which the physical relations [Step (3)] have been superposed.

The equilibrium relations [Step (2)] are (taking moments clockwise)

$$\begin{aligned}\Sigma M_1^* &= J_1\ddot{\theta}_1 + b_1\dot{\theta}_1 + k_1\theta_1 - b_3(\dot{\theta}_2 - \dot{\theta}_1) - k_3(\theta_2 - \theta_1) = 0 \quad \text{for disk 1} \\ \Sigma M_2^* &= J_2\ddot{\theta}_2 + b_2\dot{\theta}_2 + k_2\theta_2 + b_3(\dot{\theta}_2 - \dot{\theta}_1) + k_3(\theta_2 - \theta_1) = 0 \quad \text{for disk 2}\end{aligned} \quad (12.40)$$

As usual, it is assumed that the natural motions of the system are of the form e^{st}:

$$\theta_1 = \Theta_1 e^{st} \qquad \theta_2 = \Theta_2 e^{st} \quad (12.41)$$

Then Eqs. (12.40) become

$$\begin{aligned}[J_1 s^2 + (b_1 + b_3)s + (k_1 + k_3)]\Theta_1 - (b_3 s + k_3)\Theta_2 &= 0 \\ -(b_3 s + k_3)\Theta_1 + [J_2 s^2 + (b_2 + b_3)s + (k_2 + k_3)]\Theta_2 &= 0\end{aligned} \quad (12.42)$$

† Cramer's rule is given in Appendix H.

Notice that coupling is produced only by the spring k_3 and the damping b_3 which act *between* the disks.

Characteristics (eigenvalues)

From the determinant of (12.42) we obtain the *characteristic equation* of the system:

$$[J_1 s^2 + (b_1 + b_3)s + (k_1 + k_3)][J_2 s^2 + (b_2 + b_3)s + (k_2 + k_3)]$$
$$- (b_3 s + k_3)^2 = 0 \qquad (12.43)$$

When multiplied out, this equation is fourth-order in s with all terms present.

There are a number of methods for extracting the roots of a quartic. For the problem at hand, the root-locus method (Secs. 11.8 and 11.9) seems to be the best-suited by a considerable margin, because it allows the characteristics to be seen graphically as functions of the parameters of interest.

We recall that in the root-locus method we do not multiply out all the terms of the characteristic equation but, rather, leave them grouped together by subsystems and then take advantage of the fact that the char-

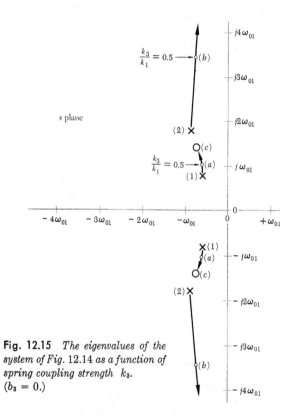

Fig. 12.15 *The eigenvalues of the system of Fig. 12.14 as a function of spring coupling strength k_3. ($b_3 = 0$.)*

[Sec. 12.8] GENERAL CASE: EIGENVALUES AND EIGENVECTORS

acteristics of the individual subsystems are already known. The first step, therefore, is to regroup Eq. (12.43) as follows:

$$\left[\left(s^2 + \frac{b_1}{J_1}s + \frac{k_1}{J_1}\right)\left(s^2 + \frac{b_2}{J_2}s + \frac{k_2}{J_2}\right)\right]$$
$$+ k_3\left(\frac{J_1 + J_2}{J_1 J_2}\right)\left[s^2 + \left(\frac{b_1 + b_2}{J_1 + J_2}\right)s + \left(\frac{k_1 + k_2}{J_1 + J_2}\right)\right]$$
$$+ b_3 s\left(\frac{J_1 + J_2}{J_1 J_2}\right)\left[s^2 + \left(\frac{b_1 + b_2}{J_1 + J_2}\right)s + \left(\frac{k_1 + k_2}{J_1 + J_2}\right)\right] = 0 \quad (12.44)$$

Notice that the characteristic equation is so arranged that it consists of the sum of three terms. The first term is the characteristic equation as it would exist if there were *no* coupling. (That is, it consists of the characteristics of the individual subsystems.) The second term involves all quantities which are added because of the spring coupling k_3, and the third term consists of the quantities which are added because of the damping term b_3.

Let us first investigate the characteristics in the absence of the joint damping; i.e., let $b_3 = 0$. This problem is the same as that studied in Sec. 12.4, except that now there is the added generality of damping on the individual elements.

The root-locus approach is ideal for this study because for the first step we need only let $b_3 = 0$ in Eq. (12.44), leaving

$$\left(s^2 + \frac{b_1}{J_1}s + \frac{k_1}{J_1}\right)\left(s^2 + \frac{b_2}{J_2}s + \frac{k_2}{J_2}\right)$$
$$+ k_3\left(\frac{J_1 + J_2}{J_1 J_2}\right)\left[s^2 + \left(\frac{b_1 + b_2}{J_1 + J_2}\right)s + \left(\frac{k_1 + k_2}{J_1 + J_2}\right)\right] = 0 \quad (12.45)$$

The roots of this equation are plotted as a function of k_3 in Fig. 12.15 for one set of physical constants.† When $k_3 = 0$, the roots are simply the characteristics of the individual elements, marked with crosses ×. When k_3 is infinitely stiff, one pair of characteristics has disappeared at infinity, leaving only one pair of roots for the combined system of two rigidly connected masses $J_1 + J_2$ on combination spring $k_1 + k_2$. This pair is denoted by circles ○ in Fig. 12.15. Roots corresponding to intermediate values of k_3 are as shown. Note the close similarity between Fig. 12.15 and Fig. 12.7, which represents the same system for the special case of no damping. Comparison of the two figures indicates that the presence of light damping has little affect on the natural frequencies of the coupled system.

Next, we select a given value of k_3 and proceed to investigate the effect of varying degrees of viscous coupling, parameter b_3. Suppose, for example, that we select the value of k_3 corresponding to the roots marked (*a*) and (*b*) in Fig. 12.15. For this value of k_3 we can rewrite Eq. (12.43) as follows:

$$(s + \sigma_a \pm j\omega_a)(s + \sigma_b \pm j\omega_b)$$
$$+ b_3 s\left(\frac{J_1 + J_2}{J_1 J_2}\right)\left[s^2 + \left(\frac{b_1 + b_2}{J_1 + J_2}\right)s + \left(\frac{k_1 + k_2}{J_1 + J_2}\right)\right] = 0 \quad (12.46)$$

† The values used in plotting Fig. 12.15 are $J_1 = J_2$, $k_2/k_1 = 4$, $b_1/J_1 = 1.2\omega_{01}$, and $b_2/J_2 = 1.7\omega_{01}$.

Compare (12.46) with (12.44) and note that what we have done is to factor the part of (12.44) which does not involve b_3, using the root-locus method. We have used the factors so obtained in (12.46), which is now in a form to which the root-locus method can be applied again, this time to study the change in characteristics as a function of b_3, as shown in Fig. 12.16.

When $b_3 = 0$, the roots of (12.46) are simply the roots (a) and (b) (from Fig. 12.15) which are marked with ×'s in Fig. 12.16. At the other extreme, when the interdisk damping is infinitely stiff, one of the roots disappears at infinity, and a pair of roots is found at the complex points marked ○, which correspond as before (see Fig. 12.15) to the single-degree-of-freedom system involving a rigidly connected combined mass $J_1 + J_2$ vibrating against the combined spring $k_1 + k_2$ and damped by $b_1 + b_2$.

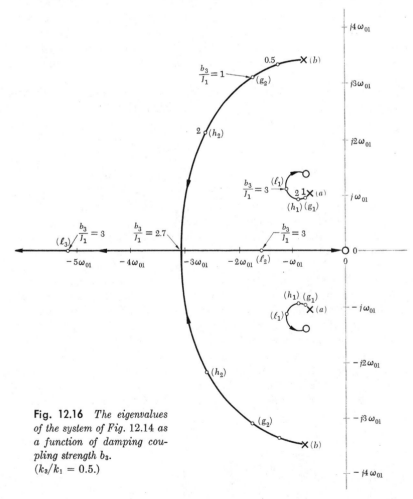

Fig. 12.16 *The eigenvalues of the system of Fig. 12.14 as a function of damping coupling strength b_3.* ($k_3/k_1 = 0.5$.)

[Sec. 12.8] GENERAL CASE: EIGENVALUES AND EIGENVECTORS

In addition, there is another real root at the origin of the s plane corresponding to a first-order motion having a very long time constant (infinitely long in the limit). The physical motion corresponding to this real root can be visualized by assuming that the two disks in Fig. 12.14 have been rotated against one another, so that shaft k_3 is twisted. Then the disks are released, and the shaft attempts to untwist them against the very high viscous resistance of b_3. The resulting motion, for finite b_3, is an exponential decay of the twist angle. In the limit when b_3 is infinite, the disks simply remain in their initial relative position, shaft k_3 remains twisted, and the only motion is the damped vibration of the combined masses on combined spring $k_1 + k_2$.

Intermediate values of b_3 may lead either to two complex pairs of roots—points (g) or (h) in Fig. 12.16—or, when b_3 becomes quite heavy, to one complex pair of roots plus two real roots, one fast and one slow—e.g., points (ℓ) in Fig. 12.16.

For specific values of k_3 (in Fig. 12.15) and b_3 (in Fig. 12.16) the roots, or *eigenvalues*, of the system's characteristic equation are the final set of points (values of s) obtained in Fig. 12.16 [e.g., points (g), (h), or (ℓ)].

Problems 12.33 through 12.40

Modes of motion: eigenvectors

We now turn to the question of what *modes* of natural motion the system of Fig. 12.14 can have. As usual, the answer to this question is found by returning to Eqs. (12.42) with the values of s obtained above. From the first of Eqs. (12.42), for example, we have the following expression for the relation between Θ_2 and Θ_1:

$$\frac{\Theta_2}{\Theta_1} = \left[\frac{s^2 + [(b_1 + b_3)/J_1]\,s + [(k_1 + k_3)/J_1]}{(b_3/J_1)\,s + (k_3/J_1)}\right]_{s=-\sigma_{g_1}\pm j\omega_{g_1}} \quad \text{or} \quad -\sigma_{g_2}\pm j\omega_{g_2} \tag{12.47}$$

in which s is to have values obtained by solving the characteristic equation as in Fig. 12.16—e.g., roots (g) in Fig. 12.16.

We recognize that the relation between Θ_2 and Θ_1 is now a complex number. More generally, the values of Θ_1 and Θ_2 in Eqs. (12.41) may be complex constants (rather than simply real ones), which we should expect, since the motions are going to be damped sine waves (see Sec. 8.3 and Fig. 8.3).

What we find, in the present case of a two-mass system with damping, is that each mass will have a damped sinusoidal motion having the natural characteristics of the *system;* and that, further, when the system moves in a natural mode, the two masses move with a definite phase relation—as well as a definite amplitude ratio—between them. This *vector* relation between Θ_2 and Θ_1 is given by Eq. (12.47). The value of Θ_2/Θ_1 for each root of the characteristic equation is known as an *eigenvector*† of the system.

† It is a vector not in physical space, but in "state space," which is a mathematical space, just as the s plane is a mathematical space in which we have vectors. (The concept of state space is discussed in Sec. 10.2.)

An eigenvector is fundamentally a set of numbers describing the natural mode of motion associated with an eigenvalue. We shall write eigenvectors in two forms. For a lumped-parameter system, the set of complex numbers Θ_2/Θ_1, Θ_3/Θ_1, and so on, evaluated for one eigenvalue (root of the characteristic equation), constitute an eigenvector because they give the relative magnitude and phase for each member when the system is moving in the natural mode associated with that eigenvalue.

Another useful form in which to describe an eigenvector is as the set of (complex) numbers which give the initial configuration *and velocities*—the initial *state*—which will cause the system to respond purely with only one of its natural motions and not of the others. We shall call this the *state* form of the eigenvector. For clarity, we may call the form Θ_2/Θ_1, Θ_3/Θ_1, and so on, the *phasor* form of the eigenvector.

There is an eigenvector corresponding to each eigenvalue or root of the system's characteristic equation. In general, each eigenvector has n components, where n is the order of the system (i.e., the order of the characteristic equation).

Whenever two eigenvalues (i.e., two roots of the characteristic equation) occur as a complex conjugate pair (e.g., $s = -\sigma_1 - j\omega_1$, $s = -\sigma_1 + j\omega_1$), indicating an underdamped natural vibration, the eigenvectors corresponding to the two members of the pair are found to be identical, indicating, of course, that there is just one natural mode of motion corresponding to the *pair* of roots.

For the special case of vibrational systems without damping, all the eigenvalues are purely imaginary (all occurring in conjugate pairs), and all the $n/2$ eigenvectors are real, indicating that each natural mode of motion can be started from the system in a *given initial configuration* and *zero initial velocity* (or alternatively, with a phase angle between elements which is either 0° or 180°). This configuration, the *mode shape* defined by the eigenvector, will then repeat itself every cycle; in fact, it will be preserved throughout the motion, changing only its magnitude. (Consider, for example, the system in Fig. 12.4 vibrating in one of its natural modes.) Eigenvectors are especially easy to contemplate when each one consists only of a mode shape, with no velocities involved, as in an undamped system like that in Fig. 12.4.

With damping present, however, as in Fig. 12.14, the roots of the characteristic equation (eigenvalues) are complex, and so therefore is each eigenvector. This means that for each root or eigenvalue a natural mode of motion is to be started by giving each mass a certain displacement *and velocity* relative to the others, or alternatively, a certain relative amplitude and relative phase as specified, for example, by Θ_2/Θ_1, Eq. (12.47).

Consider the system of Fig. 12.14. If all four eigenvalues are complex, as points (g) are in Fig. 12.16, they will be in conjugate pairs and there will be only two distinct eigenvectors, corresponding to the two oscillatory natural modes of motion. But if, for example, two of the eigenvalues are real, and two are complex, as points (ℓ) are in Fig. 12.16, then there will be three eigenvectors, one for each of the two exponential natural modes and one for the damped oscillatory natural mode. This situation is analyzed in Example 12.2.

As another situation, consider the system of Fig. 13.1. There, each (phasor form) eigenvector consists of a set of values (complex, in general) for both Θ_2/Θ_1 *and* Θ_3/Θ_1. If the damping is light, as we shall assume in Sec. 13.1, the eigenvalues may consist of three complex pairs of roots. The corresponding

[Sec. 12.8] GENERAL CASE: EIGENVALUES AND EIGENVECTORS

eigenvectors are three sets of complex values for the array

$$1 \quad \frac{\Theta_2}{\Theta_1} \quad \frac{\Theta_3}{\Theta_1}$$

Each set is one eigenvector. For the special case of zero damping, these are calculated in Sec. 13.1 and shown in Fig. 13.2. Since there is no damping, each eigenvector in the state form consists only of a mode shape, with no velocities.

If, at the other extreme, the damping in Fig. 13.1 were heavy enough to produce roots that were all real, then there would be six distinct sets of values for the array

$$1 \quad \frac{\Theta_2}{\Theta_1} \quad \frac{\Theta_3}{\Theta_1}$$

Each set would consist of numbers (real, in this case) dictating the relative magnitudes, or alternatively, the initial configuration and set of initial velocities which would produce a pure natural mode of (exponential) motion.

Example 12.2. Eigenvectors and natural modes for the system of Fig. 12.14.
Find the eigenvectors and sketch the natural modes of motion the system in Fig. 12.14 will have if $k_3/k_1 = 0.5$ and $b_3/J_1 = \omega_{01}$ [points (g) in Fig. 12.16], and also if $k_3/k_1 = 0.5$ and $b_3/J_1 = 3\omega_{01}$ [points (ℓ) in Fig. 12.16]. Also, as before, $J_1 = J_2$, $k_2/k_1 = 4$, $b_1/J_1 = 1.2\omega_{01}$, $b_2/J_2 = 1.7\omega_{01}$.

The eigenvectors for this system are given (phasor form) by (12.47), where, for the given k_3, s is given by points (g_1) and (g_2) in Fig. 12.16 for $b_3/J_1 = \omega_{01}$, and by points (ℓ_1), (ℓ_2), and (ℓ_3) for $b_3/J_1 = 3\omega_{01}$.

Equation (12.47) can be written

$$\frac{\Theta_2}{\Theta_1} = \left[\frac{\left(s^2 + \frac{b_1}{J_1}s + \frac{k_1}{J_1}\right)}{\left(\frac{b_3}{J_1}\right)\left(s + \frac{k_3/J_1}{b_3/J_1}\right)} + 1 \right]_{s = -\sigma_{g_1} + j\omega_{g_1}, -\sigma_{g_2} + j\omega_{g_2}, \cdots,} \quad (12.48)$$

or, for the numbers given,

$$\frac{\Theta_2}{\Theta_1} = \left[\frac{(s + 0.6\omega_{01} + j0.8\omega_{01})(s + 0.6\omega_{01} - j0.8\omega_{01})}{\left(\frac{b_3}{J_1}\right)\left(s + \frac{0.5\omega_{01}^2}{b_3/J_1}\right)} + 1 \right]_{s = -\sigma_{g_1} + j\omega_{g_1}, -\sigma_{g_2} + j\omega_{g_2}, \cdots,} \quad (12.49)$$

Evaluation of this expression for the values of s required [(g_1) and (g_2) or (ℓ_1), (ℓ_2), and (ℓ_3)] is easily accomplished graphically from the s plane. (It is rather tedious by any other method.) For demonstration we do the first calculation once in complete detail in Fig. 12.17. We do it for the second natural mode, $s = -\sigma_{g_2} + j\omega_{g_2}$, for which it happens that the vectors are most easily seen.

Figure 12.17 is an s-plane plot containing points $s = -\sigma_1 \pm j\omega_1$ ($\times$'s), point $s = -(k_3/J_1)/(b_3/J_1)$ (○), and points (g_1) and (g_2). The three complex vectors in (12.49) are drawn in Fig. 12.17 (vectors v_1, v_2, v_3) for the case $s = -\sigma_{g_2} + j\omega_{g_2}$. For convenience a fourth vector, v_4, is laid out with its arrowhead at point (g_2) and having length b_3/J_1 and angle 0°.

Now with a Spirule at point (g_2), a single manipulation (four settings of the log spiral and one reading) gives the following computation:

$$\frac{|v_1|\,|v_2|}{|v_3|\,|v_4|} = 3.1$$

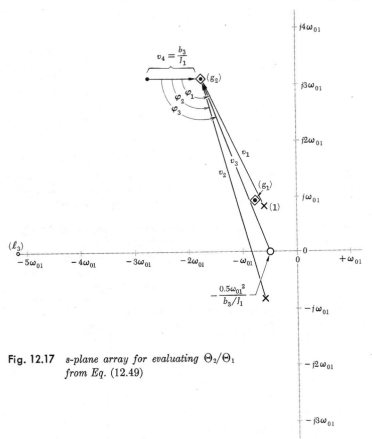

Fig. 12.17 *s-plane array for evaluating Θ_2/Θ_1 from Eq. (12.49)*

while one more manipulation (three settings and one reading with the arm and protractor) gives

$$\underline{/v_1} + \underline{/v_2} - \underline{/v_3} - \underline{/v_4} = 112°$$

from which, substituting in (12.49), the *eigenvector for the second natural mode is:*

$$\text{Phasor form} \quad \frac{\Theta_2}{\Theta_1} = 3.1 \underline{/112°} + 1 \underline{/0°} = 2.9 \underline{/93°} \quad (12.50)$$

which is in phasor form. [The summation in (12.50) was also performed with a Spirule.] With $b_3/J_1 = 1$, the *second natural mode of motion* is thus given by

$$\theta_1 = \theta_1(0)e^{-\sigma_{g_2}t} \cos(\omega_{g_2}t)$$
$$\theta_2 = 2.9\theta_1(0)e^{-\sigma_{g_2}t} \cos(\omega_{g_2}t + 93°) \quad (12.51)$$

where, from Fig. 12.16, $\sigma_{g_2} = -2.75\omega_{01}$, $\omega_{g_2} = 3.1\omega_{01}$, and $\theta_1(0)$ is any arbitrary magnitude.† This motion is plotted in Fig. 12.18b. It is produced by the initial

† Any initial time can be used, of course. The one shown is for $\Theta_1 = \theta_1(0) \underline{/0°}$ and is the simplest to write, but (for example) an appropriate change in starting time can be made so that $\theta_1(0)$ will be zero [in the state form of the eigenvector, (12.52)]. (See Problem 12.41.)

[Sec. 12.8] GENERAL CASE: EIGENVALUES AND EIGENVECTORS

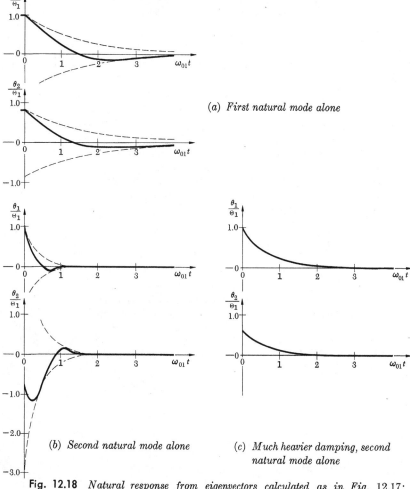

(a) First natural mode alone

(b) Second natural mode alone

(c) Much heavier damping, second natural mode alone

Fig. 12.18 Natural response from eigenvectors calculated as in Fig. 12.17; system of Fig. 12.14

conditions [from (12.51) and its derivative with $t = 0$]:

$$\begin{aligned}
&& \theta_1(0) &= \theta_1(0) \\
\text{State} && \dot{\theta}_1(0) &= -\sigma_{g_2}\theta_1(0) \\
\text{form} && \theta_2(0) &= 2.9\theta_1(0)\cos 93° \\
&& \dot{\theta}_2(0) &= 2.9\theta_1(0)(-\sigma_{g_2}\cos 93° - \omega_{g_2}\sin 93°)
\end{aligned} \quad (12.52)$$

and (12.52) is the (alternative) *state form of the eigenvector* for this natural mode. The eigenvector may be written conveniently as a column matrix:

$$\begin{bmatrix} \theta_1(0) \\ \dot{\theta}_1(0) \\ \theta_2(0) \\ \dot{\theta}_2(0) \end{bmatrix}_{\text{2nd mode}} = \begin{bmatrix} 1 \\ -\sigma_{g_2} \\ 2.9\cos 93° \\ 2.9(-\sigma_{g_2}\cos 93° - \omega_{g_2}\sin 93°) \end{bmatrix} \theta_1(0) \quad (12.53)$$

The computation of the first natural mode for $b_3/J_1 = 1$ is made, using Fig. 12.17, by drawing vectors to point (g_1).[†] The result is

$$\frac{\Theta_2}{\Theta_1} = [0.35 \underline{/127°} + 1 \underline{/0°}] = 0.84 \underline{/20°}$$

which the reader may verify. The corresponding natural motion may be written

$$\theta_1 = \Theta_1 e^{-\sigma_{g_1} t} \cos (\omega_{g_1} t)$$
$$\theta_2 = 0.84\Theta_1 e^{-\sigma_{g_1} t} \cos (\omega_{g_1} t + 20°) \qquad (12.54)$$

in which, from Fig. 12.16, $\sigma_{g_1} = 0.7\omega_{01}$, $\omega_{g_1} = 0.95\omega_{01}$, and Θ_1 is again arbitrary. This motion is sketched in Fig. 12.18a. The masses move approximately together, rather like the first natural mode of motion, Fig. 12.4a, for the similar undamped system. By comparison, in the second natural mode, Fig. 12.18b, the masses are approximately 90° out of phase, instead of 180° as for the undamped system (Fig. 12.4b). Equations (12.54) imply the alternative state form of the eigenvector:

$$\begin{bmatrix} \theta_1(0) \\ \dot{\theta}_1(0) \\ \theta_2(0) \\ \dot{\theta}_2(0) \end{bmatrix}_{\text{1st mode}} = \begin{bmatrix} 1 \\ -\sigma_{g_1} \\ 0.84 \cos 20° \\ 0.84(-\sigma_{g_1} \cos 20° - \omega_{g_1} \sin 20°) \end{bmatrix} \Theta_1 \qquad (12.55)$$

For calculation of the (three) natural modes of motion corresponding to eigenvalues (ℓ) in Fig. 12.16 (i.e., for $b_3/J_1 = 3$) another s-plane array like Fig. 12.17 is constructed. (This is left as an exercise, Problem 12.48.) From (12.49) it should contain, again, points $s = (-0.6 \pm j0.8)/\omega_{01}$ (as ×'s, say); the point $s = -(k_3/J_1)/(b_3/J_1) = -\frac{1}{3}\omega_{01}$ (as a ○); and the points (ℓ_1), (ℓ_2), and (ℓ_3) from Fig. 12.16. From such a plot the following eigenvectors are measured (see Problem 12.48):

For s at (ℓ_1): $\quad \dfrac{\Theta_2}{\Theta_1} = 0.85 \underline{/15°} \qquad (12.56)$

For s at (ℓ_2): $\quad \dfrac{\Theta_2}{\Theta_1} = 0.58 \qquad (12.57)$

For s at (ℓ_3): $\quad \dfrac{\Theta_2}{\Theta_1} = -0.50 \qquad (12.58)$

which indicate the following natural modes of motion:

For (ℓ_1): $\theta_1 = \Theta_1 e^{-1.1\omega_{01} t} \cos(1.1\omega_{01} t) \quad \theta_2 = 0.85\Theta_1 e^{-1.1\omega_{01} t} \cos(1.1\omega_{01} t + 15°)$ (12.59)

For (ℓ_2): $\theta_1 = \Theta_1 e^{-1.6\omega_{01} t} \quad \theta_2 = 0.58\Theta_1 e^{-1.6\omega_{01} t}$ (12.60)

For (ℓ_3): $\theta_1 = \Theta_1 e^{-5\omega_{01} t} \quad \theta_2 = -0.50\Theta_1 e^{-5\omega_{01} t}$ (12.61)

For practice, write the eigenvectors in state form for eigenvalues (ℓ_1), (ℓ_2), and (ℓ_3), using (12.59), (12.60), and (12.61).

The motion corresponding to real eigenvalue (ℓ_2) is sketched in Fig. 12.18c.

Problems 12.41 through 12.50

[†] The vector v_1 will be so short in this case that for Spirule calculation it is better to measure it separately and set it on the log spiral by hand, to avoid a large percentage error due to a small absolute error.

⊙Chapter 13

COUPLED MODES OF NATURAL MOTION: MANY DEGREES OF FREEDOM

The concepts introduced in Chapter 12 for linear systems having two degrees of freedom are readily extended to systems having any finite number of degrees of freedom. This will be demonstrated in Sec. 13.1.

By further extension, distributed-parameter systems may be thought of as having an infinite number of degrees of freedom. In this chapter we introduce the subject in terms of a class of one-dimensional, distributed-parameter systems. We encounter the following new concepts:

(1) The equation of motion for a distributed-parameter system is a *partial differential equation* in the independent variables time and space. Solutions to these equations which also meet both the time and space boundary conditions are very difficult to find, in general.

(2) For certain classes of distributed-parameter vibration problems, solutions are well established, however. Interesting examples are the vibration of a uniform shaft in torsion, the lateral vibration of a taut string, and the longitudinal vibration of an air column, which all happen to have the same equation of motion, known as the *wave equation*. Musical instruments of the string and wind families are in this class. The natural motions are sinusoidal vibrations in time, and the mode shapes are sinusoidal in space. Musical sounds are made from a wide variety of combinations of these natural motions. An abrupt disturbance at one end of such a system will propagate along its length as a wave.

(3) *Rayleigh's method* uses an energy relation to estimate the natural frequencies of undamped systems. It is applicable to both lumped- and distributed-parameter systems. For complicated systems whose exact

analysis would be lengthy and involved, Rayleigh's method permits a rapid, remarkably accurate estimation of natural frequency. The accuracy can be improved by iteration, if desired.

13.1 MANY DEGREES OF FREEDOM

The principles we illustrated in Chapter 12—the ideas of physical coupling and the effect of coupling strength, of natural modes of oscillation and their superposition, and more generally the concept of eigenvalues and eigenvectors in damped systems—are fully as valid for linear, lumped-parameter, constant-coefficient physical systems of any order as they are for the two-degree-of-freedom systems we used to demonstrate them.

In this section we illustrate some of the ideas briefly for a three-degree-of-freedom system. The extension to more than three degrees of freedom is perfectly straightforward.

Three-degree-of-freedom example

Consider the system shown in Fig. 13.1a, which might, for example, be a physical model (Stage I) of an axial-flow compressor for a jet engine, each disk representing a set of blades. We are interested in what natural frequencies (eigenvalues) the system has, and what the corresponding natural modes of motion (eigenvectors) are. For this purpose we consider the ends of the shaft to be stationary (clamped, in the model). Note that in this simple model, damping of the blades relative to the housing is included,

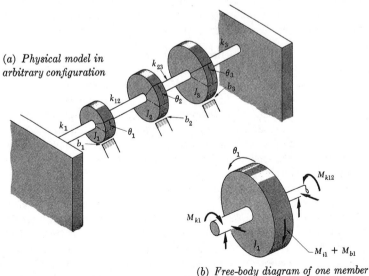

(a) Physical model in arbitrary configuration

(b) Free-body diagram of one member

Fig. 13.1 A three-degree-of-freedom system

[Sec. 13.1] MANY DEGREES OF FREEDOM

but damping of blades relative to one another is neglected. This is reasonable, since there is a stationary stage of blades between each of the rotor stages.

Next, we derive the equations of motion (Stage II) for the linear model given. We shall perform the derivation in some detail here, partly to serve as a reference for similar analysis, in Sec. 13.2, of a shaft having distributed mass.

(1) *Geometry.* In Fig. 13.1a the system is shown displaced to an arbitrary configuration. Specify, for each disk, the absolute angle of rotation: θ_1, θ_2, θ_3, as shown. The twist in shaft 1 is θ_1; the twist in shaft 2 is $\theta_2 - \theta_1$; and so on. The velocities across the three dampers are $\dot{\theta}_1$, $\dot{\theta}_2$, and $\dot{\theta}_3$.

(2) *Force equilibrium.* A free-body diagram of one of the masses (disk 1) is shown in Fig. 13.1b, and the positive directions for moments are defined there. Summing moments, we obtain

$$\Sigma \overset{\frown}{M}{}^* = M_{i1} + M_{b1} + M_{k1} - M_{k12} = 0$$

The rotary equilibrium equations for disks 2 and 3 are similar.

(3) *Physical relations.*

$$M_{i1} = J_1 \ddot{\theta}_1 \qquad M_{b1} = b_1 \dot{\theta}_1 \qquad M_{k1} = k_1 \theta_1 \qquad M_{k12} = k_{12}(\theta_2 - \theta_1)$$

for the arrow directions shown. Other physical relations are similar.

Combining the three steps, we obtain

$$\left.\begin{array}{l} J_1 \ddot{\theta}_1 + b_1 \dot{\theta}_1 + k_1 \theta_1 - k_{12}(\theta_2 - \theta_1) = 0 \\ k_{12}(\theta_2 - \theta_1) + J_2 \ddot{\theta}_2 + b_2 \dot{\theta}_2 - k_{23}(\theta_3 - \theta_2) = 0 \\ k_{23}(\theta_3 - \theta_2) + J_3 \ddot{\theta}_3 + b_3 \dot{\theta}_3 + k_3 \theta_3 = 0 \end{array}\right\} \quad (13.1)$$

Now we are ready to solve the equations of motion for the desired dynamic behavior (Stage III)—in this case for the eigenvalues and eigenvectors. We begin by assuming, as usual, that the motion is of the form $\theta = \Theta e^{st}$. Substituting assumed solutions of this form into (13.1) and canceling e^{st}, we obtain

$$\left.\begin{array}{l} (J_1 s^2 + b_1 s + k_1 + k_{12})\Theta_1 - k_{12}\Theta_2 = 0 \\ - k_{12}\Theta_1 + (J_2 s^2 + b_2 s + k_{12} + k_{23})\Theta_2 - k_{23}\Theta_3 = 0 \\ - k_{23}\Theta_2 + (J_3 s^2 + b_3 s + k_{23} + k_3)\Theta_3 = 0 \end{array}\right\} \quad (13.2)$$

Natural frequencies (eigenvalues). The determinant of Eqs. (13.2) set equal to zero is the characteristic equation of the system,

$$(J_1 s^2 + b_1 s + k_1 + k_{12})[(J_2 s^2 + b_2 s + k_{12} + k_{23})(J_3 s^2 + b_3 s + k_{23} + k_3) - k_{23}{}^2] - (J_3 s^2 + b_3 s + k_{23} + k_3)k_{12}{}^2 = 0 \quad (13.3)$$

whose roots are the eigenvalues of the system.

Often in physical systems like that of Fig. 13.1 the damping is small and has, therefore, negligible effect on the natural frequencies (see p. 449). It is thus expedient to find the natural frequencies first by neglecting damping, which, of course, is much easier because without the b's the characteristic equation has (in effect) only half the order, since only the even powers of s appear in it:

$$(J_1s^2 + k_1 + k_{12})[(J_2s^2 + k_{12} + k_{23})(J_3s^2 + k_{23} + k_3) - k_{23}^2] \\ - (J_3s^2 + k_{23} + k_3)k_{12}^2 = 0 \quad (13.4)$$

As an example, let us find the natural frequencies for the special case that all the J's are equal and all the k's are equal. Then (13.4) can be written:

$$(Js^2 + 2k)[(Js^2 + 2k)(Js^2 + 2k) - 2k^2] = 0 \quad (13.5)$$

This is merely a cubic in s^2, and it is already partly factored (we shall see the physical reason shortly), and thus its roots may be written by inspection:

$$s^2 = -\frac{2k}{J} \quad -0.586\frac{k}{J} \quad -3.414\frac{k}{J}$$

That is, the roots s (rearranged by magnitude) are

$$s = \pm j0.765\sqrt{\frac{k}{J}} \quad \pm j1.414\sqrt{\frac{k}{J}} \quad \pm j1.847\sqrt{\frac{k}{J}}$$

which means that the system has the natural frequencies

$$\omega_1 = 0.765\sqrt{\frac{k}{J}} \quad \omega_2 = 1.414\sqrt{\frac{k}{J}} \quad \omega_3 = 1.847\sqrt{\frac{k}{J}} \quad (13.6)$$

Natural modes of motion (eigenvectors). To find the natural modes of motion we simply write all but one of Eqs. (13.2) with the Θ_1 term on the right-hand side.† Again, for the assumption of zero damping,

$$-k_{12}\Theta_2 = -(J_1s^2 + b_1s + k_1 + k_{12})\Theta_1 \\ (J_2s^2 + b_2s + k_{12} + k_{23})\Theta_2 - k_{23}\Theta_3 = k_{12}\Theta_1 \quad (13.7)$$

† We may omit any one of the equations, as the reader may verify by obtaining (13.8) through the use of a different pair of equations in place of (13.7).

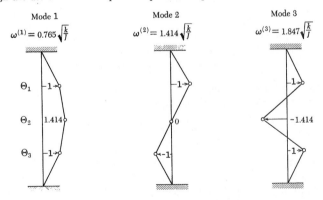

Fig. 13.2 *Natural modes of motion for the system of Fig.* 13.1, *with b's* = 0, *equal k's and equal J's.*

[Sec. 13.1] MANY DEGREES OF FREEDOM 461

These we solve for the ratios Θ_2/Θ_1 and Θ_3/Θ_1 by using Cramer's rule†:

$$\frac{\Theta_2}{\Theta_1} = \frac{J_1}{k_{12}} s^2 + \frac{b_1}{k_{12}} s + \frac{k_1}{k_{12}} + 1$$

$$\frac{\Theta_3}{\Theta_1} = -\frac{k_{12}}{k_{23}} + \left(\frac{J_2}{k_{23}} s^2 + \frac{b_2}{k_{23}} s + \frac{k_{12}}{k_{23}} + 1\right)\left(\frac{J_1}{k_{12}} s^2 + \frac{b_1}{k_{12}} s + \frac{k_1}{k_{12}} + 1\right) \quad (13.8)$$

For example, for the case of equal k's and equal J's with no damping, there is for each of the natural frequencies (13.6) a corresponding natural mode given by the ratios Θ_2/Θ_1, and Θ_3/Θ_1 shown in the following table.

	Mode 1	Mode 2	Mode 3
ω	$0.765 \sqrt{\frac{k}{J}}$	$1.414 \sqrt{\frac{k}{J}}$	$1.847 \sqrt{\frac{k}{J}}$
$\dfrac{\Theta_2}{\Theta_1}$	1.414	0	-1.414
$\dfrac{\Theta_3}{\Theta_1}$	1	-1	1

These natural modes are indicated in Fig. 13.2.

Note that the natural-mode shapes are independent of the values of k and J (when the k's are equal and the J's are equal). Note also that mode shapes at higher frequency show more twisting of shafts (slope of connecting lines), as one would expect.

For the special case considered (equal k's and equal J's), the second natural frequency can be checked with exceptional ease by noting the special character of its mode shape: the center disk does not move! Each of the other two disks can therefore be considered as a single-degree-of-freedom system having the spring constant $2k$ and inertia J, which leads at once to the result $\omega_2 = \sqrt{2k/J}$. This is why (13.5) appears partially factored.

Motions in the three natural modes can be superposed (as they were in Fig. 12.5) to describe any arbitrary natural motion of the system.

In the more general case of unequal k's and unequal J's, the modes may not be nicely symmetrical, as they are in Fig. 13.2, but their determination is just as straightforward (if cumbersome).

In the still more general case of arbitrarily large damping, the eigenvalues may be complex or real, and are much more difficult to obtain because a sixth-order characteristic equation (13.3) must be solved. Once it is solved, the eigenvalues are put back into the damped version of (13.7) to obtain the eigenvectors, which are also complex, indicating a phase as well as a magnitude. That is, a specific set of initial *velocities* $\dot{\theta}(0)$, besides

† Solution of (13.7) one at a time happens also to be easy. In general, however, use of Cramer's rule is more efficient. A systematic matrix formulation of this procedure is included in Sec. 14.6.

an initial configuration, must be employed to set in motion any one of the pure natural modes.

Extension to any number of degrees of freedom

The extension to 4, 5, . . . , n degrees of freedom is straightforward, although, of course, the more degrees of freedom there are, the more tedious and time-consuming the solution becomes. (The availability of digital computers greatly alleviates this problem.)

If one assumes first that there is no damping, then there will be as many natural frequencies as there are degrees of freedom, and this will also be (effectively) the order of the characteristic equation that must be solved for its roots. There will be a set of $n-1$ simultaneous equations like (13.7) to be solved for the natural modes of motion (e.g., using Cramer's rule).

In the more general case that damping must be considered, the characteristic equation is of order $2n$, and its roots may be complex or real. There are still just $n-1$ simultaneous equations to be solved for the eigenvectors, which are complex, in general.

Rayleigh's method

The natural frequencies of physical systems of any order, including distributed systems, can be estimated quickly and accurately by the method of Rayleigh, which is presented in Sec. 13.7. Section 13.7 may readily be studied now, without first reading the intervening sections on distributed-parameter systems, if desired.

Problems 13.1 through 13.13

°13.2 DISTRIBUTED-PARAMETER SYSTEMS: EQUATIONS OF MOTION

For the most part, this book is concerned with linear, lumped-parameter systems having constant coefficients. Occasionally, however, we go beyond these bounds in the interest of further perspective. In Chapter 10 we explored briefly the behavior of nonlinear systems and systems whose parameters vary with time. That study gave us a better appreciation for the implications of our assumptions of linearity and constant coefficients—of how much labor the assumptions save, and also of how much we may miss the true picture if we make them rashly. By our cursory study of nonlinear systems, we built a bridge to an adjacent area which we may wish to use later, en route to a careful study of that specific area.

In the same spirit, in the present section we take time to build another bridge. This time we abandon (temporarily) the assumption that our system has lumped parameters—e.g., that (like Fig. 13.1) it consists only of discrete masses and springs. We now consider the possibility that mass and flexibility are distributed continuously along a member, as in Fig. 13.3.

Our brief study of distributed-parameter systems will not be at all

A continuous shaft

Consider a long uniform shaft, Fig. 13.3, which is transmitting torque from one of its ends to the other. Such a shaft is subject to torsional moments and can vibrate in various torsional modes, and these can affect severely the system of which the shaft is a part. The drive shaft of a car, running from the engine to the rear wheels, and the drive shaft of a ship, carrying power the length of the ship from the turbines to the propellers, are common examples. An extreme example is an oil well rig, which may carry power from an engine at the surface to a drill several *miles* below. (Such shafts operate with a total steady-state twist of hundreds of revolutions.)

Figure 13.3a is a physical model (Stage I) of such a drive shaft, supported at various points along its length. We choose in this particular model to let one end of the shaft be fixed and the other torsionally free (representing, for example, a rigid engine at one end, and the comparatively light resistance of a propeller to *changes* in speed, at the other end). In the model the shaft has uniform moment of inertia γ per unit length—e.g., [(lb ft)/(rad/sec^2)]/ft—and uniform stiffness κ per unit length. That is, there is a restoring moment at any point in the shaft proportional to the angle of twist at that point. [The units of κ are (ft lb/rad)/ft.]

We now proceed (Stage II) to write the equations of motion of the system. To do so, we might think of the distributed-mass shaft as simply an extension of the many-degree-of-freedom case, like Fig. 13.1 but with n a very large number. Instead, however, we take a fundamentally different approach—that of considering an infinitesimally thin disk-shaped element dx of the shaft at an arbitrary location x, as indicated in Fig. 13.3b. This element of shaft is defined by two imaginary planes normal to the shaft axis and very close together. It will be instructive, as we proceed, to compare each step taken below with the corresponding step taken in Sec. 13.1 (where we considered finite disks).

(1) *Geometry.* To describe the configuration of the shaft, imagine that the shaft is vibrating back and forth in torsion, so that a thin line, painted originally straight along the side, now twists and contorts, undulating with time. Then imagine that we take a snapshot of the shaft at some arbitrary time t, and that Fig. 13.3a is that snapshot, the shape of the shaft being indicated by the distorted line. Designate the *location* of an element of the shaft by coordinate x, measured from a reference point to the first imaginary plane; denote the thickness of the element between planes by dx. Denote the angle of twist at location x and time t by coordinate $\theta(x,t)$.

Figure 13.3b shows an expanded view of the part of the shaft containing

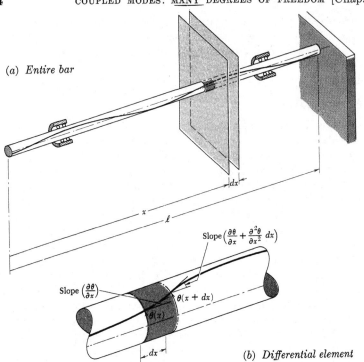

Fig. 13.3 *A distributed-mass torsion bar: geometry*

the element we have chosen to study. At a given instant the twist angle at one end of the element is $\theta(x,t)$ and at the other end it is slightly different, $\theta(x + dx, t)$.

We are interested in the slope of the twist line. At the near end of the element it is $\partial\theta/\partial x$. (Using the symbol $\partial/\partial x$ rather than d/dx indicates that we want to see the rate of change with x only, *t being frozen*. $\partial/\partial x$ is called the "partial derivative with respect to x.")

At the far end of the element the slope is slightly different. The rate at which the slope changes with location is denoted $(\partial/\partial x)(\partial\theta/\partial x)$ or simply $\partial^2\theta/\partial x^2$. Thus the slope at the far end of the element is given by

$$\frac{\partial \theta}{\partial x} + \frac{\partial^2 \theta}{\partial x^2} dx$$

(the slope at the near end plus the amount the slope has changed across dx).†

(2) *Force equilibrium.* As our free body we naturally take the shaft element we have been looking at, Fig. 13.4. There are only three moments

---

† We could get involved here also with the rate at which the rate of change changes, $\partial^3\theta/\partial x^3$, and so on: but more elaborate rigorous analysis proves that these higher-order terms may safely be neglected

about the shaft axis acting on this slice of shaft: an elastic-shear moment at each end of the slice due to distortion, and the D'Alembert inertial moment due to angular acceleration. Moment equilibrium then gives

$$M_i + M_\kappa(x) - M_\kappa(x + dx) = 0 \qquad (13.9)$$

(3) *Physical force-geometry relations.* The *inertial moment* depends merely on the total moment of inertia of the element, and (to an adequate approximation, as can be shown) the angular acceleration at x,

$$M_i = (\gamma\, dx)\frac{\partial^2 \theta}{\partial t^2} \qquad (13.10)$$

for the direction shown as positive in the picture. (The symbol $\partial/\partial t$ means derivative with respect to t for a fixed x.)

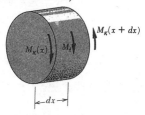

Fig. 13.4 *Free-body diagram of a thin slice of the shaft in Fig. 13.3*

The *elastic-shear moment* depends only on the slope of the twist line at each face of the element. For the arrow directions shown in Fig. 13.4:

$$M_\kappa(x) = \kappa\frac{\partial \theta}{\partial x}$$
$$M_\kappa(x + dx) = \kappa\left(\frac{\partial \theta}{\partial x} + \frac{\partial^2 \theta}{\partial x^2} dx\right) \qquad (13.11)$$

Combining [entering (13.10) and (13.11) into (13.9)], we obtain

$$\gamma\, dx \frac{\partial^2 \theta}{\partial t^2} + \kappa \frac{\partial \theta}{\partial x} - \kappa\left(\frac{\partial \theta}{\partial x} + \frac{\partial^2 \theta}{\partial x^2} dx\right) = 0$$

or

$$\boxed{\gamma \frac{\partial^2 \theta}{\partial t^2} - \kappa \frac{\partial^2 \theta}{\partial x^2} = 0} \qquad (13.12)$$

In Sec. 13.3 we shall determine the dynamic behavior (Stage III) implied by this equation of motion. First, we mention its role in physical dynamics.

The wave equation

Equation (13.12), called the wave equation, has really widespread applications in the dynamics of continuous media. As we shall note in Sec. 13.4, it describes not only the dynamics of twist in a shaft but also the lateral motions of a string and the longitudinal motions of plane pressure waves through a solid or fluid medium, e.g., along a bar or in an air column such as an organ pipe.

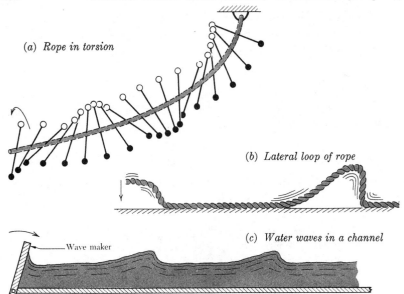

Fig. 13.5 *Demonstrations of wave propagation*

Equation (13.12) is a *partial* differential equation, and we are interested in solutions to it for the case where the physical medium has a specific length, being either clamped at both ends, free at both ends, or clamped at one end and free at the other. In this case we shall find that the system can have only simple harmonic motions, that is, motions which are sinusoidal with time; and it is these standing-sine-wave solutions we shall seek.

As an interesting digression, however, Eq. (13.12) can also be satisfied by a displacement which propagates along the shaft (or string or air column). This phenomenon is well known to freshman physics students, who are always treated to a demonstration such as that depicted in Fig. 13.5a, which shows a rope-in-torsion experiment. A simple variation is the propagation of a lateral loop of rope, Fig. 13.5b. Another involves a straight surface wave in a long narrow channel of water, which can be started at one end and seen to propagate the length of the channel, Fig. 13.5c.

For the study of such wave propagation, Eq. (13.12) is more convenient to study in the form

$$\frac{\partial^2 \theta}{\partial t^2} = c^2 \frac{\partial^2 \theta}{\partial x^2} \qquad (13.13)$$

in which c turns out to be the velocity of propagation. For a shaft in torsion, $c^2 = \kappa/\gamma$. The solution to Eq. (13.13)† is that a wave of any shape $\theta(x)$ will

† Provided that the displacements are small.

[Sec. 13.3] ONE-DIMENSIONAL DISTRIBUTED-PARAMETER SYSTEMS

travel along the shaft at the speed c until it reaches a boundary. For a string, $c^2 = F/\mu$, the ratio of tensile force to mass per unit length; for a bar or air column, $c^2 = EA/\mu$, the ratio of longitudinal stiffness per unit length to mass per unit length.

When the shaft (or string or column) is bounded at both ends, the propagating waves reflect back and forth and produce a pattern which varies *periodically* with time, and which may therefore be described, alternatively, simply as the sum of many sine waves in space which do *not* travel along the shaft but which stand in place and oscillate sinusoidally with time. These are called *standing waves*.

Since we are interested in the harmonic oscillation of shafts, strings, and columns, we shall assume such motion at the outset in solving Eq. (13.12).

○13.3 NATURAL MOTIONS OF A CLASS OF ONE-DIMENSIONAL DISTRIBUTED-PARAMETER SYSTEMS

We now turn to the study (Stage III) of the dynamic behavior of the system of Fig. 13.3, insofar as we can deduce it from Eq. (13.12).

Tactically, the approach to solving Eq. (13.12) is the same as for an *ordinary* linear differential equation: we try to guess a solution that will work. The trick is to guess a solution having sufficiently general form that it can accommodate all the special conditions it needs to.

The thing that makes Eq. (13.12) different from an ordinary differential equation is, of course, that θ varies with two different quantities, t and x (time and space), instead of with t only. It turns out that a good guess to make, as a first step toward solving (13.12), is that the time variation of θ is sinusoidal: $\theta(x,t) = \theta(x) \cos \omega t$. This is a reasonable guess, since the motion of systems similar to that in Fig. 13.3 has always turned out to be sinusoidal with time. Moreover, this guess is the key which unlocks the problem: Once we have made it, the rest of the solution comes rather quickly.

To establish that this is a correct guess, we must show that it does indeed solve Eq. (13.12) and meet all other necessary conditions. To do this, we assume that

$$\theta(x,t) = \theta(x) \cos \omega t \qquad (13.14)$$

in which $\theta(x)$ and ω are as yet unknown, and we substitute this assumed solution into (13.12):

$$-\gamma\omega^2\theta(x) - \kappa\frac{d^2\theta(x)}{dx^2} = 0$$

or

$$\frac{d^2\theta(x)}{dx^2} + \frac{\gamma}{\kappa}\omega^2\theta(x) = 0 \qquad (13.15)$$

Equation (13.15) can be written as an ordinary differential equation because $\theta(x)$ does not vary with time. We do not yet know ω.

Physically, this assumed motion (13.14) consists of the painted line in

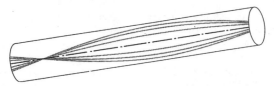

Fig. 13.6 *Strobe-light picture of an oscillating torsion bar*

Fig. 13.3a simply moving back and forth sinusoidally in time between two extreme configurations which are mirror images of one another, as indicated in Fig. 13.6. (Figure 13.6 may be thought of as a strobe-light multiple-exposure version of the snapshot in Fig. 13.3a.) Notice that all intermediate configurations are the same shape as the extreme ones, differing only in amplitude. This motion is rather easy to imagine and seems quite reasonable.

Our next move is to solve Eq. (13.15), and this should be pretty easy. It has the same form as the equations for undamped motion which we have been solving readily ever since Chapter 7, except that in (13.15) the independent variable is location x rather than time t. The solution that experience has taught us to try for an equation like (13.15) is

$$\theta(x) = C \cos(\alpha x + \psi) \tag{13.16}$$

in which α is the "space frequency" and C and ψ are constants. That is, we are guessing that the *shape* of the configuration in Fig. 13.6 is a cosine wave having a wavelength of $2\pi/\alpha$ (ft), and a maximum amplitude of C radians, which occurs ψ of a wavelength from the end. We do not yet know α, C, or ψ.

We quickly find α in terms of the other variables by substituting (13.16) into (13.15) and noting what it has to be if (13.16) is to be a correct solution to (13.15):

$$-\alpha^2 C \cos(\alpha x + \psi) + \frac{\gamma}{\kappa} \omega^2 C \cos(\alpha x + \psi) = 0$$

$$\alpha^2 = \frac{\gamma}{\kappa} \omega^2 \tag{13.17}$$

To find C and ψ we do the same kind of thing we did in Sec. 8.5, for example: We figure out what they must be in order to satisfy the "initial conditions" on θ; only now it is an initial condition in space x rather than in time t that we must consider. The conditions on the independent variable that are invoked at particular points in space are called *boundary conditions* (just as conditions invoked at particular points in time are called *initial conditions*).

The spatial boundary conditions in Fig. 13.3a are the following: at $x = 0$ there is no moment, and therefore the slope $\partial\theta/\partial x$ must be zero (the angle θ itself can have any value at $x = 0$); at $x = \ell$, the twist angle θ must be 0 (the slope $\partial\theta/\partial x$ may have any value, because the wall is presumed capable of exerting whatever moment it needs to).

[Sec. 13.3] ONE-DIMENSIONAL DISTRIBUTED-PARAMETER SYSTEMS 469

We now introduce these boundary conditions into (13.16):

$$\left[\frac{\partial \theta}{\partial x}\right]_{x=0} = -C\alpha \sin \psi = 0 \tag{13.18}$$
$$\theta(\ell) = C \cos(\alpha\ell + \psi) = 0$$

These are merely simultaneous trigonometric equations, and their solution is

and therefore
$$\left.\begin{array}{c} \psi = 0 \\ \cos \alpha\ell = 0 \\ \alpha\ell = n\pi/2 \quad n = \text{any odd integer} \\ C = \theta_{\max} \quad \text{(arbitrary)} \end{array}\right\} \tag{13.19}$$

From (13.17), $\omega = \sqrt{\kappa/\gamma}\,\alpha$, and thus the total solution for the free–fixed torsion problem is:

Solution for free-fixed uniform shaft in torsion
$$\boxed{\theta(x,t) = \left[\theta_{\max} \cos\left(\frac{n\pi}{2\ell} x\right)\right] \cos\left(\sqrt{\frac{\kappa}{\gamma}} \frac{n\pi}{2\ell} t\right)} \tag{13.20}$$
$$n = \text{any odd integer}$$

That is, for $n = 1$ the solution is a cosine wave in space which is maximum at the free end and 0 at the fixed end, as shown (to an exaggerated scale) in Fig. 13.7a. The maximum magnitude $\theta_{\max}$, which occurs at the end, can have any value, and hence depends on how the free motion was started. This shape is preserved as its amplitude changes sinusoidally with time, the time for one complete cycle being $T = 2\pi/\omega = 4\ell \sqrt{\gamma/\kappa}$.

For $n = 3$ the configuration looks like Fig. 13.7b. This shape is pre-

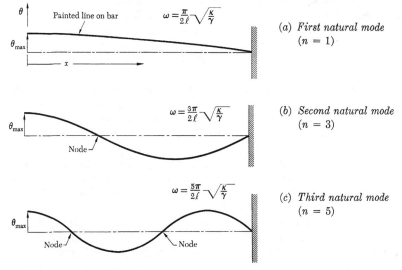

Fig. 13.7 *Natural modes of motion for a fixed-free, distributed-mass torsion bar*

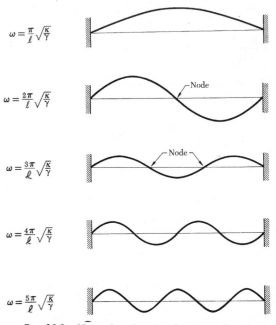

Fig. 13.8 *Natural modes of motion for a fixed-fixed, distributed-mass torsion bar*

served as the shaft vibrates back and forth, now at a higher frequency, $\omega = (3\pi/2\ell)\sqrt{\kappa/\gamma}$.

For $n = 5$ the configuration is as in Fig. 13.7c, and the frequency (still higher) is $\omega = (5\pi/2\ell)\sqrt{\kappa/\gamma}$, and so on for larger values of n.

There are, obviously, an *infinite number* of natural modes of motion. If we start a natural motion carefully by displacing the shaft into a natural-mode configuration *all along its length,* then it will oscillate in just that one of its natural modes.

More generally, however, if we start the motion by displacing the shaft into some arbitrary configuration, then the ensuing motion will involve (in general) *all* the natural modes at once; and, of course, the total motion will not be sinusoidal and may not even be periodic. Such general free motions of a distributed system are analyzed by decomposing them into natural modes—as many as are required to approximate the exact motion with sufficient accuracy. The use of superposition in this way is justified of course, because the partial differential equation of motion (13.12) is *linear.*

Consider next another, more symmetrical set of boundary conditions. Suppose *both* ends of the shaft in Fig. 13.3 to be fixed.† The solution to the problem is the same up to the point at which boundary conditions are intro-

---

† For a given physical situation (like the ship propeller drive), the two boundary-condition assumptions—free-fixed and fixed-fixed—are likely to bracket the actual conditions and hence give a good basis for estimating the actual behavior.

duced. That is, the equation of motion (13.12) holds, of course, and so do relations (13.14)–(13.17); but (13.18) must be modified. Now the situation is that

$$\begin{array}{lll}\text{Fixed-fixed} & \text{At } x = 0 & \theta = 0 \\ \text{boundary conditions} & \text{At } x = \ell & \theta = 0\end{array} \quad (13.21)$$

At each end any slope is possible, the walls being capable of sustaining whatever moment results. Substituting (13.21) into (13.16) gives

$$\begin{aligned}\theta(0) &= C \cos \psi = 0 \\ \theta(\ell) &= C \cos (\alpha \ell + \psi) = 0\end{aligned} \quad (13.22)$$

The solution of these simple trigonometric equations is†

$$\left.\begin{aligned}\psi &= -\frac{\pi}{2} \\ \alpha \ell &= n\pi \quad n \text{ any integer} \\ C &= \theta_{\max} \quad \text{(arbitrary)}\end{aligned}\right\} \quad (13.23)$$

That is, the complete solution is now

Solution for fixed-fixed uniform shaft in torsion

$$\boxed{\theta(x,t) = \left[\theta_{\max} \sin\left(\frac{n\pi}{\ell} x\right)\right] \cos\left(\frac{n\pi}{\ell}\sqrt{\frac{\kappa}{\gamma}}\, t\right) \\ n = \text{any integer}} \quad (13.24)$$

The first five natural modes of motion for this case are plotted in Fig. 13.8, with their corresponding natural frequencies noted. The points on the shaft which remain stationary during a given natural motion are called *nodes*, and, of course, they are highly significant in any physical vibration problem.

It is to be noted that the above two examples have involved a rather special situation, namely, that mass and stiffness are uniform all along the shaft. When this is not so, we may certainly expect the equation of motion to be more difficult—perhaps much more difficult—to solve.

We remind ourselves that, unlike ordinary differential equations, partial differential equations cannot generally be solved on a computer (even if they are linear).

Comparison with lumped-parameter case

Comparison of the first three natural modes in Fig. 13.8 for a distributed-parameter system with Fig. 13.2 for a simple lumped-parameter system establishes at once the image that the behavior of the two kinds of motion is very similar. The mode shapes are similar, as Fig. 13.9 shows strikingly, and even the natural frequencies have very nearly the same ratio to one another. In fact, by making $k/J = 2\pi^2 \kappa/\ell^2 \gamma$, and by choosing Θ_1

---

† Although ψ could be $m\pi/2$, with m any odd integer, this added generality would *not* add to the number of possible solutions, the shape with $\psi = 3\pi/2$ being identical to the shape with $\psi = \pi/2$, and so on.

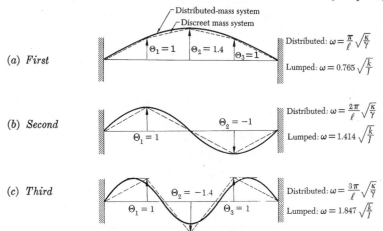

Fig. 13.9 *Comparison between natural modes of motion of distributed-mass and discrete-mass systems*

appropriately, we can reduce the differences between the two sets of results to less than 40% in both frequency and mode shape. For the first two natural modes the agreement is to 5% in frequency and 20% in mode shape.

This leads us to consider the possibility that distributed-parameter systems—ones that are obviously and purely so—can be analyzed quite successfully by means of an astute lumped-parameter model. This technique is often used. Obviously, the lumped-parameter model will have only as many natural modes as it has degrees of freedom, and hence cannot represent the distributed-parameter system at all, beyond that number of modes. Also, the approximation turns out to be poorer for the higher modes, as Fig. 13.9 indicates.

Use is also made of the reverse technique—that of analyzing a many-element lumped-parameter system as if it were a distributed-parameter system. This is the basis of the Holzer method† for analyzing vibrations in ship turbines and similar systems.

Problems 13.14 through 13.17

○13.4 BRIEF DESCRIPTION OF GENERAL DISTRIBUTED-PARAMETER SYSTEMS

One-dimensional systems

What we considered in Sec. 13.3 is a distributed system in which the medium moves in just one space dimension, and in which forces were a

---

† See J. P. Den Hartog, "Mechanical Vibrations," 4th ed., pp. 184–197, McGraw-Hill Book Company, New York, 1956.

[Sec. 13.4] GENERAL DISTRIBUTED-PARAMETER SYSTEMS 473

function of the first space derivative. There are many other physical systems which very nearly fit this mathematical description, including vibrating strings, organ pipes, and bars vibrating longitudinally. These are discussed in Sec. 13.5. (In a bar vibrating longitudinally, the geometric variable is the longitudinal *strain* as a function of location.)

Vibrating beams like Fig. 13.10a, Fig. 13.10b, and Fig. 13.10c are somewhat more complicated because it happens that the internal forces depend on the *second* spatial derivative of displacement instead of the first. This leads to a partial differential equation of the form

$$EI \frac{\partial^4 z}{\partial x^4} = \mu \frac{\partial^2 z}{\partial t^2}. \tag{13.25}$$

which is fourth-order in x instead of second-order, and hence requires the satisfaction of four boundary conditions for its solution instead of two. (EI is the bending stiffness.)† The principles and procedures are similar to those used above, however.

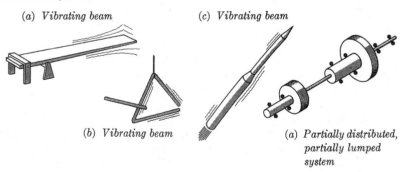

(a) *Vibrating beam*

(b) *Vibrating beam*

(c) *Vibrating beam*

(a) *Partially distributed, partially lumped system*

Fig. 13.10 *Some additional one-dimensional, distributed-parameter systems*

Frequently one encounters systems in which parameters are partially lumped and partially distributed, as in Fig. 13.10d. In an exact analysis such systems are studied piecewise, with boundary conditions to be matched at the various interfaces between a distributed segment and a lumped element. Various approximations are often used in which either the distributed parameters may be approximated by lumped ones, or alternatively, lumped elements may be thought of as being "smeared out" in an equitable way so that the whole system becomes a distributed-parameter one.

Two-dimensional systems

More generally, we may be interested in distributed-parameter systems which can distort in two spatial dimensions, as in Fig. 13.11a, b, and c. The membrane of Fig. 13.11a is the two-dimensional counterpart of the string, and has a partial differential equation of motion which is second-order

† *Ibid.*, pp. 148–155.

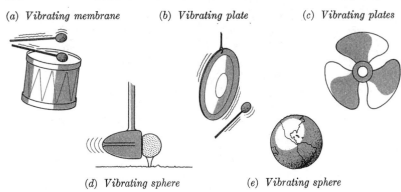

(a) *Vibrating membrane* (b) *Vibrating plate* (c) *Vibrating plates*

(d) *Vibrating sphere* (e) *Vibrating sphere*

Fig. 13.11 *Some two- and three-dimensional, distributed-parameter systems*

in both spatial locations x and y, as well as in time t. (In Fig. 13.11a air inside the drum plays a crucial role, as is evident if we remove the bottom and beat on the membrane by itself. Thus the system shown is not really well represented by a model consisting of a membrane alone.)

The vibrating plate, Fig. 13.11b and c, is the two-dimensional counterpart of the beam, and has a partial differential equation of motion which is fourth-order in x and y and second-order in t. Finding solutions to these equations and satisfying the boundary conditions is, of course, much more difficult than in one dimension.

The nodes for a given natural motion, which were points in Fig. 13.8, are *nodal lines* on a vibrating membrane or plate. Experimentally, these lines may be detected by sprinkling sawdust onto the membrane or plate and then setting it into vibration. The sawdust is undisturbed along the nodal lines.

Another kind of two-dimensional system is a plate vibrating in its own plane. This is the two-dimensional counterpart of the bar vibrating longitudinally. The coordinates that define the geometry of motion are the two components of local strain, which are a function of location.

Three-dimensional systems

Two systems whose parameters are distributed in three dimensions are shown in Fig. 13.11d and e. The geometric quantity that denotes the motion is again the strain, which is now a function of location (three coordinates) and which has itself three components. In both the golf ball and the earth, elasticity and mass are distributed with spherical symmetry.

The motion of the mass center of a golf ball has been analyzed with more or less rigor by many people. The accompanying distortions of the ball, as in Fig. 13.11d, have been given relatively little attention. The simultaneous solution of three partial differential equations with distributed boundary conditions is involved. The distortions of the earth, due principally to its spinning and to the gravitational field of the moon, are of great interest to geophysicists. The reason for the slight pear shape is not understood.

[Sec. 13.5] CERTAIN MUSICAL INSTRUMENTS 475

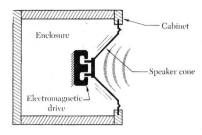

Fig. 13.12 *An audio speaker system: distributed-parameter plate coupled with distributed-parameter air*

(This out-of-roundness is extremely small, of course: The earth is about as nearly spherical as, for example, a billiard ball.) In a three-dimensional system like Fig. 13.11d or e, what is the counterpart of a nodal point or nodal line?

Audio speaker. The speaker in a sound system is essentially a vibrating plate, Fig. 13.12, with a mass at the center and with distributed-parameter air—distributed mass and distributed stiffness in three dimensions—interacting with it on both sides. It should be clear from the above discussion of such systems that, on the one hand, the exact analysis of speakers is approximately hopeless, while on the other hand, the model suggested in Problem 2.91, which ignores the mass of the diaphragm and also ignores the air, is not likely to be a very good one. The development of speaker systems has been largely empirical—that is, people who develop them try hundreds of configurations in a systematic way, and run exhaustive tests on them. The wonderful quality that often results is a tribute to their skill and patience.

○13.5 CERTAIN MUSICAL INSTRUMENTS

Certain musical instruments, namely, those utilizing vibrating strings (the violin family, the guitar, the piano, and so on) or vibrating air columns (woodwinds, brass, organs) have Eq. (13.12), very nearly, as their equation of motion. Since we have already solved Eq. (13.12) in Sec. 13.3, it is easy and instructive to draw some interesting conclusions about this whole class of musical instruments.

Vibrating strings†

Consider first the vibrating string. If the nominal tensile force F in the string is large (compared to small changes in it due to stretch of the string as it vibrates), we may consider F constant. Then the equation of motion of the string can readily be shown to be (Problem 13.14)

$$\mu \frac{\partial^2 z}{\partial t^2} = F \frac{\partial^2 z}{\partial x^2} \tag{13.26}$$

† The theory of the vibrating string was first developed by Helmholtz, following experiments he performed. His "On the Sensations of Tone," Dover Publications, Inc., New York, 1954, is a classic in the field of musical physics.

in which z is lateral displacement of the string and is a function of both location and time, $z = z(x,t)$, and μ is the mass per unit length.

The boundary conditions are that the string is fixed at both ends, and thus the solution given by (13.24) and plotted in Fig. 13.8 applies exactly! That is, the plots of Fig. 13.8 are the actual shapes of the string vibrating in several of its natural modes. The frequencies of the various natural modes of vibration are given by

$$\omega = \frac{n\pi}{\ell} \sqrt{\frac{F}{\mu}} \tag{13.27}$$

Rich sound is made by string instruments because, of course, the vibrating string sets into vibration the skillfully built light wooden box structure. Vibration of this base, in turn, sets the air in motion around it, and the resulting pressure waves carry the sound to our ears.†

If the motion of the string is started by plucking it, as for a guitar, Fig. 13.13a, then the resulting motion will consist of the *set* of natural motions shown in Fig. 13.13b‡ (plus vanishingly small amounts of infinitely many more natural modes). The natural frequency is indicated for each natural-mode shape.§ As with the vibrating shaft, the natural modes are related in an integral way, so that the shape repeats itself with a relatively short period. This order and integral relationship of the vibrations makes their collective sound *musical*. For the most part, sounds which do not have this order about them are considered by our ear-brain system to be unmusical, apparently on purely esthetic grounds. (There are notable exceptions, such as the Oriental gong.)

We have assumed no damping in the above discussion, but, of course, there is a little. Moreover, it is more effective at the higher frequencies. The result is that the sharp shape, Fig. 13.13a, of the initial "pluck" is quickly rounded out to something like Fig. 13.13c, which is made up of perhaps the first half-dozen natural modes (Fig. 13.13b and d). The result is the characteristic "twang" sound of the guitar.

> In an early keyboard instrument, the harpsichord, the strings are plucked mechanically, producing the vibration we have just analyzed. But in the piano the keys are struck with a felt hammer. The sound is somewhat different, as we know, largely because the sharp corner in Fig. 13.13a is "rounded off" by

---

† An extremely interesting discussion of the important parameters in instruments of the violin family is given in an article by Carleen Hutchins, "The Physics of Violins," *Scientific American*, vol. 207, no. 5, p. 73, 1962.

‡ The magnitudes of the various components in Fig. 13.13b and d are obtained by Fourier analysis, which will be discussed in Chapter 16. (In fact, this particular problem is to be solved in Problem 16.8.)

§ The second natural mode is called the "first harmonic" by musicians; the third natural mode is called the "second harmonic," and so on. (Musicians also number the fingers of a violinist differently from those of a pianist, which is grounds for according additional esteem to people who learn to play both instruments.)

[Sec. 13.5] CERTAIN MUSICAL INSTRUMENTS 477

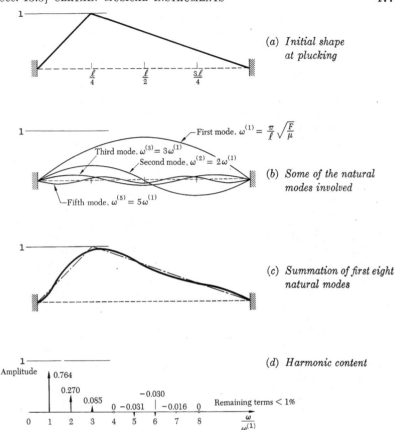

Fig. 13.13 *Typical harmonic spectrum of a plucked string*

the felt, and so therefore is the "twang." There is still the characteristic hard initial sound, however, which rapidly becomes gentler as the higher harmonics of Fig. 13.13d fade faster than the lower ones.

A bowed violin string produces a quite different sound. Here the system is put quickly into steady-state vibration, in which energy is fed steadily into the system by the "self-excited" or "negative-damping" mechanism described for the one-degree-of-freedom system in Fig. 11.9b. The steady-state wave shape is similar to Fig. 13.13c; but it is different, as we know from the sounds we hear. In fact, one of the primary skills of the accomplished violinist is his ability to bow his instrument in such a way that the pleasing "harmonics" (higher natural modes) are produced, and the unpleasing ones suppressed. The beginner is most conspicuous by his failure in this skill.

A slightly practiced ear can distinguish between the various instruments—say a violin, a clarinet, a flute, and a trumpet each playing a note of the same pitch—either in the presence of the listener or on a radio or recording, singly or in groups.† Yet the only difference between the sound made by the different instruments, obviously, is in the relative amounts of the fundamental and the various harmonics (higher modes). Thus the composer has at his disposal a tremendous variety of pleasing sounds which he can call upon by varying the mixture of instruments he employs in a given passage.

When a violinist or guitarist wishes to change the pitch of the note he is playing, he merely presses the string down at a different point on the fingerboard, thus controlling directly the length ℓ of the vibrating string, and hence, from (13.24), the frequency of the "fundamental" (first-mode) vibration and all of the higher modes (which still are integral multiples of the fundamental).

Another part of the violinist's art is that he often modulates slightly the frequency of the tone he produces, by vibrating his finger on the fingerboard. When skillfully done, this produces a most pleasing sound. It is called *vibrato*.

Still another trick, which is called for occasionally in the scores of violin concerti, is the production of a harmonic sans the fundamental and certain of the lower harmonics. The technique is known as "playing harmonics." This is accomplished by holding the finger very lightly exactly at a node (Fig. 13.8)—i.e., at the middle of the string for the elimination of the fundamental, or at the one-third point for elimination of both fundamental and first harmonic, and so on—and then bowing in just the right way. The resulting sound is thinner, more ethereal than that made by simply playing the corresponding higher-pitched note.

Wind instruments: vibrating columns

Another class of musical systems which obey an equation of the form (13.12) involves a long thin column—a bar (solid) or air column (fluid)—vibrating longitudinally.

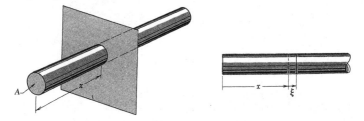

Fig. 13.14 *Geometric description of longitudinal vibration*

To describe what we mean by longitudinal vibration, Fig. 13.14, consider first that the column is at rest, and that we identify one sheet of its molecules by passing a plane normal to the column, at location x, and painting red all the molecules it intersects. Now when the column vibrates, our red sheet of molecules moves back and forth longitudinally.

Let us denote by ξ the longitudinal displacement of the red molecules at a given instant. Then ξ is a function of both x and t: $\xi = \xi(x,t)$. Then it

† On a much more subtle level, some really skilled listeners can identify the great violinists by name partly from the personal tone quality each produces.

can be shown (Problem 13.16) that, for a uniform column, the equation of motion is

$$\mu \frac{\partial^2 \xi}{\partial t^2} = AE \frac{\partial^2 \xi}{\partial x^2} \tag{13.28}$$

which is identical in form to (13.12). In (13.28), μ is the mass per unit length, A is the column's cross-sectional area, and E its elastic modulus.

Equation (13.28) is obtained in terms of the *strain*, $\partial \xi / \partial x$, at a point in the bar, which is entirely analogous to the slope of the string in Fig. 13.8, and to the slope of the twist line on the shaft in Fig. 13.3.

Since the air column (or bar) obeys Eq. (13.12), its natural frequencies and natural modes of motion will be as given by Fig. 13.7 or Fig. 13.8, depending on the boundary conditions. An air column with a closed end is like a torsional shaft with a fixed end; an air column with an open end is like a torsional shaft with a free end. Most musical instruments have one end closed and the other end open, and for these Fig. 13.7 gives the correct pattern. The first natural frequency (fundamental) is

$$\omega = (\pi/2\ell)\sqrt{AE/\mu} \tag{13.29}$$

the second is $\quad \omega = (3\pi/2\ell)\sqrt{AE/\mu} \quad$ and so on

While strings are set vibrating by plucking and bowing, air columns are set vibrating by introducing an alternating pressure at one end. In reed instruments (clarinet, oboe, and so on), the musician blows air past the reed and sets it into vibration through the mechanism of "self excitation" discussed on p. 391. The vibrating reed then supplies the pressure oscillation at one end of the column which sets the whole column in vibration. The reed actually generates *many* frequencies at once, and the column picks out from this array its own natural frequencies at which it resonates. The musician in turn controls the array of reed frequencies by pressure from his lips. Since reeds have many natural frequencies, the tones of reed instruments contain many more harmonics than those of other instruments.

In a flute or an organ (flute stop), the air column is set vibrating by blowing across a hole in the side of the cylinder containing the air column. This seems to be the best means of producing a fundamental vibration with few overtones. Spectral analysis (Fig. 13.13d) of the tones produced by flutes and by the flute stop on organs shows almost no harmonic content.[†] The resulting sound is "pure" and pleasing, but less colorful than the sound of a reed instrument, for example. Flutists add interest to their tone by modulating the *amplitude* at some frequency, perhaps 10 cps, through control of the air supply passing their larynx. The process is also called vibrato. (This is analogous to, but basically different from, the small *frequency-* or pitch-modulation vibrato achieved by the violinist oscillating his hand on the fingerboard.)

In brass instruments (trumpet, French horn, trombone, and so on), vibra-

---

[†] The early success in developing electronic organs may be partly due to the simplicity of the sound to be imitated, and therefore of the circuit required. Authentic simulation of the reed stops has proved much more difficult.

tion of the air column is sustained by the vibrating lips of the player. Because he can control his own lips in a more precise way than, for example, a clarinetist can control his reed, the skilled brass player can evoke almost any one of the first ten or so modes of natural vibration with a given valve setting, and he can get each of them in remarkably pure form.

The method of changing pitch chromatically (by small increments) in a wind instrument is to change the length of the air column. In an organ, many pipes of different fixed lengths are used. In a trombone, a slide is moved in and out. In a trumpet, valves change from one set of "plumbing" to another of different length. In the woodwinds, keys open to change the length of the closed pipe. (In the lower register the scale may be played from high pitch to low pitch merely by closing the keys in succession, beginning near the source of air.)

Pitch is varied also, with a fixed configuration of the instrument, by controlling the excitation so as to produce higher harmonics without the fundamental tone. The player controls the excitation by pinching his lips together (brass) or by clamping the reed more tightly in his lips (reed instruments).

○13.6 A NOTE ON THE MUSICAL SCALE

A brief digression regarding the scale of frequencies used in music may be of interest. It also makes an elegant demonstration of log plotting of frequency, which we introduced in Chapter 9.

As we have noted, the sounds we enjoy listening to in sequence (melody) or in concert (harmony), and which we term "musical," are made up of frequencies which are simple multiples of one another, and which therefore make a repetitive frequency pattern, as distinguished from "noise," which does not.

The simplest frequency relation—or *interval*—in music is the "octave," in which one note has twice the frequency of the other. The sound of notes played one octave apart (middle C and the next C above it on a piano, say) is not very complicated, the higher note merely enforcing what already existed as harmonics in the lower note.

When two notes are played which have the frequency ratio $\frac{3}{2}$ (G and C, for example†), the result is a little more interesting. If next, the ratio $\frac{5}{4}$ is added (E and C, for example), we have the elements of the simplest harmony. Any set of three notes having ratios $\frac{5}{4}$ and $\frac{3}{2}$ is known as a "major triad."

The frequency ratios used in various musical scales are shown in Fig. 13.15. A logarithmic scale is used, so that equal ratios are shown by equal distances, regardless of absolute frequency.‡ An octave ($\frac{2}{1}$) is shown by ○ and the ratios for a major triad are indicated by ⊙ . Indeed, the piano keyboard itself is a standard log scale, covering $6\frac{2}{3}$ octaves, or about 2.2 decades.

The ancient Chinese musical scale includes the ratios of the triad, plus two more ratios: $\frac{5}{3}$ and $\frac{9}{8}$. Simple tunes in the modern musical scale (e.g., *Yankee Doodle*) are composed with the Chinese scale plus the ratios $\frac{4}{3}$ and $\frac{15}{8}$

---

† If the C an octave lower is used, the ratio is simply 3:1. (The interval is called a "major twelfth.")

‡ Musicologists use a log-scale unit called a *cent*. There are 1200 cents per octave; 100 cents per half-tone (e.g., from C to C♯). Thus, from Fig. 9.9, u = 400 cents, and 1 db = 200 cents.

[Sec. 13.6] A NOTE ON THE MUSICAL SCALE

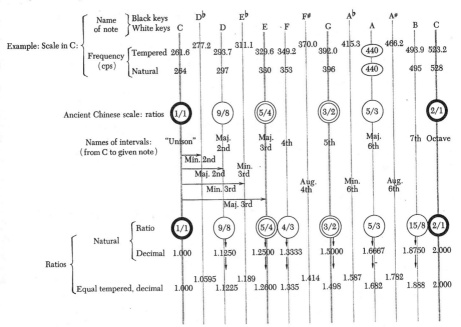

Fig. 13.15 *Frequency ratios used in musical scales*

(plus the octaves above and below each of these). Members of the total set of eight ratios are denoted by ◯ or ⊚ in Fig. 13.15. They are given names, such as "fourth," "major fifth," and so on, as shown. In particular, "octave" is from the Latin *octavus*: "eighth."

For the key of C, the eight ratios are played on the white keys on the piano. For more sophisticated music, we add the sharps and flats (black keys for the key of C), and thus each octave is divided into twelve intervals. Quartertone and microtone compositions also exist, and perhaps in future generations a scale of more than twelve notes, divided into still smaller intervals, will be used almost exclusively, and the scale familiar to us will be considered primitive.

Actually, in our modern keyed instruments (notably the piano) it is impossible to utilize in all cases exactly the simple ratios that make for "pure" harmony. If we did, then we would have to use a separate set of keys for each "key" (scale base) in which we wish to play, e.g., the key of C in the example of Fig. 13.15. For keyboard instruments that pluck or strike a string (so that harmonics die quickly, as in Fig. 13.9), J. S. Bach advanced the tempered scale wherein each octave is divided into twelve *equal* ratios (*equal intervals* on a log scale).† Each ratio must then be $(2)^{\frac{1}{12}} = 1.0595$. (This is readily checked

---

† More precisely, this is known as the tempered scale of equal temperament. The "natural" scale we have been describing is said to have "just temperament." Numerous other tempered scales have been employed. Some are described in a paper by D. D. Boyden, "Prellear, Geminiani, and Just Intonation," *J. Amer. Mus. Soc.*, vol. IV, no. 3 (1951), which reports with expert commentary on the history of tempered-scale usage.

by multiplying 1.0595 by itself 12 times.) The tempered scale is compared with the natural scale in Fig. 13.15:† the difference is never more than about one percent. (The difference is most noticeable in the major third. To a practiced ear the natural major triad chord a string quartet may play when unaccompanied sounds definitely different from the tempered version they must play when accompanied by a piano, for example.)

Musical tones, of course, are merely the building blocks of music. The genius of the composer in ordering their pitch sequence (melody), duration and intensity sequence (rhythm and dynamics), simultaneous combination (harmony), and means of generation (orchestration), in exquisite patterns of infinite variety to evoke images and ideas and emotions—this genius, while based fundamentally on physics, extends far beyond science into the realm of noble art.

Problems 13.18 through 13.21

13.7 RAYLEIGH'S METHOD

Rayleigh's method is a procedure for estimating accurately the natural frequencies of complex physical systems without having to derive and solve their differential equations. It is especially useful for making rapid estimates of the natural frequencies of many-degree-of-freedom and distributed-parameter systems, the exact analysis of which would be highly time-consuming and tedious. The method could have been presented earlier, but we chose to present it instead after study of distributed-parameter systems, in order to indicate its full power.

Lord Rayleigh (John William Strutt) was, along with Helmholtz (see p. 475) one of the nineteenth century's two dozen or so "universal scientists," who made numerous contributions of lasting importance to several branches of science. Rayleigh's *The Theory of Sound*,‡ first published in 1847, presents his extensive studies in mechanical vibration, and is a most comprehensive treatment of that field. One famous and widely used section of *The Theory of Sound* is the presentation of what is now known as "Rayleigh's method" for finding the natural frequency of complex systems.

Rayleigh's method is an energy method. It neglects damping, but is highly accurate for systems having light damping. The method is a direct extension of the energy method used on p. 229 to find the natural frequency of a one-degree-of freedom system. That is, for any system vibrating in a natural mode, Rayleigh's method consists of computing the maximum total potential energy and the maximum total kinetic energy, then equating these and solving for the natural frequency (which must appear in the expression for kinetic energy, but not in that for potential energy).

† In Fig. 13.15 the scales are made to coincide at 440 cps, the standard A used by most orchestras and piano tuners.
‡ Dover Publications, Inc., New York, 1945.

[Sec. 13.7] RAYLEIGH'S METHOD

Specifically, Rayleigh's method consists of the following:

> For an undamped physical model,
> (1) Assume an approximate shape (configuration) for a natural mode of motion.
> (2) Calculate the maximum potential energy and the maximum kinetic energy for vibration in the assumed mode.
> (3) Equate the two energy expressions, and solve for the natural frequency.

Procedure C-5
Rayleigh's method

(The three steps concern, respectively, geometric, physical, and equilibrium relations.) Rayleigh's method always gives, for the first natural mode, a frequency which is somewhat too high.

The key to why Rayleigh's method is so useful is the word *approximate* in Step (1): for Rayleigh found that the method gives a remarkably close approximation to the lowest natural frequency, *even when the mode shape assumed is a poor approximation to the true shape.* The method can be successfully applied, therefore, without knowledge of the mode shape, and even without very good physical intuition. Moreover, the mode shape can be chosen for convenience of computation without degrading the result significantly.

Rayleigh's method is equally applicable to either lumped- or distributed-parameter systems. We now give an example of each, after which we shall make some additional comments about the method.

Example 13.1. Lumped-parameter system. Use Rayleigh's method to find the lowest natural frequency for the system shown in Fig. 13.16a.

Following Procedure C-5, we (1) assume some shape for the first natural mode of vibration. Suppose we guess that the mode shape is the one shown in Fig. 13.16b. (We know better, of course, from the exact analysis of Sec. 13.1 and Fig. 13.2a; but here we intentionally make an incorrect guess, to demonstrate the versatility of the method.)

Next, we (2) calculate the maximum potential energy and the maximum kinetic energy the system would have if it were vibrating in the assumed natural mode.

Potential energy is maximum when the system is (instantaneously) stationary at the extremity of its motion. Suppose we let the extreme displacement of the central disk be θ_{2m}. Then our assumed mode shape is given by

$$\theta_{1m} = \frac{\theta_{2m}}{2} \qquad \theta_{3m} = \frac{\theta_{2m}}{2}$$

Recall from Sec. 2.3 that the (potential) energy stored in a distorted elastic member is equal to the work required to distort it. Thus, if ψ is the amount the shaft is twisted, and the torque-displacement relation is linear, the work is given by $U = \frac{1}{2}k\psi^2$, the area under the torque-displacement curve, Fig. 13.16c.

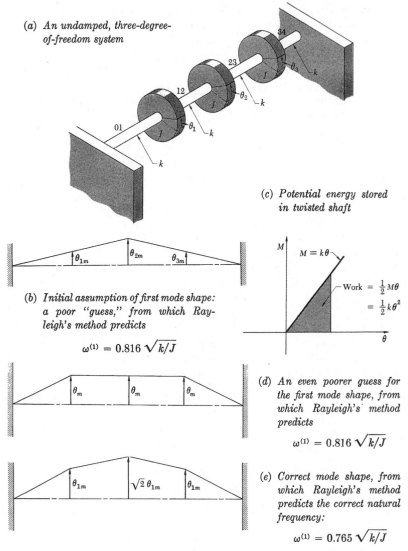

Fig. 13.16 Demonstration of Rayleigh's method for a lumped-parameter system

[Sec. 13.7] RAYLEIGH'S METHOD

For the configuration assumed, this is the twist energy in the two lengths of shaft 02 and 24, in Fig. 13.16a:

$$U_{max} = \frac{1}{2}\left(\frac{k}{2}\right)\theta_{2m}^2 + \frac{1}{2}\left(\frac{k}{2}\right)\theta_{2m}^2 = \frac{k}{2}\theta_{2m}^2$$

(The spring constant of two shafts k in series is $k/2$.)

Recall from Sec. 2.3 that the *kinetic energy* of a rotating inertia is given by

$$T = \tfrac{1}{2} J\Omega^2 \tag{13.30}$$

Kinetic energy is maximum when the system has maximum velocity. Since the shaft in Fig. 13.16 moves synchronously, this occurs as it passes through the static equilibrium position. Thus, when the system in Fig. 13.16 is vibrating in its first natural mode, the angular velocities of the masses, versus time, are

$$\dot\theta_1 = \frac{d}{dt}[\theta_{1m}\cos\omega^{(1)}t] = -\theta_{1m}\omega^{(1)}\sin\omega^{(1)}t$$

$$\dot\theta_2 = \frac{d}{dt}[\theta_{2m}\cos\omega^{(1)}t] = -\theta_{2m}\omega^{(1)}\sin\omega^{(1)}t$$

$$\dot\theta_3 = \frac{d}{dt}[\theta_{3m}\cos\omega^{(1)}t] = -\theta_{3m}\omega^{(1)}\sin\omega^{(1)}t$$

and these have (simultaneously) the following maximum values:

$$(\dot\theta_1)_{max} = -\theta_{1m}\omega^{(1)} \qquad (\dot\theta_2)_{max} = -\theta_{2m}\omega^{(1)} \qquad (\dot\theta_3)_{max} = -\theta_{3m}\omega^{(1)}$$

from which the maximum kinetic energy is, from (13.30),

$$T_{max} = \frac{1}{2}J_1\theta_{1m}^2\omega^{(1)2} + \frac{1}{2}J_2\theta_{2m}^2\omega^{(1)2} + \frac{1}{2}J_3\theta_{3m}^2\omega^{(1)2}$$

$$= \frac{1}{2}\omega^{(1)2}J\left[\left(\frac{\theta_{2m}}{2}\right)^2 + \theta_{2m}^2 + \left(\frac{\theta_{2m}}{2}\right)^2\right]$$

Finally [Step (3) of Procedure C-5], equating U_{max} and T_{max}, we obtain

$$\frac{k}{2}\theta_{2m}^2 = \frac{1}{2}\omega^{(1)2}J\left(\frac{3}{2}\theta_{2m}^2\right)$$

from which

$$\omega^{(1)}_{Rayleigh} = \sqrt{\frac{2}{3}\frac{k}{J}} = 0.816\sqrt{\frac{k}{J}} \tag{13.31}$$

This result is to be compared with the correct value (13.6)

$$\omega^{(1)}_{exact} = 0.765\sqrt{\frac{k}{J}}$$

The discrepancy is only 6% (the approximate answer being too high).

Next, as a further demonstration, consider an even poorer guess for the first-mode shape, which is even easier to calculate from. Consider the mode shape $\theta_{1m} = \theta_{2m} = \theta_{3m} = \theta_m$ (the three disks move as one), as shown in Fig. 13.16d.

Potential energy is stored only in the end shafts (the others being untwisted). The maximum potential energy is thus

$$U_{max} = 2[\tfrac{1}{2}k\theta_m^2]$$

Maximum kinetic energy is given by

$$T_{max} = 3[\tfrac{1}{2}J(\theta_m \omega^{(1)})^2]$$

and the first natural frequency is

$$\omega^{(1)} = \sqrt{\frac{2}{3}\frac{k}{J}} \tag{13.32}$$

as before! Again, the discrepancy is only 6%, even though we guessed a mode shape which is, from intuition alone, a poor approximation to the correct one.

As a final demonstration, we "guess" the correct mode shape, to verify that then Rayleigh's method gives the exact natural frequency. The correct mode shape, from Fig. 13.2, is that shown in Fig. 13.16e. Potential energy is stored in each of the four shafts, and the maximum value of the total potential energy is

$$\begin{aligned} U_{max} &= \tfrac{1}{2}k\theta_{1m}^2 + \tfrac{1}{2}k(\theta_{2m} - \theta_{1m})^2 + \tfrac{1}{2}k(\theta_{3m} - \theta_{2m})^2 + \tfrac{1}{2}k\theta_{3m}^2 \\ &= \tfrac{1}{2}k\theta_{1m}^2[1 + (0.414)^2 + (-0.414)^2 + 1] \\ &= 1.171 k\theta_{1m}^2 \end{aligned}$$

The maximum kinetic energy is

$$T_{max} = \tfrac{1}{2}J\omega^{(1)2}\theta_{1m}^2[1 + (\sqrt{2})^2 + 1] = 2J\omega^{(1)2}\theta_{1m}^2$$

Equating these gives the natural frequency:

$$\omega^{(1)} = \sqrt{\frac{1.171k}{2J}} = 0.765\sqrt{\frac{k}{J}} \tag{13.33}$$

which is the exact value.

Example 13.2. Distributed-parameter system. Use Rayleigh's method to find the natural frequency of vibration of a taut string. (Consider string tension to be constant, F_0.)

Following Procedure C-5, we (1) assume a mode shape for the first natural mode of vibration. Let us assume, for simplicity of calculation, the one given in Fig. 13.17a (which is not a very good approximation). This shape may be described by the following geometric relations:

$$\left.\begin{aligned} y_m(x) &= y_{cm}\frac{x}{\ell/2} & 0 < x < \frac{\ell}{2} \\ y_m(x) &= y_{cm}\left(\frac{\ell - x}{\ell/2}\right) & \frac{\ell}{2} < x < \ell \end{aligned}\right\} \tag{13.34}$$

(2) *To find the potential energy* stored when the string is in the extreme position, we calculate the work required to get it there. This particular configuration could be established physically by pulling laterally on the center of the string with a force f, as indicated in Fig. 13.17b. From a free-body-diagram force balance, Fig. 13.17c, the magnitude of f will be proportional to F_0 and will depend on the deflection y_c as follows: $f = 2F_0 y_c/(\ell/2)$ for small deflections. Therefore the work required to pull the string to its extreme position is

$$\begin{aligned} U_{max} &= \text{work} = \int_0^{y_{cm}} \mathbf{f} \cdot d\mathbf{y}_c = \int_0^{y_{cm}} 2F_0 \left(\frac{y_c}{\ell/2}\right) dy_c \\ U_{max} &= \frac{2F_0 y_{c\,max}^2}{\ell} \end{aligned}$$

[Sec. 13.7] RAYLEIGH'S METHOD

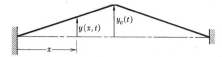

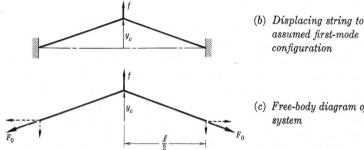

(a) *Initial assumption of first-mode shape*

(b) *Displacing string to assumed first-mode configuration*

(c) *Free-body diagram of static system*

Fig. 13.17 *Demonstration of Rayleigh's method for a distributed-parameter system*

To find the kinetic energy, we must first find the velocity of each element of mass along the string. At an arbitrary location x the displacement is $y = y_m \cos \omega^{(1)} t$, from Fig. 13.17a, and the velocity is $\dot{y} = -y_m \omega^{(1)} \sin \omega^{(1)} t$. This has the maximum value

$$\dot{y}_{\max} = -y_m \omega^{(1)} \tag{13.35}$$

All mass elements have maximum velocity at the same instant (when the string is passing through neutral); and at that instant the total kinetic energy is given by a space integration along the string:

$$T = \tfrac{1}{2} \int \dot{y}^2_{\max} \mu \, dx$$

or, using (13.34) and (13.35),

$$\begin{aligned} T &= \left[\tfrac{1}{2} \int_{x=0}^{\ell/2} \left(-\frac{x}{\ell/2} y_{cm} \omega^{(1)} \right)^2 \mu \, dx \right] \times 2 \\ &= \frac{4 y_{cm}^2}{\ell^2} \omega^{(1)2} \mu \left. \frac{x^3}{3} \right|_0^{\ell/2} \\ &= y_{cm}^2 \omega^{(1)2} \mu \frac{\ell}{6} \end{aligned} \tag{13.36}$$

(From symmetry, the kinetic energy stored in the two halves is obviously the same, hence we need evaluate only one of the integrals.)

(3) Finally, equating the expressions for maximum kinetic and potential energy, we obtain

$$y_{cm}^2 \omega^{(1)2} \mu \frac{\ell}{6} = \frac{2 F_0 y_{cm}^2}{\ell}$$

$$\omega^{(1)} = \frac{3.46}{\ell} \sqrt{\frac{F_0}{\mu}} \tag{13.37}$$

The correct answer [from (13.27)] is

$$\omega^{(1)} = \frac{\pi}{\ell}\sqrt{\frac{F_0}{\mu}} \qquad (13.38)$$

and thus the approximate answer is only 10% high.

The reader is invited to repeat the Rayleigh procedure for the exact mode shape, to show that the result agrees exactly with (13.38). (See Problem 13.36.)

Iteration to the exact natural frequencies: The Rayleigh-Ritz method

An added feature of Rayleigh's method is that subsequent iteration procedures have been developed to improve its accuracy—to any degree desired, in fact. Briefly, the iteration procedures (the best known being due to Ritz) consist of (a) guessing a mode shape and calculating an approximation to the natural frequency by Procedure C-5, (b) using that calculated frequency to improve the estimation of the mode shape, (c) applying Rayleigh's method again to the new mode shape to get a more accurate estimate of $\omega^{(1)}$, and so on. For the first natural mode the procedure converges very rapidly.

For natural modes other than the first, the simple Procedure C-5 gives a less accurate estimate of natural frequency, and the iteration method does not converge to the higher modes. Therefore when the second and higher natural frequencies are desired, a more elaborate technique is employed. The first natural frequency and natural mode are found first (using an iteration method to obtain good accuracy); and then that mode is "subtracted out" in subsequent calculation for the second mode, and so on. The iteration methods are described in textbooks on vibrations.†

Problems 13.22 through 13.37

---

† See, for example, J. P. Den Hartog, "Mechanical Vibrations," 4th ed., pp. 184–197, McGraw-Hill Book Company, New York, 1956.

PART D

TOTAL RESPONSE OF COMPOUND SYSTEMS

Chapter 14. e^{st} and Transfer Functions

Chapter 15. Forced Oscillations of Compound Systems

Chapter 16. Response to Periodic Functions: Fourier Analysis

Chapter 17. The Laplace Transform Method

Chapter 18. From Laplace Transform to Time Response by Partial Fraction Expansion

Chapter 19. Complete System Analysis: Some Case Studies

In Part C we considered the natural behavior alone of compound systems of systems. In Part D we shall consider first the forced motion alone (Chapters 14–16), and then the total motion (Chapters 17–19) of compound systems. This approach parallels the one we followed in Chapter 6 for elementary first-order systems and in Chapter 8 for elementary second-order systems.

In the study of compound systems, a key role is played by the *transfer-function* concept (which we introduced in Sec. 8.8) because of the ease of manipulation it makes possible in complicated systems and because of the considerable physical insight it provides into a system's dynamic character.

The transfer-function method is based on the fact that if an input or forcing function of the particular form e^{st} is applied to an l.c.c. (linear, constant-coefficient) dynamic system, then the *output* of the system *will also be of the form e^{st}*, although in general its magnitude and phase will be different from those of the input. That is, the system will *transfer* this input to an output of the same form, but modified in a manner described by the system's transfer function, as illustrated, for example, in Fig. 14.3b.

Because a dominant class of functions of interest in the forced motion of dynamic systems is inherently of the form e^{st} [Eq. (8.21)], the transfer-function method is used widely.

Further, as we learned in Part C, the *natural*, unforced motions of l.c.c. systems are always of the form e^{st}. Thus, when several such systems are coupled together, they have natural motions in which the output of one subsystem—and therefore the input to the next—is always of the form e^{st}. For this reason, transfer-function methods can be used advantageously to study natural motions of compound systems, as well as forced motion of the form e^{st}.

To begin Part D we consider, in Chapter 14, forced or *transferred response only* to inputs of the simple form e^{st}. We shall develop and demonstrate the transfer-function method in this context. We shall find that, in general, simple e^{st} signals go through high-order, compound, interrelated systems of systems in the same simple way they go through an elementary first- or second-order system (Sec. 8.8). The algebra may become a little laborious, but the concept remains the same: A transfer function, a purely algebraic expression, can be calculated, from the input to the output of interest, and the system's forced response can then be readily written down for any particular input of the form e^{st}.

The situation is especially simple for the special case of subsystems in cascade (Sec. 14.2). Then the e^{st} signal merely runs through the elementary transfer functions one after another, like water through a series of gates, getting its magnitude and phase changed by each transfer function.

In Chapter 15 we shall apply the method to the particular case of forced vibration and other problems in which the input is sinusoidal ($s = j\omega_f$). This will be a continuation of our discussion in Chapter 9; this time we are interested in the response of *compound* systems. (Note that study of

Chapter 15 may be deferred, if desired. Its content is not essential to the presentations that follow.)

In Chapter 16 we extend the transfer-function method to periodic forcing functions of more general form by use of the Fourier series technique. Finally, in Chapter 17, the Laplace transform method is developed, whereby all the functions involved in a dynamics problem—both *arbitrary* forcing functions and natural response to initial conditions—can be represented in the form e^{st}. It will then be possible to investigate all the motions of an l.c.c. system from a transfer-function viewpoint, with significant advantages in systematic uniformity, efficiency, and physical insight.

In Chapter 18 the useful rules for partial-fraction expansion are presented, and in Chapter 19 the methods of Part D are applied to obtain complete solutions for total system response in a variety of dynamics problems.

Chapter 14

e^{st} AND TRANSFER FUNCTIONS

In a linear system with constant coefficients, the time behavior of all of the signals—voltages, forces, displacements, and so on—can be cast in the form Ce^{st}. From this the following concepts can be developed:

(1) When such a signal passes through the components of the system, *the basic form e^{st} is preserved;* but the magnitude and phase, C, of the signal are altered.

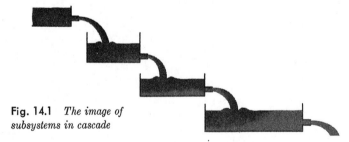

Fig. 14.1 *The image of subsystems in cascade*

(2) The *transfer function* for each subsystem, an expression in s, tells the amount by which a signal of the form Ce^{st} is altered by that subsystem. For a specific signal (e.g., $s = j\omega_f$, for a sine wave) the amount of signal alteration is obtained by substituting for s in the transfer function. The process is facilitated by graphical calculation.

(3) For a *cascaded system*† the transfer function of the overall system is merely the product of those of the subsystems.

† The word *cascade* denotes a one-way flow between subsystems, as in Fig. 14.1, where each subsystem is unaffected by the behavior of others *downstream* from it. Figure 14.3 shows a cascaded electromechanical system. (The isolation amplifiers prevent upstream flow.) In Fig. 14.1 the quantity which flows is water. In Fig. 14.3 there is a "flow" of several kinds of signal—voltage, moment, displacement. The word *tandem* is used interchangeably with the word *cascade*. Refer to Sec. 11.5.

[Sec. 14.1] REVIEW OF THE TRANSFER-FUNCTION CONCEPT

(4) More generally, two-way coupling leads to a more complicated transfer function for the overall system. For either cascaded or more generally coupled systems, the following are true:
   (a) The system's *characteristic function* is contained in the denominator of the system transfer function.
   (b) To determine system response, the poles of the transfer function—roots of the characteristic equation—must first be found. That is, the denominator must be factored (e.g., using the root-locus technique).
(5) *Matrix representation* greatly facilitates the handling of sets of many simultaneous equations. It is also important conceptually.

14.1 REVIEW OF THE TRANSFER-FUNCTION CONCEPT

We recall, from Sec. 8.9, that a transfer function for an l.c.c. system evolves from its equation of motion in the following way. First, we assume that both the forcing input and the forced part of the system's response are of the form Ce^{st}. For example, for the system of Fig. 14.2 the equation of motion is

$$J\ddot{\theta} + b\dot{\theta} + k\theta = M \tag{14.1}$$

and we assume that

$$M = \mathsf{M}e^{st} \quad \text{and} \quad \theta_f = \Theta e^{st} \tag{14.2}$$

where M and Θ may be complex numbers, and are called phasors.

Next, by substituting (14.2) into (14.1) and canceling e^{st} throughout, we obtain output Θ in terms of input M:

$$\Theta = \frac{1}{Js^2 + bs + k} \mathsf{M} \tag{14.3}$$

The ratio Θ/M is known as the *transfer function* of the system of Fig. 14.2,

$$\frac{\Theta}{\mathsf{M}} = \frac{1}{Js^2 + bs + k} \tag{14.4}$$

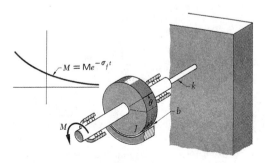

Fig. 14.2 *Forced second-order system*

because it indicates how much of input $M = Me^{st}$ will be *transferred* to output $\theta = \Theta e^{st}$, for any input of this form. That is, the general form of the forced-response solution to (14.1) is

$$\boxed{\theta_f = \left[\frac{\Theta}{M}\right] M} \tag{14.5}$$

Thus, for example, if the input to the system in Fig. 14.2 is $M = M_0 e^{-\sigma_f t}$ (that is, if $s = -\sigma_f$ and $M = M_0$), the forced portion of the response is written at once from (14.5):

$$\theta_f = \left\{\left[\frac{\Theta}{M}\right] M\right\}_{s=-\sigma_f} = \left[\frac{1}{Js^2 + bs + k}\right]_{s=-\sigma_f} M_0 e^{-\sigma_f t} = \left(\frac{M_0}{J\sigma_f^2 - b\sigma_f + k}\right) e^{-\sigma_f t}$$

If the input is sinusoidal, $M = M_0 \cos(\omega_f t + \psi_0) \stackrel{r}{=} (M_0 e^{j\psi_0}) e^{j\omega_f t}$ (that is, if $s = j\omega_f$ and $M = M_0 e^{j\psi_0}$), the forced part of the response is

$$\theta_f \stackrel{r}{=} \left[\frac{1}{Js^2 + bs + k}\right]_{s=j\omega_f} M_0 e^{j\psi_0} e^{j\omega_f t} = \left|\frac{M_0}{-J\omega_f^2 + jb\omega_f + k}\right| e^{j(\psi_0+\psi_f)} e^{j\omega_f t}$$

$$= \left|\frac{M_0}{-J\omega_f^2 + jb\omega_f + k}\right| \cos(\omega_f t + \psi_0 + \psi_f)$$

$$\psi_f = \bigg/ \frac{1}{-J\omega_f^2 + jb\omega_f + k}$$

and so on.

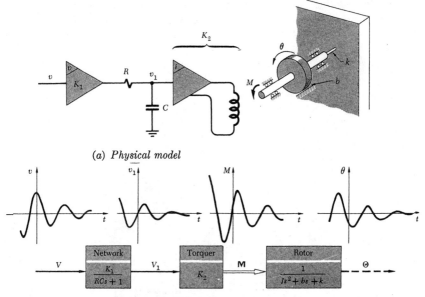

(a) *Physical model*

(b) *Mathematical block diagram representation*

Fig. 14.3 *Subsystems in cascade*

[Sec. 14.2] TRANSFERRED RESPONSE OF SUBSYSTEMS IN CASCADE 495

Notice that we are considering here only the forced part of the response, and are not concerned with the system's transient natural motions which may accompany it. For this reason, we are considering the motions to be *continuing*, rather than abruptly commencing ones. That is, we write $M = M_0 e^{-\sigma_f t}$ rather than $M = M_0 e^{-\sigma_f t} u(t)$. This is indicated graphically by the time plots of the various motions in Fig. 14.3b. (We shall take up the question of *total* response to abrupt inputs, beginning in Chapter 17.)

The general procedure (14.3) through (14.5) may seem to entail unnecessary steps in solving a simple problem; but in more involved systems the advantages of carrying a general solution are several and considerable, as will be seen. Two of the most important are (1) the characteristics of the system's motions are instantly available from its transfer function, and (2) transfer functions are readily manipulated. (Additional—and more fundamental—advantages will be expounded in Chapters 17 and 18.)

Advantage number (1) was discussed in Sec. 8.9, where we showed (for a second-order system) that the denominator of the transfer function contains the characteristic function of the system represented. That this is so—for *any* transfer function—is clear from the derivation of Eq. (14.4) from (14.1) and (14.2): When (14.2) was substituted into (14.1), the left-hand side of (14.1) became the characteristic equation, just as it always does (see, for example, Sec. 8.8). The entire left-hand side of (14.1) in turn became the denominator of (14.4). Since this procedure is the same for solving the differential equation *for any l.c.c. dynamic system, the denominator of any transfer function will always contain the characteristic function of the system.*

14.2 TRANSFERRED RESPONSE OF SUBSYSTEMS IN CASCADE

To see advantage number (2)—the ease with which transfer functions can be manipulated in studying a compound system—consider, for example, the model of two subsystems in cascade in Fig. 14.3a. The differential equations for the subsystems are

$$K_1 \frac{v}{R} = \frac{v_1}{R} + C\dot{v}_1 \qquad K_2 v_1 = M \qquad M = J\ddot{\theta} + b\dot{\theta} + k\theta \qquad (14.6)$$

To use the transfer-function method for solving Eqs. (14.6) we assume that the input is of the form $v = Ve^{st}$, and that the motions of every other variable in the system are of this same form:

$$v_1 = V_1 e^{st} \qquad M = \mathsf{M} e^{st} \qquad \theta = \Theta e^{st} \qquad (14.7)$$

Next, the appropriate transfer functions are obtained by substitution of expressions (14.7) into each of differential equations (14.6):

$$\frac{V_1}{V} = \frac{K_1}{RCs + 1} \qquad \frac{\mathsf{M}}{V_1} = K_2 \qquad \frac{\Theta}{\mathsf{M}} = \frac{1}{Js^2 + bs + k} \qquad (14.8)$$

Then, by (14.5), the behavior of each variable is given by

$$v_1 = \left[\frac{V_1}{V}\right] v \qquad M = \left[\frac{\mathsf{M}}{V_1}\right] v_1 \qquad \theta = \left[\frac{\Theta}{\mathsf{M}}\right] M \qquad (14.9)$$

Relations (14.8) are conveniently displayed in block-diagram form, as shown in Fig. 14.3b. (The block-diagram convention was introduced in Fig. 8.19.) The input and output of each block are customarily labeled with the appropriate capital letters, from (14.8). For further convenience of physical reference, electrical signals are represented in Fig. 14.3b by single lines →, forces by ⇒, and geometric displacements by --→.†

The block-diagram display has the merit of identifying each of the physical member subsystems and their interrelation to the others, and of making it easy to follow a signal—a motion, force, or voltage—through the system from one subsystem to the next, the modification of the signal by each subsystem being explicitly indicated by a transfer function.

The total signal modification produced by the entire system, from input v to output θ, is readily represented by noting the following identity:

$$\theta = \left(\frac{\theta}{M}\right)\left(\frac{M}{v_1}\right)\left(\frac{v_1}{v}\right) v$$

where the terms in parenthesis are, by (14.9),‡ simply the corresponding transfer functions:

$$\theta = \left[\frac{\Theta}{\mathsf{M}}\right]\left[\frac{\mathsf{M}}{V_1}\right]\left[\frac{V_1}{V}\right] v$$

for any specified s, and the transfer functions are, in turn, simply the contents of the blocks in Fig. 14.3b. Thus it is proper to obtain the output of the system simply by multiplying together the blocks, as follows:

$$\Theta = \left[\frac{K_1}{RCs + 1}\right][K_2]\left[\frac{1}{Js^2 + bs + k}\right] V \qquad (14.10)$$

and
$$\theta = \Theta e^{st}$$

For example, if the input is $v = Ve^{-\sigma_f t}$, then the forced response of this system is obtained at once by substituting $s = -\sigma_f$ into (14.10):

$$\theta = \left[\frac{K_1}{-RC\sigma_f + 1}\right][K_2]\left[\frac{1}{J\sigma_f^2 - b\sigma_f + k}\right] Ve^{-\sigma_f t} \qquad (14.11)$$

The general case for subsystems in cascade is given in Fig. 14.4:

$$y(t) = \{Ye^{st}\}_{s=\alpha_f} \qquad Y = \left[\prod_{i=1}^{i=n} G_i(s)\right] X \qquad (14.12)$$

where α_f is complex, in general, and $\prod_{i=1}^{i=n}$ means "product of terms 1 to n."

---

† These symbols will be especially helpful in avoiding confusion at the design stage, where they will remind us, for example, that while signals → are available to us electrically, signals --→ are not, and we must first measure them if we wish to use them.

‡ Provided that v can be expressed as $v = Ve^{st}$.

[Sec. 14.3] GRAPHICAL EVALUATION FROM POLE-ZERO DIAGRAM

The paramount feature of cascaded subsystems is that each subsystem affects only those others which are *downstream* from it. It is this property that makes possible the simple multiplication of transfer functions in (14.12).

Problems 14.1 through 14.7

14.3 GRAPHICAL EVALUATION OF TRANSFER FUNCTIONS FROM THE SYSTEM POLE-ZERO DIAGRAM

For the special case that α_f in Fig. 14.4 is real, as in (14.11), obtaining the numerical solution for the magnitude of the output is simply a matter of arithmetic with real numbers. If s is complex, however, it is often quicker and less cumbersome to evaluate the transfer function graphically. If we use the system of Eq. (14.10) as an example, the procedure is as follows:

First, the numerator and denominator of the transfer function are factored and the factors are plotted as zeros, O, and poles, ×, on the s

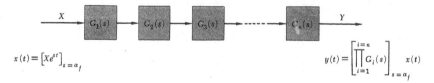

Fig. 14.4 *Forced response of subsystems in cascade: general case*

plane, as discussed on p. 227. In Eq. (14.10) there are no numerator factors, and the factors of the denominator are $[s + (1/RC)]$ and—assuming ζ is less than 1 for the spring-restrained rotor—$(s + \sigma_2 + j\omega_2)$ and $(s + \sigma_2 - j\omega_2)$, with $\sigma_2 = b/2J$, and with $\omega_2 = \sqrt{(k/J) - (b/2J)^2}$ (refer to Sec. 8.2). Thus, ×'s are plotted in Fig. 14.5 at points $s = -1/RC$, $s = -\sigma_2 - j\omega_2$, and $s = -\sigma_2 + j\omega_2$.

Next, the magnitude and phase angle of the product of square brackets in Eq. (14.10) can be evaluated graphically, for any complex value of s, from the system pole-zero diagram, as indicated in Fig. 14.5.

For example, suppose that the input voltage in Fig. 14.3 had the form

$$v = Ve^{-\sigma_f t} \cos \omega_f t \stackrel{r}{=} Ve^{(-\sigma_f + j\omega_f)t} \tag{14.13}$$

Then the output would be

$$\theta \stackrel{r}{=} \frac{K_1 K_2}{RCJ} \left[\left\{ \frac{1}{[s + (1/RC)](s + \sigma_2 + j\omega_2)(s + \sigma_2 - j\omega_2)} \right\} Ve^{st} \right]_{s = -\sigma_f + j\omega_f} \tag{14.14}$$

The point $s = -\sigma_f + j\omega_f$ can be plotted in Fig. 14.5 as shown (point P). The three vectors indicated in the transfer function—that is, in the square brackets in (14.14)—are simply vectors drawn from the poles of the transfer

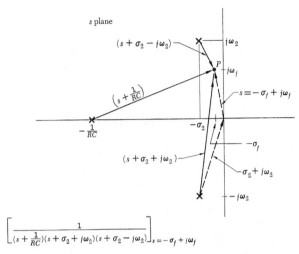

Fig. 14.5 *Graphical evaluation of a transfer function*

function to point P, as shown in Fig. 14.5, and thus (14.14) can be rewritten

$$\theta \stackrel{r}{=} \frac{K_1 K_2}{RCJ} V \left(\frac{1}{\text{product of solid vectors in Fig. 14.5}} \right) e^{(-\sigma_f + j\omega_f)t}$$

$$\stackrel{r}{=} \frac{K_1 K_2}{RCJ} V \left| \frac{1}{\text{product of solid vectors in Fig. 14.5}} \right| e^{[(-\sigma_f + j\omega_f)t + j\psi_f]}$$

$$= \frac{K_1 K_2}{RCJ} V \left| \frac{1}{\text{product of solid vectors in Fig. 14.5}} \right| e^{-\sigma_f t} \cos(\omega_f t + \psi_f) \quad (14.15)$$

where ψ_f is equal to minus the sum of the *angles* of all the vectors in Fig. 14.5. Thus the quantities in (14.14) can be quickly evaluated in two steps: (1) obtain graphically the product of the magnitudes of the vectors in Fig. 14.5; (2) obtain graphically the sum of the angles of the vectors in Fig. 14.5. As we noted in Chapter 11, these two operations can be performed rapidly with a Spirule (or, less rapidly, with a scale and a protractor).

Input (14.13) is a damped sine wave. If the input had been a simple exponential instead—$v = Ve^{-\sigma_f t}$—point P in Fig. 14.5 would have been on the real axis at location $s = -\sigma_f$. The reader is invited to show that the value of θ obtained graphically in this case is identical with (14.11).

Another special case of great interest is the simple sinusoidal input $v = V \cos \omega_f t \stackrel{r}{=} Ve^{j\omega_f t}$. For this case, point P is on the imaginary axis (at distance ω_f from the origin). This special case will be studied in considerable detail in Chapter 15.

A third special case is for an input that is a constant; then point P appears at the origin of the s plane, so that $s = 0$ and $v = Ce^{0t} = C$. Then the forced part of the system's response is, of course, also a constant:

$$\theta = \frac{K_1 K_2}{RCJ} \left[\left\{ \frac{1}{[s + (1/RC)](s + \sigma_2 \pm j\omega_2)} \right\} Ve^{st} \right]_{s=0} = \frac{K_1 K_2/J}{\omega_{02}{}^2} V$$

[Sec. 14.3] GRAPHICAL EVALUATION FROM POLE-ZERO DIAGRAM 499

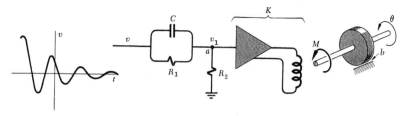

(a) *Physical model*

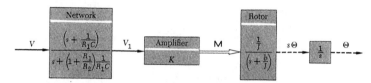

(b) *Mathematical block-diagram representation*

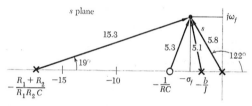

(c) *Evaluating coefficient* Θ

Fig. 14.6 *Studying the forced response of a cascaded system*

Example 14.1. Cascaded subsystems. Find the response $\theta(t)$ of the system shown in Fig. 14.6a if $R_1 C = 0.2$ sec, $R_1/R_2 = 2.5$, $b/J = 2$ sec^{-1}, and if the input is given by

$$v = V_0 e^{-3t} \cos\left(5t + \frac{\pi}{6}\right)$$

(Note that the input is not abrupt, as shown.) The equations of motion are

$$C(\dot{v} - \dot{v}_1) + \frac{v - v_1}{R_1} = \frac{v_1}{R_2} \qquad \text{for the network (KCL)}$$
$$v_1 K = M \qquad \text{for the amplifier}$$
$$M = J\ddot{\theta} + b\dot{\theta} \qquad \text{for the rotor}$$

(as the reader may show). Assume solutions of the form $v = Ve^{st}$, and so on. Then the above equations lead to the following transfer functions:

$$\frac{V_1}{V} = \left\{\frac{s + (1/R_1 C)}{s + [(R_1 + R_2)/R_1 R_2 C]}\right\} \qquad \frac{M}{V_1} = K \qquad \frac{\Theta}{M} = \frac{1}{Js^2 + bs}$$

From these a block diagram, Fig. 14.6b, can be drawn to represent the system. (Note that since $\theta = \Theta e^{st}$, we have $\dot{\theta} = s\theta$, and thus the signal $s\Theta$ in the block diagram represents $\dot{\theta}$, and the last block represents an integration from $\dot{\theta}$ to θ.)

Response $\theta(t)$ to the prescribed input is given by

$$\theta = [\{T.F.\}V]_{s=-\sigma_f+j\omega_f} \stackrel{r}{=} \left[\left\{\frac{(K/J)[s + (1/R_1C)]}{\left(s + \frac{R_1 + R_2}{R_1R_2C}\right)s\left(s + \frac{b}{J}\right)}\right\}V_0 e^{(st+j\psi_0)}\right]_{s=-\sigma_f+j\omega_f}$$

For the given input, $v = V_0 e^{-3t} \cos(5t + \pi/6)$, we have $s = -3 + j5$, and the value of $\{T.F.\}_{s=-\sigma_f+j\omega_f}$ is obtained from the s-plane picture in Fig. 14.6c (with the given physical values), from which

$$\left(s + \frac{1}{R_1C}\right) = 5.3 \ \underline{/70°} \qquad \left(s + \frac{R_1 + R_2}{R_1R_2C}\right) = 15.3 \ \underline{/19°}$$

$$s = 5.8 \ \underline{/122°} \qquad \left(s + \frac{b}{J}\right) = 5.1 \ \underline{/103°}$$

and

$$\theta \stackrel{r}{=} \frac{K}{J}\left[\frac{(5.3)}{(5.8)(15.3)(5.1)} e^{j(+70°-122°-19°-103°)}\right] V_0 e^{[(-3+j5)t+j30°]}$$

or

$$\boxed{\theta = 0.012 \frac{K}{J} V_0 e^{-3t} \cos(5t - 144°)}$$

Note that since the characteristic function $s(s + 17.5)(s + 2)$ is contained in the denominator of the transfer function,† the system is also capable of natural motions of the form

$$\theta_n = C_1 + C_2 e^{-17.5t} + C_3 e^{-2t}$$

in which the C's depend upon initial conditions and upon when input v is applied.

Problems 14.8 through 14.11

14.4 TRANSFERRED RESPONSE OF SYSTEMS WITH GENERAL COUPLING

Although one encounters many cases in which the elements in systems affect one another in a one-way, or cascaded manner, as in Fig. 14.3a or Fig. 12.1a, there are many more general cases in which coupling is two-way, and in which one element may interact with *several* other elements, as in Fig. 14.7. Moreover, inputs may occur at several locations in the system.

In this more general situation the analysis is more complicated than for cascaded subsystems, but the transfer-function technique is still of great utility because it permits us, by considering every variable (including inputs) to be of the form e^{st}, to convert the general set of *differential* equations into a set of *simultaneous algebraic equations*, which can then be solved by straightforward algebra (e.g., by Cramer's rule) for the transfer function between any two variables desired.

† Which implies the characteristic equation $s(s + 17.5)(s + 2) = 0$. See p. 277.

Fig. 14.7 *System having two-way coupling*

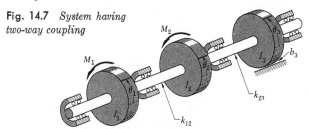

Thus, for the system of Fig. 14.7, we first write the differential equations

$$J_1\ddot{\theta}_1 + k_{12}(\theta_1 - \theta_2) = M_1$$
$$k_{12}(\theta_2 - \theta_1) + J_2\ddot{\theta}_2 + k_{23}(\theta_2 - \theta_3) = M_2 \quad (14.16)$$
$$k_{23}(\theta_3 - \theta_2) + J_3\ddot{\theta}_3 + b_3\dot{\theta}_3 = 0$$

Then we assume that the inputs have the general behavior $M_1 = \mathsf{M}_1 e^{st}$, and so on, and that the motions follow suit, $\theta_1 = \Theta_1 e^{st}$, and so on. Substituting these into (14.16) and canceling the common multiplier e^{st} gives

$$\begin{aligned} (J_1s^2 + k_{12})\Theta_1 \quad &-k_{12}\Theta_2 &&= \mathsf{M}_1 \\ -k_{12}\Theta_1 + (J_2s^2 + k_{12} + k_{23})\Theta_2 \quad &-k_{23}\Theta_3 &&= \mathsf{M}_2 \quad (14.17) \\ -k_{23}\Theta_2 + (J_3s^2 + b_3s + k_{23})\Theta_3 &&&= 0 \end{aligned}$$

a set of simultaneous linear, algebraic equations, from which any desired transfer function can be obtained by Cramer's rule (Appendix H). For example, the transfer function between M_1 and Θ_3 is obtained by solving (14.16) with $\mathsf{M}_2 = 0$:

$$\Theta_3(s) = \frac{\begin{vmatrix} (J_1s^2 + k_{12}) & -k_{12} & \mathsf{M}_1(s) \\ -k_{12} & (J_2s^2 + k_{12} + k_{23}) & 0 \\ 0 & -k_{23} & 0 \end{vmatrix}}{\begin{vmatrix} (J_1s^2 + k_{12}) & -k_{12} & 0 \\ -k_{12} & (J_2s^2 + k_{12} + k_{23}) & -k_{23} \\ 0 & -k_{23} & (J_3s^2 + b_3s + k_{23}) \end{vmatrix}} \quad (14.18)$$

$$\frac{\Theta_3}{\mathsf{M}_1}(s) = \frac{k_{12}k_{23}}{[(J_1s^2 + k_{12})(J_2s^2 + k_{12} + k_{23})(J_3s^2 + b_3s + k_{23}) - k_{23}{}^2(J_1s^2 + k_{12}) - k_{12}{}^2(J_3s^2 + b_3s + k_{23})]} \quad (14.19)$$

The transfer function between M_2 and θ_3 would be obtained, in analogous fashion, from

$$\Theta_3(s) = \frac{\begin{vmatrix} (J_1s^2 + k_{12}) & -k_{12} & 0 \\ -k_{12} & (J_2s^2 + k_{12} + k_{23}) & \mathsf{M}_2(s) \\ 0 & -k_{23} & 0 \end{vmatrix}}{\begin{vmatrix} (J_1s^2 + k_{12}) & -k_{12} & 0 \\ -k_{12} & (J_2s^2 + k_{12} + k_{23}) & -k_{23} \\ 0 & -k_{23} & (J_3s^2 + b_3s + k_{23}) \end{vmatrix}} \quad (14.20)$$

We can also, of course, readily obtain the transfer function between either of the moments, M_1 or M_2, and either of the motions, Θ_1 or Θ_2, of the other disks.

In every case the denominator is the same, and it contains the system's characteristic function† *which—as always—indicates the natural motions the system can have.* Here the denominator is not yet in factored form [as it is, for example, in (14.10)], and the natural motions the system can have are not yet evident. As in Part C, the characteristic equation must be solved for its roots, which is identical with saying that *the factors of the denominator— the poles of the transfer function—must be found.* Each of the poles connotes a possible natural motion of the system. Root-locus or other techniques may be employed to factor the denominator.

14.5 MATRIX REPRESENTATION AND STANDARD FORM

Concept

People who deal often with cumbersome sets of linear algebraic equations, like (14.17), use a special shorthand called matrix algebra, both to represent the sets of equations in compact form and to manipulate them efficiently.

For example, the set of equations (14.17) would be written in matrix form as follows:

$$[A]\underline{\Theta} = \underline{M} \tag{14.21}$$

and the manipulation of these equations to obtain the transfer function (14.18) would be indicated by

$$\frac{\Theta_3}{M_1} = \frac{a_{13}}{|A|} \tag{14.22}$$

where the notation is explained below.

Expressions (14.21) and (14.22) are certainly more compact (and more elegant) than (14.17) and (14.18), which they replace. They also save the analyst of large systems a lot of writing. (He will nevertheless still need to write out all the detailed algebra one time if he wants a particular answer.)

Notation

The meaning of the matrix notation used in (14.21) and (14.22) is almost evident from comparison with (14.17) and (14.18).

To see the logic of the notation, consider again the set of equations (14.17). Suppose that, instead of writing the variables Θ_1, Θ_2, and Θ_3 in each equation, we simply show them at the top of the appropriate column, and write the equations as follows:

$$\begin{matrix} \Theta_1 & \Theta_2 & \Theta_3 \end{matrix}$$
$$\begin{bmatrix} (J_1 s^2 + k_{12}) & -k_{12} & 0 \\ -k_{12} & (J_2 s^2 + k_{12} + k_{13}) & -k_{23} \\ 0 & -k_{23} & (J_3 s^2 + b_3 s + k_{23}) \end{bmatrix} = \begin{bmatrix} M_1 \\ M_2 \\ M_3 \end{bmatrix} \tag{14.23}$$

† Refer to Sec. 8.9.

(We have saved a little writing already.) Next, we march the row of variables Θ around to the end of the box, solely because we shall find this convenient later:

$$\begin{bmatrix} (J_1 s^2 + k_{12}) & -k_{12} & 0 \\ -k_{12} & (J_2 s^2 + k_{12} + k_{13}) & -k_{23} \\ 0 & -k_{23} & (J_3 s^2 + b_3 s + k_{23}) \end{bmatrix} \begin{bmatrix} \Theta_1 \\ \Theta_2 \\ \Theta_3 \end{bmatrix} = \begin{bmatrix} M_1 \\ M_2 \\ M_3 \end{bmatrix}$$

(14.24)

The equations are now in *matrix form*. The large box on the left is called a "3 × 3 square matrix" for obvious reasons; the slender box with Θ's in it and the one with M's in it are called "column matrices."

Finally, each of the three matrices can be replaced by a single symbol:

$$[A][\Theta] = [M] \qquad (14.25)$$

The notation in (14.22) is now self-explanatory also:

$|A|$ means "determinant of the matrix $[A]$"

a_{13} means "the *minor* of term A_{13} (row 1, column 3) in the matrix of $[A]$."

(The minor is the smaller determinant remaining when the row and the column of which A_{13} is a member are both removed. Refer to Appendix H.)

Column matrices are often also called "vectors." The idea is that for a given set of numerical values the n members of a column could be plotted on a piece of n-dimensional paper. For example, the Θ column in (14.24) might look like Fig. 14.8. Then as time goes on, and all three Θ's change, the tip of the Θ "vector" will describe a path in this *coordinate space*.

Here an obvious point must be made most emphatically: *Column "vectors" are not physical vectors* like velocity or force. They are vectors only in the mathematical sense: They are vectors not in *physical* space, but in *mathematical* space (in the same sense that phasors or complex vectors in the s plane are). That is, in general, *a column "vector" is merely a set of*

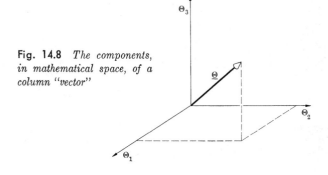

Fig. 14.8 *The components, in mathematical space, of a column "vector"*

numbers. To distinguish between physical vectors and mathematical vectors we show the former in boldface in this book,† and the latter with a bar underneath: $\underline{\Theta}$.

Column vectors (column matrices) do have a physical meaning, of course, if they represent physical variables in an equation of motion. The vector $\underline{\Theta}$ in Fig. 14.8, for example, represents at a given instant the *total configuration* (i.e., the set of shaft angles) of the system of Fig. 14.7 at that instant. But the configuration is not a physical vector, not an arrow in physical space, like velocity or force. In fact, in this case the configuration is only one-dimensional.

The set of moments in (14.24) is another column vector. A column vector might be a set of voltages (giving the configuration of a network on the node basis) or currents (loop basis), or a set of pressures or temperatures, or a combination of several of these. Any set of variables used in the equations of motion of a physical system will appear as a column matrix when the equations are written in matrix form.

State variables and standard form

An especially important modern application of matrix representation is to a set of differential equations of motion in standard form. That is, a set of state variables is selected, and used in the equations of motion, which thereupon become first order, there being as many equations as state variables. Matrix representation renders the study of such standard-form sets particularly systematic.

To illustrate, consider again Eqs. (14.16), the equations of motion of the system in Fig. 14.7. An obvious and convenient set of state variables are the coordinates θ_1, θ_2, θ_3, which already appear in (14.16), plus their derivatives, which we denote by

$$\Omega_1 \triangleq \dot{\theta}_1 \qquad \Omega_2 \triangleq \dot{\theta}_2 \qquad \Omega_3 \triangleq \dot{\theta}_3 \qquad (14.25a)$$

Substituting these into (14.16) wherever a second derivative appears, and also incorporating them as three additional equations of motion, we obtain:

$$\begin{aligned}
\dot{\theta}_1 &= \Omega_1 \\
\dot{\Omega}_1 &= -(k_{12}/J_1)\theta_1 + (k_{12}/J_1)\theta_2 & & & + M_1/J_1 \\
\dot{\theta}_2 &= \Omega_2 \\
\dot{\Omega}_2 &= (k_{12}/J_2)\theta_1 - [(k_{12}+k_{23})/J_2]\theta_2 & (k_{23}/J_3)\theta_3 & & + M_2/J_2 \\
\dot{\theta}_3 &= \Omega_3 \\
\dot{\Omega}_3 &= (k_{23}/J_3)\theta_2 & -(k_{23}/J_3)\theta_3 & -(b_3/J_3)\Omega_3 & + M_3/J_3
\end{aligned}$$
$$(14.25b)$$

† The reader may find it convenient in handwriting to use a heavy overbar: $\overline{V} \equiv \mathbf{V}$, to denote *space* vectors, thus distinguishing them from column matrices, $\underline{V}$, and from phasors and complex vectors in the s plane, for which we sometimes use a light arrow over, $\vec{V}$.

Next, we define three matrices as follows:

$$x \triangleq \begin{bmatrix} \theta_1 \\ \Omega_1 \\ \theta_2 \\ \Omega_2 \\ \theta_3 \\ \Omega_3 \end{bmatrix} \qquad [F] \triangleq \begin{bmatrix} 0 & 1 & 0 & 0 & 0 & 0 \\ -k_{12}/J_1 & 0 & k_{12}/J_1 & 0 & 0 & 0 \\ 0 & 0 & 0 & 1 & 0 & 0 \\ k_{12}/J_2 & 0 & -(k_{12}+k_{23})/J_2 & 0 & k_{23}/J_2 & 0 \\ 0 & 0 & 0 & 0 & 0 & 1 \\ 0 & 0 & k_{23}/J_3 & 0 & -k_{23}/J_3 & -b_3/J_3 \end{bmatrix}$$

$$\underline{\mu} \triangleq \begin{bmatrix} 0 \\ M_1/J_1 \\ 0 \\ M_2/J_2 \\ 0 \\ M_3/J_3 \end{bmatrix} \qquad (14.25c)$$

Then (14.25b) can be written in matrix form as

$$\text{Standard matrix form} \qquad \boxed{\underline{\dot{x}} = [F]\underline{x} + \underline{\mu}} \qquad (14.25d)$$

which is a general form much used in advanced system theory because of its compactness and generality. For transfer-function studies, we can assume all variables to be of the form $\underline{x} = \underline{X}e^{st}$ or $\underline{\mu} = \underline{\mathfrak{M}}e^{st}$ and write at once

$$s\underline{X} = [F(s)]\underline{X} + \underline{\mathfrak{M}}(s) \qquad (14.25e)$$

Column vectors sometimes represent physical vectors

In special cases the members of a column "vector" may actually be the same as the components of a physical vector. Suppose, for example, we write the differential equation for the motion of the mass center of a rigid body, and suppose we choose as independent variables the three rectangular ("cartesian") coordinates defining the location of the mass center. Then the column matrix of variables is also the set of coordinates of the *physical* position vector that locates the mass center. Moreover, the components of the *physical* force vector will also show up as the members of a column matrix in this case. Similarly, if we write the equations of motion in terms of the cartesian components of the velocity of the mass center, then the column matrix of velocity will contain the components of the physical velocity vector.

Indeed, *any physical-vector equation* can also be written in matrix form in terms of cartesian coordinates, and then the components of the physical vectors will appear as members of column-matrix "vectors."

> The test for whether the members of a column matrix represent a physical vector is the property of *invariance*. Because a physical vector is an arrow in physical space, its magnitude and direction are independent of the reference frame or coordinate space it is viewed from: they are invariant. This invariance dictates a specific pattern in which the vector's components change (transform) as the coordinate frame is rotated with respect to the vector. If the members of a column matrix do not abide by this pattern, they are not the components of a physical vector.

Rotational equations of motion for a single rigid body

A particularly interesting and important set of equations of motion are those for the rotational motions about the mass center of a rigid body, such as a spacecraft. For small angular velocities they may be approximated in the following general form (as derived in Appendix E):

$$\begin{aligned} I_x \dot{\Omega}_x - I_{xy} \dot{\Omega}_y - I_{xz} \dot{\Omega}_z &= M_x \\ -I_{xy} \dot{\Omega}_x + I_y \dot{\Omega}_y - I_{yz} \dot{\Omega}_z &= M_y \\ -I_{xz} \dot{\Omega}_x - I_{yz} \dot{\Omega}_y + I_z \dot{\Omega}_z &= M_z \end{aligned} \quad (14.26)$$

where Ω_x, Ω_y, and Ω_z are components of the angular-velocity vector along a set of three orthogonal axes *fixed in the vehicle;* M_x, M_y, M_z are components of the external vector moment acting on the vehicle along the same axes; and the I's are moments of inertia—mass-distribution properties of the rigid body. (Notice that if Ω_y and Ω_z are constrained to be 0, the first equation reduces to the familiar $J_x \dot{\Omega}_x = M_x$ for rotation about a single axis.)

Suppose the input moments are of the form $M = \mathsf{M} e^{st}$, and assume, as usual, that the motions are of the same form. Then Eqs. (14.26) (we write them directly in matrix form) become:

$$s \begin{bmatrix} I_x & -I_{xy} & -I_{xz} \\ -I_{xy} & I_y & -I_{yz} \\ -I_{xz} & -I_{yz} & I_z \end{bmatrix} \begin{bmatrix} \Omega_x \\ \Omega_y \\ \Omega_z \end{bmatrix} = \begin{bmatrix} \mathsf{M}_x \\ \mathsf{M}_y \\ \mathsf{M}_z \end{bmatrix} \quad (14.27)$$

which can be written in the compact and useful form

$$s[I]\underline{\Omega} = \underline{\mathsf{M}} \quad (14.28)$$

where the members of $\underline{\Omega}$ and $\underline{\mathsf{M}}$ are also the x, y, and z components of the physical vectors in space $\boldsymbol{\Omega}$ and $\mathbf{M}$. (Recall, for comparison, that column vector $\underline{\mathsf{M}}$ in (14.21) was not associated with any physical vector.)

The square matrix $[I]$ is called "the inertia tensor" of the body. For one particular set of body-fixed axes, x_p, y_p, z_p, the inertia tensor is a diagonal matrix, the "cross-product-of-inertia" terms I_{xy}, I_{xz}, I_{yz} being zero:

$$[I] = \begin{bmatrix} I_{x_p} & 0 & 0 \\ 0 & I_{y_p} & 0 \\ 0 & 0 & I_{z_p} \end{bmatrix} \quad (14.29)$$

The axes x_p, y_p, z_p are called the principal axes; obviously, there is great advantage in using them for writing the equations of motion, because the equations of motion will then be uncoupled and their solution made much simpler.

Conceptual aspect of matrix methods

We have indicated here only the notational ideas of matrix algebra—ideas which confer conciseness and generality. In addition, there are a number of matrix-*manipulation* techniques which permit the numerous intermediate steps in solving a large set of equations to be performed in very compact form, the terms being written in detail only at the last step. Matrix methods can contribute heavily during this kind of calculation.

But quite aside from its convenience in the handling of cumbersome equations, the matrix notation has an important conceptual utility, for it

[Sec. 14.5] MATRIX REPRESENTATION AND STANDARD FORM 507

allows us to discuss in general terms a *class* of systems of equations in the form (14.21) and to describe the procedure for their solution most succinctly, unencumbered by algebraic details, and even without specifying the exact number of such equations.

Example 14.2. A coupled electrical circuit. For the circuit represented in Fig. 14.9, find the transfer function between voltage v_1 as input and the current in resistor R_9 as output. Our procedure will be first to derive the equations of motion; then, to obtain the required transfer function we shall assume that all signals are of the form Ce^{st} and substitute into the equations of motion to obtain a set of algebraic equations, which we can then solve for the desired ratio I_9/V_1 by purely algebraic means.

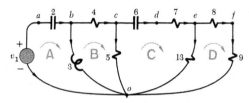

Fig. 14.9 *An electric network with general coupling*

The network shown is, of course, already a physical model (Stage I) of a real circuit. Since there are six independent nodes and only four loops, we elect to write the equations of motion (Stage II) on the loop basis, using mesh currents. The reader may show that the equations of motion are

$$\frac{1}{C_2} \int i_A \, dt + L_3 \frac{d}{dt} (i_A - i_B) = v_1$$

$$-L_3 \frac{d}{dt} (i_A - i_B) + R_4 i_B + R_5 (i_B - i_C) = 0 \quad (14.30)$$

$$-R_5 (i_B - i_C) + \frac{1}{C_6} \int i_C \, dt + R_7 i_C + R_{13} (i_C - i_D) = 0$$

$$-R_{13} (i_C - i_D) + (R_8 + R_9) i_D = 0$$

These do not contain the desired variable i_9, but it will be readily obtained from i_D by the identity $i_9 \equiv i_D$.

To begin the study of system behavior (Stage III) we assume that all voltages have the form $v = Ce^{st}$ and all currents $i = Ie^{st}$, and substitute these into (14.30). We write the result in matrix form for convenience:

$$\begin{bmatrix} \left(\frac{1}{C_2 s} + L_3 s\right) & -L_3 s & 0 & 0 \\ -L_3 s & (L_3 s + R_4 + R_5) & -R_5 & 0 \\ 0 & -R_5 & \left(R_5 + \frac{1}{C_6 s} + R_7 + R_{13}\right) & -R_{13} \\ 0 & 0 & -R_{13} & (R_{13} + R_8 + R_9) \end{bmatrix} \begin{bmatrix} I_A \\ I_B \\ I_C \\ I_D \end{bmatrix}$$

$$= \begin{bmatrix} V_1 \\ 0 \\ 0 \\ 0 \end{bmatrix} \quad (14.31)$$

This matrix equation has the form $[Z][I] = [V]$.

The desired transfer function is now obtained from (14.31) by Cramer's rule [refer to (14.22)]:

$$\frac{I_9}{V_1} = \frac{I_D}{V_1} = \frac{z_{14}}{|Z|} =$$

$$-\frac{\begin{vmatrix} -L_3 s & (L_3 s + R_4 + R_5) & -R_5 \\ 0 & -R_5 & \left(R_5 + \dfrac{1}{C_6 s} + R_7 + R_{13}\right) \\ 0 & 0 & -R_{13} \end{vmatrix}}{\begin{vmatrix} \left(\dfrac{1}{C_2 s} + L_3 s\right) & -L_3 s & 0 & 0 \\ -L_3 s & (L_3 s + R_4 + R_5) & -R_5 & 0 \\ 0 & -R_5 & \left(R_5 + \dfrac{1}{C_6 s} + R_7 + R_{13}\right) & -R_{13} \\ 0 & 0 & -R_{13} & (R_{13} + R_8 + R_9) \end{vmatrix}}$$

(14.32a)

where z_{14} is the minor of the row 1, column 4, term of $[Z]$. Multiplying out the matrices gives

$$\frac{I_9}{V_1} = \frac{+L_3 s R_5 R_{13}}{|Z|} \tag{14.32b}$$

with

$$|Z| = \left(\frac{1}{C_2 s} + L_3 s\right) \left\{ (L_3 s + R_4 + R_5) \left[\left(R_5 + \frac{1}{C_6 s} + R_7 + R_{13}\right)(R_{13} + R_8 + R_9) - R_{13}{}^2 \right] - R_5{}^2 (R_{13} + R_8 + R_9) \right\}$$
$$- L_3{}^2 s^2 \left[\left(R_5 + \frac{1}{C_6 s} + R_7 + R_{13}\right)(R_{13} + R_8 + R_9) - R_{13}{}^2 \right] \quad (14.33)$$

Thus, with suitable algebraic manipulation, transfer function (14.33) may be written in either of the forms

$$\frac{I_9}{V_1} = \frac{(r_1/L_3)s^3}{[s^4 + a_3 s^3 + a_2 s^2 + a_1 s + a_0]} \tag{14.34a}$$

$$\frac{I_9}{V_1} = \frac{(L_3 C_2 C_6/r_2) s^3}{\left[\dfrac{s^4}{a_0} + a_3 \left(\dfrac{s^3}{a_0}\right) + a_2 \left(\dfrac{s^2}{a_0}\right) + a_1 \left(\dfrac{s}{a_0}\right) + 1\right]} \tag{14.34b}$$

where the r's are dimensionless ratios of R's, and the a's are combinations of the R's, L's, and C's obtained from (14.34). (They are to be evaluated in Problem 14.31.) Are the dimensions proper in (14.34a) and (14.34b)?

Pole-zero diagram. Once we have factored (14.34) (for given numerical values of the R's, C's, and L's) to obtain the poles of (14.33), we can then draw the pole-zero diagram. Vector multiplication may be performed on this diagram to obtain the magnitude and phase of (14.33) for any given value of s.

14.6 MATRIX DESCRIPTION OF EIGENVECTOR CALCULATION

Some of the results of Chapter 12 can be summarized efficiently by application of the matrix ideas we have been discussing.

Consider, for example, the set of equations (12.5) which describe the unforced motion of the system in Fig. 12.1c. These equations can, of course, be put in matrix form:

$$\begin{bmatrix} (J_1s^2+b_1s+k_{12}) & -k_{12} & -K_1 \\ -k_{12} & (J_2s^2+b_2s+k_{12}+k_{23}) & -k_{23} \\ 0 & -k_{23} & (J_3s^2+b_3s+k_3+k_{23}) \end{bmatrix} \begin{bmatrix} \Theta_1 \\ \Theta_2 \\ \Theta_3 \end{bmatrix} = 0 \tag{14.35}$$

or $\qquad\qquad\qquad [A]\underline{\Theta} = 0 \qquad\qquad\qquad$ (14.36)

Then the characteristic equation is just the determinant of A set equal to 0:

$$\boxed{|A| = 0} \tag{14.37}$$

[which represents Eqs. (12.6) and (12.7)]. That is, the roots of (14.37) are the *eigenvalues* of the system.

To obtain the *eigenvectors* of the system, we select one of the variables—say, Θ_1—as a reference, and write all but one of the equations with that variable on the right-hand side (we can omit any equation). For example, omitting the first equation, we have

$$\begin{bmatrix} (J_2s^2 + b_2s + k_{12} + k_{23}) & -k_{23} \\ -k_{23} & (J_3s^2 + b_3s + k_3 + k_{23}) \end{bmatrix} \begin{bmatrix} \Theta_2 \\ \Theta_3 \end{bmatrix} = \begin{bmatrix} -k_{12} \\ 0 \end{bmatrix} \Theta_1 \tag{14.38}$$

which we can represent in the matrix form

$$[a_{11}]\underline{\Theta} = [c_1]\Theta_1 \tag{14.39}$$

(Note that Θ_1 is not a matrix, but a single letter.) Now the *eigenvectors are simply the transfer functions* Θ_2/Θ_1 *and* Θ_3/Θ_1, which we obtain readily from (14.38):†

$$\begin{aligned} \frac{\Theta_2}{\Theta_1} &= \frac{-k_{12}(J_3s^2 + b_3s + k_3 + k_{23})}{a_{11}} \\ \frac{\Theta_3}{\Theta_1} &= \frac{-k_{12}k_{23}}{a_{11}} \end{aligned} \tag{14.40}$$

where each of the specific natural modes is obtained by letting s equal one of the roots found in (14.37).

Problems 14.12 through 14.34

---

† The symbol a_{11} denotes the minor of term A_{11} in (14.35) and (14.36).

⊙Chapter 15

FORCED OSCILLATIONS
OF COMPOUND SYSTEMS

In this chapter we shall apply the results of Chapters 9 and 14 to study the response of compound systems to a particular class of forcing functions, namely, sinusoidal ones. Sinusoidal response has broad importance because (1) many systems are disturbed or driven primarily by sinusoidal inputs, and (2) a most convenient method of testing a system is to measure its response to sinusoidal test inputs.

In Chapter 9 we studied the response of elementary first- and second-order systems to sinusoidal disturbances. We demonstrated the rotating-vector idea, the meaning of "phase angle," and the labor-saving technique of working with the real parts of complex quantities. (We used the symbol $\stackrel{r}{=}$ to mean "equals the real part of.") We showed how sinusoidal response can be obtained rapidly by graphical calculation from the s-plane pole-zero diagram, and how it can be plotted—versus disturbing frequency—with great convenience on logarithmic coordinates, employing the straight-line asymptotes of Bode. A number of electrical- and mechanical-vibration examples were discussed, and particular attention was given the phenomenon of resonance.

In Chapter 14 we have seen how the forced response of compound systems can be studied efficiently by using transfer functions. For subsystems in cascade, the system transfer function is simply the product of the subsystem transfer functions; for more general coupling, the system transfer function is found by simultaneous solution of algebraic equations in s, and its denominator then needs to be factored.

In this chapter we specialize the results of Chapter 14 to the case of sinusoidal response (by properly setting $s = j\omega_f$). We consider first cascaded systems in Sec. 15.1, and then generally coupled systems in Sec. 15.2. Resonance in coupled systems is examined in Sec. 15.3, and some interesting examples are discussed, including the ingenious mechanical vibration absorber (Sec. 15.4).

15.1 FREQUENCY RESPONSE OF SUBSYSTEMS IN CASCADE

For a system of elements in cascade, the sinusoidal response is available from the general representation of Fig. 14.4, with $\alpha_f = j\omega_f$ as in Fig. 15.1:

$$y(t) \stackrel{r}{=} Y e^{j\omega_f t} \tag{15.1}$$

$$Y = \left[\prod_{i=1}^{i=n} G_i(s)\right]_{s=j\omega_f} X \tag{15.2}$$

$\left(\text{in which } \prod_{i=1}^{i=n} \text{ means "product of terms 1 to } n\text{"}\right).$

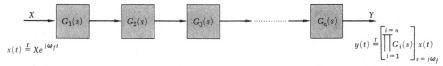

Fig. 15.1 *Subsystems in cascade with sinusoidal input: general relations*

To illustrate the procedure for cascaded systems, consider first the electromechanical system modeled in Fig. 15.2a. Using the methods of Chapter 14, we can represent the subsystems by their transfer functions, as shown in Fig. 15.2b. If the input voltage v is sinusoidal,

$$v = v_m \cos \omega_f t \stackrel{r}{=} [Ve^{st}]_{s=j\omega_f} \quad \text{with } V = v_m \tag{15.3}$$

then the steady-state output v_1 of the first element will be

$$v_1 = v_{1m} \cos (\omega_f t + \psi_1) \stackrel{r}{=} V_1 e^{st} \tag{15.4}$$

with $V_1 = v_{1m} e^{j\psi_1}$:

$$v_1 = V_1 e^{st} \stackrel{r}{=} \{[G_1(s)] V e^{st}\}_{s=j\omega_f} \stackrel{r}{=} \left\{\left[\frac{K_1}{RCs+1}\right] V e^{st}\right\}_{s=j\omega_f} \tag{15.5}$$

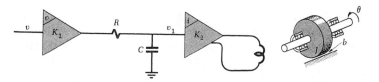

(a) *Physical model*

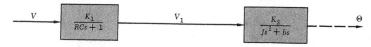

(b) *Block-diagram representation*

Fig. 15.2 *Example of cascaded subsystems*

Similarly, the output of the system can be found as a product:

$$\theta = \theta_m \cos(\omega_f t + \psi_2) \stackrel{r}{=} [\Theta e^{st}]_{s=j\omega_f} \quad (15.6)$$

$$\theta \stackrel{r}{=} \{[G_2(s)]V_1 e^{st}\}_{s=j\omega_f} \stackrel{r}{=} \{[G_2(s)][G_1(s)] V e^{st}\}_{s=j\omega_f}$$

$$\stackrel{r}{=} \left\{ \left[\frac{K_1}{RCs+1} \right] \left[\frac{K_2}{Js^2+bs} \right] V e^{st} \right\}_{s=j\omega_f} \quad (15.7)$$

That is, the frequency response is given by $\theta_m/v_m = |\Theta/V|$ and $\psi_2 = \underline{/\Theta/V}$.

Since the total frequency response from v to θ (Fig. 15.2) is the product of the individual frequency responses, v to v_1 and v_1 to θ, it is possible to obtain a plot of the total by first plotting each individual frequency response and then multiplying them together graphically, a la Bode. The advantage of the Bode method is that, on both the magnitude and phase plots, the multiplication can be accomplished graphically by adding ordinates. (The product involves the sum of the phase angles and the product of magnitudes, and the latter quantity is represented by the *sum of the logarithms*.) Use of magnitude and phase-angle asymptotes (explained in Chapter 9) is especially helpful in this process.

Example 15.1. Plotting frequency response of a cascaded system. Plot the frequency response of the system of Fig. 15.2. It is most convenient to convert transfer functions to a dimensionless form:

$$\frac{\Theta}{V} = \frac{K_1}{\left(\frac{s}{1/RC}+1\right)} \frac{K_2/b}{s\left(\frac{s}{b/J}+1\right)} \quad (15.8)$$

For definiteness, let $K_1 = 0.5$, $K_2/b = 10$, $b/J = 4$, $1/RC = 50$.

Then
$$\frac{\Theta}{V} = \left[\frac{0.5}{\left(\frac{s}{50}+1\right)} \right] \left[\frac{1}{s} \right] \left[\frac{10}{\left(\frac{s}{4}+1\right)} \right] \quad (15.9)$$

in which $s = j\omega_f$ for the frequency-response plot, of course. The steps in constructing the plot are shown in Fig. 15.3 and described below.

Transfer function (15.9) has three separate parts, denoted by the square brackets. To obtain a plot of the composite function, we first plot the frequency response of each of the individual parts separately, as shown by the solid light lines in Fig. 15.3: Magnitude is plotted at the top and phase at the bottom of the figure. The asymptote techniques described in Sec. 9.6 are used to effect rapid plotting of the individual curves. (The asymptotes are shown as light dashed lines.) Note that the low-frequency magnitude of the first term is 0.5, and that of the third term 10. For the first term the magnitude asymptote has a "break frequency" at $\omega_f = 50$, and for the third term, at $\omega_f = 4$.

Once the magnitude and phase curves for the three separate parts of (15.9) have been plotted, the curves for the complete function are obtained simply by adding the individual curves. For the magnitude curves this is most easily done by first constructing the asymptote of the product of the three terms in (15.9) (heavy dashed lines), by adding the individual asymptotes graphically. (Recall again that graphical addition of log plots is tantamount to

[Sec. 15.1] FREQUENCY RESPONSE OF SUBSYSTEMS IN CASCADE

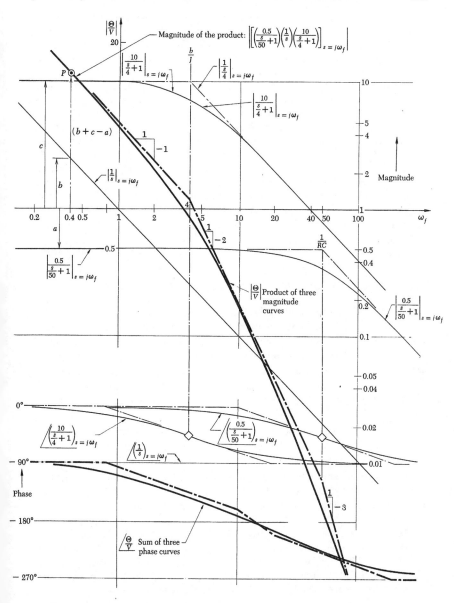

Fig. 15.3 *Constructing a Bode plot (frequency-response plot) for the function:*

$$\frac{\Theta}{V} = \left(\frac{0.5}{\frac{s}{50}+1}\right)\left(\frac{1}{s}\right)\left(\frac{10}{\frac{s}{4}+1}\right)$$

multiplication of the plotted quantities.) For example, point P, for $\omega_f = 0.4$, is obtained by adding vertically, from magnitude = 1, the distances $b + c - a$ (because a lies below 1). The reader may perform this graphical addition, laying off distances by marking on the edge of a card, for example. Some other points should also be checked. The product asymptote is obtained quickly, being a straight line between each of the break points involved. Then some key points are spotted—for example, the magnitude at each break point—after which the exact magnitude curve can be faired in readily (solid heavy curve).

Constructing the sum of phase-angle asymptotes is a little more fussy because the phase-angle asymptote for each individual term has two "break points" in it. Even so, it is a straightforward step to construct the phase-angle asymptote for the complete system (heavy dashed line at the bottom of Fig. 15.3), and it is usually convenient to do so. Then several exact phase-angle points are spotted (for example, at $\omega_f = 0.8, 10, 20,$ and 100), and the phase-angle curve for the complete system is drawn (solid heavy line at the bottom of Fig. 15.3).

Problems 15.1 through 15.8

15.2 FREQUENCY RESPONSE OF SYSTEMS WITH TWO-WAY COUPLING

We are often interested in the frequency response of compound systems in which there is more general coupling than the simple one-way cascade connections of Fig. 15.1 or Fig. 15.2. The principal feature in the more general case is that the denominator of the transfer function is not already factored. That is, the transfer function is not a product of simple functions as it is in expression (15.9), and therefore the method of adding log-magnitude curves and adding phase curves is not applicable without an additional step.

In practice, two methods are commonly used for producing a frequency-response plot from an unfactored transfer function:
(1) Factor the transfer function, and then use the method of Sec. 15.1.
(2) Evaluate the transfer function numerically, term by term, for a number of frequencies, and fair in the curve.

We demonstrate both techniques with a simple example.

Example 15.2. *A fluid clutch* is to be tested to establish its dynamic torque-transmission capability. The output shaft is connected to a limber spring, as shown in Fig. 15.4, and the input shaft is subjected to a sinusoidally varying torque, $M(t)$. The resulting response, θ_2, is measured. The test is conducted at many frequencies. The following numerical quantities are known:

$$\frac{k}{J_2} = 10{,}400 \qquad \frac{b}{J_1} = 7.7 \qquad \frac{J_1}{J_2} = 31.57 \qquad (15.10)$$

Plot the frequency response θ_{2m}/M_m and ψ to be expected from the tests. That is, if the input to the clutch is the sinusoidal moment $M = M_m \cos \omega_f t$, so that the steady-state output is $\theta_2 = \theta_{2m} \cos(\omega_f t + \psi)$, find the magnitude ratio θ_{2m}/M_m and the phase angle ψ as a function of frequency ω_f.

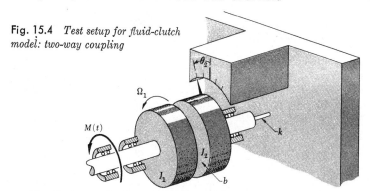

Fig. 15.4 *Test setup for fluid-clutch model: two-way coupling*

For the linear physical model shown, the *equations of motion* are

$$J_1\dot{\Omega}_1 + b(\Omega_1 - \dot{\theta}_2) = M$$
$$-b(\Omega_1 - \dot{\theta}_2) + J_2\ddot{\theta}_2 + k_2\theta_2 = 0 \quad (15.11)$$

To find the *system response* (Stage III), we note that M has the form $M = \mathsf{M}e^{st}$, and assume that therefore also $\Omega_1 = \Omega_1 e^{st}$, $\theta_2 = \Theta_2 e^{st}$, with the result

$$\begin{bmatrix} (J_1 s + b) & -bs \\ -b & (J_2 s^2 + bs + k) \end{bmatrix} \begin{bmatrix} \Omega_1 \\ \Theta_2 \end{bmatrix} = \begin{bmatrix} \mathsf{M} \\ 0 \end{bmatrix} \quad (15.12)$$

Suppose that we are interested in the transfer function from input torque M to indicator rate $\dot{\theta}_2$. We obtain it from (15.12) by Cramer's rule, and using the identity $\Omega_2 = s\Theta_2$:

$$\frac{\Omega_2}{\mathsf{M}/k} = \frac{(kb/J_1 J_2)s}{s^3 + [(b/J_1) + (b/J_2)]s^2 + (k/J_2)s + (k/J_2)(b/J_1)} \quad (15.13)$$

For the numbers given, this transfer function can be written

$$\frac{\Omega_2}{\mathsf{M}/k} = \frac{80{,}000 s}{s^3 + 250s^2 + 10{,}400s + 80{,}000} \quad (15.14)$$

which, of course, has a denominator that is not in factored form. For this particular denominator the factors are rather easily obtained by trial and error:

$$\frac{\Omega_2}{\mathsf{M}/k} = \frac{80{,}000 s}{(s+10)(s+40)(s+200)} \quad (15.15)$$

As a matter of interest, the factors of the denominator of (15.13) (roots of the system characteristic equation) are plotted in the s plane in Fig. 15.5, as a function of coupling strength b. The plot was made in the manner of Secs. 11.8 and 11.9, by first writing the characteristic equation [denominator of (15.13)] with the terms in b separated from the others:

$$s\left(s^2 + \frac{k}{J_2}\right) + \frac{b}{J_1}\left(1 + \frac{J_1}{J_2}\right)\left(s^2 + \frac{k}{J_1 + J_2}\right) = 0 \quad (15.16)$$

or, in the Evans form:

$$\frac{s\left(s^2 + \dfrac{k}{J_2}\right)}{\left(s^2 + \dfrac{k}{J_1 + J_2}\right)} = -\frac{b}{J_1}\left(1 + \frac{J_1}{J_2}\right) \quad (15.17)$$

Fig. 15.5 *Locus of roots for the fluid-clutch system of Fig. 15.4*

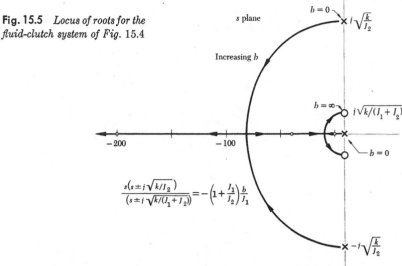

Figure 15.5 shows that for no coupling ($b = 0$ in Fig. 15.4), one pair of system roots represents the undamped vibration of mass J_2 on spring k, and the other root, at the origin, represents the unrestrained velocity of mass J_1. As coupling is increased, the system may exhibit damped oscillations or, for certain intermediate values of b (including that used in this example), may have three exponential natural motions (three real roots). For even larger values of b, there are damped oscillations again; and when b is infinite, disks J_1 and J_2 are essentially welded together and the system is an undamped single-degree-of-freedom system, as indicated by the $\bigcirc$'s in Fig. 15.5.

Plotting frequency response when denominator is factored. When the denominator is factored, we can use the same technique as for cascaded subsystems (e.g., Fig. 15.3), namely, plotting the individual terms of the product function (15.15), and then obtaining the total response by adding log magnitudes and adding phase angles. It is convenient to write all the terms in nondimensional form:

$$\frac{\Omega_2}{M/k} = \frac{s}{[(s/10) + 1][(s/40) + 1][(s/200) + 1]} \quad (15.18)$$

Fig. 15.6 shows magnitude and phase plots for the four individual functions (gray asymptotes and curves). Then the frequency response for the complete system is obtained (black curves) by adding together the four log-magnitude plots and by adding together the four phase plots.

Plotting frequency response from the unfactored form of the transfer function. It is possible to obtain the frequency response without taking time to factor the denominator, by working directly from (15.14). Again, it is convenient to put the function into a dimensionless form:

$$\frac{\Omega_2}{M/k} = \frac{s}{(s/43.1)^3 + (s/17.9)^2 + (s/7.7) + 1} \quad (15.19)$$

Fig. 15.6 *Bode plot for fluid-clutch test system of Fig. 15.4*

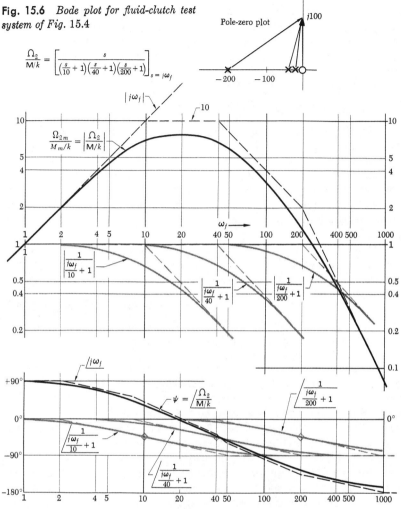

and then, letting $s = j\omega_f$, to separate the real and imaginary parts:

$$\frac{\Omega_2}{M/k} = \frac{j\omega_f}{1 - (\omega_f/17.9)^2 + j[(\omega_f/7.7) - (\omega_f/43.1)^3]} \quad (15.20)$$

The straightforward thing to do now is to substitute various values of ω_f into (15.20), calculate the magnitude and phase angle, and plot the result. Two asymptotes are helpful, however. For quite small values of ω_f—much smaller than 10—(15.20) looks essentially like $j\omega_f$ alone, so we can plot an asymptote $j\omega_f$ which will coincide with that in Fig. 15.6. At the other extreme, when ω_f is very large, relation (15.20) looks essentially like

$$\frac{j\omega_f}{(j\omega_f/43.1)^3} = \frac{43.1}{(j\omega_f/43.1)^2}$$

which will be found to coincide with the asymptote on the far right in the magnitude and phase plots of Fig. 15.6. In the region between the low-frequency asymptote and the high-frequency asymptote, three or four points may be calculated from (15.20) and plotted, which will be enough to give a good indication of the shape of the curve. Note that the calculation is especially easy for $\omega_f = 17.9$. Why?

A theorem by Bode

The above discussion illustrates an important theorem, due to H. W. Bode, which holds for *any* l.c.c. system whose poles and zeros are all in the left half of the s plane or on the imaginary axis. Such systems are called "minimum-phase" systems. (Systems having poles and/or zeros in the right-half plane are called "nonminimum-phase" systems.)

Bode's theorem states that *for any minimum-phase system, the phase-angle part of the frequency-response plot is uniquely related to the magnitude part.*

Thus, if we are given the magnitude plot for such a system, we can always deduce the phase plot, and vice versa. This is illustrated in Figs. 15.3, 15.6, and 15.9, for example, and in the problems.

> Problems 15.9 through 15.13

15.3 RESONANCE IN COUPLED SYSTEMS

In the preceding examples it happened that the poles of the coupled system [i.e., the poles of Eq. (15.9) and of Eq. (15.17)] were all real, with the result that the plots of amplitude versus frequency, Figs. 15.3 and 15.6, lay entirely below the asymptotes and did not display any "resonance peaks" such as those we noted in earlier studies of elementary second-order systems (e.g., Fig. 9.20). Resonance in coupled systems is common, however. For example, Fig. 15.5 indicates that there are values b for which there are complex roots which will result in such resonant response. [The reader is invited in Problem 15.25 to find out for what range of b this is true.]

In this section we consider two other resonant systems, one mechanical and one electrical. The first, shown in Fig. 15.7, is representative of an important class of mechanical systems in which power is being delivered through a shaft and in which, because of flexibility in the power-carrying shaft, the system may resonate—display large output motions in response to sinusoidal inputs at certain frequencies. The drive system of a ship—consisting of a turbine, long drive shaft (half the length of the ship) and propeller, an electric motor driving a load through a flexible shaft, and the rig for drilling an oil well are examples.

Sometimes a one-mass physical model like Fig. 9.18 is adequate to represent such systems dynamically, but often a two-mass (or more) model is required to predict the essential behavior. For the electric motor and for the ship's drive system, Fig. 15.7 may actually be an acceptable model.

[Sec. 15.3] RESONANCE IN COUPLED SYSTEMS

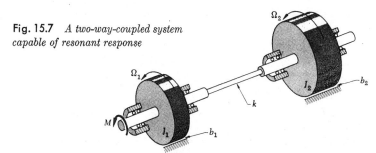

Fig. 15.7 *A two-way-coupled system capable of resonant response*

(For the oil-well rig, a distributed-mass, distributed-elasticity model should be used. An introduction to such systems was given in Secs. 13.2 and 13.3.)

Returning to Fig. 15.7, we seek the frequency response of the system. Suppose that it is the angular velocity of rotor J_1 (e.g., of the driving motor or turbine itself) in which we are interested. That is, given the input

$$M = M_m \cos \omega_f t \stackrel{r}{=} [Me^{st}]_{s=j\omega_f} \tag{15.21}$$

We expect response

$$\Omega_1 = \Omega_{1m} \cos (\omega_f t + \psi_1) \stackrel{r}{=} [\Omega_1 e^{st}]_{s=j\omega_f} \tag{15.22}$$

and we seek Ω_{1m}/M_m and ψ_1 as a function of frequency ω_f. [Expressions (15.21) and (15.22) are written also in terms of s for convenience.]

Example 15.3. Resonant mechanical system. Let Fig. 15.7 represent an electric motor driving a load, and find the motion Ω_1 of the motor rotor versus frequency for the following numerical values:

$$\sigma_1 \triangleq \frac{b_1}{J_1} = 0.1 \qquad \omega_{01} = \sqrt{\frac{k}{J_1}} = 90$$
$$\sigma_2 \triangleq \frac{b_2}{J_2} = 0.4 \qquad \omega_{02} = \sqrt{\frac{k}{J_2}} = 10 \qquad \frac{J_2}{J_1} = 81 \tag{15.23}$$

The physical meanings of these parameters should be clear: The coast-down time constant of the motor alone is $1/\sigma_1 = 10$ sec, while that of the load alone (disconnected from the motor) is $1/\sigma_2 = 2.5$ sec. The undamped natural frequency of the motor alone on the shaft is 90 rad/sec, and of the load alone 10 rad/sec.

The reader may readily show that the transfer function from M to Ω_1 is (it is helpful to use θ's, rather than Ω's in the derivation)

$$\frac{\Omega_1}{M} = \frac{s(J_2 s^2 + b_2 s + k)}{(J_1 s^2 + b_1 s + k)(J_2 s^2 + b_2 s + k) - k^2} \tag{15.24}$$

which may be written in the following useful form:

$$\frac{\Omega_1}{M/(b_1 + b_2)} = \frac{\dfrac{(b_1 + b_2)}{J_1}\left(s^2 + \dfrac{b_2}{J_2}s + \dfrac{k}{J_2}\right)}{\left[s\left(s + \dfrac{b_1}{J_1}\right)\left(s + \dfrac{b_2}{J_2}\right) + \dfrac{k}{J_2}\left(1 + \dfrac{J_2}{J_1}\right)\left(s + \dfrac{b_2 + b_1}{J_2 + J_1}\right)\right]} \tag{15.25}$$

in which the effect of shaft stiffness k appears more clearly. (The transition from one form to the other is purely algebraic, of course.)

Before proceeding, we check an important special case, namely, the case that the connecting shaft is not limber: $k \to \infty$. Then (15.25) becomes

$$\frac{\Omega_1}{M/(b_1 + b_2)} = \frac{(b_1 + b_2)/(J_1 + J_2)}{[s + (b_1 + b_2)/(J_1 + J_2)]} \tag{15.26}$$

That is, the system is then a simple first-order one involving the combined masses (now effectively clamped together by the infinitely stiff shaft connecting them) and their combined viscous resistance. The system obviously shows no sign of any resonance. Consider also the expression obtained from (15.25) for the other extreme, $k \to 0$. Does it also make complete sense physically, both as to natural motion—denominator factors—and as to frequency response?

To see the behavior of the system with varying frequency, we shall need to factor the denominator of (15.25). When we have done so, we can then write (15.25) in the form

$$\frac{\Omega_1}{M/(b_1 + b_2)} = \frac{\left(\dfrac{s^2}{\omega_{02}^2} + \dfrac{1}{Q_2}\dfrac{s}{\omega_{02}} + 1\right)}{\left(\dfrac{s}{\sigma_a} + 1\right)\left(\dfrac{s^2}{\omega_{0b}^2} + \dfrac{1}{Q_b}\dfrac{s}{\omega_{0b}} + 1\right)} \tag{15.27}$$

and from this we can plot the frequency response, using the asymptotic methods of Sec. 15.1. The constants in (15.27) are given by $\omega_{02}^2 = k/J_2$, $Q_2 = \omega_{02}J_2/b_2$, and the newly found factors of the denominator are $s = -\sigma_a$ and $s = -\sigma_b \pm j\omega_b$ with $\omega_{0b} = \sqrt{\omega_b^2 + \sigma_b^2}$ and $Q_b = \omega_{0b}/2\sigma_b$. [As one quick check on the form of (15.27), notice that when $s \to 0$, both (15.27) and (15.25) reduce to 1.]

For the numerical values given, the reader can quickly verify that the factors of the denominator of (15.25) are

$$(s + 0.4)(s^2 + 0.1s + 90.6^2) \tag{15.28}$$

so that, in the form (15.27), the transfer function is

$$\frac{\Omega_1}{M/(b_1 + b_2)} = \frac{\left(\dfrac{s^2}{10^2} + \dfrac{1}{25}\dfrac{s}{10} + 1\right)}{\left(\dfrac{s}{0.4} + 1\right)\left(\dfrac{s^2}{90.6^2} + \dfrac{1}{906}\dfrac{s}{90.6} + 1\right)} \tag{15.29}$$

The amplitude of each of the separate terms in (15.29), with $s = j\omega_f$, can be plotted versus ω_f to log coordinates, using the asymptotic technique of Sec. 9.8, and noting that the Q's—25 and 906—indicate deviation from the asymptotes at $s = j\omega_{02} = 10$ and at $s = j\omega_{0b} = 90.6$. Then the total magnitude plot can be obtained by adding the log magnitudes. The result is shown in Fig. 15.8.

The difference between the response with a flexible shaft (black curve) and that with a rigid shaft (gray curve) is negligible at very low frequencies—ω_f less than 5 or 6 rad/sec. But as the frequencies $\omega_f = 10$ and $\omega_f = 90.6$ are approached, the difference in the two curves is spectacular. When $\omega_f = 10$ the motor moves only 1/25 as much when the shaft is limber as when it is stiff (although the load moves almost exactly the same amount in either case, as will be seen in Problem 15.26). When $\omega_f = 90.6$ the motor response with the limber shaft is larger than with a rigid shaft by the factor $906 \times (90.6/10)^2$!

[Sec. 15.3] RESONANCE IN COUPLED SYSTEMS 521

(The load response will be found to be greater by 906.) That is, an extremely small electrical torque M, applied exactly at the frequency $\omega_f = 90.6$, can produce very large oscillations of the load, and especially of the rotor. It is left as a problem (Problem 15.27) to sketch the phase-angle plot accompanying the amplitude plot of Fig. 15.8. Note particularly that the phase angle changes rapidly through 180° as the frequency passes $\omega_f = 10$ and again as it passes $\omega_f = 90.6$.

The frequency at which the high amplification occurs ($\omega_f = 90.6$ in Fig. 15.8) is called the *resonant frequency*. It is the (undamped) natural fre-

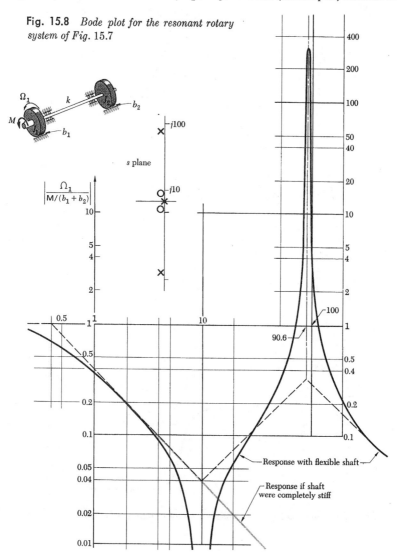

Fig. 15.8 *Bode plot for the resonant rotary system of Fig. 15.7*

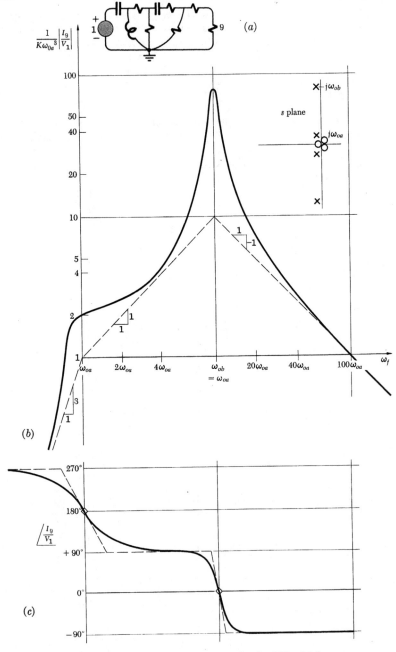

Fig. 15.9 *Bode plot for the coupled resonant circuit of Fig. 14.9*

quency of the *coupled* system. In higher-order systems, numerous resonant frequencies may occur—as many as there are oscillatory natural frequencies of the system. The resonance problem lurks wherever we try to transmit power through limber connecting members. It is kept from being serious by making the members sufficiently stiff that the natural frequencies are well above any at which input power is to be applied.

An interesting feature of the frequency-response plot in Fig. 15.8 is the large attenuation in the response at $\omega_f = 10$. A frequency at which the output is sharply attenuated, like $\omega_f = 10$ in Fig. 15.8, is called an "antiresonance." The possibility of such attenuation at a particular frequency is exploited in a fascinating manner in the dynamic vibration absorber, which will be discussed in Sec. 15.4.

Resonance can, of course, occur in any dynamic system. To illustrate, we study in the following example the frequency response of the electric circuit of Fig. 14.9.

Example 15.4. Resonance in a compound circuit. Consider again the circuit represented in Fig. 14.9, and studied in Example 14.2. Suppose now that the input signal v_1 is sinusoidal (Fig. 15.9a):

$$v_1 = v_{1m} \cos \omega_f t \tag{15.30}$$

and we are interested in how the current $i_9 = i_{9m} \cos(\omega_f t + \psi_9)$ responds to v_1. In the complex form

$$v_1 \stackrel{r}{=} V_1 e^{j\omega_f t} \qquad i_9 \stackrel{r}{=} I_9 e^{j\omega_f t} \tag{15.31}$$

in which V_1 and I_9 are the phasors: $V_1 = v_{1m} \qquad I_9 = i_{9m} e^{j\psi_9}$ (15.32)

To evaluate I_9/V_1, and thence $i_{9m}/v_{1m} = |I_9/V_1|$ and $\psi_9 = /I_9/V_1$, we turn at once to transfer function (14.32), between V_1 and I_9. We choose to factor the denominator, using method (1) of p. 514. Suppose that the result is two complex pairs of poles. Then we may write

$$\frac{I_9}{V_1} = \left[\frac{Ks^3}{\left(\dfrac{s^2}{\omega_{0a}^2} + \dfrac{1}{Q_a} \dfrac{s}{\omega_{0a}} + 1 \right)\left(\dfrac{s^2}{\omega_{0b}^2} + \dfrac{1}{Q_b} \dfrac{s}{\omega_{0b}} + 1 \right)} \right]_{s=j\omega_f} \tag{15.33}$$

where $K = L_3 C_2 C_6/r_2$. (There are two other possibilities, of course.)

The shape of the corresponding frequency response is indicated in Fig. 15.9. The following numerical values were used: $Q_a = 2$, $\omega_{0b} = 10\omega_{0a}$, $Q_b = 8$. The numerical term $[Ks^3]_{s=j\omega_f} = (L_3 C_2 C_6/r_2)(j\omega_f)^3$, in (15.33) dominates at very low frequency, giving an amplitude which increases as ω_f^3, Fig. 15.9b, and a phase angle of $+270°$, Fig. 15.9c. At very high frequency the transfer function looks like $K\omega_{0a}^2 \omega_{0b}^2/s$, and the amplitude decreases in direct proportion to frequency, Fig. 15.9b, while the phase is $-90°$, Fig. 15.9c.

For intermediate frequencies, the frequency-response plot is obtained from (15.33) in the same way that it was for elements in cascade in Sec. 15.1 (refer to Example 15.1 and Fig. 15.3). That is, the second-order transfer functions can be sketched individually and then combined as in Fig. 15.3. It is convenient to use asymptotes for this purpose, as indicated in Figs. 15.9. An alternative method of calculation, which may give amplitude and (especially) phase more quickly and with high accuracy for specific values of ω_f, is to evaluate the transfer function graphically from a pole-zero plot.

Figure 15.9 is seen to exhibit resonance at $\omega_f = \omega_{0a}$ and at $\omega_f = \omega_{0b}$; that is, at the two natural frequencies of the coupled network. The *amplification*—the amount by which the asymptotes are exceeded—is given, of course, by the appropriate Q in the denominator of (15.33), either Q_a or Q_b.

To obtain another image to help visualize the frequency-response situation

$$\frac{I_9}{V_1}(s) = \frac{Ks^3}{\left(\frac{s^2}{\omega_{oa}^2} + \frac{1}{Q_a}\frac{s}{\omega_{oa}} + 1\right)\left(\frac{s^2}{\omega_{ob}^2} + \frac{1}{Q_b}\frac{s}{\omega_{ob}} + 1\right)}$$

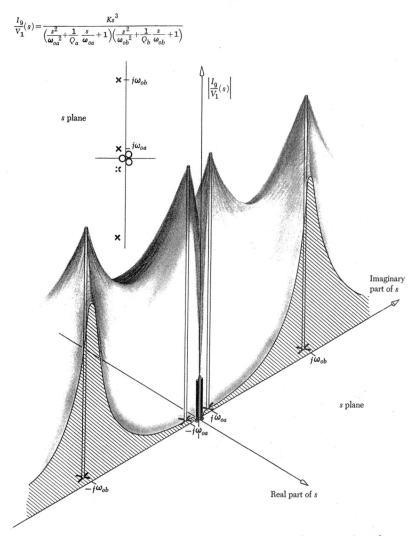

Fig. 15.10 *Three-dimensional "rubber-sheet" image of a system's pole-zero diagram. The frequency response is obtained from the imaginary-axis cross section.*

[Sec. 15.4] THE VIBRATION ABSORBER 525

from the known pole-zero configuration, we can construct the three-dimensional plot of Fig. 15.10. The base plane is the s plane. Vertically above it, we plot the magnitude of the transfer function for each (complex) value of s. The result is a surface over the s plane, as shown in Fig. 15.10b. To imagine its shape, it is helpful to note that a stretched rubber sheet with sticks inserted for the poles and zeros—poles upward to "infinity," zeros downward to zero—will deform in a shape quite close to the magnitude surface we are contemplating.

Next we freeze the rubber sheet and make a slice in the vertical plane containing the $j\omega$ axis. This gives us the cross section shown in Fig. 15.10b, and this cross section is the magnitude part of the frequency-response plot, e.g., Fig. 15.9b, because it is the value of the magnitude of transfer function (15.33) when $s = j\omega_f$. Since we customarily plot only positive values of ω_f in our frequency-response plots, we need look only at the right-hand half of the cross section. Also, in comparing Fig. 15.10b and Fig. 15.9b, we must bear in mind that the cross section in Fig. 15.10b gives magnitude to a linear scale, while Fig. 15.9b is plotted to a log scale.

15.4 DESIGN FOR A SINGLE FREQUENCY: THE VIBRATION ABSORBER

The vibration absorber is a fascinating mechanical device, invented by Frahm in 1909 as a solution to an annoying type of problem. The problem, in one form or another is a common one: Some constant-speed mechanical power device—an engine, a pump, a motor—happens to be operating at the natural resonant frequency of a nearby piece of mounted equipment, causing it to vibrate at large amplitude.

Several mechanical-resonance situations are depicted in the physical models of Fig. 15.11. In Fig. 15.11a an equipment package is "vibration isolated" from the floor, which is vibrating slightly (owing to an engine in the next room, perhaps). Unfortunately, however, the vibration happens to be at the system's resonant frequency and, the damping being relatively light, the floor vibration is magnified manyfold, resulting in an unacceptably large motion of the equipment package.

A more common situation is depicted in Fig. 15.11b, where the machine of which the power source is a part is itself set vibrating at resonance. Figure 15.11b might be a machine tool, mounted on a nonrigid floor and powered by a big electric drive motor that has an unbalanced rotor. The resulting centrifugal force has a vertical component that is sinusoidal, as shown in the physical model of Fig. 15.11c. (The magnitude of this force is readily derived, $f = mr\omega_f^2 \sin \omega_f t$.) If ω_f coincides with $\sqrt{k/m_1}$, a serious resonance may result. The rotary version of the same situation, depicted in Fig. 15.11d, is common in mounted engines (e.g., early automotive engines). In each case the resonance can be avoided, and the motion of the equipment can sometimes be reduced to the nonresonant value, by (1) balancing the rotor, (2) changing the natural frequency of the mount, (3) increasing the damping of the mount, (4) changing the operating frequency of the motor. Often, however, all four methods are quite undesirable for practical reasons.

526 FORCED OSCILLATIONS OF COMPOUND SYSTEMS [Chap. 15]

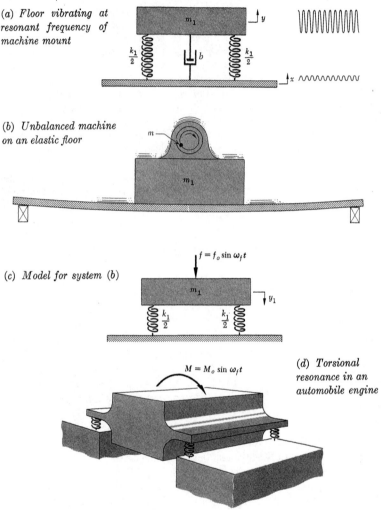

Fig. 15.11 *Some forms of vibrational resonance in machinery*

The vibration absorber approaches the problem with more finesse and, under favorable conditions, may reduce the vibration of the equipment not just to its nonresonant value, but essentially to zero. The simplest case, and the easiest to understand, is that in which all damping may be neglected, as in the idealized physical model of Fig. 15.11c. This also represents the most severe problem, of course, the response of the undoctored system being "infinite" when disturbed at exactly its resonant frequency, as shown by the dashed plot in Fig. 15.12b, which gives frequency response in the neighbor-

[Sec. 15.4] THE VIBRATION ABSORBER

hood of resonance. The transfer function, from force f to motion y_1 in Fig. 15.11c is readily derived:

$$\frac{Y_1}{F/k_1} = \frac{k_1/m_1}{s^2 + (k_1/m_1)} \tag{15.34}$$

The undamped vibration absorber. The vibration absorber consists of a small added mass and spring, Fig. 15.12a. To see how it works, we obtain the system transfer function, from force f to motion y_1 of the main equipment

Fig. 15.12 *In the absence of damping, the vibration absorber absorbs all main-mass motion at the tuned frequency (but engenders resonance at two other frequencies)*

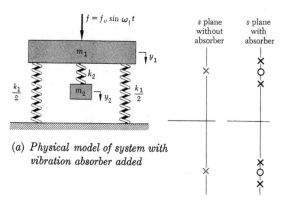

(a) *Physical model of system with vibration absorber added*

(b) *Frequency response*

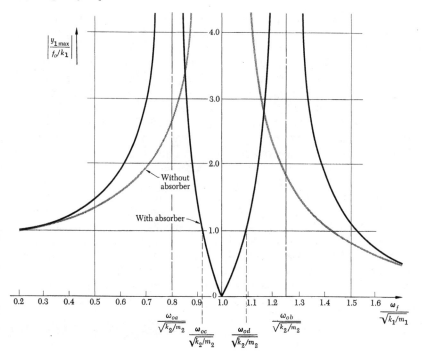

m_1 in Fig. 15.12a. The reader may show that this is

$$\frac{Y_1}{F/k_1} = \frac{(k_1/m_1)[s^2 + (k_2/m_2)]}{\{s^2 + [(k_1 + k_2)/m_1]\}[s^2 + (k_2/m_2)] - (k_2{}^2/m_1 m_2)} \quad (15.35)$$

or
$$\frac{Y_1}{F/k_1} = \frac{(k_1/m_1)[s^2 + (k_2/m_2)]}{[(s^2 + \omega_{0a}{}^2)(s^2 + \omega_{0b}{}^2)]} \quad (15.36)$$

where ω_{0a} and ω_{0b} are obtained by factoring the denominator of (15.35). The frequency response of this system is, of course, obtained by setting $s = j\omega_f$ in (15.36). It is now seen that when $\omega_f = k_2/m_2$, the result is $Y_1/F = 0$. That is, there is a frequency for which mass m_1 does not move at all! Thus, by adjusting k_2 and m_2 so that this frequency coincides with the frequency of force f, the vibration of m_1 can be completely eliminated. What happens physically is that mass m_2 is set vibrating at *its* resonant frequency, and it vibrates with just the amplitude for which the force $k_2(y_2 - y_1)$ transmitted by spring k_2 exactly equals disturbance force $f(t)$ at every instant. The net force on mass m_2 is thus zero, and it does not move. What amplitude of $(y_2 - y_1)$ is necessary to accomplish this? (Since $y_1 = 0$, this equals the amplitude of y_2 alone. The reader may easily obtain the transfer function Y_2/F to verify this amplitude.)

The denominator of (15.36) tells us that there will, however, be two *other* frequencies, $\omega_f = \omega_{0a}$ and $\omega_f = \omega_{0b}$, for which the system response will be infinite. The complete frequency response of the system of Fig. 15.12a is given in Fig. 15.12b (solid curves) for particular values of k_2 and m_2 (chosen so that $\sqrt{k_2/m_2} = \omega_1 = \sqrt{k_1/m_1}$ and $m_2/m_1 = 0.2$). It is clear from Fig. 15.12b that the vibration absorber is very much a one-frequency device: If the input frequency strays a certain small amount either way from the value $\omega_f = \sqrt{k_2/m_2}$, another resonance will occur just as bad as the original one.

The central design parameter for the undamped vibration absorber is the ratio m_2/m_1 of the absorber mass to the mass of the original equipment. Intuitively, we expect that large m_2/m_1 will lead to a smaller motion y_2 at resonance, and will also provide a wider range of frequency over which the absorber is effective (the range between ω_{0c} and ω_{0d} in Fig. 15.12b). These relationships are studied in Problems 15.28 and 15.29, where we shall find this intuition to be correct. There are, of course, practical limitations to the acceptable size of m_2, and thus a trade-off is involved in a given design.

The damped vibration absorber. It may be desirable to include damping in the vibration-absorber system m_2, k_2 (a small amount will be inevitable). Also, there will always be some damping in the original system. There are then three design parameters, m_2, k_2, b_2, and the analysis becomes quite involved. The result, however, can be a system with desirable low response over a whole range of frequencies. An elegant treatment of the problem, with rules for an optimum design, is given by Professor Den Hartog.†

Problems 15.14 through 15.30

---

† J. P. Den Hartog, "Mechanical Vibrations," 4th ed., secs. 3.2 and 3.3, McGraw-Hill Book Company, New York, 1956.

Chapter 16

RESPONSE TO PERIODIC FUNCTIONS: FOURIER ANALYSIS

It is well known that if a function is periodic—that is, if it repeats itself exactly every T seconds, where T is a constant—then that function can be represented as the sum of many pure sine waves whose frequencies are whole multiples of the frequency of the function itself, and whose magnitudes are to be determined such that when all the sine waves are added together the original function is reproduced. The set of sine waves is known as a Fourier series, and the set of coefficients, which tell the frequency content of the function, are known as its spectrum. The coefficients are often plotted versus frequency in a spectral diagram.

The most familiar form of the Fourier series is known as the "real Fourier series," given in Eq. (16.1) below. By Euler's equation this familiar relation can be rewritten in the form of a series of exponential functions known as the "complex Fourier series." This latter form will be of more interest to us because we have found in our study of linear systems that functions e^{st} (with s complex, in general) are of central importance. By way of introduction, however, the real Fourier series is reviewed in Sec. 16.1.

In Sec. 16.3 we introduce the important concept that a periodic function can be represented by its frequency spectrum.

It is expected that Sec. 16.1, if not this entire short chapter, will be review for most readers.

16.1 REAL FOURIER SERIES

If a function $x(t)$ is periodic, then it can be represented by a series of simple sine waves:†

$$x(t) = \frac{a_0}{T} + \frac{2}{T}\sum_1^\infty a_n \cos n\omega_0 t + \frac{2}{T}\sum_1^\infty b_n \sin n\omega_0 t \qquad (16.1a)$$

† The factors $2/T$ are more commonly included in the coefficients a_n and b_n, but in this book the form (16.1) will be more convenient for later extensions of the method.

where coefficients a_n and b_n depend, of course, on the shape of the function itself, as follows:

$$a_n = \int_\tau^{\tau+T} x(t) \cos n\omega_0 t \, dt \qquad b_n = \int_\tau^{\tau+T} x(t) \sin n\omega_0 t \, dt \qquad (16.1b)$$

and where a_0/T is the average value of $x(t)$.

The mechanics by which Eqs. (16.1) produces the desired result are most easily reviewed by means of a demonstration.

Example 16.1. A sum of sine waves. Suppose that a waveform of interest can be represented by the function

$$x(t) = 4 + 5 \cos \omega t + 2 \cos 2\omega t + 3 \sin \omega t + \sin 3\omega t \qquad (16.2)$$

Suppose, however, that we are not given the function in its analytical form, but rather are given simply a graphical picture of the waveform, such as Fig. 16.1. The task is to determine the coefficients in Eq. (16.1a), that is, *to determine how much of each frequency is present in the given waveform.*

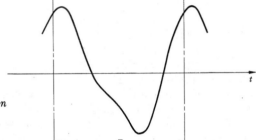

Fig. 16.1 *The periodic function represented by Eq. (16.2)*

Having no analytical information, we would proceed to find each coefficient according to Eqs. (16.1b) by means of a graphical technique. That is, to get a_1, we would multiply the given waveform, point by point, by $\cos \omega_0 t$ [where $\omega_0 = 2\pi/T = \omega$ in (16.2)], and then integrate the result over one period. The numerical value of this integral would be a_1. Similarly, to get a_2 we would multiply the waveform by $\cos 2\omega_0 t$, integrate as before, and so on [constant a_0 is a special case given by

$$a_0 = \int_\tau^{\tau+T} x(t) \, dt$$

that is, a_0/T is the average value of $x(t)$.] If this procedure is carried out for each of the a's and b's, we should get numbers which match exactly those given by (16.2). (Fig. 16.1 has been drawn carefully, and the reader is invited to perform the graphical integration upon it.)

We can prove that the above process yields the proper answers by carrying out analytically the procedure outlined. It is instructive to do so. For a_1 (with $\tau = 0$, $T = 2\pi/\omega_0$, $\omega_0 = \omega$),

$$a_1 = \int_0^{2\pi/\omega} x(t) \cos 1\omega t \, dt$$

[Sec. 16.1] REAL FOURIER SERIES

From (16.1b). Substituting the value of $x(t)$ from (16.2), we obtain

$$a_1 = \int_0^{2\pi/\omega} 4 \cos \omega t \, dt + \int_0^{2\pi/\omega} 5 \cos^2 \omega t \, dt + \int_0^{2\pi/\omega} 2 \cos 2\omega t \cos \omega t \, dt$$
$$+ \int_0^{2\pi/\omega} 3 \sin \omega t \cos \omega t \, dt + \int_0^{2\pi/\omega} \sin 3\omega t \cos \omega t \, dt$$
$$= 0 + 5 + 0 + 0 + 0$$

That is, all products $(\cos n\omega t)(\cos m\omega t)$ integrate to zero over one period unless $m = n$. Furthermore, for $m = n$, the products—$\cos^2 n\omega t$—integrate to $T/2$:

$$\left.\begin{aligned}\int_\tau^{\tau+2\pi/\omega} \cos n\omega t \cos m\omega t \, dt &= \begin{cases} 0 \text{ if } m \neq n \\ T/2 \text{ if } m = n \end{cases} \\ \int_\tau^{\tau+2\pi/\omega} \sin n\omega t \sin m\omega t \, dt &= \begin{cases} 0 \text{ if } m \neq n \\ T/2 \text{ if } m = n \end{cases} \\ \int_\tau^{\tau+2\pi/\omega} \sin n\omega t \cos m\omega t \, dt &= 0, \quad \text{whether } m = n \text{ or not}\end{aligned}\right\} \quad (16.3)$$

Pairs of functions which integrate to zero over one period, such as $\cos n\omega t$ and $\cos m\omega t$ with $m \neq n$, are known as *orthogonal* functions. *The Fourier technique works because when a periodic waveform is multiplied by a sine wave which it does not contain—one which is orthogonal to it—the resulting integral over one period is zero.*

The reader is invited to verify the remainder of Eq. (16.2) by calculating the coefficients up through, say, a_4 and b_4.

Example 16.2. Square wave. Find the Fourier series for the square wave shown in Fig. 16.2. Demonstrate the plausibility of the convergence of the Fourier series by finding the coefficients up through b_5 and then plotting the waveform given by the Fourier series first with only one term, then with three, then with five.

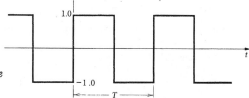

Fig. 16.2 *Periodic square wave*

We expect at the outset that all of the a's will be zero because the waveform given resembles a sine wave, rather than a cosine wave. The coefficients can be calculated analytically by using the following "piecewise" expression for $x(t)$ over one period, $t = 0$ to $t = T$:

$$\begin{aligned} x(t) &= 1 & 0 < t < \frac{T}{2} \\ x(t) &= -1 & \frac{T}{2} < t < T \end{aligned} \quad (16.4)$$

$$x(t) = \frac{4}{\pi}\left[\sin \omega t + \frac{1}{3}\sin 3\omega t + \frac{1}{5}\sin 5\omega t + \ldots\right],$$

$$\omega = \frac{2\pi}{T}$$

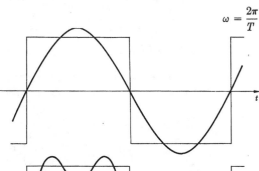

(a) *First term only*

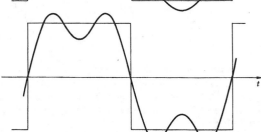

(b) *First two terms*

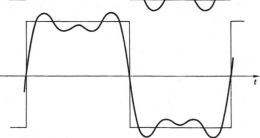

(c) *First three terms*

Fig. 16.3 *Fourier series approximation to a square wave*

Then, from (16.1b), we have

$$a_0 = \int_0^{T/2} (1)\, dt + \int_{T/2}^T (-1)\, dt = 0 \qquad (x_{\text{av}} = 0)$$

$$a_n = \int_0^{T/2} (1) \cos n\left(\frac{2\pi}{T} t\right) dt + \int_{T/2}^T (-1) \cos n\left(\frac{2\pi}{T} t\right) dt = 0$$

$$b_n = \int_0^{T/2} (1) \sin n\left(\frac{2\pi}{T} t\right) dt + \int_{T/2}^T (-1) \sin n\left(\frac{2\pi}{T} t\right) dt = \begin{cases} \dfrac{2T}{n\pi} \text{ for } n \text{ odd} \\ 0 \text{ for } n \text{ even} \end{cases}$$

The Fourier series representing Fig. 16.2 is then, from Eq. (16.1a),

$$x(t) = \frac{4}{\pi}\sin \omega t + \frac{4}{3\pi}\sin 3\omega t + \frac{4}{5\pi}\sin 5\omega t + \cdots \qquad (16.5)$$

where $\omega = 2\pi/T$.

[Sec. 16.2] COMPLEX FOURIER SERIES

Expression (16.5) is plotted several times in Fig. 16.3, each time with more terms included in the approximation. Note that quite a close approximation is obtained with only two or three terms.

16.2 COMPLEX FOURIER SERIES

Equation (16.1a) can be represented more concisely by the use of complex numbers, together with the application of Euler's equation (7.3).

The Fourier series (complex form)

$$\left. \begin{array}{l} x(t) = \dfrac{1}{T} \sum\limits_{-\infty}^{\infty} C_n e^{st} \\ C_n = \displaystyle\int_{\tau}^{T+\tau} x(t) e^{-st}\, dt \end{array} \right\} \quad s = jn\omega_0 = jn\dfrac{2\pi}{T}$$

$$\text{(In particular, } C_0 = \int_{\tau}^{T+\tau} x(t)\, dt \to x_{\text{av}} T\text{)}$$

(16.6)

Note that the complex Fourier series goes from $-\infty$ to $+\infty$. This is necessary because the Euler expressions (7.24) for $\sin \omega t$ and $\cos \omega t$ contain both $e^{+j\omega t}$ and $e^{-j\omega t}$.

Use of the complex Fourier series can be conveniently demonstrated, as in the preceding example, with a square wave.

Example 16.3. Square wave by complex form. Find the coefficients of the complex Fourier series to represent the square wave of Fig. 16.2.

As before, we represent the square wave by (16.4). Then, from (16.6):

$$C_n = \left[\int_0^{T/2} (1) e^{-st}\, dt + \int_{T/2}^{T} (-1) e^{-st}\, dt \right]_{s=jn(2\pi/T)}$$
$$= \frac{T}{jn2\pi} [1 - 2e^{-jn\pi} + e^{-jn2\pi}]$$

The values of the exponential terms are conveniently discerned from a complex-plane plot (based on Euler's equation; see Fig. 7.6):

$$C_n = \begin{cases} \dfrac{T}{jn2\pi}(1 + 2 + 1) = \dfrac{2T}{jn\pi} & \text{for } n \text{ odd} \\ \dfrac{T}{jn2\pi}(1 - 2 + 1) = 0 & \text{for } n \text{ even} \end{cases}$$

Substituting this result into (16.6), we obtain

$$x(t) = \frac{2}{j\pi}\left[\left(\frac{e^{j\omega_0 t}}{1} + \frac{e^{-j\omega_0 t}}{-1}\right) + \left(\frac{e^{j3\omega_0 t}}{3} + \frac{e^{-j3\omega_0 t}}{-3}\right) + \cdots\right]$$

$$= \frac{4}{\pi}[\sin \omega_0 t + \tfrac{1}{3}\sin 3\omega_0 t + \tfrac{1}{5}\sin 5\omega_0 t + \cdots] \qquad (16.7)$$

where $\omega_0 = 2\pi/T$. This is the same as the result obtained in Example 16.2.

16.3 SPECTRAL REPRESENTATION

We have pointed out that each of the coefficients in a Fourier series tells how much of each frequency is present in the function being represented. This is an important viewpoint and suggests representing a function alternatively by a plot of the magnitude of its coefficients versus frequency. Such a plot is known as a *spectral diagram*, and the aggregate of the coefficients is known as the *spectrum* of the function

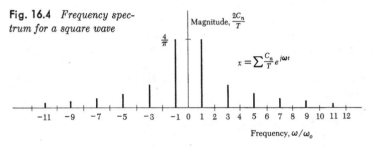

Fig. 16.4 *Frequency spectrum for a square wave*

Since a periodic function is represented by sine waves which are integral multiples of the fundamental frequency, its coefficients correspond only to discrete frequencies and therefore show up in the spectral diagram as isolated lines. For example, Fig. 16.4 shows the spectrum of the square wave of Fig. 16.2.

The representation of complicated functions by their spectral diagrams is a useful analytical tool. In Chapter 17 the spectral technique will be extended to allow representation of functions of any form, nonperiodic as well as periodic.

Problems 16.1 through 16.8

Chapter 17

THE LAPLACE TRANSFORM METHOD

In Chapter 14 we found that when the input to an l.c.c. (linear, constant-coefficient) system is expressed in the form e^{st} (where, in general, s may be a complex number), the output will then be of the same form. This important fact makes possible the clear, efficient transfer-function method of analyzing such systems described in Chapter 14. It makes it quite simple to trace mathematically the modification of any signal of this form as it flows from member to member in a system.

We have already found that most of the interesting motions of l.c.c. systems can be expressed in the form e^{st}. Specifically, (1) in Parts B and C we noted that all the possible natural motions of such a system can be represented quite simply in the form e^{st}, Fig. 17.1a, and (2) in Chapter 16 we showed that not only simple sinusoidal motions, but—by application of the Fourier series—*any periodic forcing function*, can be represented by expressions of the form e^{st}, Fig. 17.1b, where, in this case, $s = jn\omega$. Moreover, so long as these e^{st} functions are not abrupt, we can readily study the response to them, using simple transfer-function methods.

In this chapter we shall make one further extension of this technique: We shall establish what additional steps are necessary to allow e^{st} to represent almost completely arbitrary functions and functions which begin abruptly, as in Fig. 17.1c. This we shall do in Sec. 17.2. The result will be a process known as the Laplace transformation, around which an important method of analysis, the *Laplace transform method*, is built.

But, while the motivation for developing the Laplace transform method is to be able to express arbitrary functions in the form e^{st}, and thus to extend the transfer-function method to arbitrary inputs, we shall find that the tool which results is extremely handy also for elementary problems. In practice the Laplace transform method is used widely to solve problems for which its power and generality are by no means required, problems involving only simple functions which can readily be cast in the form e^{st} by less erudite means.

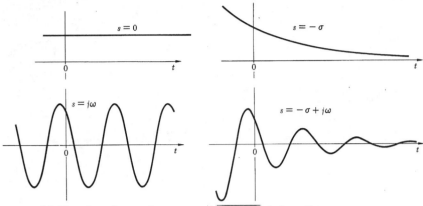

(a) *Functions that can be represented in the simple form Ce^{st}*

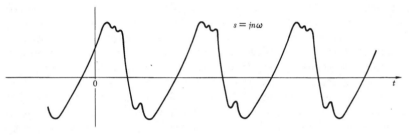

(b) *Periodic function that can be represented by a Fourier series, $\Sigma C_n e^{st}$*

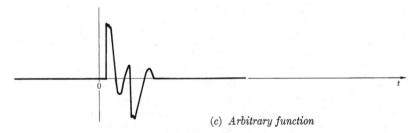

(c) *Arbitrary function*

Fig. 17.1 *Types of time functions*

The reason for this widespread usage is not the power of the Laplace transform method. The reason is rather that the method is so *convenient* and so *systematic* for setting up transfer functions, and can be applied universally to any l.c.c. system. Laplace transforms are most often used because they supply a splendid bookkeeping procedure for the solution of linear differential equations. In particular, the handling of *abrupt inputs* and *initial conditions* is made quite routine.

As a matter of fact, the Laplace transform method is so routine that it

[Sec. 17.1] DEMONSTRATION OF THE LAPLACE TRANSFORM METHOD 537

can be applied to solve most common l.c.c. differential equations of motion by someone who has never been through its derivation, or even seen the formal result.† This is possible because time functions and their transfer functions can be tabulated, and the process of finding one from the other can be reduced, in many cases, to looking in a table.

To demonstrate this, and to give a preview of the method, we shall, in Sec. 17.1, simply state the procedure and demonstrate its use in finding transient response. Hopefully, this will make more meaningful the detailed development of the method which will follow in Sec. 17.2.

17.1 DEMONSTRATION OF THE LAPLACE TRANSFORM METHOD

Note that we are concerned here only with "Stage III: from mathematical model to system behavior," in the sequence of investigation outlined in Sec. 1.1. That is, we begin with the differential equations of motion, and we are to find the time response of the system.

The Laplace transform method accomplishes this in three simple steps:

Step 1. ***Laplace transform*** the entire equation(s) of motion, term by term, using a table of Laplace transform pairs. (The resulting algebraic equations contain all of the information about the problem, including forcing functions and initial conditions.)

Step 2. ***Manipulate*** these algebraic equations to obtain:
  (a) The desired *transfer functions* and the complete *response function* for the variable of interest.
  (b) The *roots of the characteristic equation* [factors of the denominator in (a)]. Study natural behavior of the system.
  (c) *Final and initial values.* If applicable, employ the Final-Value Theorem (FVT) and the Initial-Value Theorem (IVT) as checks on the response function.

Step 3. ***Inverse transform*** the response function to obtain the complete time response of the variable of interest. Use a table of transforms, or partial fraction expansion (Procedure D-2).

Procedure D-1
Using the Laplace transform method to solve equations of motion

† For this reason many excellent texts merely state part of the result without development, and confine themselves to demonstrating its use.

The total time response of the system will always be made up of the sum of individual elementary motions which can be analyzed and plotted by the methods of Part B (e.g., Fig. 8.8).

Procedure D-1 can be stated more concisely:

Procedure D-1 *The Laplace transform method*

Step 1. **Laplace transform** the equations of motion.

Step 2. **Manipulate** to obtain:
(a) transfer functions, response function, or both;
(b) characteristics; and
(c) final and initial value (FVT and IVT).

Step 3. **Inverse transform** to get time response.

Procedure D-1 has been likened to that for multiplying numbers by means of logarithms, in which one (1) "transforms" the numbers into logarithms by means of a table, (2) manipulates the logarithms, and then (3) transforms the logarithms back into numbers ("inverse transformation"), using the same table. Table X is a short table of Laplace transform pairs. Thus, for ordinary systems, Procedure D-1 is a routine "crank-grinding" operation, as we now demonstrate for an example which we have already worked out by "classical" methods on p. 195.

The value of the Laplace transform is that it transports the routine mathematical part of a problem to another domain, where operations and manipulations are much easier to perform, and then transports it back after the work is done.

Example 17.1. Laplace analysis of simple system. A sudden step input, $M = M_0 u(t)$, is applied to the system of Fig. 6.2, p. 185. Find the response $\theta(t)$ if at $t = 0$ the value of θ happened to be θ_0. Find also the transfer function from M to θ.

The equation of motion of the system is (6.3):

$$b\dot{\theta} + k\theta = M = M_0 u(t) \tag{17.1}$$

Now we follow Procedure D-1.

Step 1. Laplace transform the equation of motion. (The one-sided Laplace transform is used.) From Table X, the transformed equation of motion is found to be

$$b[s\Theta(s) - \theta(0^-)] + k\Theta(s) = \mathsf{M}(s) = \frac{M_0}{s} \tag{17.2}$$

Transform pairs 1 and 5 of Table X were used. The term $\theta(0^-)$ denotes the value of θ "immediately *before* $t = 0$," and its value will be θ_0, as given.

Step 2. Manipulate the transformed equation algebraically to obtain:
(a) Desired *transfer function* and the *response function* for the variable of

interest. The transfer function between M and θ is obtained directly from (17.2), with $\theta(0^-) = 0$:

$$\frac{\Theta(s)}{\mathsf{M}(s)} = \frac{1}{bs + k} \qquad (17.3)$$

and this can be displayed in block-diagram form:

$$\mathsf{M}(s) \longrightarrow \boxed{\frac{1}{bs+k}} \longrightarrow \Theta(s)$$

The complete response function for $\Theta(s)$ is also obtained from (17.2):

$$\Theta(s) = \frac{\mathsf{M}(s)}{bs + k} + \frac{b\theta(0^-)}{bs + k}$$

or, with $\theta(0^-) = \theta_0$, and $\mathsf{M}(s) = \mathcal{L}[M_0 u(t)] = M_0/s$; and rearranging slightly (for convenience in using Table X), we have

$$\Theta(s) = \frac{(M_0/k)(k/b)}{s[s + (k/b)]} + \frac{\theta_0}{s + (k/b)} \qquad (17.4)$$

(b) *System natural behavior.* The denominator in (17.3) and (17.4) has come already factored. It reveals that the natural characteristic of the system is

$$s = -\frac{k}{b}$$

Step 3. Inverse transform the response function. Happily, we find in Table X both of the terms in (17.4), as transform pairs 11 and 6. The time response is thus written at once:

$$\theta(t) = u(t)\frac{M_0}{k}(1 - e^{-(k/b)t}) + u(t)\theta_0 e^{-(k/b)t} \qquad (17.5)$$

which is identical to the solution (6.14), obtained by the classical method. This response is plotted versus time in Fig. 6.7 (for $\theta_0 = 0$).

Hopefully, the above example has raised a number of questions in the reader's mind, such as (1) Why does Procedure D-1 work? (2) Under what circumstances does it work? (3) Is it somehow the same as assuming a solution of the form e^{st}? (4) What are "$\Theta(s)$" and "$\mathsf{M}(s)$?" (5) Where does Table X come from? (6) Is transfer function (17.3) like those obtained in Chapter 14? (7) What is a response function? (8) What do you do when Step 2 gives a "response function" which is not in the table? (9) What is a "one-sided" Laplace transform? (10) Why didn't we learn this simple Procedure D-1 in the first place, and save a lot of work?

The first five questions are answered in Secs. 17.2 and 17.3: The Laplace transform and its application are developed from the e^{st} idea, and the general expression for Laplace transform pairs is derived [Eqs. (17.11)]. These results are summarized in Sec. 17.4. The contents of Table X are derived in Sec. 17.5. A much more comprehensive table of Laplace transforms is provided in Appendix J.

The nature of the "transformed variables," like $\mathsf{M}(s)$ and $\Theta(s)$, is

TABLE X
Short table of Laplace transform pairs

$f(t)$		$F(s)$	
		Two-sided	One-sided
0. $Cx(t)$		$CX(s)$	$CX(s)$
1. $\dot{x}(t) = \dfrac{dx}{dt}$		$sX(s)$	$sX(s) - x(0^-)$
2. $\dfrac{d^n x}{dt^n}$		$s^n X(s)$	$s^n X(s) - s^{n-1}x(0^-) - s^{n-2}\dot{x}(0^-) - \cdots - \left[\dfrac{d^{n-1}x}{dt^{n-1}}\right]_{t=0^-}$
3. $\int x(t)\,dt$		$\dfrac{1}{s} X(s)$	$\dfrac{1}{s} X(s) + \dfrac{1}{s}\left[\int x\,dt\right]_{t=0^-}$
4. $\delta(t)$	[Area = 1 impulse at origin]	1	
5. $u(t)$	[unit step]	$\dfrac{1}{s}$	s plane
5.1. $u(t-\tau)$	[delayed unit step at τ]	$\dfrac{1}{s} e^{-\tau s}$	
5.2. $x(t-\tau)$		$e^{-\tau s} X(s)$	

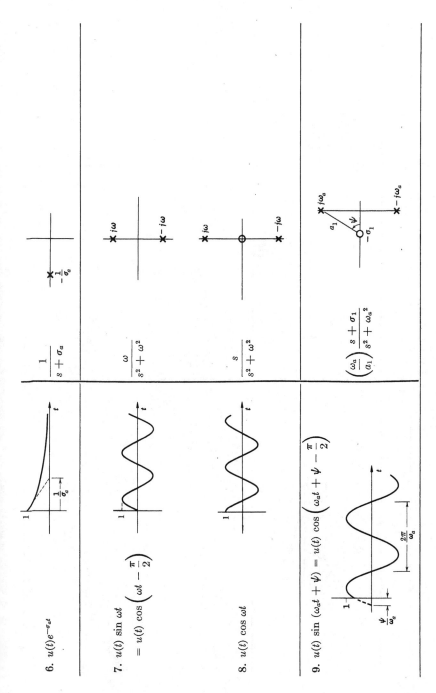

TABLE X (continued)

$f(t)$		$F(s)$
10. $u(t)e^{-\sigma_a t}\sin(\omega_a t + \psi) = u(t)e^{-\sigma_a t}\cos\left(\omega_a t + \psi - \dfrac{\pi}{2}\right)$	$\left(\dfrac{\omega_a}{a_1}\right)\dfrac{s+\sigma_1}{s+\sigma_a \mp j\omega_a}$ $= \left(\dfrac{\omega_a}{a_1}\right)\dfrac{s+\sigma_1}{s^2+2\sigma_a s + \omega_{0a}^2}$ $\zeta = \sigma_a/\omega_{0a}$	
11. $u(t)(1-e^{-\sigma_a t})$	$\dfrac{\sigma_a}{s(s+\sigma_a)}$	
12. $u(t)(e^{-\sigma_a t} - e^{-\sigma_b t})$	$\dfrac{(-\sigma_a + \sigma_b)}{(s+\sigma_a)(s+\sigma_b)}$	

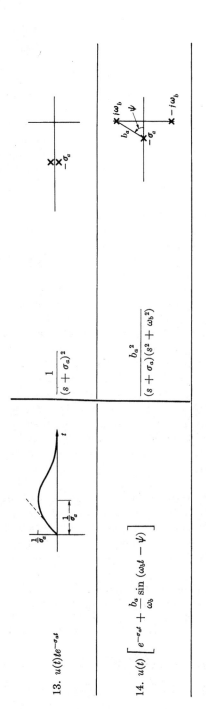

13. $u(t) t e^{-\sigma_a t}$

$$\frac{1}{(s+\sigma_a)^2}$$

14. $u(t)\left[e^{-\sigma_a t} + \dfrac{b_a}{\omega_b} \sin(\omega_b t - \psi)\right]$

$$\frac{b_a^2}{(s+\sigma_a)(s^2+\omega_b^2)}$$

developed in Sec. 17.2. In solving transient problems these may simply be regarded as alternate variables, obtained from a conversion table (e.g., Table X). They may be manipulated like any algebraic variable, and finally, they may be converted back into their time-function counterparts [e.g., $M(t)$ and $\theta(t)$] via the same table.

Question (6) is carefully answered in Sec. 17.6. Response functions [question (7)] are discussed at length in Sec. 17.7, and many examples are worked out. The problem of inverse-transforming a response function not in the table [question (8)] is the subject of Chapter 18, where the general procedure is developed, using straightforward partial fraction expansion.

The one-sided Laplace transform [question (9)] is introduced in Sec. 17.3, and its utility in handling initial conditions is demonstrated there and in Sec. 17.7.

The pedagogical reasons for delaying the introduction of the Laplace transform method [question (10)] have been discussed in the preface. There are important technical reasons as well: When we are interested *only* in the dynamic characteristics and stability of a system's natural behavior (as in Part C), *or* when we are interested *only* in the response of a system to *continuing inputs*, particularly the steady-state response to *sinusoidal inputs* (as in Chapters 9, 15, and 16), then the Laplace transform method is actually substantially *more cumbersome* than the simple e^{st} approach we used in Part C and in Chapters 9, 15, and 16. The Laplace transform method can be used in these cases, as we shall see in Secs. 17.6 and 17.7, but it is more cumbersome. Furthermore, in these cases it tends to obscure the physical understanding we have been trying to develop. We have therefore delayed introduction of the Laplace transform method until some familiarity with simple behavior had been gained, and until the method would really be useful.

Now, however, as we address the problem of finding total response to abrupt disturbances, it will be useful indeed. We therefore proceed in Sec. 17.2 to carry out a detailed (if not entirely rigorous) description of how the Laplace transformation evolves when we set out to represent an arbitrary function as a combination of terms in the form e^{st}. In doing so we hope to establish an appreciation for why the Laplace transform method works and for how it is, in fact, basically an extension, through superposition, of the technique of assuming a solution of the form e^{st}, a technique we have been using all along.

For the reader who prefers to become more familiar with the *application* of the Laplace transform (Procedure D-1) before delving into the theory of why it works, numerous examples and problems are presented in Sec. 17.7, and those can be studied before reading Secs. 17.2 through 17.6 if desired.

ⓒ 17.2 EVOLUTION OF THE LAPLACE TRANSFORM

Our goal is to cast arbitrary, nonperiodic functions as a combination of terms in the form Ce^{st}, with s a general complex number. A common and instructive demonstration is first to extend the Fourier series—which casts

[Sec. 17.2] EVOLUTION OF THE LAPLACE TRANSFORM 545

arbitrary *periodic* functions in the form Ce^{st}, with $s = j\omega$—to the Fourier integral and Fourier transform, which cast *nonperiodic* functions in the same form. Then the final step, from Fourier transform to Laplace transform, removes the restriction $s = j\omega$.

From Fourier series to Fourier integral

The task here is to represent an arbitrary nonperiodic function, such as that in Fig. 17.1c, in the form Ce^{st}, with $s = j\omega$. The strategy for doing so is to pretend that such a function *is* periodic, as in Fig. 17.2, and represent it by a Fourier series; and then to let its period grow to infinity and see how this affects the series.

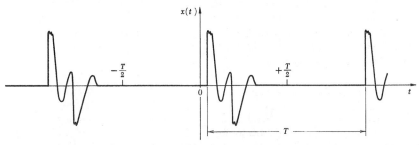

Fig. 17.2 *Pretending that Fig. 17.1c repeats periodically*

A periodic function, such as that in Fig. 17.2, can be represented by (16.6),† which we write in a slightly different, but equivalent form:

$$x(t) = \left[\frac{1}{2\pi} \sum_{n=-\infty}^{\infty} \omega_0 C(s) e^{st} \right]_{s=jn\omega_0} \quad (17.6a)$$

with
$$C(s) = \left[\int_{-\pi/\omega_0}^{\pi/\omega_0} x(t) e^{-st} dt \right]_{s=jn\omega_0} \quad \text{and} \quad \omega_0 = \frac{2\pi}{T} \quad (17.6b)$$

[We have replaced C_n by $C(s)$, which is legitimate, since for every n there is also an s, given by $s = jn\omega_0$]. Notice that ω_0 is the separation between the frequencies of succeeding terms in the series.

We now let T grow to ∞ in Fig. 17.2 and in (17.6). What happens then is that the terms in (17.6a) become infinitesimally close together in "frequency." In the limit, we write $d\omega$, or ds/j, for the infinitesimal separation ω_0, and replace the sum by an integral. Equations (17.6) then become

$$x(t) = \left[\frac{1}{2\pi} \int_{s=-\infty}^{s=\infty} ds \, C(s) e^{st} \right]_{s=j\omega} \quad (17.7a)$$

with
$$C(s) = \left[\int_{t=-\infty}^{t=\infty} x(t) e^{-st} dt \right]_{s=j\omega} \quad (17.7b)$$

† Recall, again, that the second equation of (16.6) states the orthogonality principle that a specific coefficient, C_n/T, in the series of sine waves making up $x(t)$ can be obtained by multiplying the entire function $x(t)$ by the term under study ($e^{-jn\omega_0 t}$) and integrating. Multiplication by $e^{-jn\omega_0 t}$ culls out all the content of frequencies other than $n\omega_0$.

The formal limiting process in going from relations (17.6) to relations (17.7) is somewhat involved and is omitted here. Expression (17.7a) is known as the *Fourier integral*, and expression (17.7b), which gives the coefficients for (17.7a), is called the *Fourier transform*.

The pair of relations (17.7) for a nonperiodic *function is completely analogous to the pair (17.6) [or (16.6)] for a* periodic *function.*

Expressions (17.7) state that a function $x(t)$ can be written as the sum of an extremely large number of terms, each of the form $C(\omega)e^{j\omega t}\, d\omega$, where multiplier $d\omega$ is extremely small, and coefficient C has a specific value for *every* ω. All of these myriad (sinusoidal) terms are to be summed together to produce the motion $x(t)$. The value of this sum at any specific instant t is given by (17.7a). As with the ordinary Fourier series [Eq. (17.7b) tells us], the value of the coefficient C for each ω, i.e., *the amount of each frequency present*, is to be found out by multiplying the function by, in turn, a sine wave of each frequency, and integrating. (The orthogonality culls out all of the other frequency content.)

To visualize the summation represented by integrals (17.6a) and (17.7a), it is helpful to consider the frequency spectrum—the magnitude of the coefficients $C(s)$ plotted versus frequency ω. For a *periodic* function this plot consists of a series of "spikes" at discrete frequencies, as in Fig. 16.4. For the periodic function of Fig. 17.2 the set of spikes might be as shown in Fig. 17.3a. [Since $C(s)$ is a complex number, its real and imaginary parts are shown separately.] For each spike, a rectangle has been constructed with the spike as altitude and with frequency separation ω_0 as base. Now the summation of (17.6a) can be obtained by increasing the height of each rectangle in Fig. 17.3a by the ratio $[e^{st}]_{s=jn\omega_0}$, and then finding the total area of all the rectangles. [The imaginary part always sums to zero, as it must for $x(t)$ to be a real function.] For $t = 0$, this is just the shaded area shown. For positive values of t, the rectangle at the origin remains the same, but the others grow in exponential proportion to their distance from the origin.

For a *nonperiodic* function, like that in Fig. 17.1c, the "spikes" in the frequency spectrum are infinitesimally close together, so that the spectrum is continuous, as in Fig. 17.3b. The corresponding rectangles have infinitesimal width $d\omega$, i.e., infinitesimal frequency separation. The summation of (17.7a) is obtained by multiplying the curve in Fig. 17.3b by $[e^{st}]_{s=j\omega}$ and then measuring the total area under the curve. (Again, the imaginary part sums to zero.) For $t = 0$, this is the shaded area in Fig. 17.3b. For positive t, the curve $[C(s)e^{st}]_{s=j\omega}$ is larger in exponential proportion to its distance from the origin.

In each part of Fig. 17.3 we have performed the summation or integration along the $j\omega$ axis. The σ axis† has also been included in Fig. 17.3 to indicate that we might consider these as three-dimensional plots of $C(s)$ over the whole s plane as the base. So far, however, $C(s)$ has value only above the $j\omega$ axis, because $s = j\omega$.

---

† In this section, we shall employ the symbol σ to represent the real part of s: $s = \sigma + j\omega$. (Elsewhere in this book we use the symbol for another purpose, as defined on p. 249 and discussed in the footnote there.)

[Sec. 17.2] EVOLUTION OF THE LAPLACE TRANSFORM

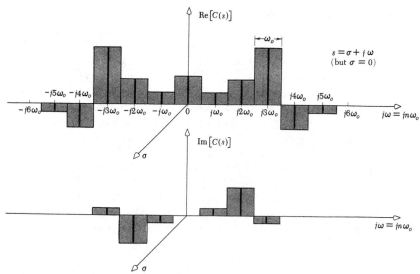

(a) Spectrum of $C(s)$ at a specific time for the periodic function $x(t)$ of Fig. 17.2

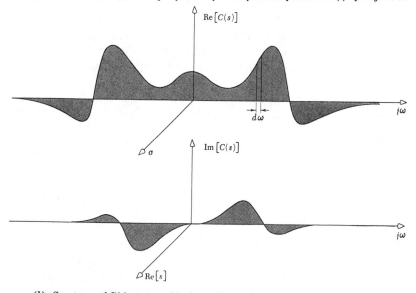

(b) Spectrum of $C(s)$ at a specific time for the nonperiodic function $x(t)$ of Fig. 17.1c

Fig. 17.3 Plots of $C(s)$ at some specific time, for the time functions of Figs. 17.2 and 17.1c

From Fourier transform to Laplace transform

We have still not quite reached our goal, which was to represent arbitrary function $x(t)$ as a combination of terms of the form Ce^{st} with s a general complex number. We are nearly there, but expressions (17.7) have still the restriction $s = j\omega$. That is, we are restricted to coefficients $C(s)$ which lie along the $j\omega$ axis. This restriction turns out to be a fundamental one because, with it, the integral (17.7b) fails to converge for many time functions of interest. For example, if $x(t)$ is the important step function $x(t) = x_o u(t)$, then $x(t)e^{-j\omega t}$ goes on and on, and its integral is unbounded. Then $C(s) = C(j\omega)$ in (17.7b) will be infinitely large.

To insure convergence of (17.7b) for a much wider class of functions, we next modify (17.7b) by letting s have a more general value than $s = j\omega$: We let it be $s = \sigma + j\omega$, where σ is large enough to make the integral (17.7b) converge. It does this because letting $s = \sigma + j\omega$ in (17.7b) is the same as multiplying by $e^{-\sigma t}$ within the original (Fourier) transform:

$$C(s) = \left[\int_{t=-\infty}^{\infty} x(t) e^{-st}\, dt \right]_{s=\sigma+j\omega}$$
$$= \int_{t=-\infty}^{\infty} x(t) e^{-\sigma t} e^{-j\omega t}\, dt \qquad (17.8)$$

We see at once that (17.8) will converge for the unit step for all $0 < \sigma$. Multiplication by $e^{-\sigma t}$ has squeezed down the time plot as $t \to \infty$ so that the area under $x_o u(t) e^{-j\omega t} e^{-\sigma t}$ is finite. More generally, (17.8) will converge for any function of the form

$$x(t) = A e^{\alpha t} u(t) \qquad (17.9)$$

so long as $\quad$ Re $[\alpha]$ < Re $[s]$ $\qquad (17.10)$

Expression (17.8) is known as the *two-sided Laplace transform* of $x(t)$ [or "the $\mathcal{L}$ transform of $x(t)$"]:

Two-sided Laplace transform
$$\boxed{\mathcal{L}[x(t)] = X(s) \triangleq \int_{t=-\infty}^{\infty} x(t) e^{-st}\, dt} \qquad (17.11b)$$

This expression [(17.8) or (17.11b)] is a modification of (17.7b) in which s is restricted no longer to the $j\omega$ axis, but to whole regions of the s plane: namely, to those ranges of values of s for which σ, the real part of s, causes (17.11b) to converge. Consider, for example, the function $x(t) = u(t)e^{-\sigma_2 t}$. For this time function, (17.11b) converges only when $-\sigma_2 <$ Re $[s]$, and therefore the Laplace transform $X(s)$ of the function (which we shall compute in Example 17.2) is defined only in the region of the s plane *to the right of the line $s = -\sigma_2$*. This is typical for functions that have a definite starting time, such as $u(t)e^{-2t}$ or $u(t-4)\sin 5t$ or $u(t+100)e^t$. (For functions that extend backward in time to $t = -\infty$, a convergence problem may arise, but

[Sec. 17.3] APPLICATION: THE ONE-SIDED LAPLACE TRANSFORM

we shall not consider such cases here. Rather, we shall restrict ourselves to functions having a definite starting time, and proceed with the discussion.)

The modification of (17.7b) to (17.11b) must be accompanied by a corresponding modification of (17.7a) to (17.11a), below, which is known as the *inverse Laplace transform*:

$$x(t) = \frac{1}{2\pi j} \int_{s=\sigma_o-j\infty}^{\sigma_o+j\infty} C(s)e^{st}\,ds \qquad (17.11a)$$

where σ_o can have any value, so long as it is large enough to make (17.11b) converge for the given $x(t)$. That is, to match the restrictions (17.10) on (17.8), the integration in (17.11a) is to be performed, not along the $j\omega$ axis as in Fig. 17.3, but along another line parallel to the $j\omega$ axis at distance σ_o away, and within the usable area of the s plane.

To visualize this completely we must note that $C(s)$, or $X(s)$ as we shall call the Laplace transform of $x(t)$ from now on, is now a function of $s = \sigma + j\omega$, and no longer of $j\omega$ alone. Therefore, the functions Re $[X(s)]$ and Im $[X(s)]$ should now be thought of as complete surfaces above the s plane. The *Fourier transform*, (17.7b), can be thought of graphically as a procedure for finding $x(t)$ from such a surface by making a vertical slice through the surface *at the jω axis*, multiplying the intercepted curve by $e^{j\omega t}$, and then measuring the area of the slice as described on p. 546. The *Laplace transform*, (17.11b), involves the same process, except that the vertical slice is made parallel to the $j\omega$ axis, but a distance σ_o out in front of it.

In practice, the inverse Laplace transform (17.11a) is almost never used. Instead, a table (like Table X) is formed, listing the needed time functions and transforms in pairs. Transformation is then performed by looking up functions in the table. The transforms in the table are always obtained by application of (17.11b) rather than (17.11a), because (17.11b) involves simple integration of the known time function under a line in a plane plot, while (17.11a) involves integration of a section under a surface, as we have just described, and is more subtle and more difficult. Therefore, restriction (17.10) is not encountered explicitly. However, it must always be borne in mind.

17.3 APPLICATION OF THE LAPLACE TRANSFORM: THE ONE-SIDED LAPLACE TRANSFORM

Some additional familiarity with the character of Laplace transforms will be obtained from applying (17.11b) to some specific functions in Examples 17.2 through 17.4, which follow. The transformation of abrupt functions in Example 17.2 will suggest the one-sided version of the $\mathcal{L}$ transform, which is presented after Example 17.3.

Example 17.2. $\mathcal{L}$ transform of an exponential. Find the Laplace transform for the time function

$$x(t) = u(t)e^{-\sigma_2 t} \qquad (17.12a)$$

For this function it is convenient to perform the integration (17.11b) in two sections, from $t = -\infty$ to 0 and from $t = 0$ to ∞:

$$\begin{aligned} X(s) &= \int_{t=-\infty}^{\infty} x(t)e^{-st}\,dt \\ &= \int_{-\infty}^{0} 0 e^{-st}\,dt + \int_{0}^{\infty} e^{-\sigma_2 t} e^{-st}\,dt \\ &= 0 + \frac{e^{-(s+\sigma_2)t}}{-(s+\sigma_2)}\bigg|_{0}^{\infty} \\ &= \frac{1}{s+\sigma_2} \qquad \text{for } -\sigma_2 < \text{Re}[s] \end{aligned} \qquad (17.12b)$$

Transform (17.12b) will be used in our algebraic manipulations whenever we are studying a differential equation containing the function $u(t)e^{-\sigma t}$. Note that (17.12b) exists all over the s plane, but recall from p. 548 that it represents (17.12a) *only* for $-\sigma_2 < \text{Re}[s]$.

The real and imaginary parts of (17.12b) could be plotted as surfaces over the s plane. Sketch them. We would, of course, find that each surface has a discontinuity at the point $s = -\sigma_2$, because (17.12b) has a pole there. A more common way of visualizing a function such as (17.12b) is to plot the *magnitude* $|X(s)|$ as a surface above the s plane. For function (17.12b) the magnitude plot is shown in Fig. 17.4. This sort of picture is appealing aesthetically because of its symmetry, and is convenient for visualizing the locations of poles and zeros of a function $X(s)$. From this image of the function (17.12b) it is clear that the value of σ_o, which we would use if we were to carry out the inverse Laplace transform (17.11a) to recover the time function, would have to be a value in front of $s = -\sigma_2$. If we used $\sigma_o = -\sigma_2$, we would encounter the pole, and if we used a value $\sigma_o < -\sigma_2$, the integral would not converge.

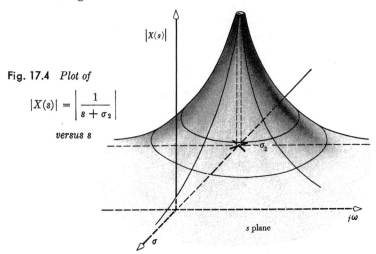

Fig. 17.4 *Plot of*

$$|X(s)| = \left|\frac{1}{s+\sigma_2}\right|$$

versus s

[Sec. 17.3] APPLICATION: THE ONE-SIDED LAPLACE TRANSFORM

The $|X(s)|$ surface in Fig. 17.4 may be produced by the convenient physical device of pushing a slender vertical pole up under a large thin horizontal rubber membrane which is held quite taut at its edges, as suggested in Fig. 17.4. If the coordinates σ and $j\omega$ are marked under the membrane, and if the stick is located at $s = -\sigma_2$, then the shape of the membrane will closely approximate the $|X(s)|$ surface. We used the same image in Fig. 15.10.

More generally, the membrane can be made to represent a compound function $X(s)$ such as

$$X(s) = \frac{(s+a)(s+b)}{s(s+\sigma_1)(s+\sigma_2+j\omega_2)(s+\sigma_2-j\omega_2)} \qquad (17.13)$$

by pushing a vertical pole up under the membrane at each of the pole locations $s = 0$, $s = -\sigma_1$, $s = -\sigma_2 - j\omega_2$, $s = -\sigma_2 + j\omega_2$, and by pinning the membrane to zero at the zero locations $s = -a$ and $s = -b$. The resulting membrane shape is a helpful image of the function. The names poles and zeros have arisen because of this image. More commonly, of course, the poles and zeros of a transfer function are represented by ×'s and ○'s on a two-dimensional s-plane plot, as we have been doing all along (e.g., Fig. 8.20). It may be helpful from now on to allow the ×'s and ○'s to call up the image of a three-dimensional $|X(s)|$ surface, as in Fig. 17.4.

Example 17.3. $\mathcal{L}$ transform of the derivative of a function: two-sided transform. Find the Laplace transform of the derivative of a function, $\dot{x}(t)$.

Substituting in (17.11b), we obtain

$$\mathcal{L}[\dot{x}(t)] = \int_{t=-\infty}^{t=\infty} \frac{dx}{dt} e^{-st}\, dt = \int_{t=-\infty}^{t=\infty} e^{-st}\, dx$$

we integrate by parts, letting

$$u = e^{-st} \qquad dv = dx \qquad du = -se^{-st}\, dt \qquad v = x(t)$$

and using the general relation

$$\int_{t=t_1}^{t=t_2} u\, dv = uv \bigg|_{t=t_1}^{t=t_2} - \int_{t=t_1}^{t=t_2} v\, du \qquad (17.14)$$

[which, in turn, is readily derived from the derivative of a product: $d(uv) = u\, dv + v\, du$]. The result of integration by parts is

$$\begin{aligned}\mathcal{L}[\dot{x}(t)] &= \int_{t=-\infty}^{t=\infty} e^{-st}\, dx \\ &= x(t)e^{-st}\bigg|_{t=-\infty}^{t=\infty} - \int_{t=-\infty}^{t=\infty} x(t)(-se^{-st}\, dt) \\ &= -x(-\infty)e^{+\infty s} + sX(s) \\ &= 0 + sX(s)\end{aligned}$$

That is, s must be restricted to values for which $x(-\infty)e^{+\infty s}$ is zero, or the transform will not represent dx/dt, as we explained in detail on p. 548.

Collecting results, we have

$$\boxed{\mathcal{L}[\dot{x}(t)] = sX(s)} \qquad (17.15a)$$

The one-sided Laplace transform

The widespread use of the Laplace transform in studying the *transient response* following abrupt inputs or disturbances has lead to the popularity of a special version, known as the one-sided Laplace transform. Notice that in Example 17.2 the Laplace integration is actually performed only from $t = 0$ to $t = \infty$, because the function $f(t)$ does not start until $t = 0$. The suggestion is thus made that, for problems in which we are not concerned with anything that happens prior to $t = 0$, an alternative Laplace transformation be defined as

$$\text{One-sided Laplace transform} \qquad \boxed{X(s) = \mathcal{L}_-[x(t)] \triangleq \int_{t=0^-}^{t=\infty} x(t)e^{-st}\, dt} \qquad (17.11c)$$

where $t = 0^-$ is defined as a time "just before" $t = 0$. The symbol $\mathcal{L}_-$ denotes this. [The specification of either "0^+" or "0^-" rather than "0" is necessary because of the discontinuity at $t = 0$ in such functions as $\delta(t)$. The reasons for choosing 0^- will be given presently. Refer also to Sec. 6.5.†]

It is clear at once that the one-sided Laplace transform of abruptly commencing functions like that of Example 17.2 is identical to the two-sided Laplace transform, and the derivation is slightly more convenient. Moreover, in studying transient response, there is another, major convenience obtainable from the one-sided Laplace transform: the *"initial conditions"* at $t = 0$ *are accounted for automatically*.‡ To see this, let us repeat Example 17.3 for the case of the one-sided Laplace transform.

Example 17.4. $\mathcal{L}_-$ **of the derivative of a function.** Find the one-sided Laplace transform of the derivative of a function $\dot{x}(t)$.

Substituting in 17.11c, we obtain

$$\mathcal{L}_-[\dot{x}(t)] = \int_{t=0^-}^{\infty} \frac{dx}{dt} e^{-st}\, dt = \int_{t=0^-}^{\infty} e^{-st}\, dx$$

Again integrating by parts, we have

$$\mathcal{L}_-[\dot{x}(t)] = x(t)e^{-st}\Big|_{t=0^-}^{t=\infty} - \int_{t=0^-}^{t=\infty} x(t)(-se^{-st}\, dt)$$

$$\boxed{\mathcal{L}_-[\dot{x}(t)] = -x(0^-) + sX(s)} \qquad (17.15b)$$

We shall find the one-sided Laplace transform, relation (17.11c), extremely useful when we are studying the response of systems to abrupt inputs, in Secs. 17.7 and 17.10, and in Chapter 19.

---

† Another popular one-sided transform integrates from $t = 0^+$, and is denoted by $\mathcal{L}_+$.
‡ Warning: The one-sided Laplace transform should be applied to equations of motion in their original physically derived form: the equations should *not* be differentiated first, because such differentiation leads to additional "initial conditions" that may not be amenable to physical interpretation and evaluation.

More generally, *the one-sided transform can be used to advantage whenever the **state** of the system (e.g., the output variable and all its derivatives) is known at one particular time, and the input is known after that time*. The procedure is to let $t = 0^-$ at the time when the state is known, and then use the one-sided Laplace transform to study the motion for all time thereafter.

To illustrate this more general situation, consider Fig. 17.5. The figure depicts four different forcing functions $M(t)$ being applied to a system, Fig. 17.5a, b, c, and d. The four forcing functions differ greatly prior to $t = 0$, the one in Fig. 17.5a being 0 for $t < 0$ and the other three having various patterns versus time. After $t = 0$, however, all four forcing functions are the same, and the initial *state* of the system $[\theta(0), \dot{\theta}(0),$ and so on] is the same in all four cases. Since the state is known at $t = 0^-$, we can find the response after $t = 0$ by using the one-sided Laplace transform. In the process we simply turn our back on whatever happened prior to $t = 0^-$, as depicted in Fig. 17.5e. That is, we recognize that it doesn't matter how the system got to the given state at $t = 0^-$: its response thereafter will be the same in all four cases.

When an abrupt input (particularly an impulse) occurs at $t = 0$, the

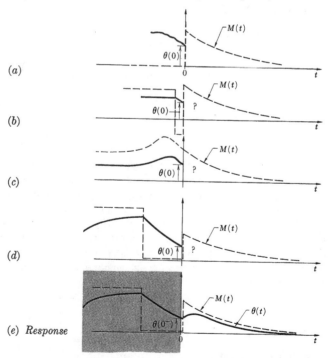

Fig. 17.5 *If input is given for $0 < t$, and state is given at $t = 0$, response for $0 < t$ may be found without concern for motions prior to $t = 0$: The response for $0 < t$ in (a) through (d) will be the same as in (e).*

initial conditions at $t = 0^+$ may differ from those at $t = 0^-$. By taking the lower limit of integration at $t = 0^-$ in (17.11c), we (1) include the entire input function in the Laplace transform,† and (2) automatically require exactly the initial conditions we have been given, namely those at $t = 0^-$. If we had taken the lower limit at $t = 0^+$ (a not uncommon practice), we would then have to (1) handle impulses at $t = 0$ as a separate problem, and (2) compute the change in initial conditions between $t = 0^-$ and $t = 0^+$ as another (related) problem. The $\mathcal{L}_-$ transform avoids completely this separate extra analysis, which is often not trivial.

17.4 SUMMARY OF BASIC LAPLACE TRANSFORM RELATIONS

The Laplace transform has been presented as a natural evolution of the technique of using e^{st} functions to represent signals passing through one element after another in a linear system, because this approach is most in keeping with its intended use for finding the forced response of systems.

What we have found is that any time function $x(t)$—abrupt functions in particular—can be written as the integral sum of many e^{st} functions by using the inverse Laplace transform, (17.11a):‡

$$x(t) = \mathcal{L}^{-1}[X(s)] = \frac{1}{2\pi j} \int_{s=\sigma_0-j\infty}^{s=\sigma_0+j\infty} X(s)e^{st}\, ds \qquad (17.11a)$$
[repeated]

in which the "coefficients" $X(s)$ of the terms in the sum are given by (17.11b), the two-sided Laplace transform:

Two-sided
Laplace transform
$$X(s) = \mathcal{L}[x(t)] \triangleq \int_{t=-\infty}^{\infty} x(t)e^{-st}\, dt \qquad (17.11b)$$
[repeated]

Expression (17.11a) represents $x(t)$ only for values of s for which (17.11b) converges, i.e., in some strip in the s plane. To use (17.11a) we must choose σ_0 to be in this strip. For time functions that commence at some definite time, the choice of σ_0 is straightforward (as in Example 17.2).

The most widespread use of the Laplace transform is in studying system response to time functions that commence abruptly at $t = 0$. In this case the limits of integration in (17.11b) may be taken from 0^- to ∞ in a version known as the one-sided Laplace transform. This version also has advantages with regard to initial conditions.

One-sided
Laplace transform
$$X(s) = \mathcal{L}_-[x(t)] \triangleq \int_{t=0^-}^{\infty} x(t)e^{-st}\, dt \qquad (17.11c)$$
[repeated]

† This is discussed further on p. 568.
‡ The symbols $\mathcal{L}[x(t)]$ and $\mathcal{L}^{-1}[X(s)]$ are commonly used to denote the Laplace transform and its inverse.

[Sec. 17.5] DERIVATION OF COMMON LAPLACE TRANSFORM PAIRS

There is now an added restriction: When $X(s)$ is obtained from the one-sided Laplace transform (17.11c), expression (17.11a) holds only for $0 < t$.

In common practice the Laplace transform is used to solve dynamics problems without the formal integration of (17.11a) ever being performed. Instead, a table is formed (like Table X), listing a number of common time functions and their corresponding Laplace transforms.† Then both transformation [(17.11b) or (17.11c)] and inverse transformation (17.11a) are readily performed by looking up functions in the table. That is, both Steps 1 and 3 of Procedure D-1 are performed completely by looking in a table of Laplace transforms, the terms in the table having been obtained "once for all" by application of (17.11b) or (17.11c) to many time functions of interest. Our next order of business, therefore, is to derive transform pairs for such a table. We shall do this in Sec. 17.5.

In cases where the response function (obtained in Step 2 of Procedure D-1) contains compound terms which do not appear explicitly in the table at hand, the use of partial fraction expansion (Procedure D-2, presented in Chapter 18) will convert these into sums of simpler terms, after which the table can then be used.

17.5 DERIVATION OF COMMON LAPLACE TRANSFORM PAIRS

In this section we shall derive, or at least indicate the derivation of the other Laplace transforms in Table X. The terms in the table itself are obtained by applying (17.11b) or (17.11c) to time functions of interest, as in Examples 17.2 to 17.4, hence (17.11a) almost never gets used.

There are two reasons for using (17.11b) or (17.11c) to build the table rather than (17.11a): (1) it is the time functions $x(t)$ that are known to begin with; (2) Eq. (17.11b) or (17.11c) involves integration under a line $x(t)e^{-st}$ in a plane plot $x(t)e^{-st}$ versus t, while (17.11a) involves integration of a section under a surface, as described on p. 549, and is more difficult.

Table X reverses the usual arrangement of a Laplace transform table. Usually, the transform is presented on the left and the time function on the right, and the multiplying factors such as ω_a/a_1 in transforms 9 and 10 are usually tabulated with the time function rather than with the transform. The reason for the more usual arrangement (which is also the one used in the longer table of Laplace transforms in Appendix J) is to facilitate inverse transformation, which is normally more involved than direct transformation.

It is immediately apparent that many of the transform pairs in the table can be obtained as special cases of others. Thus transform pairs 5 through 10 in Table X are all special cases of $\mathcal{L}[u(t)e^{-\alpha t}]$, α being a complex number; transform pair 2 in Table X can be obtained by repeated application of pair 1; and so on.

† A comprehensive table of Laplace transforms is given in Appendix J.

Exponential functions

Let us derive first the Laplace transform for the function $u(t)e^{-\alpha t}$. The formal derivation consists simply of substituting $u(t)e^{-\alpha t}$ for $f(t)$ in Eq. (17.11b) or (17.11c), and is similar to Example 17.2:

$$\begin{aligned} F(s) = \mathcal{L}[u(t)e^{-\alpha t}] &= \int_{-\infty}^{\infty} f(t)e^{-st}\,dt \\ &= \int_{-\infty}^{\infty} u(t)e^{-\alpha t}e^{-st}\,dt = \int_0^{\infty} e^{-(s+\alpha)t}\,dt \\ &= \left[-\frac{1}{s+\alpha}e^{-(s+\alpha)t} \right]_0^{\infty} \\ &= \frac{1}{s+\alpha} \end{aligned} \qquad (17.16)$$

[The two-sided transformation (17.11b) has been used, but for this function exactly the same result would, of course, have been obtained with the one-sided transform.]

Note that the Laplace transform of a function always has the units of the function, *multiplied by time*. This is, of course, because (17.11b) integrates the function over time.

Now to obtain transform 5 in Table X we simply let $\alpha = 0$ in (17.16):

$$\mathcal{L}[u(t)e^{0t}] = \frac{1}{s+0} = \frac{1}{s} \qquad (17.17)$$

To obtain transform 6 we let $\alpha = \sigma$, of course. To obtain Transform 7 we replace $\sin \omega t$ by its Euler equation equivalent, Eq. (7.24b):

$$u(t)\sin \omega t = \frac{e^{j\omega t} - e^{-j\omega t}}{2j} u(t)$$

and then apply transform (17.16) successively to the two terms:

$$\begin{aligned} \mathcal{L}[u(t)\sin \omega t] &= \frac{1}{2j}\left[\frac{1}{s-j\omega} - \frac{1}{s+j\omega} \right] \\ &= \frac{\omega}{s^2+\omega^2} \end{aligned} \qquad (17.18)$$

Result (17.18) could also have been obtained, of course, by letting $f(t) = u(t)\sin \omega t$ in Eq. (17.11b), as the reader may show. Derivation of transform pairs 8 through 14 is left to the reader in the problems.

Time delay

Delayed forcing functions, particularly the delayed step, are often encountered. The Laplace transform of a delayed step is, from (17.11b),

$$\mathcal{L}[u(t-\tau)] = \int_{-\infty}^{\infty} u(t-\tau)e^{-st}\,dt = \int_{-\infty}^{\tau}(0)e^{-st}\,dt + \int_{\tau}^{\infty} e^{-st}\,dt$$

$$= \left[\frac{e^{-st}}{-s}\right]_{\tau}^{\infty} = \frac{e^{-s\tau}}{s}$$

$$\mathcal{L}[u(t-\tau)] = \frac{e^{-s\tau}}{s} \qquad (17.19)$$

[Sec. 17.5] DERIVATION OF COMMON LAPLACE TRANSFORM PAIRS

which is transform 5.1 in Table X. [Show that the result is the same for the one-sided Laplace transform, (17.11c), provided that τ is positive.] More generally, the transform of any function which starts abruptly at time τ is

$$\mathcal{L}[u(t - \tau)f(t - \tau)] = e^{-s\tau}F(s) \qquad (17.20)$$

which is pair 5.2. The proof, again, is left as an exercise.

Unit impulse

The Laplace transform of the all-important unit impulse may be obtained from (17.17) by the same sort of limiting process used in the original conception of the unit impulse (Fig. 6.20), or much more directly, by noting that the function $\delta(t)e^{st}$ is an impulse at $t = 0$ having an *area* of unity times $e^0 = 1$, so that for the two-sided transformation we have

$$\mathcal{L}[\delta(t)] = \Delta(s) = \int_{-\infty}^{\infty} \delta(t)e^{st}\,dt = \text{Area of impulse} = 1 \qquad (17.21)$$

Note that the transform of $\delta(t)$ is dimensionless. [$\delta(t)$ has dimension τ^{-1}.]

In performing the *one-sided* Laplace transform of a unit impulse, we return briefly to Sec. 6.8 and to Fig. 6.20 where the unit impulse was defined. We note that this function $\delta(t)$ is defined as the limiting case of a finite impulse "which has unit area, and which starts at $t = 0$ and terminates an infinitesimally short time later." The key phrase is "starts at $t = 0$," because this means that when we perform the integration in (17.11c) to obtain the Laplace transform of $\delta(t)$ we find the entire function still within the integral:

$$\mathcal{L}_-[\delta(t)] = \int_{t=0^-}^{t=\infty} \delta(t)e^{-st}\,dt = 1 \qquad (17.22)$$

and the result is the same as for the two-sided Laplace transform. This is shown also in Fig. 17.6.

Notice that if we had chosen $t = 0^+$ as the lower limit of integration in defining the one-sided Laplace transform, then the impulse would not have been within the region of integration and the result of (17.22) would have been zero. The fact that choosing $t = 0^-$ as the lower limit in (17.11c) permits inclusion of impulse functions was a major reason for the choice.

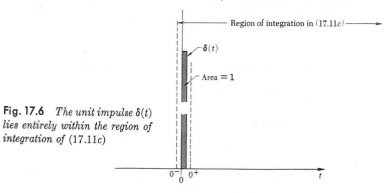

Fig. 17.6 *The unit impulse $\delta(t)$ lies entirely within the region of integration of* (17.11c)

Derivative

The two-sided and one-sided Laplace transformations of the first derivative of a function were derived respectively in Examples 17.3 and 17.4. The results, (17.15a) and (17.15b), appear as pair 1 in Table X.

The Laplace transform for a higher derivative is most readily obtained by repeated application of (17.15a) or (17.15b). For the two-sided transform,

$$\mathcal{L}[\ddot{x}] = s\mathcal{L}[\dot{x}] = s^2 X(s)$$

$$\mathcal{L}\left[\frac{d^n x}{dt^n}\right] = s^n X(s) \tag{17.23}$$

For the one-sided transform it is most straightforward to use intermediate steps. That is, for the second derivative, let $u = \dot{x}$. Then

in which
$$\mathcal{L}_-[\dot{u}] = s\mathcal{L}_-[u] - u(0^-)$$
$$\mathcal{L}_-[u] = \mathcal{L}_-[\dot{x}] = sX(s) - x(0^-)$$
or, combining,
$$\mathcal{L}_-[\ddot{x}] = s\{sX(s) - x(0^-)\} - u(0^-)$$
$$\mathcal{L}_-[\ddot{x}] = s^2 X(s) - sx(0^-) - \dot{x}(0^-) \tag{17.24}$$

Repeating the process for higher-order derivatives establishes the pattern, and for nth order,

$$\boxed{\mathcal{L}_-\left[\frac{d^n x}{dt^n}\right] = s^n X(s) - s^{n-1} x(0^-) - s^{n-2} \dot{x}(0^-) - \cdots - \left[\frac{d^{n-1} x}{dt^{n-1}}\right]_{t=0^-}} \tag{17.25a}$$

or in closed form,

$$\boxed{\mathcal{L}_-\left[\frac{d^n x}{dt^n}\right] = s^n X(s) - \sum_{k=1}^{n} s^{n-k} \left[\frac{d^{k-1} x}{dt^{k-1}}\right]_{t=0^-}} \tag{17.25b}$$

Integral

To derive the Laplace transform for the integral of a function we employ integration by parts, using relation (17.14) as we did in Examples 17.3 and 17.4 to derive the transform of a derivative. Let

$$h(t) \triangleq \int x(t) \, dt$$

For the two-sided transform, (17.11b),

$$\mathcal{L}[h(t)] = \int_{t=-\infty}^{\infty} h(t) e^{-st} \, dt$$

To integrate by parts, let

$$u = h \qquad dv = e^{-st} \, dt$$

so that
$$du = dh = x(t) \, dt \qquad v = -\frac{e^{-st}}{s}$$

[Sec. 17.6] TRANSFER FUNCTIONS FROM LAPLACE TRANSFORMATION 559

Then, using (17.14), we have

$$\mathcal{L}[h(t)] = uv \Big|_{t=-\infty}^{\infty} - \int_{t=-\infty}^{\infty} v\, du$$

$$= -\frac{he^{-st}}{s}\Big|_{t=-\infty}^{\infty} - \int_{-\infty}^{\infty} -\frac{e^{-st}}{s} x(t)\, dt$$

$$\boxed{\mathcal{L}[\textstyle\int x(t)\, dt] = 0 + \frac{X(s)}{s}} \qquad (17.26a)$$

[As in Example 17.3, s must be so restricted that the term $-h(-\infty)e^{+\infty s}/s$ is zero, otherwise the transform does not represent $\int x\, dt$.]

For the one-sided transform the steps are the same, but the limits of integration are from $t = 0^-$ to $t = \infty$:

$$\mathcal{L}_-[h(t)] = -\frac{he^{-st}}{s}\Big|_{t=0^-}^{\infty} - \int_{t=0^-}^{\infty} -\frac{e^{-st}}{s} x(t)\, dt$$

$$\boxed{\mathcal{L}_-[\textstyle\int x(t)\, dt] = \frac{h(0^-)}{s} + \frac{X(s)}{s}} \qquad (17.26b)$$

in which $h(0^-) \triangleq [\int x(t)\, dt]_{t=0^-}$

Expressions (17.26a) and (17.26b) are transform pair 3 in Table X.

The transforms of *higher-order integrals* are obtained by repeated application of (17.26a) or (17.26b). For the two-sided transform this is particularly simple: The Laplace transform of the nth integral of $x(t)$ is

$$\frac{X(s)}{s^n} \qquad (17.27)$$

For the one-sided transform the pattern is involved but straightforward. It is left to the reader in Problem 17.15.

Problems 17.1 through 17.15

17.6 TRANSFER FUNCTIONS FROM LAPLACE TRANSFORMATION

Consider a typical differential equation like Eq. (17.1) (refer to Example 17.1):

$$b\dot\theta + k\theta = M(t) \qquad (17.1)$$
[repeated]

If $M(t)$ happens to be a continuing function of the form $M(t) = Me^{st}$ (where M is a constant), then we can use the classical method of Chapter 14 to obtain a transfer function relating M and θ. We assume that $\theta(t)$ is also of the form $\theta = \Theta e^{st}$ and, by substitution,

$$(bs + k)\Theta e^{st} = Me^{st}$$

obtain a relation between the constants M and Θ:

$$\frac{\Theta}{\mathsf{M}} = \frac{1}{bs + k} \qquad (17.28)$$

Now, however, with the Laplace transform method at our disposal, we can consider a much broader class of forcing functions $M(t)$, and obtain a *transfer function* which is much more widely applicable. In fact $M(t)$ can now be almost any arbitrary function, and we can still represent it in the form e^{st} by finding its Laplace transform, $\mathsf{M}(s)$, and then writing the summation:

$$M(t) = \frac{1}{2\pi j} \int_{s=\sigma_0-j\infty}^{s=\sigma_0+j\infty} \mathsf{M}(s) e^{st}\, ds$$

If we now Laplace transform the entire equation of motion, with $\theta(0^-) = 0$,† we are, in effect, assuming that $\theta(t)$ has also the form

$$\theta(t) = \frac{1}{2\pi j} \int_{s=\sigma_0-j\infty}^{s=\sigma_0+j\infty} \Theta(s) e^{st}\, ds$$

and we are substituting these assumed solutions into the equation of motion, with the result

$$(bs + k)\Theta(s) = \mathsf{M}(s)$$

from which we obtain the transfer function

$$\frac{\Theta(s)}{\mathsf{M}(s)} = \frac{1}{bs + k} \qquad (17.29)$$

The process used is perfectly analogous to the classical method, and the resulting transfer function is the same. *Now, however, the transfer function has much broader applicability* because it relates the variables $\mathsf{M}(s)$ and $\Theta(s)$ representing "almost any function," instead of constants Θ and M which can represent certain important continuing functions (enumerated in Fig. 17.1a), but no others.

The transfer function can still be represented in block-diagram form, Fig. 17.7, which strengthens the image of cause and effect, or of transfer of a signal from input to output, and which will be found extremely helpful in studying the behavior of complex systems (as in Part E).

Fig. 17.7 *Block-diagram representation of a transfer function* $\mathsf{M}(s) \longrightarrow \boxed{\dfrac{1}{bs+k}} \dashrightarrow \Theta(s)$

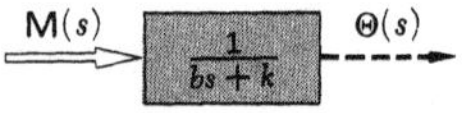

Henceforth in this book when we speak of a transfer function we shall mean that it was obtained by Laplace transformation, and when we use symbols like Θ and M we shall mean the Laplace variables $\Theta(s)$ and $\mathsf{M}(s)$.

† We are interested here only in the response to the applied torque. Natural motions, resulting from initial conditions, would be superposed later.

[Sec. 17.6] TRANSFER FUNCTIONS FROM LAPLACE TRANSFORMATION 561

Example 17.5. Ideal fluid clutch: a coupled mechanical system. For the idealized physical model of a fluid clutch shown in Fig. 17.8a, find the transfer function between M and Ω_2 and also between M and Ω_1. Show your results in a block diagram. It is assumed in the model that the coupling is purely viscous, and that all other damping is absent.

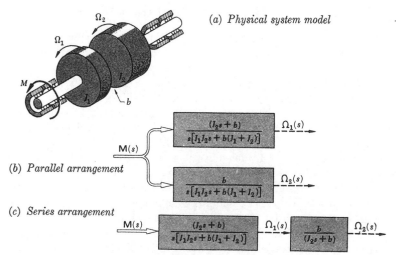

Fig. 17.8 *Transfer functions for a coupled mechanical system*

The equations of motion are (refer to Prob. 2.38 with $M_L = 0$)

$$J_1 \dot{\Omega}_1 + b(\Omega_1 - \Omega_2) = M$$
$$-b(\Omega_1 - \Omega_2) + J_2 \dot{\Omega}_2 = 0 \tag{17.30}$$

Laplace transforming these term by term, Step 1 of Procedure D-1 (and assuming no initial conditions), gives

$$(J_1 s + b)\Omega_1(s) - b\Omega_2(s) = \mathsf{M}(s)$$
$$-b\Omega_1(s) + (J_2 s + b)\Omega_2(s) = 0 \tag{17.31}$$

Solving simultaneously for $\Omega_1(s)/\mathsf{M}(s)$ first, and then for $\Omega_2(s)/\mathsf{M}(s)$ (e.g., by Cramer's rule), Step 2 of Procedure D-1, yields the desired transfer functions:

$$\frac{\Omega_1(s)}{\mathsf{M}(s)} = \frac{(J_2 s + b)}{(J_1 s + b)(J_2 s + b) - b^2} = \frac{(J_2 s + b)}{s[J_1 J_2 s + b(J_1 + J_2)]} \tag{17.32}$$

$$\frac{\Omega_2(s)}{\mathsf{M}(s)} = \frac{b}{s[J_1 J_2 s + b(J_1 + J_2)]} \tag{17.33}$$

We can also obtain the transfer function between $\Omega_1(s)$ and $\Omega_2(s)$ directly from the second of the transformed equations, (17.31):

$$\frac{\Omega_2(s)}{\Omega_1(s)} = \frac{b}{(J_2 s + b)} \tag{17.34}$$

To put these results in block-diagram form, we can either use (17.32) and (17.33) in a parallel arrangement, Fig. 17.8b, or we can use (17.32) with (17.34) in a series arrangement, Fig. 17.8c. The latter more nearly resembles the actual physical arrangement. From the block diagrams we can visualize how an input torque (for example, a sinusoidal torque) is transferred to an output motion.

Problems 17.16 through 17.25

17.7 TOTAL RESPONSE BY THE LAPLACE TRANSFORM METHOD

We have now developed the basic tools with which to find the total response of a system to abrupt inputs, using Procedure D-1 (after a set of linear equations of motion with constant coefficients has been derived). There are some "auxiliary" tools which we shall add to our "kit" in Secs. 17.8 through 17.10 and in Chapter 18. These auxiliary tools—the Final-Value and Initial-Value Theorems, impulse representation in block diagrams, and partial fraction expansion—will be most helpful in studying compound systems of several degrees of freedom.

First, however, we shall demonstrate the basic tools for some simple systems, to show the advantages of Procedure D-1 in abrupt-input problems. In all of the examples given here, the state of the system at $t = 0^-$ will be presumed known, and we shall be interested only in the behavior after $t = 0$. We shall therefore be using the one-sided Laplace transform.

In Step 1 of Procedure D-1, each of the terms in the differential equations is Laplace transformed, using a table such as Table X. The result is a set of simultaneous *algebraic* equations in s. These contain the forcing functions and the initial conditions.

In Step 2 the set of algebraic equations is manipulated to obtain the "response function of the variable of interest." If, for example, we are interested ultimately in the time response for $x_3(t)$, then we solve the algebraic equations explicitly for $X_3(s)$. *The resulting expression for $X_3(s)$ is known as the response function for $X_3(s)$.*

Finally, in Step 3 the response function is inverse transformed, term by term, to obtain the corresponding time response [e.g., $x_3(t)$].

These operations will be made more clear by some examples. In the first example only the essential steps are taken.

Example 17.6. A simple rotor. Find the response of the ideal first-order system in Fig. 17.9 to the sudden application of the moment shown, which commences at $t = 0$. The velocity is Ω_0 just before the moment is applied. We wish to know the velocity thereafter.

The complete differential equation of motion for the above system is

$$J\dot{\Omega} + b\Omega = M_f(t) = u(t)M_0(1 - e^{-\sigma_f t}) \qquad (17.35)$$

We now apply Procedure D-1.

Step 1. Laplace transform the equation of motion, term by term. (The terms are obtained from Table X.)

$$J[s\Omega(s) - \Omega(0^-)] + b\Omega(s) = \mathsf{M}(s) = M_0\left(\frac{1}{s} - \frac{1}{s + \sigma_f}\right)$$

where $\Omega(0^-) = \Omega_0$.

[Sec. 17.7] TOTAL RESPONSE BY LAPLACE TRANSFORM METHOD 563

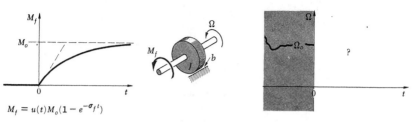

$M_f = u(t)M_o(1 - e^{-\sigma_f t})$

Fig. 17.9 *A simple problem in finding total response*

Step 2. *Manipulate* algebraically to obtain:
(a) the *response function* for $\Omega(s)$:

$$\Omega(s) = \frac{M_0}{s(Js + b)} - \frac{M_0}{(Js + b)(s + \sigma_f)} + \frac{J\Omega_0}{(Js + b)} \quad (17.36)$$

Step 3. *Inverse transform* the response function. (Table X contains the necessary pairs, 11, 12, and 6.)

$$\Omega(t) = \frac{M_0}{b} u(t)[1 - e^{-(b/J)t}] - \frac{M_0/J}{(b/J) - \sigma_f} u(t)[e^{-\sigma_f t} - e^{-(b/J)t}] + \Omega_0 u(t) e^{-(b/J)t} \quad (17.37)$$

The reader should check to be sure that this expression gives the correct value for Ω when $t = 0^+$. [Expression (17.37) does not purport to give any information for $t < 0$.]

Physical interpretation. Of the five terms in Eq. (17.37) for the time response, the first and third terms are, of course, the "forced" part of the response, having the same character as the input but different magnitude. After a long time, only $\Omega = M_0/b$ is left. This checks physical reasoning: After a long time an equilibrium is reached in which $M_0 = b\Omega$. The second and fourth terms are natural responses produced by the sudden application of the applied torque. They have just the right magnitude so that at $t = 0$ they cancel terms one and three, thus satisfying the initial conditions if $\Omega(0)$ had been 0. Finally, term five is simply the natural response to initial condition $\Omega(0^-) = \Omega_0$. Note in particular that the Laplace solution has correctly told us *automatically* that, for this problem, $\Omega(0^+) = \Omega(0^-) = \Omega_0$.

Notice how much more readily the total response is obtained by turning the Laplace transform "crank" than by the classical method of using unknown coefficients and then setting up the $t = 0$ conditions to evaluate them. Notice, in particular, that the situation between $t = 0^-$ and $t = 0^+$ is automatically handled correctly without physical reasoning.

The next example is a little more interesting, involving a simple two-degree-of-freedom system.

Example 17.7. A fluid clutch. Find the response $\Omega_2(t)$ for the fluid clutch modeled in Fig. 17.8a, if $M(t)$ is an abruptly started sinusoid: $M = M_0 u(t) \cos \omega_f t$. Both members have initial velocities, Ω_{10} and Ω_{20}.

We refer to Example 17.5, but we note that there initial conditions were taken zero. To include them it is easiest to begin again. The equations of motion are, from (17.30),

$$J_1\dot{\Omega}_1 + b\Omega_1 - b\Omega_2 = M = M_0 u(t)\cos\omega_f t$$
$$-b\Omega_1 + J_2\dot{\Omega}_2 + b\Omega_2 = 0 \tag{17.38}$$

We now follow Procedure D-1.

Step 1. Laplace transform the equations of motion:

$$(J_1 s + b)\Omega_1 - b\Omega_2 = \frac{M_0 s}{s^2 + \omega_f^2} + J_1\Omega_1(0^-)$$
$$-b\Omega_1 + (J_2 s + b)\Omega_2 = J_2\Omega_2(0^-) \tag{17.39}$$

where $\Omega_1(0^-) = \Omega_{10}$ and $\Omega_2(0^-) = \Omega_{20}$.

Step 2. (a) *Manipulate algebraically to obtain:*

(a) the *response function* of interest. Since we are interested in $\Omega_2(t)$, we solve the algebraic equations for $\Omega_2(s)$, using Cramer's rule, to obtain the response function:

$$\Omega_2(s) = \frac{\left[\dfrac{M_0 s}{s^2 + \omega_f^2} + J_1\Omega_{10}\right]b + (J_1 s + b)J_2\Omega_{20}}{[(J_1 s + b)(J_2 s + b) - b^2]} \tag{17.40}$$

[Note that the transfer function (17.33) between M and Ω_2 is readily obtained from the response function by considering only $\mathsf{M}(s) = M_0 s/(s^2 + \omega_f^2)$, and omitting the initial-condition terms.]

(b) *System natural behavior.* We factor the denominator of (17.40) to obtain the natural characteristics. This is easy in this problem; the result is

$$J_1 J_2 s \left(s + \frac{J_1 + J_2}{J_1 J_2}b\right) \tag{17.41}$$

That is, the natural characteristics are

$$s = 0 \qquad s = -[(J_1 + J_2)/J_1 J_2]b$$

The first of these implies a constant velocity, and the second implies an exponential decay whose time constant is faster than that of either disk with the other held still. Physically, this means that (for $M = 0$) if the two disks have different initial velocities there will be a transient motion of both while they adjust to a common velocity, followed by a steady state in which they rotate together at constant velocity (dissipating no energy).

Using (17.41), we can write the response function

$$\Omega_2(s) = \frac{(M_0 b/J_1 J_2)s}{(s^2 + \omega_f^2)s[s + b(J_1 + J_2)/J_1 J_2]} + \frac{(b/J_2)\Omega_{10} + [s + (b/J_1)]\Omega_{20}}{s[s + b(J_1 + J_2)/J_1 J_2]} \tag{17.42}$$

(in which the rearranging shown will make it more convenient to use Table X).

Step 3. Inverse transform the response function to obtain the time response. Again each of the terms appears in Table X (pairs 14, 11, and 6).

$$\Omega_2(t) = \frac{M_0 b}{J_1 J_2}\left[\frac{1}{c^2}e^{-\sigma t} + \frac{1}{c\omega_f}\sin(\omega_f t - \phi)\right]u(t) + \Omega_{10}\frac{J_1}{J_1 + J_2}(1 - e^{-\sigma t})u(t)$$
$$+ \Omega_{20}\left[e^{-\sigma t} + \frac{J_2}{J_1 + J_2}(1 - e^{-\sigma t})\right]u(t) \tag{17.43}$$

where $\qquad \sigma = \dfrac{b(J_1 + J_2)}{J_1 J_2} \qquad c = \sqrt{\sigma^2 + \omega_f^2} \qquad \phi = \tan^{-1}\dfrac{\omega_f}{\sigma}.$

[Sec. 17.7] TOTAL RESPONSE BY LAPLACE TRANSFORM METHOD   565

Physical interpretation. Physically, the first third of (17.43) contains the steady-state response to the applied torque, at amplitude $M_0 b / J_1 J_2 c \omega_f$ and phase angle ϕ, together with the natural transient it produces, an exponential decay having time constant $1/\sigma$, where σ is a natural characteristic of the system. (In general, a natural term in $s = 0$, a steady change in average speed, would also be expected. It just happens, for this particular forcing function, that the $s = 0$ natural motion has zero amplitude because the s's canceled in the response function. For other forcing functions this would not be so.)

The remaining two-thirds of (17.43) indicate the unforced response when the disks 1 and 2, respectively, have initial velocities. For example, if disk 1 has an initial velocity Ω_{10} and disk 2 has none, then the final velocity of disk 2 will be $[J_1/(J_1 + J_2)]\Omega_{10}$. In Problem 17.66 it will be found that disk 1 also has the same final velocity: that is, the disks "end up" spinning together, as intuition tells us they should.

When both an applied torque and initial velocities exist, then the final motion is simply a superposition of the separate responses. As a special check on the response to initial conditions, note that at $t = 0$ the value of $\Omega_2(t)$ is Ω_{20} (regardless of the value of Ω_{10} or of M_0), as it should be. This is a good check to make as soon as Step 3 is completed.

We consider next an electrical network which is mathematically identical to the system in Fig. 17.8. (The reader may show this, using the methods of Chapter 2.) To emphasize the analogy, we find that we can paraphrase the preceding example almost word for word.

Example 17.8. A network. In the network shown in Fig. 17.10, switches S_1 and S_2 are closed at $t = 0$. Find the response $v_2(t)$. Both capacitors have initial voltages, γ_1 and γ_2.

The equations of motion are obtained, using KCL:

Node a: $C_1 \dot{v}_1 + \dfrac{1}{R}(v_1 - v_2) \quad\quad = i_3 = i_{30} u(t) \cos \omega_f t$

Node b: $-\dfrac{1}{R}(v_1 - v_2) + C_2 \dot{v}_2 = 0$ (17.38')

where $v_1(0^-) = \gamma_1$ and $v_2(0^-) = \gamma_2$.

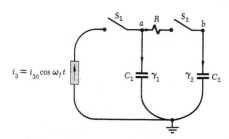

Fig. 17.10 *Network analogous to the fluid-clutch model of Fig. 17.8a*

Following Procedure D-1, the reader may now carry out Steps 1, 2, and 3, obtaining expressions precisely analogous to (17.39) through (17.43), with the final result:

$$v_2(t) = \frac{i_{30}}{RC_1C_2}\left[\frac{1}{c^2}e^{-\sigma t} + \frac{1}{c\omega_f}\sin(\omega_f t - \phi)\right]u(t) + \frac{\gamma_1 C_1}{C_1 + C_2}(1 - e^{-\sigma t})u(t)$$
$$+ \gamma_2\left[e^{-\sigma t} + \frac{C_2}{C_1 + C_2}(1 - e^{-\sigma t})\right]u(t) \qquad (17.43')$$

where $\sigma = \dfrac{1}{RC_1} + \dfrac{1}{RC_2} \qquad c = \sqrt{\sigma^2 + \omega_f^2} \qquad \phi = \tan^{-1}\dfrac{\omega_f}{\sigma}$

Physical interpretation. Physically, the first third of (17.43') contains the steady-state response to the applied current, having magnitude $i_{30}/RC_1C_2c\omega_f$ and phase angle ϕ, together with the natural transient it produces, an exponential decay having time constant $1/\sigma$, where σ is a natural characteristic of the system. (In general, a natural term in $s = 0$, an average voltage, would also be expected. It just happens, for this particular forcing function, that the $s = 0$ natural motion has zero amplitude because the s's canceled in the response function. For other forcing functions this would not be so.)

The remaining two-thirds of (17.43') indicate the unforced response due to the initial charges on the capacitors. When both the applied current and initial charges exist, the final motion is simply a superposition of the separate responses.

As a special check on the response to initial conditions, note that at $t = 0$ the value of $v_2(t)$ is γ_2 (regardless of the value of γ_1 or i_{30}), as it should be. This is a good check to make as soon as Step 3 is completed.

Response to continuing functions

When *only* the response to a *continuing* sine wave†—i.e., the steady-state sinusoidal response—is of interest, *the Laplace transform offers no advantage over simply assuming a solution of the form e^{st}* as we did in Chapters 9 and 15 (and in all chapters prior to this one). As a matter of fact, the Laplace transform method is unnecessarily cumbersome in this case. For one thing, the Laplace transform of a continuing time function is a delta function in the s plane,‡ which is awkward to work with and almost never used. (Note that Laplace transforms for no continuing functions appear in Table X —only for abrupt functions.)

An alternative and easier procedure, which is demonstrated in Example 17.9, is to find instead the response to an abruptly commencing function, and then throw away all the transient terms.

Example 17.9. Sinusoidal response of network. Use the results of Example 17.8 to find the steady-state response of the same system to a continuing sinusoidal input,

$$i_3 = i_{30}\cos\omega_f t$$

if there is no "d-c" charge on the capacitors.

† Or other continuing (nonabrupt) functions of the form e^{st}, as in Fig. 8.13a through d.
‡ In a three-coordinate plot like Fig. 17.4, for example, such functions do not have the "tent" shape above the s plane that poles produce: instead they are shaped like "flagpoles"; they are infinitely high for one single value of s, and *zero* everywhere else.

[Sec. 17.8] FINAL-VALUE THEOREM AND INITIAL-VALUE THEOREM

In the steady state, after the input has been in operation for a long time, the transient portions of time response (17.43′) will all have died out, and thus an easy way to get the response is to omit the transients and write

$$v_2(t) = \frac{i_{30}}{RC_1C_2c\omega_f} \sin(\omega_f t - \phi) \tag{17.44}$$

with c and ϕ defined as for (17.43′).

As it turns out, this method of getting the response to a continuing input is the easiest, when the equations of motion have already been Laplace transformed. If they have not, however—e.g., if Example 17.8 had not already been worked out—then some unnecessary extra steps are involved. It is quicker and more direct simply to assume a solution of the form e^{st} (where $s = j\omega_f$ for a sine-wave input of frequency ω_f, for example) and proceed in the usual way. This will be clear from Problem 17.50.

In Problem 17.51 the sinusoidal steady-state response of a fluid clutch is to be obtained by analysis analogous to that in Example 17.9.

Problems 17.26 through 17.59

17.8 THE FINAL-VALUE THEOREM AND THE INITIAL-VALUE THEOREM

As we have seen in Sec. 17.7, the response function for even relatively simple systems can become quite involved, requiring the factoring of higher-order polynominals, followed by partial fraction expansion, or the interpretation of tedious complex Laplace transforms in a table, or both. Following good engineering practice, we look for some special cases to help us get a quick initial view of the behavior and to furnish an intermediate check on our analysis. Two helpful tools for this are the Final-Value Theorem and the Initial-Value Theorem, which follow directly from the definition of the Laplace transform.

The *Final-Value Theorem* states that the time response of a variable after a long time, $x(\infty)$, can be obtained from its response function (i.e., its Laplace transform) through the relation

$$\text{Final-Value Theorem} \quad \boxed{x(\infty) = \lim_{t \to \infty} x(t) = \lim_{s \to 0} [sX(s)]} \tag{17.45}$$

provided that *all poles of $X(s)$ are in the left half-plane* or at the origin. If there are poles on the $j\omega$ axes or in the right half-plane, then of course $x(t)$ has no definite final value.

The *Initial-Value Theorem* states that the value of a variable at time $t = 0^+$ can be obtained from its response function (i.e., its one-sided Laplace transform) through the relation

$$\text{Initial-Value Theorem} \quad \boxed{x(0^+) = \lim_{t \to 0^+} x(t) = \lim_{s \to \infty} [sX(s)]} \tag{17.46}$$

In (17.46) the limit $t = 0^+$ is approached "from the right," i.e., from a time after $t = 0$.†

In writing both (17.45) and (17.46), the intermediate expressions are understood, and are usually omitted. As we shall see, both theorems are easy to apply, and the algebra involved is usually trivial. The Final-Value Theorem thus permits us to obtain the steady-state portion of the response quickly, without performing the inverse transformation. If the final value is obvious from physical reasoning, then the Final-Value Theorem serves as an easy check on the response function.

The Initial-Value Theorem is used principally as another check on the response function, since the initial conditions will be known. It must be emphasized that (17.46) gives the value at $t = 0^+$, not at $t = 0^-$. For cases in which $x(0^+)$ differs from $x(0^-)$, the Initial-Value Theorem will thus *not* give the "initial conditions." For example, if the system is disturbed by an impulse at $t = 0$, then $x(0^+)$ will differ from $x(0^-)$ by its response to the impulse. This aspect of the Initial-Value Theorem will be demonstrated in the examples which follow, particularly Example 17.11.

We shall first prove the two theorems, after which a number of examples of their application will be given. The proofs utilize integration by parts, and are similar to the derivation of the Laplace transform of a derivative.

We begin by noting that the function $sX(s)$ is available directly from the $\mathcal{L}_-$ transform of the derivative of a function $x(t)$, (17.15b):

$$sX(s) = x(0^-) + \int_{t=0^-}^{\infty} e^{-st} \dot{x}(t)\, dt \tag{17.47}$$

Proof of Final-Value Theorem

To derive the Final-Value Theorem we take the limit indicated in (17.45):

$$\lim_{s \to 0}[sX(s)] = \lim_{s \to 0}\left[x(0^-) + \int_{t=0^-}^{\infty} e^{-st}\,\dot{x}(t)\, dt \right] \tag{17.48}$$

Consider now the last term only. Since s is independent of t, the order of integrating and of taking the limit can be interchanged:

$$\lim_{s \to 0}\left[\int_{t=0^-}^{\infty} e^{-st}\dot{x}(t)\, dt\right] = \int_{t=0^-}^{\infty} \lim_{s \to 0}[e^{-st}\dot{x}(t)\, dt] = \int_{t=0^-}^{\infty} \dot{x}(t)\, dt$$

Then (because we know the form desired) we rewrite the last form as follows:

$$\int_{t=0^-}^{\infty} \dot{x}(t)\, dt = \lim_{\tau \to \infty}\left[\int_{t=0^-}^{t=\tau} \dot{x}(t)\, dt\right] = \lim_{\tau \to \infty}\left[x(t)\Big|_{t=0^-}^{t=\tau}\right] = \lim_{t \to \infty}[x(t) - x(0^-)]$$

† Recall that the lower limit used in the one-sided Laplace transform (17.11c) was $t = 0^-$, which results in the appearance of initial conditions at $t = 0^-$, e.g., $x(0^-)$. When the $\mathcal{L}_+$ transform is used (see p. 552), the Initial-Value Theorem gives a more direct check on the initial conditions, which are at 0^+ also [e.g., $x(0^+)$]. However, this advantage is not, in the author's opinion, nearly so useful as those accruing from the $\mathcal{L}_-$ transform, as noted on p. 554.

[Sec. 17.8] FINAL-VALUE THEOREM AND INITIAL-VALUE THEOREM

(Symbol τ was used as a "dummy variable" in the intermediate steps.) Finally, we substitute this form for the last term in (17.48) with the result

$$\lim_{s \to 0} [sX(s)] = \lim_{s \to 0} \{x(0^-) + \lim_{t \to \infty} [x(t) - x(0^-)]\}$$

which, since there are no s's in the brace on the right-hand side, reduces to

$$\lim_{s \to 0} [sX(s)] = \lim_{t \to \infty} [x(t)]$$

which is the desired expression (17.45), for the *Final-Value Theorem*.

The reader is invited to repeat the above proof, using the two-sided Laplace transform instead of the one-sided Laplace transform, to show that exactly the same expression, (17.45) results.

Proof of Initial-Value Theorem

To prove the Initial-Value Theorem we return to (17.47), and this time take the limit indicated in (17.46):

$$\lim_{s \to \infty} [sX(s)] = \lim_{s \to \infty} \left[x(0^-) + \int_{t=0^-}^{\infty} e^{-st} \dot{x}(t)\, dt \right] \quad (17.49)$$

In this case we must be careful about taking the limit $s \to \infty$, because within the interval between $t = 0^-$ and $t = 0^+$ the quantity $\lim_{s \to \infty} (e^{-st})$ is indeterminate. To avoid this indeterminate form, we perform the integration in two intervals, from 0^- to 0^+ and from 0^+ to ∞:

$$\lim_{s \to \infty} \left\{ \lim_{\epsilon \to 0^+} \left[\int_{t=0^-}^{t=\epsilon} e^{-st} \dot{x}(t)\, dt \right] + \lim_{\epsilon \to 0^+} \left[\int_{t=\epsilon}^{t=\infty} e^{-st} \dot{x}(t)\, dt \right] \right\} \quad (17.50)$$

In the second expression we change the order of integration and limit-taking, as we did in proving the Final-Value Theorem. In the first expression, however, we note that within the interval $0^- < t < 0^+$, the value of e^{-st} is unity. We thus write, for (17.50)

$$\lim_{s \to \infty} \left\{ \lim_{\epsilon \to 0^+} \left[\int_{t=0^-}^{t=\epsilon} \dot{x}(t)\, dt \right] \right\} + \lim_{\epsilon \to 0^+} \left\{ \int_{t=\epsilon}^{t=\infty} \lim_{s \to \infty} [e^{-st} \dot{x}(t)\, dt] \right\}$$

or $\qquad\qquad\qquad x(t) \Big|_{t=0^-}^{t=0^+} \quad + \quad 0$

or $\qquad\qquad\qquad x(0^+) - x(0^-) \quad + \quad 0$

The last term is zero because $\lim_{s \to \infty} (e^{-st})$ is zero throughout the time interval $0^+ < t < \infty$.

Substituting this result for the last term in (17.49), we obtain

$$\lim_{s \to \infty} [sX(s)] = [x(0^-) + x(0^+) - x(0^-)] = x(0^+)$$

which is the *Initial-Value Theorem* (17.46).

For the Initial-Value Theorem there is, of course, no counterpart involving the two-sided Laplace transform, since the "initial conditions" in that case occur at $t = -\infty$, and are mathematically inaccessible.

We now present some examples of the application of (17.45) and (17.46).

Example 17.10. Initial and final speed of rotor. For the system of Example 17.6 find the value of Ω a long time after application of the applied torque, $M_f(t)$. Find also the initial value of Ω.

We begin with the response function for Ω, from Example 17.6:

$$\Omega(s) = \frac{M_0}{s(Js+b)} - \frac{M_0}{(Js+b)(s+\sigma_f)} + \frac{J\Omega(0^-)}{(Js+b)}$$

To find $\Omega(\infty)$ we apply the Final-Value Theorem (17.45):

$$\Omega(\infty) = \lim_{s\to 0}[s\Omega(s)] = \frac{M_0}{b} + 0 + 0$$

That is, after a long time, when all the transient motions have died out, a steady-state equilibrium is established in which applied torque M_0 is exactly balanced by the viscous reaction torque $b\Omega$, in full accordance with physical reasoning. Notice that this important result has been obtained quickly, without carrying out any inverse transformation (Step 3), which we found to be somewhat tedious and time-consuming even in Example 17.6.

To find the initial value of Ω, we apply the Initial-Value Theorem (17.46):

$$\Omega(0^+) = \lim_{s\to\infty}[s\Omega(s)] = \Omega(0^-)$$

which we are sure is right because no impulse acts on J at $t = 0$, and J will therefore not suddenly change its velocity between $t = 0^-$ and $t = 0^+$. Here the Initial-Value Theorem is useful merely as a quick check on the validity of the response function obtained in Step 2. Notice, however, that it is a check which can be performed very quickly on the response function (Step 2) before the often-tedious inverse transformation (Step 3) has been performed: thus, if use of the Initial-Value Theorem does uncover a mistake in Step 2, a large amount of incorrect algebra is avoided.

Example 17.11. Initial rotor speed after an impulse. The system modeled in Fig. 17.11a has an initial velocity Ω_0. At $t = 0$ it is disturbed by an impulsive moment, $M(t) = (M_0\tau)\delta(t)$. Find the response $\Omega(t)$. Check your result by means of the Initial-Value Theorem.

The equation of motion is, from (17.35),

$$J\dot{\Omega} + b\Omega = M(t) = (M_0\tau)\delta(t)$$

where $M_0\tau$ is the value (area) of the moment impulse applied. We now follow Procedure D-1.

Step 1. We take the one-sided *Laplace transform* of the equation of motion:

$$(Js+b)\Omega = M_0\tau + J\Omega(0^-)$$

where $\Omega(0^-) = \Omega_0$.

Step 2. *Manipulate algebraically to obtain*
(a) *The response function*:

$$\Omega(s) = \frac{M_0\tau}{Js+b} + \frac{J\Omega(0^-)}{Js+b}$$

(a) *Physical model, with input and initial condition*

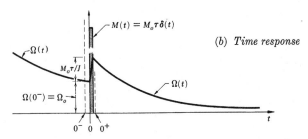

Fig. 17.11 *The response of a coasting rotor to a torque impulse*

(b) *The characteristic of natural motion:* This is $s = -b/J$.

(c) *Final value.* From (17.45), the final value of $\Omega(t)$ is given by

$$\Omega(\infty) = \lim_{s \to 0} [s\Omega(s)] = 0$$

which is obviously correct from physical reasoning. From (17.46), the "initial value" is given by

$$\Omega(0^+) = \lim_{s \to \infty} [s\Omega(s)] = \frac{M_0\tau}{J} + \Omega(0^-)$$

Physically, this result is also reasonable: the impulse which occurred between $t = 0^-$ and $t = 0^+$ has changed the velocity abruptly by the amount

$$\Delta\Omega = M_0\tau/J$$

(Refer to Example 6.5, p. 200. Does $M_0\tau/J$ have the right dimensions?)

Step 3. Inverse transform to obtain the time response:

$$\Omega(t) = \left(\frac{M_0\tau}{J} + \Omega_0\right) e^{-(b/J)t} \qquad 0 < t$$

This response is plotted in Fig. 17.11b. Note carefully the behavior between $t = 0^-$ and $t = 0^+$ and correlate this with the prediction of the Initial-Value Theorem.

Problems 17.60 through 17.66

17.9 A SYSTEM'S RESPONSE TO AN IMPULSE AND ITS ℒ TRANSFORM

Notice that if the input $M(t)$ to a system is a unit impulse (one having unit area, Fig. 17.11a),

$$M(t) = \delta(t)$$

then its Laplace transform will be unity,

$$M(s) = 1$$

and the system response, $\theta(t) = h(t)$, will simply be the inverse transform of the system's transfer function. This is a completely general and most important property of a linear system with constant coefficients:

> *The transfer function of a system is the Laplace transform of its impulse response:*
>
> $$\text{T.F.} = \mathcal{L}[h(t)] = H(s)$$
(17.51)

Although it is almost a truism, (17.51) is stated explicitly because it will be found so useful in advanced studies. We shall use it in Sec. 17.12 for example, in the discussion of convolution.

The transfer function of a system is often denoted by $H(s)$, as a reminder of (17.51).

17.10 INITIAL CONDITIONS AND IMPULSE RESPONSE: A PHYSICAL INTERPRETATION

As we have seen in Examples 17.4 and 17.6 through 17.11, the one-sided Laplace transform provides a simple, completely automatic procedure for accounting for initial conditions in calculations of system response—a procedure which is much quicker and easier than any other. It is entirely rote and requires no physical reasoning. This feature of the Laplace transform alone is responsible for much for its popularity in linear-system analysis.

In this section we wish to add physical interpretation and physical insight to the procedure, since our whole purpose in carrying out the method is to learn about physical behavior. A key to such insight is the unit impulse.

First-order system

Recall first the discussion in Sec. 6.8, where we showed that the natural response of a first-order system following an initial displacement is the same as its response to an impulse. For example, if the equation of motion of a system is

$$\dot{y} + \sigma y = \nu(t)$$

[Sec. 17.10] INITIAL CONDITIONS AND IMPULSE RESPONSE 573

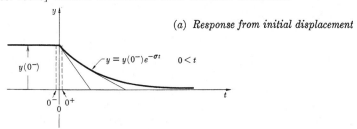

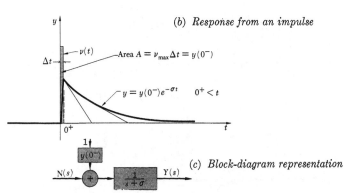

Fig. 17.12 *Initial condition generated by an impulse*

then the response to an initial condition $y(0^-)$, Fig. 17.12a, is, after $t = 0^+$, the same as the response to the impulse input $v(t) = A\,\delta(t)$, Fig. 17.12b, provided only that A is made equal to $y(0^-)$:

$$y = Ae^{-\sigma t} = y(0^-)e^{-\sigma t} \qquad 0^+ < t \qquad (6.34)$$
[repeated]

[Is it clear that all the dimensions are compatible in Fig. 17.12b? What are the dimensions of $v(t)$?] This result is obtained at once, of course, when we solve the problem by using the Laplace transform:

Step 1. Laplace transform the equation of motion ($\mathcal{L}_-$ transform):

$$(s + \sigma)Y(s) = N(s) + y(0^-)$$

Step 2. Manipulate algebraically to obtain
  (a) The response function:

$$Y(s) = \frac{N(s)}{s + \sigma} + \frac{y(0^-)}{s + \sigma}$$

or with $v(t)$ an impulse of magnitude (area) A, $v(t) = A\,\delta(t)$:

$$Y(s) = \frac{A}{s + \sigma} + \frac{y(0^-)}{s + \sigma}$$

Step 3. Inverse transform the response function:

$$y(t) = Ae^{-\sigma t} + y(0^-)e^{-\sigma t} \qquad 0^+ < t$$

That is, the response after $t = 0^+$ from an impulse of area $A = y(0^-)$ is identical to that from the initial condition itself.

We can use this fact to construct a useful artifice which is shown, in block-diagram form, in Fig. 17.12c. Given a system having an initial condition [e.g., $y(0^-)$], we represent that system by one for which the initial condi-

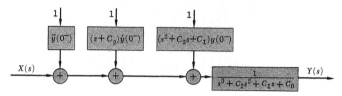

(a) *Block diagram representing Eq.* (17.53)

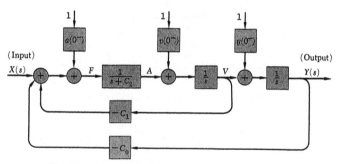

(b) *Block diagram representing Eqs.* (17.57) *(same system)*

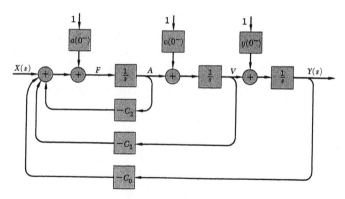

(c) *Same system using integrators only (similar to analog-computer mechanization)*

Fig. 17.13 *Block-diagram evolution for a third-order system*

[Sec. 17.10] INITIAL CONDITIONS AND IMPULSE RESPONSE   575

tion is zero, but to which we add a (fictitious) impulse of magnitude A, and we set A equal to $y(0^-)$ (Fig. 17.12c).† We know that after $t = 0^+$, the behavior of our artificial system will be identical to that of the real system. The reason for doing this is that now *we can treat all of the response of the system as forced response.* That is, in Fig. 17.12c, the total response is that due to forcing function $v(t)$ *plus* that due to "forcing function" $A\,\delta(t)$, with $A = y(0^-)$. This lends a convenient uniformity and routineness to our study of total response.

Physically, we think of the impulse in Fig. 17.12c as an input function that "jars" the system to the displacement $y = y(0^-)$ (see Fig. 17.12b).

Higher-order systems

To make the above result more general, consider next a higher-order system whose equation of motion is

$$\dddot{y} + C_2\ddot{y} + C_1\dot{y} + C_0 y = x(t) \tag{17.52}$$

We proceed to solve this equation by the Laplace transform method.

Step (1) gives

$$(s^3 + C_2 s^2 + C_1 s + C_0)Y(s) = X(s) + (s^2 + C_2 s + C_1)y(0^-) \\ + (s + C_2)\dot{y}(0^-) + \ddot{y}(0^-) \tag{17.53}$$

Step (2), the response function for $Y(s)$ is

$$Y(s) = \frac{X(s)}{s^3 + C_2 s^2 + C_1 s + C_0} \\ + \frac{(s^2 + C_2 s + C_1)y(0^-) + (s + C_2)\dot{y}(0^-) + \ddot{y}(0^-)}{s^3 + C_2 s^2 + C_1 s + C_0} \tag{17.54}$$

This result can also be represented in the block-diagram form of Fig. 17.13a. Again, the 1's are the Laplace transforms of unit impulses which are "generating" the initial conditions. Thus *all* motion is thought of as forced motion.

It is easy to interpret the "$\ddot{y}(0^-)$" block in Fig. 17.13a; we are simply saying that, for a third-order physical system, an impulse "jars" the system to an initial *acceleration*. The other blocks are a little more difficult to interpret. The "$\dot{y}(0^-)$" block implies that an initial system velocity is obtained by means of a "jar" made up of an impulse, $C_2\dot{y}(0^-)\,\delta(t)$, plus the *derivative* of an impulse. That is, the term $s\dot{y}(0^-)(1)$ implies the time function

$$\frac{d}{dt}[\dot{y}(0^-)\,\delta(t)] = \dot{y}(0^-)\frac{d}{dt}[\delta(t)]$$

The derivative of an impulse is a pair of "spikes," as shown in Fig. 17.14, of equal and opposite magnitude, one occurring at the start of the impulse and the other at the end. This is the sort of physical jarring required to get an initial velocity established in a third-order system.

† Note that, since the block diagram presents the $\mathcal{L}$ transforms of all the quantities, the impulse appears as 1. That is, $\mathcal{L}[\delta(t)] = 1$.

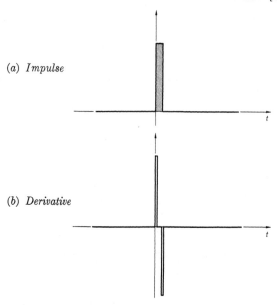

(a) *Impulse*

(b) *Derivative*

Fig. 17.14 *The derivative of an impulse*

Finally, to establish the initial position of the third-order system, the "$y(0^-)$" block in Fig. 17.13a tells us, it is necessary to perform *two* differentiations of an impulse. Such higher-order impulse differentiations become less easy to think about physically, and the concept becomes less useful.

Even so, construction, from the response function, of a block diagram of the form shown in the Fig. 17.13a is often a helpful step in gaining physical insight into a problem.

17.11 EQUATIONS OF MOTION IN STANDARD FORM: STATE VARIABLES

An alternative procedure, which returns the mathematical description of the system to a form that is very easy to contemplate physically, is to use additional variables to convert the simple nth-order differential equation into a set of n first-order differential equations. The latter form is known as the "standard form," and in addition to the use we describe below, it is also extremely useful as a purely mathematical tool in dealing with systems of equations of unspecified order, as discussed in Sec. 14.5.

To convert Eq. (17.52) into standard form we assign a new variable for each derivative, e.g.,

$$v \triangleq \dot{y}$$
$$a \triangleq \dot{v}$$ (17.55)

[Sec. 17.11] EQUATIONS OF MOTION IN STANDARD FORM

Substitution of these new variables into the original equation yields a first-order differential equation which, together with the new-variable definitions (17.55), forms a set of n first-order differential equations (where n is the order of the original equation, $n = 3$ in this example):

$$\begin{aligned} \dot{a} + C_2 a + C_1 v + C_0 y &= x(t) \\ \dot{v} &= a \\ \dot{y} &= v \end{aligned} \quad (17.56)$$

Whenever a set of variables for a physical system is so chosen that the equations of motion of the system contain those variables and their first derivatives only, the set is a set of state variables. Thus a, v, y are a set of state variables. State variables often have obvious physical meaning. For example, if the system is a mechanical one, with y a position coordinate, then v and a correspond to velocity and acceleration (the derivatives of position).

Next let us perform the one-sided Laplace transform on the set of equations in standard form:

$$\begin{aligned} (s + C_2)A &= -C_1 V - C_0 Y + X + a(0^-) \\ sV &= A \hphantom{-C_1 V - C_0 Y + X} + v(0^-) \\ sY &= V \hphantom{-C_1 V - C_0 Y + X} + y(0^-) \end{aligned} \quad (17.57)$$

We now construct a block diagram based on (17.57), Fig. 17.13b. Here the physical interpretation of the impulses is straightforward: each jars the system to an initial condition (and none is differentiated).

Analog-computer form

With a slight modification, the block diagram of Fig. 17.13b can be represented with *pure integrating elements only*. The modification is to note that a block such as that shown in Fig. 17.15a can also be drawn as shown in Fig. 17.15b. [The reader may verify that the two diagrams are equivalent by writing the algebraic Laplace transform equation represented by each— $(s + C_2)A = F$.] The complete block diagram, using integrators only, is shown in Fig. 17.13c. Notice in particular that each initial condition is now associated with one "integrator."

The arrangement shown in Fig. 17.13c is almost precisely that which would be used in mechanizing an analog computer to solve the differential equation (17.52), with a given input and given initial conditions.† An analog computer consists primarily of two building blocks, electrical integrators and electrical adders (or summers). (Refer to Fig. 8.31.) Differential equations are first put in standard form (17.56), and then each of the resulting first-order equations is "mechanized" in the form

$$\dot{y} = -\sigma y + \text{input}$$

† Actually, each initial condition is established on an analog computer by a step at the output of an integrator, rather than by an impulse at the input. This is mathematically equivalent, of course.

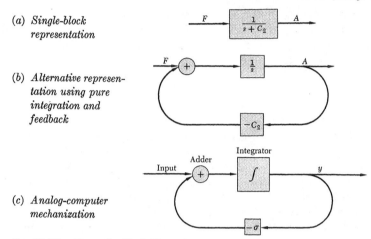

(a) *Single-block representation*

(b) *Alternative representation using pure integration and feedback*

(c) *Analog-computer mechanization*

Fig. 17.15 *Alternative block-diagram arrangement using pure integrator*

as in Fig. 17.15c. These units are in turn connected together according to the given differential equations. The reader may wish to "mechanize" differential equations (17.56) in this way and note that his resulting diagram corresponds one to one with Fig. 17.13c. The principles of analog computation are discussed more at length in Sec. 8.15.

The system we have created in Fig. 17.13c is, of course, a member of the class called "feedback" systems (because signals are fed back from the output end to the input end). The general analysis of such systems is the subject of Part E.

Matrix version of standard form

By a slight change in nomenclature, a set of equations such as (17.56) can be put into a form which is very compact and much more efficient to manipulate. The change in nomenclature consists in giving each variable the same letter with a different subscript. For example, in the system of Fig. 17.13c the following variable assignments would be made

$$y = y_0 \quad v = y_1 \quad a = y_2$$

Then the block diagram would appear as in Fig. 17.16. (Notice that for convenience of identification we have let the subscript on each variable take the order of differentiation of y.) Now the equations of motion (17.56) can be written:

$$\dot{y}_2 = -C_2 y_2 - C_1 y_1 - C_0 y_0 + x(t)$$
$$\dot{y}_1 = y_2$$
$$\dot{y}_0 = y_1$$

[Sec. 17.11] EQUATIONS OF MOTION IN STANDARD FORM

In matrix form these appear as

$$\begin{bmatrix} \dot{y}_2 \\ \dot{y}_1 \\ \dot{y}_0 \end{bmatrix} = \begin{bmatrix} -C_2 & -C_1 & -C_0 \\ 1 & 0 & 0 \\ 0 & 1 & 0 \end{bmatrix} \begin{bmatrix} y_2 \\ y_1 \\ y_0 \end{bmatrix} + \begin{bmatrix} x(t) \\ 0 \\ 0 \end{bmatrix} \quad (17.58)$$

which can be written in the compact form (refer to Sec. 14.5)

$$\underline{\dot{y}} = [C]\underline{y} + \underline{x} \quad (17.59)$$

in which $\underline{\dot{y}}$, $\underline{y}$, and $\underline{x}$ are column matrices and $[C]$ is a square matrix. This is the *standard matrix form*, which can be studied and manipulated in the general form of (17.59) with great efficiency and generality.

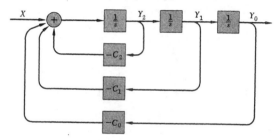

Fig. 17.16 *Block diagram corresponding to matrix form of equations of motion, Eqs.* (17.58)

Example 17.12. Representing fluid-clutch dynamics. Construct block diagrams analogous to those of Fig. 17.13 to represent the dynamics of the fluid clutch analyzed in Example 17.7, p. 563. Also, rewrite the equations of motion in standard matrix form.

One block diagram (like Fig. 17.13a) can be drawn from the system response function (17.40) obtained in Step 2 of Example 17.7:

$$\Omega_2(s) = \frac{b\{[M_0 s/(s^2 + \omega_f^2)] + J_1\Omega_1(0^-)\} + (J_1 s + b)J_2\Omega_2(0^-)}{[(J_1 s + b)(J_2 s + b) - b^2]} \quad (17.60)$$

which can be shown as in Fig. 17.17a. (Signals ⟹ are moments and ----▶ are motions.) Other block diagrams based on the overall system transfer function are given in Fig. 17.8.

A more meaningful block diagram is obtained by going back to the transformed equations of Step 1, (17.39), which are first-order:

$$\begin{aligned} (J_1 s + b)\Omega_1 \quad -b\Omega_2 &= \mathsf{M} + J_1\Omega_1(0^-) \\ -b\Omega_1 + (J_2 s + b)\Omega_2 &= J_2\Omega_2(0^-) \\ \mathsf{M} &= \frac{M_0 s}{s^2 + \omega_f^2} \end{aligned} \quad (17.61)$$

The block-diagram arrangement is quickly indicated by rewriting the first equation as an expression for Ω_1:

$$\Omega_1 = \frac{1}{J_1 s + b}[b\Omega_2 + M + J_1\Omega_1(0^-)] \quad (17.62)$$

and the second as an expression for Ω_2:

$$\Omega_2 = \frac{1}{J_2 s + b}[b\Omega_1 + J_2\Omega_2(0^-)] \quad (17.63)$$

The resulting block diagram, Fig. 17.17b, shows beautifully the coupling (and its symmetry) and all the other cause-and-effect relations: the system being "jarred" to its initial conditions by the fictitious impulses, and being driven by $M(t)$. (We imagine M also to be produced by a fictitious function generator with impulse input.)

Figure 17.17b can be quickly converted to "analog-computer form" by application of Fig. 17.15, as shown in Fig. 17.17c.

The equations of motion (17.61) are readily written in the standard matrix form of (17.59), because each equation is only first-order already:

$$\begin{bmatrix} \dot{\Omega}_1 \\ \dot{\Omega}_2 \end{bmatrix} = \begin{bmatrix} -\dfrac{b}{J_1} & +\dfrac{b}{J_1} \\ +\dfrac{b}{J_2} & -\dfrac{b}{J_2} \end{bmatrix} \begin{bmatrix} \Omega_1 \\ \Omega_2 \end{bmatrix} + \begin{bmatrix} M \\ 0 \end{bmatrix} \quad (17.64)$$

or
$$\underline{\dot{\Omega}} = [A]\underline{\Omega} + \underline{M} \quad (17.65)$$

Problems 17.67 through 17.75

17.12 CONVOLUTION AND THE LAPLACE TRANSFORM

Consider a linear system represented by a differential equation of the general form

$$\frac{d^n y}{dt^n} + A_{n-1}\frac{d^{n-1} y}{dt^{n-1}} + \cdots + A_1\dot{y} + A_0 y$$
$$= B_m \frac{d^m x}{dt^m} + B_{m-1}\frac{d^{m-1} x}{dt^{m-1}} + \cdots + B_1\dot{x} + B_0 x(t) \quad (17.66)$$

The transfer function of the system is obtained by Laplace transforming the equation and taking the ratio of output to input (initial conditions being considered zero):

$$\text{Transfer function} = \frac{Y(s)}{X(s)} = \frac{B_m s^m + \cdots + B_1 s + B_0}{s^n + A_{n-1} s^{n-1} + \cdots + A_1 s + A_0} \quad (17.67)$$

In Sec. 6.10 we showed that the response of such a linear system to any arbitrary input $x(t)$ can be calculated by means of the convolution integral

$$y(t) = x(t) * h(t)$$
or
$$y(t) = \int_{\tau=-\infty}^{\tau=t} x(\tau) h(t-\tau)\, d\tau \quad (6.54)$$
[repeated]

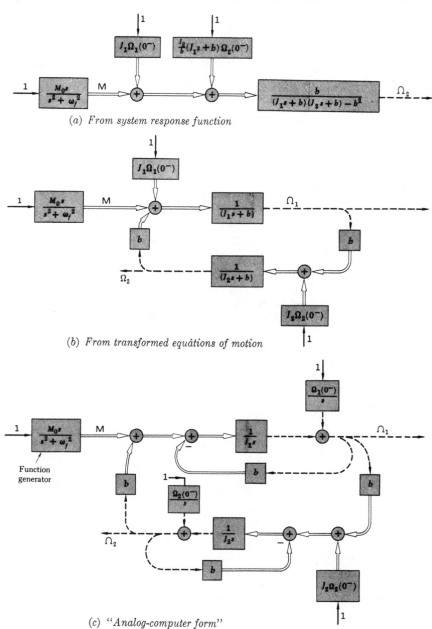

Fig. 17.17 *Block-diagram representations of fluid clutch*

where $h(t)$ is the known *response* of the system *to a unit impulse input*. We have used this result in a number of examples concerning simple systems, in which we first established the unit impulse response $h(t)$ of the system, and then used it with (6.54) to compute the system's response to other inputs.

In Sec. 17.9 we showed that the Laplace transform of the output response of a system to a unit-impulse input is identical with the transfer function of the system:

$$\mathcal{L}[h(t)] = H(s) = \frac{Y(s)}{X(s)} \qquad (17.51)$$
[repeated]

These two results can be combined to produce another relation of profound mathematical importance, particularly in the study of response to random inputs, namely

Convolution and
$\mathcal{L}^{-1}$ *transformation*

Given:	$Y(s) = H(s)X(s)$
and	$h(t)$ and $x(t)$
Then:	$y(t) = \mathcal{L}^{-1}[H(s)X(s)]$
or	$y(t) = h(t) * x(t)$

(17.68)

That is, whenever the response function for a system is known as the *product* of two functions of s, the corresponding time function of the system can be found by *convolving* the two time responses corresponding to those functions of s.†

It is possible to derive the Laplace transform itself (17.11b), beginning with the convolution integral, as an alternative to beginning with the Fourier series and Fourier transform.

Commonly, it is easiest to find system response by simply inverse transforming an algebraic product $H(s)X(s)$. Sometimes, however, an input function $x(t)$ is encountered for which the Laplace transform is not readily available. Then it may be most convenient to find the system response by direct application of the convolution integral. These two methods are demonstrated in the example that follows.

Example 17.13. Comparison of solution by $\mathcal{L}^{-1}$ and by convolution. The transfer function and the impulse response for a certain system are, respectively,

$$H(s) = \frac{Y(s)}{X(s)} = \frac{1}{(s+1)(s+2)} \quad \text{and} \quad h(t) = (e^{-t} - e^{-2t})u(t)$$

Find the response of the system to an input whose Laplace transform and time behavior are, respectively,

$$X(s) = \frac{C}{s+5} \quad \text{and} \quad x(t) = Ce^{-5t}u(t)$$

Solve by inverse Laplace transformation and also by convolution.

† Recall that convolution is commutative; $h(t) * x(t) = x(t) * h(t)$.

[Sec. 17.12] CONVOLUTION AND THE LAPLACE TRANSFORM

Steps in the two procedures are presented side by side in the columns below.

Inverse transformation (Partial fraction expansion)†	Convolution
Response function: $$Y(s) = H(s)X(s)$$ $$= \frac{C}{(s+1)(s+2)(s+5)} = \frac{A_1}{(s+1)}$$ $$+ \frac{A_2}{(s+2)} + \frac{A_3}{(s+5)}$$ $$y = [A_1 e^{-t} + A_2 e^{-2t} + A_3 e^{-5t}]u(t)$$ $$A_1 = \left[\frac{C}{(s+2)(s+5)}\right]_{s=-1} = \frac{C}{4}$$ $$A_2 = \left[\frac{C}{(s+1)(s+5)}\right]_{s=-2} = -\frac{C}{3}$$ $$A_3 = \left[\frac{C}{(s+1)(s+2)}\right]_{s=-5} = \frac{C}{12}$$ $$y = C\left(\frac{e^{-t}}{4} - \frac{e^{-2t}}{3} + \frac{e^{-5t}}{12}\right)u(t)$$	Let $$y(t) = h(t) * x(t)$$ $$= \int_{\tau=0}^{\tau=t} h(\tau)x(t-\tau)\,d\tau$$ $$y(t) = \int_{\tau=0}^{\tau=t} (e^{-\tau} - e^{-2\tau})Ce^{-5(t-\tau)}\,d\tau$$ $$= Ce^{-5t} \int_0^t (e^{4\tau} - e^{3\tau})\,d\tau$$ $$= Ce^{-5t}\left[\frac{e^{4\tau}}{4} - \frac{e^{3\tau}}{3}\right]_{\tau=0}^{\tau=t}$$ $$= Ce^{-5t}\left(\frac{e^{4t}-1}{4} - \frac{e^{3t}-1}{3}\right)u(t)$$ $$= C\left(\frac{e^{-t}}{4} - \frac{e^{-2t}}{3} + \frac{e^{-5t}}{12}\right)u(t)$$

† The details of partial fraction expansion are given in Chapter 18. At this point, the reader may easily satisfy himself that the second expression for $Y(s)$ is identical to the first, for the values given for A_1, A_2, A_3.

Problems 17.76 through 17.82

Chapter 18

FROM LAPLACE TRANSFORM TO TIME RESPONSE BY PARTIAL FRACTION EXPANSION

We now turn attention to the third (and last) step of Procedure D-1: finding the complete time response implied by a given system response function. Sometimes it is most convenient to do this by looking up the response function in a large table of Laplace transforms (like the one in Appendix J of this book), and routinely copying the inverse transform term by term.

An alternative procedure, which is often more efficient, involves performing a partial fraction expansion of the response function. In this short chapter we shall describe and demonstrate techniques for determining time response from a system response function, using partial fraction expansion. The results are summarized in Sec. 18.6 as Procedure D-2.

18.1 FORMULATION OF THE TASK

Suppose a given physical system can be represented by the following transfer function:

$$X(s) \longrightarrow \boxed{\frac{K(s + a_1)(s + a_2) \ldots}{[(s + \alpha_1)(s + \alpha_2)(s + \alpha_3) \ldots]}} \longrightarrow Y(s) \qquad (18.1)$$

where the α's may be real, or complex, or zero. (If they are complex, then—for a physical system—they must come in conjugate pairs.) We know that the denominator of the system transfer function contains the natural characteristics of the system. (Refer to Sec. 8.9.) Therefore we expect that if the system is disturbed in any way, its response will include some natural motion corresponding to each α.

[Sec. 18.2] PARTIAL FRACTION EXPANSION: CASE ONE 585

Suppose further that the response of the system to a particular disturbance is desired. To be specific, let the disturbance be applied abruptly at $t = 0$, and let the initial conditions of the system all be zero. Suppose that the Laplace transform of the disturbance is

$$X(s) = \frac{(s + a_f)x_0}{(s + \alpha_{f1})(s + \alpha_{f2})} \qquad (18.2)$$

The response of the system to this particular disturbance can be found from the corresponding *system response function:*

$$Y(s) = \frac{x_0(s + a_f)K(s + a_1)(s + a_2) \cdots}{(s + \alpha_{f1})(s + \alpha_{f2})[(s + \alpha_1)(s + \alpha_2)(s + \alpha_3) \cdots]} \qquad (18.3)$$

The $\mathcal{L}^{-1}$ transform of (18.3) gives the time response,† which will have the form

$$y(t) = [\underbrace{A_{f1}e^{-\alpha_{f1}t} + A_{f2}e^{-\alpha_{f2}t}}_{\text{Forced motion}} + \underbrace{A_1 e^{-\alpha_1 t} + A_2 e^{-\alpha_2 t} + A_3 e^{-\alpha_3 t} + \cdots}_{\text{Natural motion}}]u(t) \qquad (18.4)$$

where, in the spirit of Part B, the forced and natural components of the motion are as indicated. Notice that the numerator terms—the zeros—have no influence on the *character* of the response, only on its magnitude and phase, i.e., on the constants A.

The task before us now is to evaluate the coefficients (the A's) in (18.4). The most systematic way to do so is by means of partial fraction expansion.

18.2 PARTIAL FRACTION EXPANSION: CASE ONE

Fortunately, the task of finding the coefficient of each term in the response of a system, once its response function is known, is a straightforward one involving purely algebraic manipulations. The technique is to expand the response function (18.3) into a summation of simple fractions:

$$Y(s) = \frac{Kx_0(s + a_f)(s + a_1)(s + a_2) \cdots}{(s + \alpha_{f1})(s + \alpha_{f2})(s + \alpha_1)(s + \alpha_2)(s + \alpha_3) \cdots}$$
$$= \frac{A_{f1}}{s + \alpha_{f1}} + \frac{A_{f2}}{s + \alpha_{f2}} + \frac{A_1}{s + \alpha_1} + \frac{A_2}{s + \alpha_2} + \frac{A_3}{s + \alpha_3} + \cdots \qquad (18.5)$$

where, in general, the A's may be complex numbers. Once we have done this, the response of the system is simply the sum of the terms given in (18.4). That is, the A's required in (18.4) are identically those obtained in (18.5), as specified by (17.16) (or transform pair 6 of Table X, p. 541).

The result of a partial fraction expansion can always be checked

† The $\mathcal{L}^{-1}$ transform can be performed only when the order of the denominator *exceeds* that of the numerator. Functions not having this property can be put in proper form by dividing denominator into numerator.

directly, of course, simply by converting the expanded form (18.5) to a common denominator in the usual way, and comparing with the original expression.

The technique used to evaluate a particular coefficient in the partial fraction expansion is to eliminate all other coefficients from (18.5) by a mathematical trick. There are two procedures, depending upon whether poles $(s + \alpha_n)$ occur singly or in multiples. Let us consider first the common case that all the poles of the composite transfer function are different.

Case one: all poles different

The mathematical trick in this case is to multiply the whole equation by the factor whose coefficient is sought. To find coefficient A_1 in (18.5), for example, we multiply all of (18.5) by factor $(s + \alpha_1)$:

$$
\begin{aligned}
(s + \alpha_1)Y(s) &= \frac{Kx_0(\cancel{s+\alpha_1})(s + a_f)(s + a_1)(s + a_2)\cdots}{(s + \alpha_{f1})(s + \alpha_{f2})(\cancel{s+\alpha_1})(s + \alpha_2)(s + \alpha_3)\cdots} \\
&= (s + \alpha_1)\frac{A_{f1}}{(s + \alpha_{f1})} + (s + \alpha_1)\frac{A_{f2}}{(s + \alpha_{f2})} \\
&\quad + \cancel{(s+\alpha_1)}\frac{A_1}{\cancel{(s+\alpha_1)}} + (s + \alpha_1)\frac{A_2}{(s + \alpha_2)} \\
&\quad\quad + (s + \alpha_1)\frac{A_3}{(s + \alpha_3)} + \cdots \quad (18.6)
\end{aligned}
$$

When this is done, all the coefficients *except* A_1 are now multiplied by $(s + \alpha_1)$, and the next move is to let $(s + \alpha_1) \to 0$ in (18.6), i.e., to let $s = -\alpha_1$. This causes all the other coefficients on the right-hand side to vanish, and leaves a fine expression for A_1:

$$
\begin{aligned}
A_1 &= [(s + \alpha_1)Y(s)]_{s=-\alpha_1} \\
&= \left[\frac{Kx_0(s + a_f)(s + a_1)(s + a_2)\cdots}{(s + \alpha_{f1})(s + \alpha_{f2})(1)(s + \alpha_2)(s + \alpha_3)\cdots}\right]_{s=-\alpha_1} \quad (18.7)
\end{aligned}
$$

The pattern is the same for each of the other coefficients. This procedure is sometimes known as the *"cover-up" method*, because each coefficient A_n is evaluated by *covering up its term, $(s + \alpha_n)$, in the response function and then evaluating the remaining expression at $s = -\alpha_n$.*

The results can be stated in general form as follows:

If $\quad Y(s) = \dfrac{KN(s)}{\Pi(s + \alpha_n)}$

then $\quad y(t) = \Sigma A_n e^{-\alpha_n t} u(t)$

in which $\quad A_n = [(s + \alpha_n)Y(s)]_{s=-\alpha_n}$

provided that the α_n are all different.

Procedure D-2.1†
Partial fraction expansion, all poles different

---

† All of Procedure D-2, partial fraction expansion for all cases, is summarized in Sec. 18.6, which will be more convenient for future reference.

[Sec. 18.2] PARTIAL FRACTION EXPANSION: CASE ONE

The expressions for the coefficients are, in general, complex numbers which can be evaluated by straightforward complex algebra. (Often the process of evaluation may be speeded up by making use of a graphical plot in the s plane of the poles and zeros of the original composite response function, as will be demonstrated in Sec. 18.4.)

Example 18.1. All real poles. Find the time response of a system having the following response function:

$$Y(s) = \frac{15(s+3)}{s(s+5)(s+10)} \qquad (18.8)$$

From Procedure D-2.1, the time response is given by

$$y(t) = [A_0 e^{0t} + A_1 e^{-5t} + A_2 e^{-10t}]u(t) \qquad (18.9)$$

where the A's are given by

$$A_0 = \left[\frac{15(s+3)}{(s+5)(s+10)}\right]_{s=0} = 0.9 \qquad A_1 = \left[\frac{15(s+3)}{s(s+10)}\right]_{s=-5} = 1.2$$

$$A_2 = \left[\frac{15(s+3)}{s(s+5)}\right]_{s=-10} = -2.1 \qquad (18.10)$$

[Check this result by converting $Y(s) = \dfrac{0.9}{s} + \dfrac{1.2}{s+5} - \dfrac{2.1}{s+10}$ to a common denominator and comparing with (18.8).]

From (18.10) the time response is given by

$$y(t) = [0.9 + 1.2e^{-5t} - 2.1e^{-10t}]u(t) \qquad (18.11)$$

Notice that the numerator term, $(s+3)$, has no influence on what the exponents appearing in the response are, but does influence the magnitude and sign of the coefficient A.

Example 18.2. One complex pair. Find the time response of a system having the response function

$$Y(s) = \frac{15(s+3)}{s(s+13)(s^2+14s+113)} = \frac{15(s+3)}{s(s+13)(s+7 \mp j8)} \qquad (18.12)$$

in which $(s+7 \mp j8)$ represents the product $(s+7-j8)(s+7+j8)$.

The partial fraction expansion has the form

$$Y(s) = \frac{15(s+3)}{s(s+13)(s+7 \mp j8)} = \frac{A_0}{s} + \frac{A_1}{s+13} + \frac{\vec{A}_{21}}{s+7-j8} + \frac{\vec{A}_{22}}{s+7+j8} \qquad (18.13)$$

(where the arrows over $\vec{A}_{21}$ and $\vec{A}_{22}$ remind us that these coefficients will be complex numbers). Then the form of the time response will be

$$y(t) = [A_0 + A_1 e^{-13t} + \vec{A}_{21} e^{(-7+j8)t} + \vec{A}_{22} e^{(-7-j8)t}]u(t) \qquad (18.14)$$

in which the last two terms must combine to form a real sinusoidal function of time.

Evaluation of the coefficients is as follows (by Procedure D-2.1):

$$A_0 = \left[\frac{15(s+3)}{(s+13)(s^2+14s+113)}\right]_{s=0} = \frac{15(3)}{(13)(113)} = 0.0306 \quad (18.15a)$$

$$A_1 = \left[\frac{15(s+3)}{s(s^2+14s+113)}\right]_{s=-13} = \frac{15(-10)}{(-13)(169-182+113)} = 0.1153 \quad (18.15b)$$

$$\vec{A}_{21} = \left[\frac{15(s+3)}{s(s+13)(s+7+j8)}\right]_{s=-7+j8} = \frac{15(-4+j8)}{(-7+j8)(6+j8)(j16)} = 0.079\underline{/202°} \quad (18.15c)$$

$$\vec{A}_{22} = \left[\frac{15(s+3)}{s(s+13)(s+7-j8)}\right]_{s=-7-j8} = \frac{15(-4-j8)}{(-7-j8)(6-j8)(-j16)} = 0.079\underline{/-202°} \quad (18.15d)$$

and thus

$$y(t) = [0.0306 + 0.1153e^{-13t} + 0.079e^{-7t}(e^{j8t}e^{j3.53} + e^{-j8t}e^{-j3.53})]u(t) \quad (18.16)$$

The last two terms can be combined to give, finally

$$y(t) = [0.0306 + 0.1153e^{-13t} + 0.158e^{-7t}\cos(8t+3.53)]u(t) \quad (18.17)$$

Problems 18.1 through 18.3

18.3 SPECIAL HANDLING OF COMPLEX CONJUGATE POLES

As the above example demonstrates, the occurrence of poles in complex conjugate pairs always leads to considerable manipulation with complex numbers. Since such pairs occur frequently, it is worth-while to perform the bulk of the algebra—that involving converting from the exponential to the sinusoidal form—for the general case, once for all. This will now be done, and a simple rule will be found which can be applied to all such problems, at substantial saving of labor.

Suppose, in general, that a response function has the form

$$Y(s) = \frac{1}{(s+\sigma_1-j\omega_1)(s+\sigma_1+j\omega_1)} G(s) \quad (18.18)$$

where $G(s)$ is any ratio of polynomials in s. Then the partial fraction expansion is

$$Y(s) = \frac{\vec{A}_{11}}{s+\sigma_1-j\omega_1} + \frac{\vec{A}_{12}}{s+\sigma_1+j\omega_1} + \text{other terms} \quad (18.19)$$

where

$$\vec{A}_{11} = \left[\frac{1}{s+\sigma_1+j\omega_1}G(s)\right]_{s=-\sigma_1+j\omega_1} = |\vec{A}_{11}|e^{j\underline{/\vec{A}_{11}}} \quad (18.20a)$$

$$\vec{A}_{12} = \left[\frac{1}{s+\sigma_1-j\omega_1}G(s)\right]_{s=-\sigma_1-j\omega_1} = |\vec{A}_{12}|e^{j\underline{/\vec{A}_{12}}} \quad (18.20b)$$

Notice that $\vec{A}_{12}$ is always the *conjugate* of $\vec{A}_{11}$. (Note in particular that this was true of $\vec{A}_{22}$ and $\vec{A}_{21}$ in Example 18.2.)

That is, $\quad |\vec{A}_{12}| = |\vec{A}_{11}| \qquad \underline{/\vec{A}_{12}} = -\underline{/\vec{A}_{11}}$

The time response can therefore be written

$$\begin{aligned}
y(t) &= [\vec{A}_{11} e^{(-\sigma_1 + j\omega_1)t} + \vec{A}_{12} e^{(-\sigma_1 - j\omega_1)t}] u(t) + \text{other terms} \\
&= |\vec{A}_{11}| e^{-\sigma_1 t} [e^{j\underline{/\vec{A}_{11}}} e^{j\omega_1 t} + e^{-j\underline{/\vec{A}_{11}}} e^{-j\omega_1 t}] u(t) + \text{other terms} \\
&= |\vec{A}_{11}| e^{-\sigma_1 t} [e^{j(\omega_1 t + \underline{/\vec{A}_{11}})} + e^{-j(\omega_1 t + \underline{/\vec{A}_{11}})}] u(t) + \text{other terms}
\end{aligned}$$

or, applying (7.31)

$$\boxed{y(t) = [2|\vec{A}_{11}| e^{-\sigma_1 t} \cos(\omega_1 t + \underline{/\vec{A}_{11}})] u(t) + \text{other terms}} \quad \begin{array}{c} \text{Procedure D-2.2} \\ \text{Simplification for} \\ \text{complex poles} \end{array} \quad (18.21)$$

This useful rule states that (1) we need only evaluate one of the coefficients of a complex pair, and (2) we can write the real sinusoidal solution at once. The reader is invited to solve Example 18.2 by this technique and notice the reduction in labor. Further streamlining is facilitated by calculating the coefficient $\vec{A}_{11}$ graphically in the s plane, as described in Sec. 18.4.

18.4 USE OF THE s-PLANE POLE-ZERO ARRAY TO COMPUTE RESPONSE COEFFICIENTS

Algebraic computation of response coefficients such as (18.15c) can sometimes be tedious and cumbersome, particularly when a number of complex poles are involved. Alternatively, each coefficient can be rapidly computed (or, even more rapidly, estimated by eye) by a simple graphical procedure using an s-plane plot of the poles and zeros of the system response function. The procedure was developed in Sec. 14.3 and is illustrated in Figure 18.1 for a rather general response function:

$$\begin{aligned}
Y(s) &= \frac{K(s + \sigma_a)}{s(s + \sigma_1)(s + \sigma_2 \mp j\omega_2)(s \mp j\omega_f)} \\
&= \frac{A_0}{s} + \frac{A_1}{s + \sigma_1} + \frac{\vec{A}_{21}}{s + \sigma_2 - j\omega_2} + \frac{\vec{A}_{22}}{s + \sigma_2 + j\omega_2} + \frac{\vec{A}_{f1}}{s - j\omega_f} + \frac{\vec{A}_{f2}}{s + j\omega_f}
\end{aligned} \quad (18.22)$$

First, the poles and zeros of the response function are plotted in the s plane in Fig. 18.1. Next, the expression for each coefficient is written in algebraic form. For example, the expression for $\vec{A}_{21}$ is

$$\vec{A}_{21} = \left[\frac{K(s + \sigma_a)}{s(s + \sigma_1)(s + \sigma_2 + j\omega_2)(s \mp j\omega_f)} \right]_{s = -\sigma_2 + j\omega_2} \quad (18.23)$$

Each of the terms in the denominator of (18.23) is a vector in the s plane which begins at a pole of the original response function, (18.22), and *termi-*

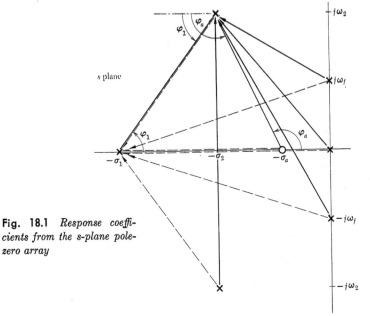

Fig. 18.1 *Response coefficients from the s-plane pole-zero array*

nates at the pole whose coefficient is being determined, i.e., at $s = -\sigma_2 + j\omega_2$. The vectors shown solid in Fig. 18.1 are for $s = -\sigma_2 + j\omega_2$.

Because the vectors terminate at a single point, they can be multiplied together rapidly by graphical means—particularly if a Spirule is available. The magnitude $|\vec{A}_{21}|$ is K times the product of the lengths of the (solid) vectors from the zeros divided by the product of the lengths of the (solid) vectors from the poles. The phase angle $\angle \vec{A}_{21}$ is the sum of the vector angles from the zeros minus the sum of the vector angles from the poles. (Notice that all the angles are available at the single point $s = -\sigma_2 + j\omega_2$. Two of these angles, ϕ_a and ϕ_1, are indicated for illustration.)

Calculating coefficient A_1 for pole $(s + \sigma_1)$ involves the dashed set of vectors in Fig. 18.1—each of which, again, terminates at the pole whose coefficient is being calculated, i.e., at $s = -\sigma_1$. The same procedure is applied to each of the poles in turn until all the coefficients have been computed. (For the poles on the real axis, algebraic calculation may be quicker.)

Example 18.3. Graphical calculation. Use the s plane to calculate the response in Example 18.2. It is given that

$$Y(s) = \frac{15(s + 3)}{s(s + 13)(s + 7 \mp j8)} = \frac{A_0}{s} + \frac{A_1}{s + 13} + \frac{\vec{A}_{21}}{s + 7 - j8} + \frac{\vec{A}_{22}}{s + 7 + j8}$$
(18.24)

[Sec. 18.4] USING POLE-ZERO ARRAY TO COMPUTE COEFFICIENTS

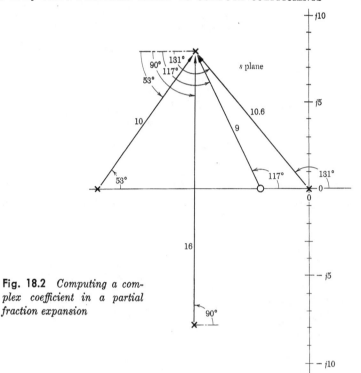

Fig. 18.2 *Computing a complex coefficient in a partial fraction expansion*

By Procedures D-2.1 and D-2.2, we have

$$y(t) = [A_0 + A_1 e^{-13t} + 2|\vec{A}_{21}|e^{-7t} \cos(8t + \underline{/\vec{A}_{21}})]u(t) \quad (18.25)$$

in which $\vec{A}_{21} = \left[\dfrac{15(s+3)}{s(s+13)(s+7+j8)}\right]_{s=-7+j8}$ (18.26)

The poles (×) and zeros (○) of this expression are plotted in Fig. 18.2. The magnitudes and angles of the vectors in the above expression for $\vec{A}_{21}$ are drawn on the plot, from which

$$|\vec{A}_{21}| = \frac{15(9)}{(10.6)(10)(16)} = 0.08 \quad (18.27)$$

and $\underline{/\vec{A}_{21}} = 117° - (131° + 90° + 53°) = -157°$, or 202° or 3.53 radians. (With a Spirule, the individual magnitudes and angles would not be measured or noted. Instead, the totals would be read directly from the Spirule. See Appendix I.) Thus the last term in the time response is

$$0.16 e^{-7t} \cos(8t + 3.53)u(t) \quad (18.28)$$

which checks the result of Example 18.2.

Problems 11.10 through 11.15

18.5 THE CASE OF REPEATED POLES: CASE TWO

If a pole is repeated, then the evaluation of coefficients is slightly more complicated.

Consider first the case of a double pole:

$$Y(s) = \frac{K(s + a_1) \cdots}{(s + \alpha_1) \cdots (s + \alpha_4)^2(s + \alpha_5) \cdots} \qquad (18.29)$$

The proper partial fraction expansion in this case is†

$$Y(s) = \frac{K(s + a_1) \cdots}{(s + \alpha_1) \cdots (s + \alpha_4)^2(s + \alpha_5) \cdots} = \frac{A_1}{s + \alpha_1} + \cdots$$
$$+ \frac{A_{42}}{(s + \alpha_4)^2} + \frac{A_{41}}{s + \alpha_4} + \frac{A_5}{s + \alpha_5} + \cdots \qquad (18.30)$$

which leads to response of the form

$$y(t) = [A_1 e^{-\alpha_1 t} + \cdots + A_{42} t e^{-\alpha_4 t} + A_{41} e^{-\alpha_4 t}$$
$$+ A_5 e^{-\alpha_5 t} + \cdots]u(t) \qquad (18.31)$$

Coefficient A_{42} can be obtained by eliminating all other A's by the method of case 1; i.e., by multiplying (18.30) by $(s + \alpha_4)^2$ and then letting $s + \alpha_4 = 0$, or $s = -\alpha_4$:

$$(s + \alpha_4)^2 Y(s) = \frac{K(s + a_1) \cdots}{(s + \alpha_1) \cdots (1)^2(s + \alpha_5) \cdots}$$
$$= \frac{(s + \alpha_4)^2 A_1}{(s + \alpha_1)} + \cdots + A_{42} + (s + \alpha_4)A_{41}$$
$$+ \frac{(s + \alpha_4)^2 A_5}{(s + \alpha_5)} + \cdots \qquad (18.32)$$

$$A_{42} = [(s + \alpha_4)^2 Y(s)]_{s=-\alpha_4} \qquad (18.33)$$

It is found, however, that this method will not work to isolate A_{41}; instead, another mathematical trick must be used to eliminate A_{42} but retain A_{41} in (18.30). In this case, the trick is to differentiate, with respect to s, both sides of (18.32):

$$\frac{d}{ds}[(s + \alpha_4)^2 Y(s)] = \frac{d}{ds}\left[\frac{K(s + a_1) \cdots}{(s + \alpha_1) \cdots (1)^2(s + \alpha_5) \cdots}\right]$$
$$= \frac{2(s + \alpha_4)(s + \alpha_1) - (s + \alpha_4)^2}{(s + \alpha_1)^2} A_1 + \cdots$$
$$+ 0 + (1)A_{41} + \frac{2(s + \alpha_4)(s + \alpha_5) - (s + \alpha_4)^2}{(s + \alpha_5)^2} A_5 + \cdots \qquad (18.34)$$

which eliminates A_{42}. Then, as before, $(s + \alpha_4)$ is made zero ($s = -\alpha_4$), whereby all unwanted terms on the right-hand side are eliminated, leaving

† This is easily proved by reconverting the right-hand side to a common denominator.

[Sec. 18.5] THE CASE OF REPEATED POLES: CASE TWO

an expression for A_{41}:

Partial fraction expansion for double pole
$$A_{41} = \left\{\frac{d}{ds}[(s+\alpha_4)^2 Y(s)]\right\}_{s=-\alpha_4} \quad (18.35)$$

In the more general case that a pole is repeated m times, the technique of differentiating is repeated $(m-1)$ times in order to obtain all the coefficients. That is, if

$$Y(s) = \frac{KN(s)}{(s+\alpha_r)^m(s+\alpha_2)\cdots} \quad (18.36)$$

then the proper partial fraction expansion is

$$Y(s) = \frac{A_{rm}}{(s+\alpha_r)^m} + \frac{A_{r(m-1)}}{(s+\alpha_r)^{m-1}} + \cdots + \frac{A_{ri}}{(s+\alpha_r)^i} + \cdots$$
$$+ \frac{A_{r1}}{s+\alpha_r} + \frac{A_2}{s+\alpha_2} + \cdots$$
$$= \sum_{i=1}^{m} \frac{A_{ri}}{(s+\alpha_r)^i} + \frac{A_2}{s+\alpha_2} + \cdots \quad (18.37)$$

which leads to the response

$$y(t) = \left\{\left[A_{rm}\frac{t^{m-1}}{(m-1)!} + A_{r(m-1)}\frac{t^{m-2}}{(m-2)!} + \cdots \right.\right.$$
$$\left.\left. + \frac{A_{ri}t^{i-1}}{(i-1)!} + \cdots + A_{r1}\right]e^{-\alpha_r t} + A_2 e^{-\alpha_2 t} + \cdots\right\} u(t)$$

which is displayed in closed form in (18.38) below. By repeated application of the differentiation "trick" of (18.35) each of the coefficients can be evaluated in turn. The general result is (18.39), below.

$$y(t) = \left\{\sum_{i=1}^{m} A_{ri}\frac{t^{(i-1)}}{(i-1)!}e^{-\alpha_r t} + A_2 e^{-\alpha_2 t} + \cdots\right\} u(t) \quad (18.38)$$

in which
$$A_{ri} = \frac{1}{(m-i)!}\left\{\frac{d^{(m-i)}}{ds^{(m-i)}}[(s+\alpha_r)^m Y(s)]\right\}_{s=-\alpha_r} \quad (18.39)$$

Example 18.4. Repeated pole. Find the time response of a system whose response function is

$$Y(s) = \frac{y_0}{s^4(s+3)} \quad (18.40)$$

The proper partial fraction expression is, by (18.39)

$$Y(s) = y_0\left[\frac{1}{s^4(s+3)}\right] = y_0\left[\frac{A_{14}}{s^4} + \frac{A_{13}}{s^3} + \frac{A_{12}}{s^2} + \frac{A_{11}}{s} + \frac{A_2}{s+3}\right] \quad (18.41)$$

from which, by (18.38), the time response is

$$y(t) = y_0\left[\left(A_{14}\frac{t^3}{3!} + A_{13}\frac{t^2}{2!} + A_{12}t + A_{11}\right)e^{0t} + A_2 e^{-3t}\right] u(t) \quad (18.42)$$

(In the above expressions the common term y_0 has been factored out, purely for convenience.)

594   TIME RESPONSE BY PARTIAL FRACTION EXPANSION [Chap. 18]

The A's are given by (18.39), as follows:

$$A_{14} = \left[\frac{1}{s+3}\right]_{s=0} = \frac{1}{3}$$

$$A_{13} = \left[\frac{d}{ds}\left(\frac{1}{s+3}\right)\right]_{s=0} = \left[\frac{-1}{(s+3)^2}\right]_{s=0} = -\frac{1}{9} \quad (18.43)$$

$$A_{12} = \frac{1}{2!}\left[\frac{d^2}{ds^2}\left(\frac{1}{s+3}\right)\right]_{s=0} = \frac{1}{2!}\left\{\frac{d}{ds}\left[\frac{-1}{(s+3)^2}\right]\right\}_{s=0} = \frac{1}{2!}\left[\frac{2}{(s+3)^3}\right]_{s=0}$$
$$= +\frac{1}{27}$$

$$A_{11} = \frac{1}{3!}\left[\frac{d^3}{ds^3}\left(\frac{1}{s+3}\right)\right]_{s=0} = \frac{1}{3!}\left\{\frac{d}{ds}\left[\frac{2}{(s+3)^3}\right]\right\}_{s=0} = \frac{1}{6}\left[\frac{-2(3)}{(s+3)^4}\right]_{s=0} = -\frac{1}{81}$$

$$A_2 = \left[\frac{1}{s^4}\right]_{s=-3} = \frac{1}{81}$$

Thus
$$y(t) = y_0\left[\left(\frac{t^3}{18} - \frac{t^2}{18} + \frac{t}{27} - \frac{1}{81}\right) + \frac{1}{81}e^{-3t}\right]u(t) \quad (18.44)$$

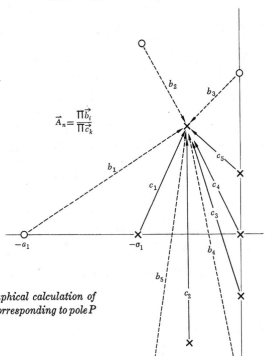

Fig. 18.3 *Graphical calculation of the coefficient corresponding to pole* P

[Sec. 18.6] SUMMARY OF PROCEDURE D-2

Check this result by substituting into the right-hand side of (18.41), then converting to a common denominator and comparing with (18.40).

Note that less labor is involved in carrying out the differentiations required by (18.39) than we might have feared, because each coefficient A_{ri} can be obtained by just a *single* differentiation of the previous coefficient, $A_{r(i-1)}$.

Problems 18.6 through 18.9

18.6 SUMMARY OF PROCEDURE D-2: PARTIAL FRACTION EXPANSION

The complete procedure for partial fraction expansion is summarized in the accompanying Procedure D-2.

The coefficients A in (18.46) through (18.51) can be rapidly calculated by graphical means from the pole-zero plot of $Y(s)$ in the s plane, as illustrated in Fig. 18.2 and Fig. 18.3. In many cases an acceptable estimate of each coefficient can be made by eye.

**Procedure D-2
Partial fraction expansion**

Given a response function in factored form,

$$Y(s) = \frac{K(s + a_1)(s + a_2) \cdots}{(s + \sigma_1)(s + \sigma_2) \cdots (s + \sigma_r)^m \cdots (s + \sigma_3 \mp j\omega_3) \cdots} \quad (18.45)$$

with the order of the denominator greater than the order of the numerator (see footnote, p. 585). The corresponding time response will contain terms of the form Ae^{st} for each pole or set of poles (denominator factors):

$$y(t) = \left\{ A_1 e^{-\sigma_1 t} + A_2 e^{-\sigma_2 t} + \cdots + \left[A_{rm} \frac{t^{m-1}}{(m-1)!} + \cdots + A_{ri} \frac{t^{i-1}}{(i-1)!} + \cdots + A_{r1} \right] e^{-\sigma_r t} \right.$$
$$\left. + 2|\vec{A}_{31}| e^{-\sigma_3 t} \cos(\omega_3 t + \underline{/\vec{A}_{31}}) + \cdots \right\} u(t) \quad (18.46)$$

in which the coefficients are given as follows:

(Procedure D-2.1)   For single poles (case one),

$$A_1 = [(s + \sigma_1) Y(s)]_{s=-\sigma_1} \quad (18.47)$$

(Procedure D-2.3) For multiple poles (case two),†

$$A_{ri} = \frac{1}{(m-i)!} \left\{ \frac{d^{(m-i)}}{ds^{(m-i)}} [(s+\sigma_r)^m Y(s)] \right\}_{s=-\sigma_r} \quad (18.48)$$

(Procedure D-2.2) For a complex pair of poles,

$$\vec{A}_{31} = [(s+\sigma_3-j\omega_3)Y(s)]_{s=-\sigma_3+j\omega_3}$$

$$= \frac{\Pi(\vec{b_i}'s)}{\Pi(\vec{c_k}'s)} \text{ in Fig. 18.3} \quad (18.49)$$

Multiple complex pairs are handled by the obvious combination of (18.48) and (18.49).
Common special cases include

double poles: $y(t) = (A_{r2}t + A_{r1})e^{-\sigma_r t}u(t)$
in which $\quad A_{r2} = [(s+\sigma_r)^2 Y(s)]_{s=-\sigma_r}$
$$A_{r1} = \left\{ \frac{d}{ds}[(s+\sigma_r)^2 Y(s)] \right\}_{s=-\sigma_r} \quad (18.50)$$

Procedure D-2
(Cont.)

imaginary complex pair:

$$y(t) = 2|\vec{A}_{31}| \cos(\omega_3 t + \underline{/\vec{A}_{31}})u(t) \quad (18.51)$$

in which $\quad \vec{A}_{31} = [(s-j\omega_3)Y(s)]_{s=j\omega_3}$

The response corresponding to two nearly equal real poles can be closely approximated by considering the poles to be equal, and applying (18.50).

† For computer programming it is often easier to clear fractions in Eq. (18.30), equate coefficients of like powers of s, and solve simultaneously the resulting set of algebraic equations.

Problems 18.10 and 18.11

Chapter 19

COMPLETE SYSTEM ANALYSIS: SOME CASE STUDIES

We have now completed our study of the separate stages involved in predicting comprehensively the dynamic behavior of a given physical system that can be represented (with sufficient accuracy) by ordinary, linear differential equations with constant coefficients (l.c.c. equations). The detailed procedures involved in each stage have been developed carefully in Parts A through D of this book. At this point let us summarize these procedures, after which we shall demonstrate their collective application in a number of examples.

The major stages of a dynamic investigation were first delineated in Sec. 1.2 and in Fig. 1.2. These are summarized in Fig. 19.1. Stage I is to imagine a physical model which is amenable to mathematical analysis, and whose dynamic behavior will match closely enough that of the actual system. Stage II is to derive the differential equations of motion—the mathematical model—for the chosen physical model. Further, whenever it is possible to do so with reasonable fidelity, the equations of motion will be approximated by a set of ordinary l.c.c. differential equations. Stages I and II were the principal concern of Part A.

Parts B, C, and D are concerned primarily with Stage III: studying the dynamic behavior of the mathematical model, that is, studying the system time response implied by the differential equations of motion. As indicated in Fig. 19.1, the specific analytical path we choose to follow will depend upon certain features of the problem at hand.

If it is sufficient to have a thorough understanding of the *stability of the system* and of the *character of the natural motions* it is capable of, without actually knowing the precise time response of the system, then the relatively simple procedures of Part C will give the desired information most directly, as suggested in Fig. 19.1. One first assumes that the natural motions of the system are of the form Ce^{st}, and then obtains the characteristic equation of the system by introducing this assumption into the differential equations of motion. (The assumption of Ce^{st} converts the differential equations into algebraic equations, from which the characteristic equation is readily obtained.) The characteristic equation can then be factored, either numerically (using trial and error if necessary), or by using the root-locus technique (Procedure C-2). The factors, or roots, of the characteristic equation then tell precisely the character of all possible natural motions of the system. Alternatively, if it is sufficient merely to know under what circumstances the system will be stable, without knowing the frequencies or decay rates of the natural motions, then the Routh stability criterion (Procedure C-4) may be applied. The *modes of natural motion* ("eigenvectors") of the system can also be learned simply by returning to the equations of motion in which solutions Ce^{st} have been substituted, and solving for the values C, as described in Chapters 12 and 13.

If it is required, on the other hand, to know the *response of the system to simple continuing inputs* of the form e^{st}, such as *continuing sinusoidal inputs*, then one may obtain the response by assuming that the *forced* motions are also of the form e^{st}, and then substituting this assumption into the equations of motion, as described in Chapters 14 and 15. The use of transfer functions greatly facilitates the algebra in such calculations. For the particular case of response to sinusoidal inputs, the results can be plotted efficiently by using logarithmic techniques (Chapters 9 and 15).

If, however, it is desired to predict completely the *time response* of a system to specific initial conditions and to abruptly commencing inputs (and if we are not interested in behavior before time $t = 0$), then the most efficient procedure is the one-sided Laplace transform method. The Laplace transform method systematically and automatically yields also all the other information of interest: the characteristic equation, the natural motions, and the response to continuing inputs. Many people therefore use it to the exclusion of all other analytical methods, even though it is often more elaborate than necessary for a given problem.

The complete procedure for a dynamic investigation using the *Laplace transform method* may be summarized as follows:

STAGE I.	Establish a PHYSICAL MODEL.	
STAGE II.	Derive the EQUATIONS OF MOTION (Procedures A).	**Procedure D-3** *Complete investigation using the Laplace transform method*
STAGE III.	Study the DYNAMIC BEHAVIOR (Procedure D-1, p. 537):	
	Step 1. **Laplace transform** the equations of motion.†	
	Step 2. **Manipulate** the resulting algebraic equations to obtain:	

† Transform equations of motion in their original form: do not differentiate first. (Footnote, p. 552.)

[Chap. 19] COMPLETE SYSTEM ANALYSIS: SOME CASE STUDIES 599

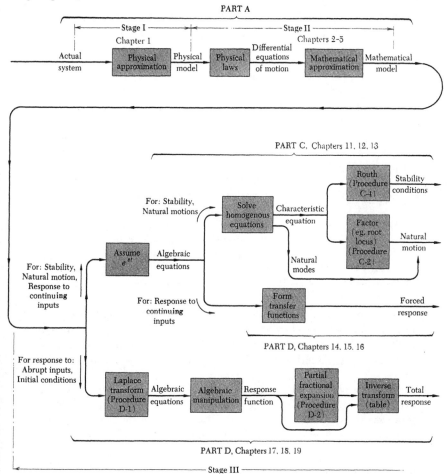

Fig. 19.1 *The stages of dynamic analysis of a physical system*

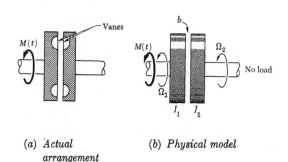

(a) Actual arrangement

(b) Physical model

Fig. 19.2 *Dynamical model of a fluid clutch*

(a) *Transfer functions* and *response functions* (block diagrams). (b) *The roots of the characteristic equation*; study *natural behavior* [e.g., using the root-locus technique, Procedure C-2 (p. 401), or Routh's method, Procedure C-4 (p. 410)]. (c) *Final and initial values* (FVT and IVT). Step 3. **Inverse transform** to obtain the time response [using a table of Laplace transform pairs or partial fraction expansion, Procedure D-2 p. 595)].	**Procedure D-3** *(Cont.)*
STAGE IV. Carry out SYSTEM DESIGN.	

Each element of the expression for time response will be either a simple exponential or a damped sinusoid (or, in singular cases, a power function of t). The response can therefore be readily plotted, term by term, using the methods of Fig. 6.3 for exponentials and Procedure B-3, p. 260, for damped sinusoids.

The remainder of this chapter will illustrate Procedure D-3 with a variety of "case-study" examples, several of which may act as bridges to more specialized studies. To concentrate attention on the sequence of steps, we begin with a "rerun" of a familiar problem.

19.1 A FLUID CLUTCH

The fluid clutch shown in Fig. 19.2a has a torque between its members which is proportional to their relative velocity. If an external torque $M = M_0 e^{-\sigma_f t} u(t)$ is applied abruptly to member 1, find the resulting motion of member 2. The members have initial velocities Ω_{10} and Ω_{20} at the instant just before the torque is applied. Bearing friction is negligibly small.

STAGE I. PHYSICAL MODEL. If bearing friction is neglected and clutch friction assumed ideally viscous, we have the linear physical model shown in Fig. 19.2b. This model is quite accurate so long as Ω_1 and Ω_2 are nearly equal and the observing time is relatively short. (Eventually, of course, small bearing friction will slow down the disk assembly.)

[Sec. 19.1] A FLUID CLUTCH

STAGE II. EQUATIONS OF MOTION. The equations of motion may be readily derived for this physical model (e.g., using Procedure A-m). They are

$$\begin{aligned} J_1\dot{\Omega}_1 + b(\Omega_1 - \Omega_2) &= M(t) \\ -b(\Omega_1 - \Omega_2) + J_2\dot{\Omega}_2 &= 0 \end{aligned} \quad (19.1)$$

in which we are given $M(t) = M_0 e^{-\sigma_f t} u(t)$.

STAGE III. DYNAMIC BEHAVIOR. (Part of what follows was also done in Example 17.7.)

Step 1. Laplace transform Eq. (19.1):

$$\begin{aligned} J_1[s\Omega_1 - \Omega_1(0^-)] + b(\Omega_1 - \Omega_2) &= \mathsf{M}(s) \\ -b(\Omega_1 - \Omega_2) + J_2[s\Omega_2 - \Omega_2(0^-)] &= 0 \end{aligned}$$

with $\quad \mathsf{M}(s) = M_0 \dfrac{1}{s + \sigma_f} \quad \Omega_1(0^-) = \Omega_{10} \quad \Omega_2(0^-) = \Omega_{20}$

Combining, and rearranging for convenience:

$$\begin{aligned} (J_1s + b)\Omega_1 - b\Omega_2 &= \frac{M_0}{s + \sigma_f} + J_1\Omega_{10} \\ -b\Omega_1 + (J_2s + b)\Omega_2 &= J_2\Omega_{20} \end{aligned} \quad (19.2)$$

Step 2. Manipulate to obtain:
(a) *The response function of interest.*
We use Cramer's rule to solve algebraically for Ω_2

$$\Omega_2 = \frac{b\left[\dfrac{M_0}{s + \sigma_f} + J_1\Omega_{10}\right] + (J_1s + b)J_2\Omega_{20}}{(J_1s + b)(J_2s + b) - b^2}$$

$$\Omega_2 = \frac{\dfrac{M_0 b}{J_1 J_2}\dfrac{1}{(s + \sigma_f)} + \dfrac{b}{J_2}\Omega_{10} + \left(s + \dfrac{b}{J_1}\right)\Omega_{20}}{s^2 + \left(\dfrac{b}{J_1} + \dfrac{b}{J_2}\right)s} \quad (19.3)$$

(Block diagrams for this system are given in Fig. 17.8.)

(b) *Natural characteristics.* The characteristic equation can be obtained from the denominator of response function (19.3) or the determinant of (19.2). In factored form, it is

$$s\left[s + \left(\frac{b}{J_1} + \frac{b}{J_2}\right)\right] = 0 \quad (19.4)$$

Since b is by nature a positive quantity, the system is always stable. Its possible natural motions include a steady angular velocity ($s = 0$) and an exponential decay with time constant $1/[(b/J_1) + (b/J_2)]$.

(c) *Final and initial values.* We first note that (19.3) has no poles on the $j\omega$ axis. We may therefore apply the Final-Value Theorem (17.45). This yields

$$\Omega_2(\infty) = \lim_{s\to 0} [s\Omega_2(s)] = \frac{(M_0 b/J_1 J_2 \sigma_f) + (b/J_2)\Omega_{10} + (b/J_1)\Omega_{20}}{(b/J_1) + (b/J_2)} \quad (19.5)$$

The second and third terms in (19.5) give the (common) equilibrium speed to which the two disks would coast if they had different initial speeds and were undisturbed. The first term gives the final (common) speed the disks would have if they had been at rest when $M(t)$ was applied. [We know both disks will have the same "final" speed; otherwise any relative speed would prevent steady state from obtaining. The reader may verify this by solving (19.2) also for $\Omega_1(s)$ and applying (17.45).]

Applying the Initial-Value Theorem (17.46) to response function (19.3) yields

$$\Omega_2(0^+) = \lim_{t\to 0^+} \Omega_2(t) = \lim_{s\to\infty} s\Omega_2(s) = \Omega_{20} = \Omega_2(0^-) \quad (19.6)$$

We expect the result $\Omega_2(0^+) = \Omega_2(0^-)$ because no impulsive torques act on disk 2, and an inertia will not otherwise change its speed instantaneously. [As a special case that is physically obvious, suppose the disks had equal initial velocities and were undisturbed. Then their velocities would not change, and $\Omega_2(\infty)$ should equal $\Omega_2(0^-)$. Equation (19.5) is in agreement with this conclusion also.]

Step 3. Inverse transform the response function to obtain the time response. The first term in (19.3) may be written

$$G_1(s) = \left(\frac{M_0}{J_1}\frac{b}{J_2}\right)\frac{1}{s(s+\sigma)(s+\sigma_f)}$$

in which $\sigma = (b/J_1) + (b/J_2)$. This function appears in most tables of Laplace transforms (e.g., the table in Appendix J), but let us use partial fraction expansion (Procedure D-2.1) here for demonstration purposes. The time response corresponding to $G_1(s)$ [refer to (18.46)] is

$$\frac{M_0}{J_1}\frac{b}{J_2}(A_0 + A_1 e^{-\sigma t} + A_f e^{-\sigma_f t})$$

in which, applying (18.47), we have

$$A_0 = [sG_1(s)]_{s=0} = \frac{1}{\sigma\sigma_f} \qquad A_1 = [(s+\sigma)G_1(s)]_{s=-\sigma} = \frac{1}{-\sigma(-\sigma+\sigma_f)}$$

$$A_f = [(s+\sigma_f)G_1(s)]_{s=-\sigma_f} = \frac{1}{-\sigma_f(-\sigma_f+\sigma)}$$

The corresponding response for this part of (19.3) is therefore

$$\frac{M_0 b}{J_1 J_2}\left[\frac{1}{\sigma\sigma_f} + \frac{1}{\sigma(\sigma-\sigma_f)}e^{-\sigma t} - \frac{1}{\sigma_f(\sigma-\sigma_f)}e^{-\sigma_f t}\right]u(t) \quad (19.7)$$

The other two terms in (19.3) are also readily inverse transformed. [They

[Sec. 19.1] A FLUID CLUTCH

both appear in Table X, for example, if the numerator term $s + (b/J_1)$ is separated into s and (b/J_1).] The corresponding time functions are

$$\left\{\frac{b}{J_2}\Omega_{10}\frac{1-e^{-\sigma t}}{(b/J_1)+(b/J_2)} + \Omega_{20}\left[e^{-\sigma t} + \frac{(b/J_1)(1-e^{-\sigma t})}{(b/J_1)+(b/J_2)}\right]\right\} u(t) \quad (19.8)$$

Combining (19.7) and (19.8) gives the total response

$$\Omega_2(t) = \frac{M_0 b}{J_1 J_2}\left[\frac{1}{\sigma\sigma_f} + \frac{1}{\sigma-\sigma_f}\left(\frac{e^{-\sigma t}}{\sigma} - \frac{e^{-\sigma_f t}}{\sigma_f}\right)\right]$$
$$+ \frac{b}{J_2\sigma}\Omega_{10}(1-e^{-\sigma t}) + \Omega_{20}\left(\frac{b}{J_1\sigma} + \frac{b}{J_2\sigma}e^{-\sigma t}\right) \qquad 0 < t \quad (19.9)$$

where $\sigma = (b/J_1) + (b/J_2)$. The terms in the [ ] brackets constitute the total response to $M(t)$, while the other terms, of course, give the response following from initial wheel speeds Ω_{10} and Ω_{20}.

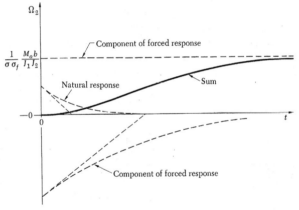

(a) *Response to applied torque $M(t)$*

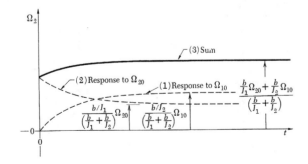

(b) *Response to initial wheel speeds $\Omega_2(0)$ and $\Omega_1(0)$*

Fig. 19.3 *Total response of the system of Fig. 19.2*

To spot-check (19.9), we note first the value of Ω_2 at $t = 0$, as given by (19.9) [this will be $\Omega_2(0^+)$ because (19.9) holds only for $0 < t$]:

$$\Omega_2(0^+) = [0] + 0 + \Omega_{20}$$

which checks the known initial condition and (19.6). At the other extreme, we note that as $t \to \infty$, (19.9) gives

$$\Omega_2(\infty) = \frac{M_0 b}{J_1 J_2 \sigma \sigma_f} + \frac{(b/J_2)\Omega_{10} + (b/J_1)\Omega_{20}}{(b/J_1) + (b/J_2)} \tag{19.10}$$

which checks (19.5).

If numerical values are known, the various terms in (19.9) can be quite readily plotted, as in Fig. 19.3. (There are only three kinds of response in this problem: constant terms, $e^{-\sigma_f t}$ terms, and $e^{-\sigma t}$ terms.) The total response to $M(t)$ [the first line of (19.9)], and the elements of which it is composed, are plotted in Fig. 19.3a. A ratio of $\sigma = 3\sigma_f$ was used for illustration. This would be the total response if neither disk had an initial velocity. Note that $\Omega_2(0)$ is zero in Fig. 19.3a, as it should be. [So is $\dot{\Omega}_2(0)$ in this particular case.]

In Fig. 19.3b are plotted (1) the response of disk 2 to an initial velocity of disk 1, (2) the response of disk 2 to an initial velocity of disk 2, and (3) the sum (superposition) of responses (1) and (2) [second line of (19.9)]. Ratios of $\Omega_{10} = 1.5\Omega_{20}$ and $J_1 = J_2$ were used for illustration. If $\Omega_2(0) = 0$ and

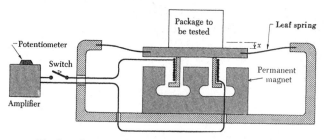

(a) *Actual arrangement*

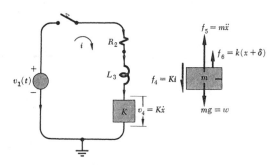

(b) *Physical model*

Fig. 19.4 *Dynamical model of an electromechanical shaker*

[Sec. 19.2] AN ELECTROMECHANICAL SHAKER 605

$\Omega_1(0) \neq 0$, then disk 2 speeds up and disk 1 slows down until the two disks are rotating at the same speed, and vice versa. The speeding up and slowing down occurs with time constant $1/\sigma$.

The total response to $M(t)$ plus initial conditions is of course the superposition of the heavy curves in Figs. 19.3a and b. Note that no response is plotted before $t = 0$, because (19.9) holds only for $0 < t$.

19.2 AN ELECTROMECHANICAL SHAKER

Figure 19.4a shows schematically a shaker used in the vibration testing of flight hardware. The system is capable of providing a precisely known vibratory input to a test package mounted on it, as shown in the figure. In performing a test, a sinusoidal signal (from a sine-wave generator) is amplified and applied to the driver coil. Typically, the operator turns "down" the potentiometer, closes the switch, sets the frequency of the input signal, and then turns the potentiometer "up" gradually until the vibration level (as measured, for example, by an accelerometer on the package) has the desired value. Then he records his data.

Suppose, however, that the operator fails to turn the potentiometer down before closing the switch. Then when the switch is closed the system will be subjected to abrupt transient motion as well as the intended continuing sinusoidal motion. We are asked to compute the total motion in this case, to see whether damage may result.

Specifically, then, we assume that the driving circuit is subjected (in the worst case) to the abruptly commencing input voltage

$$v_1 = v_{10}u(t) \cos \omega_f t \qquad (19.11)$$

Suppose that the following approximate numerical values are known:

Natural frequency of mass on leaf spring (with switch open),

$$\sqrt{\frac{k}{m}} = 300 \text{ rad/sec}$$

Natural time constant of coil circuit (with mass clamped),

$$\frac{L}{R} = \frac{1}{600} \text{ sec}$$

Electromechanical coupling coefficient of coil, $K^2/mL = 4 \times 10^5 \text{ sec}^{-2}$
Input frequency, $\omega_f = 150$ rad/sec
Input voltage level, such that $(Kv_{10}/kR) = 10^{-3}$ meter $= 0.1$ cm

The last combination of variables is significant because it is the steady displacement of the shaker produced by a constant voltage v_{10}, as is to be proved in Problem 19.28. (The reader should check to see that this term has the dimensions of length.)

To perform the analysis, we follow Procedure D-3.

COMPLETE SYSTEM ANALYSIS: SOME CASE STUDIES [Chap. 19]

STAGE I. PHYSICAL MODEL. Following the discussion of Sec. 2.9 (page 79) we adopt the physical model shown in Fig. 19.4b, representing the coil as an inductor and resistor in series, with an additional voltage drop v_4 due to velocity of the current-carrying coil moving through the magnetic field of the permanent magnet, and representing the mechanical subsystem as just a mass and spring driven by a force proportional to current i. As we shall see, the total system has substantial damping due to the resistor, and thus the small mechanical damping due to windage and internal friction in the leaf spring may be neglected by comparison. (If, at the end of our analysis, we were to find that the overall system had negligible damping, then we would have to return and amend our physical model to include the mechanical damping.)†

STAGE II. EQUATIONS OF MOTION. We follow Procedure A-em (p. 82).

(1) *Variables and geometry.* The system is shown in an arbitrary configuration in Fig. 19.4a. For the network, we select current i as the independent variable, and plan to use a loop analysis. The mechanical variable of interest is displacement x (and its derivatives). We measure x from the position of static equilibrium where there will be a static displacement δ. Thus the total displacement of the leaf springs from their zero-strain condition is $x + \delta$. The electrical constraint is that $v_1(t)$ is prescribed by (19.11).

(2) *Kirchhoff's laws and force equilibrium.* (Refer to Fig. 19.4b.)

For loop shown, KVL: $\qquad \Sigma v = 0 = v_2 + v_3 + v_4 - v_1$
For free body shown, (2.1): $\quad -\Sigma f_x^* = 0 = -f_4 + f_5 + f_6 - w$ $\qquad$ (19.12)

(3) *Constitutive physical relations.* These are shown in Fig. 19.4b. Substituting them into (19.12), and dropping the term $k\delta\text{-}w = 0$, we obtain

$$\begin{aligned} R_2 i + L_3 \frac{di}{dt} + K\dot{x} &= v_1(t) \\ -Ki + m\ddot{x} + kx &= 0 \end{aligned} \qquad (19.13)$$

STAGE III. STUDY DYNAMIC BEHAVIOR. We follow Procedure D-1.

Step 1. Laplace transform the equations of motion. (Subscripts on R and L are dropped here.) In the matrix form:

$$\begin{bmatrix} (Ls + R) & Ks \\ -K & (ms^2 + k) \end{bmatrix} \begin{bmatrix} I \\ X \end{bmatrix} = \begin{bmatrix} V_1 + Li(0^-) + Kx(0^-) \\ msx(0^-) + m\dot{x}(0^-) \end{bmatrix} \qquad (19.14)$$

Step 2. Manipulate algebraically to obtain:
(a) *Transfer functions and response function.*

† The system of Fig. 19.4a and b could also represent, qualitatively and approximately, a speaker such as Fig. 2.23, and the analysis that follows is correspondingly applicable to a preliminary study of a speaker. As we have noted elsewhere, however, experience shows that a good speaker relies heavily on the dynamics of its air enclosure, so that a more complicated model than Fig. 19.4b would be required for its accurate study.

[Sec. 19.2] AN ELECTROMECHANICAL SHAKER

If initial conditions are all zero (as they will be for the problem given), the transfer function and response function for x are

transfer function $\quad \dfrac{X}{V_1} = \dfrac{K}{[(ms^2 + k)(Ls + R) + K^2 s]} \quad$ (19.15)

response function $\quad X = \dfrac{K v_{10} s}{(s^2 + \omega_f{}^2)[(ms^2 + k)(Ls + R) + K^2 s]} \quad$ (19.16)

To find time response from (19.16) we shall first have to factor its denominator, which we do in our study of natural characteristics.

(b) *Natural behavior.* The characteristic equation of the system [from the determinant of (19.14) or from the denominator of (19.15)] is

$$\left(s^2 + \frac{k}{m}\right)\left(s + \frac{R}{L}\right) + \frac{K^2}{mL} s = 0 \quad (19.17)$$

Since we have values for k/m, R/L, and K^2/mL, we can solve for the roots of (19.17), to obtain the natural dynamic characteristics of the system.

The quick way to obtain the roots of (19.17) (quicker to do than to describe) is to sketch the locus of roots, following Evans' method, Procedure C-2. In the process we can also learn some interesting things about the design of a shaker. It is convenient to use K^2/mL, the physical coupling, as a variable. We write (19.17) in the Evans form:

$$C(s) = \frac{\left(s^2 + \dfrac{k}{m}\right)\left(s + \dfrac{R}{L}\right)}{s} = -\frac{K^2}{mL} \quad (19.18)$$

Next, we plot the numerator values of $C(s)$ as ×'s and the denominator terms as ○'s in Fig. 19.5. Then, considering first only the angle condition $\underline{/C(s)} = 180°$ in (19.18), we apply Procedure C-3, p. 405, to sketch the locus of roots. From Rule 1 there is a real-axis locus between $s = -R/L$ and the origin. From Rule 2 there are loci asymptotic to a vertical line through the point $s = -R/2L$. From Rule 3 the loci depart the complex ×'s in the directions shown. The loci can be sketched rather accurately without further information. (Actually, to improve the accuracy of the locus in Fig. 19.5, one point on it, point P, was computed by "trial and error," using a Spirule to add the angles. Point P, where the angles sum to 180°, was found in two trials.)

Finally, to find the points on the loci where the *magnitude* condition $|C(s)| = K^2/mL = 4 \times 10^5$ in (19.18) is satisfied, we measure the constituent vector lengths and multiply [as indicated by (19.18)] for several points on each locus until, by trial and error, we find the points (1) and (2) where the magnitude of $|C(s)|$ is 4×10^5. (Three trials each should suffice.)

The result, shown in Fig. 19.5, tells the following story: (i) If there is no electromechanical coupling ($K = 0$), then the mechanical subsystem behaves as if the switch were open ($s = \pm j \sqrt{k/m}$) and the electrical sub-

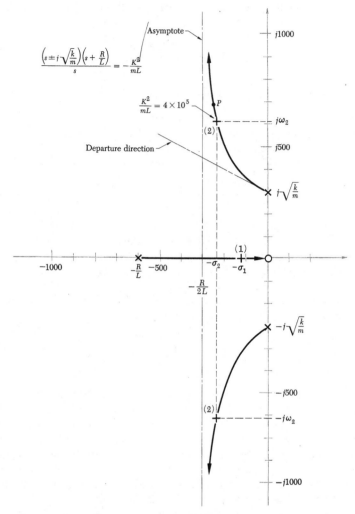

Fig. 19.5 *Locus of the roots of Eq.* (19.17): *The dynamic characteristics of the electromechanical shaker*

system behaves as if the mass were clamped ($s = -R/L$). (ii) With coupling present, one natural motion of the system is a *damped* sinusoid having frequency higher than $\sqrt{k/m}$, and the other natural motion is an exponential decay having a time constant slower than L/R. The stronger the coupling, the higher the frequency of the oscillatory natural motion and the slower the time constant of the exponential natural motion. (iii) Resistor R is directly responsible for damping the oscillatory motion of the coupled system. (What would the locus look like, in Fig. 19.5, if $R = 0$?) The reason is, of course,

[Sec. 19.2] AN ELECTROMECHANICAL SHAKER

that resistor R is the only energy-dissipation element in the entire physical system model, Fig. 19.4b.

Figure 19.5 would be most helpful in designing a shaker for a given job—i.e., in choosing the values K and R. If the shaker is built, and K is already fixed, but R is still at the designer's disposal, then it will be more useful to plot the locus of roots versus R, as is suggested in Problem 19.29.

(c) *Final and initial values.* Before proceeding to obtain the complete time response from (19.16), we see whether quick application of the Final- and Initial-Value Theorems can tell us anything useful. Since the denominator of (19.16) contains the term $s^2 + \omega_f^2$, the final value of $x(t)$ will not be defined (it will be oscillatory). The Initial-Value Theorem gives the following:

$$x(0^+) = \lim_{s \to \infty} [sX(s)] = \left\{ \frac{Kv_{10}s^2}{(s^2)[mLs^3]} \right\}_{s=\infty} = 0 \qquad (19.19)$$

Physically, the abrupt application of voltage v_1 produces a force on mass m which is finite, and therefore m does not displace instantaneously. What can you deduce about the initial value of $\dot{x}$ by applying the IVT to a modified version of (19.16)? What about $\ddot{x}$? What physical role is played by the inductor during the starting transient?

Step 3. Inverse transform to obtain the time response. We insert the roots of the characteristic equation [obtained in Step 2(b) above] as the denominator factors of (19.16):

$$X = \frac{(Kv_{10}/mL)s}{(s \mp j\omega_f)[(s + \sigma_1)(s + \sigma_2 \mp j\omega_2)]} \qquad (19.20)$$

Then we use partial fraction expansion, Procedure D-2, to obtain the following expression for the total time response of the system to the given input:

$$x(t) = [2|\vec{A}_{f1}| \cos (\omega_f t + \underline{/\vec{A}_{f1}}) + A_1 e^{-\sigma_1 t} + 2|\vec{A}_{21}| e^{-\sigma_2 t} \cos (\omega_2 t + \underline{/\vec{A}_{21}})]u(t) \qquad (19.21)$$

in which the A's are given by

$$2\vec{A}_{f1} = \frac{Kv_{10}}{kR} \left[\frac{2s\sigma_1\omega_{02}^2}{(s + j\omega_f)(s + \sigma_1)(s + \sigma_2 \mp j\omega_2)} \right]_{s=j\omega_f}$$

$$= \frac{Kv_{10}}{kR} \left[\frac{\sigma_1\omega_{02}^2}{bac} \underline{/-\phi_b - \phi_a - \phi_c} \right] = 0.054 \underline{/298°} \text{ cm}$$

$$A_1 = \frac{Kv_{10}}{kR} \left[\frac{8s\sigma_1\omega_{02}^2}{(s \mp j\omega_f)(s + \sigma_2 \mp j\omega_2)} \right]_{s=-\sigma_1}$$

$$= \frac{Kv_{10}}{kR} \left[\frac{-\sigma_1^2\omega_{02}^2}{b^2d^2} \right] = -0.037 \text{ cm}$$

$$2\vec{A}_{21} = \frac{Kv_{10}}{kR} \left[\frac{2s\sigma_1\omega_{02}^2}{(s \mp j\omega_f)(s + \sigma_1)(s + \sigma_2 + j\omega_2)} \right]_{s=-\sigma_2+j\omega_2}$$

$$= \frac{Kv_{10}}{kR} \left[\frac{\sigma_1\omega_{02}^3}{aed\omega_2} \underline{/\phi_2 - \phi_a' - \phi_e - \phi_d - \pi/2} \right]$$

$$= 0.022 \underline{/58°} \text{ cm}$$

Pole-zero array for $\vec{A}_{f1}$ (also A_1, A_2)

where the numerical values are obtained by using distances and angles measured from poles (1) and (2) in Fig. 19.5. Substituting these values into (19.21), we have

$$x(t) = [0.054 \cos (150t + 298°) - 0.036e^{-120t} + 0.022e^{-230t} \cos (610t + 58°)]u(t) \text{ cm} \quad (19.22)$$

The steady-state motion is given by the first term, and the remaining two terms make up the starting transient due to closing the switch. [As a check, show that (19.22) gives $x(0^+) = 0$.] The starting transient somewhat resembles Fig. 8.25, and the reader is invited to plot it. But even without plotting the response, we now see that the total motion might be at most about twice the intended steady-state motion [the first term in (19.22)], and this is the report we would make in answer to the question posed in the statement of the problem.

A more comprehensive analysis would determine x_{max}/x_{allow} over the whole spectrum of frequencies.

19.3 A THERMAL QUENCHING OPERATION

One process in the production of steel involves the quenching (sudden cooling) of cherry-red-hot ingots in a bath that is initially at about room temperature, as depicted in Fig. 19.6. The ingots are, typically, about 8 feet tall and 2 feet square, and the bath is perhaps 12 feet deep and 6 feet square.

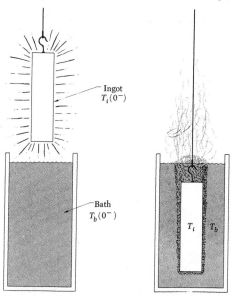

Fig. 19.6 *The dynamics of ingot quenching*

[Sec. 19.3] A THERMAL QUENCHING OPERATION   611

The rate of cooling is, of course, crucial to the quality of the steel. It is therefore desired to estimate the time history of ingot temperature.

STAGE I. PHYSICAL MODEL. Any precise model of the system would have to include the distribution of temperature within the ingot and within the bath. That is, the capacity of the ingot to store heat is distributed throughout the ingot. A distributed-heat-capacity model would lead, of course, to partial differential equations just as, for example, the distributed-mass (distributed-kinetic-energy-storage) model of a vibrating string leads to partial differential equations (Sec. 13.2). At some point in the detailed investigation of the ingot-quenching process, such a model must be developed and studied if the effect of the quenching is to be understood at the microscopic level.

In this study, however, our task is to make a first quick approximate estimate of the time history of the *average* ingot temperature, and for this we use a much simpler lumped-capacity model. That is, we assume that at any instant the temperature is the same everywhere in the ingot. Similarly, we assume that at any instant a single (different) uniform temperature obtains everywhere in the bath. Heat transfer between the bath and the room is neglected.

We realize that our study with this model will give us only an approximate picture of the system dynamic behavior, but this lumped-parameter analysis is simple and quick, and would normally precede any more elaborate analysis. A somewhat closer approximation to the actual situation can be made by modeling the ingot as several lumped heat capacities instead of one. These would be separated from one another by imaginary conducting cylindrical layers, as in Fig. 3.3. The bath would be modeled in a similar way. A model of this type is suggested in Problem 3.3.

In our model we assume that, during the quenching operation, circulation and convection in the bath are neglected. (It is as if the bath were made of jelly, so far as heat transfer is concerned.) When we have obtained an estimate of the quenching time, we can compare it with the time for circulation and convection to see whether we have made a good approximation.

As we derive the equations of motion of the system, we shall see that we need certain physical constants to complete our physical model.

STAGE II. EQUATION OF MOTION

(1) *Geometry*. We first select thermal variables. For the lumped-parameter model chosen, the choice is simple:† We choose as variables the temperatures T_i of the ingot and T_b of the bath, the rate of heat flow q from ingot to bath (we arbitrarily take q positive when flow is from ingot to bath), and the total heat (in calories or Btu's) $C_i T_i$ and $C_b T_b$ contained in the ingot and in the bath, respectively, at a given instant. There are no geometric constraints.

† In a distributed-parameter model this selection would have to be done with much more care: A coordinate system would be defined, and temperature would be a continuous function of the coordinates.

(2) *Equilibrium.* (i) Consider the ingot and bath as subsystems. (ii) Conservation of thermal energy requires that heat flow q decrease the heat in the ingot at rate q and increase the heat in the bath at rate q:

$$-C_i \dot{T}_i = q = C_b \dot{T}_b \tag{19.23}$$

(3) *Physical relations.* It is well established (by exhaustive experiment) that the rate of heat flow between two (lumped) thermal storage elements is simply proportional to their temperature difference:

$$q = \frac{1}{R}(T_i - T_b) \tag{19.24}$$

The heat-transfer resistance R depends on the area and nature of the interface where the ingot surface is exposed to the bath. This constant would be obtained empirically or estimated from the storehouse of measured heat-transfer coefficients to be found in the heat-transfer literature.

Combining relations, we obtain

$$\begin{aligned} C_i \dot{T}_i &= -\frac{1}{R}(T_i - T_b) \\ C_b \dot{T}_b &= \frac{1}{R}(T_i - T_b) \end{aligned} \tag{19.25}$$

STAGE III. DYNAMIC BEHAVIOR. We follow Procedure D-1.

Step 1. Laplace transform the equations of motion.

$$\begin{aligned} \left(s + \frac{1}{RC_i}\right) \mathsf{T}_i - \frac{1}{RC_i} \mathsf{T}_b &= T_i(0^-) \\ -\frac{1}{RC_b} \mathsf{T}_i + \left(s + \frac{1}{RC_b}\right) \mathsf{T}_b &= T_b(0^-) \end{aligned} \tag{19.26}$$

Step 2. Manipulate algebraically to obtain:

(b) *Natural behavior.*† The characteristic equation is obtained from the determinant of the left-hand sides of Eqs. (19.26):

$$\left(s + \frac{1}{RC_i}\right)\left(s + \frac{1}{RC_b}\right) - \frac{1}{R^2 C_i C_b} = 0$$

or
$$s\left(s + \frac{1}{RC_i} + \frac{1}{RC_b}\right) = 0 \tag{19.27}$$

from which the characteristics of natural behavior are

$$s = 0 \qquad s = -\frac{1}{RC_i} - \frac{1}{RC_b} \triangleq -\sigma \tag{19.28}$$

That is, the possible natural behavior of either temperature T_i or T_b is that it may remain constant or that it may change exponentially with the time constant $1/\sigma$.

(a) The *response function* for T_i is obtained from (19.26) by Cramer's rule:

$$\mathsf{T}_i = \frac{[s + (1/RC_b)]T_i(0^-) + (1/RC_i)T_b(0^-)}{s[s + (1/RC_i) + (1/RC_b)]} \tag{19.29}$$

† It is sometimes more convenient to invert the order of steps.

[Sec. 19.3] A THERMAL QUENCHING OPERATION

(c) *Initial and final values.* Before obtaining the complete time response, it is interesting to apply the *Initial- and Final-Value Theorems.* The result is

$$T_i(0^+) = \lim_{s \to \infty} [sT_i(s)] = T_i(0^-) \tag{19.30}$$

$$T_i(\infty) = \lim_{s \to 0} [sT_i(s)] = \frac{C_i T_i(0^-) + C_b T_b(0^-)}{C_i + C_b} \tag{19.31}$$

That is, the initial value of the temperature of the ingot does not change between $t = 0^-$ and $t = 0^+$, and the final value of the ingot temperature is a weighted average of the initial temperatures of ingot and bath, which is intuitively clear. (In particular, if the ingot and the bath happened to have the same thermal capacity, $C_i = C_b$, the final ingot temperature would simply be the average of the initial temperatures of ingot and bath.)

Step 3. Inverse transform to obtain the time response. The response is, from (19.29),

$$\boxed{T_i = \left[\frac{C_i T_i(0^-) + C_b T_b(0^-)}{C_i + C_b} + \frac{C_b T_i(0^-) - C_b T_b(0^-)}{C_i + C_b} e^{-\sigma t} \right] u(t)} \tag{19.32}$$

which is plotted in Fig. 19.7. It is left as an exercise to show that the bath temperature varies in the following way:

$$T_b = \left[\frac{C_i T_i(0^-) + C_b T_b(0^-)}{C_i + C_b} - \frac{C_i T_i(0^-) - C_i T_b(0^-)}{C_i + C_b} e^{-\sigma t} \right] u(t) \tag{19.33}$$

which is also plotted in Fig. 19.7. That is, the temperatures of the two heat-storage elements approach one another exponentially. In practice the ingot would doubtless be withdrawn from the bath after just a few time constants, and the bath would be rapidly recirculated to reduce its temperature again to a level near room temperature, in preparation for the next ingot.

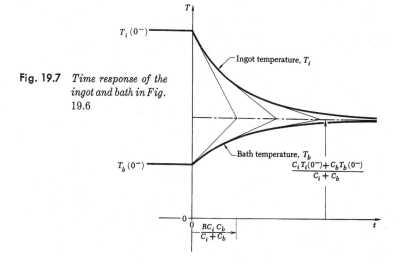

Fig. 19.7 *Time response of the ingot and bath in Fig. 19.6*

The quantity of primary interest is, of course, the time constant

$$\frac{1}{\sigma} = \frac{RC_i C_b}{C_i + C_b} \qquad (19.34)$$

It is physically logical that the time constant will be made shorter by making the thermal capacity of the two elements smaller and by making the heat-transfer resistance R smaller (by improving thermal contact and increasing the area of contact). On a much longer time scale, of course, the "final" temperature of ingot-plus-bath in Fig. 19.7 will slowly decrease toward that of the surroundings as heat leaks through the resistance of the bath walls.

The reader may note with some interest that the dynamic behavior of the thermal system just analyzed is identical to that of the fluid-clutch system of Sec. 19.1, and also to the network of Example 17.8 (p. 565). Figure 19.7 depicts the unforced behavior of all three systems. (The correspondence of physical variables is clear.)

19.4 AN AIRCRAFT HYDRAULIC SERVO

The hydraulic servomechanism for controlling the position of a rocket engine or an aerodynamic control surface has been studied earlier (see pp. 118 and 394). In this section we present a complete analysis, following Procedure D-3.

A hydraulic elevator-control system is shown in Fig. 19.8. This system is operated by the pilot, responding to displacement x of his stick, and magnifying his effort many-fold. It serves the same function as power steering in an automobile. Suppose, for illustration, that we wish to find the response to a step input from the stick, $x = x_0 u(t)$, and that the initial elevator deflection is $\epsilon(0^-) = 0$.

STAGE I. PHYSICAL MODEL. We assume that the oil is incompressible. (The compressible-fluid case is discussed in Problems 3.23 and 22.8.) We assume that linkages are rigid, that there is no leakage, and that flow through

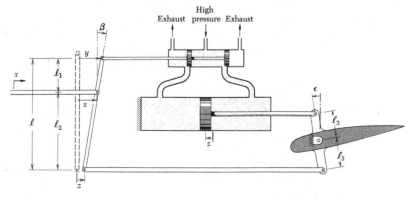

Fig. 19.8 *Aircraft manual "boost" hydraulic servo*

[Sec. 19.4] AN AIRCRAFT HYDRAULIC SERVO 615

the valve is simply proportional to valve displacement y. (In practice the orifices in the valve can be so shaped that this proportional relationship is quite faithfully followed until the valve is completely open.)

STAGE II. EQUATIONS OF MOTION. We follow Procedure A-f.

(1) *Geometry.* The configuration of the system can be defined with two coordinates, such as control-stick deflection x and elevator position ϵ. That is, the system has two degrees of freedom, of which one will be specified as a function of time. Then with the system in the arbitrary configuration shown (much exaggerated) in Fig. 19.8, the following geometric identities can be written:

$$z \equiv \ell_3 \sin \epsilon \approx \ell_3 \epsilon \tag{19.35}$$

$$y \equiv \frac{\ell}{\ell_2} x - \frac{\ell_1}{\ell_2} z \tag{19.36}$$

(The latter relation can be deduced, for example, by noting that when $z = 0$, $\sin \beta = y/\ell = x/\ell_2$; and when $x = 0$, $\sin \beta = y/\ell_1 = -z/\ell_2$.) In addition, continuity of flow q requires the following, for the incompressible case [from Eq. (3.50) of Example 3.4, p. 120].

$$A\dot{z} = q = Cy \tag{19.37}$$

or

$$\dot{\epsilon} \cong \frac{C}{A\ell_3} y \tag{19.38}$$

where A is the piston area and C is the valve constant.

(2) *Equilibrium.* No relation is required because, for the model assumed, the motion of the system is independent of the forces exerted.

(3) *Physical force-geometry relations.* None are required: Since forces are not involved in determination of the motion, force-geometry relations are not required either.

Combining the above relations, we have

$$\dot{\epsilon} = \frac{C}{A\ell_3} \left(\frac{\ell}{\ell_2} x - \frac{\ell_1}{\ell_2} z \right)$$

or

$$\dot{\epsilon} + \left(\frac{C\ell_1}{A\ell_2} \right) \epsilon = \frac{C\ell}{A\ell_2 \ell_3} x \tag{19.39}$$

STAGE III. DYNAMIC BEHAVIOR. We determine the response of the system, using Procedure D-1.

Step 1. Laplace transform the equation of motion:

$$(s + \sigma)\mathrm{E} = \frac{C\ell}{A\ell_2\ell_3} X + \epsilon(0^-) \tag{19.40}$$

with

$$X = \frac{x_0}{s} \qquad \sigma = \frac{C\ell_1}{A\ell_2} \qquad \epsilon(0^-) = 0$$

where $\mathrm{E}(s)$ is the Laplace transform of $\epsilon(t)$.

Step 2. Manipulate to obtain:

(a) *The desired response function:*

$$\mathrm{E} = \frac{C\ell/A\ell_2\ell_3}{s + \sigma} X \tag{19.41}$$

$$= \frac{(C\ell/A\ell_2\ell_3)x_0}{(s + \sigma)s} \tag{19.42}$$

Step 3. Inverse transform to obtain the response of the system:

$$\epsilon(t) = \frac{C\ell}{A\ell_2\ell_3\sigma} x_0(1 - e^{-\sigma t})u(t) = \frac{\ell}{\ell_3\ell_1} x_0(1 - e^{-\sigma t})u(t) \quad (19.43)$$

This response is plotted in Fig. 19.9.

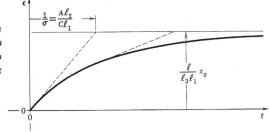

Fig. 19.9 *Response of the hydraulic servo of Fig. 19.8 to a step change in pilot's stick position*

STAGE IV. SYSTEM DESIGN

Obviously, the crucial term in the dynamic response (19.43) is the time constant

$$\frac{1}{\sigma} = \frac{A\ell_2}{C\ell_1}$$

Typically, the linkage lengths are pretty much dictated by the mechanical design, and thus the speed of response depends upon the valve characteristic C. For a manual-control arrangement, as shown here, the valve is carefully arranged to produce a characteristic time response that is easy for the pilot to work with. Commonly, for manual control, a line of slightly staggered holes in the valve cylinder is uncovered in sequence by the valve, so that the characteristic can be "shaped" to the pilot's liking. In automatic applications (e.g., in a missile control system) the emphasis has been on the fastest possible speed of response; the design of valves to give a high rate of flow for a small displacement, with a variation that is nearly linear, has received a great deal of technological attention.

Often it is convenient to test hydraulic systems with sinusoidal inputs—even though such inputs are unlikely to be encountered in flight. The frequency response of a hydraulic servo is readily calculated from the transfer function obtained in Step 2 of Stage III, e.g., the transfer function implicit in Eq. (19.41), for the physical model considered here. When oil compressibility, linkage flexibility and backlash, or valve nonlinearity, is important, the situation is, of course, much more complicated than (19.41) suggests. (See Problem 22.8.)

19.5 THE TWO-AXIS GYROSCOPE

Few devices have dynamic behavior as intriguing as that of the two-axis gyroscope. Moreover, the device is of great practical modern interest as a sensor in control and guidance systems for air, sea, and space vehicles.

Gyroscopes, or gyros, as they are more commonly called, come in many forms and sizes. One common arrangement is shown in Fig. 19.10a, where the gyro rotor is mounted to the vehicle frame via two gimbals in such a way that the vehicle and rotor can move freely through small angles about any axis without exerting any torque on one another. In other forms the gyro rotor may be spun about a spherical bearing, giving it the same complete freedom from angular motion of the vehicle; or the rotor may itself be a sphere which is supported electrically, magnetically, or by gas film, so as to be free rotationally from the vehicle.

The usual application of the free gyro is as a directional reference. The spinning mass has, as we shall show presently, an extremely strong tendency to maintain its spin direction in space. When it is supported free of the vehicle (as described above), it therefore continues to point in a fixed spatial direction despite rotation of the vehicle around it. But a free gyro may have interesting dynamic motions of its own, independent of the vehicle, and it is these we wish to study here. We use the configuration of Fig. 19.10a because a convenient coordinate system is defined by the gimbals, making it particularly easy to keep track of what is happening.

In this study, then, let us *compute the time response of the free gyro of Fig. 19.10a following initial conditions (displacement and velocity), and also the response resulting from an impulsive- or a step-disturbance torque on the gimbals*, the base being held fixed throughout.

STAGE I. PHYSICAL MODEL. Assume in this first analysis that in Fig. 19.10a the axially symmetrical rotor is carefully mounted so that its spin axis is its major axis of inertia. Assume also that the gimbals are carefully balanced, that the rotor is driven at constant speed with respect to the inner gimbal by a precision motor, and that damping in the gimbal bearings can be neglected. The effect of gravity acting on a gimbal unbalance will be considered explicitly as a steady disturbing torque on the gimbal. A study of the effect of damping is outlined in Problem 19.30. Finally, we make the influential assumption that only small angles are involved in the motion. This is essential if the equations of motion are to be linear. (The study of large-angle motions of a free gyro is an advanced topic.)

STAGE II. EQUATIONS OF MOTION. The equations of motion for the above physical model were derived in Example 5.7, p. 159. A brief review follows.

(1) *Geometry.* The system has two unconstrained degrees of freedom. In the arbitrary-configuration sketch of Fig. 19.10a the variables ψ and θ serve

to define the system configuration: ψ is rotation of gimbal o with respect to a fixed reference frame, and θ is rotation of gimbal g with respect to gimbal o. Both ψ and θ can be measured directly, e.g., using electrical gimbal potentiometers. Rotor speed n, with respect to gimbal g, is prescribed constant. (The rotor's position about its axle is of no concern when the rotor is assumed to have perfect axial symmetry.)

(2) *Equilibrium.* Two free-body diagrams are utilized, Fig. 19.10b and c. For system 1 we write (2.2) about axis y, and for system 2, about axis Z.

(3) *Physical relations.* These involve inertial moments of the form $M_i = J\ddot{\theta}$ and $M_i = h\dot{\theta}$, which are substituted into the equilibrium relations. The result, for small motions θ and ψ, and for $\dot{\theta} \ll n$ and $\dot{\psi} \ll n$, is the pair of equations (5.63) and (5.64):

$$J_y\ddot{\theta} + h\dot{\psi} = M_y \qquad (19.44)$$
$$[(5.63) \text{ repeated}]$$
$$-h\dot{\theta} + J_z\ddot{\psi} = M_z \qquad (19.45)$$
$$[(5.64) \text{ repeated}]$$

in which J_y is the moment of inertia of system 1 (Fig. 19.10b) about axis y, J_z is the moment of inertia of system 2 (Fig. 19.10c) about axis Z,† and h is defined by $h = I_{rx}n$, where I_{rx} is the moment of inertia of the rotor about its axis of symmetry [refer to (2.7)].

STAGE III. DYNAMIC BEHAVIOR. We follow Procedure D-1.

Step 1. Laplace transform the equations of motion:

$$\begin{bmatrix} s^2 & \dfrac{h}{J_y}s \\ -\dfrac{h}{J_z}s & s^2 \end{bmatrix} \begin{bmatrix} \Theta \\ \Psi \end{bmatrix} = \begin{bmatrix} \dfrac{M_y(s)}{J_y} + s\theta(0^-) + \dot{\theta}(0^-) + \dfrac{h}{J_y}\psi(0^-) \\ \dfrac{M_z(s)}{J_z} - \dfrac{h}{J_z}\theta(0^-) + s\psi(0^-) + \dot{\psi}(0^-) \end{bmatrix} \qquad (19.46)$$

Step 2. Manipulate algebraically to obtain:

(b) *Natural behavior.* The characteristic equation is given, from the left-hand side of (19.46), by

$$s^2\left(s^2 + \frac{h^2}{J_yJ_z}\right) = 0 \qquad (19.47)$$

the roots of which are $s = 0, 0$, and $s = \pm jh/\sqrt{J_yJ_z}$. The zero roots imply two natural "motions" in which the motion is a constant. We shall soon see that these are a constant displacement θ and a constant displacement ψ. The imaginary roots imply an oscillation at frequency $\omega = h/\sqrt{J_yJ_z}$. To get an idea of the magnitude of ω, consider the limiting case that the gimbals are massless in Fig. 19.10a. Then J_y and J_z are merely the diametral moments of inertia of the rotor disk, and they are equal to one-half its polar moment of inertia, $J_y = J_z = I_{rx}/2$. Then, since $h = I_{rx}n$,

$$\omega = \frac{I_{rx}n}{\sqrt{I_{rx}^2/4}} = 2n \qquad (19.48)$$

† The small variation in J_z with θ is second-order small, and is neglected in linear Eq. (19.46), as discussed in Example 5.7.

[Sec. 19.5] THE TWO-AXIS GYROSCOPE

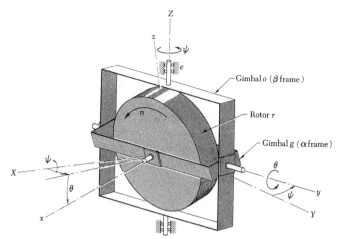

(a) *Arrangement and coordinates*

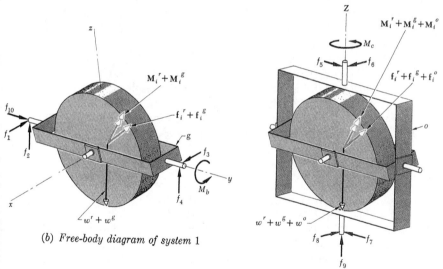

(b) *Free-body diagram of system 1*

Fig. 19.10 *The two-gimbal gyro*

(c) *Free-body diagram of system 2*

The oscillation frequency is twice the spin speed. Since real gimbals are not massless, a real gyro has an oscillation frequency somewhat less than $2n$—but typically of the same order of magnitude as n.

Another important point is evident from the characteristic equation: the oscillatory motion is undamped! This may be slightly surprising, since rate-dependent terms appear in both the equations of motion ($h\dot\theta$ and $h\dot\psi$).

But these terms, known as "gyroscopic coupling terms," appear in a special way, and they do not produce energy dissipation. The system is in fact conservative, as is clear from physical considerations, for the zero-damping model assumed in Stage I above.

(a) *Response functions* for θ and ψ, from relations (19.37). These response functions will be found to be symmetrical with respect to one another. Therefore, for demonstration purposes we consider only one moment—say, M_y—and one initial velocity—say, $\dot\theta(0^-)$. Then the equations can be written:

$$\Theta = \frac{\left[\dfrac{M_y}{J_y} + \dot\theta(0^-)\right]}{s^2 + \omega^2} + \frac{\theta(0^-)}{s}$$

$$\Psi = \sqrt{\frac{J_y}{J_z}} \frac{\left[\dfrac{M_y}{J_y} + \dot\theta(0^-)\right]\omega}{s(s^2 + \omega^2)} + \frac{\psi(0^-)}{s} \tag{19.49}$$

(c) *Final and initial values.* Let $M_y(t)$ be an impulse, $M_y(t) = \mu\delta(t)$. What does one then learn by applying FVT and IVT to (19.49)? Multiply (19.49) by s to obtain the Laplace transforms of $\dot\theta$ and $\dot\psi$; then apply IVT. Justify all these results on physical grounds.

Step 3. Inverse transform response functions (19.49) to obtain the time response. Consider first the initial conditions $\theta(0^-)$ and $\psi(0^-)$. Equations (19.49) show that if the system is given an initial displacement θ and ψ, but is otherwise undisturbed (and has no initial velocities), it will simply remain in that position indefinitely. This is physically obvious from Fig. 19.10a. That is, from (19.49), $\theta(t) = \theta(0^-)$ and $\psi(t) = \psi(0^-)$ for all $0 < t$.

Impulse response: coning

Consider next the response to an impulse, $M_y(t) = \mu\delta(t)$. The Laplace transform of M_y will then be $\mathsf{M}_y(s) = \mu$. The response of the system to this input will be *identical to its response to an initial velocity* $\dot\theta(0^-)$, provided only that $\mu = J_y\dot\theta(0^-)$. The terms in Eqs. (19.49) have been arranged to emphasize this equivalence. This is another way of saying that an initial velocity can be established in a gyro by jarring the gyro with a moment impulse. [Is this also borne out by the results of Step 2(c)?] Now let us see what motion follows. For simplicity, define the quantity Ω_{yo} by

$$\Omega_{yo} \triangleq \frac{\mu}{J_y} \tag{19.50}$$

(Ω_{yo} is recognized as the initial velocity that μ will produce.) Then (19.49) becomes (for response to the impulse only)

$$\Theta = \frac{\Omega_{yo}}{s^2 + \omega^2} \qquad \Psi = \sqrt{\frac{J_y}{J_z}}\,\Omega_{yo}\frac{\omega}{s(s^2 + \omega^2)} \tag{19.51}$$

[Sec. 19.5] THE TWO-AXIS GYROSCOPE

from which the time response to an impulse is (using, e.g., Table X)

$$\theta(t) = \frac{\Omega_{yo}}{\omega}(\sin \omega t)u(t) \qquad \psi(t) = \sqrt{\frac{J_y}{J_z}}\frac{\Omega_{yo}}{\omega}(1 - \cos \omega t)u(t) \qquad (19.52)$$

A very useful way of plotting time response (19.52) is shown in Fig. 19.11. In one time plot, plot (a), $\theta(t)$ is plotted positive downward to correspond with the direction in which the spin axis moves when θ is positive in Fig. 19.10a. In the other time plot, plot (b), $\psi(t)$ is plotted positive to the right, to correspond with the direction in which the spin axis moves when ψ is positive in Fig. 19.10a (i.e., when the spin axis is viewed from the front of the gyro). Time is plotted downward in plot (b) for convenience. Now from plots (a) and (b) we can conveniently make a cross plot of θ versus ψ, plot (c), which (for small angles) is a picture of the way the motion of the spin axis appears when viewed from the front of Fig. 19.10a. The motion in Fig. 19.11 is called *coning*. (The spin axis generates a cone.) The reader should check a number of points on plot (c), which is drawn for J_Z slightly larger than J_y, and is elliptical. [Note from (19.52) that if $J_Z = J_y$ the path of the spin axis is circular.]

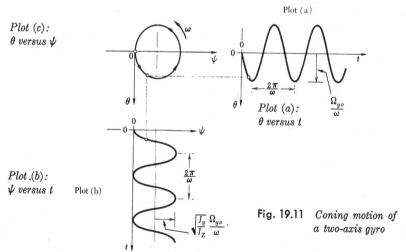

Fig. 19.11 *Coning motion of a two-axis gyro*

To get an idea of the import of this result, consider what the corresponding response to a moment impulse $\mu\delta(t)$ would be if the rotor were not spinning. Then, instead of tracing out the small ellipse of Fig. 19.11, plot (c), the spin axis would *tumble* about the y axis at velocity Ω_{yo}. Numerically, for example, suppose that the moments of inertia were

$$I_{rx} = 1000 \qquad J_y = 1200 \qquad J_Z = 1300$$

in cgs units [dn cm/(rad/sec), or gm cm²]. These represent a typical rotor of mass one-half kilogram. Then an impulse about y of

$$1000 \text{ dn cm sec} = 10^{-3} \text{ in. lb sec}$$

(one in. lb for 0.001 sec) would produce an Ω_{yo} of 0.83 rad/sec = 0.13 rev/sec. Next, suppose for comparison that the rotor is spinning at 2000 rad/sec (approximately 20,000 rpm) when the same impulse is applied. Then, instead of the spin axis *tumbling* at 0.13 revolution per second, its only motion will be a tiny coning motion of magnitude 0.52×10^{-3} rad. With typical instrument damping present, the *nonspinning* rotor will come to rest after perhaps a few *revolutions*, while the *spinning* rotor will move only half a *milliradian* (0.03 degree). The tremendous capability of a gyro to maintain its spatial orientation is thus dramatically demonstrated.

As the above calculations show, the coning motion of a typical instrument gyro is normally too small to be seen. Demonstration gyros have been developed which spin very slowly—e.g., at two or three revolutions per second ($\omega = 12$ to 20). Then a large, slow coning motion can be produced with a reasonable impulse, and the results shown in Fig. 19.11 can readily be checked quantitatively.†

In almost any real gyro there will be some damping about both axes—although in several gyro designs (notably the electrostatically suspended gyro) it is made unbelievably small. The effect of damping on the response in Fig. 19.11 is to cause both $\theta(t)$ and $\psi(t)$ [plots (a) and (b)] to decay exponentially (typically with a very long time constant) and thus the elliptical path of the spin axis [plot (c)] is seen to spiral in to the point $\theta = 0$, $\psi = \sqrt{(J_y/J_Z)}\,(\Omega_{yo}/\omega)$. The reader will have the opportunity to calculate this response in Problem 19.30.

Response to constant moment: precession and nutation

Let us study next the response of a free gyro to a constant moment M_{yo} acting on the inner gimbal about its y axis. The most common source of such a moment is that the center of support of the rotor does not quite coincide with the center of mass, and thus gravity acts to produce an unbalance moment equal to the rotor weight times the misalignment of support. Other sources of uncertainty torque are occasioned by stray magnetic and electric fields, bearing anomalies, and the like.

We calculate first the response when the system is released from rest without initial conditions, but with the constant moment $M_y(t) = M_{yo}u(t)$

---

† A precisely known impulse can be applied to such a demonstration gyro in the following way: a known mass is tied to one end of a string; the other end of the string is tied to one end of the spin axle of the gyro. The mass is then released from a position at about the height of the axle and allowed to drop to the end of the string. It is caught on first bounce. It is found that the bounce velocity of the mass is negligible so that all of its linear momentum just before it reached the end of the string has been transferred as an impulse to the gyro; the magnitude of this impulse can be accurately calculated from the mass and the known drop height.

[Sec. 19.5] THE TWO-AXIS GYROSCOPE

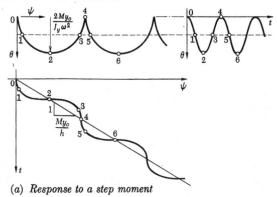

(a) Response to a step moment

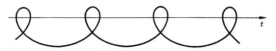

(b) Response to a combination of step and impulse moment

Fig. 19.12 *Precession and nutation of a two-axis gyro*

applied about the y axis. The Laplace transform of M_y is $\mathsf{M}_y(s) = M_{yo}/s$, and the response functions (19.49) become

$$\Theta = \frac{M_{yo}/J_y}{s(s^2 + \omega^2)} \qquad \Psi = \sqrt{\frac{J_y}{J_z}} \frac{(M_{yo}/J_y)\omega}{s^2(s^2 + \omega^2)} \qquad (19.53)$$

The corresponding time response is

$$\boxed{\theta(t) = \frac{M_{yo}}{J_y\omega^2}(1 - \cos \omega t)u(t) \qquad \psi(t) = \frac{M_{yo}}{h}\left(t - \frac{\sin \omega t}{\omega}\right)u(t)} \qquad (19.54)$$

[In the expression for $\psi(t)$, ω has been replaced by $h/\sqrt{J_yJ_z}$.] Figure 19.12a shows this response, plotted in the manner of Fig. 19.11. The path made by the spin axis is called a *cycloid*. (It is the pattern traced by a point on the rim of a rolling wheel, as is readily shown.) The steady part, $\psi(t)_{\text{steady}} = (M_{yo}/h)t$ is called *precession* of the gyro, and the nodding up and down that is superposed on it is called *nutation*. The precession $\psi(t)$ is steady about an axis perpendicular to the axis about which the torque is applied, which is somewhat spectacular the first time it is experienced.† For the no-damping model assumed (which is closely approximated by many real

---

† An easy demonstration is to spin up a free bicycle wheel, for example, and then try to rotate it about a diameter. Motion (19.54) can be obtained with the bicycle wheel by hanging it from a string fastened to one end of its axle. Gravity furnishes $M_y(t)$. (The sinusoidal part of the motion is quickly damped.)

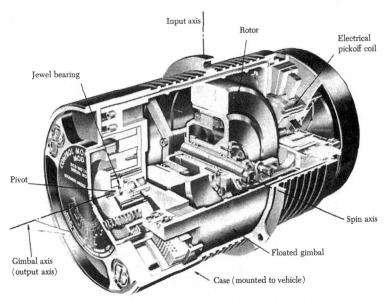

(a) *Floated, single-axis gyro (Courtesy Northrop Nortronics, Inc.)*

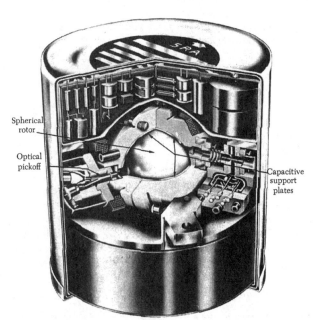

(b) *Spherical rotor supported electrically in a vacuum (Courtesy Honeywell, Inc.)*

Fig. 19.13 *Low-friction gyro designs*

[Sec. 19.5] THE TWO-AXIS GYROSCOPE

gyros) the gyro may precess around in a complete circle many times, nodding (nutating) all the way.

As with the gyro's impulse response, shown in Fig. 19.11 (which is also commonly referred to as "nutation"), the response of Fig. 19.12a is large enough to observe only in a very slowly spinning gyro having large unbalance. Other fascinating patterns such as that in Fig. 19.12b can be created by superposing some of the motion of Fig. 19.11 on that of Fig. 19.12a, by applying an impulse after the motion of Fig. 19.12a has been started, for example.

The presence of damping causes the nutation motion to decay away eventually, and also results in precession with a small vertical component (the spin axis drops slowly). See Problem 19.30.

Precession

In the design of instrument gyros the crucial quantity is steady-state gyro precession. From (19.54) and Fig. 19.12a (lower plot) this is given by

$$\text{Gyro drift} \quad \boxed{\dot{\psi}(t) = \frac{M_{yo}}{h}} \quad (19.55)$$

or, in vector form,

$$\mathbf{\Omega} \times \mathbf{h} = \mathbf{M} \quad (19.56)$$

where $\mathbf{\Omega} = \mathbf{1}_z \dot{\psi}$, $\mathbf{h} = \mathbf{1}_x h$, $\mathbf{M} = \mathbf{1}_y M_{yo}$, the precession being about an axis perpendicular to the axis of the applied torque. This is sometimes referred to as "the law of the gyro." Indeed, in most instrument gyros h is very large and there is some damping, so that the coning motions are tiny and short-lived. Then the steady precession (19.55) makes itself known over a period of time, and its magnitude determines the worth of the gyro as an inertial reference. But, as we have shown, there is much more to the dynamic behavior of a gyro than (19.55) foretells.

The accuracy of gyros

Gyroscopes, both free and single-axis, may be classified by random drift rates into three broad groups. Gyros used to stabilize short-range missiles, torpedoes, and fire-control antennas have random drift rates measured in degrees per minute. The free gyros used for attitude reference in aircraft have typical unsupervised drifts measured in degrees per hour. The gyros used to stabilize inertial navigation equipment must be made to have drifts measured in degrees per year.

Gyro drift can be controlled only by extremely careful balancing, by reducing bearing friction to the smallest possible value, and by preventing temperature and magnetic effects from causing drift-producing torques. Control of these effects has been the subject of extensive fundamental research.

An important factor in reducing the drift of gimbaled gyros is the prevalent use of the floatation technique, whereby the friction level in the gimbal bearings is greatly reduced. A typical construction is illustrated in Fig. 19.13a for a single-axis (single-gimbal) gyro. The gimbal is a sealed float which houses the rotor. The float is suspended in a liquid whose density is just sufficient to relieve the gimbal bearings of the rotor weight; therefore, friction at these bearings is greatly reduced—perhaps by a factor of 1000. Floatation may also be used with two-gimbal gyros.

An alternative technique is to use gas for support, either of a cylindrical gimbal like that in Fig. 19.13a, or of a spherical free rotor in a cup-shaped bearing.

A more fundamental approach to the bearing-friction problem, and one that has led to the most accurate gyros yet produced, employs a spherical rotor suspended electrically or magnetically in an extremely good vacuum, as in Fig. 19.13b, which shows an electrical suspension. Support is effected by controlling the plate voltages with position-feedback signals, as indicated in Fig. 2.27f. Location of the spin axis (corresponding to angles ψ and θ in Fig. 19.10a) is sensed optically.

As an extreme case, a research program at Stanford University includes plans to operate a spherical-rotor gyro having *no* support forces, by orbiting it in a special satellite controlled to follow it without touching. The purpose is to measure two general relativity effects, and the design goal is a drift rate of about one degree per 100,000 years.†

Problems 19.1 through 19.30

---

† R. H. Cannon, Jr., "Requirements and Design for a Special Gyro for Measuring General Relativity Effects from an Astronomical Satellite," *Gyrodynamics*, H. Ziegler, ed., Springer Verlag, Berlin, 1963.

PART E

FUNDAMENTALS OF CONTROL-SYSTEM ANALYSIS

Chapter 20. *Feedback Control*
Chapter 21. *Evans' Root-Locus Method*
Chapter 22. *Some Case Studies in Automatic Control*

Part E, the last (and shortest) part of this book, introduces formally a branch of system dynamics which is of great importance throughout physical technology, and which is rapidly establishing its importance also in the biological, physiological, economic, and sociological sciences. This is the field of automatic control, using feedback.

In the arena of dynamic systems, feedback systems are a special case involving a specific form of coupling (Sec. 11.6 and Fig. 12.1), and such systems have appeared naturally and often as examples in our discussions, along with other kinds of systems. Now, in Part E, we turn our full attention to this special case. We do so because it is a most important branch of dynamics to which we want to build a bridge, because it will provide excellent additional comprehensive demonstration of the analysis concepts and techniques we have developed, and because, by example, it will give us a chance to look briefly at the synthesis and design process, "Stage IV."

Not only are feedback systems numerous; they are, fortunately, inherently amenable to certain special analytical techniques, which have therefore been highly developed. Historically, these techniques had their early development in the field of communications, where feedback amplifier design and feedback circuitry became highly sophisticated at an early date. Because of the elaborateness called for in communication systems, and the relative ease with which electrical components can be assembled, these systems soon became highly complicated, and the need for rapidly analyzing them became acute.

The basic signal in communication systems is the continuing sine wave, and so analysis was based on the sinusoidal response, versus frequency, of systems and their subsystems—i.e., on their *frequency response* (Chapters 9 and 15). Exploiting the fact that a subsystem's frequency response is indicated directly by its transfer function, Nyquist[†] and Bode[‡] developed highly effective algebraic and graphical techniques for assessing "closed-loop" (feedback-coupled) system stability and response from the open-loop set of subsystem transfer functions.

By contrast, early mechanical feedback systems—engine governors, temperature regulators, antenna servos, automatic pilots, and so on—were often designed using Routh's method (Procedure C-4) as the only analytical tool. Later, as these systems became more complicated, and as electrical engineers became involved with them, the frequency-response methods of analysis also became widely used (even though some of these systems would never see a sinusoidal signal in their operating lives). The important work under wartime pressure of James, Nichols, and Phillips and their associates in the radar antenna servo group at the MIT Radiation Laboratory stands out in this era of automatic control. In particular, Nichols developed a new graphical arrangement combining the best features of the Bode and Nyquist methods.§

[†] H. Nyquist, Regeneration Theory, *Bell System Tech. J.*, vol. 11, pp. 126–147, Jan. 1932.
[‡] H. W. Bode, "Network Analysis and Feedback Amplifier Design," D. Van Nostrand Company, Princeton, N.J., 1945.

The frequency-response techniques did the design job, and they did it well. Pedagogically, however, they left much to be desired. They were indirect and complicated—unnecessarily so, in retrospect. Moreover, the only response they predict directly is steady-state response to a sinusoidal input: They indicate natural behavior and response to abrupt inputs only by inference.

In 1948 W. R. Evans provided the central analytical structure for control work with his invention and development of the root-locus method, which we described and demonstrated in Chapter 11 and subsequent chapters for problems of a general nature. For feedback control systems the root-locus method provides a direct display of system stability and natural characteristics and, even more important, shows graphically *precisely* how these qualities are influenced by changes in design parameters. The method is thus tailor-made for design studies where transient behavior is of prime interest. Graphical techniques for rapid sketching of system root loci versus a parameter are part of the Evans method.

Thus pedagogically, and often operationally as well, the root-locus method plays the central role in the study of feedback control analysis. But this is not to say that frequency-response methods have been superseded. On the contrary, they are sometimes the more rapid way of studying systems of high order and of deciding on design values for parameters. Further, it is often the case, in mechanical as well as electrical systems, that the most reliable characterization of a plant is not its imagined equations of motion but its measured frequency response. Then analysis by frequency response is the most direct procedure. Finally, a modification of frequency-response procedures is useful for analyzing nonlinear systems as well.

In practice the control engineer typically uses both the root-locus and frequency-response methods in combination, in whatever proportion will lead most directly to a good design. There is not space in this book to present both methods. We therefore present our introduction to automatic control in terms only of the root-locus method, because it provides a more direct introduction to automatic-control philosophy, because it is already familiar to us and is simpler and shorter to learn, and because it will make the frequency-response viewpoint much easier to grasp in subsequent study. It is thus best for building a bridge to the broad field of automatic control.

In Chapter 20 we introduce the philosophy and objectives of automatic control using feedback, and derive the basic algebraic relations for linear feedback systems. In Chapter 21 we present the root-locus method more formally than we did in Chapter 11, and we derive and demonstrate the sketching techniques that make it so utilitarian. Finally, in Chapter 22 we present, as we did in Chapter 19, a short collection of interesting case studies to demonstrate further the concepts and techniques of dynamic analysis in general, and of feedback-system analysis in particular.

---

§ H. M. James, N. B. Nichols, and R. S. Phillips, "Theory of Servomechanisms," MIT Rad. Lab. Series No. 25, McGraw-Hill Book Company, New York, 1947.

Chapter 20

FEEDBACK CONTROL

In this chapter we introduce the philosophy of feedback control and describe the performance objectives which motivate the design of control systems. We also derive and demonstrate the general form taken by the characteristic equation of a feedback system. Finally, in Procedure E-1, we specialize Procedure D-3 for use in the dynamic investigation of control systems.

Basic concepts

(1) In a feedback system for controlling a "plant," the plant motion (output) is compared with the desired motion (input), and the error is used to instigate corrective action. (See Fig. 20.1.)
(2) The benefits and objectives of feedback control include:
  (a) the ability to control plant motions with acceptably small error;
  (b) improvement in system stability and speed of response;
  (c) performance insensitivity to disturbances and to changes in plant parameters.
(3) The dynamic coupling in a simple feedback control system has the general form of Fig. 20.5, and the closed-loop transfer function from command (input X) to response (output Y) has the general form:

$$\frac{Y}{X} = \frac{G}{1 + GH} \qquad (20.1)$$

where $G(s)$ and $H(s)$ are the transfer functions of the forward and feedback loops, respectively.
(4) The poles (denominator factors) of (20.1) are the roots of the closed-loop system characteristic equation, which determine the character of system natural behavior.

20.1 THE PHILOSOPHY OF FEEDBACK CONTROL

There is a large class of dynamic problems in which the objective is to control a physical subsystem automatically so that it follows closely some prescribed motion.† Some now-common examples of such automatic control are the autopilot that steers a missile in accordance with radioed instructions, the automatic machine-tool control that shapes a part according to a given program, the thermostat system that holds the temperature in a building nearly constant despite large changes in outside temperature, the chemical-process system that controls temperatures and rates of flow according to a prescribed program, and the gun-fire-control system that automatically keeps artillery trained on a target despite maneuvers of the target and disturbing motions of the gun-carrying vehicle. In each case, the subsystem to be controlled is called, in control jargon, "the plant."

In nearly all cases the basic scheme used to accomplish control to the required accuracy is the *feedback* technique, depicted in Fig. 20.1, in which actual motion of the plant is measured and compared with desired motion, and the difference, or *error* signal, used to instigate corrective action. That is, if missile heading differs from desired heading, the rudder is moved appropriately; if the temperature in the building is too low, the furnace is turned on; and so on.

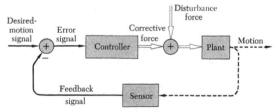

Fig. 20.1 *The basic arrangement of a feedback control system*

A typical (and by now classical) example of feedback control is the system for aiming large shipborne antiaircraft guns, as indicated in some detail in Fig. 20.2. A target aircraft—a guided missile, say—is approaching the ship. The gun turret is to be controlled so that it is always pointing in the same direction as a remote telescope device, called a director, which, in turn, is aimed optically at the target by a skilled man, called a chief gunner's mate.‡ The control problem is to ensure that the turret points always in the same direction as the director.

The turret is driven by a motor,§ and the scheme is to invoke a motor

† Here, as usual, we use "motion" in the general sense of change in any physical variable, such as voltage or temperature, as well as mechanical motion.
‡ In more elaborate systems the director may also be aimed automatically by means of a target-tracking radar. (In either case a computer in the director will automatically cause it to point ahead of the target by the proper "lead" angle.)
§ An electric motor is postulated in this example. Hydraulic drive is more effective.

torque, $K(\theta_d - \theta)$, proportional to the difference between the director angle and the turret angle, such that the turret is always torqued in the proper direction to reduce the error. (When there is no error, there is no torque.) The director angle and turret angle are measured electrically and compared in a simple circuit. The difference, or *error*, signal is then used to control the large current into the motor windings. If "tight" control is desired, then K in Fig. 20.2 must be made large, so that the corrective action is strong. K is commonly called the "gain" of the control system.

Figure 20.2a indicates the actual system, while Fig. 20.2b shows a simplified physical model of the system that will be amenable to mathematical analysis. As always, the process of contriving a model, Stage I, is crucial. In Fig. 20.2b the gears have been assumed to be completely stiff and without backlash, mechanical damping has been assumed viscous, motor torque has been assumed proportional to controlling voltage, and circuit dynamics have been neglected (i.e., circuit time constants have been assumed to be very short compared to mechanical time constants). In a later, more refined analysis we may want to study the validity of these assumptions by choosing a more elaborate model in which gear limberness, motor characteristics and torque-limiting,† circuitry, and the like are included in varying degrees. In fact, we may wish later to improve performance by *adding* circuitry with certain dynamic properties (see Sec. 22.2). For a first analysis, however, we use a simple model to develop a quick, approximate understanding of the situation. Compare Fig. 20.2b with the general representation in Fig. 20.1.

We expect the open-loop dynamics of the model of Fig. 20.2b to be fairly simple. In fact, the plant transfer function from Θ_e to Θ has the form

$$\frac{\Theta}{\Theta_e} = \frac{C}{s(s + \sigma)}$$

so that the natural response of this subsystem, without feedback, will consist of a constant displacement plus a simple exponential motion. But the complete system includes feedback coupling, and we learned in Part C that a coupled system may have natural motions that are very different from those of its constituent subsystems. Thus, if we make K very large to achieve tight control, the system will have oscillatory motions which may be poorly damped. (In fact, if a circuit lag is included in the model, for example, the oscillations may grow unstably, as we learned in Part C: refer to Figs. 11.14 and 11.15.)

The problem of choosing system parameters to achieve both the desired degree of control *and* good system dynamic characteristics is the central design problem in automatic control. To solve it, techniques are required for determining rapidly the characteristics and response of a feedback system.

---

† Certainly for large error angles the torquer will be saturated and the control torque will be constant, rather than proportional to error. Thus, the linear analysis discussed here is valid only for behavior in the vicinity of zero error.

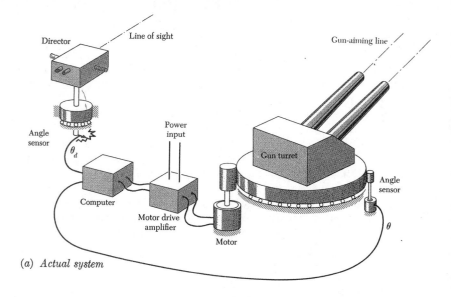

(a) *Actual system*

(b) *Physical model*

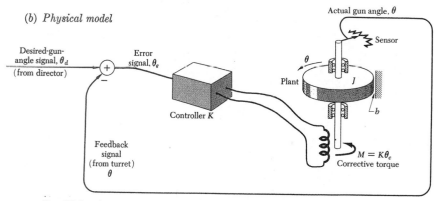

Fig. 20.2 *A gun-control system using feedback*

Fig. 20.3 *An all-human feedback control system*

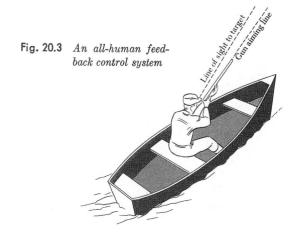

The purpose of Part E is to present the primary techniques used, and to develop both some facility with their use and an understanding of the underlying dynamic philosophy.

One further example may help strengthen the concept of feedback control in a personal way. Consider the man hunting pheasant in Fig. 20.3. In the role of hunter he is in nearly every way analogous to the gun-control system in Fig. 20.2. His arms and trunk form the gun turret, which his muscles rotate. His *eye* is both the *director*, tracking the target optically, and the gun-position *sensor*, and his brain is the *controller* which compares the target line-of-sight angle with the gun-aiming line (with the help of the gun sight) and, from the resulting error angle, controls corrective torques in his trunk muscles. The block diagram in Fig. 20.1 can represent the human

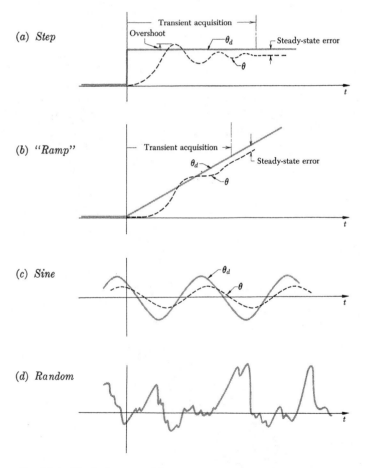

Fig. 20.4 *Typical command inputs*

control system of Fig. 20.3 just as well as it represents the electromechanical control system of Fig. 20.2a.

Because we have often the experience of *being* part of an automatic control system, the concepts of feedback control are, on reflection, especially easy and appealing for us to grasp. Indeed, our every physical activity—catching a ball, picking up a pencil, driving a car—is controlled by marvelous feedback control systems. Some are considered in Problems 20.1 through 20.10. The study of biological control systems is one of the most exciting areas of current research.

Problems 20.1 through 20.10

20.2 PERFORMANCE OBJECTIVES

Put qualitatively, the task of the control system in Fig. 20.2 is to cause the turret to follow the director promptly, with sufficient accuracy. To proceed with a quantitative design we need to know four things in quantitative terms: (1) What director motion (inputs) must be followed? (2) How quickly must the system follow? (3) What accuracy is sufficient? (4) What disturbances will be trying to force the system off target?†

In this example several kinds of input may be involved, as depicted in Fig. 20.4. Three of these—the step, ramp, and sinusoid—are especially common, and we take time here to indicate how they arise physically.

In the initial acquisition of a target, the director may be moved suddenly to a new direction, Fig. 20.4a. The system sees this as a "step input," to which it will respond as shown. Naturally, the system cannot remove such a large error instantaneously, because the turret has inertial and frictional resistance to motion. (Moreover, in the real system the motor will be velocity- and torque-limited.) But the system will respond as fast as it can, perhaps in the manner shown in Fig. 20.4a, and thus in a short time the error will be reduced as indicated. Preferably this time should be shorter than the time for the target to get within range of the guns. It must certainly be short compared to the time for the target to reach the ship. (Long response time will make the gunner's mate understandably impatient.)

Formally, the design specification might include an item such as "The system shall reduce to 1°, in less than 3 seconds, the error due to a step change in director angle of 10°; transient motions shall be 0.4 critically damped. The steady-state error shall be less than 0.1°." Note that both transient-response time and steady-state error are specified.

In the present example, as in many cases, the acquisition problem may be thought of either as a step-input problem (as above) or, alternatively, as

---

† At a later, more sophisticated stage of the design we shall want to know also how much the system's physical constants may change—e.g., how much b will change with temperature, or J with number of gun crewmen. The control system must be made adequately *insensitive* to such changes.

an initial-condition problem. Then the above specification could be stated: "The system shall reduce to within 1°, in less than 3 seconds, an initial error of 10°; transient motions shall be 0.4 critically damped. The steady-state error shall be less than 0.1°." More generally, maximum times and maximum "overshoots" may be specified for the system's transient response from its various possible initial conditions (displacements, velocities, voltages, etc.)

In addition to acceptable acquisition capability, the system must, of course, track the target closely enough that hits will be scored (as determined by the lethal range of the shells). It must do so despite target motions, which also appear as system inputs. Two cases are shown in Fig. 20.4. If the target is crossing the field of view, then the director angle (input θ_d) will be changing at a high rate, as depicted in Fig. 20.4b. Such constant-rate inputs are commonly known as "ramp" inputs. The director system will of course arrange automatically for the director angle to lead the line of sight geometrically. Here again, the system cannot instantaneously assume the proper rate; after a short transient, however, the guns will be following the target with the correct rate but, in general, with a small tracking error, as shown. The design specification might therefore say, formally, "The system shall follow an abrupt ramp input of 5°/sec with less than 0.1° steady-state tracking error, and shall reduce transient error to 1° in less than 1 second." Again note that both steady-state error and transient-response time are specified.

In some cases, control systems may be required to keep the steady-state error small during acceleration inputs, $x = ct^2$, or during inputs characterized by even higher powers of t. More generally, this class of "steady" inputs has the form $x = ct^n$, the step and ramp representing the first two levels ($n = 0, n = 1$).

The third important form of control-system input is the sinusoid, shown in Fig. 20.4c. In many cases the system inputs actually are pure sine waves of various frequencies, as in communication systems. Even when they are not, the sinusoid may be a sufficiently good approximation for design purposes. Suppose, for example, that the target in Fig. 20.2 is flying directly toward the gun, and is taking violent evasive action. Then the director angle (system input) will be more or less random, as shown in Fig. 20.4d. But even here, though the target may *wish* to make maneuvers which are *entirely* random, its motions will in fact be limited to turning frequencies determined by its own dynamic characteristics—perhaps to 0.5 cps for large turns. Then a gun-control system which follows sinusoidal inputs closely up to 0.5 cps will track the maneuvering target adequately.

Thus, if it were known that sizable target maneuvers would be restricted to frequencies below 0.5 cps, the design specifications might include an item reading "The peak error shall not exceed 0.1°, nor the phase angle 0.1 radian, for sinusoidal inputs of magnitude 1° and frequency from 0 to 0.5 cps." (Since no transient is involved here, only the steady-state error is specified.) Design is based on sinusoidal inputs whenever they are a reasonable approximation, because the response to a continuing sine wave is especially easy to compute and to measure experimentally. (Recall Sec. 9.3.)

As we have learned in Chapters 9 and 15, the steady-state response of a linear system to a sinusoidal input is also a sine wave of the same frequency, Fig. 20.4c, having a peak amplitude and a phase angle, with respect to the input, which depend upon the frequency (as in Fig. 9.3 and Table IX).

The response of a system to truly random inputs, as in Fig. 20.4d, can also be computed on a statistical basis, if necessary. (These will not be treated in this book.)

In addition to command inputs such as the above, control systems generally must deal with unwanted disturbance inputs that attempt to force the controlled element off target. The desired response to disturbances is *zero*, and the control system must provide this to within specified tolerances. The ratio of disturbance to system error is called the *stiffness* of the control system. In some systems, such as the temperature-control system in a building or the stable platform in a guidance system, there is no command input, and stiffness is the principal system specification.

For the gun controller of Fig. 20.2 the important disturbances might be (*i*) a steady wind or unbalance torque; and, (*ii*) a sinusoidal torque produced by inertial reaction to rolling motions of the ship. (The latter are especially severe for guns mounted in an aircraft rather than a ship.) Thus for the system in Fig. 20.2 a specification item might read: "Aiming errors shall not exceed 0.3° under steady or sinusoidal torque disturbances of magnitude 400 ft-lb and frequency 0 to 0.2 cps."

The various design requirements described above for the gun-control example are quite typical of those involved in common control problems of all kinds. Actually, the gun-control problem is unusual in the number of different specifications it must meet. Often one or another is predominant, and the others are of less importance.

To summarize: It is convenient, for analysis and design purposes, to classify control system specifications into the following categories:

(*a*) System characteristics: stability and transient behavior.
(*b*) Steady-state error for steady inputs (step, ramp, and so on).
(*c*) Total response to inputs and disturbances.

20.3 THE SEQUENCE OF CONTROL-SYSTEM ANALYSIS

The order in which the categories (*a*), (*b*), and (*c*) are presented above is usually the most efficient sequence in which to study a control-system problem. That is, it makes little sense to calculate the time response of a system which will be unstable to begin with, so it is usually most efficient to base the first design on studies of stability and system characteristics. Then the relatively simple study and adjustment of steady-state performance are made, followed by the more involved study of total response.

An important exception to the above sequence may occur when the response to steady sinusoidal inputs is of paramount importance, as it is in many cases. For example, the principal disturbance to a stable platform is the vibratory motion of the vehicle in which it is mounted, while the inputs to communication equipment are nearly all sinusoidal. Even for control systems—such as the rudder-position servo for a missile—whose actual inputs will not be sinusoidal, it is sometimes so much easier to perform tests using sinusoidal inputs, that specifications are written in terms of frequency-response performance. (For any linear system there exists a frequency-response plot, like Fig. 9.20, corresponding to a given set of transient characteristics.)

As noted in the introduction to Part E, sinusoidal-response techniques have been developed whereby the closed-loop stability of a system can be inferred, *as a function of control gain* (e.g., K in Fig. 20.2), directly from its open-loop sinusoidal response. These techniques, due to Nyquist and Bode and Nichols, are presented in textbooks on feedback control-system design.

In this introductory presentation we shall concentrate mainly on category (a), system characteristics. In the remainder of this chapter we shall study the general form in which, inherently, the characteristic equation of a feedback system occurs. In Chapter 21 we present the root-locus method for determining and displaying the roots of characteristic equations in this form, and thus of probing their stability and natural-response characteristics.

20.4 REVIEW OF DYNAMIC COUPLING

The objective in analyzing any coupled system is to obtain the *system transfer function*, from input to output. The natural characteristics of the system will then be evident, and the total response of the system to any specific input can be calculated (e.g., by inverse Laplace transformation). For example, in Fig. 20.2b the objective is to obtain the transfer function $(\Theta/\Theta_d)(s)$, from which to compute system response to θ_d commands.

As we emphasized in Chapter 12, feedback systems are a special class of coupled dynamic systems. In the general case of dynamic coupling (such as in Figs. 12.1c and 12.2), the motion of every subsystem in the system may, in general, influence directly every other subsystem. Moreover, such influence is commonly two-way: The motion of subsystem A causes a force on subsystem B, and the motion of B causes a force on A. But in the special case of a feedback system, such as those in Fig. 12.1b and Fig. 20.7 (p. 641), the subsystems are in "cascade," and thus A influences B, B influences C, and so on; but the motion of B has no direct effect on A, C has no effect on B, and so on, except that (typically) the control system "feeds back" the motion of the last subsystem in the chain to influence the first.

Because of the special nature of the coupling in a feedback system, such systems are especially amenable to a number of short-cut and special-case methods of analysis which greatly simplify and speed up their study. Because feedback systems are so common and so important, it is worth-while developing these methods in considerable detail. This we do next.

20.5 THE ALGEBRA OF LOOP CLOSING

The basic feedback system is shown in general form in Fig. 20.5. Quantity X is the input to the system, Y is the output of interest, and E is a difference signal produced at the feedback summing point.† Function $G(s)$ is the product of all transfer functions in the *forward* path, from E to Y. Function $H(s)$ is the product of all transfer functions in the feedback path, from Y to the summing point.

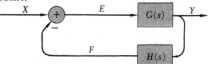

Fig. 20.5 *The basic feedback loop*

The system transfer function is readily derived:

$$Y = GE = G(X - HY)$$

$$\boxed{\frac{Y}{X} = \frac{G}{1 + GH}} \qquad (20.1)$$

Note that the output is fed back with a negative sign in Fig. 20.5. This is usual for a control system. If the feedback had been positive, (20.1) would have been $Y/X = G/(1 - GH)$. The quantity $GH = F/E$ is known as the system's *open-loop transfer function*.

Some alternative forms of (20.1) are often more efficient. Dividing by G, we obtain

$$\boxed{\frac{Y}{X} = \frac{1}{(1/G) + H}} \qquad (20.2)$$

Separating G and H into their numerators and denominators, $G = G_N/G_D$ and $H = H_N/H_D$, we obtain

$$\boxed{\frac{Y}{X} = \frac{G_N H_D}{G_D H_D + G_N H_N}} \qquad (20.3)$$

The *characteristic equation* for the system is given by the denominator of (20.1), (20.2), or (20.3):

$$1 + GH = 0$$

or $$G_D H_D + G_N H_N = 0 \qquad (20.4)$$

(We call the denominator itself, $1 + GH$, the system's *characteristic function*.)

Example 20.1. Find the transfer function, from input command to output, for the system of Fig. 20.2, if the "controller" box contains the physical elements shown in Fig. 20.6a.

The transfer function of the RC network in Fig. 20.6a has been derived on a number of previous occasions [e.g., Eq. (14.8)]: $\mathsf{M}_c/\Theta_e = K/(RCs + 1)$. The

† Quantities X, Y, and E are the Laplace transforms of time functions $x(t)$, $y(t)$, and $e(t)$.

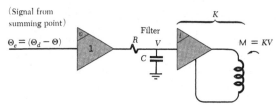

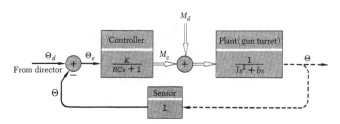

Fig. 20.6 *A mathematical model for the gun-control system of Fig. 20.2*

block diagram representing the complete system of Fig. 20.2 is thus as given in Fig. 20.6b.

The closed-loop transfer function of the system can now be obtained by direct application of Eq. (20.3):

$$\frac{\Theta}{\Theta_d} = \frac{K}{(RCs + 1)(Js^2 + bs) + K} \quad (20.5)$$

Example 20.2. Figure 20.7b constitutes a linear mathematical model of the physical system of Fig. 20.7a.† The purpose of this particular feedback system is to control the angle θ of the rotor in block C to follow the input signal θ_d. This system is somewhat more general mathematically than the one in Fig. 20.6b, in that there are a dynamic subsystem in the feedback path and numerator terms in both the forward and feedback transfer functions. Find the system transfer function, from Θ_d to Θ.

By direct application of Eq. (20.3), the system transfer function is

$$\frac{\Theta}{\Theta_d} = \frac{[K_1(s + a)(K_2/J)](s + d)}{(s + b)(s + d)(s^2 + 2\sigma s + \omega_0^2) + (K_1 K_2 K_3/J)(s + a)(s + c)} \quad (20.6)$$

The closed-loop transfer function is often shown as a single block, Fig. 20.7c, which is convenient if the system is part of a larger system.

Dynamic characteristics

The denominator of (20.6) gives the *characteristic equation* of the system of Fig. 20.7:

$$(s + b)(s + d)(s^2 + 2\sigma s + \omega_0^2) + \frac{K_1 K_2 K_3}{J}(s + a)(s + c) = 0 \quad (20.7)$$

† Determining the values of a, b, c, d, σ, ω_0, and the K's in Fig. 20.7b from the physical parameters in Fig. 20.7a is left as an exercise, Problem 20.26.

[Sec. 20.5] THE ALGEBRA OF LOOP CLOSING 641

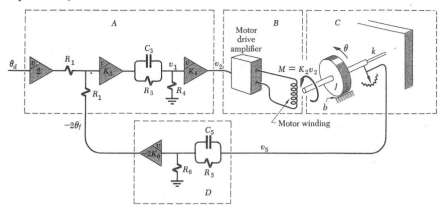

Notes:

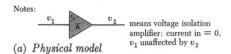

means voltage isolation amplifier: current in $= 0$, v_1 unaffected by v_2

(a) *Physical model*

Motor drive amplifier is considered a current source, so that the torque M produced on the rotor is $M = K_m i = K_2 v_2$

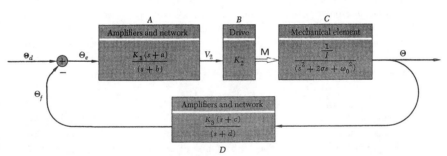

(b) *Block diagram of subsystem arrangement: mathematical model is in transfer-function form*

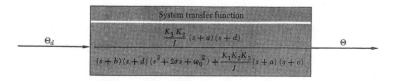

(c) *Closed-loop transfer function shown as a one-block system*

Fig. 20.7 *Steps in block-diagram analysis*

the *roots* of which are the dynamic characteristics of the system. Let us call the (unknown) roots of (20.7) $s = -\sigma_1$, $s = -\sigma_2$, and $s = -\sigma_3 \pm j\omega_3$. (There must be four roots because (20.7) is fourth order. We have assumed, for demonstration, that two are real and two are complex.) Then (20.6) can be written, alternatively, with its denominator in factored form:

$$\frac{\Theta}{\Theta_d} = \frac{[K_1(s+a)(K_2/J)](s+d)}{C(s+\sigma_1)(s+\sigma_2)(s^2+2\sigma_3 s + \omega_{03}^2)} \tag{20.8}$$

where $\omega_{03}^2 = \sigma_3^2 + \omega_3^2$. In a similar way, the reader may rewrite expression (20.5) with its denominator in factored form (using letters to represent the unknown factors).

The form (20.8) is, of course, a more useful form than (20.6), because the possible natural motions are evident at once from the denominator factors, and because the total response of the system, for any specified $\Theta_d(s)$ can be obtained by routine partial fraction expansion (Procedure D-2). Because it is normally necessary to factor the denominator of the system transfer function to see fully the system behavior, a major portion of Part E will be devoted to techniques for solving the characteristic equation of a feedback system for its roots (which *are* the desired denominator factors).

It should be pointed out, however, that some important information about the system can be deduced without actually finding its characteristic roots. First, the *form* of the denominator of (20.8) can be compared with the denominator of (20.6). For example, we can imagine that each of the two denominators is expanded by powers of s, as:

$$A_n s^n + \cdots + A_3 s^3 + A_2 s^2 + A_1 s + A_0 \tag{20.9a}$$

(In Fig. 20.7, n happens to be 4.) Then, since the two denominators are equivalent, their coefficients of like powers of s must be equal. In particular, the coefficients of s^0 and of the highest power, s^n, can be written by inspection and equated. Thus, for equal A_4's in (20.6) and (20.8):

$$C = 1 \tag{20.9b}$$

and, for equal A_0's,

$$bd\omega_0^2 + \frac{K_1 K_2 K_3}{J} ac = \sigma_1 \sigma_2 \omega_{03}^2 \tag{20.9c}$$

which will be useful to know (e.g., as a check on the roots actually calculated).

Next, we might actually multiply out the denominator of (20.6) completely to obtain the form (20.9a), and then use Routh's method (Procedure C-4) to inspect the stability of the system. (Alternatively, one of the frequency-response techniques mentioned in Sec. 20.3 might be used to study system characteristics and stability indirectly from sinusoidal response. In some problems this indirect procedure is the most efficient.)

But to obtain direct knowledge of a feedback system's dynamic characteristics, and thereby of its response to many kinds of inputs—particularly

[Sec. 20.5] THE ALGEBRA OF LOOP CLOSING 643

abrupt ones—the most straightforward procedure is first to factor the denominator of the closed-loop transfer function, i.e., to find the roots of the system's characteristic equation.

To summarize, for an l.c.c. feedback system whose open-loop transfer functions are given (as in Fig. 20.7), the dynamic behavior of the closed-loop system may be analyzed, using Procedure D-3, with the steps described in the more specific manner of Procedure E-1.

Procedure E-1
Dynamic investigation of feedback systems (special version of Procedure D-3)

STAGE I. Establish a PHYSICAL MODEL.
STAGE II. Derive the EQUATIONS OF MOTION (Procedures A).
STAGE III. Study the DYNAMIC BEHAVIOR (Procedure D-1):
  Step 1. ***Laplace transform*** the equations of motion.
  Step 2. ***Manipulate*** the resulting algebraic equations to obtain:
   (a) *Transfer functions* and *response functions:*
    (i) Components of open-loop transfer function (block diagram).
    (ii) Closed-loop system transfer function (denominator is system characteristic function).
    (iii) Factored form of (ii) (factors are unknown). Coefficients of s^0 and s^n in (iii) and (ii) must match.
    (iv) Response functions of interest.
   (b) *Roots of the characteristic equation* (20.4). Plot versus feedback gain (e.g., using Procedure C-2 or E-2). Study system *natural behavior.*
   (c) *Final and initial values* (FVT and IVT).
  Step 3. ***Inverse transform*** (for chosen gain) to obtain system time response of interest (Procedure D-2).
STAGE IV. Carry out SYSTEM DESIGN, on the basis of results of Stage III.

Once the analysis of Stage III has been made, the system *performance* can be evaluated, usually in two or three of the categories:

(a) System characteristics.
(b) Steady-state error.
(c) Total response.

Problems 20.11 through 20.26

Chapter 21

EVANS' ROOT-LOCUS METHOD

A number of algebraic and graphical techniques have been developed to facilitate the study of dynamic systems. Among them the root-locus method of Evans occupies a central position, because it deals directly, efficiently, and visually with a problem that is central to all studies of system dynamics: the determination of the characteristic roots for linear-system models—the eigenvalues which govern stability and natural behavior. The root-locus method was invented by W. R. Evans around 1947, and was first described in his paper, "Control System Synthesis by the Root Locus Method" and in his book, *Control-System Dynamics*.†

Evans' method is used almost universally in control-system analysis, and is valuable in the study of many other dynamic systems, as illustrated in Chapters 11 and 12, for example. It is emphasized that while control-system analysis, as introduced in Chapter 20, provides a strong motivation for learning the root-locus method (and did for its invention), the method has much broader applicability. Fundamentally, it is a method for finding the roots of a linear algebraic equation *such as the characteristic equation of a dynamic system*.

Basic concepts

(1) When the characteristic equation of a system can be written in the form (21.1)

$$(s + \sigma_1)(s + \sigma_2) \cdots (s + \sigma_n) + K(s + a_1)(s + a_2) \cdots (s + a_m) = 0 \quad (21.1)$$

with K a parameter of interest, the root-locus method provides an efficient graphical means for finding the roots, and presents them as a locus, versus K, in the s plane. The root-locus method merely solves the characteristic equation.

† References are given in footnote, p. 401.

[Sec. 21.1] THE BASIC PRINCIPLE   645

(2) The basic process of finding the roots is divided into two parts:
  ① Finding the loci of roots in the s plane for *all* possible values of K by satisfying a vector-*angle* relation.
  ② Determining K for specific points on the loci by satisfying a vector-*magnitude* relation.
(3) Part ① is greatly facilitated by a few sketching rules which are easy to learn and to use, and which give remarkably accurate loci.
(4) More accurate calculation for particular points is facilitated by the Spirule, a simple plastic device for adding angles in Part ① and for multiplying lengths in Part ②.
(5) When the loci cross the imaginary axis, indicating the condition of neutral stability, Routh's method provides a cross-check with Evans' which may be quite helpful.

21.1 THE BASIC PRINCIPLE

An essential step in the study of any dynamic system is finding the roots of its characteristic equation. Often in dynamic systems of all kinds, and almost always for feedback control systems, the characteristic equation can be written in the form

$$P(s) + KZ(s) = 0 \qquad (21.1a)$$

where $P(s)$ and $Z(s)$ are already factored. For example, we might have

$$(s + \sigma_1)(s + \sigma_2 - j\omega_2)(s + \sigma_2 + j\omega_2)(s + \sigma_4) \cdots (s + \sigma_n)$$
$$+ K(s + a_1)(s + a_2) \cdots (s + a_m) = 0 \qquad (21.1b)$$

where K is a parameter of the system. The *roots* of (21.1) are the poles of transfer functions of the overall system. For a feedback system, in particular, they are the poles of the closed-loop transfer function from input to output. The roots of (21.1) therefore indicate the character of the system's natural behavior.

Often it is of special interest to know how the roots are affected by the value of K, as in a control system where K is the loop gain. For example, the characteristic equation for the system of Fig. 11.14 (p. 398) is, from (11.26),

$$s\left(s + \frac{b}{J}\right)\left(s + \frac{1}{RC}\right) + \frac{K}{JRC} = 0$$

and we are interested in how the roots vary with K (Fig. 11.15). For the more elaborate control system of Fig. 20.7 the characteristic equation is

$$(s + b)(s + d)(s + \sigma - j\omega)(s + \sigma + j\omega) + \frac{K_1 K_2 K_3}{J}(s + a)(s + c) = 0$$

and we are interested in how the roots vary with K_1. For other systems, we may wish to study variations with some other parameter. Thus, for the mechanical clutch of Fig. 11.16 the characteristic equation can be written

$$s\left(s^2 + \frac{k}{J_2}\right) + b\left(\frac{J_1 + J_2}{J_1 J_2}\right)\left(s^2 + \frac{k}{J_1 + J_2}\right) = 0$$

and we are interested in how the roots vary with b.

For a control system (such as the first two examples above), when the characteristic equation is in the form (21.1), the first set of factors, $P(s)$, are the poles of the open-loop transfer function, and the second set, $Z(s)$, are the zeros of the open-loop transfer function, i.e., from v_1 to $K_3\theta$ in Fig. 11.14, and from Θ_e to Θ_f in Fig. 20.7. [This is, of course, the reason for the designation $P(s)$ and $Z(s)$.] Using the symbols of Sec. 20.5 and Fig. 20.5, we can write (21.1) also in the form of (20.4):

$$G_D(s)H_D(s) + G_N(s)H_N(s) = 0$$

That is,

$$P(s) = G_D(s)H_D(s) \qquad Z(s) = G_N(s)H_N(s)/K$$

The root-locus method is a graphical procedure for finding the roots of a characteristic equation like (21.1) by taking advantage of the fact that it is already partially factored. In general, the roots of any numerical algebraic equation of more than second order which can be put in the form (21.1) can be found more quickly by the root-locus method (to about one per cent) than by any other nonmachine method. Moreover, the roots are obtained for *all* values of K.

Procedure C-2 is given again here in more general form.

The object is to find values of s which satisfy (21.1)—to find the *roots* of (21.1). To begin the root-locus method, the characteristic equation is rewritten in the Evans form: $$C(s) \triangleq \frac{P(s)}{Z(s)} = \frac{(s + \sigma_1)(s + \sigma_2 - j\omega_2)(s + \sigma_2 + j\omega_2)(s + \sigma_4) \cdots (s + \sigma_n)}{(s + a_1)(s + a_2) \cdots (s + a_m)}$$ $$= -K = |K|\,\underline{/180°} \qquad (21.2)$$ wherein the left-hand side is totally factored, and $0 < K < \infty$ is of interest. We call $C(s)$ the system's *Evans function*. Then, restated, the object is to find values of s that make $C(s)$ equal to $-K$: That is, the *magnitude* of $C(s)$ is to be K and its *phase angle* is to be 180°.	**Procedure C-2** *Root-locus method: basic construction* (repeated)

Note that, for feedback systems, $C(s)$ is simply the reciprocal of the open-loop transfer function with $K = 1$. That is, since $G(s)H(s) = KZ(s)/P(s)$ we have

$$C(s) = \frac{1}{G(s)H(s)/K}$$

where K is the overall loop gain.

[Sec. 21.1] THE BASIC PRINCIPLE 647

When using (21.2) to study control systems it is customary to refer to factors $P(s)$ in (21.2) as the "system poles" (or sometimes less carefully as "the poles"), and to refer to factors $Z(s)$ as "the system zeros" (or "the zeros"), because factors $P(s)$ are the poles of the system's open-loop transfer function, and factors $Z(s)$ are its zeros. Moreover, we plot factors $P(s)$ as $\times$'s, and factors $Z(s)$ as $\bigcirc$'s.

Factors $P(s)$ happen not to be the poles of $C(s)$. [They appear, in fact, as zeros in $C(s)$.] Nevertheless, they are still the *open-loop system* poles, and there is no reason why they cannot maintain this identity, even though we happen to have written them "upstairs" in (21.2). Confusion is easily avoided by referring to factors $P(s)$ *either* as "the *open-loop system* poles," or "the zeros of $C(s)$." Both are correct titles. The analogous statement for factors $Z(s)$ is evident. Further possible confusion is circumvented by the pronunciation "crosses" and "circles" for the s-plane symbols.

The question arises, of course, why not write $C(s)$ in the form

$$\frac{Z(s)}{P(s)} = -\frac{1}{K}$$

The answer is that (21.2) is a more convenient form for the algebraic manipulations involved in actual root-locus plotting and interpretation (as will be evident in all that follows). We therefore choose the form (21.2).

Many people do choose to use the above-suggested form, however, trading some computational convenience for the satisfaction of seeing factors $P(s)$ downstairs, and $Z(s)$ upstairs, "where they belong."

Continuing the root-locus method, the roots of (21.1) are next plotted in the s plane for the two extreme cases $K = 0$ and $K = \infty$. For $K = 0$, the roots of (21.1) are the numerator factors in (21.2) $[s = -\sigma_1, s = -\sigma_2 + j\omega_2, \text{etc.}]$; these are plotted as $\times$'s. [For a feedback system, they are the poles of $G(s)H(s)$.] For $K = \infty$, the roots of (21.1) are the denominator factors of (21.2) $[s = -a_1, s = -a_2, \text{etc.}]$; these are plotted as $\bigcirc$'s. [For a feedback system, they are the zeros of $G(s)H(s)$.] A root locus departs from each $\times$ as K increases from zero; a root locus arrives at each $\bigcirc$ as K approaches ∞.†

The roots for values of K between 0 and ∞ are found in two parts:

Part ① Find the loci (in the s plane) of all points s for which $\underline{/C(s)} = 180°$.‡

Part ② Evaluate $|C(s)| = K$ at selected points on the loci of Part ①.

Graphical techniques for performing each part quickly are given in Procedure E-2.

Procedure C-2
Root-locus method: basic construction (continued)

Procedure C-2 is best explained by example.

† Other loci may approach certain asymptotes as $K \to \infty$, as will be seen.
‡ These are called "180° loci." (The companion case of $-\infty < K < 0$ involves 0° loci, and is considered later.)

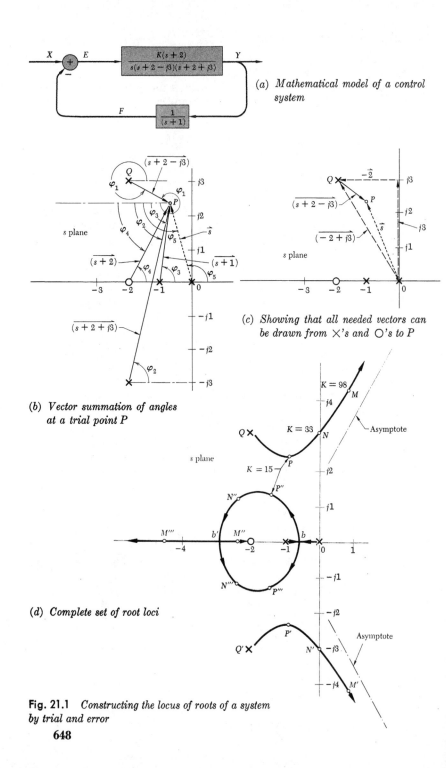

(a) *Mathematical model of a control system*

(b) *Vector summation of angles at a trial point P*

(c) *Showing that all needed vectors can be drawn from ×'s and ○'s to P*

(d) *Complete set of root loci*

Fig. 21.1 *Constructing the locus of roots of a system by trial and error*

[Sec. 21.1] THE BASIC PRINCIPLE 649

Example 21.1. A control system. For the feedback control system shown in Fig. 21.1a, find the roots of the characteristic equation for several values of K. Use the result to select a reasonable value for K, and discuss the corresponding system natural behavior.

Using (20.4), we find that the characteristic equation of the system is

$$s(s + 2 - j3)(s + 2 + j3)(s + 1) + K(s + 2) = 0$$

which we put in the Evans form for root-locus analysis:

$$C(s) = \frac{s(s + 2 - j3)(s + 2 + j3)(s + 1)}{(s + 2)} = -K \quad (21.3)$$

Next, the numerator factors are plotted as ×'s, Fig. 21.1b, and the denominator factor as a ○. We note that for $s = 0$ or $s = -2 + j3$ or $s = -2 - j3$ or $s = -1$, we have $K = 0$; we note that for $s = -2$, we have $K = \infty$.

Now we proceed to find the loci of s for intermediate values of K.

Part ①. To find the loci of roots we begin a search for all possible locations of the tip of the s vector which will make the *phase angle* of (21.3) be 180°:

$$\underline{/C(s)} = \underline{/\frac{s(s + 2 - j3)(s + 2 + j3)(s + 1)}{(s + 2)}} = 180° \pm k(360°)$$

which can also be written [by Eq. (7.26c)]

$$\underline{/s} + \underline{/(s + 2 - j3)} + \underline{/(s + 2 + j3)} + \underline{/(s + 1)} - \underline{/(s + 2)}$$
$$= 180° \pm k(360°) \quad (21.4)$$

where k is any integer.

Consider a trial point P in Fig. 21.1b, for the tip of the $\vec{s}$ vector.† Note that each of the vectors in (21.4)—$\vec{s}$, $\overrightarrow{(s + 2 - j3)}$, $\overrightarrow{(s + 2)}$, and so on—also appears on the s plane and, moreover, each of them terminates at point P. The details are shown in Fig. 21.1c for one of the vectors, namely, the vector $\overrightarrow{(s + 2 - j3)}$. The dashed vector from the origin to the upper × (marked Q) is made up of $+j3$ and -2, and is therefore $\overrightarrow{(-2 + j3)}$. The dotted vector from the origin to point P is $\vec{s}$. Therefore the vector from Q to P is $\overrightarrow{s - (-2 + j3)}$ or $\overrightarrow{(s + 2 - j3)}$. In like manner, each of the other terms in (21.4)—s, $(s + 2 + j3)$, $(s + 1)$, $(s + 2)$—is a vector from a × or ○ to point P. Thus the angles to be added in (21.4) are, respectively, the angles ϕ_1, ϕ_2, ϕ_3, ϕ_4, and ϕ_5 in Fig. 21.1b.

As a practical matter it is an important fact that, alternatively, these same angles can *all* be measured *at point* P, as shown in Fig. 21.1b. This makes it possible to measure *and add* all the angles with a single location of a simple protractor device known as a Spirule. (See footnote, p. 241. Very rapid addition of the angles, using the movable arm of the Spirule and reading only the final sum, is explained in Appendix J.) Thus we can establish quickly whether or not (21.4) is satisfied at point P. If it is, we proceed to look for another point. If

---
† In this explanation we write arrows over the terms, merely as a reminder that each is a vector in the s plane.

it is not, we make another trial. Eventually, we obtain loci of *all* the values of vector s which satisfy (21.4). The loci for this example are drawn in Fig. 21.1d.

The above process may sound slow and tedious, but in fact it is usually quick, for two reasons. First, the loci can be quite closely approximated and sketched in "by eye," using a set of sketching rules to be presented in Sec. 21.2. Second, for the three or four trial-and-error points which are typically required, trials can be made quickly with a Spirule. (Note that the required angles can all be measured *and added*, using the Spirule arm, without actually drawing the vectors, or even noting the individual angles.)

Notice that each of the root loci in Fig. 21.1d emanates from a ×. Recall that for $K = 0$ there is always one root *at each* ×; thus we infer that *a locus of roots begins at each* ×. Notice also that one locus terminates at the ○ as K becomes extremely large. This corresponds to the fact that if $s = -2$, the denominator in (21.3) is 0 and therefore K must be extremely large. We infer that, in the general case, *a locus of roots terminates at each* ○. Certain other values of s satisfy (21.3) when K is "infinite." These s's have infinite magnitude, and in addition they satisfy the phase condition (21.4). They are thus found to lie far out along certain asymptotes, which are drawn in Fig. 21.1d. Construction of the asymptotes will be explained in Secs. 21.3 and 21.5.

Part ②. To evaluate K at any given point s on one of the loci, we could merely draw the vectors—s, $(s + 2 - j3)$, $(s + 2 + j3)$, $(s + 1)$, $(s + 2)$—to that point, measure the *lengths only* of the vectors, and multiply them together as follows [refer to (21.3)]:

$$\frac{|s| \, |s + 2 - j3| \, |s + 2 + j3| \, |s + 1|}{|s + 2|} = K \qquad (21.5)$$

For example, to find the value of K at point P in Fig. 21.1d, the appropriate vectors are those drawn in Fig. 21.1b. When we multiply the lengths of these vectors, as in (21.5), we find that $K = 15$.

Again, the required vector lengths can of course be measured without actually drawing the vectors. For example, the pivoting arm of a Spirule located at P can be swung to each × and ○ in turn, and the vector length read on its linear scale. (Alternatively, the logarithmic spiral can be employed, as explained in Appendix J, to obtain directly the desired product of lengths without noting the individual lengths.) Using (21.5), the reader is invited to verify that the values of K corresponding to points N and M (on the same locus) are 33 and 98.

The values of K along the other two loci are found in the same way. In particular, points P', P'', P''', and N', N'', N''', and M', M'', M''' have the same values of K, respectively, as do points P, N, and M. (For P', N', and M' this is obvious by symmetry. K at the other points must be calculated separately.)

To summarize, in Part ① we constructed the locus of all roots of (21.3), and in Part ② we determined the value of K at specific points. The result, Fig. 21.1d, is a set of roots of (21.3) for each value of K.

System natural behavior versus K can now be inferred directly and precisely from the root-locus plot.† From Fig. 21.1d it is clear, for example, that K should

---

† Because the roots on the root-locus plot are the roots of the overall system's characteristic equation, (21.3): They *are* the system's natural characteristics (its eigenvalues).

[Sec. 21.2] ROOT-LOCUS SKETCHING PROCEDURE

be kept well under 33, to be sure to avoid unstable oscillations. A reasonable value might be $K = 15$, for which that oscillatory motion would be 0.35 critically damped.† Then the system response would be characterized by two damped natural oscillations, one having frequency 2.4 rad/sec, with $\zeta = 0.35$, and the other having natural frequency 1.3 rad/sec, with $\zeta = 0.7$. (If faster response is required, with the same stability, compensation techniques described in Sec. 22.2 may be employed.)

The above example illustrates the basic root-locus idea. But the method is made useful, in practice, by the fact that approximate loci can be sketched quickly following some simple rules. Indeed, such sketches are often adequate for preliminary design decisions. These sketching rules will now be derived and demonstrated. The rules will be more complete than in Chapter 11, and each will be proved.

Problems 21.1 through 21.5

21.2 ROOT-LOCUS SKETCHING PROCEDURE

There are several "sketching rules" which make it possible to construct approximate root loci quickly and accurately before commencing the trial-and-error procedure indicated in Sec. 21.1. These are almost always employed first, after which a few points in regions of interest can be checked by Spirule if greater accuracy is required. Often the sketched loci are adequate for the task at hand. (Sketched loci are usually accurate to about ten percent, and Spirule calculations to one percent.)

By way of outline, a sequence of steps for sketching the loci of roots of a characteristic equation will simply be listed first, without elaboration. After that, the corresponding detailed rules for sketching 180° loci will also be listed concisely (Sec. 21.3), again without elaboration. Then in the following sections each rule will be developed rigorously and demonstrated in detail. The reader may wish to read the "rules" in Sec. 21.3 rather quickly the first time, and then return to them as a summary after studying carefully the development and application of each one.

Summary of sketching procedure

Suppose we are given a characteristic equation of the general form (21.2)

$$C(s) = \frac{P(s)}{Z(s)} = \frac{(s + \sigma_1)(s + \sigma_2 + j\omega_2)(s + \sigma_2 - j\omega_2)(s + \sigma_4) \cdots (s + \sigma_n)}{(s + a_1)(s + a_2) \cdots (s + a_m)}$$
$$= -K \quad (21.2)$$
[repeated]

where n is the number of factors of $P(s)$ [poles of $G(s)H(s)$, for a feedback

† Damping ratio ζ can be read directly from one of the Spirule scales.

system] and m is the number of factors of $Z(s)$ [zeros of $G(s)H(s)$], and $m \leq n$. The graphical process for finding the roots may be separated into two parts, as described in Sec. 21.1:

Part ① Find the loci of all values of s for which $\underline{/C(s)} = 180° + k360°$ (where k is any integer). *There will be n loci.*

Part ② Find $K = |C(s)|$ for specific points on the loci.

To carry out Part ①, first plot as ×'s the factors of $P(s)$ and plot as ○'s the factors of $Z(s)$. A locus of roots will emanate from each × and a locus will terminate at each ○ (when $K \to \infty$). In addition, $n - m$ loci will proceed out to ∞ along certain asymptotes as $K \to \infty$.

The following sequence of steps usually leads to the most rapid construction of the root loci. Often the first two or three steps suffice for sketching loci with adequate accuracy (as in Chapter 11).

After plotting the ×'s and ○'s, sketch elements of the loci in the following steps (refer to Figs. 21.19 and 21.18, for example):

Step 1. Plot loci segments on the *real axis*.
Step 2. Plot high-K *asymptotes*
    (a) from the proper centroid;
    (b) in the proper directions.
Step 3. Sketch the directions in which loci
    (a) *depart* from ×'s;
    (b) *arrive* at ○'s.

Procedure E-2
Part ①:
locus of roots

Step 4. Find the points where loci *break away* from (or into) the real axis.
Step 5. (If desired) use *Routh's method* to find *j-axis crossings*.

In most problems the first two or three steps provide an ample guide. After the above steps (or the first two or three of them) have been taken, the complete loci can usually be sketched quite accurately. Spot trial-and-error checks by Spirule of a few intermediate points in critical regions will yield loci of greater accuracy, if desired. As a qualitative guide, the path of a locus is attracted by nearby ○'s and repelled by nearby ×'s.

Step 5 above is often omitted because it is time-consuming. However, it does give not only the exact locations of j-axis crossings, but also the precise value of K there, so that it is sometimes very handy. Step 4 is also sometimes rather involved, and is often omitted when a reasonable approximation to breakaway points is adequate.

[Sec. 21.3] SKETCHING RULES FOR 180° LOCI

> After loci have been drawn, evaluate K at interesting points by multiplication, as in (21.5). Utilize symmetry.

Two special cases are helpful:

Step 5. (If desired) use *Routh's method* to evaluate K at j-axis crossings.

Procedure E-2
Part ②:
values of K

> Step 6. Use the *root-centroid* rule if $n \geq m + 2$

Step 6 is quick. Step 5 is time-consuming, and so is used only when high accuracy is desired. Rules for sketching loci for the particular, but most usual, case of 180° loci are stated below. The rule numbers correspond to the above step numbers. Again, the first two or three rules often suffice for sketching a locus of adequate accuracy.

21.3 SKETCHING RULES FOR 180° LOCI

The rules which follow apply to finding the roots of any characteristic equation of the form (21.1) and (21.2) (with real constants) for the common case that K is positive, and therefore $/C(s)$ is to be made 180°. (The companion rules for K negative—i.e., for 0° loci—are readily deduced, Problems 21.49 through 21.61.)

The following is a more comprehensive presentation of Procedure C-3, p. 405. It may be helpful to refer to Figs. 21.18 and 21.19, pp. 680 and 681, for example.

Special rule: Symmetry. The pattern of the set of loci is symmetrical with respect to the real axis.†

Rule 1: Real-axis segments. There is a locus segment in each real-axis interval to the right of which the number of ×'s plus ○'s on the real axis is odd.

Rule 2: Asymptotes. As $K \to \infty$, $(n - m)$ loci will approach ∞ along asymptotes determined as follows:

(a) The asymptotes emanate from the centroid of the ×-○ array, ○'s being given negative weight. That is, the centroid is the location $s = -c$ for which

$$c = \frac{(\sigma_1 + \sigma_2 + \cdots + \sigma_n) - (a_1 + a_2 + \cdots + a_m)}{n - m} \quad (21.6)$$

(b) The $(n - m)$ asymptotes have the directions

$$\phi = \pm (2r + 1) \frac{180°}{n - m} \quad (21.7)$$

† Provided that the coefficients of the characteristic equation are real.

where r is each integer, from 0 until ϕ repeats. The array of asymptotes is symmetrical with respect to the real axis.

Rule 3: Directions of departure and arrival.

(a) *Single* ×'s (or ○'s). The direction in which a locus departs from a × at $s = -\vec{\beta}$ (or arrives at a ○ at $s = -\vec{\beta}$) is given by

$$\phi_{dep} = 180° - \underline{/\left[\frac{1}{(s+\vec{\beta})} C(s)\right]_{s=-\vec{\beta}}} \quad \text{for departure from a ×} \tag{21.8}$$

(or $\quad \phi_{arr} = 180° + \underline{/[(s+\vec{\beta})C(s)]_{s=-\vec{\beta}}}$ for arrival at a ○)

That is, if $\underline{/[C'(s)]_{s=-\vec{\beta}}}$ is the net phase angle, measured *at* $s = -\vec{\beta}$ of all terms in $C(s)$ *except* $s = -\vec{\beta}$, then ϕ is the *additional* angle required to make 180.°

Directions ϕ_{dep} and ϕ_{arr} can be quickly determined with a Spirule without measuring any angles, as follows:

For departure from a single ×, place Spirule over the ×, and add in angles to all other ×'s and subtract angles to all ○'s. Then set arm horizontal to the left, and mark departure direction at disk 0° index.

For arrival at a single ○, place Spirule over the ○, and add in angles to all ×'s and subtract angles from all other ○'s. Then set disk 0° index on 180° reference line (horizontal to the left) and mark arrival direction under the Spirule arm.

(b) *Multiple* ×'s (or ○'s). If there are k ×'s (or ○'s) in a single place, then the k departure (or arrival) directions are given by

$$\phi_{dep} = \frac{1}{k}\left\{180° - \underline{/\left[\frac{1}{(s+\vec{\beta})^k} C(s)\right]_{s=-\vec{\beta}}}\right\} + \frac{360r}{k} \tag{21.9}$$

$$\left(\text{or} \quad \phi_{arr} = \frac{1}{k}\{180° + \underline{/[(s+\vec{\beta})^k C(s)]_{s=-\vec{\beta}}}\} + \frac{360r}{k}\right)$$

where r is any k consecutive integers, such as 0 to $\pm k/2$ or $\pm (k-1)/2$. The k directions will be disposed at equal angles from one another.

(c) *Special case: Multiple ×'s (or ○'s) located on real axis.* If k ×'s (or ○'s) are located at a single point on the real axis, then, of the k directions of departure (or arrival), zero, one, or two will be along the real axis, according to Rule 1; and the k directions will be disposed at equal angles of $360°/k$ from one another, and symmetrically with respect to the real axis.

Rule 4: Breakaway from the real axis.

(a) *All ×'s and ○'s on real axis.* Loci will break away from (or into) the real axis at points $s = -b$, given by the equation

$$\sum_{k=1}^{n} \frac{1}{\sigma_k - b} - \sum_{\ell=1}^{m} \frac{1}{a_\ell - b} = 0 \tag{21.10a}$$

[Sec. 21.4] RULE 1: REAL-AXIS SEGMENTS 655

Alternatively, points $-b$ can be located as the roots of

$$\frac{d}{ds}[C(s)] = 0 \qquad (21.11)$$

[Form (21.10a) is quicker for trial-and-error evaluation.]

(b) *General case involving complex ×'s or ○'s.* For each pair of ×'s (or ○'s) at locations $s = -\sigma_q - j\omega_q,\ -\sigma_q + j\omega_q$, a term

$$\frac{2(\sigma_q - b)}{(\sigma_q - b)^2 + \omega^2} \qquad (21.10b)$$

must be added (or subtracted) in (21.10a). [Equation (21.11) may be used verbatim.]

When two loci meet on the real axis, they leave (or approach) the real axis at $\pm 90°$. In singular cases where more than two loci depart from (or arrive at) the same point on the real axis, then all the arriving and departing loci branches form a symmetrical, equal-angle array at the point.

As a qualitative guide, a breakaway point will be "repelled" by nearby ×'s and "attracted" by nearby ○'s.

Rule 5: j-Axis crossings by Routh (refer to Sec. 11.11).

If the characteristic equation is cast in a Routh array, then each value of K that produces a *double* sign change in the first column corresponds to the j-axis crossing of a pair of root loci. The crossing point is given by Routh's auxiliary equation. A value of K that produces a *single* sign change in the first column corresponds to the j-axis crossing of a single locus along the real axis.

Rule 6: Root-centroid invariance.

If $m + 2 \leq n$ (i.e., at least two more ×'s than ○'s), then as K is varied, the centroid of *all* the roots remains fixed at the centroid $c_\times$ of the ×'s:

$$c_\times = \frac{(\sigma_1 + \sigma_2 + \cdots + \sigma_n)}{n - m}$$

21.4 RULE 1: REAL-AXIS SEGMENTS

Proof of the rule

Consider a characteristic equation whose ×'s and ○'s are plotted in the s plane, as shown in Fig. 21.2, and consider a point P on the real axis. We ask "is point P on a locus of roots of the characteristic equation?" To answer, we draw a vector to P from each × and ○, and then apply (21.4). If the sum of all the vector angles is 180°, then P is on the locus; otherwise, it is not.

To show that this leads directly to Rule 1 (Sec. 21.3), first separate all the ×'s and ○'s into three categories: (1) complex pairs not on the real axis,

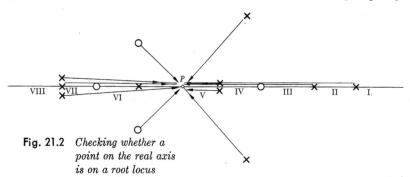

Fig. 21.2 *Checking whether a point on the real axis is on a root locus*

(2) ×'s and ○'s on the real axis to the left of P, and (3) ×'s and ○'s on the real axis to the right of P. Then note the following:

(1) The vectors from a complex pair of ×'s or ○'s to point P always have equal but opposite phase angles, and thus each pair contributes a net phase angle of 0°.

(2) The vector from any × or ○ on the real axis to the left of P has a phase angle of 0° and thus the contribution is again 0°.

(3) The vector from any × or ○ on the real axis to the right of P has a phase angle of 180°. Thus, if the total number of ×'s plus ○'s is odd, then their total phase angle is $180° + k360°$ (where k is some integer), and P will thus be on a locus of roots, by (21.4). If, however, the total number of ×'s plus ○'s is even, then their total phase angle is $k360°$, and P will thus not be on a locus.

Application

Thus, applying Rule 1 to the ×-○ array of Fig. 21.2, for example, we determine the following location for loci on the real axis:

Interval	Number of real-axis ×'s plus ○'s to the right	Locus?
I	none	no
II	1	yes
III	2	no
IV	3	yes
V	5	yes
VI	6	no
VII	7	yes
VIII	10	no

[Sec. 21.5] RULE 2: ASYMPTOTES

Example 21.2. Real-axis segments. Verify the real-axis root-locus locations in Fig. 21.7, p. 663. Solution:

Interval	Number of real-axis X's plus O's to the right	Locus?
I	0	no
II	1	yes
III	4	no
IV	8	no
V	9	yes

Problem 21.6

21.5 RULE 2: ASYMPTOTES

(b) Directions

As $K \to \infty$, those loci that do not approach O's must proceed to points far from the origin, for the magnitude condition, (21.5), to be satisfied. While they are doing so, however, they must continue to satisfy the phase condition, (21.4).

Suppose a characteristic equation is represented by an array having n X's and m O's, as in Fig. 21.3a. From a point P a large distance from the array, Fig. 21.3b, the vectors from the X's and O's to point P will *seem all to come from a single point* quite near the origin. The net number of vectors will be $n - m$, because the m vectors from O's will just cancel m of the vectors from X's. To find out whether point P is on a root locus, we need only add up the angles ϕ of the $n - m$ vectors to see whether they total 180°. In the limit, as P becomes very far from the origin, the distances between the X's and O's vanish compared to the length of the vectors to P, and all the angles ϕ become equal. For this case (21.4) can be written

$$(n - m)\phi = 180° + r360°$$

or, *the $(n - m)$ asymptotes have the directions*†

$$\phi = \pm(2r + 1)\frac{180°}{n - m} \qquad m < n \qquad \begin{array}{c}(21.7)\\ \text{[repeated]}\end{array}$$

where, for convenience, $r = 0, \pm 1, \ldots, \pm(n - m)/2$, or $\pm(n - m - 1)/2$. That is, the loci of the roots that go to infinity approach asymptotes which

† If $m = n$ no loci go to infinity. For $n < m$ the present methods can be applied by inverting $C(s)$ at the outset.

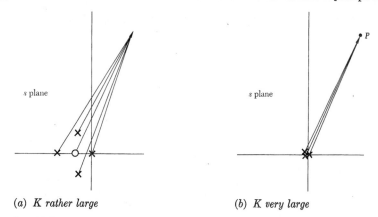

(a) K rather large   (b) K very large

Fig. 21.3 *The special case of roots located far from the pole-zero array*

are an array of straight lines having directions ϕ given by (21.7), and emanating from a common point.

(a) Point of emanation

By symmetry, the point from which the asymptotes emanate is on the real axis. To find its location, $s = -c$, we consider the characteristic equation of a mythical system having $n - m$ ×'s all located at point $s = -c$:

$$(s + c)^{n-m} = -K \qquad (21.6a)$$

Next we determine the value of c that will make the mythical system have the same asymptotes as the real one in which we are interested. To do this we perform a binomial expansion on both the mythical and the real characteristic equations, and then drop small terms produced when s becomes large, until we obtain expressions having the same form in both cases. Details of the expansion are left as an exercise. The result is that expansion of (21.6a) for the mythical system leads to

$$s^{n-m}\left[1 + (n-m)\frac{c}{s} + \frac{(n-m)(n-m-1)}{2!}\left(\frac{c}{s}\right)^2 + \cdots\right] = -K \qquad (21.6b)$$

while, for the real system represented by†

$$C(s) = \frac{(s+\sigma_1)(s+\sigma_2-j\omega_2)(s+\sigma_2+j\omega_2)(s+\sigma_4)\cdots(s+\sigma_n)}{(s+a_1)(s+a_2)\cdots(s+a_m)} = -K$$

appropriate double expansion leads to

$$s^{n-m}\left[1 + \frac{(\sigma_1 + \sigma_2 + \cdots + \sigma_n) - (a_1 + a_2 + \cdots + a_m)}{s} - \cdots\right] = -K \qquad (21.6c)$$

† A complex pair is included to show formally that the point $s = -c$ is on the real axis; the $j\omega_2$'s cancel at once in the binomial expansion.

[Sec. 21.5] RULE 2: ASYMPTOTES

Comparison of the second term in (21.6c) with the second term in (21.6b) gives the desired result:

$$c = \frac{(\sigma_1 + \sigma_2 + \cdots + \sigma_n) - (a_1 + a_2 + \cdots + a_m)}{n - m} \quad (21.6) \text{ [repeated]}$$

That is, *the asymptotes emanate from the point $s = -c$, which is the centroid of the $\times$, $\bigcirc$ array, the $\bigcirc$'s being given negative weight.*

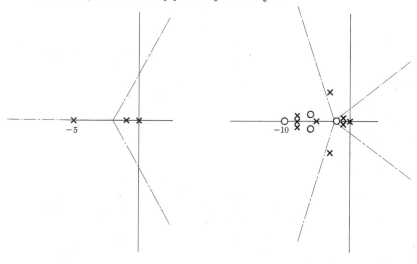

(a) *Simple system*   (b) *More complicated system*

Fig. 21.4 *Examples of root-locus asymptotes*

Example 21.3. Asymptotes. If the characteristic equation of a system is

$$s(s + 1)(s + 5) = -K$$

find the asymptotes and centroid of the root loci. The $\times$'s for this system are plotted in Fig. 21.4a. By (21.6) the centroid is at $s = -c$, where

$$s = -c = -\frac{0 + 1 + 5}{3 - 0} = -2$$

and by (21.7) the $3 - 0 = 3$ asymptote directions are

$$\phi = \left[\pm(2r + 1)\frac{180°}{3 - 0}\right]_{r=0,1} = \pm 60°, 180°$$

These asymptotes are drawn in Fig. 21.4a.

Example 21.4. Asymptotes. Find the asymptotes and centroid for the characteristic equation

$$\frac{s(s + 1)^2(s + 5)(s + 3 \mp j4)(s + 8)^3}{(s + 2)(s + 10)(s + 6 \mp j1)} = -K$$

The ×'s and ○'s for this system are plotted in Fig. 21.4b. By (21.6) the centroid is at

$$s = -c = -\frac{[0 + 2(1) + 5 + 2(3) + 3(8)] - [2 + 10 + 2(6)]}{9 - 4} = -2.6$$

and by (21.7) the $9 - 4 = 5$ asymptote directions are

$$\phi = \left[\pm(2r + 1)\frac{180°}{5}\right]_{r=0,1,2} = \pm 36°,\ \pm 108°,\ 180°$$

These asymptotes are drawn in Fig. 21.4b.

Problems 21.7 and 21.8

21.6 RULE 3: DIRECTIONS OF DEPARTURE AND ARRIVAL

(a) Single ×'s or ○'s

Consider the characteristic equation

$$\frac{s(s + 1 - j2)(s + 1 + j2)}{s + 3} = -K$$

the ×'s and ○'s of which are plotted in Fig. 21.5. We know that a locus emanates from the upper ×, marked Q, at $s = -1 + j2$, and we wish to determine its direction of departure.

Let us draw a small circle around the × at point Q and determine the location of a point P, on the circle, which is also on the locus. Then we can let the circle shrink to an arbitrarily small size and note the limiting direction from Q to P.

First, select any arbitrary point P on the small circle and draw all the vectors from ×'s and ○'s to point P, *except the vector from Q*. Next, add the angles of all the vectors from ×'s and subtract those from ○'s, and deduct

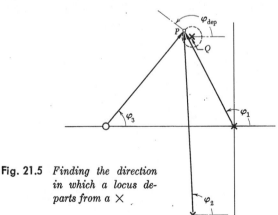

Fig. 21.5 *Finding the direction in which a locus departs from a ×.*

[Sec. 21.6] RULE 3: DIRECTIONS OF DEPARTURE AND ARRIVAL   661

the total from 180°. The result, by (21.4), is the angle ϕ_{dep} must have if P is on the locus *departing from the* $\times$:

$$\phi_{dep} + \phi_1 + \phi_2 - \phi_3 = 180° + k360°$$
or $$\phi_{dep} = 180° - (\phi_1 + \phi_2 - \phi_3) + k360°$$
or, more generally, $$\phi_{dep} = 180° - \Sigma\phi_{other} \quad (21.8a)$$

where $\Sigma\phi_{other}$ is the sum of the phase angles of the vectors from all the $\times$'s minus those from $\bigcirc$'s, and excepting the one from point Q. (The term $k360°$ is dropped because it does not affect ϕ_{dep}.) In the limit, as the circle is made vanishingly small, the summation of angles is made *at* point Q, and (21.8a) gives the direction of departure.

For arrival at a zero the procedure is exactly the same, except that the angle ϕ_{arr} will be *subtracted* from the other angles to make the required $180° + k360°$. The result is that, for a locus *arriving at a* $\bigcirc$:

$$\phi_{arr} = 180° + \Sigma\phi_{other} \quad (21.8b)$$

(Note that either $+$ or $-$ 180° gives the same answer.)

If a Spirule is used, the addition and subtraction of all the vectors from the other $\times$'s and $\bigcirc$'s to point Q can, of course, be performed without reading any of them individually, using a single placement of the Spirule at point Q. After this is done, the departure (or arrival) direction is the angle ϕ *between the Spirule arm R and the 180° line on the disk*. It is easy to show that, for departure from a $\times$, if the Spirule assembly is now rotated until arm R is horizontal to the left, the 0° index on the disk will point in the departure direction (or, for arrival at a $\bigcirc$, if the assembly is rotated until the disk 0° index points horizontal to the left, the arm will indicate the arrival direction).

As a check, one may mark the departure direction, then move the Spirule center (without rotation) to any point along the departure-direction line, and add the final angle by moving the arm from horizontal to the $\times$ at Q. The Spirule should now read 180°. A similar check may be made on an arrival direction.

Example 21.5. Departure from a $\times$. Find the direction of departure from the $\times$'s at points Q and Q' in Fig. 21.6.

Summation of the angles at Q, as in (21.8a), gives 156°, and thus

$$\phi_{dep} = 180 - 156 = 24°$$

which is plotted as shown. (Departure from Q' is plotted by symmetry.) The reader should check this calculation with a Spirule.

Example 21.6. Arrival at a $\bigcirc$. Find the direction of arrival at the $\bigcirc$ at point R in Fig. 21.6.

Summation of the angles, as in (21.8b), gives 126°, and thus

$$\phi_{arr} = 180° + 126° = 306°,$$

which is plotted as shown. Again, a Spirule check should be made.

(b) Multiple ×'s or ○'s

Consider next a characteristic equation such as

$$\frac{s[(s+1-j2)(s+1+j2)]^k}{s+3} = -K$$

It is left as an exercise to show that the angle of departure from such a multiple pole of order k is given by

$$\phi_{\text{dep}} = \frac{1}{k}[180° - \Sigma\phi_{\text{other}} + r(360°)] \quad (21.9a)$$

where r is each integer from 0 to $\pm k/2$ or $(k-1)/2$. (Quantity r may have any two consecutive integer values; these are usually the most convenient.) The angle between adjacent directions of departure is $360°/k$.

The companion expression for arrival at a ○ of order k is

$$\phi_{\text{arr}} = \frac{1}{k}[180° + \Sigma\phi_{\text{other}} + r(360°)] \quad (21.9b)$$

Again, the angle between adjacent directions of arrival is $360°/k$.

Example 21.7. Departure from a double cross. Suppose the ×'s at Q and Q' in Fig. 21.6 to be double ×'s. Then what will be the angles of departure from Q?

Summation of the angles, as in (21.9a), gives 246°. (This is 90° more than in Example 21.5 because there are now two vectors from Q' to Q.) Then, by (21.9a),

$$\phi_{\text{dep}} = \tfrac{1}{2}[180° - 246° + r(360°)]_{r=0,\pm 1}$$
$$= -33°, 147°$$

Note that the two directions of departure are 180° apart. Why?

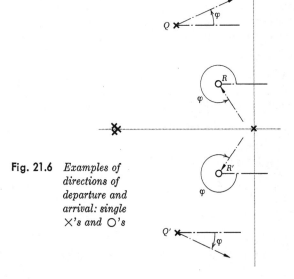

Fig. 21.6 *Examples of directions of departure and arrival: single ×'s and ○'s*

Example 21.8. Arrival at a triple circle.
Suppose the ○'s at R and R' in Fig. 21.6 to be triple ○'s. Then what will be the angles of arrival at R?

Using (21.9b), we obtain

$$\phi_{arr} = \tfrac{1}{3}[180° + 306° + r(360°)]_{r=0,\pm 1}$$
$$= 162°, 282°, 42°$$

Note that the three directions of arrival are 120° apart. Why?

(c) Special case: multiple ×'s or ○'s on the real axis

When the multiple ×'s and ○'s are on the real axis only, the directions of arrival and departure can be determined at once, by Rule 1 and by symmetry, without resorting to the calculation of (21.9). Loci will or will not depart and arrive along the real axis *in accordance with Rule 1*. Furthermore, by (21.9) the remaining departure or arrival directions will be symmetrically disposed with respect to the real axis, the angle between them being $360°/k$.

Example 21.9. Departure from a multiple ×.
Find the angles of departure from the triple × on the real axis in Fig. 21.7.

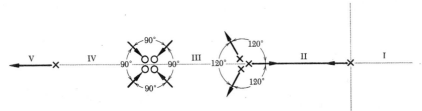

Fig. 21.7 *Rule 1 helps establish directions of departure from multiple ×'s and arrival at multiple ○'s on the real axis*

By Rule 1, there is a locus on the real axis to the right of the triple × (and none to the left), as shown. Altogether there will be three loci departing, so the angles between them must be $360°/3 = 120°$, and the other two loci must be at 120° and 240°, as shown.

Example 21.10. Arrival at a multiple ○.
Find the angles of arrival at the quadruple ○ in Fig. 21.7.

By Rule 1 there is no locus on the real axis either immediately to the right or immediately to the left of the ○'s (i.e., there is no locus in interval III or IV). Altogether there must be four loci arriving at the ○'s, so the loci must be separated by 90°. Then, by symmetry, the loci must arrive at ±45°, ±135°.

Some additional arrangements of multiple ×'s and ○'s on the real axis are cataloged in Fig. 21.8. The reader should be sure he agrees with the directions of departure or arrival in each case.

Problems 21.9 through 21.16

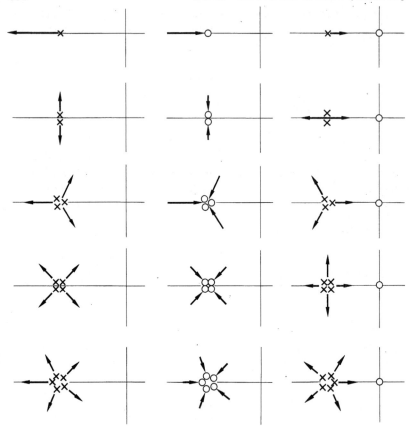

Fig. 21.8 *Root-locus departure from multiple ×'s and arrival at multiple ○'s on the real axis: some typical situations*

21.7 RULE 4: BREAKAWAY FROM THE REAL AXIS†

In many problems the exact point at which a pair of loci break away from (or into) the real axis (as points b, b' in Fig. 21.1d) is not crucial, so a quick, educated guess is made and the solution carried forward. However, in cases where the exact location of a breakaway point (or a break-in point) is important, it may be calculated exactly as follows. Two forms—(21.10) and (21.11)—are given for the calculation. One or the other may be easiest to apply in a given situation: Form (21.11) is more direct, but is often quite cumbersome to carry out. In such cases a trial-and-error solution will be much quicker than direct solution, and form (21.10) is more rapid for trial-and-error solution. Both forms will be derived. Form (21.10) is most easily seen graphically, and will be derived on geometric grounds. Form (21.11) is

† As noted, Rules 4 and 5 are seldom used. Accordingly, Secs. 21.7, 21.8, 21.9 may be omitted, if desired, without loss in continuity.

[Sec. 21.7] RULE 4: BREAKAWAY FROM THE REAL AXIS

derived algebraically and is more general, locating any other "saddle points" which exist on the loci in addition to those that are breakaways from (or break-ins to) the real axis.

(a) ×'s and ○'s on the real axis only

In deriving form (21.10) we begin with the case in which all ×'s and ○'s are on the real axis, which is a little easier to see. Consider the ×−○ array shown in Fig. 21.9a. By Rule 1, there will be real-axis loci in intervals II, IV, V, and VI. The loci in interval II will clearly have to break away from the axis because they cannot terminate there. Similarly, there must be a break *into* the real axis in interval IV, since no locus originates there. In general there may also be "breakaways" and "break-ins" at any other real-axis interval containing a locus.

To set up relation (21.10), consider a point P *on the locus* a short distance δ *above* the real axis in one of the intervals, say, interval IV, Fig. 21.9b. Point P is at $s = -b + j\delta$, and *we wish to find b when δ is extremely small*. To do so, we apply (21.4) to the set of vectors terminating at P, as shown in

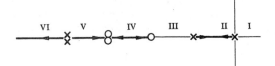

(a) *Real-axis intervals are noted*

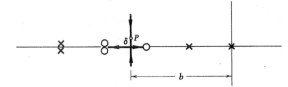

(b) *It is determined that loci must break-in in interval IV; a test point P is considered very near the real axis*

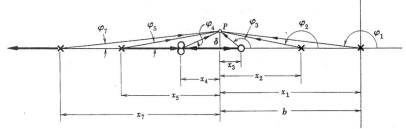

(c) *Vectors are drawn to point P*

Fig. 21.9 *Geometrical determination of a break-in point: all ×'s and ○'s on real axis*

Fig. 21.9c. Each vector is the hypotenuse of a triangle, and the triangles have side δ in common. We now make the approximation that $\tan \phi = \phi$, if ϕ is extremely small, and write (21.4) as

$$\phi_7 + \phi_5 - 2\phi_4 - \phi_3 + \phi_2 + \phi_1 = 180°$$

$$\frac{\delta}{x_7} + \frac{\delta}{x_5} - 2\frac{\delta}{x_4} - \left(180° - \frac{\delta}{x_3}\right) + \left(180° - \frac{\delta}{x_2}\right) + \left(180° - \frac{\delta}{x_1}\right) = 180°$$

or, canceling the 180°'s,† and then canceling δ's, we obtain

$$\frac{1}{x_7} + \frac{1}{x_5} - \frac{2}{x_4} + \frac{1}{x_3} - \frac{1}{x_2} - \frac{1}{x_1} = 0$$

From the above rather general example we may deduce that (21.10) could be written

$$\left[\sum\left(\frac{1}{x}\right)_\times - \sum\left(\frac{1}{x}\right)_\circ\right]_{\text{to the left}} - \left[\sum\left(\frac{1}{x}\right)_\times - \sum\left(\frac{1}{x}\right)_\circ\right]_{\text{to the right}}$$
$$= 0 \quad (21.10c)$$

where x is the distance from breakaway point $-b$ to a $\times$ or $\circ$.

The more compact form is obtained by noting that for $\times$'s or $\circ$'s to the left, $x = \sigma - b$ or $x = a - b$, while for $\times$'s or $\circ$'s to the right, $x = b - \sigma$ or $x = b - a$. This substitution converts (21.10c) to (21.10a):

$$\sum_{k=1}^{n} \frac{1}{\sigma_k - b} - \sum_{\ell=1}^{m} \frac{1}{a_\ell - b} = 0 \qquad (21.10a)$$
[repeated]

In applying (21.10) it is often convenient to recall the construction of Fig. 21.9c. Equation (21.10) will have as many solutions as there are breakaway points on the real axis. In using (21.10) for trial-and-error solution, one selects initial trials at real-axis locations where Rule 1 predicts that loci exist.

(b) General case

When there are also $\times$'s and $\circ$'s not on the real axis, the computation of Fig. 21.9c must be modified, as shown in Fig. 21.10. Now, when we write (21.4), the tangent approximation cannot be used for the vectors from the complex pairs of $\times$'s or $\circ$'s, and we must turn to Pythagoras.

First, applying (21.4) at point P, we obtain

$$\underbrace{(\phi_6 + \alpha)}_{Q'} - \underbrace{(\phi_6 - \alpha)}_{Q} + \phi_5 - 2\phi_4 - \phi_3 + \phi_2 + \phi_1 = 180°$$
$$\underbrace{}_{2\alpha}$$

(The two α's are equal in the limit as $\delta \to 0$.) To find α, construct a per-

† They must sum to $r360°$ or there is no locus.

[Sec. 21.7] RULE 4: BREAKAWAY FROM THE REAL AXIS

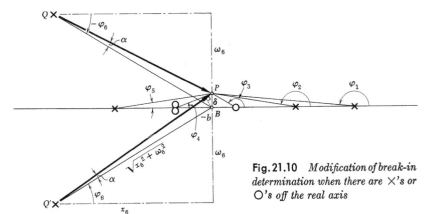

Fig. 21.10 *Modification of break-in determination when there are ×'s or ○'s off the real axis*

pendicular from point B to line $Q'P$. Then, for small δ,

$$\alpha \approx \sin \alpha = \frac{\delta \cos \phi_6}{\sqrt{x_6^2 + \omega_6^2}} = \frac{\delta x_6}{x_6^2 + \omega_6^2}$$

and thus (21.10a) is modified to (21.10b), p. 655, e.g.,

$$\frac{2x_6}{x_6^2 + \omega_6^2} + \frac{1}{x_5} - \cdots = 0$$

Trial-and-error solution

In pursuing a trial-and-error solution of (21.10) it is found that the ×'s or ○'s closest to the breakaway point have the predominant effect on its location. That is, in Fig. 21.9c it is evident that a small change in b may produce a substantial percent change in x_2 and x_4, but little change in x_7. Thus a rapid first approximation may be made by *assuming the distance to distant ×'s and ○'s to be constant* as various trials are made.

It can also be deduced from (21.9) that the breakaway (break-in) location is "attracted" by ○'s (×'s) and repelled by ×'s (○'s). Consider, for example, the sequence of arrays in Fig. 21.11. In the first plot the breakaway is midway between the ×'s, by symmetry. In the second plot the "pressure" of an additional × at $-\sigma_3$ (some distance away) moves the breakaway slightly to the right (the breakaway is repelled by the additional ×). In the third plot the added × is closer, and the breakaway is repelled further. In the limit, as $-\sigma_3$ approaches $-\sigma_2$, point b moves to a point $\tfrac{2}{3}$ the distance from $-\sigma_2$ to $-\sigma_1$. (Can you deduce where the loci go after breaking away?)

In the fourth plot of Fig. 21.11, a ○ at $-a_3$ is seen to "attract" the breakaway location. In the limit, as $-a_3$ approaches $-\sigma_2$, point b moves all the way to $-\sigma_2$.

Example 21.11. Loci that break away from the real axis. Find the location(s) where loci break away from the real axis in the array shown in Fig. 21.12a.

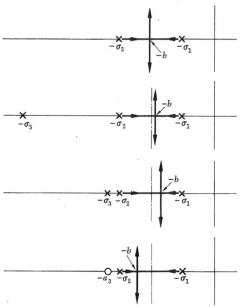

Fig. 21.11 *Attraction and repulsion of breakaway locations*

By Rule 1, there are loci in intervals II and V on the real axis. There must be a breakaway in interval II (since no loci terminate there). Moreover, since two loci must terminate at the ○'s, it is suspected that no breakaway (or break-in) will occur in interval V.

Consider a trial point P in interval II. Because we expect the two ○'s to "attract" the breakaway, we make a first estimate of $b = 0.7$, rather than 0.5.

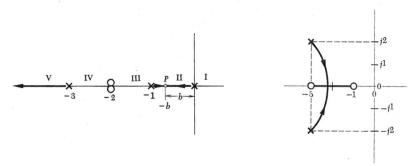

(a) *Breakaway from the real axis (all poles and zeros real)*

(b) *Break-in to the real axis (one pair of poles complex)*

Fig. 21.12 *Examples of trial-and-error calculation [Eqs. (21.10)] of the point where loci break away from or into the real axis*

[Sec. 21.8] RULE 4 (CONTINUED): THE SADDLE-POINT CONCEPT 669

(the midpoint). Then, starting from the right, (21.10a) gives (instead of zero)

$$-\frac{1}{0.7} + \frac{1}{0.3} - \frac{2}{1.3} + \frac{1}{2.3} = -1.43 + 3.333 - 1.54 + 0.435 = +0.80$$

The first two terms are most influenced by a change in b, so we split the error approximately between them. Since the first term's magnitude needs to be increased, and the second's decreased, we move to the right—to $b = 0.55$, say. Then

$$-\frac{1}{0.55} + \frac{1}{0.45} - \frac{2}{1.45} + \frac{1}{2.45} = -0.47$$

We have overshot, but linear interpolation versus b gives $b = 0.60$, which should provide a close estimate in the next trial,

$$-\frac{1}{0.6} + \frac{1}{0.4} - \frac{2}{1.4} + \frac{1}{2.4} = -0.08$$

which is quite close. For many problems our initial estimate of $b = 0.7$ would have been adequate, and, of course, it involved no computation time.

Example 21.12. Loci that break into the real axis. Find the point where loci break into the real axis for the array shown in Fig. 21.12b.

By Rule 1 there are loci on the real axis only in interval II. Since no loci originate in that interval, they must break in at some point $s = -b$. As a first trial, let $b = 2$. Then, applying (21.10b) starting from the right, we have

$$-\frac{1}{-1} - \frac{1}{1} + \frac{2(1)}{1^2 + 2^2} = +0.4$$

we must move to the left. (The break-in point is attracted by the $\times$'s.)
For the next trial let $b = 2.2$:

$$-\frac{1}{0.8} - \frac{1}{-1.2} + \frac{2(0.8)}{(0.8)^2 + 2^2} = -0.08$$

This is probably close enough.

Problems 21.17 through 21.21

⊙21.8 RULE 4 (continued): THE SADDLE-POINT CONCEPT

Consider again the Evans function

$$C(s) = \frac{(s + \sigma_1)(s + \sigma_2 - j\omega_2)(s + \sigma_2 + j\omega_2)(s + \sigma_4) \cdots (s + \sigma_n)}{(s + a_1)(s + a_2) \cdots (s + a_m)} = -K$$

and suppose we make a plot of the magnitude of $C(s)$ versus s (where $s = \sigma + j\omega$). This will be a three-dimensional plot which we may think of as a space curve over the s plane. Such a plot is shown in Fig. 21.13a, for example, for the function

$$C(s) = s(s + 1)(s + 2) = -K$$

670 EVANS' ROOT LOCUS METHOD [Chap. 21]

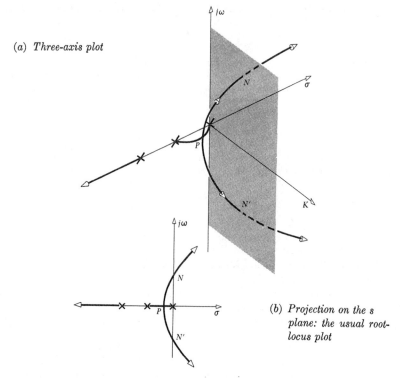

(a) *Three-axis plot*

(b) *Projection on the s plane: the usual root-locus plot*

Fig. 21.13 *Plotting K as a third coordinate of the characteristic equation: the nature of a saddle point*

The value of s is plotted to the σ and $j\omega$ coordinates, as usual, and the third coordinate is $|C(s)| = K$. The loci start at the ×'s, where $K = 0$. As we proceed along a locus, the value of K is increasing and s is changing; the usual root-locus plot, Fig. 21.13b, is merely a projection on the s plane of the plot of Fig. 21.13a. Two of the loci come together at point P, and then separate and proceed outward and to the right. At points N and N' they cross the K, $j\omega$ plane (corresponding to the $j\omega$ axis in Fig. 21.13b), and proceed out over the right half of the s plane. It is important to note that the curves in Fig. 21.13a are smooth, continuous curves, *without discontinuities in position or in slope*, because $C(s)$ is a continuous function of s.

Let us focus attention on point P, where the loci meet. The only way for the curve to have *no discontinuity in slope at this particular point* is for the *slope to be zero* there. That is, as the loci approach and leave point P they must be *parallel to the s plane*. Therefore P is what is called a *"saddle point,"* because the three-dimensional plot resembles a saddle in the vicinity of P. Mathematically, the existence of a saddle point at P indicates the following

[Sec. 21.8] RULE 4 (CONTINUED): THE SADDLE-POINT CONCEPT 671

important relation:

$$\left[\frac{d(-K)}{ds}\right]_{s=s_P} = \left[\frac{d}{ds}C(s)\right]_{s=s_P} = 0 \quad (21.11)$$
[repeated]

which can be solved for the location, in the s plane, of point P. That is, *saddle points can be located by differentiating $C(s)$ with respect to s and setting the result equal to zero.*

For the system illustrated,

$$\frac{d}{ds}[s(s+1)(s+2)] = 0$$

or
$$3s^2 + 6s + 2 = 0$$

and saddle points are located at $s = -0.42, -1.58$.

The solution $s = -0.42$ corresponds to the breakaway point P. The other point, $s = -1.58$, is in an interval where there is no 180°-locus, and so is not of interest here. In fact, point $s = -1.58$ is a point on a "0°-locus," corresponding to negative values of K. This can be quickly verified, for this simple example, by calculating the value of K corresponding to each of the saddle-point solutions. For $s = -0.42$, we have

$$K = -(-0.42)(-0.42 + 1)(-0.42 + 2) = +0.39$$

while for $s = -1.58$, we have

$$K = -(-1.58)(-1.58 + 1)(-1.58 + 2) = -0.39$$

Off-axis saddle points

Saddle points occasionally occur at locations other than the real axis. In Fig. 21.14, for example, the loci are as shown, as can be deduced by determining the asymptotes and the departure directions from the ×'s, and by noting the symmetry which must obtain.

Let us use (21.11) to find the location of points P and P':

$$C(s) = (s^2 + 2s + 2)(s^2 + 2s + 5) = -K$$
or
$$C(s) = s^4 + 4s^3 + 11s^2 + 14s + 10 = -K$$
from which
$$\frac{dC(s)}{ds} = 4s^3 + 12s^2 + 22s + 14 = 0$$

This equation has three roots, and can be solved by the trial-and-error method. Two of the roots will be points P and P'. The third root will correspond to a negative-gain point B which, by symmetry, will be at $s = -1$. Extracting $s = -1$ as one of the roots quickly solves this particular cubic:†

$$(s+1)(s + 1 \pm j1.58) = 0$$

and thus points P and P' are at $s = -1 + j1.58$ and $s = -1 - j1.58$.

† Cubics are generally more difficult to solve. Here we were helped by symmetry.

The reader may enjoy visualizing the three-dimensional plot for Fig. 21.14, which is analogous to the one in Fig. 21.13a.

Example 21.13. Off-axis saddle points. Find the points where loci break into the real axis, and also any other saddle points, for the characteristic equation

$$C(s) = s(s + 2)(s^2 + 2s + 5) = -K \qquad (21.12)$$

The X's for this Evans' function $C(s)$ are plotted in Fig. 21.15b. By Rule 1, there is a 180° locus only in interval II. By Rule 2 and symmetry, the asymptotes are as drawn, and by Rule 3 the starting directions are as shown. It is evident that three saddle points are possible.

Applying (21.11), we obtain

$$\frac{d}{ds}(s^4 + 4s^3 + 9s^2 + 10s) = 0$$

$$4s^3 + 12s^2 + 18s + 10 = 0$$

Guessing (from symmetry) that $s = -1$ is a likely breakaway location,† we write

$$(s + 1)(s + 1 \mp j1.23) = 0$$

We find our guess correct, and there are (1) A breakaway at $s = -1$ and (2) Saddle points at $s = -1 \pm j1.23$. The complete locus is shown in Fig. 21.15b.

The occurrence of saddle points off the real axis, as in Fig. 21.15b, actually constitutes a singular case, as can be recognized by comparing Fig. 21.15b with the adjacent cases in Fig. 21.15a and Fig. 21.15c. In Fig. 21.15b it is clear from symmetry that the loci branches from the real-axis X's (R_0, R_1) will have to meet those from the complex pair (Q, Q') at some point. But if Q, Q' are displaced slightly to the left, as in Fig. 21.15a, the root loci from Q, Q' move off to the left, while those from R_0, R_1 move off to the right; the two pairs of loci never meet anywhere.

Similarly, in Fig. 21.15c Q, Q' are displaced slightly to the right of the symmetrical location. Now the loci from Q, Q' move off to the right, while those from R_0, R_1 move to the left, and again the pairs of loci do not meet. Thus, a given array of X's and O's may produce loci having saddle points off the real axis only for *certain singular arrangements* out of many possible.

In Fig. 21.15b the singular arrangement is the symmetrical one that $\sigma_Q = (\frac{1}{2})\sigma_{R_1}$. For σ_Q greater or less than this amount, there are no off-axis saddle points. The required arrangement is not always a symmetrical one, however, and is not always evident at once. (See Problem 21.24.)

It should be noted that the existence of a solution to 21.11—i.e., the determination of a saddle-point location—*does not insure that the* 180° *loci pass through that saddle point*. It only insures that *some* set of loci pass through it. In Fig. 21.14, for example, we found three possible saddle-point locations, P, P', and B, but only two of these, P and P', occurred on 180° loci. The other point, B, occurred on a pair of 0° loci.

† Any other method of extracting the roots of the cubic, including the root-locus method itself may, of course, be used here.

[Sec. 21.9] RULE 5: ROUTH AND EVANS

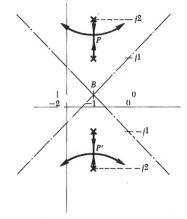

Fig. 21.14 *Saddle points not on the real axis*

Multiple saddle points

Occasionally, a saddle point will occur which involves the conjunction of more than two loci. An example is shown in Fig. 21.16b and discussed in Problem 21.25. Again, this is a singular case, as Figs. 21.16a and c indicate. (Visualize the three-dimensional K, σ, ω picture corresponding to Fig. 21.16.)

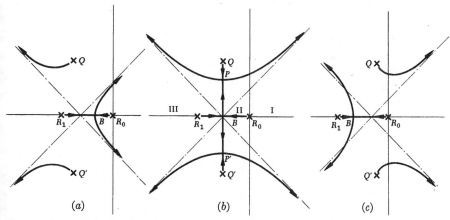

Fig. 21.15 *Saddle points are a singular case*

Problems 21.22 through 21.26

21.9 RULE 5: ROUTH AND EVANS

We have seen earlier (Sec. 11.11) that Routh's method, Procedure C-4, can be used to learn the following properties of a characteristic equation:

(a) How many roots there are in the right half-plane.

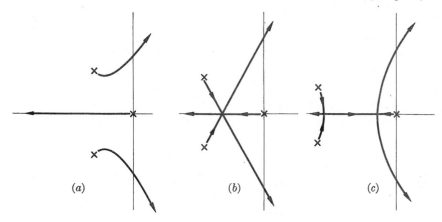

Fig. 21.16 *Example of a higher-order saddle point*

(b) Whether there are roots on the imaginary axis, and how many.
(c) The location of roots on the imaginary axis.

In using Routh's method to augment Evans' we apply the Routh technique with K as an unknown. Then from the Routh conditions we can determine the particular value of K that will produce roots on the imaginary

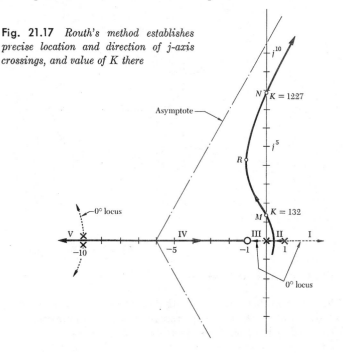

Fig. 21.17 *Routh's method establishes precise location and direction of j-axis crossings, and value of K there*

[Sec. 21.9] RULE 5: ROUTH AND EVANS

axis,† and then find the location of those roots. An example will best demonstrate the procedure.

Example 21.14. Locus crossings of the imaginary axis. Construct the loci of roots, and find carefully the locations of j-axis crossings for the following characteristic equation:

$$\frac{s(s-1)(s+10)^2}{s+1} = -K \qquad (21.13)$$

Find also the values of K at the crossing points.

The root-locus plot for characteristic equation (21.13) is constructed in Fig. 21.17. There exist 180° loci (Rule 1) in real-axis intervals II, IV, and V. The asymptotes (Rule 2) emanate from point $s = -6$, and have directions $\pm 60°$, 180° There must be a breakaway from the real axis in interval II.

At this point we need a clue to the path of the loci that eventually follow the $\pm 60°$ asymptotes. There are two possibilities: (1) the loci breaking away from interval II proceed at once to the right, or (2) the loci from II proceed as shown in Fig. 21.17, crossing the axis once to the left and then back across to the right. (In the second case the loci could break into the real axis in interval IV and then break out again.)

We could obtain our clue by trial and error, adding angles [as in (21.4)] for various points along the j axis to see whether we get 180°. This is often the quickest thing to do. In this example, however, we elect instead to apply Routh's method to (21.13) to see whether there are j-axis crossings.

To use Routh's method, the characteristic equation is first multiplied out to the form

$$s^4 + 19s^3 + 80s^2 + (K-100)s + K = 0$$

The first step in Procedure C-4 ("all coefficients must be positive") tells us that K must be at least 100 or there cannot be stability. We therefore write down the tentative conclusion:

(i) $100 < K$ for stability,

and proceed with the Routh array (Step 2):

s^4	1	80	K
s^3	19	$(K-100)$	0
s^2	$\dfrac{1520 - (K-100)}{19}$	K	0
s	$\dfrac{(1620 - K)(K-100) - (19)^2 K}{(1620 - K)}$	0	
s^0	K	0	

from which, after some algebra, we arrive at the following further conditions for stability, by Step 5:

(ii) $K < 1620$ from s^2 row
(iii) $132 < K < 1227$ from s row   (21.14)
(iv) $0 < K$ from s^0 row

† More sophisticated applications of Routh's method can be developed whereby other boundaries may be studied—e.g., the constant-damping-time line $s = -\sigma_0$.

Condition (iii) transcends all the others, and is therefore sufficient to prescribe completely the stability boundaries of the system. If K is positive, but less than 132 or greater than 1227, there will be two sign changes in the first column (Step 5) and therefore two roots in the right half-plane. If K is negative, there will be only one sign change, and one root in the right half-plane. In particular, the system will be *marginally stable* when $K = 132$ and when $K = 1227$.

To find out where the j-axis crossings are that correspond to these two values of K, we next insert the values $K = 132$ and $K = 1227$, in turn, into the Routhian array. For $K = 132$, we have

s^4	1	80	132
s^3	19	32	0
s^2	78.2	132	0
s	0	0	

The occurrence of a zero row leads us to Step 7 and the auxiliary equation

$$78.2s^2 + 132 = 0$$

whose roots are the j-axis crossings at M in Fig. 21.17:

$$s = \pm j1.30 \quad \text{for } K = 132 \tag{21.15a}$$

These are crossings *from* the right half-plane (instability) *to* the left half-plane (stability) by condition (iii) above [Eq. (21.14)]. The Routhian array is completed without further event, by differentiating the auxiliary equation (Step 7). For $K = 1227$ the array is

s^4	1	80	1227
s^3	19	1127	0
s^2	20.7	1227	0
s	0	0	

Again we proceed to the auxiliary equation:

$$20.7s^2 + 1227 = 0$$
$$s = \pm j7.80 \quad \text{for } K = 1227 \tag{21.15b}$$

These are crossings at N in Fig. 21.17, *from* the left half-plane *to* the right, by (21.14). With these strong clues, and guided by the asymptotes and approximate breakaway location, we can make an accurate sketch of the loci, except that we don't know exactly where points near R are. We will certainly want to know these most accurately, for they are in the region of useful design. We therefore find, by trial and error, a single point on the locus in the vicinity of R (by making angle trials along a horizontal line through $s = j4.5$, for example, until the location is found where $\Sigma\phi = 180°$). Usually three trials will locate a point on the locus in this vicinity with ample precision.

If there had been no j-axis crossings in Fig. 21.17—if the loci from interval I had turned off to the right at once—the Routh array would have told us so by

[Sec. 21.10] RULE 6: FIXED CENTROID

showing two changes of sign in the first column for *all* positive values of K. The reader may wish to observe this by repeating the Routh analysis with the ○ at -5, say, instead of at -1.

The Routh result that for *negative* values of K there is only one sign change in the first column (p. 675) is also of interest. When K is negative, the loci are 0° loci, which start as indicated by the dotted lines in Fig. 21.17. The asymptotes are at 0° and $\pm 120°$. Thus there is a single root in interval I for all negative values of K, as the Routh array predicts (one sign change in the first column).

The decision whether to use Routh's method to augment the root-locus analysis depends upon the complexity of the characteristic equation and the importance of determining precisely the location or value of K, or both, at the j-axis crossings. Commonly, the root loci are sketched first, using Rules 1 to 3, and then the decision is made.

It is noted in passing that the viewpoint may be reversed in some problems, and the bulk of the analysis based on Routh's method. This is advantageous, for example, when the existence of stability is of prime interest (and the actual characteristics of less importance), and when several parameters are unknown, as in general studies without numerical values. In such cases Evans' method may often be helpful as an auxiliary to Routh's.

Problems 21.27 through 21.35

21.10 RULE 6: FIXED CENTROID

When, for a given value of K, all but one root (or all but one complex pair of roots) of a set have been located in the s plane, Rule 6 can often serve as a short cut for locating the final root (or pair of roots) of the set. The rule states that if $m + 2 \leq n$ in a characteristic equation [in the form of (21.2)], then the centroid of all the roots remains fixed at the centroid of the ×'s.

To prove the rule, (21.2) is multiplied out, assuming $m + 2 \leq n$. The first two terms are

$$s^n + s^{n-1}(\sigma_1 + \sigma_2 + \cdots + \sigma_n) + \cdots = 0 \qquad (21.16)$$

Next we assume a *completely factored* form of the characteristic equation:

$$(s + \alpha_1)(s + \alpha_2 - j\omega_2) \cdots (s + \alpha_n) = 0 \qquad (21.17)$$

in which $s = -\alpha_1$, $s = -\alpha_2 + j\omega_2$, ..., are the n *roots* of (21.2) for some particular value of K. Now we multiply out also the product in (21.17). The first two terms are

$$s^n + s^{n-1}(\alpha_1 + \alpha_2 + \cdots + \alpha_n) + \cdots \qquad (21.18)$$

By comparing the coefficient of s^{n-1} in (21.16) and (21.18) we can write

$$\alpha_1 + \alpha_2 + \cdots + \alpha_n = \sigma_1 + \sigma_2 + \cdots + \sigma_n \qquad (21.19)$$

because (21.16) and (21.18) are the same equation in different form. The left-hand side of (21.19) is the centroid of the roots of (21.2) for some *arbitrary* value of K, while the right-hand side of (21.19) is the centroid of the ×'s [or the centroid of the roots of (21.2) when $K = 0$], which proves Rule 6.

If n is less than $m + 2$, then the coefficient of s^{n-1} in (21.16) will contain a's as well as σ's, and (21.19) will no longer be true.

Example 21.15. Utilizing the fixed-centroid rule. On the root-locus diagram of Fig. 11.15, p. 398, find the location of the third root when the complex pair has $\zeta = 0$.

This problem is a natural for the application of Rule 6. If the complex roots move a distance x to the right, the third root must move $2x$ to the left for the centroid to remain fixed. Thus, when the right-hand loci are at PP' ($\zeta = 0$ in Fig. 11.15), their *centroid* has moved to the right by $b/2J$ (from $-b/2J$ to 0), and the third root will have to move to the left by b/J, to point P''.

Notice that this problem was solved *without finding the actual values of K*, a considerable saving in labor for studies where K might not otherwise be needed.

> Problems 21.36 through 21.38

21.11 A TYPICAL CONSTRUCTION OF ROOT LOCI

We now consolidate the individual rules of the preceding sections into a single, rather comprehensive root-locus construction. Figure 21.18 shows the locus of roots for the characteristic equation

$$C(s) = \frac{s(s + 1)(s + 2)^3(s + 4)(s + 5 \pm j2)}{(s + 3)(s + 1 \pm j3)} = -K \quad (21.20)$$

and Fig. 21.19 shows the individual steps in its construction.

In Fig. 21.19a, (1) Rule 1 is applied to determine the real-axis intervals in which there are locus segments, and (2) Rule 2 is used to calculate the centroid of the ×'s and ○'s,

$$c = \frac{[0 + 1 + 3(2) + 4 + 2(5)] - [3 + 2(1)]}{8 - 3} = 3.2 \quad (21.21)$$

from which the asymptotes emanate, and the directions of the asymptotes,

$$\phi = \pm (2r + 1) \frac{180°}{8 - 3} \bigg|_{r=0,1,2} = \pm 36°, \pm 108°, 180° \quad (21.22)$$

In Fig. 21.19b, (3) the directions of loci departure from the off-axis ×'s are obtained, as are the directions of arrival at the off-axis ○'s. These are computed, using a Spirule and the principle of Rule 3(a). Departure from the triple × on the real axis, at $s = -2$, is calculated by Rule 3(c): One locus is on the real axis in interval IV, in accordance with Rule 1; the three loci are $360°/3 = 120°$ apart, and therefore the other two have directions $\pm 60°$, by symmetry.

[Sec. 21.11] A TYPICAL CONSTRUCTION OF ROOT LOCI

In Fig. 21.19b, also, (4) the point $s = -b$ is found where the loci in interval II break away from the real axis. By Rule 4,

$$\frac{1}{-b} + \frac{1}{1-b} + \frac{3}{2-b} - \frac{1}{3-b} + \frac{1}{4-b} + \frac{2(5-b)}{(5-b)^2 + (2)^2} \\ - \frac{2(1-b)}{(1-b)^2 + (3)^2} = 0 \quad (21.23)$$

This equation has the solution $b = 0.37$, found by trial and error. Unless we really wanted to know b to two or three significant figures, we would surely not bother with this laborious trial-and-error solution. The last statement in Rule 4 would cause us to estimate that, because of the triple $\times$ at $s = -2$, point b is between 0.3 and 0.4—somewhat to the right of the midpoint between $s = 0$ and $s = -1$. We would be unlikely to pursue the saddle-point technique in this problem because, for Eq. (21.20), the differentiation of (21.11) would be quite laborious, and would only result in a seventh-order algebraic expression whose roots would have to be found. Clearly, trial and error using (21.23) is much easier.

In Fig. 21.19c, (5) the j-axis crossings are found. In this problem the Routh array will be of the eighth order, and K will enter the terms in a highly involved way. We therefore abandon Routh, in this case, and find crossing R_1 by an angle search along the j axis, which takes a minute or two. The search area is well localized by the breakaway in real-axis interval II and the asymptotes at $\pm 36°$.

In Fig. 21.19c we also locate two points, R_2 and R_3, by trial and error. The approximate location of R_2 is indicated by the arrival direction at Q, the departure directions at T, and the probability that a locus connects the triple $\times$ at T with the $\bigcirc$ at Q. Note that we merely want to find a point— any point—on a locus about halfway between T and Q. We might, for example, search along a horizontal line midway between Q and T.

Similarly, a point R_3 is located approximately by the departure direction at P and the asymptote at 108°. Here we might search, for example, along a ray at 70° from P to find one point, R_3, through which to draw the locus from P to the asymptote. Useful points at R_2 and R_3 should be found by three trials each.

Finally, in Fig. 22.19d, we sketch in the complete root loci, using all the previous constructions we have made. The result is Fig. 21.18. After a little practice, a set of loci like those in Fig. 21.18 can be sketched in less time than it takes to read the above description of the process.

After the loci have been constructed, the value of K for several points of interest can be found by graphical measurement (e.g., by Spirule multiplication). In this problem we shall be most interested in the roots on the right-hand pair of loci, *because they are the roots which determine stability and dominate system behavior.*

Therefore, we first calculate K for point R_1, Fig. 21.18, where the system

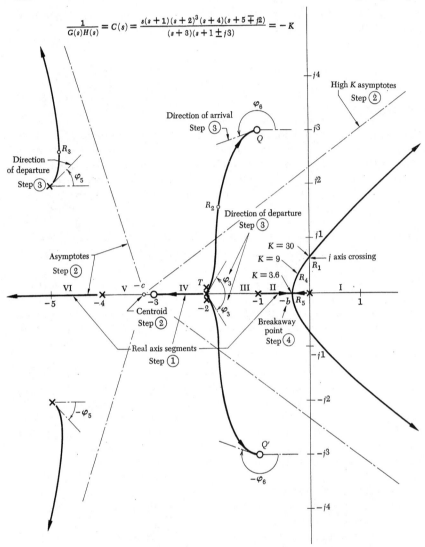

Fig. 21.18 *Complete set of root loci for the characteristic equation:*

$$C(s) = \frac{s(s+1)(s+2)^3(s+4)(s+5 \mp j2)}{(s+3)(s+1 \pm j3)} = -K$$

[Sec. 21.11] A TYPICAL CONSTRUCTION OF ROOT LOCI 681

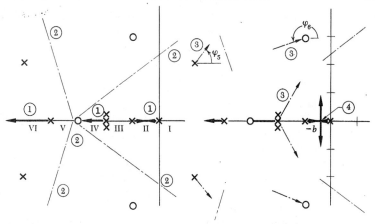

(a) Step ① real-axis segments
    Step ② asymptotes

(b) Step ③ directions of departure and arrival
    Step ④ breakaway locations

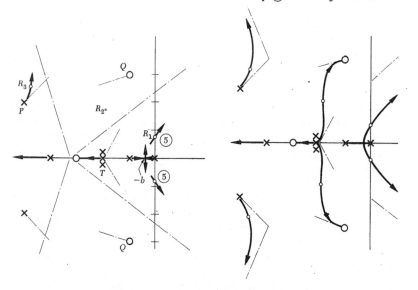

(c) Step ⑤ j-axis crossing
    Trial-and-error location of
    R_1, R_2

(d) Fill in loci
    Determine K at interesting points
    Step ⑤ check K at j-axis crossing
    Step ⑥ check K by centroid rule

Fig. 21.19 Steps in constructing the root loci of Fig. 21.18

goes unstable, and also for points R_4 and R_5, which probably represent the design range of interest. These are

$$K = 30 \quad \text{for point } R_1$$
$$K = 9 \quad \text{for point } R_4$$
$$K = 3.6 \quad \text{for point } R_5$$

These are marked on the locus.

If we had used Routh's method (Rule 5) in this problem, K at point R_1 would have been available directly. (Since we didn't, it wasn't.)

If the locations of the other roots for $K = 3.6$, 9, and 30, are of interest then they must be determined by trial and error. A more direct method is simply to calculate K directly at various points along the loci, as shown, and then interpolate between them. This is usually accurate enough for loci which are not especially critical.

Rule 6 is not especially useful in this problem. However, if points on all but one of the branches have been located for some interesting value of K, then a little work is saved by using Rule 6 to find the corresponding point on the last branch—e.g., the branch in interval IV on the real axis.

21.12 SUMMARY

As we have noted repeatedly, all that the root-locus method does is to find the roots of the system characteristic equation, and to present them in a convenient s-plane graphical display as a function of a system or control parameter of interest. This, however, is a key thing to do, because the roots of the characteristic equation tell the character of the system's natural motions—its degree of stability, its response speed, and its natural frequencies. The roots are poles of system transfer functions (e.g., the closed-loop transfer function from input to output, if we are studying a feedback control system). With the roots so displayed, versus a parameter that we can govern (e.g., for a control system, the control gain), we can make design decisions regarding natural response most directly.

Facility with, and a thorough understanding of root-locus techniques can be gained only by working a number and variety of analysis and design problems, such as those suggested (Problems 21.39 through 21.48). By way of demonstrating to the reader the breadth, scope, and utility of the concepts and techniques he has learned, and of introducing him briefly to Stage IV (p. 7), the synthesis and design of systems, we present in Chapter 22 several complete control-system design problems.

Problems 21.39 through 21.62

Chapter 22

SOME CASE STUDIES IN AUTOMATIC CONTROL

The purpose of this chapter is to demonstrate, with a variety of interesting case studies, some essential features of the process of analyzing and designing control systems. Procedure E-1 provides a convenient framework in each example, but concepts and methods from throughout the book will appear prominently, as they do in most dynamic investigations.

The classical indicator servo is analyzed in Sec. 22.1, with a typical set of specifications to be met. The passive lead-compensation network is introduced in Sec. 22.2, where design decisions are to be made.

The autopilot design problem of Sec. 22.3 demonstrates the synthesis of multiloop control systems. Section 22.4 introduces the problem of controlling an inherently unstable system and describes a machine for balancing sticks.

22.1 ANALYSIS OF AN ELECTROMECHANICAL REMOTE-INDICATOR SERVO

There are many applications for the remote position indicator, or "repeater," which consists essentially of a motor with a pointer on it, Fig. 22.1a, so driven by an amplifier that its angular position corresponds at all times to that of a reference pointer some distance (perhaps miles) away, the only connection between the two being electrical. The desired matching of reference and repeater is achieved by measuring the angular positions of both (e.g., with potentiometers, as shown), comparing them electrically, and driving the motor with the difference signal. A typical application is the remote "gyro repeater" used at many stations on a ship to display the output of the ship's master gyrocompass. An angle pickoff on the master gyro

684   SOME CASE STUDIES IN AUTOMATIC CONTROL   [Chap. 22]

(a) *Actual system*

(b) *Physical model*

(c) *Mathematical model*

furnishes the electric input to all the repeaters.† Conceptually, the scheme is like that in Fig. 20.2 for controlling the gun turret to follow the director. We shall find that the mathematical models and design specifications of the two systems also have the same form.

Suppose that it is desired to have a repeater that will meet the following specifications: (1) respond to a step change in θ_d of 8° by reducing and holding error to less than 1° within 0.3 sec, and with damping that is greater than 0.4 critical; (2) have a steady-state position error less than 0.3° for either the above step input or a ramp input of 5°/sec. We are asked to find out whether the given system will meet these specifications with simple proportional feedback, i.e., with just a constant gain in the control amplifier, and if not, to see how close it can come to meeting the specifications. That is, our

† Actually, ship's gyro repeaters use a-c motors and pickoffs, rather than the d-c devices described here. The d-c system is more straightforward to explain, and the principle of operation is the same.

(d) Root-locus plot

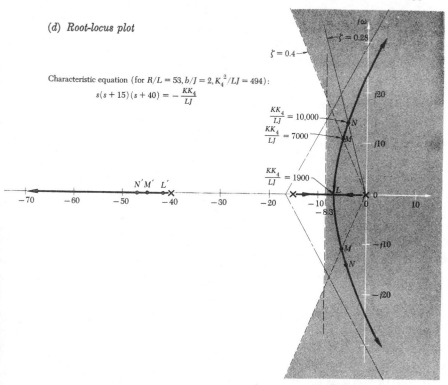

Characteristic equation (for $R/L = 53$, $b/J = 2$, $K_4^2/LJ = 494$):

$$s(s+15)(s+40) = -\frac{KK_4}{LJ}$$

task is one of analysis only, to determine the system behavior. (In Sec. 22.2 we shall have also a design, or synthesis, task to perform.) Suppose, further, that we are given the following numerical values:

$$b/J = 1 \text{ sec}^{-1} \qquad R/L = 54 \text{ sec}^{-1} \qquad K_4^2/LJ = 546 \text{ sec}^{-2} \qquad (22.1)$$

In performing the analysis, it is convenient to follow Procedure E-1.

STAGE I. PHYSICAL MODEL. Let us suppose that, upon examining the motor, we decide that the physical model shown in Fig. 2.24b, and described on p. 84, is an appropriate one. (The mass and the damping of the display dial, if appreciable, can be lumped with those of the rotor.) We neglect electrical losses in this simple reference circuit. The resulting physical model is shown in Fig. 22.1b.

STAGE II. EQUATIONS OF MOTION (MATHEMATICAL MODEL). The equations of motion for the plant consisting of the motor, amplifier, and circuit of Fig. 22.1b were derived in Sec. 2.9 (Example 2.6), Eq. (2.77); to wit:

$$\text{KVL:} \quad L\frac{di}{dt} + Ri + K_4\Omega = v \qquad (22.2)$$
$$\Sigma M^* = 0: \quad -K_4 i + J\dot{\Omega} + b\Omega = 0 \qquad \text{[(2.77) repeated]}$$

(where the load torque is zero, and subscripts have been omitted, there being no ambiguity). The only other equation required is a trivial one for the summing network (the ground is not shown):

$$2v_e \triangleq \theta_e = \theta_d - \theta \tag{22.3}$$

where the units are volts. (The potentiometer scale factor has been taken as -1 volt/radian, for convenience.)

STAGE III. DYNAMIC BEHAVIOR.

Step 1. Laplace transform the equations of motion. (Initial conditions are assumed zero, and the identity $\Omega \equiv \dot\theta$ is employed):

$$\begin{bmatrix} (Ls + R) & K_4 s \\ -K_4 & (Js + b)s \end{bmatrix} \begin{bmatrix} I \\ \Theta \end{bmatrix} = \begin{bmatrix} V \\ 0 \end{bmatrix} \tag{22.4}$$

$$\Theta_e = \Theta_d - \Theta \tag{22.5}$$

Step 2(a). Manipulate the resulting algebraic equations to obtain transfer functions.

(i) The open-loop transfer function from V to Θ is, from (22.4),

$$\frac{\Theta}{V} = \frac{K_4}{s[(Ls + R)(Js + b) + K_4^2]} \tag{22.6}$$

Using the numerical values given, and factoring the denominator, this can be written

$$\frac{\Theta}{V} = \frac{K_4/LJ}{s(s + 15)(s + 40)} \tag{22.7}$$

The block diagram for the system is as shown in Fig. 22.1c.

(ii) The closed-loop system transfer function is obtained by applying (20.3) to Fig. 22.1c:

$$\frac{\Theta}{\Theta_d} = \frac{KK_4/LJ}{[s(s + 15)(s + 40) + KK_4/LJ]} \tag{22.8}$$

the denominator of which gives the characteristic equation of the system. This is displayed in Fig. 22.1d.

(iii) We can rewrite (22.8) with the denominator represented in factored form as follows:

either
$$\frac{\Theta}{\Theta_d} = \frac{KK_4/LJ}{(s + \sigma_1)(s + \sigma_2)(s + \sigma_3)} \tag{22.9a}$$

or
$$\frac{\Theta}{\Theta_d} = \frac{KK_4/LJ}{(s^2 + 2\sigma_1 s + \omega_{01}^2)(s + \sigma_3)} \tag{22.9b}$$

in which the factors σ_1, σ_2, σ_3 or ω_{01}^2 are as yet unknown. We do know, however, that the product $\sigma_1\sigma_2\sigma_3$ or $\omega_{01}^2\sigma_3$ must equal KK_4/LJ [so that the coefficient of s^0 in the denominator of (22.9) will equal the coefficient of s^0 in the denominator of (22.8)].

(iv) The response functions in which we are interested are those giving system output θ and also system error θ_e in response to the two inputs called

[Sec. 22.1] ELECTROMECHANICAL REMOTE-INDICATOR SERVO

out in the specifications, namely, a step, $\theta_d = \theta_o u(t)$ (with $\theta_o = 8°$), and a ramp, $\theta_d = \Omega_o t u(t)$ (with $\Omega_o = 5°/\text{sec}$). The corresponding response functions are [using (22.9b) for illustration, and employing (22.5) and (22.8) in obtaining Θ_e]:

for the step input

$$\Theta = \frac{(KK_4/LJ)\theta_o}{s[(s^2 + 2\sigma_1 s + \omega_{01}^2)(s + \sigma_3)]} \tag{22.10}$$

$$\Theta_e = \frac{s(s + 15)(s + 40)\theta_o}{s[(s^2 + 2\sigma_1 s + \omega_{01}^2)(s + \sigma_3)]} \tag{22.11}$$

for the ramp input

$$\Theta = \frac{(KK_4/LJ)\Omega_o}{s^2[(s^2 + 2\sigma_1 s + \omega_{01}^2)(s + \sigma_3)]} \tag{22.12}$$

$$\Theta_e = \frac{s(s + 15)(s + 40)\Omega_o}{s^2[(s^2 + 2\sigma_1 s + \omega_{01}^2)(s + \sigma_3)]} \tag{22.13}$$

Step 2(b). The roots of the characteristic equation. These are sketched, versus KK_4/LJ, in the s plane of Fig. 22.1d, using only Rules 1 and 2 of Sec. 21.3, plus a Spirule check (two trials) to find the j-axis crossing. (The breakaway point doesn't take too long to find; however, knowing from Rule 4 that it will be somewhat to the right of the midpoint between 0 and -15, and that the loci leave the axis vertically, we can make an acceptably accurate estimate without any calculation.) Thus, for the simple proportional controller postulated, we can have any system natural characteristics on the locus of Fig. 22.1d by proper choice of gain K. For the moment, let us leave the exact choice of K open, but postulate that we shall choose it to have a complex pair of roots, to meet the fast-response specification. That is, we postulate form (22.9b) for the closed-loop transfer function. We shall return to Fig. 22.1d shortly.

Step 2(c). Final-Value Theorem. In this problem the Final-Value Theorem gives directly useful information regarding the steady-state error specification—Specification (2). (This is often the case in control system studies.) For the step input, application of the Final-Value Theorem to (22.11) gives $\theta_e(\infty) = 0$, and thus the position-error specification is met regardless of K. That is, the step response looks approximately like that in Fig. 20.4a, but the final position matches the input with *no* error. This can be shown to be a direct result of the open-loop transfer function's having a pole at the origin—i.e., to one factor of the denominator of (22.7) being $s = 0$. Physically, there can be no steady-state error because $1/s$ means an integration, and thus so long as there is any voltage θ_e in Fig. 22.1c there must be a rate $\dot\theta$, and steady state cannot obtain. [Show that application of the Final-Value Theorem to (22.10) indicates the same result. You will need the relation $\omega_{01}^2 \sigma_3 = KK_4/LJ$.]

For the ramp input, application of the Final-Value Theorem to (22.13) gives

$$\theta_e(\infty) = \frac{(15)(40)}{(KK_4/LJ)} \Omega_o \tag{22.14}$$

The system's ramp response looks like that of Fig. 20.4b, the output following the input with a constant error proportional to Ω_o. [Can this be seen directly also by applying the Final-Value Theorem to (22.12)? Why?] From (22.14) it is obvious that steady-state error can be reduced by increasing gain K—by "tightening" the control. Specifically, to meet Specification (2) we require $(KK_4/LJ) = 10,000$. However, K cannot be increased beyond a certain value without encountering poor transient-response characteristics and, eventually, instability, as we shall see presently (Fig. 22.2a). We shall therefore make K as large as transient requirements allow us to, and then see what θ_e this gives us.

Step 3. Inverse transform to obtain time response. To see whether (or how nearly) the system can be made to meet Specification (1), we want to obtain its time response to a step input. Inverse transforming (22.10) by partial fraction expansion (Procedure D-2) gives

$$\theta(t) = \theta_0[A_0 + 2|\vec{A}_{11}|e^{-\sigma_1 t} \cos(\omega_1 t + \underline{/\vec{A}_{11}}) + A_3 e^{-\sigma_3 t}]u(t) \quad (22.15)$$

in which, as the reader may show,

$$A_0 = 1$$

$$2\vec{A}_{11} = \frac{2\omega_{01}^2 \sigma_3}{(\omega_{01})(2\omega_1)(c)} \underline{/-\left(\frac{\pi}{2} + \phi_1\right) - \frac{\pi}{2} - \phi_3}$$

$$= \frac{\omega_{01}\sigma_3}{\omega_1 c} \underline{/-\phi_3 - \phi_1 - \pi}$$

$$A_3 = \frac{\omega_{01}^2 \sigma_3}{-\sigma_3 c^2} = -\frac{\omega_{01}^2}{c^2} \quad (22.16)$$

where the quantities involved will be available directly from Fig. 22.1d when a value for K is chosen. Hence

$$\theta(t) = \theta_o \left[1 - \frac{\omega_{01}\sigma_3}{\omega_1 c} e^{-\sigma_1 t} \cos(\omega_1 t - \phi_3 - \phi_1) - \frac{\omega_{01}^2}{c^2} e^{-\sigma_3 t}\right] u(t) \quad (22.17)$$

From the relative magnitudes of distances in Fig. 22.1d, it is clear that the third term in (22.17) is quite small compared to the second, and thus we have approximately the second-order step response indicated in Fig. 22.2a.

Before introducing numerical values into (22.17), let us turn now to Specification (1), which deals with transient response.† The implications of

† Recall from Step 2(c) that this system automatically meets the step-input part of Specification (2), the steady-state requirement.

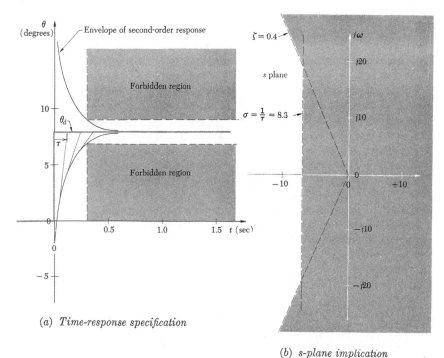

(a) *Time-response specification*

(b) *s-plane implication*

Fig. 22.2 *Specified transient time response implies characteristic root locations in the s plane*

Specification (1) are also displayed in Fig. 22.2a: The system's response is not to transgress the shaded region. For the (approximately) second-order system, we can convert this specification into a region in the s plane by the following reasoning: The (exponential) *envelope* of the second-order damped sinusoidal response in Fig. 22.2a must miss the corner of the shaded region. Assuming conservatively that factor $\omega_{01}\sigma_3/\omega_1 c$ in (22.17) is perhaps 1.25 or so, we can sketch a few trial exponential envelopes (for different time constants) in Fig. 22.2a until we find the one that just nicks the corner, as shown.† This one happens to have $\tau = 0.12$ sec., and thus Specification (1) requires that $|\sigma_1| > 1/0.12 \cong 8.3$ in Fig. 22.2b, i.e., that the system roots lie to the left of the vertical line $s = -8.3$ in Fig. 22.2b. Further, since Specification (1) states that $0.4 < \zeta$, the roots must lie to the left of the radial lines $\zeta = 0.4$. Combining these requirements, we see that the roots must lie to the left of the shaded region shown in Fig. 22.2b if Specification (1) is to be met.

† Alternatively, an exact analytical expression can, of course, be derived for this time constant.

We now transfer this region to Fig. 22.1d and find that, alas, the locus of roots does not pass into the unshaded region at any point. We must therefore answer the first part of the given question by saying, "No, the system cannot, with purely proportional control, meet Specification (1)." We now look to see how close it can come.

Recalling from (22.14) that we need a large K to achieve the ramp steady-state error of Specification (2), we are impelled to move up along the locus. In fact, at point N the gain is $KK_4/LJ = 10{,}000$ and the ramp-input steady-state specification, Specification (2), is met [Eq. (22.14)]. On the other hand, as soon as the loci have broken away from the real axis (point L, $KK_4/LJ = 1900$) they start to move to the right, thus failing increasingly to meet Specification (1) on the time-response envelope (Fig. 22.2a). Further, for $7000 < KK_4/LJ$ (point M) they no longer meet the $0.4 < \zeta$ specification either. It is clear that we shall not want KK_4/LJ greater than 10,000 nor less than 1900. But in the region $1900 < KK_4/LJ < 10{,}000$, some compromise between Specifications (1) and (2) must be made. For example, if steady-state error for the ramp input is most important (the more usual situation), we shall tend toward point N and accept a transient response that is less than ideal (i.e., that has $\zeta = 0.28$ and cuts slightly into the shaded region in Fig. 22.2a).

The third root corresponding to a given gain, points L', M', N' in Fig. 22.1d, is readily found, using the fixed-centroid rule (Rule 6 of Sec. 21.3). This root plays a minor role in system performance.

Suppose we choose point N in Fig. 22.1d to meet the ramp steady-state

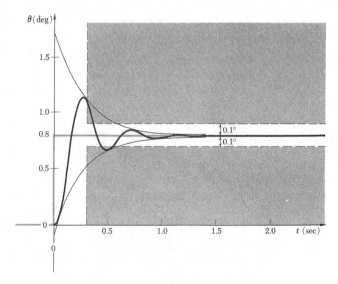

Fig. 22.3 *The indicator servo of Fig. 22.1 does not quite meet the response specifications*

[Sec. 22.2] INDICATOR SERVO USING NETWORK COMPENSATION 691

error specification. Then, obtaining values for (22.16) from Fig. 22.1d, we can write (22.17) as follows:

$$\begin{aligned} \theta(t) &= \theta_0 \left[1 - \frac{15 \times 47}{14 \times 45} e^{-4t} \cos(14t - 22° - 16°) - \frac{15^2}{47^2} e^{-47t} \right] u(t) \\ &= 8° \left[1 - 1.12 e^{-4t} \cos(14t - 38°) - 0.10 e^{-47t} \right] u(t) \end{aligned}$$ (22.18)

This response has been plotted in Fig. 22.3 (using Procedure B-3). The response envelope is seen to miss the time limitation of Specification (1) by 0.3 sec (although the actual response is within 0.1° slightly sooner). The response is, of course, less well damped than was called for.

Problems 22.1 through 22.9

22.2 SYNTHESIS OF INDICATOR SERVO USING NETWORK COMPENSATION TO IMPROVE PERFORMANCE

Physically, the reason the simple proportional control system of Fig. 22.1b is unable to meet Specification (1), described in Sec. 22.1 and illustrated in Fig. 22.2a, is that it does not get started soon enough. The controller needs to apply full torque to the motor sooner after the step command is received. It is well known that a control system can be "speeded up" by furnishing it with the *derivative* of the error signal, as well as with the error signal itself, so that the system has an early *prediction* of error, and can act sooner.

Sometimes in control problems it is possible to obtain a direct measurement of the derivative of the error. For example, an autopilot for controlling aircraft motions may use the signals from rate gyros (Sec. 22.3). In other cases, rate sensors may be unavailable or undesirably expensive. Often in such cases approximate differentiation of the error signal can be effected by means of a simple passive circuit, such as that represented by the network (model) in Fig. 22.4a. Such a network is called a "lead compensation network" because when its input is a continuing sinusoidal signal, the phase angle of its output leads that of the input, for some range of frequencies, thus compensating for lags in the system. (Refer to Problems 2.85 and 15.30.)

Suppose that we are given such a network, and asked to specify its parameters (R_1, R_2, and C) so as to enable the system of Sec. 22.1 to meet the specifications given there, subject to the constraint that the ratio $R_2/(R_1 + R_2) < 3.5$. (The physical significance of this constraint will be evident presently.) We analyze the system, following Procedure E-1. Here we are interested in synthesis—in deciding what parameters the system should have, as well as in analyzing it. We shall therefore be superposing

692 SOME CASE STUDIES IN AUTOMATIC CONTROL [Chap. 22]

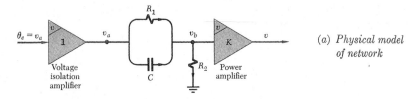

(a) *Physical model of network*

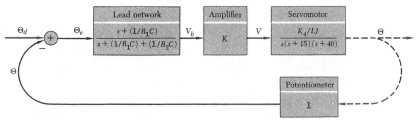

(b) *Mathematical model of system including network*

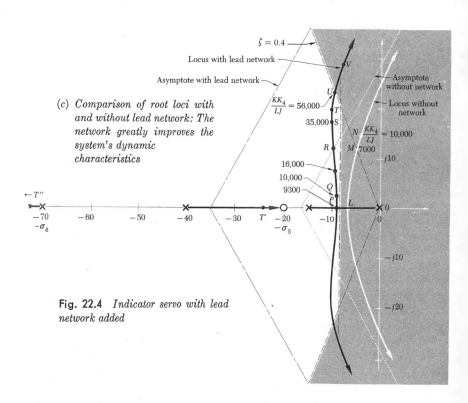

(c) *Comparison of root loci with and without lead network: The network greatly improves the system's dynamic characteristics*

Fig. 22.4 *Indicator servo with lead network added*

[Sec. 22.2] INDICATOR SERVO USING NETWORK COMPENSATION 693

some synthesis steps, Stage IV, upon the analysis steps of Stage III, as we go along.

STAGE I. PHYSICAL MODEL. This is the same as for Fig. 22.1b with the network (circuit model) of Fig. 22.4 added.

STAGE II. MATHEMATICAL MODEL. In addition to the equations of motion (22.2) and (22.3) for the motor subsystem and the summer, we need the equation of motion for the network. Writing KCL for node b (and noting that the current into isolation amplifier K is zero, Sec. 2.8) we obtain

$$i_1 + i_3 = i_2$$

or, with physical relations inserted,

$$\frac{1}{R_1}(v_a - v_b) + C(\dot{v}_a - \dot{v}_b) = \frac{1}{R_2}v_b \qquad (22.19)$$

STAGE III. DYNAMIC BEHAVIOR

Step 1. Laplace transform the equations of motion. To (22.4) and (22.5) we add the Laplace transform of (22.19) (with zero initial conditions):

$$\left(Cs + \frac{1}{R_1} + \frac{1}{R_2}\right)V_b = \left(Cs + \frac{1}{R_1}\right)V_a \qquad (22.20)$$

Step 2(a). Manipulate the resulting algebraic equations to obtain transfer functions.

(i) To the open-loop transfer functions of Fig. 22.1c we must add that of the network, which we obtain from (22.20) and display in Fig. 22.4b:

$$\frac{V_b}{V_a} = \frac{V_b}{\Theta_e} = \frac{s + \dfrac{1}{R_1 C}}{s + \dfrac{1}{R_1 C} + \dfrac{1}{R_2 C}} \qquad (22.21)$$

To see the effect this network has on the control system, it is helpful to rewrite (22.21) as

$$\frac{V_b}{\Theta_e} = \frac{s}{s + (1/R_1 C) + (1/R_2 C)} + \frac{1/R_1 C}{s + (1/R_1 C) + (1/R_2 C)} \qquad (22.22)$$

The first term is approximately a differentiator—a predictor—but with filtering of rapid signals, while the second term is approximately proportional, but with filtering. The benefits of this prediction will first be seen quantitatively in Step 2(b), the study of system characteristics. The filtering feature is important: the network attenuates slowly changing signals, by the ratio $(R_2 + R_1)/R_2$ [$s = 0$ in (22.21)], but does not attenuate rapidly changing signals [s large in (22.21)], including noise, in particular, and thus the circuit reduces the ratio of valid low-frequency signal to extraneous high-frequency noise by the amount $(R_2 + R_1)/R_2$. This is the practical reason why that ratio must be restricted to some low value—typically between 3 and 20, depending on the noise characteristics of the system.

We designate the following parameters in (22.21):

$$\sigma_5 \triangleq 1/R_1C \qquad \sigma_6 \triangleq (1/R_1C) + (1/R_2C) \qquad (22.23)$$

These are design parameters which we are free to choose, so long as we maintain $\sigma_5/\sigma_6 < 3.5$.

(ii) We now obtain the closed-loop transfer function for the system of Fig. 22.4b by application of (20.3):

$$\frac{\Theta}{\Theta_d} = \frac{(KK_4/LJ)(s + \sigma_5)}{[s(s + 15)(s + 40)(s + \sigma_6) + (KK_4/LJ)(s + \sigma_5)]} \qquad (22.24)$$

(iii) We can rewrite this transfer function with the denominator represented in factored form. Recalling Fig. 22.1d, we estimate that, for gains of interest, two of the roots will be complex and the other two real (the denominator is fourth-order):

$$\frac{\Theta}{\Theta_d} = \frac{(KK_4/LJ)(s + \sigma_5)}{[(s^2 + 2\sigma_1 s + \omega_{01}^2)(s + \sigma_3)(s + \sigma_4)]} \qquad (22.25)$$

where $\omega_{01}^2 \sigma_3 \sigma_4$ must equal $(KK_4/LJ)\sigma_5$ in order that the coefficient of s^0 in the denominator of (22.25) equal the coefficient of s^0 in the denominator of (22.24).

(iv) We are interested in the same response functions as we were in Sec. 22.1. For the step input [again, using (22.5) to obtain Θ_e],

$$\Theta = \frac{(KK_4/LJ)(s + \sigma_5)\theta_0}{s[(s^2 + 2\sigma_1 s + \omega_{01}^2)(s + \sigma_3)(s + \sigma_4)]} \qquad (22.26)$$

$$\Theta_e = \frac{s(s + 15)(s + 40)(s + \sigma_6)\theta_0}{s[(s^2 + 2\sigma_1 s + \omega_{01}^2)(s + \sigma_3)(s + \sigma_4)]} \qquad (22.27)$$

For the ramp input,

$$\Theta = \frac{(KK_4/LJ)(s + \sigma_5)\Omega_0}{s^2[(s^2 + 2\sigma_1 s + \omega_{01}^2)(s + \sigma_3)(s + \sigma_4)]} \qquad (22.28)$$

$$\Theta_e = \frac{s(s + 15)(s + 40)(s + \sigma_6)\Omega_0}{s^2[(s^2 + 2\sigma_1 s + \omega_{01}^2)(s + \sigma_3)(s + \sigma_4)]} \qquad (22.29)$$

Step 2(b). Find the roots of the characteristic equation. These are sketched, versus KK_4/LJ, in the s-plane plot of Fig. 22.4c. First, Rules 1, 2, and 4 of Sec. 21.3 were applied to obtain the real-axis intervals, the high-K asymptotes, and the breakaway points. Then several points on the complex locus were found by trial and error.

The root-locus plot of Fig. 22.1d (*no* prediction network) is also shown (in white) in Fig. 22.4c. We can now see the role played by the network, by comparing the black and white loci. For small values of gain (locus on or near the real axis) the denominator term $(s + \sigma_6)$ of the network transfer function ($\times$ at $s = -\sigma_6$) has a minor effect, but the numerator term $(s + \sigma_5)$ (○ at $s = -\sigma_5$) has a dominant effect: It attracts the breakaway point leftward from L to P (Rule 4), and causes the locus above the axis to move leftward of vertical (points R, S, T, U), instead of proceeding steadily to the right as the white locus does. The ○ accomplishes this because it

[Sec. 22.2] INDICATOR SERVO USING NETWORK COMPENSATION 695

supplies a positive angle to the summation of angles at points in the s plane in the vicinity of points Q, R, S; thus the locus goes through points Q, R, S instead of through point M. For larger distances from the real axis (e.g., point U) the angle summation contains an increasing contribution from the $\times$ at $s = -\sigma_6$, and the locus swings to the right again. (At very large distances from the origin the effect of the network's O-$\times$ combination is simply to displace the asymptotes as shown, black asymptote versus gray-white asymptote.)

Step 2(c). Final-Value Theorem. By application of the Final-Value Theorem to (22.27) and (22.29) we see that, just as in Sec. 22.1, the error following a step input is zero, and the steady-state error for a ramp input is

$$\theta_e = \frac{(15)(40)(\sigma_6)}{(KK_4/LJ)\sigma_5} \Omega_o \qquad (22.30)$$

In particular, for $\sigma_6 = 70$ and $\sigma_5 = 20$ as in Fig. 22.4c, we shall need

$$(KK_4/LJ) = \frac{(15)(40)(70)}{20} \frac{\Omega_o}{\theta_e} = 35{,}000 \qquad (22.31)$$

to make $\theta_e = 0.3°$ and meet Specification (2) (p. 684) when $\Omega_0 = 5°/\text{sec}$. That is, the gain must be higher by $(R_2 + R_1)/R_2$, compared with (22.14), to make up for the lead network's attenuation of d-c signals.

STAGE IV. SYSTEM DESIGN. Before proceeding with Step (3) of Stage III, we superpose some additional design discussion at this point. The root locus in Fig. 22.4c has a substantial portion in the unshaded region, indicating that the transient-response specification (1) can be met over a large range of K. Moreover, the gains computed and labeled on the locus show that steady-state-error specification (2), from (22.31), will be met for any point beyond point S. Thus we have the "luxury" of a range of design over which the specifications can all be met. A logical choice might be point T, where both the steady-state error and the damping ratio will be better than required. What may be even more important, the system parameters can now suffer reasonable changes (e.g., due to temperature variation) and still meet the specifications.

We note that no attempt has been made yet to select the *best* values for σ_5 (and for $\sigma_6 = 3.5\sigma_5$): We have merely made an educated guess at a useful value, and then found that it is sufficient to meet the given specifications with a comfortable margin. The guess was based on getting the O in Fig. 22.4c into the neighborhood of the locus, so as to attract the locus to the left, out of the shaded region. Presumably a better location for σ_5 can be found, and the reader is invited in Problem 22.11 to carry out a study to "optimize" the network design on the basis of a particular criterion of performance. [To bracket the study, one would expect that if the O were moved too far out— say to -40— it would have negligible affect on the locus near point S. At the other extreme, if it were moved in too close—e.g., to -8—a root would be trapped on the real axis near the origin, resulting in a slow exponential response that would fail to meet specification (1).]

696 SOME CASE STUDIES IN AUTOMATIC CONTROL [Chap. 22]

Step 3. Inverse transform to obtain the time response of interest. This step is left as an exercise (Problem 22.10). Note that there is now a relatively slow real root (T' in Fig. 22.4c), which has resulted from the presence of the O and which will have a substantial influence on the system response.

> Problems 22.10 through 22.18

22.3 ROLL-CONTROL AUTOPILOT: A MULTILOOP SYSTEM

We consider here the design of the section of an autopilot that controls the angle of roll (bank angle) of an aircraft, Fig. 22.5a. The ailerons are con-

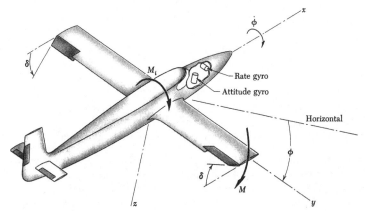

(a) *Physical arrangement of system components*

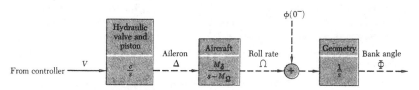

(b) *Block diagram of the physical plant to be controlled*

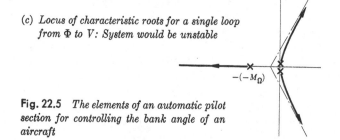

(c) *Locus of characteristic roots for a single loop from Φ to V: System would be unstable*

Fig. 22.5 *The elements of an automatic pilot section for controlling the bank angle of an aircraft*

[Sec. 22.3] ROLL CONTROL AUTOPILOT: A MULTILOOP SYSTEM 697

trolled by hydraulic pistons (Fig. 3.12 or Fig. 19.8 without the mechanical linkage), and both a rate gyro and an attitude gyro (called a "vertical gyro," Sec. 19.5) are available to measure aircraft motion. We are to design a feedback control system that will effect any desired bank angle in response to an electrical command signal from either a radio command, an on-board guidance system, or a dial set by the human pilot. Suppose that the system is to respond to a step command input with (1) a transient motion characterized by a damping-envelope time constant $\tau = 0.36$ sec and 0.7 critical damping, and (2) zero steady-state error. These are the system specifications. Suppose further that, for this particular aircraft, aerodynamic parameter M_Ω (defined below) has the value -1 sec^{-1}.

We analyze the system, following Procedure E-1, and superposing synthesis of the control system (Stage IV) at the appropriate points.

STAGE I. PHYSICAL MODEL. In the interest of simplicity we make an important assumption about the aircraft: We assume that roll motion ϕ (about the vehicle x axis, Fig. 22.5a) can be considered independent of pitch-plane motions (rotation about y axis and translation in xz plane) and of yaw-plane motions (rotation about z axis and translation in xy plane). With regard to pitch motion, this assumption is valid to a high degree for nearly all aircraft because of aircraft symmetry about the vehicle xz plane. With regard to yaw motion, the assumption is valid for aircraft having axial symmetry, such as cruciform missiles. For conventional aircraft it is not, as any light-plane pilot will attest: Roll and yaw motion are coupled, yawing motion producing a rolling moment because of the unsymmetrical rudder, and so on. It is a useful initial assumption, however, in that it is found to lead to good preliminary roll-autopilot design, which can later be modified on the basis of studies with a more accurate physical model.

We assume that each aileron hydraulic actuator, Fig. 22.5b, can be represented as a pure integrator, Eq. (3.50), and that the time for the hydraulic valve to open in response to an electric voltage δ_d can be neglected (the valve is extremely light and small). We assume that rate-gyro response (Problem 8.61) is extremely fast, compared to aircraft response, so that the rate-gyro signal is simply proportional to vehicle roll rate.

STAGE II. EQUATION OF MOTION. We follow Procedure A-m.

(1) *Geometry.* When an aircraft is in an arbitrary orientation, its bank angle is defined as the angle ϕ back through which it must roll about its longitudinal axis (x axis in Fig. 22.5a) to make the wings level. The vertical gyro is gimbaled so that its output signal (from a gimbal-angle potentiometer) gives precisely this quantity, regardless of aircraft orientation.† Let Ω represent

---

† Specifically, the attitude gyro is a two-gimbal gyro like that in Fig. 19.10, but with its spin axis vertical and its outer gimbal bearing along the aircraft's longitudinal axis. For this arrangement, a pickoff on the outer gimbal gives angle ϕ of Fig. 22.5a, regardless of aircraft orientation, as the reader may convince himself. (Another pickoff between the inner and outer gimbals gives the pitch angle—the angle between the x vehicle axis and a horizontal plane.) The terms "bank angle" and "roll attitude" are synonymous.

angular velocity (roll rate) about the x vehicle axis. Then we have the identity

$$\Omega \equiv \dot{\phi} \qquad (22.32)$$

and the ideal rate gyro will measure precisely this quantity. We define aileron deflection δ as shown, so that positive δ will produce positive rolling moment. For the physical model shown for the hydraulic system, aileron rate $\dot{\delta}$ is proportional to valve input voltage v:

$$\dot{\delta} = cv \qquad (22.33)$$

(2) *Force equilibrium.* Applying (2.2) to the rigid aircraft, Fig. 22.5a, about the x axis through the mass center gives simply

$$\Sigma M^* = M_i + M = 0 \qquad (22.34)$$

(3) *Physical relations.* For the direction defined as positive in Fig. 22.5a, we have $M_i = -J\dot{\Omega}$. For the physical model assumed (with no yawing motion), aerodynamic moment M depends only on aileron position and roll rate. For small values of δ and Ω the physical relations between M and δ and between M and Ω will be nearly linear, and we may write

$$M = \frac{\partial M}{\partial \delta}\delta + \frac{\partial M}{\partial \Omega}\Omega \qquad (22.35)$$

where $\partial M/\partial \delta$ is the slope of the plot of experimental measurements of M versus δ, with $\Omega = 0$, and $\partial M/\partial \Omega$ is the slope of the plot of experimental measurements of M versus Ω, with $\delta = 0$. Physical intuition tells us, from Fig. 22.5a, that $\partial M/\partial \delta$ will always be positive and $\partial M/\partial \Omega$ will always be negative.

Combining relations, we can rewrite (22.34):

$$-J\dot{\Omega} + \frac{\partial M}{\partial \delta}\delta + \frac{\partial M}{\partial \Omega}\Omega = 0 \qquad (22.36)$$

where J is, of course, moment of inertia about the x axis. It is convenient to define

$$M_\delta \triangleq \frac{1}{J}\frac{\partial M}{\partial \delta} \qquad M_\Omega \triangleq \frac{1}{J}\frac{\partial M}{\partial \Omega} \qquad (22.37)$$

and then to write (22.36) as follows:

$$\dot{\Omega} - M_\Omega \Omega = M_\delta \delta \qquad (22.38)$$

where we expect M_Ω to be always negative and M_δ to be always positive.

STAGE III. DYNAMIC BEHAVIOR

Step 1. Laplace transform the equations of motion. From (22.38), (22.32), and (22.33), with all initial conditions zero except ϕ (for simplicity),

$$\begin{aligned}(s - M_\Omega)\Omega &= M_\delta \Delta \\ s\Phi &= \Omega + \phi(0^-) \\ s\Delta &= cV\end{aligned} \qquad (22.39)$$

Step 2(a). Manipulate the resulting algebraic expressions to obtain transfer functions.

[Sec. 22.3] ROLL CONTROL AUTOPILOT: A MULTILOOP SYSTEM 699

(i) The components of the open-loop system transfer functions, obtained from (22.39), are shown in block-diagram form in Fig. 22.5b. At this point we begin considering the synthesis of a feedback control system (Stage IV). We might, for example, contemplate simply closing a single loop around the given system, from Φ to V, using the attitude gyro to measure ϕ. This would be rather foolish, however, as the root-locus picture in Fig. 22.5c indicates: The two ×'s at the origin engender a system which is unstable for any value of loop gain. We will be better advised to take advantage also of the other sensors available to us—the rate gyro and the aileron-position pickoff.

We can assemble a system as shown in Fig. 22.6a, on the basis of the following straightforward conceptual reasoning (which is indicated in Fig. 22.6a). Error in roll attitude, $\Phi_e = \Phi_d - \Phi$, should call for a (desired) corrective roll rate Ω_d proportional to that error: $\Omega_d = K_\phi \Phi_e$. This desired roll rate Ω_d should be compared in turn with actual roll rate Ω to determine the roll-rate error, $\Omega_e = \Omega_d - \Omega$. Roll-rate error should then call for a (desired) proportional corrective aileron deflection: $\Delta_d = K_\Omega \Omega_e$. Finally, desired aileron deflection should be compared with actual aileron deflection Δ, and the aileron error, $\Delta_e = \Delta_d - \Delta$, should apply a proportional corrective voltage V to the valve: $V = K_\delta \Delta_e$. In the arrangement of Fig. 22.6a we have three system parameters at our disposal—K_ϕ, K_Ω, and K_δ—and we shall find that these give us ample freedom to design the system as specified. We have not yet shown this, however. We continue with the analysis.

The system of Fig. 22.6a is a multiloop system—there are three "nested" loops to be closed. This is a common configuration. A systematic and effective procedure is to close them one at a time, beginning with the innermost and proceeding outward, as outlined in Fig. 22.6b, c, d, carrying out Step 2 of Procedure E-1 for each loop in turn: 2(a) (ii) writing the closed-loop system transfer function, (iii) rewriting (ii) in factored form, and 2(b) studying the roots of the characteristic equation. Finally, when we reach the outermost loop, we obtain the roots of the complete system and (iv) the system response functions of interest.

First, then, the hydraulic-servo loop in Fig. 22.6a is closed, giving the closed-loop transfer function

$$\frac{\Delta}{\Delta_d} = \frac{K_\delta c}{s + K_\delta c} \qquad (22.40)$$

This is a subsystem in its own right, similar conceptually to the remote-indicator servo of Sec. 22.1 and, of course, also to the hydraulic servo with mechanical feedback we studied in Sec. 19.4. While Eq. (22.40) oversimplifies the servo considerably, it is sufficiently accurate for the present study. A more elaborate study would include the effects of valve dynamics and perhaps of fluid compressibility (Problem 22.8). The design and testing of hydraulic servos has been a major project for many engineers.

Next we turn to the roll-rate loop, which is shown by itself in Fig. 22.6b, with the hydraulic servo represented by its transfer function (22.40) (box with double border). It can be shown (Problem 22.21) that aircraft roll attitude can be controlled without rate-gyro feedback. However, only a relatively weak feedback can be used without the system becoming too lightly damped, and the speed of response is only about one-tenth as fast as the aircraft is capable of achieving. A set of tight rate loops (one for each vehicle axis) is the heart of high-performance autopilots.

The characteristic equation for the roll-rate loop alone is [by (20.4)]

$$(s + K_\delta c)(s - M_\Omega) + K_\Omega K_\delta c M_\delta = 0 \qquad (22.41)$$

The locus of roots for the closed-loop roll-rate system of Fig. 22.6b is sketched beside the block diagram in Fig. 22.6b. Since the servo gain K_δ can be made as high as desired (within reason), the roots of the roll-rate loop can be located almost anywhere in the s plane. (One limitation is aileron maximum deflection, but typically ailerons are tremendously effective, producing vehicle rates as high as one *revolution* per sec.) As a first design, let us choose to make the two rate-loop roots critically damped, at location $s = -\sigma_r$, as indicated in Fig. 22.6b. The closed-loop transfer function of the roll-rate loop can then be written [by (20.3)]

$$\frac{\Omega}{\Omega_d} = \frac{K_\Omega K_\delta c M_\delta}{(s + K_\delta c)(s - M_\Omega) + K_\Omega K_\delta c M_\delta} = \frac{K_\Omega K_\delta c M_\delta}{(s + \sigma_r)^2} \qquad (22.42)$$

in which $\sigma_r = (K_\delta c - M_\Omega)/2$.

Finally, we turn to the attitude loop, Fig. 22.6c, including the rate-loop transfer function, Eq. (22.42) (box with double border). Its closed-loop transfer function (ii) can be written, by (20.3),

$$\frac{\Phi}{\Phi_d} = \frac{K_\phi K_\Omega K_\delta c M_\delta}{s(s + \sigma_r)^2 + K_\phi K_\Omega K_\delta c M_\delta} \qquad (22.43)$$

the denominator of which gives the system characteristic equation

$$s(s + \sigma_r)^2 = -K_\phi K_\Omega K_\delta c M_\delta \triangleq -K_c \qquad (22.44)$$

For the rate loop chosen, the locus of roots for the attitude loop (i.e., for the complete system) has the shape shown in Fig. 22.6c. It is clear, however, that by other choices of rate-loop characteristics in Fig. 22.6b, loci can be obtained in Fig. 22.6c which pass through any desired location in the s plane. In Problem 22.22, for example, all three loci are to be made to meet at one specified point on the real axis.

STAGE IV. SYSTEM DESIGN. We can do the system design corresponding to Specification (1) at this point in the analysis. First, we choose K_ϕ to put the roots in Fig. 22.6c just on the $\zeta = 0.7$ line, points (1). [The third root, point (3), is readily located by the fixed-centroid rule, Rule 6.] By the specification, we wish to have $\sigma_1 = 1/\tau = 2.8$. Now from the simple geometry of

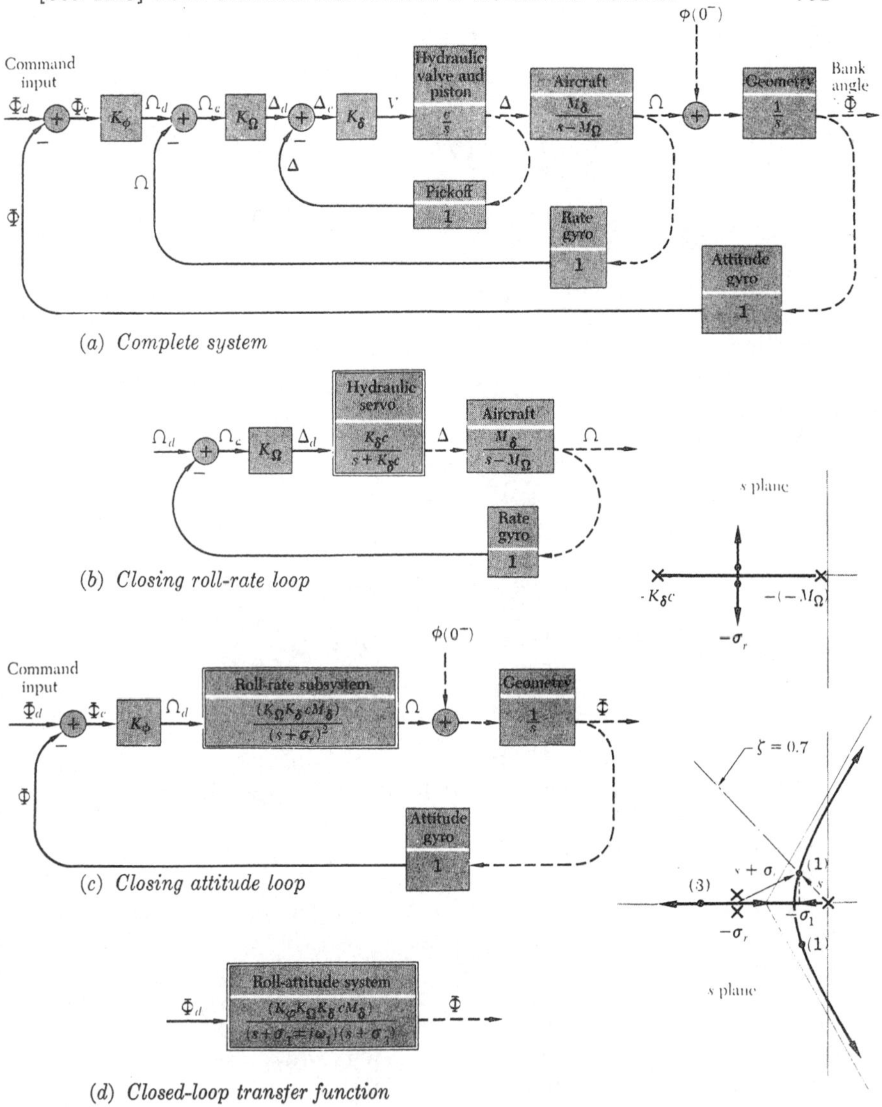

[Sec. 22.3] ROLL-CONTROL AUTOPILOT: A MULTILOOP SYSTEM 701

(a) Complete system

(b) Closing roll-rate loop

(c) Closing attitude loop

(d) Closed-loop transfer function

Fig. 22.6 *Steps in analyzing the multiloop roll autopilot system*

the root-locus drawings in Fig. 22.6b and c we can see what values to give K_δ, K_Ω, and K_ϕ so that this will be true. First, by estimating distances on the s plane in Fig. 22.6c we see that $\sigma_r \approx 3.6\sigma_1 = 10$, and [using (22.44)] $K_c \approx (0.4\sigma_r)(0.75\sigma_r)^2 = 0.23\sigma_r^3 = 230$. From Fig. 22.6b,

$$\sigma_r = [K_\delta c + (-M_\Omega)]/2$$

and thus $K_\delta c = 2\sigma_r + M_\Omega = 20 - 1 = 19$, and, using (22.41),

$$K_\Omega M_\delta = -\frac{(-\sigma_r + K_\delta c)(-\sigma_r - M_\Omega)}{K_\delta c} = -\frac{(-10 + 19)(-10 + 1)}{19} = 4.3$$

The complete-system transfer function, from ϕ_d to ϕ, may now be written, from (22.43) and Fig. 22.6c,

$$\frac{\Phi}{\Phi_d} = \frac{K_\phi K_\Omega K_\delta c M_\delta}{(s^2 + 2\sigma_1 s + \omega_{01}^2)(s + \sigma_3)} = \frac{230}{(s^2 + 5.6s + 16)(s + 14)} \quad (22.45)$$

where $\omega_{01}^2 \sigma_3$ must equal $K_\phi K_\Omega K_\delta c M_\delta = K_c$ by comparison of (22.45) with (22.43), and σ_1, ω_{01}, and σ_3 are the chosen roots in the root-locus plot of Fig. 22.6c. This result is displayed in block-diagram form in Fig. 22.6d to emphasize its equivalence to the block diagrams of Fig. 22.6a and c.

Having met Specification (1), we now turn to Specification (2) for steady-state error. For this we need the response function for ϕ when ϕ_d is a step, $\phi_d = \phi_{do} u(t)$. From (22.45) we have

$$\Phi = \frac{K_c \phi_{do}}{s(s^2 + 2\sigma_1 s + \omega_{01}^2)(s + \sigma_3)} \quad (22.46)$$

where the numerical constants have been combined, $K_c \triangleq K_\phi K_\Omega K_\delta c M_\delta$.

Step 2(c). Final value. We can now check the steady-state error. Applying the Final-Value Theorem to (22.46) gives

$$\phi(\infty) = \phi_{do}$$
or
$$\phi_e(\infty) = \phi_{do} - \phi(\infty) = 0 \quad (22.47)$$

(because $\omega_{01}^2 \sigma_3 = K_c$). That is, there is no steady-state error. As in Sec. 22.1 (p. 687), this is true because the attitude loop contains an integration— a $\times$ at the origin. Thus Specification (2) has been met.

Certainly, it just "turned out" that the specification of zero steady-state error was met by the system; we did not design specifically for it. Had it not been so, then a control technique known as *integral control* could have been employed to achieve zero steady-state error. The technique is discussed in books on automatic control.†

The result of our synthesis is an autopilot design arrangement which provides fast, well-damped control of aircraft bank angle in response to commands introduced electrically, with zero steady-state error (so long as the vertical gyro is accurate).

Our specifications have been so written that it is not necessary to perform Step 3 to obtain a specific time response. This is often the case in preliminary design of control systems. Response to the step input can, of course, be readily obtained from (22.46) (Problem 22.24). The most desirable value for ζ is discussed in Problem 22.23. Another response pattern that might well be of interest is the response to an initial bank angle, with zero called-for

† See J. G. Truxal, "Control System Synthesis," pp. 252–254, McGraw-Hill Book Company, New York, 1955.

[Sec. 22.4] UNSTABLE MECHANICAL SYSTEM: STICK BALANCER

angle, as when the autopilot is first engaged from an arbitrary aircraft orientation. The pilot will want this response also to be quick and smooth. From Fig. 22.6c, what is the appropriate response function for calculating this? (See Problem 22.25.)

Problems 22.19 through 22.28

22.4 CONTROL OF AN UNSTABLE MECHANICAL SYSTEM: THE STICK BALANCER

Consider the system shown in Fig. 22.7a. It consists of a motor-driven cart on which a stick—an inverted pendulum—is mounted with a ball-bearing pivot having almost no friction: The stick is free to fall over about the (horizontal) pivot axis. This system is of considerable interest because it represents a useful laboratory idealization of unstable mechanical systems which are encountered from time to time and are to be controlled.

The position θ of the stick is measured as an electrical signal. The objective is to use this signal to drive the cart so as to balance the stick. We may also be interested in the dynamic behavior of the controlled system as it recovers from an initial tilt angle. Specifically, suppose we are given the task of designing a control to make the system stable, with $0.4 < \zeta$, and of predicting time response in recovery from a small initial tilt angle.

To analyze the system and synthesize a control for it we follow Procedure E-1.

STAGE I. PHYSICAL MODEL. For a first analysis, we assume that the pivot is completely frictionless, and that the dynamic response of the motor electric circuit is sufficiently fast that it can be considered instantaneous (i.e., we assume that an ideal current amplifier applies torque to the motor without delay). We assume that the wheels of the cart do not slip. We assume, for a first analysis, that the motor-torque limit is not encountered, which may be true when θ is sufficiently small.

STAGE II. EQUATIONS OF MOTION. We follow Procedure A-m.

(1) *Geometry.* The system is shown in an arbitrary position in Fig. 22.7a, and coordinates x and θ are defined. We elect to use the method of D'Alembert; therefore we shall need the accelerations of the mass centers C of the cart and Q of the stick.† For the cart the acceleration is, obviously,

$$\mathbf{a}^C = \dot{\mathbf{v}}^C = \mathbf{1}_x \ddot{x} \qquad (22.48)$$

For the stick we can obtain the required acceleration with only scalar differentiation by working from the fixed, nonrotating coordinate frame xy shown in Fig. 22.7a. First we write the vector position $\mathbf{r}$ of the stick's mass center Q in

† The reader may prefer to use an energy method; then only the magnitudes of the velocities will be needed.

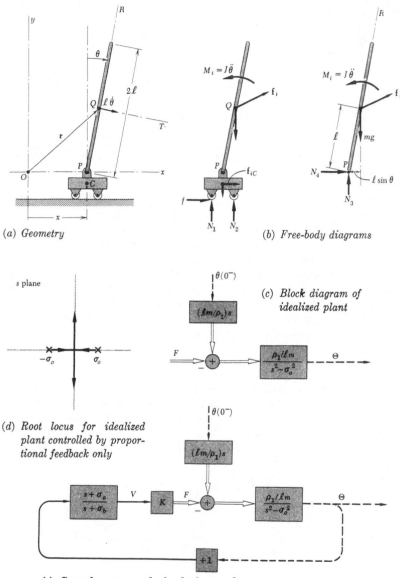

(a) Geometry

(b) Free-body diagrams

(c) Block diagram of idealized plant

(d) Root locus for idealized plant controlled by proportional feedback only

(e) Control system employing lead network

Fig. 22.7 *A system for controlling a cart to balance an inverted pendulum*

[Sec. 22.4] UNSTABLE MECHANICAL SYSTEM: STICK BALANCER

the xy coordinate frame:

$$\mathbf{r}^Q = \mathbf{r}^P + \mathbf{r}^{Q/P}$$
$$= \mathbf{1}_x x + (\mathbf{1}_x \ell \sin\theta + \mathbf{1}_y \ell \cos\theta)$$

Since the xy frame is nonrotating, we can differentiate (22.48) by components to obtain $\mathbf{v}^Q$:

$$\mathbf{v}^Q = \dot{\mathbf{r}}^Q = \mathbf{1}_x(\dot{x} + \ell\dot\theta \cos\theta) + \mathbf{1}_y(-\ell\dot\theta \sin\theta) \qquad (22.49)$$

and again to obtain $\mathbf{a}^Q$:

$$\mathbf{a}^Q = \dot{\mathbf{v}}^Q = \mathbf{1}_x(\ddot{x} + \ell\ddot\theta \cos\theta - \ell\dot\theta^2 \sin\theta) + \mathbf{1}_y(-\ell\ddot\theta \sin\theta - \ell\dot\theta^2 \cos\theta) \qquad (22.50)$$

The components can then be regrouped conveniently, using also the RT coordinate frame in Fig. 22.7a, to obtain a simpler expression whose components are each readily checked physically:

$$\mathbf{a}^Q = \mathbf{1}_x(\ddot{x}) + \mathbf{1}_T(\ell\ddot\theta) + \mathbf{1}_R(-\ell\dot\theta^2) \qquad (22.51)$$

(2) *Force equilibrium.* (i) Define system. The two free-body diagrams shown in Fig. 22.7b are the most efficient for obtaining the equations of motion, using D'Alembert's principle. Force f is the traction force, due to the motor, between the drive wheel and the floor. (ii) Force balance. In the first free-body diagram we apply (2.1) in the horizontal direction (to avoid unknown forces N_1 and N_2):

$$\Sigma f_x{}^* = 0 = (\mathbf{f}_{iC})_x + (\mathbf{f}_i)_x + f \qquad (22.52)$$

In the second free-body diagram we apply (2.2) about pivot point P (to avoid unknown forces N_3 and N_4); the z axis is out of the paper:

$$\overset{\frown}{\Sigma M_p}{}^* = 0 = M_i - [(\mathbf{1}_R \ell) \times \mathbf{f}_i]_z - mg\ell \sin\theta \qquad (22.53)$$

(3) *Physical Force-Geometry Relations.* From (2.15) and (2.18) we can write

$$\begin{aligned} f_{iC} &= -m_C \ddot{x} \qquad M_i = J\ddot\theta \\ \mathbf{f}_i &= -m\mathbf{a}^Q = -m[\mathbf{1}_x(\ddot{x}) + \mathbf{1}_T(\ell\ddot\theta) + \mathbf{1}_R(-\ell\dot\theta^2)] \\ (\mathbf{f}_i)_x &= -m(\mathbf{a}^Q)_x = -m[\ddot{x} + \ell\ddot\theta \cos\theta - \ell\dot\theta^2 \sin\theta] \\ -[\mathbf{1}_R(\ell) \times (\mathbf{f}_i)]_z &= +\ell m[\ddot{x} \cos\theta + \ell\ddot\theta] \end{aligned} \qquad (22.54)$$

where J is the moment of inertia of the stick about its mass center: $J = m\ell^2/3$ for a stick of length 2ℓ. Substituting these relations into (22.52) and (22.53):

$$\begin{aligned} m_C \ddot{x} + m\ddot{x} + m\ell\ddot\theta \cos\theta - m\ell\dot\theta^2 \sin\theta &= f \\ J\ddot\theta + m\ell(\ddot{x} \cos\theta + \ell\ddot\theta) - mg\ell \sin\theta &= 0 \end{aligned} \qquad (22.55)$$

At this point we confine our study to small angles θ and postulate that the angular velocity $\dot\theta$ will also be small, and we therefore drop products of θ and $\dot\theta$ from the equations of motion (including product terms in the series expansions for $\sin\theta$ and $\cos\theta$). The result is the pair of equations

$$\begin{aligned} (m_C + m)\ddot{x} + m\ell\ddot\theta &= f \\ m\ell\ddot{x} + (J + m\ell^2)\ddot\theta - mg\ell\theta &= 0 \end{aligned} \qquad (22.56)$$

In addition, we are given that the force f can be considered directly proportional to input voltage v to the motor-drive amplifier:

$$f = cv \tag{22.57}$$

STAGE III. DYNAMIC BEHAVIOR

Step 1. Laplace transform the equations of motion. Let the only initial condition be tilt angle, $\theta(0^-)$.

$$\begin{bmatrix} s^2 & \left(\dfrac{\ell}{1+\rho}\right)s^2 \\ \left(\dfrac{3}{4\ell}\right)s^2 & \left(s^2 - \dfrac{3}{4}\dfrac{g}{\ell}\right) \end{bmatrix} \begin{bmatrix} X \\ \Theta \end{bmatrix} = \begin{bmatrix} \dfrac{F}{(1+\rho)m} + \dfrac{\ell}{1+\rho} s\theta(0^-) \\ s\theta(0^-) \end{bmatrix} \tag{22.58}$$

$$F = cV \tag{22.59}$$

where $\rho \triangleq m_C/m$, and we have let $J = m\ell^2/3$.

Step 2(a). Manipulate the resulting algebraic expressions to obtain transfer functions and response functions.

(i) Obtain components of open-loop transfer function (draw block diagrams). Because (22.58) involves two inputs, the most convenient thing to do is to write the response function for the output Θ of the subsystem. The reader may verify the following:

$$\Theta = \dfrac{-\left(\dfrac{3}{1+4\rho}\right)\dfrac{F}{\ell m}}{s^2 - \dfrac{3(1+\rho)}{1+4\rho}\dfrac{g}{\ell}} + \dfrac{s\theta(0^-)}{s^2 - \dfrac{3(1+\rho)}{1+4\rho}\dfrac{g}{\ell}} \tag{22.60}$$

or, using the shorthand notation, $\sigma_o \triangleq \sqrt{[3(1+\rho)/(1+4\rho)](g/\ell)}$, $\rho_1 = 3/(1+4\rho)$, relation (22.60) can be written

$$\Theta = \dfrac{-(\rho_1/\ell m)F}{s^2 - \sigma_o^2} + \dfrac{s\theta(0^-)}{s^2 - \sigma_o^2} \tag{22.61}$$

This relation is shown in block-diagram form in Fig. 22.7c. [Do the dimensions check in (22.61)?]

Step 2(b). Study natural characteristics. Equation (22.61) substantiates the physically obvious fact that the uncontrolled system of Fig. 22.7a is unstable, and its motion will "diverge" (the stick will fall over) following any initial tilt angle $\theta(0^-)$ or any disturbance f. The initial divergent motion will have the time constant $1/\sigma_o$. As the angle becomes large, Eqs. (22.56) no longer hold, but we expect that the stick will continue to fall over.

STAGE IV. CONTROL-SYSTEM SYNTHESIS. The open-loop poles of the system are plotted in the s plane of Fig. 22.7d. If a purely proportional feedback loop is closed around this system, the root loci will come toward each other from the two poles on the real axis, meet at the origin, and then separate vertically along the $\pm j\omega$ axes, as shown. Thus, for small feedback

[Sec. 22.4] UNSTABLE MECHANICAL SYSTEM: STICK BALANCER 707

gains there would be an unstable divergence (root on the real axis in the right half-plane), and for large gain the system would display, ideally, neutrally stable oscillations. However, since no real system is free from small time lags (additional dynamic terms), we may expect the roots of a real system to move into the right half-plane, although this will be mitigated somewhat by the presence of friction in the bearing, as will be seen in Problem 22.31. Thus the stability is at best highly tenuous: It is more likely that the system will be thoroughly unstable.

To effect control that provides solid stability to the system, we turn to the lead-network technique introduced in Sec. 22.2. There it was helpful in improving performance; here it is essential to system stability. We use the network physical model of Fig. 22.3a and borrow transfer function (22.21), which we add to the block diagram, Fig. 22.7e. Quantities σ_a and σ_b are design parameters which we can deploy as we wish to effect a good design. Let us take $\sigma_b/\sigma_a = 4$ as a reasonable design choice (refer to p. 693).

Now we continue with the analysis of the system characteristics, Stage III, Step 2(b). The characteristic equation for the complete system of Fig. 22.7e is, by (20.4),

$$(s + \sigma_b)(s + \sigma_o)(s - \sigma_o) + \frac{K\rho_1}{\ell m}(s + \sigma_a) = 0 \quad (22.62)$$

(Does it check dimensionally?), or, in the Evans form,

$$\frac{(s + \sigma_b)(s + \sigma_o)(s - \sigma_o)}{(s + \sigma_a)} = -\frac{K\rho_1}{\ell m} \quad (22.63)$$

The open-loop $\times$'s and $\bigcirc$'s are shown in Fig. 22.8 for a particular choice of σ_a and σ_b. The root loci are plotted versus $K\rho_1/\ell m$ in Fig. 22.8, and we see that the system is stable (ideally) for all values of $K\rho_1/\ell m$ greater than $(\sigma_b/\sigma_a)\sigma_o^2 = 4\sigma_o^2$. [Verify this value from the geometry of Fig. 22.8. Verify it independently by using Routh's method on (22.63).] For $0.4 < \zeta$ we may use any $K\rho_1/\ell m < 8.5\sigma_o^2$, as shown.

Putting the $\bigcirc$ to the right of the pole at $-\sigma_0$ is, in fact, a conservative thing to do: It insures that the root from the right half-plane pole will be drawn into the left half-plane and will be real, as shown. But its time constant is limited, and the possibility of fast response is sacrificed. (The alternative of placing the $\bigcirc$ to the left of $-\sigma_0$ is to be explored.)

From the locus in Fig. 22.8 we choose gain K to give a desirable combination of speed of response ω and relative damping ζ. We might choose $K\rho_1/\ell m = 8.5\sigma_o^2$, for example, to make the slow real root as fast as possible without violating the specification $0.4 < \zeta$. The resulting roots are as indicated. (Better characteristics than these may be obtained by exploring other possible values for σ_a; see Problem 22.32.)

We return to Step 2(a). (iv) To find the time response of the complete system, we need the response function from the initial-condition "input" to

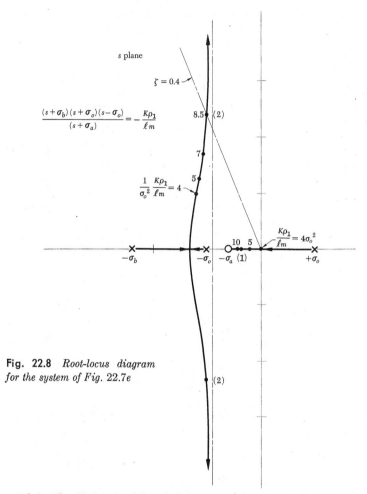

Fig. 22.8 *Root-locus diagram for the system of Fig. 22.7e*

output Θ in Fig. 22.7e. Applying (20.3), we obtain

$$\Theta = \frac{s(s + \sigma_b)\theta(0^-)}{[(s + \sigma_b)(s^2 - \sigma_o^2) + (K\rho_1/\ell m)(s + \sigma_a)]}$$

$$= \frac{s(s + \sigma_b)\theta(0^-)}{(s + \sigma_1)(s^2 + 2\sigma_2 s + \omega_{02}^2)} \quad (22.64)$$

in which $\sigma_1\omega_{02}^2$ must equal $(K\rho_1/\ell m)\sigma_a$, by comparison of the two denominators.

Step 2(c). IVT and FVT. Before calculating time response, result (22.64) may be checked conveniently by the Initial-Value Theorem [Eq. (17.46)]:

$$\theta(0^+) = \lim_{s \to \infty} \left[\frac{s^3}{s^3} \theta(0^-)\right] = \theta(0^-) \quad (22.65)$$

[Sec. 22.4] UNSTABLE MECHANICAL SYSTEM: STICK BALANCER 709

as we expect from physical reasoning, since the masses will not be moved instantaneously. Applying the Final-Value Theorem [Eq. (17.45)] gives $\theta(\infty) = 0$; physically, we would be surprised by any other answer.

Step 3. Inverse transform to obtain desired time response. From (22.64) we obtain

$$\theta(t) = \theta(0^-)[A_1 e^{-\sigma_1 t} + 2|\vec{A}_{21}|e^{-\sigma_2 t}\cos(\omega_2 t + \underline{/\vec{A}_{21}})]u(t) \quad (22.66)$$

where the A's are given by

$$A_1 = \frac{(-\sigma_1)(\sigma_b - \sigma_1)}{x^2}$$

$$2\vec{A}_{21} = \frac{\omega_{02} y}{x \omega_2} \Big/ \phi_2 + \phi_b - \phi_1 - \frac{\pi}{2}$$

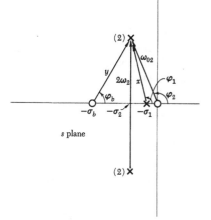

For the numerical values available directly from points (1) and (2) in Fig. 22.8, the response (22.66) is

$$\boxed{\theta(t) = \theta(0^-)[-0.14 e^{-0.4\sigma_o t} + 1.24 e^{-\sigma_o t}\cos(2.3\sigma_o t - 23°)]u(t)} \quad (22.67)$$

This dynamic behavior during recovery from an initial tilt angle is plotted in Fig. 22.9. (Procedure B-3 was used.) The stick is brought smoothly to a stable upright position. [We must remember, of course, that our linearized equations of motion (22.56), and therefore our result, hold only for $\theta(0^-)$ small—e.g., less than 0.3 radian.]

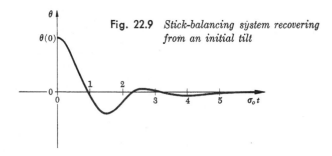

Fig. 22.9 *Stick-balancing system recovering from an initial tilt*

The behavior of coordinate x (cart position) during the controlled recovery is also of interest: We shall see in Problem 22.33 that simple additional control is necessary to keep the cart on the premises.

Some more general studies of the automatic stick-balancing problem have been carried out in a small project at Stanford University. Experimental results have included the successful balancing of two sticks side by side on the same cart, of one stick atop another, and of an extremely limber stick. In each case, there was a low limit on motor torque. Several papers and reports discuss some new analytical techniques, applicable to an arbitrary number of unstable elements, that were developed while synthesizing the control systems.†

The reader is invited, in Problems 22.34 and 22.35, to study the problem of balancing two sticks side by side, and of balancing one stick on top of another.

Problems 22.29 through 22.35

† D. T. Higdon and R. H. Cannon, Jr., "On the Control of Unstable Multiple-Output Mechanical Systems," presented at the ASME Winter Annual Meeting, Philadelphia, Pa., November 17–22, 1963, paper no. 63-WA-148.
D. T. Higdon, "Automatic Control of Inherently Unstable Systems With Bounded Control Inputs," Ph.D. Dissertation, Department of Aeronautics and Astronautics, Stanford University, December 1963.
J. F. Schaefer and R. H. Cannon, Jr., "On the Control of Unstable Mechanical Systems," presented at the 1966 International Federation of Automatic Control, London, England.
J. F. Schaefer, "On the Bounded Control of Some Unstable Mechanical Systems," Ph.D. Dissertation, Department of Electrical Engineering, Stanford University, April 1965.

Appendix A

PHYSICAL CONVERSION FACTORS TO EIGHT SIGNIFICANT FIGURES[†]

Quantity	Constant or unit-conversion factor
Length ℓ	(m) (ft) (m/ft) $\ell = \ell \times 0.30480000$
Force $\mathbf{f}$	(n) (lb) (n/lb) $\mathbf{f} = \mathbf{f} \times 4.4482216$
Moment $\mathbf{M}$	(n m) (ft lb) (n m/ft lb) $\mathbf{M} = \mathbf{M} \times 1.3558179$
Angular displacement θ	(rad) (deg) $\theta = \theta /57.295780$ (deg/rad)
Mass m	[n/(m/sec^2)] [lb/(ft/sec^2)] or (kg) or (slugs) (kg/slug) $m = m \times 14.593903$
Acceleration of gravity, standard value g_0 (sea level)	Nominal weight conversion: 32.174000 ft/sec^2 9.8100000 m/sec^2
Amount of heat Q	(n m) (jl) (kcal) (jl/kcal) $Q = Q = Q \times 4186.7400$[‡] (ft lb) (Btu) (ft lb/Btu) $Q = Q \times 778.15760$
Power P	(watts) (hp) (watts/hp) $P = P \times 745.69987$

[†] Refer to Table II, p. 22. After E. A. Mechtly, *The International System of Units*, National Aeronautics and Space Administration SP-7012, 1964. [Based on the 1960 Eleventh General Conference on Weights and Measures, which defined an international system of units (designated as SI).]

[‡] Based on International Steam Table.

Appendix B

VECTOR DOT PRODUCT AND CROSS PRODUCT

Dot product. The dot product of two vectors, say, **A** and **B** in Fig. B.1a, is defined as the product of their magnitudes times the cosine of the angle θ between them. Equivalently, it is the product of the magnitude of one of the vectors (say, B) times the projection ($A \cos \theta$) of the other upon it:

$$\mathbf{A} \cdot \mathbf{B} \triangleq AB \cos \theta \tag{B.1}$$

An example is the work done by a constant force **A** acting on a point P as it moves through a displacement **B**. The dot product is a scalar, and it is commutative: $\mathbf{B} \cdot \mathbf{A} \equiv \mathbf{A} \cdot \mathbf{B}$.

Resolved into *components* along a given orthogonal set of axes x, y, z, the vectors and their dot product are given by

$$\left. \begin{aligned} \mathbf{A} &= \mathbf{1}_x A_x + \mathbf{1}_y A_y + \mathbf{1}_z A_z \\ \mathbf{B} &= \mathbf{1}_x B_x + \mathbf{1}_y B_y + \mathbf{1}_z B_z \end{aligned} \right\} \tag{B.2}$$

$$\mathbf{A} \cdot \mathbf{B} = A_x B_x + A_y B_y + A_z B_z \tag{B.3}$$

where $\mathbf{1}_x, \mathbf{1}_y, \mathbf{1}_z$ are unit vectors along x, y, z. That is, the dot product of two vectors is simply the sum of the products of corresponding components of the vectors.

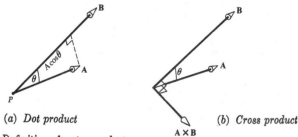

(a) *Dot product* (b) *Cross product*

Fig. B.1 *Definition of vector products*

VECTOR DOT PRODUCT AND CROSS PRODUCT

Cross product. The cross product of two vectors, say, **A** and **B** in Fig. B.1b, is another vector whose magnitude is the product of their magnitudes times the sine of the angle θ between them, and whose direction is perpendicular to the common plane of the vectors, the sense being given by the right-hand rule,† as the first is rotated into the second:

$$\mathbf{A} \times \mathbf{B} = \mathbf{1}_\perp AB \sin \theta \tag{B.4}$$

An example is the moment of a force **A** about a point O, Fig. B.2, if the force's point of application Q is located, with respect to O, by a position vector **B**.

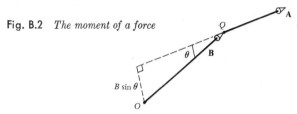

Fig. B.2 *The moment of a force*

The cross product is not commutative: by the right-hand rule (e.g., Fig. B.1b), $\mathbf{B} \times \mathbf{A} = -\mathbf{A} \times \mathbf{B}$.

In terms of their components [given in Eqs. (B.2)] along a given set of axes x, y, z, the cross product of **A** and **B** is given by

$$\mathbf{A} \times \mathbf{B} = \mathbf{1}_x(A_y B_z - A_z B_y) + \mathbf{1}_y(A_z B_x - A_x B_z) + \mathbf{1}_z(A_x B_y - A_y B_x) \tag{B.5a}$$

which is conveniently cast as a determinant:‡

$$\mathbf{A} \times \mathbf{B} = \begin{vmatrix} \mathbf{1}_x & \mathbf{1}_y & \mathbf{1}_z \\ A_x & A_y & A_z \\ B_x & B_y & B_z \end{vmatrix} \tag{B.5b}$$

[Use (B.5) to prove algebraically that $\mathbf{B} \times \mathbf{A} = -\mathbf{A} \times \mathbf{B}$.]

† Right-hand rule: Put tails of **A** and **B** together (e.g., by sliding **A** parallel to itself), and push **A** toward **B** with fingertips of right hand: then right thumb gives direction of $\mathbf{A} \times \mathbf{B}$.
‡ Refer to Appendix H.

Appendix C

VECTOR DIFFERENTIATION IN A ROTATING REFERENCE FRAME

Figure C.1 shows a vector **A** which is changing—both in magnitude and in direction—so that after an (infinitesimal) elapsed time dt, the vector tip has moved from a to b. If $\dot{\mathbf{A}}$ is the absolute rate of change of **A**, then vector **ab** has the value $\dot{\mathbf{A}}\,dt$, as shown, and this would be correctly observed from a stationary reference frame.

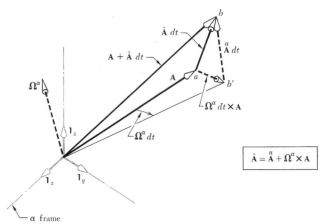

Fig. C.1 *Rate of change (derivative) of a vector*

Now consider a reference frame α which is rotating with angular velocity $\mathbf{\Omega}^\alpha$. A point which is *fixed to the α reference frame*, and which was initially at a, will itself move to point b' during time interval dt. Thus, to an observer in the (rotating) α reference frame, the vector tip *appears* to have moved along **b'b**, and the apparent rate of change of **A**, which we shall denote by $\overset{\alpha}{\mathbf{A}}$, is given by $\overset{\alpha}{\mathbf{A}} = \mathbf{b'b}/dt$. The differ-

VECTOR DIFFERENTIATION

ence between *absolute* change $\mathbf{ab} = \dot{\mathbf{A}}\,dt$ and *apparent* change $\mathbf{b'b} = \overset{\alpha}{\dot{\mathbf{A}}}\,dt$ is vector $\mathbf{ab'}$ which, from Fig. C.1, has the value $\mathbf{ab'} = (\mathbf{\Omega}^\alpha\,dt) \times \mathbf{A}$. From this we can write down the very important relation

$$\dot{\mathbf{A}} = \overset{\alpha}{\dot{\mathbf{A}}} + \mathbf{\Omega}^\alpha \times \mathbf{A} \tag{C.1}$$

or, in words: *The absolute rate of change $\dot{\mathbf{A}}$, of any vector $\mathbf{A}$, is equal to its apparent rate of change $\overset{\alpha}{\dot{\mathbf{A}}}$—as observed from a reference frame having angular velocity $\mathbf{\Omega}^\alpha$—plus a correction term $\mathbf{\Omega}^\alpha \times \mathbf{A}$.* It is helpful to visualize Fig. C.1 for the special case $\dot{\mathbf{A}} = 0$ (i.e., the case that $\mathbf{A}$ does not actually move). Then it is obvious physically that $\overset{\alpha}{\dot{\mathbf{A}}} = -\mathbf{\Omega}^\alpha \times \mathbf{A}$.

Relation (C.1) holds also if the reference frame is translating, and if its origin is displaced from the vector, because the location of the vector is not involved—only its magnitude and direction.

To prove Eq. (C.1) algebraically, designate unit vectors $\mathbf{1}_x$, $\mathbf{1}_y$, $\mathbf{1}_z$ to define an orthogonal set of axes fixed in reference frame α. At a given instant, vectors $\mathbf{\Omega}^\alpha$ and $\mathbf{A}$ can be resolved into components along these axes:

$$\begin{aligned}\mathbf{\Omega}^\alpha &= \mathbf{1}_x \Omega_x^\alpha + \mathbf{1}_y \Omega_y^\alpha + \mathbf{1}_z \Omega_z^\alpha \\ \mathbf{A} &= \mathbf{1}_x A_x + \mathbf{1}_y A_y + \mathbf{1}_z A_z\end{aligned} \tag{C.2}$$

To obtain $\dot{\mathbf{A}}$, we perform a total differentiation of $\mathbf{A}$, noting that the unit vectors $\mathbf{1}_x, \mathbf{1}_y, \mathbf{1}_z$, have nonzero derivatives because they are rotating:

$$\dot{\mathbf{A}} = \underbrace{[\mathbf{1}_x \dot{A}_x + \mathbf{1}_y \dot{A}_y + \mathbf{1}_z \dot{A}_z]}_{\overset{\alpha}{\dot{\mathbf{A}}}} + \underbrace{[\dot{\mathbf{1}}_x A_x + \dot{\mathbf{1}}_y A_y + \dot{\mathbf{1}}_z A_z]}_{\mathbf{\Omega}^\alpha \times \mathbf{A}} \tag{C.3}$$

in which the equivalence of the component and the vector forms of (C.3) is readily shown by applying (B.5) to the vectors of (C.2), and noting from Fig. C.1 that $\dot{\mathbf{1}}_x\,dt = (\mathbf{\Omega}^\alpha\,dt) \times \mathbf{1}_x = \mathbf{1}_y \Omega_z^\alpha\,dt - \mathbf{1}_z \Omega_y^\alpha\,dt$, and so on.

Appendix D

NEWTON'S LAWS OF MOTION

Newton's three laws of motion are given in his *Principia*† as follows:

Lex I (in edition of 1726). Corpus omne perseverare in statu suo quiescendi vel movendi uniformiter in directum, nisi quatenus illud a viribus impressis cogitur statum suum mutare.

Lex II. Mutationem motis proportionalem esse vi motrici impressae, et fieri secondum lineam rectam qua vis illa imprimitur.

Lex III. Actioni contrariam semper et aequalem esse reactionem: sive corporum duorum actiones in se mutuo semper esse aequales et in partes contrarias dirigi.

These are translated by Cajorian,‡ together with Newton's elaborations on them (which add to our understanding of his meaning), as follows:

Law I. Every body continues in its state of rest, or of uniform motion in a straight line, unless it is compelled to change that state by forces impressed upon it.

Projectiles continue in their motions, so far as they are not retarded by the resistance of the air, or impelled downwards by the force of gravity. A top, whose parts by their cohesion are continually drawn aside from rectilinear motions, does not cease its rotation, otherwise than as it is retarded by the air. The greater bodies of the planets and comets, meeting with less resistance in freer spaces, preserve their motions both progressive and circular for a much longer time.

Law II. The change of motion is proportional to the motive force impressed; and is made in the direction of the right line in which that force is impressed.

If any force generates a motion, a double force will generate double the motion, a triple force triple the motion, whether that force be impressed altogether and at once, or gradually and successively. And this motion (being always directed the same way with the generating force), if the body moved before, is added to or subtracted from the former motion, according as they directly conspire with or are directly contrary to each other; or obliquely joined, when they are oblique, so as to produce a new motion compounded from the determination of both.

† I. Newton, *Principia Mathematica Philosophiae Naturalis*, 1686.
‡ F. Cajorian, "Sir Isaac Newton's Mathematical Principles of Natural Philosophy and His System of the World," University of California Press, Berkeley, 1960.

NEWTON'S LAWS OF MOTION

Law III. To every action there is always opposed an equal reaction: or, the mutual actions of two bodies upon each other are always equal, and directed to contrary parts.

Whatever draws or presses another is as much drawn or pressed by that other. If you press a stone with your finger, the finger is also pressed by the stone. If a horse draws a stone tied to a rope, the horse (if I may so say) will be equally drawn back towards the stone; for the distended rope, by the same endeavor to relax or unbend itself, will draw the horse as much towards the stone as it does the stone towards the horse, and will obstruct the progress of the one as much as it advances that of the other. If a body impinge upon another, and by its force change the motion of the other, that body also (because of the equality of the mutual pressure) will undergo an equal change, in its own motion, towards the contrary part. The changes made by these actions are equal, not in the velocities but in the motions of bodies; that is to say, if the bodies are not hindered by any other impediments. For, because the motions are equally changed, the changes of the velocities made towards contrary parts are inversely proportional to the bodies. This law takes place also in attractions, as will be proved in the next Scholium.†

It is clear from the above that Newton used the word *motion* to mean precisely what we call *momentum* ($m\mathbf{v}$, for a point mass). By "inversely proportional to the bodies," he meant inversely proportional to their masses. The term *vector* makes a concise substitute for "in a right line." Finally, we find it helpful to specify explicitly a point mass or particle. The extension to rigid bodies and other mass systems is straightforward but not trivial. [For example, the laws may be shown to apply verbatim to the mass centers of rigid bodies.]

Thus, in modern parlance, the laws may be stated as follows:

Law I. (*Conservation of momentum*). The momentum of a particle is constant in the absence of external forces.

Corollary: The linear and angular momentum of systems of particles (such as rigid bodies) is thus also constant in the absence of external forces.

Law II. ("$\mathbf{f} = m\mathbf{a}$"). The vector change in the momentum of a particle due to an applied force is proportional to the impulse of that force [where impulse $\triangleq \int \mathbf{f}\, dt$].

Law III. ("*Action equals reaction*"). To every action there is always opposed an equal reaction; or, if body A is exerting a force on body B, then body B is exerting an equal and (vectorially) opposite force on body A.

Corollary: The collision of two particles results in equal but opposite changes in their momenta (and thus in velocity changes inversely proportional to their masses).

Actually, we commonly use a derivative of Law II. That is, differentiating $m(\mathbf{v} - \mathbf{v}_{\text{initial}}) = \int \mathbf{f}\, dt$, we obtain $(d/dt)(m\mathbf{v}) = \mathbf{f}$, or for a particle (constant m),

$$\text{Law II} \qquad m\dot{\mathbf{v}} = \mathbf{f} \qquad (D.1)$$

or $\mathbf{f} = m\mathbf{a}$, as it is commonly phrased. Equations (2.1) and (2.2) are obtained for a rigid body by space integration of (D.1) over all its mass particles, invoking Law III to cancel internal forces properly.‡

---

† The presentation to this point is from a compilation by J. M. Slater.
‡ See, for example, G. W. Housner and D. E. Hudson, "Applied Mechanics–Dynamics," Chaps. 6 and 7, Second Edition, D. Van Nostrand Company, Inc., Princeton, N.J., 1959.

Appendix E

ANGULAR MOMENTUM AND ITS RATE OF CHANGE FOR A RIGID BODY; MOMENTS OF INERTIA

Angular momentum. Let $\mathbf{H}$ denote the angular momentum of a rigid body about a specified point P fixed in the body (e.g., its mass center). Then $\mathbf{H}$ is defined as the sum (integral) over the body of the moments, about P, of all the linear momenta of its infinitesimal mass constituents:

$$\mathbf{H} \triangleq \int_{\text{body}} \mathbf{r} \times (dm\, \mathbf{v}) \tag{E.1}$$

where $\mathbf{r}$ locates constituent dm, with respect to P, and $\mathbf{v}$ is the instantaneous linear velocity of dm with respect to P. Thus $\mathbf{v}$ is given by

$$\mathbf{v} = \boldsymbol{\Omega} \times \mathbf{r} \tag{E.2}$$

where $\boldsymbol{\Omega}$ is the angular velocity of the rigid body.

Let unit vectors $\mathbf{1}_x, \mathbf{1}_y, \mathbf{1}_z$ define an orthogonal set of axes x, y, z, whose origin is P. (In general, the xyz frame rotates in space and also with respect to the rigid body.) Denote this axis set as the α reference frame. Write the components, along the x, y, z axes, of $\boldsymbol{\Omega}, \mathbf{r}, \mathbf{v}$ and $\mathbf{H}$, as follows:

$$\boldsymbol{\Omega} = \mathbf{1}_x \Omega_x + \mathbf{1}_y \Omega_y + \mathbf{1}_z \Omega_z$$
$$\mathbf{r} = \mathbf{1}_x x + \mathbf{1}_y y + \mathbf{1}_z z$$

From (E.2)†

$$\mathbf{v} = \boldsymbol{\Omega} \times \mathbf{r} = \mathbf{1}_x(z\Omega_y - y\Omega_z) + \mathbf{1}_y(x\Omega_z - z\Omega_x) + \mathbf{1}_z(y\Omega_x - x\Omega_y)$$

Then, from (E.1), after some rearranging

$$\begin{aligned}\mathbf{H} = \mathbf{1}_x \{ & [\int (y^2 + z^2)\, dm]\Omega_x - [\int xy\, dm]\Omega_y - [\int xz\, dm]\Omega_z \} \\ + \mathbf{1}_y \{ & -[\int yx\, dm]\Omega_x + [\int (z^2 + x^2)\, dm]\Omega_y - [\int yz\, dm]\Omega_z \} \\ + \mathbf{1}_z \{ & -[\int zx\, dm]\Omega_x - [\int zy\, dm]\Omega_y + [\int (x^2 + y^2)\, dm]\Omega_z \} \end{aligned} \tag{E.3}$$

† The vector cross product is defined in Appendix B.

ANGULAR MOMENTUM; MOMENTS OF INERTIA   719

where all integrals are taken over the rigid body. The integrals on the diagonal of the array in (E.3) are designated *moments of inertia*, and the others, *products of inertia*. The following notation is common:†

$$\mathbf{H} = \mathbf{1}_x\{\ J_x\Omega_x - J_{xy}\Omega_y - J_{xz}\Omega_z\}$$
$$+ \mathbf{1}_y\{-J_{xy}\Omega_x + J_y\Omega_y - J_{yz}\Omega_z\}$$
$$+ \mathbf{1}_z\{-J_{xz}\Omega_x - J_{yz}\Omega_y + J_z\Omega_z\} \quad (E.4)$$

where the definitions of the J's are clear by comparison of (E.4) with (E.3), and $J_{xy} = J_{yx}$, and so on. Expression (2.7) derives from (E.4) and (E.3).

As a special case, for *plane motion about the z axis*,

$$\mathbf{H} = \mathbf{1}_x(-J_{xz}\Omega_z) + \mathbf{1}_y(-J_{yz}\Omega_z) + \mathbf{1}_z(J_z\Omega_z) \quad (E.5)$$

Rate of change of angular momentum. The rate of change of $\mathbf{H}$ is, of course, simply $\dot{\mathbf{H}}$. In terms of components, if the axes x, y, z (α frame) into which $\mathbf{H}$ is resolved are nonrotating, then $\dot{\mathbf{H}}$ is obtained by differentiating the nine scalar terms of (E.4), the unit vectors being constant. More generally, if the α frame *is* rotating,

$$\mathbf{\Omega}^\alpha = \mathbf{1}_x\Omega_x{}^\alpha + \mathbf{1}_y\Omega_y{}^\alpha + \mathbf{1}_z\Omega_z{}^\alpha \quad (E.6)$$

then Eq. (C.1) must be used to calculate $\dot{\mathbf{H}}$. The resulting expression has 36 terms:

$$\dot{\mathbf{H}} = \overset{\alpha}{\dot{\mathbf{H}}} + \mathbf{\Omega}^\alpha \times \mathbf{H}$$
$$= \mathbf{1}_x\{\dot{J}_x\Omega_x + J_x\dot{\Omega}_x - \dot{J}_{xy}\Omega_y - J_{xy}\dot{\Omega}_y - \dot{J}_{xz}\Omega_z - J_{xz}\dot{\Omega}_z$$
$$+ (-J_{xz}\Omega_x - J_{yz}\Omega_y + J_z\Omega_z)\Omega_y{}^\alpha - (-J_{xy}\Omega_x + J_y\Omega_y - J_{yz}\Omega_z)\Omega_z{}^\alpha\}$$
$$+ \mathbf{1}_y\{-\dot{J}_{xy}\Omega_x - J_{xy}\dot{\Omega}_x + \dot{J}_y\Omega_y + J_y\dot{\Omega}_y - \dot{J}_{yz}\Omega_z - J_{yz}\dot{\Omega}_z$$
$$+ (J_x\Omega_x - J_{xy}\Omega_y - J_{xz}\Omega_z)\Omega_z{}^\alpha - (-J_{xz}\Omega_x - J_{yz}\Omega_y + J_z\Omega_z)\Omega_x{}^\alpha\}$$
$$+ \mathbf{1}_z\{-\dot{J}_{xz}\Omega_x - J_{xz}\dot{\Omega}_x - \dot{J}_{yz}\Omega_y - J_{yz}\dot{\Omega}_y + \dot{J}_z\Omega_z + J_z\dot{\Omega}_z$$
$$+ (-J_{xy}\Omega_x + J_y\Omega_y - J_{yz}\Omega_z)\Omega_x{}^\alpha - (J_x\Omega_x - J_{xy}\Omega_y - J_{xz}\Omega_z)\Omega_y{}^\alpha\} \quad (E.7)$$

where the second line following each unit vector gives the component of $\mathbf{\Omega}^\alpha \times \mathbf{H}$.

To avoid the problem of nonconstant J's, it is common to *fix the α frame to the body*. Then $\mathbf{\Omega}^\alpha = \mathbf{\Omega}$, all $\dot{J}$'s $= 0$, and (E.7) is reduced to 27 terms.

For the special case of *plane motion about the z axis* (and again, with the α frame fixed in the body), (E.7) reduces further to

$$\dot{\mathbf{H}} = \mathbf{1}_x\{-J_{xz}\dot{\Omega}_z + J_{yz}\Omega_z{}^2\} + \mathbf{1}_y\{-J_{yz}\dot{\Omega}_z - J_{xz}\Omega_z{}^2\} + \mathbf{1}_z\{J_z\dot{\Omega}_z\} \quad (E.8)$$

in which commonly only the z component is of interest. This leads to Eqs. (2.2) and (5.37).

Principal axes. For a given origin P of axes, there is one set of axis directions, fixed in the body, for which all products of inertia are zero. These are called principal axes. For principal axes, (E-4) becomes

$$\mathbf{H} = \mathbf{1}_x(J_x\Omega_x) + \mathbf{1}_y(J_y\Omega_y) + \mathbf{1}_z(J_z\Omega_z) \quad (E.9)$$

(E.7) becomes

$$\dot{\mathbf{H}} = \mathbf{1}_x[J_x\dot{\Omega}_x + (J_z - J_y)\Omega_y\Omega_z] + \mathbf{1}_y[J_y\dot{\Omega}_y + (J_x - J_z)\Omega_z\Omega_x]$$
$$+ \mathbf{1}_z[J_z\dot{\Omega}_z + (J_y - J_x)\Omega_x\Omega_y] \quad (E.10)$$

---

† Sometimes J_{xx} is used for J_x, and so on, for symmetry. (Also, I is often used instead of J.) For additional properties of moments and products of inertia see, for example, G. W. Housner and D. E. Hudson, "Applied Mechanics–Dynamics," Second Edition, D. Van Nostrand Company, Inc. Princeton, N.J., 1959.

and (E.8), for plane motion, becomes

$$\dot{\mathbf{H}} = \mathbf{1}_z(J_z\dot{\Omega}_z) \tag{E.11}$$

If x, y, z are principal axes, then J_x, J_y, J_z are called the principal moments of inertia.

For the special case that two principal moments of inertia are equal (as for a coin), any axis in the plane of the corresponding two axes is a principal axis.

For the special case that three principal moments of inertia are equal (as for a sphere or cube), every axis is a principal axis.

The process of obtaining moments of inertia in one set of axes from those in another is discussed in texts on rigid-body dynamics.[†]

[†] For example, R. L. Halfman, "Dynamics," vol. I, Addison-Wesley, Reading, Mass., 1962.

Appendix F

FLUID FRICTION FOR FLOW THROUGH LONG TUBES AND PIPES

For flow of a fluid (liquid or gas) in a long tube or pipe, there are two regimes of flow, as indicated in Fig. F.1. The velocity at which transition occurs from one regime to the other depends on Reynolds number, a dimensionless quantity defined by

$$\mathcal{R} = \frac{vD\rho}{\mu} \tag{F.1}$$

in which v is average velocity, D is pipe diameter, ρ is fluid mass-density (mass per unit volume) [dimensions $(f/\ell^3)/(\ell/\tau^2) = f\tau^2\ell^{-4}$], and μ is fluid viscosity [dimensions

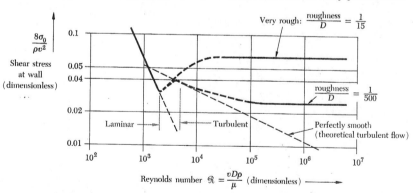

Fig. F.1 *Friction force for fluid flow in long tubes or pipes*[†]

of shear stress/velocity gradient: $(f/\ell^2)/(\ell\tau^{-1}/\ell) = f\tau\ell^{-2}$]. Figure F.1 is remarkable in that it gives the necessary relationships for wall shear stress σ_0, and hence for

[†] After L. F. Moody, "Friction Factors in Pipe Flow," *Trans. ASME*, vol. 66, pp. 671–684, 1944.

total friction f_f, for all fluids and for pipes of any roughness, from perfectly smooth to extremely rough (roughness-to-diameter ratio of one to fifteen).

It is instructive to replot the data for a specific fluid. This is done in Fig. F.2 for water flowing in pipes of diameters from 0.01 ft. to 1 ft., with a roughness range from perfectly smooth (dashed lines) to very rough: roughness/diameter = 1/15.

In the low-speed laminar regime, stress is simply proportional to velocity: $\sigma_0 = (8\mu/D)v_{av}$ [refer to Eq. (3.20)]. At high speed, it is proportional to v^2 (rough pipe), or to $v^{1.75}$ (smooth pipe). Coefficients b in (3.20) and b_α in (3.22) can be obtained from Fig. F.2 for water, or from Fig. F.1 for any fluid. [Note that friction force $f_f = \sigma_0 \times$ (wetted area).] Refer to p. 111.

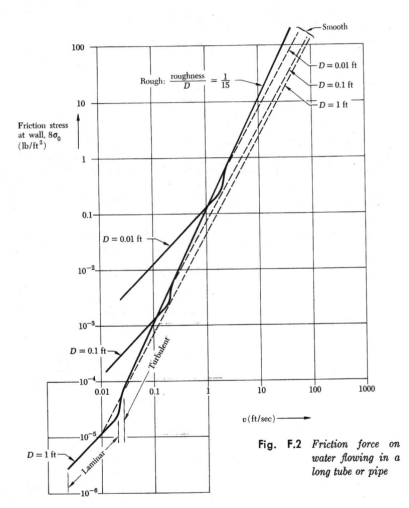

Fig. F.2 *Friction force on water flowing in a long tube or pipe*

Appendix G

DUALS OF ELECTRICAL NETWORKS

For every ideal planar network, there exists a perfect dual in which node voltages and loop currents are interchanged, and in which each element is replaced by a corresponding dual element such that, as a result, the equations of motion of the dual network in terms of node voltages are identical to the equations of motion of the original network in terms of loop currents, and vice versa.

To obtain the dual of a given network, the following exchanges are made:

Table XI

Exchanges made in constructing the dual of a network

$$i \to v$$
$$R \to \frac{1}{R}$$
$$L \to C$$
$$C \to L$$
current source $\to$ voltage source
node $\to$ mesh

The symbol $L \to C$ is to be read "every inductance is to be exchanged for a capacitance," and so on.

The key to obtaining the dual of a network graphically is the relation node $\to$ mesh. This is most readily explained by example. Consider the network of Fig. 2.13, which is drawn again in Fig. G.1a. The structure of the dual of this network is established by first marking a new "node" in the interior of each of the meshes of the original network (points A, B, C, D) and a reference node at some point *outside* the original network (point O), and then indicating a branch, shown dashed gray, which passes through each of the original elements and connects the new nodes. The resulting network is shown in Fig. G.1b, where each element has been replaced by its dual, each of the original meshes has become a node, and each of the original nodes

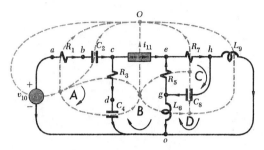

(a) Original network

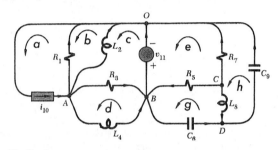

(b) Dual

Fig. G.1 *Constructing the dual of a network*

has become a mesh. (Element-arrow directions in Fig. G.1b are conveniently chosen consistently by rotating the arrows emerging from the elements of Fig. G.1a counterclockwise 90°.) Figure G.1b is the dual of Fig. G.1a. Complete dualism may now be demonstrated as follows:

Write the equations of motion (i) of the original network on the loop basis, and (ii) of the dual on the node basis; show that the individual terms of (i) and (ii) correspond precisely according to Table XI.

Write the equations of motion (iii) of the original network on the node basis and (iv) of the dual on the loop basis; show that the individual terms of (iii) and (iv) correspond precisely according to Table XI.

To illustrate, the first pair of dual equations are

(i) KVL for loop A in Fig. G.1a:

$$-v_{10} + R_1 i_A + \frac{1}{C_2} \int i_A \, dt + R_3(i_A - i_B) + \frac{1}{C_4} \int (i_A - i_B) \, dt = 0$$

(ii) KCL for node A in Fig. G.1b:

$$-i_{10} + \frac{1}{R_1} v_A + \frac{1}{L_2} \int v_A \, dt + \frac{1}{R_3} (v_A - v_B) + \frac{1}{L_4} \int (v_A - v_B) \, dt = 0$$

The reader may complete the set, and may also carry out the alternative procedure (iii), (iv).

Appendix H

DETERMINANTS AND CRAMER'S RULE

Determinants. A determinant is a square array of numbers or symbols:

$$D = \begin{vmatrix} D_{11} & D_{12} & D_{13} & \cdots & \cdots & D_{1n} \\ D_{21} & D_{22} & D_{23} & & & \\ \cdot & \cdot & \cdot & & & \\ \cdot & \cdot & \cdot & & & \\ \cdot & & & D_{ij} & & \\ \cdot & & & & & \\ \cdot & & & & & \\ D_{n1} & & & & & D_{nn} \end{vmatrix} \qquad (H.1)$$

To find the *value* of a determinant, one may expand it by the *minors* of any row or any column in the following manner: (i) using the ith row,

$$D = \pm(D_{i1}d_{i1} - D_{i2}d_{i2} + \cdots \pm D_{ij}d_{ij} \mp \cdots \pm D_{in}d_{in}) \qquad (H.2a)$$

or (ii) using the jth column,

$$D = \pm(D_{1j}d_{1j} - D_{2j}d_{2j} + \cdots \pm D_{ij}d_{ij} \mp \cdots \pm D_{nj}d_{nj}) \qquad (H.2b)$$

where d_{ij} is the *minor* of term D_{ij}—the determinant formed by removing the ith row and jth column of D. The sign rule is as follows: If $i + j$ is even, the sign is $+$; if $i + j$ is odd, the sign is $-$. (Thus the signs alternate.) Since the d's are themselves determinants, they must be expanded in turn, the process being repeated until no determinants remain.

Example H.1. Evaluate the determinant

$$D = \begin{vmatrix} A & B & C \\ E & F & G \\ H & I & J \end{vmatrix} \qquad (H.3)$$

Expanding by minors of column 2, for example, yields

$$D = -B\begin{vmatrix} E & G \\ H & J \end{vmatrix} + F\begin{vmatrix} A & C \\ H & J \end{vmatrix} - I\begin{vmatrix} A & C \\ E & G \end{vmatrix}$$

Then expanding these 2×2 determinants leads to the final answer:

$$D = -B(EJ - HG) + F(AJ - HC) - I(AG - EC) \quad (H.4)$$

Cramer's rule. Given a set of simultaneous, linear, algebraic equations,

$$\begin{aligned}
A_{11}x_1 + A_{12}x_2 + A_{13}x_3 + \cdots + A_{1j}x_j + \cdots + A_{1n}x_n &= R_1 \\
A_{21}x_1 + A_{22}x_2 + A_{23}x_3 + \cdots + A_{2j}x_j + \cdots + A_{2n}x_n &= R_2 \\
&\vdots \\
A_{i1}x_1 + A_{i2}x_2 + A_{i3}x_3 + \cdots + A_{ij}x_j + \cdots + A_{in}x_n &= R_i \\
&\vdots \\
A_{n1}x_1 + A_{n2}x_2 + A_{n3}x_3 + \cdots + A_{nj}x_j + \cdots + A_{nn}x_n &= R_n
\end{aligned} \quad (H.5)$$

the value of the variable x_j is given by

$$x_j = \frac{D_j}{D} \quad (H.6)$$

where D is the determinant formed from the coefficients A_{ij},† and D_j is the same determinant, except that its jth column is replaced by the column of R's. Cramer's rule is merely a highly systematic procedure for solving the set of equations simultaneously for one of its variables.

Example H.2. Find x and y from the set of linear algebraic equations

$$\begin{aligned}
Ax + By + Cz &= M \\
Ex + Fy + Gz &= N \\
Hx + Iy + Jz &= L
\end{aligned} \quad (H.7)$$

Applying (H.6) with $j = 1$, for x, and with $j = 2$, for y, yields

$$x = \frac{\begin{vmatrix} M & B & C \\ N & F & G \\ L & I & J \end{vmatrix}}{D} \qquad y = \frac{\begin{vmatrix} A & M & C \\ E & N & G \\ H & L & J \end{vmatrix}}{D} \quad (H.8)$$

where

$$D = \begin{vmatrix} A & B & C \\ E & F & G \\ H & I & J \end{vmatrix}$$

Determinant D was evaluated in (H.4). The numerators of (H.8) are also to be evaluated by application of (H.2).

The reader may want to make up and work some numerical practice problems.

† It is assured that this will be a square array if there are as many equations as variables. (If there are fewer, the system is unsolvable; if there are more, some are redundant.)

Appendix I

COMPUTATION WITH A SPIRULE†

Adding the angles of vectors. The sum of the angles of a set of vectors in a plane, whose heads are at a common point P, as in Fig. I.1, can be obtained rapidly with a Spirule, Fig. I.2, using only the red *angle* scale on the disk (0° through 359°) and the black reference line R on the arm.‡ The steps are as follows.

Preliminaries. On the plot, mark a 180° (horizontal) fixed reference line F through point P (see Fig. I.1). Place Spirule center (hole) over point P, and line up R (on arm) with 0° (on disk).

To add angles of vectors. (a) Holding center button to paper, but with disk free, rotate assembly (arm and disk together) to make R coincide with F. Then (b) clamp disk to paper (with your fingers) and rotate arm alone to align R with first vector, $\vec{V}_1$ say. Angle scale now reads ϕ_1 (i.e., R crosses disk scale at ϕ_1). For example, in Fig. I.2 the angle scale reads 30°. Repeat (a) and then (b) for each vector in turn. Final angle scale reading is $\Sigma\phi$, the sum of the angles.

To add in minus the angle of a vector. (c) Holding center button to paper, but with disk free, rotate assembly together to make R coincide with given vector, $\vec{V}_5$, say. Then (d) clamp disk to paper and rotate arm alone to align R with F. Angle scale now reads: previous angle $-\phi_5$.

The angles [or minus angles] from any set of vectors can be added in any sequence by following steps (a) and then (b) [or (c) and then (d)] for each in turn. Commonly, one does not read the separate angles, only the final sum.

Multiplying the lengths of vectors. The product of the lengths of a set of vectors in a plane, whose heads are at a common point P, as in Fig. I.1, can be obtained rapidly with a Spirule, using the black magnitude scale, the black reference line R, and the black logarithmic spiral S (all on the arm).

Algebraic structure. Suppose that we desire the following product of vector magnitudes from Fig. I.1:

$$\frac{|\vec{V}_1| \cdot |\vec{V}_3| \cdot |\vec{V}_4| \cdot |\vec{V}_6|}{|\vec{V}_2| \cdot |\vec{V}_5|} = K \tag{I.1}$$

† Available in bookstores or from The Spirule Company, 9728 El Venado, Whittier, Calif.

‡ The other scales and curves on a Spirule, and their many uses, are described in the instruction pamphlet that comes with the Spirule.

(where K is just a number, the desired answer). To avoid scaling problems, it is helpful to rewrite (I.1) in the equivalent form

$$\frac{|\vec{V}_1| \cdot |\vec{V}_3| \cdot |\vec{V}_4| \cdot |\vec{V}_6|}{1 \cdot 1 \cdot |\vec{V}_2| \cdot |\vec{V}_5|} = K \tag{I.2}$$

where unit vectors are included to make the same number of terms in numerator and denominator. If the scale to which the vectors are plotted is such that a unit vector is less than $\frac{1}{2}$ in. long, then the following equivalent algebraic form should be used:

$$(10 \cdot 10) \frac{|\vec{V}_1| \cdot |\vec{V}_3| \cdot |\vec{V}_4| \cdot |\vec{V}_6|}{10 \cdot 10 \cdot |\vec{V}_2| \cdot |\vec{V}_5|} = K \tag{I.3}$$

where the simple multiplication by 10^2 will be performed last. (If the plot scale is such that 10 is also less than $\frac{1}{2}$ in. long, factors of 100 should be used, and so on.) This will become clear when examples are worked (e.g., Example I.1).

Entering a numerator length on the Spirule. With the Spirule center over point P, align arm line R with disk angle 0°. Then (i) with disk free, rotate assembly (arm and disk together) to align R with the vector, $\vec{V}_1$, say. Then (ii) clamp disk to paper and rotate arm alone until logarithmic spiral S intersects the (tail) end of the vector. The magnitude scale now reads the length of $\vec{V}_1$ times a scale factor. (Reading the magnitude is discussed below.)

Next, to multiply $|\vec{V}_1|$ by another length $|\vec{V}_3|$, say, repeat steps (i) and then (ii) for $\vec{V}_3$. The magnitude scale now reads the product $|\vec{V}_1| \cdot |\vec{V}_3|$ times a scale factor.

Entering a denominator length on the Spirule. (iii) With disk free, rotate assembly together until spiral S intersects the (tail) end of the vector, $\vec{V}_2$, say. Then (iv) clamp disk to paper and rotate arm alone until line R is aligned with $\vec{V}_2$. The magnitude scale now reads: (previous magnitude) $\div |\vec{V}_2|$, again, times a scale factor.

Avoiding scale-factor algebra. The scale factors involved in Spirule multiplication are actually not difficult to account for algebraically. They are a nuisance, however, and a frequent source of error. They are more simply accounted for graphically: All we need do is use form (I.2) [or (I.3)] so that there are as many divisions as multiplications. Then no scale factor is involved (the final answer is a pure *ratio*, and is independent of the plotting scale).

Procedure summary. To perform (I.3), say, first place the Spirule center at the origin of a *scale* for the vectors to be used (e.g., at the origin of the s plane) and enter $\frac{1}{10}$ twice, using steps (iii) and then (iv). Then place the Spirule center at P and enter— in any convenient order†—numerator lengths via steps (i) and then (ii), and remaining denominator lengths via steps (iii) and then (iv). Finally, read the magnitude and multiply it by $10 \cdot 10$, as in (I.3), to obtain K.

Reading the magnitude. For a given position of the arm relative to the disk, one of the four red arrows on the disk points to a number (from .1 to .99) on the magnitude scale. Three of the arrows are marked with *index multipliers:* ×.1, ×1, ×10. The index multiplier of the fourth arrow is either ×100 or ×.01. The magnitude reading, corresponding to a given position of arm relative to disk, is the product of

† A systematic sequence—e.g., taking vectors in turn, 1 through 6—will help keep track of what has been done and what has not.

COMPUTATION WITH A SPIRULE   729

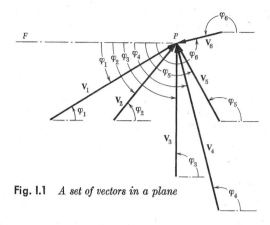

Fig. I.1 *A set of vectors in a plane*

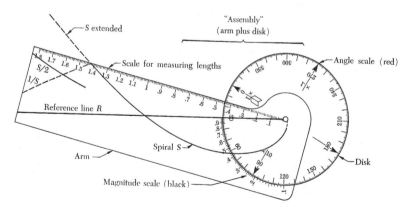

Fig. I.2 *Spirule (only magnitude and angle scales are shown). Courtesy W. R. Evans (inventor) and North American Aviation, Inc. (copyright 1959).*

the number pointed to by a red arrow times that arrow's index multiplier. Thus in Fig. I.2, for example, the magnitude reading is

$$\begin{aligned}\text{magnitude} &= (\text{number pointed to}) \times (\text{arrow's index multiplier}) \\ &= (.213) \times (10) = 2.13\end{aligned}$$

Procedure when a vector length exceeds 7.1 *in.* To enter a numerator length $|\vec{V}|$ on the Spirule when 7.1 in. $< |\vec{V}| <$ 9.1 in., use the $S/2$ curve and perform steps (i), (ii) *twice*. [This is the same as performing them once with the S curve (extended). That is, the total angle moved by the arm relative to the disk is the same.] To enter a denominator length $|\vec{V}|$ on the Spirule when 7.1 in. $< |\vec{V}| <$ 9.1 in., use the $S/2$ curve and perform steps (iii), (iv) *twice*.

Example I.1. Draw the vectors of Fig. I.1 to a scale of $\frac{1}{2}$ in. $= 10$, with $\vec{V}_1 = 70\underline{/30°}$, $\vec{V}_2 = 45\underline{/50°}$, $\vec{V}_3 = 60\underline{/90°}$, $\vec{V}_4 = 80\underline{/105°}$, $\vec{V}_5 = 39\underline{/120°}$, $\vec{V}_6 = 20\underline{/195°}$. Then compute the following (magnitude and angle), using a Spirule: (a) $\vec{V}_2/\vec{V}_1$, (b) $\vec{V}_2\vec{V}_4/\vec{V}_1\vec{V}_3$, (c) $\vec{V}_2\vec{V}_4/\vec{V}_1$, (d) $\vec{V}_2\vec{V}_4\vec{V}_6/\vec{V}_1$.

Example I.2. Redraw the above vectors to the scale 1.1 in. $= 10$, and repeat Example I.1.

After many practice problems have been worked, use of the Spirule will become "second nature."

Appendix J

TABLE OF LAPLACE TRANSFORM PAIRS

PART 1: GENERAL RELATIONSHIPS

Operation	One-sided Laplace transform $F(s) = \mathcal{L}_-[f(t)] \triangleq \int_{0^-}^{\infty} f(t)e^{-st}\, dt$	Time function $f(t)$
	$X(s)$	$x(t)$
Derivative	$sX(s) - x(0^-)$	$\dot{x}(t)$
	$s^2 X(s) - sx(0^-) - \dot{x}(0^-)$	$\ddot{x}(t)$
	$s^n X(s) - s^{n-1}x(0^-) - s^{n-2}\dot{x}(0^-) - \cdots$ $- \left[\dfrac{d^{n-1}x}{dt^{n-1}}\right]_{t=0^-}$	$\dfrac{d^n x}{dt^n}$
Integral	$\dfrac{1}{s}X(s) + \dfrac{1}{s}\left[\int x\, dt\right]_{t=0^-}$	$\int x(t)\, dt$
	$\dfrac{1}{s^2}X(s) + \dfrac{1}{s^2}\left[\int x\, dt\right]_{t=0^-}$ $+ \dfrac{1}{s}\left[\iint x\, dt\, dt\right]_{t=0^-}$	$\iint x(t)\, dt\, dt$
Time delay	$e^{-\tau s} X(s)$	$x(t - \tau)$
Convolution	$X(s)Y(s)$	$x(t) * y(t)$
Final value	$\lim\limits_{s \to 0} sX(s)$	$\lim\limits_{t \to \infty} x(t)$
Initial value	$\lim\limits_{s \to \infty} sX(s)$	$\lim\limits_{t \to 0} x(t)$

APPENDIX J

PART 2: SPECIFIC TIME FUNCTIONS

Numbering system

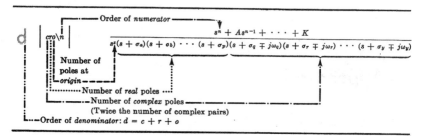

Example of numbering system

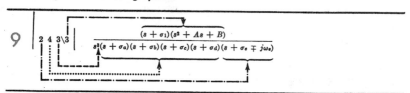

Order of denominator	Pair number $cro\backslash n$	$F(s)$	[s-plane plot]	$f(t)$ $0 < t$ (All time functions zero for $t < 0$)	[Time plot]
0	000\0	1		$\delta(t)$	Area = 1; 0^- 0 0^+
1	001\0	$\dfrac{1}{s}$	×	$u(t)$	1; 0
		$\dfrac{e^{-\tau s}}{s}$		$u(t - \tau)$	1; 0 ←τ→
1	010\0	$\dfrac{1}{s + \sigma_a}$	× $-\sigma_a$	$e^{-\sigma_a t}$	1; 0 $\dfrac{1}{\sigma_a}$

TABLE OF LAPLACE TRANSFORM PAIRS

Order of denominator	Pair number $cro\backslash n$	$F(s)$	[s-plane plot]	$f(t)$ $0 < t$ (All time functions zero for $t < 0$)	[Time plot]
2	002\0	$\dfrac{1}{s^2}$		t	
	002\1	$\dfrac{s+A}{s^2}$	$-A$	$1 + At$	
	011\0	$\dfrac{1}{s(s+\sigma_a)}$	$-\sigma_a$	$\dfrac{1}{\sigma_a}(1 - e^{-\sigma_a t})$	
	011\1	$\dfrac{s+A}{s(s+\sigma_a)}$	$-A\ -\sigma_a$	$\dfrac{A}{\sigma_a} + \left(1 - \dfrac{A}{\sigma_a}\right)e^{-\sigma_a t}$	
	020\0	$\dfrac{1}{(s+\sigma_a)(s+\sigma_b)}$	$-\sigma_b\ -\sigma_a$	$\dfrac{1}{\sigma_b - \sigma_a}(e^{-\sigma_a t} - e^{-\sigma_b t})$	
		$\dfrac{1}{(s+\sigma_a)^2}$	$-\sigma_a$	$te^{-\sigma_a t}$	
	020\1	$\dfrac{s+A}{(s+\sigma_a)(s+\sigma_b)}$	$-\sigma_b\ -A\ -\sigma_a$	$\dfrac{A - \sigma_a}{\sigma_b - \sigma_a}e^{-\sigma_a t} + \dfrac{\sigma_b - A}{\sigma_b - \sigma_a}e^{-\sigma_b t}$	

Order of denominator	Pair number $cro\backslash n$	$F(s)$	[s-plane plot]	$f(t)\quad 0 < t$ (All time functions zero for $t < 0$)	[Time plot]
2 (Cont'd)	020\1 (Cont'd)	$\dfrac{s + A}{(s + \sigma_a)^2}$		$e^{-\sigma_a t} + (A - \sigma_a)t e^{-\sigma_a t}$	
	200\0	$\dfrac{1}{s^2 + \omega_a^2}$		$\dfrac{1}{\omega_a}\sin(\omega_a t)$	
		$\dfrac{1}{(s + \sigma_a \mp j\omega_a)}$		$\dfrac{1}{\omega_a}e^{-\sigma_a t}\sin(\omega_a t)$	
	200\1	$\dfrac{s + \sigma_1}{s^2 + \omega_a^2}$		$\dfrac{a_1}{\omega_a}\sin(\omega_a t + \alpha_1)$	
2		$\dfrac{s + \sigma_1}{(s + \sigma_a \mp j\omega_a)}$		$\dfrac{a_1}{\omega_a}e^{-\sigma_a t}\sin(\omega_a t + \alpha_1)$	

TABLE OF LAPLACE TRANSFORM PAIRS

Order of denominator	Pair number $cro\backslash n$	$F(s)$	[s-plane plot]	$f(t) \quad 0 < t$ (All time functions zero for $t < 0$)
3	003\0	$\dfrac{1}{s^3}$		$\dfrac{t^2}{2}$
	003\1	$\dfrac{s+A}{s^3}$		$t + \tfrac{1}{2}At^2$
	003\2	$\dfrac{s^2 + As + B}{s^3}$		$1 + At + \tfrac{1}{2}Bt^2$
	012\0	$\dfrac{1}{s^2(s+\sigma_a)}$		$\dfrac{1}{\sigma_a}t - \dfrac{1}{\sigma_a{}^2}(1 - e^{-\sigma_a t})$
	012\1	$\dfrac{s+A}{s^2(s+\sigma_a)}$		$\dfrac{A-\sigma_a}{\sigma_a{}^2}(1 - e^{-\sigma_a t}) + \dfrac{A}{\sigma_a}t$
	012\2	$\dfrac{s^2 + As + B}{s^2(s+\sigma_a)}$		$\dfrac{1}{\sigma_a}\left(A - \dfrac{B}{\sigma_a}\right) + \dfrac{B}{\sigma_a}t + \left(1 - \dfrac{A}{\sigma_a} + \dfrac{B}{\sigma_a{}^2}\right)e^{-\sigma_a t}$
	021\0	$\dfrac{1}{s(s+\sigma_a)(s+\sigma_b)}$		$\dfrac{1}{\sigma_a \sigma_b} + \dfrac{1}{\sigma_a(\sigma_a - \sigma_b)}e^{-\sigma_a t} + \dfrac{1}{\sigma_b(\sigma_b - \sigma_a)}e^{-\sigma_b t}$
		$\dfrac{1}{s(s+\sigma_a)^2}$		$\dfrac{1}{\sigma_a{}^2}(1 - e^{-\sigma_a t}) - \dfrac{1}{\sigma_a}te^{-\sigma_a t}$
	021\1	$\dfrac{s+A}{s(s+\sigma_a)(s+\sigma_b)}$		$\dfrac{A}{\sigma_a \sigma_b} + \dfrac{A - \sigma_a}{\sigma_a(\sigma_a - \sigma_b)}e^{-\sigma_a t} + \dfrac{A - \sigma_b}{\sigma_b(\sigma_b - \sigma_a)}e^{-\sigma_b t}$
		$\dfrac{s+A}{s(s+\sigma_a)^2}$		$\dfrac{A}{\sigma_a{}^2} - \dfrac{\sigma_1}{\sigma_a{}^2}e^{-\sigma_a t} + \left(\dfrac{A}{\sigma_a} - 1\right)te^{-\sigma_a t}$
	021\2	$\dfrac{s^2 + As + B}{s(s+\sigma_a)(s+\sigma_b)}$		$\dfrac{B}{\sigma_a \sigma_b} + \dfrac{\sigma_a{}^2 - A\sigma_a + B}{\sigma_a(\sigma_a - \sigma_b)}e^{-\sigma_a t} + \dfrac{\sigma_b{}^2 - A\sigma_b + B}{\sigma_b(\sigma_b - \sigma_a)}e^{-\sigma_b t}$
		$\dfrac{s^2 + As + B}{s(s+\sigma_a)^2}$		$\dfrac{B}{\sigma_a{}^2} + \left(1 - \dfrac{B}{\sigma_a{}^2}\right)e^{-\sigma_a t} + \left(\sigma_a - A + \dfrac{B}{\sigma_a}\right)te^{-\sigma_a t}$
	030\0	$\dfrac{1}{(s+\sigma_a)(s+\sigma_b)(s+\sigma_c)}$		$\dfrac{1}{(\sigma_a - \sigma_b)(\sigma_a - \sigma_c)}e^{-\sigma_a t} + \dfrac{1}{(\sigma_b - \sigma_a)(\sigma_b - \sigma_c)}e^{-\sigma_b t} + \dfrac{1}{(\sigma_c - \sigma_a)(\sigma_c - \sigma_b)}e^{-\sigma_c t}$
		$\dfrac{1}{(s+\sigma_a)(s+\sigma_b)^2}$		$\dfrac{1}{(\sigma_a - \sigma_b)^2}(e^{-\sigma_a t} - e^{-\sigma_b t}) + \dfrac{1}{\sigma_a - \sigma_b}te^{-\sigma_b t}$
		$\dfrac{1}{(s+\sigma_a)^3}$		$\tfrac{1}{2}t^2 e^{-\sigma_a t}$

Order of denominator	Pair number $cro\backslash n$	$F(s)$	[s-plane plot]	$f(t) \quad 0 < t$ (All time functions zero for $t < 0$)
3 (Cont'd)	030\1	$\dfrac{s + A}{(s + \sigma_a)(s + \sigma_b)(s + \sigma_c)}$		$\dfrac{A - \sigma_a}{(\sigma_a - \sigma_b)(\sigma_a - \sigma_c)} e^{-\sigma_a t}$ $+ \dfrac{A - \sigma_b}{(\sigma_b - \sigma_a)(\sigma_b - \sigma_c)} e^{-\sigma_b t}$ $+ \dfrac{A - \sigma_c}{(\sigma_c - \sigma_b)(\sigma_c - \sigma_a)} e^{-\sigma_c t}$
		$\dfrac{s + A}{(s + \sigma_a)(s + \sigma_b)^2}$		$\dfrac{A - \sigma_a}{(\sigma_a - \sigma_b)^2}(e^{-\sigma_a t} - e^{-\sigma_b t}) + \dfrac{A - \sigma_b}{\sigma_a - \sigma_b} t e^{-\sigma_b t}$
		$\dfrac{s + A}{(s + \sigma_a)^3}$		$t e^{-\sigma_a t} + \dfrac{A - \sigma_a}{2} t^2 e^{-\sigma_a t}$
	030\2	$\dfrac{s^2 + As + B}{(s + \sigma_a)(s + \sigma_b)(s + \sigma_c)}$		$\dfrac{\sigma_a^2 - A\sigma_a + B}{(\sigma_a - \sigma_b)(\sigma_a - \sigma_c)} e^{-\sigma_a t}$ $+ \dfrac{\sigma_b^2 - A\sigma_b + B}{(\sigma_b - \sigma_a)(\sigma_b - \sigma_c)} e^{-\sigma_b t}$ $+ \dfrac{\sigma_c^2 - A\sigma_c + B}{(\sigma_c - \sigma_a)(\sigma_c - \sigma_b)} e^{-\sigma_c t}$
		$\dfrac{s^2 + As + B}{(s + \sigma_a)(s + \sigma_b)^2}$		$\dfrac{\sigma_a^2 - A\sigma_a + B}{(\sigma_a - \sigma_b)^2} e^{-\sigma_a t}$ $+ \dfrac{\sigma_b^2 - 2\sigma_a\sigma_b + A\sigma_a - B}{(\sigma_a - \sigma_b)^2} e^{-\sigma_b t}$ $+ \dfrac{\sigma_b^2 - A\sigma_b + B}{\sigma_a - \sigma_b} t e^{-\sigma_b t}$
		$\dfrac{s^2 + As + B}{(s + \sigma_a)^3}$		$e^{-\sigma_a t} + (A - 2\sigma_a) t e^{-\sigma_a t}$ $+ \dfrac{\sigma_a^2 - A\sigma_a + B}{2} t^2 e^{-\sigma_a t}$
	201\0	$\dfrac{1}{s(s^2 + \omega_a^2)}$		$\dfrac{1}{\omega_a^2}[1 - \cos(\omega_a t)]$
		$\dfrac{1}{s(s + \sigma_a \mp j\omega_a)}$		$\dfrac{1}{\omega_{0a}^2} + \dfrac{1}{\omega_a \omega_{0a}} e^{-\sigma_a t} \sin(\omega_a t - \alpha)$
	201\1	$\dfrac{s + \sigma_1}{s(s^2 + \omega_a^2)}$		$\dfrac{\sigma_1}{\omega_a^2} - \dfrac{a_1}{\omega_a^2} \cos(\omega_a t + \alpha_1)$

TABLE OF LAPLACE TRANSFORM PAIRS

Order of denominator	Pair number $cro\backslash n$	$F(s)$	[s-plane plot]	$f(t) \quad 0 < t$ (All time functions zero for $t < 0$)
3 (Cont'd)	201\1 (Cont'd)	$\dfrac{s + \sigma_1}{s(s + \sigma_a \mp j\omega_a)}$		$\dfrac{\sigma_1}{\omega_{0a}{}^2} + \dfrac{a_1}{\omega_a \omega_{0a}} e^{-\sigma_a t} \sin(\omega_a t + \alpha_1 - \alpha)$
	201\2	$\dfrac{(s + \sigma_1)(s + \sigma_2)}{s(s + \sigma_a \mp j\omega_a)}$		$\dfrac{\sigma_1 \sigma_2}{\omega_{0a}{}^2} - \dfrac{a_1 a_2}{\omega_a \omega_{0a}} e^{-\sigma_a t} \sin(\omega_a t + \alpha_1 + \alpha_2 - \alpha)$
		$\dfrac{(s + \sigma_1 \mp j\omega_1)}{s(s + \sigma_a \mp j\omega_a)}$		$\dfrac{\omega_{01}{}^2}{\omega_{0a}{}^2} + \dfrac{a_1 \bar{a}_1}{\omega_a \omega_{0a}} e^{-\sigma_a t} \sin(\omega_a t + \alpha_1 + \bar{\alpha}_1 - \alpha)$
	210\0	$\dfrac{1}{(s + \sigma_a)(s^2 + \omega_b{}^2)}$		$\dfrac{1}{a^2} e^{-\sigma_a t} + \dfrac{1}{a\omega_b} \sin(\omega_b t - \alpha)$

APPENDIX J

Order of denominator	Pair number $cro\backslash n$	$F(s)$ [s-plane plot]	$f(t)$ $0 < t$ (All time functions zero for $t < 0$)
3 (Cont'd)	210\0 (Cont'd)	$\dfrac{1}{(s+\sigma_a)(s+\sigma_b \mp j\omega_b)}$	$\dfrac{1}{a^2}e^{-\sigma_a t} + \dfrac{1}{a\omega_b}e^{-\sigma_b t}\sin(\omega_b t - \beta)$
	210\1	$\dfrac{s+\sigma_1}{(s+\sigma_a)(s+\sigma_b \mp j\omega_b)}$	$\dfrac{\sigma_1 - \sigma_a}{b_a{}^2}e^{-\sigma_a t}$ $+ \dfrac{b_1}{b_a\omega_b}e^{-\sigma_b t}\sin(\omega_b t + \beta_1 - \beta_a)$
	210\2	$\dfrac{(s+\sigma_1)(s+\sigma_2)}{(s+\sigma_a)(s+\sigma_b \mp j\omega_b)}$	$\dfrac{(\sigma_1-\sigma_a)(\sigma_2-\sigma_a)}{b_a{}^2}e^{-\sigma_a t}$ $+ \dfrac{b_1 b_2}{b_a\omega_b}e^{-\sigma_b t}\sin(\omega_b t + \beta_1 + \beta_2 - \beta_a)$
3		$\dfrac{s+\sigma_1 \mp j\omega_1}{(s+\sigma_a)(s+\sigma_b \mp j\omega_b)}$	$\dfrac{a_1{}^2}{b_a{}^2}e^{-\sigma_a t}$ $+ \dfrac{b_1 \bar{b}_1}{b_a \omega_b}e^{-\sigma_b t}\sin(\omega_b t + \beta_1 + \bar{\beta}_1 - \beta_a)$
4	004\0	$\dfrac{1}{s^4}$	$\dfrac{t^3}{3!}$
	004\1	$\dfrac{s+A}{s^4}$	$\dfrac{1}{2}t^2 + \dfrac{A}{6}t^3$
	004\2	$\dfrac{s^2 + As + B}{s^4}$	$t + \dfrac{A}{2}t^2 + \dfrac{B}{6}t^3$
	004\3	$\dfrac{s^3 + As^2 + Bs + C}{s^4}$	$1 + At + \dfrac{B}{2}t^2 + \dfrac{C}{6}t^3$
	013\0	$\dfrac{1}{s^3(s+\sigma_a)}$	$\dfrac{1}{\sigma_a{}^3}(1 - e^{-\sigma_a t}) - \dfrac{1}{\sigma_a{}^2}t + \dfrac{1}{2\sigma_a}t^2$

TABLE OF LAPLACE TRANSFORM PAIRS

Order of denominator	Pair number $cro\backslash n$	$F(s)$	[s-plane plot]	$f(t)$ $\quad 0 < t$ (All time functions zero for $t < 0$)
4 (Cont'd)	013\1	$\dfrac{s + A}{s^3(s + \sigma_a)}$		$\dfrac{1}{\sigma_a^2}\left(\dfrac{A}{\sigma_a} - 1\right)(1 - e^{-\sigma_a t})$ $\quad + \dfrac{\sigma_a - A}{\sigma_a^2} t + \dfrac{A}{2\sigma_a} t^2$
	013\2	$\dfrac{s^2 + As + B}{s^3(s + \sigma_a)}$		$\dfrac{1}{\sigma_a}\left(1 - \dfrac{A}{\sigma_a} + \dfrac{B}{\sigma_a^2}\right)(1 - e^{-\sigma_a t})$ $\quad + \dfrac{1}{\sigma_a}\left(\dfrac{A}{\sigma_a} - B\right) t + \dfrac{B}{2\sigma_a} t^2$
	013\3	$\dfrac{s^3 + As^2 + Bs + C}{s^3(s + \sigma_a)}$		$\left(\dfrac{C}{\sigma_a^3} - \dfrac{B}{\sigma_a^2} + \dfrac{A}{\sigma_a} - 1\right)(1 - e^{-\sigma_a t})$ $\quad + \dfrac{1}{\sigma_a}\left(B - \dfrac{C}{\sigma_a}\right) t + \dfrac{C}{2\sigma_a} t^2$
	022\0	$\dfrac{1}{s^2(s + \sigma_a)(s + \sigma_b)}$		$-\dfrac{\sigma_a + \sigma_b}{\sigma_a^2 \sigma_b^2} + \dfrac{1}{\sigma_a \sigma_b} t$ $\quad + \dfrac{1}{\sigma_b^2(\sigma_b - \sigma_a)} e^{-\sigma_a t} + \dfrac{1}{\sigma_a^2(\sigma_a - \sigma_b)} e^{-\sigma_b t}$
		$\dfrac{1}{s^2(s + \sigma_a)^2}$		$-\dfrac{2}{\sigma_a^3}(1 - e^{-\sigma_a t}) + \dfrac{1}{\sigma_a^2} t(1 + e^{-\sigma_a t})$
	022\1	$\dfrac{s + A}{s^2(s + \sigma_a)(s + \sigma_b)}$		$\dfrac{1}{\sigma_a \sigma_b}\left(1 - A \dfrac{\sigma_a + \sigma_b}{\sigma_a \sigma_b}\right) + \dfrac{A}{\sigma_a \sigma_b} t$ $\quad + \dfrac{A - \sigma_a}{\sigma_a^2(\sigma_b - \sigma_a)} e^{-\sigma_a t} + \dfrac{A - \sigma_b}{\sigma_b^2(\sigma_a - \sigma_b)} e^{-\sigma_b t}$
		$\dfrac{s + A}{s^2(s + \sigma_a)^2}$		$\dfrac{1}{\sigma_a^2}\left(1 - \dfrac{2A}{\sigma_a}\right)(1 - e^{-\sigma_a t}) + \dfrac{A}{\sigma_a^2} t$ $\quad + \dfrac{1}{\sigma_a}\left(\dfrac{A}{\sigma_a} - 1\right) t e^{-\sigma_a t}$
	022\2	$\dfrac{s^2 + As + B}{s^2(s + \sigma_a)(s + \sigma_b)}$		$\dfrac{1}{\sigma_a \sigma_b}\left(A - B \dfrac{\sigma_a + \sigma_b}{\sigma_a \sigma_b}\right) + \dfrac{B}{\sigma_a \sigma_b} t$ $\quad + \dfrac{\sigma_a^2 - A\sigma_a + B}{\sigma_a^2(\sigma_b - \sigma_a)} e^{-\sigma_a t}$ $\quad + \dfrac{\sigma_b^2 - A\sigma_b + B}{\sigma_b^2(\sigma_a - \sigma_b)} e^{-\sigma_b t}$
		$\dfrac{s^2 + As + B}{s^2(s + \sigma_a)^2}$		$\dfrac{1}{\sigma_a^2}\left(A - \dfrac{2B}{\sigma_a}\right)(1 - e^{-\sigma_a t}) + \dfrac{B}{\sigma_a^2} t$ $\quad + \left(1 - \dfrac{A}{\sigma_a} + \dfrac{B}{\sigma_a^2}\right) t e^{-\sigma_a t}$
	022\3	$\dfrac{s^3 + As^2 + Bs + C}{s^2(s + \sigma_a)(s + \sigma_b)}$		$\dfrac{1}{\sigma_a \sigma_b}\left(B - C \dfrac{\sigma_a + \sigma_b}{\sigma_a \sigma_b}\right) + \dfrac{C}{\sigma_a \sigma_b} t$ $\quad + \dfrac{\sigma_a^3 - A\sigma_a^2 + B\sigma_a - C}{\sigma_a^2(\sigma_a - \sigma_b)} e^{-\sigma_a t}$ $\quad + \dfrac{\sigma_b^3 - A\sigma_b^2 + B\sigma_b - C}{\sigma_b^2(\sigma_b - \sigma_a)} e^{-\sigma_b t}$
		$\dfrac{s^3 + As^2 + Bs + C}{s^2(s + \sigma_a)^2}$		$\dfrac{1}{\sigma_a^2}\left(B - \dfrac{2C}{\sigma_a}\right) + \dfrac{C}{\sigma_a^2} t$ $\quad + \dfrac{1}{\sigma_a^2}\left(\dfrac{3C}{\sigma_a} - B\right) e^{-\sigma_a t}$ $\quad + \left(\sigma_a - A + \dfrac{B\sigma_a - C}{\sigma_a^2}\right) t e^{-\sigma_a t}$

APPENDIX J

Order of denominator	Pair number $cro\backslash n$	$F(s)$	[s-plane plot]	$f(t) \quad 0 < t$ (All time functions zero for $t < 0$)
4 (Cont'd)	031\0	$\dfrac{1}{s(s+\sigma_a)(s+\sigma_b)(s+\sigma_c)}$		$\dfrac{1}{\sigma_a \sigma_b \sigma_c} - \dfrac{1}{\sigma_a(\sigma_a - \sigma_b)(\sigma_a - \sigma_c)} e^{-\sigma_a t} - \dfrac{1}{\sigma_b(\sigma_b - \sigma_a)(\sigma_b - \sigma_c)} e^{-\sigma_b t} - \dfrac{1}{\sigma_c(\sigma_c - \sigma_a)(\sigma_c - \sigma_b)} e^{-\sigma_c t}$
		$\dfrac{1}{s(s+\sigma_a)(s+\sigma_b)^2}$		$\dfrac{1}{\sigma_a \sigma_b^2} - \dfrac{1}{\sigma_a(\sigma_b - \sigma_a)^2} e^{-\sigma_a t} + \dfrac{2\sigma_b - \sigma_a}{\sigma_b^2(\sigma_b - \sigma_a)^2} e^{-\sigma_b t} + \dfrac{1}{\sigma_b(\sigma_b - \sigma_a)} t e^{-\sigma_b t}$
		$\dfrac{1}{s(s+\sigma_a)^3}$		$\dfrac{1}{\sigma_a^3}(1 - e^{-\sigma_a t}) - \dfrac{1}{\sigma_a^2} t e^{-\sigma_a t} - \dfrac{1}{2\sigma_a} t^2 e^{-\sigma_a t}$
	031\1	$\dfrac{s+A}{s(s+\sigma_a)(s+\sigma_b)(s+\sigma_c)}$		$\dfrac{A}{\sigma_a \sigma_b \sigma_c} + \dfrac{\sigma_a - A}{\sigma_a(\sigma_a - \sigma_b)(\sigma_a - \sigma_c)} e^{-\sigma_a t} + \dfrac{\sigma_b - A}{\sigma_b(\sigma_b - \sigma_a)(\sigma_b - \sigma_c)} e^{-\sigma_b t} + \dfrac{\sigma_c - A}{\sigma_c(\sigma_c - \sigma_a)(\sigma_c - \sigma_b)} e^{-\sigma_c t}$
		$\dfrac{s+A}{s(s+\sigma_a)(s+\sigma_b)^2}$		$\dfrac{A}{\sigma_a \sigma_b^2} + \dfrac{\sigma_a - A}{\sigma_a(\sigma_a - \sigma_b)^2} e^{-\sigma_a t} + \dfrac{\sigma_b(\sigma_b - 2A) + A\sigma_a}{\sigma_b^2(\sigma_b - \sigma_a)^2} e^{-\sigma_b t} + \dfrac{A - \sigma_b}{\sigma_b(\sigma_b - \sigma_a)} t e^{-\sigma_b t}$
		$\dfrac{s+A}{s(s+\sigma_a)^3}$		$\dfrac{A}{\sigma_a^3}(1 - e^{-\sigma_a t}) + \dfrac{1}{\sigma_a}\left(2 - \dfrac{A}{\sigma_a}\right) t e^{-\sigma_a t} + \dfrac{1}{2}\left(1 - \dfrac{A}{\sigma_a}\right) t^2 e^{-\sigma_a t}$
	031\2	$\dfrac{s^2 + As + B}{s(s+\sigma_a)(s+\sigma_b)(s+\sigma_c)}$		$\dfrac{B}{\sigma_a \sigma_b \sigma_c} - \dfrac{\sigma_a^2 - A\sigma_a + B}{\sigma_a(\sigma_a - \sigma_b)(\sigma_a - \sigma_c)} e^{-\sigma_a t} - \dfrac{\sigma_b^2 - A\sigma_b + B}{\sigma_b(\sigma_b - \sigma_a)(\sigma_b - \sigma_c)} e^{-\sigma_b t} - \dfrac{\sigma_c^2 - A\sigma_c + B}{\sigma_c(\sigma_c - \sigma_a)(\sigma_c - \sigma_b)} e^{-\sigma_c t}$
		$\dfrac{s^2 + As + B}{s(s+\sigma_a)(s+\sigma_b)^2}$		$\dfrac{B}{\sigma_a \sigma_b^2} - \dfrac{\sigma_a^2 - A\sigma_a + B}{\sigma_a(\sigma_a - \sigma_b)^2} e^{-\sigma_a t} + \dfrac{B(2\sigma_b - \sigma_a) + \sigma_b^2(\sigma_a - A)}{\sigma_b^2(\sigma_b - \sigma_a)^2} e^{-\sigma_b t} + \dfrac{\sigma_b^2 - A\sigma_b + B}{\sigma_b(\sigma_b - \sigma_a)} t e^{-\sigma_b t}$
		$\dfrac{s^2 + As + B}{s(s+\sigma_a)^3}$		$\dfrac{B}{\sigma_a^3}(1 - e^{-\sigma_a t}) + \left(1 - \dfrac{B}{\sigma_a^2}\right) t e^{-\sigma_a t} + \dfrac{1}{2}\left(\sigma_a - A + \dfrac{B}{\sigma_a}\right) t^2 e^{-\sigma_a t}$

TABLE OF LAPLACE TRANSFORM PAIRS

Order of denominator	Pair number $cro\backslash n$	$F(s)$	$f(t)$ $0 < t$ (All time functions zero for $t < 0$)
4 (Cont'd)	031\3	$\dfrac{s^3 + As^2 + Bs + C}{s(s + \sigma_a)(s + \sigma_b)(s + \sigma_c)}$	$\dfrac{C}{\sigma_a \sigma_b \sigma_c} + \dfrac{\sigma_a^3 - A\sigma_a^2 + B\sigma_a - C}{\sigma_a(\sigma_a - \sigma_b)(\sigma_a - \sigma_c)} e^{-\sigma_a t}$ $+ \dfrac{\sigma_b^3 - A\sigma_b^2 + B\sigma_b - C}{\sigma_b(\sigma_b - \sigma_a)(\sigma_b - \sigma_c)} e^{-\sigma_b t}$ $+ \dfrac{\sigma_c^3 - A\sigma_c^2 + B\sigma_c - C}{\sigma_c(\sigma_c - \sigma_a)(\sigma_c - \sigma_b)} e^{-\sigma_c t}$
		$\dfrac{s^3 + As^2 + Bs + C}{s(s + \sigma_a)(s + \sigma_b)^2}$	$\dfrac{C}{\sigma_a \sigma_b^2} + \dfrac{\sigma_a^3 - A\sigma_a^2 + B\sigma_a - C}{\sigma_a(\sigma_a - \sigma_b)^2} e^{-\sigma_a t}$ $+ \dfrac{(C + \sigma_b^3)(2\sigma_a - \sigma_b) + \sigma_b^2(B - \sigma_a A)}{\sigma_a^2(\sigma_b - \sigma_a)^2} e^{-\sigma_b t}$ $+ \dfrac{\sigma_b^3 - A\sigma_b^2 + B\sigma_b - C}{\sigma_a(\sigma_b - \sigma_a)} t e^{-\sigma_b t}$
		$\dfrac{s^3 + As^2 + Bs + C}{s(s + \sigma_a)^3}$	$\dfrac{C}{\sigma_a^3} + \left(1 - \dfrac{C}{\sigma_a^3}\right) e^{-\sigma_a t}$ $- \left(2\sigma_a - A + \dfrac{C}{\sigma_a^2}\right) t e^{-\sigma_a t}$ $+ \dfrac{1}{2}\left(\sigma_a^2 - A\sigma_a + B - \dfrac{C}{\sigma_a}\right) t^2 e^{-\sigma_a t}$
	040\0	$\dfrac{1}{(s + \sigma_a)(s + \sigma_b)(s + \sigma_c)(s + \sigma_d)}$	$\dfrac{1}{(\sigma_b - \sigma_a)(\sigma_c - \sigma_a)(\sigma_d - \sigma_a)} e^{-\sigma_a t}$ $+ \dfrac{1}{(\sigma_a - \sigma_b)(\sigma_c - \sigma_b)(\sigma_d - \sigma_b)} e^{-\sigma_b t}$ $+ \dfrac{1}{(\sigma_a - \sigma_c)(\sigma_b - \sigma_c)(\sigma_d - \sigma_c)} e^{-\sigma_c t}$ $+ \dfrac{1}{(\sigma_a - \sigma_d)(\sigma_b - \sigma_d)(\sigma_c - \sigma_d)} e^{-\sigma_d t}$
		$\dfrac{1}{(s + \sigma_a)(s + \sigma_b)(s + \sigma_c)^2}$	$\dfrac{1}{(\sigma_b - \sigma_a)(\sigma_c - \sigma_a)^2} e^{-\sigma_a t}$ $+ \dfrac{1}{(\sigma_a - \sigma_b)(\sigma_c - \sigma_b)^2} e^{-\sigma_b t}$ $+ \dfrac{2\sigma_c - \sigma_a - \sigma_b}{(\sigma_a - \sigma_c)^2(\sigma_b - \sigma_c)^2} e^{-\sigma_c t}$ $+ \dfrac{1}{(\sigma_a - \sigma_c)(\sigma_b - \sigma_c)} t e^{-\sigma_c t}$
		$\dfrac{1}{(s + \sigma_a)^2(s + \sigma_b)^2}$	$\dfrac{2}{(\sigma_a - \sigma_b)^3}(e^{-\sigma_a t} - e^{-\sigma_b t})$ $+ \dfrac{1}{(\sigma_a - \sigma_b)^2} t(e^{-\sigma_a t} + e^{-\sigma_b t})$
		$\dfrac{1}{(s + \sigma_a)(s + \sigma_b)^3}$	$\dfrac{1}{(\sigma_a - \sigma_b)^3}(-e^{-\sigma_a t} + e^{-\sigma_b t})$ $+ \dfrac{1}{(\sigma_a - \sigma_b)^2} t e^{-\sigma_b t} + \dfrac{1}{2(\sigma_a - \sigma_b)} t^2 e^{-\sigma_b t}$
		$\dfrac{1}{(s + \sigma_a)^4}$	$\dfrac{1}{3!} t^3 e^{-\sigma_a t}$

Order of denominator	Pair number $cro\backslash n$	$F(s)$	[s-plane plot]	$f(t)$ $0 < t$ (All time functions zero for $t < 0$)
4 (Cont'd)	040\1	$\dfrac{s + A}{(s + \sigma_a)(s + \sigma_b)(s + \sigma_c)(s + \sigma_d)}$		$\dfrac{A - \sigma_a}{(\sigma_b - \sigma_a)(\sigma_c - \sigma_a)(\sigma_d - \sigma_a)} e^{-\sigma_a t}$ $+ \dfrac{A - \sigma_b}{(\sigma_a - \sigma_b)(\sigma_c - \sigma_b)(\sigma_d - \sigma_b)} e^{-\sigma_b t}$ $+ \dfrac{A - \sigma_c}{(\sigma_a - \sigma_c)(\sigma_b - \sigma_c)(\sigma_d - \sigma_c)} e^{-\sigma_c t}$ $+ \dfrac{A - \sigma_d}{(\sigma_a - \sigma_d)(\sigma_b - \sigma_d)(\sigma_c - \sigma_d)} e^{-\sigma_d t}$
		$\dfrac{s + A}{(s + \sigma_a)(s + \sigma_b)(s + \sigma_c)^2}$		$\dfrac{A - \sigma_a}{(\sigma_b - \sigma_a)(\sigma_c - \sigma_a)^2} e^{-\sigma_a t}$ $+ \dfrac{A - \sigma_b}{(\sigma_a - \sigma_b)(\sigma_c - \sigma_b)^2} e^{-\sigma_b t}$ $+ \dfrac{A(2\sigma_c - \sigma_a - \sigma_b) + (\sigma_a \sigma_b - \sigma_c^2)}{(\sigma_a - \sigma_c)^2(\sigma_b - \sigma_c)^2} e^{-\sigma_c t}$ $+ \dfrac{A - \sigma_c}{(\sigma_a - \sigma_c)(\sigma_b - \sigma_c)} t e^{-\sigma_c t}$
		$\dfrac{s + A}{(s + \sigma_a)^2 (s + \sigma_b)^2}$		$\dfrac{2A - \sigma_a - \sigma_b}{(\sigma_a - \sigma_b)^3} (e^{-\sigma_a t} - e^{-\sigma_b t})$ $+ \dfrac{A - \sigma_a}{(\sigma_a - \sigma_b)^2} t e^{-\sigma_a t} + \dfrac{A - \sigma_b}{(\sigma_a - \sigma_b)^2} t e^{-\sigma_b t}$
		$\dfrac{s + A}{(s + \sigma_a)(s + \sigma_b)^3}$		$\dfrac{A - \sigma_a}{(\sigma_a - \sigma_b)^3} (-e^{-\sigma_a t} + e^{-\sigma_b t})$ $- \dfrac{A - \sigma_a}{(\sigma_a - \sigma_b)^2} t e^{-\sigma_b t} + \dfrac{A - \sigma_b}{2(\sigma_a - \sigma_b)} t^2 e^{-\sigma_b t}$
		$\dfrac{s + A}{(s + \sigma_a)^4}$		$\dfrac{1}{2} t^2 e^{-\sigma_a t} + \dfrac{A - \sigma_a}{6} t^3 e^{-\sigma_a t}$
	040\2	$\dfrac{s^2 + As + B}{(s + \sigma_a)(s + \sigma_b)(s + \sigma_c)(s + \sigma_d)}$		$\dfrac{\sigma_a^2 - A\sigma_a + B}{(\sigma_b - \sigma_a)(\sigma_c - \sigma_a)(\sigma_d - \sigma_a)} e^{-\sigma_a t}$ $+ \dfrac{\sigma_b^2 - A\sigma_b + B}{(\sigma_a - \sigma_b)(\sigma_c - \sigma_b)(\sigma_d - \sigma_b)} e^{-\sigma_b t}$ $+ \dfrac{\sigma_c^2 - A\sigma_c + B}{(\sigma_a - \sigma_c)(\sigma_b - \sigma_c)(\sigma_d - \sigma_c)} e^{-\sigma_c t}$ $+ \dfrac{\sigma_d^2 - A\sigma_d + B}{(\sigma_a - \sigma_d)(\sigma_b - \sigma_d)(\sigma_c - \sigma_d)} e^{-\sigma_d t}$
		$\dfrac{s^2 + As + B}{(s + \sigma_a)(s + \sigma_b)(s + \sigma_c)^2}$		$\dfrac{\sigma_a^2 - A\sigma_a + B}{(\sigma_b - \sigma_a)(\sigma_c - \sigma_a)^2} e^{-\sigma_a t}$ $+ \dfrac{\sigma_b^2 - A\sigma_b + B}{(\sigma_a - \sigma_b)(\sigma_c - \sigma_b)^2} e^{-\sigma_b t}$ $+ \dfrac{A(\sigma_a \sigma_b - \sigma_c^2) + B(2\sigma_c - \sigma_a - \sigma_b) + \sigma_c^2(\sigma_a + \sigma_b) - 2\sigma_a \sigma_b \sigma_c}{(\sigma_a - \sigma_c)^2 (\sigma_b - \sigma_c)^2} e^{-\sigma_c t}$ $+ \dfrac{\sigma_c^2 - A\sigma_c + B}{(\sigma_a - \sigma_c)(\sigma_b - \sigma_c)} t e^{-\sigma_c t}$
		$\dfrac{s^2 + As + B}{(s + \sigma_a)^2 (s + \sigma_b)^2}$		$\dfrac{2B - A(\sigma_a + \sigma_b) + 2\sigma_a \sigma_b}{(\sigma_a - \sigma_b)^3} (e^{-\sigma_a t} - e^{-\sigma_b t})$ $+ \dfrac{\sigma_a^2 - A\sigma_a + B}{(\sigma_a - \sigma_b)^2} t e^{-\sigma_a t}$ $+ \dfrac{\sigma_b^2 - A\sigma_b + B}{(\sigma_b - \sigma_a)^2} t e^{-\sigma_b t}$

TABLE OF LAPLACE TRANSFORM PAIRS

Order of denominator	Pair number $cro\backslash n$	$F(s)$	[s-plane plot]	$f(t) \quad 0 < t$ (All time functions zero for $t < 0$)
4 (Cont'd)	040\2 (Cont'd)	$\dfrac{s^2 + As + B}{(s+\sigma_a)(s+\sigma_b)^3}$		$\dfrac{\sigma_a^2 - A\sigma_a + B}{(\sigma_a - \sigma_b)^3}(-e^{-\sigma_a t} + e^{-\sigma_b t})$ $+ \dfrac{\sigma_b(\sigma_b - 2\sigma_a) + B\sigma_a - C}{(\sigma_a - \sigma_b)^2} te^{-\sigma_b t}$ $+ \dfrac{\sigma_b^2 - A\sigma_b + B}{2(\sigma_a - \sigma_b)} t^2 e^{-\sigma_b t}$
		$\dfrac{s^2 + As + B}{(s+\sigma_a)^4}$		$te^{-\sigma_a t} + \left(\dfrac{A}{2} - \sigma_a\right) t^2 e^{-\sigma_a t}$ $+ \dfrac{\sigma_a^2 - A\sigma_a + B}{6} t^3 e^{-\sigma_a t}$
	040\3	$\dfrac{s^3 + As^2 + Bs + C}{(s+\sigma_a)(s+\sigma_b)(s+\sigma_c)(s+\sigma_d)}$		$\dfrac{\sigma_a^3 - A\sigma_a^2 + B\sigma_a - C}{(\sigma_a - \sigma_b)(\sigma_a - \sigma_c)(\sigma_a - \sigma_d)} e^{-\sigma_a t}$ $+ \dfrac{\sigma_b^3 - A\sigma_b^2 + B\sigma_b - C}{(\sigma_b - \sigma_a)(\sigma_b - \sigma_c)(\sigma_b - \sigma_d)} e^{-\sigma_b t}$ $+ \dfrac{\sigma_c^3 - A\sigma_c^2 + B\sigma_c - C}{(\sigma_c - \sigma_a)(\sigma_c - \sigma_b)(\sigma_c - \sigma_d)} e^{-\sigma_c t}$ $+ \dfrac{\sigma_d^3 - A\sigma_d^2 + B\sigma_d - C}{(\sigma_d - \sigma_a)(\sigma_d - \sigma_b)(\sigma_d - \sigma_c)} e^{-\sigma_d t}$
		$\dfrac{s^3 + As^2 + Bs + C}{(s+\sigma_a)(s+\sigma_b)(s+\sigma_c)^2}$		$\dfrac{\sigma_a^3 - A\sigma_a^2 + B\sigma_a - C}{(\sigma_a - \sigma_b)(\sigma_a - \sigma_c)^2} e^{-\sigma_a t}$ $+ \dfrac{\sigma_b^3 - A\sigma_b^2 + B\sigma_b - C}{(\sigma_b - \sigma_a)(\sigma_b - \sigma_c)^2} e^{-\sigma_b t}$ $+ \dfrac{A\sigma_c[\sigma_c(\sigma_a + \sigma_b) - 2\sigma_a\sigma_b] + B(\sigma_a\sigma_b - \sigma_c^2) + C(2\sigma_c - \sigma_b - \sigma_a) + \sigma_c^2[\sigma_c^2 - 2\sigma_c(\sigma_a + \sigma_b) + 3\sigma_a\sigma_b]}{(\sigma_c - \sigma_a)^2(\sigma_c - \sigma_b)^2} e^{-\sigma_c t}$ $+ \dfrac{\sigma_c^3 - A\sigma_c^2 + B\sigma_c - C}{(\sigma_c - \sigma_a)(\sigma_c - \sigma_b)} te^{-\sigma_c t}$
		$\dfrac{s^3 + As^2 + Bs + C}{(s+\sigma_a)^2(s+\sigma_b)^2}$		$\dfrac{2A\sigma_a\sigma_b - B(\sigma_a + \sigma_b) + 2C + \sigma_a^2(\sigma_a - 3\sigma_b)}{(\sigma_a - \sigma_b)^3} e^{-\sigma_a t}$ $- \dfrac{\sigma_a^3 - A\sigma_a^2 + B\sigma_a - C}{(\sigma_a - \sigma_b)^2} te^{-\sigma_a t}$ $+ \dfrac{2A\sigma_a\sigma_b - B(\sigma_a + \sigma_b) + 2C + \sigma_b^2(\sigma_b - 3\sigma_a)}{(\sigma_b - \sigma_a)^3} e^{-\sigma_b t}$ $- \dfrac{\sigma_b^3 - A\sigma_b^2 + B\sigma_b - C}{(\sigma_b - \sigma_a)^2} te^{-\sigma_b t}$
		$\dfrac{s^3 + As^2 + Bs + C}{(s+\sigma_a)(s+\sigma_b)^3}$		$\dfrac{\sigma_a^3 - A\sigma_a^2 + B\sigma_a - C}{(\sigma_a - \sigma_b)^3} e^{-\sigma_a t}$ $+ \dfrac{\sigma_b^3 - 3\sigma_a\sigma_b^2 + \sigma_a\sigma_b(A + 3\sigma_a) - A\sigma_a^2 + B\sigma_a - C}{(\sigma_b - \sigma_a)^3} e^{-\sigma_b t}$ $- \dfrac{2\sigma_b^3 - \sigma_b^2(3\sigma_a + 2A) + 2A\sigma_a\sigma_b + B\sigma_a - C}{(\sigma_b - \sigma_a)^2} te^{-\sigma_b t}$ $+ \dfrac{\sigma_b^3 - A\sigma_b^2 + B\sigma_b - C}{2(\sigma_b - \sigma_a)} t^2 e^{-\sigma_b t}$

Order of denominator	Pair number $cro\backslash n$	$F(s)$	[s-plane plot]	$f(t)$ $0 < t$ (All time functions zero for $t < 0$)
4 (Cont'd)	040\3 (Cont'd)	$\dfrac{s^3 + As^2 + Bs + C}{(s + \sigma_a)^4}$		$e^{-\sigma_a t} + (A - 3\sigma_a)te^{-\sigma_a t}$ $+ \dfrac{3\sigma_a^2 - 2A\sigma_a + B}{2} t^2 e^{-\sigma_a t}$ $- \dfrac{\sigma_a^3 - A\sigma_a^2 + B\sigma_a - C}{6} t^3 e^{-\sigma_a t}$
	202\0	$\dfrac{1}{s^2(s^2 + \omega_a^2)}$		$\dfrac{1}{\omega_a^3}[\omega_a t - \sin(\omega_a t)]$
		$\dfrac{1}{s^2(s + \sigma_a \mp j\omega_a)}$		$\dfrac{1}{\omega_a^2} t + \dfrac{1}{\omega_a \omega_{0a}^2} e^{-\sigma_a t} \sin(\omega_a t - 2\alpha)$
	202\1	$\dfrac{s + \sigma_1}{s^2(s^2 + \omega_a^2)}$		$\dfrac{1}{\omega_a^2} + \dfrac{\sigma_1}{\omega_a^2} t - \dfrac{a_1}{\omega_a^3} \sin(\omega_a t + \alpha_1)$
		$\dfrac{s + \sigma_1}{s^2(s + \sigma_a \mp j\omega_a)}$		$\dfrac{1}{\omega_{0a}^2}\left(1 - \dfrac{2\sigma_1 \sigma_a}{\omega_{0a}^2}\right) + \dfrac{\sigma_1}{\omega_{0a}^2} t$ $+ \dfrac{a_1}{\omega_a \omega_{0a}^2} e^{-\sigma_a t} \sin(\omega_a t + \alpha_1 - 2\alpha)$
	202\2	$\dfrac{(s + \sigma_1)(s + \sigma_2)}{s^2(s + \sigma_a \mp j\omega_a)}$		$\dfrac{1}{\omega_{0a}^2}\left(\sigma_1 + \sigma_2 - \dfrac{2\sigma_1 \sigma_2 \sigma_a}{\omega_{0a}^2}\right) + \dfrac{\sigma_1 \sigma_2}{\omega_{0a}^2} t$ $+ \dfrac{a_1 a_2}{\omega_a \omega_{0a}^2} e^{-\sigma_a t} \sin(\omega_a t + \alpha_1 + \alpha_2 - 2\alpha)$

TABLE OF LAPLACE TRANSFORM PAIRS

Order of denominator	Pair number $cro\backslash n$	$F(s)$ [s-plane plot]	$f(t)$ $\quad 0 < t$ (All time functions zero for $t < 0$)
4 (Cont'd)	202\2 (Cont'd)	$\dfrac{(s + \sigma_1 \mp j\omega_1)}{s^2(s + \sigma_a \mp j\omega_a)}$	$\dfrac{1}{\omega_{0a}{}^2}\left(2\sigma_1 - \dfrac{2\sigma_a\omega_{01}{}^2}{\omega_{0a}{}^2}\right) + \dfrac{\omega_{01}{}^2}{\omega_{0a}{}^2}t$ $+ \dfrac{a_1\bar{a}_1}{\omega_a\omega_{0a}{}^2}e^{-\sigma_a t}\sin(\omega_a t + \alpha_1 + \bar{\alpha}_1 - 2\alpha)$
	202\3	$\dfrac{(s + \sigma_1)(s + \sigma_2)(s + \sigma_3)}{s^2(s + \sigma_a \mp j\omega_a)}$	$\dfrac{1}{\omega_{0a}{}^2}\left(\sigma_1\sigma_2 + \sigma_1\sigma_3 + \sigma_2\sigma_3 - \dfrac{2\sigma_1\sigma_2\sigma_3\sigma_a}{\omega_{0a}{}^2}\right)$ $+ \dfrac{\sigma_1\sigma_2\sigma_3}{\omega_{0a}{}^2}t + \dfrac{a_1 a_2 a_3}{\omega_a\omega_{0a}{}^2}e^{-\sigma_a t}\sin(\omega_a t + \alpha_1 + \alpha_2 + \alpha_3 - 2\alpha)$
		$\dfrac{(s + \sigma_1)(s + \sigma_2 \mp j\omega_2)}{s^2(s + \sigma_a \mp j\omega_a)}$	$\dfrac{1}{\omega_{0a}{}^2}\left(\omega_{02}{}^2 + 2\sigma_1\sigma_2 - \dfrac{2\sigma_1\sigma_a\omega_{02}{}^2}{\omega_{0a}{}^2}\right) + \dfrac{\sigma_1\omega_{02}{}^2}{\omega_{0a}{}^2}t$ $+ \dfrac{a_1 a_2 \bar{a}_2}{\omega_a\omega_{0a}{}^2}e^{-\sigma_a t}\sin(\omega_a t + \alpha_1 + \alpha_2 + \bar{\alpha}_2 - 2\alpha)$
	211\0	$\dfrac{1}{s(s + \sigma_a)(s^2 + \omega_b{}^2)}$	$\dfrac{1}{\sigma_a\omega_b{}^2} - \dfrac{1}{\sigma_a b_a{}^2}e^{-\sigma_a t} - \dfrac{1}{b_a\omega_b{}^2}\cos(\omega_b t - \beta_a)$
		$\dfrac{1}{s(s + \sigma_a)(s + \sigma_b \mp j\omega_b)}$	$\dfrac{1}{\sigma_a\omega_{0b}{}^2} - \dfrac{1}{\sigma_a b_a{}^2}e^{-\sigma_a t}$ $+ \dfrac{1}{b_a\omega_b\omega_{0b}}e^{-\sigma_b t}\sin(\omega_b t - \beta_a - \beta)$

Order of denominator	Pair number $cro\backslash n$	$F(s)$ [s-plane plot]	$f(t)$ $0 < t$ (All time functions zero for $t < 0$)
4 (Cont'd)	211\1	$\dfrac{s + \sigma_1}{s(s + \sigma_a)(s^2 + \omega_b^2)}$	$\dfrac{\sigma_1}{\sigma_a \omega_b^2} + \dfrac{1}{b_a^2}\left(1 - \dfrac{\sigma_1}{\sigma_a}\right) e^{-\sigma_a t}$ $- \dfrac{b_1}{b_a \omega_b^2} \cos(\omega_b t + \beta_1 - \beta_a)$
		$\dfrac{s + \sigma_1}{s(s + \sigma_a)(s + \sigma_b \mp j\omega_b)}$	$\dfrac{\sigma_1}{\sigma_a \omega_{0b}^2} + \dfrac{1}{b_a^2}\left(1 - \dfrac{\sigma_1}{\sigma_a}\right) e^{-\sigma_a t}$ $+ \dfrac{b_1}{b_a \omega_b \omega_{0b}} e^{-\sigma_b t} \sin(\omega_b t + \beta_1 - \beta_a - \beta)$
	211\2	$\dfrac{(s + \sigma_1)(s + \sigma_2)}{s(s + \sigma_a)(s + \sigma_b \mp j\omega_b)}$	$\dfrac{\sigma_1 \sigma_2}{\sigma_a \omega_{0b}^2} - \dfrac{(\sigma_a - \sigma_1)(\sigma_a - \sigma_2)}{\sigma_a b_a^2} e^{-\sigma_a t}$ $+ \dfrac{b_1 b_2}{b_a \omega_b \omega_{0b}} e^{-\sigma_b t} \sin(\omega_b t + \beta_1$ $+ \beta_2 - \beta_a - \beta)$
		$\dfrac{(s + \sigma_1 \mp j\omega_1)}{s(s + \sigma_a)(s + \sigma_b \mp j\omega_b)}$	$\dfrac{\omega_{01}^2}{\sigma_a \omega_{0b}^2} - \dfrac{a_1^2}{\sigma_a b_a^2} e^{-\sigma_a t}$ $+ \dfrac{b_1 \bar{b}_1}{b_a \omega_b \omega_{0b}} e^{-\sigma_b t} \sin(\omega_b t + \beta_1$ $+ \bar{\beta}_1 - \beta_a - \beta)$
	211\3	$\dfrac{(s + \sigma_1)(s + \sigma_2)(s + \sigma_3)}{s(s + \sigma_a)(s + \sigma_b \mp j\omega_b)}$	$\dfrac{\sigma_1 \sigma_2 \sigma_3}{\sigma_a \omega_{0b}^2} + \dfrac{(\sigma_a - \sigma_1)(\sigma_a - \sigma_2)(\sigma_a - \sigma_3)}{\sigma_a b_a^2} e^{-\sigma_a t}$ $+ \dfrac{b_1 b_2 b_3}{b_a \omega_b \omega_{0b}} e^{-\sigma_b t} \sin(\omega_b t + \beta_1 + \beta_2$ $+ \beta_3 - \beta_a - \beta)$

TABLE OF LAPLACE TRANSFORM PAIRS

Order of denominator	Pair number $cro\backslash n$	$F(s)$ [s-plane plot]	$f(t)$ $0 < t$ (All time functions zero for $t < 0$)
4 (Cont'd)	211\3 (Cont'd)	$\dfrac{(s + \sigma_1)(s + \sigma_2 \mp j\omega_2)}{s(s + \sigma_a)(s + \sigma_b \mp j\omega_b)}$	$\dfrac{\sigma_1 \omega_{02}^2}{\sigma_a \omega_{0b}^2} + \dfrac{a_2^2}{b_a^2}\left(1 - \dfrac{\sigma_1}{\sigma_a}\right)e^{-\sigma_a t}$ $+ \dfrac{b_1 b_2 \bar{b}_2}{b_a \omega_b \omega_{0b}} e^{-\sigma_b t} \sin(\omega_b t + \beta_1 + \beta_2 + \bar{\beta}_2 - \beta_a - \beta)$
	220\0	$\dfrac{1}{(s + \sigma_a)(s + \sigma_b)(s^2 + \omega_c^2)}$	$\dfrac{1}{c_a^2(\sigma_b - \sigma_a)} e^{-\sigma_a t} + \dfrac{1}{c_b^2(\sigma_a - \sigma_b)} e^{-\sigma_b t}$ $+ \dfrac{1}{c_a c_b \omega_c} \sin(\omega_c t - \gamma_a - \gamma_b)$
		$\dfrac{1}{(s + \sigma_a)(s + \sigma_b)(s + \sigma_c \mp j\omega_c)}$	$\dfrac{1}{c_a^2(\sigma_b - \sigma_a)} e^{-\sigma_a t} + \dfrac{1}{c_b^2(\sigma_a - \sigma_b)} e^{-\sigma_b t}$ $+ \dfrac{1}{c_a c_b \omega_c} e^{-\sigma_c t} \sin(\omega_c t - \gamma_a - \gamma_b)$
		$\dfrac{1}{(s + \sigma_a)^2(s^2 + \omega_b^2)}$	$\dfrac{2\sigma_a}{b_a^4} e^{-\sigma_a t} + \dfrac{1}{b_a^2} t e^{-\sigma_a t}$ $+ \dfrac{1}{b_a^2 \omega_b} \sin(\omega_b t - 2\beta_a)$
		$\dfrac{1}{(s + \sigma_a)^2(s + \sigma_b \mp j\omega_b)}$	$\dfrac{2(\sigma_a - \sigma_b)}{b_a^4} e^{-\sigma_a t} + \dfrac{1}{b_a^2} t e^{-\sigma_a t}$ $+ \dfrac{1}{b_a^2 \omega_b} e^{-\sigma_b t} \sin(\omega_b t - 2\beta_a)$
	220\1	$\dfrac{s + \sigma_1}{(s + \sigma_a)(s + \sigma_b)(s + \sigma_c \mp j\omega_c)}$	$\dfrac{\sigma_a - \sigma_1}{c_a^2(\sigma_a - \sigma_b)} e^{-\sigma_a t} + \dfrac{\sigma_b - \sigma_1}{c_b^2(\sigma_b - \sigma_a)} e^{-\sigma_b t}$ $+ \dfrac{c_1}{c_a c_b \omega_c} e^{-\sigma_c t} \sin(\omega_c t + \gamma_1 - \gamma_a - \gamma_b)$

APPENDIX J

Order of denominator	Pair number $cro\backslash n$	$F(s)$	[s-plane plot]	$f(t) \quad 0 < t$ (All time functions zero for $t < 0$)
4 (Cont'd)	220\1 (Cont'd)	$\dfrac{s + \sigma_1}{(s + \sigma_a)^2(s + \sigma_b \mp j\omega_b)}$		$\dfrac{2\sigma_1(\sigma_a - \sigma_b) + \omega_{0b}^2 - \sigma_a^2}{b_a^4} e^{-\sigma_a t}$ $+ \dfrac{\sigma_1 - \sigma_a}{b_a^2} t e^{-\sigma_a t}$ $+ \dfrac{b_1}{b_a^2 \omega_b} e^{-\sigma_b t} \sin(\omega_b t + \beta_1 - 2\beta_a)$
	220\2	$\dfrac{(s + \sigma_1)(s + \sigma_2)}{(s + \sigma_a)(s + \sigma_b)(s + \sigma_c \mp j\omega_c)}$		$\dfrac{(\sigma_1 - \sigma_a)(\sigma_2 - \sigma_a)}{c_a^2(\sigma_b - \sigma_a)} e^{-\sigma_a t}$ $+ \dfrac{(\sigma_1 - \sigma_b)(\sigma_2 - \sigma_b)}{c_b^2(\sigma_a - \sigma_b)} e^{-\sigma_b t}$ $+ \dfrac{c_1 c_2}{c_a c_b \omega_c} e^{-\sigma_c t} \sin(\omega_c t + \gamma_1 + \gamma_2 - \gamma_a - \gamma_b)$
		$\dfrac{(s + \sigma_1 \mp j\omega_1)}{(s + \sigma_a)(s + \sigma_b)(s + \sigma_c \mp j\omega_c)}$		$\dfrac{a_1^2}{c_a^2(\sigma_b - \sigma_a)} e^{-\sigma_a t}$ $+ \dfrac{b_1^2}{c_b^2(\sigma_a - \sigma_b)} e^{-\sigma_b t}$ $+ \dfrac{c_1 \bar{c}_1}{c_a c_b \omega_c} e^{-\sigma_c t} \sin(\omega_c t + \gamma_1 + \bar{\gamma}_1 - \gamma_a - \gamma_b)$
		$\dfrac{(s + \sigma_1)(s + \sigma_2)}{(s + \sigma_a)^2(s + \sigma_b \mp j\omega_b)}$		$\dfrac{2\sigma_1\sigma_2(\sigma_a - \sigma_b) + (\sigma_1 + \sigma_2)(\omega_{0b}^2 - \sigma_a^2) - 2\sigma_a(\omega_{0b}^2 - \sigma_a\sigma_b)}{b_a^4} e^{-\sigma_a t}$ $+ \dfrac{(\sigma_1 - \sigma_a)(\sigma_2 - \sigma_a)}{b_a^2} t e^{-\sigma_a t}$ $+ \dfrac{b_1 b_2}{b_a^2 \omega_b} e^{-\sigma_b t} \sin(\omega_b t + \beta_1 + \beta_2 - 2\beta_a)$
		$\dfrac{(s + \sigma_1 \mp j\omega_1)}{(s + \sigma_a)^2(s + \sigma_b \mp j\omega_b)}$		$\dfrac{2\omega_{01}^2(\sigma_a - \sigma_b) + 2\sigma_1(\omega_{0b}^2 - \sigma_a^2) - 2\sigma_a(\omega_{0b}^2 - \sigma_a\sigma_b)}{b_a^4} e^{-\sigma_a t}$ $+ \dfrac{a_1^2}{b_a^2} t e^{-\sigma_a t}$ $+ \dfrac{b_1 \bar{b}_1}{b_a^2 \omega_b} e^{-\sigma_b t} \sin(\omega_b t + \beta_1 + \bar{\beta}_1 - 2\beta_a)$

TABLE OF LAPLACE TRANSFORM PAIRS

Order of denominator	Pair number $cro\backslash n$	$F(s)$ [s-plane plot]	$f(t)$ $0 < t$ (All time functions zero for $t < 0$)
4 (Cont'd)	220\3	$\dfrac{(s+\sigma_1)(s+\sigma_2)(s+\sigma_3)}{(s+\sigma_a)(s+\sigma_b)(s+\sigma_c \mp j\omega_c)}$	$\dfrac{(\sigma_a-\sigma_1)(\sigma_a-\sigma_2)(\sigma_a-\sigma_3)}{c_a{}^2(\sigma_a-\sigma_b)}e^{-\sigma_a t}$ $+ \dfrac{(\sigma_b-\sigma_1)(\sigma_b-\sigma_2)(\sigma_b-\sigma_3)}{c_b{}^2(\sigma_b-\sigma_a)}e^{-\sigma_b t}$ $+ \dfrac{c_1 c_2 c_3}{c_a c_b \omega_c}e^{-\sigma_c t}\sin(\omega_c t + \gamma_1 + \gamma_2 + \gamma_3 - \gamma_a - \gamma_b)$
		$\dfrac{(s+\sigma_1)(s+\sigma_2 \mp j\omega_2)}{(s+\sigma_a)(s+\sigma_b)(s+\sigma_c \mp j\omega_c)}$	$\dfrac{a_2{}^2(\sigma_a-\sigma_1)}{c_a{}^2(\sigma_a-\sigma_b)}e^{-\sigma_a t} + \dfrac{b_2{}^2(\sigma_b-\sigma_1)}{c_b{}^2(\sigma_b-\sigma_a)}e^{-\sigma_b t}$ $+ \dfrac{c_1 c_2 \bar{c}_2}{c_a c_b \omega_c}e^{-\sigma_c t}\sin(\omega_c t + \gamma_1 + \gamma_2 + \bar{\gamma}_2 - \gamma_a - \gamma_b)$
		$\dfrac{(s+\sigma_1)(s+\sigma_2)(s+\sigma_3)}{(s+\sigma_a)^2(s+\sigma_b \mp j\omega_b)}$	$\dfrac{\begin{Bmatrix} 2\sigma_1\sigma_2\sigma_3(\sigma_a-\sigma_b) \\ - 2\sigma_a(\omega_0 b^2 - \sigma_a\sigma_b)(\sigma_1+\sigma_2+\sigma_3) \\ + (\omega_0 b^2 - \sigma_a{}^2)(\sigma_1\sigma_2+\sigma_1\sigma_3+\sigma_2\sigma_3) \\ + \sigma_a{}^2(\sigma_a{}^2 - 4\sigma_a\sigma_b + 3\omega_0 b^2) \end{Bmatrix}}{b_a{}^4}e^{-\sigma_a t}$ $+ \dfrac{(\sigma_1-\sigma_a)(\sigma_2-\sigma_a)(\sigma_3-\sigma_a)}{b_a{}^2}te^{-\sigma_a t}$ $+ \dfrac{b_1 b_2 b_3}{b_a{}^2 \omega_b}e^{-\sigma_b t}\sin(\omega_b t + \beta_1 + \beta_2 + \beta_3 - 2\beta_a)$
		$\dfrac{(s+\sigma_1)(s+\sigma_2 \mp j\omega_2)}{(s+\sigma_a)^2(s+\sigma_b \mp j\omega_b)}$	$\dfrac{\begin{Bmatrix} 2\sigma_1\omega_{02}{}^2(\sigma_a-\sigma_b) \\ + (\omega_0 b^2 - \sigma_a{}^2)(2\sigma_1\sigma_2+\omega_{02}{}^2) \\ + 2\sigma_a(\sigma_1+2\sigma_2)(\sigma_a\sigma_b - \omega_0 b^2) \\ + \sigma_a{}^2(\sigma_a{}^2 - 4\sigma_a\sigma_b + 3\omega_0 b^2) \end{Bmatrix}}{b_a{}^4}e^{-\sigma_a t}$ $+ \dfrac{a_2{}^2(\sigma_1-\sigma_a)}{b_a{}^2}te^{-\sigma_a t}$ $+ \dfrac{b_1 b_2 \bar{b}_2}{b_a{}^2 \omega_b}e^{-\sigma_b t}\sin(\omega_b t + \beta_1 + \beta_2 + \bar{\beta}_2 - 2\beta_a)$
	400\0	$\dfrac{1}{(s^2+\omega_a{}^2)(s^2+\omega_b{}^2)}$	$\dfrac{1}{\omega_a(\omega_b{}^2-\omega_a{}^2)}\sin(\omega_a t) + \dfrac{1}{\omega_b(\omega_a{}^2-\omega_b{}^2)}\sin(\omega_b t)$

750 APPENDIX J

Order of denominator	Pair number $cro\backslash n$	$F(s)$	[s-plane plot]	$f(t)$ $0 < t$ (All time functions zero for $t < 0$)
4 (Cont'd)	400\0 (Cont'd)	$\dfrac{1}{(s^2 + \omega_a^2)(s + \sigma_b \mp j\omega_b)}$		$\dfrac{-1}{b_a \bar{b}_a \omega_a} \sin(\omega_a t - \beta_a - \alpha_b)$ $+ \dfrac{1}{b_a \bar{b}_a \omega_b} e^{-\sigma_b t} \sin(\omega_b t - \beta_a - \bar{\beta}_a)$
		$\dfrac{1}{(s + \sigma_a \mp j\omega_a)(s + \sigma_b \mp j\omega_b)}$		$\dfrac{1}{a_b \bar{a}_b \omega_a} e^{-\sigma_a t} \sin(\omega_a t - \alpha_b - \bar{\alpha}_b)$ $- \dfrac{1}{a_b \bar{a}_b \omega_b} e^{-\sigma_b t} \sin(\omega_b t - \alpha_b - \beta_a)$
		$\dfrac{1}{(s^2 + \omega_a^2)^2}$		$\dfrac{1}{2\omega_a^3} \sin(\omega_a t) - \dfrac{1}{2\omega_a^2} t \cos(\omega_a t)$
		$\dfrac{1}{(s + \sigma_a \mp j\omega_a)^2}$		$\dfrac{1}{2\omega_a^3} e^{-\sigma_a t} \sin(\omega_a t) - \dfrac{1}{2\omega_a^2} t e^{-\sigma_a t} \cos(\omega_a t)$
	400\1	$\dfrac{s + \sigma_1}{(s + \sigma_a \mp j\omega_a)(s + \sigma_b \mp j\omega_b)}$		$\dfrac{a_1}{a_b \bar{a}_b \omega_a} e^{-\sigma_a t} \sin(\omega_a t + \alpha_1 - \alpha_b - \bar{\alpha}_b)$ $- \dfrac{b_1}{a_b \bar{a}_b \omega_b} e^{-\sigma_b t} \sin(\omega_b t + \beta_1 - \beta_a - \alpha_b)$
		$\dfrac{s + \sigma_1}{(s + \sigma_a \mp j\omega_a)^2}$		$\dfrac{\sigma_1 - \sigma_a}{2\omega_a^3} e^{-\sigma_a t} \sin(\omega_a t)$ $- \dfrac{a_1}{2\omega_a^2} t e^{-\sigma_a t} \cos(\omega_a t + \alpha_1)$

TABLE OF LAPLACE TRANSFORM PAIRS

Order of denominator	Pair number $cro\backslash n$	$F(s)$ [s-plane plot]	$f(t) \quad 0 < t$ (All time functions zero for $t < 0$)
4 (Cont'd)	400\2	$\dfrac{(s + \sigma_1)(s + \sigma_2)}{(s + \sigma_a \mp j\omega_a)(s + \sigma_b \mp j\omega_b)}$	$\dfrac{a_1 a_2}{a_b \bar{a}_b \omega_a} e^{-\sigma_a t} \sin(\omega_a t + \alpha_1 + \alpha_2 - \alpha_b - \bar{\alpha}_b)$ $- \dfrac{b_1 b_2}{a_b \bar{a}_b \omega_b} e^{-\sigma_b t} \sin(\omega_b t + \beta_1 + \beta_2 - \beta_a - \alpha_b)$
		$\dfrac{(s + \sigma_1 \mp j\omega_1)}{(s + \sigma_a \mp j\omega_a)(s + \sigma_b \mp j\omega_b)}$	$\dfrac{a_1 \bar{a}_1}{a_b \bar{a}_b \omega_a} e^{-\sigma_a t} \sin(\omega_a t + \alpha_1 + \bar{\alpha}_1 - \alpha_b - \bar{\alpha}_b)$ $- \dfrac{b_1 \bar{b}_1}{a_b \bar{a}_b \omega_b} e^{-\sigma_b t} \sin(\omega_b t + \beta_1 + \bar{\beta}_1 - \beta_a - \alpha_b)$
		$\dfrac{(s + \sigma_1)(s + \sigma_2)}{(s + \sigma_a \mp j\omega_a)^2}$	$\dfrac{\sigma_1 \sigma_2 - \sigma_a(\sigma_1 + \sigma_2) + \omega_{0a}^2}{2\omega_a^3} e^{-\sigma_a t} \sin(\omega_a t)$ $- \dfrac{a_1 a_2}{2\omega_a^2} t e^{-\sigma_a t} \cos(\omega_a t + \alpha_1 + \bar{\alpha}_1)$
		$\dfrac{(s + \sigma_1 \mp j\omega_1)}{(s + \sigma_a \mp j\omega_a)^2}$	$\dfrac{\omega_{01}^2 - 2\sigma_1 \sigma_a + \omega_{0a}^2}{2\omega_a^3} e^{-\sigma_a t} \sin(\omega_a t)$ $- \dfrac{a_1 \bar{a}_1}{2\omega_a^2} t e^{-\sigma_a t} \cos(\omega_a t + \alpha_1 + \bar{\alpha}_1)$

APPENDIX J

Order of denominator	Pair number $cro\backslash n$	$F(s)$	[s-plane plot]	$f(t)$ $0 < t$ (All time functions zero for $t < 0$)
4 (Cont'd)	400\3	$\dfrac{(s+\sigma_1)(s+\sigma_2)(s+\sigma_3)}{(s+\sigma_a \mp j\omega_a)(s+\sigma_b \mp j\omega_b)}$		$\dfrac{a_1 a_2 a_3}{a_b \bar{a}_b \omega_a} e^{-\sigma_a t} \sin(\omega_a t + \alpha_1 + \alpha_2 \\ \qquad\qquad + \alpha_3 - \alpha_b - \bar{\alpha}_b)$ $- \dfrac{b_1 b_2 b_3}{a_b \bar{a}_b \omega_b} e^{-\sigma_b t} \sin(\omega_b t + \beta_1 + \beta_2 \\ \qquad\qquad + \beta_3 - \beta_a - \alpha_b)$
		$\dfrac{(s+\sigma_1)(s+\sigma_2 \mp j\omega_2)}{(s+\sigma_a \mp j\omega_a)(s+\sigma_b \mp j\omega_b)}$		$\dfrac{a_1 a_2 \bar{a}_2}{a_b \bar{a}_b \omega_a} e^{-\sigma_a t} \sin(\omega_a t + \alpha_1 + \alpha_2 \\ \qquad\qquad + \bar{\alpha}_2 - \alpha_b - \bar{\alpha}_b)$ $- \dfrac{b_1 b_2 \bar{b}_2}{a_b \bar{a}_b \omega_b} e^{-\sigma_b t} \sin(\omega_b t + \beta_1 + \beta_2 \\ \qquad\qquad + \bar{\beta}_2 - \beta_a - \alpha_b)$
		$\dfrac{(s+\sigma_1)(s+\sigma_2)(s+\sigma_3)}{(s+\sigma_a \mp j\omega_a)^2}$		$e^{-\sigma_a t} \cos(\omega_a t)$ $- \dfrac{a_1 a_2 a_3}{2 \omega_a^2} t e^{-\sigma_a t} \cos(\omega_a t + \alpha_1 + \alpha_2 + \alpha_3)$ $+ \{[\sigma_1 \sigma_2 \sigma_3 - \sigma_a(\sigma_1 \sigma_2 + \sigma_1 \sigma_3 + \sigma_2 \sigma_3) \\ \qquad + \omega_{0a}^2(\sigma_1 + \sigma_2 + \sigma_3) \\ \qquad - \sigma_a(\omega_{0a}^2 + 2\omega_a^2)]/2\omega_a^3\} e^{-\sigma_a t} \sin(\omega_a t)$
		$\dfrac{(s+\sigma_1)(s+\sigma_2 \mp j\omega_2)}{(s+\sigma_a \mp j\omega_a)^2}$		$e^{-\sigma_a t} \cos(\omega_a t)$ $- \dfrac{a_1 a_2 \bar{a}_2}{2 \omega_a^2} t e^{-\sigma_a t} \cos(\omega_a t + \alpha_1 + \alpha_2 + \bar{\alpha}_2)$ $+ \{[\sigma_1 \omega_{02}^2 - \sigma_a(\omega_{02}^2 + 2\sigma_1 \sigma_2) \\ \qquad + \omega_{0a}^2(\sigma_1 + 2\sigma_2) \\ \qquad - \sigma_a(\omega_{0a}^2 + 2\omega_a^2)]/2\omega_a^3\} e^{-\sigma_a t} \sin(\omega_a t)$

TABLE OF LAPLACE TRANSFORM PAIRS

SELECTED TRANSFORM PAIRS WHOSE DENOMINATOR IS FIFTH-ORDER

Order of denominator	Pair number $cro\backslash n$	$F(s)$	[s-plane plot]	$f(t)$ $0 < t$ (All time functions zero for $t < 0$)
5	023\4	$\dfrac{As^4 + Bs^3 + Cs^2 + Ds + E}{s^3(s + \sigma_a)(s + \sigma_b)}$		$\dfrac{C\sigma_a\sigma_b - D(\sigma_a + \sigma_b) - E}{\sigma_a\sigma_b}$ $+ \dfrac{D}{\sigma_a\sigma_b}t + \dfrac{E}{2\sigma_a\sigma_b}t^2$ $+ \dfrac{A\sigma_a^4 - B\sigma_a^3 + C\sigma_a^2 - D\sigma_a + E}{\sigma_a^3(\sigma_a - \sigma_b)}e^{-\sigma_a t}$ $+ \dfrac{A\sigma_b^4 - B\sigma_b^3 + C\sigma_b^2 - D\sigma_b + E}{\sigma_b^3(\sigma_b - \sigma_a)}e^{-\sigma_b t}$
		$\dfrac{As^4 + Bs^3 + Cs^2 + Ds + E}{s^3(s + \sigma_a)^2}$		$\dfrac{C\sigma_a^2 - 2D\sigma_a - E}{\sigma_a^2} + \dfrac{D}{\sigma_a^2}t + \dfrac{E}{2\sigma_a^2}t^2$ $+ \dfrac{A\sigma_a^4 - C\sigma_a^2 - 2D\sigma_a - 3E}{\sigma_a^4}e^{-\sigma_a t}$ $- \dfrac{A\sigma_a^4 - B\sigma_a^3 + C\sigma_a^2 - D\sigma_a + E}{\sigma_a^3}te^{-\sigma_a t}$
	032\4	$\dfrac{As^4 + Bs^3 + Cs^2 + Ds + E}{s^2(s + \sigma_a)(s + \sigma_b)(s + \sigma_c)}$		$\left[\dfrac{D}{\sigma_a\sigma_b\sigma_c} - \dfrac{E}{\sigma_a\sigma_b\sigma_c}\left(\dfrac{1}{\sigma_a} + \dfrac{1}{\sigma_b} + \dfrac{1}{\sigma_c}\right)\right]$ $+ \dfrac{E}{\sigma_a\sigma_b\sigma_c}t$ $+ \dfrac{A\sigma_a^4 - B\sigma_a^3 + C\sigma_a^2 - D\sigma_a + E}{\sigma_a^2(\sigma_a - \sigma_b)(\sigma_a - \sigma_c)}e^{-\sigma_a t}$ $+ \dfrac{A\sigma_b^4 - B\sigma_b^3 + C\sigma_b^2 - D\sigma_b + E}{\sigma_b^2(\sigma_b - \sigma_a)(\sigma_b - \sigma_c)}e^{-\sigma_b t}$ $+ \dfrac{A\sigma_c^4 - B\sigma_c^3 + C\sigma_c^2 - D\sigma_c + E}{\sigma_c^2(\sigma_c - \sigma_a)(\sigma_c - \sigma_b)}e^{-\sigma_c t}$
		$\dfrac{As^4 + Bs^3 + Cs^2 + Ds + E}{s^2(s + \sigma_a)(s + \sigma_b)^2}$		$\dfrac{\sigma_a\sigma_b D - (\sigma_b + 2\sigma_a)E}{\sigma_a^2\sigma_b^3} + \dfrac{E}{\sigma_a\sigma_b^2}t$ $+ \dfrac{A\sigma_a^4 - B\sigma_a^3 + C\sigma_a^2 - D\sigma_a + E}{\sigma_a^2(\sigma_b - \sigma_a)^2}e^{-\sigma_a t}$ $+ \dfrac{A\sigma_b^4 - B\sigma_b^3 + C\sigma_b^2 - D\sigma_b + E}{\sigma_b^2(\sigma_a - \sigma_b)}te^{-\sigma_b t}$ $+ \dfrac{\left[A\sigma_b^5 - C\sigma_b^3 + 2D\sigma_b^2 - 3E\sigma_b\right.}{\sigma_b^3(\sigma_b - \sigma_a)^2}$ $\left.+ \sigma_a(-2A\sigma_b^4 + B\sigma_b^3 - D\sigma_b + 2E)\right]$ $\cdot e^{-\sigma_b t}$
		$\dfrac{As^4 + Bs^3 + Cs^2 + Ds + E}{s^2(s + \sigma_a)^3}$		$\dfrac{D\sigma_a - 3E}{\sigma_a^4} + \dfrac{E}{\sigma_a^3}t + \dfrac{A\sigma_a^4 + D\sigma_a - E}{\sigma_a^4}e^{-\sigma_a t}$ $- \dfrac{2A\sigma_a^4 - B\sigma_a^3 + D\sigma_a - 2E}{\sigma_a^3}te^{-\sigma_a t}$ $+ \dfrac{A\sigma_a^4 - B\sigma_a^3 + C\sigma_a^2 - D\sigma_a + E}{2\sigma_a^2}t^2e^{-\sigma_a t}$
	041\4	$\dfrac{As^4 + Bs^3 + Cs^2 + Ds + E}{s(s + \sigma_a)^2(s + \sigma_b)^2}$		$\dfrac{E}{\sigma_a^2\sigma_b^2}$ $- \dfrac{A\sigma_a^4 - B\sigma_a^3 + C\sigma_a^2 - D\sigma_a + E}{\sigma_a(\sigma_b - \sigma_a)^2}te^{-\sigma_a t}$ $- \dfrac{A\sigma_b^4 - B\sigma_b^3 + C\sigma_b^2 - D\sigma_b + E}{\sigma_b(\sigma_a - \sigma_b)^2}te^{-\sigma_b t}$ $+ \dfrac{\sigma_a(A\sigma_a^4 - C\sigma_a^2 + 2D\sigma_a - 3E)}{\sigma_a^2(\sigma_b - \sigma_a)^3}e^{-\sigma_a t}$ $\dfrac{-\sigma_b(3A\sigma_a^4 - 2B\sigma_a^3 + C\sigma_a^2 + E)}{\sigma_a^2(\sigma_b - \sigma_a)^3}e^{-\sigma_a t}$ $+ \dfrac{\sigma_b(A\sigma_b^4 - C\sigma_b^2 + 2D\sigma_b - 3E)}{\sigma_b^2(\sigma_a - \sigma_b)^3}e^{-\sigma_b t}$ $\dfrac{-\sigma_a(3A\sigma_b^4 - 2B\sigma_b^3 + C\sigma_b^2 + E)}{\sigma_b^2(\sigma_a - \sigma_b)^3}e^{-\sigma_b t}$

APPENDIX J

Order of denominator	Pair number $cro\backslash n$	$F(s)$ [s-plane plot]	$f(t) \quad 0 < t$ (All time functions zero for $t < 0$)
5 (Cont'd)	041\4 (Cont'd)	$\dfrac{As^4 + Bs^3 + Cs^2 + Ds + E}{s(s + \sigma_a)(s + \sigma_b)^3}$	$\dfrac{E}{\sigma_a \sigma_b{}^3}$ $+ \dfrac{A\sigma_a{}^4 - B\sigma_a{}^3 + C\sigma_a{}^2 - D\sigma_a + E}{\sigma_a(\sigma_a - \sigma_b)^3} e^{-\sigma_a t}$ $+ \dfrac{A\sigma_b{}^4 - B\sigma_b{}^3 + C\sigma_b{}^2 - D\sigma_b + E}{2\sigma_b(\sigma_b - \sigma_a)} t^2 e^{-\sigma_b t}$ $+ \dfrac{\sigma_a(3A\sigma_b{}^4 - 2B\sigma_b{}^3 + C\sigma_b{}^2 - E) - (2A\sigma_b{}^5 + B\sigma_b{}^4 - D\sigma_b{}^2 - 2E\sigma_b)}{\sigma_b{}^2(\sigma_b - \sigma_a)^2} t e^{-\sigma_b t}$ $- \left\{ \begin{array}{l} (A\sigma_b{}^7 + D\sigma_b{}^4 - 3E\sigma_b{}^3) \\ + \sigma_a(A\sigma_b{}^6 - D\sigma_b{}^3 + 6E\sigma_b{}^2) \\ - \sigma_a{}^2(11A\sigma_b{}^5 - 3B\sigma_b{}^4 \\ + C\sigma_b{}^3 - 2D\sigma_b{}^2) \\ - \sigma_a{}^3(9A\sigma_b{}^3 - 5B\sigma_b{}^2 + 3C\sigma_b) \\ \hline \sigma_b{}^3(\sigma_b - \sigma_a)^4 \end{array} \right.$ $\left. + \dfrac{E\sigma_a{}^3}{\sigma_b{}^4(\sigma_b - \sigma_a)^4} \right\} e^{-\sigma_b t}$
	203\4	$\dfrac{As^4 + Bs^3 + Cs^2 + Ds + E}{s^2(s + \sigma_a \mp j\omega_a)}$ *Zeros shown in arbitrary configuration*	$\left(\dfrac{C}{\omega_{a0}{}^2} - \dfrac{2D\sigma_a + E}{\omega_{a0}{}^4} + \dfrac{2E\sigma_a{}^2}{\omega_{a0}{}^6} \right)$ $+ \left(\dfrac{D}{\omega_{a0}{}^2} - \dfrac{2\sigma_a E}{\omega_{a0}{}^4} \right) t + \dfrac{E}{2\omega_{a0}{}^2} t^2$ $+ \dfrac{a_1 a_2 \bar{a}_2 a_3}{\omega_a \omega_{a0}{}^3} e^{-\sigma_a t} \sin(\omega_a t + \alpha_1 + \alpha_2 + \bar{\alpha}_2 + \alpha_3 - 3\alpha)$
	212\4	$\dfrac{As^4 + Bs^3 + Cs^2 + Ds + E}{s^2(s + \sigma_a)(s + \sigma_b \mp j\omega_b)}$ *Zeros shown in arbitrary configuration*	$\left(\dfrac{D\sigma_a - E}{\sigma_a{}^2 \omega_{b0}{}^2} - \dfrac{2E\sigma_b}{\sigma_a \omega_{b0}{}^4} \right) + \dfrac{E}{\sigma_a \omega_{b0}{}^2} t$ $+ \dfrac{A\sigma_a{}^4 - B\sigma_a{}^3 + C\sigma_a{}^2 - D\sigma_a + E}{\sigma_a{}^2 b_a{}^2} e^{-\sigma_a t}$ $+ \dfrac{b_1 b_2 \bar{b}_2 b_3}{\omega_b \omega_{b0}{}^2 b_a} e^{-\sigma_b t} \sin(\omega_b t + \beta_1 + \beta_2 + \bar{\beta}_2 + \beta_3 - 2\beta - \beta_a)$

TABLE OF LAPLACE TRANSFORM PAIRS 755

Order of denominator	Pair number $cro\backslash n$	$F(s)$ [s-plane plot]	$f(t)$ $\quad 0 < t$ (All time functions zero for $t < 0$)
5 (Cont'd)	221\4	$\dfrac{As^4 + Bs^3 + Cs^2 + Ds + E}{s(s + \sigma_a)^2(s + \sigma_b \mp j\omega_b)}$ *Zeros shown in arbitrary configuration*	$\dfrac{E}{\sigma_a{}^2 \omega_{b0}{}^2} + \dfrac{b_1 b_2 \bar{b}_2 b_3}{\omega_b \omega_{b0} b_a{}^2} e^{-\sigma_b t} \sin(\omega_b t$ $+ \beta_1 + \beta_2 + \bar{\beta}_2 + \beta_3 - \beta - 2\beta_a)$ $- \dfrac{A\sigma_a{}^4 - B\sigma_a{}^3 + C\sigma_a{}^2 - D\sigma_a + E}{\sigma_a b_a{}^2} t e^{-\sigma_a t}$ $+ \dfrac{b_a{}^2(3A\sigma_a{}^4 - 2B\sigma_a{}^3 + C\sigma_a{}^2 - E) + 2\sigma_a(\sigma_b - \sigma_a)(A\sigma_a{}^4 - B\sigma_a{}^3 + C\sigma_a{}^2 - D\sigma_a + E)}{\sigma_a{}^2 b_a{}^4} e^{-\sigma_a t}$
	401\4	$\dfrac{As^4 + Bs^3 + Cs^2 + Ds + E}{s(s + \sigma_a \mp j\omega_a)^2}$ *Zeros shown in arbitrary configuration*	$\dfrac{E}{\omega_a{}^4}$ $+ \dfrac{1}{\omega_a}\left(B - 2\sigma_a A - \dfrac{2\sigma_a E}{\omega_a{}^4}\right) e^{-\sigma_a t} \sin(\omega_a t)$ $+ \dfrac{1}{\omega_a}\left(A - \dfrac{E}{\omega_a{}^4}\right) e^{-\sigma_a t} \sin(\omega_a t - \alpha)$ $+ \dfrac{a_1 a_2 \bar{a}_2 a_3}{\omega_a \omega_{0a}{}^2} t e^{-\sigma_a t} \sin(\omega_a t + \alpha_1 + \alpha_2 + \bar{\alpha}_2 + \alpha_3 - \alpha)$

PROBLEMS

CHAPTER 1: DYNAMIC INVESTIGATION

In Probs. 1.1 through 1.4 draw a functional block diagram that is, in an approximate, qualitative way, descriptive of each system named. Use a box to represent each major subsystem, and lines between boxes to represent the interactions between subsystems. (Do not show physical details—only functions.)

1.1 A thermostat-controlled home heating system.

1.2 An automatic beverage dispenser with syrup, carbonated water, and ice dispensed in that order.

1.3 A system (including the motormen) for actuating signal lights to keep subway trains safely spaced.

1.4 The cooling system for a water-cooled automobile engine.

1.5 The following are some assumptions commonly made in modeling physical systems prior to analyzing them mathematically. For each assumption, state what sort of analytical simplifications may result. (Refer to Table I, p. 18.) (i) The gravity torque on a pendulum (figure), which is actually proportional to the sine of pendulum angle θ, is approximated as being proportional to the angle itself. (ii) The pendulum hinge moment is assumed proportional to pendulum angular velocity.

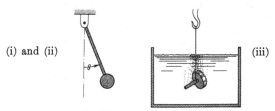

Prob. 1.5

(iii) In predicting the time for a heat-treated part to cool (figure) the bath temperature is assumed constant. (iv) Temperature throughout the part is assumed uniform.

(v) The output of an amplifier is assumed to depend only on the input, and to be proportional to the input. (vi) Wind drag is ignored in studying the motion of an ocean liner. (vii) *Only* wind forces and gravity are considered in studying the motion of an aircraft. (viii) A temperature sensor is assumed to report temperature exactly. (ix) The pressure in a pneumatic cylinder is assumed uniform throughout the cylinder. (x) In studying the response of a sound system to a test tone, the tone is assumed to be a pure sine wave.

1.6 State whether each of the following ordinary differential equations is (i) non-linear; (ii) linear, with time-varying coefficients; or (iii) l.c.c. (linear with constant coefficients, i.e., linear and time-invariant). For those marked (i), state precisely why, and suggest a linear version (which might be a viable approximation, given additional information).

(a) $\dfrac{d^2x}{dt^2} + 3\dfrac{dx}{dt} + 7x = 0$ (b) $\dfrac{d^2x}{dt^2} + 3\dfrac{dx}{dt} + 7x = 5$

(c) $\dfrac{d^2x}{dt^2} + 3\dfrac{dx}{dt} + 7x = 5tx$ (d) $\dfrac{d^2x}{dt^2} + 3\dfrac{dx}{dt} + 7x = 5t^2$

(e) $\dfrac{d^2x}{dt^2} + 3x\dfrac{dx}{dt} + 7x = 5t$ (f) $\dfrac{d^2x}{dt^2} + 3t\dfrac{dx}{dt} + 7x = 5t$

(g) $\begin{cases} \dfrac{d^2x}{dt^2} + 3t\dfrac{dy}{dt} + 7x = 5t \\ 4\dfrac{dx}{dt} + 7ty = 0 \end{cases}$ (h) $\begin{cases} \dfrac{d^2x}{dt^2} + 3t\dfrac{dy}{dt} + 7xy = 5t \\ 4y\dfrac{dx}{dt} + 7\,ty = 0 \end{cases}$

(i) $\dfrac{d^2x}{dy^2} + 3\dfrac{dx}{dy} + 7x = 5y$ (j) $\dfrac{d^2x}{dt^2} + 3\dfrac{dx}{dt} + 7x^2 = 5\cos t$

[Note that y is the independent variable in (i).]

(k) $\dfrac{d^3x}{dt^3} + 4t\dfrac{d^2x}{dt^2} + 3\left(\dfrac{dx}{dt}\right)^2 + 7xy = 5t$

In Probs. 1.7 through 1.10, involving physical modeling, notice that a propitious model of a system can be contemplated only after stating the model's analytical purpose.

1.7 Sketch and describe a physical model for an elevator, from which an estimate of transit times versus load can be computed.

1.8 Sketch a physical model for a d-c electric iron with on-off thermostat, the model to be used for estimating current usage versus time. List your assumptions.

1.9 Sketch and describe a gross physical model for a home forced-air heating system, with thermostat control, suitable for a first, *very approximate* analysis of the temperature variation during 24 hours on a typical cold day.

1.10 The figure for Prob. 5.5 (p. 776) shows a physical model for an automobile. The model is to be used to study pitching motions over a rough road. List the assumptions that have been made.

[Chap. 1] PROBLEMS

1.11 Complete the following table:

Physical quantity	Given value			
	mks units	cgs units	British units	Combination or other units
Height of the Eiffel tower	300 m			
Diameter of an oxygen molecule		10^{-8} cm		
Distance to Polaris				1086 light years†
Thrust of a rocket engine			10^6 lb	
Torque in a drive shaft	10^4 n m			
Velocity of a sailboat				12 kt‡ due west
Stiffness of a torsion bar		12 dn cm/rad		
Mass of a spacecraft	10^4 n/(m/sec^2)			
Momentum of a bullet		2×10^5 dn sec		
Moment of inertia of a gyro rotor		2×10^{12} dn cm/sec^{-2}		
Angular momentum of a gyro rotor		2×10^5 dn cm sec		
Pressure 36 $\times 10^3$ ft under the sea			2.26×10^6 lb/ft^2	_____ atmospheres
Mass flow rate of blood through the heart	0.1 kg/sec			
Viscosity of mercury at 50°K			3×10^{-5} lb sec/ft^2.	
Lowest recorded temperature in Montana			−78°F	
Heat-loss rate of a lightbulb	95 watts			
Specific heat of a re-entry heat shield		0.1 cal/gm		
Current in an electric toaster				9.1 amp
Voltage of a flashlight battery		1.5×10^{10} dn cm/cl		
Resonant frequency, $\sqrt{LC}$, of a circuit				1×10^3 h$^{1/2}$ f$^{1/2}$
The resistance of a 100-ohm resistor				101 ohms
Electric field in a TV picture tube	1×10^4 n/cl			
Magnetic induction of a cyclotron				2×10^4 gauss

† One light year is the distance light travels in one year. (The speed of light is 1.86×10^5 miles/sec.)
‡ One knot = 1 nautical mile per hour = 6080 ft/hour.

1.12 Check the following equations of motion dimensionally, and report any inconsistencies. (Symbols have the meanings given in Table II, p. 21.)

(a) Electrical: $\dfrac{1}{C}\dfrac{dv}{dt} + \dfrac{v}{R} = i$ (b) Mechanical: $b\dfrac{d\theta}{dt} + k\theta^2 + J\dfrac{d^2\dot{\theta}}{dt^2} = M$

(c) Mechanical: $m\left(\dfrac{dx}{dt}\right)^2 + kx^2 = 2fx/t$ (d) Thermal: $\dfrac{1}{C}\int q\,dt + Rq = 5T$

1.13 An important parameter in fluid flow is Reynolds number, defined by $\mathcal{R} \triangleq v\ell\rho/\mu$. Show that it is dimensionless.

CHAPTER 2: EQUATIONS OF MOTION FOR SIMPLE PHYSICAL SYSTEMS: MECHANICAL, ELECTRICAL, AND ELECTROMECHANICAL

For each system in Probs. 2.1 through 2.9, (i) *sketch* a reference configuration and an arbitrary configuration, and (ii) define an independent set of *coordinates*. Indicate the positive direction for each coordinate. (iii) In each case, state the number of *degrees of freedom*. (iv) Write an equation to describe each geometric *constraint* that is not already implied precisely by (ii). (v) Where indicated, define at least one alternate set of coordinates, and relate it to the set in (ii) by writing *identity* equations.

2.1 A model for a crank and piston (figure). The crank rotates at constant speed. Include instruction (v).

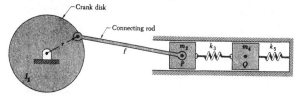

Prob. 2.1

2.2 A lever system (figure). Assume small rotations only. Include instruction (v).

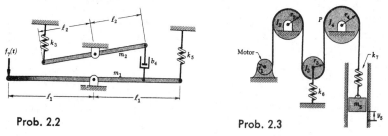

Prob. 2.2   Prob. 2.3

2.3 A belt-drive system (figure). The purpose of pulley J_3 is to take up slack in the belt. Assume the belt to be inextensible, and assume no belt slippage. The motor runs at constant speed. Include instruction (v).

2.4 A jeep driving at constant angle α with respect to the fore-and-aft line of the deck of a moving aircraft carrier. Assume the carrier does not roll or pitch.

2.5 A cam and follower (figure). The cam is flat at the ends, and has a sinusoidal shape $y = h[1 - \cos(2\pi x/\ell)]$ between points A and B. The follower is constrained to the slot.

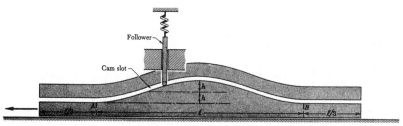

Prob. 2.5

[Chap. 2] PROBLEMS

2.6 A mile-long oil-well drill. (They twist.)
2.7 A vibrating violin string.
2.8 A rotary cam and follower (figure) (such as the valve-opening mechanism in an automobile engine). The cam rotates at constant speed and is shaped to have the following relation between r and θ:

$$r = r_o \qquad\qquad -\frac{\pi}{2} \leqslant \theta \leqslant \frac{\pi}{2}$$

$$r = r_o - \frac{h}{4}(3\cos\theta + \cos 3\theta) \qquad \frac{\pi}{2} < \theta < \frac{3\pi}{2}$$

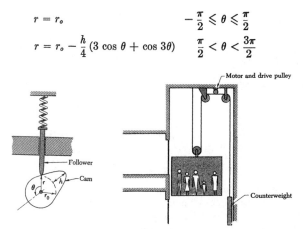

Prob. 2.8   Prob. 2.9

2.9 A simplified model of an elevator system (figure). The cable makes several turns around the drive pulley. Assume that the cable stretches *uniformly* (except that the stretch around pulleys and between the fixed pulleys at the roof may be neglected). Consider motor speed to be prescribed.

2.10 In Prob. 2.1, write expressions for the velocity and acceleration of points P and Q in terms of the coordinates you chose. Write two possible sets of state variables for the system. Give explicitly the velocity and acceleration of point P if the crank disk rotates at constant speed ω.

2.11 Find an expression for the stretch in spring k_7 in Prob. 2.3, and also for the acceleration of m_5, in terms of the coordinates you chose there. Define a set of state variables.

2.12 Write an identity expression for the velocity "across" the dashpot in Prob. 2.2, in terms of the angular velocities $\dot\theta_1$ and $\dot\theta_2$ of the bars.

2.13 Assume the cam in Prob. 2.5 moves from end to end at constant velocity, and write explicit expressions for the velocity and acceleration of the follower.

2.14 Assume the cam in Prob. 2.8 rotates at constant speed ω, and find an explicit expression for the velocity and acceleration of the follower. In particular, show that there is no discontinuity (abrupt change) in velocity at $\theta = \pi/2$.

2.15 Write a system compatibility relation for the crank system in Prob. 2.1. [An example is Eq. (2.28), p. 52.]

2.16 Write a system compatibility relation for the lever system in Prob. 2.2.

2.17 Write a system compatibility relation for the belt-drive system in Prob. 2.3.

In Probs. 2.18 through 2.28 (i) each member named is to be sketched in a free-body diagram showing (and labeling, as f_1, M_3, f_i, and so on) all forces and moments acting, with their positive directions, as in Fig. 2.3, p. 41. Inertial forces and moments are to be included. Then (ii) write Eq. (2.1) or (2.2), or both, as indicated. [Examples are Eqs. (2.3) and (2.4).]

2.18 Mass m_1 (figure). Write $\Sigma f^*_{\text{horiz}}$. (Assume arbitrary friction.)

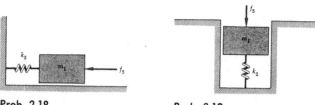

Prob. 2.18   Prob. 2.19

2.19 Mass m_1 (figure). Write Σf^*_{vert}. (Assume arbitrary friction.)

2.20 Mass m_5 in Prob. 2.3 (vertical motion only). Write Σf^*_{vert}.

2.21 Members P and Q of Prob. 2.1. Write $\Sigma f^*_{\text{horiz}}$. (Assume viscous friction.)

2.22 Each bar in Prob. 2.2. Write $\Sigma M^*_{\text{pivot}}$ and $\Sigma M^*_{\text{one end}}$.

2.23 Pulley J_4 (with part of the belt) in Prob. 2.3. Write ΣM^*_{axle} and ΣM^*_P.

2.24 From Prob. 2.3, pulley J_3 plus pulley J_4 plus the belt connecting them, taken as a single free body. Write Σf^*_{vert} and $\Sigma M^*_{\text{axle 4}}$.

2.25 From Prob. 2.3, the system consisting of pulley J_3 plus pulley J_4 plus mass m_5, together with the belt connecting them, as a single free body. Write $\Sigma M^*_{\text{axle 4}}$ and Σf^*_{vert}.

2.26 Obtain the result in Prob. 2.25 by combining the results of Probs. 2.20 and 2.24.

2.27 Disks J_1, J_2, J_3 (figure) with k_5 missing. No slippage. Write ΣM^* for the shaft axis of each disk.

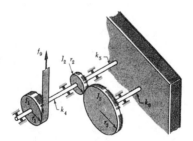

Probs. 2.27 and 2.28

2.28 Same as Prob. 2.27, but with k_5 present.

2.29 In static tests, various weights are hung on a vertical spring, and the spring length measured. The results are

w (lb)	0	2	4	6	8
x (in.)	10.1	14.1	18.1	21.6	22.7

[Chap. 2] PROBLEMS 763

The spring is also inverted and compressed by different weights:

w (lb)	0	2	4	6
x (in.)	9.9	5.9	4.5	4.5

Discuss the spring "constant" k and the circumstances under which its use is advisable in analyzing a system containing this spring. (Use an appròpriate plot.)

In Probs. 2.30 through 2.33 (i) write the physical force-geometry relation for each element named, on the basis of the physical model described, and then (ii) substitute these into the equations of dynamic equilibrium previously derived.

2.30 (i) The inertia, elastic spring, and dry friction in Prob. 2.18. (ii) Refer to Prob. 2.18.

2.31 (i) Each rotary inertia and elastic shaft in Fig. 2.2b, p. 36, assuming viscous friction. (ii) Refer to Eqs. (2.4).

2.32 (i) Each of the masses, the springs, and each friction element in Prob. 2.1. Assume linear springs, viscous damping on mass P, and ideal dry friction on mass Q. (ii) See Prob. 2.21.

2.33 (i) In Prob. 2.2, rotary inertias $J_1 = m_1\ell_1^2/3$ and $J_2 = m_2\ell_2^2/3$, the viscous dashpot, and the elastic springs. (ii) See Prob. 2.22, pivots only.

In Probs. 2.34 through 2.39 use Procedure A-m to derive a complete set of equations of motion. List the unknown variables in your set of equations, and verify that you have as many equations as unknown variables.

2.34 Blocks in a slot (figure). Viscous friction.

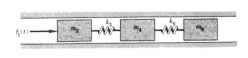

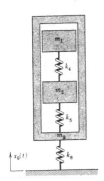

Prob. 2.34   Prob. 2.35

2.35 Model for seismic instruments in a frame (figure). Motion of the earth is the prescribed input. Viscous friction, elastic springs.

2.36 Turbine model (figure). Square-law rim damping and elastic shafts.

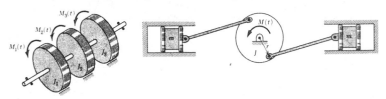

Prob. 2.36   Prob. 2.37

2.37 Model of an air compressor (or automobile engine without ignition) (figure). Gas in the chambers is elastic and doesn't leak. Dry friction at cylinders, viscous at flywheel pivot. Torque $M(t)$ is prescribed, from the electric driver (or starter motor). Make a reasonable assumption about f_{gas}.

2.38 Model of a fluid clutch (figure). Viscous friction between disks. Drive torque M_D and load torque M_L are prescribed.

Prob. 2.38

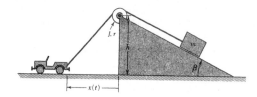

Prob. 2.39

2.39 Model of a loading system (figure). Elastic cables (stretch *per unit length* proportional to tension; cable mass negligible), heavy pulley, dry sliding friction, viscous pivot. Jeep motion $x(t)$ is prescribed.

2.40 Consider again the model in Fig. 2.7d, p. 48, of an aircraft landing on a carrier. This time, obtain one possible equation of motion from aircraft-plus-sandbag taken as a single free body. [This equation plus either of (2.27) form an independent set.] Check your result by adding (2.27) together.

2.41 Model for a ship propulsion system (figure). Only the turbine rotor has appreciable inertia. It also has viscous damping. The (massless) shaft is elastic in torsion. The load torque on the (massless) propeller is proportional to propeller speed squared.

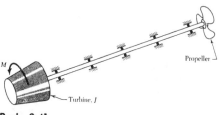

Prob. 2.41

In Probs. 2.42 through 2.46 (i) write a set of system compatibility equations for each system specified, and (ii) substitute physical relations to obtain a set of differential equations in terms of internal forces, as specified. (Refer to Example 2.3.)

2.42 The massless system model shown (figure).

Prob. 2.42

2.43 The system shown (figure), with viscous dampers. Neglect the mass of the levers, and friction in the guides, for m_4. Define variables precisely. Obtain final equations in terms of the forces in the two springs. (Hint: Compatibility equations must be written for each of two "loops," because the *absolute* motion x_4 of the mass—i.e., with respect to "ground"—is involved.)

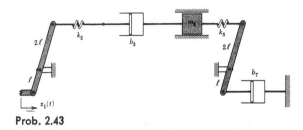

Prob. 2.43

2.44 The boat-drive system shown (figure), with friction in the clutch and propeller, viscous and other friction negligible, and the motor *speed* prescribed (not a very realistic assumption). Neglect propeller inertia. Obtain final equations in terms of shaft moments M_2 and M_7. (Hint: Compatibility must be written for two loops, because the absolute motion θ_6 of inertia J_6 must appear.)

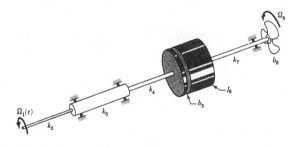

Prob. 2.44

2.45 The seismic instrument system of Prob. 2.35, with friction neglected. Obtain equations in terms of forces in the springs. (Hint: Three loops—see Prob. 2.43.)

2.46 The elevator model of Prob. 2.9. Neglect friction, and assume drive-pulley motion is prescribed. Obtain final equations in terms of forces in the long cable segments (which, again, are elastic: their stretch per unit length is proportional to the tension in them).

2.47 A torque $M_1(t)$ is applied electrically to the rotor of the motor (figure), whose moment of inertia is J_r. The radii and moments of inertia of the gears are as labeled. The load has moment of inertia J_6 and is subjected to an independent load torque $M_7(t)$. The shafts are elastic, as indicated. Write the equations of motion, neglecting gear friction. (How many degrees of freedom has the system?)

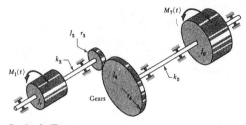

Prob. 2.47

2.48 Write the equation of motion for the gear train shown (figure) if all members are rigid and friction is negligible.

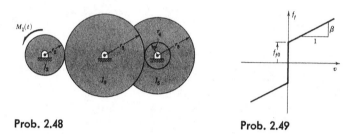

Prob. 2.48 **Prob. 2.49**

2.49* Augment your result in Prob. 2.48 to include gear friction assuming that, for each gear, the *tangential* component of total friction *force* on its teeth varies with linear rim speed v as shown (figure).

2.50* Augment your result in Prob. 2.47 to include gear friction modeled as in Prob. 2.49.

2.51 Write the equations of motion for the system in Prob. 2.27, with f_9 prescribed, versus time. No slippage.

2.52 Write the equation of motion for the system shown (figure), with a piece of lever of length ℓ having mass m. Neglect pin friction; $f_1(t)$ is prescribed. (Note: The moment of inertia of a bar of mass m and length ℓ *about one end* is $m\ell^2/3$. Moments of inertia are additive, of course. A bar's moment of inertia about an arbitrary point is $m_1\ell_1^2/3 + m_2\ell_2^2/3$, where 1 and 2 are the parts into which the point divides it.)

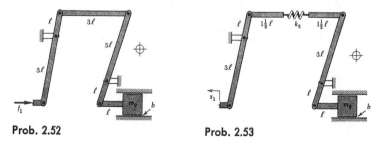

Prob. 2.52   **Prob. 2.53**

*Some of the more difficult problems are marked with an asterisk.

[Chap. 2] PROBLEMS 767

2.53 Write the equations of motion for the system shown, which is the same as for Prob. 2.52, except that a spring has been added and the input displacement $x_1(t)$ is now prescribed.

2.54 Augment Prob. 2.53 to include viscous pin friction.

In Probs. 2.55 through 2.60 (i) determine the number of independent voltages and the number of independent currents. Then (ii) select and label an independent set of node voltages and an independent set of loop currents. Define positive directions for loop currents, branch voltages, and branch currents on the drawings. Next, (iii) write identities relating every branch voltage drop to the selected node voltages, and relating every branch current to the selected loop currents. Finally, (iv) write all the constraint relations.

2.55 The network of Fig. 2.12b, p. 58 (which is a physical model for the circuit of Fig. 2.12a).

2.56 The network shown.

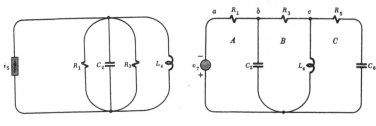

Prob. 2.56   Prob. 2.57

2.57 The network shown.
2.58 The network of Fig. 2.13, p. 62.
2.59 The network shown (which is not mappable onto a plane).

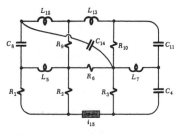 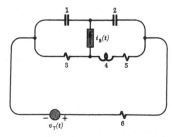

Prob. 2.59   Prob. 2.60

2.60 The network shown.

In Probs. 2.61 through 2.66 write a full set of KCL relations in terms of branch currents and a full set of KVL relations in terms of branch voltage drops.

2.61 The network of Fig. 2.12b, p. 58.
2.62 The network of Prob. 2.56.
2.63 The network of Prob. 2.57.
2.64 The network of Fig. 2.13, p. 62.
2.65 The network of Prob. 2.59.
2.66 The network of Prob. 2.60.

2.67 Suppose the capacitor in Fig. 2.17, p. 68, has capacitance $C_2 = 4$ microfarads, and the input current is given be

$$i_1(t) = 2t \text{ milliamps} \quad 0 < t < 2 \text{ sec}$$
$$i_1(t) = 4 \text{ milliamps} \quad 2 \text{ sec} < t < 5 \text{ sec}$$
$$i_1(t) = 0 \quad 5 \text{ sec} < t$$

Plot $v_2(t)$ versus t to scale, and label the plot numerically.

2.68 Suppose the capacitor in Fig. 2.17 has capacitance $C_2 = 4$ microfarads, and the input current is given by

$$i_1(t) = t^2 \text{ milliamps} \quad 0 < t < 2 \text{ sec}$$
$$i_1(t) = 2t \text{ milliamps} \quad 2 \text{ sec} < t < 5 \text{ sec}$$
$$i_1(t) = 10 \text{ milliamps} \quad 5 \text{ sec} < t$$

Plot $v_2(t)$ versus t to scale, and label the plot numerically.

2.69 Repeat Prob. 2.68, with the capacitor replaced by a resistor $R_2 = 200$ ohms.

2.70 Repeat Prob. 2.68, with the capacitor replaced by an inductor $L_2 = 2$ millihenrys.

2.71 Suppose the capacitor in Fig. 2.17 has capacitance $C_2 = 4$ microfarads, and the current source is replaced by a voltage source which is given by

$$v_1(t) = \frac{t^2}{3} \text{ volts} \quad 0 < t < 3 \text{ sec}$$
$$v_1(t) = t \text{ volts} \quad 3 \text{ sec} < t < 5 \text{ sec}$$
$$v_1(t) = 5 \text{ volts} \quad 5 \text{ sec} < t$$

Plot $i_2(t)$ versus t to scale, and label the plot numerically.

2.72 Repeat Prob. 2.71, with the capacitor replaced by a resistor $R_2 = 200$ ohms.
2.73 Repeat Prob. 2.71, with the capacitor replaced by an inductor $L_2 = 2$ millihenrys.

In Probs. 2.74 through 2.79 (i) use KCL to derive a complete set of equations of motion in terms of an independent set of node voltages (by first writing the KCL equations in terms of branch currents and then substituting in appropriate physical voltage-current relations, using identities as needed). Also, (ii) use KVL to derive a complete set of equations of motion in terms of an independent set of loop currents (by first writing the KVL equations in terms of branch voltage drops and then substituting in appropriate physical relations, using identities as needed). (iii) For the KCL set and for the KVL set, list the independent variables, and show that the number of equations equals the number of independent variables, and that this number agrees with (2.46) and (2.47). Finally, (iv) remove an equation from the set wherever a constraint relation makes this possible by removing an independent variable from the list of unknowns. *Show that the final set of equations contains the same number of equations as unknown variables.*

2.74 The network of Fig. 2.12b, p. 58. (Use the results of Probs. 2.55 and 2.61.) [Note: The KVL equation was obtained in Example 2.4, Eq. (2.64), p. 72.]
2.75 The network of Prob. 2.56. (Use the results of Probs. 2.56 and 2.62.)
2.76 The network of Prob. 2.57. (Use the results of Probs. 2.57 and 2.63.)
2.77 The network of Fig. 2.13, p. 62. (Use the results of Prob. 2.64.)
2.78 The network of Prob. 2.59. (Use the results of Probs. 2.59 and 2.65.)
2.79 The network of Prob. 2.60. (Use the results of Probs. 2.60 and 2.66.)

In Probs. 2.80 through 2.83 obtain a set of independent equations of motion in which the number of equations is equal to the number of unknowns. (Use Procedure A-e with either KCL or KVL. Draw equivalent networks, if it is helpful.)
2.80 The network shown.

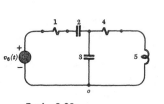

Prob. 2.80   Prob. 2.81

2.81 The network shown. (This is known as a "bridged-T" network.)
2.82 The network shown in Fig. 2.19, p. 73, with mutual inductance M_{67} between elements L_6 and L_7. [Augment Eqs. (2.65) appropriately.] With v_6 and v_7 defined positive by the arrows shown there, assume that v_7 is positive when di_6/dt is positive and that v_6 is positive when di_7/dt is positive.
2.83 The network of Prob. 2.59, with mutual inductance M_{57} between elements L_5 and L_7. [Augment the results of Prob. 2.78 appropriately.] With v_5 and v_7 defined with arrows going from left to right, assume that v_7 is positive when di_5/dt is positive and that v_5 is positive when di_7/dt is positive.

In Probs. 2.84 through 2.97 obtain a set of independent equations of motion in which the number of equations is equal to the number of unknowns. (Use Procedure A-e with either KCL or KVL. Draw equivalent networks if it is helpful.)
2.84 The system of Fig. 2.20, p. 76.
2.85 The network shown. Subsystem R_5, C_6, R_7 is called a "lead network" [because, for steady-state sinusoidal signals, $v_g(t)$ is found to precede, or "lead," $v_b(t)$].

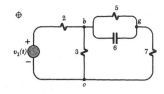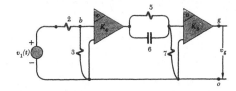

Prob. 2.85   Prob. 2.86

2.86 The network shown. The lead network is isolated by voltage isolation amplifiers.

2.87 The network shown.

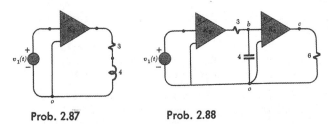

Prob. 2.87   Prob. 2.88

2.88 The network shown. The section between the isolation amplifiers is called a "low-pass" filter (because $v_b \approx v_a$ for low-frequency sinusoidal signals, but $v_b \ll v_a$ for high-frequency signals).

2.89 The network of Prob. 2.88 with R_3 and C_4 interchanged. The section between amplifiers is now called a "high-pass" filter (because $v_b \approx v_a$ for high-frequency signals, but $v_b \ll v_a$ for low-frequency signals).

2.90 The network shown. [Assume that the variables used represent the amplitude of an a-c carrier. Consider the output current from the transformer to be simply related to the input by Eqs. (2.66), p. 78. That is, neglect the inductances of the transformer coils in dealing with the low-frequency variables of interest here.]

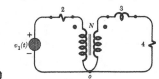

Prob. 2.90

2.91 The speaker of Fig. 2.23, p. 79, is driven by a circuit which can be modeled like the one in Fig. 2.24b. Derive the equations of motion for the system.

2.92 A network is shown below for converting motion x to voltage v, using the potentiometer of Fig. 2.26a, p. 86. (An "RC filter" is included.) Write the equation of motion for the system.

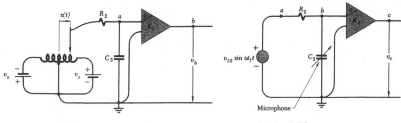

Prob. 2.92   Prob. 2.93

2.93 The figure represents the capacitive microphone of Fig. 2.26b connected in an a-c circuit. Write the equation of motion for the system if v_{10} is constant and the sound pressure on the microphone, $p(t)$, is variable. Define parameters as required.

2.94 A strain gauge cemented to an aircraft wing, Fig. 2.26c, is connected in place of the potentiometer in the circuit of Prob. 2.92. The strain, $\epsilon(t)$, at the gauge varies

[Chap. 3] PROBLEMS 771

sinusoidally, $\epsilon(t) = \epsilon_{\max} \sin \omega t$. Write the system equation of motion. Define parameters as needed.

2.95 Write the equation of motion for the D'Arsonval meter, Fig. 2.27e, where i is a prescribed time function $i(t)$. State your assumptions.

2.96 The rotor in Fig. 2.24a, p. 80, is driven as a generator by the belt, which is connected to a steam turbine. The applied moment $M_7(t) = r[f_7(t) - f_8(t)]$ is a constant: $M_7 = -C$ ft lb. With the voltage source v_1 replaced by a resistor R_1, write the system equations of motion.

2.97 The speaker of Prob. 2.91 is to be used as a dynamic microphone. With the source v_1 replaced by a resistor R_1, write the equations of motion between sound pressure $p(t)$ and current $i(t)$. Define quantities as needed.

2.98 The electrostatic cathode ray tube of Fig. 2.28a, p. 92, has a specified time-varying voltage $v_1(t)$ applied to its plates. Using letters for plate dimensions, and assuming electron-gun "muzzle velocity" u_o, find $\theta(t)$.

CHAPTER 3: EQUATIONS OF MOTION FOR SIMPLE HEAT-CONDUCTION AND FLUID SYSTEMS

3.1 Describe a one-lump physical model for the ingot in Fig. 3.1a, p, 95, and derive its equation of motion, considering bath temperature T_b to be constant. Assume resistance to heat flow to be concentrated at the interface between ingot and bath.

3.2 Describe a two-lump physical model for the system of Fig. 3.1a, the ingot being one heat-storage element and the bath another, with the outside temperature, T_o, constant. (In this approximate model, heat transfer is attributed entirely to conduction.) Write the equations of motion. Sketch qualitatively the temperature distribution at several times, as in Fig. 3.3a.

3.3 Describe a three-lump physical model for the system of Fig. 3.1a by dividing the ingot arbitrarily into two concentric cylinders of given heat-storage capacity, with the bath as the third lump. Let outside temperature T_o be constant. Let heat transfer be attributed entirely to conduction, and considered to take place at the interfaces between lumps. Write the equations of motion. Sketch qualitatively the temperature distribution at several times, as in Fig. 3.3a, p. 100.

3.4 A package for making marine measurements is falling at constant speed v through the ocean. The ocean temperature T_o varies with depth in a known way. Describe a one-lump thermal model, and write an equation of motion from which temperature T inside the package can be estimated as a function of time. Is the prescribed (constrained) variable in this problem the through variable or the across variable?

3.5 When a satellite vehicle re-enters the earth's atmosphere, heat is generated (converted from kinetic energy) on its nose at a rate proportional to air density and to vehicle velocity squared. With both given, versus time, write the differential equation for temperature T inside the vehicle, based on a crude one-lump model. Make a sketch of your physical model. Is the prescribed (constrained) variable in this problem the through or the across variable?

3.6 Electronic equipment is assembled in modular packages, and these are mounted to a base plate for installation in a vehicle, as shown in the figure. A number of base plates are connected to a single refrigerated heat sink, so that heat generated within the electronic packages will be carried away adequately. For the assembly shown, with only one package P, describe a two-lump model (storage elements P and B), and write the equations of motion for temperature T_P, following turning on of the electric power. Let the rate of heat generation within P be given as q_P, and the temperature of S be constrained to T_S. Assume frame F to be a perfect insulator, and ignore heat transfer to the atmosphere.

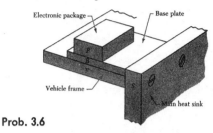

Prob. 3.6

3.7 Repeat Prob. 3.6, with four packages mounted on the base B (but not in contact with one another).

3.8 Repeat Prob. 3.6, with a small fraction of the heat flowing through frame F to the heat sink. Use a three-lump model (storage elements P, B, and S).

3.9 Consider a simple, very approximate model for the forced-air heating of a one-room building: Let the room be a single heat-storage element of capacity C (within which air convection effects uniform temperature distribution relatively rapidly). When the furnace is on, heat is introduced at a prescribed constant rate. Write the equation of motion for temperature T in the room, with temperature external to the room uniform and constant, and the thermal resistance of each of the walls and roof known. Assume the floor to be a perfect insulator.

3.10 Repeat Prob. 3.9 for a room that is 12 ft square by 8 ft high, where the heat transfer coefficient of the wall and ceiling material is α [(Btu/hour)/ft² °F].

3.11 A building consists of four cubical rooms in a row, as shown. They are heated by a single furnace which delivers equal quantities of heat to the four. The outside temperature T_o is constant, and the resistance of each external wall and of the roof over each room is R_e, while the resistance of each internal wall is R_i. For simplicity, consider the floor and ducts to be perfect insulators. Using the model of Prob. 3.9 for each room, derive a set of equations of motion from which the temperatures in the four rooms can be found, versus time, after the furnace is turned on. In the manner of Fig. 3.3a, sketch qualitatively the temperature distribution you expect at various times. Can symmetry be used to reduce the number of variables and the number of equations of motion?

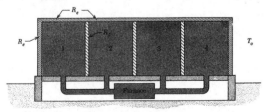

Prob. 3.11

3.12 Repeat Prob. 3.11 for a furnace that supplies heat to only one room, the right-hand inner one.

[Chap. 3] PROBLEMS   773

3.13 Heat q_r is generated at the rotor of the gyro in Fig. 3.1d, p. 95, and Fig. 19.13a, p. 625, by both electrical and mechanical resistance. Describe a concentric four-lump thermal model for the system and write the thermal equations of motion. Take, as the four lumps, (1) the rotor, (2) the floated gimbal (a can in which the rotor is mounted), (3) the liquid between float and case, (4) the case. External temperature is constant, and rotor heat during a 1-minute startup is constant at $6q_{ro}$, after which it has the operating value q_{ro}.

3.14 Derive a set of linearized equations of motion for the water supply system of Fig. 3.10, p. 115, by using the perturbation technique. Refer to the last paragraph of Example 3.2. Use variables as defined in Fig. 3.10, and neglect the inertial force due to acceleration of water in the pipe, so that (3.28) may be used. [Obtain R' in terms of w_{03} and $(h_{01} - h_{02})$.] Compare your results with Eqs. (3.34).

3.15 Repeat Prob. 3.14, with a small additional reservoir midway between the original two. The new reservoir is used for storage only, and has no flow in or out except from the pipe.

3.16 Derive exact equations of motion for the system shown, assuming the flow through the orifices obeys (3.17), p. 108. Assume dimensions and coefficients as needed.

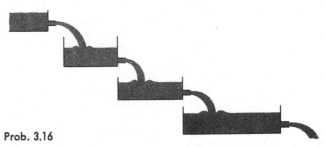

Prob. 3.16

3.17 A simple bellows element is affixed to an air line as shown. Derive the equations of motion from which one could obtain its response x to prescribed changes in p_1, with the air taken as compressible but uniform throughout chamber 2, and orifice flow obeying (3.17), p. 108.

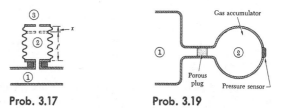

Prob. 3.17   Prob. 3.19

3.18 Repeat Prob. 3.17, using perturbation to obtain approximate linear equations of motion. Assume a steady-state pressure differential and bellows deflection x_0.

3.19 A segment of a fluid computer is shown. Its purpose is to register slow changes in the pressure p_1 in vessel 1 but to ignore rapid changes. The output of the device is p_2 (which is sensed by a diaphragm and capacitive pickup, as in Fig. 2.26d). Assume that pressure p_1 is prescribed (and is unaffected by the relatively small flow in and out of accumulator 2). That is, vessel 1 is a *pressure source*. Write the equation of motion for p_2.

3.20 Apply perturbation to Prob. 3.19 to obtain a linearized (approximate) equation of motion, valid for small changes in p_1 and p_2. Refer to Eqs. (3.12), (3.13), and (3.14), p. 106. From your result, write an expression for capacitance C_2.

3.21 A certain gas system can be represented as shown. With p_1 prescribed, versus time, derive equations of motion from which one can find x versus time. Assume that flow is laminar in the long lines, and that body forces can be ignored there, so that Eq. (3.21) is valid. Neglect resistance entering chambers 3 and 5.

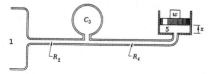

Prob. 3.21

3.22 Modify the results of Prob. 3.21 for the case in which flow in the lines is found to be turbulent.

3.23 Repeat the analysis of the hydraulic force-amplification system in Example 3.4, p. 118, with elastic compressibility of the oil in chambers 1 and 2 considered: $\rho_{01} + \rho_1$ and $\rho_{02} + \rho_2$, with $\rho_1 = p_1/K$, $\rho_2 = p_2/K$. Also, allow the connecting rod (from piston to engine) to be elastic, with stiffness k_3. Compare your results with those of Example 3.4 by allowing K and k_3 to become infinitely stiff.

3.24 In closed hydraulic systems, such as the one of which the actuator in Fig. 3.12 (p. 119) is a part, oil is recirculated by a constant-displacement pump. Thus the flow rate out of the pump is prescribed; i.e., the pump is a flow source. To make oil available on demand, an accumulator is provided. Consider the system shown. Assume that w_1 is constant and that p_5 is prescribed, versus time. Assume that the only resistance anywhere is through the valve, which has a constant opening. The bladder supports no pressure differential. Write the equation(s) of motion.

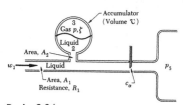

Prob. 3.24

3.25 Repeat Prob. 3.24, with the valve opening also varied in a prescribed way, versus time. Assume that c_α in Eq. (3.17) (p. 108) is simply proportional to valve opening.

CHAPTER 4: ANALOGIES

In Probs. 4.1 through 4.12 draw an electrical network which is an analog of each thermal system cited. Use Fig. 4.1 as an example. Write the equations of motion for

[Chap. 4] PROBLEMS   775

your network, and compare with those obtained earlier for the thermal system (in the problem cited). (If you have difficulty, work backward from the thermal equations to contrive a network which obeys them.)

4.1 The system of Prob. 3.1.   **4.7** The system of Prob. 3.7.
4.2 The system of Prob. 3.2.   **4.8** The system of Prob. 3.8.
4.3 The system of Prob. 3.3.   **4.9** The system of Prob. 3.9.
4.4 The system of Prob. 3.4.   **4.10** The system of Prob. 3.11.
4.5 The system of Prob. 3.5.   **4.11** The system of Prob. 3.12.
4.6 The system of Prob. 3.6.   **4.12** The system of Prob. 3.13.

In Probs. 4.13 through 4.20 draw an electrical network which is the analog of the fluid system cited. Write the equations of motion for your network, and compare with those obtained earlier for the fluid system (in the problem or example cited). If you have difficulty, work backward from the fluid-system equations to contrive a network which obeys them.

4.13 The system of Prob. 3.21.   **4.17** The system of Prob. 3.18.
4.14 The system of Fig. 3.10.   **4.18** The system of Prob. 3.20.
4.15 The system of Prob. 3.15.   **4.19** The system of Prob. 3.23.
4.16 The system of Example 3.3, p. 115.   **4.20** The system of Prob. 3.25.

4.21 Write the remaining equations of motion in Example 4.2, p. 129.

In Probs. 4.22 through 4.33 draw an electrical network which is an analog of the mechanical system cited. Write the equations of motion for your network, and compare with those obtained directly from the mechanical system. (Use KCL, except as noted.)

4.22 The torsional system of Fig. 2.2b, p. 36.
4.23 The Scotch Yoke system of Fig. 2.9, p. 52. (Use KVL.) Let ω be a prescribed constant.
4.24 The mass-spring system of Prob. 2.18, p. 71.
4.25 The vertical mass-spring system of Prob. 2.19.
4.26 The three-mass system of Prob. 2.34.
4.27 The seismic system of Prob. 2.35. (Use both KCL and KVL.)
4.28 The turbine model of Prob. 2.36.
4.29 The fluid-clutch model of Prob. 2.38.
4.30 The lever system of Prob. 2.2.
4.31 The system of Prob. 2.42. (Use both KVL and KCL.)
4.32 The system of Prob. 2.43. (Use KVL.)
4.33 The boat-drive system of Prob. 2.44. (Use KVL.)
4.34 Draw an all-electric analog of the electromechanical system of Fig. 2.24, p. 80. (This technique is commonly employed for electromechanical systems.) Write the equations of motion (using KVL for the drive system and KCL for the rotor), and compare with Eqs. (2.77).
4.35 Contrive an all-electrical analog to model the system of Fig. 2.23, p. 79. (Hint: A transformer will be useful.)

4.36 Develop the relations for the "force-voltage" mechanical-electrical analogy, and add them to Fig. 4.2, p. 126. Use them to construct another electrical analog to the mechanical system of Fig. 4.3a.

4.37 Use the relations of Prob. 4.36 to construct another electrical analog for the mechanical system of Fig. 4.4a.

4.38* Show that the network obtained in answer to Prob. 4.37 is the dual of the network in Fig. 4.4b. (Refer to Appendix G.)

4.39 Write "KCL" for node b in Fig. 4.7b, p. 138, and compare your result with Eq. (4.8).

In Probs. 4.40 through 4.49, draw a "fluid network" or a "mechanical network," as appropriate, using the symbols of Table V. Then write "KCL" or "KVL" as indicated (Procedure A-n).

4.40 The system of Prob. 3.21. Write KCL, and compare results with Prob. 4.13.

4.41 The system of Fig. 3.10, p. 115. Write KCL. [Refer to Prob. 3.14, and compare results with those of Prob. 4.14 and with Eqs. (3.34).]

4.42 The system of Prob. 3.20. Write KCL. (Compare with the results of Prob. 4.18.)

4.43 The torsional system of Fig. 2.2b, p. 36. Write KCL. (Compare with the results of Prob. 2.31.)

4.44 The Scotch Yoke system of Fig. 2.9, p. 52. Write KVL. Let ω be a prescribed constant. (Compare with the results of Prob. 4.23.)

4.45 The three-mass system of Prob. 2.34. Write KCL. (Compare with the results of Prob. 4.26.)

4.46 The seismic system of Prob. 2.35. Write both KCL and KVL. (Compare with the results of Probs. 4.27, 2.35, and 2.45.)

4.47 The fluid-clutch system of Prob. 2.38. Write KCL. (Compare with the results of Prob. 4.29.)

4.48 The system of Prob. 2.42. Write KVL. (Compare with the results of Prob. 2.42.)

4.49 The system of Prob. 2.43. Write KVL. (Compare with the results of Prob. 2.43.)

CHAPTER 5: EQUATIONS OF MOTION FOR MECHANICAL SYSTEMS IN TWO AND THREE DIMENSIONS

For each system in Probs. 5.1 through 5.15 (i) *sketch* a reference configuration and an arbitrary configuration, and (ii) define an independent set of *coordinates*. Indicate the positive direction for each coordinate. (iii) In each case, state the number of *degrees of freedom*. (iv) Write an equation to describe each geometric *constraint* that is not already implied precisely in (ii). (v) Where indicated, define at least one alternate set of coordinates and relate it to the set in (ii) by *identity* relations.

5.1 A hockey puck, flat on the ice. (Paint a line on it.) Include (v).

5.2 A steam shovel. (Simplify to just a few degrees of freedom.)

5.3 A child's swing, if the separate ropes remain straight.

5.4 A truck with swinging tailgate, if the truck does not turn.

5.5 A model for studying only the "pitch plane" motions of an automobile on its suspension system (figure). Include (v).

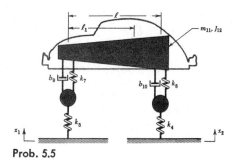

Prob. 5.5

Prob. 5.6

5.6 The mass center of a ladder which slips, but stays in contact with the wall (figure).
5.7 A cat running up the ladder in Prob. 5.6.
5.8 A "spherical" pendulum (figure).

Prob. 5.8   Prob. 5.9

5.9 A flyball governor (figure).
5.10 A yo-yo. Assume that the string remains taut and the plane of the yo-yo always stays vertical, but that otherwise it has complete freedom.
5.11* A rigid aircraft in maneuvering flight.
5.12* The tip of a propeller blade on the aircraft in Prob. 5.11.
5.13* A rigid satellite in earth orbit (not just a point mass).
5.14* A uniform vibrating string.
5.15* A chain, swinging pendulously.
5.16* Consider a coin which is constrained to roll without slipping on a flat surface. It may tilt, however. (i) Sketch the coin in an arbitrary configuration, and define sufficient coordinates to prescribe the configuration completely. Show that $c = 5$ coordinates are required. (ii) Write a purely geometric relation which must hold between coordinate rates of change because of the (nonholonomic) no-slipping constraint. Since there is $n = 1$ such nonholonomic relation, the number of degrees of freedom is $d = c - n = 5 - 1 = 4$.
5.17* Repeat Prob. 5.16, part (i) with the coin free to slip.
5.18* Repeat Prob. 5.16 for a billiard ball, on which you may paint convenient markings.
5.19* Repeat Prob. 5.18, part (i), with the billiard ball free to slip.
5.20* What are c, n, and d for a maneuvering automobile whose tires do not slip? Assume springs and tires infinitely stiff.

5.21 For small angles, a pendulum is known to swing in pure simple harmonic motion. In Fig. 2.2c, p. 36, for example, the angle θ is known to vary as $\theta = \theta_m \cos \omega t$. Find the horizontal and vertical components of the velocity and the acceleration of the bob.

5.22 Find the instantaneous velocity and acceleration of an arbitrary point on the rim of a wheel which is rolling in a straight line at constant speed.

5.23 An undisturbed projectile has constant horizontal velocity, $v_x = v_{x0}$, and its upward vertical velocity is $v_z = v_{z0} - gt$, where v_{x0} and v_{z0} are the components of the launch (muzzle) velocity, and $t = 0$ is launch time. Find an expression for the projectile's path by integrating the velocity expressions and eliminating t between them. Sketch a typical flight.

5.24 A car races at constant speed γ along (relative to) a track whose path can be described by $y = cx^3$. Find the x and y components of the car's acceleration at an arbitrary point x_1.

5.25 The bottom end of a ladder is pulled away from the wall at constant speed v_{20} (figure). Write expressions for the velocity and acceleration of the top end, assuming it stays in contact with the wall.

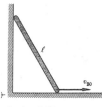

Prob. 5.25

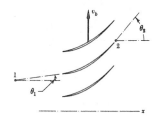

Prob. 5.26

5.26 The figure shows particles of air entering and leaving one moving stage of the axial-flow compressor in a jet engine. Flow is tangential to the blade at both 1 and 2. By continuity, v_{2x} must equal v_{1x}. What other velocity relation may be written?

5.27 The figure represents one type of grinding mill, consisting of a fixed sole, a vertical driving shaft, and two conical rollers A and B. The driving shaft turns at angular speed n, and there is no slippage between the rollers and the sole. Find the (total) angular velocity of roller A.

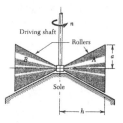

Prob. 5.27

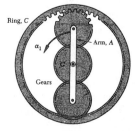

Prob. 5.28

[Chap. 5] PROBLEMS

5.28 In the figure, the arm A carries two gears which are free to rotate, and which mesh with an inner gear G and an outer ring C. The arm A and the ring C are free to rotate about the axis of gear G. The pitch diameter of the three gears is the same. At time $t = 0$, the arm A is given an angular acceleration of α_1 rad/sec^2, counterclockwise. Determine (a) the motion that gear G must be given so that ring C remains stationary; (b) the final motion of ring C if the motion from part (a) for gear G is not applied until $t = t_1$; and (c) the angle through which ring C has turned when $t = 2t_1$, with the conditions of part (b). (The intermediate gears are called *planetary* gears.)

5.29 Derive the velocity and acceleration of the pendulum bob in Fig. 5.3, p. 150, in terms of θ, using the xy reference frame only. Compare with the derivation of (5.17) and (5.18).

5.30 The horses on a merry-go-round move up and down approximately sinusoidally ($z = h \sin \omega t$). Meanwhile, the platform rotates at constant speed n. Find the total velocity and acceleration of a child on one of the horses.

5.31 Find the total velocity and acceleration of the father of the child in Prob. 5.30 if he walks radially outward on the platform at constant speed c relative to the platform.

5.32 Repeat Prob. 5.31, with the father walking at constant speed c around the rim of the merry-go-round.

5.33 Repeat Prob. 5.31, with the father walking at constant speed c along a chord of the platform.

5.34 Find the total velocity and acceleration of the end of the spherical pendulum shown in Prob. 5.8, if the base has prescribed angular velocity $\dot{\psi}(t)$, and the pendulum has prescribed motion $\phi(t)$ relative to the base. (Obtain your answer in terms of ψ and ϕ.)

5.35 The outboard motor has struck a rock. The boat continues with velocity v_o, and the propeller continues to spin at constant speed n, but the motor is caused to rotate about its pivot with angular velocity $\dot{\phi}$ and acceleration $\ddot{\phi}$. Write an expression for the absolute velocity and acceleration of a blade tip when the blade is (instantaneously) horizontal.

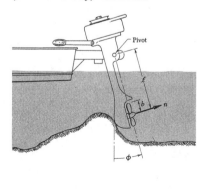

Prob. 5.35

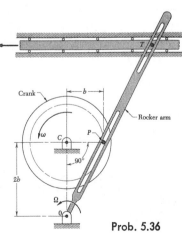

Prob. 5.36

5.36* The mechanism shown is used on shapers to return the cutting tool to position rapidly after each working stroke. As the crank C rotates at constant angular velocity ω, the crank pin P, which slides in a groove on the rocker arm, causes this arm to oscillate about its pivot O, driving the table T. For the dimensions shown in the schematic sketch, and *for the instantaneous position shown*, calculate the angular velocity $\dot{\Omega}$ of the rocker arm OPT. (This device is called a "quick-return mechanism.")

5.37 A cat runs up the ladder in Prob. 5.25 with constant speed u_0 relative to the ladder. Find the cat's total velocity and acceleration.

5.38 The gyro rotor shown is driven at constant angular speed n relative to its gimbal. The test table on which it sits is driven at constant speed N relative to the earth, about a vertical axis. The latitude is λ. The gyro gimbal angle $\theta(t)$ varies in some prescribed way. Find the total angular acceleration of the rotor. (Include earth rotation.)

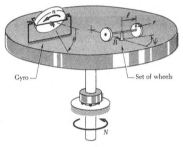

Probs. 5.38 and 5.39

5.39 If the wheels roll without slipping, find the angular velocity of one of them as a function of $\dot{\psi}$ and N. Neglect earth rotation.

5.40 The air-cushion vehicle of Fig. 5.5 is tethered as shown below, so that it cannot move. If $f_1 = 10$ lb, what does the scale read? (Neglect friction.)

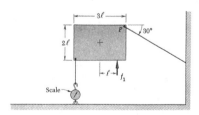

Prob. 5.40

5.41 A circular racing track of radius R is banked at angle β. For what car velocity will the driver be sitting perpendicular to the seat?

5.42 If the system shown is stationary (motor braked), find the force f_1 in the motor cable. (Use the free-body diagrams.)

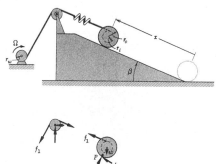

Prob. 5.42

5.43 A subway train has constant acceleration a. At what angle do the straps hang?

5.44 Determine precisely the angle between a plumb bob at 30° N latitude and the local direction to the center of the earth.

5.45* In what general direction must the father in Prob. 5.31 lean?

5.46 A manned space station 100 meters in diameter is shaped like a bicycle wheel. How fast should it spin to give an "artificial gravity" of $\tfrac{1}{4}g$? A man climbs a ladder connecting the rim to the hub. His speed relative to the ladder is 0.4 m/sec. What is the resultant (vector) force exerted on him by the ladder?

5.47 Write the equation of motion for the system shown below, assuming the cylinder does not slip.

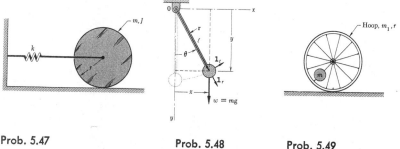

Prob. 5.47   Prob. 5.48   Prob. 5.49

5.48 Write the equation of motion for the simple pendulum shown. Assume viscous friction.

5.49 Write the equation of motion for the unbalanced bicycle wheel shown, assuming it does not slip. (Consider only motion in the plane of the page.) The wheel alone has mass m_1 and radius r. The unbalance mass is at a distance ℓ from the center.

5.50 Someone gives the subway strap in Prob. 5.43 a flick which sets it oscillating. Write the differential equation that governs the ensuing motion. (The moment of inertia of a rod of length ℓ about its mass center is $m\ell^2/12$.)

5.51 Suppose the motor in Prob. 5.42 is run at constant speed. Derive the equation of motion for the drum, assuming it does not slip. The spring represents flexibility in the cable. The free-body diagram shown below may be helpful.

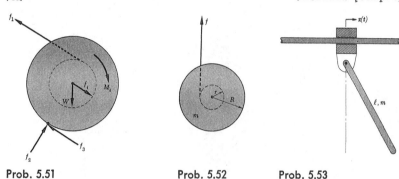

Prob. 5.51   Prob. 5.52   Prob. 5.53

5.52 What vertical motion must the top of a yo-yo string be given to make its center stay fixed in space? What will be the force in the string? (The polar moment of inertia of a cylinder about its mass center is $mR^2/2$. For simplicity, neglect the mass of the hub.)

5.53 The pendulum support is moved with a prescribed, time-varying displacement $x(t)$ (figure). Derive the equation of motion. Neglect friction. (Note: $J_c = m\ell^2/12$ about the mass center.)

5.54 Derive the equation of motion for the flyball governor of Prob. 5.9, with $\Omega(t)$ prescribed and time-varying. Assume viscous sliding friction, and ignore hinge friction.

5.55 The table (figure) turns at constant speed. Write the equation of motion for the point mass, (a) with $\ell = 0$, and (b) with $\ell \neq 0$. Assume dry friction.

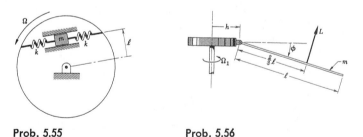

Prob. 5.55   Prob. 5.56

5.56 In the highly simplified model of a helicopter rotor shown (figure), the hub rotates at constant speed Ω_1. Derive the equation of motion for one blade. The resultant lift force L passes through the point $\frac{2}{3}\ell$, as shown, and is proportional to Ω_1^2 and to $\alpha_0 + \dot\phi$, where α_0 is a constant. (For a blade, $J_c = m\ell^2/12$ about the mass center.) Let $\phi = 0$ be the position of static equilibrium at speed Ω_1.

5.57 As an exercise in force equilibrium, write the equations of *static* equilibrium for the double pendulum constrained by a string (figure). Determine the static angles θ_1 and θ_2, with $x_2 = \ell$ given (determined by the length of the string).

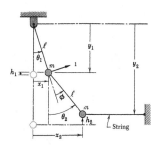

Probs. 5.57 and 5.58

5.58 Use D'Alembert's method to write the equations of motion of the double pendulum after the string has been cut. Neglect friction. (Note: Internal forces can be completely avoided by astute use of free-body diagrams.)

In Probs. 5.59 through 5.63, derive a complete set of equations of motion for each system of rigid bodies shown.

5.59 Simplified model of an automobile. (Pitch plane only.)

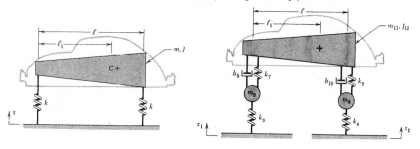

Prob. 5.59   **Prob. 5.60**

5.60 More realistic model of automobile, including tires. (Again, pitch plane only. Specialize your result for comparison with Prob. 5.59.)

5.61 Block on sliding wedge. (Viscous friction.)

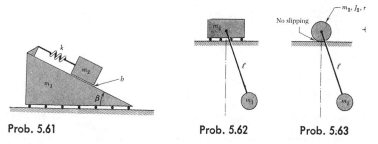

Prob. 5.61   **Prob. 5.62**   **Prob. 5.63**

5.62 Pendulum and sliding block. (Let $m_2 \to \infty$ for comparison with Prob. 5.48.)
5.63 Pendulum and rolling cylinder.

5.64 For the system shown, write the equations of motion for small motions only in the (vertical) plane of the page. (Note that the center of the right-hand bar is free to move. For a bar, $J_c = m\ell^2/12$ *about the mass center.*)

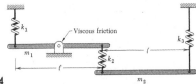

Prob. 5.64

5.65 For the system shown, write the equations of motion, assuming no slippage.

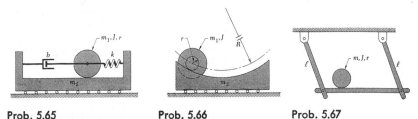

Prob. 5.65   **Prob. 5.66**   **Prob. 5.67**

5.66 For the system shown, write the equations of motion, assuming no slippage.
5.67 For the system shown, write the equations of motion, assuming no slippage. Assume viscous bearing friction everywhere. Assume the bars are massless.
5.68 Write the equations of motion for the system shown, assuming there is no friction.

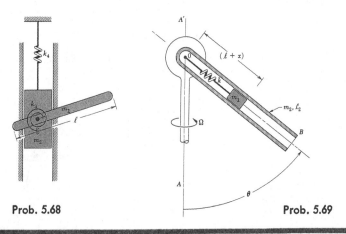

Prob. 5.68   **Prob. 5.69**

5.69 The frictionless system shown consists of a fixed vertical shaft AA' having negligible mass and rotating around AA' at constant angular velocity Ω. Tube OB is pivoted at O around an axle perpendicular to AA' and OB. The particle of mass m_1 can move along OB only. The spring has unstressed length ℓ. (*a*) Derive the equa-

[Chap. 5] PROBLEMS 785

tions of motion. (b) Find the equations from which steady-state value of θ could be found.

5.70 Consider again the model of a helicopter rotor in Prob. 5.56. Assume now, however, that Ω_1 is a variable, and that the engine delivers constant torque M_0 to the shaft. Assume also that the resultant drag force on each of the two blades is $D = D_0 + K\dot{\phi}^2\Omega^2$ located also at $\tfrac{2}{3}\ell$. Assuming that the hub moment of inertia is J_h, write the equations of motion of the system (hub plus two blades).

In Probs. 5.71 through 5.75 obtain the equation of motion by simply equating the maximum kinetic energy (when potential energy is minimum) to the maximum change in potential energy (using the zero-kinetic-energy point), for the conservative system indicated.

5.71 The simple pendulum of Prob. 5.48. Neglect friction.
5.72 The rolling-cylinder system of Prob. 5.47.
5.73 Mercury in a U tube (figure). Neglect friction.

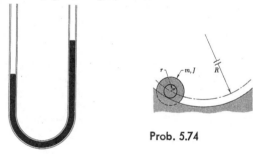

Prob. 5.73   Prob. 5.74

5.74 The system shown (pure rolling without energy loss).
5.75 The flyball governor of Prob. 5.9. Assume no friction, constant shaft speed Ω, equilibrium angle θ_o.

In Probs. 5.76 through 5.99 use the method of Lagrange to derive a complete set of equations of motion. Compare your results with those obtained previously by using Newton-D'Alembert methods.

5.76 The simple pendulum of Prob. 5.48.
5.77 The rolling-cylinder system of Prob. 5.47.
5.78 The two-mass system of Fig. 5.10, p. 166. (Compare your analysis with that attempted on p. 166.)
5.79 The rotary-mass system of Fig. 2.2b, p. 36, if the angle θ_9 is held to zero.
5.80 The flyball governor of Prob. 5.9. Assume no friction, constant shaft speed Ω.
5.81 The system of Prob. 5.55. Neglect friction.
5.82 The very approximate model of an automobile in Prob. 5.59.
5.83 The more realistic model of an automobile in Prob. 5.60. Neglect damping.
5.84 The cylinder-and-sliding-wedge system of Example 5.6, p. 157.
5.85 The moving-pivot pendulum of Prob. 5.62.
5.86 The double pendulum of Prob. 5.58. (No friction.)
5.87 The system of Prob. 5.64, with $b = 0$.

5.88 The system of Prob. 5.66.
5.89 The gyro system of Fig. 5.8, p. 160. Derive for arbitrarily large angles. Compare your result with Eqs. (5.61), (5.62). Then make small-angle approximations, and compare with Eqs. (5.63), (5.64).

5.90 The system of Fig. 2.2a, p. 36.
5.91 The system of Fig. 2.2b, with $\theta_9(t)$ prescribed.
5.92 The simple pendulum of Fig. 2.2c, with friction included. (Compare with Prob. 5.48.)
5.93 The propeller-drive system of Prob. 2.41.
5.94 The seismic instrument system of Prob. 2.35.
5.95 The fluid clutch of Prob. 2.38.
5.96 The sliding pendulum of Prob. 5.53, with $x(t)$ prescribed.
5.97 The pendulum system of Prob. 5.63.
5.98 The automobile model of Prob. 5.60, with $x_1(t)$ and $x_2(t)$ prescribed.
5.99 The air-cushion vehicle of Example 5.4, p. 154.

CHAPTER 6: FIRST-ORDER SYSTEMS

6.1 Write the complete equation of motion for the torque motor in Fig. 6.2a, p. 185, assuming that the moment of inertia, J, is *not* negligible. Then obtain Eq. (6.3) by letting J go to zero. (Refer to Example 2.1, p. 46.)

6.2 Make a time plot, for $0 < t$, of the following functions, and label on the plot the value at $t = 0$, the final value, and the time constant.

(a) $5e^{-3t}$ (b) $-3e^{-t/2}$ (c) $4(1 - e^{-t/4})$

6.3 How many time constants elapse before the free motion of a first-order system decays to 10% of its original displacement? To 5%? 1%? 0.001%? How far has it decayed in 10 time constants?

6.4 What time function is plotted below?

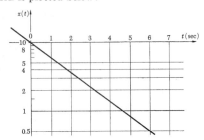

Prob. 6.4

6.5 The circuit in Fig. 6.1c, p. 184, is initially connected to the battery for a long time. At $t = 0$ the switch is suddenly switched to ground. Derive the equation of motion for $0 < t$. (Procedure A-e may be helpful.) Then determine and plot the time history of voltage v_3 for $0 < t$. (Hint: Find the magnitude of the current through the inductor just prior to switching, and note that the current through the inductor does not change instantaneously.)

[Chap. 6] PROBLEMS 787

6.6 In Fig. 6.1g, a pebble has been dropped into a deep bucket (or a depth charge into the ocean). After some time the pebble reaches a terminal velocity at which hydrodynamic drag just balances the force of gravity. Assume the pebble is dropped from an initial height h above the water, and make a plot of its velocity versus time in terms of the necessary physical constants. Assume the drag force on the pebble to be linearly proportional to velocity. [This is tantamount to assuming laminar flow around the pebble, as we did for Eq. (3.20).]

6.7 In Fig. 6.1e, a chunk of copper having thermal capacitance C is heated to an initial temperature T_0 and then immersed suddenly in a very large vat of water at lower temperature T_1. Obtain the equation of motion and the characteristic equation. Find the temperature of the copper as a function of time. Assume that the temperature of the water remains always at T_1 and that the temperature throughout the copper is uniform at any given time. [The equation of motion is readily obtained by using (3.1) and (3.2), as in Prob. 3.1. Let the copper-water interface have total thermal resistance R.]

6.8 It would be convenient to assume that liquid flows—because of gravity—out of the bottom of a vessel at a rate proportional to the height, h, of the remaining liquid. Suppose this to be true, and make a plot of h versus time.

6.9 Fig. 6.1f, p. 184, shows a simple hydraulic feedback system for positioning the angle of an aerodynamic control surface. When the pilot moves his control stick, he displaces point P on the linkage through a distance x. Since the bottom of the linkage is initially held fixed by the control surface, the top end of the linkage must move through a distance y, and the hydraulic valve is opened by this amount. This allows oil to flow from the high-pressure inlet into the right side of the piston. The piston then moves to the left, forcing low-pressure oil out the exhaust port. In so doing, the piston rotates the control surface through an angle θ. As the control surface rotates, it moves the linkage about point P as a pivot, thus closing the valve. The system finally comes to equilibrium with the valve closed and the control surface at some new angle proportional to x. The hydraulic valve and piston constitute a nearly perfect integrator (if the oil can be considered incompressible, and if there is no leakage), as discussed in Example 3.4, p. 118. That is, the piston will move at a rate $\dot{z}$ which is proportional to valve opening y: $\dot{z} = K_1 y$ [refer to Eq. (3.50)]. Write the equation of motion of the system in Fig. 6.1f—θ as a function of x and t—in terms of K_1 and the linkage lengths $\ell_1, \ell_2, \ell_3, \ell_4$. Obtain the system's characteristic equation. Plot the response $\theta(t)$, with x always zero, the valve initially open an amount y_0, and the control surface initially in the corresponding position θ_0. (Assume $y = 0$ to be the valve neutral position, and let $\theta = 0$ when $x = 0$ and $y = 0$.) θ always small.

6.10 Repeat Prob. 6.9, with link SQ extended and pivot point P (fixed for these studies) located a distance ℓ_5 *below* point Q (instead of between Q and the valve as shown in Fig. 6.1f). Explain your result thoroughly on both mathematical and physical grounds.

6.11 The system shown represents a rudimentary feedback amplifier arrangement. The section between the amplifiers is an ideal electrical integrator. Before $t = 0$, the switch is open, and the voltage due to charge on C_2 is γ. The switch is closed at $t = 0$. Plot and label appropriately the subsequent time history of v_a (i) if K_1 is positive and K_3 is negative, (ii) if K_1 and K_3 are both positive. Explain your results on both mathematical and physical grounds. (The simple properties of ideal amplifiers are given in Sec. 2.8.)

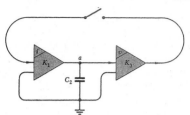

Prob. 6.11

6.12 Obtain the characteristic equation for the pressure-rate sensor of Example 3.3, p. 115. If the bellows is displaced by x_0 and released, and line pressure p_1 is constant, find and sketch the subsequent motion $x(t)$. Label your sketch accurately. [Hint: Eliminate p_2 between the two equations of motion, (3.43a) and (3.43b).]

6.13ᵉ Place a good thermometer in a liquid bath at other than room temperature, and plot the thermometer reading versus time. Repeat for several bath temperatures. Compare this arrangement and results with those of Prob. 6.7.

6.14ᵉ Construct the circuits of Figs. 6.1b and c from electrical engineering or physics laboratory components, and check their time response, using voltmeter or ammeter and stopwatch. (Choose components to give time constants of several seconds, if possible.)

6.15ᵉ Fill a tall graduate with a clear, highly viscous fluid. Drop a small spherical object (ball bearing, BB, marble, or the like) into the graduate from a measured height. Mark the object's location each second, using a ticking clock as a time base (or verbal "marks" from an assistant with a sweep-second-hand watch). Compare your measured results with those predicted by Prob. 6.6. Explain any discrepancies.

6.16 The equation of motion of a system is $\dot{y} + y = x$. Find $y(t)$, with $x(t) = e^{-2t}u(t)$ and $y(0) = 2$. Make an accurate plot of $x(t)$ and $y(t)$ from $t = -1$ to $t = 3$.

6.17 If $v_1(t) = v_{1m}e^{-\sigma_f t}u(t)$ for the network shown, find only the forced portion of the response $v_3(t)$.

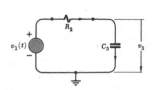

Prob. 6.17

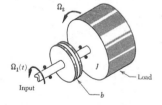

Prob. 6.18

6.18 A fluid clutch connects a load (e.g., the wheels of an automobile) to an engine. Assume that the torque transmitted through the clutch, when it is engaged, is proportional to the difference in speed of the input and output shafts (proportionality constant b). Assume also that the speed Ω_1 of the input shaft is constant, $\Omega_1 = \Omega_{10}$, unaffected by the load. If the load is a pure inertia J, find the time history of angular velocity $\Omega_2(t)$ of the output shaft after the clutch is suddenly engaged. Suppose the load to be stationary before engagement.

ᵉ Experimental problem.

[Chap. 6] PROBLEMS

6.19 A box sits on a conveyor belt. There is *viscous* friction between the box and the belt. The belt and box are initially moving along together at constant velocity $v_0 = r\Omega_0$. Suddenly the belt motor is turned off and the belt slows down exponentially: $\Omega = \Omega_0 u(t) e^{-\sigma_f t}$. Obtain a complete expression for the resulting displacement of the box with respect to the floor. Sketch displacement versus time for the box and for a point on the belt, with $\sigma_f = 0.2b/m$.

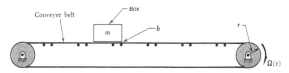

Prob. 6.19

6.20 The pilot effects a step input x (with his stick) into the hydraulic "boost" system of Fig. 6.1f. Find the resulting motion θ of the elevator. (Refer to Prob. 6.9.) Assume that before the input the system is stationary, with $y = 0$, $\theta = 0$.

6.21 The Mariner spacecraft which made the first Venus fly-by mission performed magnetometer measurements to determine whether a "cosmic wind" exists in space. For isolation, the magnetometer was extended from the spacecraft at the end of a long telescoping boom, modeled in the figure. The boom was activated by an explosive squib which filled the chamber with high-pressure gas and thus forced the magnetometer outward. (i) Discuss the assumptions under which the system may be represented by the equation of motion $m\ddot{x} + b\dot{x} = pA$. (ii) With $p(t)$ given by $p = p_0 e^{+7t} u(t)$, and $x(0) = 0$, $\dot{x}(0) = 0$, find and plot the total motion of the magnetometer. Let $b/m = 2$ sec^{-1}.

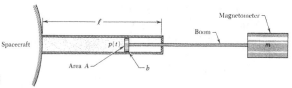

Prob. 6.21

6.22 Find and plot the total response $v_3(t)$ of the network in Prob. 6.17, for an input of (i) $v_1(t) = Au(t)$, and (ii) $v_1(t) = Be^{-\sigma_f t} u(t)$, with $\sigma_f = 1/(5R_2C_3)$. In both cases, assume $v_3 = 0$ at $t = 0$.

6.23 Use the results of Prob. 6.22 to sketch immediately the response $v_3(t)$ of the network in Prob. 6.17 to the input $v_1(t) = [4 + 3e^{-5t}]u(t)$ volts, if $v_3(0) = 0$. Let $R_2C_3 = 0.04$ sec.

6.24 Use the results of Prob. 6.22 to write down the total response $v_3(t)$ of the network to the input $v_1(t) = [Ae^{-\sigma_{f1} t} + Be^{-\sigma_{f2} t}]u(t)$, if A and B are constants, $\sigma_{f1} \neq \sigma_{f2} \neq 1/R_2C_3$, and $v_3(0) = 0$.

6.25 A container of liquid at room temperature T_r is immersed in a large vat of another liquid at a higher temperature T_v. A short time t_2 later, a thermometer whose time constant is τ_1 is inserted into the container. Plot the reading of the thermometer versus time. Assume that all of the liquid in the container warms up uniformly, that the container does not affect the temperature of the vat, and that the thermometer does not affect the temperature of the container. The initial reading on the thermometer is $T(0) = T_r$. [The basic thermal relations are (3.1) and (3.2).]

(Hint: First find the time history of the temperature T_c in the container. Then use this as the input to the thermometer. The container has thermal capacity C.)

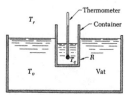

Prob. 6.25

6.26 The systems of Prob. 6.17 and Fig. 6.1a are connected in tandem, as indicated. [The simple model of a motor driven by a current source is discussed at the end of Sec. 2.9: Fig. 2.25 and Eqs. (2.78).] A tachometer, represented by the sleeve, and an ideal voltage amplifier, are used to convert motion Ω to an electrical signal. Let the tachometer scale factor be 1 volt/(rad/sec), and the amplifier gain be 1 volt/volt. For $v_1(t) = v_{1m}u(t)$, find $v_a(t)$. Then, using this as an input to the current amplifier, find $\Omega(t)$ and thence $v_8(t)$. Let $v_3 = 0$ and $\Omega = 0$ at $t = 0$.

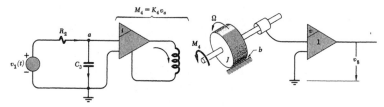

Prob. 6.26

6.27 The subsystems of Prob. 6.26 are interchanged. For $v_1(t) = v_{1m}u(t)$, find $\Omega(t)$. Then, using this as an input to the voltage amplifier, find $v_3(t)$. Show that the system output is identical for Probs. 6.26 and 6.27, demonstrating an important property of linear systems stated in Sec. 6.4.

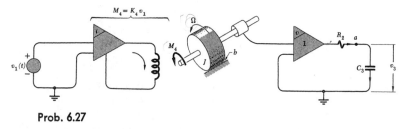

Prob. 6.27

6.28 Repeat Prob. 6.22, using γ to denote the voltage across C_3 before $t = 0$. That is, $v_3(0^-) = \gamma$. [Use Table VI to determine $v_3(0^+)$.]

6.29 Repeat Prob. 6.18 for $\Omega_2(0^-) = 3$ rad/sec. [Use Table VI to determine $\Omega_2(0^+)$.]

6.30 Repeat Prob. 6.18 for $\Omega_2(0^-) = 3$ rad/sec and $\dot{\Omega}_2(0^-) = 2$ rad/sec^2.

6.31 Repeat Prob. 6.20 for $y(0^-) = 0.01$ in. What must $\dot{\theta}(0^-)$ be? Why?

6.32 Repeat Prob. 6.26 for $v_3(0^-) = \gamma$ and $\Omega(0^-) = \rho$.

6.33 Repeat Prob. 6.27 for $v_3(0^-) = \gamma$ and $\Omega(0^-) = \rho$. Compare with Prob. 6.32.

[Chap. 6] PROBLEMS   791

6.34 Consider a pure rotary inertia J, as in Fig. 6.1a, p. 184, on which friction may be neglected. If a moment $M = M_{max}e^{-\sigma_f t}u(t)$ is applied to the rotor, plot is subsequent angular velocity versus time (i) if it was initially at rest, (ii) if it had initial angular velocity $\Omega = 3$ rad/sec. Refer to Eq. (6.19).

6.35 Find $v_2(t)$ in Fig. 6.15, p. 205, with $v_1(t) = v_{1m}(1 - e^{-\sigma_f t})u(t)$, and with an initial voltage drop across C_2 of $v_2(0^-) = \gamma$. Refer to Eq. (6.21). Plot both $v_1(t)$ and $v_2(t)$.

6.36 If the applied moment in Prob. 6.34 is as shown below (left), plot Ω versus t and label appropriately. Assume $\Omega(0^-) = 0$.

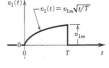

Prob. 6.36   Prob. 6.37

6.37 If the input in Prob. 6.35 is as shown above (right), plot and label v_2 versus t. Let $v_2(0^-) = 0.3v_{1m}$.

6.38 Let the link QR be removed in Fig. 6.1f, and point Q be fixed. Show that the motion $\theta(t)$ will then be simply the integral of input displacement $x(t)$, and find the pertinent physical constants. For $x(t) = x_0 e^{-\sigma_f t}u(t)$, find and plot $\theta(t)$ with labels. Let $\theta(0^-) = 0.01$ rad.

6.39 Repeat part (ii) of Prob. 6.22, with $\sigma_f = 1/R_2 C_3$.

6.40 Find the response of the fluid clutch of Prob. 6.18, with $\Omega_1 = \Omega_{1m}e^{-\sigma_f t}u(t)$ and $\sigma_f = b/J$. Assume $\Omega_2 = 0$ prior to $t = 0$. Repeat with $\Omega_2(0^-) = 0.3\Omega_{1m}$.

6.41 Repeat Prob. 6.25 for the special case $\tau_1 = RC$.

6.42 Repeat Prob. 6.26 for the special case $R_2 C_3 = J/b$.

6.43 Repeat part (i) of Prob. 6.34 for an applied moment of $M = M_m u(t)$. Show that your result is in accord with Eq. (6.28) and Fig. 6.18.

6.44 Show that your solution to Prob. 6.35 is in accord with Eq. (6.28) and Fig. 6.18.

6.45 Use Eq. (6.28) to plot the response of a first-order system $\dot{y} + 2y = \nu(t)$ in. to a near-resonant abrupt input $\nu(t) = 3\ u(t)e^{-1.90t}$ in./sec. Plot also the resonant approximation, i.e., the first term only in (6.28). Finally, plot the difference.

6.46 An ocean liner weighs 64,400 tons. At very low velocity, its resistance to motion is approximately proportional to velocity, the constant being 2 tons/(ft/sec). The liner is initially at rest in the harbor. A tugboat pushes on its stern with a force of 3 tons for 10 sec, and then backs away. Find the resulting motion (i) by assuming the input to be an impulse, and (ii) exactly. Compare the two results. For what initial conditions would the time response be the same as for (i)?

6.47 Repeat Example 6.6, p. 215 (telegraph circuit) with R_3 included. Compare your result with that of Example 6.6.

In Probs. 6.48 through 6.58 find the response of the designated system to the impulse input specified.

6.48 The rotor in Fig. 6.1a, p. 184, for $M(t) = M_0 \tau \delta(t)$, where τ is a constant. Repeat with damping $b = 0$.

6.49 The network of Prob. 6.17, for $v_1(t) = v_{1m}\tau\delta(t)$.

6.50 The small copper object in Fig. 6.1e, p. 184, if it is dipped into the tank for an extremely short time τ, but the tank temperature T_1 is very high.

6.51 The hydraulic servo of Fig. 6.1f if the input motion is a "flick," $x = x_0\tau\delta(t)$.

6.52 Repeat Prob. 6.51, with link QR removed and point Q fixed.

6.53 Repeat Prob. 6.51, with link SQ extended and point P located a distance ℓ_5 *below* point Q. (Refer to Prob. 6.10.) Let $y(0^-) = 0$ and $\theta(0^-) = 0$.

6.54 Find the response $\Omega_2(t)$ of the clutch system of Prob. 6.18 if Ω_1 has a large constant value Ω_{10}, $\Omega_2(0^-) = 0$, and the clutch is engaged only momentarily (nervous driver).

6.55 Find the response of the telegraph circuit of Example 6.6 if the input is a series of dots spaced 0.1 sec apart.

6.56 Find the response $v_8(t)$ of the tandem electromechanical system of Prob. 6.26 if $v_1(t) = v_{1m}\tau\delta(t)$, where τ is very short compared to either R_2C_3 or J/b. [First find $v_a(t)$; then use that as the input to the rotor drive.]

6.57 Repeat Prob. 6.56 with the subsystems interchanged, as in Prob. 6.27.

6.58 Find the response of the electrical integrator of Fig. 6.15, p. 205, to an impulsive input, $v_1(t) = v_{1m}\tau\delta(t)$.

6.59 Find the response $\theta(t)$ in Example 6.8, p. 218, by method (ii) of p. 220. That is, use (6.43) to obtain $b_2\ddot{x}(t)$ in (6.45) for $0^- < t$. (The result has two terms.) Substitute this for the right-hand side of (6.46). Then solve for the forced motion. Finally, choose the coefficient of the natural motion to match the initial condition at $t = 0^-$ (rather than at $t = 0^+$). Compare your total solution with (6.50).

6.60 For the circuit shown, find the response $v_3(t)$ if voltage $v_1(t)$ varies with time as plotted. There is initially no charge on the capacitor.

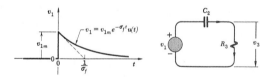

Prob. 6.60

6.61 Repeat Prob. 6.60 if the capacitor has initial voltage drop $v_2(0^-) = \gamma$ across it. Explain your mathematical result carefully on physical grounds. (Note that it is the voltage *drop across* the capacitor that cannot be changed abruptly.)

6.62 Find the response $x(t)$ of the pressure-rate sensor of Example 3.3, p. 115, to a step change in line pressure. Refer to Prob. 6.12.

In Probs. 6.63 through 6.72 first find the unit-impulse response of the system. Then use convolution to find the response requested.

6.63 The response of the torque motor model of Fig. 6.2, p. 185, to the step input of Fig. 6.7, p. 193. Compare your result with Eq. (6.14). [The unit-impulse response $h(t - \tau)$ for this system is given by Eq. (6.36).]

6.64 Repeat Prob. 6.63, with the step terminated after one time constant.

[Chap. 7] PROBLEMS 793

6.65 Repeat Prob. 6.63, with the input given by Eq. (6.15). Compare your result with Eq. (6.17).

6.66 The response of the network in Fig. 6.15, p. 205, to the input given below. Justify your answer by physical reasoning.

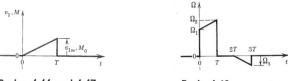

Probs. 6.66 and 6.67   Prob. 6.68

6.67 The torque-motor model system of Prob. 6.63 to the above input. (Show that if $R_2 = 0$, the result has the same form as for Prob. 6.66.)

6.68 The box on the conveyor belt in Prob. 6.19, with the belt stopping and starting as shown.

6.69 The hydraulic servo system of Fig. 6.1f, p. 184, to a step input. (Refer to Prob. 6.20.)

6.70 The tandem system of Prob. 6.26 to the triangular input given for Prob. 6.66.

6.71 Continue the convolution analysis of Example 6.10, p. 223, to obtain an expression for the response during time interval $0 < t < T$ and during time interval $2T < t$.

6.72 With the ocean liner of Prob. 6.46 again at rest, the tugboat comes back and applies force in the following program. Find the motion of the ship by convolution.

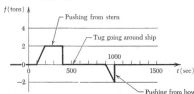

Prob. 6.72

CHAPTER 7: UNDAMPED SECOND-ORDER SYSTEMS: FREE VIBRATIONS

For Probs. 7.1 through 7.19, find the natural frequency of *small* oscillations. It is recommended that both the direct method of writing the equation of motion and the energy method be used in several of the problems, to gain familiarity with both.

7.1 The mass in Fig. 7.2a, p. 230, stretches the spring an amount δ when it hangs motionless. What is the natural frequency (in terms of δ) when the mass vibrates up and down on the spring? (The equation of motion may be written with the variable x representing either displacement from the static position or displacement from the point where the spring is unstretched. Discuss.)

7.2 In Fig. 7.2b the middle rigid member has negligible mass.

7.3 In Fig. 7.2c the rotating member is a single solid piece which rotates about a fixed axis.

7.4 The piston in Fig. 7.2d rides up and down on the air column trapped below it in the cylinder. There is no leakage around the piston. (Assume that the air is a perfect gas, and that no heat transfer takes place.)

7.5 The cylindrical buoy in Fig. 7.2e bobs up and down in the water. Assume that it does not tilt. (Neglect acceleration of the water mass.)

7.6 Find the frequency of vibration of mercury in the U tube in Fig. 7.2f. (The energy method may be easier for this problem.)

7.7 The switch in Fig. 7.2g is initially connected to the battery for a long time until there is no current flow. (The battery has some internal resistance.) Then the switch is connected to ground. The current oscillates back and forth in the inductor and capacitor.

7.8 Figure 7.2h is a two-stage hydraulic servo system. The position of the "pilot piston" is controlled by moving pivot point P (which requires negligible force). When the pilot valve is displaced an amount x, the slave valve moves at a proportional rate: $\dot{y} = -K_1 x$. Similarly, displacement of the slave valve causes the piston to move at a proportional rate: $\dot{z} = -K_2 y$. Show that if point P is held fixed, with the pilot valve initially open, the system will have an undamped oscillation. Find the frequency of oscillation in terms of K_1, K_2, and the linkage ratio ℓ_1/ℓ_2. Discuss the energy situation in the system. Let θ be small so that $\sin \theta \approx \theta$.

7.9 In Fig. 7.2i a uniform plank lies across two rotating cylinders. The coefficient of *dry* friction between the plank and a cylinder is β ($f_{\text{friction}} = \beta f_{\text{normal}}$). If the cylinders rotate toward one another at high constant speed Ω, the plank is seen to oscillate back and forth. Find the frequency of oscillation in terms of whatever parameters are required. What happens if both wheels are rotated in the directions opposite those shown?

7.10 The upside-down pendulum in Fig. 7.2k is held in place by two springs, as shown. What is the frequency of small oscillations? For what values of k will the pendulum fall over?

7.11 The pendulum in Fig. 7.2l consists of a rod with a disc of diametral moment of inertia J welded onto one end.

7.12 In Fig. 7.2m a stiff horizontal rod of negligible mass carries a concentrated weight at its end. The rod is pivoted freely at the other end and is supported by a string as shown.

7.13 In Fig. 7.2n a concentrated weight is supported at the end of a cantilever beam of stiffness EI. Find the natural frequency in the position shown, and also when the cantilever is vertically up and when it is vertically down.

7.14 In Fig. 7.2o the central rod spins on its axis and the mass is held out by centrifugal force at some equilibrium angle. If the mass is disturbed slightly from this equilibrium, it will oscillate up and down.

7.15 In Fig. 7.2q the cylinders roll without slipping. The two rods pivot freely on the cylinders.

7.16 The half cylinder in Fig. 7.2r rocks back and forth without slipping.

7.17 In Fig. 7.2s the railroad wheel rotates without slipping on the track, which has a radius of curvature R in the *vertical* plane.

7.18 The wheel axle in Fig. 7.2t makes an angle β with respect to the vertical. The assembly is made pendulous by the extra concentrated mass m (neglect bearing friction).

7.19* The table in Fig. 7.2u has a hollow ring welded to it. The internal radius of the ring is c, and the farthest point on its inside surface is a distance b from the axis of the table. A small solid disk of radius r rolls without slipping inside the ring, but has negligible friction with respect to the table. The table rotates at constant speed Ω. Find the frequency of oscillation of the small disk.

[Chap. 8] PROBLEMS

In Probs. 7.20 through 7.24 evaluate the quantities indicated, obtaining an answer in terms of a magnitude and an angle. (Graphical techniques may be used wherever desired.) All angle quantities are in radians.

7.20 $10e^{j1} + 5e^{j0.5}$

7.21 $\dfrac{(1+j2)(2-j3)}{(2+j1)(1+j3)}$

7.22 $\text{Re}\left[\dfrac{1+j7}{2-j3} + \dfrac{1-j4}{5+j2}\right]$

7.23 $\exp\left(2 + j\dfrac{\pi}{4}\right) + \exp\left(5 - j\dfrac{\pi}{2}\right)$

7.24 $\sqrt{j}$ (two roots)

7.25 $e^{j\pi}$

7.26 $\sqrt[5]{-1}$ (five roots)

7.27 $\ln(-1)$

7.28 $(1+j3)(2-j)(1+j1)(3-j2)$ (Use two methods, direct algebra and polar plot.)

7.29 Show that $\dfrac{d}{dt}\text{Re}\,[\vec{A}] = \text{Re}\left[\dfrac{d\vec{A}}{dt}\right]$, where $\vec{A}$ is a time-varying vector.

7.30 Verify Eq. (7.24a) by showing that $\sin\phi = \text{Im}\,[e^{j\phi}] = \dfrac{e^{j\phi} - e^{-j\phi}}{2j}$

7.31 Use Eqs. (7.24a) and (7.24b) to verify the identities
(a) $\sin\theta\sin\phi = \tfrac{1}{2}[\sin(\theta+\phi) + \sin(\theta-\phi)]$
(b) $\cos 2\phi = 1 - 2\sin^2\phi$

7.32 Use Eq. (7.21) to show that the right-hand sides of (7.27a) and (7.27b) are identical.

CHAPTER 8: DAMPED SECOND-ORDER SYSTEMS

8.1 Plot in the s plane the roots of Eq. (8.3) that are tabulated in Table VII. Label each point with the corresponding values of b from Example 8.1. For what value of b is the system critically damped, when J and k have the values of Example 8.1?

8.2 Obtain the roots of Eq. (8.3) and plot and label them in the s plane for $J = 3$ (in. lb)/(rad/sec^2), $b = 24$ (in. lb)/(rad/sec), and for k having the following values: (a) 0, (b) 10, (c) 100, (d) 200, (e) 500, (f) -100 (in. lb/rad). From your plot, estimate the value of k for which the system is critically damped. Estimate the value for which $= 0.4$. (Note the trigonometry of Fig. 8.2b). Discuss the physical significance of the roots you obtain for case (a).

8.3 For the network shown, with the switch at a, derive the equation of motion for $i(t)$ and obtain the characteristic equation. How should σ and ω_0 be defined to correspond to Eq. (8.1)?

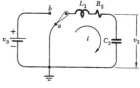

Prob. 8.3

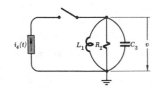

Prob. 8.4

8.4 For the network shown, with the switch open, derive the equation of motion for $v(t)$ and obtain the characteristic equation. How should σ and ω_0 be defined to correspond to Eq. (8.1)?

8.5 Plot the locus of roots of the characteristic equation for Prob. 8.3 as R_2 is varied from 0 to ∞. Let $L_1 = 0.02$ henry, $C_3 = 10^{-5}$ farad. What is the physical meaning of your result for $R_2 = \infty$? [Note that a root at the origin corresponds to a motion of the form $(Ce^{st})_{s=0} = Ce^{0t}$.] For what value of R_2 (in terms of L_1 and C_3) will $\zeta = 0.4$? [Refer to Eq. (8.5) and Figs. 8.2 and 8.5.] For what value of R_2 is the system critically damped ($\zeta = 1$)?

8.6 Repeat Prob. 8.5 for the system of Prob. 8.4. In what ways do your results differ from those of Prob. 8.5? Explain on physical grounds.

8.7 For the system of Prob. 8.3 plot the locus of roots as C_3 is varied from 0 to ∞. Explain your results on physical grounds, particularly the results for $C_3 = 0$ and $C_3 = \infty$. Let $L_1 = 0.02$ henry, $R_2 = 1000$ ohms. For what value of C_3 (in terms of L_1 and R_2) will $\zeta = 1$? Compare with Prob. 8.5.

8.8 Repeat Prob. 8.7, but vary L_1 instead of C_3. Let $C_3 = 10$ microfarads, $R_2 = 1$ kilohm. How does your result compare with that obtained when C_3 is varied? Explain on mathematical and physical grounds.

In Probs. 8.9 through 8.14, obtain the equation of motion and the characteristic equation. Then define σ and ω_0 to correspond to Eq. (8.1).

8.9 The translational system of Prob. 2.19, with viscous friction.

8.10 The simple pendulum of Fig. 2.2c, p. 36, with viscous hinge damping.

8.11 The two-stage hydraulic servo of Prob. 7.8.

8.12 The plank on two rotating cylinders of Prob. 7.9.

8.13 The flyball governor of Probs. 5.9 and 5.75. Linearize the equations of motion.

8.14 The helicopter rotor of Prob. 5.56, p. 782, with Ω_1 constant.

8.15 Calculate and sketch accurately the motion of the system in Example 8.2, p. 256, if it is released with the initial conditions $x(0) = 2$ in., $\dot{x}(0) = -2$ in./sec.

8.16 The network of Prob. 8.3 has the following physical constants: $L_1 = 0.02$ henry, $R_2/L_1 = 2000$ sec^{-1}, $1/L_1C_3 = 40{,}000$ sec^{-2}, $v_0 = 1.6$ volts. The switch is in position b a long time. Suddenly it is switched to position a. Use KCL to obtain $v_3(t)$. Sketch and label carefully the subsequent behavior of voltage v_3. Use KVL to obtain $i_3(t) = i(t)$. Sketch and label carefully the behavior of i_3. (Refer to Prob. 8.3.) Consider the physical relation between v_3 and i_3. Is it verified at every instant by your results?

8.17 The pendulum of Prob. 8.10 has an undamped period of 0.5 sec. Its pivot has extremely heavy damping, such that $\zeta = 3$. What relation does this imply between b, ℓ, and m? [The units of b are (n m)/(rad/sec), for example.] (a) The pendulum is released without initial velocity from an angle of 0.06 rad. Plot its motion. (b) Find the minimum initial velocity required for the pendulum to *overshoot* before coming to rest. (Hint: The condition of overshooting can be represented mathematically as achieving zero displacement in *finite* time.)

8.18 Derive Eq. (8.11a) from the relations preceding it in Sec. 8.5.

8.19 Derive Eq. (8.11b) without employing complex numbers, as suggested in the text following Eq. (8.11b).

8.20 Show analytically that the period of a damped cosine wave, such as Fig. 8.8, Step (3), can be obtained by measuring the time between maxima (points Δ).

[Chap. 8] PROBLEMS

In Probs. 8.21 and 8.22 use Procedure B-3 to sketch accurately the time response of the underdamped second-order system described. In each sketch label the magnitudes of key quantities: T, τ, ψ, x_{max}, and so on. Accompany each time plot with an s-plane picture of the characteristics.

8.21 The mass-spring-damper system of Fig. 8.7a with the following physical values:

	$\dfrac{b}{2m}$ (sec^{-1})	$\dfrac{k}{m}$ (sec^{-2})	$x(0)$ (in.)	$\dot{x}(0)$ (in./sec)
(a)	1	10	1	0
(b)	3	5	1	0
(c)	1	10	1	2
(d)	3	5	1	2

8.22 Voltage v in the network shown, if the switch is switched from a to b at $t = 0$, and $2R_2C_3 = 1$ sec, $L_1C_3 = 0.1$ sec^2. (Hint: The equation of motion is most readily obtained by KCL.) The initial current in L_1 connotes an initial value of $\int v\, dt$, which is readily found from the physical voltage-current relation for element L_1.

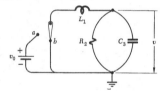

Prob. 8.22

8.23 Consider the following idealization. A railroad car approaches, at constant velocity 260 in./sec, a massless bumper (figure) and couples with the bumper without backlash and without energy loss. The bumper position is recorded versus time, as shown. Suppose the viscous dashpot is removed and the coupling repeated: What will be the maximum bumper excursion? What will be the frequency of oscillation after coupling?

Prob. 8.23

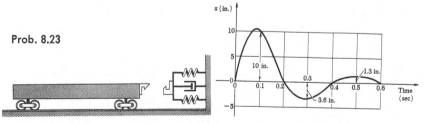

8.24 The bumper damping in Prob. 8.23 is increased 20-fold. Sketch the resulting motion.

8.25 By what ratio should the damping in Prob. 8.23 be increased to produce critical damping ($\zeta = 1$)? Assume that this has been done, and repeat Prob. 8.23.

8.26 For what value of R_2/L_1 will the network of Prob. 8.3 have critical damping ($\zeta = 1$), if $1/L_1C_3 = 40{,}000$ sec^{-2}? Assume this value for R_2/L_1, and repeat Prob. 8.3.

8.27 (a) For what relation between initial conditions $[\dot{\theta}(0)$ and $\theta(0)]$ will a critically damped system "overshoot" (cross $\theta = 0$)? Refer to Eq. (8.17). At what time T_c will the crossing occur? (b) After overshooting, the system will reverse direction at some time T_s, then return slowly toward $\theta = 0$. Find T_s. [Hint: $\theta(t)$ has zero slope where $\dot{\theta}(t) = 0$.]

8.28 Consider the system of Eq. (8.20). Utilizing Eqs. (8.23) and Eqs. (7.24), show rigorously, by carrying all terms, that if $M(t) = M_0 \cos \omega_f t \equiv M_0(e^{i\omega_f t} + e^{-i\omega_f t})/2$, then $\theta_f(t) = \theta_{fm} \cos(\omega_f t + \psi_f)$, with $\theta_{fm} = |\Theta|$ and $\psi_f = \underline{/\Theta}$, where Θ is given by (8.23a). Show that this result is identical to (8.23b). Note the generality of this result.

8.29 Make a careful plot of the response worked out in Example 8.5, p. 270, using the following numerical values: $2\pi/\omega_f = 0.4$ sec, $1/\sigma_f = 0.3$ sec, $k/J = 20$ sec^{-2}, $b/2J = 4$ sec^{-1}, $\psi_0 = 0$.

In Probs. 8.30 through 8.34, (i) obtain the transfer function between the quantities indicated; (ii) state the poles and zeros of the transfer function, and plot them in an s plane; (iii) state the system characteristic equation; (iv) obtain the forced-response-only to the given input.

8.30 For the translational system shown, obtain the transfer function from x to y. Assuming that the motion x of the upper end of the spring is $x = 0.2e^{-3t}$ in., plot the resulting forced-motion-only y of the mass, with $k = 9$ lb/in., $b = 0.3$ lb/(in./sec), and $m = 1$ lb/(in./sec^2). If the input were sinusoidal, $x = x_m \cos \omega_f t$, for what frequency ω_f would resonance occur?

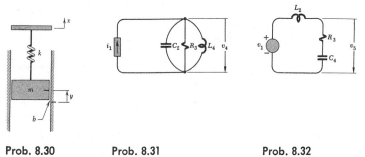

Prob. 8.30   Prob. 8.31   Prob. 8.32

8.31 For the parallel network shown, obtain the transfer function from i_1 to v_4. (See preceding instructions.) If the current source produces $i_1 = 2e^{-300t}$ amps, plot the resulting forced-motion-only of v_4 for $C_2 = 100$ microfarads, $R_3 = 100$ ohms, and $L_4 = 10$ millihenrys.

8.32 For the series network shown, obtain the transfer function from v_1 to i. (See preceding instructions.) The voltage generator produces a voltage $v_1 = 2e^{+10t}$ volts. Make a plot of the forced-motion-only of i versus time if $C_4 = 100$ microfarads, $R_3 = 100$ ohms, and $L_2 = 10$ millihenrys. If v_1 were sinusoidal, $v_1 = v_{1m} \cos \omega_f t$, for what value of ω_f would resonance occur?

8.33 A model of a ship's drive system is shown. The speed Ω_1 of the turbine is presumed given, versus time. Find the transfer function from Ω_1 to propeller speed

[Chap. 8] PROBLEMS 799

Ω_2. (See preceding instructions.) The propeller has inertia J and damping b, and the long flexible shaft has stiffness k. When the turbine is running down, its motion is given by $\Omega_1 = \Omega_{10} e^{-\sigma t}$. Find the resulting forced-motion-only of $\Omega_2(t)$. If Ω_1 were sinusoidal, $\Omega_1 = \Omega_{1m} \cos \omega_f t$, for what frequency ω_f would resonance occur?

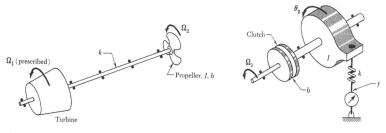

Prob. 8.33   Prob. 8.34

8.34 The fluid clutch system of Prob. 6.18, p. 788, is connected to a "prony brake" through a spring, for testing purposes, as shown. Find the transfer function from Ω_1 to f, the force read by the scale. (See preceding instructions.) If Ω_1 is given by $\Omega_1 = \Omega_{1m} e^{-\sigma t}$, find the resulting forced-motion-only of force f. Also, obtain the transfer function from Ω_{10} to θ_2. The radius to the scale is r, the shafts very stiff.

In Probs. 8.35 through 8.45 obtain the transfer function indicated, state its poles and zeros and plot them on an s plane, and deduce the system characteristic equation. Note specifically that although several transfer functions may be obtained for a given system, there is just one characteristic equation. Where ζ is not known or specified, assume that it is less than 1.

8.35 The network of Example 8.7, p. 272, if $1 < \zeta$. Compare with Fig. 8.20d.

8.36 From input Ω_1 to the moment M_k in the shaft in Prob. 8.33. Note that M_k is an important quantity to know. (Compare this transfer function with that from Ω_1 to Ω_2 obtained in Prob. 8.33. The system characteristic equation is, of course, the same.)

8.37 From input Ω_1 to the moment M_b transmitted by the clutch in Prob. 8.34. (Compare with the transfer function from Ω_1 to f obtained in Prob. 8.34.)

8.38 From i_1 to i_4 in Prob. 8.31. Also, from i_1 to i_2 and from i_1 to i_3. Check your result by adding the three transfer functions. (To what should they sum?)

8.39 From v_1 to v_5 in Prob. 8.32; also, from v_1 to v_3 and from v_1 to v_4.

8.40 From Ω_1 to θ_2 in the system of Prob. 6.18, where θ_2 is the output shaft angle. Compare with the transfer function from Ω_1 to θ_2 for the same system with prony brake in Prob. 8.34.

8.41 From container temperature T_c to thermometer reading T in the system of Prob. 6.25.

8.42 From line pressure p_1 to indicator angle θ in the sensor of Example 3.3, p. 115. (Refer to Prob. 6.12.)

8.43 From pilot stick input x to control-surface position θ in the hydraulic "boost" system of Prob. 6.9.

8.44 From voltage input v_1 to moment M_4 in the system of Prob. 6.26, and also from M_4 to output v_8. Display as two boxes (like those in Fig. 8.19, p. 276) in tandem.

8.45 Repeat Prob. 8.44 for the system of Prob. 6.27, and compare results. Note that the tandem boxes are merely interchanged. This feature, which derives directly from superposition, will be the basis for methods of studying compound l.c.c. systems in Chapter 14.

In Probs. 8.46 through 8.50 obtain (i) the driving-point impedance or admittance and (ii) the transfer impedance or admittance indicated for the given network. Simultaneous solution of algebraic equations may be required. Make use of transfer functions obtained previously, where possible.

8.46 For the network shown, driving-point admittance and transfer admittances I_5/V_1 and I_4/V_1.

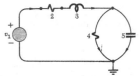 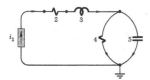

Prob. 8.46   Prob. 8.47

8.47 For the network shown, driving-point impedance and transfer impedance V_5/I_1.

8.48 For the network of Prob. 6.17, driving-point admittance and transfer admittance I_3/V_1. (Why are these the same?)

8.49 For the system of Fig. 8.1, driving-point impedance Θ/M.

8.50 For the system of Prob. 8.34, transfer admittance F/Ω_1.

8.51 Use Procedure B-3, p. 260, to sketch accurately the time responses in Fig. 8.23, p. 285, for $\zeta = 0$ and 0.2. Use expression (8.52).

8.52 Obtain and sketch the response $v_4(t)$ of the network of Fig. 2.12, p. 58, and Fig. 2.18 if the battery is switched in at $t = 0$ and the initial conditions at $t = 0^-$ were $v_3(0^-) = \gamma$, $i_2(0^-) = 0$. The equation of motion is (2.64). Use the following numerical values: $R_1 = R_4 = 10$ ohms, $R_6 = 4$ ohms, $L_2 = 1$ millihenry, $C_3 = 10^{-3}$ farad, $v_{bat} = 1.5$ volts, $\gamma = 0.5$ volt.

8.53 Derive equation of motion (8.54).

8.54 Find the total response of the system of Example 8.8, p. 285, to a step input of 1 ft lb, if the initial velocity is zero, and $k = 2.25$ ft lb/rad, $J = 0.25$ ft lb/(rad/sec²), and (i) $b = 0.1$ ft lb/(rad/sec), (ii) $b = 2.5$ ft lb/(rad/sec).

8.55 Find the total response of the system in Prob. 8.30, assuming the motion of x starts abruptly at $t = 0$: $x = 0.2e^{-3t}u(t)$ in. Repeat again with $b = 0$.

8.56 Repeat Prob. 8.31, assuming the flow of current commences abruptly at $t = 0$: current $i = 2e^{-3t}u(t)$ amp.

8.57 Repeat Prob. 8.30, with $x = 0.2 \cos 0.5t u(t)$. Make a careful plot. (Procedure B-3 will be helpful.)

8.58* The switch (figure) is moved from a to b at $t = 0$. Plot the resulting variation in v_5 (after $t = 0$), if $i_1 = 2e^{-3t}$ amps, $R_4 = 0.5$ ohm, $R_2 = 3.0$ ohms, $L_5 = 0.5$ henrys, and $C_3 = 1.0$ farad.

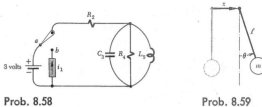

Prob. 8.58   Prob. 8.59

8.59 The top end of a 2-lb simple pendulum, of length $\ell = 2$ ft, is moved abruptly from one location to another, as follows: $x = 2u(t)$ in. Plot x and the resulting swing, θ, of the pendulum. (Refer to Prob. 5.53.) A good approximation to the initial conditions is that the bob remains stationary between $t = 0^-$ and $t = 0^+$.

8.60 A loudspeaker [Fig. 2.23 and Eq. (2.78)] is driven in a magnetic field by varying the current in its windings. The coil is 2 cm in diameter and has 20 turns. The field strength of the permanent magnet is $B = 0.1$ weber/cm^2 = 0.1 dn/amp cm, and the current, produced by a current source, is a suddenly applied 60-cycle sinusoid: $i = 0.3 \cos(120 \pi t) u(t)$ amps. Find the responding motion of the speaker, neglecting (in this problem) the effect of surrounding air, if the effective mass is 10 gm and the measured spring constant is $6.25/^5$ dn/cm.

8.61 The equation of motion for a single-axis gyro is given by (5.66). In the "rate-gyro" version, both damping and a restoring spring are present. Suppose the gyro is mounted to sense the roll rate of an aircraft, and assume uncertainty torques to be negligible. The roll rate of the aircraft has the time history $\dot{\psi} = \Omega_0(1 - e^{-\sigma_f t}) u(t)$. Find an expression for the responding gyro output signal.

8.62 Repeat Prob. 8.61 for a "rate-integrating" gyro, obtained by making $k = 0$, $J \ll b$. Discuss the appropriateness of the name.

8.63 Suppose the Scotch Yoke of Example 2.3, p. 52, initially at rest, is suddenly started in rotation at constant speed ω_f. Find the resulting time history of the *force* f in the link k_1.

8.64* Find the response of the shock-mounted instrument package in Fig. 1.1 to a step motion $x = x_0 u(t)$ of the vehicle frame. (The discussion in Example 8.12 may be helpful.) Find also an expression for the maximum value of acceleration of the package. If this must be made less than a specified value, $a_{\text{allow.}}$, and if m is given, indicate the allowable design ranges of k and b. The *static deflection* ξ_0 of the springs under the weight of the package is also limited by space in the vehicle. If the maximum allowable value is $\xi_{0\,\text{allow.}}$, how are the available design ranges of k and b affected? What is the lowest permissible system natural frequency? [Note: If the specified maximum vehicle acceleration is to be n g's (and this may be vertically upward), then the maximum spring deflection will be $(n+1)\xi_0$, and this is, of course, the basis on which $\xi_{0\,\text{allow.}}$ is actually specified.] Take $\zeta = 0.5$.

In Probs. 8.65 through 8.71 use Eq. (8.68) to write down the response of the system named to the specified short-impulse input. Sketch the response very approximately (using letters rather than numbers) to make sure that changes between $t = 0^-$ and $t = 0^+$ are in accord with physical reasoning.

8.65 Displacement θ in Fig. 8.1, with $M = (M_0 \tau)\delta(t)$, for $\zeta < 1$, $\zeta = 1$, and $1 < \zeta$. [Compare (8.20) with the equation preceding (8.68), and apply (8.68).]

8.66 Voltage v in the network of Prob. 8.22, with the switch initially at b, and an impulse $(v_0\tau)$ produced by switching it momentarily to a and then back to b. Let $1 < \zeta$. To obtain an equation of motion analogous to that preceding (8.68), use the variable $\lambda_1 \triangleq \int v_1\, dt$. (Physically, λ_1 is the amount of magnetic flux in the inductor.)

8.67 The mass-spring system of Fig. 8.24, p. 286, to the impulse $x = (x_0\tau)\delta(t)$. [Refer to Eq. (8.54).] Assume $\zeta < 1$.

8.68 Specialize Prob. 8.67 to the case $b = 0$.

8.69 Current i_4 in the network of Prob. 8.31, with $i_1 = (i_{1m}\tau)\delta(t)$. Take $1 < \zeta$. Use variable λ_4, as in Prob. 8.66.

8.70 Voltage v_4 in the network of Prob. 8.32, if $v_1 = (v_{1m}\tau)\delta(t)$. Take $1 < \zeta$. Use charge $q_4 = \int i_4\, dt$ as the unknown variable.

8.71 Specialize Prob. 8.69 to the case $R_3 = \infty$. Discuss the response and accompanying energy interchange on physical grounds.

In Probs. 8.72 through 8.78 use convolution, and the impulse response obtained previously, to find the response of the variable specified to the given input.

8.72 Displacement θ in Fig. 8.1, p. 247, if M is a step, $M = M_0 u(t)$, and $\zeta < 1$. Compare your result with Eq. (8.52). (Refer to Prob. 8.65.)

8.73 Repeat Prob. 8.72, assuming the applied torque has been $M = M_0$ (constant) for a long time prior to $t = 0$, then is removed at $t = 0$. Compare the result with that of Prob. 8.72.

8.74 Repeat Prob. 8.72, with the step terminating at time T.

8.75 Repeat Prob. 8.72, with $\zeta = 1$.

8.76 Displacement y in Fig. 8.24, p. 286, with $x(t)$ looking like $v_1(t)$ in Prob. 6.66.

8.77 The response of voltage v_4 in the network of Prob. 8.32, with $v_1 = v_{10}e^{-\sigma t}u(t)$. Refer to Prob. 8.69.

8.78 Repeat Prob. 8.77, with $v_1(t)$ given as in Prob. 6.66.

CHAPTER 9: FORCED OSCILLATIONS OF ELEMENTARY SYSTEMS

9.1 For the system of Fig. 9.1a, p. 310, with $R_2 = 5000$ ohms and $R_2C_3 = 0.5$ sec, plot to scale the phasor diagram of Fig. 9.1c for three values of ω_f: 0.4, 2, and 10 rad/sec. On each diagram, show also the phasor construction for driving-point admittance Y_{11}, as in Fig. 9.2. [Note from Eq. (9.9) that $1/Y_{11} = R + (1/Cj\omega_f)$.]

9.2 For the model shown of the rotor of an electric motor (with damping taken to be viscous), obtain the transfer function from M to Ω. This is sometimes called the mechanical admittance. Suppose $J/b = 10$ sec (the coast-down time constant). If $M = M_m \cos \omega_f t \stackrel{\mathrm{r}}{=} \mathsf{M}e^{j\omega_f t}$ then Ω will follow: $\Omega = \Omega_m \cos(\omega_f t + \psi) \stackrel{\mathrm{r}}{=} \Omega e^{j\omega_f t}$ (where $\mathsf{M} = M_m e^{j0}$ and $\Omega = \Omega_m e^{j\psi}$). Construct a phasor diagram of $\mathsf{M}/J = (M_m/J)e^{j0}$ and $\Omega = \Omega_m e^{j\psi}$, for $\omega_f = 0.2b/J$, b/J, and $5b/J$.

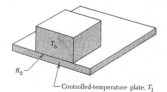

Prob. 9.2   Prob. 9.3

[Chap. 9] PROBLEMS 803

9.3 A box containing sensing equipment is mounted to a controlled-temperature plate, as shown, to test its performance during temperature cycling. The other five faces of the box are insulated from the room air. The temperature T_1 of the plate is controlled to be $T_1 = T_{1m} \cos \omega_f t$. Derive an expression for the time variation of T_b. Consider the box to be a single lump with thermal capacity C_3, and let the interface resistance between box and plate be R_2. If $R_2 C_3 = 360$ sec, construct a phasor diagram for T_1 and T_2 for cycling *periods* of 72 seconds, 6 minutes, and 1 hour. Obtain also an expression for the thermal driving-point admittance, transfer function $Q_1/T_1 = Q_2/T_1$.

9.4 Derive the driving-point admittance and impedance for the network shown. Are they reciprocal? Construct a phasor diagram for each, à la Fig. 9.2a, p. 314, with $R_2 = 500$ ohms, $R_3 = 1000$ ohms, $C_4 = 10$ microfarads $1\backslash^{-5}$ farad), and a forcing frequency of 100 *cycles* per sec. Obtain also an expression for the "transfer impedance" (transfer function) V_4/I_1.

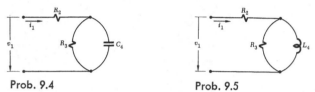

Prob. 9.4   Prob. 9.5

9.5 Derive the driving-point admittance and impedance for the network shown. Are they reciprocal? Construct a phasor diagram for each, a la Fig. 9.2a, p. 314, with $R_2 = 500$ ohms, $R_3 = 1000$ ohms, $L_4 = 10$ henrys, and a forcing frequency of 100 *cycles* per sec. Obtain also an expression for the "transfer impedance" (transfer function) V_4/I_1.

9.6 From your results in Prob. 9.1, plot the quantities $|I_1/V_1|$ and $/I_1/V_1$ [i.e., the magnitude and phase of (9.9)] versus driving frequency ω_f for $\omega_f = 0.4, 2,$ and 10. Use a scale like that in Fig. 9.5, p. 318. Fair in the curves for $|I_1/V_1|$ versus ω_f and $/I_1/V_1$ versus ω_f.

9.7 For the network of Fig. 9.1, p. 310, we find that the driving-point impedance Z_{11}, given by (9.10), is the reciprocal of driving-point admittance Y_{11}, given by (9.9) and plotted in Prob. 9.6. Plot the quantities $|Z_{11}|$ and $/Z_{11}$ versus driving frequency ω_f, for $\omega_f = 0.4, 2,$ and 10. Use the same plot as in Prob. 9.6, and compare. Fair in the curves of $|Z_{11}|$ versus ω_f and $/Z_{11}$ versus ω_f. Note that the $|Z_{11}|$ curve is the mirror image, about the unity-magnitude line, of $|Y_{11}| = |I_1/V_1|$, while $/Z_{11}$ is the mirror image, about the 0° line, of $/Y_{11} = /I_1/V_1$. Explain.

9.8 Show that, with the coordinates appropriately relabeled, the plot of Fig. 9.5 correctly gives the frequency response of the rotor of Prob. 9.2. Redraw Fig. 9.5 with the correct labels.

9.9 Show that, with the coordinates appropriately relabeled, the plot of Fig. 9.5 correctly gives the frequency response of the box in Prob. 9.3 to sinusoidal variations in temperature T_1 of the plate. Redraw Fig. 9.5 with the correct labels.

9.10 The voltage $x = 3 \cos \omega t$ is applied to the horizontal plates of a cathode ray tube, and the voltage $y = 2 \cos(\omega t + 30°)$ is applied to the vertical plates. Sketch carefully the Lissajous pattern that will appear on the crt face.

9.11 Generalize the preceding problem by letting $x = x_m \cos \omega t$ and $y = y_m \cos(\omega t + \psi)$. Show what measurements to make on the Lissajous pattern, and what relations to use them in, to obtain x_m, y_m, and ψ. Test your method by applying it to the Lissajous pattern you drew in Prob. 9.10.

9.12 For the low-pass filter of Fig. 9.6, p. 318, let v_1 be applied to the horizontal plates of a crt and v_3 to the vertical plates. Refer to Fig. 9.5 to sketch accurately the resulting Lissajous pattern for $\omega_f = 0$, 0.2, 2, 4, 20, and 10,000 rad/sec.

9.13 A Lissajous pattern produced by two variables, one of which has twice the frequency of the other, is commonly encountered. The pattern appears generally as a distorted figure 8. Sketch the Lissajous pattern produced by $x = 3 \cos \omega t$, $y = 2 \cos(2\omega t + 30°)$.

9.14 Consider again the model of a ship's power system from Prob. 8.33. Assume that the very long flexible drive shaft is massless, that damping on the propeller is viscous, and that the propeller inertia is negligible. The system output is, of course, propeller speed Ω_2. The turbine has a vibration $\Omega_1 = \Omega_{1m} \cos \omega_f t$ whose effect, by superposition, may be studied independently of the high-speed steady rotation. For $b/k = 0.1$ sec, sketch a Bode plot of the frequency response of the system. Use the asymptote techniques of Fig. 9.10, p. 325.

9.15 Assuming that the network of Prob. 9.4 is connected to a voltage source $v_1 = v_{1m} \cos \omega_f t$, and output voltage v_4 is measured, sketch a Bode plot of the system frequency response, with $R_2 = 500$ ohms, $R_3 = 1000$ ohms, $C_4 = 10$ μf. Use the asymptote techniques of Fig. 9.10.

9.16 In a test, a sinusoidal valve motion x, of varying frequency, is applied to the hydraulic boost system of Fig. 3.12, p. 119. The system output is, of course, rocket-engine motion θ. Sketch a Bode plot of the system frequency response, with Eq. (3.50) holding, and $C = 100$ (in./sec)/in., $\ell_1 = 6$ in.

9.17 Sketch and label a Bode plot of the frequency response of the heat-conduction system of Prob. 9.3.

9.18 The pneumatic computing element of Fig. 3.11, p. 116, is tested by connecting it to a line in which the pressure can be varied sinusoidally. Sketch a Bode plot of the resulting frequency response, using combined literal coefficients. [Hint: Eliminate p_2 between Eqs. (3.43a) and (3.43b).]

9.19* The electric network shown plays an important role in the synthesis of feedback control systems. It is called a *lead network*. Use asymptote techniques to sketch its frequency response to suitable dimensionless scales, with $R_3 C_2 = 0.01$ sec and $R_3/R_4 = 4$. (Note: The product of two frequency-response plots may be obtained by adding graphically their log magnitudes at each frequency and adding their phase angles at each frequency.) From your phase-angle plot, explain the meaning of the term *lead network*.

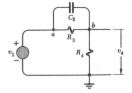

Prob. 9.19

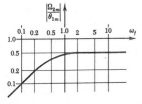

Prob. 9.20

[Chap. 9] PROBLEMS 805

9.20 In the fluid-clutch system of Prob. 6.18, the input disk *displacement* is driven sinusoidally (by a crank mechanism not shown): $\theta_1 = \theta_{1m} \cos \omega_f t$. Output *velocity* Ω_2 is measured, and the ratio $|\Omega_{2m}/\theta_{1m}|$ is measured and plotted as shown above. With $J = 5$ in. lb/(rad/sec^2), find b. Sketch the phase-angle plot that would accompany the given magnitude plot.

9.21 The instrument shown is for measuring velocity changes. Spring k is for centering only. It is extremely light, and *may be ignored* here. In a test, the case of the instrument is given a continuing sinusoidal motion: $x = x_m \cos \omega_f t$. The relative *velocity*, $\dot{\xi}$, is measured, and $|\dot{\xi}|$ is plotted versus ω_f below. If m weighs 0.384 lb, (i) find b. (Specify units.) (ii) Find x_m. (iii) Sketch the phase plot $/\dot{\xi}/x$.

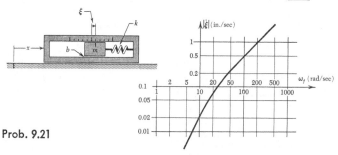

Prob. 9.21

9.22 An unbalanced machine, running on a slightly flexible mounting—e.g., a nonrigid floor—may vibrate severely under certain conditions. (For an extreme demonstration, ask the clerk in the hardware store to shake a can of paint for you in his mechanical shaker.) The situation, depicted in the first figure, may be modeled in one dimension as shown in the second figure, where f represents the vertical component of centrifugal force due to the unbalance. (If the unbalance is $m\ell$, what is f?) Obtain the transfer function from f to y, and show that the frequency response is as in Fig. 9.16, p. 333. How should the coordinates be labeled? State the conditions for resonance.

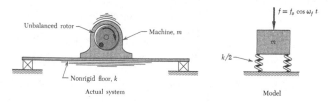

Prob. 9.22

9.23 A rotary tool for working in hard-to-reach places is driven by a long flexible shaft which connects it to the motor. A simple model for rotary motion is shown. If the motor speed contains a small sinusoidal ripple, the tool-shaft system will respond. By superposition, this sinusoidal variation—call it θ—may be studied separately from the steady angular motion. Calling the tool inertia J and the shaft torsional flexibility k, and neglecting shaft inertia and damping, obtain the transfer function from θ_1 to θ_2. Show that it leads to a form like (9.19), and that Fig. 9.16

therefore applies. How should the coordinates in Fig. 9.16 be relabeled? Explain the behavior physically at low frequency, at high frequency, and especially at resonance.

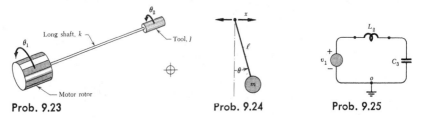

Prob. 9.23 **Prob. 9.24** **Prob. 9.25**

9.24 The top of a simple pendulum is given sinusoidal motion x in a horizontal straight line. Obtain the transfer function from x to θ. (Refer to Prob. 5.53.) For what coordinate labels will Fig. 9.16 represent this system? What is the condition for resonance? Demonstrate experimentally. (A key chain makes a good model. Steady the top end on the edge of a table to effect the desired tiny horizontal motions.)

9.25 For the resonant oscillator circuit shown, find the transfer function from v_1 to v_3, and state the condition for resonance. How should the coordinates be relabeled in Fig. 9.16 to make it applicable? Explain physically the behavior at low frequency, at high frequency, and at resonance.

9.26 For the network of Prob. 9.25, find the transfer function from v_1 to i_1 (the driving-point admittance). Sketch the system's frequency response in plots similar to Fig. 9.16. Again, explain physically the behavior (both magnitude and phase) at low frequency, at high frequency, and at resonance.

9.27 Consider again the two-stage hydraulic boost system of Fig. 7.2h, p. 231, in which, ideally, slave-valve motion $y = C_1 \int x\, dt$ and elevator motion $\theta = C_2 \int y\, dt$. Let the elevator linkages have length ℓ_3. In a test, the pilot's stick input (motion of point P) is made sinusoidal $x_P = x_{P0} \cos \omega_f t$. Obtain the transfer function from x_P to θ, and indicate what relabeling will make Fig. 9.16 applicable to this system. Explain physically the behavior at low frequency, at high frequency, and at resonance.

9.28 The Frahm tachometer consists of a set of reed-like cantilever springs of different lengths mounted to a common base, as shown. The base is held against a machine whose frequency of vibration, and therefore whose speed, is to be determined. Each separate reed has a frequency response like Fig. 9.16, but each has a different resonant frequency. The resonant frequency of each reed is printed on the base. Thus, when the base is held against a machine, the frequency of vibration is estimated by noting which reed is responding with large-amplitude vibrations. Modeling the reeds as massless cantilevers with a mass m concentrated at the end, find the driving frequency ω_f at which a reed of length ℓ will resonate. The deflection ξ of a cantilever beam of length ℓ under a static end-load f is $\xi = f\ell^3/3EI$, where EI is the beam's cross-sectional stiffness.

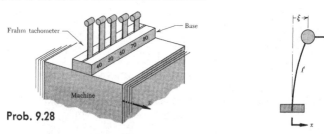

Prob. 9.28

[Chap. 9] PROBLEMS   807

9.29 Another version of the resonance tachometer of Prob. 9.28 consists of a single reed whose length can be adjusted, while it is in contact with a vibrating machine, until resonance is observed. The natural frequencies corresponding to various lengths are printed on the reed. Use the cantilever model of Prob. 9.28 to design such an instrument for the frequency range 10–1000 cps, and draw a picture to scale showing the frequency markings.

9.30 It is known that the frequency response of a certain system may be represented by Eqs. (9.21), p. 337. It is known further that $\zeta = 0.125$ for the system. Use Fig. 9.20 to fill in the following table.

Nondimensional frequency, β:	0.4	0.5	0.707	0.8	$\beta_{\alpha\max} = ?$	1.0	1.325	1.414	2	4	10
Nondimensional amplitude, α:											
Phase angle, ψ_f:											

9.31 The asymptotes shown in Fig. 9.20b are obtained by considering alone the terms 1 and $-\beta^2$ in the denominator of (9.21). A third "asymptote" may be obtained by considering the term $j(2\zeta\beta)$ alone. Draw it in Fig. 9.20b. Show that the amplitude plot actually touches it at point R, where $\beta^2 = 1$.

9.32 For the physical parameters listed in the table below, use Fig. 9.20 to find the steady sinusoidal response of the system of Fig. 9.18 to the specified input.

Physical parameters			Input M (in. lb)	Output θ (rad)
k (in. lb/rad)	b [in. lb/(rad/sec)]	J [in. lb/(rad/sec^2)]		
500	10	5	$2 \cos 5t$	
500	50	5	$2 \cos 5t$	
125	10	5	$2 \cos 5t$	
125	10	5	$2 \cos (5t + 0.1)$	
5000	10	5	$2 \sin 5t$	

9.33 Use the asymptotic plotting techniques of Sec. 9.9 to sketch accurately a Bode plot (amplitude and phase) for the system of Fig. 9.18, with $k/J = 100$ and $b/2J = 2$. Repeat for $b/2J = 10$. Label carefully. Compare your results point by point with the appropriate values from Fig. 9.19. In each case, what is the maximum value of α, and what is its value at resonance?

In Probs. 9.34 through 9.36 use the asymptotic plotting techniques of Sec. 9.9 to sketch accurately a Bode plot (magnitude and phase) for the function specified, with $\zeta = 0.1$.

9.34 $\alpha = \dfrac{j\beta}{1 - \beta^2 + j2\zeta\beta}$ **9.35** $\alpha = \dfrac{-\beta^2}{1 - \beta^2 + j2\zeta\beta}$

9.36 $\alpha = 1 - \beta^2 + j2\zeta\beta$

9.37 Use the phasor construction of Fig. 9.21c, p. 342, to check all the values obtained in Prob. 9.30. (Draw to a large convenient scale, and proceed systematically.)

9.38 Use the s-plane construction of Fig. 9.22, p. 344, to check all the values obtained in Prob. 9.30. (Use a Spirule, if available. See Appendix I, p. 727. Compare with Prob. 9.37.)

9.39 Use the phasor construction of Fig. 9.21c to check the amplitude and phase given in Fig. 9.19 for the following values: $\beta = 0.5$ and $\zeta = 0, 0.2, 0.5, 1.0$; $\beta = 1$ and $\zeta = 0.2, 0.5, 1$; $\beta = 2$ and $\zeta = 0, 0.2, 0.5, 1.0$.

9.40 Modify the phasor construction of Fig. 9.21c appropriately, and use it to check the three points obtained in Prob. 9.34 for $\beta = 0.5, 1, 5$.

9.41 Modify the s-plane construction of Fig. 9.22 appropriately, and use it to check the three points obtained in Prob. 9.34 for $\beta = 0.5, 1, 5$. (Compare with Prob. 9.40.)

9.42 Find the frequency ω_f for which the reaction on the wall is a maximum (figure).

Prob. 9.42

9.43 In Prob. 9.42 add a spring to the dashpot between the wall and the mass. Again find the frequency for which the total reaction on the wall is a maximum.

9.44 Construct an amplitude and phase plot, versus ω_f, for the system of Prob. 9.43 from a measurement of the magnitudes and angles of appropriate vectors in the s plane, with $\zeta = 0.2$.

9.45 For the network shown, voltages v_1 and v_2 will both vary as $v = v_m \cos(\omega_f t + \psi_f)$. Find and construct a Bode plot of v_{cm} and ψ_c as a function of ω_f and v_m, if $v_{1m} = v_{2m} = v_m$, and $\psi_f = 0$.

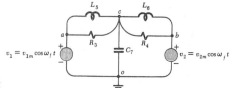

Prob. 9.45

9.46 The frame m_1 (figure) rides on an air cushion whose friction may be neglected. Uniform rod m_2 is pivoted to the frame with viscous friction. A horizontal force $f = f_m \cos \omega_f t$ is applied to the frame. Obtain an expression for the magnitude and phase angle of rod angle θ versus ω_f, for small θ. Sketch the Bode plot for light damping. Discuss the case $b \to 0$. Discuss the case $k \to 0$.

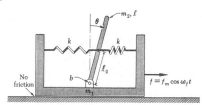

Prob. 9.46

9.47 A rocket vehicle has a steady vertical acceleration of 64.4 ft/sec². What will be reported by a longitudinal on-board accelerometer?

9.48 An accelerometer is calibrated on a static tilt table: Angle α of Fig. 9.23, p. 347, is set at various values, from 0 to 2π, and the instrument's output is recorded. Plot the output of an ideal accelerometer versus α. If it is desired to measure the error at null (zero output) with an accuracy of 10 micro g's (10 millionths of a g), how accurately must α be known? If it is desired to measure the error at one g input, to an accuracy of 10 micro g's, how accurately must α be known?

9.49 An accelerometer, having a natural frequency of 100 cps and 0.5 critical damping, is mounted horizontally at the nose of a vibrating rocket vehicle during a test firing. It reports an acceleration given by $a = 16.1 \cos (12.56t + \pi/6)$ ft/sec². (Time t is taken with respect to some arbitrary reference.) What is the actual acceleration? At a later time, the accelerometer reports $a = 3.22 \cos (555.2t + \pi/4)$ ft/sec². What is the actual acceleration?

9.50 Use the techniques of Secs. 9.9 and 9.10 to construct the phase plot that accompanies Fig. 9.25, p. 349, for a seismometer having $\zeta = 0.05$.

9.51 The seismometer of Fig. 9.25 reports a vertical motion of the laboratory consisting essentially of three sine waves: $\xi = [0.002 \cos (3t + \pi/12) + 0.01 \cos (12t + \pi/6) + 0.02 \cos 10t]$ in. (Time t is measured with respect to an arbitrary reference.) The seismometer has a natural frequency of 1 cps and $\zeta = 0.05$. What is the true motion of the laboratory, insofar as you can deduce it? (Use the results of Prob. 9.50 to compute phase shift.)

9.52 The true vertical motion of an aircraft wing undergoing vibration testing is $x = [6 \cos 4t + 0.6 \cos 40t + 0.006 \cos 400t]$ in. Mounted on the wing are a seismograph having natural frequency $\omega_0 = 6$ rad/sec and $\zeta = 0.05$, a velocimeter having $\omega_0 = 60$ rad/sec and $\zeta = 10$, and an accelerometer having $\omega_0 = 600$ rad/sec and $\zeta = 0.5$. Find the outputs of the three instruments (if they have no bias or scale-factor errors). The outputs of the three instruments are recorded with the following scale factors: seismograph 1 in. represents 1 in., velocimeter 1 in. represents 1 in./sec, accelerometer 1 in. represents 1 g. In the light of your results, discuss a reasonable vibration-test procedure for obtaining frequency response, with regard to quantities to be recorded in various frequency regimes.

9.53* Consider again the vibration-isolation system of Fig. 1.1. Obtain the transfer function from x to y, and construct the corresponding Bode plot. From your Bode plot, discuss quantitatively how well the isolation system protects the package (a) from high-frequency vibration, (b) from low-frequency vibration. As discussed in Prob. 8.64, the lower design boundary on system natural frequency is established by the allowable deflection of the springs under the weight of the package. Is this an important restriction? What are the factors in selection of damper size?

9.54 Develop analytical expressions for the asymptotes in Fig. 9.28b. Prove that $\alpha_{max} = Q$.

9.55 For the system of Fig. 9.27 (shown again below), with $L = 0.1$ millihenry, $C = 100\mu_f$, $R = 20$ ohms, use Fig. 9.28, p. 353, to find the response $i(t)$, with $v(t) = 100 \sin \omega_f t$ volts, and $\omega_f = 5000$ rad/sec, 10,000 rad/sec, 20,000 rad/sec, and 100,000 rad/sec.

9.56 Use a phasor construction similar to Fig. 9.21c to check your first answer in Prob. 9.55.

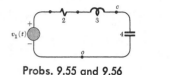

Probs. 9.55 and 9.56

Prob. 9.57

9.57 Expression (8.35), p. 274, gives the driving-point admittance for the network in Fig. 9.27, and this is plotted in Fig. 9.28. By replacing the voltage source with a current source, as shown above, obtain an expression for the driving-point impedance. Relabel the coordinates in Fig. 9.28 so that that figure becomes precisely the Bode plot for driving-point impedance. What special properties of the Bode plot are exploited?

9.58 Referring to Prob. 9.57, obtain an expression for the *transfer* impedance V_c/I_1. Sketch an accurate Bode plot for this function.

9.59 Obtain the transfer function from V_1 to V_c in the system of Fig. 9.27 and Prob. 9.55, and show that a certain Bode plot already printed in Chapter 9 applies also to this function, with appropriate relabeling of coordinates. State precisely how the coordinates are to be relabeled.

In Probs. **9.60**, **9.61**, and **9.62** (figures), find (i) the driving-point admittance and impedance, (ii) the transfer admittance I_4/V_1, (iii) the transfer function I_4/I_1. (Imagine the appropriate source in each case.)

Prob. 9.60

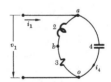

Prob. 9.61

Prob. 9.62

9.63 For the network in Prob. 9.60, with $Q = 5$, sketch an accurate Bode plot of (i) the driving-point admittance, (ii) the transfer admittance I_4/V_1, and (iii) the transfer function I_4/I_1. Label coordinates and key values carefully.

CHAPTER 10: NATURAL MOTIONS OF NONLINEAR SYSTEMS AND TIME-VARYING SYSTEMS

10.1 The figure shows a typical measured physical force-displacement relation for a nonlinear spring. Select the equivalent linear-spring constant k you would use to make an approximate analysis of motions about null having a peak amplitude of (i) 0.2 in., (ii) 1.0 in., (iii) 1.4 in.

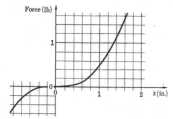

Prob. 10.1

10.2 For the spring in Prob. 10.1, estimate by eye the appropriate k to use for very small motions about the position $x = x_0 = 1$ in.

10.3 Repeat Prob. 10.2, with motions occurring between $x = +0.6$ in. and $+1.4$ in.

10.4 The curve in Prob. 10.1 is actually a cubic. Determine its parameters and do Prob. 10.2 analytically, using a series expansion about $x = x_0 = 1$ in. Compare with your graphical estimate.

10.5 Evaluate and plot the error and the per cent error in approximating $\sin \theta$ by θ for $0 < \theta < 30°$.

10.6 In the network shown the initial voltage across C_1 is γ, and the initial current in L_2 is zero. We wish to study the behavior after the switch is closed. Choose as state variables v_1 and Ri_2, and write the equations of motion in standard form (see p. 361). Construct a phase-plane plot of the behavior. Let $L_2 C_1 = 253$ sec^{-2}, and indicate the time at several points on your plot.

Prob. 10.6

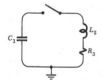
Prob. 10.7

10.7 Repeat the analysis of Prob. 10.6 with resistance added, as shown. The method of isoclines is convenient. Let $\zeta = R_3/(2\sqrt{L_2/C_1}) = 0.1$.

10.8 Make a phase-plane plot of the vertical motion of a projectile fired upward with initial velocity v_0. Ignore air drag. First obtain the equations of motion in standard form.

10.9 Consider a system consisting of a mass and a piecewise-linear spring like that in Fig. 10.7, but with $k = 0$ for the weaker spring, as shown below. There is no damping. Obtain the equations of motion in standard form. Then make a phase-plane study of the motion following an initial state in which $x(0) = x_0 > x_d$ and $v(0) = 0$.

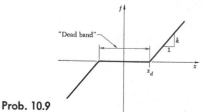

Prob. 10.9

10.10 Make a purely analytical study of the motion in Prob. 10.9 by solving the usual equation of motion several times, as the displacement is first within and then without the dead band of the spring, and matching initial conditions each time. Compare your solution with that of Prob. 10.9.

10.11° Mechanize the exact equation of motion for a pendulum, Eq. (10.4) or (10.21) (which are equivalent), on an analog computer. Arrange to have the computer plot $x = \theta$ versus time, and duplicate the plots of Fig. 10.5c for the values of $\theta(0)$ indicated there. Arrange also to plot y versus x [Eqs. (10.21a) and (10.21b)], and duplicate the phase-plane trajectories of Fig. 10.5b for several sets of initial conditions. Also, arrange to switch to a linear mechanization ($\sin \theta$ replaced with θ), and compare the behavior of the linear and nonlinear systems for $\theta(0) = 5°$, $\theta(0) = 10°$, $\theta(0) = 20°$, $\theta(0) = 60°$, and so on.

10.12° Repeat Prob. 10.11, with pendulum damping included. That is, mechanize Eqs. (10.30). Study the behavior of $x = \theta$ versus time, and then duplicate the phase-plane trajectories in Fig. 10.6. [Use Eq. (10.32) to deduce the value of ζ that was used in Fig. 10.6.] Switch to a linear mechanization, and compare the linear-system trajectories with the corresponding nonlinear ones, for small and large initial conditions.

10.13° Mechanize on an analog or digital computer a mass-spring system, $m\ddot{x} + f(x) = 0$, in which the spring $f(x)$ obeys (closely) the relation of Prob. 10.1. (A curve-follower function generator or just a "tapped" potentiometer may be used.) For simplicity, let $m = 1$ lb/(in./sec^2). Set the system oscillating with $x_{max} = 1$ in., and record x versus time. Then switch to a linear system, $m\ddot{x} + kx = 0$, and repeat the recording several times, adjusting k until the linear and nonlinear time responses match most nearly, in your judgment. Then check your value of k with the one you estimated in Prob. 10.1(ii). Repeat for $x_{max} = 0.2$ in. and $x_{max} = 1.4$ in.

10.14° Mechanize on a computer the piecewise-linear system of Prob. 10.9, and check the results you obtained in Probs. 10.9 and 10.10.

10.15° Mechanize on a computer the parametrically excited system represented in Fig. 10.8, p. 371. Study the natural motions for various values of $\omega_1/\sqrt{g/\ell_0}$, and discuss your result. Then build the system, and compare the behavior of the real system with its computer simulation. (A simple hand-operated lever to move the top of the string should be adequate. A motor drive can be excellent.)

CHAPTER 11: DYNAMIC STABILITY

11.1 The philosophy of stability is often illustrated using the image of a marble free to roll on a surface, as shown. The marble may be in static equilibrium (gravity and support forces exactly balance) at any one of locations A, B, C. Discuss the *stability* of the equilibrium at each location, using the concept of stability introduced at the beginning of Sec. 11.1.

Prob. 11.1

° Problem to be solved by using an analog or digital computer.

11.2 If the equation of the surface in Prob. 11.1 may be written $y = a(x - \ell_1)^2$ near A, $(y - h_2) = -b(x - \ell_2)^2$ near B, and $y = c = $ const. near C, develop analytical support for the conclusions you reached in Prob. 11.1.

11.3 A man and his wife sleep under an electric blanket, the two halves of which are controlled by separate dials ("his" and "hers"). One night the control dials become interchanged. Discuss the stability of the situation.

11.4 For the system of Fig. 11.2a, p. 378, make a plot like Fig. 11.2b for the case $b/2J = 3$, where k/J is a variable between -9 and $+100$.

11.5 Plot the roots of characteristic equation (11.3) versus inertia J. Label your plot appropriately.

11.6 Suppose the inverted pendulum of Example 11.1 to be (i) on the earth, (ii) on the moon, (iii) in free fall. Plot the roots of characteristic equation (11.9) for each case. Recall that $b/2m\ell^2 = 1.0$ sec^{-1} and $\ell = 10.7$ ft.

11.7 A feedback amplifier may be arranged as shown. (Refer to Sec. 2.8.) Plot the roots of the system characteristic equation in the s plane, versus K, with $RC = 100$. Let K have both positive and negative values.

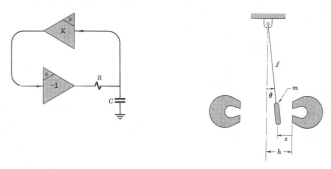

Prob. 11.7 Prob. 11.8

11.8 A pendulum with a metal bob hangs between two magnets, as shown. The force one magnet exerts on the bob is given by $f = c/x^2$. Write the exact equation of motion. Then linearize for small motions ($\ell\theta \ll h - x$) about the position $\theta = 0$. Obtain the characteristic equation and discuss the effect of c and h upon stability.

11.9 In the two-stage hydraulic boost system of Fig. 7.2h, p. 231, and Prob. 7.8, the lines connecting the slave valve to the piston are accidentally connected opposite from the way shown. With point P fixed, obtain the system characteristic equation, and discuss the system stability. Assume $\dot{y} = c_1 x$ and $\dot{\theta} = c_2 y$, and let $\ell_1 = \ell_3 = \ell$. Plot the locus of roots versus ℓ.

11.10 Use the construction of Fig. 11.6 to prove that the locus in Fig. 11.2, from points (c) onward, is the perpendicular bisector of segment (a)(a).

11.11 Make a plot like Fig. 11.7 for the case $k/J = 9$, where $b/2J$ is a variable from -1 to $+5$.

11.12 Let $k/J = 25$ sec^{-2}, and use the graphical construction illustrated in Fig. 11.6, p. 386, to check points (g), (h), (j), (f), (k) in Fig. 11.7. That is, draw to scale the vectors indicated in Eq. (11.16), measure their magnitude and phase, and show that they satisfy (11.16). (The value of b/J corresponding to each point can be computed from the corresponding ζ, which is noted in Fig. 11.8.)

11.13 Derive the equation of motion for small motions of the system of Fig. 11.9b about static equilibrium. Let the local slope of the curve be $-b$ there. (What *is* the location of the static equilibrium point, relative to the unstretched spring point?)

11.14 Pursuing Prob. 11.13 further, let $k/m = 144$ and $-b/m = -2$. If the mass is displaced from the static equilibrium point by 1 in. and released, plot the ensuing motion carefully, and label important quantities. (Procedure B-3 is helpful.)

11.15* Assuming that the curve in Fig. 11.9b can be represented closely by the function $f = f_0(1 - 0.5 \sqrt{v_{\text{rel}}/v_0})$, where f_0 and v_0 are constants, obtain a linearized equation of motion for small perturbations about the static equilibrium point. (Use the series-expansion method described in Sec. 10.1.)

In Probs. 11.16 through 11.18 obtain the characteristic equation of the system named. Assign letters to physical parameters as required.

11.16 The system models in Fig. 11.12a, b, c, d.
11.17 The system modeled in Fig. 11.12e.
11.18 The system modeled in Fig. 11.12f.

In Probs. 11.19 through 11.24 derive the equations of motion for the system in terms of the variables named, in the manner of Sec. 11.7. Then use Procedure C-1 to obtain the characteristic equation, and sketch the shape of a locus of the roots versus the parameter specified. (Include both positive and negative values.) Assign letters to coefficients as needed, and use appropriate combinations of letters as labels for key quantities in your s-plane plots.

11.19 The system of Fig. 11.13a: variables M, θ, v, with $M = -Kv$, $v = 1\dot{\theta}$. Plot roots versus K.

11.20 The system of Fig. 11.13b: variables M, θ, v, with $M = -bv$, $v = 1\dot{\theta}$. Plot roots versus b.

11.21 The system of Fig. 11.13c: variables x, y, v, with $\dot{y} = cx$, $v = 1y$. Plot roots versus K. Be careful of the sign.

11.22 The system of Fig. 11.13c': variables x, y, with $\dot{y} = cx$. Plot roots versus ℓ_1/ℓ.

11.23 The system of Fig. 11.13d: variables x, y, v, with $\dot{y} = cx$, $v = 1y$. Plot roots versus K.

11.24 The system of Fig. 11.13e: variables M, θ, v_a, v_b, with $M = -Kv_b$, $v_a = 1\dot{\theta}$. Plot roots versus K.

11.25 (i) Construct to scale the vectors indicated by the characteristic equation displayed in Fig. 11.17, p. 403, for point P and for another point on the locus, and show that the angles of the vectors sum to 180°, if $1/RC = 3b/J$. (ii) With $b/J = 1$ and $1/RC = 3$, compute the value of K/JRC at point P by multiplying the vector lengths.

11.26 A certain system consists essentially of three simple identical networks connected in series, as shown. There is a voltage-isolation amplifier between each. The output of the third is connected to the input of the first. Derive the characteristic equation of the system, and plot the locus of roots versus $K = K_1K_2K_3$, the product of all the amplifier constants. (Consider both positive and negative values of K.)

Compare your result with Fig. 11.15. For what range of values of K is this system unstable?

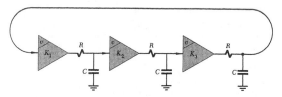

Prob. 11.26

In Probs. 11.27 through 11.39 rewrite the given characteristic equation in the Evans form [Eq. (21.2), p. 405], for example, Eq. (11.27), with $-K$ on the right-hand side. Then perform the following 180° root-locus operations on a carefully drawn s-plane plot: (i) Determine and mark real-axis intervals; (ii) determine the directions of the asymptotes and their emanation point, and draw them lightly; (iii) determine directions of departure or arrival, or both, if appropriate; (iv) sketch the 180° loci of roots as accurately as steps (i), (ii), and (iii) permit. Estimate by eye any points where loci leave or enter the real axis. (v) Use a Spirule to check the angle summation for one convenient point on each locus.

11.27 $s(s+1)(s+3) + K = 0$. (Refer to Figs. 11.15 and 11.17.)

11.28 $(s+1)^3 + K = 0$.
11.29 $s(s+1)^2 + K = 0$.
11.30 $s^2(s+1) + K = 0$.
11.31 $(s^2+1) + K = 0$.
11.32 $s(s^2+2s+2) + K = 0$.
11.33 $(s^2+1) + Ks = 0$. (Refer to Fig. 11.7.)
11.34 $s^2 + K(s+1) = 0$.
11.35 $s^2 + K(s+1)^2 = 0$.
11.36 $s^2 + K(s+1)(s+2) = 0$.
11.37 $s(s+1) + K(s^2+2s+2) = 0$.
11.38 $(s-1)^2(s+10)^2 + Ks = 0$.
11.39 $s(s^2+1) + K(s^2+2s+5) = 0$.
11.40 $(s+1)^5 + K = 0$.

11.41 Construct a root-locus plot, with b variable, for the characteristic equation accompanying Fig. 11.16. Show that the system is stable for all (positive) values of b. Explain the meaning of the roots when $b \to \infty$.

11.42 For Prob. 11.27 find the value of K where the locus crosses the $j\omega$ axis.

11.43 For Prob. 11.28 find the value of K where the locus crosses the $j\omega$ axis. Find also the value of K where the locus crosses the $\zeta = 0.707$ line.

11.44 For Prob. 11.29 find the value of K where the locus crosses the $j\omega$ axis.

11.45 For Prob. 11.30 find the value of K where the locus crosses the $j\omega$ axis.

11.46 For Prob. 11.32 find the value of K where the locus crosses the $j\omega$ axis.

11.47 For Prob. 11.34 find the value of K where the roots are critically damped.

11.48 For Prob. 11.39 find the value of K where the locus crosses the $\zeta = 0.1$ line.

11.49 Rewrite the root-locus sketching rules (Procedure C-3) for the case in which the 0° loci are to be sketched.

In Probs. 11.50 through 11.60 sketch the loci for negative values of K—i.e., the 0° loci—for the characteristic equation specified. (Employ the rules you developed in Prob. 11.49.)

11.50 The characteristic equation of Prob. 11.27.

11.51 The characteristic equation of Prob. 11.28.
11.52 The characteristic equation of Prob. 11.30.
11.53 The characteristic equation of Prob. 11.31.
11.54 The characteristic equation of Prob. 11.32.
11.55 The characteristic equation of Prob. 11.33.
11.56 The characteristic equation of Prob. 11.34.
11.57 The characteristic equation of Prob. 11.36.
11.58 The characteristic equation of Prob. 11.37.
11.59* The characteristic equation of Prob. 11.38.
11.60* The characteristic equation of Prob. 11.39.

11.61 Use Routh's method to prove formally Eqs. (11.37) and (11.38).
11.62 The root loci in Fig. 11.17, p. 403, are drawn for the particular case $(1/RC) = 3(b/J)$. For this case, obtain the value of $(K/JRC)/(b/J)^3$ corresponding to point P in Fig. 11.15 in two independent ways: (a) by applying Routh's criterion, Eq. (11.45g), and (b) by measuring and multiplying lengths in Fig. 11.15. Compare your results. Use Routh's method to find the range of values of K for instability, and the frequency at point P in Fig. 11.17, if $b = 10$ n/(m/sec), $J = 3$ n/(m/sec^2), $RC = 0.1$ sec. What does Eq. (11.41) imply about the locus in Fig. 11.17 for negative values of K? Compare with the results of Prob. 11.50.

In Probs. 11.63 through 11.68 use Routh's method to determine, for the given characteristic equation, the number of roots having positive real parts. If there are roots *on* the imaginary axis, state their frequency.

11.63 $s^4 + 5s^3 + 13s^2 + 19s + 10 = 0$.

11.64 $s^5 + 4s^4 + 7s^3 + 8s^2 + 6s + 4 = 0$.
11.65 $s^5 + s^4 + 2s^3 + 2s^2 + 3s + 15 = 0$.
11.66 $s^4 + 2s^3 + 11s^2 + 18s + 18 = 0$.
11.67 $s^7 + 2s^6 + 4s^5 + 4s^4 + 3s^3 + 2s^2 + s + 1 = 0$.
11.68 $s^5 + s^4 + 4s + 4 = 0$.
11.69 Develop the general expression, similar to (11.37), for any fourth-order characteristic equation.
11.70 Develop the general expression, similar to (11.37), for any fifth-order characteristic equation.

In Probs. 11.71 through 11.78 use Routh's method, Step 6, to determine the frequency at which the indicated locus crosses the $j\omega$ axis, and to find the value of K there.
11.71 The locus obtained in Prob. 11.27. (Compare with result of Prob. 11.42.)
11.72 The locus obtained in Prob. 11.28.
11.73 The locus obtained in Prob. 11.51. (0° locus.)
11.74 The locus obtained in Prob. 11.29. (Compare with result of Prob. 11.44.)
11.75 The locus obtained in Prob. 11.30. (Compare with result of Prob. 11.45.)
11.76 The locus obtained in Prob. 11.52. (0° locus.)
11.77* The locus obtained in Prob. 11.38.
11.78* The locus obtained in Prob. 11.59. (0° locus.)

[Chap. 11] PROBLEMS

11.79 The characteristic equation for a certain system is

$$(s + 10)^2(s - 1)^2 + K(s + 1)^2(s - 10)^2 = 0$$

The loci of roots of this equation are plotted, versus K, below. (a) Find K_{n1} and K_{n2}. (b) Describe *in detail* the possible motions of the system for $K = 0$, for $K = K_{n1}$, for $K = K_{n2}$, and for $K > K_{n1}$ and K_{n2}. (c) Use Routh's method to verify as much as you can of the root locus and the results of (a) and (b).

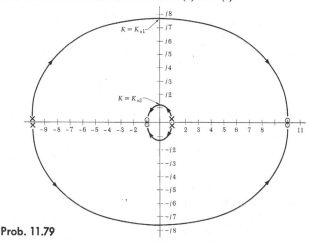

Prob. 11.79

11.80* A single-wheel trailer is attached to a car with a vertical pin connector as shown in the sketch. To approximate the lateral flexibility of the drawbar, replace it with a rigid drawbar and a torsion-spring connection at the center of mass of the trailer. The car moves at velocity V, and the trailer has mass m and moment of inertia J about its center of mass; the mass of the drawbar and wheel are negligible. We assume that *the wheel does not skid* and that there is a small drag force D, due to rolling friction which acts in the plane of the wheel and depends only on V. Investigate the stability of the trailer motion. (Hint: What does the no-skid constraint dictate about the direction of $\mathbf{v}^W$? What is such a constraint called? See p. 144.)

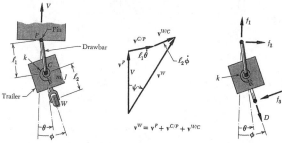

Prob. 11.80

11.81* Write the equation of motion for the marble in Probs. 11.1 and 11.2 for small motions in the vicinity of each point, A, B, C. (Let its mass be m, its radius r, and its moment of inertia J.) Obtain the characteristic equation and plot its roots in the s plane for each case. Label distances in the s plane.

CHAPTER 12: COUPLED MODES OF NATURAL MOTION: TWO DEGREES OF FREEDOM

In Probs. 12.1 through 12.8 (i) obtain the equations of motion for the unforced system specified; (ii) assume that all motions are of the form $x = Xe^{st}$, and obtain an algebraic set of equations like Eqs. (12.3); (iii) obtain the system characteristic equation (in unfactored form).

12.1 The two-pendulum system shown. The spring is fastened at $\frac{2}{3}\ell$.

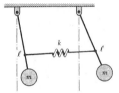

Prob. 12.1

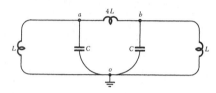

Prob. 12.2

12.2 The electrical network shown.
12.3 The three-mass system shown.

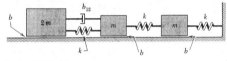

Prob. 12.3

12.4 The electrical network shown.

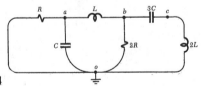

Prob. 12.4

12.5 The system of Fig. 4.4a, p. 129, with $f_7 = 0$.
12.6 The system of Fig. 4.4b, with $i_7 = 0$.
12.7 The speaker drive system shown, with zero input voltage. (Later we shall be interested in the decay transient after the input is suddenly shorted.) Refer to Sec. 2.9 and to Example 2.6 there.

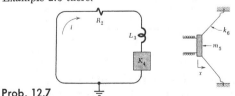

Prob. 12.7

[Chap. 12] PROBLEMS   819

12.8 The hydraulic reservoir system of Fig. 3.9a, p. 113, with $\alpha = 1$ in Eq. (3.38), and with $w_2 = 0$. Retain the $\dot{w}_3$ term. (Later we shall be interested in the natural oscillations of the system following a sudden shutoff of w_2, for example.)

12.9 Two beads are fixed to a string having high tension F_0 as shown. Derive the equations of motion, for small lateral displacements in the plane of the paper and normal to the string, and show that they are analogous to Eqs. (12.21). Changes in F_0 may be ignored, and gravity is not involved. Assume solutions of the form $y = Ye^{st}$, and obtain a pair of algebraic equations analogous to (12.22). Then obtain the characteristic equation and the natural frequencies and natural modes of motion of the system. Finally specialize your result for the case $m_1 = m_2$.

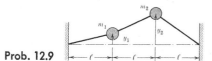

Prob. 12.9

12.10 For the special case $m_1 = m_2$, find the total motion of the system in Prob. 12.9 if the system is released from rest from the initial configuration $y_1(0) = y_{1m}$, $y_2(0) = 0$.

12.11 Repeat Prob. 12.10, with both beads having zero initial displacement, but m_1 having an initial velocity u_0 (e.g., following an impulsive tap).

12.12 Study carefully the behavior of Eqs. (12.27) and (12.28) as $k_3 \to \infty$, and show that ω for the first natural mode approaches $\sqrt{k_3[(1/J_1) + (1/J_2)]}$ and $\Theta_2/\Theta_1 \to -0.5$, as indicated in Fig. 12.9.

In Probs. 12.13 through 12.17 do the following, for the undamped system described: (i) Derive the equations of motion; (ii) assume solutions of the form $x = Xe^{st}$ for both variables, and obtain simultaneous algebraic equations like (12.22); (iii) obtain the characteristic equation and the system's natural frequencies (eigenvalues); (iv) determine the natural modes of motion (eigenvectors); (v) state the amount of each natural mode that will be present if motion is initiated as specified.

12.13 The system shown. (Friction is negligible.) For (v), take $x_1(0) = 1$, $x_2(0) = 0$.

Prob. 12.13   Prob. 12.14

12.14 The system shown. For (v), the left-hand capacitor has charge γ, and all other voltages and currents are zero.

12.15 The two-mass system shown. (Friction is negligible.) What is the physical meaning of the fact that one root is $s = 0$? For (v), take $x_1(0) = 1$, $x_2(0) = 0$.

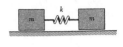
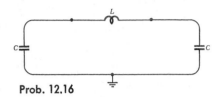

Prob. 12.15 **Prob. 12.16**

12.16 The network shown. What is the physical meaning of the fact that one root is $s = 0$?

12.17 The point-mass system shown: small displacements in the plane of the page. Note that the single point mass has two coupled equations of motion.

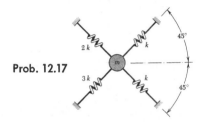

Prob. 12.17

12.18* The two-axis gyro described in Sec. 5.5. The equations of motion are given, (5.63) and (5.64). Describe the physical path followed by the tip of the spin axis as the system vibrates in each of its natural modes.

12.19 Prove Eq. (12.29) by starting with the first equation in Sec. 12.5 and employing suitable trigonometric identities.

12.20 Consider the system of Fig. 12.12a, p. 440. Let $k/m\ell^2 = 0.0016\, g/\ell$, and find expressions for the total motion if one of the pendulums is held vertical and the other released with zero velocity from a displaced position θ_0. In how many cycles will one complete beat pattern occur? Sketch the motion, using Fig. 12.11 as a guide. (Refer to Prob. 12.1 for the equations of motion.)

12.21° Build the system studied in Prob. 12.20, and test your results. (The accurate observation of beat period is especially easy to make.) Good ball bearings or knife edges will be required to achieve the necessary low friction. The spring may be made from piano wire.

12.22 Consider again the system of Prob. 12.14. Now let the central capacitance be 400 C. Determine the number of cycles of rapid oscillation that will occur during one beat period. Assuming the initial voltage across one of the small capacitors is γ, and all other voltages and currents are initially zero, obtain an expression for the resulting time response.

12.23 Derive the equations of motion for the Wilberforce spring, Fig. 12.12b. (Consider only vertical motion and twist, and assume physical constants as needed.) Obtain expressions for its natural frequencies and natural modes of motion.

12.24° Build (or borrow) a Wilberforce spring assembly, and tune it to produce a long beat period.

[Chap. 12] PROBLEMS   821

12.25 Define normal (natural) coordinates for the system in Fig. 12.4, p. 428 (equal springs, equal inertias), and rederive the system equations in terms of them. Then obtain expressions for the time response displayed in Fig. 12.5, but again, in terms of your normal coordinates. Finally, use appropriate geometric relations ("transformation relations") to obtain time expressions for $\theta_1(t)$ and $\theta_2(t)$, and compare these with the response shown in Fig. 12.5b.

12.26 Repeat Prob. 12.25 for the system of Fig. 12.6.

12.27 Define normal (natural) coordinates for the system of Prob. 12.13, and rederive the equations of motion in terms of them. Find the system natural frequencies, and compare with the results of Prob. 12.13. With a constant force f_0 applied leftward to mass $2m$, the system allowed to come to static equilibrium, and then f_0 suddenly removed, determine the ensuing motion, first in terms of the normal coordinates, and then in terms of x_1 and x_2.

In Probs. 12.28 through 12.32 do the following for the (undamped) system depicted: (i) Derive the equations of motion; (ii) find the natural frequencies and natural modes of motion; (iii) sketch the natural modes, and define a set of normal (natural) coordinates; (iv) rewrite the equations of motion in terms of the normal coordinates in such a way as to obtain two uncoupled equations of motion.

12.28 The system shown: small vertical-plane motions only. (Hint: For (iv), take moments about each of the two nodes in turn.)

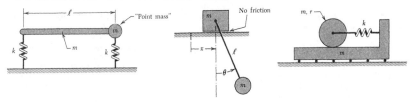

Prob. 12.28   Prob. 12.29   Prob. 12.30

12.29 The system shown: small θ. (Refer to Prob. 5.62.)
12.30 The system shown: cylinder rolling without slipping. (Refer to Prob. 5.65.)
12.31 The double pendulum of Prob. 5.58.
12.32 The system of Prob. 5.68.

In Probs. 12.33 through 12.40 obtain the characteristic equation and plot its roots in the s plane, versus the specified coupling parameter, for values from 0 to ∞. (The networks are analogies of the mechanical systems.)

12.33 The coupling parameter is b, and $k_1 = k$, $k_2 = 10k$. (Scale your s plane in multiples of $\sqrt{k/m}$.)

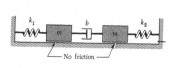   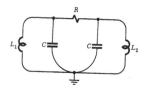

Prob. 12.33   Prob. 12.34

12.34 The coupling parameter is R, and $L_1 = L$, $L_2 = 0.1 L$. (Scale your s plane in multiples of $\sqrt{1/LC}$.)

12.35 Repeat Prob. 12.33, with $k_1 = 0$. Explain physically the meaning of the root $s = 0$.

12.36 Repeat Prob. 12.34, with $L_1 = \infty$ (open circuit). Explain physically the meaning of the root $s = 0$.

12.37 The coupling parameter is k_1.

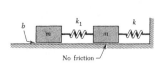

Prob. 12.37   Prob. 12.38

12.38 The coupling parameter is L_1.
12.39 Repeat Prob. 12.37, with $k = 0$.
12.40 Repeat Prob. 12.38, with $L = \infty$ (open circuit).

12.41 Select the starting time in (12.51) to make $\dot{\theta}_1(0)$ equal to zero. Then determine the components of the state form of the eigenvector (12.53) for this case.

For Probs. 12.42 through 12.47 find the eigenvectors and plot the response, assuming the system is initially stationary with the initial conditions specified.

12.42 The system of Prob. 12.35, with $b/m = 0.2\sqrt{k/m}$, and with spring k_2 initially compressed by ξ_0.

12.43 The system of Prob. 12.36, with ρ the initial current in L_2, and $RC = 5\sqrt{LC}$.

12.44 The system of Prob. 12.39, with ξ_0 the initial spring compression.

12.45 The system of Prob. 12.40, with ρ the initial current in L_1. Let $RC = 2\sqrt{L_1C}/3$.

12.46 The system of Prob. 12.37, with ξ_0 the initial spring compression in k_1 and with no initial spring compression in k. Let $b \to 0$.

12.47 The system of Prob. 12.38, with ρ the initial current in L_1, and with no initial current in L. Let $R \to \infty$.

12.48 (i) Verify results (12.56) through (12.61) of Example 12.2, p. 453, by computing the eigenvectors (in phasor form) corresponding to eigenvalues (ℓ_1), (ℓ_2), and (ℓ_3) in Fig. 12.16. Make use of an s-plane array like Fig. 12.17. (ii) Convert the eigenvectors to state form. (iii) Sketch carefully the time response for each natural mode, à la Fig. 12.18.

12.49° Mechanize on a computer the equations of motion (12.8) for the pair of disks of Fig. 12.4 (equal J's and equal k's). Arrange to have plotted, versus time, the physical coordinates θ_1 and θ_2, and also the normal coordinates $q_1 \triangleq \theta_1 + \theta_2$ and $q_2 \triangleq \theta_1 - \theta_2$ (refer to Fig. 12.5a). Duplicate the general time traces in Fig. 12.5b. Then demonstrate that if the system is started in either of its *natural modes*, $\theta_1(0) = \theta_2(0)$ or $\theta_1(0) = -\theta_2(0)$, a purely sinusoidal motion ensues.

12.50° Repeat the study of Prob. 12.49° for the case of unequal J's and unequal k's. Specifically, verify some of the results of Fig. 12.9 for the case $J_2 = 2J_1$, $k_2 = 8k_1$, with k_3 having various values.

CHAPTER 13: COUPLED MODES OF NATURAL MOTION: MANY DEGREES OF FREEDOM

In Probs. 13.1 through 13.12, for each system specified, (i) obtain the system characteristic equation, (ii) obtain the natural frequencies (eigenvalues), (iii) determine and indicate by modal sketches (like Fig. 13.2) the natural modes of motion.

13.1 Three beads of equal mass on a (massless) string. Initial tension F_0 is very high. Consider only motion in the plane of the page. (This is an approximate lumped model for a continuous string.) Use physical intuition and symmetry to deduce one natural mode and thence one natural frequency. (Small motions.)

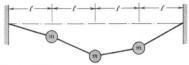

Prob. 13.1

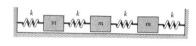

Prob. 13.2

13.2 Three equal masses moving laterally only, without friction. (This is an approximate lumped model for a bar or air column vibrating longitudinally.) Use symmetry as in Prob. 13.1. Compare the form of your results for Probs. 13.1 and 13.2.

13.3 The same system as in Prob. 13.2, but with one end "open."

Prob. 13.3   Prob. 13.4

13.4 The same system as in Prob. 13.2, but with both ends "open." (Make use of symmetry.)

13.5 Electrical analog of the system of Probs. 13.1 and 13.2. (Use node voltages as the variables.) Make use of symmetry to deduce one natural mode and its natural frequency.

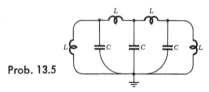

Prob. 13.5

13.6 A four-mass system. (Extend the analysis of Prob. 13.4.)

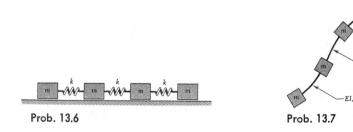

Prob. 13.6   Prob. 13.7

13.7* A lumped model for a flexible rocket vehicle: four point-masses connected by massless beams. Consider small motions in the plane of the picture only. The bending deflection of a cantilever beam is given by $\xi = f\ell^3/3EI$ and its end slope by $\varphi = f\ell^2/2EI$ (where f is force applied at the end, ℓ is length, E is the modulus of elasticity of the material, and I is the area-moment of the cross section; assume EI and ℓ to be given constants). Do not forget rigid-body motions.

13.8 A system of pendulums: consider small rotations only. Discuss the conditions for stability, and find natural modes for both a stable and an unstable value of $k/(mg/\ell)$.

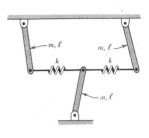

Prob. 13.8   **Prob. 13.9**

13.9* A physical model for a Volkswagen, including suspension springs and tires, but neglecting damping of the shock absorbers. Consider only vertical-plane motion and small angles of rotation. Use coordinates y_C and θ for the body. (The equations of motion were derived in Prob. 5.5.)

13.10 The system of Fig. 13.1, with the J's all equal, the k's all equal, and with $b_1 = 0$, $b_3 = 0$, but $b_2/J = 0.2 \sqrt{k/J}$.

13.11 The system of Prob. 13.5, with the right-hand and left-hand L's replaced by R's, with $RC = 0.0025$ sec, and with $LC = 10^{-4}$ sec². Do parts (i) and (ii) only.

13.12 The system of Prob. 13.5, with only the right-hand L replaced by an R. Do part (i) only.

13.13* In order to see how the analysis develops for lumped systems having a large number of degrees of freedom, consider an axial-flow compressor for a jet engine represented as in Fig. 13.1, but with ten stages. Write the equations of motion for the first three stages and for the last two. Then indicate the form of the determinant from which the characteristic equation would be obtained, by filling in enough of the terms to establish the pattern. [Consider Eqs. (13.2) as the special case $n = 3$.] What order will the characteristic equation have? How many roots will it have? (Digital computers are capable of extracting the roots from equations this large in a remarkably short time.)

13.14 Derive the exact equation of motion, (13.26), for small lateral displacements of a uniform taut string. The keys to success will be carefully made geometric definitions analogous to those in Fig. 13.3, plus a carefully drawn free-body diagram, analogous to Fig. 13.4, of an infinitesimally short string segment. Modify Eq. (13.24) so that it yields solutions for natural vibrations of a string. Compare your results with those obtained in Prob. 13.1 for a lumped-model approximation, using sketches like Fig. 13.9.

[Chap. 13] PROBLEMS 825

13.15 A violin string is 32 cm long. Calculate the tension in a properly tuned A string (440 cps), if $\mu = 0.0075$ gm/cm.

13.16 Derive the exact equation of motion, (13.28), for small *longitudinal* displacements of a metal bar or an air column (e.g., in an organ pipe). The keys to the derivation will be precise definition of geometric quantities, as stated in Fig. 13.14 and described on p. 478, and a free-body diagram, analogous to Fig. 13.4, of an infinitesimally short slice of the column. Modify Eqs. (13.23) and (13.24) so that they yield solutions for natural vibrations of an organ pipe which is (i) closed at one end and open at the other, and (ii) closed at both ends. Compare your results with those of Probs. 13.3 and 13.2, respectively.

13.17* Calculate the length of an organ pipe whose pitch is 16 cps (about the lower audible limit for human ears), if the cross-sectional area of the pipe is uniformly 40 cm^2.

13.18 Use the information in Eq. (13.27) and Fig. 13.15, p. 481, to determine the finger locations a violinist should use to play (part of) a natural scale in C on his A string, if the string is tuned to vibrate at 440 cps when it is open (not fingered). The length of a violin string is 32 cm.

13.19 Repeat Prob. 13.18 for an equal-tempered scale in C. What is the largest difference in finger placement (in cm) between one scale and the other?

13.20 Repeat Prob. 13.18 for the natural scale in A.

13.21° Mark the fingerboard of a violin with the locations you calculated in Probs. 13.18 and 13.19. Then, "stopping" the string *accurately* (e.g., with the thumb nail), pluck the notes of the scales calculated. Can you hear the difference between the natural and equal-tempered scales? The A–C interval is easy to check because A is open for both scales.

In Probs. 13.22 through 13.37 use Rayleigh's method to calculate the approximate first natural frequency of the specified system, without deriving the equations of motion. In each case, assume an approximate shape that is *convenient*. Compare your answer with the exact answer obtained in previous work.

13.22 The torsional system of Fig. 12.4, p. 428.

13.23 The torsional system of Fig. 12.8, p. 435.

13.24 The translational system of Prob. 12.15. Use symmetry to make the problem trivial.

13.25 Repeat Prob. 13.24 for a right-hand mass of $5m$.

13.26 The translational system of Prob. 12.13.

13.27 The electrical system of Prob. 12.14. The energy stored in a linear capacitor is given by $U = \int v\, dq = \frac{1}{2}Cv^2$ (where v is the voltage drop across the capacitor), while the energy stored in a linear inductor is given by

$$T = \int (\text{power})\, dt = \int iv\, dt = \int i\left[L\frac{di}{dt}\right] dt = \tfrac{1}{2}Li^2$$

13.28 The beads-on-a-string system of Prob. 13.1.

13.29 The electrical system of Prob. 13.5. (Refer to Prob. 13.27.)

13.30 The double-pendulum system of Prob. 12.20.

13.31 The two-dimensional system of Fig. 12.13, p. 442.

13.32 The Wilberforce spring of Prob. 12.23.
13.33 The two-dimensional system of Prob. 12.28.
13.34 The sliding pendulum system of Prob. 12.29.
13.35* The flexible-rocket model of Prob. 13.7.
13.36* The distributed string of Prob. 13.14. Consider the string tension constant, so that the stored potential energy is obtained from $U = \int_{\xi_1}^{\xi_2} F_0 \, d\xi$, where ξ is the amount of string stretch.
13.37* The distributed air column of Prob. 13.16.

CHAPTER 14: e^{st} AND TRANSFER FUNCTIONS

In Probs. 14.1 through 14.7 (i) derive the equations of motion; (ii) obtain the transfer function for each of the subsystems, and show their interconnection in a block diagram like Fig. 14.3b; (iii) write by inspection the forced part of the system response to the input specified.

14.1 The system of Fig. 12.1a, if a disturbing moment $M_1(t)$ is applied to disk J_1. For part (iii) let $M_1 = M_{10}e^{-\sigma_f t}$.

14.2 A cascaded electrical system. For part (iii), $v_1 = v_{10}e^{-\sigma_f t}$.

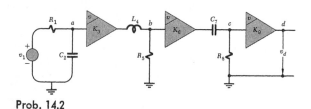

Prob. 14.2

14.3 The system of reservoirs in Fig. 14.1. Consider small *changes* from a steady-flow condition, so that linear equations of motion may be used. (Refer to Example 3.2.) For part (iii) let the *change* in flow rate into the first reservoir be given by $w_1 = w_{10}e^{-\sigma_f t}$.

14.4 The system shown: a setup for vibration testing of an instrument-panel mount like that in Fig. 1.1. Input motion $x(t)$ is prescribed. The displacement of the package is sensed by an inductive voltage-modulation device whose output can be assumed perfectly proportional to displacement: $v_a = K_i y$. This electrical signal is filtered, as shown, and then used to drive a pen recorder, via a current amplifier ($M = K_2 v_b$). For part (iii) of this problem, let $x = x_0 e^{-\sigma_f t}$. (The system will be studied further in Probs. 14.11 and 15.5.)

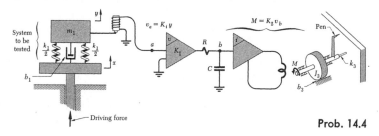

Prob. 14.4

14.5 A hydraulic boost system. The torque motor of Fig. 6.2 operates the valve, and the piston output drives the large inertia of a rocket engine (Fig. 3.12) through a flexible link. In parts (i), (ii), and (iii) assume the oil to be incompressible, so that $\dot{y} = c\phi$. For part (iii) let $v_a = v_{a0}e^{-\sigma_f t}$.

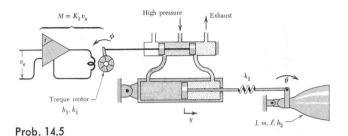

Prob. 14.5

14.6 The mechanical system shown, with the influence of motion x_3 on motion x_2 ignored (because k_2 is extremely small). In part (iii) let $f_1 = f_{10}e^{-\sigma_f t}$, and find both $x_1(t)$ and $x_2(t)$.

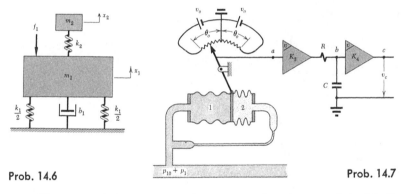

Prob. 14.6 Prob. 14.7

14.7 The pneumatic sensor shown. (Its equations of motion were derived in Example 3.3.) In part (iii) find $v_c(t)$, with $p_1(t) = p_{1m}e^{-\sigma_f t}$.

14.8 For the system of Fig. 14.3 assume the following numerical values, and calculate the forced part of the response, with $v = 3 \cos 20t$ volts: $K_1 = 50$, $RC = 0.02$ sec, $k = 1000$ in. lb/rad, $K_2 = 2$ in. lb/volt, $J = 10$ in. lb/(rad/sec^2), $b = 40$ in. lb/(rad/sec). (The technique of Fig. 14.5 may be helpful.) Also, determine by inspection the steady output angle of the system if v has the constant value of 3 volts.

14.9 Repeat Prob. 14.8, with an input of $v = 3e^{-4t} \cos 20t$ volts.

14.10 For the system of Prob. 14.2, assume the following numerical values and calculate the forced part of the response, with $v_1 = 3e^{-40t} \cos 200t$: $R_1C_2 = 0.01$ sec, $L_4/R_5 = 0.05$ sec, $R_8C_7 = 0.004$ sec, $K_3 = 3$ volts/volt, $K_6 = 6$ volts/volt, $K_9 = 9$ volts/volt.

14.11 For the system of Prob. 14.4, calculate the forced part of the response $\theta(t)$, with $x(t) = 0.1 \cos 30t$ in., and $k_1 = 5$ lb/in., $\sqrt{k_1/m_1} = 30$ sec^{-1}, $b_1/2m_1 = 0.5$ sec^{-1},

$RC = 0.005$ sec, $k_3 = 0.1$ in. lb/rad, $\sqrt{k_3/J_3} = 600$ sec^{-1}, $b_3/2J_3 = 200$ sec^{-1}, $K_i = 100$ v/in., $K_1 = 1$, and $K_2 = 0.001$ in. lb/v.

In Probs. 14.12 through 14.26 (i) write the equations of motion (they will usually be available from earlier work); (ii) assume solutions of the form $x = Xe^{st}$, and obtain a set of algebraic equations in matrix form, as Eqs. (14.24) and (14.31); (iii) use Cramer's rule to solve these for the transfer function specified, as (14.33) and (14.34). (Do not factor the denominator.)

14.12 A translational system (refer to Prob. 12.13). In part (iii) find X_3/F_1 and X_2/F_1.

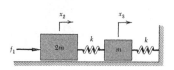

Prob. 14.12

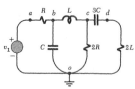

Prob. 14.13

14.13 A network (refer to Prob. 12.4). In part (iii) find V_c/V_1 and also the driving-point impedance I_1/V_1.

14.14 The network of Fig. 2.19. [Refer to Eqs. (2.65).] In part (iii) find V_c/I_1.

14.15 An electromechanical speaker system (refer to Sec. 2.9 and to Prob. 12.7). In part (iii) find X/V_1.

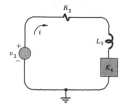 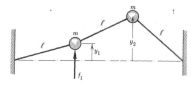

Prob. 14.15   **Prob. 14.16**

14.16 Beads on a taut string, one of which is being forced (refer to Prob. 12.9). In part (iii) find Y_1/F_1 and Y_2/F_1. (Assume y's $\ll \ell$.)

14.17 The system of Prob. 14.6, but now with two-way influence. In part (iii) find X_1/F_1 and also X_2/F_1. Compare with the results obtained by using a cascaded model in Prob. 14.6.

14.18 The translational system of Fig. 4.4a.

14.19 The network of Fig. 4.4b (the electrical analog for Prob. 14.18).

14.20 A six-degree-of-freedom system, with one coordinate (x) constrained. The only damping is b_4. In part (iii) find Y_5/X and Y_3/X.

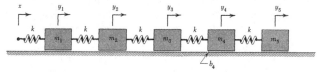

Prob. 14.20

[Chap. 14] PROBLEMS

14.21* The Wilberforce spring of Fig. 12.12b, with the top given a vertical displacement $x(t)$ (refer to Prob. 12.23).

14.22* The Volkswagen model of Prob. 13.9. The ground "displacement" under the front wheels is now $x_1(t)$ and that under the rear wheels $x_2(t)$. In part (iii) find Y_C/X_1 and Θ/X_1.

14.23* With the road shape in Prob. 14.22 approximated by $x = x_0 \cos 2\pi z/L$, where z is horizontal location along the road and L is the distance between bumps, and with a car length of ℓ, derive appropriate expressions for $x_1(t)$ and $x_2(t)$. (A road with regular bumpiness might be approximated more closely by a Fourier series of such inputs.)

14.24 The two-axis gyro, Eqs. (5.63), (5.64). In part (iii) find Ψ/M_b.

14.25* The lumped model of a flexible rocket vehicle of Prob. 13.7, with a lateral force $f_1(t)$ applied at one end.

14.26 The network shown, which has mutual inductance. Assume that positive di_3/dt produces positive drop v_5, and vice versa. To avoid confusion, it will be helpful to follow each separate step of Procedure A-e. Then the assumption that all motions have the form $v = Ve^{st}$ or $i = Ie^{st}$ can be made, and the resulting equations combined algebraically.

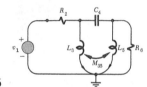

Prob. 14.26

In Probs. 14.27 through 14.30 define a propitious set of state variables for the system specified, and rewrite in standard form the *algebraic* version of its matrix equations of motion, that is, in the form (14.25e), with all matrices defined explicitly as in (14.25c).

14.27 The two-mass system of Prob. 14.12.

14.28 The network of Prob. 14.14.

14.29 The speaker system of Prob. 14.15.

14.30 The two-axis gyro of Prob. 14.24.

14.31 Find expressions for the coefficients a_0, a_1, a_2, a_3 in Eqs. (14.34) in terms of the R's, L's, and C's of Eq. (14.33).

In Probs. 14.32 through 14.34 find (i) the driving-point impedance Z_{11}, (ii) the driving-point admittance Y_{11}, and (iii) the transfer function specified.

14.32 A bridge network. In part (iii) find transfer impedance V_9/I_1 and also transfer function V_9/V_1.

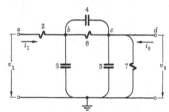

Prob. 14.32   Prob. 14.33

14.33 A five-node network. In part (iii) find the transfer function V_9/V_1. Assume that positive di_4/dt produces positive drop v_7, and vice versa.

14.34 The network of Fig. 2.19. [For part (ii) the current source must be replaced by a voltage source, of course.]

CHAPTER 15: FORCED OSCILLATIONS OF COMPOUND SYSTEMS

Note: In the problems which follow, a convenient scale for log plotting will usually be $\frac{3}{4}$ in. = 2, for which ordinary $\frac{1}{4}$-in. "quadrille" paper is quite satisfactory. (When precision is required, the same scale with fine subdivisions can be found on the edge of a Spirule.)

15.1 Sketch the frequency-response (Bode) plot on logarithmic coordinates for the system shown below, with $b_1 = 0.1$ in. lb/(rad/sec), $b_1/J_1 = 0.1$ sec^{-1}, $K_1 K_2 = 0.02$ in. lb/(rad/sec), $b_2 = 0.3$ rad/sec, $b_2/J_2 = 1$ sec^{-1}.

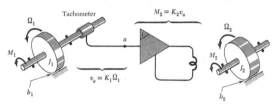

Prob. 15.1

15.2 Repeat Prob. 15.1, using measured wheel *position* θ_1, rather than the speed Ω_1, as the input to the isolation amplifier. Can you infer from this the qualitative effect of a single integration on the logarithmic frequency-response plot?

15.3 Sketch the frequency-response (Bode) plot for the system of Fig. 15.2, using the following values for the parameters: $K_1 = 10$, $K_2/b = 10$ (rad/sec)/v, $b/J = 4$ sec^{-1}, and $RC = 0.25$ sec.

15.4 Sketch the Bode plot for the uncoupled bandpass filter arrangement shown, with $K_1 = K_2 = K_3 = 2$ and $R_1 C_1 = 1$ sec, $R_2 C_2 = 0.2$ sec.

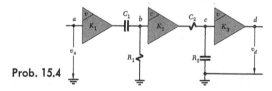

Prob. 15.4

15.5 Sketch the frequency response for the system of Probs. 14.11 and 14.4.

15.6 The figure shows a physical model of part of a servo system, including a network, current amplifier, and rotor. It is understood that this is a reasonably good model. The values of C, R_1, R_2, and J are well known; but K and b are in doubt. To determine what they are, a frequency-response test is run in which $v_5 = v_{5m} \cos \omega_f t$ is supplied from a calibrated voltage source, and shaft velocity Ω is measured. The resulting data are plotted below. Determine the values of K and b, with $C = 1$ μf = $1^{\backslash -6}$ farad, $R_1 = 50K = 5^{\backslash 4}$ ohms, $R_2 = 12.5K = 1.25^{\backslash 4}$ ohms, $J = 0.5$ in. lb/(rad/sec^2). Suppose it is known qualitatively that the mechanical "break frequency" b/J is much lower than the electrical break frequencies. After a system transfer function has been derived, all the desired information can be deduced from the frequency-response plot by fairing in various asymptotes. The steps are indicated in numerical sequence in the plot below. Figure out the meaning of each construction line, and deduce the numerical values of K and b.†

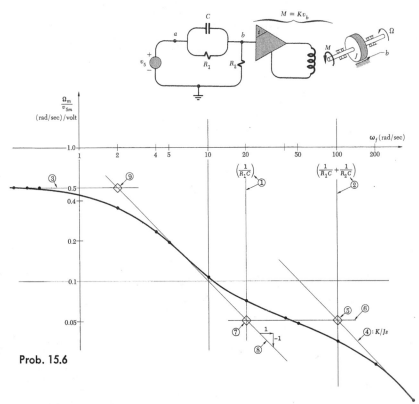

Prob. 15.6

15.7 The transfer function for the system in Prob. 15.6 has no poles or zeros in the right half-plane. Therefore, by Bode's theorem, p. 518, the magnitude plot given defines completely the corresponding phase plot. Sketch it.

15.8 A frequency-response test was run on the high-pass filter shown in Fig. 9.12. A plot of v_{2m}/v_{1m} and the phase shift versus ω_f was made, as shown. Does the plotted

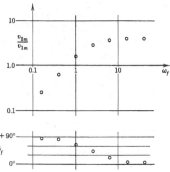

Prob. 15.8

† See footnote, page 20.

data abide by Bode's theorem, p. 518? The resistor's value was found to be 10 kilohms. Find the value of the capacitor. Assume that the resistor is replaced by a one-megohm resistor, and plot the frequency response that results.

15.9 A certain system can be represented by two cylinders, arranged as shown, with a damping fluid carrying a torque between them. (This damping, associated with the *relative* motion between cylinders, is the only damping existing in the system.) In order to determine the damping strength b, a test is devised in which a sinusoidal moment, $M = M_m \sin \omega_f t$, is applied to the first disk, and the resulting angular acceleration $\ddot{\theta}_1$ of the *same* disk is measured. The test results are shown plotted on log coordinates below. What is the value of b? (Specify units.) Does the plot obey Bode's theorem?

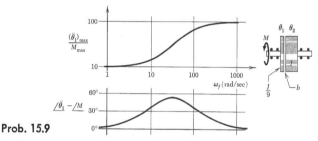

Prob. 15.9

15.10 For the network shown, sketch carefully a Bode plot for the function V_b/V_1, with $R_2C_3 = 1\backslash^{-3}$ sec, $R_5C_4 = 1\backslash^{-4}$ sec, $R_2C_4 = 1\backslash^{-6}$ sec.†

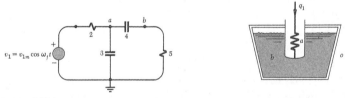

Prob. 15.10 **Prob. 15.11**

15.11 A heating element is inserted into a bath and cycled at various frequencies, from once per day to once per minute. Assume that the water in the bath is well stirred, so that a two-lump approximation for the system is reasonable. The time constant of the heating element alone has been measured previously, with the heat turned off, and is only 1 minute (because of its low thermal resistance and low thermal capacity). The thermal capacity of the bath is 100 times that of the heater, and the thermal resistance of the insulated outer enclosure of the bath is the same as the thermal resistance of the heater case, despite its much greater area. Suppose the heat input is $q_1 = q_{1m}(1 - \cos \omega_f t)$, and T_b is measured. Sketch carefully the resulting Bode plot $\overline{T}_b/Q_1$. For convenience, measure T_a and T_b with respect to T_o.

15.12 Consider again the water-supply system of Fig. 3.10, and consider small changes from a steady-flow condition (in which $w_{20} = w_{10} = $ const.). The input is constant ($w_1 = 0$), but the output is more or less periodic, and can be approximated by a summation of sinusoids (a Fourier series), the dominant ones having periods of

† See footnote, page 20.

one-half day (some citizens shower in the morning, some at night), one day, one week, and one year. To help see the situation, sketch a Bode plot of the function H_2/W_2, with $\rho A_2 R'$ (the transient time constant of the small reservoir when the large one is maintained at constant depth) as one day, and $\rho A_1 R'$ (the time constant of the large reservoir when the small one has constant depth) as one month.

15.13 For the network shown, sketch carefully the Bode plot for V_b/V_1 and also V_a/V_1, with $C_3 = C_5 = C_7 = 1\backslash^{-5}$ farad, $R_2 = 2\backslash^4$ ohms, $R_4 = 1\backslash^5$ ohms, $R_6 = 1\backslash^6$ ohms.

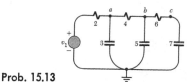

Prob. 15.13

In Probs. 15.14 through 15.20 sketch carefully a Bode plot (magnitude and phase) for the system specified, using the numerical values given or suitable dimensionless ratios. The appropriate transfer function is available from previous work.

15.14 $\Theta_1/(M_1/k)$ for the system shown. Plot using suitable dimensionless ratios.

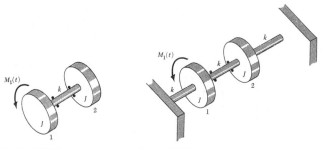

Prob. 15.14   **Prob. 15.15**

15.15 $\Theta_1/(M_1/k)$ for the system shown. (Sec. 12.3 may be helpful.) Plot, using suitable dimensionless ratios.

15.16 Repeat Prob. 15.15, for damping b on each disk, with $b/2J = 0.1\sqrt{k/J}$.

15.17 Y_1/F_1 for the beads on a string in Prob. 14.16. Again, use suitable dimensionless ratios. Compare with results of Prob. 15.15.

15.18 X_3/F_1 for the translational system of Prob. 14.12. (No damping.)

15.19 X/V_1 for the speaker system of Prob. 14.15, with $L/R = 10^{-5}$ sec, $\sqrt{k/m} = 6\backslash^3$ sec^{-1}, $R_2 = 10^4$ ohms, $K_4 = 0.01$ n/amp, $k = 30$ n/m, and $v_1 = 3 \cos \omega_f t$ volts.

15.20 I_1/V_1 for the network of Prob. 14.32, with the R's equal and the C's equal.

15.21 From Fig. 15.5, determine the ranges of values of b/J_1 over which the system of Example 15.2 (Fig. 15.4) will exhibit resonance in its frequency response. Use numerical values from Example 15.2. Explain physically why there are two such ranges. Sketch very carefully a Bode plot of $\Omega_2/(\dot{M}/k)$ for $b/J_1 = 0.6$ sec^{-1}. Then, for comparison, copy to the same scale the plot of Fig. 15.6 for $b/J_1 = 7.7$ sec^{-1}.

15.22 Repeat Prob. 15.21 for $b/J_1 = 88$ sec^{-1}.

15.23 Consider again the undamped three-disk system of Fig. 13.1, with all J's equal and all k's equal (figure below). Its natural frequencies and natural modes of motion are as given in Fig. 13.2; we expect that if the system is forced sinusoidally, it will resonate at any of these frequencies. To verify this, assume that a moment $M_1 = M_{1m} \cos \omega_f t$ is applied to the first disk, as shown, and sketch the resulting Bode plots for $\Theta_1/(\mathsf{M}_1/k)$, $\Theta_2/(\mathsf{M}_1/k)$, and $\Theta_3/(\mathsf{M}_1/k)$. (The phase plots are, of course, very easy to sketch—though not trivial. Include them.)

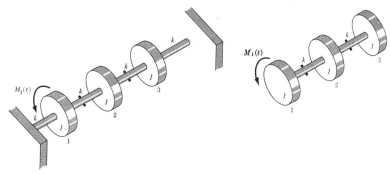

Prob. 15.23

15.24 Repeat Prob. 15.23 for the case in which the torsional springs at the ends are removed, as shown.
15.25 For what range of values will the fluid-clutch system of Example 15.2, p. 514, have underdamped roots? (Refer to Fig. 15.5.)
15.26 Obtain the transfer function for Ω_2/M for the resonant torsional system of Example 15.3, p. 519. Sketch the Bode plot, and compare with Fig. 15.8.
15.27 Sketch the phase-angle plot to accompany Fig. 15.8.
15.28 Make a design study of the undamped vibration absorber, Sec. 15.4, and produce a plot of ω_{0d} and ω_{0c} versus m_2/m_1. (Refer to Fig. 15.12.) What value for m_2/m_1 seems to you to be practical and useful in a typical application?
15.29 For the undamped vibration absorber, Sec. 15.4, obtain the transfer function for $Y_2/(F/k_1)$, and make a plot of the value of $y_{2\text{max}}$ at resonance versus m_2/m_1.
15.30 Sketch carefully a Bode plot (magnitude and phase) of v_g/v_b for the lead network of Prob. 2.85 with $R_6/R_7 = 9$. Show that, for a steady sinusoidal input, v_g "leads" v_b; i.e., $\psi = /V_g/V_b$ is positive. For what driving frequency is ψ maximum? What is its value there?

CHAPTER 16: RESPONSE TO PERIODIC FUNCTIONS: FOURIER ANALYSIS

16.1 Find the coefficients for a complex Fourier series to represent the function below. Find the first seven coefficients. Represent your results in a spectral diagram.

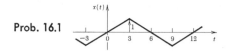

Prob. 16.1

[Chap. 17] PROBLEMS 835

16.2 Plot the time function obtained in Prob. 16.1 for one, two, and three terms (e.g., for $n = \pm 1, \pm 2, \pm 3$).

16.3 Repeat Prob. 16.1 for the following partial cosine wave.

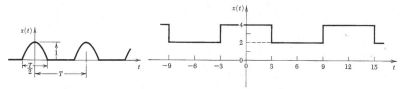

Prob. 16.3   Prob. 16.4

16.4 Repeat Prob. 16.1 for the function shown above.

16.5 Repeat Prob. 16.1 for the sawtooth wave shown below.

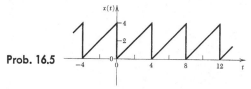

Prob. 16.5

16.6 Repeat Prob. 16.1 for the train of cosine-squared pulses shown below.

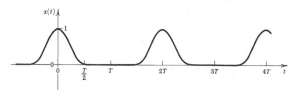

Prob. 16.6

16.7 Calculate the terms in a Fourier series to represent the evenly spaced train of impulses shown, and plot the spectrum of the series. Let the area of each impulse be A and the spacing T, and perform the integration over a period beginning *just before* an impulse.

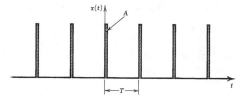

Prob. 16.7

16.8 Develop the complex Fourier series to represent the shape of a plucked string, as represented in Fig. 13.13a, p. 477, and verify the spectral diagram of Fig. 13.13d.

CHAPTER 17: THE LAPLACE TRANSFORM METHOD

In Probs. 17.1 through 17.9 derive the $\mathcal{L}_-$ transform (which in these cases will be identical to the $\mathcal{L}_+$ transform) of the given function, and compare your result with Table X. In each case, verify the pole-zero array given in Table X.

17.1 $u(t) \cos \omega t$.
17.2 $u(t) \sin (\omega t + \psi)$.
17.3 $u(t) \cos (\omega t + \psi)$. (Check against Table X by redefining ψ.)
17.4 $u(t)e^{-\sigma_a t} \sin (\omega_a t + \psi)$.
17.5 $u(t)(1 - e^{-\sigma_a t})$.
17.6 $u(t)(e^{-\sigma_a t} - e^{-\sigma_b t})$.
17.7 $u(t)te^{-\sigma_a t}$.
17.8 $u(t)\left[e^{-\sigma_a t} + \dfrac{b_a}{\omega_b} \sin (\omega_b t - \psi)\right]$.
17.9 $u(t - \tau)f(t - \tau)$ (where f is any arbitrary function of time).
17.10 Derive the $\mathcal{L}_-$ transform of $u(t)te^{-\sigma_a t} \sin \omega_a t$.
17.11 Derive the $\mathcal{L}_-$ transform of the time function shown.

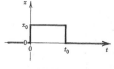
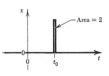

Prob. 17.11 Prob. 17.12

17.12 Derive the $\mathcal{L}_-$ transform of the time function shown, first by direct integration in Eq. (17.11c), and second by application of transform pair 5.2 of Table X.
17.13 Derive the Laplace transform for the function shown by the solid lines.

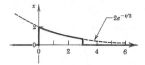
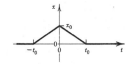

Prob. 17.13 Prob. 17.14

17.14 Derive both the $\mathcal{L}$ transform and the $\mathcal{L}_-$ transform for the function shown.
17.15 Derive an expression for the $\mathcal{L}_-$ transform of the nth integral of a function $x(t)$. [Refer to the procedure employed in Eq. (17.26).]

In Probs. 17.16 through 17.25 (i) take the $\mathcal{L}_-$ transform of the indicated equations of motion. [This is Step 1 of Procedure D-1.] Then (ii) assume all initial conditions to be zero, and write down the specified transfer function. [This is part of Step 2 of Procedure D-1.] In each case, check the dimensions of your answer, both from term to term, and overall.

17.16 Equation (2.21) for a "torque motor," p. 49. Find $\Theta(s)/\mathsf{M}_1(s)$.
17.17 Equations (2.65) for a network, p. 73. Find $V_c(s)/I_1(s)$.
17.18 Equations (2.77) for an electric motor, p. 84. Find $\Omega(s)/V_1(s)$, with $M_7 = 0$.
17.19 Equations (3.7) for heat flow through the walls of a rocket engine, p. 101. Find $\mathsf{T}_b(s)/\mathsf{T}_i(s)$, with $T_o = 0$ (i.e., if all other T's are considered measured with respect to T_o).
17.20 Equations (3.43) for a pneumatic computer element, p. 117. Find $X(s)/P_1(s)$.
17.21 Equations (5.63) and (5.64) for a two-axis gyro, p. 163. Find $\Theta(s)/\mathsf{M}_b(s)$, with $M_c = 0$.

[Chap. 17] PROBLEMS   837

17.22 The translational system of Prob. 14.12. Find $X_2(s)/F_1(s)$, and compare with your earlier result.

17.23 The speaker system of Prob. 14.15. Find $X(s)/V_1(s)$, and compare with your earlier result.

17.24 The three-disk system of Prob. 15.23. Find $\Theta_2(s)/(M_1(s)/k)$, and compare with your earlier result.

17.25* The Volkswagen model of Prob. 14.22. Find Y_C/X_1 and Θ/X_1, and compare with your earlier result.

17.26 through **17.34.** For these problem numbers, (i) write the *complete response function* for the first-named variable in each of Probs. 17.16 through 17.24, respectively, for arbitrary values of all initial conditions. [This is, of course, part of Step 2(a), Procedure D-1.] Also, (ii) write explicitly the system characteristic equation. Obtain its roots. [This is Step 2(b).] In Prob. 17.33 let $R_2/L_3 = 1.4\backslash^3$, $k_6/m_5 = 5.71\backslash^4$, $k_4{}^2/L_3 m_5 = 4.23\backslash^5$.

17.35 In Eq. (17.4), is the s which appears in the denominator of the first term a characteristic of the system? If it is, why does it not appear also in the second term? If it is not, where does it come from?

In Probs. 17.36 through 17.49, use Table X to obtain the explicit time function $x(t)$ corresponding to the given $\mathcal{L}$ transform expression, i.e., the given response function $X(s)$.

17.36 $X(s) = \dfrac{2}{s + 3}$

17.37 $X(s) = \dfrac{6}{s^2 + 9}$

17.38 $X(s) = \dfrac{6s}{s^2 + 9}$

17.39 $X(s) = \dfrac{6(s + 1)}{s^2 + 9}$

17.40 $X(s) = \dfrac{6}{s(s + 9)}$

17.41 $X(s) = \dfrac{6}{(s + 1)(s + 9)}$

17.42 $X(s) = \dfrac{6}{(s + 9)^2}$

17.43 $X(s) = \dfrac{6}{(s + 1)(s^2 + 9)}$

17.44 $X(s) = \dfrac{s + 6}{s^2 + 10s + 9}$

17.45 $X(s) = \dfrac{s + 6}{s^2 + 2s + 9}$

17.46 $X(s) = \dfrac{s + 6}{s^2 + 6s + 9}$

17.47 $X(s) = \dfrac{e^{-0.5s}}{s}$

17.48 $X(s) = \dfrac{2e^{-0.5s}}{s + 3}$

17.49 $X(s) = \dfrac{6e^{-0.5s}}{s^2 + 9}$

17.50 Use your results from Prob. 17.43 to write down the steady-state part only of the response to a continuing sine wave, $3 \sin 3t$, of a system whose transfer function is $2/(s + 1)$.

17.51 From Eq. (17.43), write down the steady-state sinusoidal response of the fluid clutch analyzed in Example 17.7. Compare with Example 17.9.

In Probs. 17.52 through 17.58 you are to obtain an explicit expression for the time response of the system named, for the given disturbance and initial conditions. [This is Step 3 of Procedure D-1.]

17.52 The torque motor of Fig. 2.5 and Prob. 17.16: Find $\theta(t)$, with

$$M_1(t) = M_{10}u(t)e^{-\sigma_f t} \quad \text{and} \quad \theta(0^-) = \theta_0, \quad \dot{\theta}(0^-) = 0$$

Let $k/J = 1\backslash^4 \text{ sec}^{-2}$, $b/2J = 200 \text{ sec}^{-1}$, $\sigma_f = 10 \text{ sec}^{-1}$.

17.53 The network of Fig. 2.19 and Prob. 17.17: Find $v_c(t)$, with $v_3(0^-) = \gamma$ and $i_6(0^-) = \rho$, $i_7(0^-) = 0$, and $i_1(t) = i_{1m}u(t) \cos \omega_f t$. (Current ρ is, of course, the value of $(1/L_6)\int v_c \, dt$ at $t = 0^-$, by the physical relation for an inductor.) Let $C_3R_4 = C_3R_5 = 0.1$ sec, $L_6/R_5 = L_7/R_5 = 0.02$ sec, $C_3R_2 = 0.2$ sec.

17.54 The electric motor of Fig. 2.24 and Prob. 17.18: Find $\Omega(t)$, with $v_1(t) = v_{10}u(t)$, and $i(0^-) = \rho$, $\Omega(0^-) = \Omega_0$, $M_7 = 0$. Let $J_5/b_6 = 10$ sec, $L_3/R_2 = 0.1$ sec, $K_4^2/J_5L_3 = (4.45)^2 \sec^{-2}$.

17.55 The rocket-engine heat-flow model of Fig. 3.3 and Prob. 17.19: Find $T_b(t)$, with $T_b(0^-) = T_c(0^-) = T_i(0^-) = T_o = 0$ (T_o is taken arbitrarily as the datum), and $T_i(t) = T_{i0}u(t)$. Let $R_2C_3 = 5$ min, $R_4C_3 = 1$ min, $R_4C_5 = 0.2$ min, $R_6C_5 = 1$ min.

17.56 The pneumatic computer element of Fig. 3.11 and Prob. 17.20: Find $x(t)$, if $p_1(t) = p_{10}u(t)e^{-\sigma_f t}$ and $p_2(0^-) = 0$, $x(0^-) = x_0$. Let $A\ell R/K_2 = 0.2$ sec, $A^2\rho_0 R/k_3 = 0.02$ sec.

17.57 The torque motor of Prob. 17.52 again, this time with an impulsive input, $M_1(t) = (M_{10}\tau)\,\delta(t)$ and $\theta(0^-) = 0$, $\dot\theta(0^-) = \Omega_0$. Find also the response $\Omega(t)$, by first multiplying the response function by s—because $\Omega(s) \equiv s\Theta(s)$, Pair 1 of Table X—and then inverse transforming. As a check, differentiate your expression for $\theta(t)$.

17.58 The network of Prob. 17.53 again, this time with an impulsive input $i_1 = (i_{10}\tau)\,\delta(t)$.

17.59 Use your results from Prob. 17.53 to write down the steady-state response only of the system to a sinusoidal input $i(t) = i_{10} \cos \omega_f t$. Compare the total work required with that of letting $s = j\omega_f$ in the transfer function obtained in Prob. 14.14.

17.60 through 17.65. For these problem numbers, apply the Final-Value Theorem and the Initial-Value Theorem to the response functions you used in Probs. 17.52 through 17.57, respectively. Then, for the corresponding time functions, set $t = \infty$ and then set $t = 0$ and check the FVT and IVT predictions. When FVT is not applicable, explain this on physical, as well as mathematical, grounds. When you find that a variable has $x(0^+) \neq x(0^-)$, as in Prob. 17.64, explain on physical grounds.

17.66 For the fluid clutch of Example 17.7, p. 563, obtain the response function for $\Omega_1(s)$. Then apply FVT to show at once that $\Omega_1(\infty) = \Omega_2(\infty)$. [$\Omega_2(s)$ is given by (17.42).]

In each of Probs. 17.67 through 17.75, a set of equations of motion is specified. (i) Rewrite the equations of motion in standard form [as Eqs. (17.64)], selecting additional variables as needed (and including their definitions as additional equations of motion). (ii) $\mathcal{L}_-$ transform the equations, assuming arbitrary forcing functions [e.g., $\mathsf{M}_1(s)$] and arbitrary initial conditions [e.g., $\Omega_1(0^-)$]. (iii) Construct a block diagram of the result, like Fig. 17.13b or c, with initial conditions represented as being established by unit impulses.

17.67 Equation (2.21) for a torque motor, p. 49.
17.68 Equations (2.65) for a network, p. 73.
17.69 Equations (2.77) for an electric motor system, p. 84.
17.70 Equations (3.7) for heat flow through the walls of a rocket engine, p. 101.
17.71 Equations (3.43) for a pneumatic computer element, p. 117.
17.72 Equations (5.63) and (5.64) for a two-axis gyro [both $\mathsf{M}_b(s)$ and $\mathsf{M}_c(s)$ present].

[Chap. 18] PROBLEMS 839

17.73 The equations obtained in Fig. 4.4a, p. 129, for a three-mass system.
17.74 The equations obtained in Fig. 4.4b, p. 129, for a three-capacitor system.
17.75 The equations obtained in Prob. 14.22 for a Volkswagen.

In Probs. 17.76 through 17.82 obtain, in two ways, the output time response $y(t)$ corresponding to the given system transfer function $H(s)$ [which is always also the $\mathcal{L}$ transform of the system's impulse response $h(t)$] and the given input $x(t)$, having $\mathcal{L}$ transform $X(s)$: (i) by taking the $\mathcal{L}^{-1}$ transform of the product,

$$y(t) = \mathcal{L}^{-1}[Y(s)] = \mathcal{L}^{-1}[H(s)X(s)]$$

and making use of Table X; (ii) by convolving the corresponding time functions, $y(t) = h(t) * x(t)$.

17.76 $h(t) = u(t)e^{-t}$ $\quad H(s) = \dfrac{1}{s+1}$ $\quad x(t) = 3u(t)$ $\quad X(s) = \dfrac{3}{s}$

17.77 $h(t) = u(t)e^{-t}$ $\quad H(s) = \dfrac{1}{s+1}$ $\quad x(t) = 8u(t)e^{-5t}$ $\quad X(s) = \dfrac{8}{s+5}$

17.78 $h(t) = 0.5u(t)(e^{-t} - e^{-5t})$ $\quad H(s) = \dfrac{2}{(s+1)(s+5)}$
$x(t) = 4\delta(t)$ $\quad X(s) = 4$

17.79 $h(t) = u(t)\sin 2t$ $\quad H(s) = \dfrac{2}{s^2+4}$
$x(t) = 3u(t)e^{-0.1t}$ $\quad X(s) = \dfrac{3}{s+0.1}$

17.80 $h(t) = u(t)e^{-0.1t}$ $\quad H(s) = \dfrac{1}{s+0.1}$
$x(t) = 3u(t)\sin 2t$ $\quad X(s) = \dfrac{6}{s^2+4}$

17.81 $h(t) = u(t)e^{-2t}$ $\quad H(s) = \dfrac{1}{s+2}$ $\quad x(t) = 4u(t)e^{-2t}$ $\quad X(s) = \dfrac{4}{s+2}$

17.82 $h(t) = u(t)te^{-2t}$ $\quad H(s) = \dfrac{1}{(s+2)^2}$ $\quad x(t) = 4\delta(t)$ $\quad X(s) = 4$

CHAPTER 18: FROM LAPLACE TRANSFORM TO TIME RESPONSE BY PARTIAL FRACTION EXPANSION

In Probs. 18.1 through 18.10 find expressions for time response $y(t)$ from the given transfer functions and given input time functions $x(t)$.

18.1 $\dfrac{Y(s)}{X(s)} = \dfrac{1}{(s+1)(s+2)}$ with $x(t) = e^{-3t}u(t)$

18.2 $\dfrac{Y(s)}{X(s)} = \dfrac{1}{(s^2+9)}$ with $x(t) = 3u(t)$

18.3 $\dfrac{Y(s)}{X(s)} = \dfrac{s+2}{s(s+1)}$ with $x(t) = 2e^{-2t}u(t)$

18.4 $\dfrac{Y(s)}{X(s)} = \dfrac{8}{(s^2+4)}$ with $x(t) = \sin t\, u(t)$

18.5 $\dfrac{Y(s)}{X(s)} = \dfrac{s+2}{(s+1)(s^2+3s+9)}$ with $x(t) = e^{-3t}u(t)$

18.6 $\dfrac{Y(s)}{X(s)} = \dfrac{8}{(s+1)^2}$ with $x(t) = 2e^{-2t}u(t)$

18.7 $\dfrac{Y(s)}{X(s)} = \dfrac{1}{(s+2)^3}$ with $x(t) = 2u(t)$

18.8 $\dfrac{Y(s)}{X(s)} = \dfrac{(\tau_1 s + 1)}{(\tau_a s + 1)}$ with $x(t) = x_m\,\delta(t)$ and with $x(t) = x_m t^2 u(t)$

18.9 $\dfrac{Y(s)}{X(s)} = \dfrac{1}{(s+\sigma_a)^5}$ with $x(t) = x_m u(t)$

18.10 $\dfrac{Y(s)}{X(s)} = \dfrac{(s^2 + 2\sigma_1 s + w_{01}^2)}{s(s+\delta_a)^2}$ with $x(t) = x_m e^{-\sigma_f t} \sin(\omega_f t + \psi_f) u(t)$.

18.11 When a certain linear system is disturbed by the input $x(t)$ shown below, it has the following response: $y(t) = u(t)[3 - 6 \sin 2t - 3e^{-4t}]$. Find the transfer function, $Y(s)/X(s)$.

$x(t) = 3u(t) \sin 2t$ → Linear system → $y(t)$

CHAPTER 19: COMPLETE SYSTEM ANALYSIS: SOME CASE STUDIES

In Probs. 19.1 through 19.27, (i) use Procedure D-3 as an outline to obtain the required information. Include an explicit statement of assumptions made in your physical model (Stage I), precise definition of variables, and full use of Laplace transform techniques and partial-fraction expansion where useful. (ii) Make checks wherever possible (e.g., using FVT and IVT). (iii) Discuss and defend your results on physical grounds.

19.1 Repeat Prob. 6.19 (the box on a conveyor belt), now using Procedure D-3.

19.2 Find the response of the circuit shown to the signal $v_1(t) = v_{1m} \cos \omega_f t\, u(t)$, with $v_4(0^-) = 0$, and $R_2 = 3$ kilohms, $R_3 = 12$ kilohms, $C_4 = 10$ μf, $v_{1m} = 100$ volts, $\omega_f = 2$ rad/sec.

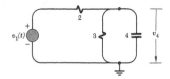

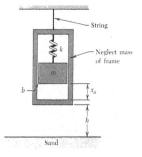

Prob. 19.2   Prob. 19.3

[Chap. 19] PROBLEMS 841

19.3 The string is cut and the assembly drops to the sand. Sketch *carefully* the value of x versus time, assuming the frame is essentially massless. Let $m = 2$ lb/(in./sec^2), $b = 2.5$ lb/(in./sec), $k = 20$ lb/in., $h = 1.5$ ft.

19.4 A cylindrically symmetrical space vehicle whose moment of inertia is 3 ft lb/(rad/sec^2) is in orbit, and is initially rotating about its mass center with constant angular velocity $\Omega_0 = 12$ rad/sec. A malfunction occurs, causing gas to flow from the pair of jets. This produces a reaction moment which slowly decreases with time as shown (the gas supply is being exhausted). Sketch carefully the vehicle's angular velocity versus time. (Include $t < 0$.) Is your answer reasonable physically?

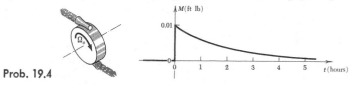

Prob. 19.4

19.5 For the circuit shown, let $v_1(t) = v_{1m}u(t) \sin \omega_f t$, and let the initial voltage across C_3 be γ, while the initial current in L_4 is ρ. Find the total response $v_4(t)$ after $t = 0$. Let $L_4 C_3 = 10^{-8}$ sec^2, $L_4/R_2 = 10^{-3}$ sec.

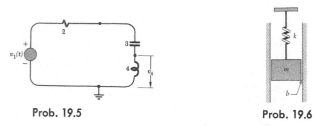

Prob. 19.5   Prob. 19.6

19.6 The system hangs initially at rest (figure). Suddenly the upper end of the spring is given the motion $x = x_0 \cos \omega_f t$. Find the resulting motion of m. Friction between m and the guiding walls is approximately viscous. Compare your results with Eq. (8.59).

19.7 Repeat Probs. 6.26 and 6.27, using Procedure D-3. For generality, let $v_3(0^-) = \gamma$, $\Omega(0^-) = \rho$.

19.8 In the network shown, the switch is in position x for a long time, after which it is suddenly switched to position y. Find the resulting time variation in the voltage drop through resistor R_5, with $v_1(t) = [v_{10} + v_{11} \cos \omega_f t]u(t)$. Let $R_3 = R_5$, $L_2/R_3 = 1.11\backslash^{-3}$ sec, $R_5 C_4 = 0.09\backslash^{-3}$ sec, $v_{11} = 0.5 v_{10}$, $\gamma = 0.2 v_{10}$, $\omega_f = 3000$ rad/sec.

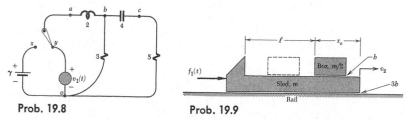

Prob. 19.8   Prob. 19.9

19.9 A rocket sled of mass m carries a piece of heavy equipment (shown as a box) of mass $m/2$ which (unfortunately) is free to slide. The friction between box and sled, and between sled and track, is viscous (proportional to relative velocity, as indicated). The sled is initially at rest ($v = 0$) with the box at location $x(0) = x_0$. At $t = 0$, the rocket engine is ignited, producing constant thrust $f_1(t) = f_{10}u(t)$. Calculate the magnitude f_{10} which will cause the box finally just to reach the wall at the back of the sled with zero relative velocity. (Is it necessary to find the time response in this problem?)

19.10 Repeat the analysis of the vibration-isolation system of Fig. 1.1, p. 6, as described in Prob. 8.71, p. 802. This time use Procedure D-3.

19.11 The study of space-vehicle dynamics is particularly elegant because of the very high degree of momentum conservation. The attitude of the Orbiting Astronomical Observatory is controlled by rotating a reaction wheel within it, as shown. Show that a proper equation of motion is $J_{\text{veh}}\ddot{\theta} + J_{\text{wheel}}(\dot{\Omega}_{\text{rel}} + \ddot{\theta}) = M(t)$, where Ω_{rel} is wheel angular velocity relative to the vehicle and M represents moments *external* to the vehicle-wheel system. To rotate the vehicle precisely from one pointing direction to another, i.e., by $\Delta\theta$, the wheel is rotated ("indexed") through the correct number of revolutions and then stopped. Suppose that, with $M = 0$ and the system initially motionless at $\theta(0^-) = \theta_0$, we have Ω_{rel} rotated according to $\Omega_{\text{rel}}(t) = \Omega_0[u(t) - u(t - \tau)]$. Plot $\theta(t)$, and obtain the relation between $\Delta\theta$ and wheel angle $\Omega_0\tau$.

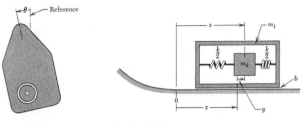

Prob. 19.11   **Prob. 19.12**

19.12 A box of mass m_1 comes down a chute and slides across a greasy floor. Inside the box a weight of mass m_2 is suspended by two springs, as shown. The equations of motion for this system can be written in terms of x and y (motion of m_2 relative to m_1) or in terms of x and z (motion of m_2 with respect to fixed reference). Show that, for a physical model which you are to describe, the resulting equations of motion are $m_1\ddot{x} + b\dot{x} - ky = 0$ and $m_2\ddot{x} + m_2\ddot{y} + ky = 0$, or $m_1\ddot{x} + b\dot{x} + kx - kz = 0$ and $-kx + m_2\ddot{z} + kz = 0$. (a) Discuss the kinds of natural motion of which the system is capable, and discuss the stability of the system. Suppose the box happens to have constant velocity v_0 as it passes point 0, and m_2 happens to be in the middle with no velocity relative to the box. (b) Given m_1 negligible, $m_2 = 0.1$ lb/(ft/sec^2), $k = 10$ lb/ft, $b = 0.833$ lb/(ft/sec), $v_0 = 5$ ft/sec, find the time response $x(t)$ after the box passes point 0, and sketch, indicating important quantities. (c) Discuss the initial value of $x(t)$ and $\dot{x}(t)$ indicated in part (b). Apply physical reasoning.

19.13 Consider again the Scotch Yoke mechanism of Fig. 2.9. Find the time history of force f_3 in the right-hand connecting rod, if the system is initially at rest in static equilibrium with $\theta = \pi/2$, and the disk is then suddenly started in rotation at constant speed ω. (Note: Never differentiate an equation of motion before $\mathcal{L}_-$ transforming it.) Assume the system to be overdamped.

19.14 Augment the results of Prob. 19.13 for the new initial conditions $x_a(0^-) = x_b(0^-) = r$; $x_c(0^-) = x_d(0^-) = 0$.

19.15 Two unequal masses are connected by a single spring, as shown. Initially they are connected also by a string which compresses the spring by an amount ξ. There is no friction. The string breaks at $t = 0$. (a) Derive the equations of motion, and use the Laplace transform method to obtain expressions for the subsequent lateral displacement, versus time, of each of the two masses. (b) Sketch this motion carefully. (c) Explain on physical grounds as many aspects of your results as you can. (d) What are the two natural modes of motion? How much of each is present in the motion plotted in (b)?

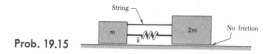

Prob. 19.15

19.16 Postulate an approximate linear physical model for the carrier landing of an aircraft, Example 2.2, p. 49, stating the additional assumptions required. Then use Procedure D-3 to obtain a general expression for the distance the aircraft moves across the deck before its relative velocity is reduced to zero. Suppose the airplane weighs 3000 lb, the sandbags together weigh 400 lb, and k_6 for the cable is 40,000 lb/ft. The ship is making 24 knots (40.6 ft/sec) upwind, the wind speed is 9 knots, and the airplane's indicated air speed (i.e., relative to the wind) at touchdown is 48 knots. Take the average values of the *linear* approximate damping coefficients to be 4 lb/(ft/sec) for the airplane and 15 lb/(ft/sec) for the sandbags. To approximately what values of β_5 and f_{40} do these correspond in Eqs. (2.27)? (Use graphs of f versus $\dot{x}$.) Show that the constant component of wind force on the aircraft (independent of $\dot{x}_1$) is implied to be 110 lb; i.e., the value of $\beta_5 = 0.0363$ lb/(ft/sec^2). From this linear model, how long a deck is advisable for this particular landing? Use engineering judgment. How good an indication do you think this really is? (Discuss carefully.)

19.17^c Mechanize the nonlinear carrier-landing equations of motion, Eqs. (2.27), on a computer and obtain the time history of $x_1(t)$. Use $\beta_5 = 0.0363$ lb/(ft/sec)^2 and $f_{40} = 200$ lb. Compare your results with those of Prob. 19.16. In particular, how long should the deck be for this landing? Would your answer to this question from the linear analysis of Prob. 19.16 have been a good one? Discuss.

19.18 A single-axis rate gyro, p. 163, is mounted to measure the roll angular velocity of an aircraft. During a maneuver, the aircraft angular velocity is as plotted below. Obtain an expression for the resulting signal to be expected from the instrument, if the physical constants of the gyro (as defined on p. 163) are such that $\sqrt{k/J} = 600$, if damping is critical, and if the electrical pickoff scale factor is known without error. Assume uncertainty torque M_u to be zero. Sketch the *error* in the instrument's report.

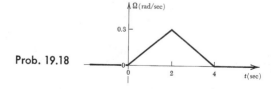

Prob. 19.18

19.19 Another version of the single-axis gyro, called a rate-integrating gyro, is constructed like the rate gyro of Prob. 19.18, except that the restoring spring is omitted. The ideal output of such a gyro is a gimbal angle which is proportional to the integral of rate or, for rotation about a single axis only, the angle of rotation itself: $\psi = \int \Omega\, dt$. Plot the value of actual rotation ψ corresponding to the time input shown above. Then, to the same scale, plot the value of ψ that would be reported by a rate-integrating gyro having $b/h = 2$ (dimensionless) and $J/b = 0.1$ sec. [Refer to Eq. (5.66).] Again, assume no scale-factor error and assume uncertainty torque M_u to be zero.

19.20 The potentiometer is nulled when the crank is vertical. The crank is initially at rest in this position: Suddenly it commences to rotate at constant speed ω. (a) Write all differential equations. (b) Obtain the transfer functions and draw a block diagram. (c) Obtain the complete expression for $v_8(t)$. Assume in (c) that ℓ/r is large and thus that it is reasonable to approximate the motion x as a pure sine wave. (d) Set up the response function $V_8(s)$ corresponding to an *exact* analysis (but do not inverse transform).

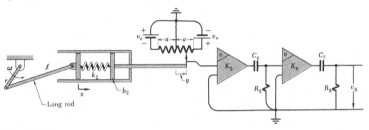

Prob. 19.20

19.21 Mass m is initially at rest with the potentiometer nulled (figure). Element No. 2 has initial velocity Ω_0. Suppose that x is suddenly moved upward a distance x_0. (a) Write all the differential equations involved. (b) Obtain transfer functions, and draw a block diagram. (c) Obtain expressions for displacement y of element No. 1 and angular velocity Ω of element No. 2 after $t = 0$, using letters for constants, and also using the following values: $b_1/2m_1 = 1$, $k/m_1 = 17$, $b_2/J_2 = 3$,

$$kK_2v_0/J_2am_1 = 10$$

(d) Check final and initial values by FVT and IVT.

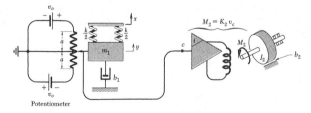

Prob. 19.21

19.22 The water-supply system modeled in Fig. 3.10, p. 115, has been in steady-state equilibrium for some time, with $w_{10} = w_{20}$, and h_2 and h_1 varying slightly as w_2 varies. One day about noon, h_2 happens to be down slightly, $h_2 = -2$ ft, when someone notices it and starts a rumor to the effect that there is a water shortage. Then everyone in town decides to fill his bathtub and other containers, as a precaution. This causes a large temporary increase in w_2, approximately as plotted. Find and plot the resulting time variation in h_2 (which everyone is watching, to see whether there is any truth to the rumor). Will h_2 have a chance to recover substantially before the daily cooking and bathing transient starts at around 6 P.M.? (Note: Derivation of the linearized equations of motion is outlined in Prob. 3.14.) For numerical values, let $h_{10} - h_{20} = 100$ ft, $w_{10} = w_{20} = w_{30} = 2$ slugs/sec $= 2$ lb sec/ft (about 8 gal/sec), and $A_2 = 1000$ ft^2 (a water tank). Show from (3.25) and (3.26) that R' in (3.28) is given by $R' = 2(h_{10} - h_{20})/w_{30}$. That is, we can infer pipe friction from measured flow. Consider two values for A_1, $A_1 = 1000\ A_2$ (A_1 is a small lake), and $A_1 = A_2$ (A_1 is another water tank).

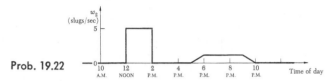

Prob. 19.22

19.23 Repeat the thermal analysis of the ingot-quenching operation described in Sec. 19.3, but use a two-lump model for the ingot—an inner solid cylinder surrounded by an outer one of equal volume (and therefore equal heat capacity). A reasonable division of resistance [R in (19.24)] might be 20% internal and 80% at the solid-liquid interface. Plot the (more realistic) time responses for $T_i(t)$, $T_o(t)$, and $T_b(t)$ (inner cylinder, outer cylinder, bath), and compare with the plot of Fig. 19.7. Comment on the suitability of the one-lump model. Would you expect a three-lump model to produce a result differing significantly from that of a two-lump model? Estimate and plot the distribution of temperature across a diameter of the ingot at $t = 0$, $t = RC_iC_b/(C_i + C_b)$, and $t = \infty$. How much would your results change if the allocation of resistance between internal and interface were inverted? To plot, let $c_b = c_i$.

19.24 A hi-fi speaker and part of its accompanying circuitry are modeled as shown. (Again, the omission, in the model, of enclosed air is serious.) Source $v_1(t)$ represents the broadcast signal after it has been processed by the receiver. A hi-fi speaker is distinguished by its reproduction of high-frequency sounds (e.g., words beginning with s) and percussive sounds, as from plucked strings, castanets, and the like. Suppose that a good approximation to the signal generated when the string of a guitar is plucked is the following: $v_1 = v_{10}[0.8e^{-t/0.01}\cos 6000t + 0.2e^{-t/2}\sin 60{,}000t]u(t)$. Find the response of the system modeled, and comment on how well it reproduces the character of the original signal. Physical constants are $\sqrt{k_7/m_6} = 20{,}000$ sec^{-1}, $b/2m_6 = 20{,}000$ sec^{-1}, $R_3C_2 = 1/20{,}000$ sec, $R_4C_2 = 1/80{,}000$ sec.

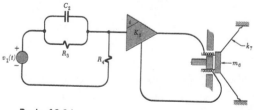

Prob. 19.24

19.25° Mechanize on a computer Eqs. (3.34) for the behavior of the reservoir system of Fig. 3.9; these equations have not been linearized. Take $\alpha = 2$ (rough pipe), and add a term $+\omega_1$ to the right-hand side of the first equation to describe water being supplied to the first reservoir. Study the behavior of the system for the situation described in Prob. 19.22. Compare your results with those of the linear perturbation analysis of Prob. 19.22.

19.26° A tugboat is towing a barge on a flexible cable, as shown. After towing steadily at constant speed for some time, the tug engine thrust is suddenly increased to a higher constant level. Find an expression for the subsequent velocity $v_2(t)$ of the barge. The following data are given: mass of tug $m_1 = 600$ lb/(ft/sec^2), mass of barge $m_2 = 6000$ lb/(ft/sec^2), original propeller thrust $f = 12{,}000$ lb, new propeller thrust $f = 24{,}000$ lb. The stretch characteristic of the tow cable, and drag curves of tug and barge are shown.

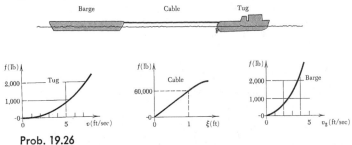

Prob. 19.26

19.27* The air-cushion vehicle of Example 5.4 is initially located at x_0, y_0, and rotating with angular velocity Ω_0. Its initial translational velocity is zero. Also, $\theta(0) = 0$. At $t = 0$, the jet is turned full on: $f_1 = f_0$. Make a plot of the subsequent path of point C, with $b/m = 1$ sec^{-1}, $b\ell^2/J = 2$ sec^{-1}, $f_0/b\ell = 0.5$ sec^{-1}, $\ell_2 = 1.5\ell$. Assume $\theta \ll 1$ and linearize Eqs. (5.43).

19.28 Show that the steady-state displacement of the shaker in Fig. 19.4 produced by constant voltage v_{10} is $x_{ss} = Kv_{10}/kR$.

19.29 Suppose that K in Fig. 19.4 is fixed, such that $K^2/mL = 4^{\backslash 5}$ sec^{-2}. Rearrange Eq. (19.17) appropriately and plot the locus of roots vs. R/L. (Take $\sqrt{k/m} = 300$ sec^{-1} as before.) Select a reasonable value for R/L.

19.30 Repeat the analysis of the two-axis gyro of Sec. 19.5 assuming that light viscous damping b_z is present on the outer gimbal. Show that your result reduces to Eq. (19.52) for $b_z \to 0$. Sketch Fig. 19.11 for the numerical case $b_z/J_z = 0.05h/\sqrt{J_y J_z}$.

CHAPTER 20: FEEDBACK CONTROL

20.1 Draw a nonmathematical cause-and-effect block diagram to indicate the interrelation between the various elements of the human system of Fig. 20.3, particularly the feedback feature.

20.2 Draw a nonmathematical cause-and-effect block diagram for the steering system consisting of automobile plus driver on a winding road.

20.3 Make an estimate, in the form of a transfer function, for the dynamic character of each of the blocks in Prob. 20.2.

20.4 Establish reasonable error criteria for acceptable performance of the driver-automobile system of Prob. 20.2, using test inputs which you think are appropriate and practicable. (Refer to Fig. 20.4.)

20.5 Repeat Probs. 20.2 and 20.3 for the man-automobile *speed*-control system. Include estimates (based on your experience) of the time lags in the various blocks. Establish reasonable error criteria for safe driving.

20.6 Draw a block diagram to represent the temperature-regulation system for a home. Include variation in outside temperature as a system input, and note that the thermostat is capable of on-off control only, with the "on" and "off" temperatures different. (Note also that there is a built-in time delay in the system. Why?)

20.7 Draw a block diagram, with appropriate transfer functions, to represent the aircraft hydraulic control unit of Fig. 6.1*f*, including particularly a correct mathematical representation of the feedback. (Assume incompressible oil.)

20.8 In an experiment, liquid is being removed from the bottom of a container at a variable rate. A technician is supposed to maintain the liquid level constant in the container by controlling the flow into it with a valve. Draw a block diagram of this level-controlling system.

20.9 Draw a block diagram of the control system you use for catching a ball.

20.10 Draw a block diagram of the control system you use for balancing a broom.

In Probs. **20.11** through **20.18** (figures) write expressions for the transfer functions Y/X and E/X. (Do not factor the denominator: leave it as it comes—the sum of polynomials in s.) Whenever possible, specialize your answer to an earlier one by letting the proper quantities be zero.

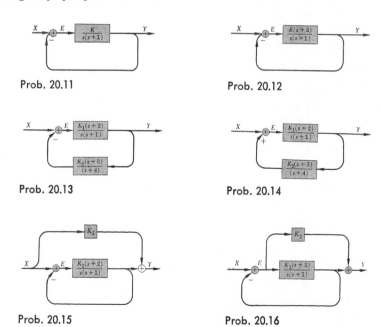

Prob. 20.11

Prob. 20.12

Prob. 20.13

Prob. 20.14

Prob. 20.15

Prob. 20.16

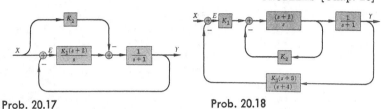

Prob. 20.17   Prob. 20.18

20.19 Find the transfer function Y/X for the block diagram shown.

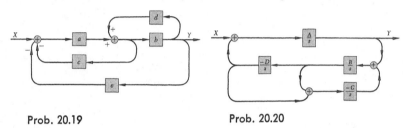

Prob. 20.19   Prob. 20.20

20.20 Write the transfer function Y/X for the system shown. Discuss the stability of the system on the basis of Routh's method.

For Probs. **20.21** through **20.24** (figures) *write first the characteristic equation*. Then obtain $Y(s)$ in the form

$$Y(s) = X(s)\frac{C_1(s+a)(s+b)\cdots}{(s+\sigma_1)(s+\sigma_2)\cdots + C_2(s+c)\cdots} + y(0)\frac{C_3(\ )\cdots}{(\ )(\ )\cdots + (\ )\cdots} + \cdots$$

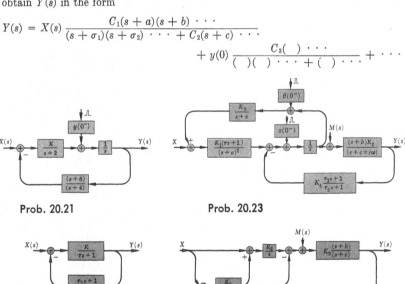

Prob. 20.21   Prob. 20.23

Prob. 20.22   Prob. 20.24

20.25 For the system shown, (a) write the response function $Y(s) = ?$, given $X(s)$ is zero and M is an impulse: $M(t) = M_0\delta(t)$. Note that there are no minus signs in the diagram. (b) Write the characteristic equation of the system.

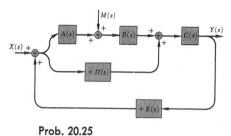

Prob. 20.25

20.26 Determine the values of a, b, c, d, σ, ω, and the K's in Fig. 20.7b, p. 641, in terms of the physical parameters in Fig. 20.7a.

CHAPTER 21: EVANS' ROOT-LOCUS METHOD

In Probs. 21.1 and 21.2 obtain the system characteristic equation in the Evans form (21.2). Plot as ×'s in an s plane the numerator factors $P(s)$ of the Evans function $C(s)$, and plot the denominator terms $Z(s)$ as ○'s. Sketch, in a qualitative way, the major features of the locus of roots. (Note that the characteristic equations for Probs. 21.1 and 21.2 are available from the denominators of the transfer functions obtained in Probs. 20.11 and 20.12, respectively.) Recall that a locus must begin at every × and a locus must terminate at every ○.

21.1 The system of Prob. 20.11.   **21.2** The system of Prob. 20.12.

21.3 For the feedback system at left below, show that the characteristic equation is $s(s + 1)(s + 3) + K = 0$, or in Evans form, $C(s) = s(s + 1)(s + 3) = -K$. The correct locus of roots (for K positive) is as shown at right. Copy the locus carefully onto a full-sized sheet of paper, and verify by measurements with a Spirule (or protractor) that points P, Q, R, and S are indeed points at which the 180° angle condition of (21.2) is met. Compute the value of K at each of the points.

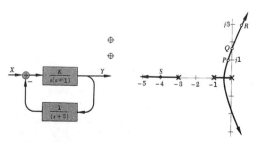

Prob. 21.3

21.4 Prove that the correct 180° root locus for the system shown at left below is as shown at right. That is, prove analytically that the angle condition of Eq. (21.2) is satisfied. Find the value of K where $\zeta = 0$ ($j\omega$ axis crossing), and also where $\zeta = 0.3$.

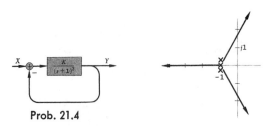

Prob. 21.4

21.5 Verify that Eq. (21.3) is the correct characteristic equation for the system of Fig. 21.1a. Copy Fig. 21.1d carefully, and verify, using a Spirule, that points P, N, M, P'', N'' all satisfy the 180° condition of Eq. (21.4). Then verify, again by actual measurement and multiplication, the values shown in Fig. 21.1d for K at those points.

21.6 Real-axis segments. For each of the s-plane plots given, (i) identify the real-axis segments for 180° root loci (by copying the figure and sketching the loci); (ii) state in which cases the entire locus is on the real axis. Are there cases in which there is *no* locus on the real axis?

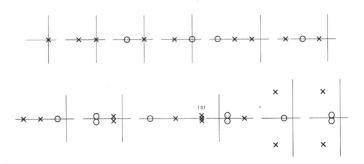

Prob. 21.6

21.7 Asymptotes. For each of the s-plane plots given, copy the plot, locate the centroid of the ×–○ array (the point from which the asymptotes emanate), and draw the asymptotes for 180° root loci. Also fill in the real-axis segments for 180° loci, and sketch in the loci themselves very approximately. (Recall that a locus must emanate from every × and a locus must terminate at every ○.)

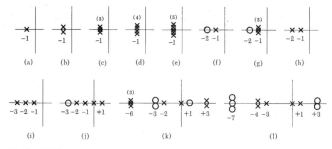

Prob. 21.7

21.8 For each of the loci sketched in parts b, c, g, i, j of Prob. 21.7 use a Spirule to locate *accurately*, by trial and error, one interesting point on the locus—for example, the point where a locus crosses the $j\omega$ axis.

21.9 Departure and arrival directions. For each of the s-plane arrays given, (i) copy the array accurately on a full-sized sheet of paper, (ii) determine the direction(s) of departure of 180° root loci from each × and the direction(s) of arrival at each ○, using Eqs. (21.8), p. 661, for single ×'s and ○'s, or Eqs. (21.9) for multiple ×'s and ○'s, and employing a Spirule (or protractor). Also (iii) fill in real-axis segments and construct the asymptotes for 180° loci. Finally, sketch in—approximately—the complete set of 180° root loci.

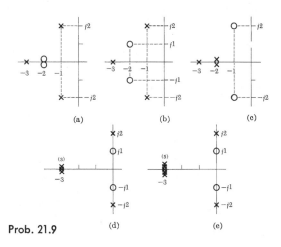

Prob. 21.9

In Probs. **21.10** through **21.14** (figures) use sketching rules 1 through 3 of Sec. 21.3 to produce the most accurate sketches you can of the loci of roots for positive K. (Do not use a Spirule in these problems.) [Recall from p. 415 that $(s + \sigma \mp j\omega)$ means $(s + \sigma - j\omega)(s + \sigma + j\omega)$.]

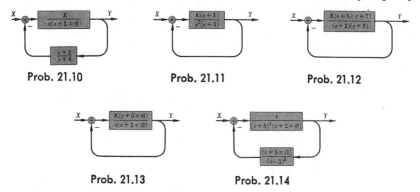

Prob. 21.10 Prob. 21.11 Prob. 21.12

Prob. 21.13 Prob. 21.14

21.15 In Prob. 21.10 find the value of K for which there are roots on the $j\omega$ axis. (Use a Spirule for this, if desired.)

21.16 In Prob. 21.14 find the values of K for which there are roots on the $j\omega$ axis. (Use a Spirule, if desired.)

21.17 Real-axis breakaway. For each of the s-plane arrays given, find each location where 180° root loci break away from or break into the real axis. Use three methods: (i) Estimation by eye (refer to Fig. 21.11); (ii) Eq. (21.10a); (iii) Eq. (21.11). [Equations (21.10a) and (21.11) appear on p. 654. Equation (21.10a) is discussed in Sec. 21.7, and Eq. (21.11) in Sec. 2.8.] Also, fill in real-axis segments, construct asymptotes, and sketch 180° root loci approximately.

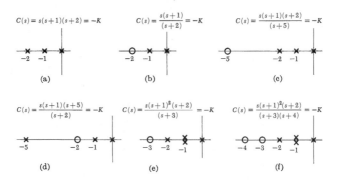

Prob. 21.17

21.18 Real-axis breakaway in the presence of complex ×'s and ○'s. For each s-plane array given, locate all points where loci break away from or into the real axis, using Eq. (21.10b), p. 654.

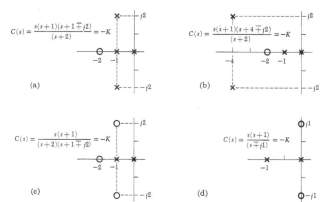

Prob. 21.18

21.19 Use Rule 4 to improve the loci you sketched in Prob. 21.3.
21.20 Use Rule 4 to improve the loci you sketched in Prob. 21.12.
21.21 Use Rule 4 to improve the loci you sketched in Prob. 21.11.

21.22 Saddle points. Find the value of K at the off-axis saddle point, point P, in Fig. 21.14.
21.23 Find the value of K at points P and B in Fig. 21.15b, with $C(s) = s(s+2)(s+1 \mp j3) = -K$.
21.24* Given the Evans function $C(s) = \dfrac{s(s+1)(s+1 \mp j2)}{(s+b)} = -K$, find by trial and error the value of b for which there is an off-axis saddle point. Sketch the 180° root loci for this case, and also for the cases in which b is twice or half that value.
21.25 Find the value of σ_a and ω_a for which the root locus for $C(s) = s(s + \sigma_a \mp j\omega_a) = -K$ has a triple saddle point at -1. (Refer to Fig. 21.16.) Sketch a three-dimensional picture, analogous to Fig. 21.13, of the situation.
21.26 Find by trial and error the value of ω_a for which the root locus for $C(s) = s(s+2)(s+1 \mp j\omega_a)$ has a quadruple saddle point. Where is it? Sketch the loci of roots for that case, and also for the cases in which ω has half or twice that value. (You should obtain a set of locus pictures analogous to those in Fig. 21.16a, b, c.)

In Probs. 21.27 through 21.32 find the location of each $j\omega$-axis crossing of a 180° root locus, and also the value of K there, in two ways: (i) by sketching the root loci, using sketching rules 1 through 4 (Sec. 21.3), and then using a Spirule to establish the crossing point accurately by trial and error, and to measure K there; (ii) by application of Routh's method, Procedure C-4, p. 410, as demonstrated in Sec. 2.9. (iii) In each case, state the range of values of K for which the system is stable.
21.27 The system of Example 21.14, p. 674. [Part (ii) is completed in the example. The results are Eqs. (21.15).]
21.28 A system for which $C(s) = (s+1)^3 = -K$.

21.29 A system for which $C(s) = s(s+1)^2 = -K$.

21.30 A system for which $C(s) = \dfrac{s(s-1)}{(s+1)} = -K$.

21.31 A system whose characteristic equation is $s(s-1)(s+4)^2 + K(s+1) = 0$.

21.32 A system for which $C(s) = s(s+2)(s+1 \mp j2) = -K$.

21.33 Use Routh's method to improve further the loci you sketched in Prob. 21.10, and to check the value of K where the loci cross the $j\omega$ axis.

21.34 Use Routh's method to improve further the loci you sketched in Prob. 21.14, and to check the values of K where loci cross the $j\omega$ axis. For what range of values of K is the system stable?

21.35 For the system represented, *first* find the range of values of K for neutral stability by Routh's method. Then find the location of the undamped roots, also by Routh. Then sketch the locus of roots (get the main features of the locus carefully, but don't be fussy about intermediate points), and verify the Routh results.

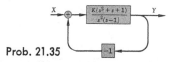

Prob. 21.35

21.36 Use the fixed-centroid rule (Rule 6, Sec. 21.3) to prove by inspection that the value of K at points P'' and P''' in Fig. 21.1d is the same as it is at points P, and that the gain at points N'' and N''' is the same as at point N.

21.37 Use Rule 6 to locate, in Fig. 21.16b, the real-axis root which has the same value of K as the imaginary-axis roots. (Refer to Prob. 21.25.)

21.38 For each of the s-plane arrays below, the 180° root loci have two branches which cross the $j\omega$ axis and one which proceeds out along the negative real axis. In each case, determine by inspection the location of the real root when the gain is such that the other two roots are on the $j\omega$ axis.

Prob. 21.38

In Probs. **21.39** through **21.48** (figures) (i) sketch accurately all the loci of roots of the characteristic equation for the system shown. Use sketching rules (in sequence) to construct the loci as accurately as possible. Then use a Spirule to check a few key points. (ii) Label the value of K at several useful points on the loci, including all imaginary-axis crossings. (iii) State the range of positive K for which the system is stable. (Check your answer by Routh's method for systems of sixth order or less.)

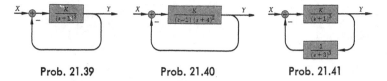

Prob. 21.39   Prob. 21.40   Prob. 21.41

[Chap. 22] PROBLEMS 855

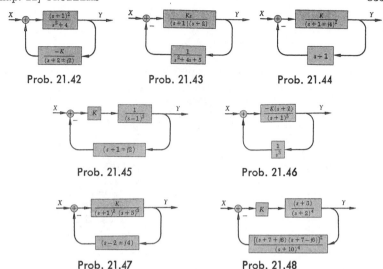

Prob. 21.42   Prob. 21.43   Prob. 21.44

Prob. 21.45   Prob. 21.46

Prob. 21.47   Prob. 21.48

21.49 Rewrite sketching rules 1, 2, 3 of Sec. 21.3 for 0° loci—i.e., for the case in which K in Eq. (21.2) has only negative values.

In Probs. 21.50 through 21.61, use the results of Prob. 21.49 to sketch the 0° loci for the system named. (Superimpose the 0° loci on the 180° loci you obtained earlier.)

21.50	The system of Prob. 21.3.	**21.57**	The system of Prob. 21.12.
21.51	The system of Prob. 21.4.	**21.58**	The system of Prob. 21.13.
21.52	The system of Prob. 21.6.	**21.59**	The system of Prob. 21.14.
21.53	The system of Prob. 21.7.	**21.60**	The system of Prob. 21.31.
21.54	The system of Prob. 21.9.	**21.61**	The system of Prob. 21.32.
21.55	The system of Prob. 21.10.		
21.56	The system of Prob. 21.11.		

21.62* Plot the loci of roots for the equation $s(s + j) + K = 0$ for positive values of K.

CHAPTER 22: SOME CASE STUDIES IN AUTOMATIC CONTROL

In each of Probs. 22.1 through 22.9 a physical model (Stage I) is described. Perform a system analysis and design, following Procedure E-1, p. 643. Present the following results: (i) the complete set of differential equations of motion (Stage II); (ii) a system block diagram, based on the Laplace transformed equations of motion, and including specified initial conditions shown as impulses; (iii) a locus of roots of the system characteristic equation versus the parameter specified in the problem statement [Step 2(b)]; (iv) the complete system response function for the variables specified in the problem statement [Step 2(a)], including specified initial conditions as well as inputs and disturbances; (v) time-response indices as required [Steps 2(c) and 3]; (vi) design selections as required (Stage IV).

22.1 Machine-tool control. The figure shows a system for the precise positioning of a machine tool (e.g., a milling cutter) in one coordinate, using a hydraulic servo. (Two similar systems, not shown, position the tool in the other two coordinates.) An optical pickup (not shown) provides a voltage proportional to tool displacement y. (Let the constant be 1 volt per cm.) This is compared, by the controller, with specified-displacement voltage y_d from the cutting program stored on tape; the controller then applies electromagnetically a proportional force f to the valve: $f = K(y_d - y)$. In (i) assume incompressible oil. Assume that the inertia of the valve can be neglected compared to viscous and spring forces $b\dot{x}$ and kx. (The valve is spring-centered.) In (ii) and (iv) include both $y(0^-)$ and $x(0^-)$. In (iii) plot roots versus K. In (iv) find the response function for the position error $y_e = y_d - y$. In (v) find $y_e(0^+)$ and $y_e(\infty)$ first for $y_d = 0$; then, with $y(0^-) = 0$ and $x(0^-) = 0$, let $y_d = y_{d0}u(t)$ (a step), and $y_d = v_{d0}tu(t)$ (a ramp). That is, find the steady-state position error following a step and a ramp command. Explain each result carefully on physical grounds. Finally, sketch $y(t)$ with $y_d = 0$, $x(0^-) = 0$, $y(0^-) \neq 0$, $KC/b = (k/b)^2$, where C is a combination of physical constants. (This is the response after turning on the controller.) In (vi), if $b/k = 0.02$ sec, choose KC/b to make the ramp error $y_e < |v_{d0}/40|$, and state whether ζ for the system will then meet the desirable criterion $0.5 < \zeta$.

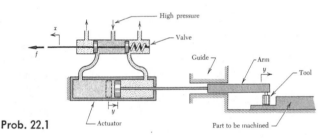

Prob. 22.1

22.2 Temperature control. A control system is shown for maintaining the temperature T of a liquid bath at a desired value T_d, which may be set manually. (External ambient air temperature T_a may vary in a random way.) Bath temperature T is sensed by a thermocouple and compared electrically with T_d. A current amplifier supplies to a heater (resistor) a current proportional to the square root of temperature error, and thus the heat q_1 generated is given by $q_1 = K(T_d - T)$. The total resistance of the insulating tank is R_2, and thus heat lost to the air is given by $q_2 = (T - T_a)/R_2$. Assume that (because of convection) the bath temperature is uniform, and thus that the bath can be considered a single heat-storage element having heat capacity C. This same model may be used to represent the heating of a room if proportional control of furnace heat can be presumed. In (ii), (iv), and (v) include the initial value $T(0^-)$ of bath temperature. In (iii) plot the locus of roots versus K. In (v) let the reference and air temperatures be constants, $T_d(t) = T_{d0}u(t)$ and $T_a(t) = T_{a0}u(t)$, and find $T(\infty)$ and $T(0^+)$. Check with physical reasoning. Then find and plot T versus time t, letting the initial bath temperature equal air temperature equal $0.8T_{d0}$, $R_2C = 100$ sec, and $K = 20/R_2$, and otherwise using letters instead of numbers. Is the heater called upon to deliver negative heat in this time response? (vi) For what range of values of K (in terms of R_2) is the system stable? Select a value of K such that in steady state $T(\infty) \geq 0.95T_{d0}$ when $T_{a0} = 0.8T_{d0}$.

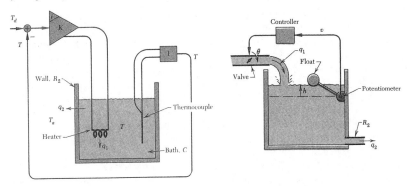

Prob. 22.2   Prob. 22.3

22.3 Liquid-level control. The level of fluid in the mixing tank is to be controlled by varying the flow q_1 in one of the input lines, as shown in the figure. The operation is described in terms of variation from a nominal flow condition as follows (refer to Fig. 3.10): The change in input flow rate q_1 is proportional to the valve angle setting: $q_1 = K_1 \theta$. The change in output flow is proportional to the change in liquid level: $q_2 = (1/R_2)h$. The float and potentiometer are arranged so that the output voltage is given by $v = K_3 h$. The tank diameter is such that one kilogram of liquid raises the level one millimeter. Analyze the system, assuming a proportional control, i.e., $\theta = Kv$. In (iii) sketch loci versus K. Carry the analysis through Stage III only, and then comment on its similarity to Prob. 22.2. Use convenient magnitudes in sketching loci.

22.4 Simple feedback amplifier. In the first figure a "voltage amplifier," of the kind we have idealized frequently (Sec. 2.8), is made more realistic by the inclusion of an output resistance R_3 (representing, for example, the collector resistance of the last transistor in the amplifier). Now the output voltage v_ℓ will be reduced when current is drawn by a load R_5. Ideally the amplifier would have $v_\ell = -Kv_a$, independent of R_5. Show that actually, for the first model, $v_\ell = -Kv_a/[1 + (R_3/R_5)]$, and v_ℓ is highly *dependent* on load R_5. For example, with $K = 1$, if $R_5/R_3 = \infty$, then $v_\ell = -v_a$; but if $R_5 = R_3$, then $v_\ell = -v_a/2$ instead of $v_\ell = -v_a$. If R_3 is greater or R_5 is less, the deterioration is worse. In the second figure, simple feedback is used to make v_ℓ vastly more independent of R_3. In (i) derive the simple equations of motion. In (ii) draw a block diagram, and show that by making K very large, v_ℓ/v_a can be made arbitrarily close to unity, even for low R_5. For example, for $K = 10^6$, we see that v_ℓ/v_a differs from unity by only four parts in 10^6 when $R_5 = R_3$ (compared with 50 percent in the nonfeedback model). Omit parts (iii)-(vi); we have no dynamics in this idealized system. (In actual feedback amplifier design, a capacitor must be included in parallel with the feedback resistor to compensate for lags within the amplifier. Otherwise the extremely high gain would produce instability.)

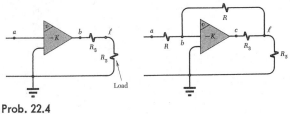

Prob. 22.4

22.5 Space-vehicle attitude control. Derivative control is sometimes added to proportional control to improve a system's dynamic characteristics. In fact some systems cannot be made stable with proportional control only; then derivative control is one of several (similar) means for achieving stability. Consider the space vehicle shown, whose attitude is to be controlled by gas jets. In (i) show that, if the only external torque M acting is independent of vehicle attitude, application of Eq. (2.2) about the mass center gives $J\dot{\Omega} = f_c \ell_1 + M$, where Ω is angular velocity and f_c is the thrust of a control jet. Suppose that angle θ can be measured by a star-tracking telescope, but angular velocity Ω cannot be measured directly. Show in (iii) that if proportional control $f_c = -K\theta_e$ is used, where $\theta_e \triangleq \theta_d - \theta$, then even for the ideal case the system will have $\zeta = 0$ for all positive K (and will be unstable for negative K). Next, consider an ideal proportional-plus-derivative controller: $f_c = -K_2\dot{\theta}_e - K_1\theta_e$. Show that then any desired dynamic characteristics can be obtained, and select K_1/J and K_2/J to produce equal real roots at $s = -0.5$ sec^{-1}. In (iii) and (iv) include both $\Omega(0^-)$ and $\theta(0^-)$. (Heed scrupulously the warning in the footnote on p. 552.) In (v) let $\theta_d = 0$, $\Omega(0^-) = 0$, $\theta(0^-) = 0$, and let M be the result of a speeding meteorite (mass m, relative velocity v) embedding itself in the vehicle a distance ℓ_2 from the mass center: $M = mv\ell_2 \, \delta(t)$. In (vi) find the minimum value of K_1/J required to keep $\theta_{max} < (\sqrt{K_1 J} e)^{-1} mv\ell_2$ throughout the transient motions (given $\zeta = 1$, as above).

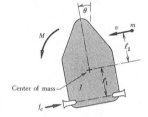

Prob. 22.5

22.6 Shaft-speed control. The figure shows a linear physical model for a speed-control system involving a fluid clutch and an electrical filter. In constructing the block diagram (ii) it will be convenient to consider the clutch subsystem, from M_6 to Ω_9, as a single block (because of the two-way coupling within this subsystem). Let $R_2 = R_3 = 10,000$ ohms, $b/J_7 = b/J_9 = 0.5$ sec^{-1}, $K_9 = 0.6$ volt/(rad/sec), $R_3 C_4 = \frac{1}{6}$ sec. In (iii) select $K_5 K_6$ in terms of the other parameters to produce $\zeta = 0.275$. In (v) find and sketch the time response if the system is initially in steady state with $\Omega_9 = \Omega_7 = \Omega_{9d}$, and then at $t = 0$, Ω_{9d} is made zero. (Check initial conditions.) Also give a brief physical description of what happens.

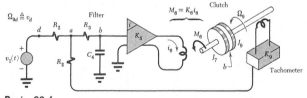

Prob. 22.6

22.7 Automatic pilot. Many of the features of designing an autopilot for airplanes, ships, and especially rockets, can be seen from the dynamically simpler model of

[Chap. 22] PROBLEMS 859

an air-cushion vehicle in Fig. 5.5, p. 154, which is discussed in Example 5.4. Add another jet symmetrical with the original one, as shown below, and suppose that each jet has a constant thrust $f_0/2$ plus a variable increment $f/2$ for steering: $f_1 = (f_0/2) + (f/2)$, $f_2 = (f_0/2) - (f/2)$. Assume $\dot{y} = \text{const.} = v_0$, and therefore $f_0 = bv_0$. (i) Modify Eqs. (5.43) accordingly, and *linearize for small angles* θ. In (ii) show that then the transfer function from control force f to heading angle θ is $\dfrac{\Theta}{F} = \dfrac{\ell_2}{J}\left(s + \dfrac{b}{m}\right) \Big/ s\left[s^2 + \left(\dfrac{b}{m} + \dfrac{b\ell^2}{J}\right)s + \dfrac{bv_0\ell}{J}\right]$. (iii) Is the vehicle stable or unstable in steady travel in the y direction without control (i.e., powered by thrust f_0, with $f = 0$)? Is this result predicted by intuition? Discuss four special cases on mathematical and physical grounds: First, $b = 0$, second $v_0 = 0$, third, $\ell = 0$, and fourth, ℓ negative (center of friction forward of mass center). In (vi), for the following numerical values, design a stable control system $[f = f(\theta)]$ for controlling vehicle heading θ, assuming that $\dot{\theta}$ and θ are available as electrical signals, $m/b = 2$ sec, $J/(b\ell^2) = 5$ sec, $v_0 = 4.9$ ft/sec, $\ell = 2$ ft, $\ell_2 = 3$ ft. Can the system be made stable without $\dot{\theta}$ feedback? Compare the closed-loop system dynamic characteristics for the two cases.

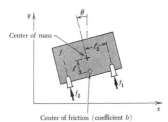

Prob. 22.7

22.8 *Hydraulic servo.* A realistic model for a single-stage hydraulic servo with electrical feedback (refer to Prob. 22.1) should include representation of the valve dynamics and of oil compressibility. (Entrained air in the oil can make the oil more compressible than normal.) For the valve, viscous friction and the force in a centering spring are typically large compared to valve inertia, and thus the valve equation of motion can be written $b\dot{x} + kx = f$, where f is the force applied electromechanically by the controller. Thus $X/F = (1/b)/[s + (k/b)]$. To account for oil compressibility, let the density in chambers 1 and 2 of the actuator (Fig. 3.12) be $\rho_0 + \rho_1$ and $\rho_0 + \rho_2$, respectively, and represent oil elasticity by $\rho_1 = p_1/\Gamma$, $\rho_2 = p_2/\Gamma$. (See Fig. 3.5.) (Assume p and ρ uniform throughout each chamber.) Then the total mass of oil in chamber 1 at a given instant is $A(\rho_0 + \rho_1)(\ell + y)$, the rate of change of which must be w_1: $w_1 = A\rho_0\dot{y} + A\ell\dot{\rho}_1$ plus a term which may be neglected in a small-change perturbation study. Similarly, $w_2 = A\rho_0\dot{y} - A\ell\dot{\rho}_2$ (for the definition of positive w_2 shown in Fig. 3.12). These replace Eqs. (3.45) of Example 3.4. As shown in Example 3.4, we have $w_1 = Cx$, $w_2 = Cx$, approximately. The other equations of motion are (3.46), to which we add a damping term $b_3\dot{\theta}$. (i) Show that, with these assumptions, the transfer function from valve displacement x to engine rotation θ has the form $\Theta/X = C'/s(s^2 + 2\sigma_3 s + \omega_{03}^2)$, where ω_{03} is the natural frequency of the engine vibrating against the "spring" of the compressible oil. Discuss physically. What is the response to an impulse in valve position: $x = v_0 u(t)$? Show that if elasticity $\Gamma \to \infty$ (incompressible oil), this transfer function goes to $\Theta/X = (C/A\rho_0^2\ell_1^2)/s$, as given by (3.50) for the incompressible case. (ii) With the numerical values $\omega_{03} = 600\ \text{sec}^{-1}$, $\sigma_3 = 20\ \text{sec}^{-1}$, $b/k = 0.03$ sec, design a simple electrical feedback control system using the above valve and actuator. Discuss the control-system stability, dynamic characteristics, and steady-state error following a step input $\theta_d = \theta_{d0} u(t)$.

22.9* *On-off control of temperature.* Consider again the temperature control system of Prob. 22.2 (for a chemical bath or for a room in a house). Suppose now that control is to be effected by having the current either full on, $q_1 = q_{10}$, or off $q_1 = 0$ (or the furnace full on or off): The system is to be turned on whenever the temperature is, say, 2° or more below the desired temperature, and is to be turned off whenever the temperature is, say, 1° or more above the desired temperature. Assume the ambient temperature outside the bath (or outside the house) to be constant. Then there are actually five reference temperatures of interest: ambient outside temperature T_a, desired temperature T_d, turn-on temperature $T_{on} = T_d - 2°$, turn-off temperature $T_{off} = T_d + 1°$, and the equilibrium temperature that would be reached if the heater stayed on indefinitely. Show that the latter is given by $R_2 q_{10} + T_a$. Suppose the control system has been turned on when the temperature T is slightly larger than ambient: $T_a < T(0^-)$. Using two versions of the differential equations—one for heater on and one for heater off—and the corresponding response functions, obtain and plot T versus time t. Notice that after the system turns off the first time, the time behavior simply repeats, exhibiting what is called a "limit cycle." Make a plot of the same behavior in a "phase plane" of $R_2 C \dot{T}$ versus T. The plot is deceptively simple. The vertical lines T_{on} and T_{off} are called "switching lines." Compare the response found here with that one would obtain with the linear model of Prob. 22.2 under the same conditions. (Use superimposed time plots or phase-plane plots.)

22.10 Obtain and plot the response of the rate-network-controlled indicator servo of Fig. 22.4 to a step input $\theta_d = 0.8° u(t)$, and compare the results with the response of Fig. 22.3 for a system controlled without a network. Let K be set to obtain system roots T, T', T'' in Fig. 22.4c. Are the design criteria for transient response now met completely?

22.11 Consider other possible values for σ_5 in Fig. 22.4c, including very small and very large values. (Assume $\sigma_6 = 3.5 \sigma_5$, as before.) Find, by trial, the value of σ_5 for which the system natural frequency ω_{01} is largest, with the restrictions that $\zeta \geq 0.4$ and, simultaneously, that the slow real root σ_3 must satisfy $\omega_{01} < \sigma_3$.

22.12 Return to the development of Eq. (22.20), and assume now that there is an initial charge $v(0^-) = v_a(0^-) - v_b(0^-) = \gamma$ on capacitor C. Alter Fig. 22.4b appropriately to represent this added initial condition.

22.13 A certain system has the plant transfer function $Y/F = 1/s(s + a)$. Determine the controller lead network $(s + \sigma_1)/(s + \alpha \sigma_1)$ having the smallest α which will still put the closed-loop poles in the shaded region shown.

Prob. 22.13

22.14 To improve the speed of response of the hydraulic machine-tool control system studied in Prob. 22.1, it is proposed that a lead network like that in Fig. 22.4a be added. That is, the control logic, which was $f = K(y_d - y)$ in Prob. 22.1, will now have the equation of motion $\dot{f} + \alpha\sigma_1 f = K\dot{y}_e + K\sigma_1 y_e$, where $y_e \triangleq y_d - y$. Carry out three design studies: (i) With $\alpha = 4$ and $\sigma_1 = 1.25k/b$, select K so that all three system roots (two complex, one real) have the same real part (same damping envelope). (ii) For this K, evaluate the steady-state error y_e incurred in following a ramp input: $y_d = v_{d0} t u(t)$. Compare with the corresponding error in Prob. 22.1. (iii) With $\alpha = 4$, find by trial the combination of σ_1 and K that will produce three equal real roots (a triple saddle point).

22.15 As a limiting case, proportional-plus-pure-derivative control may be imagined for the machine-tool control system. Show that this is tantamount to employing a lead network with $\alpha = \infty$. Plot the locus of roots for this case, with $\sigma_1 = 1.25k/b$. Compare with the result of Prob. 22.14.

22.16 The figure shows schematically one axis of a stable platform to be controlled by using a single-axis rate-integrating gyro [Eq. (5.66) with $k = 0$, J negligible]. The angle ϕ is to be kept at zero. For a first analysis neglect platform bearing friction, so that the equations of motion are, for the platform, $J\ddot{\phi} = M_c + M_{\text{dist}}$, and for the gyro, $b\dot{\theta} = h\dot{\phi} + M_u$ and $v = K_1\theta$, where M_{dist} is the disturbance torque acting on the platform, M_u is uncertainty torque on the gyro, and M_c is the control torque exerted magnetically by the controller between the platform and support. Supposing that the controller includes an ideal current amplifier, (i) design a set of control logic, $M_c = M_c(v)$ to produce a stable system. Use a lead network to achieve stability. (ii) Draw a complete block diagram, including M_{dist}, M_u, and initial conditions. (iii) Obtain an expression for steady-state system response, versus frequency, to a sinusoidal disturbance $M_{\text{dist}} = M_{d0} \cos \omega_f t$. Sketch the corresponding Bode plot. (iv) Discuss the steady-state effect on platform angle ϕ of a constant gyro torque $M_u = M_{u0} u(t)$.

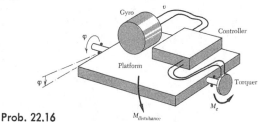

Prob. 22.16

22.17 Sometimes stable platforms are controlled by using a single-axis rate-*double*-integrating gyro [Eq. (5.66) with $k = 0$ *and* $b = 0$]. That is, the gyro equation of motion is $I\ddot{\theta} = h\dot{\phi} + M_u$. Repeat Prob. 22.16 for this case. A *triple* lead network, $V_{\text{out}}/V_{\text{in}} = [(s + \sigma_1)/(s + \alpha\sigma_1)]^3$ will be required. In part (i) (of Prob. 22.16), first let $\alpha = 3.5$ and find the range of stable values of control gain; second, select α so that the locus of roots has a triple saddle point. (The situation is somewhat analogous to Fig. 21.15.) This choice of α provides the maximum attainable ζ for the complex roots. For this case, complete parts (ii) through (iv) as in Prob. 22.16.

22.18 A flexible rocket vehicle is to be controlled by an autopilot, as indicated below. (See also Fig. 1.4.) The vehicle transfer function—from engine swivel position E to vehicle angular velocity Ω—is represented in the box marked "Vehicle": The poles and zeros on the imaginary axis are caused by coupling between engine dynamics and bending of the flexible vehicle body. So long as the closed-loop roots near these poles are *stable*, they may have very low damping ratios without adverse effect. To help keep them stable, a compensation network is furnished, as shown. (The engine-angle servo is taken here to be very fast, and the entire inner loop in Fig. 1.4b is represented simply by 1.) Select values of σ_1, α, and K such that the low-frequency roots have 2 sec^{-1} < ω_0 and 0.4 < ζ, while the high-frequency roots are stable.

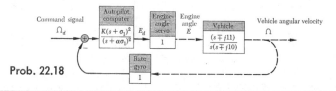

Prob. 22.18

22.19 The characteristic speed of response of the machine-tool control system of Prob. 22.1 needs to be improved. As an alternative to the lead network of Prob. 22.14, whose limiting case is the proportional-plus-pure-derivative control of Prob. 22.15, it is proposed to measure arm velocity as well as position, and to employ an inner rate loop, as was done in Fig. 22.6a. Add such a loop to your block diagram for Prob. 22.1 (working carefully from the proper equations of motion), obtain the system characteristic equation, and plot the loci of roots. Compare the dynamic characteristics achievable with those obtained in Prob. 22.1 (no prediction), Prob. 22.14 (lead network), and Prob. 22.15 (proportional-plus-derivative control). Show that, in fact, the two-loop arrangement using rate feedback is mathematically equivalent to the proportional-plus-pure-derivative arrangement. Discuss the physical limitations on components which mitigate each of the latter ideal results.

22.20 Discuss the dynamic characteristics and probable behavior of the roll autopilot system of Fig. 22.6 if the design of Sec. 22.3 (Stage IV) is being flown, but the rate gyro has just broken (middle loop in Fig. 22.6a is open).

22.21 Design a roll autopilot system like that in Fig. 22.6 which will operate without a rate gyro. Take $M_\Omega = -1$ and $K_\delta c = 19$, as before, and set $K_\Omega = 1$ for convenience. Then choose K_φ to achieve $\zeta = 0.7$ for the oscillatory roots. Now compare the value of ω_{01} for this system with that achievable with a rate gyro, Eq. (22.45).

22.22 Carry out the following "optimization" of the roll autopilot design in Fig. 22.6: Beginning with the values $M_\Omega = -1$, $K_\delta c = 19$, as in Fig. 22.6b, select K_Ω and K_φ so that the final system roots are all equal and real. (Hint: The rate-loop roots will have to be complex. Refer to Fig. 21.16.) Show that this real-root location can have any desired value if $K_\delta c$ may also be chosen freely.

22.23 We have seen in Sec. 22.3 and in Prob. 22.22 that the roll autopilot can provide virtually any system characteristics desired. Discuss the factors to be considered in specifying performance characteristics—ω_{01} and ζ—(i) for a manned aircraft, and (ii) for an unmanned aircraft (a guided missile).

22.24 Find and sketch the response of the roll autopilot system of Fig. 22.6d to a step command in bank angle of 60°. (Use the numerical values of p. 701.) Assume zero initial conditions.

22.25 Find and sketch the response of the roll autopilot system of Fig. 22.6d, where the command bank angle is 0°, and the aircraft has an initial bank angle of 60° (and no angular velocity) when the autopilot is turned on. Do you think the pilot will approve of this response?

[Chap. 22] PROBLEMS

22.26 Sketch, as an addition to Fig. 22.1b, an arrangement of components which would provide an inner rate feedback loop for the remote-indicator servo, and with which the system dynamic characteristics could be improved to be comparable with those in Fig. 22.4c. Sketch the appropriate root loci for comparison with Fig. 22.4c, to prove the point.

22.27 The figure shows part of a two-stage hydraulic servo for controlling a large rocket engine (refer to Prob. 22.1). The additional stage is required to realize the necessary increase in force level. Neglect oil compressibility throughout. The valve dynamics for the first stage, called the *pilot valve*, are as in Prob. 22.8: That is, the transfer function from electromechanical force on the valve to valve displacement is $X/F = 1/(bs + k)$, where k represents a centering spring. The intermediate valve, called the *slave valve*, is a pure integrator, as is the main actuator itself: $\dot{y} = C_1 x$; $\dot{z} = C_2 y$. Slave-valve position y may be sensed by an electrical pickoff. Design the control-system logic, first without and then with a lead network. Discuss qualitatively the performance improvement to be gained from the network.

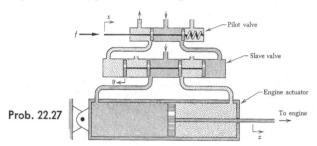

Prob. 22.27

22.28 Design a system to control the lateral *displacement* x of the air-cushion vehicle of Prob. 22.7. Begin with the control system of Prob. 22.7. Next, show from the original equations of motion that the transfer function from θ to x is

$$\frac{X}{\Theta} = \left(\frac{b\ell}{m}\right)\left(s + \frac{v_0}{\ell}\right) \bigg/ s\left(s + \frac{b}{m}\right).$$

If rate ($\dot{x}$) feedback and lead compensation are not used, what is the highest loop gain that can be used without incurring instability? What will be the time constant or natural frequency of the slowest mode of natural motion in this case?

22.29 It is desired to suspend a metal sphere magnetically at a fixed position in a low-density wind tunnel (to avoid struts or wires). Part of the suspension system is shown. The magnetic force on the sphere is proportional to current in the electromagnet windings and (approximately) inversely proportional to the square of the sphere's distance from the magnet: $f = C(i_0 + i)/(\ell + z)^2$, where i_0 and ℓ are the current and distance, respectively, for a nominal condition, and i and z are small changes in current and distance from the nominal. The nominal condition is so chosen that the ball is supported against gravity: $f_0 = Ci_0/\ell^2 = mg$. For constant current, the system is unstable: If the ball is raised, the upward force is *increased*, and so on. For small motions about the nominal, the magnetic force on the sphere can be approximated by the linear relation $f = f_0[1 - (2z/\ell) + (i/i_0)]$. To provide a stable suspension, sphere position is sensed optically, and current i is controlled. You are to design suitable control logic to provide rapid, well-damped recovery from disturbances. As a design goal, all roots should be at least $2\sqrt{g/\ell}$ from the origin, with $0.5 \leq \zeta$. A lead network may be used, with $\alpha \leq 4$.

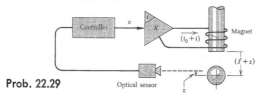

Prob. 22.29

22.30 Repeat the design of an autopilot to steer the air-cushion vehicle of Prob. 22.7 for the case $\ell = -1$ ft—the center of friction is forward of the center of mass, and the uncontrolled vehicle is thus unstable. Let other numerical values remain as in Prob. 22.7.

22.31 Repeat the analysis of the stick-balancing system of Sec. 22.4 under the more realistic assumption that light viscous friction is present in the hinge. From your new root-locus plot, comment on whether the friction makes the system easier or more difficult to stabilize.

22.32 Consider further the idealized stick-balancing problem of Sec. 22.4. Study two or three other possible locations for σ_a in Fig. 22.8, keeping σ_a/σ_b constant, and comment on where the optimum location seems to be, from the stability viewpoint.

22.33 Design a simple auxiliary loop, to be added to the stick-balancing system of Fig. 22.7, whose purpose is simply to control cart position x to be zero with a leisurely speed of response (i.e., merely to keep the cart on the premises). The transfer function from θ to x is available from the second of Eqs. (22.56), and the cart position may be assumed to be measured and available as an electrical signal. The effectiveness of the control may be demonstrated merely by (i) showing that its characteristic equation has all stable roots, and (ii) writing the overall system response function for an initial x, then using FVT to show that $x(\infty)$ is 0. (iii) As an additional feature, IVT may be used (shrewdly) to show that the initial *velocity* $\dot{x}$ will be *negative*. That is, the cart corrects an x error by first *backing up*. Explain physically.

22.34* Extend the analysis of Sec. 22.4 to design a system for controlling a cart on which *two* sticks (of different lengths) are pivoted side by side without friction, just as the stick in Fig. 22.7a is. Both sticks are to be kept balanced in the inverted position with a single cart drive. What does physical intuition tell you about the possibility of balancing two sticks in this manner if they are of equal length? Does your analysis bear this out? Do you see any fundamental reason why three or more unequal sticks cannot be balanced on a single cart in the same manner?

22.35* Design a cart-control system for balancing the inverted pendulum of Fig. 22.7a with another inverted pendulum pivoted without friction on top of it. (The equations of motion must, of course, be derived meticulously.)

ANSWERS TO ODD-NUMBERED PROBLEMS

CHAPTER 1

1.5 (i) d (ii) d (iii) b (iv) c (v) b, d, f (vi) a (vii) a (viii) f (ix) c (x) f

CHAPTER 2

2.1 (iii) 2 (iv) $r + \ell \equiv x_2 + r\cos\theta + \ell\cos\psi$ where $\ell\sin\psi \equiv r\sin\theta$
2.3 (iii) 3 (iv) Ω_1 given **2.5** (iii) 1 (iv) $y \equiv h[1 - \cos(2\pi x/\ell)]$, $\ell/3 < x < 4\ell/3$ $y = 0, x < \ell/3, 4\ell/3 < x$ **2.7** (iii) infinite **2.9** (iii) 4
(iv) Ω_m given **2.11** $\xi_7 = r_4\theta - y_5$, $a_5 = \ddot{y}_5$, six state variables **2.13**
$\dot{y} = 2\pi h v_c \sin(2\pi v_c t/\ell)/\ell$, $\ddot{y} = 4\pi^2 h v_c{}^2 \cos(2\pi v_c t/\ell)/\ell^2$ $\ell/3v_c < t < 4\ell/3v_c$
2.15 $r(1 - \cos\theta) + \ell(1 - \cos\psi) - \xi_3 - \xi_5 = 0$ where $\ell\sin\psi = r\sin\theta$
2.17 $r\theta_1 - 2y_3 - \xi_7 - y_5 = 0$ **2.19** $\Sigma f^* = 0 = f_5 - f_2 - f_i - f_{\text{friction}} + f_{\text{gravity}}$
2.21 $\Sigma f_P^* = 0 = f_\ell \cos\psi - f_3 - f_{2i} - f_{2\text{friction}}$ $\Sigma f_Q^* = 0 = f_3 - f_5 - f_{4i} - f_{4\text{friction}}$
2.23 $\Sigma M_{\text{axle}}^* = 0 = r_4(f_7 - f_{34}) + M_{4i}$ $\Sigma M_P^* = 0 = r_4(2f_7 + f_{4\text{gravity}} - f_{4n}) + M_{4i}$
2.25 $\Sigma M_{\text{axle 4}}^* = 0 = f_{23}(2r_3 + r_4) - (f_{3i} + f_6 + f_{3\text{gravity}})(r_3 + r_4) + M_{4i} + r_4(f_{5i} + f_{5\text{gravity}})$ $\Sigma f_{\text{vertical}}^* = 0 = f_{23} - f_{3i} - f_6 - f_{3\text{gravity}} + f_{4n} - f_{4\text{gravity}} - f_{5i} - f_{5\text{gravity}}$
2.27 $\Sigma M_{\text{axle 1}}^* = 0 = f_9 r_1 - M_{1i} - M_4$ $\Sigma M_{\text{axle 2}}^* = 0 = M_4 - M_{2i} - f_{23} r_2$
$\Sigma M_{\text{axle 3}}^* = 0 = f_{23} r_3 - M_{3i} - M_6$ **2.29** $k = 2$ lbs/in. $6 < x < 18$ in.
2.31 $J_1\ddot{\theta}_1 + b_3\dot{\theta}_1 + k_5(\theta_1 - \theta_2) - k_8(\theta_9 - \theta_1) = 0$ $J_2\ddot{\theta}_2 + b_4\dot{\theta}_2 - k_5(\theta_1 - \theta_2) + k_6\theta_2 = 0$ **2.33** $(m_1\ell_1{}^2/3)\ddot{\theta}_1 + (b_4\ell_1\ell_2/2)(\dot{\theta}_1 - \dot{\theta}_2) + k_5\ell_1\theta_1 - f_7\ell_1 = 0$
$(m_2\ell_2{}^2/3)\ddot{\theta}_2 + b_4\ell_2{}^2(\dot{\theta}_2 - \dot{\theta}_1) + k_3\ell_2\theta_2 = 0$ **2.35** $m_1\ddot{x}_1 + b_{13}(\dot{x}_1 - \dot{x}_3) + k_4(x_1 - x_2) = 0$ $m_2\ddot{x}_2 + b_{23}(\dot{x}_2 - \dot{x}_3) - k_4(x_1 - x_2) + k_5(x_2 - x_3) = 0$
$m_3\ddot{x}_3 + b_{13}(\dot{x}_3 - \dot{x}_1) + b_{23}(\dot{x}_3 - \dot{x}_2) + k_5(x_3 - x_2) + k_6(x_3 - x_9) = 0$ **2.37**
$m\ddot{x} + f_0 \, \text{sgn}\, \dot{x} + kx - f_\ell \cos\psi = 0$ $M(t) - 2f_\ell r(\sin\psi\sin\theta + \cos\psi\cos\theta) - J\ddot{\theta} - b\dot{\theta} = 0$ where $x = \ell\cos\psi + r\sin\theta - \ell\cos\psi_0$, $\ell\sin\psi = r\cos\theta$, $\sin\psi_0 = r/\ell$
2.39 $J\ddot{\theta} + b\dot{\theta} + [kr(r\theta - y)/(h\,\text{cosec}\,\beta - y)] - [kr/\sqrt{x^2 + (h + r)^2}][\sqrt{x^2 + (h + r)^2} - (h + r) - r\theta] = 0$ $m\ddot{y} + f_0 \, \text{sgn}\, \dot{y} + mg\sin\beta - [k(r\theta - y)/(h\,\text{cosec}\,\beta - y)] = 0$
2.41 $J\ddot{\theta}_1 + b_1\dot{\theta}_1 + k(\theta_1 - \theta_2) = M$ $k(\theta_1 - \theta_2) - b\dot{\theta}_2{}^2 = 0$ **2.43** $2x_1 = $

$(f_2/k_2) + (1/b_3)\int f_2\, dt + (1/m_4)\int\int(f_2 - f_5)\, dt\, dt = 0$ $\quad (1/m_4)\int\int(f_2 - f_5)\, dt\, dt - (f_5/k_5) = (2/b_7)\int f_5\, dt$ **2.45** $(1/m_1)\int\int f_4\, dt\, dt - (f_4/k_4) - (1/m_2)\int\int(f_4 - f_5)\, dt\, dt = 0$ $\quad (1/m_2)\int\int(f_4 - f_5)\, dt\, dt - (f_5/k_5) - (1/m_3)\int\int(f_5 - f_6)\, dt\, dt = 0$ $\quad (1/m_3)\int\int(f_5 - f_6)\, dt\, dt - (f_6/k_6) - x_9 = 0$ **2.47** $J_r\ddot{\theta}_1 + k_2(\theta_1 - \theta_3) = M_1(t)$ $J_6\ddot{\theta}_6 + k_5(\theta_6 - \theta_4) = -M_7(t)$ $\quad [J_3 + (r_3{}^2 J_4/r_4{}^2)]\ddot{\theta}_3 + [k_2 + (r_3{}^2 k_5/r_4{}^2)]\theta_3 - k_2\theta_1 - (r_3 k_5\theta_6/r_4) = 0$ Three **2.49** $M_1(t) - J_2\dot{\Omega}_2 - r(f_{23} + f_f) = 0$ $\quad J_3\dot{\Omega}_3 - r_3(f_{23} - f_{34} - 2f_f) = 0$ $\quad J_5\dot{\Omega}_5 - r_4(f_{34} - f_f) = 0$ where $f_f = f_{f0}\,\mathrm{sqn}\,v + \beta v$, $v = r_2\Omega_2 = r_3\Omega_3 = r_4\Omega_4$ **2.51** $J_1\ddot{\theta}_1 + k_4(\theta_1 - \theta_2) = r_1 f_9(t)$ $\quad [J_2 + (r_2{}^2 J_3/r_3{}^2)]\ddot{\theta}_2 + [k_4 + (r_2{}^2 k_6/r_3{}^2)]\theta_2 - k_4\theta_1 = 0$ **2.53** $[m_9 + (143m/6)]\ddot{x}_9 + b\dot{x}_9 + 9k_3 x_9 = k_3 x_1(t)$ **2.55** (i) 5 voltages, 1 current (iii) $v_6 \equiv v_f - v_a$ $\quad v_1 \equiv v_a - v_b$ $v_2 \equiv v_b - v_c$ $\quad v_3 \equiv v_c - v_d$ $\quad v_4 \equiv v_d$ $\quad i_6 \equiv i_1 \equiv i_2 \equiv i_3 \equiv i_4 \equiv i_A$ (iv) $v_f \equiv v_5$ **2.57** (i) 4 voltages, 3 currents (iii) $v_1 \equiv v_a - v_b$ $\quad v_2 \equiv v_b$ $v_3 \equiv v_b - v_c$ $\quad v_4 \equiv v_c$ $\quad v_5 \equiv v_c - v_d$ $\quad v_6 \equiv v_d$ $\quad i_1 \equiv i_A \equiv i_7$ $\quad i_2 \equiv i_A - i_B$ $\quad i_3 \equiv i_B$ $\quad i_4 \equiv i_B - i_C$ $\quad i_5 \equiv i_6 \equiv i_C$ (iv) $v_a \equiv -v_7$ **2.59** (i) 8 voltages, 7 currents (iii) $v_1 \equiv v_d - v_h$ etc., $v_{14} \equiv v_a - v_f$ $\quad i_1 \equiv i_D$ $\quad i_2 \equiv i_D - i_E$ etc., $i_{14} \equiv i_G$ $\quad i_6 \equiv i_E - i_B - i_G$ $\quad i_5 \equiv i_D - i_A - i_G$ $\quad i_8 \equiv i_A + i_G$ (iv) $i_E \equiv i_{15}$ **2.61** $v_6 + v_1 + v_2 + v_3 + v_4 = v_5$ $\quad i_6 = i_1 = i_2 = i_3 = i_4 = i_A$ **2.63** $i_1 - i_2 - i_3 = 0$ $\quad i_3 - i_4 - i_5 = 0$ $\quad i_5 - i_6 = 0$ $\quad i_7 - i_1 = 0$ $\quad v_7 + v_1 + v_2 = 0$ $\quad -v_2 + v_3 + v_4 = 0$ $\quad -v_4 + v_5 + v_6 = 0$ **2.65** $i_8 - i_{12} - i_{14} = 0$ $\quad i_{12} - i_{13} - i_9 = 0$ $\quad i_{13} - i_{11} - i_{10} = 0$ $\quad i_1 - i_8 - i_5 = 0$ $\quad i_5 + i_9 - i_6 - i_2 = 0$ $\quad i_6 + i_{14} + i_{10} - i_7 - i_3 = 0$ $\quad i_7 + i_{11} - i_{14} = 0$ $\quad i_{15} - i_1 + i_2 = 0$ $\quad v_{12} + v_9 - v_5 + v_8 = 0$ $\quad v_{13} + v_{10} - v_6 - v_9 = 0$ $\quad v_{11} - v_7 - v_{10} = 0$ $\quad v_1 + v_5 + v_2 = 0$ $\quad -v_2 + v_6 + v_3 + v_{15} = 0$ $\quad -v_3 + v_7 + v_4 = 0$ $\quad v_{14} - v_6 - v_5 + v_8 = 0$ **2.67** $v_2 = 250t^2$ volts $\quad 0 < t < 2$ sec $\quad v_2 = 1000(t-1)$ volts $\quad 2 < t < 5$ sec $\quad v_2 = 4000$ volts $\quad 5 < t$ **2.69** $v_2 = 0.200t^2$ volts $\quad 0 < t < 2$ sec $\quad v_2 = 0.400t$ volts $\quad 2 < t < 5$ sec $\quad v_2 = 2$ volts $\quad 5 < t$ **2.71** $i_2 = 2.67 \times 10^{-6}t$ amps $0 < t < 3$ sec $\quad i_2 = 4.0 \times 10^{-6}$ amps $\quad 3 < t < 5$ sec $\quad i_2 = 0$ $\quad 5 < t$ **2.73** $i_2 = 500t^3/9$ amps $\quad 0 < t < 3$ sec $\quad i_2 = -750 + 250t^2$ amps $\quad 3 < t < 5$ sec $\quad i_2 = -7000 + 2500t$ amps $\quad 5 < t$ **2.75** (i) $-i_5 + (v_a/R_1) + (v_a/R_3) + (1/L_4)\int v_a\, dt + C_2\, dv_a/dt = 0$ (ii) $v_5 + (i_A - i_B)R_1 = 0$ $\quad (i_B - i_A)R_1 + (1/C_2)\int(i_B - i_C)\, dt = 0$ $\quad (1/C_2)\int(i_C - i_B)\, dt + (i_C - i_D)R_3 = 0$ $\quad (i_D - i_C)R_3 + L_4(di_D/dt) = 0$ (iii) v_a $\quad i_A, i_B, i_C, i_D$ (iv) $i_A = i_5$ remove $v_5 + (i_A - i_B)R_1 = 0$ **2.77** (i) $i_{10} + (1/R_1)(v_a - v_b) = 0$ $\quad -(1/R_1)(v_a - v_b) + C_2 d(v_b - v_c)/dt = 0$ $\quad C_2 d(v_b - v_c)/dt - (1/R_3)(v_c - v_d) - i_{11} = 0$ $\quad i_{11} + (1/R_5)(v_g - v_e) - (1/R_7)(v_e - v_f) = 0$ $\quad -(1/R_3)(v_c - v_d) + C_4\, dv_d/dt = 0$ $\quad (1/R_5)(v_g - v_e) + (1/L_6)\int v_g\, dt + C_8 d(v_g - v_f)/dt = 0$ $\quad (1/R_7)(v_e - v_f) - (1/L_9)\int v_f\, dt - C_8 d(v_f - v_g)/dt = 0$ (ii) $i_A R_1 + (1/C_2)\int i_A\, dt + (i_A - i_B)R_3 + (1/C_4)\int(i_A - i_B)\, dt = v_{10}$ $\quad -(i_A - i_B)R_3 - (1/C_4)\int(i_A - i_B)\, dt + v_{11} - (i_B - i_C)R_5 + L_6 d(i_B - i_D)/dt = 0$ $\quad (i_C - i_B)R_5 + i_C R_7 + (1/C_8)\int(i_C - i_D)\, dt = 0$ $\quad -L_6 d(i_B - i_D)/dt - (1/C_8)\int(i_C - i_D)\, dt + L_9\, di_D/dt = 0$ (iii) $v_a, v_b, v_c, v_d, v_e, v_f, v_g$ $\quad i_A, i_B, i_C, i_D$ (iv) $v_a \equiv v_{10}$ remove $i_{10} + (1/R_1)(v_a - v_b) = 0$ $\quad i_B \equiv i_{11}$ remove $-(i_A - i_B)R_3 - (1/C_4)\int(i_A - i_B)\, dt + v_{11} - (i_B - i_C)R_5 + L_6 d(i_B - i_D)/dt = 0$ **2.79** (i) $(1/R_3)(v_a - v_c) + C_1 d(v_a - v_b)/dt - i_7 = 0$ $\quad C_1 d(v_b - v_a)/dt + C_2 d(v_b - v_e)/dt - i_8 = 0$ $\quad (1/R_3)(v_c - v_a) + (1/L_4)\int(v_c - v_d)\, dt + i_8 = 0$ $\quad (1/L_4)\int(v_d - v_c)\, dt + (1/R_5)(v_d - v_e) = 0$ $\quad C_2 d(v_e - v_b)/dt + (1/R_5)(v_e - v_d) + (v_e/R_6) = 0$ (ii) $(i_C - i_A)R_3 + L_4 d(i_C - i_B)/dt + (i_C - i_B)R_5 + i_C R_6 + v_7 = 0$ $\quad (1/C_1)\int i_A\, dt - v_8 + (i_A - i_C)R_3 = 0$ $\quad (1/C_2)\int i_B\, dt + v_8 + (i_B - i_C)R_5 + L_4 d(i_B - i_C)/dt = 0$ (iii) v_a, v_b, v_c, v_d, v_e $\quad i_A, i_B, i_C$

ANSWERS TO ODD-NUMBERED PROBLEMS   867

(iv) $v_a \equiv v_7$ remove $(1/R_3)(v_a - v_c) + C_1 d(v_a - v_b)/dt - i_7 = 0$ $i_A \equiv i_B - i_8$ combine two equations containing v_8 to obtain $(1/C_1)\int(i_B - i_8)\,dt + (i_B - i_8 - i_C)R_3 + (1/C_2)\int i_B\,dt + (i_B - i_C)R_5 + L_4 d(i_B - i_C)/dt = 0$ **2.81** KCL: $(1/R_4)(v_c - v_b) + (1/R_5)(v_c - v_d) + C_7\,dv_c/dt = 0$ $(1/R_1)(v_b - v_8) + (1/R_4)(v_b - v_c) + (1/R_3)(v_b - v_d) + (1/L_2)\int(v_b - v_d)\,dt = 0$ $(1/R_5)(v_d - v_c) + (1/R_3)(v_d - v_b) + (1/L_2)\int(v_d - v_b)\,dt + (v_d/R_6) = 0$ KVL: $i_C R_1 + (i_C - i_B)R_4 + (1/C_7)\int(i_C - i_D)\,dt = v_8$ $(1/C_7)\int(i_D - i_C)\,dt + (i_D - i_B)R_5 + i_D R_6 = 0$ $(i_A - i_B)R_3 + L_2\,di_A/dt = 0$ $(i_B - i_C)R_4 + (i_B - i_A)R_3 + (i_B - i_D)R_5 = 0$
2.83 KCL: $C_8 d(v_a - v_d)/dt + (1/L_{12})\int(v_a - v_b)\,dt + C_{14} d(v_a - v_f)/dt = 0$ $(1/L_{12})\int(v_b - v_a)\,dt + (1/L_{13})\int(v_b - v_c)\,dt + (1/R_9)(v_b - v_e) = 0$ $(1/L_{13})\int(v_c - v_b)\,dt + (1/R_{10})(v_c - v_f) + C_{11} d(v_c - v_g)/dt = 0$ $C_8 d(v_d - v_g)/dt + (1/R_1)(v_d - v_h) + (1/L_5)\int(v_d - v_e)\,dt + (1/M_{57})\int(v_f - v_g)\,dt = 0$ $(1/L_5)\int(v_e - v_d)\,dt - (1/M_{57})\int(v_f - v_g)\,dt + (1/R_9)(v_e - v_b) + (1/R_2)(v_e - v_h) + (1/R_6)(v_e - v_f) = 0$ $(1/L_7)\int(v_f - v_g)\,dt + (1/M_{57})\int(v_d - v_e)\,dt - C_{14} d(v_a - v_f)/dt + (1/R_6)(v_f - v_e) + (1/R_{10})(v_f - v_c) + (v_f/R_3) = 0$ $(1/L_7)\int(v_g - v_f)\,dt - (1/M_{57})\int(v_d - v_e)\,dt + C_{11} d(v_g - v_c)/dt + C_4\,dv_g/dt = 0$ $(1/R_1)(v_d - v_h) + (1/R_2)(v_e - v_h) + i_{15} = 0$ KVL: $L_{12}\,di_A/dt + (i_A - i_B)R_9 + L_5 d(i_A + i_G - i_D)/dt + (1/C_8)\int(i_A + i_G)\,dt + (L_7/M_{57})(i_C - i_F) = 0$ $L_{13}\,di_B/dt + (i_B - i_A)R_9 + (i_B - i_C)R_{10} + (i_B + i_G - i_{15})R_6 = 0$ $R_{10}(i_C - i_B) + (1/C_{11})\int i_C\,dt + L_7 d(i_C - i_F)/dt = 0$ $R_1 i_D + L_5 d(i_D - i_A - i_G)/dt + R_2(i_D - i_{15}) = 0$ $R_3(i_F - i_{15}) + L_7 d(i_F - i_C)/dt + (1/C_4)\int i_F\,dt = 0$ $(1/C_{14})\int i_G\,dt + R_6(i_G + i_B - i_{15}) + L_5 d(i_G + i_A - i_D)/dt + (1/C_8)\int(i_A + i_G)\,dt = 0$
2.85 KCL: $(1/R_2)(v_b - v_1) + (v_b/R_3) + (1/R_5)(v_b - v_g) + C_6 d(v_b - v_g)/dt = 0$ $C_6 d(v_g - v_b)/dt + (v_g/R_7) + (1/R_5)(v_g - v_b) = 0$ KVL: $i_A(R_2 + R_3) - i_B R_3 = v_1$ $i_B(R_3 + R_7) - i_A R_3 + (1/C_6)\int(i_B - i_C)\,dt = 0$ $i_C R_5 + (1/C_6)\int(i_C - i_B)\,dt = 0$ **2.87** Constraint $i_A = K_2 v_1$ KCL: $(v_a - v_b)/R_3 = i_A$ $(1/L_4)\int v_b\,dt = i_A$
2.89 $v_a = K_2 v_1$ $v_b/R_3 = C_4 d(v_a - v_b)/dt$ $v_c = K_5 v_b$ $v_c/R_6 = i_6$
2.91 $iR_2 + L_3\,di/dt + K_4 \dot{x} = v_1$ $K_4 i = m\ddot{x} + kx$ **2.93** $v_c = K_4[(1/C_{30}) + K_P p(t)]\int i\,dt$ $iR_2 + [(1/C_{30}) + K_P p(t)]\int i\,dt = v_{10}\sin\omega_f t$ **2.95** $J\ddot{\theta} + b\dot{\theta} + k\theta = Ki(t)$ **2.97** $L_3\,di/dt + (R_1 + R_3)i + K_4\dot{x} = 0$ $m\ddot{x} + kx - K_4 i = -p(t)A$

CHAPTER 3

3.1 $C_i\,dT_i/dt = (T_b - T_i)/R_{ib}$ **3.3** $C_1\,dT_1/dt = (T_2 - T_1)/R_{12}$ $C_2\,dT_2/dt = (1/R_{12})(T_1 - T_2) + (1/R_{2b})(T_b - T_2)$ $C_b\,dT_b/dt = (1/R_{2b})(T_2 - T_b) + (1/R_{bo})(T_o - T_b)$ **3.5** $C_v\,dT/dt = K\rho(t)v^2(t)$ (through variable q is constrained)
3.7 $C_n\,dT_n/dt = (1/R_{nB})(T_B - T_n)$ $n = 1, 2, 3, 4$ $C_B\,dT_B/dt = \sum_{n=1}^{4}(1/R_{nB})(T_n - T_B) + (1/R_{BS})(T_S - T_B)$ **3.9** $C\,dT/dt = [(4/R_{W_o}) + (1/R_{R_o})](T_o - T) + q(t)$ **3.11** $C\,dT_1/dt = (4/R_e)(T_o - T_1) + (1/R_i)(T_2 - T_1) + q(t)$ $C\,dT_4/dt = (4/R_e)(T_o - T_4) + (1/R_i)(T_3 - T_4) + q(t)$ $C\,dT_2/dt = (3/R_e)(T_o - T_2) + (1/R_i)(T_1 - T_2) + (1/R_i)(T_3 - T_2) + q(t)$ $C\,dT_3/dt = (3/R_e)(T_o - T_3) + (1/R_i)(T_2 - T_3) + (1/R_i)(T_4 - T_3) + q(t)$ **3.13** $C_r\,dT_r/dt = (1/R_{rg})(T_g - T_r) + q_r(t)$ $C_g\,dT_g/dt = (1/R_{gl})(T_l - T_g) + (1/R_{rg})(T_r - T_g)$ $C_l\,dT_l/dt = (1/R_{lc})(T_c - T_l) + (1/R_{gl})(T_g - T_l)$ $C_c\,dT_c/dt = (1/R_c)(T_e - T_c) + (1/R_{lc})(T_l - T_c)$ **3.15** $\rho A_1\,dh_1/dt + [w_{03}(h_1 - h_3)/(h_{01} - h_{03} + z_1 - z_3)] = 0$ $\rho A_2\,dh_2/dt - [w_{04}(h_3 - h_2)/(h_{03} - h_{02} + z_3 - z_2)] + w_2 = 0$ $\rho A_3\,dh_3/dt - [w_{03}(h_1 - h_3)/(h_{01} - h_{03} + z_1 - z_3)] + [w_{04}(h_3 - h_2)/(h_{03} - h_{02} + z_3 - z_2)] = 0$

3.17 $c_{12}\sqrt{p_1 - p_2} - c_{23}\sqrt{p_2 - p_3} = A\dot{\rho}_2(\ell + x) + A\rho_2\dot{x}$ $\qquad \rho_2 = (p_2/K)^{1/\gamma}$
$(p_2 - p_3)A = kx$ **3.19** $dp_2/dt = \gamma K^{1/\gamma} p_2^{(\gamma-1)/\gamma}(p_1 - p_2)/V_2 R_{12}$ **3.21** $\dot{\rho}_3 C_3$
$= (1/R_2)(p_1 - p_3) - (1/R_4)(p_3 - p_5)$ $\qquad \dot{\rho}_5 Ax + \rho_5 A\dot{x} = (1/R_4)(p_3 - p_5)$
$[(w/g) + m]\ddot{x} + b\dot{x} = p_5 A$ $\qquad p_3 = K_3 \rho_3^\gamma$ $\qquad p_5 = K_5 \rho_5^\gamma$ **3.23** $(J + m\ell_2^2)d^2\theta/dt^2$
$- k_3\ell_1(y - \ell_1\theta) = 0$ $\qquad k_3(y - \ell_1\theta) = A(p_1 - p_2)$ $\qquad c\sqrt{p_s - p_1}\, x =$
$[A(\ell + y)/K]\, dp_1/dt + A[(p_1/K) + \rho_{01}]\, dy/dt$ $\qquad c\sqrt{p_2 - p_s}\, x =$
$[A(\ell + y)/K]\, dp_2/dt + A[\rho_{02} - (p_2/K)]\, dy/dt$ **3.25** $dp_2/dt =$
$K^{(\gamma+1)/\gamma} p_2^{(\gamma-1)/\gamma} \gamma(w_1 - K_0 x \sqrt{p_2 - p_5})/V_2$

CHAPTER 4

4.1 $C_i\, dv_i/dt = (v_b - v_i)/R_{ib}$ **4.3** $C_1\, dv_1/dt = (v_2 - v_1)/R_{12}$ $\qquad C_2\, dv_2/dt$
$= (1/R_{12})(v_1 - v_2) + (1/R_{2b})(v_b - v_2)$ $\qquad C_b\, dv_b/dt = (1/R_{2b})(v_2 - v_b) +$
$(1/R_{bo})(v_o - v_b)$ **4.5** $C_v\, dv_v/dt = K\rho(t)v^2(t)$
4.7 $C_n\, dv_n/dt = (1/R_{nB})(v_B - v_n) + i_n(t)$ $\qquad n = 1, 2, 3, 4$ $\qquad C_B\, dv_B/dt =$
$\sum_{n=1}^{4} (1/R_{nB})(v_n - v_B) + (1/R_{BS})(v_S - v_B)$ **4.9** $C\, dv/dt = [(4/R_{Wo}) +$
$(1/R_{Ro})](v_o - v) + i(t)$ **4.11** $C_1\, dv_1/dt = (4/R_e)(v_0 - v_1) + (1/R_i)(v_2 - v_1)$
$C_2\, dv_2/dt = (3/R_e)(v_0 - v_2) + (1/R_i)(v_1 - v_2) + (1/R_i)(v_3 - v_2)$ $\qquad C_3\, dv_3/dt =$
$(3/R_e)(v_0 - v_3) + (1/R_i)(v_2 - v_3) + (1/R_i)(v_4 - v_3) + i(t)$ $\qquad C_4\, dv_4/dt =$
$(4/R_e)(v_0 - v_4) + (1/R_i)(v_3 - v_4)$ **4.13** $i_3 = (1/R_2)(v_1 - v_3) - (1/R_4)(v_3 - v_5)$
$i_5 = (1/R_4)(v_3 - v_5)$ $\qquad C\, dv_4/dt + v_4/R = v_5/R_A$ $\qquad v_1 = p_1$ $\qquad v_3 = p_3$ $\qquad v_4 = \dot{x}$
$v_5 = p_5$ $\qquad i_3 = C_3\, d\rho_3/dt$ $\qquad i_5 = dA\rho_5 x/dt$ $\qquad C = (w/g) + m$ $\qquad R = 1/b$
$R_A = 1/A$ **4.15** $C_1\, dv_1/dt + [i_{03}(v_1 - v_3)/(v_{01} - v_{03} + V_1 - V_3)] = 0$
$C_2\, dv_2/dt - [i_{04}(v_3 - v_2)/(v_{03} - v_{02} + V_3 - V_2)] + i_2 = 0$ $\qquad C_3\, dv_3/dt$
$- [i_{03}(v_1 - v_3)/(v_{01} - v_{03} + V_1 - V_3)] + [i_{04}(v_3 - v_2)/(v_{03} - v_{02} + V_3 - V_2)] = 0$
4.17 $v_1/R_1 = (v_2/R_2) - (1/L)\int v_1\, dt$ $\qquad 1/R_1 = A[(K_b x_0 + p_3 A)/K_g A]^{1/\gamma}[1 +$
$(K_b x_0 + K_b \ell)/\gamma(K_b x_0 + p_3 A)]$ $\qquad 1/R_2 = c_{12}/2\sqrt{p_{10} - p_{30} - (K_b x_0/A)}$ $\qquad 1/L =$
$c_{12} K_b/2A\sqrt{p_{10} - p_{30} - (K_b x_0/A)} + c_{23}\sqrt{K_b/A x_0/2}$ $\qquad v_1 = dx/dt$ $\qquad v_2 = p_3$
4.19 $(1/L_1)\int v_x\, dt = (1/R)v_y$ $\qquad C\, dv_8/dt + (1/L_2)\int v_8\, dt = (1/L_2)\int v_y\, dt$ $\qquad (1/L_1)$
$= c\sqrt{p_s - p_{10}}$ $\qquad (1/R) = A\rho_0$ $\qquad C = (J + m\ell_2^2)/\ell_1$ $\qquad (1/L_2) = k_3\ell_1$
$v_8 = \ell_1\, d\theta/dt$ **4.21** $C_1\, dv_a/dt + (1/L_{12})\int(v_a - v_b)\, dt + (1/R_{13})(v_a - v_c) = 0$
$(1/R_{13})(v_c - v_a) + (1/R_{23})(v_c - v_b) + C_3\, dv_c/dt + (1/R_4)v_c + (1/R_6)v_c = 0$
4.23 $L_1 i_3 + R_2\int i_3\, dt + L_3 i_3 + (1/C_4)\int\int i_3\, dt = (v_r/\omega)\sin\omega t$ **4.25** $i_5 -$
$(1/L_2)\int v\, dt - C_2\, dv/dt - i_{\text{friction}} + i_{\text{gravity}} = 0$ **4.27** $C_1\, dv_1/dt + (1/R_1)(v_1 - v_3)$
$+ (1/L_4)\int(v_1 - v_2 - v_3)\, dt = 0$ $\qquad C_2\, dv_2/dt + (1/R_2)(v_2 - v_3) -$
$(1/L_4)\int(v_1 - v_2 - v_3)\, dt + (1/L_5)\int(v_3 - v_1)\, dt = 0$ $\qquad C_3\, dv_3/dt -$
$(1/R_1)(v_1 - v_3) - (1/R_2)(v_2 - v_3) + (1/L_5)\int(v_3 - v_1)\, dt + (1/L_6)\int(v_3 - v_9)\, dt = 0$
4.29 $C_1\, dv_1/dt + (1/R)(v_1 - v_2) - i_D = 0$ $\qquad C_2\, dv_2/dt -$
$(1/R)(v_1 - v_2) - i_L = 0$ **4.31** $\int v\, dt = L_2 i_n + R_3\int i_n\, dt + L_4 i_n$ $\qquad n = 2, 3, 4$
$\int i\, dt = C_2 v_n + (1/R_3)\int v_n\, dt + C_4 v_n$ $\qquad n = 2, 3, 4$ **4.33** $\int(i_1 - i_6)\, dt =$
$(C_2 + C_4 + C_6)(L_6\, di/dt + R_8 i_8) + (1/R_5)\int(L_6\, di/dt + R_8 i_8)\, dt$ $\qquad R_8 i_8 +$
$(1/C_7)\int i_8\, dt = (1/C_7)\int i_6\, dt$ **4.35** $K_1 i = C_3\, dv_a/dt + (1/R_3)v_a + (1/L_3)\int v_a\, dt$
$iR + L\, di/dt + K_1 v_a = v_1(t)$ **4.37** $L_1\, di_a/dt + (1/C_{12})\int(i_a - i_b)\, dt$
$+ R_{13}(i_a - i_c) = 0$ $\qquad v_7 = (1/C_{12})\int(i_b - i_a)\, dt + L_2\, di_b/dt + R_{23}(i_b - i_c)$
$+ (1/C_5)\int i_b\, dt$ $\qquad R_{23}(i_c - i_b) + R_{13}(i_c - i_a) + R_4 i_c + R_6 i_c + L_3\, di_c/dt = 0$
4.39 $i_7 = (1/L_{12})\int(v_b - v_a)\, dt + C_2\, dv_b/dt + (1/R_{23})(v_b - v_c) + (1/L_5)\int v_b\, dt$
4.41 $C_1\, dh_1/dt + (h_1 - h_2)/R_3 = 0$ $\qquad (1/R_3)(h_1 - h_2) - C_2\, dh_2/dt - w_2 = 0$

ANSWERS TO ODD-NUMBERED PROBLEMS 869

$C_1 = \rho A_1$ $C_2 = \rho A_2$ $1/R_3 = w_{30}/(h_{10} - h_{20} + z_1 - z_2)$ **4.43**
$k_4 \int (\Omega_9 - \Omega_1) \, dt - J_1 \dot{\Omega}_1 - k_5 \int (\Omega_1 - \Omega_2) \, dt = 0$ $k_5 \int (\Omega_1 - \Omega_2) \, dt - J_2 \dot{\Omega}_2 - k_6 \int \Omega_2 \, dt = 0$ **4.45** $f_1 - m_2 \dot{u}_2 - k_3 \int (u_2 - u_4) \, dt = 0$ $k_3 \int (u_2 - u_4) \, dt - m_4 \dot{u}_4 - k_5 \int (u_4 - u_6) \, dt = 0$ $k_5 \int (u_4 - u_6) \, dt - m_6 \dot{u}_6 = 0$ $u_2 = \dot{x}_2$, $u_4 = \dot{x}_4$, $u_6 = \dot{x}_6$ **4.47** $M_D - J_1 \dot{\Omega}_1 - b(\Omega_1 - \Omega_2) = 0$ $b(\Omega_1 - \Omega_2) - J_2 \dot{\Omega}_2 + M_L = 0$ **4.49** $2u_1 - (1/k_2) \dot{f}_A - b_3 f_A - (1/m_4) \int (f_A - f_B) \, dt = 0$ $(1/m_4) \int (f_A - f_B) \, dt - (1/k_5) \dot{f}_B - (b_7/2) f_B = 0$ $u_1 = \int x_1 \, dt$

CHAPTER 5

5.1 (iii) 3   **5.3** (iii) 3   **5.5** (iii) 4   **5.7** (iii) 2   **5.9** (iii) 2
5.11 (iii) 6   **5.13** (iii) 6   **5.15** (iii) infinite   **5.17** 2 position, 3 orientation   **5.19** 2 position, 3 orientation   **5.21** $dx/dt = -\ell \cos(\theta_m \cos \omega t) \theta_m \omega \sin \omega t$ $dy/dt = -\ell \sin(\theta_m \cos \omega t) \theta_m \omega \sin \omega t$ $d^2x/dt^2 = -\ell \sin(\theta_m \cos \omega t) \theta_m^2 \omega^2 \sin^2 \omega t - \ell \cos(\theta_m \cos \omega t) \omega^2 \theta_m \cos \omega t$ $d^2y/dt^2 = \ell \cos(\theta_m \cos \omega t) \theta_m^2 \omega^2 \sin^2 \omega t - \ell \sin(\theta_m \cos \omega t) \omega^2 \theta_m \cos \omega t$ **5.23** $z = (v_{z0} x/v_{x0}) - (gx^2/2v_{x0}^2)$ **5.25** $dy/dt = -v_{20}^2 t/(\ell^2 - v_{20}^2 t^2)^{\frac{1}{2}}$ $d^2y/dt^2 = -v_{20}^2 \ell^2/(\ell^2 - v_{20}^2 t^2)^{\frac{3}{2}}$ **5.27** $1_z n + 1_r(nh/a)$ **5.29** $\mathbf{v} = 1_x \ell \dot{\theta} \cos \theta + 1_y \ell \dot{\theta} \sin \theta$ $\mathbf{a} = 1_x \ell (\ddot{\theta} \cos \theta - \dot{\theta}^2 \sin \theta) + 1_y \ell (\ddot{\theta} \sin \theta + \dot{\theta}^2 \cos \theta)$ **5.31** $\mathbf{v} = 1_r c + 1_\theta n r$ $\mathbf{a} = 1_r(-n^2 r) + 1_\theta 2nc$ **5.33** $\mathbf{v} = 1_x(-ny) + 1_y(c + nx)$ $\mathbf{a} = 1_x(-n^2 x - 2nc) + 1_y(-n^2 y)$ center of chord located at $(x,0)$ in rotating xy frame
5.35 $\mathbf{v} = 1_x(v_0 - \ell \dot{\phi} \cos \phi - bn \sin \phi) + 1_z(\ell \dot{\phi} \sin \phi - bn \cos \phi)$ $\mathbf{a} = 1_x(-\ell \ddot{\phi} \cos \phi + \ell \dot{\phi}^2 \sin \phi - bn \dot{\phi} \cos \phi) + 1_z(\ell \ddot{\phi} \sin \phi + \ell \dot{\phi}^2 \cos \phi + bn \dot{\phi} \sin \phi)$ **5.37** $\mathbf{v} = 1_x v_{20}[1 - (2u_0 t/\ell)] + 1_y[u_0(\ell^2 - 2v_{20}^2 t^2)/\ell(\ell^2 - v_{20}^2 t^2)^{\frac{1}{2}}]$ $\mathbf{a} = 1_x(-2u_0 v_{20}/\ell) + 1_y[-u_0 v_{20}^2 t(3\ell^2 - 2v_{20}^2 t^2)/\ell(\ell^2 - v_{20}^2 t^2)^{\frac{3}{2}}]$ **5.39** $1_\ell (\dot{\psi}/r) + 1_z(N + \dot{\psi})$ **5.41** $(gR \tan \beta)^{\frac{1}{2}}$ **5.43** $\arctan(a/g)$ **5.45**   backwards and in the direction of rotation   **5.47** $[(mr^2 + J)\ddot{x}/r^2] + kx = 0$ **5.49** $\{2m_1 - [2m\ell \cos \theta/r] + [m(r^2 + \ell^2)/r^2]\} \ddot{\theta} + mg\ell \sin \theta + (m\ell \dot{\theta}^2 \sin \theta)/r = 0$ **5.51** $[m + (J/r_o^2)]\ddot{x} + k(r_o - r_i)[(r_o - r_i)(x/r_o) - r_w \Omega t]/r_o + mg \sin \beta = 0$ **5.53** $m(\ell \cos \theta \ddot{x} + \frac{2}{3} \ell^2 \ddot{\theta}) + mg\ell \sin \theta = 0$ **5.55** $m\ddot{x} + b \operatorname{sgn} \dot{x} + 2kx - m\Omega^2 x = 0$
5.57 $T_1 \cos \theta_1 - T_2 \cos \theta_2 - mg = 0$ $T_1 \sin \theta_1 - T_2 \sin \theta_2 = 0$ $T_2 \cos \theta_2 - mg = 0$ $T_2 \sin \theta_2 - T_3 = 0$ $\theta_1 = 21°55'$, $\theta_2 = 38°50'$ **5.59** $m \ddot{y} + 2ky - k(\ell - 2\ell_1)\theta = 0$ $J \ddot{\theta} + k(\ell^2 + 2\ell \ell_1 + 2\ell_1^2)\theta - (\ell - 2\ell_1) y = 0$ **5.61** $m_1 \ddot{x} + m_2(\ddot{x} + \ddot{y} \cos \beta) = 0$ $m_2(\ddot{x} \cos \beta + \ddot{y}) + b\dot{y} + ky + mg \sin \beta = 0$ **5.63** $[(m_1 + m_2)r^2 + m_1(\ell^2 - 2r\ell \cos \theta) + J_2] \ddot{\theta} + m_1 r \ell \sin \theta \dot{\theta}^2 + m_1 g \ell \sin \theta = 0$
5.65 $m(\ddot{x}_1 + \ddot{x}_2) + (J \ddot{x}_1/r^2) + b\dot{x}_1 + kx_1 = 0$ $m_2 \ddot{x}_2 + m(\ddot{x}_1 + \ddot{x}_2) = 0$
5.67 $[m + (J/r^2)]\ddot{x} + m\ell \ddot{\psi} \cos \psi - m\ell \dot{\psi}^2 \sin \psi = 0$ $m\ddot{x}\ell \cos \psi + m\ell^2 \ddot{\psi} + mg\ell \sin \psi + 4b\dot{\psi} = 0$ **5.69**   (a) $m_1 \ddot{x} - m_1 \Omega^2 (\ell + x) \sin^2 \theta - m_1 (\ell + x) \dot{\theta}^2 - m_1 g \cos \theta + kx = 0$ $m_1 (\ell + x)^2 \ddot{\theta} + (m_2 \ell_2^2 \dot{\theta}/3) + 2m_1(\ell + x)\dot{x}\dot{\theta} - [m_1(\ell + x)^2 + (m_2 \ell_2^2/3)] \Omega^2 \sin \theta \cos \theta + [(m_2 \ell_2 g/2) + m_1 g(\ell + x)] \sin \theta = 0$ (b) $-m_1 \Omega^2 (\ell + x) \sin^2 \theta - m_1 g \cos \theta + kx = 0$ $-[m_1 (\ell + x)^2 + (m_2 \ell_2^2/3)] \Omega^2 \sin \theta \cos \theta + [(m_2 \ell_2 g/2) + m_1 g(\ell + x)] \sin \theta = 0$ **5.71** $\ddot{\theta} + (g/\ell) \sin \theta = 0$ **5.73** $m \ddot{h} + 2 \rho g h = 0$ **5.75** $m \ell^2 \ddot{\theta}(1 + 4 \sin^2 \theta_0) + 3 mg \ell \cos \theta_0 \sin \theta - (m \ell^2 \Omega^2/2) \cos 2\theta_0 \sin 2\theta = 0$ **5.79** $J_1 \dot{\theta}_1 + k_5(\theta_1 - \theta_2) + k_8 \theta_1 = 0$ $J_2 \dot{\theta}_2 - k_5(\theta_1 - \theta_2) + k_6 \theta_2 = 0$ **5.83** $m_5 \ddot{x}_5 + k_3(x_5 - x_1) - k_7(x_{1:} - \ell_1 \theta - x_5) = 0$ $m_6 \ddot{x}_6 + k_4(x_6 - x_2) - k_8[x_{11} - (\ell - \ell_1)\theta - x_6] = 0$ $m_{11} \ddot{x}_{11} + k_8[x_{11} - (\ell - \ell_1)\theta - x_6] + k_7(x_{11} + \ell_1 \theta - x_5) = 0$ $J_{12} \ddot{\theta} - k_8(\ell - \ell_1)[x_{11} - (\ell - \ell_1)\theta - x_6] + k_7 \ell_1(x_{11} + \ell_1 \theta - x_5) = 0$ **5.85** $\ddot{\theta} + (g/\ell) \sin \theta + (\ddot{x}/\ell) \cos \theta = 0$ $(m_1 + m_2) \ddot{x} + m_1 \ell \ddot{\theta} \cos \theta - m_1 \ell^2 \dot{\theta}^2 \sin \theta = 0$
5.87 $(m_1 \ell \ddot{\theta}_1/6) + (k_1 \ell \theta_1/2) + k_2[(\ell \theta_1/2) + (\ell \theta_2/2) - x] = 0$ $(m_2 \ell \ddot{\theta}_2/6) +

$k_2[(\ell\theta_1/2) + (\ell\theta_2/2) - x] + k_3[(\ell\theta_2/2) + x] = 0 \qquad m_2\ddot{x} + k_3[(\ell\theta_2/2) + x] - k_2[(\ell\theta_1/2) + (\ell\theta_2/2) - x] = 0$ **5.91** $J_1\ddot{\theta}_1 + b_3\dot{\theta}_1 + k_5(\theta_1 - \theta_2) - k_8(\theta_9 - \theta_1) = 0 \qquad J_2\ddot{\theta}_2 + b_4\dot{\theta}_2 - k_5(\theta_1 - \theta_2) + k_6\theta_2 = 0$ **5.95** $J_1\ddot{\theta}_1 + b(\dot{\theta}_1 - \dot{\theta}_2) - M_D = 0 \qquad J_2\ddot{\theta}_2 - b(\dot{\theta}_1 - \dot{\theta}_2) + M_L = 0$

CHAPTER 6

6.3 2.3, 3.0, 4.6, 11.5, 0.0045% **6.5** $v_3(t) = -[v_0R_2/(R_1+R_2)]\exp[-R_1R_2t/L_3(R_1+R_2)] \quad 0 < t$ **6.7** $T(t) = T_1 + (T_0 - T_1)\exp(-t/RC) \quad 0 < t$ **6.9** $s + (K_1\ell_1\ell_4/\ell_2\ell_3) = 0 \qquad \theta(t) = \theta_0 \exp(-K_1\ell_1\ell_4t/\ell_2\ell_3) \quad 0 < t$ **6.11** $v_a(t) = \gamma \exp(K_1K_3t/C_2) \quad 0 < t$ **6.17** $v_{3f}(t) = [v_{1m}/(1 - \sigma_fR_2C_3)] \exp(-\sigma_ft) \quad 0 < t$ **6.19** $x(t) = x_0 + \dfrac{r\Omega_0}{\sigma_f}\left[1 - \left(\dfrac{b}{m}e^{-\sigma_ft} - \sigma_f e^{-\frac{b}{m}t}\right)\Big/\left(\dfrac{b}{m} - \sigma_f\right)\right] + (mr\Omega_0/b)[1 - e^{-(b/m)t}] \quad 0 < t$ **6.21** (ii) $x(t) = (Ap_0/m)(-\tfrac{1}{14} + \tfrac{1}{63}e^{7t} + \tfrac{1}{18}e^{-2t}) \quad 0 < t$ **6.25** $T(t) = T_r e^{(t_2-t)/\tau_1} + T_v(1 - e^{(t_2-t_1)/\tau_1}) + (T_r - T_v)(e^{-t/\tau_1} - e^{-t/RC})/\tau_1[(1/RC) - (1/\tau_1)] \quad t_2 < t$ **6.27** $\Omega(t) = (K_4v_{1m}/b)[1 - \exp(-bt/J)] \qquad v_3(t) = (K_4v_{1m}/b) + [K_4v_{1m}/(R_2C_3b - J)][(J/b)\exp(-bt/J) - R_2C_3\exp(-t/R_2C_3)] \quad 0 < t$ **6.29** $\Omega_2(t) = \Omega_{10} + (3 - \Omega_{10})\exp(-bt/J) \quad 0 < t$ **6.31** $\theta(t) = x_0[(\ell_1 + \ell_2)/\ell_1\ell_4][1 - \exp(-K_1\ell_1\ell_4t/\ell_2\ell_3)] - 0.01(\ell_2/\ell_1\ell_4)\exp(-K_1\ell_1\ell_4t/\ell_2\ell_3) \quad 0 < t$ **6.33** $v_3(t) = (K_4v_{1m}/b) + [K_4v_{1m}/(R_2C_3b - J)][(J/b)\exp(-bt/J) - R_2C_3\exp(-t/R_2C_3)] + \gamma + \rho[1 - \exp(-t/R_2C_3)] \quad 0 < t$ **6.35** $v_2(t) = \gamma + (Kv_{1m}t/C_2) + \{(Kv_{1m}/C_2\sigma_f)[1 - \exp(-\sigma_ft)]\} \quad 0 < t$ **6.37** $v_2(t) = 0.3v_{1m} + (2Kv_{1m}/3C_2\sqrt{T})t^{3/2} \quad 0 < t < T \qquad v_2(t) = v_{1m}[0.3 + (2KT/3C_2)] \quad T < t$ **6.39** $v_3(t) = (Bt/R_2C_3)\exp(-t/R_2C_3) \quad 0 < t$ **6.41** $T(t) = T_r e^{(t_2-t)/\tau_1} + T_v(1 - e^{(t_2-t_1)/\tau_1}) + (T_r - T_v)(t - t_2)e^{-t/RC}/RC \quad t_2 < t$ **6.47** $i(t) = [v_0R_3 \Delta t/L_2(R_1 + R_3)]\exp[-R_1R_3t/L_2(R_1 + R_3)]u(t)$ **6.49** $v_3(t) = (v_{1m}\tau/R_2C_3)\exp(-t/R_2C_3)u(t)$ **6.51** $\theta(t) = [K_1x_0\tau(\ell_1 + \ell_2)/\ell_2\ell_3]\exp(-K_1\ell_1\ell_4t/\ell_2\ell_3)u(t)$ **6.53** $\theta(t) = [-K_1x_0\tau(\ell_1 + \ell_2)/\ell_3\ell_5]\exp[K_1\ell_4(\ell_1 + \ell_2 + \ell_5)t/\ell_3\ell_5]u(t)$ **6.55** $i(t) = 0.30\exp(-10t) \quad 0 < t < 0.1 \qquad i(t) = 0.30\{\exp(-10t) + \exp[-10(t - 0.1)]\} \quad 0.1 < t < 0.2 \qquad i(t) = 0.30\{\exp(-10t) + \exp[-10(t - 0.1)] + \exp[-10(t - 0.2)]\} \quad 0.2 < t < 0.3 \ldots$ **6.57** $v_3(t) = [K_4v_{1m}\tau/(R_2C_3b - J)][\exp(-t/R_2C_3) - \exp(-bt/J)]u(t)$ **6.61** $v_3(t) = [v_{1m}/(R_3C_2\sigma_f - 1)][R_3C_2\sigma_f \exp(-\sigma_ft) - \exp(-t/R_3C_2)] - \gamma \exp(-t/R_3C_2) \quad 0 < t$ **6.67** $\theta(t) = (M_0/kT)\{t - (b/k)[1 - \exp(-kt/b)]\} \quad 0 < t < T \qquad \theta(t) = (M_0/kT)\{T\exp(kT/b) - (b/k)[\exp(kT/b) - 1]\}\exp(-kt/b) \quad T < t$ **6.69** $\theta(t) = x_0[(\ell_1 + \ell_2)/\ell_1\ell_4][1 - \exp(-K_1\ell_1\ell_4t/\ell_2\ell_3)] \quad 0 < t$ **6.71** $\theta(t) = (M_0/kT)\{t - (b/k)[1 - \exp(-kt/b)]\} \quad 0 < t < T \qquad \theta(t) = (M_0b/k^2T)[1 - 2\exp(kT/b) + \exp(2kT/b)]\exp(-kt/b) \quad 2T < t$

CHAPTER 7

7.1 $\omega^2 = g/\delta$ **7.3** $\omega^2 = kr_1^2/(J + mr_2^2)$ **7.5** $\omega^2 = A\rho_w g/m$ **7.7** $\omega^2 = 1/LC$ **7.9** $\omega^2 = 2\beta g/d$ **7.11** $\omega^2 = [m + (m_{\text{rod}}/2)]g\ell/[J + m\ell^2 + (m_{\text{rod}}\ell^2/3)]$ **7.13** $\omega^2 = 3EI/m\ell^3$ **7.15** $\omega^2 = mgr/[2J + (2M + m)R^2 + m(R - r)^2]$ **7.17** $\omega^2 = mgr^2/(R - r)(J + mr^2)$ **7.19** $\omega^2 = \Omega^2(b - c)/2(c - r)$ **7.21** $-1.14 \exp(j \arctan 57.0)$ **7.23** $144 \exp(j \arctan -27.6)$ **7.25** -1 **7.27** $j\pi$

ANSWERS TO ODD-NUMBERED PROBLEMS   871

CHAPTER 8

8.1 $b = 48$ **8.3** $s^2 + (R_2 s/L_1) + (1/L_1 C_3) = 0$ $\omega_0^2 = (1/L_1 C_3)$ $\sigma = (R_2/2L_1)$ **8.5** $R_2 = 0.8 \sqrt{L_1/C_3}$, $R_2 = 2\sqrt{L_1/C_3}$ **8.7** $C_3 = 4L_1/R_2^2$ **8.9** $m_1 \ddot{x} + b\dot{x} + k_2 x = f_5 + m_1 g$ $s^2 + (bs/m_1) + (k_2/m_1) = 0$ $\sigma = b/2m_1$ $\omega_0^2 = k_2/m_1$ **8.11** $\dot{z} + (K_1 K_2 \ell_1/\ell_2) \int z\, dt = 0$ $s^2 + (K_1 K_2 \ell_1/\ell_2) = 0$ $\omega_0^2 = (K_1 K_2 \ell_1/\ell_2)$ $\sigma = 0$ **8.13** $m\ell^2 \ddot{\theta}(1 + 4\sin^2\theta_0) + 3mg\ell \cos\theta_0 \theta - m\ell^2 \Omega^2 \cos 2\theta_0 \theta = 0$ $m\ell^2(1 + 4\sin^2\theta_0) s^2 + 3mg\ell \cos\theta_0 - m\ell^2 \Omega^2 \cos 2\theta_0 = 0$ $\omega_0^2 = (3g\cos\theta_0 - \ell\Omega^2 \cos 2\theta_0)/\ell(1 + 4\sin^2\theta_0)$ $\sigma = 0$ **8.15** $x = 2e^{-t}$ **8.17** $b = 24\pi m\ell^2$ (a) $\theta(t) = (0.06/16\sqrt{2})[(8\sqrt{2} - 12)e^{-(12+8\sqrt{2})t} + (8\sqrt{2} + 12)e^{-(12-8\sqrt{2})t}]$ (b) $\dot{\theta}(0) < -\sigma_1 \theta(0)$ for $0 < \theta(0)$, $\sigma_2 < \sigma_1$ $\dot{\theta}(0) < -\sigma_2 \theta(0)$ for $0 < \theta(0)$, $\sigma_1 < \sigma_2$ **8.23** $f = 2.62$ cps $x_{max} = 15.76$ in. **8.25** 3.296 $x_{max} = 5.80$ in **8.27** (a) $\dot{\theta}(0) < -\theta(0)\sigma$ $0 < \dot{\theta}(0)$ $T_c = -\theta(0)/[\dot{\theta}(0) + \sigma\theta(0)]$ (b) $T_s = \theta(0)/\sigma[\dot{\theta}(0) + \sigma\theta(0)]$ **8.31** (i) $R_3 L_4 s/(C_2 R_3 L_4 s^2 + L_4 s + R_3)$ (ii) zero at origin, poles at $-50 \pm j999$ (iii) $C_2 R_3 L_4 s^2 + L_4 s + R_3 = 0$ (iv) $-5.7 e^{-300t}$ **8.33** (i) $k/(Js^2 + bs + k)$ (iii) $Js^2 + bs + k = 0$ (iv) $[\Omega_{10} k/(J\sigma^2 - b\sigma + k)]e^{-\sigma t}$ $\omega = \sqrt{k/J}$ **8.35** $Cs/(LCs^2 + RCs + 1)$ zero at origin, poles at $-(R/2L) \pm \sqrt{(R/2L)^2 - (1/LC)}$ $LCs^2 + RCs + 1 = 0$ **8.37** $b(Js^2 + kr^2)/(Js^2 + bs + kr^2)$ zeros at $\pm j\sqrt{kr/J}$, poles at $-(b/2J) \pm j\sqrt{(kr/J) - (b/2J)^2}$ $Js^2 + bs + kr = 0$ **8.39** $(R_3 C_4 s + 1)/(L_2 C_4 s^2 + R_3 C_4 s + 1)$ zero at $-(1/R_3 C_4)$, poles at $-(R_3/2L_2) \pm j\sqrt{(1/L_2 C_4) - (R_3/2L_2)^2}$ $L_2 C_4 s^2 + R_3 C_4 s + 1 = 0$ $R_3 C_4 s/(L_2 C_4 s^2 + R_3 C_4 s + 1)$ $1/(L_2 C_4 s^2 + R_3 C_4 s + 1)$ **8.41** $1/(\tau_1 s + 1)$ pole at $-(1/\tau_1)$ $\tau_1 s + 1 = 0$ **8.43** $K_1(\ell_1 + \ell_2)/(\ell_2 \ell_3 s + K_1 \ell_1 \ell_4)$ pole at $-(K_1 \ell_1 \ell_4/\ell_2 \ell_3)$ $\ell_2 \ell_3 s + K_1 \ell_1 \ell_4 = 0$ **8.45** $K_4/(Js + b)$ $1/(R_2 C_3 s + 1)$ poles at $-(b/J)$, $-(1/R_2 C_3)$ $Js + b = 0$ $R_2 C_3 s + 1 = 0$ **8.47** $R_2 + sL_3 + [R_4/(R_4 C_5 s + 1)]$ $R_4/(R_4 C_5 s + 1)$ **8.49** $1/(Js^2 + bs + k)$ **8.55** $0.1052 e^{-3t} - 0.1460 e^{-0.15t} \cos(2.995t - 43.6°)$ $0.1000 e^{-3t} - 0.1414 \cos(3.0t - 45°)$ **8.57** $0.114[\cos(0.5t + \arctan 0.017) - e^{-0.15t} \cos(3t - \arctan 0.053)]$ **8.59** $-0.083 \cos(4.01t)$ **8.61** $\theta(t) = -(h\Omega_0/k) + [h\Omega_0/(k - b\sigma_f + J\sigma_f^2)]e^{-\sigma_f t} + [h\Omega_0 \sigma_f/J(\sigma_1 - \sigma_2)]\{[e^{-\sigma_2 t}/\sigma_2(\sigma_f - \sigma_2)] - [e^{-\sigma_1 t}/\sigma_1(\sigma_f - \sigma_1)]\}$ $\sigma_1 = (b/2J) - \sqrt{(b/2J)^2 - (k/J)}$ $\sigma_2 = (b/2J) + \sqrt{(b/2J)^2 - (k/J)}$ **8.63** $\dfrac{rk_1 k_3 \omega_f}{k_1 + k_3}\left[\dfrac{1}{\sigma_2 - \sigma_1}\left(\dfrac{\sigma_1^2 e^{-\sigma_1 t}}{\omega_f^2 + \sigma_1^2} - \dfrac{\sigma_2^2 e^{-\sigma_2 t}}{\omega_f^2 + \sigma_2^2}\right) + \dfrac{\omega_f \sin(\omega_f t + \alpha)}{(\omega_f^2 + \sigma_1^2)^{\frac{1}{2}}(\omega_f^2 + \sigma_2^2)^{\frac{1}{2}}}\right]$ $\alpha = \arctan(\omega_f/\sigma_1) + \arctan(\omega_f/\sigma_2)$ $\sigma_1, \sigma_2 = [k_1 k_2/2b_2(k_1 + k_2)] \mp \sqrt{[k_1 k_2/2b_2(k_1 + k_2)]^2 - k_1 k_2/m_4(k_1 + k_2)}$ **8.65** $(M_0 \tau/J\omega)e^{-\sigma t}\sin\omega t$, $(M_0 \tau/J)te^{-\sigma t}$, $(M_0 \tau/J)(e^{-\sigma_1 t} - e^{-\sigma_2 t})/(\sigma_2 - \sigma_1)$ $\sigma = b/2J$ $\omega = \sqrt{(k/J) - (b/2J)^2}$ $\sigma_1, \sigma_2 = (b/2J)[1 \mp \sqrt{1 - (4kJ/b^2)}]$ **8.67** $(x_0 \tau/m\omega)e^{-\sigma t}\sin\omega t$ $\sigma = b/2m$ $\omega = \sqrt{(k/m) - (b/2m)^2}$ **8.69** $[i_{1m}\tau/L_4 C_2(\sigma_2 - \sigma_1)](e^{-\sigma_1 t} - e^{-\sigma_2 t})$ $\sigma_1, \sigma_2 = (1/2R_3 C_2)[1 \mp \sqrt{1 - (4R_3^2 C_2/L_4)}]$ **8.71** $(i_{1m}\tau/\sqrt{L_4 C_2}) \sin\sqrt{1/L_4 C_2}\, t$ **8.73** $(M_0 \sqrt{\omega^2 + \sigma^2}/k\omega)e^{-\sigma t}\cos[\omega t - \arctan(\sigma/\omega)]$ $\omega = \sqrt{(k/J) - (b/2J)^2}$ $\sigma = b/2J$ **8.75** $(M_0/k)[1 - e^{-bt/2J} - (2k/b)te^{-bt/2J}]$ **8.77** $v_{10}\{[10^3/(10^6 - 10^4 \sigma + \sigma^2)]e^{-\sigma t} + [1/4\sqrt{6}(\sigma + 2000\sqrt{6} - 5000)]e^{-(5000 - 2000\sqrt{6})t} - [1/4\sqrt{6}(\sigma - 2000\sqrt{6} - 5000)]e^{-(5000 + 2000\sqrt{6})t}\}10^3$

CHAPTER 9

9.3 $T_b(t)_f = [T_{1m}/(\omega_f^2 R_2^2 C_3^2 + 1)^{\frac{1}{2}}] \cos [\omega_f t - \arctan (\omega_f R_2 C_3)]$
$sC_3/(R_2C_3s + 1)$ **9.5** $[R_2R_3 + j\omega L_4(R_2 + R_3)]/(R_3 + j\omega L_4)$ $(R_3 + j\omega L_4)/$
$[R_2R_3 + j\omega L_4(R_2 + R_3)]$ $j\omega R_3 L_4/(R_3 + j\omega L_4)$ **9.21** (i) 0.020 lb-sec/in.
(ii) 0.005 in. **9.25** $1/(s^2 L_2 C_3 + 1)$ **9.27** $\ell_2 C_1 C_2/\ell_3(\ell_2 s^2 + C_1 C_2 \ell_1)$
9.43 $\omega = \sqrt{(k/m) - (b^2/2m^2)}$ **9.45** $V_c/V_m = [R_3R_4(L_5 + L_6) + sL_5L_6(R_3 + R_4)]/[R_3R_4(L_5 + L_6) + sL_5L_6(R_3 + R_4) + s^2R_3R_4L_5L_6C_7]$ **9.47**
96.6 ft/sec² **9.49** 16.1 cos [12.56t + ($\pi/6$) + 0.02], 2.92 cos [555.2t +
$(3\pi/4) - 0.23$] **9.51** $-0.0067 \cos [3t + (\pi/20)] + 0.0071 \cos [12t + (7\pi/36)]$
$+ 0.0125 \cos [10t + (\pi/30)]$ **9.53** $(4bs + 4k)/(ms^2 + 4bs + 4k)$ **9.55**
5 sin (5000t + 0.075), 5 sin 10,000t, 5 sin (20,000t − 0.075), 4.46 sin (100,000t −
arctan 0.5) amps **9.57** $R_2 + sL_3 + (1/sC_4)$ **9.59** $1/(L_3C_4s^2 + R_2C_4s + 1)$
9.61 (i) $(L_2C_4s^2 + R_3C_4s + 1)/(R_3 + L_2s), (R_3 + L_2s)/(L_2C_4s^2 + R_3C_4s + 1)$
(ii) sC_4 (iii) $(R_3C_4s + L_2C_4s^2)/(L_2C_4s^2 + R_3C_4s + 1)$

CHAPTER 10

10.1 (i) 0 (ii) 0.3 (iii) 0.5 **10.3** 1.2 **10.9** $m\ddot{x} + k(x + x_d)$
$= 0$ $x \leq -x_d$ $m\ddot{x} = 0$ $-x_d \leq x \leq x_d$ $m\ddot{x} + k(x - x_d)$
$= 0$ $x_d \leq x$

CHAPTER 11

11.1 A stable B unstable C neutrally stable **11.9** $s^2 - (c_1c_2\ell^2/\ell_2) = 0$
11.13 $m\ddot{x} - b\dot{x} + kx = 0$ $x_0 = (f_{\text{friction}|\dot{x}=0})/k$ **11.15** $m\ddot{x} - (f_0/4v_0)\dot{x}$
$+ kx = 0$ **11.17** $J_1\dot{\Omega}_1 + b(\Omega_1 - \Omega_2) = 0$ $J_2\dot{\Omega}_2 + b(\Omega_2 - \Omega_1) = 0$
$s\{s + [b(J_1 + J_2)/J_1J_2]\} = 0$ **11.19** $J\ddot{\theta} + b\dot{\theta} + K\theta = 0$ $Js^2 + bs + K$
$= 0$ **11.21** $\dot{y} + Kcy = 0$ $s + Kc = 0$ **11.23** $\dot{y} = cx$ $v = 1y$
$(1/R)(v_2 - v) + C\,dv_2/dt = 0$ $x = -Kv_2$ $s^2 + (1/RC)s + (Kc/RC) = 0$
11.25 (ii) 12.0 **11.43** 8, 0.392 **11.45** 0 **11.47** 4 **11.63**
none **11.65** 2 **11.67** 2 **11.69** $A_1(A_2A_3 - A_1) > A_0A_3^2 > 0$, all
coefficients > 0 **11.71** $\omega = \sqrt{3}, K = 12$ **11.73** $\omega = 0, K = -1$
11.75 $\omega = K = 0$ **11.77** $\omega = 1.3, K = 288$ $\omega = 7.7, K = 1250$
11.79 $K_{n1} = K_{n2} = 1$ **11.81** A $[m + (J/r^2)](1 - 2ar)\ddot{x} + 2mgax = 0$
$[m + (J/r^2)](1 - 2ar)s^2 + 2mga = 0$ B $[m + (J/r^2)](1 + 2br)\ddot{x} - 2mgbx = 0$
$[m + (J/r^2)](1 + 2br)s^2 - 2mgb = 0$ C $[m + (J/r^2)]\ddot{x} = 0$
$[m + (J/r^2)]s^2 = 0$

CHAPTER 12

12.1 (i) $m\ell\ddot{\theta}_1 + mg\theta_1 + (4k\ell\theta_1/9) - (4k\ell\theta_2/9) = 0$ $m\ell\ddot{\theta}_2 + mg\theta_2 +$
$(4k\ell\theta_2/9) - (4k\ell\theta_1/9) = 0$ (ii) $[m\ell s^2 + mg + (4k\ell/9)]\Theta_1 - (4k\ell/9)\Theta_2 = 0$
$-(4k\ell/9)\Theta_1 + [m\ell s^2 + mg + (4k\ell/9)]\Theta_2 = 0$ (iii) $[m\ell s^2 + mg + (4k\ell/9)]^2 -$
$(4k\ell/9)^2 = 0$ **12.3** (i) $m\ddot{x}_1 + b\dot{x}_1 + 2kx_1 - kx_2 = 0$ $m\ddot{x}_2 + (b + b_{12})\dot{x}_2$
$+ 2kx_2 - kx_1 - b_{12}\dot{x}_3 - kx_3 = 0$ $2m\ddot{x}_3 + (b + b_{12})\dot{x}_3 + kx_3 - b_{12}\dot{x}_2 - kx_2 = 0$
(ii) $(ms^2 + bs + 2k)X_1 - kX_2 = 0$ $-kX_1 + [ms^2 + (b + b_{12})s + 2k]X_2 -$
$(b_{12}s + k)X_3 = 0$ $-(b_{12}s + k)X_2 + [2ms^2 + (b + b_{12})s + k] = 0$
(iii) $(ms^2 + bs + k)[ms^2 + (b + b_{12})s + 2k][2ms^2 + (b + b_{12})s + k] -$
$(b_{12}s + k)^2[m s^2 + bs + 2k] - k^2[2ms^2 + (b + b_{12})s + k] = 0$ **12.5** (i) $m_1\dot{u}_1$
$+ b_{13}(u_1 - u_3) + k_{12}\int(u_1 - u_2)\,dt = 0$ $m_2\dot{u}_2 + b_{23}(u_2 - u_3) + k_{12}\int(u_2 - u_1)\,dt$

$+ k_5 \int u_2 \, dt = 0$ $\qquad m_3\dot{u}_3 + (b_4 + b_6)u_3 + b_{13}(u_3 - u_1) + b_{23}(u_3 - u_2) = 0$
(ii) $[m_1s + b_{13} + (k_{12}/s)]U_1 - (k_{12}/s)U_2 - b_{13}U_3 = 0$ $\qquad -(k_{12}/s)U_1 + [m_2s + b_{23} + (k_{12} + k_5)/s]U_2 - b_{23}U_3 = 0$ $\qquad -b_{13}U_1 - b_{23}U_2 + (m_3s + b_4 + b_6 + b_{13} + b_{23})U_3 = 0$ (iii) $[m_1s + b_{13} + (k_{12}/s)][m_2s + b_{23} + (k_{12} + k_5)/s](m_3s + b_4 + b_6 + b_{13} + b_{23}) - (2k_{12}b_{13}b_{23}/s) - b_{23}^2[m_1s + b_{13} + (k_{12}/s)] - b_{13}^2[m_2s + b_{23} + (k_{12} + k_5)/s] - [k_{12}^2(m_3s + b_4 + b_6 + b_{13} + b_{23})/s^2] = 0$ **12.7** (i) $iR_2 + L_3 \, di/dt + K_4\dot{x} = 0$ $\qquad K_4 i = m_5\ddot{x} + k_6 x$ (ii) $(R_2 + sL_3)I + K_4 sX = 0$ $-K_4 I + (m_5 s^2 + k_6)X = 0$ (iii) $(m_5 s^2 + k_6)(sL_3 + R_2) + K_4^2 s = 0$
12.9 $m_1\ddot{y}_1 + (2F_0 y_1/\ell) - (F_0 y_2/\ell) = 0$ $\qquad m_2\ddot{y}_2 + (2F_0 y_2/\ell) - (F_0 y_1/\ell) = 0$
$[m_1 s^2 + (2F_0/\ell)]Y_1 - (F_0/\ell)Y_2 = 0$ $\qquad -(F_0/\ell)Y_1 + [m_2 s^2 + (2F_0/\ell)]Y_2 = 0$
$m_1 m_2 s^4 + [2F_0(m_1 + m_2)s^2/\ell] + (3F_0^2/\ell^2) = 0$ $\qquad \omega^2 = [F_0(m_1 + m_2)/\ell m_1 m_2]$
$\pm (F_0/\ell m_1 m_2)\sqrt{m_1^2 - m_1 m_2 + m_2^2}$ $\qquad Y_2/Y_1 = 2 - [(m_1 + m_2)/m_2] \mp$
$(1/m_2)\sqrt{m_1^2 - m_1 m_2 + m_2^2}$ $\qquad \omega^2 = F_0(2 \pm 1)/m\ell$ $\qquad Y_2/Y_1 = \mp 1$
12.11 $y_1 = (u_0/2)\sqrt{m\ell/F_0}[\sin\sqrt{F_0/m\ell}\,t + (1/\sqrt{3})\sin\sqrt{3F_0/m\ell}\,t]$ $\qquad y_2 = (u_0/2)\sqrt{m\ell/F_0}[\sin\sqrt{F_0/m\ell}\,t - (1/\sqrt{3})\sin\sqrt{3F_0/m\ell}\,t]$ **12.13** (i) $m\ddot{x}_1 + 2kx_1 - kx_2 = 0$ $\qquad -kx_1 + 2m\ddot{x}_2 + kx_2 = 0$ (ii) $(ms^2 + 2k)X_1 - kX_2 = 0$
$-kX_1 + (2ms^2 + k)X_2 = 0$ (iii) $2m^2 s^4 + 5kms^2 + k^2 = 0$ $\qquad \omega^{(1)}, \omega^{(2)} = (1.25 \mp \sqrt{1.0625})^{\frac{1}{2}}\sqrt{k/m}$ (iv) $X_2/X_1 = 0.75 \pm \sqrt{1.0625}$ for $\omega = \omega^{(1)}, \omega^{(2)}$
(v) $(\sqrt{1.0625} \mp 0.75)/2\sqrt{1.0625}$ for $\omega = \omega^{(1)}, \omega^{(2)}$ **12.15** (i) $m\ddot{x}_1 + kx_1 - kx_2 = 0$ $\qquad -kx_1 + m\ddot{x}_2 + kx_2 = 0$ (ii) $(ms^2 + k)X_1 - kX_2 = 0$
$-kX_1 + (ms^2 + k)X_2 = 0$ (iii) $ms^2(ms^2 + 2k) = 0$ $\qquad \omega^{(1)}, \omega^{(2)} = 0, \sqrt{2k/m}$
(iv) $X_2/X_1 = \pm 1$ for $\omega = \omega^{(1)}, \omega^{(2)}$ **12.17** (i) $m\ddot{x} + (7k/\sqrt{2})x + (k/\sqrt{2})y = 0$ $\qquad (k/\sqrt{2})x + m\ddot{y} + (7k/\sqrt{2})y = 0$ (ii) $[ms^2 + (7k/\sqrt{2})]X + (k/\sqrt{2})Y = 0$ $\qquad (k/\sqrt{2})X + [ms^2 + (7k/\sqrt{2})]Y = 0$ (iii) $m^2 s^4 + (7km/\sqrt{2})s^2 + 12k^2 = 0$ $\qquad \omega^{(1)}, \omega^{(2)} = 3\sqrt{k/m}, 4\sqrt{k/m}$ (iv) $Y/X = \mp 1$
for $\omega = \omega^{(1)}, \omega^{(2)}$ **12.23** $m\ddot{x} + k_1 x + k_3(x - \ell\theta) = 0$ $\qquad J\ddot{\theta} + k_2\theta + k_3\ell(\ell\theta - x) = 0$ $\qquad \omega^2 = [(k_1 J + k_2 m + k_3 J + k_3 m\ell^2)/2Jm] \pm \{[(k_1 J + k_2 m + k_3 J + k_3 m\ell^2)^2/4J^2 m^2] - [(k_1 k_2 + k_2 k_3 + k_1 k_3\ell^2)/Jm]\}^{\frac{1}{2}}$ $\qquad \Theta/X = [(Jk_1 + Jk_3 - k_2 m - k_3 m\ell^2)/2Jk_3\ell] \mp (m/k_3\ell)\{[(k_1 J + k_2 m + k_3 J + k_3 m\ell^2)^2/4J^2 m^2] - [(k_1 k_2 + k_2 k_3 + k_1 k_3\ell^2)/Jm]\}^{\frac{1}{2}}$ **12.25** $q_1: \theta_1 = \theta_2 = 0.1$ rad
$q_2: \theta_1 = -\theta_2 = 0.1$ rad $\qquad J\ddot{q}_1 + kq_1 = 0$ $\qquad J\ddot{q}_2 + 3kq_2 = 0$ $\qquad q_1(t) = 0.1 \cos\sqrt{k/J}\,t$ $\qquad q_2(t) = 0.1 \cos\sqrt{3k/J}\,t$ **12.27** $q_1: x_1 = 1$ ft,
$x_2 = (3 - \sqrt{17})/4$ ft $\qquad q_2: x_1 = 1$ ft, $x_2 = (3 + \sqrt{17})/4$ ft $\qquad m\ddot{q}_1 + [(5 - \sqrt{17})/2]kq_1 = 0$ $\qquad m\ddot{q}_2 + [(5 + \sqrt{17})/2]kq_2 = 0$ $\qquad q_1(t) = -[(5 - \sqrt{17})f_0/2k]$
$\cos\sqrt{(5 - \sqrt{17})k/2m}\,t$ $\qquad q_2(t) = -[(5 + \sqrt{17})f_0/2k]\cos\sqrt{(5 + \sqrt{17})k/2m}\,t$
$x_1(t) = [(5 + \sqrt{17})f_0/4\sqrt{17}\,k]\cos\sqrt{(5 + \sqrt{17})k/2m}\,t - [(5 - \sqrt{17})f_0/$
$4\sqrt{17}\,k]\cos\sqrt{(5 - \sqrt{17})k/2m}\,t$ $\qquad x_2(t) = [(4 + \sqrt{17})f_0/2\sqrt{17}\,k]$
$\cos\sqrt{(5 + \sqrt{17})k/2m}\,t - [(4 - \sqrt{17})f_0/2\sqrt{17}\,k]\cos\sqrt{(5 - \sqrt{17})k/2m}\,t$
12.29 (i) $2m\ddot{x} + m\ell\ddot{\theta} = 0$ $\qquad m\ddot{x} + m\ell\ddot{\theta} + mg\theta = 0$ (ii) $\omega = 0, \sqrt{2g/\ell}$
(iii) $q_1 = x + (\ell\theta/2)$ $\qquad q_2 = -\ell\theta/2$ (iv) $\ddot{q}_1 = 0$ $\qquad \ddot{q}_2 + (2g/\ell)q_2 = 0$
q_1 consists of $x = 1$ inch, $\theta = 0$ $\qquad q_2$ consists of $x = 1$ inch, $\theta = -(2/\ell)$ in.
12.31 (i) $2m\ell^2\ddot{\theta}_1 + 2mg\ell\theta_1 + m\ell^2\ddot{\theta}_2 = 0$ $\qquad m\ell^2\ddot{\theta}_1 + m\ell^2\ddot{\theta}_2 + mg\ell\theta_2 = 0$
(ii) $\omega^2 = (2 \pm \sqrt{2})g/\ell$ $\qquad \Theta_2/\Theta_1 = \mp\sqrt{2}$ (iii) $q_1: \theta_1 = 0.1, \theta_2 = -\sqrt{2}/10$ rad $\qquad q_2: \theta_1 = 0.1, \theta_2 = \sqrt{2}/10$ rad $\qquad q_1 = [\theta_1 + (\theta_2/\sqrt{2})]/2$

$q_2 = [\theta_1 - (\theta_2/\sqrt{2})]/2$ (iv) $(2 + \sqrt{2})m\ell^2\ddot{q}_1 + 2mg\ell q_1 = 0$
$(2 - \sqrt{2})m\ell^2\ddot{q}_2 + 2mg\ell q_2 = 0$ **12.33** $m^2s^4 + 2bms^3 + 11kms^2 + 11bks + 10k^2 = 0$ **12.35** $s(m^2s^3 + 2bms^2 + 10kms + 10bk) = 0$ **12.37** $m^2s^4 + bms^3 + m(k + 2k_1)s^2 + b(k + k_1)s + kk_1 = 0$ **12.39** $s(m^2s^3 + bms^2 + 2k_1ms + bk_1) = 0$ **12.41** $t = -(1/\omega_{g2}) \arctan (\sigma_{g2}/\omega_{g2})$ $E\omega_{g2}/(\sigma_{g2}^2 + \omega_{g2}^2)^{\frac{1}{2}}$, 0, $2.9E \cos (104° - \alpha)$, $2.9E[-\sigma_{g2} \cos (104° - \alpha) - \omega_{g2} \sin (104° - \alpha)]$ $E = \exp (\sigma_{g2}\alpha/\omega_{g2})$ $\alpha = \arctan (\sigma_{g2}/\omega_{g2})$
12.43 $V_a/V_b = 1/(1 + RCs) = R[Cs + (1/R) + (10/sL)]$
$v_a = -0.200\rho\{e^{-0.2008t/\sqrt{LC}} + \sin [\sqrt{10/LC}\,t - \arctan (\sqrt{10}/0.1008)]\}$
$v_b = -\rho\{0.0008e^{-0.2008t/\sqrt{LC}} + \sqrt{10} \sin [\sqrt{10/LC}\,t - \arctan (0.0008/\sqrt{10})]\}$
12.45 $V_a/V_b = 1 + s^2L_1C = 1/[1 + (sL_1/R) + s^2L_1C]$
$v_a = \rho\sqrt{L_1/C}\{(\frac{1}{2})e^{-t/\sqrt{L_1C}} - (6/\sqrt{70}) \cos [\sqrt{\frac{5}{4}}t + \arctan (1/\sqrt{20}) - \arctan (\sqrt{20}/3)]\}$ $v_b = \rho\sqrt{L_1/C}\{(\frac{1}{4})e^{-t/\sqrt{L_1C}} + (\sqrt{\frac{12}{7}}) \sin [\sqrt{\frac{5}{4}}t + \arctan (\sqrt{20}/5) - \arctan (\sqrt{20}/3)]\}$ **12.47** $V_a/V_b = 1/(L_1Cs^2 + 1) = (1/L)(LL_1Cs^2 + L_1 + L)$ $v_a = \{-\rho/[CL_1(\omega_b^2 - \omega_a^2)]\}\{[\omega_b - (L_1 + L)/\omega_b CLL_1] \sin \omega_b t - [\omega_a - (L_1 + L)/\omega_a CLL_1] \sin \omega_b t\}$
$v_b = [-\rho/L_1{}^2C^2(\omega_b^2 - \omega_a^2)][(1/\omega_a) \sin \omega_a t - (1/\omega_b) \sin \omega_b t]$ $\omega_a{}^2, \omega_b{}^2 = (2L + L_1 \mp \sqrt{4L^2 + L_1{}^2})/2LL_1C$

CHAPTER 13

13.1 (i) $[ms^2 + (2F_0/\ell)]\{[ms^2 + (2F_0/\ell)]^2 - 2(F_0/\ell)^2\} = 0$ (ii) $\omega^2 = (2 - \sqrt{2})F_0/m\ell$, $2F_0/m\ell$, $(2 + \sqrt{2})F_0/m\ell$ (iii) $Y_2/Y_1 = \sqrt{2}, 0, -\sqrt{2}$ $Y_3/Y_1 = 1, -1, 1$ **13.3** (i) $m^3s^6 + 5m^2ks^4 + 6mk^2s^2 + k^3 = 0$ (ii) $\omega^2 = 0.198k/m, 1.556k/m, 3.246k/m$ (iii) $X_2/X_1 = 1.802, 0.444, -1.246$ $X_3/X_1 = 2.25, -0.796, 0.553$ **13.5** (i) $[sC + (2/sL)]\{[sC + (2/sL)]^2 - 2(1/sL)^2\} = 0$ (ii) $\omega^2 = (2 - \sqrt{2})/LC, 2/LC, (2 + \sqrt{2})/LC$ (iii) $V_b/V_a = \sqrt{2}, 0, -\sqrt{2}$ $V_c/V_a = 1, -1, 1$ **13.7** for a three-mass model, (i) $s^4[s^2 + (9EI/m\ell^3)] = 0$ (ii) $\omega^2 = 0, 0, 9EI/m\ell^3$ (iii) $\xi_1/Y = \xi_2/Y = 0, 0, \frac{3}{2}$ Θ/Y indeterminate **13.9** (i) $1500m^4s^8 + 18{,}250m^3ks^6 + 56{,}759m^2k^2s^4 + 7590mk^3s^2 + 225k^4 = 0$ (ii) $\omega^2 = 0.045k/m, 0.095k/m, 6.009k/m, 6.018k/m$ **13.11** (i) $(s^2 + 400s + 10^4)(s^3 + 400s^2 + 3 \times 10^4s + 8 \times 10^6) = 0$ (ii) roots $-26.8, -373, -377, -11.7 \pm j145$ **13.13** $J_1\ddot{\theta}_1 + b_1\dot{\theta}_1 + k_1\theta_1 - k_{12}(\theta_2 - \theta_1) = 0$ $-k_{(n-1)n}(\theta_{n-1} - \theta_n) + J_n\ddot{\theta}_n + b_n\dot{\theta}_n - k_{n(n+1)}(\theta_{n+1} - \theta_n) = 0$ $n = 2, 3 \ldots 9$ $J_{10}\ddot{\theta}_{10} + b_{10}\dot{\theta}_{10} + k_{10}\theta_{10} - k_{910}(\theta_9 - \theta_{10}) = 0$ 10th order in s^2, 10 pairs of roots **13.15** 5.94×10^6 dynes **13.17** 5.34 m **13.19** 0.27 cm **13.25** $\omega_{\text{exact}}^2 = 6k/5m$ **13.27** $\omega^2 = (6 \pm 1)/5LC$ **13.33** $\omega^2 = (3 \pm \sqrt{5})k/m$ **13.37** (i) $\omega = [(2n - 1)\pi/2\ell]\sqrt{AE/\mu}$ (ii) $\omega = (n\pi/\ell)\sqrt{AE/\mu}$

CHAPTER 14

14.1 (i) $J_1\ddot{\theta}_1 + b_1\dot{\theta}_1 = M_1$ $J_2\ddot{\theta}_2 + b_2\dot{\theta}_2 = K_2\theta_1$ $J_3\ddot{\theta}_3 + b_3\dot{\theta}_3 + k_3\theta_3 = K_3\theta_2$ (ii) $\Theta_1/M_1 = 1/(J_1s^2 + b_1s)$ $\Theta_2/\Theta_1 = K_2/(J_2s^2 + b_2s)$ Θ_3/Θ_2

ANSWERS TO ODD-NUMBERED PROBLEMS

$= K_3/(J_3 s^2 + b_3 s + k_3)$ (iii) $\theta_{1f} = M_0 e^{-\sigma_f t}/(J_1 \sigma_f^2 - b\sigma_f)$ $\theta_{2f} = K_2 M_0 e^{-\sigma_f t}/(J_1 \sigma_f^2 - b_1 \sigma_f)(J_2 \sigma_f^2 - b_2 \sigma_f)$ $\theta_{3f} = K_2 K_3 M_0 e^{-\sigma_f t}/(J_1 \sigma_f^2 - b_1 \sigma_f)(J_2 \sigma_f^2 - b_2 \sigma_f)(J_3 \sigma_f^2 - b_3 \sigma_f + k_3)$ **14.3** (i) $A_1 \dot{h}_1 + (h_1/R_1) = w_1$ $A_2 \dot{h}_2 + (h_2/R_2) = (h_1/R_1)$ $A_3 \dot{h}_3 + (h_3/R_3) = (h_2/R_2)$ (ii) $H_1/W_1 = R_1/(A_1 R_1 s + 1)$ $H_2/H_1 = R_2/[R_1(A_2 R_2 s + 1)]$ $H_3/H_2 = R_3/[R_2(A_3 R_3 s + 1)]$ (iii) $h_{1f} = R_1 w_{10} e^{-\sigma_f t}/(1 - \sigma_f A_1 R_1)$ $h_{2f} = R_2 w_{10} e^{-\sigma_f t}/(1 - \sigma_f A_1 R_1)(1 - \sigma_f A_2 R_2)$ $h_{3f} = R_3 w_{10} e^{-\sigma_f t}/(1 - \sigma_f A_1 R_1)(1 - \sigma_f A_2 R_2)(1 - \sigma_f A_3 R_3)$ **14.5** (i) $b_1 \dot{\phi} + k_1 \phi = K_1 v_a$ $\dot{y} = c\phi$ $J\ddot{\theta} + b_2 \dot{\theta} + k_2 \ell^2 \theta = k_2 y \ell$ (ii) $\Phi/V_a = K_1/(b_1 s + k_1)$ $Y/\Phi = c/s$ $\Theta/Y = \ell k_2/(Js^2 + b_2 s + k_2 \ell^2)$ (iii) $\phi_f = K_1 v_{a0} e^{-\sigma_f t}/(k_1 - b_1 \sigma_f)$ $y_f = c K_1 v_{a0} e^{-\sigma_f t}/\sigma_f(b_1 \sigma_f - k_1)$ $\theta_f = k_2 c K_1 v_{a0} e^{-\sigma_f t}/\sigma_f(b_1 \sigma_f - k_1)(J\sigma_f^2 - b_2 \sigma_f + k_2 \ell)$ **14.7** (ii) $X/P_1 = A\ell s/[(\ell k_3 + AK_2 \rho_0)s + (K_2 k_3/RA)]$ $V_a/X = v_0/\theta_0 \ell_2$ $V_b/V_a = K_3/(R_7 Cs + 1)$ $V_c/V_b = K_4$ (iii) $v_{cf} = p_{1m} K_3 K_4 v_0 A \ell \sigma_f e^{-\sigma_f t}/\theta_0 \ell_2 (1 - R_7 C \sigma_f)[\sigma_f(\ell k_3 + AK_2 \rho_0) - (K_2 k_3/RA)]$
14.9 $-0.0974 e^{-4t} \cos(20t - 38.5°)$ **14.11** $2.97 \cos(30t - 98.04°)$
14.13 (i) $(1/L) \int (v_b - v_c) \, dt + v_b/R + C \, dv_b/dt = v_1/R$ $(1/L) \int (v_c - v_b) \, dt + v_c/2R + 3Cd(v_c - v_a)/dt = 0$ $(1/2L) \int v_d \, dt + 3Cd(v_d - v_c)/dt = 0$
(ii) $\begin{vmatrix} [(1/sL) + (1/R) + Cs] & -(1/sL) & 0 \\ -(1/sL) & [(1/sL) + (1/2R) + 3Cs] & -3Cs \\ 0 & -3Cs & [(1/2sL) + 3Cs] \end{vmatrix} \begin{vmatrix} V_b \\ V_c \\ V_d \end{vmatrix}$

$= \begin{vmatrix} V_1/R \\ 0 \\ 0 \end{vmatrix}$

(iii) $V_c/V_1 = (1/sRL)[3sC + (1/2sL)]/\Delta$ $I_1/V_1 = (1/R)\{1 - [3sC + (1/2R) + (1/sL)][3sC + (1/2sL)]/R\Delta\} - (9s^2 C^2/R^2 \Delta)$ $\Delta = \{[sC + (1/R) + (1/sL)][3sC + (1/2R) + (1/sL)][3sC + (1/2sL)] - 9s^2 C^2 [sC + (1/R) + (1/sL)] - (1/sL)^2 [3sC + (1/2sL)]\}$ **14.15** (iii) $X/V_1 = K_4/[(m_5 s^2 + k_6)(R_2 + sL_3) + K_4^2 s]$ **14.17** (i) $m_1 \ddot{x}_1 + b_1 \dot{x}_1 + k_1 x + k_2(x_1 - x_2) = -f_1$ $-k_2 x_1 + m_2 \ddot{x}_2 + k_2 x_2 = 0$
(ii) $\begin{vmatrix} [s^2 + (b_1/m_1)s + (1/m_1)(k_1 + k_2)] & -(k_2/m_1) \\ -(k_2/m_2) & [s^2 + (k_2/m_2)] \end{vmatrix} \begin{vmatrix} X_1 \\ X_2 \end{vmatrix} = \begin{vmatrix} -(1/m_1)F_1 \\ 0 \end{vmatrix}$
(iii) $X_1/F_1 = -(m_2 s^2 + k_2)/[(m_2 s^2 + k_2)(m_1 s^2 + b_1 s + k_1 + k_2) - k_2^2]$ $X_2/F_2 = -k_2/[(m_2 s^2 + k_2)(m_1 s^2 + b_1 s + k_1 + k_2) - k_2^2]$
14.19 (i) $(1/L_{12}) \int (v_a - v_b) \, dt + (1/L_{13})(v_a - v_c) + C_1 \, dv_a/dt = 0$
$(1/L_{12}) \int (v_b - v_a) \, dt + (1/L_5) \int v_b \, dt + (1/R_{23})(v_b - v_c) + C_2 \, dv_b/dt = i_7$
$(1/R_{13})(v_c - v_a) + (1/R_{23})(v_c - v_b) + [(1/R_4) + (1/R_6)]v_c + C_3 \, dv_c/dt = 0$
(ii) $\begin{vmatrix} [(1/sL_{12}) + (1/R_{13}) + C_1 s] & -(1/sL_{12}) & -(1/R_{13}) \\ -(1/sL_{12}) & [(1/sL_{12}) + (1/sL_5) + (1/R_{23}) + C_2 s] & -(1/R_{23}) \\ -(1/R_{13}) & -(1/R_{23}) & [(1/R_{13}) + (1/R_{23}) + (1/R_4) + (1/R_6) + C_3 s] \end{vmatrix} \begin{vmatrix} V_a \\ V_b \\ V_c \end{vmatrix} = \begin{vmatrix} 0 \\ I_7 \\ 0 \end{vmatrix}$
14.21 (i) $m\ddot{y} + k_1(y - x) + k_3(y - x - \ell\theta) = 0$ $J\ddot{\theta} + k_2\theta - k_3\ell(y - x - \ell\theta) = 0$
(ii) $\begin{vmatrix} [s^2 + (1/m)(k_1 + k_3)] & -(k_3 \ell/m) \\ -(k_3 \ell/J) & [s^2 + (1/J)(k_2 + k_3 \ell^2)] \end{vmatrix} \begin{vmatrix} Y \\ \Theta \end{vmatrix} = \begin{vmatrix} (1/m)(k_1 + k_3)X \\ -(k_3 \ell/J)X \end{vmatrix}$
(iii) $Y/X = [(Js^2 + k_2 + k_3 \ell^2)(k_1 + k_3) - k_3^2 \ell^2]/[(Js^2 + k_2 + k_3 \ell^2)(ms^2 + k_1 + k_3) - k_3^2 \ell^2]$ $\Theta/X = -k_3 \ell m s^2/[(Js^2 + k_2 + k_3 \ell^2)(ms^2 + k_1 + k_3) - k_3^2 \ell^2]$

14.23 $x_1(t) = x_0 \cos[(2\pi/L)\int v(t)\,dt]$ $x_2(t) = x_0 \cos[(2\pi/L)\int v(t)\,dt - (6\pi\ell/L)]$

14.25 For a three-mass model (i) $m(\ddot{\xi}_1 + \ddot{y} + \ell\ddot{\theta}) + (3EI/\ell^3)\xi_1 - f_1 = 0$
$m(\ddot{\xi}_2 + \ddot{y} - \ell\ddot{\theta}) + (3EI/\ell^3)\xi_2 = 0$ $m(3\ddot{y} + \ddot{\xi}_1 + \ddot{\xi}_2) - f_1 = 0$
$m\ddot{y}\ell + m(\ddot{y} + \ddot{\xi}_2 - \ell\ddot{\theta})2\ell = 0$

(ii) $\begin{vmatrix} (s^2+\omega^2) & 0 & s^2 & s^2 \\ 0 & (s^2+\omega^2) & s^2 & -s^2 \\ s^2 & s^2 & 3s^2 & 0 \\ 0 & 2s^2 & 3s^2 & -2s^2 \end{vmatrix} \begin{vmatrix} \Xi_1 \\ \Xi_2 \\ Y \\ \ell\Theta \end{vmatrix} = \begin{vmatrix} (1/m)F_1 \\ 0 \\ (1/m)F_1 \\ 0 \end{vmatrix}$ where $\omega^2 = (3EI/m\ell^3)$

(iii) $\xi_1/F_1 = \xi_2/F_1 = 1/2m(s^2+9EI/m\ell^3)$ $Y/F_1 = \omega^2/ms^2(s^2+9EI/m\ell^3)$
$\ell\Theta/F_1 = 1/2ms^2$

14.27 $s\begin{vmatrix} X_2 \\ V_2 \\ X_3 \\ V_3 \end{vmatrix} = \begin{vmatrix} 0 & 1 & 0 & 0 \\ -(k/2m) & 0 & (k/2m) & 0 \\ 0 & 0 & 0 & 1 \\ (k/m) & 0 & -(2k/m) & 0 \end{vmatrix} \begin{vmatrix} X_2 \\ V_2 \\ X_3 \\ V_3 \end{vmatrix} + F_1(s)\begin{vmatrix} 0 \\ (1/2m) \\ 0 \\ 0 \end{vmatrix}$

14.29 $s\begin{vmatrix} I \\ X \\ V \end{vmatrix} = \begin{vmatrix} -(R_2/L_3) & 0 & -(K_4/L_3) \\ 0 & 0 & 1 \\ (K_4/m_5) & -(K_6/m_5) & 0 \end{vmatrix}\begin{vmatrix} I \\ X \\ V \end{vmatrix} + V_1(s)\begin{vmatrix} (1/L_3) \\ 0 \\ 0 \end{vmatrix}$

14.31 $a_3 = (1/2R_A^2)[(R_B^3/L_3) + (R_D/C_6)]$ $a_2 = (1/2R_A^2L_3)[(R_A^2/C_2) + (R_C^2/C_6)]$ $a_1 = R_B^3/2R_A^2C_2L_3^2$ $a_0 = R_C^2/2R_A^2C_2C_6L_3^2$ $R_A^2 = (R_5 + R_7 + R_{13})(R_8 + R_9 + R_{13}) - R_{13}^2$ $R_B^3 = R_D[(L_3/C_6) + (R_4 + R_5)(R_7 + R_{13}) + R_4R_5] - R_{13}^2(R_4 + R_5)$ $R_C^2 = (R_4 + R_5)(R_8 + R_9 + R_{13})$ $R_D = (R_8 + R_9 + R_{13})$ **14.33** $Z_{IN} = R_2\Delta/[\Delta + \{[R_8 + R_2(1/sC_{12})][sL_7 + R_6 + (1/sC_5) + (1/sC_{12})] - (1/sC_{12})^2\}]$ $Y_{IN} = (1/R_2) + \{[R_8 + (1/sC_{12})][sL_7 + R_6 + (1/sC_5) + (1/sC_{12})] - (1/sC_{12})^2\}/\Delta$ $V_9/V_1 = -(1/sC_{12})[M_{47}s - (1/sC_{12})]/\Delta$ $\Delta = [sL_4 + R_3 + (1/sC_5)]\{[R_8 + (1/sC_{12})][sL_7 + R_6 + (1/sC_5) + (1/sC_{12})] - (1/sC_{12})^2\} - M_{47}s - (1/sC_5)]^2[R_3 + (1/sC_{12})]$

CHAPTER 15

15.9 $b = 0.9$, units in.-lb-sec if M is in in.-lbs and θ_1 in radians, yes   **15.21** $0 < (b/J_1) < 9.23$, $45.8 < (b/J_1)$ **15.25** $0 < (b/J_1) < 9.23$, $45.8 < (b/J_1)$
15.29 $Y_2/(F/k_1) = k_1k_2/[(m_1s^2 + k_1 + k_2)(m_2s^2 + k_2) - k_2^2]$

CHAPTER 16

16.1 $C_n = -24(e^{jn\pi/2} - e^{-jn\pi/2})/n^2\pi^2$ **16.3** $C_n = -T(e^{jn\pi/2} + e^{-jn\pi/2})/[2\pi(n^2-1)]$ **16.5** $C_0 = 8$, $C_n = -(8e^{-2jn\pi}/jn\pi) + 4(e^{-2jn\pi} - 1)/n^2\pi^2$
16.7 $C_n = A$

CHAPTER 17

17.11 $X(s) = x_0(1 - e^{-t_0s})/s$ **17.13** $X(s) = 6[1 - e^{-(3s+1)}]/(3s+1)$
17.15 $(1/s^n)X(s) + (1/s^n)\int x\,dt\Big|_{t=0^-} + (1/s^{n-1})\iint x\,dt\,dt\Big|_{t=0^-} + \cdots + (1/s)\underbrace{\iint \cdots \int}_{n} x\,dt\,dt \cdots dt\Big|_{t=0^-}$ **17.17** (i) $[(1/R_4) + (1/R_5) + C_3s]V_b - (1/R_5)V_c = I_1 + C_3v_b(0^-)$ $-(1/R_5)V_b + [(1/R_5) + (1/sL_6) + (1/sL_7)]V_c = -(1/sL_6)\int v_c\,dt\Big|_{t=0^-} - (1/sL_7)\int v_c\,dt\Big|_{t=0^-}$ (ii) $sL_6L_7R_4R_5/[(R_4 + R_5 + R_4R_5C_3s)(R_5L_6 + R_5L_7 + sL_6L_7) - R_4L_6L_7s]$ **17.19** (i) $[C_3s + (1/R_2) + $

ANSWERS TO ODD-NUMBERED PROBLEMS   877

$(1/R_4)]\mathsf{T}_b - (1/R_4)\mathsf{T}_c = (1/R_2)\mathsf{T}_i + C_3 T_b(0^-)$
$+ (1/R_6)]\mathsf{T}_c = (1/R_6)\mathsf{T}_0 + C_5 T_c(0^-)$
$[(R_2R_4C_3s + R_2 + R_4)(R_4R_6C_5s + R_4 + R_6) - R_2R_6]$
$= \mathsf{M}_b + J_{ys}\theta(0^-) + J_y\dot\theta(0^-) + h\psi(0^-)$
$J_{zs}\psi(0^-) + J_z\dot\psi(0^-)$ $J_z/(J_yJ_zs^2 + h^2)$
$+ 2k\ell\Theta = 5kX_1 + msy_1(0^-) + m\dot y_1(0^-)$
$5kX_2 + msy_2(0^-) + m\dot y_2(0^-)$
$30msy_c(0^-) + 30m\dot y_c(0^-)$
$50m\ell^2 s\theta(0^-) + 50m\ell^2\dot\theta(0^-)$
(ii) $Y_c/X_1 = 5k^2\ell[(ms^2 + 6k)(50ms^2 + 5k) - 2k(ms^2 + 6k) - 3k^2]/\Delta$
$\Theta/X_1 = -5k^2[2(ms^2 + 6k)(30ms^2 + 2k) - k(ms^2 + 6k) - 3k^2]/\Delta$
$\Delta = \ell[1500m^4s^8 + 18{,}250m^3ks^6 + 56{,}759m^2k^2s^4 +$
$7590mk^3s^2 + 225k^4]$ **17.27** (i) $V_c(s) = \{sL_6L_7R_4R_5[I_1(s) + C_3v_b(0^-)] -$
$R_5(L_6 + L_7)(R_4 + R_5 + R_4R_5C_3s)\int v_c \, dt\Big|_{t=0^-}\}/[(R_4 + R_5 + R_4R_5C_3s)(R_5L_6 +$
$R_5L_7 + sL_6L_7) - R_4L_6L_7s]$ (ii) $(-b \pm \sqrt{b^2 - 4ac})/2a$ $a = C_3R_4R_5L_6L_7$, $b =$
$[R_4R_5{}^2C_3(L_6 + L_7) + L_6L_7(R_4 + R_5)]$, $c = R_5(R_4 + R_5)(L_6 + L_7)$ **17.29** (i) $\mathsf{T}_b(s)$
$= \{(R_4R_6C_5s + R_4 + R_6)[R_4\mathsf{T}_i(s) + R_2R_4C_3T_b(0^-)] + R_2[R_4\mathsf{T}_0(s) + R_6R_4C_5T_c(0^-)]\}/$
$[(R_4R_6C_5s + R_4 + R_6)(R_2R_4C_3s + R_2 + R_4) - R_2R_6]$ (ii) $(-b \pm \sqrt{b^2 - 4ac})/2a$
$a = R_2R_4R_6C_3C_5$, $b = [R_2C_3(R_4 + R_6) + C_5R_6(R_2 + R_4)]$, $c = (R_2 + R_4 + R_6)$
17.31 (i) $\Theta(s) = \{J_{zs}[\mathsf{M}_b(s) + J_{ys}\theta(0^-) + J_y\dot\theta(0^-) + h\psi(0^-)] - h[\mathsf{M}_c(s) - h\theta(0^-)$
$+ J_{zs}\psi(0^-) + J_z\dot\psi(0^-)]\}/s(J_yJ_zs^2 + h^2)$ (ii) $0, \pm jh/\sqrt{J_yJ_z}$ **17.33**
(i) $X(s) = \{K_4[V_1(s) + L_3i(0^-) + K_4x(0^-)] + (R_2 + sL_3)[m_5sx(0^-) + m_5\dot x(0^-)]\}/$
$[(m_5s^2 + k_6)(sL_3 + R_2) + K_4{}^2s]$ (ii) $-1000, -200 \pm j200$ **17.37**
$2u(t) \sin 3t$ **17.39** $2\sqrt{10}\, u(t) \sin(3t + \arctan 3)$ **17.41**
$3u(t)[e^{-t} - e^{-9t}]/4$ **17.43** $\tfrac{3}{8}u(t)[e^{-t} + (\sqrt{10}/3)\sin(3t - \arctan 3)]$ **17.45**
$\sqrt{33/8}\, u(t)e^{-t}\sin[\sqrt{8}\, t + \arctan(\sqrt{8}/5)]$ **17.47** $u(t - 0.5)$ **17.49**
$2u(t - 0.5)\sin[3(t - 0.5)]$ **17.51** $(M_0b/J_1J_2c\omega_f)\sin(\omega_f t - \phi)$ **17.53**
$(i_{1m}/C_3)[(8.26/\omega_a{}^2)e^{-23t} - (118.2/\omega_b{}^2)e^{-87t} - (\omega_f/\omega_a\omega_b)\sin(\omega_f t - \phi)] -$
$(\gamma/64)(23e^{-23t} - 87e^{-87t}) - (R_5\rho/64)(67e^{-87t} - 3e^{-23t})$ $\omega_a{}^2 = \omega_f{}^2 + 529$,
$\omega_b{}^2 = \omega_f{}^2 + 7569$, $\phi = \arctan(\omega_f/23) + \arctan(\omega_f/87)$
17.55 $(0.5455 - .5416e^{-0.32t} - .0039e^{-0.88t})T_{i0}$ t in minutes
17.57 $\Omega(t) = [(M_{10}\tau + J\Omega_0)/200\sqrt{3}\, J][e^{-(200-100\sqrt{3})t} - e^{-(200+100\sqrt{3})t}]$
17.59 $-(i_{1m}\omega_f/C_3\omega_a\omega_b)\sin(\omega_f t - \phi)$ **17.61** $v_c(0^+) = \gamma - \rho R_5$
17.63 $T_b(0^+) = 0$ $T_b(\infty) = 0.5455T_{i0}$ **17.65** $\theta(0^+) = \theta(\infty) = 0$
$\Omega(0^+) = \Omega_0 + (M_{10}\tau/J)$ $\Omega(\infty) = 0$
17.67   (ii) $s\begin{vmatrix}\Omega(s)\\ \Theta(s)\end{vmatrix} = \begin{vmatrix}-(b/J) & -(k/J)\\ 1 & 0\end{vmatrix}\begin{vmatrix}\Omega(s)\\ \Theta(s)\end{vmatrix} + \begin{vmatrix}(1/J)\\ 0\end{vmatrix}\mathsf{M}_1(s) + \begin{vmatrix}\Omega(0^-)\\ \theta(0^-)\end{vmatrix}$
17.69   (ii) $s\begin{vmatrix}I(s)\\ \Omega(s)\end{vmatrix} = \begin{vmatrix}-(R_2/L_3) & -(K_4/L_3)\\ (K_4/J_5) & -(b_6/J_5)\end{vmatrix}\begin{vmatrix}I(s)\\ \Omega(s)\end{vmatrix}$
$+ \begin{vmatrix}(1/L_3)\\ 0\end{vmatrix}V_1(s) + \begin{vmatrix}0\\ -(1/J_5)\end{vmatrix}\mathsf{M}_7(s) + \begin{vmatrix}i(0^-)\\ \Omega(0^-)\end{vmatrix}$
17.71   (ii) $s\begin{vmatrix}P_2(s)\\ X_3(s)\end{vmatrix} = \dfrac{K_2}{R(K_2A\rho_0 + \ell k_3)}\begin{vmatrix}0 & (k_3{}^2/A^2)\\ 0 & -(k_3/A)\end{vmatrix}\begin{vmatrix}P_2(s)\\ X_3(s)\end{vmatrix}$
$+ \dfrac{AK_2s}{(K_2A\rho_0 + \ell k_3)}\begin{vmatrix}\rho_0\\ (\ell/K_2)\end{vmatrix}P_1(s) + \begin{vmatrix}p_2(0^-)\\ x_3(0^-)\end{vmatrix}$

17.73 (ii)

$$s\begin{vmatrix}X_1(s)\\U_1(s)\\X_2(s)\\U_2(s)\\X_3(s)\\U_3(s)\end{vmatrix} = \begin{vmatrix}0 & 1 & 0 & 0 & 0 & 0\\ -(k_{12}/m_1) & -(b_{12}/m_1) & (k_{12}/m_1) & (b_{12}/m_1) & 0 & 0\\ 0 & 0 & 0 & 1 & 0 & 0\\ (k_{12}/m_2) & (b_{12}/m_2) & -[(k_{12}+k_3)/m_2] & -(b_{23}/m_2) & 0 & (b_{23}/m_2)\\ 0 & 0 & 0 & 0 & 0 & 1\\ 0 & 0 & (b_{23}/m_3) & (b_{23}/m_3) & -[(b_{13}+b_{23}+b_4+b_6)/m_3] & 1\end{vmatrix}\begin{vmatrix}X_1(s)\\U_1(s)\\X_2(s)\\U_2(s)\\X_3(s)\\U_3(s)\end{vmatrix} + \begin{vmatrix}0\\0\\0\\0\\(1/m_2)\\0\\0\end{vmatrix}F_7(s) + \begin{vmatrix}x_1(0^-)\\u_1(0^-)\\x_2(0^-)\\u_2(0^-)\\x_3(0^-)\\u_3(0^-)\end{vmatrix}$$

17.75 (ii)

$$s\begin{vmatrix}Y_1(s)\\V_1(s)\\Y_2(s)\\V_2(s)\\Y_C(s)\\V_C(s)\\\Theta(s)\\\Omega(s)\end{vmatrix} = \begin{vmatrix}0 & 1 & 0 & 0 & 0 & 0 & 0 & 0\\ -(6k/m) & -(6k/m) & 0 & 0 & (k/m) & 0 & -(2k\ell/m) & 0\\ 0 & 0 & 0 & 1 & 0 & 0 & 0 & 0\\ 0 & 0 & 0 & 0 & (k/m) & 0 & (k\ell/m) & 0\\ 0 & 0 & 0 & 0 & 0 & 1 & 0 & 0\\ (k/30m) & (k/30m) & -(k/15m) & 0 & 0 & 0 & (k\ell/30m) & 0\\ 0 & 0 & 0 & 0 & 0 & 0 & 0 & 1\\ -(k/25m\ell) & 0 & (k/50m\ell) & 0 & (k/50m\ell) & 0 & -(k/10m) & 0\end{vmatrix}\begin{vmatrix}Y_1(s)\\V_1(s)\\Y_2(s)\\V_2(s)\\Y_C(s)\\V_C(s)\\\Theta(s)\\\Omega(s)\end{vmatrix} + \begin{vmatrix}0\\(5k/m)\\0\\0\\0\\0\\0\\0\end{vmatrix}X_1(s) + \begin{vmatrix}0\\0\\0\\(5k/m)\\0\\0\\0\\0\end{vmatrix}X_2(s) + \begin{vmatrix}y_1(0^-)\\v_1(0^-)\\y_2(0^-)\\v_2(0^-)\\y_C(0^-)\\v_C(0^-)\\\theta(0^-)\\\Omega(0^-)\end{vmatrix}$$

ANSWERS TO ODD-NUMBERED PROBLEMS

17.77 $2u(t)[e^{-t} - e^{-5t}]$ **17.79** $1.496u(t)[e^{-0.1t} + 1.0012 \sin(2t - \arctan 20)]$
17.81 $4u(t)te^{-2t}$

CHAPTER 18

18.1 $u(t)[e^{-t} - 2e^{-2t} + e^{-3t}]/2$ **18.3** $2u(t)[2t + e^{-t} - 1]$
18.5 $u(t)\{(\frac{1}{14})e^{-t} + (\frac{1}{15})e^{-3t} - (1/3\sqrt{6.75})e^{-1.5t}\cos[\sqrt{6.75}\,t + \arctan(\sqrt{6.75}/0.5) - \arctan(\sqrt{6.75}/1.5) - \arctan(0.5/\sqrt{6.75})]\}$ **18.7** $u(t)[1 - e^{-2t} - 2t(1+t)e^{-2t}]/4$ **18.9** $u(t)x_m\{[(1 - e^{-\sigma_a t})/\sigma_a^5] - (1/\sigma_a^4)te^{-\sigma_a t} - (1/2\sigma_a^3)t^2e^{-\sigma_a t} - (1/6\sigma_a^2)t^3e^{-\sigma_a t} - (1/24\sigma_a)t^4e^{-\sigma_a t}\}$ **18.11** $8(1-s)/s(s+4)$

CHAPTER 19

19.3 $x = 3.22(1 - e^{-8t})$ $0 < t < 0.072$ sec $x = 1.49e^{-.625(t-.072)}\cos(3.1t - 0.22 + \arctan 0.33)$ $0.072 < t$ **19.5** $-[v_{1m}\omega_f/\sqrt{(500^2 + \omega_a^2)(500^2 + \omega_b^2)}][\omega_f \sin(\omega_f t - \phi_1 + \phi_2) - 10^4 e^{-500t}\sin(10^4 t - 2\arctan 0.05 - \phi_1 - \phi_2)] - \gamma e^{-500t}\cos(10^4 t + \arctan 0.05) - 10R_2\rho e^{-500t}\sin(10^4 t + \arctan 0.1005)$ $\omega_a = \omega_f - 10^4$, $\omega_b = \omega_f + 10^4$, $\phi_1 = \arctan(500/\omega_a)$ $\phi_2 = \arctan(500/\omega_b)$
19.9 $f_{10} = 6b^2\ell/m$ **19.11** $\theta = \theta_0 - J_{\text{wheel}}\Omega_{10}t/(J_{\text{wheel}} + J_{\text{veh}})$ $0 < t < \tau$ $\theta = \theta_0 - J_{\text{wheel}}\Omega_{10}\tau/(J_{\text{wheel}} + J_{\text{veh}})$ $\tau < t$
19.13 $-\{kr\omega^2/\sqrt{[(k/m_4) - \omega^2]^2 + (k^2\omega^2/b_2^2)]}\}\sin\{\omega t - \arctan[k\omega m_4/b_2(k - m_4\omega^2)]\} + [kr\omega/2\sqrt{(k^2/4b_2^2) - (k/m_4)}]\{[a^2/(a^2 + \omega^2)]e^{-at} - [b^2/(b^2 + \omega^2)]e^{-bt}\}$ $a,b = -(k/2b) \mp \sqrt{(k^2/4b_2^2) - (k/m_4)}$ $k = k_1k_3/(k_1 + k_3)$
19.15 (a) $x_1 = -(2\xi/3)\cos(\sqrt{3k/2m}\,t)$ $x_2 = (\xi/3)\cos(\sqrt{3k/2m}\,t)$
(d) natural modes $x_2/x_1 = -\frac{1}{2}, 1$ **19.19** $-\theta = 0.0375t^2 - 0.0075t + 0.00075(1 - e^{-10t})$ $0 < t < 2$ sec $-\theta = -0.33075 + 0.00075(2e^{20} - 1)e^{-10t} + 0.3075t - 0.0375t^2$ $2 < t < 4$ sec $-\theta = 0.30000 - .00075(e^{40} - 2e^{20} + 1)e^{-10t}$ $4 < t$ **19.21** (a) $m_1\ddot{y} + b_1\dot{y} + ky = kx$ $v_c = v_0y/a$ $M_2 = K_2v_c$ $J_2\dot{\Omega} + b_2\Omega = M_2$ (b) $Y(s)/X(s) = k/(m_1s^2 + b_1s + k)$ $V_c(s)/Y(s) = v_0/a$ $M_2(s)/V_c(s) = K_2$ $\Omega(s)/M(s) = 1/(J_2s + b_2)$ (c) $y = x_0\{1 - (\sqrt{17}/4)e^{-t}\cos[4t - \arctan(\frac{1}{4})]\}$ $\Omega = 10x_0\{(\frac{1}{51}) - (\frac{1}{60})e^{-3t} - (1/8\sqrt{85})e^{-t}\cos[4t - \arctan(\frac{1}{4}) - \arctan 2]\}$
19.23 $T_i = [T_i(0^-) - T_b(0^-)][0.398E_1 - 0.065E_2] + \frac{1}{3}[2T_i(0^-) + T_b(0^-)]$
$T_o = [T_i(0^-) - T_b(0^-)][0.259E_1 + 0.074E_2] + \frac{1}{3}[2T_i(0^-) + T_b(0^-)]$
$T_b = [T_b(0^-) - T_i(0^-)][0.656E_1 + 0.010E_2] + \frac{1}{3}[2T_i(0^-) + T_b(0^-)]$
$E_1 = e^{-1.75t/RC}$ $E_2 = e^{-10.75t/RC}$ **19.27** $x = x_0 - \ell\Omega_0(0.07e^{-3.3t} + 2.43e^{+0.3t} - 0.5t - 2.5) - 1.5\ell(8.02e^{+0.3t} - 0.02e^{-3.3t} + 0.25t^2 - 2.5t - 8)$
$y = y_0 + 0.5\ell(e^{-t} + t - 1)$ for $\theta(t) \ll 1$

CHAPTER 20

20.11 $Y/X = K/[s(s+1) + K]$ $E/X = s(s+1)/[s(s+1) + K]$
20.13 $Y/X = K_1(s+2)(s+4)/[s(s+1)(s+4) + K_1K_2(s+2)(s+3)]$
$E/X = s(s+1)(s+4)/[s(s+1)(s+4) + K_1K_2(s+2)(s+3)]$
20.15 $Y/X = K_2 + \{K_1(s+2)/[s(s+1) + K_1(s+2)]\}$ $E/X = s(s+1)/[s(s+1) + K_1(s+2)]$ **20.17** $Y/X = [K_1(s+2) - K_2s]/[s(s+1) + K_1(s+2)]$ $E/X = [s(s+1) + K_2s]/[s(s+1) + K_1(s+2)]$
20.19 $Y/X = ab/(1 + ac - bd + abe)$

20.21 $Y(s) = X(s)\{K(s+4)/[s(s+2)(s+4) + K(s+8)]\} + y(0^-)\{(s+2)(s+4)/[s(s+2)(s+4) + K(s+8)]\}$ **20.23** $Y(s) = \{X(s)[K_1K_2(s+b)(s+e)(\tau s+1)(\tau_2 s+1)] + \theta(0^-)[K_1K_2K_3(s+b)(\tau s+1)(\tau_2 s+1)] + z(0^-)[K_2(s+b)(s+e)(s+a)^2(\tau_2 s+1) + M(s)[K_2 s(s+b)(s+e)(s+a)^2(\tau_2 s+1)]\}/[s(s+a)^2(s+c \mp j\omega)(\tau_2 s+1)(s+e) - K_1K_3(\tau s+1)(\tau_2 s+1)(s+c \mp j\omega) + K_2K_4(s+b)(s+e)(s+a)^2(\tau_1 s+1)]$
20.25 (a) $Y(s) = M_0 B(s)C(s)/[1 - A(s)B(s)C(s)E(s) - C(s)D(s)E(s)]$
(b) $1 - A(s)B(s)C(s)E(s) - C(s)D(s)E(s) = 0$

CHAPTER 21

21.1 $s(s+1) = -K$ **21.3** $K_P = 3.7$ $K_Q = K_S = 12$ $K_R = 49$
21.7 Centroids at (a), (b), (c), (d), (e) -1 (f) none (g) $-\frac{1}{2}$
(h) $-\frac{3}{2}$ (i) -2 (j) $\frac{1}{3}$ (k) $-\frac{9}{4}$ (l) none **21.9**
(a) $\phi_{\text{dep}} = 172°, 180°, 188°$ $\phi_{\text{arr}} = 90°, -90°$ (c) $\phi_{\text{dep}} = 90°, 180°, -90°$ $\phi_{\text{arr}} = 98°, -98°$ **21.15** 80 **21.23** $K_P = 25$
$K_B = 9$ **21.25** $\sigma_a = 1.5, \omega_a = 0.866$ **21.29** $\omega = 1, K = 2$ (iii) $0 < K < 2$ **21.31** unstable for all K **21.33** $\omega = 4, K = 80$ **21.35** $\omega = \sqrt{2}, K = 2$ **21.37** -3 **21.39** (ii) $\omega = \sqrt{3}, K = 8$ (iii) $0 < K < 8$ **21.41** (ii) $\omega = 0.926, K = 78.4$ (iii) $0 < K < 78.4$
21.43 (ii) $\omega = 4.3, K = 106$ (iii) $0 < K < 106$ **21.45** (ii) $\omega = 3.56, K = 4.8$ (iii) $4.8 < K$ **21.47** (ii) $\omega = 1.47, K = 23.5$ $\omega = 13.1, K = 2{,}980$ (iii) $0 < K < 23.5$

CHAPTER 22

22.1 (i) $b\dot{x} + kx + Ky = Ky_d$ $\dot{y} = Cx$ (iii) $s(bs+k) + KC = 0$
(iv) $Y_e(s) = Y_d(s)\{s(bs+k)/[s(bs+k) + KC]\} - x(0^-)\{bC/[s(bs+k) + KC]\} - y(0^-)\{(bs+k)/[s(bs+k) + KC]\}$ (v) $y_e(0^+) = -y(0^-)$ $y_e(\infty) = 0$
$y_{e\text{ step}}(\infty) = 0$ $y_{e\text{ ramp}}(\infty) = v_{d0}k/KC$ $y(t) = [4y(0^-)/\sqrt{3}]e^{-kt/2b}\sin[(\sqrt{3}k/2b)t + \arctan\sqrt{3}]$ (vi) $2000 < (KC/b) < 2500$ for $y_e < (v_{d0}/40)$ and $0.5 < \zeta$ **22.3** (i) $A\dot{h} + [(1/R_2) - KK_1K_3]h = 0$ (iii) $s + (1/R_2) - KK_1K_3 = 0$ (iv) $H(s) = h(0^-)/[s + (1/R_2) - KK_1K_3]$ (v) $h(t) = h(0^-)\exp\{[(-1/R_2) + KK_1K_3]t\}$ (vi) unstable for $(1/R_2K_1K_3) < K$
22.5 (iv) $\Theta_e(s) = \{M(s) + 0.25\Theta_d(s) + \Omega(0^-) + (s+1)\theta(0^-)\}/J(s^2+s+0.25)$
(v) $\theta_e(t) = (mv\ell_2/J)te^{-0.5t}$ (vi) $(1/\ell_1) < (K_1/J)$ **22.7** (i) $m\ddot{x} + b(\dot{x} + \ell\dot{\theta}) + bv_0\theta = 0$ $J\ddot{\theta} + b\ell\dot{x} + b\ell v_0\theta + b\ell^2\dot{\theta} = f\ell_2$ (iii) $s\{s^2 + [(b/m) + (b\ell^2/J)]s + (bv_0\ell/J)\} = 0$ Neutrally stable, but unstable for all special cases (vi) yes **22.9** $T(t) = T_a(1 - e^{-t/R_2C}) + T(0^-)e^{-t/R_2C} + q_{10}R_2(1 - e^{-t/R_2C})$ first "on" cycle $T(t) = T_a(1 - e^{-t/R_2C}) + (T_d + 1°)e^{-t/R_2C}$ first "off" cycle and subsequent "off" cycles $T(t) = T_a(1 - e^{-t/R_2C}) + (T_d - 2°)e^{-t/R_2C} + q_{10}R_2(1 - e^{-t/R_2C})$ subsequent "on" cycles with $t = 0$ at beginning of each cycle **22.11** $\sigma_5 = 17\text{ sec}^{-1}$ **22.13** $\sigma_1 = 2.29a$ $\alpha = 1.92$ **22.17** (i) $19.8 < (K_1h/IJ\sigma^3) < 96.5$ $\alpha = 4$
(iii) $\phi = (M_{d0}/J)a\sin(\omega_f t + \alpha)$ $a = -\omega_f(16\sigma_1^2 + \omega_f^2)^{\frac{1}{2}}/\{[(28 + 8\sqrt{12})\sigma_1^2 + \omega_f^2]^{\frac{1}{2}}[(28 - 8\sqrt{12})\sigma_1^2 + \omega_f^2]^{\frac{1}{2}}[(4\sigma_1^2 + \omega_f^2)^2 + 4\sigma_1^2\omega_f^2]\}$
$\alpha = \arctan[\omega_f/(4 + \sqrt{12})\sigma_1] + \arctan[\omega_f/(4 - \sqrt{12})\sigma_1] + 2\arctan[\sigma_1\omega_f/(4\sigma_1^2 + \omega_f^2)] - 3\arctan(\omega_f/4\sigma_1)$

SELECTED REFERENCES

PART A: PHYSICAL LAWS AND EQUATIONS OF MOTION

Physical Laws

D. Halliday and R. Resnick, "Physics for Students of Science and Engineering," combined ed., chaps. 1–37, appendixes A, G, H; Wiley, New York, 1962.

R. P. Feynman, "Lectures on Physics," vols. I and II, chaps. 1–17; Addison-Wesley, Reading, Mass., 1964.

Equations of Motion—Mechanical

ONE-DIMENSIONAL (Chapter 2)

J. L. Meriam, "Dynamics," 3d ed., chap. 3; Wiley, New York, 1966.

G. W. Housner and D. E. Hudson, "Applied Mechanics—Dynamics," 2d ed., chaps. 1, 3; Van Nostrand, Princeton, N.J., 1959.

THREE-DIMENSIONAL (Chapter 5)

G. W. Housner and D. E. Hudson, "Applied Mechanics—Dynamics," 2d ed., chaps. 1–7, 9; Van Nostrand, Princeton, N.J., 1959.

R. L. Halfman, "Dynamics (vol. 1)," chaps. 1–6; Addison-Wesley, Reading, Mass., 1962.

A. Higdon and W. B. Stiles, "Engineering Mechanics (vol. 2: Dynamics)," chaps. 8–11; Prentice-Hall, Englewood Cliffs, N.J., 1962.

J. L. Meriam, "Dynamics," 3d ed., chaps. 1–6; Wiley, New York, 1966.

H. Goldstein, "Classical Mechanics," chaps. 1–5; Addison-Wesley, Reading, Mass., 1950.

R. F. Deimel, "Mechanics of the Gyroscope," Dover, New York, 1952.

R. N. Arnold and L. Maunder, "Gyrodynamics and Its Engineering Applications." Academic, New York, 1961.

Equations of Motion—Electrical (Chapter 2)

E. A. Guillemin, "Introductory Circuit Theory," chaps. 1, 2; Wiley, New York, 1958.

R. E. Scott, "Elements of Linear Circuits," chaps. 1–3; Addison-Wesley, Reading, Mass., 1965.
S. B. Hammond, "Electrical Engineering," McGraw-Hill, New York, 1961.
R. J. Smith, "Circuits, Devices, and Systems," chaps. 9–11; Wiley, New York, 1966.
F. E. Terman, "Electronic and Radio Engineering," 4th ed., chap. 2; McGraw-Hill, New York, 1955.

Equations of Motion—Electromechanical (Chapter 2)

H. H. Skilling, "Electromechanics," chaps. 1–5; Wiley, New York, 1962.
D. C. White and H. H. Woodson, "Electromechanical Energy Conversion," chap. 1; Wiley, New York, 1959.
D. Halliday and R. Resnick, "Physics for Students of Science and Engineering," combined ed., chaps. 26–37; Wiley, New York, 1962.
W. W. Harman and D. W. Lytle, "Electrical and Mechanical Networks," chap. 11; McGraw-Hill, New York, 1962.

Equations of Motion—Heat Conduction (Chapter 3)

J. P. Holman, "Heat Transfer," McGraw-Hill, New York, 1963.
W. H. Giedt, Jr., "Principles of Engineering Heat Transfer," Van Nostrand, Princeton, N.J., 1957.
E. R. G. Eckert, "Heat and Mass Transfer," translated by J. F. Gross, McGraw-Hill, New York, 1963.

Equations of Motion—Fluid (Chapter 3)

J. C. Hunsaker and B. G. Rightmire, "Engineering Applications of Fluid Mechanics," chaps. 1–8; McGraw-Hill, New York, 1947.
J. K. Vennard, "Elementary Fluid Mechanics," 4th ed., chaps. 1–4, 7–9; Wiley, New York, 1961.
A. H. Shapiro, "The Dynamics and Thermodynamics of Compressible Fluid Flow," Ronald, New York, 1953.

Analogies (Chapter 4)

W. A. Lynch and J. G. Truxal, "Introductory System Analysis," chap. 3; McGraw-Hill, New York, 1961.

PART B: DYNAMIC RESPONSE OF ELEMENTARY SYSTEMS (Chapters 6–10)

F. B. Hildebrand, "Advanced Calculus for Applications," chap. 1; Prentice-Hall, Englewood Cliffs, N.J., 1962.
H. B. Phillips, "Differential Equations," 3d ed.; Wiley, New York, 1951.
G. B. Thomas, "Calculus," 2d ed., chap. 15; Addison-Wesley, Reading, Mass., 1961.
J. P. Den Hartog, "Mechanical Vibrations," 4th ed., chaps. 1, 2, 8; McGraw-Hill, New York, 1956.
L. A. Manning, "Electrical Circuits," chaps. 3–7; McGraw-Hill, New York, 1966.
W. W. Harman and D. W. Lytle, "Electrical and Mechanical Networks," chaps. 2–4; McGraw-Hill, New York, 1962.
W. A. Lynch and J. G. Truxal, "Introductory System Analysis," McGraw-Hill, New York, 1961.

SELECTED REFERENCES 883

W. R. Evans, "Control System Dynamics," chaps. 3–6, 11, 12; McGraw-Hill, New York, 1954.

F. E. Terman, "Electronic and Radio Engineering," 4th ed., chap. 3; McGraw-Hill, New York, 1955.

C. Hayashi, "Nonlinear Oscillations in Physical Systems," McGraw-Hill, New York, 1964.

D. Graham and D. T. McRuer, "Analysis of Nonlinear Control Systems," Wiley, New York, 1961.

PART C: NATURAL BEHAVIOR OF COMPOUND SYSTEMS (Chapters 11–13)

J. P. Den Hartog, "Mechanical Vibrations," 4th ed., chaps. 3, 4, 7; McGraw-Hill, New York, 1956.

E. A. Guillemin, "The Mathematics of Circuit Analysis," pp. 395–409; Wiley, New York, 1956.

E. J. Routh, "A Treatise on the Dynamics of a System of Rigid Bodies," 6th rev. ed., part II, arts. 290–307; Dover, New York, 1955.

W. R. Evans, "Control System Dynamics," chaps. 1, 7, appendix D; McGraw-Hill, New York, 1954.

M. F. Gardner and J. L. Barnes, "Transients in Linear Systems," pp. 197–201; Wiley, New York, 1954.

J. W. S. Rayleigh, "Theory of Sound," vol. I, Dover, New York, 1945.

H. L. F. v. Helmholz, "On the Sensations of Tone as a Physiological Basis for the Theory of Music," translated by A. J. Ellis, 6th ed.; Peter Smith Publisher, Gloucester, Mass., 1954.

P. M. Morse, "Vibration and Sound," 2d ed., chaps. 1–4; McGraw-Hill, New York, 1948.

F. E. Terman, "Electronic and Radio Engineering," 4th ed., chap. 4; McGraw-Hill, New York, 1955.

PART D: TOTAL RESPONSE OF COMPOUND SYSTEMS (Chapters 14–19)

W. A. Lynch and J. G. Truxal, "Introductory System Analysis," McGraw-Hill, New York, 1961.

W. R. Evans, "Control System Dynamics," chap. 10; McGraw-Hill, New York, 1954.

R. N. Clark, "Introduction to Automatic Control Systems," chaps. 2–4, 8; Wiley, New York, 1962.

J. P. Den Hartog, "Mechanical Vibrations," 4th ed., chaps. 3, 4; McGraw-Hill, New York, 1956.

H. W. Bode, "Network Analysis and Feedback Amplifier Design," sec. 12.6; Van Nostrand, Princeton, N.J., 1945.

R. Bracewell, "The Fourier Transform and Its Applications," chaps. 1–3, 11; McGraw-Hill, New York, 1965.

R. V. Churchill, "Modern Operational Methods in Engineering," McGraw-Hill, New York, 1944.

M. F. Gardner and J. L. Barnes, "Transients in Linear Systems," chaps. 4–7; Wiley, New York, 1954.

J. A. Aseltine, "Transform Method in Linear System Analysis," chaps. 1–13; McGraw-Hill, New York, 1958.

R. G. Brown and J. W. Nilsson, "Introduction to Linear Systems Analysis," chaps. 1-9; Wiley, New York, 1962.

P. M. DeRusso, R. J. Roy, and C. M. Close, "State Variables for Engineers," Wiley, New York, 1965 (an advanced text).

R. N. Arnold and L. Maunder, "Gyrodynamics and Its Engineering Applications," Academic, New York, 1961.

G. R. Pitman (ed.), "Inertial Guidance," chap. 2 by J. M. Slater, chap. 3 by J. S. Ausman; Wiley, New York, 1962.

PART E: FUNDAMENTALS OF CONTROL-SYSTEM ANALYSIS
(Chapters 20-22)

Scientific American, "Automatic Control," Simon and Schuster, New York, 1955.

W. R. Evans, "Control System Dynamics," McGraw-Hill, New York, 1954.

W. R. Evans, Graphical Analysis of Control Systems, *Trans. AIAA*, vol. 67, pp. 547-551, 1948.

R. N. Clark, "Introduction to Automatic Control Systems," Wiley, New York, 1962.

J. G. Truxal, "Automatic Feedback Control System Synthesis," McGraw-Hill, New York, 1955.

H. L. Hazen, Theory of Servomechanisms, *J. Franklin Inst.*, vol. 218, pp. 279-331, Sept., 1934.

H. W. Bode, "Network Analysis and Feedback Amplifier Design," Van Nostrand, Princeton, N.J., 1945.

H. Nyquist, Regeneration Theory, *Bell System Tech. J.*, vol. 11, pp. 126-147, Jan., 1932.

H. M. James, N. B. Nichols, R. S. Phillips, "Theory of Servomechanisms," McGraw-Hill, New York, 1947.

W. E. Bollay, Aerodynamic Stability and Automatic Control, *J. Aeron. Sci.*, Sept., 1951.

B. Etkin, "Dynamics of Flight," Wiley, New York, 1959.

INDEX

A **boldface** number indicates the page on which a term is defined.

Absolute velocity and acceleration, 38, 44n., 50, 146–149
a-c carrier (electrical), 770
Acceleration, 23, **38**, 44n., **146–151**
  absolute and relative, 38, 44n., 146–151
  angular, 23, 38, 147, 779, 780
  of gravity, 24
  positive direction for, 38, 227
Accelerometer, 8, 345–348, 809
  equation of motion, 346
  frequency response, 336, 809
  sensing gravity, 347
Accumulator, fluid, 774
Admittance (*see* Impedance and admittance)
Aerodynamic coefficients, 697
Aerodynamic control surface, 614, 696, 787
Aerodynamic flutter, 345, 391
Aerodynamic moment, 697
Aileron-control servo (*see* Hydraulic servo)
Aileron position pickoff, 699
Air column, 230, 793, 826
  vibrating, 478–480
Air compressor, 764
Aircraft, 13, 49, 102, 442, 696, 764, 777, 843
  automatic control (*see* Automatic pilot)
  carrier landing, 48–51, 764, 843
  equations of motion in roll, 696, 697
  hydraulic servo (*see* Hydraulic servo)
Air-cushion vehicle, 154, 780, 786, 846, 859
Algebra of loop closing, 639–643
Ampére, 20, 25, 59
Amplifiers, 75–78
  current, 77, 420, 790
  equivalent network for, 77
  feedback, 628, 787, 813
  voltage isolation, **75–77**, 393, 420
    symbol for, 76, 393, 420

Analog computer, 133, **299–308**, 398
  arrangement, 302–305
  components used, 300, 301
  experiments to do, 305, 306
  form for equations, 303, 577, 580
  power of, 307
Analog simulation (*see* Simulation)
Analogies, 121–142, 774–776
  benefits and limitations, 133
  classification of relations, 132–136
  fluid-electrical, 59, 123–125, 134
  heat-electrical, 121–123, 134
  mechanical-electrical, 125–132, 134
    construction procedure, 130–131
    force-voltage analogy, 776
  "mobility" analog, 126
  network approach, 137–142
Analogs, physical (*see* Analogies)
Analysis, 4, 66
Angular velocity, acceleration, 23, **38**, 44, 147, 718, 779, 780
Angular momentum, 24, 151, 718–720
Antiresonance, 523
Approximations, **11–18**, 356–360, 757, 758
  in fluid flow, 102, 106, 112, 115, 120
Arbitrary configuration, **36, 37**, 47, 52, **148**, 154, 161, 171, 760, 776
Armature, 82
Ash, R., 61n.
Asymptotes
  for plotting exponentials, 189
  for plotting frequency response, 324–326, 334, 339–341, 512
    first-order systems, 326
    higher-order systems, 512, 830
    second-order systems, 334, 339–341, 807
  in root-locus sketching (*see* Root-locus sketching rules)

885

Asymptotic stability, 377
Attitude control (see Space vehicle)
Attitude reference gyro, 617, 696
Audio speaker (see Speaker)
Automatic control, 628–710, 846–864
  benefits and objectives, 630
  concepts, 630
  philosophy, 631–634, 846, 847
Automatic control systems, 628–710
  analysis of, 637–643
    methods, 639–643
    sequence, 637, 638
    summary procedure, 643
  characteristics of, inspecting, 642
  compensation in, 691–696, 860, 861
  derivative (prediction) control, 691, 693, 858, 861
  design case studies, 683–710, 855–864
  design criteria, 632, 635–637
  feedback amplifier, 857
    insensitivity of, 634
  human, 632, 634, 846, 847
  insensitivity of, 634
  integral control, 702
  multiple loop (multiloop) 696–703, 862–864
  on-off, 860
  optimization of, 862
  performance objectives, 634–637, 643
  remote indicator servo (see Remote indicator servo)
  steady-state error, 687, 702
    position, 687
    velocity (ramp), 688
  stiffness of, 637
  synthesis of (design) to control specific physical quantities, 690–710, 855–864
    aircraft motions (see Autopilot)
    air-cushion-vehicle motion, 859, 863, 864
    inverted-pendulum position, 703–710, 864
    liquid level, 857
    machine-tool position, 856, 860–862
    rocket-engine or aileron motion (see Hydraulic servo)
    space-vehicle attitude, 3, 842, 852, 858
    speed of a shaft, 858
    stable-platform orientation, 861
    temperature, 847–856
      on-off control, 860
    use of compensation in, 691–696
  for unstable plants, 703–710
Automatic pilot (autopilot), 1, **8, 9,** 691, **696–703,** 861–864
  design, 701–703, 858, 859, 861–864
  multiloop synthesis, 699, 700, 863, 864
  specifications for, 696

Automobile
  clutch, 400, 514–518, 788
  engine, 764
  engine cooling, 95, 757
  engine resonance, 526
  pitching motion, 442–445, 758, 776, 783, 785, 786, 829, 837, 839
  power steering, 614
  racing, 778, 780
  speed control, 847
  steering control, 846
  traveling, 150
Autonetics, 302
Autopilot (see Automatic pilot)

Bach, J. S., 481
Back emf (electromotive force), 83, 189
Bank-angle control (aircraft), 696–703
Bar, swinging, 231, 784
  moment of inertia, 766, 784
Battery (electrical), 58, 71
Beads on a string, 823, 828
Beam, flexible, 824, 826, 829
  vibration, 473, 824, 826
Beat generation, 437–441
  phenomena, 420
Beats, 438, 820
Bellows (pneumatic), 102, **115–118,** 773, 788, 804, 827, 836, 838
Belt drive, 80, 760, 761
Bernoulli effect, 383
Bernoulli's formula, $105n$.
β, frequency ratio, 332, 337
Bicycle wheel, 623, 781
Billiard ball, 144, 777
Biology, 3
Block diagram, **8, 9,** 275, 276, **495,** 496, 499, 560, 573–575, 578–581, 633, 634, 639–643, 826–829, 838
  analysis (control systems), 639–643
  recommended symbols, $496n.$
  representation of
    control systems, 633, **640,** 846–849, 855–864
    equations of motion, 578–581, 838
    initial-condition generation, 573–575, 578–581, 838
    transfer functions, 275, 499, 560
Boat drive, 765, 775
Bode, H. W., 322, 518, 628
  plotting method, 322–330, 512–518, 802–810
    for compound systems, 512, 513, 516–518
      general coupling, 516–518
      subsystems in cascade, 512, 513
    reciprocal property, 354
    for RLC circuit, 353

INDEX

Bode, plotting method, for second-order systems, 337–342, 807
  theorem, 518, 831, 832
Boundaries of a system, 10, 33, 105, 165, 170
Boundary conditions, 468
  spatial, 468
Boundary layer, 112
Boyden, D. D., 481n.
Branch (electrical), 62
Break frequency (corner frequency), 326, 830
Bridged-T network, 769
British system of units, 20–28
Brush Company, 320
Bucket and pulley, 156
Buoy, floating, 230, 794

Cajorian, F., 716
Cam and follower, 760, 761
Cannon, R. H., Jr., 626n., 710n.
Cantilever, 14, 17, 231, 794
Capacitance (electrical), 25, 68
  microphone, 88, 770
  phonograph pickup, 88
Capacitor ("condenser"), 68, 69, 768
  energy stored in, 825
  parallel-plate, 88, 91
  variable (radio), 88
Capacity, electrical, 68, 122
  fluid, 107, 112, 125
  thermal, 96, 98, 122, 787
Cascaded (tandem) systems, **393**, 395, 420, 492, 499, 790, 792, 793, 800, 826–828
  frequency response of, 512–514, 830–832
Case studies, in automatic control, 683–710, 855–864
  in general system response, 597–626, 840–846
Cathode ray tube (crt), 90–92, 304, 320
  electromagnetic, 90
  electrostatic, 91, 771, 803
Cause-effect relations, 8, 275
Cent (musical), 480
cgs system of units, 20–28
Characteristic equation, **183**, 188, 236, 247, 248, 396, 401, 427, 459, 639, 645, 814
  compound systems, 396, 427, 459, 814
  Evans form, 405, 646
  feedback system, 639
  first-order system, 188
  roots of (eigenvalues), 183, 188, **247,** 248, 379–393, **399–401,** 639, 645, 646
    (*See also* Locus of roots)
  second-order system, 248, 795–798
  from a *set* of l.c.c. equations, 396
  symbols (physical meaning), 248
  from transfer function, 276

Characteristic function, **276,** 494, 502, 639
Characteristics, dynamic (eigenvalues), 188, **247,** 248, 276, 374, 375, 394, **399–**401, 419, 427, 432, 448, 795–798, 819–824
  feedback system, 640–643, 848, 849
  physical meaning, 190, 248
  plotting in the s plane, 399–406, 448
    versus coupling strength, 448
  (*See also* Root-locus method)
  (*See also* Roots of characteristic equation)
Charge, electric, 19, 24, **58,** 59
Checking, dimensional, 20, 759
Chinese scale (musical), 480
Clutch, fluid (*see* Fluid clutch)
Coin, rolling, 144, 777
Command inputs, 636
Communication systems, 1
Commutator, 80
Compatibility relations, 33, 39, 64, 98, 122–142
  electrical (Kirchhoff's voltage law), 64
  geometric, **39,** 52, 127, 761, 764, 765
  thermal, 98, 122
Complex numbers, definition, manipulation, utilization, 235–243, 795
Complex plane (*see* s plane)
Complex vectors, 238
  manipulation of, 238–241, 727–730
  sine-wave generation with, 242–243
  Spirule calculations with, 727–730
Compressibility of fluid, 774
Compressor (air), 764
  axial flow, 778, 824
Compound dynamic systems, 374–376
  characteristics of natural motion and stability, 376–417, 814–825
  forced oscillation of, 510–528
  forced response alone of, 492–508
  modes of natural motion, (eigenvectors), 419, 428–488, 818–824
    distributed-parameter systems, 462–476, 824, 825
    many degrees of freedom, 458–462
    two degrees of freedom, 426–456
  obtaining characteristic equation (Procedure C-1), 396, 427, 814–825
  subsystems lose identity, 374, 400, 419
  total response of, 490, 491, 562–566
    (Procedure D-3), 599
  vibrating (undamped), 426–446, 458–488
    Rayleigh's method for (Procedure C-5), 482–488, 825, 826
Compound pendulum, 231
Computers, 1, 8, 299–308, 355
  analog, digital, 299–308

Condensor (*see* Capacitor)
Conduction, of electricity, 57–74
 of heat, 94–102
  electrical analog of, 122–124, 134
Conductivity, electrical, 69
 thermal, 95
Configuration, 36, **144,** 145, 173
 reference, arbitrary, **36, 37,** 47, 49, 52, **148,** 154, 161, 171, 760, 776, 777
Connectedness relations, 33, 39, 64
Conservation, of energy, 106, 163, 229
 of momentum, 106, 717
Conservative system, 165, 169, 229
Consolidated Engineering Co., 320
Constitutive physical relations (*see* Physical relations)
Constraint relations, electrical, 63, 72
 fluid, 104
 geometric, 37, **144,** 760, 776, 777
 holonomic, nonholonomic, 144, **777**
 thermal, 97
 (*See also* Identity and constraint relations)
Continuity relations, 33, **105,** 114, 124, 134
Control, automatic (*see* Automatic control)
 of unstable systems, 703–710
Control surface (in a fluid), 33, **105,** 614
Control system analysis methods, 628, 629, 639–643
Control systems (*see* Automatic control systems)
Conversion factors, physical units of measure, 21–28, 711, 759
Convolution, 183, **220–224,** 298, 299, 792, 793, 802
 integral, 221
 and Laplace transform, 580, 731, 839
Coordinate frame, 146–151, 718, 719
 (*See also* Frame of reference)
Coordinates, 36
 Cartesian, 168, 505
 generalized, 168
 independent, 39, 144, 167
 mechanical (position), **36,** 49, **144**
 normal (natural), 445, 446, 821
 rectangular, 38, 168, 505
Coriolis, G., 151
Coulomb's law, 58
Coupling, 420
 compound systems, 392–401, **420–488,** 499–502, 514–525, 638, 814–824, 828
  cascaded subsystems, 499, 814
  general, 396–401, **420–426,** 500–502, 638
   frequency response, 514–525, 832–834
  feedback, 393, 397, 638, 639
  first-order subsystems, 392, 395

Coupling, physical forms, 420–423
 inertial, 422, 442–445
 strength, effect of, 431–437
Cramer's rule, 446, 461n., 501, 601, **725**
Crandall, S. H., 35n.
Crank and piston, 760, 761, 844
Critical damping ($\zeta = 1$), 263, 797, 798
crt (*see* Cathode ray tube)
Current, electrical, 20, 25, 59
 source, 77
Current amplifier, 75, 77, 84
Cycloid (path of gyro), 623
Cylinder on wedge, 148, 157–159, 176

Dahl, N. C., 35n.
D'Alembert's principle (D'Alembert's method), 40, **44–46,** 152–159, 465
 advantages of, 159
Damper, 23, 809
 (*See also* Damping)
Damping, 43, 46, 174, 251, 388–392, 422
 critical ($\zeta = 1$), 263, 797, 798
 dry-friction, 388, 391
 negative, 388–392
 as a variable, 386, 450
 viscous, 43, 46, 422
Damping ratio ζ, 251
Damping time constant $\tau = 1/\sigma$, 187, 249
D'Arsonval meter, 90, 320, 771
Dashpot, 5, 761
DeBra, D. B., 341n.
Decade, 323
Decibel scale, 324, 326
Degrees of freedom, 37, 39, **144,** 145, 462
Den Hartog, J. P., 388n., 472n., 473n., 488n., 528
Density, 24, 104, 106, 110
Departure and arrival of root loci, 405
Design, 4, 632
Determinants, 398, 425, **725,** 726
Deterministic system, 18
Dielectric constant, 68, 91
Differential equations, 13–18, 182n., 465–471, 758
 constant parameters, 17, 182n.
 conversion to algebraic equations, 500
 linear approximation, 15–17, 106, 112, 356–360
 (*See also* Perturbation technique)
 linear constant-coefficient (l.c.c.), 16, 182n.
 linear time-invariant (l.t.i.), 16, 182n.
 linearity of, 15
 nonlinear, 16, 114, 115, 365–369
  phase-plane solution, 365–369
 partial, 15, 457, 464–474
  solution to, 467–472

INDEX 889

Differential equations, simultaneous, linear, 500, 501
  (*See also* Matrix representation)
  solution of, formal (Procedure B-1), 197
    homogeneous (complementary) solution, 180, 182, 186, 197
    particular solution, 180, 182, 197
    standard form, 502
  time-varying coefficients, 17, 370, 371
Differentiating systems, step response of, 218, 792
Digital computer, 307
Dimensional checking, 20, 759
Dimensions, fundamental, 19
  table of, 21–28
Director (fire control), 631
Displacement, 23, 36, 38, 144, 145, 173
  virtual, 173
Dissipation of energy, electrical, 69
  mechanical, 164
Distributed-parameter systems, 14, 15, 457–488
  natural motion, 467, 824–826
Double cross, 662
Drag, fluid, 787
Drive shaft, flexible, 805
  (*See also* Ship drive system)
Drum (vibrating membrane), 474
Duals, network, 63, 121n., **723**, 724, 776
Dynamic analysis, 2, 10
Dynamic behavior (*see* Response; Natural motion; Stability; Characteristics; Natural modes of motion)
Dynamic characteristics (*see* Characteristics)
Dynamic equilibrium (*see* Equilibrium)
Dynamic investigation, 2
  pattern, scope, stages, 3–7, 599
Dynamic response (*see* Response)
Dynamic stability (*see* Stability)
Dynamic systems (*see* Systems)
Dynamics, 3

e^{st}, 181, 187
  and arbitrary functions, 535
  in continuing-function response, 266–282, 490–508
  and Fourier series, 529, 533
  in frequency response, 313
  and transfer functions, 273–282, 492–508
  to unify time functions, 235, 266
Eckert, E. R. G., 102
Eigenvalues (*see* Characteristics)
Eigenvector, 419, 451–456, 460, 819–824
  definition of, 452
  by matrix method, 509
  phasor form, 452, 454
  state form, 452, 455, 822

Elastic shear energy, shaft, 465, 483–487
Electric blanket, 813
Electric field **E**, 26
Electric generator, 81, 83
Electric isolators, 75, 135
  (*See also* Amplifiers, isolation)
Electric motor (*see* Motor)
Electrical systems, 57–79
  analogs (*see* Analogies)
  capacitance, 25, 68
  charge, 19, 24, **58**, 59
  circuits, 58
  conductivity, 69
  constraint relations, 72
  current, 20, 25, 59
  force, 58, 91
  identity relations, 72
  network (*see* Networks)
  node (junction), 59
  potential, 59
  voltage, 59
Electromechanical elements, 85–92
  energy-conversion devices, 89–92
  fundamental relations, 90, 93
  modulation devices, 87–89
Electromechanical systems, 79–93
Electromotive force (emf), 25, 70, 83, 90
Electron, charge on, 59
  flow of, 68–70
Electronic organ, 479
Electronic tube, 393
Elevator, 758, 761, 765
emf (*see* Electromotive force)
Energy
  from an amplifier, 400
  balance, 35, 163
  conservation of, 106, 163
  dimensions, 28
  dissipation, 164
  electromechanical, 83n.
  internal, 95, 98
  kinetic, 54, **164**, 177, 485–487, 785
  mechanical, 53, 164
  into a mechanical system, 332
  methods of analysis, 163, 229, 793
    (*See also* Lagrange's method *and* Rayleigh's method)
  migration in a system, 438
  potential (*see* Potential energy)
  relations, 35, 163, 785
  in a spring, 54
  and state, 39, 64
  stored, 38, 64, 134, 183, 226, 825
    A-type and T-type elements, 133–135
    in a capacitor, 69
    electrically, 64, **69**, **70**, 134, 233, 352, 825
    by gravity, 54
    in an inductor, 70

Energy
  stored, kinetically, 54, **164**, 170, 172, 177, 485–487
    mechanically, 53, 134, 164, 483–487, 826
    by a pendulum, 172
    in a shaft, 483, 484
    in a spring, 53, 170, 173
    in a string, 486, 487
  thermal (heat), 95, 134
Energy conversion, electrical-electrical, 78, 136
  electrical-mechanical, 83, 136
  mechanical-mechanical, 56, 136
Energy-conversion devices, 89–92
  electromagnetic, 89, 90
  electrostatic, 91, 92
Engine, automobile, 95, 525, 757, 764
  rocket (*see* Rocket engine)
  turbojet, 330, 458, 763, 778, 824
Engineering, 2
  approximations, 12–18
  judgment, 11
  system of units, 21–28
Equations of motion, 4, **31–178**, 760–786
  analogies between media (*see* Analogies)
  basic kinds of relations in deriving, 33, 74
  cancellation of static terms in, 115, 271
    steady-flow terms, 115
    weight and static spring deflection, 271
  classification of relations, 134–136
  considerations in deriving, 33, 74
  electrical systems, 57–79, 767–770
  electromechanical systems, 79–93, 770
  energy methods, 163–178, 482, 785, 786
  fluid systems, 102–120, 773, 774
  heat-conduction systems, 94–102, 771–773
  mechanical systems, 35–57, 143–178
    by Lagrange's method, 167–178
    in one dimension, 35–57, 74, 762–767
    plane motion, 153, 720
    in three dimensions, 143–178, 782–786
  pattern of solution (Procedures B-1), **193–197**
  for a single rigid body (general), 152–156
  solution of (*see* Characteristics; Differential equations; Natural modes of motion; Natural motion; Response)
  standard form (*see* Standard form)
  for systems of rigid bodies, 156–163
Equilibrium, dynamic, **32–35**, 40, **74**, 104, 105, 114, **151**, 163
  electrical networks (*see* Kirchhoff's laws)
  energy, 35, 163
  fluid system, 104, 105, 114
  force, 35, **40**, 47, 50, 105, 114, **151**
  relations (node), 32, 35, 74, 134
  thermal, 97
Euler reference frame, 104

Euler's equation for complex numbers, 226, **236–238**, 251, 385, 795
Euler's equation for fluid flow, 105
Evans, W. R., 376, 401*n*., 629, 644, 729
Evans form, 405, 646, 849
Evans function, 405, 407, 646, 849
Evans method (*see* Root-locus method of Evans)
Experiments suggested
  aircraft carrier-landing (simulation), 843
  analog simulation, 305–307, 812, 822, 843
  beats, 439, 820
  gyro, 622
  resonance, 272, 334, 806
  rotor time constant, 191
  thermal time constant, 788
  vibrating devices, 234, 825
  violin, 825
  water-supply reservoir system (simulation), 846
Exponential function e^{st} (*see* e^{st})
Exponential input, abrupt, 195, 288–291
Exponential time response, 180, 188, 235
  growing, 192
  sketching accurately, 189

"$\mathbf{f} = m\mathbf{a}$ and $M = J\alpha$," 153
Faraday, 59
Feedback, 7, 393, 397, 628, 638
  algebra of loop closing, 639–643
  control systems, 628–710
    (*See also* Automatic control systems)
  definition, 638
  (*See also* Automatic control)
Feedback amplifier, 628, 787, 857
Filter, 316, 317, 328, 330, 693
  band-pass, 830
  bridged-T, 769
  high-pass (RC), 328, 770, 831
  high-Q (LC), 330
  low-pass (RC), 316, 770, 804, 858
Final-value theorem (FVT), 567–569, 602
  in control system studies, 687
First law of thermodynamics, 96–98, 122, 163
First-order systems, 183–186, 786–793
  behavior of, 182–224, 786–793
    mathematical solution, 187
    natural response, 188, 786–788
  by physical reasoning, 186
Fixed-centroid rule (*see* Root-locus)
Flexible shaft (*see* Shaft)
Flow
  analogies between media, 121–125, 774–776
  of electricity, 25, 59
  of fluid (*see* Fluid flow)
  of heat, 27, 95
  relations, 33, 64

INDEX

Fluid clutch, 400, 514–518, 563–565, 579, **600–605,** 764, 775, 776, 786, 788, 799, 805, 834, 858
  characteristics, 400
  complete analysis, 600–605
  frequency response, 514–518
  Laplace analysis, 563–565
  total response, 563, 603, 788, 791, 792, 838
    block diagram, 579
    physical interpretation, 565, 604
Fluid compressibility, 774
Fluid computer element, 102, 115–118, 773
Fluid flow, **107–114,** 721, 722
  electrical analog of, 123–125, 134
  laminar, 110, 722
    Newton's relation for, 111
  through pipes or tubes, 109, 111, 721
  through short constrictions, 108
  source, 774
  transition, 721
  turbulent, 111, 114, 722
Fluid properties, capacitance, 107
  mass-density, 110, 721
  resistance (friction), 107–113, 721
  shear stress, 721
  viscosity, 110, 721
Fluid systems, 102–120, 773–775
  electrical analog for, 123–125
Flutter, aerodynamic, 345, 391
Flyball governor, 777, 782, 785, 796
Force, 3, 19, 22, 35, **40,** 47, 711
  balance (*see* Force equilibrium)
  body, 45
  electric, 58, 91
  electromagnetic, 83, 90
  equilibrium, 35, **40,** 47, 50, 105, 114, **151**
  friction, 40, 43, 46, 174
  generalized, 173
  gravity, 40, 43, 46, 171
  inertial, 40, **45,** 46, **152,** 762
  and mass, 19, 711
  spring, 40, 43, 46, 271, 465
  -transmission system, 52
  units, 22, 711
Force-balance construction for frequency response, 341
Force sensor, 88
Forced motion (*see* Motion)
Forced motion alone, 266–282, 490–508
Forced oscillation (*see* Frequency response)
Forced vibration, 309–354, 510–528
  (*See also* Frequency response)
Forcing functions, 266, 267, 635, 636
  abrupt, 193, 195
  arbitrary nonperiodic, 545, 636
  continuing, form e^{st}, 266, 267, 494
  impulse, 183, 210–217
  parabolic, 267

Forcing functions, periodic, 267, 529–534, 635
  ramp, 267, 635, 636, 688
  random, 267, 545, 636
  step, 193, 283, 635, 636
Fourier integral, 545, 546
Fourier series, 529–534, 834, 835
  average-value term, 530
  complex, 533, 834, 835
  real, 529
  for square wave, 531–534
  waveform analysis, 530
Fourier transform, 545, 546, 549
Frahm tachometer, 806
Frahm vibration absorber, 525
Frame of reference (coordinate frame), 37, 38, **146–151,** 714, 715, 718, 719
  rotating, 149, 714, 715, 718, 719
  translating, 38, 147
Frankel, F., 302
Franklin, B., 59, 124
Free-body diagram, 33, **40,** 47, 51, 84, 114, 155, 161, 761, 762, 781–785, 825
Free vibrations, 225–234
  of coupled systems, 419–488
Frequency response (forced vibration), 309, 314–354, 511–528, 628, 802–810
  amplitude, 313, 316, 318, 324, 339
  compound systems, 510–528
    cascaded subsystems, 511–514
    general coupling, 515–528
  computing, 321–330, 337–345
    Bode's method (*see* Bode plotting method)
    phasor construction, 341
    from the s plane, 321, 322, 342–345
  definition, 316
  demonstration, 314–318
  first-order system (low-pass filter), 324
  high-pass filter, 329
  importance of, 309, 510
  from Laplace analysis, 566
  low-pass filter, 316, 325
  measurement, demonstration, 311–318
  methods in feedback-system analysis, 628, 629, 638
  phase angle, 312, 316, 318, 326, 340
  phasor construction for, 341
  plotting, 316
  recording, 319, 320
  of RLC network, 352
  from the s plane, 321, 322, 342–345
  second-order systems, 331–377
    damped, 335–377
      family of curves, 336, 353
      nondimensional form, 337
    undamped, 331–335
Friction, 40, 43, 46, 49, 164, 174, 388
  dissipation through, 43, 54, 164

Friction, dry, 43, 49, 388, 763
  electrical, 69
  in fluid flow, 107–113
  in Lagrange's equations, 174, 176
  square-law, 49, 763
Friedland, B., 61n.
Function generator, 303

Gain (control system), 631, 645
Galloping transmission line, 388
Galvanometer, mirror, 320
Gas in a cylinder, 15
Gas flow (see Fluid flow)
Gas system, 774
  (See also Pneumatic computing element)
Gears, 54–56, 765, 766
  planetary, 779
General relativity experiment gyro, 626
Generalized coordinates, 168n.
Generalized forces, 173, 175
Generator (electric), 81, 83
Geometric compatibility, **39**, 52, 127
Geometric constraint relations, 37, **144**
Geometric identities, 37, 50
Geometry, one-dimensional, 35–39
  three-dimensional, **143–151**, 167, 168
Governor, flyball, 777, 782, 785, 796
Graham, D., 356n.
Gravity, acceleration of (g), 24, 28, 711
  energy stored by, 54, 172
  sensing, 347
  static spring force cancels weight, 271
  torque on a pendulum, 364, 757
Grid (electronic tube), 393
Gross, J. F., 102
Guillemin, E. A., 63, 66, 409n.
Guitar, 476
Gun-aiming system, 631–637
  automatic, 631–633
  human, 632, 634
Gyro (gyroscope), 9, **159–163**, 442, **617–626**, 691, 696, 697, 780, 786, 801, 829, 836, 838, 843, 844, **846**, 861
  accuracy of, 625, 626
  attitude reference, 617, 696
  directional, 617
  drift, 625
  dynamic behavior, 617–626
    effect of damping, 622, 846
    impulse response (coning), 620–622
    natural motions, 619, 820
    precession and nutation, 622–625, 846
  electrical suspension for, 91
  equations of motion (two-axis), 159–163, 617–619, 780, 786
    linearized, 163
    single-axis, 163, 691, 696, 801

Gyro (gyroscope), floatation, 626
    heating, 773
  gas support, 626
  law of, 625
  low-friction designs, 624–626
  rate, **163**, 624–626, 691, 696, 700, 801, 843, 861
    rate-integrating, 624–626, 801, 844, 861
    relativity experiment, 626
    repeater, 683, 684
    research on, 625, 626
    rigidity, 622
    single-axis, **163**, 624–626, 691, 696, 801, 843, 844, 861
    spin momentum, 162
    two-gimbal (two-axis), **159–163, 617–626**, 697, 829, 836, 838, **846**
    unbalance, 622, 623
    unsupported, 626
    uses of, 617
    vertical, 696
Gyro repeater, 683, 684
Gyroscopic coupling, 620

Halfman, R. L., 720n.
Halliday, D., 67n., 93n.
Harman, W. W., 83n., 253n.
Harmonic (musical), 476n., 478
Harmonic oscillators, 228, 365, 778
Harmony, 480
Harpsichord, 476
Hayashi, C., 356n.
Heat, 94–102
  capacity, 27, 96, 98, 122, 787
  conduction, **94–102**, 771–773, 787–792, 803, 832, 836, 838, 845, 847, 856, 860
    electrical analog of, 122–124, 134
  conductivity, 95, 98
  flow rate, 27, 95
  generation, electrical, 69
  resistance, 95, 98
  specific, 27, 99
  -transfer coefficient, 99
  units of, 27, 711
Heating system, 758, 772, 832
Helicopter rotor blade, 782, 785
Helmholtz, H. L. F. v., 475n.
Henry, 59
Higdon, D. T., 710n.
Holonomic constraints, 144, 777
Holzer method, 472
Homogeneous (complementary) solution, 180, 182, 186, 197
Honeywell, Inc., 624
Housner, G. W., 153n., 167n., 717n., 719n.
Hudson, D. E., 153n., 167n., 717n., 719n.
Hutchins, Carleen, 476n.
Hybrid simulation, 308

INDEX   893

Hydraulic boost system (*see* Hydraulic servo)
Hydraulic force-amplification (*see* Hydraulic servo)
Hydraulic piston (*see* Hydraulic servo)
Hydraulic reservoir system, 102, **113–115**, 773, 819, 826, 832, 844, 846
Hydraulic servo, 1, 9, **118–120**, 205, 230, 394, 397, **614–616**, 774, 787, 789–794, 799, 804, 827, 847, 856, **859**, 860, 863
   complete time response, 616, 787, 789–793
   equations of motion, 118–120, 615, 859
   manual control, 616, 787, 799
   testing, 616, 804
   two-stage, 230, 794, 796, 806, 813, 863
   valve characteristic, 119, 120, 616
   with compressible oil, 859
Hydraulic valve (*see* Hydraulic servo)
Hydraulics (*see* Fluid systems)

Identity and constraint relations, 37
   electrical, 63, 72, 83, 767
   fluid, 104
   geometric, 37, 50, **144**, 760, 776, 777
   thermal, 97, 101, 122
Impedance and admittance, 280–282, 313–314, 800, 828, 829
   driving-point, 280, 281, 314, 802–810
   electrical, 280, 313, 314, 800, 828, 829
   mechanical, 281, 800
   transfer, 280, 800
Impulse, 183, 210–217
   derivative of, 575, 576
   magnitude of (definition), 210
   representation of initial conditions, 211, 213, 572–580
   response to, 210–217, 291–297
   symbol for, 214
   and $t = 0^+$, 557
   unit, 183, 211
Independence of translation and rotation, 152
Independent environment, 13
Independent variables, 39, 63, 167
   electrical, 63
   mechanical, 39, 167
Inductance, electrical, 70
   mutual, 70, 769, 829
Inductive microphone, 79
Inductor, 70
   energy stored in, 825
Inertia, moment of, 23, **44**, 153, 162, 506, **718, 719**
   principal axes of, 153, 506, 719
   product of, 506
   tensor, 506
Inertial coupling, 442–445

Inertial force and moment, 40, **45**, 46, **152**, 762
Inertial guidance, 159, 347
Inertial instruments, 1, 159, 345–352, 617–626
   (*See also* Gyro *and* Accelerometer)
Ingot quenching (*see* Quenching an ingot)
Initial conditions, 188, 192, **200–204**, 229, 237, **255–265**, 552, 568, **572–580**, 790
   in block diagrams, 213, **572–577**, 581
   generated by impulses ("jarring"), 211, 572–580, 838
   Laplace handling of, 201, 552–554
   procedure, 203
   for second-order systems, 237, 255–265
      critically damped ($\zeta = 1$), 263
      overdamped ($1 < \zeta$), 255
      underdamped ($\zeta < 1$), 257
   sudden change, 218
   $t = 0^-$ vs. $t = 0^+$, 200, 201, 554, 557, 568
Initial-value theorem (IVT), 567, 569–571, 602, 609, 838
Integral control, 702
Integration by parts, 551
Integrators, physical, 204–206, 394, 787
   impulse response, 216
   natural motion, 206
   "resonance" in, 210
Interchangeability of subsystems, 198
Internal energy, 95, 98
Interval (musical), 480
Inverse Laplace transform, 549
   using partial fraction expansion (*see* Partial fraction expansion)
Isocline, 362–369, 811
Isolators, electrical, 75–78, 769, 770, 790
   (*See also* Amplifiers, voltage isolation)

$j = \sqrt{-1}$, 237, 238
James, H. M., 628

KCL (Kirchhoff's current law) (*see* Kirchhoff's laws)
Kinetic energy of a rigid body, 54, **164**, 170, 172, 177, 485–487
Kirchhoff's laws, 61, **64**–66, 71–75, 122–142
KVL (Kirchhoff's voltage law) (*see* Kirchhoff's laws)

Ladder, 777, 778
Lagrange, J. L., 104, 166, 178, 229, 785
Lagrange's equations, 168
Lagrange's method, 166–178, 785, 786
   advantages of, 178
   conservative systems, 169–173

Lagrange's method, nonconservative systems, 173–177
  systems with friction, 174, 176
Lagrangian, 168
Laminar flow, 110, 722
Laplace, P. S. de, 181
Laplace transform, 181, 201, 535–583
  convergence, 548
  and convolution, 580
  general relationships, 731
  graphical visualization, 549
  and impulse response, 572
  inverse, 549
  one-sided, 552, 554
    lower limit, 554, 557, 569
    product of two functions (convolution), 582, 731
  region of definition, 548
  summary, 554
  symbol for ($\mathcal{L}_-$ and $\mathcal{L}_+$), 552, 554
  and system state, 553
  two-sided, 548, 554
Laplace transform method, 181, 201n., 535–583, 835–839
  application of, 549–555
  in control-system analysis, 643
  demonstration, 537–544
  evolution of, 544–549
  initial-condition handling, 181, 201, 552–554, 837, 838
  procedure, 537, 538
  questions about, 539
  reasons for delaying in text, 544
  total response using, 562–567, 599, 600, 643, 837, 838, 840–846, 855–864
  and transfer functions, 559, 836, 837
  use of partial fraction expansion in (*see* Partial fraction expansion)
  using standard form, 577
  when most useful, 535–537
Laplace transform pairs, 538
  comprehensive table of, 731–755
  derivation of, 555–559, 835, 836
    derivative, 551, 552, 558
    exponential, 549, 556
    integral, 558
    time delay, 556
    unit impulse, 557
  dimensions of, 556
  general relationships, 731
  notation, 561
  pole-zero diagrams for, 550
  short table, 540–543
Laws of motion (Newton), **40, 152,** 111, 156, 716, 717
LC circuit, 230, 330
  coupled pair, 440
l.c.c. (linear, constant-coefficient) equations, 16, 182n.

l.c.c. systems, 182n.
Lead angle (gun), 635
Lead compensation, 691–696, 704, 707, 769
  (*See also* Network, lead)
Levers, **56,** 760, 761, 765–767, 775
Limit cycle, 860
Linear approximation, 15–17, 199, 356–360
  by "equivalent" linear relations, 357
  by perturbation, 94, 104–106, 112, 358–360
  by series expansion, 234, 358
Linear characteristics, 17, 42–46, 67, 197–200
Linear, constant-coefficient equations (l.c.c. equations), 16, 182n.
Linear differential equations (*see* Differential equations)
Linear systems, 15, 182, 197–200
  properties of, 198–200
Linear, time-invariant equations, (l.t.i. equations), 182n.
Linearity, 15, 197–200, 758
  importance of, 16, 199, 355
Liquid depth sensor, 88
Lissajous pattern, 320, 321, 803, 805
Locus of roots, 254, 376, 399, 401, 433, 796, 813–816
  graphical construction, 385–388, 401–404, 645–651
  saddle points in, 670–673
  sketching, 404–407, 650–682
  (*See also* Root-locus method)
  180° loci, 647
  0° loci, 647, 815
Log scale, 323, 324
  for frequency response, 318, 830
    construction, 323, 324
    relation to decibel scale, 324
  piano keyboard, 480
  for time response, 187, 189
Loop, electrical, 62
  equations (Kirchhoff's voltage law), 66
  feedback (*see* Feedback)
  relations, 33, 64, 765, 767
Loop-closing algebra, 639–643
Loop gain, 631, 645
Loudspeaker (*see* Speaker)
Low-pass filter (*see* Filter)
l.t.i. equations (linear, time-invariant equations), 182n.
Lumped physical parameters, 13–15, 18, 97, 102, 114, 471
Lytle, D. W., 83n., 253n.

"$M = J\alpha$ and $\mathbf{f} = m\mathbf{a}$," 153
Machine-tool chatter, 388
Machine-tool control system, 856, 860–862

INDEX 895

McRuer, D. T., 356n.
Magnetic field, 70, 90
Magnetic flux, 70, 90
Magnetic induction **B**, 26, 90
Many degrees of freedom, 458–462, 823
Marble (stability demonstration), 812, 817
Mass, 19, 23, 28, 711
  attraction, 28
  density, 24
  dimensions, 28, 711
  flow rate, 24, 104
  and force, 19n.
Mass-spring-damper system, 256, 270, 286–288, 797
  static spring force cancels weight, 271
Mass-spring system (no damping), 165, 169, 288, 330, 775, 802, 812
Mathematical model, 10, 32, 597
Matrix representation, 502–509, 828–830
  column, 503
  concepts, 506
  for eigenvector calculation, 509
  manipulation, 506
    electric circuit example, 507
  square, 503
  standard form (state form), 505, 578
Maxwell, 59, 93
Mechanical energy and power, 28, 53–57, 711
  (*See also* Energy)
Mechanical systems, 34–57, 143–178
Mechtly, E. A., 711n.
Membrane, 473
Mercury in U-tube, 231, 785, 794
Merry-go-round, 779
Mesh (electrical), 63
Microphone, carbon, 88
  crystal, 92
  dynamic (inductance), 83, 85, 771
  variable-capacitance, 88, 770
Minimum-phase systems, 518
Missile control system (*see* Hydraulic servo)
MIT Radiation Laboratory, 628
mks system of units, 20–28
Mobility analog, 126
Modal sketches, **434–436**, 442, 460, 469, 470, **472**, 477, 484, 487
Model, physical (*see* Physical model)
  mathematical (*see* Mathematical model)
Modes of motion (*see* Natural modes of motion)
Modulation devices, 87–89
Moment, 22, 711
  of inertia, 23, **44**, 153, 162, 506, **719**
  inertial, 45, 152
  of momentum (*see* Momentum, angular)

Momentum, angular, 24, 151, 718, 719
  rate of change of, 719
    for plane motion, 719, 720
  conservation of, 106, 717
  translational, 24
Moody, L. F., 721n.
Morse, P. M., 266n.
Motion, 4, 32, 180
  equations of (*see* Equations of motion)
  exponential, 180, 188
  forced, 180, 182, 193, 266–282, 490–508
  natural (*see* Natural motion)
  plane, 143, 147, 153, 719
  sinusoidal, 180
    damped, 244–308
    overdamped, 253
    undamped, 225–243, 309, 793–795
    (*See also* Sinusoidal oscillations *and* Frequency response)
  total, 180, 193–224, 282–299, 490, 562
  the two elementary patterns of, 180, 235, 245
  (*See also* Response)
Motion, Newton's laws of, 40, 44–46, 152, 156, 716, 717
  and D'Alembert's principle, 44–46
Motion sensor, 88
Motor (electric), 80–85, 765, 775, 836, 838
  (*See also* Torque motor *and* Rotor)
Multiloop control system, 696–703
Multiloop network equations, 73, 74
Murphy, A. T., 133n.
Musical instruments, 475–482
  strings, 475–478
  wind, 478–480
    brass, 479, 480
    flute and organ, 479
    reed, 479
Musical scales, 480–482
Mutual inductance, 70, 769, 829

n degrees of freedom, 462
Natural coordinates, 445, 446, 821
Natural frequency ω, 228, **249**, 821–825
  of compound (coupled) systems, 432, 459, 523, 821–825
  damped (ω) and undamped (ω_0), 249–251, 333
Natural modes of motion, 419, 420, **428–430**, 433–436, 441, 451, 455, 460–462, 469–472, 477, 818–826
  distributed systems, 469–472, 477
  eigenvectors, 451–456, 460, 822
    definition, 452
  many degrees of freedom, 456–462, 823–826
  two degrees of freedom, 426–456, 818–822

Natural motion, 180, 183, 186, 225, 246, 374, 786–788, 796–798
  compound (higher-order) systems, 374–488
  distributed-parameter systems, 467
  first-order systems, 183, 186–193
  many-degree-of-freedom systems, 457–462
  second-order systems, 225–233, 246–265, 796–798
  treated as forced response, 575
  two-degree-of-freedom systems, 419–456
  (*See also* Characteristics)
Navier-Stokes equations, 105*n.*
Negative spring constant, 381
Network, 62–75, 767–770
  approach to analysis, 121, 137–142
  branch, 62
  bridge, 829
  bridged-T, 769
  compound example, sinusoidal response, 566
    total response, 565
  duals, 723, 724, 776
  electrical, definition, 60
  first-order (*see RC* network)
  LC, 230, 330, 440
  lead (lead-compensation), **691–696**, 704, 707, 769, 804, 834, 860, 861–863
    design selection, 695, 707
    effect, on characteristics, 694
      on steady-state error, 695
    frequency response, 804
    physical limitations, 691
    triple, 861
  loop, 62
  mesh, 63
  nonplanar, 767
  for other physical systems, 137–142
  planar and nonplanar, 63, 723, 767
  RC (*see RC* network)
  RLC (*see RLC* network)
  variables, 62–64
Neutral stability, 193, 251, 388
Newton, Sir Isaac, 38, 40, 44, 111, 716
  second law, 40, 44–46, 151, 716, 717
    and D'Alembert's principle, 44–46
  unit of measure, 20
  viscous-fluid relation, 111
Nichols, N. B., 628
Nodal lines (vibrational), 474
Node, electrical, 59, 63, 64, 66, 767
  (*See also* Kirchhoff's laws)
  vibrational, 444
Node relations, 33, 53
  (*See also* Equilibrium)
Noise, 18, 347
  filtering, 316, 693
  nonmusical, 480

Nomenclatural conventions defined
  $\overset{\alpha}{\mathbf{A}}$, 714, 715
  $\triangleq$, 44
  $\overset{r}{=}$, 243
  $4^{\backslash 3}$ (slant notation), 20*n.*
  $j = \sqrt{-1}$, 237, 238
  $\mathcal{L}_-$ and $\mathcal{L}_+$, 552
  $<$, 191
  matrix notation, 502
  $t = 0^+$, $t = 0^-$, 200, 214
  vectors (physical and mathematical), 503*n.*
Nonlinear systems, 16, 114, 355–370
  analysis techniques, 356, 365–370
  linear approximation, 356–360
    (*See also* Perturbation technique)
  pendulum, 365–369
  phase-plane study of, 365–369
  piecewise-linear systems, 369
  resistor, 200
Nonplanar network, 767
Normal (natural) coordinates, 445, 446, 821
North American Aviation, 729
Northrup Nortronics, 624
Notation, scientific (standard), 20*n.*
  slant, 20*n.*
Nutation and precession (gyro), 623, 625
Nyquist, H., 628

Octave, 323, 480
Ohm's law, 67, 124
Oil-well rig, 463, 518, 761
One-dimensional system, 34, 49
One-wheel trailer (stability), 817
Open system, 105
Orbiting Astronomical Observatory, 842
Order of a system, 39, 183, 226, 244
Organ (musical), 479, 825
Orientation, 144
Orthogonal functions, 531, 545
Orthogonality principle, 531, 545, 546
Oscillations
  self-excited, 391, 477
  sinusoidal (*see* Sinusoidal oscillations)
  unstable, 383
  (*See also* Frequency response)
Oscillator circuit, 232, 806
Oscilloscope (*see* Cathode ray tube)
Overdamped system, 249, 253, 255–257

Parameters (constant or variable), 17
  distributed (*see* Distributed-parameter systems)
  lumped, 13–15, 18, 97, 102, 114, 471
Parametric excitation, 371, 812

INDEX   897

Partial differential equations, 14, 15, 457, 464–474
 solution to, 467–472
Partial fraction expansion, 584–596
 checking result of, 585
 examples of use, 602, 609, 839, 840
 formulation, 584
 graphical, pole-zero array, 589–591
 repeated poles, 592
 simplification for complex poles, 589
 summary, **595**
Particular solution, 180, 182, 197
Passive elements, 66, 134
Path relations (*see* Compatibility relations)
Pen recorder, 304, 310, 311
Pendulum, compound, 231, 782–785
 simple, 38, 172, 234, 368, 703–710
  analog-computer experiment, 306
  coupled pair, 440, 818, 820, 821, 825
  equation of motion (exact), 172, 234
   linear approximation, 234
  inverted, 230, 383, 703–710, 794, 813
   automatic control of, 703–710
   equations of motion, 703–706
  large-motion solutions, 364–**369**, 812
   with damping, 367–369
  natural frequency, 234, 794
  restoring torque, 364, 757
  on rolling cylinder, 783
  with time-varying length, 371
  velocity and acceleration of, 149, 150, 779
 spherical, 777, 779
Period T of sinusoidal motion, 228, 249
 how to measure, 260
Perturbation technique, 94, 104, 358
 fluid systems, 94, 104–106, 112, 120
 mechanical systems, 358–360, 814
Phase angle, 243, 259, **268**, 312
 forced motion, 268, 312
 natural motion, 259
Phase plane, 360–369, 811, 812, 860
 importance, 367
 solution for pendulum, 368
 (*See also* State space)
Phasor, 238, 242, 243, 312, 314, 438
 diagram, 310, 314, 341, 802, 803
 form of eigenvector, 452
 in transfer functions, 493
 use in frequency response, 341, 808
Phillips, R. S., 628
Phonograph pickup, crystal, 92
 variable-capacitance, 88
 variable-reluctance, 88
Physical constants, 21–28, 711
Physical laws, 32
 First law of thermodynamics, 97
 Maxwell's equations, 93
 Newton's laws, 40, 151, 166, 716, 717
Physical laws, (*See also* Physical relations)
Physical media, 31
Physical model, 4, 5, **10**
Physical modeling, 4, 5, 10–18, 34, 60
 approximations, 18
 electrical, 60
 fluid, 102
 mechanical, 11, 13, 49, 154
 thermal, 96, 100, 772
Physical properties, 104
Physical relations (constitutive relations), 32, 134
 classification of, 134, 135
 electrical, 66–71
  voltage-current, 66–70, 768
 electromechanical, 83, 84, 87
 energy-geometry, 165
 fluid mechanics, 106–114
  pressure-density, 106–114
 heat-temperature, 9, 98, 99
 mechanical, 43–46
  force-acceleration, 44–46
  force-displacement, 43
  force-geometry, 43, 47, 763
  force-velocity, 43
Physical variables (*see* Variables)
Physical vibrations, 225–235, 309, 793–795
 (*See also* Sinusoidal oscillations *and* Frequency response)
Piano, 480
Piecewise-linear elements, 369, 811
Piezoelectric devices, 92
Pipes or tubes, flow through, 109, 111, 721, 722
Piston, on air column, 230, 793
 hydraulic (*see* Hydraulic servo)
Planar networks, 63, 723, 767
Plane motion, 143, 147, 153–159, 719, 720
Plank on rollers, 230
Plant (control system), 631
 unstable, 704
Plate, vibrating, 474
Plotting
 characteristic roots in the s plane, 249–255, 378, 380, 387, 399
 (*See also* Locus of roots)
 frequency response, 316, 324
  amplitude, 324, 333, 339
   unity amplitude, 339
  break frequency, 326, 830
  phase angle, 326, 334, 340
  phasor construction, 341
 (*See also* Sketching)
Plumb bob, 781
Pneumatic computing element (bellows), 102, 115–118, 773, 788, 804, 827, 836, 838
Pneumatic sensor (*see* Pneumatic computing element)

Poise (measure of viscosity), 24
Poles and zeros, 245, 273, 276–279, 502
　genesis of names, 551
　open-loop system, 647
　and response of systems, 295
Pole-zero diagrams (s-plane plot), 276–279, 342, 497–500, 508, 799, 800, 835, 836
　to evaluate transfer functions, 497–500
　"rubber-sheet" image, 524, 525, 550
　of Laplace transforms, 550, 551
Porous plug, flow through, 108
Position, 36, 49, 144, 145
Positive directions, definition of, **37**
　displacement, 37
　force, 40
　velocity and acceleration, 38, 227
　voltage and current, 62
Potential energy, **53**, 165, 170–177, 483–487
　in a bar, 483–485
　in a gravity field, 54, 171, 176
　in a pendulum, 171
　in a spring, 54, 175, 483–485
　in a string, 486
Potentiometer, 88, 770, 812
Pound (force), 22
Power, dimensions of, 28, 711
　electrical, 69, 78
　mechanical, 54, 56
Precession and nutation, 623, 625
Pressure, 4, 24, 104
Pressure-density relation, 106
Pressure-rate sensor, 788, 792
Principia, 716
Principle axes, 153, 162, 506, 719
Procedures (Suggested "check lists")
　A. Equations of motion
　　A-e　　Electrical, 61, 75
　　A-em　Electromechanical, 82
　　A-f　　Fluid, 104
　　A-h　　Heat conduction, 97
　　A-m　　Mechanical, 35, 74
　　A-mL　Mechanical, Lagrange, 168
　B. Elementary response
　　B-1　General form, 182
　　B-2　Initial conditions, 203
　　B-3　Sketching natural response, 260
　C. Natural motion, compound systems
　　C-1　Obtaining characteristic equation, 398
　　C-2　Root-locus method, 401, 646
　　C-3　Root-locus sketching, 405, 652
　　C-4　Routh's method, 410
　　C-5　Rayleigh's method, 483
　D. Total response, compound systems
　　D-1　Laplace transform method, 537
　　D-2　Partial fraction expansion, 595
　　D-3　Total investigation, 599

Procedures (Suggested "check list")
　E. Control system analysis
　　E-1　Total investigation, 643
　　E-2　Root-locus sketching, 652
　summary diagram, 599
　(*See also* front end paper)
Product of inertia, 506, 719
Projectile, 778
Propeller, 765, 777, 779, 786, 799, 804
Properties (pressure, etc.), 104
Proportionality, law of, 198
Pump, constant-displacement, 774

Q, resonance parameter (network quality coefficient), 337–340, 354, 520, 524
　relation to ζ, 354
Quenching an ingot, 95–97, 610–614, 771
　equations of motion, 611, 771
　lumped-capacity model, 611
　time response, 613, 845
Quick-return mechanism, 780

Railroad car, 797
Ramp input, 635
Random signals, 18, 267, 635
　statistical treatment, 18
Rate gyro, **163**, 624–626, 691, 696, 699, 700, 801, 843, 861
Rate-integrating gyro, 624–626, 801, 844
Rate sensors, 691
　rate gyro (*see* Rate gyro)
　tachometer, 393, 790, 806, 807
Rayleigh, Lord, J. W. S., 229
Rayleigh's method, 457, 462, **482–488**, 825
　distributed parameters, 486–488
　lumped parameters, 483–486
　with Ritz iteration method, 488
RC network (filter), 191, 313, **316**, 317, 328–330, 770, 789–793, 804
　high-pass, 328, 770
　low-pass, 316, 770, 804
Reactance, 280
Reaction-wheel attitude control, 307, 842
Real-axis breakaway (root locus), 664–669
Real-axis segments (root locus), 655–657
Recorders, 304, 305, 319–321
　cathode-ray oscilloscope, 304, 320
　for frequency response, 319–321
　mirror galvanometer, 320
　pen, 304, **310–312**, 319, 335, 826
　x-y plotter (input-output), 305
Reference configuration, **36**, 37, 49, 52, **148**, 154, 161, 171, 760, 776, 777
Reference frame, 37, 38, **146–151**, 718, 719
　Eulerian (fluid), 104
　rotating, 149, 162, 714, 715, 718, 719
　translating, 147

Relative damping ζ, 251
Relative motion, 38, 146–149, 778–780
Relative velocity, acceleration, 38, 146–149
Relativity, 46n., 626
Relativity experiment using gyro, 626
Remote-indicator servo (remote-position indicator), 683–696, 860
  design, 683, 689, 690, 695
  network compensation, 691–696, 860
Repeater (*see* Remote-indicator servo)
Reservoir (fluid), 107, 108, 773
  gravity, 107, 108, 114, 773
  static relations for, 112
  system, 102, **113–115**, 773, 819, 826, 832, 844, 846
Resistance, electrical, 25, 69, 122–124
  fluid, 107–114, 124, 721, 722
    (*See also* Fluid flow)
  thermal, 27, 95, 98, 122
Resistor, 67, 69
  nonlinear, 200
  variable (potentiometer), 88, 770, 812
Resnick, R., 67n., 93n.
Resonance, 206–209, 271–273, **333–335**, **337**, 518–528, 791, 805, 833, 834
  in higher-order systems, 518–528, 834
    electrical circuit, 523–525
    mechanical systems, 519–528, 834
  near-(plotting), 209
  Q index of (*see* Q)
  in second-order systems, 271, 333, 354
    damped, 337–340, 354, 806
    undamped, 330–335
Resonant frequency, 272, 333, 337–340, 354, 521
Response (dynamic), 3, 180, 182, 183
  forced, 180, 182, 193–200
    arbitrary input (convolution), 220–224
    to continuing functions, 266–282, 492–528
      by Laplace, 566
    by convolution (*see* Convolution)
    to damped sine wave, 270, 271
    to periodic functions, 529–534
  frequency (*see* Frequency response)
  general pattern (Procedure B-1), 197
  impulse, 183, 210–217, 791, 792, 801
  natural (*see* Natural motion)
    compound systems (*see* Natural modes)
  sinusoidal (*see* Sinusoidal response *and* Frequency response)
  resonant (*see* Resonance)
  total, to abrupt inputs, 193–224, 282–299, 490, 536, 562–626
    by Laplace method, 562–583
Response function, 537, **562–564**, 855–864
Reynolds number, 109, 721, 759

Richardson, H. H., 133n.
Right-hand rule, 713n.
Rigid-body equations of motion, 152, 156, 168, 506, 782–786
RLC network, **261–263**, 272–275, 278–282, 352–354
  equation of motion and natural response, 261–263, 796–798
  forced response and transfer function, 272–275, 278–282, 798, 799, 808
  frequency response (family of curves), **352–354**, 808, 810
  impedance and admittance, 280–282, 800, 810
  phase-plane study, 811
  resonance in, 272, 806
  resonant oscillator, 806
  sinusoidal response, 272, 806
  some arrangements, 354
Rocket engine, 1, 8, 9
  control of, 1, 8, 9, 118, 614
    (*See also* Hydraulic servo)
  heat transfer, 95, 99, 836, 838
    electrical analogy, 123
Rocket sled, 841
Rocket vehicle, 331, 809, 826
  flexible, 826, 829, 861
Roll control system (aircraft), 696–703
Roll-rate loop (autopilot), 699
Rolling cylinder on sliding wedge, 148, 157, 176, 785
Root-locus method of Evans, 376, **401–407**, 629, **644–682**, 849–855
  basic construction, 401–404, 644–647
  comparison with Routh, 413, 673–677
  for coupled systems, 433, 434, 437, 448–451, 502, 515, 516, 607, 608
  general case, 404
  repeated application, 449, 450
  saddle points, 669–673, 853
  sketching procedure (sequence), 404–407, 651–653, 814
  sketching rules, 404–407, 650, 653–655
    accuracy, 651
    demonstration, 407, 648–651, 678–682
    derivations, with examples, 655–678
      Rule 1: real-axis segments, 655–657
      Rule 2: asymptotes 657–660, 850
      Rule 3: departure and arrival, 660–664, 851
      Rule 4: real-axis breakaway, 664–669, 852
      Rule 4: saddle points, 669–673
      Rule 5: Routh application, 673–677
      Rule 6: fixed centroid, 677, 678
  for 0° loci, 815, 816, 855
  problems, 815, 816, 850–855
  summary, 405, 651–653, 814, 854
  utility, 682

Roots of a characteristic equation (eigenvalues), 183, 188, 247, 248, 795–798
  equal, 265
  Evans' method (*see* Root-locus method of Evans)
  locus of (*see* Locus of roots)
  Routh's method (*see* Routh's method)
  (*See also* Characteristics)
Rotating reference frames, 149–151, 161, 162
  differentiation in, 149, 162, 714, 778
Rotating vectors (mathematical), 235, 236, 242
Rotation and translation, independence of, 152
Rotor
  spring-restrained, 46, 47, 184, 225, 246
    first-order model (*see* Torque motor)
    second-order model, damped (*see* Second-order systems)
    undamped, 225–229
      equation of motion, 225
      forced oscillation, 330–335
      natural vibration, 226–229
  unrestrained (free), 189–191
    coastdown, 191
    impulse response, 214–217
    without damping (pure integrator), 205, 216, 217
Routh, E. J., 376, 409
Routh's method, 376, 400, **406–418**, 628, 642, 653, 655, 673–677, 816, 817, 848, 853
  comparison with root-locus method, 413, 673–677, 816, 817
  general stability criteria, 410, 411, 415, 417
  third-order example, 406–409
  zero row, 416, 417
  zero term in first column, 415
Rumford, Count, 96n.

s plane, 244, 249, 252, 376, 795
  correspondence with time response, 249–253, 380
  locus of roots in (*see* Locus of roots)
  system pole-zero diagram, 277–279
    use in constructing frequency response, 321, 342, 808
Saddle points in root loci, 669–673
  three-dimensional picture, 670
Sanborn Company, 320
Satellite vehicle, 3, 442, 771, 777, 842
Scales, musical, 480–482
Schaefer, J. F., 710n.
Science, 2
Scientific (standard) notation, 20n.
Scotch yoke, 52, 53, 775, 776, 801, 842

Second-order systems, 225–308, 793–802
  analog computer experiments, 305, 306
    forced motion, 306
    natural motion (variable ζ), 305, 306
  characteristic roots (eigenvalues), 248, 249, 795, 796
  s-plane location and time response, 249–253
  damped, 244–308, 795–802
    equation of motion, 244, 245
    natural behavior of, 246–265
      deduced physically, 246
      critically damped, 263–265, 797, 798
      mathematical solution, 247
      overdamped, 249, 255–257, 796
      underdamped, 249, 257–263, 797
  forced motion alone of, 266–282
  forced oscillation of, 335–345, 802, 803
  impulse response of, 291–297
  stability of, 379
  total response to abrupt inputs, 282–290
    step response, 283–285
  transfer functions of, 273–282
  undamped, 225–235, 793–795
    equation of motion, 225
    examples of, 230, 231, 793–795
    forced oscillations of, 330–335
    natural behavior of, 227–229
      deduced physically, 227
      mathematical solution, 227
Seismic instruments, 345–352, 763, 765, 775, 776, 786
  accelerometer, 345–348, 809
  seismograph, 348–350, 809
  velocity meter, 350–352, 809
Seismograph (seismometer), 348–350, 809
Self-excited oscillations, 391, 477
Sensor (*see specific type, as* Pressure-rate sensor)
Separatrix, 366
Series expansion, 106, 112, 120, 234, 358
Servo, 683–696, 830
  hydraulic (*see* Hydraulic servo)
  remote-indicator, 683–696
  with compensation, 693–696
Shaft, continuous, 463–472
  drive, 463
  equation of motion, 465
    solution, 467–472
      fixed-fixed, 471
      free-fixed, 469
  speed control, 858
Shaft drive (lumped model), 518–521
Shaker, electromechanical, 605–610, 846
  dynamic response, 610
  equations of motion, 606
Shearer, J. L., 133n.
Ship drive system, 463, 518, 764, 798, 804

INDEX   901

Shock absorbers, 442
Shock mounting (see Vibration isolation)
σ, definition of, 249, 546
Sign conventions, 37, 38, 62, 227
   electrical voltage, current, 62
   in equations of motion, 227
   position, 37
   velocity and acceleration, 38, 227
Signal (system context), 496
Simple harmonic motion, 228
Simulation, 133, 299–308, 398
   analog, 133, 299–302
   digital, 307, 308
   fixed-base, 307
   hybrid, 308
Sine wave, 228, 259
   damped, 259
      maxima, 260
      period, 260
   generator, 303
   undamped, 228
      complex-vector representation, 242
      representation by e^{st}, 237
      rotating-vector generation, 236
Single-wheel trailer (stability), 817
Sinusoidal oscillations (natural), 225–243, 426–488, 793–797, 819–826
   of compound systems, 426–488, 819–826
   in nature, 309
   of second-order systems, 225–243, 793–795
      damped, 244–255, 796, 797
   state-space picture, 363, 812
Sinusoidal response (forced), 268–273, 309–354, 510–528, 802–810
   by Laplace, 566
   nature of, 311
   (See also Frequency response)
Sketching accurately and rapidly
   damped sinusoidal time response, 259–263, 796, 797
   exponential time response, 188, 189, 786–789
   frequency response, 324–327, 333, 337–342, 802–810
      compound systems, 512–524, 830–834
      elementary systems, 324–327, 337–345
   near-resonant time response, 209, 791
   root loci (see Root-locus sketching)
Skilling, H. H., 83n.
Slant notation, 20n.
Sleter, J. M., 717n.
Slave valve (hydraulic), 863
Sliding mass, 203, 328
Sliding wedge, 783
Small effects, 12, 13, 18
Sound, musical, 475–482
   via audio speaker, 79
Source, voltage, current, 63, 70, 76, 77

Space station, 781
Space vehicle, 3, 442, 789, 841, 842, 858
   attitude control for, 3, 842, 858
Speaker (audio), 79, 83, 422, 475, 605, 606n., 770, 771, 801, 818, 828, 829, 837, 845
Specific-force sensor, 348
Specific heat, 27, 99
Specific volume, 99
Spectral diagram, 534, 834, 835
Spectrum, 477, 529, **534**, 547
Spin momentum (gyro), 162
Spirule, 241, 343, 403n., 498, 590, 645, 649–654, 661, 678, **727–730**
   accuracy, 651
   company, 241n., 727n.
   manipulation (procedure, examples), 727–730
   picture of, 729
Spring, 5, 23, 40, **42, 43,** 46, 381, 465
   cantilever, 14, 17, 231, 794
   characteristic, 16, 762
   constant, 23, 43
   energy stored in, 54, 175, 483–485
   equivalent linear, 810
   via feedback, 397
   with negative constant, 381
   piecewise-linear, 811
   static deflection under weight, 801
   time-variable, 17
   torsional, 42, 43
   Wilberforce, 440, 820, 826, 829
Stability (dynamic), 192, 193, 251, 377–417, 812–818
   definition, 377, 812
      for linear systems, 378
      for nonlinear systems, 377
   feedback systems, 394
   index of, ζ, 251, 379
   marble demonstration, 812, 817
   neutral (definition), 193, 251
   Routh's method (see Routh's method)
   second-order systems, 379–396
Stable platform, 330, 861
Stages of dynamic investigation, 4, 597–600
   detailed outline, 597–600
Standard form of equations of motion (state form), 303, 356, 361, 502, 504, **505,** 576, 577, 838
   matrix version, 505, 578
Standing waves, 467
Stanford University, 626, 710
Star-tracking telescope, 858
State, definition, 39, 197, 577
   electrical, 62
   equations, 362
   form of eigenvector, 452, 822
   and the Laplace transform, 553

State, mechanical, 38, 39
  relation (thermodynamic), 99
  space, 356, 451
    analysis, 360–369, 811, 812
    in system studies, 197, 419n., 504, 577
    variables, **39**, 64, 99, 303, 356, 360, **504**, 576, **577**
Steady state, 290, 291
Step function, 193, 194, 283, 635, 636
  derivative of, 217
  integral of, 210
Stick-balancer, 703–710, 864
  control-system synthesis, 703–707, 864
  performance, 709, 710
Strain gauge, 88, 770
String, vibrating, 475–478, 486–488, 761, 777, 823–826
  energy stored, 486; 826
  exact equation, 824
  Fourier series for shape, 835
  violin, 388, 391, 476–478, 761, 825
Stochastic processes, 18
Strobe light, 468
Strutt, J. W. (*see* Rayleigh, Lord)
Subway train, 781
Superposition, 16, 182, 195, **197–200**, 220
  of natural modes, 419, 429, 430, 441, 446, 477, 819–822
  utilization of, 199
Swinging bar (*see* Bar, swinging)
Switching, 192, 203, 605, 800–802, 841
Switching lines, 860
Symmetry of root loci, 405, 653
System, 1, **33**, 40, 105, 164
  boundaries, 33
  cascaded (tandem) (*see* Cascaded systems)
  closed, open, 105
  compound (*see* Compound dynamic systems)
  conservative, 165, 169
  defining, 33, 40, 164
  distributed-parameter, 462–488
  electrical, 57–79
  electromechanical, 79–93
  feedback, 393
    (*See also* Automatic control systems)
  first-order (*see* First-order systems)
  many-degree-of-freedom, 458–462
  mechanical, 34–56, 143–178
  relations, 32
  second-order (*see* Second-order systems *and* Undamped second-order systems)
  six-degree-of-freedom, 828
  third-order, 396
  two-degree-of-freedom (*see* Two-degree-of-freedom systems)
  of units, 20–28

$t = 0$, $t = 0^+$, $t = 0^-$, definition of, 200, 214
  (*See also* Initial conditions)
Tacoma Narrows Bridge, 390, 391
Tachometer, 393, 790
  resonance (Frahm), 806, 807
Tandem systems (*see* Cascaded systems)
τ, time constant (*see* Time constant)
Telegraph circuit, 215, 792
Telescope (automatic tracking), 858
Temperature, 19, 27, 95, 104
  control, 847, 856, 860
  electronic packages, 771, 803
  source, 122
Test pad, isolated, 294
Thermal, capacity, 27, 96, 98, 122, 787
  constraint relations, 97
  lumped-element model, 97
  resistance (conductivity), 95, 98
Thermal quenching (*see* Quenching an ingot)
Thermocouple, 422
Thermometer, response of, 788–791, 799
Thermostat, 757, 758
Time, fundamental dimension, 19, 22
  -invariant systems, 16
  -variable systems, 17, 370, 371
  -varying coefficients, 370, 371
Time constant, $\tau = 1/\sigma$, 187, 251, 786
  for damped sinusoidal motion, 251
  for exponential motion, 187, 786
Time response (*see* Response)
  natural (*see* Natural motion)
Torque motor, 46, 47, 49, **184–189, 193–197**, 245, 393, 786, 792, 793, 827
  first-order physical model, 185, 317
  forced response of, 193–197, 317
    by convolution, 222–224, 192, 793
  frequency response of, 317, 318
  impulse response of, 211, 212, 792
  natural motion of, 186–188
  second-order physical model, 245
  response of (*see* Second-order systems)
  transfer function of, 273, 274, 836
Torsion bar (*see* Shaft)
Total response (*see* Response, total)
Trajectories, mathematical, 356, 362
Transfer functions, 245, 273–279, 826–829
  block representation, 275, 826–829
  for common physical systems, 274, 275
  in compound systems, **495–502**, 508, 511–528, 559–562, 826–829
  concept, 273, 493
  feedback system, general, 639, 645, 847
  open-loop, 639
  from Laplace transform, 559–562, 836, 837
  graphical evaluation, 497
  manipulation of, 495
  modification of signals, 495

INDEX
903

Transformers, 78, 770, 775
Transient and steady state, 290, 291
Translation, 144–147, 152
Transmission line, electrical, 15, 331
  galloping, 388
Truxal, J. G., 702n.
Tugboat, 791, 793, 846
Tuning "condensor" (radio), 88
Turbine, 330, 458, 763
Turbulent flow, 111, 114, 722
Two- and three-dimensional motion, 57, 143–178, 776–779
Two-degree-of-freedom systems, 426–456, 818–822
  characteristic equation, 427, 432
  eigenvectors, 446–456
  locus of roots, 433–437
  natural frequencies, 432
  natural modes, 428, 433–437
    physical meaning, 428

U-tube containing mercury, 230, 794
Unbalanced rotor, 526, 805
Uncertainty, 18
Undamped second-order systems (*see* Second-order systems, undamped)
Underdamped system, 249, 257–263
Unit impulse, 183, 211
  evolution of, 211, 212
  importance of, 210
  Laplace transform of, 557
Unit-impulse response $h(t)$, 183, 213, **214**, 293
  first-order system, 214
  second-order system, 293
Unit step function, 194
Unit vectors, 146, 712, 715, 718
Units of measure, 19, 759
  conversion factors, 21–28, 711
  table of, 21–28
Unstable behavior, 193, 251, 383, 703
Unstable mechanical system, control of, 703–710

Valve, fuel injection, 382
  hydraulic, 118–120, 616
  slave, 863
Variables (physical), 32, 74, 90
  electrical, 61–63, 767
  electromechanical, 82
  fluid, 104
  independent, 39, 63, 167, 767
  mechanical (selection of), 36–39, 144–151
  state, **39**, 62, 64, 99, 303, 356, 360, **504**, 576, 577, 761
  that cannot change instantaneously, 202
  thermal, 97

Variables (physical), through (flow) and across, 32, 36, 59, 74, 90, 97, 104, 122–142
Variational principles, 173n.
Vector differentiation in rotating reference frame, 149, 162, 714, 715
Vectors
  mathematical, 238
    column vector (column matrix), 503
    complex (*see* Complex vectors)
    distinguished from physical vectors, 503
    eigenvectors (*see* Eigenvectors)
    phasor (*see* Phasor)
    rotating vectors, 235
  nomenclature for, 503n.
  physical, 145–165, 503, 505, 712, 713
    dot, cross product, 54, 164, 712
    invariance of, 505
    unit vectors, 146, 712, 715, 718
Vehicle, on earth, 150
  space, 3
Velocity, 23, **38**, 50, **145–151**, 761, 778–780
  absolute and relative, 38, 146–151
  angular, 23, 38, 44, 147, 779
  of light, 46n.
  positive direction for, 38, 227
Velocity meter, 350–352, 805, 809
Vertical gyro, 696
Vextex relations, 33, 64
  (*See also* Equilibrium)
Vibrating systems, 225–235, 245–254, 257–263, 268–273, 309–371, 426–488, 510–528
  air columns, 478–480
  coupled, 426–488
  distributed (shaft, etc.), 467–488
    three-dimensional, 474
  many degrees of freedom, 458–462
  one degree of freedom, 225–273, 309–371
  strings, 475–478
    (*See also* String, vibrating)
  two degrees of freedom, 426–456
Vibration absorber, 525–528, 834
Vibration isolation (shock mounting), 2, 5, 8, 11, **270, 271**, 330, 338, 525, 801, **809**
  static deflection under weight, 801
Vibration testing, 335–337, 345–352, 804–809, 826
Vibration-testing machine (*see* Shaker, electromechanical)
Vibrations, forced (*see* Forced vibration)
  free, 225–235, 363, 793–795
    damped, 244–255, 257–263, 796, 797
    physical occurrence, 225, 230, 309
Vibrato (musical), 478
Violin, 388, 391, 476–478, 761, 825
  fingerboard, 825

Virtual displacement, 173
Virtual work, 173
Viscosity, 24, 109, 110, 721
Viscous damping, 43, 46, 422, 763–765
　drag, 787
Volkswagen, 824, 839
　(*See also* Automobile)
Volt, 20, 24, 59
Voltage, 20, 25, 59
　branch, 62
　node, 62
Voltage divider, 204
Voltage-isolation amplifier, 75, 420
Voltage source, 76

Water-supply system, 102, **113–115,** 773, 819, 826, 832, 844, 846
Wave equation, 457, 465–467
Wave propagation, 466
　velocity of, 466, 467

Waveform, Fourier analysis of, 530–534
Weight, 24
　and static spring force cancel, 271
White, D. C., 83n.
Wilberforce spring, 440, 820, 826, 829
Wind-tunnel suspension system, 863
Wing, O., 61n.
Woodson, H. H., 83n.
Work, 19, 28, 164
　virtual, 173

Yo-yo, 777, 782

Zadeh, L., 39
Zeros, and poles (*see* Poles and zeros)
　response of systems having, 295
0° root loci, 402n., 405n., 815, 816, 855
ζ, damping ratio (relative damping), 251
Ziegler, H., 626n.

ROBERT H. CANNON, JR.

As Charles Lee Powell Professor, Robert H. Cannon, Jr. chaired Stanford's Department of Aeronautics and Astronautics through the 1980s. Within that decade, he was able to assemble a faculty of peerless colleagues that makes Stanford a premier center for teaching tomorrow's aerospace engineering leaders. In 1983 he founded the Stanford Aerospace Robotics Laboratory, which has graduated 49 Ph.D.'s to date. Currently he is second research adviser to 10 more of its remarkable Ph.D. students who explore new concepts for human-like graceful robots, including free-flying space robots and underwater robots. "Being with such creative students is a daily delight," he notes. He is also co-initiator of Stanford's space gyroscope test of Einstein's General Theory of Relativity, now nearing launch. (The specified gyro drift rate is one degree every million years.)

Dr. Cannon's degrees are from the University of Rochester and MIT. During the summers, he helped develop a hydrofoil sailboat that was clocked at 31 knots in a 13-knot wind. At North American Aviation, he was a supervisor in flight control and inertial navigation systems, helping develop gyros and stable platforms for Navajo and Minuteman missiles and for the Nautilus and Skate submarines to make man's first voyages under the North Pole. After rejoining the MIT faculty for two years, he then came to Stanford in 1959, and founded its Guidance and Control Laboratory. Early projects included the first drag-free control of a satellite (the Navy's Transit satellite).

During the years 1966–74 Cannon served two tours in Washington—as Chief Scientist of the Air Force, and as Assistant Secretary of Transportation. He was close to the Apollo program in those years. From 1974–79 he chaired the Division of Engineering and Applied Science at the California Institute of Technology, where he added sixteen talented young faculty members. He then returned to Stanford to head the Department of Aeronautics and Astronautics, a position he held through 1990.

Throughout his career, Cannon has served frequently in Washington, chairing the Assembly of Engineering of the National Academy of Engineering, chairing the Energy Engineering Board, serving on the National Research Council's Governing Board, its Aeronautics and Space Engineering Board and its Ocean Studies Board, and chairing the President's Committee on the National Medal of Science. He has authored or co-authored some 60 technical papers and 11 encyclopedia articles. His textbook, *Dynamics of Physical Systems,* of which this Dover edition is the latest version, is widely used. In 1988 he was awarded the Oldenberger Medal in Automatic Control.

A CATALOG OF SELECTED
DOVER BOOKS
IN ALL FIELDS OF INTEREST

A CATALOG OF SELECTED DOVER BOOKS IN ALL FIELDS OF INTEREST

CONCERNING THE SPIRITUAL IN ART, Wassily Kandinsky. Pioneering work by father of abstract art. Thoughts on color theory, nature of art. Analysis of earlier masters. 12 illustrations. 80pp. of text. 5⅜ x 8½. 23411-8

ANIMALS: 1,419 Copyright-Free Illustrations of Mammals, Birds, Fish, Insects, etc., Jim Harter (ed.). Clear wood engravings present, in extremely lifelike poses, over 1,000 species of animals. One of the most extensive pictorial sourcebooks of its kind. Captions. Index. 284pp. 9 x 12. 23766-4

CELTIC ART: The Methods of Construction, George Bain. Simple geometric techniques for making Celtic interlacements, spirals, Kells-type initials, animals, humans, etc. Over 500 illustrations. 160pp. 9 x 12. (Available in U.S. only.) 22923-8

AN ATLAS OF ANATOMY FOR ARTISTS, Fritz Schider. Most thorough reference work on art anatomy in the world. Hundreds of illustrations, including selections from works by Vesalius, Leonardo, Goya, Ingres, Michelangelo, others. 593 illustrations. 192pp. 7⅛ x 10¼. 20241-0

CELTIC HAND STROKE-BY-STROKE (Irish Half-Uncial from "The Book of Kells"): An Arthur Baker Calligraphy Manual, Arthur Baker. Complete guide to creating each letter of the alphabet in distinctive Celtic manner. Covers hand position, strokes, pens, inks, paper, more. Illustrated. 48pp. 8¼ x 11. 24336-2

EASY ORIGAMI, John Montroll. Charming collection of 32 projects (hat, cup, pelican, piano, swan, many more) specially designed for the novice origami hobbyist. Clearly illustrated easy-to-follow instructions insure that even beginning papercrafters will achieve successful results. 48pp. 8¼ x 11. 27298-2

THE COMPLETE BOOK OF BIRDHOUSE CONSTRUCTION FOR WOODWORKERS, Scott D. Campbell. Detailed instructions, illustrations, tables. Also data on bird habitat and instinct patterns. Bibliography. 3 tables. 63 illustrations in 15 figures. 48pp. 5¼ x 8½. 24407-5

BLOOMINGDALE'S ILLUSTRATED 1886 CATALOG: Fashions, Dry Goods and Housewares, Bloomingdale Brothers. Famed merchants' extremely rare catalog depicting about 1,700 products: clothing, housewares, firearms, dry goods, jewelry, more. Invaluable for dating, identifying vintage items. Also, copyright-free graphics for artists, designers. Co-published with Henry Ford Museum & Greenfield Village. 160pp. 8¼ x 11. 25780-0

HISTORIC COSTUME IN PICTURES, Braun & Schneider. Over 1,450 costumed figures in clearly detailed engravings–from dawn of civilization to end of 19th century. Captions. Many folk costumes. 256pp. 8⅜ x 11¾. 23150-X

CATALOG OF DOVER BOOKS

STICKLEY CRAFTSMAN FURNITURE CATALOGS, Gustav Stickley and L. & J. G. Stickley. Beautiful, functional furniture in two authentic catalogs from 1910. 594 illustrations, including 277 photos, show settles, rockers, armchairs, reclining chairs, bookcases, desks, tables. 183pp. 6½ x 9¼. 23838-5

AMERICAN LOCOMOTIVES IN HISTORIC PHOTOGRAPHS: 1858 to 1949, Ron Ziel (ed.). A rare collection of 126 meticulously detailed official photographs, called "builder portraits," of American locomotives that majestically chronicle the rise of steam locomotive power in America. Introduction. Detailed captions. xi+ 129pp. 9 x 12. 27393-8

AMERICA'S LIGHTHOUSES: An Illustrated History, Francis Ross Holland, Jr. Delightfully written, profusely illustrated fact-filled survey of over 200 American lighthouses since 1716. History, anecdotes, technological advances, more. 240pp. 8 x 10¾. 25576-X

TOWARDS A NEW ARCHITECTURE, Le Corbusier. Pioneering manifesto by founder of "International School." Technical and aesthetic theories, views of industry, economics, relation of form to function, "mass-production split" and much more. Profusely illustrated. 320pp. 6⅛ x 9¼. (Available in U.S. only.) 25023-7

HOW THE OTHER HALF LIVES, Jacob Riis. Famous journalistic record, exposing poverty and degradation of New York slums around 1900, by major social reformer. 100 striking and influential photographs. 233pp. 10 x 7⅞. 22012-5

FRUIT KEY AND TWIG KEY TO TREES AND SHRUBS, William M. Harlow. One of the handiest and most widely used identification aids. Fruit key covers 120 deciduous and evergreen species; twig key 160 deciduous species. Easily used. Over 300 photographs. 126pp. 5⅜ x 8½. 20511-8

COMMON BIRD SONGS, Dr. Donald J. Borror. Songs of 60 most common U.S. birds: robins, sparrows, cardinals, bluejays, finches, more–arranged in order of increasing complexity. Up to 9 variations of songs of each species.
Cassette and manual 99911-4

ORCHIDS AS HOUSE PLANTS, Rebecca Tyson Northen. Grow cattleyas and many other kinds of orchids–in a window, in a case, or under artificial light. 63 illustrations. 148pp. 5⅜ x 8½. 23261-1

MONSTER MAZES, Dave Phillips. Masterful mazes at four levels of difficulty. Avoid deadly perils and evil creatures to find magical treasures. Solutions for all 32 exciting illustrated puzzles. 48pp. 8¼ x 11. 26005-4

MOZART'S DON GIOVANNI (DOVER OPERA LIBRETTO SERIES), Wolfgang Amadeus Mozart. Introduced and translated by Ellen H. Bleiler. Standard Italian libretto, with complete English translation. Convenient and thoroughly portable–an ideal companion for reading along with a recording or the performance itself. Introduction. List of characters. Plot summary. 121pp. 5¼ x 8½. 24944-1

TECHNICAL MANUAL AND DICTIONARY OF CLASSICAL BALLET, Gail Grant. Defines, explains, comments on steps, movements, poses and concepts. 15-page pictorial section. Basic book for student, viewer. 127pp. 5⅜ x 8½. 21843-0

CATALOG OF DOVER BOOKS

THE CLARINET AND CLARINET PLAYING, David Pino. Lively, comprehensive work features suggestions about technique, musicianship, and musical interpretation, as well as guidelines for teaching, making your own reeds, and preparing for public performance. Includes an intriguing look at clarinet history. "A godsend," *The Clarinet,* Journal of the International Clarinet Society. Appendixes. 7 illus. 320pp. 5⅜ x 8½. 40270-3

HOLLYWOOD GLAMOR PORTRAITS, John Kobal (ed.). 145 photos from 1926-49. Harlow, Gable, Bogart, Bacall; 94 stars in all. Full background on photographers, technical aspects. 160pp. 8⅜ x 11¼. 23352-9

THE ANNOTATED CASEY AT THE BAT: A Collection of Ballads about the Mighty Casey/Third, Revised Edition, Martin Gardner (ed.). Amusing sequels and parodies of one of America's best-loved poems: Casey's Revenge, Why Casey Whiffed, Casey's Sister at the Bat, others. 256pp. 5⅜ x 8½. 28598-7

THE RAVEN AND OTHER FAVORITE POEMS, Edgar Allan Poe. Over 40 of the author's most memorable poems: "The Bells," "Ulalume," "Israfel," "To Helen," "The Conqueror Worm," "Eldorado," "Annabel Lee," many more. Alphabetic lists of titles and first lines. 64pp. 5³⁄₁₆ x 8¼. 26685-0

PERSONAL MEMOIRS OF U. S. GRANT, Ulysses Simpson Grant. Intelligent, deeply moving firsthand account of Civil War campaigns, considered by many the finest military memoirs ever written. Includes letters, historic photographs, maps and more. 528pp. 6⅛ x 9¼. 28587-1

ANCIENT EGYPTIAN MATERIALS AND INDUSTRIES, A. Lucas and J. Harris. Fascinating, comprehensive, thoroughly documented text describes this ancient civilization's vast resources and the processes that incorporated them in daily life, including the use of animal products, building materials, cosmetics, perfumes and incense, fibers, glazed ware, glass and its manufacture, materials used in the mummification process, and much more. 544pp. 6⅛ x 9¼. (Available in U.S. only.) 40446-3

RUSSIAN STORIES/RUSSKIE RASSKAZY: A Dual-Language Book, edited by Gleb Struve. Twelve tales by such masters as Chekhov, Tolstoy, Dostoevsky, Pushkin, others. Excellent word-for-word English translations on facing pages, plus teaching and study aids, Russian/English vocabulary, biographical/critical introductions, more. 416pp. 5⅜ x 8½. 26244-8

PHILADELPHIA THEN AND NOW: 60 Sites Photographed in the Past and Present, Kenneth Finkel and Susan Oyama. Rare photographs of City Hall, Logan Square, Independence Hall, Betsy Ross House, other landmarks juxtaposed with contemporary views. Captures changing face of historic city. Introduction. Captions. 128pp. 8¼ x 11. 25790-8

AIA ARCHITECTURAL GUIDE TO NASSAU AND SUFFOLK COUNTIES, LONG ISLAND, The American Institute of Architects, Long Island Chapter, and the Society for the Preservation of Long Island Antiquities. Comprehensive, well-researched and generously illustrated volume brings to life over three centuries of Long Island's great architectural heritage. More than 240 photographs with authoritative, extensively detailed captions. 176pp. 8¼ x 11. 26946-9

NORTH AMERICAN INDIAN LIFE: Customs and Traditions of 23 Tribes, Elsie Clews Parsons (ed.). 27 fictionalized essays by noted anthropologists examine religion, customs, government, additional facets of life among the Winnebago, Crow, Zuni, Eskimo, other tribes. 480pp. 6⅛ x 9¼. 27377-6

FRANK LLOYD WRIGHT'S DANA HOUSE, Donald Hoffmann. Pictorial essay of residential masterpiece with over 160 interior and exterior photos, plans, elevations, sketches and studies. 128pp. 9¼ x 10¾. 29120-0

THE MALE AND FEMALE FIGURE IN MOTION: 60 Classic Photographic Sequences, Eadweard Muybridge. 60 true-action photographs of men and women walking, running, climbing, bending, turning, etc., reproduced from rare 19th-century masterpiece. vi + 121pp. 9 x 12. 24745-7

1001 QUESTIONS ANSWERED ABOUT THE SEASHORE, N. J. Berrill and Jacquelyn Berrill. Queries answered about dolphins, sea snails, sponges, starfish, fishes, shore birds, many others. Covers appearance, breeding, growth, feeding, much more. 305pp. 5¼ x 8¼. 23366-9

ATTRACTING BIRDS TO YOUR YARD, William J. Weber. Easy-to-follow guide offers advice on how to attract the greatest diversity of birds: birdhouses, feeders, water and waterers, much more. 96pp. 5³⁄₁₆ x 8¼. 28927-3

MEDICINAL AND OTHER USES OF NORTH AMERICAN PLANTS: A Historical Survey with Special Reference to the Eastern Indian Tribes, Charlotte Erichsen-Brown. Chronological historical citations document 500 years of usage of plants, trees, shrubs native to eastern Canada, northeastern U.S. Also complete identifying information. 343 illustrations. 544pp. 6½ x 9¼. 25951-X

STORYBOOK MAZES, Dave Phillips. 23 stories and mazes on two-page spreads: Wizard of Oz, Treasure Island, Robin Hood, etc. Solutions. 64pp. 8¼ x 11. 23628-5

AMERICAN NEGRO SONGS: 230 Folk Songs and Spirituals, Religious and Secular, John W. Work. This authoritative study traces the African influences of songs sung and played by black Americans at work, in church, and as entertainment. The author discusses the lyric significance of such songs as "Swing Low, Sweet Chariot," "John Henry," and others and offers the words and music for 230 songs. Bibliography. Index of Song Titles. 272pp. 6½ x 9¼. 40271-1

MOVIE-STAR PORTRAITS OF THE FORTIES, John Kobal (ed.). 163 glamor, studio photos of 106 stars of the 1940s: Rita Hayworth, Ava Gardner, Marlon Brando, Clark Gable, many more. 176pp. 8⅜ x 11¼. 23546-7

BENCHLEY LOST AND FOUND, Robert Benchley. Finest humor from early 30s, about pet peeves, child psychologists, post office and others. Mostly unavailable elsewhere. 73 illustrations by Peter Arno and others. 183pp. 5⅜ x 8½. 22410-4

YEKL and THE IMPORTED BRIDEGROOM AND OTHER STORIES OF YIDDISH NEW YORK, Abraham Cahan. Film Hester Street based on *Yekl* (1896). Novel, other stories among first about Jewish immigrants on N.Y.'s East Side. 240pp. 5⅜ x 8½. 22427-9

SELECTED POEMS, Walt Whitman. Generous sampling from *Leaves of Grass*. Twenty-four poems include "I Hear America Singing," "Song of the Open Road," "I Sing the Body Electric," "When Lilacs Last in the Dooryard Bloom'd," "O Captain! My Captain!"–all reprinted from an authoritative edition. Lists of titles and first lines. 128pp. 5³⁄₁₆ x 8¼. 26878-0

THE BEST TALES OF HOFFMANN, E. T. A. Hoffmann. 10 of Hoffmann's most important stories: "Nutcracker and the King of Mice," "The Golden Flowerpot," etc. 458pp. 5⅜ x 8½. 21793-0

FROM FETISH TO GOD IN ANCIENT EGYPT, E. A. Wallis Budge. Rich detailed survey of Egyptian conception of "God" and gods, magic, cult of animals, Osiris, more. Also, superb English translations of hymns and legends. 240 illustrations. 545pp. 5⅜ x 8½. 25803-3

FRENCH STORIES/CONTES FRANÇAIS: A Dual-Language Book, Wallace Fowlie. Ten stories by French masters, Voltaire to Camus: "Micromegas" by Voltaire; "The Atheist's Mass" by Balzac; "Minuet" by de Maupassant; "The Guest" by Camus, six more. Excellent English translations on facing pages. Also French-English vocabulary list, exercises, more. 352pp. 5⅜ x 8½. 26443-2

CHICAGO AT THE TURN OF THE CENTURY IN PHOTOGRAPHS: 122 Historic Views from the Collections of the Chicago Historical Society, Larry A. Viskochil. Rare large-format prints offer detailed views of City Hall, State Street, the Loop, Hull House, Union Station, many other landmarks, circa 1904-1913. Introduction. Captions. Maps. 144pp. 9⅜ x 12¼. 24656-6

OLD BROOKLYN IN EARLY PHOTOGRAPHS, 1865-1929, William Lee Younger. Luna Park, Gravesend race track, construction of Grand Army Plaza, moving of Hotel Brighton, etc. 157 previously unpublished photographs. 165pp. 8⅞ x 11¾. 23587-4

THE MYTHS OF THE NORTH AMERICAN INDIANS, Lewis Spence. Rich anthology of the myths and legends of the Algonquins, Iroquois, Pawnees and Sioux, prefaced by an extensive historical and ethnological commentary. 36 illustrations. 480pp. 5⅜ x 8½. 25967-6

AN ENCYCLOPEDIA OF BATTLES: Accounts of Over 1,560 Battles from 1479 B.C. to the Present, David Eggenberger. Essential details of every major battle in recorded history from the first battle of Megiddo in 1479 B.C. to Grenada in 1984. List of Battle Maps. New Appendix covering the years 1967-1984. Index. 99 illustrations. 544pp. 6½ x 9¼. 24913-1

SAILING ALONE AROUND THE WORLD, Captain Joshua Slocum. First man to sail around the world, alone, in small boat. One of great feats of seamanship told in delightful manner. 67 illustrations. 294pp. 5⅜ x 8½. 20326-3

ANARCHISM AND OTHER ESSAYS, Emma Goldman. Powerful, penetrating, prophetic essays on direct action, role of minorities, prison reform, puritan hypocrisy, violence, etc. 271pp. 5⅜ x 8½. 22484-8

MYTHS OF THE HINDUS AND BUDDHISTS, Ananda K. Coomaraswamy and Sister Nivedita. Great stories of the epics; deeds of Krishna, Shiva, taken from puranas, Vedas, folk tales; etc. 32 illustrations. 400pp. 5⅜ x 8½. 21759-0

THE TRAUMA OF BIRTH, Otto Rank. Rank's controversial thesis that anxiety neurosis is caused by profound psychological trauma which occurs at birth. 256pp. 5⅜ x 8½. 27974-X

A THEOLOGICO-POLITICAL TREATISE, Benedict Spinoza. Also contains unfinished Political Treatise. Great classic on religious liberty, theory of government on common consent. R. Elwes translation. Total of 421pp. 5⅜ x 8½. 20249-6

CATALOG OF DOVER BOOKS

MY BONDAGE AND MY FREEDOM, Frederick Douglass. Born a slave, Douglass became outspoken force in antislavery movement. The best of Douglass' autobiographies. Graphic description of slave life. 464pp. 5⅜ x 8½. 22457-0

FOLLOWING THE EQUATOR: A Journey Around the World, Mark Twain. Fascinating humorous account of 1897 voyage to Hawaii, Australia, India, New Zealand, etc. Ironic, bemused reports on peoples, customs, climate, flora and fauna, politics, much more. 197 illustrations. 720pp. 5⅜ x 8½. 26113-1

THE PEOPLE CALLED SHAKERS, Edward D. Andrews. Definitive study of Shakers: origins, beliefs, practices, dances, social organization, furniture and crafts, etc. 33 illustrations. 351pp. 5⅜ x 8½. 21081-2

THE MYTHS OF GREECE AND ROME, H. A. Guerber. A classic of mythology, generously illustrated, long prized for its simple, graphic, accurate retelling of the principal myths of Greece and Rome, and for its commentary on their origins and significance. With 64 illustrations by Michelangelo, Raphael, Titian, Rubens, Canova, Bernini and others. 480pp. 5⅜ x 8½. 27584-1

PSYCHOLOGY OF MUSIC, Carl E. Seashore. Classic work discusses music as a medium from psychological viewpoint. Clear treatment of physical acoustics, auditory apparatus, sound perception, development of musical skills, nature of musical feeling, host of other topics. 88 figures. 408pp. 5⅜ x 8½. 21851-1

THE PHILOSOPHY OF HISTORY, Georg W. Hegel. Great classic of Western thought develops concept that history is not chance but rational process, the evolution of freedom. 457pp. 5⅜ x 8½. 20112-0

THE BOOK OF TEA, Kakuzo Okakura. Minor classic of the Orient: entertaining, charming explanation, interpretation of traditional Japanese culture in terms of tea ceremony. 94pp. 5⅜ x 8½. 20070-1

LIFE IN ANCIENT EGYPT, Adolf Erman. Fullest, most thorough, detailed older account with much not in more recent books, domestic life, religion, magic, medicine, commerce, much more. Many illustrations reproduce tomb paintings, carvings, hieroglyphs, etc. 597pp. 5⅜ x 8½. 22632-8

SUNDIALS, Their Theory and Construction, Albert Waugh. Far and away the best, most thorough coverage of ideas, mathematics concerned, types, construction, adjusting anywhere. Simple, nontechnical treatment allows even children to build several of these dials. Over 100 illustrations. 230pp. 5⅜ x 8½. 22947-5

THEORETICAL HYDRODYNAMICS, L. M. Milne-Thomson. Classic exposition of the mathematical theory of fluid motion, applicable to both hydrodynamics and aerodynamics. Over 600 exercises. 768pp. 6⅛ x 9¼. 68970-0

SONGS OF EXPERIENCE: Facsimile Reproduction with 26 Plates in Full Color, William Blake. 26 full-color plates from a rare 1826 edition. Includes "The Tyger," "London," "Holy Thursday," and other poems. Printed text of poems. 48pp. 5¼ x 7. 24636-1

OLD-TIME VIGNETTES IN FULL COLOR, Carol Belanger Grafton (ed.). Over 390 charming, often sentimental illustrations, selected from archives of Victorian graphics–pretty women posing, children playing, food, flowers, kittens and puppies, smiling cherubs, birds and butterflies, much more. All copyright-free. 48pp. 9¼ x 12¼. 27269-9

PERSPECTIVE FOR ARTISTS, Rex Vicat Cole. Depth, perspective of sky and sea, shadows, much more, not usually covered. 391 diagrams, 81 reproductions of drawings and paintings. 279pp. 5⅜ x 8½. 22487-2

DRAWING THE LIVING FIGURE, Joseph Sheppard. Innovative approach to artistic anatomy focuses on specifics of surface anatomy, rather than muscles and bones. Over 170 drawings of live models in front, back and side views, and in widely varying poses. Accompanying diagrams. 177 illustrations. Introduction. Index. 144pp. 8⅜ x11¼. 26723-7

GOTHIC AND OLD ENGLISH ALPHABETS: 100 Complete Fonts, Dan X. Solo. Add power, elegance to posters, signs, other graphics with 100 stunning copyright-free alphabets: Blackstone, Dolbey, Germania, 97 more–including many lower-case, numerals, punctuation marks. 104pp. 8⅛ x 11. 24695-7

HOW TO DO BEADWORK, Mary White. Fundamental book on craft from simple projects to five-bead chains and woven works. 106 illustrations. 142pp. 5⅜ x 8. 20697-1

THE BOOK OF WOOD CARVING, Charles Marshall Sayers. Finest book for beginners discusses fundamentals and offers 34 designs. "Absolutely first rate . . . well thought out and well executed."–E. J. Tangerman. 118pp. 7¾ x 10⅝. 23654-4

ILLUSTRATED CATALOG OF CIVIL WAR MILITARY GOODS: Union Army Weapons, Insignia, Uniform Accessories, and Other Equipment, Schuyler, Hartley, and Graham. Rare, profusely illustrated 1846 catalog includes Union Army uniform and dress regulations, arms and ammunition, coats, insignia, flags, swords, rifles, etc. 226 illustrations. 160pp. 9 x 12. 24939-5

WOMEN'S FASHIONS OF THE EARLY 1900s: An Unabridged Republication of "New York Fashions, 1909," National Cloak & Suit Co. Rare catalog of mail-order fashions documents women's and children's clothing styles shortly after the turn of the century. Captions offer full descriptions, prices. Invaluable resource for fashion, costume historians. Approximately 725 illustrations. 128pp. 8⅜ x 11¼. 27276-1

THE 1912 AND 1915 GUSTAV STICKLEY FURNITURE CATALOGS, Gustav Stickley. With over 200 detailed illustrations and descriptions, these two catalogs are essential reading and reference materials and identification guides for Stickley furniture. Captions cite materials, dimensions and prices. 112pp. 6½ x 9¼. 26676-1

EARLY AMERICAN LOCOMOTIVES, John H. White, Jr. Finest locomotive engravings from early 19th century: historical (1804–74), main-line (after 1870), special, foreign, etc. 147 plates. 142pp. 11⅜ x 8¼. 22772-3

THE TALL SHIPS OF TODAY IN PHOTOGRAPHS, Frank O. Braynard. Lavishly illustrated tribute to nearly 100 majestic contemporary sailing vessels: Amerigo Vespucci, Clearwater, Constitution, Eagle, Mayflower, Sea Cloud, Victory, many more. Authoritative captions provide statistics, background on each ship. 190 black-and-white photographs and illustrations. Introduction. 128pp. 8⅞ x 11¾. 27163-3

CATALOG OF DOVER BOOKS

LITTLE BOOK OF EARLY AMERICAN CRAFTS AND TRADES, Peter Stockham (ed.). 1807 children's book explains crafts and trades: baker, hatter, cooper, potter, and many others. 23 copperplate illustrations. 140pp. $4^{5}/_{8}$ x 6.   23336-7

VICTORIAN FASHIONS AND COSTUMES FROM HARPER'S BAZAR, 1867–1898, Stella Blum (ed.). Day costumes, evening wear, sports clothes, shoes, hats, other accessories in over 1,000 detailed engravings. 320pp. $9^{3}/_{8}$ x $12^{1}/_{4}$.   22990-4

GUSTAV STICKLEY, THE CRAFTSMAN, Mary Ann Smith. Superb study surveys broad scope of Stickley's achievement, especially in architecture. Design philosophy, rise and fall of the Craftsman empire, descriptions and floor plans for many Craftsman houses, more. 86 black-and-white halftones. 31 line illustrations. Introduction 208pp. $6^{1}/_{2}$ x $9^{1}/_{4}$.   27210-9

THE LONG ISLAND RAIL ROAD IN EARLY PHOTOGRAPHS, Ron Ziel. Over 220 rare photos, informative text document origin (1844) and development of rail service on Long Island. Vintage views of early trains, locomotives, stations, passengers, crews, much more. Captions. $8^{7}/_{8}$ x $11^{3}/_{4}$.   26301-0

VOYAGE OF THE LIBERDADE, Joshua Slocum. Great 19th-century mariner's thrilling, first-hand account of the wreck of his ship off South America, the 35-foot boat he built from the wreckage, and its remarkable voyage home. 128pp. $5^{3}/_{8}$ x $8^{1}/_{2}$.   40022-0

TEN BOOKS ON ARCHITECTURE, Vitruvius. The most important book ever written on architecture. Early Roman aesthetics, technology, classical orders, site selection, all other aspects. Morgan translation. 331pp. $5^{3}/_{8}$ x $8^{1}/_{2}$.   20645-9

THE HUMAN FIGURE IN MOTION, Eadweard Muybridge. More than 4,500 stopped-action photos, in action series, showing undraped men, women, children jumping, lying down, throwing, sitting, wrestling, carrying, etc. 390pp. $7^{7}/_{8}$ x $10^{5}/_{8}$.
20204-6 Clothbd.

TREES OF THE EASTERN AND CENTRAL UNITED STATES AND CANADA, William M. Harlow. Best one-volume guide to 140 trees. Full descriptions, woodlore, range, etc. Over 600 illustrations. Handy size. 288pp. $4^{1}/_{2}$ x $6^{5}/_{8}$.   20395-6

SONGS OF WESTERN BIRDS, Dr. Donald J. Borror. Complete song and call repertoire of 60 western species, including flycatchers, juncoes, cactus wrens, many more–includes fully illustrated booklet.   Cassette and manual 99913-0

GROWING AND USING HERBS AND SPICES, Milo Miloradovich. Versatile handbook provides all the information needed for cultivation and use of all the herbs and spices available in North America. 4 illustrations. Index. Glossary. 236pp. $5^{3}/_{8}$ x $8^{1}/_{2}$.   25058-X

BIG BOOK OF MAZES AND LABYRINTHS, Walter Shepherd. 50 mazes and labyrinths in all–classical, solid, ripple, and more–in one great volume. Perfect inexpensive puzzler for clever youngsters. Full solutions. 112pp. $8^{1}/_{8}$ x 11.   22951-3

CATALOG OF DOVER BOOKS

PIANO TUNING, J. Cree Fischer. Clearest, best book for beginner, amateur. Simple repairs, raising dropped notes, tuning by easy method of flattened fifths. No previous skills needed. 4 illustrations. 201pp. 5⅜ x 8½. 23267-0

HINTS TO SINGERS, Lillian Nordica. Selecting the right teacher, developing confidence, overcoming stage fright, and many other important skills receive thoughtful discussion in this indispensible guide, written by a world-famous diva of four decades' experience. 96pp. 5⅜ x 8½. 40094-8

THE COMPLETE NONSENSE OF EDWARD LEAR, Edward Lear. All nonsense limericks, zany alphabets, Owl and Pussycat, songs, nonsense botany, etc., illustrated by Lear. Total of 320pp. 5⅜ x 8½. (Available in U.S. only.) 20167-8

VICTORIAN PARLOUR POETRY: An Annotated Anthology, Michael R. Turner. 117 gems by Longfellow, Tennyson, Browning, many lesser-known poets. "The Village Blacksmith," "Curfew Must Not Ring Tonight," "Only a Baby Small," dozens more, often difficult to find elsewhere. Index of poets, titles, first lines. xxiii + 325pp. 5⅜ x 8¼. 27044-0

DUBLINERS, James Joyce. Fifteen stories offer vivid, tightly focused observations of the lives of Dublin's poorer classes. At least one, "The Dead," is considered a masterpiece. Reprinted complete and unabridged from standard edition. 160pp. 5³⁄₁₆ x 8¼. 26870-5

GREAT WEIRD TALES: 14 Stories by Lovecraft, Blackwood, Machen and Others, S. T. Joshi (ed.). 14 spellbinding tales, including "The Sin Eater," by Fiona McLeod, "The Eye Above the Mantel," by Frank Belknap Long, as well as renowned works by R. H. Barlow, Lord Dunsany, Arthur Machen, W. C. Morrow and eight other masters of the genre. 256pp. 5⅜ x 8½. (Available in U.S. only.) 40436-6

THE BOOK OF THE SACRED MAGIC OF ABRAMELIN THE MAGE, translated by S. MacGregor Mathers. Medieval manuscript of ceremonial magic. Basic document in Aleister Crowley, Golden Dawn groups. 268pp. 5⅜ x 8½. 23211-5

NEW RUSSIAN-ENGLISH AND ENGLISH-RUSSIAN DICTIONARY, M. A. O'Brien. This is a remarkably handy Russian dictionary, containing a surprising amount of information, including over 70,000 entries. 366pp. 4½ x 6⅛. 20208-9

HISTORIC HOMES OF THE AMERICAN PRESIDENTS, Second, Revised Edition, Irvin Haas. A traveler's guide to American Presidential homes, most open to the public, depicting and describing homes occupied by every American President from George Washington to George Bush. With visiting hours, admission charges, travel routes. 175 photographs. Index. 160pp. 8¼ x 11. 26751-2

NEW YORK IN THE FORTIES, Andreas Feininger. 162 brilliant photographs by the well-known photographer, formerly with *Life* magazine. Commuters, shoppers, Times Square at night, much else from city at its peak. Captions by John von Hartz. 181pp. 9¼ x 10¾. 23585-8

INDIAN SIGN LANGUAGE, William Tomkins. Over 525 signs developed by Sioux and other tribes. Written instructions and diagrams. Also 290 pictographs. 111pp. 6⅛ x 9¼. 22029-X

CATALOG OF DOVER BOOKS

ANATOMY: A Complete Guide for Artists, Joseph Sheppard. A master of figure drawing shows artists how to render human anatomy convincingly. Over 460 illustrations. 224pp. 8⅜ x 11¼. 27279-6

MEDIEVAL CALLIGRAPHY: Its History and Technique, Marc Drogin. Spirited history, comprehensive instruction manual covers 13 styles (ca. 4th century through 15th). Excellent photographs; directions for duplicating medieval techniques with modern tools. 224pp. 8⅜ x 11¼. 26142-5

DRIED FLOWERS: How to Prepare Them, Sarah Whitlock and Martha Rankin. Complete instructions on how to use silica gel, meal and borax, perlite aggregate, sand and borax, glycerine and water to create attractive permanent flower arrangements. 12 illustrations. 32pp. 5⅜ x 8½. 21802-3

EASY-TO-MAKE BIRD FEEDERS FOR WOODWORKERS, Scott D. Campbell. Detailed, simple-to-use guide for designing, constructing, caring for and using feeders. Text, illustrations for 12 classic and contemporary designs. 96pp. 5⅜ x 8½. 25847-5

SCOTTISH WONDER TALES FROM MYTH AND LEGEND, Donald A. Mackenzie. 16 lively tales tell of giants rumbling down mountainsides, of a magic wand that turns stone pillars into warriors, of gods and goddesses, evil hags, powerful forces and more. 240pp. 5⅜ x 8½. 29677-6

THE HISTORY OF UNDERCLOTHES, C. Willett Cunnington and Phyllis Cunnington. Fascinating, well-documented survey covering six centuries of English undergarments, enhanced with over 100 illustrations: 12th-century laced-up bodice, footed long drawers (1795), 19th-century bustles, 19th-century corsets for men, Victorian "bust improvers," much more. 272pp. 5⅜ x 8¼. 27124-2

ARTS AND CRAFTS FURNITURE: The Complete Brooks Catalog of 1912, Brooks Manufacturing Co. Photos and detailed descriptions of more than 150 now very collectible furniture designs from the Arts and Crafts movement depict davenports, settees, buffets, desks, tables, chairs, bedsteads, dressers and more, all built of solid, quarter-sawed oak. Invaluable for students and enthusiasts of antiques, Americana and the decorative arts. 80pp. 6½ x 9¼. 27471-3

WILBUR AND ORVILLE: A Biography of the Wright Brothers, Fred Howard. Definitive, crisply written study tells the full story of the brothers' lives and work. A vividly written biography, unparalleled in scope and color, that also captures the spirit of an extraordinary era. 560pp. 6⅛ x 9¼. 40297-5

THE ARTS OF THE SAILOR: Knotting, Splicing and Ropework, Hervey Garrett Smith. Indispensable shipboard reference covers tools, basic knots and useful hitches; handsewing and canvas work, more. Over 100 illustrations. Delightful reading for sea lovers. 256pp. 5⅜ x 8½. 26440-8

FRANK LLOYD WRIGHT'S FALLINGWATER: The House and Its History, Second, Revised Edition, Donald Hoffmann. A total revision–both in text and illustrations–of the standard document on Fallingwater, the boldest, most personal architectural statement of Wright's mature years, updated with valuable new material from the recently opened Frank Lloyd Wright Archives. "Fascinating"–*The New York Times*. 116 illustrations. 128pp. 9¼ x 10¾. 27430-6

PHOTOGRAPHIC SKETCHBOOK OF THE CIVIL WAR, Alexander Gardner. 100 photos taken on field during the Civil War. Famous shots of Manassas Harper's Ferry, Lincoln, Richmond, slave pens, etc. 244pp. 10⅝ x 8¼. 22731-6

FIVE ACRES AND INDEPENDENCE, Maurice G. Kains. Great back-to-the-land classic explains basics of self-sufficient farming. The one book to get. 95 illustrations. 397pp. 5⅜ x 8½. 20974-1

SONGS OF EASTERN BIRDS, Dr. Donald J. Borror. Songs and calls of 60 species most common to eastern U.S.: warblers, woodpeckers, flycatchers, thrushes, larks, many more in high-quality recording. Cassette and manual 99912-2

A MODERN HERBAL, Margaret Grieve. Much the fullest, most exact, most useful compilation of herbal material. Gigantic alphabetical encyclopedia, from aconite to zedoary, gives botanical information, medical properties, folklore, economic uses, much else. Indispensable to serious reader. 161 illustrations. 888pp. 6½ x 9¼. 2-vol. set. (Available in U.S. only.) Vol. I: 22798-7
Vol. II: 22799-5

HIDDEN TREASURE MAZE BOOK, Dave Phillips. Solve 34 challenging mazes accompanied by heroic tales of adventure. Evil dragons, people-eating plants, blood-thirsty giants, many more dangerous adversaries lurk at every twist and turn. 34 mazes, stories, solutions. 48pp. 8¼ x 11. 24566-7

LETTERS OF W. A. MOZART, Wolfgang A. Mozart. Remarkable letters show bawdy wit, humor, imagination, musical insights, contemporary musical world; includes some letters from Leopold Mozart. 276pp. 5⅜ x 8½. 22859-2

BASIC PRINCIPLES OF CLASSICAL BALLET, Agrippina Vaganova. Great Russian theoretician, teacher explains methods for teaching classical ballet. 118 illustrations. 175pp. 5⅜ x 8½. 22036-2

THE JUMPING FROG, Mark Twain. Revenge edition. The original story of The Celebrated Jumping Frog of Calaveras County, a hapless French translation, and Twain's hilarious "retranslation" from the French. 12 illustrations. 66pp. 5⅜ x 8½. 22686-7

BEST REMEMBERED POEMS, Martin Gardner (ed.). The 126 poems in this superb collection of 19th- and 20th-century British and American verse range from Shelley's "To a Skylark" to the impassioned "Renascence" of Edna St. Vincent Millay and to Edward Lear's whimsical "The Owl and the Pussycat." 224pp. 5⅜ x 8½. 27165-X

COMPLETE SONNETS, William Shakespeare. Over 150 exquisite poems deal with love, friendship, the tyranny of time, beauty's evanescence, death and other themes in language of remarkable power, precision and beauty. Glossary of archaic terms. 80pp. 5³⁄₁₆ x 8¼. 26686-9

THE BATTLES THAT CHANGED HISTORY, Fletcher Pratt. Eminent historian profiles 16 crucial conflicts, ancient to modern, that changed the course of civilization. 352pp. 5⅜ x 8½. 41129-X

CATALOG OF DOVER BOOKS

THE WIT AND HUMOR OF OSCAR WILDE, Alvin Redman (ed.). More than 1,000 ripostes, paradoxes, wisecracks: Work is the curse of the drinking classes; I can resist everything except temptation; etc. 258pp. 5⅜ x 8½. 20602-5

SHAKESPEARE LEXICON AND QUOTATION DICTIONARY, Alexander Schmidt. Full definitions, locations, shades of meaning in every word in plays and poems. More than 50,000 exact quotations. 1,485pp. 6½ x 9¼. 2-vol. set.
Vol. 1: 22726-X
Vol. 2: 22727-8

SELECTED POEMS, Emily Dickinson. Over 100 best-known, best-loved poems by one of America's foremost poets, reprinted from authoritative early editions. No comparable edition at this price. Index of first lines. 64pp. 5³⁄₁₆ x 8¼. 26466-1

THE INSIDIOUS DR. FU-MANCHU, Sax Rohmer. The first of the popular mystery series introduces a pair of English detectives to their archnemesis, the diabolical Dr. Fu-Manchu. Flavorful atmosphere, fast-paced action, and colorful characters enliven this classic of the genre. 208pp. 5³⁄₁₆ x 8¼. 29898-1

THE MALLEUS MALEFICARUM OF KRAMER AND SPRENGER, translated by Montague Summers. Full text of most important witchhunter's "bible," used by both Catholics and Protestants. 278pp. 6⅝ x 10. 22802-9

SPANISH STORIES/CUENTOS ESPAÑOLES: A Dual-Language Book, Angel Flores (ed.). Unique format offers 13 great stories in Spanish by Cervantes, Borges, others. Faithful English translations on facing pages. 352pp. 5⅜ x 8½. 25399-6

GARDEN CITY, LONG ISLAND, IN EARLY PHOTOGRAPHS, 1869–1919, Mildred H. Smith. Handsome treasury of 118 vintage pictures, accompanied by carefully researched captions, document the Garden City Hotel fire (1899), the Vanderbilt Cup Race (1908), the first airmail flight departing from the Nassau Boulevard Aerodrome (1911), and much more. 96pp. 8⅞ x 11¾. 40669-5

OLD QUEENS, N.Y., IN EARLY PHOTOGRAPHS, Vincent F. Seyfried and William Asadorian. Over 160 rare photographs of Maspeth, Jamaica, Jackson Heights, and other areas. Vintage views of DeWitt Clinton mansion, 1939 World's Fair and more. Captions. 192pp. 8⅞ x 11. 26358-4

CAPTURED BY THE INDIANS: 15 Firsthand Accounts, 1750-1870, Frederick Drimmer. Astounding true historical accounts of grisly torture, bloody conflicts, relentless pursuits, miraculous escapes and more, by people who lived to tell the tale. 384pp. 5⅜ x 8½. 24901-8

THE WORLD'S GREAT SPEECHES (Fourth Enlarged Edition), Lewis Copeland, Lawrence W. Lamm, and Stephen J. McKenna. Nearly 300 speeches provide public speakers with a wealth of updated quotes and inspiration–from Pericles' funeral oration and William Jennings Bryan's "Cross of Gold Speech" to Malcolm X's powerful words on the Black Revolution and Earl of Spenser's tribute to his sister, Diana, Princess of Wales. 944pp. 5⅜ x 8⅜. 40903-1

THE BOOK OF THE SWORD, Sir Richard F. Burton. Great Victorian scholar/adventurer's eloquent, erudite history of the "queen of weapons"–from prehistory to early Roman Empire. Evolution and development of early swords, variations (sabre, broadsword, cutlass, scimitar, etc.), much more. 336pp. 6⅛ x 9¼.
25434-8

CATALOG OF DOVER BOOKS

AUTOBIOGRAPHY: The Story of My Experiments with Truth, Mohandas K. Gandhi. Boyhood, legal studies, purification, the growth of the Satyagraha (nonviolent protest) movement. Critical, inspiring work of the man responsible for the freedom of India. 480pp. 5⅜ x 8½. (Available in U.S. only.) 24593-4

CELTIC MYTHS AND LEGENDS, T. W. Rolleston. Masterful retelling of Irish and Welsh stories and tales. Cuchulain, King Arthur, Deirdre, the Grail, many more. First paperback edition. 58 full-page illustrations. 512pp. 5⅜ x 8½. 26507-2

THE PRINCIPLES OF PSYCHOLOGY, William James. Famous long course complete, unabridged. Stream of thought, time perception, memory, experimental methods; great work decades ahead of its time. 94 figures. 1,391pp. 5⅜ x 8½. 2-vol. set.
Vol. I: 20381-6   Vol. II: 20382-4

THE WORLD AS WILL AND REPRESENTATION, Arthur Schopenhauer. Definitive English translation of Schopenhauer's life work, correcting more than 1,000 errors, omissions in earlier translations. Translated by E. F. J. Payne. Total of 1,269pp. 5⅜ x 8½. 2-vol. set.   Vol. 1: 21761-2   Vol. 2: 21762-0

MAGIC AND MYSTERY IN TIBET, Madame Alexandra David-Neel. Experiences among lamas, magicians, sages, sorcerers, Bonpa wizards. A true psychic discovery. 32 illustrations. 321pp. 5⅜ x 8½. (Available in U.S. only.) 22682-4

THE EGYPTIAN BOOK OF THE DEAD, E. A. Wallis Budge. Complete reproduction of Ani's papyrus, finest ever found. Full hieroglyphic text, interlinear transliteration, word-for-word translation, smooth translation. 533pp. 6½ x 9¼. 21866-X

MATHEMATICS FOR THE NONMATHEMATICIAN, Morris Kline. Detailed, college-level treatment of mathematics in cultural and historical context, with numerous exercises. Recommended Reading Lists. Tables. Numerous figures. 641pp. 5⅜ x 8½.
24823-2

PROBABILISTIC METHODS IN THE THEORY OF STRUCTURES, Isaac Elishakoff. Well-written introduction covers the elements of the theory of probability from two or more random variables, the reliability of such multivariable structures, the theory of random function, Monte Carlo methods of treating problems incapable of exact solution, and more. Examples. 502pp. 5⅜ x 8½. 40691-1

THE RIME OF THE ANCIENT MARINER, Gustave Doré, S. T. Coleridge. Doré's finest work; 34 plates capture moods, subtleties of poem. Flawless full-size reproductions printed on facing pages with authoritative text of poem. "Beautiful. Simply beautiful."–*Publisher's Weekly*. 77pp. 9¼ x 12. 22305-1

NORTH AMERICAN INDIAN DESIGNS FOR ARTISTS AND CRAFTSPEOPLE, Eva Wilson. Over 360 authentic copyright-free designs adapted from Navajo blankets, Hopi pottery, Sioux buffalo hides, more. Geometrics, symbolic figures, plant and animal motifs, etc. 128pp. 8⅜ x 11. (Not for sale in the United Kingdom.) 25341-4

SCULPTURE: Principles and Practice, Louis Slobodkin. Step-by-step approach to clay, plaster, metals, stone; classical and modern. 253 drawings, photos. 255pp. 8⅛ x 11.
22960-2

THE INFLUENCE OF SEA POWER UPON HISTORY, 1660–1783, A. T. Mahan. Influential classic of naval history and tactics still used as text in war colleges. First paperback edition. 4 maps. 24 battle plans. 640pp. 5⅜ x 8½. 25509-3

CATALOG OF DOVER BOOKS

THE STORY OF THE TITANIC AS TOLD BY ITS SURVIVORS, Jack Winocour (ed.). What it was really like. Panic, despair, shocking inefficiency, and a little heroism. More thrilling than any fictional account. 26 illustrations. 320pp. 5⅜ x 8½. 20610-6

FAIRY AND FOLK TALES OF THE IRISH PEASANTRY, William Butler Yeats (ed.). Treasury of 64 tales from the twilight world of Celtic myth and legend: "The Soul Cages," "The Kildare Pooka," "King O'Toole and his Goose," many more. Introduction and Notes by W. B. Yeats. 352pp. 5⅜ x 8½. 26941-8

BUDDHIST MAHAYANA TEXTS, E. B. Cowell and others (eds.). Superb, accurate translations of basic documents in Mahayana Buddhism, highly important in history of religions. The Buddha-karita of Asvaghosha, Larger Sukhavativyuha, more. 448pp. 5⅜ x 8½. 25552-2

ONE TWO THREE . . . INFINITY: Facts and Speculations of Science, George Gamow. Great physicist's fascinating, readable overview of contemporary science: number theory, relativity, fourth dimension, entropy, genes, atomic structure, much more. 128 illustrations. Index. 352pp. 5⅜ x 8½. 25664-2

EXPERIMENTATION AND MEASUREMENT, W. J. Youden. Introductory manual explains laws of measurement in simple terms and offers tips for achieving accuracy and minimizing errors. Mathematics of measurement, use of instruments, experimenting with machines. 1994 edition. Foreword. Preface. Introduction. Epilogue. Selected Readings. Glossary. Index. Tables and figures. 128pp. 5⅜ x 8½. 40451-X

DALÍ ON MODERN ART: The Cuckolds of Antiquated Modern Art, Salvador Dalí. Influential painter skewers modern art and its practitioners. Outrageous evaluations of Picasso, Cézanne, Turner, more. 15 renderings of paintings discussed. 44 calligraphic decorations by Dalí. 96pp. 5⅜ x 8½. (Available in U.S. only.) 29220-7

ANTIQUE PLAYING CARDS: A Pictorial History, Henry René D'Allemagne. Over 900 elaborate, decorative images from rare playing cards (14th–20th centuries): Bacchus, death, dancing dogs, hunting scenes, royal coats of arms, players cheating, much more. 96pp. 9¼ x 12¼. 29265-7

MAKING FURNITURE MASTERPIECES: 30 Projects with Measured Drawings, Franklin H. Gottshall. Step-by-step instructions, illustrations for constructing handsome, useful pieces, among them a Sheraton desk, Chippendale chair, Spanish desk, Queen Anne table and a William and Mary dressing mirror. 224pp. 8⅛ x 11¼. 29338-6

THE FOSSIL BOOK: A Record of Prehistoric Life, Patricia V. Rich et al. Profusely illustrated definitive guide covers everything from single-celled organisms and dinosaurs to birds and mammals and the interplay between climate and man. Over 1,500 illustrations. 760pp. 7½ x 10⅛. 29371-8

Paperbound unless otherwise indicated. Available at your book dealer, online at www.doverpublications.com, or by writing to Dept. GI, Dover Publications, Inc., 31 East 2nd Street, Mineola, NY 11501. For current price information or for free catalogues (please indicate field of interest), write to Dover Publications or log on to www.doverpublications.com and see every Dover book in print. Dover publishes more than 500 books each year on science, elementary and advanced mathematics, biology, music, art, literary history, social sciences, and other areas.